AUTHORS OF PLANT NAMES

AUTHORS OF PLANT NAMES

A list of authors of scientific names of plants,
with recommended standard forms of their names,
including abbreviations

Edited by

R.K. BRUMMITT AND C.E. POWELL

Principal compilers:

C.E. Powell   (overall cumulation, and Spermatophytes after R.D. Meikle)
P.M. Kirk   (Fungi and Lichens)
P.C. Silva   (Algae)
R.E.G. Pichi Sermolli   (Pteridophytes)
M.R. Crosby   (Bryophytes)
H.T. Clifford   (Fossils)

ROYAL BOTANIC GARDENS, KEW
1992

First published 1992
Reprinted 1996
Reprinted 2001

General Editor of Series J.M. Lock

Cover Design by Media Resources, Royal Botanic Gardens, Kew

The *International Working Group on Taxonomic Databases for Plant Sciences* (TDWG) has endorsed this list as a standard for the citation of authors of plant names. TDWG is an international organization established to further collaboration among plant taxonomic database projects by promoting common use and interpretation of terminology, data fields, data dictionaries, and common logical rules and data relationships. For more information about TDWG, contact the TDWG Secretariat, c/o James L. Zarucchi, Missouri Botanical Garden, P.O. Box 299, St. Louis, Missouri 63366 (USA).

ISBN 1 842460 85 4

Typeset at the Royal Botanic Gardens, Kew by Pam Arnold, Christine Beard, Brenda Carey, Margaret Newman, Pam Rosen and Helen Ward

PREFACE

From before the time of Linneaus, botanists have sought to clarify their usage of plant names by citing the original author after the name of the plant. In the eighteenth century there were comparatively few botanists who had published such names, and simple traditions were soon established. Thus, the citation of the abbreviation L. was sufficient to indicate that the name was originally published by Linnaeus. This practice has persisted and grown up to the present time as an integral part of plant nomenclature, and the International Code now contains sections which allow for formal citation of author names. In modern times the need has arisen for a comprehensive list of such authors, with a standardised unique way of citing each one. This volume, including nearly 30,000 authors, aims to fulfil this need.

Historically the principal need was for a standard way of *abbreviating* author names to give a conveniently short way of indicating an earlier author. In the modern computer age there are those who see such abbreviations as an archaic irrelevance, and some publications and databases insist on giving surnames in full, just as journal titles sometimes now are given in full in bibliographic references. Many others, however, preferring to follow well established tradition, or perhaps because they understand the problems of accommodating author names on herbarium labels or determinavit slips, are happy to use abbreviations. This is a matter of personal preference, but, whichever view one takes, the need for a standard list remains. The rate of publication of new names for plants is probably as high now as it has ever been, and the list of authors of such names is expanding rapidly every year. The main problem these days is not how to abbreviate names but how many initials or other devices need to be *added* to distinguish between authors with the same surname. The standard forms of names offered here do include many abbreviations of surnames, but those who prefer not to abbreviate can simply give the surname in full and still have a unique form for every author.

Because a majority of names now have to be expanded to include initials, we have tried to avoid referring to this volume as offering a list of 'author abbreviations', preferring to call them 'standard forms' of names. (Is there a need for a simple term to indicate the standard form of the name of an author of plant names in botanical works — 'botaniconym' abbreviated to 'botanym' perhaps?). We also plead for the use of the word 'author' rather than 'authority' in this context.

Numerous lists of authors with 'standard forms', including 'abbreviations', have been published in the past, but none has been comprehensive. Details are given in the Introduction below of such lists for Bryophytes published by Sayre, Bonner & Culberson in 1964, for Pteridophytes by Pichi Sermolli in 1965, for Lichens by Laundon in 1979, for Fungi and Lichens by Hawksworth in 1980, and for flowering plants by the Royal Botanic Gardens, Kew (supervised by R.D. Meikle) in 1980 and 1984. The seven volume second edition of *Taxonomic Literature* by F.A. Stafleu & R.S. Cowan, 1976–1988, (commonly known as *TL-2*), provided recommended forms of author names in all major plant groups and fossils, but only for those who were listed as authors of books included in the work. Many Floras and other taxonomic works have produced their own lists of 'abbreviations' of those authors included within their own context. There has, however, been no standardisation between any of these lists, and the need for one consolidated list has become urgent.

In aiming for a comprehensive list of standard forms of authors in all plant groups, the compilers have tried as far as possible to maintain established traditions. Personal feelings, logical or illogical, can run surprisingly high in such matters, and many botanists have their own principles or their own preferred lists which they regard as sacrosanct. We know well that it is impossible to please everybody in every instance, and probably everybody will find something here which they do not like. In producing this consolidated list we have found it imperative to adopt many compromises and changes from previously published lists in order to achieve reasonable consistency and to avoid

1

duplication or ambiguity. The present recommendations have been produced after wide consultation with a large international Working Party in the hope of producing the most acceptable consensus. We are very glad that the list has been adopted as a standard by the International Working Group on Taxonomic Databases for Plant Sciences (TDWG) (see reverse of title page), which we hope will promote its acceptance in the botanical community.

Some users who are concerned mainly with authors of names of flowering plants may be disconcerted to find that a considerable number of changes have been made from the *Draft Index* produced at Kew in 1980. The present list, covering cryptogamic groups as well as flowering plants, is approximately three times as big as the *Draft Index* (about 30,000 names compared with about 10,000), and many changes were inevitable. The Working Party considered that the best option for maintaining traditional abbreviations was to follow Stafleu & Cowan's *TL-2*, which has the advantage of covering the cryptogamic groups as well as flowering plants. A major source of difference also arises from the Working Party's decision usually to include all initials when initials are needed, instead of only single initials given in the *Draft Index*. In view of the much greater coverage now achieved, and the very rapid current rate of growth of the author list, this decision was imperative. It also has the advantage of moving away from many anomalous single-initial standard forms previously recommended, such as J. Howell for Tom Howell, A. Leveillé for Hector Leveillé, J. Boivin for Bernard Boivin, C. Hitchc. for Leo Hitchcock, F. Lewis for Harlan Lewis, and W. Clayton for Derek Clayton (of Kew!). The opportunity has also been taken to change anomalies such as Ruiz Lopez for Ruiz of Ruiz & Pavon fame, Vahl for Jens Vahl instead of Martin Vahl, and others which had caused adverse comment previously.

Many people have contributed data to the present publication. Our thanks go firstly to those who have supplied the lists of authors for major plant groups, who are listed on the title page. Without their kind assistance and hard work this list could never have had any pretensions to giving comprehensive coverage. We are grateful also to all those botanists who took part in the correspondence of the Working Party between 1985 and 1991, who are listed in the Acknowledgements below. And finally we wish to express our particular gratitude to a small group of people who have commented very extensively on the list throughout the operation, and provided enormous amounts of information. Their enthusiasm and devotion to the project has been a great spur to us to complete the task. They are: Alfred Hansen (Copenhagen) who has supplied very many biographical details and references from his own records over a period of ten years; D.H. Kent (London) who has added hundreds of new names to the list from his extensive bibliographical and biographical researches; E.G. Voss (Ann Arbor) who has provided many additions and corrections; and G. Buchheim (Providence, R.I.), R.W. Kiger (Hunt Library, Pittsburgh), D.H. Nicolson (Smithsonian Institution, Washington) and Gea Zijlstra (Utrecht) who have all devoted their energies to assisting in making the data as complete and accurate as possible. Our sincere thanks go to all of these.

Any list such as this is bound to be out of date as soon as it is published. Every effort will be made to keep the database up to date and to augment the data already included there. It is expected that new authors, publishing for the first time after 1991, will be added automatically through appropriate indexing services. Information concerning omissions or additional data relating to earlier authors will be gratefully received, and should be sent either to the principal compiler for the cryptogamic group if appropriate, or to the Royal Botanic Gardens, Kew, or to both.

Dick Brummitt
Emma Powell

The Herbarium
Royal Botanic Gardens
Kew

December 1991

INTRODUCTION

Recommended 'standard forms', including 'abbreviations', of the names of authors of plant names have been published previously for some of the major groups of plants. For the Bryophyta such a list was first published by G. Sayre, C.E.B. Bonner and W.L. Culberson in *The Bryologist* 67: 113–135 (1964); for the Pteridophyta R.E.G. Pichi Sermolli included a list in *Index Filicum, Suppl.* 4: 351–361 (1965); for the Lichens a list was published by J.R. Laundon in *The Lichenologist* 11: 1–26 (1979); for the Fungi and Lichens D.L. Hawksworth published a list in *Review of Plant Pathology* 59: 474–480 (1980); and for the Flowering Plants the *Draft Index of Author Abbreviations* compiled at the Herbarium, Royal Botanic Gardens, Kew, under the supervision of R.D. Meikle, was published for Kew by HMSO in 1980, and reprinted with a supplement in 1984.

In 1985 a letter was received by the Director of the Royal Botanic Gardens at Kew from the Director of the Hunt Botanical Library, Pittsburgh, commenting that while Kew's *Draft Index* was extremely valuable and useful it could well be expanded and revised. It was noted that some of the 'abbreviations' adopted seemed idiosyncratic rather than based on any clear principles, and encouragement was given to Kew to produce a more definitive version.

Reaction at Kew was that this letter was very positive and helpful and drew attention to a clear demand felt by the botanical community. The title '*Draft Index...*' had indicated that the 1980 publication had not been intended as a final product, as emphasised in the foreword. While it was felt that a degree of idiosyncrasy is an essential aspect of sound botanical tradition, it was agreed that the question of standard forms for citation of botanical authors should be considered in a broad context. After correspondence with other interested institutions, particularly those dealing with cryptogamic groups, it was decided that a revised version should cover the whole plant kingdom and not just the flowering plants.

WORKING PARTY 1985–1991

In late 1985 a broadly based Working Party was established, with R.K. Brummitt acting as convener and secretary at Kew. The aim was to try to co-ordinate efforts and opinions of as wide a range of interested parties as possible, including those working on cryptogamic groups. Anybody who expressed an interest and a wish to participate in correspondence was invited to join, but it was made clear that those who did not respond to any letter would not receive further circulars. The Working Party operated in this way for six years, with new members joining and others being eliminated throughout the period, and a small core of dedicated collaborators continuing throughout. A list of those who have contributed is given in the Acknowledgements below.

The task of the Working Party was firstly to agree a set of principles to be followed in compiling the list, particularly in determining how to arrive at the recommended standard forms of author names, and secondly to scrutinise the details of the list. In the early stages only data on flowering plant authors were circulated for comment, but later, as authors in other groups were added to the database, more complete lists were circulated. It is hoped that in this way every opportunity has been given for data to be made as complete and accurate as possible and for as wide a body of people as possible to have a say in formulation of the standard forms recommended.

SCOPE

The intention is to include all authors who have published names of plants covered by the *International Code of Botanical Nomenclature*, which excludes the Bacteria. However, coverage of authors of names of fossil plants is very restricted (see notes under Compilation below). Authors who have published names only invalidly (but effectively) are included, since they may appear in an 'ex' citation. Authors who have proposed only names which have never been effectively published (such as manuscript notes on herbarium sheets), and which have not been taken up effectively by other authors, are excluded. All authors in F.A. Stafleu & R.S. Cowan's second edition of *Taxonomic Literature*, 1976-1988, (here subsequently referred to by its popular abbreviation *TL-2*) have also been included even though some of them may never have published any plant names. Since standard forms were proposed for all of them, it seemed a pity not to include them all, and anyway it was impossible to be sure which had published new names and which had not.

COMPILATION

The work has been compiled in sections according to major plant groups. Responsibility for Spermatophyta has been based at RBG Kew, where Emma Powell, who started full time work on the project in 1989, has augmented the list included in the Kew *Draft Index* (see below for further comments). For the Pteridophyta R.E.G. Pichi Sermolli (Montagnana) has continued to add to his 1965 *Index Filicum* list, assisted by P. Bizzari and with further information provided by B.S. Parris (Kew, now in Auckland, New Zealand). Authors of names in Bryophyta and some lichens were provided by M.R. Crosby (St. Louis) from his own database (for further details, see below). For the Algae P.C. Silva (Berkeley) and R. Moe (now at Waterville, Maine) produced lists, which were supplemented by data from D. John and D. Williams (Natural History Museum, London). For the Fungi, including Lichens, P.M. Kirk (International Mycological Institute, Kew) has greatly augmented Hawksworth's 1980 list with information from the *Index of Fungi* database, and has incorporated new information from D.N. Pegler and B.M. Spooner (RBG, Kew) (see further note below).

The inclusion of some authors of names of fossil plants was a late decision. A complete list for such plants, including fossil pollen, would be very large and have rather little overlap with the list of authors for extant plants. Initially palaeobotanists showed little enthusiasm for including authors of names of fossil plants in the present list, and it was found to be completely impractical and not necessarily even desirable to attempt to be comprehensive in this coverage. Nonetheless, we were very pleased to accept an offer from H.T. Clifford (Brisbane), to contribute a list of over 500 names of the better known authors for fossil plants. Some other authors of names of fossils were added from *TL-2*. In addition, authors of names of fossil fungi, including fungal spores, were included in the mycological authors (and so are coded as mycologists).

Overall cumulation into one database at RBG Kew was achieved by Emma Powell. The lists received of authors of names in Pteridophyta, Bryophyta and Fungi and Lichens were accompanied by suggested 'standard forms' of their names, but for the Algae and Fossils no suggestions were made by the contributors. Editing of the consolidated list, including final decisions on the recommended 'standard forms' of all names, has been done by R.K. Brummitt and Emma Powell, together with P.M. Kirk for mycological authors, in accordance with the principles agreed by the Working Party.

The major plant group or groups in which each author is known to have published names are recorded in the database, to allow selective output in future. They are also indicated in the present publication by single-letter codes after the author's dates. Where an author has published in more than one group, all appropriate codes are listed alphabetically. For some names taken from *TL-2* the precise group is not always specified and only 'cryptogamic group' is indicated. A further code is added for another group of authors taken from *TL-2* when their publications were before 1753, and they are referred to as 'pre-Linnaean'. The numbers of authors coded in these ways in the present list are as follows, listed in order of magnitude:

Code	Group	Number
S	Spermatophytes (flowering plants and gymnosperms)	16354
M	Fungi and Lichens (Mycology)	9196
A	Algae	4525
P	Pteridophytes	1862
B	Bryophytes	1609
F	Fossils	570
L	Pre-Linnaean (from *TL-2*)	120
C	Cryptogamic, unspecified (from *TL-2*)	67

Total authors included in list	29694

SOURCES OF INFORMATION FOR AUTHORS IN SPERMATOPHYTA

R.D. Meikle's foreword to the 1980 *Draft Index of Author Abbreviations* notes that it was derived largely from a card index kept by successive compilers of the *Index Kewensis*. This included only authors of generic and specific names in the flowering plants. A first typescript of the first draft was widely circulated, and additional names were added before publication.

The *Index Kewensis* database has now been scanned for additional authors publishing in the period 1981-91. The structure of the database is such that it is not easy to search the earlier references for authors, and in the time available it has not been possible to do this.

A number of other works have been checked for additional entries and data. These include:

Correll, D.S. & Johnston, M.C., Manual of the Vascular Plants of Texas (1970)
Dore, W.G. & McNeill, J., Grasses of Ontario (1980)
Encke, F., Buchheim, G. & Seybold, S., ed. 13 of Zander, R., Handwörterbuch der Pflanzennamen (1984)
Flora Europaea, vol. 5 (1980), and ed. 2, vol. 1 (in press)
Flora Iberica, vol. 1 (1986), vol. 2 (1990)
Flora of Iraq, vol. 1 (1966)
Flora Reipublicae Popularis Romanicae, vol. 13 (1975)
How Foon-Chew, Dictionary of Families and Genera of Chinese Seed Plants (1982)
Liberty Hyde Bailey Hortorium, Hortus Third (1976)
Mabberley, D.J., The Plant Book (1987)
Munz, P.A., California Flora (1959)
Ohwi, J., Flora of Japan (1959).

COMPILATION OF DATA ON MYCOLOGISTS
(by P.M. Kirk, IMI, Kew)

For the Fungi, including Lichens, P.M. Kirk assisted by A.E. Ansell (International Mycological Institute, Kew) has greatly augmented Hawksworth's 1980 list with information from the *Index of Fungi* database (1981-1991), the printed volumes of the *Index of Fungi* (1940-1980, vol. 1-4), *Petrak's Lists* (1920-1939, parts 1-8 and supplement, 1969) and *Saccardo's Omissions* (1985). Other sources include the register of yeast names in

Barnett, Payne & Yarrow, *The Yeasts* (1983), and those cited by Hawksworth (1980). D.N. Pegler and B.M. Spooner provided information from Lindau & Sydow's *Thesaurus litteraturae mycologicae et lichenologicae* (1908–1917) and the supplement to this work by Ciferri (1957–1960) and formulated abbreviations for a large number of previously unlisted names.

COMPILATION OF DATA ON BRYOLOGISTS
(by M.R. Crosby, St. Louis)

The entries for the authors of names of bryophytes make use of several previously published listings but include many new entries. Initially, the list of authors published in *The Bryologist* by Sayre, Bonner & Culberson (1964) was very useful as a basic list. Additional names were gleaned from Sayre's list of authors of bryophyte names and the whereabouts of their herbaria (1977). The list of author abbreviations for the *Index Muscorum* (Margadant & Terken, 1981) was also useful, though this list is not a list of names *per se*, but rather a list of the abbreviations of names used in the *Index Muscorum*. Hence, it contains some entries that are actually mistakes; therefore, these names are not included. Cheryl Bauer, an editorial assistant at the Missouri Botanical Garden from 1981–1986, worked to collate information from these various lists, producing the initial draft of a list of names of authors of bryophytes. The files of *TL-2* were searched personally while they were still in Utrecht for information for some authors, and Bernadette Callery, Research Librarian, The New York Botanical Garden, subsequently carried out additional searches at my request. During the ongoing accumulation of information for the *Index of Mosses*, I have routinely sent postcards to new authors asking for biographical information, and I thank all of those who responded with this information, which is included in the listing.

Margadant, W.D. & L. Terken. *Index of author abbreviations in Index Muscorum.* I–III, 1–28. Utrecht. 1981.

Sayre, G., C.E.B. Bonner & W.L. Culberson. The authorties for the epithets of mosses, hepatics, and lichens. *Bryologist* 67: 113–135. 1964.

Sayre, G. Authors of names of bryophytes and the present location of their herbaria. *Bryologist* 80: 502–521. 1977.

BIOGRAPHICAL DATA

Dates of birth and death have been obtained from a wide variety of sources. Particularly valuable have been J.H. Barnhardt's *Biographical Notes upon Botanists*, 3 vols. (1965), *Flora Europaea*, vol. 5 (1980), Stafleu & Cowan's *TL-2* (1976–1988), and R. Zander's *Handwörterbuch*, ed. 13 by F. Encke, G. Buchheim & S. Seybold (1984). We are very grateful to Pat Holmgren (New York) for extracting dates of birth from the *Index Herbariorum* database prior to publication of ed. 8 (*Regnum Vegetabile* 120, 1990), and to Marshall Crosby for assistance in relating these to entries in the present list. As noted elsewhere, we are also very grateful to Alfred Hansen (Copenhagen) for sending many details from his personal files.

SURNAME PROBLEMS

Even before one considers whether to abbreviate names or not, there are certain recurring problems concerning what constitutes the operative surname or how it should be spelled.

Cyrillic names

We have tried to adopt the transliteration which we believe the author himself would prefer. This is not always easy to determine, and we have an example of one Russian author whose name was transliterated in four different ways in four journals in the same year. Nonetheless, it seems undesirable to follow rigidly any one of the several available transliteration systems, which inevitably sometimes produce a version of a name which is markedly different from that usually adopted by the author. Although the Foreword to the Kew *Draft Index* states that the British Standards Institution system of transliteration was adopted there (which would give Bochantsev for Botschantzev, Goryaninov for Horaninov, etc.), in fact this is not so and the more traditional versions were actually accepted.

We have adopted as a standard the transliterations given in Komarov's *Flora of URSS, Indices Alphabetici* I–XXX: 229–259 (1964) and, for Bulgarians, those given in *Flora Reipublicae Popularis Bulgaricae* (1963–1982). For botanists not given in those Floras we have gladly accepted advice from A.E. Bobrov and D.V. Geltman of the St. Petersburg herbarium; to them, and to C. Jeffrey (Kew), our sincere thanks are due for checking lists of Russian names. Beyond this, we have tried to introduce some consistency in some situations, such as standardising the ending of many Georgian names as -schvili (rather than -shvili). It seems, however, dangerous to apply similar standardisations in other cases, such as -ovski and -ovsky, where different authors seem to prefer different versions.

Alternative transliterations which have actually been encountered in the literature have been included with a cross-reference to the accepted version. No attempt has been made to include all the possible transliterations, which in some cases could be very numerous.

Chinese names

A major complication arises from the conflict between the modern Pin-yin and the generally older and more traditional Wade-Giles conventions. These give alternative spellings, such as Zhang and Chung, which are often not recognisable as the same name to most western readers. In current literature Chinese authors will frequently use the Pin-yin spelling for the authorship of a paper, and within it use the Wade-Giles spelling for authorship of a new name. We have here adopted the Wade-Giles version, which seems to be generally still favoured in China in this context. Cross-references have been made where both spellings have been found in the literature.

The part of the full name other than the operative surname commonly consists of two single syllables. In Pin-yin these are joined by a hyphen, with the second not capitalised, but they are given as two separate words, both with capital initial, in Wade-Giles. Again, we have consistently adopted the latter version.

The last two decades have seen an enormous increase in the number of Chinese authors publishing new names, and the rather small range of names available in China has meant great duplication of surnames in the present list. Addition of initials to distinguish authors is essential in nearly all cases, but even then frequent duplications occur. Considerable trouble has therefore been taken to record full names in as many cases as possible, and we are most grateful to Lisa von Schlippe (Kew) for checking back to original publications of the very many Chinese authors quoted in *Index Kewensis* since 1975. In doing this she has also established the equivalence in many cases between the Pin-yin and Wade-Giles spellings. Her painstaking assistance is much appreciated. We are also indebted to C. Jeffrey (Kew) for much advice on Chinese names, and to Chen Chia-jui (PE, Beijing) for helpful comments and advice. Nonetheless, there may well be considerable problems remaining in distinguishing different Chinese authors, or in merging apparently different names as referable to the same author, particularly where data have been obtained from sources relating to more than one major plant group. When in doubt about two names being synonymous, we have left both without a cross-reference.

Compound names of Spanish or Portuguese origin

In Spanish or Spanish-American compound names the first part of the name is usually the more important and may sometimes stand on its own, as in Ruiz Lopez commonly known as Ruiz. In Portuguese and many Brazilian compound names, however, the second part is usually the more important and may often stand on its own, as in Amaral Franco usually known as Franco. In many cases, however, it is unclear to us whether authors with such names wish both parts or only one part to be included in their effective surname. We find the compromise situation where one part is reduced to its initial letter only, as in Gomez P. (which could be Gomez Pompa or Gomez Pignataro) to be unacceptable. To distinguish authors with compound names where the operative part is the first part and is the same, we have added initials of forenames rather than the second part of the name, e.g. C. Sánchez (rather than Sánchez Villaverde) to distinquish Carlos Sánchez Villaverde from others named Sánchez. Rarely we have also had to add the second half of the name (or an abbreviation of it) when both the operative surname and forename initials are the same, as in M. Fernández and M. Fernández Zeq.

We have sought authoritative advice from the countries concerned. A list of such compound names was extracted and submitted to Prof. B. Valdés (Sevilla, Spain), Dr. J.A.R. Paiva (Coimbra, Portugal), Dr. L. Rico (Mexico and Kew), A.L. Toscano de Brito (Brazil and Kew) and C.V. Mendonça Filho (Brazil). We are very grateful to all of them for their advice on what should be treated as the operative surname. The abbreviations of these, however, are the responsibility of the editors. Guidance on Spanish names has also been sought from *Flora Iberica* (vol. 1 1986, vol. 2 1990), and some hundreds have been checked in their original publications.

Prefixes in Dutch, Belgian, South African and other names

Conflicting views were originally expressed within the Working Party over whether the prefixes 'van', 'van der' and 'de' (variously capitalised) should be included in standard forms of names or not. Several Dutch botanists were consulted, but each seemed to differ in one way or another from the others. Two submitted copies of Dutch telephone directories to show how names should be alphabetised, but the two copies differed from each other! When Dutch botanists whose names actually included a prefix were consulted, it emerged that those with 'van' or 'van der' slightly preferred to drop the prefix — or at least did not object strongly if this was suggested. On the other hand, most of those with 'de' objected strongly when it was suggested that their prefixes might be omitted. Advice was also sought from Belgian and South African botanists, who generally agreed with each other. Eventually the following principles emerged:

Dutch — 'van' and 'van der' to be dropped, 'de' to be retained, all spelled with lower case initial.

Belgian — 'Van', 'Van der' and 'De' to be retained, all spelled with upper case first initial.

South African — 'van', 'van der' and 'de' to be retained, spelled with a capital first initial if no forename initial precedes them but with lower case first initial if forename initials are included.

For other nationalities also having these prefixes, these were usually retained but no general rule about capitalisation emerged. Each case should be considered separately.

The problem then remained of deciding which of the names listed with 'van' or 'de' were Dutch, which were Belgian, which were South African, and which were others. Print-outs of all such names were sent to the herbaria at Leiden, Utrecht, Wageningen, Brussels (Meise), Pretoria and Kirstenbosch, and to R.E.G. Pichi Sermolli in Italy. Sympathetic replies were received from all, with many constructive comments and corrections. Our thanks are due to all, and to B. de Winter (Pretoria), A.E. van Wyk (Witwatersrand), J.J.A. van der Walt (Stellenbosch) and A. Nicholas (Pretoria, formerly Liaison Officer at Kew) for advice on South African authors.

The prefix has been retained in the Dutch names 't Hart and 't Mannetje. The German prefix 'von' has been dropped from the surname consistently.

Ethiopian names

Ethiopians do not have a family name or 'surname' which is inherited from one generation to another as in most western nationalities. Instead the father's name is appended after his child's name which is usually mononomial in character. Thus Sebsebe Demissew is Sebsebe, the son of Demissew, and in the next generation the name Demissew will be lost. The operative name for Sebsebe Demissew is Sebsebe, which is to be used in the context of authorship of plant names. Such authors are listed here without a separating comma.

'STANDARD FORMS' OF NAMES

The 'standard form' of a name is a surname, or an abbreviation of it (e.g. Adans. for Adanson), or rarely a contraction of it (e.g. Michx. for Michaux), with or without initials or other distinguishing appendages (see principles 7–9 below). In establishing a set of principles for deriving standard forms of names, the Working Party has aimed above all for practicality and has been largely sympathetic to well established traditions. Very few absolute rules have been adopted, since experience shows that sooner or later it becomes inconvenient or undesirable to follow them. However, clear general principles have been laid down which are followed except when there is a good reason not to do so. The standard form adopted is the product of the interaction of all the principles, of which one may be of paramount importance in one circumstance but less so in another. The principles were agreed by the Working Party by a vote where necessary, but in the majority of cases the vote was overwhelmingly in favour of the practice adopted here.

The first four 'principles' are applied rigidly as absolute rules:

1. *Script.* Names are given in Roman characters (see also notes on transliteration and similar problems under 'Surname Problems' above).

2. *Uniqueness.* Every 'standard form' must be unique to one person. The Working Party considered a suggestion that this should apply only within the context of each major plant group, but voted heavily against it. (For comments on the converse situation, concerning the same author always having the same 'standard form', see 10 and 11 below.)

3. *Uniform treatment of names.* The same surname (i.e. identical spelling) must always be given in the same form (e.g. Miller is always abbreviated to Mill.), unless it is part of a compound name (see 14 below), and different surnames must not be given in the same form (if Brown is abbreviated to Br. then Browne must not also be abbreviated to Br.). Unlike the Kew *Draft Index*, we have treated masculine and feminine forms of one name as different names. Thus, for example, Botschantzev and Botschantzeva are not given the same standard form.

4. *Full-stops and accents.* All abbreviations and contractions are terminated by a full-stop (e.g. Adans., Walt.Jones, Michx.) but the full-stop does not make a standard form different from the same spelling without a full stop (e.g. Lam. for Lamarck and Lam for H.J. Lam would be treated as homonymous, and initials are required for the latter's standard form). Similarly names differing only by presence or absence of an accent (e.g. Love and Löve, Leonard and Léonard) or apostrophe (Ohara and O'Hara) are treated as homonymous.

The remaining principles are not considered absolutely binding except for 12. The first, no. 5, over-rides all the subsequent ones. Those from 6 to 11 refer to full surnames and when to include initials and other appendages, while 12 to 14 refer to when and how to abbreviate surnames.

5. *TL-2 as a standard.* The Working Party strongly favoured the retention of traditional abbreviations and other standard forms, and considered that the best way to ensure this was to accept the standard forms recommended in Stafleu & Cowan's *TL-2*. This has been done in the great majority of cases, but in a small number *TL-2* has not been followed. Reasons for departing from *TL-2* included: a) contravention of principles 2 and 3 above (rare, e.g. Stef.); b) serious conflict with the Kew *Draft Index* involving transfer of one standard form from one author to another; c) failure to abbreviate very long names, such as Schlagdenhauffen; d) rather excessive abbreviation of a few names of six letters or fewer where no outstandingly strong precedent seems to exist (e.g. Wilson is given in full rather than abbreviated to Wils.); e) failure to give any of a large of number of authors with the same surname a standard form without initials, e.g. Moore; and f) occasionally, conflict with particularly well established abbreviations used elsewhere, such as Copel. which is very widely adopted in pteridological literature for Copeland.

The following principles are applied only to names of authors not appearing in *TL-2*:

6. *Surname only.* A surname alone, or its abbreviation or contraction, is adopted as the standard form if it is applicable to only one author in the list. A surname alone, or its abbreviation or contraction, is usually also adopted for one of a number of authors with the same surname (see 7 below).

7. *Initials.* Persons with identical surnames are distinguished by use of intials of forenames, except as in 8 and 9 below. Usually the earliest born is given without initials and all later ones with initials, but in some cases a better known later author may be given without and all others, including the earliest, with initials. Occasionally, when all having the same surname are more or less contemporary and equally well known, all may be given initials. Where initials are required: a) if the author has one forename, then the one initial is given; b) if the author has two forenames, both initials are given, except in occasional cases where an author consistently omits one initial in authorship of books or papers (as distinct from authorship of plant names), e.g. P.Taylor, H.Rob., B.Nord. for P.G. Taylor, H.E. Robinson and R.B. Nordenstam who consistently publish their work using only one of their forenames; c) if the author has three forenames, three initials are given unless he or she has a clear preference for using only one and no ambiguity arises (L.A.S.Johnson is given all three initials to avoid confusion with Lennart Johnson); if the author has more than three forenames, an *ad hoc* decision is taken. We recommend that no spaces be left after full stops.

8. *Abbreviated or full forenames.* When two authors have identical surnames and initials, full or abbreviated forenames may be used. One author may be given with only surname, or initial(s) plus surname, and the other with fuller name, or both may be given fuller names, e.g. Thomas Hogg (1777–1855) and Thomas Hogg (1820–1892) are distinguished as Hogg and T.Hogg respectively, while Walter Jones and William Jones have standard forms Walt.Jones and Wm.Jones respectively.

9. *Suffixes.* In a few cases persons with identical surnames may be distinguished by a suffix instead of, or in addition to, initials. In well known cases of father and son, the son may be distinguished by 'f.', an abbreviation of 'filius' (a narrow majority of the Working Party preferred 'f.' to 'fil.'). Where this follows an abbreviated name with a full-stop, we recommend that no space be left between full-stop and f., e.g. Rech.f. for Rechinger the son. Decisions on when to use 'f.' and when to use initials were usually decided by a vote of the Working Party, which voted for Aiton and W.T.Aiton, but Baker and Baker f., for examples. Tradition may also allow a different suffix, such as 'Arg.' for 'Argoviensis' in

Müll.Arg. (Müller of Aargau). In a few cases where different authors have identical surname and forename we have used the suffixes 'bis' for the second and 'ter' for the third, as in R.Br., R.Br. bis and R.Br. ter for the three Robert Browns.

10. *Variant names for the same person.* Except as noted in 11 below, one person is always given the same standard form, even though he or she may have modified the spelling of their surname during their lifetime, or different transliterations may exist, or a compounding form may have been adopted during their lifetime. For example, Meisner adopted the spelling Meissner later in life but is always given the same standard form, Meisn. The variant transliterations Tsvelev, Tsvelov, Tzvelev and Tzvelov have all appeared for the same person, whose recommended standard form Tzvelev should be used in all cases. Alan Radcliffe Smith originally published plant names as A.R. Smith, but later modified his name to A. Radcliffe-Smith; the standard form Radcl.-Sm. should be used for both.

11. *Different names for the same person.* When a person has published under completely different names, different standard forms may be used for the same person. For example, Inger Nordal has published under both her married name Bjørnstad and her maiden name Nordal, and both may be used in standard forms, as in *Scadoxus pseudocaulus* (I.Bjørnstad & Friis) Friis & Nordal. However, where one person has simultaneously used alternative names, such as Brother Alain who also uses the names Liogier, Enrique Eugenio, and others, we have adopted a single name (in this case Alain, as agreed with him in correspondence).

The following principles concern when and how to abbreviate surnames. It should be noted again that these have been applied only to names not given in *TL-2* (see 5 above).

12. *Where to abbreviate.* Names are never abbreviated before a consonant.

13. *How many letters to save.*
a)   Names are usually not abbreviated unless more than two letters are eliminated and replaced by a full-stop.
b)   Names of authors living before the 20th century are more likely to be abbreviated than later ones, and names of authors in the later 20th century tend to be given in full.
c)   Where no strong tradition exists, names of 8 letters or fewer are not abbreviated, names of 9 letters are abbreviated if more than 3 letters are eliminated (e.g. Verdc. for Verdcourt), and names of 10 letters or more are usually abbreviated. (In the suggested abbreviations submitted by the compiler for Bryology there was a tendency for short names (7 letters or fewer) to be abbreviated where they probably would not have been abbreviated in the list for other groups. In the final editing some of these have been expanded to the full name to make them more comparable with others, but the tendency to greater abbreviation for bryological authors may still be noted.)
d)   Other things being equal, if an abbreviation is made, 2-syllable names are abbreviated to one syllable, 3-syllable names are most likely to be abbreviated to one (Pennington to Penn.), and names of 4 or more syllables are most often abbreviated to two syllables (Tirvengadum to Tirveng.). Many Japanese names have four short syllables, and we have tried to be consistent in abbreviating them to two syllables (Kitagawa to Kitag., Hashimoto to Hashim., etc) except where ambiguity would occur, or except where the name has only seven letters.
e)   Application of these principles may depend on whether initials are also needed in standard forms of well known authors. Decisions may also be affected when an author is known to be commonly cited as a joint author with somebody else.

14. *Compound names.* Compound names, whether hyphenated or not, may be treated as special cases, and some principles given above may be discounted. In order to keep the

standard form as short as possible, the principles of how many letters should be saved (see 13 above) may often be broken. Similarly the principle that the same name must always be given in the same way (see 3 above) is over-ridden in compound names, so that, for example, although Gonzales is given in full when it stands on its own, Gonzales Albo is abbreviated to Gonz.Albo. We recommend, again that no space be left after the full-stop.

THE DATABASE

The present list has been printed from a database held on a PRIME computer at the Royal Botanic Gardens, Kew. More information is held in the database than is printed here. In addition to the standard form recommended here, the recommendations of *TL-2* and of the Kew *Draft Index* are recorded where applicable. Some very brief biographical data in addition to dates of birth and death are sometimes given, such as maiden names and married names or titles. Such alternative names are given in the printed version only when the author has published new names under both. The source of the entry, and sometimes of the biographical data if they were received separately, are also recorded. For some authors in Spermatophyta (especially less well-known ones) a reference is given to where they have published one name. It is intended that copies of this database will be sold in electronic form, and enquiries should be directed to the Publications Department, Royal Botanic Gardens, Kew.

ACKNOWLEDGEMENTS

The Working Party was established in 1985 and anybody who asked to join was included on the understanding that if they did not reply to one circular they would not be sent any more. The following participated in the correspondence at one time or another: R. Allkin (Kew), F.A. Bisby (Southampton), G. Buchheim* (Providence, R.I.), H.M. Burdet (Geneva), J.F.M. Cannon (BM, London), P.F. Cannon (IMI, Kew), W.G. Chaloner (London), A.D. Chapman (Canberra), A.O. Chater* (BM, London), H.T. Clifford* (Brisbane), M.J.E. Coode (Kew), M.R. Crosby* (St. Louis), W.G. D'Arcy (St. Louis), H. Eichler (Canberra), E. Forero (St. Louis), I. Friis (Copenhagen), G.E. Gibbs Russell (Pretoria), W. Greuter (Berlin), A. Hansen* (Copenhagen), D.L. Hawksworth* (IMI, Kew), R.J.F. Henderson (Brisbane), V.H. Heywood (Reading), S. Hill (Beltsville), J. Jansonius (Calgary), D.H. Kent* (London), R.W. Kiger* (Pittsburgh), P.M. Kirk* (IMI, Kew), R.P. Korf (Ithaca), H.-W. Lack (Berlin), O.A. Leistner (Pretoria), R. Magill (St. Louis), J. McNeill (Edinburgh, later Toronto), L. Morse (Arlington), D.H. Nicolson* (Washington), B.S. Parris* (Kew, later Auckland), D.N. Pegler* (Kew), R.E.G. Pichi Sermolli* (Montagnana), G.T. Prance (New York, later Kew), P.H. Raven (St. Louis), J.L. Reveal (College Park), A.Y. Rossman (Beltsville), P.C. Silva* (Berkeley), B.M. Spooner* (Kew), C.A. Stace (Leicester), F.A. Stafleu (Utrecht), D.A. Sutton (BM, London), N.P. Taylor (Kew), E.E. Terrell (Beltsville), J. Tolsma (Utrecht), A. Traverse (Pennsylvania), E.G. Voss* (Ann Arbor), O. Wijnands (Wageningen), G. Zijlstra* (Utrecht). Thanks are extended to all of these for their participation. Especial acknowledgement is made to those marked with an asterisk (*) for contribution of much original data, including in some cases the complete lists for authors of names of cryptogamic groups or fossils.

Others who have kindly responded to specific requests for information or opinions are P. Bamps (Meise, Brussels), A.E. Bobrov (St. Petersburg), B. de Winter (Pretoria), J.R. Edmondson (Liverpool), E. Farr (Washington), D.V. Geltman (St. Petersburg), A.R.A. Görts-van Rijn (Utrecht), P. Holmgren (New York), C. Kalkman (Leiden), J. van der Maesen (Wageningen), C.V. Mendonça Filho (Brazil), R.R. Mill (Edinburgh), R.L. Moe

(Waterville, Maine), J.A.R. Paiva (Coimbra), A.L. Toscano de Brito (Herbarium Bradeanum, Rio de Janeiro), B. Valdés (Sevilla), J.J.A. van der Walt (Stellenbosch), A.E. van Wyk (Witwatersrand). Sincere thanks are due to all.

At Kew Veronica Marx and Margaret Beyer have given invaluable service in keyboarding information; Rosemary Davies and Katherine Lloyd provided information from *Index Kewensis*, Lisa von Schlippe painstakingly compiled data on Chinese authors; C. Jeffrey also advised on Chinese and Russian authors; W.N. Loader, M. Jackson, J. Farnon and Sarah Edwards have provided the computing services; G.E. Wickens took charge of data after the retirement of R.D. Meikle; R. Milne extracted data from *TL-2*; Lulu Rico advised on Latin American names; J.M. Lock supervised publication; and the team of typesetters acknowledged individually elsewhere produced the camera-ready copy. The editors are also very grateful to P.M. Kirk of IMI, Kew, for much assistance in checking proofs. Again, sincere thanks are due to all.

EXPLANATION OF THE LIST

Authors are listed alphabetically by their surname in the left hand column, followed by their forenames where known or initials when full names are not known. Alternative transliterations of cyrillic names, and alternative spellings of Chinese names, with cross reference to accepted names, are given in brackets when these alternative forms have been found in the literature; no attempt is made to give all possible transliteration or spelling variants. Where doubt may exist as to what constitutes the surname, such as when prefixes occur or when compound names are involved, alternative forms are listed in brackets with cross reference to the accepted form. In making such cross references we have tried to bear in mind problems which may arise in use of the database in electronic form as well as in the present printed copy. When the same author has published under two completely different names (see p.11 above), both include cross references to the other, and separate abbreviations are given for each (see, for example S. Soenarko and S. Dransfield). If only one name has been used for publication of new plant names, others may be held in the database but not printed.

Alphabetisation is by computer, with accents ignored, upper and lower case treated the same, apostrophes and hyphens sorted before letters, Mc not expanded to Mac and St not expanded to Saint. The position of a comma is ignored, and for compound names the second part of a name may be treated as though it were the first forename. It should be noted, however, that a hyphen may make a big difference to the alphabetic placement of a name. Thus, for example, while Fernández Carvajal, M., follows after Fernández, C.F., the alternative form Fernández-Carvajal, M., is treated as one word and follows after Fernández Zequeira.

After the forename or initials is the date of birth and/or death, as applicable and as far as is known, in brackets. Where neither date of birth nor of death is known, any one date when the author is known to have published a name is given after the abbreviation 'fl.' ('floruit'). Where no dates are known the brackets are omitted; this happens especially frequently for algologists, for whom, in general, biographical data have been provided less comprehensively than for other groups.

After the dates, or after a space where no date is given, is an indication of the group in which the author is known to have published names. Codes are as follows:

A   Algae
B   Bryophyta
C   Cryptogamic, unspecified
F   Fossils
L   pre-Linnaean
M   Fungi and Lichens (Mycology)
P   Pteridophyta
S   Spermatophyta (Flowering Plants and Gymnosperms)

In the right hand column is the recommended 'standard form' of the name.

Aa, H.A.van der (1935–) MS   **Aa**
Aalders, Lewis Eldon (1933–) S   **Aalders**
Aalto, Marjatta (1939–) M   **Aalto**
Aaronsohn, Aaron (1876–1919) S   **Aarons.**
Aas, O. (fl. 1978) M   **Aas**
Aasamaa, Heinrich (fl. 1956) M   **Aasamaa**
Aase, Hannah Caroline (1883–) S   **Aase**
Abadie, F. (fl. 1954) M   **Abadie**
Abadjieff, V. (fl. 1971) M   **Abadj.**
Abalo, J.E. (fl. 1981) S   **Abalo**
Abarca, Loudes (fl. 1990) M   **Abarca**
Abassi Maaf, L. (fl. 1984) M   **Abassi**
Abat (fl. 1792) S   **Abat**
Abawi, George S. (fl. 1971) M   **Abawi**
Abbareddy, N.R. (fl. 1983) S   **Abbar.**
Abbas, Syed Qaiser (fl. 1972) M   **Abbas**
Abbasi, P. (fl. 1987) M   **Abbasi**
Abbayes, Henri Robert Nicollon des (1898–1974) MS   **Abbayes**
Abbe, Ernst Cleveland (1905–) PS   **Abbe**
Abbiatti, Delia (1918–) PS   **Abbiatti**
Abbot, Charles (1761–1817) S   **C.Abbot**
Abbot, John (1751–c.1840) S   **Abbot**
Abbot, Maxine L. (fl. 1968) F   **M.L.Abbot**
Abbott, Ernest Victor (1899–1980) M **E.V.Abbott**
Abbott, George (fl. 1927) S   **Abbott**
Abbott, Isabella Aiona (1919–) A   **I.A.Abbott**
Abbott, L.K. (fl. 1982) M   **L.K.Abbott**
Abbott, Sean P. (fl. 1988) M   **S.P.Abbott**
Abdallah, Moustafa Sayed–Ahmed (1918–) S   **Abdallah**
(Abdel Bari, Ekhlas Mohammed Mahmud)
  see Bari, Ekhlas M.M.Abdel   **Bari**
Abdel Rahman, Mahmoud Hafez A   **Abdel Rahman**
Abdel–Fattah, H.M. (fl. 1978) M   **Abdel–Fattah**
Abdel–Hafez (fl. 1977) M   **Abdel–Hafez**
Abdul–Majeed, M. A   **Abdul–Majeed**
Abdul–Wahid, O.A. (fl. 1989) M   **Abdul–Wahid**
Abdulina, S.A. S   **Abdulina**
Abdulkadder, Maithum A. (fl. 1989) M   **Abdulk.**
Abdullaeva, H.N. (1923–) S   **Abdull.**
Abdullaeva, M.N. (fl. 1981) S   **M.N.Abdull.**
Abdullah, N. (fl. 1990) M   **N.Abdullah**
Abdullah, Samir K. (fl. 1977) M   **Abdullah**
Abdus Salam, A.M. A   **Abdus Salam**
Abdusalyamova, L.N. (fl. 1970) S   **Abdusal.**
Abe, H. (fl. 1986) M   **H.Abe**
Abe, S. (fl. 1956–89) MS   **S.Abe**
Abé, Tohru Hidemiti (1899–1971) A   **T.H.Abé**
Abe, Yasuhisa (1951–) M   **Y.Abe**
(Abebe, Dawit)
  see Dawit Abebe   **Dawit**

Abedin, Sultanul (fl. 1986) S	**Abedin**
Abel, Clarke (1789–1826) S	**C.Abel**
Abel, Gottlieb Friedrich (1763–) S	**Abel**
Abélard, Christiane A	**Abélard**
Abeleven, Theodor (Theodoor) Hendrik Arnoldus Jacobus (1822–1904) B	**Abeleven**
Abels, Jürgen (fl. 1973) S	**Abels**
Aberconway, Henry Duncan McLaren, Lord (1879–1953) S	**Aberc.**
Aberdeen, J.E.C. (fl. 1962) M	**Aberdeen**
Aberdeen, V. (fl. 1987) M	**V.Aberdeen**
Åberg, A.Ewart (1909–) S	**A.E.Åberg**
Åberg, Johann Gerhard ('Goran') (1868–1940) B	**Åberg**
Åberg, Mette A	**M.Åberg**
Abeywickrama, Bartholomeusz Aristides (1920–) PS	**Abeyw.**
Abich, (Otto) Hermann Wilhelm (von) (1806–1886) F	**Abich**
Abiko, K. (fl. 1970) M	**Abiko**
Abildgaard, Peder Christian (1740–1801) S	**Abildg.**
Ablajev, A.G. S	**Ablajev**
Ablakatova, A.A. (fl. 1960) M	**Ablak.**
Abolin, Robert Ivanovich (1886–) S	**Abolin**
Abraham, Rajender A	**R.Abraham**
Abraham, S.P. (fl. 1980) M	**S.P.Abraham**
Abraham, V. (fl. 1981) S	**V.Abraham**
Abraham, W.R. (fl. 1987) S	**W.R.Abraham**
(Abramoff, Ivan N.)	
see Abramov, Ivan N.	**Abramov**
Abramov, Ivan Ivanovich (1912–1990) BS	**I.I.Abramov**
Abramov, Ivan N. (1884–) MS	**Abramov**
Abramov, N.V. (1942–) S	**N.V.Abramov**
Abramova, Anastasia Laurentievna (1915–1976) B	**Abramova**
Abramova, L.I. (1940–) S	**L.I.Abramova**
Abramova, S.N. (1937–) S	**S.N.Abramova**
Abrams, LeRoy (1874–1956) PS	**Abrams**
Abramyan, D.G. (fl. 1969) M	**Abramyan**
Abreu, Cordélia Luiza (fl. 1982) S	**Abreu**
Abromeit, Johannes (1857–1946) S	**Abrom.**
Accordi, Bruno A	**Accordi**
Acebes Ginovés, J.R. (1950–) S	**Acebes**
Acedo, C. (fl. 1990) S	**Acedo**
Acevedo de Vargas, Rebecca (1903–) S	**Acevedo**
(Acevedo, Pedro)	
see Acevedo–Rodriguez, Pedro	**Acev.-Rodr.**
(Acevedo San Martin, Pedro)	
see Acevedo–Rodriguez, Pedro	**Acev.-Rodr.**
Acevedo–Rodriguez, Pedro (1954–) S	**Acev.-Rodr.**
Acharius, Erik (1757–1819) AMPS	**Ach.**
Achepohl, Ludwig (–1902) FP	**Achepohl**
Achey, Daisy Bird (1906–) S	**Achey**
Achintre, Joseph Frédéric (fl. 1867–68) S	**Achintre**
Achmetova, N.I. A	**Achmetova**

(Achtar, Kazmi Sayed Ali)
  see Akhtar, Kazmi Sayed Ali   **Akhtar**
(Achtaroff, Boris T.)
  see Achtarov, Boris T.   **Acht.**
Achtarov, Boris T. (1885–1959) S   **Acht.**
Achundov, G.F. (1914–) S   **Achundov**
Achundov, T.M. (fl. 1956) M   **T.M.Achundov**
Achutan, V. (fl. 1968) F   **Achutan**
Achuthan, Manjapra Variath A   **Achuthan**
Achverdov, Agazi Asaturovich (1907–) S   **Achv.**
(Aciole de Queiroz, Lusinete)
  see Queiroz, Lusinete Aciole de   **L.A.Queiroz**
Ackerman, James D. (1950–) S   **Ackerman**
Ackley, Alma Bernice (1893–) A   **Ackley**
(Ackverdov, Agazi Asaturovich)
  see Achverdov, Agazi Asaturovich   **Achv.**
(Acleto, César)
  see Acleto Osorio, César   **Acleto**
Acleto Osorio, César (1937–) A   **Acleto**
Acloque, Alexandre Noël Charles (1871–1908) MS   **Acloque**
Acosta Castellanos, Salvador (1957–) S   **Acosta**
Acosta, Christobal (1512–1580) L   **C.Acosta**
Acton, Elizabeth (–1923) A   **E.Acton**
Acton, H.W. (fl. 1919) M   **Acton**
Acuña Gale, Julián Baldomero (1900–) S   **Acuña**
Adachi, Rokuro (1931–) A   **Adachi**
Adam, David Bonar (fl. 1924) M   **D.B.Adam**
Adam, J.I. (fl. 1805) S   **Adam**
Adam, Jacques–Georges (1909–1980) S   **J.-G.Adam**
(Adam, Johannes Michael Friedrich)
  see Adams, Johannes Michael Friedrich   **Adams**
Adam, Jumaat Haji (1956–) S   **J.H.Adam**
Adametz, L. (fl. 1886) M   **Adametz**
Adamiker, Dieter A   **Adamiker**
Adamov, Vladimir Vladimirowitsch (1875–1932) S   **Adamov**
Adamović, Lujo (Lulji) (Lucian) (1864–1935) S   **Adamović**
Adams, Bryan Roger (1942–) S   **B.R.Adams**
Adams, Charles Dennis (1920–) PS   **C.D.Adams**
Adams, James Fowler (1888–) M   **J.F.Adams**
Adams, Johannes Michael Friedrich (1780–1838) S   **Adams**
Adams, Joseph Edison (1904–1981) S   **J.E.Adams**
Adams, Joseph William (1906–) S   **J.W.Adams**
Adams, Laurence George (1929–) S   **L.G.Adams**
Adams, Nancy M. (1926–) A   **N.M.Adams**
Adams, P.B. (fl. 1978) M   **P.B.Adams**
Adams, R.M. (fl. 1984) S   **R.M.Adams**
Adams, Robert Perry (1886–) S   **R.P.Adams**
Adams, William Preston (1930–) S   **W.P.Adams**
Adamska, I. (fl. 1983) S   **Adamska**

Adamson, Alastair Martin (fl. 1933) A	A.M.Adamson
Adamson, Robert Stephen (1885–1965) S	Adamson
Adanson, Michel (1727–1806) AMPS	Adans.
Adaskaveg, J.E. (fl. 1988) M	Adask.
Ade, Alfred (1876–1968) MS	Ade
Adediran, S.A. A	Adediran
Adelbert, Albert George Ludwig (1914–1972) PS	Adelb.
Adema, Fredericus Arnoldus Constantin Basil (1938–) S	Adema
Aderhold, Rudolph Ferdinand Theodor (1865–1907) M	Aderh.
Adey, Patricia J. (1936–) A	P.J.Adey
Adey, Walter H. (1934–) A	W.H.Adey
Adhikari, Mahesh Kumar (fl. 1986) M	Adhikari
Adhya, T.K. A	Adhya
Adjanohoun, Edouard (1928–) S	Adjan.
Adlakha, K.L. (fl. 1958) M	Adlakha
Adler, Mónica T. (fl. 1987) M	Adler
Adlerz, Ernst (1854–1918) B	Adlerz
Adolf, N.A. (fl. 1930) S	Adolf
Adsersen, H. (fl. 1980–) S	Adsersen
Adylov, T.A. (1920–) S	Adylov
Adzet, José–Maria (fl. 1958) M	Adzet
Aebe, Bertha (fl. 1972) M	Aebe
Aeberhardt, H. (fl. 1983) M	Aeberh.
Aebi, B. (fl. 1972) M	Aebi
Aellen, Paul (1896–1973) S	Aellen
Aepnelaeus, C. S	Aepnel.
(Afanasiev, C.S.)	
see Afanassiev, C.S.	Afan.
Afanassiev, C.S. (1905–1960) S	Afan.
Affolter, J.M. (fl. 1985) S	Affolter
(Afonso, Maria da Luz de Oliviera Tavares Monteiro da)	
see Rocha Afonso, Maria da Luz de Oliveira Tavares Monteiro	Rocha Afonso
Afonso–Carrillo, Julio (1954–) A	Afonso–Carr.
Afrikyan, R. (fl. 1933) M	Afrikyan
Afzelius, Adam (1750–1837) MPS	Afzel.
Afzelius, Karl Rudolf (1887–1971) S	K.Afzel.
Afzelius, Pehr Conrad (1817–1850) S	P.Afzel.
Agababjan, M.V. (1964–) S	M.V.Agab.
Agababjan, V.S. (1932–) S	Agab.
Agadi, V.V. A	Agadi
(Agadzhanof, S.D.)	
see Agadzhanov, S.D.	Agadzh.
Agadzhanov, S.D. S	Agadzh.
Agajeva, G.B. (fl. 1971) M	Agajeva
Agajeva, O.D. S	O.D.Agajeva
Agapova, N.D. (1932–) S	Agapova
Agardh, Carl Adolf (1785–1859) AMPS	C.Agardh
Agardh, Jakob Georg (1813–1901) AMPS	J.Agardh
Agarkar, D.S. A	D.S.Agarkar

Agarkar, M.S. A	**M.S.Agarkar**
Agarwal, A.K. (fl. 1989) M	**A.K.Agarwal**
Agarwal, D.K. (1945–) M	**D.K.Agarwal**
Agarwal, G.P. (fl. 1960) M	**G.P.Agarwal**
Agarwal, Manjoo Rani (fl. 1968) BM	**M.R.Agarwal**
Agarwal, S.C. (fl. 1969) M	**S.C.Agarwal**
(Agarwal, Sunita)	
see Agrawal, Sunita	**S.Agrawal**
Agassiz, Jean Louis Rodolphe (1807–1873) AS	**Agassiz**
Agelopoulos, Johann A	**Agelop.**
Agerer, Reinhard (1947–) MS	**Agerer**
(Agerer-Kirchhoff, Christina Hedwig)	
see Kirchhoff, Christina Hedwig	**Kirchhoff**
Aggeenko, Wladimir Naumowitsch (1860–1907) S	**Aggeenko**
Aggéry, Anna Berthe Emma (1892–) M	**Aggéry**
Aggéry, N. (fl. 1935) M	**N.Aggéry**
(Aggéyenko, Vladimir Naumovitch)	
see Aggeenko, Wladimir Naumowitsch	**Aggeenko**
(Agguéenko, Vladimir Naumovitch)	
see Aggeenko, Wladimir Naumowitsch	**Aggeenko**
Agnew, Andrew David Quentin (1929–) PS	**Agnew**
Agnihothrudu, V. (fl. 1956) M	**Agnihothr.**
Agnihotri, V.P. (fl. 1967) M	**Agnihotri**
Agosti, Guiseppe (1715–1785) S	**Agosti**
Agostini, Angela (fl. 1926) MS	**A.Agostini**
Agostini, Getulio (1943–) S	**G.Agostini**
(Agrasar, Sulma (Zulma) E.Rúgolo)	
see Rúgolo de Agrasar, Sulma (Zulma) E.	**Rúgolo**
Agrawal, P.D. (fl. 1972) M	**P.D.Agrawal**
(Agrawal, S.C.)	
see Agarwal, S.C.	**S.C.Agarwal**
Agrawal, Sunita (later Garg, S.) (fl. 1982) S	**S.Agrawal**
Agsteribbe, Etienne (1901–1964) B	**Agst.**
Aguiar, Izonete de Jesus Araújo (1947–) M	**I.A.Aguiar**
Aguiar, Joaquim Macedo de (1854—1882) S	**Aguiar**
Aguilella, Antoni (1957–) S	**Aguil.**
(Aguirre Ceballos, Jaime)	
see Aguirre, Jaime	**J.Aguirre**
Aguirre Garcia, Maria Begoña (fl. 1990) M	**M.B.Aguirre**
Aguirre, Jaime (1951–) B	**J.Aguirre**
Aguirre León, Ernesto (1951–) S	**E.Aguirre**
(Aguirre-Hudson, Maria Begoña)	
see Aguirre García, Maria Begõna	**M.B.Aguirre**
Aguirre-Olavarrieta, Ignacio (1955–) S	**Aguirre-Olav.**
Ahearn, Donald G. (fl. 1970) M	**Ahearn**
Ahlburg, Hermann (–1878) MS	**Ahlb.**
Ahles, Harry E. (1924–1981) S	**H.E.Ahles**
Ahles, Wilhelm Elias von (1829–1900) S	**Ahles**
Ahlfvengren, Frederik Elias (1862–1921) S	**Ahlfv.**

Ahlner, Klas (1845–1908/1932) A	**K.Ahlner**
Ahlner, Sten Gustav Edverd (1905–) M	**Ahlner**
Ahlquist, Abraham (1794–1844) S	**Ahlq.**
Ahlstrom, Elbert Halvor (1910–) A	**Ahlstrom**
Ahmad, A. (fl. 1971) A	**A.Ahmad**
Ahmad, Javed (fl. 1967) M	**J.Ahmad**
Ahmad, Nasim (fl. 1987) M	**N.Ahmad**
Ahmad, Shamsuddin (fl. 1942) B	**Ahmad**
Ahmad, Sultan (fl. 1951) MS	**S.Ahmad**
Ahmadjian, Vernon (1930–) AM	**Ahmadjian**
Ahmed Kunju, T.U. (fl. 1987) S	**Ahmed Kunju**
Ahmed, Mokter (fl. 1989) S	**M.Ahmed**
Ahmed, S.Iktikhar (fl. 1969) M	**S.I.Ahmed**
Ahmed, Z.U. (fl. 1970) M	**Z.U.Ahmed**
Ahmedunnisa (no forenames) (fl. 1971) M	**Ahmedunn.**
Ahnfelt, Nicolaus ('Nils') Otto (1801–1837) B	**Ahnf.**
Aho, P.E. (fl. 1979) M	**Aho**
Ahokas, Hannu (1947–) S	**Ahokas**
Ahrendt, Leslie Walter Allen (1903–1969) S	**Ahrendt**
(Ahti, Lena Hämet)	
see Hämet-Ahti, Raija–Lena	**Hämet-Ahti**
Ahti, Teuvo Tapio (1934–) MS	**Ahti**
Ahuja, S.C. (fl. 1991) M	**Ahuja**
Ahumada, Luisa Zulema (1930–) S	**Ahumada**
Ai, Tie Min (1946–) S	**Ai**
Aichele, Dietmar Ernst (1928–) S	**Aichele**
Aiello, A. (fl. 1979) S	**Aiello**
Aigret, Louis Clément Joseph (1856–1921) BCM	**Aigret**
Aiken, M. A	**Aiken**
Aikman, A. (fl. 1794) S	**Aikman**
Ainslie, Whitelaw (1767–1837) S	**Ainslie**
Ainsworth, Geoffrey Clough (1905–) M	**Ainsw.**
Airoldi, Marco (1900–1937) A	**Airoldi**
Airy Shaw, Herbert Kenneth (1902–1985) S	**Airy Shaw**
Aitchison, James Edward Tierney (1836–1898) S	**Aitch.**
Aitken, John B. (–1939) S	**Aitken**
Aiton, William (1731–1793) PS	**Aiton**
Aiton, William Townsend (1766–1849) S	**W.T.Aiton**
Aiyengar, N.K.N. (fl. 1952) F	**Aiyengar**
Aizpuru Oiarbide, Iñaki (1956–) PS	**Aizpuru**
Ajello, Libero (1916–) M	**Ajello**
Ajita Sen (fl. 1977) S	**Ajita Sen**
Ajrekar, S.L. (fl. 1938) M	**Ajrekar**
Akagi, S. (fl. 1929) M	**Akagi**
Akahori, Akira (fl. 1963) S	**Akahori**
Akasawa, Yoshyuki (1915–) PS	**Akasawa**
Akatsuka, Isamu A	**Akatsuka**
Aké Assi, Laurent (1931–) S	**Aké Assi**
Akechi, Koichiro (fl. 1965) M	**Akechi**

Aken, Mark E.   A   **Aken**
Åkermark, Sophia (1817–1882) C   **Åkermark**
Akeroyd, John Robert (1952–   ) S   **Akeroyd**
Akers, John F. (1906–   ) S   **Akers**
Akhani, Kh. (fl. 1989) S   **Akhani**
Akhatar, P.A. (fl. 1979) M   **Akhatar**
Akhtar, Kazmi Sayed Ali (1899–   ) S   **Akhtar**
(Akhundov, T.M.)
    see Achundov, T.M.   **T.M.Achundov**
(Akhverdov, Agazi Asaturovich)
    see Achverdov, Agazi Asaturovich   **Achv.**
Akiba, Fumio   A   **Akiba**
Akinfiev, Ivan Jakovlevic (fl. 1880) S   **Akinf.**
(Akinfiew, Jwan Jakovlewitsch)
    see Akinfiev, Ivan Jakovlevic   **Akinf.**
(Akinfiyev, Ivan Yakovlevic)
    see Akinfiev, Ivan Jakovlevic   **Akinf.**
Akioka, Hidetsugu (1937–   ) A   **Akioka**
Akiyama, Hiroyuki (1956–   ) B   **H.Akiyama**
Akiyama, Masaru (1928–   ) A   **M.Akiyama**
Akiyama, Shigeo (1906–   ) S   **Akiyama**
Akkermans, R.W.A.P. (fl. 1982) S   **Akkermans**
Aksel, M.J. (fl. 1953) M   **Aksel**
Akselman, Rut   A   **Akselman**
Aksenov, E.S. (fl. 1972) S   **Aksenov**
Akulova, Z.V. (1952–   ) S   **Akulova**
Al Musawi, A.H.E. (fl. 1986) S   **Al Musawi**
Al–Achmed, M.A. (fl. 1977) M   **Al–Achmed**
Al–Bader, Salah M. (fl. 1980s) M   **Al–Bader**
Al–Bermani, A.K.K.A. (fl. 1986) S   **Al–Bermani**
Al–Eisawi, D.M. (fl. 1982) S   **Al–Eisawi**
Al–Helfi, M.A. (fl. 1981) M   **Al–Helfi**
(al–'Izzi, Husham Ayub)
    see Alizzi, Husham Ayub   **Alizzi**
Al–Khakani, M.K. (fl. 1983) S   **Al–Khakani**
Al–Khayat, Abdul–Hussain M.A. (1932–   ) S   **Al–Khayat**
Al–Musallam, A. (fl. 1989) M   **Al–Musallam**
(al–Rawi, Ali)
    see Al–Rawi, Ali   **Al–Rawi**
Al–Rawi, Ali (fl. 1966) S   **Al–Rawi**
Al–Sarraf, M.A.A. (fl. 1981) S   **Al–Sarraf**
Al–Shehbaz, Ihsan Ali (1939–   ) S   **Al–Shehbaz**
Al–Sohaily, Ibrahim A. (fl. 1963) M   **Al–Sohaily**
Al–Taey, R.A. (fl. 1984) S   **Al–Taey**
Alain (Brother) (1916–   ) PS   **Alain**
Alam, M.K. (fl. 1985) S   **Alam**
Alasoadura, S.O. (fl. 1968) M   **Alas.**
Alava, Reino Olavi (1915–   ) S   **Alava**
Alavi, S.A. (1934–   ) S   **Alavi**
Albequerque, I.L.d' (fl. 1959) M   **Albeq.**

Albers, F. (fl. 1979) S	**F.Albers**
Albers, M.A. (fl. 1986) M	**M.A.Albers**
Albert, Abel (1836–1909) PS	**Albert**
Albert, L. (fl. 1986) M	**L.Albert**
Albert, Nairn R. A	**N.R.Albert**
Alberti, Antonio ('Florentino Rial') (1785–1861) MS	**Alberti**
(Alberti, Florentino Rial)	
see Alberti, Antonio	**Alberti**
Alberti, Gerhard AF	**G.Alberti**
Albertini, Johannes Baptista von (1769–1831) MS	**Alb.**
Albertson, Nils (1909–1956) B	**Alberts.**
(Albkron Benz, Robert von)	
see Benz, Robert von Albkron	**Benz**
Albo, Giacomo (fl. 1905–1953) S	**Albo**
(Alboff, Nikolai Michailovich)	
see Albov, Nikolai Michailovich	**Albov**
Albov, Nikolai Michailovich (1866–1897) PS	**Albov**
(Albow, Nikolai Michailovich)	
see Albov, Nikolai Michailovich	**Albov**
Albrecht, David Edward (1962–) S	**Albr.**
Albrecht Rohner, J.Huldrich (1900–1971) B	**Albr.Rohn.**
Albuquerque, Byron Wilson Pereira da (1932–) S	**Albuq.**
Albuquerque, Fernando Carneiro de (1932–) M	**F.C.Albuq.**
(Alcala, Juan Varo)	
see Varo Alcala, Juan	**Varo**
Alcalde, Maria Bausá (fl. 1945) M	**Alcalde**
Alcaraz Ariza, Francisco (1958–) S	**Alcaraz**
Alcocer (fl. 1911) S	**Alcocer**
Alcock, Charles Raymond (1921–) M	**Alcock**
Alcorn, John Leonard (1937–) M	**Alcorn**
Aldave, Augusto (1931–) A	**Aldave**
(Aldave Pajares, Augusto)	
see Aldave, Augusto	**Aldave**
Aldén, Björn G. (1948–) S	**Aldén**
Alderman, D.J. (fl. 1971) M	**Alderman**
Alderwerelt van Rosenburgh, Cornelis Rugier Willem Karel van (1863–1936) PS	**Alderw.**
Aldrovandi, Ulisse (1522–1605) L	**Aldrovandi**
Alé-Agha, Nosratollah (fl. 1971) M	**Alé-Agha**
Alechin, Vasiliǐ Vasilievich (1884–1946) S	**Alechin**
Alecrim, Ivan C. (fl. 1955) M	**Alecrim**
Aleem, Anwar Abdel (fl. 1953) AM	**Aleem**
Alefeld, Friedrich Georg Christoph (1820–1872) S	**Alef.**
Alejandre, Juan Antonio (1947–) S	**Alejandre**
(Alejandre Sáenz, Juan Antonio)	
see Alejandre, Juan Antonio	**Alejandre**
(Alekseenko, N.)	
see Alexeenko, N.	**N.Alex.**
(Alekseev, A.G.)	
see Alexeev, A.G.	**A.G.Alexeev**

(Alekseev, A.M.?)	
see Alexeev, A.M.?	**A.M.Alexeev**
(Alekseev, E.B.)	
see Alexeev, E.B.	**E.B.Alexeev**
(Aleksenko, M.A.)	
see Alexeenko, M.A.	**Alex.**
Alessio, Carlo Luciano (fl. 1975) M	**Alessio**
Alexander, Arthur (fl. 1928) M	**A.Alexander**
Alexander, Edward Johnston (1901–1985) S	**Alexander**
Alexander, James Crinan Murray (1944–) S	**C.Alexander**
Alexandrescu, Liviu (1899–1980) S	**Alexandr.**
Alexandri, A.V. (fl. 1932) M	**Alexandri**
Alexandrov, L.P. (1958–1922) S	**Alexandrov**
Alexeenko, F.N. S	**F.N.Alex.**
Alexeenko, M.A. (1861–1919) A	**Alex.**
Alexeenko, M.I. (1905–) S	**M.I.Alex.**
Alexeenko, N. A	**N.Alex.**
Alexeev, A. (fl. 1911) M	**Alexeev**
Alexeev, A.G. A	**A.G.Alexeev**
Alexeev, A.M.? (1896–) S	**A.M.Alexeev**
Alexeev, E.B. (1946–1976) S	**E.B.Alexeev**
(Alexejenko, F.N.)	
see Alexeenko, F.N.	**F.N.Alex.**
(Alexejenko, M.I.)	
see Alexeenko, M.I.	**M.I.Alex.**
(Alexenko, M.A.)	
see Alexeenko, M.A.	**Alex.**
(Alexieff, A.)	
see Alexeev, A.	**Alexeev**
(Alexiev, A.)	
see Alexeev, A.	**Alexeev**
Alexopoulos, Constantine John (1907–1986) AM	**Alexop.**
Alfenas, A.C. (fl. 1987) M	**Alfenas**
Alfieri, S.A. (fl. 1970) M	**Alfieri**
Alfinito, Silvia (1951–) A	**Alfinito**
Alfonseca, J.D. (fl. 1931) M	**Alfons.**
Ali, S.J. (fl. 1991) S	**S.J.Ali**
Ali, Syed Irtifaq (1930–) S	**Ali**
Ali-Shtayeh, M.S. (fl. 1985) M	**Ali-Shtayeh**
Ali-Zade, A.V. (fl. 1978) S	**Ali-Zade**
Aliev, D.A. (1913–) S	**Aliev**
Alieva, A.A. S	**Alieva**
Alippi, H.E. (fl. 1960) M	**Alippi**
Alişan, C. (fl. 1989) M	**Alişan**
Alissova-Klobukova, Eugenija Nikolaevna (1889–1962) PS	**Aliss.**
Alizzi, Husham Ayub S	**Alizzi**
Allam, Bensayah A	**Allam**
Allamand, Frédérique (Frédéric) Louis (1735–1803) S	**F.Allam.**
Allamand, Jean Nicolas Sébastien (1731–1793) S	**J.Allam.**
Allan, Harry Howard Barton (1882–1957) PS	**Allan**

Allander, Helge (1894–) S	**Allander**
Allard, S.T. (fl. 1986) S	**Allard**
Allegre, Charles Frederick (1911–) A	**Allegre**
Alleizette, Aymar Charles d' (1884–1967) S	**Alleiz.**
Allem, Antonio Costa (1949–) S	**Allem**
Allemann, Franz A	**Allemann**
Allemão e Cysneiro, Francisco Freire (1797–1874) S	**Allemão**
Allemão, Manoel (-1863) S	**M.Allemão**
Allen, Betty Eleanor Gosset Molesworth (1913–) P	**B.M.Allen**
Allen, Bruce H. (1952–) B	**B.H.Allen**
Allen, Caroline Kathryn (1904–1975) S	**C.K.Allen**
Allen, David Elliston (1932–) S	**D.E.Allen**
Allen, Geraldine A. (1950–) S	**G.A.Allen**
Allen, Guy Oldfield (1883–1963) A	**G.O.Allen**
Allen, J.G. (fl. 1991) M	**J.G.Allen**
Allen, James (c.1830–1906) S	**J.Allen**
Allen, John Alphaeus (1863–1916) B	**J.A.Allen**
Allen, John Fisk (1807–1876) S	**J.F.Allen**
Allen, Mary Belle (1922–1973) A	**M.B.Allen**
Allen, Paul Hamilton (1911–1963) S	**P.H.Allen**
Allen, Thomas Cort (1899–) M	**T.C.Allen**
Allen, Timothy Field (1837–1902) A	**Allen**
Allender, Bruce M. A	**Allender**
Allescher, Andreas (1828–1903) FMS	**Allesch.**
Allington, William B. (1912–) M	**Allington**
Allioni, Carlo (1728–1804) PS	**All.**
Allison, Kenneth Willway (1894–1976) B	**Allison**
Allison, Patricia (fl. 1969) M	**P.Allison**
Allman, George James (1812–1898) AS	**G.J.Allman**
Allman, William (1776–1846) S	**Allman**
Allorge, Pierre (1891–1944) AS	**P.Allorge**
Allorge, Valia (Valentine) Selitsky (1888–1977) AB	**V.Allorge**
Allred, Kelly W. (1949–) S	**Allred**
Alm, Carl Gustav (1888–) MS	**Alm**
Almagia, Roberto (fl. 1930s) S	**Almagia**
Almborn, Ove (1914–) M	**Almb.**
Almeda, Frank (1946–) S	**Almeda**
Almeida, Floriano Paulo de (fl. 1933) M	**F.P.Almeida**
Almeida, J.F.R.d' (1891–1949) PS	**J.F.R.Almeida**
Almeida, Joaquin de Almeida Pinto (1870–) S	**Almeida**
Almeida, José Veríssimo (1834–1915) M	**J.V.Almeida**
Almeida, M.R. (fl. 1977–1986) PS	**M.R.Almeida**
Almeida, Maria Teresa de (1940–) PS	**M.T.Almeida**
Almeida, Rogério T. (fl. 1991) M	**R.T.Almeida**
Almeida, Sarah M. (1940–) PS	**S.M.Almeida**
Almgren, Knut A	**Almgren**
Almquist, Erik Gustaf (1892–1974) S	**E.G.Almq.**
Almquist, Ernst Bernhard (1852–1946) S	**E.B.Almq.**
Almquist, Johan Magnus (fl. 1920s) S	**J.M.Almq.**
Almquist, Sigfried Osker Immanuel (1844–1923) MS	**Almq.**

Aloj, B. (fl. 1987) M	**Aloj**
Alonso Paz, Eduardo (1954–) S	**Alonso Paz**
Alpers, Ferdinand (1841–1912) S	**Alpers**
Alphand, Jean Charles Adolphe (1817–1891) S	**Alphand**
Alpinar, Kerim (1954–) S	**Alpinar**
(Alpini, Prospero)	
see Alpinio, Prospero	**Alpinio**
Alpinio, Prospero (1553–1617) L	**Alpinio**
Alschinger, Andreas (1791–1864) S	**Alsch.**
Alsina Aser, Mercedes (fl. 1987) S	**Alsina**
Alsop, R. (fl. 1983) S	**Alsop**
Alston, Arthur Hugh Garfit (1902–1958) PS	**Alston**
Alston, M.Meyer (fl. 1981–89) S	**M.M.Alston**
(Alstroemer, Clas)	
see Alströmer, Clas	**Alstr.**
Alströmer, Clas (1736–1794) S	**Alstr.**
Alstrup, Vagn (fl. 1986) M	**Alstrup**
Alt, Karen S. (later Grant, K.A.) (1936–) PS	**K.S.Alt**
Altamirano, Fernando S	**Altam.**
(Altamirano y Rose, Fernando)	
see Altamirano, Fernando	**Altam.**
Altamura, Luciano (1950–) S	**Altamura**
Alten, Hermann von A	**H.Alten**
Alten, Johann Wilhelm von (1770–) S	**J.Alten**
Alteraş, Ion (fl. 1959) M	**Alteraş**
Alth, Alojzy (Alois) (1819–1886) A	**Alth**
Altmann, Paul (fl. 1894) S	**Altmann**
Altschul, Siri Sylvia Patricia von Reis (1931–) S	**Altschul**
Altstatt, George E. (1905–) M	**Altstatt**
Álvarez de Zayas, Alberto (fl. 1981) S	**A.Álvarez**
Álvarez García, Luis A. (1903–) M	**Álv.García**
Álvarez Martínez, M.J. (fl. 1986) S	**Álv.Mart.**
Alveal, Krisler A	**Alveal**
(Alveal V., Krisler)	
see Alveal, Krisler	**Alveal**
Alverson, Andrew Halstead (1845–1916) S	**Alverson**
Alverson, Edward R. (fl. 1989) P	**E.R.Alverson**
Alverson, William S. (fl. 1981) S	**W.S.Alverson**
Alvik, Gunnar A	**Alvik**
Alvin, K.L. (1927–) M	**Alvin**
Alwan, A.-R.A. (fl. 1985) S	**Alwan**
Alyon, Pierre Phillipe (1746–1816) S	**Alyon**
Alziar, Gabriel (1948–) S	**Alziar**
Amadori, L. (fl. 1927) M	**Amadori**
Amakawa, Tairoko (1917–) B	**Amakawa**
Amann, Jean Jules (1859–1939) B	**J.J.Amann**
Amann, Johann (see also Kurz, W.S.) (1814–1878) P	**Amann**
Amano, N. (fl. 1979) M	**N.Amano**

Amano, Tetsuo (fl. 1967) P	**T.Amano**
Amano, Y. (fl. 1975) M	**Y.Amano**
(Amans, Jean Florimond Boudon de Saint)	
see Saint Amans, Jean Florimond Boudon de	**St.Amans**
Amaral, Ayrton (1948–) S	**Amaral**
(Amaral Franco, João Manuel Antonio)	
see Franco, João Manuel Antonio	**Franco**
Amaral, Maria C.E. (fl. 1991) S	**M.C.E.Amaral**
(Amatao, R.F.d')	
see D'Amatao, R.F.	**D'Amatao**
(Amato, Giovanni Frederico d')	
see D'Amato, Giovanni Frederico	**D'Amato**
Amberg, Otto (1875–1920) A	**Amberg**
Ambronn, Hermann (1856–1927) A	**Ambronn**
Ambrosi, Francesco (1821–1897) S	**Ambrosi**
Ambrosioni, P. (fl. 1938) M	**Ambros.**
Ambroz, Josef (1886–1950) B	**Ambroz**
Ambwani, K. (fl. 1971) FM	**Ambwani**
(Amechin, O.A.)	
see Amekhin, O.A.	**Amekhin**
Amekhin, O.A. (fl. 1988) S	**Amekhin**
(Amelchenko, V.P.)	
see Ameljczenko, V.P.	**Ameljcz.**
Ameljczenko, V.P. (1948–) S	**Ameljcz.**
Amelung, R.M. (fl. 1971) M	**Amelung**
Ames, Adeline (1879–19??) M	**A.Ames**
Ames, Lawrence Marion (1900–1966) MS	**L.M.Ames**
Ames, Oakes (1874–1950) S	**Ames**
Ames, Robert N. (fl. 1976) M	**R.N.Ames**
Amich García, Francisco (1953–) S	**Amich**
Amici, Giovanni Battista (1786–1863) A	**Amici**
Amin, Amal (1929–) S	**Amin**
Amin, K.S. (fl. 1978) M	**K.S.Amin**
(Amirchanov, A.M.)	
see Amirkhanov, A.M.	**Amirkh.**
Amirkhanov, A.M. S	**Amirkh.**
Amman, Johann (1707–1741) S	**Amman**
Ammirati, Joseph F. (1942–) M	**Ammirati**
Ammon, H.U. (fl. 1963) M	**Ammon**
Amo, N. (fl. 1983) M	**N.Amo**
Amo y Mora, Mariano del (1809–1896) MS	**Amo**
Amon, J.P. (fl. 1984) M	**Amon**
Amos, Raymond E. (fl. 1966) M	**Amos**
Amossé, Auguste A	**Amossé**
Amshoff, Gerda Jane Hillegonda (1913–1985) S	**Amshoff**
Amsler, Charles D. (1959–) A	**Amsler**
Amstutz, Erika (fl. 1957) PS	**Amstutz**
An, Cheng Hsi (fl. 1970s) S	**C.H.An**
An, Qing Fu (fl. 1986) S	**Q.F.An**

(An, Zhang Xi)
  see An, Cheng Hsi   **C.H.An**
(Ana Magán, F.J.)
  see Fernández de Ana Magan, F.J.   **Fern.Magán**
(Anachin, J.K.)
  see Anakhin, J.K.   **Anakhin**
Anagnostakis, Sandra L. (1939–) M   **Anagnost.**
Anagnostidis, Konstantinos Th. (1924–) A   **Anagn.**
Anahosur, K.H. (fl. 1968) M   **Anahosur**
Anakhin, J.K.   A   **Anakhin**
Anand, G.P.S. (fl. 1956) M   **Anand**
Anand Kumar (c.1955–) S   **Anand Kumar**
Anand, Pyare Lal   A   **P.L.Anand**
Anand–Prakash   A   **Anand–Prak.**
Ananev, A.R. (1911–) AF   **Ananev**
(Ananiev, A.R.)
  see Ananev, A.R.   **Ananev**
(Ananjyev, A.R.)
  see Ananev, A.R.   **Ananev**
Ananthanarayanan, S. (fl. 1962) M   **Ananthan.**
Ananthapadmanaban, D. (fl. 1988) M   **Ananthap.**
Anashchenko, A.V. (fl. 1970) S   **Anashch.**
(Anashczenko, A.V.)
  see Anashchenko, A.V.   **Anashch.**
Anastasia, Giuseppe Emilio (1870–1934) M   **Anastasia**
Anastasiou, C.J. (fl. 1962) M   **Anastasiou**
Ančev, M.E. (fl. 1982) S   **Ančev**
Anchabadze, T. (fl. 1963) M   **Anchab.**
(Anchev, M.E.)
  see Ančev. M.E.   **Ančev**
Ancona, Ignacio   A   **Ancona**
Andary, Claude (fl. 1983) M   **Andary**
Anderberg, Arne A. (1954–) S   **Anderb.**
Anders, Joseph (1863–1936) M   **Anders**
Andersen, Johannes Carl (1873–1962) S   **Andersen**
Andersen, Robert A.   A   **R.A.Andersen**
Anderson, Alexander (–1811) S   **A.Anderson**
Anderson, Anthony Benett (1950–) S   **A.B.Anderson**
Anderson, Charles Lewis (1827–1919) A   **C.L.Anderson**
Anderson, Christiane Eva (née Seidenschnur, C.E.) (1944–) S   **C.E.Anderson**
Anderson, Daryl J.   A   **D.J.Anderson**
Anderson, Denis Elmo (1934–   ) S   **D.E.Anderson**
Anderson, Edgar Shannon (1897–1969) S   **E.S.Anderson**
Anderson, Edward Frederick (1932–) S   **E.F.Anderson**
Anderson, Frederick William (1866–1891) AM   **F.W.Anderson**
Anderson, George W. (fl. 1800–1817) S   **G.Anderson**
Anderson, Gregory Joseph (1944–) S   **G.J.Anderson**
Anderson, Harvey Warren (1885–1971) MS   **H.W.Anderson**
Anderson, J.F. (fl. 1979) M   **J.F.Anderson**
Anderson, J.L. (fl. 1987) S   **J.L.Anderson**

Anderson, Jacob Peter (1874–1953) PS	J.P.Anderson
Anderson, James (1739–1809) S	Anderson
Anderson, James (fl. 1868) S	J.Anderson
Anderson, John Graham (1926–1970) S	J.G.Anderson
Anderson, Lewis Edward (1912–) B	L.E.Anderson
Anderson, Loran Crittendon (1936–) S	L.C.Anderson
Anderson, Lucia F. (fl. 1948) M	Lu.E.Anderson
Anderson, Paul Johnson (1884–1971) MS	P.J.Anderson
Anderson, R.Y. S	R.Y.Anderson
Anderson, Redvers B. (fl. 1924) M	R.B.Anderson
Anderson, Robert Henry (1899–1969) S	R.H.Anderson
Anderson, Robert James (1952–) A	R.J.Anderson
Anderson, Roger Arthur (1935–) M	R.A.Anderson
Anderson, T.F. (fl. 1969) M	T.F.Anderson
Anderson, Thomas (1832–1870) S	T.Anderson
Anderson, Trante-Heidi (fl. 1976) M	T.-H.Anderson
Anderson, William (1750–1778) S	W.Anderson
Anderson, William Russell (1942–) PS	W.R.Anderson
Andersson, Bengt Lennart (1948–) S	I..Andersson
Andersson, Carl Filip Gunnar (1865–1928) BF	C.Andersson
Andersson, Olof (fl. 1989) M	O.Andersson
Andersson, Gustaf Oskar (later Malme, G.O.A.(N.)) (1864–1967) M	G.O.Andersson
Andersson, I.A. (1959–) S	I.A.Andersson
Andersson, Nils Johan (1821–1880) PS	Andersson
Andersson, Oskar Fredrik (1862–1938) A	O.F.Andersson
Andianova, T.V. (fl. 1989) M	Andianova
Ando, Hisatsugu (1922–) B	Ando
Ando, Katsuhiko (fl. 1984) M	K.Ando
Ando, Kazuo A	Kaz.Ando
(Andrä, Carl Justus)	
see Andrae, Carl Justus	Andrae
Andrade, Aydil Grave de (1930–) S	A.G.Andrade
(Andrade Chiappeta, Alda de)	
see de Andrade Chiappeta	de Andrade
(Andrade, Edmundo Navarro de)	
see Navarro, Edmundo	Navarro
Andrade, José Cardoso de (fl. 1981) S	J.C.Andrade
Andrade, M.H.de A. A	M.H.A.Andrade
Andrade-Lima, Artura Dardano de (1919–1981) PS	Andrade-Lima
Andrae, Carl Justus (1816–1885) F	Andrae
Andrásovszky, Jószef (Josef) (1889–1943) PS	András.
(Andrászovszky, Jószef (Josef))	
see Andrásovszky, Jószef (Josef)	András.
André, Éduard-François (1840–1911) PS	André
André, L. (fl. 1955) M	L.André
André, L.E. S	L.E.André
Andréanszky, Gábor (Gabriel) (1895–1967) FS	Andr.
Andreas, Charlotte Henriette (1898–) S	Andreas
Andreata, Regina Helena Potsch (fl. 1981) S	Andreata
Andreetta, A. (fl. 1978) S	Andreetta

(Andreev, Vladimir Nikolaevich)	
see Andrejev, Vladimir Nikolaevič	**Andrejev**
(Andreew, Wladimir Nikolaewitsch)	
see Andrejev, Vladimir Nikolaevič	**Andrejev**
Andrejev, Vladimir Nikolaevič (1907–1987) S	**Andrejev**
Andrejeva, E.I. (fl. 1979) M	**E.I.Andrejeva**
Andrejeva, R.V. (fl. 1979) M	**R.V.Andrejeva**
Andreoli, C. A	**Andreoli**
Andres, Heinrich (1883–1970) CMS	**Andres**
Andrés, Jaime (1937–) S	**J.Andrés**
(Andrés y Tubilla, Tomás)	
see Tubilla, Tomás Andrés y	**Tubilla**
Andresen, John William (1925–) S	**Andresen**
Andreucci, A. (fl. 1926) M	**Andreucci**
Andrews, Albert LeRoy (1878–1961) B	**A.L.Andrews**
Andrews, Cecil Rollo Payton (1870–1951) S	**C.R.P.Andrews**
Andrews, Charles William (1866–1924) S	**C.W.Andrews**
Andrews, Darwin Maxson (1869–1938) S	**D.M.Andrews**
Andrews, Ebenezer Baldwin (1821–1880) F	**E.B.Andrews**
Andrews, Frederick William (–1961) S	**F.W.Andrews**
Andrews, George W. (1929–) A	**G.W.Andrews**
Andrews, Henry Charles (1794–1830) S	**Andrews**
Andrews, Henry Nathaniel (1910–) F	**H.N.Andrews**
Andrews, S.Brian (fl. 1977) P	**S.B.Andrews**
Andrews, Susyn M. (1953–) S	**S.Andrews**
Andrews, Theodore Francis (1917–) A	**T.F.Andrews**
(Andreyev, Vladimir Nikolaevič)	
see Andrejev, Vladimir Nikolaevič	**Andrejev**
Andriamampandry, A.V. A	**Andriam.**
Andrianova, Tatjana V.(B.) (1960–) M	**Andrian.**
Andrieu, Bernard A	**Andrieu**
Andronov, Nikolaï Matveevich (fl. 1955) S	**Andronov**
(Androsov, N.V.)	
see Androssov, N.V.	**Androssov**
Androssov, N.V. (1870–1941) S	**Androssov**
Androssova, E.Ja. A	**Androssova**
(Androssow, N.V.)	
see Androssov, N.V.	**Androssov**
Andrus, Charles Frederick (1906–) M	**Andrus**
Andrus, Richard E. (1941–) B	**R.E.Andrus**
Andrusov, Nikolaj (Nicolaus) Ivanovich (1861–) A	**Andrusov**
(Andrussow, Nikolaj (Nicolaus) Ivanovich)	
see Andrusov, Nikolaj (Nicolaus) Ivanovich	**Andrusov**
(Andrzeiovski, Antoni Lukianowicz)	
see Andrzejowski, Antoni Lukianowicz	**Andrz.**
(Andrzeiowski, Antoni Lukianowicz)	
see Andrzejowski, Antoni Lukianowicz	**Andrz.**
(Andrzeiowsky, Antoni Lukianowicz)	
see Andrzejowski, Antoni Lukianowicz	**Andrz.**

Andrzejowski, Antoni Lukianowicz (1785–1868) S	**Andrz.**
Anfalov, P.A. (fl. 1943) S	**Anfalov**
Ang, Put (1955–) A	**Ang**
(Angeli, João Alberto)	
see Angely, João Alberto	**Angely**
Angelis, Moritz (1805–1894) S	**Angelis**
Angelis, P. (fl. 1844) S	**P.Angelis**
Angely, João Alberto (1917–) S	**Angely**
Anghel, Gheorghe (1910–) S	**Anghel**
Angst, Ernest C. (–1930) A	**Angst**
Ångström, Johan (1813–1879) BP	**Ångstr.**
Anguli, V.C. (fl. 1965) M	**Anguli**
Anikster, J. (fl. 1967) M	**J.Anikster**
Anikster, Y. (fl. 1987) M	**Y.Anikster**
Anisimova, N.W. A	**Anisimova**
Anisomova, O.I. (fl. 1973) F	**Anisomova**
(Anissimowa, N.W.)	
see Anisimova, N.W.	**Anisimova**
Anjum, Ghazala A	**Anjum**
Ann, P.J. (fl. 1980) M	**Ann**
Annable, C.R. (fl. 1988) S	**Annable**
(Annaliev, S.A.)	
see Annalijev, S.A.	**Annal.**
Annalijev, S.A. (fl. 1960) M	**Annal.**
Annenkov, Nicolas Ivanovich (1819–1889) C	**Annenkov**
Añon, Delia C.Suarez de Cullen (1917–) S	**Añon**
Ansari, A.A. (fl. 1985) S	**A.A.Ansari**
Ansari, M.A. (fl. 1972) S	**M.A.Ansari**
Ansari, M.Y. (1929–) S	**Ansari**
Ansari, Muhammad Atiqur Rahman A	**M.A.R.Ansari**
Ansari, R. (1948–) S	**R.Ansari**
Ansberque, E. (fl. 1866–1873) S	**Ansb.**
Ansel, M. (fl. 1971) M	**Ansel**
Anselmino, Elizabeth (fl. 1932) S	**Anselmino**
Ansorge, Car. (1849–1915) S	**Ansorge**
Antarikanonda, Pongtep A	**Antarik.**
Anters, E. (fl. 1913) F	**Anters**
Anthony, Emilia Crane (–1904) B	**Anthony**
Anthony, John (1891–1972) S	**J.Anthony**
Anthony, Margery Stuart (1924–) S	**M.S.Anthony**
Anthony, Nicola C. (fl. 1982–86) PS	**N.C.Anthony**
Antikajian, Grace (fl. 1947) M	**Antik.**
Antinori, Orazio (1811–1882) S	**Antinori**
Antipova, N. (fl. 1961) M	**Antipova**
Antipova, N.A. A	**N.A.Antipova**
Antipova, N.L. A	**N.L.Antipova**
Antoine, E. (fl. 1917) M	**E.Antoine**
Antoine, Franz (1815–1886) S	**Antoine**
Antokolskaya, M.P. (fl. 1922) M	**Antok.**
Anton, Ana Maria (1942–) S	**Anton**

Antonescu, Emanuel A	**Antonescu**
Antonín, Vladimír (1955–) M	**Antonín**
Antonio O., Rachel (fl. 1991) S	**R.Antonio**
Antonio, T.M. (fl. 1981) S	**Antonio**
Antoniu–Murgoci, Adriana (1909–1987) M	**Ant.-Murg.**
Antonopoulos, Antonios A. (fl. 1976) M	**Antonop.**
Antropov, I.A. A	**I.A.Antropov**
Antropov, Vasiliĭ Ivanovich (1889–1942) S	**Antropov**
Antropova, Varvara Filipovna (1890–) S	**Antropova**
Antsupova, T.P. (1938–) S	**Antsupova**
Antz, Carl Cäsar (1805–1859) S	**Antz**
Anwar, Nadia (1946–) M	**Anwar**
Anway, J.C. (fl. 1969) S	**Anway**
Anzalone, Bruno (1921–) S	**Anzal.**
Anzalone, Louis (fl. 1957) M	**L.Anzal.**
Anzi, Martino (1812–1883) ABCM	**Anzi**
Anzotegin, Luisa Matilde (fl. 1968) F	**Anzotegin**
Ao, Balakrishna (fl. 1974) M	**Ao**
Aoki, F. (fl. 1957) M	**Aoki**
Aoki, Takayuki (fl. 1986) M	**T.Aoki**
Aoshima, Kiyowo (1925–) M	**Aoshima**
Aparicio Martínez, Abelardo (1956–) S	**Aparicio**
Aparina, T. (fl. 1966) S	**Aparina**
Apel, K. A	**Apel**
(Aphanasiev, C.S.)	
see Afanassiev, C.S.	**Afan.**
Apinis, Arvids Eduards (1907–1979) M	**Apinis**
Apolinnaire, Antoine Laurent S	**Apol.**
Apostolides, C.A. (fl. 1952) M	**Apostol.**
Appan, Subramanian G. (1937–) S	**Appan**
Appel, Friedrich Carl Louis Otto (1867–1952) MS	**Appel**
Appel, K.R. (fl. 1980) M	**K.R.Appel**
Appent, Otto (fl. 1973) F	**Appent**
Appenzeller (fl. 1986) S	**Appenz.**
Applegate, Elmer Ivan (1867–1949) S	**Applegate**
Apstein, Carl (1862–1950) AM	**Apstein**
Apt, Kirk E. (1956–) A	**Apt**
Apte, V.V (fl. 1942) AB	**Apte**
Aptekar, Esfir' Markovich (1892–) A	**Aptekar**
Aptroot, Andre (fl. 1988) M	**Aptroot**
Ara, Rownak (fl. 1964) M	**Ara**
Aragaki, Minoru (fl. 1979) M	**Aragaki**
Aragão, Henrique de Beaurepaire A	**Aragão**
Arakawa, Kesatoshi (fl. 1974) S	**Arakawa**
Araki, T. (fl. 1986) M	**T.Araki**
Araki, Yeiichi (1904–1955) S	**Araki**
Aralbaev, N.K. (fl. 1990) S	**Aralbaev**
Arambarri, Angélica M. (1945–) M	**Aramb.**
Aránega Jiménez, Raquel (fl. 1990) S	**Aránega**

(Arango, A.Posada–)	
see Posada–Arango, Andres	Posada–Ar.
Arasaki, Seibin (1912–) A	Arasaki
(Araujo Aguiar, Izonete de Jesus)	
see Araujo, Izonete de Jesus	I.J.Araujo
Araujo, Izonete de Jesus (fl. 1983) M	I.J.Araujo
Araujo, Paulo Agostinho de Matos S	Araujo
(Araujo Schwarz, E.de)	
see Schwartz, Elizabeth de Araujo	E.A.Schwartz
(Arbeláez, Enrique Pérez)	
see Pérez Arbeláez, Enrique	Pérez Arbel.
Arber, Agnes (1879–1960) S	A.Arber
Arber, Edward Alexander Newell (1870–1918) F	E.Arber
Arbo, Maria Mercedes (1945–) S	Arbo
Arcangeli, Giovanni (1840–1921) AMPS	Arcang.
(Arcangelsky, Sergio)	
see Archangelsky, Sergio	S.Archang.
Arce, Gina (1929–) A	Arce
Archambault, D. (fl. 1985) S	Archamb.
Archangelsky, Sergio (fl. 1963–76) AFM	S.Archang.
Archer, Alan W. (1930–) M	A.W.Archer
Archer, Mary Ellinor Lucy (fl. 1917) S	M.E.L.Archer
Archer, R.H. (fl. 1987) S	R.H.Archer
Archer, S.A. (fl. 1973) M	S.A.Archer
Archer, Thomas Croxen (1817–1885) S	Archer
Archer, William (1827–1897) A	W.Archer
Archer, William (1820–1874) B	W.Archer bis
Archer, William Andrew (1894–1973) M	W.A.Archer
Archer–Hind, Thomas H. (fl. 1880) S	Archer–Hind
Archibald, C.G.M. A	C.G.M.Archibald
Archibald, Eily Edith Agnes (1916–) S	Archibald
Archibald, Patricia A. (1934–) A	P.A.Archibald
Archibald, R.E.M. (1940–) A	R.E.M.Archibald
Archibald, R.G. (fl. 1916) M	R.G.Archibald
Arde, Walker Robert (1902–) M	Arde
Ardissone, Francesco (1837–1910) A	Ardiss.
Ardoino, Honoré Jean Baptiste (1819–1874) S	Ardoino
Ardré, Françoise (1931–) A	Ardré
Arduino, Luigi (1759–1834) S	L.Ard.
Arduino, Pietro (1728–1805) S	Ard.
(Arêa Leão, A.E.de)	
see Leão, A.E. de Arêa	Leão
Areces Mallea, L.A.E. (fl. 1976) S	Areces
Arechavaleta, José (1838–1912) S	Arechav.
(Arechavaleta y Balpardo, José)	
see Arechavaleta, José	Arechav.
Areger, F.W. (fl. 1917) M	Areger
Aregood, Carol Croley A	Aregood
Arekal, Govindappa D. (fl. 1972) S	Arekal
Arenberg, Florence F. (fl. 1941) M	Arenb.

Arendholz, Wolf–Rüdiger (fl. 1983) M	**Arendh.**
Arends, Georg (1862–1952) S	**Arends**
Arends, Johan Coenraad (1940–) S	**J.C.Arends**
Arendt, Johann Josef Franz (fl. 1828–1840) S	**Arendt**
Arènes, Jean (1898–1960) S	**Arènes**
Arens, Pedro Martin José (1884–1954) B	**Arens**
Areschoug, Frederic Wilhelm Christian (1830–1908) S	**F.Aresch.**
Areschoug, John Erhard (1811–1887) A	**Aresch.**
Argent, Graham Charles George (1941–) BS	**Argent**
(Argentelle, Louis Marc Antoine Robillard d')	
see Robillard d'Argentelle, Louis Marc Antoine	**Robill.**
Argus, George William (1929–) S	**Argus**
Arias, Ileana (1958–) S	**I.Arias**
Arias y Costa, Antonio Sandalio de (1764–1839) S	**Arias**
Arietti, Nino (Giovanni) (1902–1979) S	**Arietti**
Arillaga de Maffei, Blanca Renée (1917–) S	**B.R.Arill.**
Arillaga, Jaime Guiscafre (fl. 1955) M	**Arill.**
Arima, K. (fl. 1956) M	**Arima**
Aristeguieta, Leandro (1923–) S	**Aristeg.**
Ariza Espinar, Luis (1933–) S	**Ariza**
Ark, Peter Alexander (1899–) M	**Ark**
Arkhangelskij, Andrei Dmetrievich (1879–1940) A	**Arkhang.**
(Armas, Matilde de la Luz de)	
see de la Luz de Armas, Matilde	**de la Luz**
Armbruster, W.Scott (1951–) S	**Armbr.**
Armenise-Porcelli, Vittoria (1926–1979) S	**Arm.-Porc.**
Armitage, Edward (1822–1906) S	**Arm.**
Armitage, Eleanora (1865–1961) B	**E.Arm.**
Armstrong, C.A. (fl. 1934) S	**C.A.Armstr.**
Armstrong, C.W. (–1950) S	**C.W.Armstr.**
Armstrong, George Miller (1893–) M	**G.M.Armstr.**
Armstrong, James Andrew (1950–) S	**J.A.Armstr.**
Armstrong, Joanne K. (fl. 1966) M	**J.K.Armstr.**
Armstrong, John Francis (1820–1902) S	**Armstr.**
Armstrong, Joseph Beattie (1850–1926) PS	**J.B.Armstr.**
Armstrong, Margaret (1867–) S	**M.Armstr.**
Armstrong, P.A. (fl. 1986) M	**P.A.Armstr.**
Armstrong, P.M. (fl. 1983) M	**P.M.Armstr.**
Arnaiz, C. (fl. 1980–) S	**Arnaiz**
(Arnaoudov, Nicolas N.)	
see Arnautov, Nicolas N.	**Arnautov**
Arnaud, Ch. (fl. 1883) S	**C.Arnaud**
Arnaud, Gabriel (1882–1957) M	**G.Arnaud**
Arnaud, Jean André Michel (1760–1831) MS	**Arnaud**
Arnaud, Madeleine A. (fl. 1924) M	**M.A.Arnaud**
Arnaud, P. A	**P.Arnaud**
Arnaudi, C. (fl. 1927) M	**Arnaudi**
(Arnaudow, Nicolas N.)	
see Arnautov, Nicolas N.	**Arnautov**

Arnautov, Nicolas N. (1887–1961) MS   **Arnautov**
Arndt, Charles Homer (1892–) A   **Arndt**
Arnell, Hampus Wilhelm (1848–1932) S   **Arnell**
Arnell, Sigfrid Wilhelm (1895–1970) S   **S.W.Arnell**
Arneson, Ronald D. A   **Arneson**
Arnold, Chester Arthur (fl. 1947–56) F   **C.A.Arnold**
Arnold, Ferdinand Christian Gustav (1829–1901) BM   **Arnold**
Arnold, Günter Rudolph Walter (fl. 1970) M   **G.R.W.Arnold**
Arnold, J.D. S   **J.D.Arnold**
Arnold, Johann Franz Xaver (fl. 1785) S   **J.F.Arnold**
Arnold, Lillian Eleanore (1895–) S   **L.E.Arnold**
Arnold, Ralph Edward (1891–) S   **R.E.Arnold**
Arnold, Ruth Horner (fl. 1961) M   **R.H.Arnold**
Arnold, Trevor Henry (1947–) S   **T.H.Arnold**
Arnold, W. (fl. 1834) S   **W.Arnold**
Arnoldi, Wladimir Mitrofanowitsch (1871–1924) AS   **Arnoldi**
Arnolds, Eef J.M. (fl. 1974) M   **Arnolds**
Arnott, George Arnott Walker (1799–1868) AMPS   **Arn.**
Arnott, Howard J. (1928–) A   **H.J.Arn.**
Arnott, Samuel (1852–1930) S   **S.Arn.**
Arnould, Léon (fl. 1893) M   **Arnould**
Arny, D.C. (fl. 1971) M   **Arny**
Aroche, Regla Maria (fl. 1986) M   **Aroche**
Aronsen, A. (fl. 1989) M   **Aronsen**
Arora, C.M. (fl. 1969) S   **Arora**
(Aroza Montes, Pilar)
  see Aroza, Pilar   **Aroza**
Aroza, Pilar (1951–) S   **Aroza**
Arp, Gerald Kench (1947–) S   **Arp**
Arpin, M. (fl. 1963) M   **Arpin**
Arrábida, D.Francisco Antonio de (1771–1850) S   **Arráb.**
Arrando, Oscar Guillermo (fl. 1974) F   **Arrando**
Arreguín, María da la Luz (1950–) S   **Arreguín**
(Arreguin Sanchez, María de la Luz)
  see Arreguín, María da la Luz   **Arreguín**
Arrhenius, Johan Israel Axel (1858–1950) S   **A.Arrh.**
Arrhenius, Johan Peter (Pehr, Petter) (1811–1889) MS   **Arrh.**
Arrigoni, Pier Virgilio (Virgilis) (1932–) S   **Arrigoni**
(Arrillaga de Maffei, Blanca Renée)
  see Arillaga, Blanca Renée   **B.R.Arill.**
(Arrillaga Oronoz de Maffei, Blanca Renée)
  see Arillaga, Blanca Renée   **B.R.Arill.**
Arrondeau, Étienne Théodore (–1882) S   **Arrond.**
Arroyo, Irene (fl. 1988) M   **I.Arroyo**
Arroyo, Mary Therese Kalin (1944–) S   **Arroyo**
Arruda da Cámara, Manoel (1752–1810) S   **Arruda**
Arsène, Gustave (Gerfroy) (1867–1938) S   **Arsène**
Arshad, M. (fl. 1972) M   **Arshad**
Årsvoll, K. (fl. 1965) M   **Årsvoll**
Artagaveytia–Allende, R.C. (fl. 1970) M   **Artag.–All.**

Artari, Alexander Petrovich (1862–1919) AM	**Artari**
Artault, St. (fl. 1893) M	**Artault**
Artelari, P. (fl. 1984) S	**P.Artelari**
Artelari, Rea (1948–) S	**R.Artelari**
(Artemchuk, Ivan Vlasovich)	
see Artemczuk, Ivan Vlasovich	**Artemczuk**
(Artemchuk, N.J. (Ya.))	
see Artemczuk, N.J. (Ya.)	**N.J.Artemczuk**
Artemczuk, Ivan Vlasovich (1898–1973) S	**Artemczuk**
Artemczuk, N.J. (Ya.) (fl. 1968–) M	**N.J.Artemczuk**
(Artemieff, G.V.)	
see Artemiev, G.V.	**Artemiev**
Artemiev, G.V. (fl. 1935) M	**Artemiev**
(Artemtchuk, N.J. (Ya.))	
see Artemczuk, N.J. (Ya.)	**N.J.Artemczuk**
Arthaud–Berthet, Julio Loão (1875–1941) M	**Arth.-Berth.**
Arthur, John Morris (1893–) M	**J.M.Arthur**
Arthur, Joseph Charles (1850–1942) MPS	**Arthur**
Artis, Edmund Tyrell (1789–1847) F	**Artis**
(Artjuschenko, Z.T.)	
see Artjushenko, Z.T.	**Artjush.**
Artjushenko, Z.T. (1916–) S	**Artjush.**
Artraria, Ferdinando Augusto (1853–1929) B	**Artr.**
Artsikhovskij, Vladimir Martinovich (1876–1931) A	**Artsikh.**
(Artyushenko, Z.T.)	
see Artjushenko, Z.T.	**Artjush.**
Artzner, Darrah G. A	**Artzner**
Aruch, E. (fl. 1895) M	**Aruch**
Aruna (fl. 1979) M	**Aruna**
Arup, Ulf (1959–) M	**Arup**
Aruta, C. (fl. 1974) M	**Aruta**
(Aruta M., C.)	
see Aruta, C.	**Aruta**
Arutyunyan, M.G. (fl. 1987) S	**Arut.**
Arvat, A. (1890–1950) S	**Arvat**
Arvet-Touvet, Jean Maurice Casimir (1841–1913) S	**Arv.-Touv.**
Arvidsson, Lars (1949–) M	**Arv.**
Arvy, Lucie (fl. 1969) M	**Arvy**
Arwidsson, Thorsten (1904–1948) AMS	**Arw.**
Arx, Josef Adolph von (1922–1988) M	**Arx**
Arya, A. (fl. 1981) M	**A.Arya**
Arya, B.S. (fl. 1971) P	**B.S.Arya**
Arya, Harish C. (1925–) MS	**Arya**
Arya, S. (fl. 1990) M	**S.Arya**
Arystangaliev, S.A. (fl. 1961) S	**Aryst.**
(Arystangalijew, S.A.)	
see Arystangaliev, S.A.	**Aryst.**
Arzeni, Charles Basel (1925–) B	**Arzeni**
Arzt, L. (fl. 1934) M	**Arzt**
Asad, Fatima (fl. 1969) M	**Asad**

Asahina, Yasuhiko (Jasuhiko) (1881-1975) M	**Asahina**
Asai, Yasuhiro (1933-) MS	**Asai**
Asama, K. S	**Asama**
Asami, Yoshichi (1894-) S	**Asami**
Asano, Iasmu (fl. 1965) M	**I.Asano**
Asano, Kazno (fl. 1972) S	**K.Asano**
Asche, W.W. (1872-1932) S	**Asche**
Ascherson, Ferdinand Moritz (1798-1879) S	**F.Asch.**
Ascherson, Paul Friedrich August (1834-1913) MPS	**Asch.**
Aschieri, Eugenia (fl. 1931) M	**Aschieri**
Aseginolaza, Carlos (1956-) S	**Aseg.**
(Aseginolaza Iparragirre, Carlos)	
see Aseginolaza, Carlos	**Aseg.**
(Asenov, Vulevi Ivan)	
see Assenov, Vulevi Ivan	**Assenov**
Asensi, Aldo Oscar A	**Asensi**
Asfaw Hunde (1940-) S	**Asfaw**
Ash, S.R. (fl. 1969-1979) F	**Ash**
Ashburn, L.L. (fl. 1945) M	**Ashburn**
Ashburner, K. (fl. 1985) S	**Ashburner**
Ashby, Alison Marjorie (1901-) S	**A.M.Ashby**
Ashby, Edwin (1861-1941) S	**Ashby**
Ashby, Marshall Rhodes (1923-) S	**M.R.Ashby**
Ashby, Sidney Francis (1874-1954) M	**S.F.Ashby**
Ashe, William Willard (1872-1932) S	**Ashe**
Asher, James H. (1941-) S	**Asher**
Ashford, Bailey Kelly (fl. 1929) AM	**Ashford**
Ashkan, M. (fl. 1981) M	**Ashkan**
Ashlee (fl. 1932) F	**Ashlee**
Ashok Kumar, C.K. A	**Ashok Kumar**
Ashtekar, P.V. A	**Ashtekar**
Ashton, Peter Shaw (1934-) S	**P.S.Ashton**
Ashton, Ruth Elizabeth (1896-) S	**Ashton**
Ashwin, Margot Bernice (1935-) S	**Ashwin**
Askelöf, Johan Christoffer (1787-1848) S	**Askelöf**
Askenasy, Eugen (1845-1903) A	**Askenasy**
Askerov, Ajdyn Musaogly (1948-) P	**Askerov**
Askerova, Rosa K. (1929-) S	**Askerova**
Asmund, Berit Charlotte (1904-1985) A	**Asmund**
Asmunt, Rudolphi? (-1831) M	**Asmunt**
Asokan Nair, R. (fl. 1988) S	**Asokan Nair**
Asonganyi, Joseph Nchendia (1945-) S	**Asong.**
Aspegren, Georg Carsten (1791-1828) S	**Aspegren**
Asperges, M.G.I. (fl. 1983) M	**Asperges**
Asplund, Erik (1888-1974) PS	**Aspl.**
Assadi, Mostafa (1950-) S	**Assadi**
Assawah, M.W. (fl.1962) M	**Assawah**
(Assem, J.van den)	
see Van den Assem, J.	**Van den Assem**
Assenov, Vulevi Ivan (fl. 1982) S	**Assenov**

(Assi, Laurent Aké)
  see Aké Assi, Laurent   **Aké Assi**
Assis-Lopes, L. (fl. 1953) M   **Assis-Lopes**
Asso y del Rio, Ignacio Jordán de (1742–1814) S   **Asso**
(Assunção Diniz, Manuel)
  see Diniz, Manuel (de) Assunção   **Diniz**
Ast, Suzanne (later Jovet-Ast, S.) (1914–) S **Ast**
Asta, J. (fl. 1973) M   **Asta**
Asta-Giacometti, Julliet (fl. 1973) M   **Asta-Giac.**
Astanova, A. (fl. 1971) S   **A.Astan.**
Astanova, S.B. (1941–) S   **S.B.Astan.**
Astegiano, Marta E. (1944–) S   **Asteg.**
Asthana, Deep Kumar A   **D.K.Asthana**
Asthana, R.P. (fl. 1944) M   **Asthana**
Astier, Joseph (fl. 1973) M   **Astier**
Aston, Helen Isabel (1934–) S   **Aston**
Aswal, B.S. (1948–) PS   **Aswal**
Ataeva, A.A. (1948–) S   **Ataeva**
Atehortúa, Lucia G. (1949–) P   **Atehortúa**
Athanasiadis, A. A   **Athanas.**
Athow, Kirk L. (fl. 1954) M   **Athow**
Atienza, Juan D. (fl. 1931) M   **Atienza**
Atkins, D. (fl. 1955) M   **D.Atkins**
Atkins, G.A. (fl. 1958) M   **G.A.Atkins**
Atkins, R.J. (fl. 1983–) S   **R.J.Atkins**
Atkins, Sandy (1948–) S   **S.Atkins**
Atkinson, George Francis (1854–1918) AMS **G.F.Atk.**
Atkinson, J.D. (fl. 1940) M   **J.D.Atk.**
Atkinson, R.G. (fl. 1952) M   **R.G.Atk.**
Atkinson, T.A. (1950–) S   **T.A.Atk.**
Atkinson, William Sackston (1821–1875) PS **Atk.**
Atri, N.S. (fl. 1982) M   **Atri**
Atthey, Th. (fl. 1869) M   **Atthey**
Attinger, E. (fl. 1965) PS   **Attinger**
Attolini, L. (fl. 1933) M   **Attolini**
Atwood, John T. (1946–) S   **J.T.Atwood**
Atwood, N.Duane (1938–) S **N.D.Atwood**
Atzei, Aldo Domenico (1932–) S   **Atzei**
Aubert, Alfred Bellamy (1853–1912) A **A.B.Aubert**
(Aubert du Petit-Thouars, Abel)
  see Thouars, Abel Aubert du Petit– **A.Thouars**
(Aubert du Petit-Thouars, Louis Marie)
  see Thouars, Louis Marie Aubert du Petit **Thouars**
Aubert, Gustave (1829–1902) B   **Aubert**
Aubin, P. (fl. 1986–) S   **Aubin**
Aublet, Jean Baptiste Christophore Fusée (1720–1778) PS **Aubl.**
Aubréville, André (1897–1982) S   **Aubrév.**
Aubriet, Claude (1665?–1742) BL   **Aubriet**
Aubriot, L.J. (fl. 1885) S   **Aubriot**

Aucher-Éloy, Pierre Martin Rémi (1792–1838) S	**Aucher**
Auclair, Firmin A	**Auclair**
Audas, James Wales Claredon (1872–1959) S	**Audas**
Audibert, Urbain (1791–1846) S	**Audib.**
Audot, N. (fl. 1845) S	**Audot**
Audouin, Jean Victor (1797–1841) AS	**Audouin**
Audubon, John James (1785–1851) S	**Audubon**
Aue, Regina (fl. 1967) M	**Aue**
Auerswald, Bernhard (1818–1870) M	**Auersw.**
Augias (fl. 1988) M	**Augias**
Augier, H. A	**H.Augier**
Augier, Jean (1909–) B	**Augier**
Augusta (fl. 1934) F	**Augusta**
Augustin, K. (fl. 1854) S	**Augustin**
Auquier, Paul Henri (1939–1980) S	**Auquier**
Aur, Chih Wen (1932–) B	**Aur**
Auret, Theodora B. (fl. 1930) M	**Auret**
Auriault, R. (fl. 1983) S	**Auriault**
(Aurifici, Michelangelo)	
see Ucria, Bernardino da	**Ucria**
Aurivillius, Carl W.S. (1854–1899) A	**Auriv.**
Ausfeld, J.G.von (fl. 1864) S	**Ausfeld**
Ausserdorfer, Anton (1836–1885) S	**Ausserd.**
Austin, Coe Finch (1831–1880) B	**Austin**
Austin, Daniel Frank (1943–) S	**D.F.Austin**
Austwick, P.K.C. (fl. 1965) M	**Austwick**
Autran, Eugène John Benjamin (1855–1912) S	**Autran**
Avakjan, K.G. (fl. 1957) M	**Avakjan**
Avazneli, A.A. (1981–) S	**Avazneli**
Avdeev, G.V. A	**G.V.Avdeev**
Avdeev, V.I. (1908–) S	**Avdeev**
Avdulov, Nikolai Pavlović (1899–) S	**Avdulov**
Avé-Lallemant, Julius Léopold Edouard (1803–1867) S	**Avé-Lall.**
Avellaneda, Ramón (fl. 1932) M	**Avell.**
(Avellar Brotero, Felix de (Silva))	
see Brotero, Felix de (Silva) Avellar	**Brot.**
Averett, John Earl (1943–) S	**Averett**
Averintsev, S. A	**Averintsev**
Averna–Saccá, Rosario (1883–1951) M	**Av.-Saccá**
Avery, Charles (1880–1960) S	**Avery**
Averyanov, Leonid V. (1955–) S	**Aver.**
(Avetisian, Evgenia M.)	
see Avetissjan, Evgenia M.	**Avet.**
(Avetisian, Vanda E.)	
see Avetissjan, Vanda E.	**V.E.Avet.**
(Avetissian, Evgenia M.)	
see Avetissjan, Evgenia M.	**Avet.**
(Avetissian, Vanda E.)	
see Avetissjan, Vanda E.	**V.E.Avet.**
Avetissjan, Evgenia M. (1923–) S	**Avet.**

Avetissjan, Vanda E. (1928–) S	V.E.Avet.
(Avetisyan, Evgenia M.)	
see Avetissjan, Evgenia M.	Avet.
(Avetisyan, Vanda E.)	
see Avetissjan, Vanda E.	V.E.Avet.
Avetta, Carlo (1861–1941) AP	Avetta
Avetta, P. (fl. 1880s) M	P.Avetta
Avidor, A. (fl. 1974) M	Avidor
Avizohar-Hershenzan, Zehara (fl. 1959) M	Aviz.-Hersh.
Avrorin, N.A. (1906–) S	Avrorin
Awakura, T. (fl. 1977) M	Awakura
Awao, Takeyoshi (fl. 1973) M	Awao
Awasthi, Dharani Dhar (1922–) M	D.D.Awasthi
Awasthi, Garima (fl. 1984) M	G.Awasthi
Awasthi, N. (fl. 1982) F	N.Awasthi
Awasthi, Priti A	P.Awasthi
Awasthi, U.S (fl. 1979) B	U.S.Awasthi
Awati, J.B. (fl. 1972) M	Awati
(Aweke, Getachew)	
see Getachew Aweke	Getachew
Axelius, Barbro (1952–) S	Axelius
Axelrod, Daniel I. (1910–) F	Axelrod
(Axenov, E.S.)	
see Aksenov, E.S.	Aksenov
Ayala, Franklin (fl. 1981) S	Ayala
Ayala, Stephen C. A	S.C.Ayala
Ayasligil, Y. (fl. 1984) S	Ayasligil
Ayensu, Edward Solomon (1935–) S	Ayensu
Ayer, F. (fl. 1974) M	Ayer
Ayers, Theodore Thomas (1900–1967) M	Ayers
Ayers, Tina J. (1957–) S	T.J.Ayers
Ayers, W.A. (fl. 1978) M	W.A.Ayers
Aymard C., Gerardo (fl. 1992) S	Aymard
Aymonin, Gérard Guy (1934–) S	Aymonin
(Aymonin, Monique Keraudren)	
see Keraudren, Monique	Keraudren
Ayobangira, François-Xavier (fl. 1987) S	Ayob.
Ayouty, E.E.L. A	Ayouty
Ayres, William Port (1815–1875) S	Ayres
Ayyangar, C.R. (fl. 1928) M	Ayyangar
Azambuja, David de (fl. 1948) S	Azambuja
(Azancot de Meneses, Oscar J.)	
see Meneses, Oscar J. Azancot de	Meneses
Azaola, Inigo Gonzalez y (fl. 1845) S	Azaola
Azbukina, Z.M. (fl. 1956) M	Azbukina
Azcuy, C.L. (fl. 1984–83) F	Azcuy
Azema, R.C. (fl. 1967) M	Azema
Azevedo, A.M.G.de (fl. 1982) S	A.M.G.Azevedo
(Azevedo de Menezes, Carlos)	
see Menezes, Carlos Azevedo de	Menezes

Azevedo, Maria Teresa de P. A	M.T.P.Azevedo
Azevedo, Nearch da Silveira S	N.S.Azevedo
Azevedo, P.C.de (fl. 1956) M	P.C.Azevedo
Azevedo Penna, Ivo de (fl. 1983) S	I.A.Penna
Azevedo Penna, Leonam de (1903–1979) S	L.A.Penna
Aziz, A. A	A.Aziz
Aziz Khan, M.A. (fl. 1989) S	Aziz Khan
Aziz, Khwaja Muhammad Sultanul (1936–) A	Aziz
Aznavour, Georges V. (1861–1920) S	Azn.
Azpeitia Moros, Florentina (1859–) A	Azpeitia
(Azumbuja, David de)	
see Azambuja, David de	Azambuja
Baad, Michael Francis (1941–) S	Baad
Baagøe, Jette (1946–) S	J.Baagøe
Baagøe, Johannes Schønberg (1838–1905) S	Baagøe
(Baalen, Chase Van)	
see Van Baalen, Chase	Van Baalen
Baard, S.W. (fl. 1985) M	Baard
Baardseth, Egil (1912–) A	Baardseth
(Baare Hellquist, C.)	
see Hellquist, C.Baare	Hellq.
Baas, Pieter (1944–) S	Baas
Baas-Becking, L.G.M. (1894–1963) A	Baas-Beck.
Bååth, E. (fl. 1979) M	Bååth
Babacauh, K.D. (fl. 1983) M	Babacauh
Babajan, A.A. (fl. 1949) M	Babajan
Babajan, D.N.Teterenvnikova (fl. 1964) M	D.N.Babajan
Babalonas, Dimitrios (1944–) S	Babal.
Babcock, C.E. (fl. 1987) M	C.E.Babc.
Babcock, Ernest Brown (1877–1954) S	Babc.
Babel, William Keith (1917–) S	Babel
Babenko (fl. 1925) M	Babenko
Babeva, I.P. (fl. 1974) M	Babeva
Babey, Claude Marie Philibert (1786–1848) S	Babey
Babington, Charles Cardale (1808–1895) AMPS	Bab.
Babington, Churchill (1821–1889) M	C.Bab.
(Babjeva, I.P.)	
see Babeva, I.P.	Babeva
Babos, Margit G. (1931–) M	Babos
Babu, A. (fl. 1984) S	A.Babu
Babu, Cherukuri Raghavendra (1940–) S	Babu
Babu, Kochu (fl. 1977) M	K.Babu
Baburina, A.A. (1895–) S	Baburina
Babushkina, I.N. (fl. 1989) M	Babushk.
Baccarini, Pasquale (1858–1919) M	Bacc.
Bacelar, J.J.A.H.de (1934–) S	Bacelar
Bach, Michael (1808–1878) S	Bach
Bachelot de la Pylaie, Auguste Jean Marie (1786–1856) ABPS	Bach.Pyl.
(Bachinskaya, A.A.)	
see Bachinskaya-Raichenko, A.A.	Bach.-Raich.

Bachinskaya–Raichenko, A.A. (fl. 1911) M	**Bach.-Raich.**
Bachlechner, Gregor (1808–1873) S	**Bachl.**
Bachmann, Alfred A	**A.Bachm.**
Bachmann, Christine (fl. 1963) M	**C.Bachm.**
Bachmann, Franz Ewald Theodor (1850–1937) BM	**Bachm.**
Bachmann, Hans (1866–1940) A	**H.Bachm.**
Bachteev, Fatikh Khafizovich (1905–) S	**Bachteev**
(Bachtin, V.S.)	
see Bakhtin, V.S.	**Bakhtin**
Bacigalupi, Rimo Carlo Felice (1901–) S	**Bacig.**
Bacigalupo, Nélida María (1924–) S	**Bacigalupo**
Bäck, Abraham (1713–1795) S	**Bäck**
Bäck, Ragnar Gottfrid (1904–) S	**R.G.Bäck**
Back, Saara B	**S.Back**
Backeberg, Curt (1894–1966) S	**Backeb.**
Backer, Cornelis Andries B. (1874–1963) PS	**Backer**
Backhouse, James (1794–1869) PS	**Backh.**
Backhouse, James (1825–1890) PS	**Backh.f.**
Backman, Albin Lennart (1880–1967) A	**Backman**
Backus, M.P. (fl. 1961) M	**Backus**
Bacle, César Hippolyte (1794–1838) S	**Bacle**
Bacon, J.D. (fl. 1978) S	**J.D.Bacon**
Bacon, P.S. (fl. 1982) S	**P.S.Bacon**
Badarò, Giovanni Battista (1793–1831) S	**Badarò**
Badea, Mircea (fl. 1935) M	**Badea**
Baden, Claus (1952–) S	**Baden**
Baden Powell, C. (fl. 1893) S	**Baden Pow.**
Badham, Charles David (1806–1857) M	**Badham**
Badhwar, Rattan Lal (fl. 1931) B	**Badhw.**
(Badillo F., Victor Manuel)	
see Badillo, Victor Manuel	**V.M.Badillo**
Badillo, Franceri (1920–) S	**F.Badillo**
Badillo, Victor Manuel (1920–) S	**V.M.Badillo**
Badini, J. (fl. 1977) S	**Badini**
Badoux, Henri (1871–1951) S	**Badoux**
Badré, Frederic Jean (1937–) PS	**Badré**
Badura, L. (fl. 1954) M	**Badura**
Badurowa, Maria (fl. 1964) M	**Badurowa**
Baechler (fl. 1926) M	**Baechler**
Baehni, Charles (1906–1964) BS	**Baehni**
Baenitz, Karl (Carl) Gabriel (1837–1913) BPS	**Baen.**
Baer, K.M. S	**K.M.Baer**
Baer, Karl Reinhold Ernst von (1792–1876) S	**Baer**
Baerts, F. (fl. 1925) M	**Baerts**
Baesecke, Paul (fl. 1903) P	**Baesecke**
Baeza, Victor Manuel von (1889–1944) M	**Baeza**
Bagchee, Krishnadas D. (fl. 1929) M	**Bagchee**
Baghdadi, V.K.H. (fl. 1968) M	**Baghd.**
Baglietto, Francesco (1826–1916) AMS	**Bagl.**
Bagnall, James Eustace (1830–1918) B	**Bagn.**

Bagnall, R.G. (fl. 1972) M	R.G.Bagn.
Bagnis, Carlo (1854–1879) M	Bagnis
Bagshawe, Arthur William Gerrard (1871–1950) S	Bagsh.
Bagyanarayana, G. (fl. 1986) M	Bagyan.
Bahadur, Kunwar Naresh (1935–1984) S	Bahadur
Bahadur, P. (fl. 1957) M	P.Bahadur
Baharaeen, S. (fl. 1982) M	Baharaeen
Bahariev, D.B. (fl. 1984) M	Bahariev
Bahekar, V.S. (fl. 1964) M	Bahekar
Bahl, N. (fl. 1976) M	Bahl
Bahnweg, G. (fl. 1972) M	Bahnweg
Bai, C.K. (fl. 1984) M	C.K.Bai
Bai, En Zhong (1945–) S	E.Z.Bai
Bai, H.C. (fl. 1985) M	H.C.Bai
Bai, Jin Kai (1925–) M	J.K.Bai
Bai, P.Y. (fl. 1983) S	P.Y.Bai
Bai, Xue (Xie) Liang (1950–) B	X.L.Bai
Baibulatova, N.E. (fl. 1988) M	Baibul.
Baici, A. (fl. 1984) M	Baici
Baijal, Usha (fl. 1962) M	Baijal
Baijnath, Himansu (1943–) S	Baijnath
Bail, Carl Adolf Emmo Theodor (1833–1922) M	Bail
Bailey, Alan (1937–) A	A.Bailey
Bailey, Charles (1838–1924) S	C.Bailey
Bailey, Dana K. (1916–) S	D.K.Bailey
Bailey, Ethel Zoe (1889–1983) S	E.Z.Bailey
Bailey, Floyd Douglas (fl. 1935) M	F.D.Bailey
Bailey, Frederick Manson (1827–1915) BMPS	F.M.Bailey
Bailey, Glenn P. A	G.P.Bailey
Bailey, Harold E. (1906–) M	H.E.Bailey
Bailey, Irving Widmer (1884–1967) FS	I.W.Bailey
Bailey, Jacob Whitman (1811–1857) A	Bailey
Bailey, John Frederick (1866–1938) S	J.F.Bailey
Bailey, John Paul (1951–) S	J.P.Bailey
Bailey, John William (1870–1933) B	J.W.Bailey
Bailey, Liberty Hyde (1858–1954) PS	L.H.Bailey
Bailey, Loring Woart (1839–1925) A	L.W.Bailey
Bailey, R.E. (fl. 1967) F	R.E.Bailey
Bailey, Virginia (Edith) Long (1908–) S	V.L.Bailey
Bailey-Watts, A.E. A	Bailey-Watts
Baillargeon, G. (1953–) S	Baillarg.
Baillet, Casimir Celestin (1820–1900) S	Baillet
Baillon, Henri Ernest (1827–1895) BMPS	Baill.
Bailly, A. (fl. 1921) M	A.Bailly
Bailly, Émile (1829–1894) S	Bailly
Bailly, William Hellier (1819–1888) S	W.H.Bailly
Bain, Douglas Cogburn (1908–) M	D.C.Bain
Bain, Henry Franklin (1893–) M	H.F.Bain
Bain, John F. (fl. 1988) S	J.F.Bain
Bain, Samuel McCutcheon (1869–1919) M	Bain

Baines, Henry (1798–1878) S	**Baines**
Baines, Olive B. S	**O.B.Baines**
Baines, Richard Cecil (1905–) M	**R.C.Baines**
Bainier, Georges (18??–1920) M	**Bainier**
Baird, Ralph O. (fl. 1931) S	**Baird**
Baird, Richard E. (fl. 1986) M	**R.E.Baird**
Bairiganjan, G.C. (fl. 1986) S	**Bairig.**
(Baitenov, M.S.)	
see Bajtenov, M.S.	**Bajtenov**
Bajaj, B.S. (fl. 1957) M	**Bajaj**
Bajmuchambetova, Zh.U. S	**Bajmuch.**
Bajtenov, M.S. (1927–) S	**Bajtenov**
Bakaĭ, S.M. (fl. 1967) M	**Bakai**
Bakalova, Ganka G. (fl. 1976) M	**Bakalova**
Bakare, V.B. (fl. 1973) M	**Bakare**
Baker, A.C. (fl. 1923) M	**A.C.Baker**
Baker, Ailsie F. A	**A.F.Baker**
Baker, C.P. (fl. 1918) M	**C.P.Baker**
Baker, Charles Fuller (1872–1927) MS	**C.F.Baker**
Baker, Edmund Gilbert (1864–1949) PS	**Baker f.**
Baker, Gladys Elizabeth (1908–) M	**G.E.Baker**
Baker, Herbert George (1920–) S	**H.G.Baker**
Baker, Hugh Arthur (1896–1976) S	**H.A.Baker**
Baker, Irene (1918–1989) S	**I.Baker**
Baker, J.M. (fl. 1964) M	**J.M.Baker**
Baker, John Gilbert (1834–1920) MPS	**Baker**
Baker, K.K. (fl. 1979) M	**K.K.Baker**
Baker, Kenneth Frank (1908–) M	**K.F.Baker**
Baker, Milo Samuel (1868–1961) S	**M.S.Baker**
Baker, Richard Eric Defoe (1908–1954) MS	**R.E.D.Baker**
Baker, Richard Thomas (1854–1941) FS	**R.T.Baker**
Baker, Sarah Martha (1887–1917) A	**S.M.Baker**
Baker, Shirley D. (fl. 1955) M	**S.D.Baker**
Baker, William Hudson (1911–1985) S	**W.H.Baker**
Bakhariev, D. (fl. 1984) M	**Bakhariev**
(Bakhteev, Fatikh Chafizovich)	
see Bachteev, Fatikh Khafizovich	**Bachteev**
Bakhtin, V.S. (fl. 1925) M	**Bakhtin**
Bakhuizen van den Brink, Reinier Cornelis (1911–1987) S	**Bakh.f.**
Bakhuizen van den Brink, Reinier Cornelis (1881–1945) S	**Bakh.**
Bakke, Arthur Lawrence (1886–) M	**Bakke**
Bakker, Dingeman (1917–) S	**D.Bakker**
Bakker, K. (1931–) S	**K.Bakker**
Baksay, Leona (1915–) S	**Baksay**
Bakshi, Bimal Kumar (fl. 1950) M	**B.K.Bakshi**
Bakshi, Subhendu Kumar (1931–1981) AS	**S.K.Bakshi**
Bal, S.N. (fl. 1921) M	**Bal**
(Balabanoff, Vassil Al.)	
see Balabanov, Vassil Al.	**Balab.**
Balabanov, Vassil Al. (fl. 1965) M	**Balab.**

Balachandran, Indu (fl. 1983) S	**Balach.**
Balakrishnan, M.S. (fl. 1948) AM	**M.S.Balakr.**
Balakrishnan, Nambiyath Puthansurayil (1935–) PS	**N.P.Balakr.**
Balamani, G.V.A. (fl. 1989) S	**Balamani**
Balamuth, William (1914–1981) A	**Balamuth**
Balandin, S.A. (1952–) S	**Balandin**
Balansa, Benedict (Benjamin) (1825–1892) MS	**Balansa**
Balasubramanian, K.A. (fl. 1966) M	**Balas.**
Balasundaram, M.S. A	**Balasund.**
Balátová–Tuláčková, E. (fl. 1981) S	**Bál.-Tul.**
Balayer, M. (fl. 1986) S	**Balayer**
Balazs, F. (1913–) S	**Balazs**
Balazuc, Jean (fl. 1970) M	**Balazuc**
Bałazy, Stanisław (fl. 1973) M	**Bałazy**
Balbiani, E.G. (fl. 1889) M	**Balbiani**
Balbis, Gioanni (Giovanni) Battista (1765–1831) BMPS	**Balb.**
Baldacci, Antonio (1867–1950) S	**Bald.**
Baldacci, Elio (fl. 1940) M	**E.Bald.**
Baldauf, Jack C. A	**Baldauf**
Baldev, B. (fl. 1978) M	**Baldev**
Baldinger, Ernst Gottfried (1738–1804) S	**Baldinger**
Baldock, R.N. A	**Baldock**
Baldoni, Alicia Marta (fl. 1977–85) F	**Baldoni**
Baldwin, John Thomas (1910–1974) S	**J.T.Baldwin**
Baldwin, William (1779–1819) S	**Baldwin**
Baldwin, William Kirwan Willcocks (1910–1979) S	**W.Baldwin**
Bale, V.S. (fl. 1983) M	**Bale**
Bale, W.M. A	**W.M.Bale**
Balech, Enrique A	**Balech**
Balfour, Isaac Bayley (1853–1922) PS	**Balf.f.**
Balfour, John Hutton (1808–1884) BS	**Balf.**
Balfour–Browne, Frances L. (fl. 1951) M	**Balf.-Browne**
Balgooy, Max Michael Josephus van (1932–) S	**Balgooy**
Balick, Michael Jeffrey (1952–) S	**Balick**
Balik, Michael Jeffrey (1952–) S	**Balik**
Balkovsky, B.E. (1899–) S	**Balk.**
Balkwill, Kevin (1958–) S	**K.Balkwill**
Balkwill, Mandy-Jane (née Cadman, M.-J.) (1964–) S	**M.Balkwill**
Ball, Carleton Roy (1873–1958) S	**C.R.Ball**
Ball, John (1818–1889) S	**Ball**
Ball, Oscar Melville (1868–1942) B	**O.M.Ball**
Ball, Peter William (1932–) S	**P.W.Ball**
Ballal, V.N. (fl. 1984) M	**Ballal**
Ballantine, David L. (1947–) A	**D.L.Ballant.**
Ballantine, Dorothy A	**D.Ballant.**
Ballantine, Henry (1833–1929) S	**Ballant.**
Ballard, Ernest (1871–1952) S	**Ballard**
Ballard, Francis (1896–1976) PS	**F.Ballard**
Ballard, Harvey E. (fl. 1983) PS	**H.E.Ballard**
Ballard, Robert E. (1944–) S	**R.E.Ballard**

Balle, Simone (1906–   ) S — **Balle**
Ballero, Mauro (fl. 1987) M — **Ballero**
Ballesteros, Enric   A — **Ballest.**
(Ballesteros i Sagarra, Enric)
    see Ballesteros, Enric — **Ballest.**
Ballet, Jules (1825–   ) S — **Ballet**
Balletto, Cesare (1912–   ) M — **Balletto**
Ballment, E.R. (fl. 1989) S — **Ballment**
Balloni, Waldemaro (fl. 1971) M — **Balloni**
Balls, Edward Kent (1892–1984) S — **Balls**
Bally, Peter René Oscar (1895–1980) S — **P.R.O.Bally**
Bally, Walter (1882–   ) M — **Bally**
Balme, Basil Eric (1923–   ) B — **Balme**
Balodi, B. (1960–   ) S — **Balodi**
Balogh, Miklós (fl. 1970) M — **Balogh**
(Balogh, Pamela)
    see Burns–Balogh, Pamela — **Burns–Bal.**
Balonov, I.M.   A — **Balonov**
Balsamo, Francesco (1850–1922) A — **Balsamo**
Balsamo-Crivelli, Giuseppe Gabriel (1800–1874) ABMS — **Bals.-Criv.**
Balslev, Henrik (1951–   ) S — **Balslev**
Balţatu, G. (fl. 1939) M — **Balţatu**
Baltes, Nicolae   A — **Baltes**
Baltet, Charles (1830–1908) S — **Baltet**
Baltisberger, Matthias (1951–   ) S — **Baltisb.**
Baluswami, M.   A — **Baluswami**
Balzer, F. (fl. 1912) M — **Balzer**
(Bambeke, Charles Eugène Marie Van)
    see Van Bambeke, Charles Eugène Marie — **Van Bamb.**
Bamber, Charles (1855–   ) S — **Bamber**
Bamberger, Johann Georg (1821–1872) B — **Bamb.**
Bamps, Paul Rodolphe Joseph (1932–   ) S — **Bamps**
Bân, Nguyên Tiên (fl. 1973) S — **Bân**
Bañares Baudet, Angel (fl. 1980) MS — **Bañares**
Bancroft, Claude Keith (1885–1919) M — **C.K.Bancr.**
Bancroft, Edward Nathaniel (1772–1842) S — **Bancr.**
Bancroft, Helen Holme (1887–   ) S — **H.H.Bancr.**
(Bandala–Muñoz, Victor M.)
    see Bandala, Victor M. — **Bandala**
Bandala, Victor M (fl. 1987) M — **Bandala**
Bandet, E.A.R.F. (fl. 1930) M — **Bandet**
Bandoni, Robert J. (fl. 1955) M — **Bandoni**
Banerjee, Dhiraj M.   A — **D.M.Banerjee**
Banerjee, M. (fl. 1969–84) F — **M.Banerjee**
Banerjee, Rabindra Nath (1935–   ) S — **R.N.Banerjee**
Banerjee, S. (fl. 1942) AM — **S.Banerjee**
Banerjee, S.P. (1934–   ) S — **S.P.Banerjee**
Banerji, Jogendra Chandra   A — **J.C.Banerji**
Banerji, M.L. (fl. 1965) S — **Banerji**
Bang, H. (fl. 1910) M — **H.Bang**

Bang, Miguel (1853–) S	M.Bang
Bang, Niels (1776–1855) AS	Bang
(Bang-Hofman, Niels)	
see Bang, Niels	Bang
Bange, A.-J. (1896–1950) PS	Bange
Bange, G.G.J. (fl. 1952) S	G.G.J.Bange
Bánhegi, J. (1911–) M	Bánhegi
(Bánhegyi, J.)	
see Bánhegi, J.	Bánhegi
Banister, John (1650–1692) L	Banister
Banker, Howard James (1866–1940) M	Banker
Banks, Donald Jack (1930–) S	D.J.Banks
Banks, George (fl. 1823–1832) S	G.Banks
Banks, Harlan Parker (1913–) AF	H.P.Banks
Banks, Joseph (1743–1820) S	Banks
Banning, Mary Elizabeth (1832–1901) M	Banning
Banno, Isao (fl. 1987) M	I.Banno
Banno, T.J. (fl. 1967) M	Banno
Banu, Zakia (fl. 1962) M	Banu
Banwell, A.David (fl. 1951) B	Banwell
Bányai, János (1886–1971) A	Bányai
Banyard, B.J. (fl. 1975) S	Banyard
Bao, Shi Ying (1935–) S	S.Y.Bao
Bao, Zhou Ran (fl. 1984) S	Z.R.Bao
Bapna, Karan Raj (1925–) B	Bapna
Baptist, J.N. (fl. 1976) M	Baptist
Baquar, Syed Riaz (fl. 1970) S	Baquar
Baquis, E. (fl. 1905) M	Baquis
Bär, Johannes (1877–) P	Bär
Barad, Gerald S. (fl. 1980) S	Barad
Baral, Hans Otto (1954–) M	Baral
Barale, G. S	Barale
Baranec, T. (fl. 1983) S	Baranec
Baranetzky, Josep (Osip) Wasilijewitsch (1843–1905) M	Baran.
Baranov, A.I. (fl. 1965) S	A.I.Baranov
Baranov, Paul Alexandrovich (1892–1962) S	P.A.Baranov
Baranov, Vladimir Isakovich (1889–1967) S	V.I.Baranov
Baranova, J.P. (fl. 1976) S	J.P.Baranova
Baranova, J.(Yu.) V. (fl. 1979) S	J.V.Baranova
Baranova, M.V. (1932–) S	Baranova
Baranyay, J.A. (fl. 1966) M	Baranyay
(Baratte, Jean François Gustave)	
see Barratte, Jean François Gustave	Barratte
Baratti, Giacomo S	Baratti
Barbaini, Maria (1891–1929) M	Barbaini
(Barbarich, Andrej Ivanovich)	
see Barbaricz, Andrej Ivanovič	Barbar.
Barbaricz, Andrej Ivanovič (1903–) S	Barbar.
Barbarin, I.E. (fl. 1924) M	Barbarin

Barbazita, Francesco (–1847) S	**Barbaz.**
Barbe, J. (fl. 1985) M	**Barbe**
Barber, Antonio (fl. 1991) S	**A.Barber**
Barber, Charles Alfred (1860–1933) A	**C.A.Barber**
Barber, Edwin Atlee (1851–1916) B	**Barber**
Barber, Emil (1857–1917) S	**E.Barber**
Barber, H.G. A	**H.G.Barber**
Barber, Horace Newton (1914–1971) P	**H.N.Barber**
Barber, Susan C. (1952–) S	**S.C.Barber**
Barberis, Guiseppina (1950–) S	**Barberis**
Barbero, Andres S	**A.Barbero**
Barbero, Mercedes (1940–) MS	**M.Barbero**
(Barbey, Caroline)	
see Barbey–Boissier, Caroline	**Barb.-Boiss.**
Barbey, William (1842–1914) PS	**Barbey**
Barbey–Boissier, Caroline (1847–1918) S	**Barb.-Boiss.**
(Barbey–Boissier, William)	
see Barbey, William	**Barbey**
Barbey–Gampert, M. S	**Barb.-Gamp.**
Barbier, E. (fl. 1973) S	**E.Barbier**
Barbier, M. (fl. 1904) M	**Barbier**
(Barbieri, Esther de)	
see De Barbieri, Esther	**De Barbieri**
Barbieri, Paolo (1789–1875) S	**Barbieri**
Barbosa, (Fevereiro) Vanio Perazzo (fl. 1970) S	**V.P.Barbosa**
Barbosa, A.F. (fl. 1971) S	**A.F.Barbosa**
Barbosa, Luis Agosto Grandvaux (1914–1983) S	**Barbosa**
Barbosa, M.A.de Freitas (fl. 1965) M	**M.A.F.Barbosa**
Barbosa, Maria A.de J. (fl. 1941) M	**M.A.J.Barbosa**
Barbosa, Octavio A	**O.Barbosa**
Barbosa Pereira, A. S	**Barb.Per.**
Barbosa Rodrigues, João (1842–1909) PS	**Barb.Rodr.**
Barboza, Gloria E. (fl. 1987) S	**Barboza**
Barbour, W.J. (fl. 1941) M	**Barbour**
Barcellos, Ana Maria Pinto (fl. 1973) S	**Barcellos**
Barceló y Combis, Francisco (1820–1889) S	**Barceló**
Bárcena, Mariano de la (1842–1899) S	**Bárcena**
(Barchalov, Shaban Omer-Ogly)	
see Barkhalov, Shaban Omer-Ogly	**Barkh.**
Barckhausen, Gottlieb (fl. 1775) S	**Barckh.**
Barclay, Arthur (1852–1891) M	**Barclay**
Barclay, Arthur S. (1932–) S	**A.S.Barclay**
Barclay, Frederic White (1874–1943) S	**F.W.Barclay**
Barclay, Harriet George (1901–1990) B	**H.G.Barclay**
Barclay, William R. A	**W.R.Barclay**
Barde, A.K. (fl. 1984) M	**Barde**
(Bardessi, E.De)	
see De Bardessi, E.	**De Bardessi**
Bardunov, Leonid Vladimirovich (1932–) B	**Bard.**

Baretti, Amalia   A   **Baretti**
Barfod, A. (1957–   ) S   **Barfod**
Bargagli–Petrucci, Gino (1878–1945) S   **Barg.–Petr.**
Barghoorn, Elso Sternenberg (1915–1984) AFM   **Bargh.**
Barghoorn, F.G. (fl. 1955) F   **F.G.Bargh.**
Bargoni, E.   A   **Bargoni**
Barham, Henry (1670–1726) L   **Barham**
Bari, Ekhlas Mohammed Mahmoud Abdel   **Bari**
Barkalov, V.Yu.(Ju.) (1954–   ) S   **Barkalov**
Barkas, Thomas Pallister (1819–1891) A   **Barkas**
Barker, Berthie Thomas Percival (fl. 1900) M   **B.T.P.Barker**
Barker, D.   A   **D.Barker**
Barker, Frank (fl. 1934) S   **F.Barker**
Barker, George (1776–1845) S   **Barker**
Barker, J.S.F. (fl. 1978) M   **J.S.F.Barker**
Barker, John (1838–1907) AB   **J.Barker**
Barker, John William (1877–1948) A   **J.W.Barker**
Barker, Robyn Mary (1948–   ) S   **R.M.Barker**
Barker, William Robert (1948–   ) S   **W.R.Barker**
Barker, Winsome Fanny (1907–   ) S   **W.F.Barker**
Barkhalov, Shaban Omer-Ogly (1938–   ) M   **Barkh.**
Barkley, Elizabeth Anne (1908–   ) S   **E.A.Barkley**
Barkley, Fred Alexander (1908–1989) S   **F.A.Barkley**
Barkley, Theodore Mitchell (1934–   ) S   **T.M.Barkley**
Barkman, Jan Johannes (1922–1990) BM   **Barkman**
Barkoudah, Youssef Ibrahim (1933–   ) S   **Barkoudah**
Barkworth, Mary Elizabeth (1941–   ) S   **Barkworth**
Barla, Joseph Hieronymus (Jérome) Jean Baptiste (Giambattista)
   (1817–1896) MS   **Barla**
Barlinge, S.G. (fl. 1979) M   **Barlinge**
Bärlocher, Felix (–1988) M   **Bärl.**
Barlow, Bryan Alwyn (1933–   ) S   **Barlow**
(Barnades, Miguel)
   see Barnadez, Miguel   **Barnadez**
Barnadez, Miguel (c.1717–1771) S   **Barnadez**
Barnard, E.L. (fl. 1986) M   **E.L.Barnard**
Barnard, P.D.W. (fl. 1975) F   **P.D.W.Barnard**
Barnard, Thomas Theodore (1898–1983) AS   **Barnard**
Barneby, Rupert Charles (1911–   ) S   **Barneby**
Barnéoud, François Marius (1821–   ) S   **Barnéoud**
Barnes, B.V. (fl. 1985) S   **B.V.Barnes**
Barnes, Barbara   A   **B.Barnes**
Barnes, Charles Reid (1858–1910) B   **Barnes**
Barnes, Edward (1892–1941) S   **E.Barnes**
Barnes, P.E. (fl. 1931) S   **P.E.Barnes**
Barnett, Euphemia Cowan (1890–1970) S   **Barnett**
Barnett, Horace L. (fl. 1958) M   **H.L.Barnett**
Barnett, J.A. (fl. 1973) M   **J.A.Barnett**
Barnett, Lisa C. (1959–   ) S   **L.C.Barnett**
Barney, E.E. (fl. 1877–79) S   **Barney**

Barnhardt, E.A. (fl. 1985) M	**Barnhardt**
Barnhart, John Hendley (1871–1949) S	**Barnhart**
Barnola, Joaquín Maria de (1870–1925) P	**Barnola**
Barocio de las Fuentes, Ilcana (fl. 1973) S	**Barocio**
Baron, P.Alexis (1754–) S	**Baron**
Baroni, Eugenio (1865–1943) MPS	**Baroni**
Baroni, Timothy J. (fl. 1983) M	**T.J.Baroni**
Barquin Diez, Eduardo (1941–) S	**Barquin**
Barr, Donald J.S. (1937–) M	**D.J.S.Barr**
Barr, F.S. (fl. 1956) M	**F.S.Barr**
Barr, Margaret E. (fl. 1955) M	**M.E.Barr**
Barr, Peter (1826–1909) S	**Barr**
Barr, Peter Rudolph (1862–1944) S	**P.R.Barr**
Barra, Alfredo (fl. 1984) S	**Barra**
(Barra Lázaro, Alfredo)	
see Barra, Alfredo	**Barra**
Barrande, Joachim (1799–1883) F	**Barrande**
Barrandon, Auguste (1814–1897) S	**Barrandon**
Barras, Francisco (1869–1955) S	**Barras**
Barras, Stanley J. (fl. 1971) M	**S.J.Barras**
(Barras y de Aragón, Francisco)	
see Barras, Francisco	**Barras**
Barrasa, José Maria (fl. 1977) M	**Barrasa**
Barratt, Joseph (1796–1882) S	**Barratt**
Barratte, Jean François Gustave (1857–1920) S	**Barratte**
Barreiros, Humberto de Souza (1922–) S	**Barreiros**
Barreno, Eva (1950–) M	**Barreno**
Barrera Martínez, Ildefonso (1951–)	**Barrera**
Barrère, Pierre (1690–1755) S	**Barrère**
Barret, A. (fl. 1955) M	**Barret**
Barreto, Henrique Lamahyer de Mello (1892–1962) MS	**Barreto**
Barreto, I.L. (fl. 1983) S	**I.L.Barreto**
Barreto, R.W. (fl. 1960) M	**R.W.Barreto**
Barreto Valdés, A. (fl. 1987) S	**A.Barreto**
Barrett, D.K. (fl. 1981) M	**D.K.Barrett**
Barrett, James Theophilus (1876–) M	**Barrett**
Barrett, Mary Franklin (1879–) S	**M.F.Barrett**
Barretto, G.D'A. (fl. 1976) S	**Barretto**
Barrie, Fred Rogers (1948–) S	**Barrie**
Barringer, Kerry A. (1954–) S	**Barringer**
Barrington, David S. (1948–) FP	**Barrington**
Barrios, Miguel Angel (1953–) S	**Barrios**
(Barrios R., Miguel Angel)	
see Barrios, Miguel Angel	**Barrios**
Barron, George L. (fl. 1961) M	**G.L.Barron**
Barron, John A. (1947–) A	**J.A.Barron**
Barron, William (fl. 1852–1880) S	**Barron**
Barros, Fábio de (1956–) S	**F.Barros**
Barros, Manuel (1880–1973) S	**Barros**
Barros, Maud L.de (fl. 1973) M	**M.Barros**

Barros, Oridio (fl. 1966) M	O.Barros
Barros, T.T. (fl. 1970) M	T.T.Barros
Barroso, Graziela Maciel (1912–) S	G.M.Barroso
Barroso, Liberato Joaquim (1900–1949) S	Barroso
Barrow, John (1764–1848) S	Barrow
Barrows, Charles (fl. 1979) M	Barrows
Barrus, Mortier Franklin (1879–) M	Barrus
Barry, J.P. (fl. 1949) M	J.P.Barry
Barry, Patrick (1816–1890) S	P.Barry
Barry, Redmond (1813–1880) S	Barry
Barsakoff, B. (fl. 1939) M	Barsak.
Barsali, Egidio (1876–1945) B	Bars.
Barsanti, Leopoldo (fl. 1903) F	Barsanti
Barsegyan, A.M. (fl. 1983) S	Barsegyan
Barss, Howard Philipps (1885–) M	Barss
Barta-Calmus, Sylvie A	Barta-Calmus
Bartalini, Biagio (1746–1822) S	Bartal.
Bartel, J.A. (fl. 1983) S	Bartel
Bartelli, Ingrid (fl. 1965) M	Bartelli
Barth, József (Joseph) (1833–1915) MS	Barth
Barth, O.M. (fl. 1971) S	O.M.Barth
Bartha, Zsuzsa A	Bartha
Barthelet, Jean Jules (fl. 1933) M	Barthelet
Barthlott, Wilhelm A. (1946–) S	Barthlott
Bartholomew, Bruce Monroe (1946–) S	B.M.Barthol.
Bartholomew, Elam (1852–1934) M	Barthol.
Bartholomew, Elbert Thomas (1878–) M	E.T.Barthol.
Bartik, A. (fl. 1885) S	Bartik
Bartlett, A.W. (fl. 1926) M	A.W.Bartlett
Bartlett, B. (fl. 1984) M	B.Bartlett
Bartlett, Francis Alonzo (1882–) S	F.A.Bartlett
Bartlett, Harley Harris (1886–1960) PS	Bartlett
Bartlett, John Kenneth (1945–1986) BM	J.K.Bartlett
Bartley, Floyd (1888–1974) S	Bartley
Bartling, Friedrich Gottlieb (1798–1875) S	Bartl.
Bartoli, Antonella (1943–) M	Bartoli
Bartolini, B. (fl. 1966) S	B.Bartol.
Bartolini, Carlo A	C.Bartol.
(Bartolo Brullo, Giuseppina)	
see Bartolo, Giuseppina	Bartolo
Bartolo, Giuseppina (1948–) S	Bartolo
Barton, Benjamin Smith (1766–1815) S	Barton
Barton, Ethel Sarel (later Gepp, E.S.B.) (1864–1922) A	E.S.Barton
Barton, William Charles (–1955) S	W.C.Barton
Barton, William Paul Crillon (1786–1856) S	W.P.C.Barton
Bartram, Edwin Bunting (1878–1964) B	E.B.Bartram
Bartram, John (1699–1777) S	Bartram
Bartram, William (1739–1823) S	W.Bartram
Bartsch, Alfred Frank (1913–) M	A.F.Bartsch
Bartsch, Gustav (fl. 1903) S	G.Bartsch

Bartsch, Johann (1709–1738) S	**Bartsch**
Barua, G.C.S. (fl. 1957) M	**G.C.S.Barua**
Barua, K.C. (fl. 1969) M	**K.C.Barua**
Barua, P.K. (fl. 1956) S	**P.K.Barua**
Barulina, E.I. (1895–1957) S	**Barulina**
(Bary, Heinrich Anton de)	
see de Bary, Heinrich Anton	**de Bary**
Bas, Cornelius (1928–) M	**Bas**
Basak, M. (fl. 1964) M	**Basak**
Basappa, G.P. (fl. 1983) S	**Basappa**
(Basargin, D.D.)	
see Bassargin, D.D.	**Bassargin**
Baschnagel, Raymond A. A	**Baschn.**
Başer, Kemel Hüsnü Can (1949–) S	**Başer**
Basex, A. (fl. 1954) M	**Basex**
Basgal, W. (fl. 1931) M	**Basgal**
(Basilevskaia, Nina Alexandrovna)	
see Basilevskaja, Nina Alexandrovna	**Basil.**
Basilevskaja, Nina Alexandrovna (1902–) S	**Basil.**
Basiner, Theodor Friedrich Julius (1817–1862) S	**Basiner**
Basinger, James F. (fl. 1979) BF	**Basinger**
Bassargin, D.D. (1933–) S	**Bassargin**
Basse de Ménorval, Élaine (1899–) A	**Basse Mén.**
Bassett, Ivan John (1929–) S	**Bassett**
Bassi, Agostino (1772–1856) S	**A.Bassi**
Bassi, Ferdinando (1710–1774) S	**Bassi**
Bassino, J.-P. (fl. 1966) M	**Bassino**
Bässler, Manfred (1935–) S	**Bässler**
Basson, P.W. A	**Basson**
Bastard, Thom. (–1815) S	**T.Bastard**
Bastard, Toussaint (1784–1846) PS	**Bastard**
(Bastelaer, Désiré Alexandre (Henri) Van)	
see Van Bastelaer, Désiré Alexandre (Henri)	**Van Bast.**
Baster, Job (1711–1775) S	**Baster**
Bastian, Henry Charlton (1837–1915) M	**Bastian**
Bastos, Antonia Rangel (1929–) S	**Bastos**
Bastos, C.N. (fl. 1981) M	**C.N.Bastos**
Bastow, R.Fraser A	**R.F.Bastow**
Bastow, Richard Austin (1839–1920) BM	**Bastow**
Basu, C.C. (fl. 1943) M	**Basu**
(Basu Chaudhary, Kailash Chandra)	
see Chaudhary, Kailash Chandra Basu	**K.C.B.Chaudhary**
Basu, D. (1943–) S	**D.Basu**
Basu, Monica (fl. 1976) M	**M.Basu**
Basu, Nupur (fl. 1976) B	**N.Basu**
Basu, P. (fl. 1985) S	**P.Basu**
Basu, Rabindra Krishna (fl. 1970) S	**R.K.Basu**
Basu, S.K. (1934–) P	**S.K.Basu**
Basuki, T. (fl. 1982) M	**Basuki**
Bata, Jovanka (fl. 1964) M	**Bata**

Bataille, Frédéric (1850–1946) M	**Bataille**
Bataille, J. (fl. 1958) M	**J.Bataille**
Batalin, Alexander Theodorowicz (1847–1896) S	**Batalin**
Batalla i Xatruch, Enuli (1889–1979) S	**Batalla**
Bataller, J.R. A	**Bataller**
(Batard, Toussaint)	
see Bastard, Toussaint	**T.Bastard**
(Batarda Fernandes, Rosette Mercedes Saraiva)	
see Fernandes, Rosette Mercedes Saraiva Batarda	**R.Fern.**
Batchelder, Frederick William (1838–1911) S	**Batch.**
Batcheller, Frances N. (fl. 1978) S	**Batcheller**
Bateman, H.R. A	**H.R.Bateman**
Bateman, James (1811–1897) S	**Bateman**
Bateman, Richard M. (fl. 1983) S	**R.M.Bateman**
Batenburg-van der Vegte, W.H. (fl. 1986) M	**Bat.Vegte**
Bates, David Martin (1935–) S	**D.M.Bates**
Bates, John Mallory (1846–1930) S	**Bates**
Bates, Robert J. (1946–) S	**R.J.Bates**
Bates, Vernon M. (fl. 1984) S	**V.M.Bates**
(Bâthie, Joseph Marie Henry Alfred Perrier de la)	
see Perrier de la Bâthie, Joseph Marie Henry Alfred	**H.Perrier**
(Bâthie, Pierre Eugène Perrier de la)	
see Perrier de la Bâthie, Pierre Eugène	**P.E.Perrier**
Baticados, Cecilia L. (fl. 1980) M	**Batic.**
Baticados, Ma.C. A	**M.C.Batic.**
Batikyan, S.G. (fl. 1969) M	**Batikyan**
Batista, Augusto Chaves (1916–1967) M	**Bat.**
Batka, Johann Baptista (fl. 1823–1872) S	**Batka**
Batko, Stanislaw Albert Antony Benedict (1904–) AMS	**Batko**
Batra, Lekh R. (1929–) M	**L.R.Batra**
Batra, Suzanne W.T. (fl. 1963) M	**S.W.T.Batra**
Batsch, August Johann Georg Karl (1761–1802) BMPS	**Batsch**
(Batschinskaja, A.A.)	
see Bachinskaya-Raichenko, A.A.	**Bach.-Raich.**
Batta, J. (fl. 1980) S	**Batta**
Battacharjee (fl. 1986) M	**Battach.**
(Battacharjee, M.)	
see Bhattacharjee, M.	**M.Bhattach.**
Battacharjee, R. (1937–) S	**R.Battach.**
Battacharyya, U.C. (fl. 1967) S	**Battacharyya**
Battandier, Jules Aimé (1848–1922) BPS	**Batt.**
Battarra, Giovanni Antonio (1714–1789) MS	**Battarra**
Batten, D.J. A	**Batten**
Batters, Edward Arthur Lionel (1860–1907) A	**Batters**
Battetta, Victor (fl. 1938) M	**Battetta**
Baturina, V.N. A	**Baturina**
Batyrova, G.Š. (fl. 1983) M	**Batyrova**
Bau, Y.S. (fl. 1984) M	**Y.S.Bau**
Bauch, R. (fl. 1926) M	**Bauch**
Bauchman, Teresa Cecil (later Rickett, T.C.) (1902–) S	**Bauchman**

Bauchop, T. (fl. 1988) M	**Bauchop**
Baudet, Jean C. (fl. 1970) S	**Baudet**
Baudière, André (1932–) S	**Baudière**
Baudo, Firmin (fl. 1843) S	**Baudo**
Baudon, Alfred François (Gaston) (1875–) S	**Baudon**
Baudrimont, Roland A	**Baudr.**
Baudyš, Eduard (1886–1968) M	**Baudyš**
Bauer, Ernst (1860–1942) B	**E.Bauer**
Bauer, Ferdinand Lukas (1760–1826) S	**F.L.Bauer**
Bauer, Franz (Francis) Andreas (1758–1840) APS	**F.A.Bauer**
Bauer, Gustav Heinrich (1794–1888) B	**G.H.Bauer**
Bauer, Robert (fl. 1990) M	**R.Bauer**
Bauhin, Casper (Gaspard) (1560–1624) LM	**C.Bauhin**
Bauhin, Jean Johannes (1541–1613) LM	**J.Bauhin**
Baum, Bernard René (1937–) S	**B.R.Baum**
Baum, Hugo (1866–1950) MS	**Baum**
Baum, Vicki M. (fl. 1982) S	**V.M.Baum**
Baumann, B. (fl. 1980–) S	**B.Baumann**
Baumann, Constantin Auguste Napoléon (1804–1884) S	**Baumann**
Baumann, Émile Napoléon (1835–1910) S	**É.N.Baumann**
Baumann, Eugen (1868–1933) S	**E.Baumann**
Baumann, Helmut (1937–) S	**H.Baumann**
Baumann, Karlheinz (fl. 1989) M	**K.Baumann**
Baumann, Paul A	**P.Baumann**
Baumann–Bodenheim, G. (fl. 1953) S	**Baum.–Bod.**
Baumeister, Willy (1904–) A	**Baumeister**
Baumgardt, Ernst (fl. 1856) S	**Baumgardt**
Baumgarten, Johann Christian Gottlob (1765–1843) MPS	**Baumg.**
Baumgartner, Julius (1870–1955) MS	**Baumgartner**
Baumgartner, R. (fl. 1923) M	**R.Baumgartner**
Baumgratz, José Fernando Andrade (fl. 1990) S	**Baumgratz**
Bäumler, Johann Andreas (1847–1926) M	**Bäumler**
Baur, Wilhelm (1839–1920) B	**Baur**
Bausch, Jan (1917–) S	**J.Bausch**
Bausch, Wilhelm (1804–1873) M	**Bausch**
Bautista, Hortensia Pousada (1949–) P	**Bautista**
Bawcutt, R.A. (fl. 1983) M	**Bawcutt**
Baxter, Dow Vauter (1898–1965) MS	**D.V.Baxter**
Baxter, Edgar Martin (1903–1967) S	**E.M.Baxter**
Baxter, John Wallace (1918–) M	**J.W.Baxter**
Baxter, Robert W. (fl. 1955–1967) FM	**R.W.Baxter**
Baxter, W.T. (fl. 1864) S	**W.T.Baxter**
Baxter, William (1787–1871) S	**Baxter**
Baxter, William (–before 1836) S	**W.Baxter**
Baxter, William Hart (1816–1890) S	**W.H.Baxter**
Bay, J.C. (fl. 1893) M	**Bay**
Bayer, A. (1882–1941) M	**A.Bayer**
Bayer, Edvin (1862–1921) FMS	**E.Bayer**
Bayer, Ehrentraud (fl. 1982) S	**Ehr.Bayer**
Bayer, Johann Nepomuk (1802–1870) S	**Bayer**

Bayer, Manfred (fl. 1986) S	**M.Bayer**
Bayer, Martin Bruce (1935–) S	**M.B.Bayer**
Bayer, Randall J. (1955–) S	**R.J.Bayer**
Bayle–Barelle, Guiseppe (1768–1811) S	**Bayle–Bar.**
Baylet, J. (fl. 1959) M	**Baylet**
Baylis, Edward (fl. 1791–1794) S	**Baylis**
Baylis, Geoffrey Thomas Sandford (1913–) MS	**G.T.S.Baylis**
Bayliss Elliott, Jessie Sproat (fl. 1920) M	**Bayl.Ell.**
Bayon, Eva (fl. 1984) S	**Bayon**
Bayrhoffer, Johann Daniel Wilhelm (1793–1868) BCM	**Bayrh.**
Baytop, Asuman (1920–) S	**A.Baytop**
Baytop, Turhan (fl. 1982) S	**T.Baytop**
(Bazilevskaja, Nina Alexandrovna)	
see Basilevskaja, Nina Alexandrovna	**Basil.**
Bazzalo, María E. (fl. 1982) M	**Bazzalo**
Bazzichelli, Giorgio (1924–) S	**Bazzich.**
Bé, Allan W.H. (1931–) A	**Bé**
Beach, Walter Spurgeon (1890–) M	**Beach**
Beadle, Chauncey Delos (1866–1950) S	**Beadle**
Beadle, Noel Charles William (1914–) PS	**N.C.W.Beadle**
Beal, Ernest Oscar (1928–1980) S	**E.O.Beal**
Beal, William James (1833–1924) S	**Beal**
Beaman, John Homer (1929–) S	**Beaman**
Beaman, Reed Schiele (1961–) S	**R.S.Beaman**
Beamish, Katherine I. (fl. 1972) S	**Beamish**
Bean, A.R. (fl. 1987) S	**A.R.Bean**
Bean, V. (fl. 1970) S	**V.Bean**
Bean, William Jackson (1863–1947) S	**Bean**
Beane, Lawrence (1901–) S	**Beane**
Beaney, William D. A	**Beaney**
Beard, John Stanley (1916–) S	**Beard**
Beard, Luther Stanford (1929–) S	**L.S.Beard**
Beardslee, Henry Curtis (1865–1948) M	**Beardslee**
Beatley, Janice Carson (1919–1987) S	**Beatley**
Beaton, Donald (1802–1963) S	**Beaton**
Beaton, Gordon William (1911–) M	**G.W.Beaton**
Beatson, Alexander (1759–1833) S	**Beatson**
Beattie, Rolla Kent (1875–1960) S	**Beattie**
Beauchamp, C.J. (fl. 1986) M	**C.J.Beauch.**
Beauchamp, Paul de (1883–) A	**Beauch.**
Beauchamp, R.Mitchel (fl. 1974) S	**R.M.Beauch.**
Beaudry, Jean Romuald (1917–) S	**Beaudry**
Beauge, 'André' (fl. 1973) S	**Beauge**
Beaumont, Albert (1901–) M	**Beaumont**
Beauseigneur, A. (fl. 1926) M	**Beauseign.**
Beauvais, P. (fl. 1975) M	**Beauvais**
Beauverd, Gustave (1867–1942) BPS	**Beauverd**
Beauverie, Jean Jules (1874–1938) MS	**Beauverie**
Beauvisage, Georges Eugène Charles (1852–1925) S	**Beauvis.**

Beauvisage, L. (fl. 1959–) B   **L.Beauvis.**
(Beauvois, Ambroise Marie François Joseph Palisot de)
    see Palisot de Beauvois, Ambroise Marie François Joseph   **P.Beauv.**
Bebb, Michael Schuck (1833–1895) S   **Bebb**
Bebout, J.W.   A   **Bebout**
Beccari, Odoardo (1843–1920) AMPS   **Becc.**
Becerescu, Dumitru (1923–   ) M   **Becer.**
Bechenau   B   **Bechenau**
Becherer, Alfred (1897–1977) PS   **Bech.**
Bechet, Ion (fl. 1982) M   **I.Bechet**
Bechet, Maria (1928–   ) M   **M.Bechet**
Bechstein, Johann Matthaeus (1757–1822) S   **Bechst.**
Beck, (Carl) Richard (1858–1919) F   **R.Beck**
Beck, Charles B. (fl. 1958–1967) F   **C.B.Beck**
Beck, Günther von Mannagetta und Lërchenau (1856–1931) ABMPS   **Beck**
Beck, Lewis Caleb (1798–1853) BPS   **L.C.Beck**
Beck, O. (fl. 19??) M   **O.Beck**
(Beck von Mannagetta und Lerchenau, Günther)
    see Beck, Günther von Mannagetta und Lërchenau   **Beck**
Becker, Alexander K. (1818–1901) S   **A.K.Becker**
Becker, Clarence D. (1930–   ) A   **C.D.Becker**
Becker, Elery Ronald (1896–1962) A   **E.R.Becker**
Becker, G. (fl. 1954) M   **G.Becker**
Becker, Johannes (1769–1833) S   **Becker**
Becker, Kenneth M. (fl. 1970) S   **K.M.Becker**
Becker, W.N. (fl. 1976) M   **W.N.Becker**
Becker, Wilhelm (1874–1928) S   **W.Becker**
Beckett, K.A. (fl. 1960) S   **K.A.Beckett**
Beckett, Thomas W.Naylor (1838–1906) B   **Beckett**
Beckhaus, Konrad (Conrad) Friedrich Ludwig (1821–1890) BM   **Beckh.**
Becking, R.W. (fl. 1986) S   **Becking**
Beckmann, Carl(Karl) Ludwig (1845–1898) BS   **C.L.Beckm.**
Beckmann, Jean–Pierre   A   **J.-P.Beckm.**
Beckmann, Johann (1739–1811) MS   **Beckm.**
Beckmann, Paul (1881–   ) MS   **P.Beckm.**
Beckmann, Rosemarie   A   **R.Beckm.**
Beckner, John (fl. 1968) S   **Beckner**
Beckwith, Theodore Day (1879–1946) A   **Beckwith**
Bedarff, Ute (fl. 1989) S   **Bedarff**
Beddome, Richard Henry (1830–1911) PS   **Bedd.**
Beddows, Arthur Rhys (1896–   ) S   **Beddows**
Bedell, Hollis G. (fl. 1989) S   **Bedell**
Bedford, C.L. (fl. 1942) M   **Bedford**
Bedford, David John (1952–   ) S   **D.J.Bedford**
Bedi, S.J. (1935–   ) S   **Bedi**
Bednarz, Teresa   A   **Bednarz**
Beeby, William Haddon (1849–1910) S   **Beeby**
Beech, F.W. (fl. 1965) M   **Beech**
Beech, Peter Luke   A   **P.L.Beech**

Beechey, Frederick William (1796–1856) S	**Beechey**
Beede, Joshua William (1871–1940) A	**Beede**
Beek, G. (fl. 1890) S	**Beek**
Beek, A.van de (fl. 1981) S	**A.Beek**
Beekman, W.L. (fl. 1924) P	**Beekman**
Beeley, Fred (fl. 1930) M	**Beeley**
Beeli, Maurice (1879–) M	**Beeli**
(Beem, A.P.Van)	
see Van Beem, A.P.	**Van Beem**
Beentje, Henk Jaap (1951–) S	**Beentje**
Beer, Eva S	**E.Beer**
Beer, Johann Georg (1803–1873) S	**Beer**
Beer, Rudolph (1873–1940) MS	**R.Beer**
Beers, Alma Holland (fl. 1951) M	**Beers**
Beetle, Alan Ackerman (1913–) S	**Beetle**
Beetle, Dorothy Erna (née Schoof) (1916–) S	**D.E.Beetle**
Beger, Herbert K.E. (1889–1955) AS	**Beger**
Béguinot, Augusto (1875–1940) AS	**Bég.**
Begum, Mubina A	**M.Begum**
Begum, Rehana (fl. 1978) M	**R.Begum**
Begum, S. (fl. 1978) M	**S.Begum**
Begum, Zeenatunnessa Tahmida A	**Z.T.Begum**
Behera, N. (fl. 1958) M	**Behera**
Béhéré, Jean Baptiste Joseph (1763–1840) B	**Béhéré**
Behlau, Joachim A	**Behlau**
Behlen, Stephan (1784–1847) M	**Behlen**
Behm, Johan Eric Florentin (1838–1915) S	**Behm**
Behning, Arvid Liborecič (1890–) A	**Behning**
Behr, Ernst (1903–1957) S	**E.Behr**
Behr, Hans Hermann (1818–1904) S	**Behr**
Behr, Otto (1901–1957) M	**O.Behr**
Behre, Karl (1901–1972) A	**Behre**
Behrend, Gustav (fl. 1890) M	**Behrend**
Behrendsen, Werner (–1923) S	**Behrendsen**
Beigel, H. A	**Beigel**
Beijerinck, Martinus Willem (1861–1931) AM	**Beij.**
Beijerinck, Willem (1891–1960) S	**W.Beij.**
Beille, Lucien (1862–1946) S	**Beille**
Beilschmied, Carl (Karl) Traugott (1793–1848) BS	**Beilschm.**
Beintema, K. (fl. 1934) M	**Beintema**
Beirnacka, I. A	**Beirnacka**
Beissner, Ludwig (1843–1927) S	**Beissn.**
Beitel, Joseph M. (1952–1991) P	**Beitel**
Bejlin, J. (fl. 1924) M	**Bejlin**
Beju, Dan A	**Beju**
(Beketov, Andrej Nikolaevich)	
see Beketow, Andrej Nikolaevich	**Bek.**
Beketow, Andrej Nikolaevich (1825–1902) S	**Bek.**
Bélanger, C. (fl. 1983) M	**C.Bél.**

Bélanger, Charles Paulus (1805–1881) BMPS	**Bél.**
Bělař, Karl (1895–1931) A	**Bělař**
Belcher, J. (fl. 1961) A	**J.Belcher**
Belcher, J.H. A	**J.H.Belcher**
Belcher, Robert Orange (1918–) S	**Belcher**
Beldie, Alexandru (1912–) S	**Beldie**
Beliakova, L.A. (fl. 1954) M	**Beliakova**
Belianina, Nina B. (1932–) S	**Belianina**
Beliram, R. (fl. 1960) M	**Beliram**
Belisario, Alessandra (fl. 1991) M	**Belisario**
(Beljanina, Nina B.)	
see Belianina, Nina B.	**Belianina**
Bell, Ann E. (1941–) M	**A.E.Bell**
Bell, Bruce Graham (1942–) B	**B.G.Bell**
Bell, Clyde Ritchie (1921–) S	**C.R.Bell**
Bell, D.K. (fl. 1966) M	**D.K.Bell**
Bell, Hugh Philip (1889–) M	**H.P.Bell**
Bell, Klaus P. (fl. 1970) S	**K.P.Bell**
Bell, Peter Robert (1920–) PS	**P.R.Bell**
Bell, S. (fl. 1852) S	**Bell**
Bell, Shona M. (fl. 1956) F	**S.M.Bell**
Bell, T.A. (fl. 1950) M	**T.A.Bell**
Bell, Walter Andrew (1889–) A	**W.A.Bell**
Bellair, Georges Adolphe (1860–1939) PS	**Bellair**
Bellamy, R.E. (fl. 1985) M	**Bellamy**
Belland, René Jean (1954–) B	**Belland**
Bellardi, Carlo Antonio Lodovico (1741–1826) ABMPS	**Bellardi**
Bellemère, A. (fl. 1960) M	**Bellem.**
(Bellenden Ker, John)	
see Ker Gawler, John Bellenden	**Ker Gawl.**
Beller, J. (fl. 1967) M	**Beller**
(Belleval, Pierre Richer de)	
see Richer de Belleval, Pierre	**Rich.Bell.**
Belli, Carlo Saverio (1852–1919) BMS	**Belli**
Bellia, G.G. (fl. 1971) S	**Bellia**
Belliard, L. (fl. 1973) M	**Belliard**
Bellis, Vincent J. (1938–) A	**Bellis**
Bellisari, G. (fl. 1904) M	**Bellisari**
Bello y Espinosa, Domingo (1817–1884) S	**Bello**
Bellot Rodríguez, Francisco (1911–1983) S	**Bellot**
Bellú, Francesco (fl. 1988) M	**Bellú**
Bellynck, Auguste Alexis Adolphe Alexandre (1814–1877) CMPS	**Bellynck**
Belmonte, Dolores (fl. 1989) S	**Belmonte**
Belo–Correia, A.L. (fl. 1986) S	**Belo–Corr.**
Beloserky, R.N. S	**Beloserky**
Below, Raimond A	**Below**
Belozor, N.I. (1937–) S	**Belozor**
Belsky, M. (fl. 1964) M	**Belsky**
Beltramini de Casati, Francesco (1828–1903) M	**Beltr.**

(Beltrán Bigorra, Francisco)	
see Beltrán, Francisco	**Beltrán**
Beltrán, Enrique (1903–) S	**E.Beltrán**
Beltrán, Francisco (1886–1962) MS	**Beltrán**
(Beltrán Tejera, Esperanza)	
see Beltrán-Tejera, Esperanza	**Beltrán-Tej.**
Beltrán-Tejera, Esperanza (1917–) MS	**Beltrán-Tej.**
Beltrani, Vito (fl. 1874) M	**Beltrani**
Belval, Henri (1880–) S	**Belval**
Belyaeva, V.A. (1946–) S	**Belyaeva**
(Ben, Dick Van der)	
see Van der Ben, Dick	**Van der Ben**
Ben-Sasson, Rivka (1946–) B	**Ben-Sasson**
Ben-Ze'ev, Israel S. (fl. 1986) M	**Ben-Ze'ev**
Bena, Mathew (fl. 1900) B	**Bena**
Benabid, A. (fl. 1982) S	**Benabid**
Benary, Ernst (1819–1892) S	**Benary**
Benatar, Rubens (1910–) M	**Benatar**
Bencini, A. (fl. 1928) M	**Bencini**
Benda, Irmgard (fl. 1962) M	**Benda**
Benda, L. A	**L.Benda**
Bendak, O. (fl. 1981) S	**Bendak**
Bender, Harold (Bohn) (1902–) M	**Bender**
Benderliev, K.M. A	**Benderl.**
Bendiksen, Egil (1955–) M	**Bendiksen**
Bendre, A.M. A	**Bendre**
Benecke, Franz (1857–1903) S	**Benecke**
Benecke, Wilhelm (1868–1946) AS	**W.Benecke**
Benedek, Paul Nikolaus von A	**P.N.Benedek**
Benedek, Tibor (1892–1974) M	**Benedek**
Benedí, Carles (1958–) S	**Benedí**
(Benedí González, Carles)	
see Benedí, Carles	**Benedí**
(Benedí i González, Carles)	
see Benedí, Carles	**Benedí**
(Benedí-Gonzàlez, Carles)	
see Benedí, Carles	**Benedí**
Benedict Ayers (Brother) (fl. 1944) S	**Ben.Ayers**
(Benedict, Charlotte Gilg–)	
see Gilg-Benedict, Charlotte	**Gilg-Ben.**
Benedict, James Everard (1885–) P	**J.E.Benedict**
Benedict, R.G. (fl. 1970) M	**R.G.Benedict**
Benedict, Ralph Curtiss (1883–1965) PS	**Benedict**
Benedix, Erich Heinz (1914–1983) BM	**Benedix**
Beneke, E.S. (fl. 1954) M	**Beneke**
Beneken, Ferdinand (1800–1859) S	**Beneken**
Beneš, R. (fl. 1934) M	**Beneš**
(Benevides de Abreu, Cordélia Luiza)	
see Abreu, Cordélia Luiza	**Abreu**

Benham, Dale Maurice (1957–) P	**D.M.Benham**
Benham, Rhoda Williams (fl. 1952) M	**Benham**
Benitez de Rojas, Carmen E. (1937–) S	**Benitez**
Benjamin, Chester R. (fl. 1955) M	**C.R.Benj.**
Benjamin, Dimitri Sucre (fl.c.1945–) S	**D.S.Benj.**
Benjamin, Ludwig (1825–1848) S	**Benj.**
Benjamin, Richard K. (1922–) M	**R.K.Benj.**
Benke, Hermann Conrad (1869–1946) S	**Benke**
Benkert, Dieter (1933–) M	**Benkert**
Benkö, J. (1740–1814) S	**Benkö**
Benl, Gerhard (1910–) PS	**Benl**
Bennell, A.P. (fl. 1983) S	**Bennell**
Benner, Walter Mackinett (1888–1970) S	**Benner**
Bennert, H.Wilfried (1945–) P	**Bennert**
Bennet, S.S.R. (1940–) S	**Bennet**
Bennett, Alfred William (1833–1902) AMS	**A.W.Benn.**
Bennett, Arthur (1843–1929) S	**A.Benn.**
Bennett, Carlyle Wilson (1895–) M	**C.W.Benn.**
Bennett, David E. (fl. 1989) S	**D.E.Benn.**
Bennett, Eleanor Marion (1942–) S	**E.M.Benn.**
Bennett, Frederick Thomas (fl. 1928) M	**F.T.Benn.**
Bennett, George (1804–1893) S	**G.Benn.**
Bennett, James Lawrence (1832–1904) M	**J.L.Benn.**
Bennett, John Hughes (fl. 1842) M	**J.H.Benn.**
Bennett, John (Johannes) Joseph (1801–1876) PS	**Benn.**
Bennett, W.E. (fl. 1986) M	**W.E.Benn.**
Bennetts, William James (1865–1920) S	**Bennetts**
Benny, Gerald Leonard (1942–) M	**Benny**
Benoist, Raymond (1881–1970) S	**Benoist**
Benoit, M.A. (fl. 1974) M	**Benoit**
Bensemann, Hermann (1858–) S	**Bensemann**
Bensin, Basil Mitrofanovich (1881–) S	**Bensin**
Benson, Bernard W. (fl. 1957) S	**B.W.Benson**
Benson, D.G. A	**D.G.Benson**
Benson, Gilbert Thereon (1896–1928) S	**G.T.Benson**
Benson, Lyman David (1909–) S	**L.D.Benson**
Benson, Margaret J. (fl. 1904) F	**M.J.Benson**
Benson, Robson (1822–1894) S	**Benson**
Bentham, George (1800–1884) MPS	**Benth.**
Benthem, J.van (fl. 1981) S	**Benthem**
Bentivenga, Stephen P. (fl. 1991) M	**Bentiv.**
Bentley, Robert (1821–1893) S	**Bentley**
Bentvelzen, P.A.J. (fl. 1962) S	**Bentv.**
Bentzer, Bengt (1942–) S	**Bentzer**
Benvenuto, Evangelina (1947–) S	**Benv.**
Benz, Robert von Albkron (1863–1921) S	**Benz**
Benzoni, C. (fl. 1930) M	**Benzoni**
Beppu, M. (fl. 1982) P	**M.Beppu**
Beppu, T. (fl. 1985) S	**T.Beppu**
Bequaert, J. (1886–) A	**Bequaert**

Beraha, Louis (fl. 1975) M	**Beraha**
Beránek, J. (fl. 1976) M	**Beránek**
Bérard, Edoardo (1825–1889) S	**Bérard**
Bérard–Therriault, L. A	**Bér.-Therr.**
(Berazaín Iturralde, Rosalina)	
see Berazaín, Rosalina	**Berazaín**
Berazaín, Rosalina (1947–) S	**Berazaín**
Berbee, Mary (fl. 1988) M	**Berbee**
Berch, F. (fl. 1983) M	**F.Berch**
Berch, Shannon M. (fl. 1982) M	**S.M.Berch**
Berchenko, O.I. A	**Berchenko**
Bercht, C.A.L. (fl. 1984) S	**C.A.L.Bercht**
Berchtold, Bedřicha (Friedrich) Wssemjra von (1781–1876) PS	**Bercht.**
Berckmans, Prosper Jules Alphonse (1830–1910) S	**Berckm.**
Berczi, Laszló (fl. 1942) M	**Berczi**
Berdan, Helen Berenice (1901–) M	**Berdan**
Berdau, Feliks (Félix) I. (1826–1888) MS	**Berdau**
Berde, K.von (fl. 1937) M	**Berde**
(Berdiev, B.B.)	
see Berdyev, B.B.	**Berdyev**
Berdyev, B.B. (1938–) S	**Berdyev**
Berendsen, W. B	**Berendsen**
Berendt, Georg Carl (1790–1850) BFM	**Berendt**
Berenger, Giuseppe Adolpho de (1815–1895) M	**Berenger**
Bereriguer, J. (fl. 1980) M	**Berer.**
Berestnev, N. (fl. 1941) M	**Berestnev**
(Berestnew, N.)	
see Berestnev, N.	**Berestnev**
Berg, Åke A	**Åke Berg**
Berg, Alexander (fl. 1840) S	**A.Berg**
Berg, Anthony (1888–1948) M	**Anth.Berg**
Berg, Cornelius C. (1934–) S	**C.C.Berg**
Berg, Ernst von (1782–1855) S	**Berg**
Berg, G. (fl. 1990) M	**G.Berg**
Berg, Maria Elizabeth van den (fl. 1970) S	**M.E.Berg**
Berg, Marion P. S	**M.P.Berg**
Berg, Otto Karl (Carl) (1815–1866) S	**O.Berg**
Bergamaschi, Giuseppe (1785–1867) M	**Bergam.**
Bergdolt, Ernst (1902–1948) PS	**Bergdolt**
Berge, Ernst (1836–1897) S	**Berge**
Bergen, Karl August von (1704–1759) S	**Bergen**
Berger, Alwin (1871–1931) S	**A.Berger**
Berger, Ernst Friedrich (1814–1853) S	**Berger**
Berger, G. (fl. 1938) M	**G.Berger**
Berger, H.A.C. (fl. 1838) F	**H.Berger**
Berger, Reinholdus (1824–1850) F	**R.Berger**
Berger, S. A	**S.Berger**
Berger–Perrot, Yvette A	**Berg.-Perr.**
Bergeret, Gaston (fl. 1909) S	**G.Bergeret**

Bergeret, Jean (1751–1813) S	**Bergeret**
Bergeret, Jean Pierre (1751–1813) S	**J.P.Bergeret**
Bergey, David Hendricks (1860–1937) M	**Bergey**
Berggren, Jacob (1790–1868) S	**J.Berggr.**
Berggren, Sven (1837–1917) ABS	**Berggr.**
Bergh, Rudolph Sophus (1859–1924) A	**Bergh**
(Berghen, Constant Vanden)	
see Vanden Berghen, Constant	**Vanden Berghen**
Bergius, Bengt (Benedictus) (1723–1784) MPS	**Bergius**
Bergius, Karl Heinrich (1790–1818) S	**K.Bergius**
Bergius, Peter Jonas (1730–1790) PS	**P.J.Bergius**
Bergman, Ferdinand (1826–1899) S	**Bergman**
Bergman, Herbert Floyd (1883–) S	**H.Bergman**
Bergman, O. (fl. 1966) M	**O.Bergman**
Bergman, Phyllis S. (fl. 1958) M	**P.S.Bergman**
Bergmans, Johannes (John) Baptista (1892–1980) S	**Bergmans**
Bergner, S. (fl. 1989) S	**Bergner**
Bergon, Paul (1863–1912) AS	**Bergon**
Bergquist, R.R. (fl. 1969) M	**Bergq.**
Berhaut, Jean (1902–1977) S	**Berhaut**
Berher, E. (fl. 1987) S	**Berher**
Berjak, P. (fl. 1988) M	**Berjak**
Berkeley, Emeric Streatfield (1823–1898) S	**E.S.Berk.**
Berkeley, Garven Hugh (1894–) M	**G.H.Berk.**
Berkeley, Miles Joseph (1803–1889) BCMS	**Berk.**
Berkenhout, John (1730–1791) S	**Berkenh.**
Berkhout, Christine Marie (1893–1932) M	**Berkhout**
Berkutenko, A.N. (1950–) S	**Berkut.**
Berlandier, Jean Louis (1805–1851) S	**Berland.**
Berlese, Antonio (1863–1927) M	**A.Berl.**
Berlese, Augusto Napoleone (1864–1903) M	**Berl.**
Berlin, Johan August (1851–1910) S	**Berlin**
Berliner, E. A	**Berliner**
Berman, J.D. (fl. 1973) S	**Berman**
Bernal, Henry Yesid (1953–) S	**Bernal**
Bernal, Mercè (1959–) S	**M.Bernal**
Bernal–González, Rodrigo (fl. 1985) S	**R.Bernal**
Bernalier, Annick (fl. 1989) M	**Bernalier**
Bernard, Charles Jean (1876–1967) AMS	**C.Bernard**
Bernard, Ernest C. (fl. 1978) M	**E.C.Bernard**
Bernard, F.Guy (fl. 1968) S	**F.G.Bernard**
Bernard, Francis A	**F.Bernard**
Bernard, Georges Eugène (18??–1925) M	**G.E.Bernard**
Bernard, Jean–Paul (1921–) S	**J.-P.Bernard**
Bernard, Pierre Frédéric (1749–1825) S	**Bernard**
Bernardello, Luis M. (1953–) S	**Bernardello**
Bernardi, Luciano (1920–) S	**Bernardi**
Bernardino da Bologna S	**Bernardino**
Bernatowicz, Albert John (1920–) A	**Bernat.**

Bernátsky, J. (1873–1944) S	**Bernátsky**
Bernaux, P. (fl. 1949) M	**Bernaux**
Berndes, Wilhelm Eugene (1844–1882) B	**Berndes**
Bernelot Moens, Jacob Carel (1837–1886) S	**Bern.Moens**
Bernet, Henri (1850–1904) B	**Bernet**
Bernet, Martin (1815–1887) B	**M.Bernet**
Bernhardi, Johann Jakob (1774–1850) BMPS	**Bernh.**
Bernhardt, E.A. (fl. 1985) M	**Bernhardt**
Bernheim, M. A	**Bernheim**
Bernicchia, Annarosa (fl. 1988) M	**Bernicchia**
Bernis, Francisco (1916–) S	**Bernis**
(Bernis Madrazo, Francisco)	
see Bernis, Francisco	**Bernis**
Bernoulli, Karl (Carl) Gustav (1834–1878) PS	**Bernoulli**
Bernsteil, Otto (fl. 1912) S	**Bernsteil**
Bernstein, Heinrich Agathon (1822–1865) MS	**Bernstein**
Bernstein, Tamara A	**T.Bernstein**
Beroqui de Martínez, Martha (fl. 1969) M	**Beroqui**
Berrie, Geoffrey K. (1928–) B	**Berrie**
Berry, Andrew (fl. 1780–1810s) S	**Berry**
Berry, C.S. (fl. 1983) M	**C.S.Berry**
Berry, Edward Cain (1898–) M	**E.C.Berry**
Berry, Edward Wilber (1875–1945) ABF	**E.W.Berry**
Berry, L.A. (fl. 1991) M	**L.A.Berry**
Berry, Paul Edward (1952–) S	**P.E.Berry**
Bert, Jean-Jacques (1939–) A	**Bert**
Bertault, R. (fl. 1955) M	**Bertault**
Bertéa, Paul (fl. 1988) M	**Bertéa**
Bertero, Carlo Luigi Guiseppe (1789–1831) MPS	**Bertero**
Berthault, François (1857–1916) S	**Berthault**
Berthault, Pierre (fl. 1953/54) S	**P.Berthault**
Berthelot, Sabin (1794–1880) BPS	**Berthel.**
(Berthet, Julio Loão Arthaud–)	
see Arthaud–Berthet, Julio Loão	**Arth.-Berth.**
Berthet, Paul (1933–) MP	**Berthet**
Berthier, Jacques (fl. 1963) M	**Berthier**
Berthold, Gottfried Dietrich Wilhelm (1854–1937) ABM	**Berthold**
Berthoumieu, G.Victor (1840–1916) B	**Berthoum.**
(Berti, Luis Marcano)	
see Marcano–Berti, Luis	**Marc.-Berti**
Bertin, Pierre (1800–1891) S	**Bertin**
Bertini, S. (fl. 1957) M	**Bertini**
(Bertoldi, Marco De)	
see De Bertoldi, Marco	**De Bert.**
Bertolino, V. A	**Bertolino**
Bertoloni, Antonio (1775–1869) ABMPS	**Bertol.**
Bertoloni, Giuseppe (1804–1879) MS	**G.Bertol.**
Berton, M.C. (fl. 1979) M	**Berton**
Bertoni, M.D. (fl. 1991) M	**M.D.Bertoni**
Bertoni, Moisés de Santiago (1857–1929) S	**Bertoni**

Bertová, Lýdia (1931–) S	**Bertová**
Bertram, Ferdinand Wilhelm Werner (1835–1899) S	**Bertram**
Bertram, L. (fl. 1937) M	**L.Bertram**
Bertrand, André (fl. 1956) S	**A.Bertrand**
Bertrand, Charles Eugène (1851–1917) AF	**C.E.Bertrand**
Bertrand, G. (fl. 1913) M	**G.Bertrand**
Bertrand, Marcel C. S	**Bertrand**
Bertrand, Marie–Paule (1943–) M	**M.–P.Bertrand**
Bertrand, Paul (fl. 1911–26) F	**P.Bertrand**
Bertsch, Franz (1910–1944) S	**F.Bertsch**
Bertsch, Karl (1878–1965) BPS	**Bertsch**
Bertuch, Friedrich Justin (fl. 1801) S	**Bertuch**
Bertullo, V.H. (fl. 1970) M	**Bertullo**
Bertus, A.L. (fl. 1971) M	**Bertus**
Bérubé, J.A. (fl. 1988) M	**Bérubé**
Berwald, Johann Gottfried (fl. 1778) S	**Berwald**
Besada, Waheeb H. (fl. 1969) M	**Besada**
Besant, John William (1878–1944) S	**Besant**
Beschel, Ronald Ernst (1928–1971) M	**Beschel**
Bescherelle, Émile (1828–1903) B	**Besch.**
Besl, H.B. (fl. 1979) M	**Besl**
Besler, Basilius (1561–1629) L	**Besler**
Bessa, Pancrace (1772–1835) S	**Bessa**
Besse, François Maurice (1864–1924) S	**Besse**
Besser, Wilibald Swibert Joseph Gottlieb von (1784–1842) S	**Besser**
Bessey, Charles Edwin (1845–1915) AMS	**Bessey**
Bessey, Ernst Athearn (1877–1957) M	**E.A.Bessey**
Besson, M. (fl. 1965) M	**Besson**
Best, George Newton (1846–1926) BS	**Best**
Bestagno, Giuseppe (fl. 1959) M	**Bestagno**
Betche, Ernst (1851–1913) PS	**Betche**
Betcke, Ernst Friedrich (1815–1865) S	**Betcke**
Bethel, Ellsworth (1863–1925) M	**Bethel**
Bethge, Hans A	**Bethge**
Betsche, I. (fl. 1984) S	**Betsche**
Bettfreund, Karl (fl. 1887–1901) S	**Bettfr.**
Betts, Annie D. (fl. 1912) M	**Betts**
Beug, M.W. (fl. 1980) M	**Beug**
Beukes, G.J. (fl. 1982) S	**Beukes**
Beumée, Johan Gotlieb Benjamin (1888–1966) S	**Beumée**
Beunet, A.W. S	**Beunet**
Beurlin, A. S	**Beurlin**
Beurling, Pehr Johan (1800–1866) S	**Beurl.**
Beurmann, Charles Lucien de (1851–) M	**Beurm.**
Beurton, Christa (1945–) S	**Beurton**
Beusekom, C.F.van (1940–) S	**Beusekom**
Beusekom–Osinga, R.J.van (fl. 1970) S	**Beus.–Osinga**
(Beutelspacher B., Carlos R.)	
see Beutelspacher, Carlos R.	**Beutelsp.**

Beutelspacher, Carlos R. (fl. 1971) S	**Beutelsp.**
Beuzeville, Wilfred Alexander Watt de (1884–1954) S	**Beuzev.**
Beverwijk, A.L.van (1907–1963) M	**Beverw.**
Bevis, I. (fl.c.1839) P	**Bevis**
Bewerunge, W. (fl. 1948) S	**Bewer.**
Bews, John William (1884–1938) S	**Bews**
Beyer, Liselotte (fl. 1965) M	**L.Beyer**
Beyer, Rudolf (1852–1932) S	**Beyer**
Beyerle, C.Richard (fl. 1938) S	**Beyerle**
Beyma, J.F.H.van (1885–1966) M	**J.F.H.Beyma**
Beyma, T.H.van (–1945) M	**T.H.Beyma**
(Beyma Thoe Kingma, J.F.H.van)	
see Beyma, J.F.H.van	**J.F.H.Beyma**
(Beyma Thoe Kingma, T.H.van)	
see Beyma, T.H.van	**T.H.Beyma**
Beyrich, Heinrich Karl (1796–1834) BMS	**Beyr.**
Beyschlag, Franz Heinrich August (1856–) S	**Beyschl.**
Bezerra, José Luiz (fl. 1970) M	**J.L.Bezerra**
Bezerra, Prisco (fl. 1979) S	**P.Bezerra**
Bezold, F. (fl. 1870) M	**Bezold**
Bezuchova, Z.P. (fl. 1988) M	**Bezuch.**
Bhagyanarayana, G. (fl. 1976) M	**Bhagyan.**
Bhairavanath, D. (fl. 1985) M	**Bhairav.**
Bhandari, L.L. A	**L.L.Bhandari**
Bhandari, Madan Mal (1929–) S	**Bhandari**
Bhandary, H.R. (fl. 1982) M	**Bhandary**
Bhansali, A.K. (fl. 1984) S	**Bhansali**
Bharadwaj, S.D. (fl. 1971) M	**Bharadwaj**
Bharadwaja, R.C. (fl. 1957) S	**R.C.Bharadwaja**
Bharadwaja, Yajnyavalkya (1895–1963) A	**Bharadwaja**
Bharati, S.G. A	**Bharati**
Bhardwaj, Dinesh C. (1923–) ABF	**D.C.Bhardwaj**
Bhardwaj, L.N. (fl. 1987) M	**L.N.Bhardwaj**
Bhardwaja, Triloki Nath (1933–) P	**Bhardwaja**
Bhargava, K.S. (fl. 1977) M	**Bhargava**
Bhargavan, P. (1939–) S	**Bhargavan**
Bharucha, Faridunji (Faridoon) Rustomji (1904–) S	**Bharucha**
Bhaskar Rao, T.Susan A	**Bhaskar Rao**
Bhaskar, V. (fl. 1983) S	**Bhaskar**
Bhat, D. Jayarama (fl. 1977) M	**Bhat**
Bhat, K.G. (fl. 1985) S	**K.G.Bhat**
Bhatia, B.L. A	**B.L.Bhatia**
Bhatia, Madhu (fl. 1986) M	**Bhatia**
Bhatnagar, G.C. (fl. 1960) M	**G.C.Bhatn.**
Bhatnagar, G.S. (fl. 1960) S	**G.S.Bhatn.**
Bhatnagar, M.K. (fl. 1966) M	**M.K.Bhatn.**
Bhatt, G.C. (fl. 1965) M	**G.C.Bhatt**
Bhatt, R.P. (1935–) S	**Bhatt**
Bhatt, V.V. (fl. 1956) M	**V.V.Bhatt**

Bhattacharjee, M. (fl. 1966) M   **Bhattacharjee**
Bhattacharjee, Rera (fl. 1985) S   **R.Bhattacharjee**
(Bhattacharya, S.C.)
    see Bhattacharyya, S.C.   **S.C.Bhattach.**
Bhattacharya, Sunanda (fl. 1980–) S   **Bhattacharya**
Bhattacharyya, Prasanta Kumar (1947–) S   **P.K.Bhattach.**
Bhattacharyya, S.C. (fl. 1951) S   **S.C.Bhattach.**
Bhattacharyya, Upendra Chandra (1933–) S   **U.C.Bhattach.**
Bhatty, Shaheen F. (fl. 1980) M   **Bhatty**
Bhide, R.K. (fl. 1911) S   **Bhide**
Bhide, V.P. (fl. 1949) M   **V.P.Bhide**
Bhoite, A.S. (fl. 1991) M   **Bhoite**
Bi, C.S. (fl. 1982) M   **C.S.Bi**
Bi, C.Z. (fl. 1984) M   **C.Z.Bi**
Bi, Lie Jue (fl. 1990) A   **L.J.Bi**
Bi, Zhi Shu (1925–) M   **Z.S.Bi**
Biagi, J.A. (fl. 1988) S   **Biagi**
Bian, B.Z. (fl. 1979) M   **Bian**
Bianca, Giuseppe (1801–1883) S   **Bianca**
Bianchinotti, María V. (fl. 1990) M   **Bianchin.**
Bianco, Pasqua (1927–) S   **Bianco**
Bianor, (Frère) (1859–1920) S   **Bianor**
Biasoletto, Bartolomeo (1793–1859) AMS   **Biasol.**
Biatzovsky, Johann (c.1802–1863) S   **Biatz.**
Bibby, B.T.   A   **B.T.Bibby**
Bibby, Patrick Noel Sumner (1907–1955) BM   **Bibby**
Bibra, Ernst von (1806–1878) S   **Bibra**
Bicalho, Hamilton Dias (fl. 1964) S   **Bicalho**
Bicchi, Cesare (1822–1906) AS   **Bicchi**
Bicheno, James Ebenezer (1785–1851) S   **Bicheno**
Bickis, J.   S   **Bickis**
Bicknell, Clarence (1842–1918) PS   **C.Bicknell**
Bicknell, Eugene Pintard (1859–1925) S   **E.P.Bicknell**
Bicudo, Carlos Eduardo de Mattos (1937–) A   **C.E.M.Bicudo**
Bicudo, Denise de Campos (1955–) A   **D.C.Bicudo**
Bicudo, Rosa Maria Teixeira (–1980) A   **R.M.T.Bicudo**
Bidan, P. (fl. 1955) M   **Bidan**
Bidaud, A. (fl. 1991) M   **Bidaud**
Bidault, M.   S   **Bidault**
Bidder, George Palmer (1863–1953) A   **Bidder**
Biddulph, Susanna Fry (fl. 1790–1808) S   **Biddulph**
Bidgood, Gillian Sally (1948–) S   **Bidgood**
Bidin, Abdul Aziz (1948–) P   **Bidin**
Bidwill, John Carne (1815–1853) S   **Bidwill**
Bieberdorf, F.W. (fl. 1955) M   **Bieberd.**
(Bieberstein, Friedrich August Marschall von)
    see Marschall von Bieberstein, Friedrich August   **M.Bieb.**
Biebl, Richard (1908–) A   **Biebl**
Biecheler, Berthe   A   **Biecheler**

Biedenfeld, Ferdinand Leopold Karl von (1788–1862) S	**Biedenf.**
Biedenkopf, Hermann (1870–) M	**Biedenk.**
Biehler, Johann Friedrich Theodor (c.1785–18??) S	**Biehler**
Bielawski, J.B.M. S	**Biel.**
Bielawski, Wincenty S	**W.Biel.**
Bielefield, Rudolf (fl. 1900) S	**Bielef.**
Bielz, E.A. (1827–1898) S	**Bielz**
Bienert, Theophil (–1873) S	**Bien.**
Bier, John Ertel (1909–) M	**Bier**
Bierhorst, David William (1924–1966) FPS	**Bierh.**
Bierkander, Clas (1735–1795) S	**Bierk.**
Bierner, Mark William (1946–) S	**Bierner**
Biers, P.–M. (fl. 1924) M	**Biers**
Biffi, Ulderico A	**Biffi**
Biga, M.L.Bestagno (fl. 1955) M	**Biga**
Bigeard, René (1840–1917) M	**Bigeard**
Bigelow, Howard E. (fl. 1958) M	**H.E.Bigelow**
Bigelow, Jacob (1787–1879) S	**Bigelow**
Bigelow, John Milton (1804–1878) S	**J.M.Bigelow**
Bigg, W.L. (fl. 1988) M	**Bigg**
Biggs, Rosemary Peyton (1912–) M	**Biggs**
Bigler, H. A	**Bigler**
Bigsby, John Jeremiah (1792–1881) S	**Bigsby**
Bigwood, Anthony J. (fl. 1987) F	**A.J.Bigwood**
Bigwood, J.E. (fl. 1980) M	**Bigwood**
Bihari, Gyula (1889–) S	**Bihari**
Bijhouwer, Jan Tijs Pieter (1898–) S	**Bijh.**
(Bijl, Paul Andries Van der)	
see Van der Byl, Paul Andries	**Van der Byl**
Bijlsma, R.J. (fl. 1986) S	**Bijlsma**
Biju, S.D. (fl. 1991) S	**Biju**
Bilai, Vera Iosifovna (1908–) M	**Bilai**
Bilaidi, A.S. (fl. 1971) S	**Bilaidi**
Bilewsky, Felix (1902–1979) B	**F.Bilewsky**
Bilewsky, H. (fl. 1911) M	**Bilewsky**
Bilge, Emine (1926–) S	**Bilge**
Bilgram, H. (fl. 1905) M	**Bilgram**
Bilgrami, K.S. (fl. 1973) M	**Bilgrami**
Bilgütay, Utarit A	**Bilgütay**
Bilik, Maria (fl. 1960) S	**Bilik**
Billard, Chantal (1947–) A	**Billard**
(Billardière, Jacques Julien Houtton de la)	
see Labillardière, Jacques Julien Houtton de	**Labill.**
Billberg, Gustaf Johan (1772–1844) S	**Billb.**
Billberg, Johan Immanuel (1799–1845) S	**J.I.Billb.**
Billet, A. (fl. 1898) P	**Billet**
Billiard, L.C. (fl. 1861–1870) S	**Billiard**
Billings, Elkanagh (1820–1876) AF	**Billings**
Billings, John Shaw (1838–1913) M	**J.S.Billings**

Billon–Grand, Genevieve (fl. 1989) M	**Billon-Grand**
Billore, K.V. (fl. 1970) S	**Billore**
Billot, Paul Constant (1796–1863) S	**Billot**
Bills, Gerald F. (fl. 1984) M	**Bills**
Bilý, Julius (1895–1970) A	**J.Bílý**
Bilyk, Gavriel Ivanovich (1904–) S	**Bilyk**
(Binert, Theophil)	
see Bienert, Theophil	**Bien.**
Binford, C.H. (fl. 1944) M	**Binford**
Binnendijk, Simon (1821–1883) PS	**Binn.**
Binney, Edward William (1812–1881) F	**Binney**
Binstead, Charles Herbert (1862–1941) B	**Binst.**
Bint, A.M. A	**Bint**
Binyamini, Nissan (fl. 1972) M	**Binyam.**
Binz, August (1870–1963) AS	**Binz**
Bioletti, Frederic Theodore (1865–1939) S	**Bioletti**
Biourge, Philibert Melchior Joseph Ehi (1864–19??) M	**Biourge**
Bir, Sarmukh Singh (1929–) PS	**Bir**
Biradar, N.V. (fl. 1976) M	**Biradar**
Biraghi, A. (fl. 1947) M	**Biraghi**
Birari, S.P. (fl. 1973) S	**Birari**
Birch–Andersen, Peter A	**Birch–And.**
Birchfield, Wray (fl. 1960) M	**Birchf.**
Bird, C.G. (fl. 1934) S	**Bird**
Bird, Carolyn J. (1947–) A	**C.J.Bird**
Bird, Charles Durham (1932–) BM	**C.D.Bird**
Birdsey, Monroe Roberts (1922–) S	**Birdsey**
Birdwood, George Christopher Molesworth (1832–1917) S	**Birdw.**
Biria, J.A.J. (1789–) S	**Biria**
Birina, L.M. A	**Birina**
Birk, L.A. (fl. 1980) S	**Birk**
Birkinshaw, J.H. (fl. 1942) M	**Birkinshaw**
Biroli, Giovanni (1772–1825) MS	**Biroli**
Birula, J. (fl. 1928) M	**Birula**
Bisby, Frank Ainley (1945–) S	**F.A.Bisby**
Bisby, Guy Richard (1889–1958) M	**Bisby**
Bischler, Helene (1932–) B	**Bischl.**
Bischoff, Bernhardt A	**B.Bisch.**
Bischoff, Gottlieb Wilhelm T.G. (1797–1854) BMPS	**Bisch.**
Bischoff, Harry William (1922–) A	**H.W.Bisch.**
Bishop, Ann A	**A.Bishop**
Bishop, David (1788–1849) S	**Bishop**
Bishop, L.Earl (fl. 1973) PS	**L.E.Bishop**
Bisht, B.S. (fl. 1987) S	**B.S.Bisht**
Bisht, S. (fl. 1983) M	**S.Bisht**
Bisse, Johannes (1935–1984) S	**Bisse**
Bissell, Charles Humphrey (1857–1925) P	**Bissell**
Bisset, James (1843–1911) AS	**Bisset**
Bissett, John (fl. 1979) M	**Bissett**

Biswas, A. (fl. 1986–1989) P	**A.Biswas**
Biswas, Kalipada P. (1899–1969) APS	**Biswas**
Biswas, M.C. (fl. 1982–83) P	**M.C.Biswas**
Biswas, Samarendra Nath (1939–) AS	**S.N.Biswas**
Bitancourt, Agesilau Antonio (1894?–) M	**Bitanc.**
Bitter, Friedrich August Georg (1873–1927) MPS	**Bitter**
Bittrich, Volker (fl. 1984–) S	**Bittrich**
Bivona–Bernardi, Antonius de (1774–1837) ABMS	**Biv.**
Bizot, Maurice (1905–1979) B	**Bizot**
Bizzarri, Maria Paola (1937–) PS	**Bizzarri**
Bizzozero, Giacomo (1852–1885) ABMS	**Bizz.**
Bjaerke, T. A	**Bjaerke**
Bjelcić, Zeljka (fl. 1973–) S	**Bjelcić**
Bjerkander, Clas (1735–1795) M	**Bjerk.**
Björkman, Sven Oscar (1920–1956) S	**Björkman**
Björkqvist, Ingemar (1931–) S	**Björkqv.**
Björling, Karl (1910–) M	**Björl.**
Björnstad, Anders (fl. 1970) S	**A.Björnstad**
Björnstad, Inger (later Nordal, I.) (1942–) S	**I.Björnstad**
(Björnström, Fredrik Johan)	
see Bjørnstrøm, Fredrik Johan	**Bjørnstr.**
Bjørnstrøm, Fredrik Johan (1833–1889)BS	**Bjørnstr.**
Bjornstrom, Fredrik Johan (1944–) S	**F.Bjornstr.**
Blacic, Jan M. A	**Blacic**
Black, Allan A. (1832–1865) S	**Black**
Black, George (fl. 1957) S	**G.Black**
Black, George Alexander (1916–1957) S	**G.A.Black**
Black, John McConnell (1855–1951) PS	**J.M.Black**
Black, Maurice (1904–1973) A	**M.Black**
Black, Raleigh Adelbert (1880–1963) S	**R.A.Black**
Blackburn, Benjamin Coleman (1908–) S	**Blackburn**
Blackburn, M.D. (fl. 1986) M	**M.D.Blackburn**
Blackman, A.L. A	**A.L.Blackman**
Blackman, Frederick Frost (1866–1947) A	**F.F.Blackman**
Blackman, Vernon Herbert (1872–1967) A	**V.H.Blackman**
Blackmon, C.W. (fl. 1955) M	**Blackmon**
Blackmore, John A.P. (fl. 1960) S	**Blackmore**
Blackmore, Stephen (1952–) S	**S.Blackmore**
Blackwell, Elizabeth (c.1700–1758) S	**Blackw.**
Blackwell, Kay P. (fl. 1975) S	**K.P.Blackw.**
Blackwell, Meredith (1940–) M	**M.Blackw.**
Blackwell, William Hoyle (1939–) AMS	**W.H.Blackw.**
Blada, Ion (fl. 1961) M	**Blada**
Bladh, Peter Johan (1746–1816) S	**Bladh**
Blagodatskaja, Valentina M. (fl. 1973) M	**Blagod.**
(Blagodatskaya, Valentina M.)	
see Blagodatskaja, Valentina M.	**Blagod.**
(Blagoveshchenskij, Andrey Vassilievich)	
see Blagovestschensky, Andrey Vassilievich	**Blagov.**

Blagovestschensky, Andrey Vassilievich (1889–1982) S	**Blagov.**
Blainville, Henri Marie Ducrotay de (1777–1850) AF	**Blainv.**
Blaise, S. (fl. 1970) S	**Blaise**
Blaisten, Raúl (1913–) A	**Blaisten**
Blake, Joseph (1814–1888) S	**Blake**
Blake, Sidney Fay (1892–1959) BPS	**S.F.Blake**
Blake, Stanley Thatcher (1910–1973) S	**S.T.Blake**
Blakeley, William Faris (1875–1941) S	**Blakeley**
Blakelock, Ralph Anthony (1915–1963) S	**Blakelock**
Blakeman, J.P. (fl. 1964) M	**Blakeman**
Blakeslee, Albert Francis (1874–1954) MS	**Blakeslee**
Blakiston, Th.W. (fl. 1862) P	**Blakiston**
Blanc, Albert A. (1850–1928) S	**Blanc**
Blanca, Gabriel (1954–) S	**Blanca**
(Blanca López, Gabriel)	
see Blanca, Gabriel	**Blanca**
Blanchard, Frank N. (1888–) A	**F.N.Blanch.**
Blanchard, O.J. (fl. 1979–) S	**O.J.Blanch.**
Blanchard, Raphaël (1857–1919) AM	**R.Blanch.**
Blanchard, William Henry (1850–1922) PS	**Blanch.**
Blanché i Vergés, Cesar (1958–) S	**C.Blanché**
Blanche, Charles Isodore (1823–1887) S	**C.I.Blanche**
Blanche, Emanuel (1824–1908) S	**Blanche**
Blanche, F. S	**F.Blanche**
(Blanché i Vergés, Cesar)	
see Blanché, Cesar	**C.Blanché**
Blanchet, Jacques Samuel (1807–1875) S	**Blanchet**
Blanchette, R.A. (fl. 1991) M	**Blanchette**
Blanck, A. (fl. 1884) S	**Blanck**
Blanco, Cenobio E. S	**C.E.Blanco**
Blanco, Francisco Manuel (1778–1845) APS	**Blanco**
Blanco, M.N. (fl. 1989) M	**M.N.Blanco**
Blanco, Manuel (1780–1848) M	**M.Blanco**
Blanco, Paloma (1950–) M	**P.Blanco**
Bland, C.E. (fl. 1985) M	**Bland**
Bland, Roger G. A	**R.G.Bland**
Blandow, Otto Christian (1778–1810) B	**Blandow**
Blanford, H.F. (1834–1893) PS	**Blanf.**
Blank, Paul (1936–) M	**P.Blank**
Blank, Rudolf J. A	**R.J.Blank**
Blankinship, Joseph William (1862–1938) S	**Blank.**
Blaschke, F. (fl. 1964) S	**Blaschke**
Blasdale, Walter Charles (1871–) MS	**Blasdale**
Blasdell, Robert Ferris (1929–) P	**Blasdell**
Blaser, Paul (fl. 1975) M	**Blaser**
Blastein, Raúl (fl. 1941) M	**Blastein**
Błaszkowski, Janusz (–1987) M	**Błaszk.**
Blatter, Ethelbert (1877–1934) BPS	**Blatt.**
Blattny, T. (fl. 1913) S	**Blattny**

Blaxall, Frank R. (fl. 1896) M   **Blaxall**
Blaxell, Donald Frederick (1934–) S   **Blaxell**
Blazé, Kevin L.   A   **K.L.Blazé**
Blečić, Vilotije (1911–1981) S   **Blečić**
Bleck, John   A   **J.Bleck**
Bleck, M.B. (fl. 1984–) S   **Bleck**
Bleicher, Joseph (fl. 1899) S   **J.Bleicher**
Bleicher, Marie Gustav (1838–1901) S   **Bleicher**
Bleisch   A   **Bleisch**
Blevins, Anne (fl. 1945) M   **Blevins**
Bliding, Carl Vilhelm (1891–) A   **Bliding**
Blinks, Anne Catherine Hof (1903–) A   **A.C.H.Blinks**
Blinks, Lawrence Rogers (1900–1989) A   **L.R.Blinks**
Blinn, Dean W. (1941–) A   **Blinn**
Blinova, E.I.   A   **Blinova**
Blinovskij, K.V. (1903–) S   **Blin.**
Bliss, Donald Everett (1903–1951) M   **Bliss**
Blium, Oleg B. (fl. 1970) M   **Blium**
Bljumina, L.S.   A   **Bljumina**
Bloch, Bruno (fl. 1911) M   **Bloch**
Bloch, Jean-Paul   A   **J.-P.Bloch**
Blochmann, Friedrich (1858–1931) A   **Blochm.**
Blochwitz, Adalbert von (fl. 1929) M   **Blochwitz**
Blockeel, T.L. (fl. 1982) B   **Blockeel**
Błocki, Bronislaw (1857–1919) S   **Błocki**
Bloembergen, Siebe (1905–) S   **Bloemb.**
Blok, Ida (fl. 1978) M   **I.Blok**
Blok, W.J. (fl. 1988) M   **W.J.Blok**
Blom, Carl Hilding (1888–1972) S   **C.H.Blom**
Blom, Carl Magnus (1737–1815) S   **Blom**
Blom, Hans Haavardsholm (1955–) B   **H.H.Blom**
Blomberg, Olof Gotthard (1838–1901) M   **Blomb.**
Blomgren, Nils Harald (1901–1926) S   **Blomgr.**
Blomquist, Hugo Leander (1888–1964) ABPS   **H.L.Blomq.**
Blomquist, Sven Gustaf Krister Gustafson (1882–1953) S   **Blomq.**
Bloński, Franciszek(Franz) Kzawery (1867–1910) MS   **Bloński**
Bloss, H.E. (fl. 1970) M   **Bloss**
Blossfeld, Harry (1913–1986) S   **H.Blossf.**
Blossfeld, Robert (1882–) S   **Blossf.**
Bloxam, Andrew (1801–1878) MS   **A.Bloxam**
Bloxam, Richard Rowland (1798–1877) M   **R.R.Bloxam**
Bluff, Mathias Joseph (1805–1837) BS   **Bluff**
Bluket, N.A.   S   **Bluket**
Blum, F.K.   S   **F.K.Blum**
Blum, Jean (fl. 1952) M   **J.Blum**
Blum, John Leo (1917–) A   **Blum**
Blume, Carl(Karl) Ludwig von (1796–1862) BMPS   **Blume**
Blumenfeld, S.N. (fl. 1984) M   **Blumenf.**
Blumenschein, Almiro (fl. 1960) S   **Blumensch.**

Blumer, Jacob Corwin (1872–1948) S	**Blumer**
Blumer, Samuel von (fl. 1926) M	**S.Blumer**
Blumrich, Josef (1865–1949) B	**Blumr.**
Blunden, Gerald (1939–) S	**Blunden**
Blytt, Axel Gudbrand (1843–1898) BFMPS	**A.Blytt**
Blytt, Mathias Numsen (1789–1862) BS	**Blytt**
(Blyumina, L.S.)	
see Bljumina, L.S.	**Bljumina**
Boalch, G.T. (1933–) A	**Boalch**
Boas, Friedrich (fl. 1913) S	**F.Boas**
Boas, Johan Erik Vesti (1855–1935) S	**Boas**
Boberski, Wladyslaw (1846–1891) M	**Boberski**
Bobrov, Andrej Evgenievich (1936–) PS	**A.E.Bobrov**
Bobrov, Evgenij Grigorievicz (1902–1983) FS	**Bobrov**
Bocchieri, Emanuele (1941–) S	**Bocchieri**
Boccone, Paolo (Silvio) (1633–1704) LM	**Boccone**
(Bochantsev, Victor Petrovič)	
see Botschantzev, Victor Petrovič	**Botsch.**
(Bochantseva, Vera Viktorovna)	
see Botschantzeva, Vera Viktorovna	**V.V.Botschantz.**
(Bochantseva, Zinaida Petrovna)	
see Botschantzeva, Zinaida Petrovna	**Botschantz.**
Böcher, Tyge Wittrock (1909–1983) AS	**Böcher**
Bochicchio, N. (fl. 1894) M	**Bochicchio**
Bochkarnikova, N.M. (fl. 1967) S	**Bochkarn.**
Bock, Hieronymous (Tragus) (1498–1554) L	**H.Bock**
Bock, I. (fl. 1986) S	**I.Bock**
Bock, Otto A	**O.Bock**
Bock, Thor Methven (fl. 1941) S	**T.M.Bock**
Bock, Walter A	**W.Bock**
Bock, Wilhelm (fl. 1952–1962) F	**Wilh.Bock**
Bock, Wolfgang von (fl. 1908) B	**Bock**
Böckel, Godwin (fl. 1853) P	**Böckel**
(Böckeler, Johann Otto)	
see Boeckeler, Johann Otto	**Boeck.**
(Böckler, Johann Otto)	
see Boeckeler, Johann Otto	**Boeck.**
(Bockstal, Liliane Van)	
see Van Bockstal, Liliane	**Van Bockstal**
Bocquet, Gilbert François (1927–1986) S	**Bocquet**
Bocquillon, Henri Théophile (1834–1883) S	**Bocq.**
Bocquillon-Limousin, Henri (fl. 1891–1914) S	**Bocq.-Lim.**
(Boczantzev, Victor Petrovič)	
see Botschantzev, Victor Petrovič	**Botsch.**
(Boczantzeva, Vera Viktorovna)	
see Botschantzeva, Vera Viktorovna	**V.V.Botschantz.**
(Boczantzeva, Zinaida Petrovna)	
see Botschantzeva, Zinaida Petrovna	**Botschantz.**
(Boczkarnikova, N.M.)	
see Bochkarnikova, N.M.	**Bochkarn.**

Bodard, Marcel J. (1927–1988) AS	**M.Bodard**
Bodard, Pierre Henri Hippolyte (fl. 1798–1810) S	**Bodard**
(Bödecker, Friedrich)	
see Boedeker, Friedrich	**Boed.**
Boden, Brian Peter (1921–) A	**Boden**
Bodenbender, Wilhelm (1857–1941) S	**Bodenb.**
Bodin, E. (fl. 1912) M	**E.Bodin**
Bodin, Nicolaus Gustavus (fl. 1798) S	**Bodin**
Bodkin, Norlyn L. (1937–) S	**Bodkin**
Bodman, Mary Cecilia (fl. 1953) M	**Bodman**
Bodrogközy, Gyöorgy A	**Bodrogk.**
Boechat, Sonja de Castro (fl. 1990) S	**Boechat**
Boeck, Christian Peter Bianco (1798–1877) BS	**Boeck**
Boeck, William C. A	**W.C.Boeck**
Boeckeler, Johann Otto (1803–1899) AS	**Boeck.**
Boedeker, Friedrich (1867–1937) S	**Boed.**
Boedijn, Karel Bernard (1893–1964) AMS	**Boedijn**
Boehmer, Georg Rudolf (1723–1803) ABMPS	**Boehm.**
Boehmer, Louis B. (1822–82 or 1896–1908) S	**L.B.Boehm.**
(Boekel, Norma M.da Costa Van)	
see Van Boekel, Norma M.da Costa	**Van Boekel**
Boekhout, Teun (1955–) M	**Boekhout**
Boelcke, Osvaldo (Oswaldo) (1920–) S	**Boelcke**
Boele, C. (fl. 1989) S	**Boele**
Boelens, W. (fl. 1943) M	**Boelens**
Boenninghausen, Clemens Maria Griedrich von (1785–1864) S	**Boenn.**
(Boer, Hendrik Wijbrand de)	
see de Boer, Hendrik Wijbrand	**de Boer**
(Boer, Jan Gerard Wessels)	
see Wessels Boer, Jan Gerard	**Wess.Boer**
Boerema, Gerhard H. (fl. 1962) M	**Boerema**
(Boergesen, Fredrik Christian Emil)	
see Børgesen, Fredrik Christian Emil	**Børgesen**
Boerhaave, Herman (1668–1739) LS	**Boerh.**
Boerlage, Jacob Gijsbert (1849–1900) S	**Boerl.**
(Boerner, Carl (Karl) Julius Bernhard)	
see Börner, Carl (Karl) Julius Bernhard	**Börner**
Boertmann, David (fl. 1990) M	**Boertm.**
Boesewinkel, H.J. (fl. 1976) M	**Boesew.**
Boewe, Gideon Herman (1895–1970) M	**Boewe**
Bogart, Fred van de (fl. 1975) M	**Bogart**
Bogdan, Alexis V. (fl. 1949) S	**Bogdan**
Bogenhard, Carl (1811–?1853) S	**Bogenh.**
Boggiani, Oliviero (fl. 1912) P	**Boggiani**
Bogin, Clifford (1920–) S	**Bogin**
Bogle, A.Linn (1931–) S	**Bogle**
Boglioli, L. (fl. 1951) M	**Boglioli**
Bogner, Josef (1939–) S	**Bogner**
Bogoyavlensky, N. (fl. 1922) M	**Bogoyavl.**
Bogusch, Edwin Robert (1905–) S	**Bogusch**

Boguslavskii, R.L. (fl. 1982) S	**Bogusl.**
Boguslaw, I.A. (fl. 1846) S	**Boguslaw**
Bohlen, Ann (fl. 1950) B	**Bohlen**
Bøhler, H.C. (fl. 1974) M	**H.C.Bøhler**
Bohler, John (1797–1872) M	**Bohler**
Bohlin, Berger (fl. 1971) F	**B.Bohlin**
Bohlin, Jan Erik (fl. 1988) S	**J.E.Bohlin**
Bohlin, Knut Harald (1869–) A	**Bohlin**
Bohling, Maude H. A	**Bohling**
Böhm, Anton A	**Böhm**
Bohm, B.A. (fl. 1985) S	**Bohm**
Böhme, Hannelore (fl. 1967) M	**H.Böhme**
Böhme, Olga (fl. 1931–1942) S	**O.Böhme**
Böhme, Paul (1860–1935) S	**Böhme**
(Böhmer, Georg Rudolf)	
see Boehmer, Georg Rudolf	**Boehm.**
Bohnen (? née Kies, P.) (fl. 1959) S	**Bohnen**
Bohnstedt, Alexander Reinhold (1839–1903) S	**Bohnst.**
Bohovik, I.V. (fl. 1936) M	**Bohovik**
Bohra, A. (fl. 1972) M	**Bohra**
Bohus, G. (1914–) M	**Bohus**
Bohuslav (fl. 1845) S	**Bohuslav**
Boidin, Jacques (1893–) M	**Boidin**
Boidol, Michael (1936–) M	**Boidol**
Boiffard, J. (fl. 1972) M	**Boiffard**
Boiko, E.V. (fl. 1985) S	**Boiko**
Boira, H. (1943–) S	**Boira**
Bois, Désiré Georges Jean Marie (1856–1946) S	**Bois**
Boisduval, Jean Baptiste Alphonse Déchauffour(e) de (1801/1799–1879) S	**Boisd.**
Boise, Jean R. (1952–) M	**Boise**
Boissevain, Charles Hercules (1893–1946) S	**Boissev.**
(Boissier de la Croix Sauvages, François)	
see Sauvages, François Boissier de la Croix de	**Sauvages**
Boissier, Pierre Edmond (1810–1885) MPS	**Boiss.**
Boissière, J.C. (fl. 1980) M	**Boissière**
Boissieu, Claude Victor (1784–1868) S	**Boissieu**
(Boissieu de la Martinière, Claude Victor)	
see Boissieu, Claude Victor	**Boissieu**
Boissieu, Henri de (1871–1912) S	**H.Boissieu**
Boisson S	**Boisson**
Boissonade, J. (1831–1897) S	**Boissonade**
Boistel, Alphonse Barthèlè (1836–1908) M	**Boistel**
Boiteau, Pierre L. (1911–) S	**Boiteau**
Boivin, André S	**A.Boivin**
Boivin, Joseph Robert Bernard (1916–1985) PS	**B.Boivin**
Boivin, Louis Hyacinthe (1808–1852) S	**Boivin**
Bojanovská, A. (fl. 1966) M	**Bojan.**
Bojer, Wenceslas(Wenzel) (1797–1856) PS	**Bojer**
Bojke, H. (1892–) S	**Bojke**

Bok Choon S	**Bok Choon**
Bok, Marie B. (fl. 1936) S	**Bok**
Boke, Norman Hill (1913–) S	**Boke**
Bokhari, Mumtaz Hussain (1937–) S	**Bokhari**
Boland, Douglas John (1947–) S	**Boland**
Bolander, Henry Nicholas (1832–1897) S	**Bol.**
Bolay, A. (fl. 1967) M	**Bolay**
Bold, Harold C. (1909–1987) A	**H.C.Bold**
(Bolding, Isaäc)	
see Boldingh, Isaäc	**Bold.**
Boldingh, Isaäc (1879–1938) PS	**Bold.**
Boldrini, Ilsi Job (Iob) (fl. 1970) S	**Boldrini**
Boldt, (Johan Georg) Robert (1861–1923) A	**Boldt**
Bole, Pritamial Vijoyshanker (1920–) PS	**Bole**
Bolick, Margaret R. (1950–) S	**Bolick**
Bolkhovitina, N.A. (fl. 1959) B	**Bolkh.**
Bolkhovskikh S	**Bolkhovsk.**
Boll, Ernst Friedrich August (1817–1868) S	**Boll**
Bolla, Johann (1806–1881) S	**Bolla**
Bollard, E.G. (fl. 1950) M	**Bollard**
Bolle, Carl (Karl) August (1821–1909) BPS	**Bolle**
Bolle, Friedrich Franz August Albrecht (1905–) S	**F.Bolle**
Bolle, Pierette Cornelie (1893–1945) M	**P.C.Bolle**
Bollen, G.J. (fl. 1975) M	**G.J.Bollen**
Bollen, Walter Beno (1896–) M	**Bollen**
Bolleter, Eugen (1873–1922) B	**Bolleter**
Bolley, Henry Luke (1865–1956) M	**Bolley**
Bolliger, Markus (1951–) S	**Bolliger**
Bolochonzew, Evgenij Nikolaovič (1879–1909) A	**Boloch.**
(Bolohonzcev, Evgenij Nikolaovič)	
see Bolochonzew, Evgenij Nikolaovič	**Boloch.**
(Bolòs i Capdevila, Oriol de)	
see Bolòs, Oriol de	**O.Bolòs**
Bolòs, Oriol de (1924–) PS	**O.Bolòs**
Bolòs y Vaireda, Antonio de (1889–1975) S	**A.Bolòs**
Bolschakov, N.M. (1951–) S	**Bolsch.**
Bolton, A.T. (fl. 1963) M	**A.T.Bolton**
Bolton, Ethel (fl. 1938) S	**E.Bolton**
Bolton, J.J. (1952–) A	**J.J.Bolton**
Bolton, James (1758–1799) MPS	**Bolton**
Boltovskoy, Andres (fl. 1984) AM	**Boltovskoy**
Boltshauser, Heinrich (1852?–1899) M	**Boltsh.**
Bolus, Frank (1870–1945) S	**F.Bolus**
Bolus, Harriet Margaret Louisa (née Kensit, H.M.L.) (1877–1970) S	**L.Bolus**
Bolus, Harry (1834–1911) S	**Bolus**
Bolzon, Pio (1867–1940) PS	**Bolzon**
Bomansson, Johan Oskar (1838–1906) B	**Bom.**
Bomhard, Miriam Lucile (1898–1952) S	**Bomhard**
Bommer, Elisa Caroline (1832–1910) M	**E.Bommer**

Bommer, Joseph (Jean) Édouard (1829–1895) MPS   **J.Bommer**
Bompard, H.   A   **Bompard**
Bon, Marcel (fl. 1973) M   **Bon**
Bóna, József   A   **Bóna**
(Bonada, Josep Vigo)
    see Vigo Bonada, Josep   **Vigo**
Bonadonna, Francesco Paolo   A   **Bonad.**
(Bonafé Barceló, Francesc)
    see Bonafè, Francesc   **Bonafé**
Bonafè, Francesc (fl. 1977–1989) S   **Bonafé**
Bonafous, Matthieu (1793–1852) S   **Bonaf.**
Bonamo, P.M. (fl. 1968) F   **Bonamo**
Bonamy, François (1710–1786) S   **Bonamy**
Bonaparte, Roland Napoléon (1858–1924) PS   **Bonap.**
Bonar, Lee (1891–1977) M   **Bonar**
Bonati, Gustave Henri (1873–1927) PS   **Bonati**
Bonato, Giuseppe Antonio (1753–1836) S   **Bonato**
Bonavia, Emanuel (1826–1908) S   **Bonavia**
Bond, H.A. (fl. 1985) M   **H.A.Bond**
Bond, Pauline (1917–) S   **Bond**
Bond, T.A.   A   **T.A.Bond**
Bondam, Rutger (1817–1896) S   **Bondam**
Bondar, Gregório Gregorievich (1881–1959) MS   **Bondar**
(Bondarcev, Appollinaris Semenovich)
    see Bondartsev, Appollinaris Semenovich   **Bondartsev**
(Bondarceva, Margarita Apollinavievra)
    see Bondartseva, Margarita Apollinavievra   **Bondartseva**
(Bondarceva–Monteverde, Vera Nikolaevna)
    see Bondartseva–Monteverde, Vera Nikolaevna   **Bond.-Mont.**
Bondarenko, O.N. (1923–) S   **Bondarenko**
Bondareva, N.A. (fl. 1989) S   **Bondareva**
Bondartsev, Appollinaris Semenovich (1877–1968) M   **Bondartsev**
Bondartseva, Margarita Apollinavievra (1935–) M   **Bondartseva**
Bondartseva–Monteverde, Vera Nikolaevna (1889–1944) M   **Bond.-Mont.**
(Bondartzeva–Monteverde, Vera Nikolaevna)
    see Bondartseva–Monteverde, Vera Nikolaevna   **Bond.-Mont.**
Bonde, S.D. (1950–) M   **Bonde**
Bondesen, E.   A   **Bondesen**
Bondev, I. (fl. 1966) S   **Bondev**
Bondoux, P. (fl. 1978) M   **Bondoux**
Bondt, Nicolaas (1765–1796) S   **Bondt**
Bone, E.C.P.   A   **Bone**
Bonelli, Giorgio (1724–1782) S   **Bonelli**
Bonequet, Pierre August (1882–) M   **Bonequet**
Bonet, F.   A   **Bonet**
Bonetti, Maria I.R. (fl. 1966–68) F   **Bonetti**
Bongale, U.D.   A   **Bongale**
Bongard, August (Gustav) Heinrich von (1786–1839) PS   **Bong.**
Bongini, V. (fl. 1932) M   **Bongini**

Bonhomme, J. A	**Bonhomme**
Böning, Karl (fl. 1933) M	**Böning**
Bonjean, Joseph (1780–1846) S	**Bonjean**
Bonker, Frances (fl. 1932) S	**Bonker**
Bonnard, J. (fl. 1987) M	**Bonnard**
Bonnemaison, Théophile (1774–1829) AS	**Bonnem.**
Bonnemoy, Frédérique (fl. 1989) M	**Bonnemoy**
Bonner, Charles Edmond Bradlaugh (1915–1976) CPS	**Bonner**
Bonnet, Edmond (1848–1922) S	**Bonnet**
Bonnier, Gaston Eugène Marie (1851–1922) S	**Bonnier**
(Bönninghäusen, Clemens Maria Griedrich von)	
see Boenninghausen, Clemens Maria Griedrich von	**Boenn.**
Bono, Guiseppe (1927–) S	**Bono**
Bononi, Vera Lúcia R. (1944–) M	**Bononi**
Bonorden, Hermann Friedrich (1801–1884) M	**Bonord.**
Bonpland, Aimé Jacques Alexandre (né Goujaud, A.J.A.) (1773–1858) APS	**Bonpl.**
Bonsdorff, Ernst Jakob Waldemar (1842–1936) B	**Bonsd.**
Bonstedt, Carl (1866–1953) S	**Bonstedt**
Bontea, Vera (fl. 1948) M	**Bontea**
Boo, Sung Min A	**S.M.Boo**
Booberg, Karl Gunnar (1892–1944) S	**Booberg**
Boodle, Leonard Alfred (1865–1941) AS	**Boodle**
Boom, Boudewijn Karel (1903–1980) S	**Boom**
Boom, Brian Morley (1954–) S	**B.M.Boom**
Boomsma, Clifford David (1915–) S	**Boomsma**
Boonyamalik, K. (fl. 1986) S	**Boonyam.**
Boos, G.V. (1920–) S	**G.V.Boos**
Boos, Joseph (1794–1879) S	**Boos**
Booth, Beatrice C. A	**B.C.Booth**
Booth, Colin (fl. 1957) M	**C.Booth**
Booth, John (1836–1908) S	**J.Booth**
Booth, John Richmond (1801–1847) S	**J.R.Booth**
Booth, T. (fl. 1969) M	**T.Booth**
Booth, Thomas Jonas (1829–post 1861) S	**T.J.Booth**
Booth, William Beattie (c.1804–1874) S	**Booth**
Booth, William Edwin (1909–) AS	**W.E.Booth**
Boothman, H.Stuart (fl. 1934) S	**Boothman**
Boothroyd, C.W. (fl. 1973) M	**Boothr.**
Boott, Francis M.B. (1792–1863) S	**Boott**
Boott, William (1805–1887) S	**W.Boott**
Boquiren, Daisy T. (fl. 1971) M	**Boquiren**
Bor, Norman Loftus (1893–1972) S	**Bor**
Boraiah, A. (fl. 1972) S	**A.Boraiah**
Boraiah, G. (fl. 1964) S	**G.Boraiah**
Boral, L.L. (fl. 1964) M	**Boral**
Borbás, Vinczé (Vincent, Vince) von (1844–1905) PS	**Borbás**
Borchers–Kolb, E. (fl. 1985) S	**Borch.-Kolb**
Borchsenius, Finn (fl. 1989) S	**Borchs.**
(Borckhausen, Moritz (Moriz) Balthasar)	
see Borkhausen, Moritz (Moriz) Balthasar	**Borkh.**

Bordakov, Leonid Petrovich (1885–1940) S   **Bordakov**
Borden, Carol Ann   A   **Borden**
Bordère, Henri (1825–1889) S   **Bordère**
Bordères, O. (fl. 1939–1968) S   **Bordères**
(Bordères–Rey, O.)
  see Bordères, O.   **Bordères**
Bordzilowski, Eugen Iwanowitsch (1875–1949) S   **Bordz.**
Boreau, Alexandre (1803–1875) PS   **Boreau**
Borelli, Dante (fl. 1955) M   **Borelli**
Boresch, Karl (1886–1947) A   **Boresch**
Boretti, G. (fl. 1973) M   **Boretti**
Borg, John (1873–1945) PS   **Borg**
Borgen, Liv (1943–) S   **L.Borgen**
Borgen, Torbjørn (fl. 1983) M   **T.Borgen**
Borgert, Adolf Hermann Constant (1868–1954) A   **Borgert**
Børgesen, Fredrik Christian Emil (1866–1956) A   **Børgesen**
Borgvall, Torsten Alvin (1884–1975) S   **Borgv.**
(Borgwall, Torsten Alvin)
  see Borgvall, Torsten Alvin   **Borgv.**
Borhidi, Attila L. (1932–) S   **Borhidi**
(Borisov, G.I.)
  see Borissov, G.I.   **Borissov**
(Borisova, Antonina Georgievna)
  see Borissova, Antonina Georgievna   **Boriss.**
(Borisova–Bekrjaševa, Antonina Georgievna)
  see Borissova, Antonina Georgievna   **Boriss.**
(Borissoff, G.I.)
  see Borissov, G.I.   **Borissov**
Borissov, G.I. (1901–) MS   **Borissov**
Borissova, Antonina Georgievna (1903–1970) S   **Boriss.**
(Borissova–Bekrjaševa, Antonina Georgievna)
  see Borissova, Antonina Georgievna   **Boriss.**
Borissova, V.N. (fl. 1971) M   **V.N.Boriss.**
Borja Carbonell, José (1903–) S   **Borja**
Borkhausen, Moritz (Moriz) Balthasar (1760–1806) BPS   **Borkh.**
Borkowski, R. (fl. 1913) P   **Bork.**
Borlaza, P.B. (fl. 1964) M   **Borlaza**
Born, Gerald L. (fl. 1972) M   **Born**
Bornemann, Felix (fl. 1887) A   **F.Bornem.**
Bornemann, Johann Georg (1831–1896) AF   **Bornem.**
Börner, Carl (Karl) Julius Bernhard (1880–1953) PS   **Börner**
Bornet, Jean–Baptiste Édouard (1828–1911) AMS   **Bornet**
Bornmüller, Joseph Friedrich Nicolaus (1862–1948) BMS   **Bornm.**
Borodina, A.E. (later Grabovskaja, A.E.) (1953–) S   **Borodina**
Boros, Ádám (1900–1973) BS   **Boros**
Borowitzka, Michael Armin (1948–) A   **Borow.**
Borowska, Alicja (1940–) M   **Borowska**
Borrel, Amédéee (1867–1936) A   **Borrel**
Borrer, William J. (1781–1862) MS   **Borrer**
Borrill, Martin (1924–) S   **Borrill**

Borse, B.D. (fl. 1984) M	**Borse**
Borsetti, Anna Maria A	**Borsetti**
Borshchow, Grigori Grigorievicz (fl. 1856) B	**G.G.Borshch.**
Borshchow, Iljia Grigorievich (1833–1878) BMS	**I.G.Borshch.**
Borsini, Olga Helena (1916–1981) S	**Borsini**
Borsos, Olga (1926–) S	**Borsos**
Borssum Waalkes, Jan van (1922–1985) PS	**Borss.Waalk.**
(Borszczow, Grigori Grigorievicz)	
see Borshchow, Grigori Grigorievicz	**G.G.Borshch.**
(Borszczow, Ilya (Elia) Grigorievicz)	
see Borshchow, Iljia Grigorievich	**I.G.Borshch.**
Bort, Katherine Stephens (1870–) S	**Bort**
Bortels, Hermann (1902–) A	**Bortels**
Borut, S. (fl. 1958) M	**Borut**
(Bory de Saint-Vincent, Jean Baptiste Georges Geneviève Marcellin)	
see Bory, Jean Baptiste Georges Geneviève Marcellin	**Bory**
Bory, Jean Baptiste Georges Geneviève Marcellin (1778–1846) AMPS	**Bory**
Borza, Alexandru (1887–1971) BS	**Borza**
(Borzczow, J.G.)	
see Borshchow, Iljia Grigorievich	**I.G.Borshch.**
Borzęcki, Konstantin Marian (1883–) A	**Borzęcki**
Borzí, Antonino (1852–1921) AMPS	**Borzí**
Bos, Jan Justus (1939–) S	**Bos**
(Bos, Jan Ritzema)	
see Ritzema Bos, Jan	**Ritz.Bos**
Bos, P. (fl. 1969) M	**P.Bos**
Bosc, G. (fl. 1986) S	**G.Bosc**
Bosc, J.A. S	**J.A.Bosc**
Bosc, Louis Augustin Guillaume (1759–1828) MPS	**Bosc**
Boşcaiu, Nicolae (1925–) S	**Boşcaiu**
Bosch, Roelof Benjamin van den (1810–1862) BMPS	**Bosch**
Boschma, Hilbrand (1893–1976) A	**Boschma**
Bosco, Roberto (1902–) PS	**Bosco**
Bose, M.N. (fl. 1952–61) F	**M.N.Bose**
Bose, S.R. (1888–1970) M	**Bose**
Bose, Sunil Kumar (1922–1983) M	**S.K.Bose**
Bosman, M.T.M. (1958–) S	**Bosman**
Bosnjak, Karlo (1866–1953) S	**Bosnjak**
Bosse, Georg G. (1887–1972) S	**G.G.Bosse**
Bosse, Julius Friedrich Wilhelm (1788–1864) S	**Bosse**
Bosselaers, J.P. (fl. 1984) M	**Bossel.**
Bosser, Jean M. (1922–) S	**Bosser**
Bosserdet, Pierre (fl. 1970) S	**Bosser.**
Bostick, P.E. (fl. 1969) S	**Bostick**
Bostock, P.D. (1949–) PS	**Bostock**
Boswell, Henry (1837–1897) B	**Bosw.**
(Boswell-Syme, John Thomas Irvine)	
see Syme, John Thomas Irvine Boswell	**Syme**
Both, E.E. (fl. 1977) M	**Both**
Bothmer, Roland von (1943–) S	**Bothmer**

Botnen, Astri (fl. 1990) M	**Botnen**
Botschantzev, Victor Petrovič (1910–1990) S	**Botsch.**
Botschantzeva, Vera Viktorovna (1946–) S	**V.V.Botschantz.**
Botschantzeva, Zinaida Petrovna (1907–1973) S	**Botschantz.**
Botta, Silvia Margarita (1942–) S	**Botta**
Botte, F. (fl. 1987) S	**Botte**
Botteron, Germain A	**Botteron** ·
Bottini, Antonio (1850–1931) B	**Bott.**
Bottler, Max (fl. 1882) S	**Bottler**
Bottomley, A.M. (fl. 1948) M	**Bottomley**
Boucaud–Camou, Eve A	**Bouc.-Camou**
Bouček, Bedřich (1904–1975) A	**Bouček**
Bouchard, Jean S	**Bouchard**
Boucharlet (1807–1893) S	**Bouch.**
Bouchat, Alain (1956–) S	**Bouchat**
Bouché, Carl David (1809–1881) S	**C.D.Bouché**
Bouché, Peter Carl (1783–1856) S	**Bouché**
Bouché, Pierre M. A	**P.M.Bouché**
Boucher, Charles (1944–) S	**C.Boucher**
Boucher de Crèvecoeur, Jules Armand Guillaume (1757–1844) S	**Boucher**
Boucher, Humbert (fl. 1923) M	**H.Boucher**
Bouchet, Dominique (Doumenq) (1770–1845) MS	**Bouchet**
Bouchet, Pierre (fl. 1964) M	**P.Bouchet**
Bouchinot (fl. 1904) M	**Bouchinot**
Boudier, Jean Louis Émile (1828–1920) MS	**Boud.**
Boudouresque, Charles–François A	**Boudour.**
Boudreaux, Joseph E. A	**Boudreaux**
Boudrie, Michel (1950–) P	**Boudrie**
Bouet, G. A	**Bouet**
Boufford, David E. (1941–) PS	**Boufford**
Bouget, Joseph (1867–1953) S	**Bouget**
Bougher, N.L. (fl. 1985) M	**Bougher**
Boughey, Arthur Stanley (fl. 1938) MS	**Boughey**
Boughton, Valerie H. (fl. 1978) M	**Boughton**
Bougon A	**Bougon**
Bouhot, D. (fl. 1966) M	**Bouhot**
Bouix, G. (fl. 1972) M	**Bouix**
Boulanger, Émile (fl. 1894) M	**Boulanger**
Boulay, Jean Nicolas (1837–1905) BFS	**Boulay**
Boulenger, George Albert (1858–1937) S	**Boulenger**
Boulger, George Edward Simmonds (1853–1922) S	**Boulger**
Boullu, Antoine Étienne (1813–1904) S	**Boullu**
Boulos, Loutfy (1932–) S	**Boulos**
Bouloumoy, Louis (–1926) S	**Boul.**
Boulter, Michael Charles (1942–) BF	**Boulter**
(Bouly de Lesdain, Maurice)	
see de Lesdain, Maurice Bouly	**de Lesd.**
Bouquaheux, Françoise A	**Bouq.**
Bourchier, R.J. (fl. 1957) M	**Bourch.**

Bourdillon, Thomas Fulton (1849–1930) S	**Bourd.**
Bourdon, M. (fl. 1986) S	**Bourdon**
Bourdot, Hubert (1861–1937) M	**Bourdot**
Bourdu, Robert (fl. 1957) S	**Bourdu**
Boureau, Edouard Léon François (1913–) AFMS	**Boureau**
Bourgeau, Eugène (1813–1877) S	**Bourg.**
Bourgeois, C. (fl. 1929) M	**Bourgeois**
Bouriquet, Gilbert (fl. 1939) M	**Bouriquet**
Bourquenoud, François (1785–1837) B	**Bourquet**
Bourreil, Pierre (fl. 1969) S	**Bourreil**
Bourrelly, Pierre (1910–) A	**Bourr.**
Bourret, J.A. (fl. 1967) M	**Bourret**
Boursault, Henri (fl. 1889) F	**Boursault**
Boursier, J. (fl. 1928) M	**Boursier**
Bousset, M. (fl. 1939) M	**Bousset**
Bouteille, J. (fl. 1946) M	**Bouteille**
Boutelje, Julius B. (fl. 1954) S	**Boutelje**
Boutelou, Claudio (1774–1842) S	**C.Boutelou**
Boutelou, Estéban (1823–1883) S	**E.Boutelou**
Boutelou, Estéban (1776–1813) S	**Boutelou**
(Boutelou y Soldevilla, Claudio)	
see Boutelou, Claudio	**C.Boutelou**
(Boutelou y Soldevilla, Estéban)	
see Boutelou, Estéban	**E.Boutelou**
(Boutelou y Soldevilla, Estéban)	
see Boutelou, Estéban	**Boutelou**
Bouthilet, Robert J. (fl. 1951) M	**Bouth.**
Boutigny, Jean François Désiré (1820–1884) S	**Boutigny**
Boutique, Raymond (1906–1985) S	**Boutique**
Bouton, Louis (1800–1878) S	**Bouton**
Boutroux, A. (fl. 1883) M	**Boutroux**
Bouvet, Georges (1850–1929) BS	**Bouvet**
Bouvier, Jean–Louis (1819–1908) S	**Bouvier**
Bové, Nicolas (1812–1841) S	**Bové**
Bovee, Eugene Cleveland (1915–) A	**Bovee**
Bowden, Lorna Frances (1941–) S	**L.F.Bowden**
Bowden, Wray Merrill (1914–) S	**Bowden**
Bowdich, Sarah (1791–1877) PS	**Bowdich**
Bowdich, Thomas Edward (1791–1824) PS	**T.E.Bowdich**
Bower, Frederick Orpen (1855–1948) PS	**Bower**
Bowerbank, James Scott (1797–1877) S	**Bowerb.**
Bowerman, Constance A. (fl. 1955) M	**Bowerman**
Bowers, Clement Gray (1893–1973) S	**Bowers**
Bowers, Frank Dana (1936–) BS	**F.D.Bowers**
Bowers, Maynard C. (1930–) B	**M.C.Bowers**
Bowie, James (1789–1869) S	**Bowie**
Bowler, Peter A. (1948–) M	**Bowler**
Bowles, Edward Augustus (1865–1954) S	**Bowles**
Bowles, M.L. (fl. 1986) S	**M.L.Bowles**

Bowman, John Eddowes (1785–1841) MS	**Bowman**
Bowman, R.N. (fl. 1979) S	**R.N.Bowman**
Box, Harold Edmund (1898–) PS	**Box**
Boxall, Richard (fl. 1880) S	**Boxall**
Boyce, John Shaw (1889–) M	**Boyce**
Boyce, Peter Charles (1964–) S	**P.C.Boyce**
Boyd, Daniel Alexander (1855–1928) M	**Boyd**
Boyd, Emma Sophia (fl. 1934) M	**E.S.Boyd**
Boydston, Kathryn E. (fl. 1954–1978) P	**Boydston**
Boyer, Charles S. (1856–1928) A	**C.S.Boyer**
Boyer, Gaston (fl. 1891) M	**G.Boyer**
Boyer, Léon (fl. 1890) M	**Boyer**
Boyetchko, S.M. (fl. 1986) M	**Boyetchko**
Boykin, Samuel (1786–1848) S	**Boykin**
Boykins, William T. A	**Boykins**
Boylan, Brendan V. (fl. 1970) M	**Boylan**
Boyland, Desmond Ernest (1941–) S	**Boyland**
Boyle, John Samuel (1917–) M	**J.S.Boyle**
Boyle, William Sidney (1915–) S	**Boyle**
Boynton, Frank Ellis (1859–post 1917) S	**F.E.Boynton**
Boynton, Kenneth Rowland (1891–) S	**Boynton**
Boza, P. (fl. 1986) S	**Boza**
Bozeman, John R. (1935–) S	**Bozeman**
Bozonnet, J. (fl. 1985) M	**Bozonnet**
Braarud, Trygve (1903–1985) A	**Braarud**
Braas, Lothar A. (fl. 1977) S	**Braas**
Brabez, Rosalia A	**Brabez**
Bracelin, Nina Floy (1890–) S	**Bracelin**
Bracey, B.O. (fl. 1984) S	**Bracey**
Brack, P. (fl. 1987) S	**P.Brack**
Brack, S. (fl. 1986) S	**S.Brack**
Brackenridge, William Dunlop (1810–1893) PS	**Brack.**
Brackett, Amelia Ellen (1896–1926) S	**Brackett**
Bradbury, J.Platt A	**Bradbury**
Brade, Alexander Curt (1881–1971) PS	**Brade**
Brader, L. (fl. 1964) M	**Brader**
Bradford, Edward (fl. 1845) S	**Bradford**
Bradford, Martin R. A	**M.R.Bradford**
Bradley, D.E. A	**D.E.Bradley**
Bradley, Richard (1688–1732) L	**Bradley**
Bradley, Ted Ray (1940–) S	**T.R.Bradley**
Bradley, Wilmot H. (1899–) AM	**W.H.Bradley**
Bradshaw, Margaret Elizabeth (1926–) S	**M.E.Bradshaw**
Bradshaw, Robert Vernon (1896–) S	**Bradshaw**
Brady, B.L. (fl. 1984) M	**B.L.Brady**
Brady, Henry Bowman (1835–1891) A	**Brady**
Braem, Guido J. (fl. 1980) S	**Braem**
Braga, Maria do Rosário de Almeida (1955–) A	**M.Braga**
Braga, Pedro Ivo Soares (1950–) S	**Braga**

Braga, Ruby (fl. 1964) S	R.Braga
Braggio Morucchio, Giulia (1939–) S	Braggio
(Braggio–Morucchio, Giulia)	
see Braggio Morucchio, Giulia	Braggio
Brahmam, M. (fl. 1981) S	Brahmam
Braid, Kenneth William (1897–1984) S	Braid
Brain (fl. 1923) M	Brain
Brainerd, Ezra (1844–1924) S	Brainerd
Braird, H. (fl. 1888) M	H.Braird
Braithwaite, Anthony Forester (1936–) P	A.F.Braithw.
Braithwaite, Robert (1824–1917) B	Braithw.
Brako, Lois (fl. 1985) M	Brako
Bräm, Heinrich A	Bräm
Bramlette, Milton Nunn (1896–) A	Bramlette
Bramwell, David (1942–) S	Bramwell
Brand, August (1863–1930) S	Brand
Brand, Friedrich (1842–1924) A	F.Brand
Brand, O. S	O.Brand
Brandbyge, John S	Brandbyge
Brandegee, Mary Katharine (formerly Curran, M.K.) (1844–1920) S	K.Brandegee
Brandegee, Townshend Stith (1843–1925) S	Brandegee
Brandenburger, W. (fl. 1961) M	Brandenb.
Brandes, Edvard (fl. 1845) S	Brandes
Brandes, Elmer Walker (1891–) MS	E.W.Brandes
Brandes, Wilhelm (–1916) S	W.Brandes
Brandham, Peter Edward (1937–) AS	Brandham
Brandis, Dietrich (1824–1907) S	Brandis
Brandis, E. (1834–1921) S	E.Brandis
Brandner, R. A	Brandner
Brandrud, Tor Erik (fl. 1977) M	Brandrud
Brandsberg, John W. (fl. 1971) M	Brandsb.
Brandt, Fred H. (1908–) S	F.H.Brandt
Brandt, Jean–Pierre (1921–1963) S	J.-P.Brandt
Brandt, Johann Friedrich (von) (1802–1879) S	Brandt
Brandt, Johann Theodor Hubert (1877–1939) M	J.T.H.Brandt
Brandt, Karl (1854–1931) A	K.Brandt
Brandt, Max (1884–1914) S	M.Brandt
Brandt, T. (1877–1939) M	T.Brandt
Brandt, Wilhelm (–1929) S	W.Brandt
Brandt-Pedersen, T. (fl. 1980) M	Brandt-Ped.
Brândză, Dimitrie (1846–1895) S	D.Brândză
Brândză, Marcel Alex (1868–1934) M	Brândză
Branger, Chr. (fl. 1970) S	Branger
Branson, Carl Colton (1906–1975) A	Branson
Branth, Jakob Severin Deichmann (1831–1917) BM	Branth
Brasfield, Travis W. (fl. 1938) M	Brasf.
Brasier, C.M. (fl. 1979) M	Brasier
Braslawsky-Spectorowa, H. (1890–) A	Brasl.-Spect.
Brassai, S. (1798–1898) S	Brassai

Brassard, Guy Raymond (1943–) B — **Brassard**
Bratteng, A.Stephen (fl. 1975) M — **Bratteng**
Brault, J. (fl. 1914) M — **Brault**
Braun, Alexander Karl (Carl) Heinrich (1805–1877) ABFMPS — **A.Braun**
Braun, Bernhard (fl. 1973) S — **B.Braun**
Braun, Carl Friedrich Wilhelm (1800–1864) FM — **Braun**
Braun, Emma Lucy (1889–1971) PS — **E.L.Braun**
Braun, Friederich von (fl. 1820) S — **F.Braun**
Braun, Gottlieb (1821–1882) S — **G.Braun**
Braun, Hans ('Harry') (1895/6–) M — **Hans Braun**
Braun, Heinrich (1851–1920) S — **Heinr.Braun**
(Braun, Josias)
   see Braun–Blanquet, Josias — **Braun-Blanq.**
Braun, P.J. (fl. 1985) S — **P.J.Braun**
Braun, Uwe (1953–) M — **U.Braun**
Braun–Blanquet, Josias (1884–1980) S — **Braun-Blanq.**
Braune, Franz Anton Alexander von (1766–1853) S — **Braune**
Braune, Robert A. (fl. 1913) M — **R.A.Braune**
Brauner, John F. A — **Brauner**
Brause, Guido Georg Wilhelm (1847–1922) PS — **Brause**
Bräutigam, Siegfried (1944–) S — **S.Bräut.**
Bräutigam, Volker (1939–) S — **V.Bräut.**
Braverman, Samuel W. (fl. 1960) M — **Braverman**
Bravo Hollis, Helia (1903–) S — **Bravo**
Bravo Hollis, Margarita (1903–) S — **M.Bravo**
Bravo, Lilia Dora (1945?–1986) S — **L.Bravo**
(Bravo–Hollis, Helia)
   see Bravo Hollis, Helia — **Bravo**
Brawley, Susan H. (1951–) A — **Brawley**
Braxton, H.H. (fl. 1926) M — **Braxton**
Bray, Franz Gabriel von (1765–1832) S — **Bray**
Bray, William L. (1865–1953) S — **W.L.Bray**
Brayford, David (1957–) M — **Brayford**
Brayshaw, Thomas Christopher (1919–) BS — **Brayshaw**
Brebeck, C. (fl. 1894) M — **Brebeck**
Brébisson, Louis Alphonse de (1798–1872) ABMS — **Bréb.**
Brecerescu, D. (fl. 1983) M — **Brecer.**
Brecher, G. (fl. 1941) S — **Brecher**
Breckon, Gary J. (1940–) S — **Breckon**
Breda de Haan, Jacob van (1866–1917) M — **Breda de Haan**
Breda, Jacob Gijsbert Samuel van (1788–1867) FMS — **Breda**
Bredell, H.C. S — **Bredell**
Bredell, Ingrid H. (fl. 1974) M — **I.H.Bredell**
Bredemeyer, Franz (1758–1839) S — **Bredem.**
Brederoo, A.J. ('Nol') (fl. 1971) S — **Brederoo**
Bredkina, L.I. (1938–) M — **Bredkina**
Bree, William Thomas (1787–1868) S — **Bree**
Breedlove, Dennis E. (1939–) S — **Breedlove**
Breemen, Pieter Johan van A — **P.J.Breemen**

Breen, Ruth Olive Schornhurst (1905–1987) B	**Breen**
Brefeld, Julius Oscar (1839–1925) MS	**Bref.**
Bregadze, N.N.　S	**Bregadze**
Bregmann, R. (fl. 1986) S	**Bregmann**
Brehmer, Wilhelm Georg Baptist Alexander von (1883–　) S	**Brehmer**
Breidenstein, W. (fl. 1856) S	**Breid.**
Breidler, Johann (1828–1913) BMS	**Breidl.**
Breiner, E. (fl. 1981) S	**E.Breiner**
Breiner, R. (fl. 1981) S	**R.Breiner**
Breinig, Peter　S	**Breinig**
Breistroffer, Maurice A.F. (1910–　) S	**Breistr.**
Breitenbach, Josef (fl. 1973) M	**J.Breitenb.**
Breitenbach, Wilhelm (1856–　) S	**Breitenb.**
Breiter, Christian August (1776–1840) S	**Breiter**
Breitfeld, Charlotte (1902–　) S	**Breitf.**
Breitung, August Johann Julius (1913–1987) S	**Breitung**
Breitwieser, I. (fl. 1986) S	**Breitw.**
Bremekamp, Cornelis Eliza Bertus (1888–1984) PS	**Bremek.**
(Bremekamp, Neeltje Elizabeth)	
see Nannenga-Bremekamp, Neeltje Elizabeth	**Nann.-Bremek.**
Bremer, Birgitta (1950–　) BS	**B.Bremer**
Bremer, Hans (fl. 1947) M	**Bremer**
Bremer, Kåre (Kaare) (1948–　) AS	**K.Bremer**
Brenan, John Patrick Micklethwait (1917–1985) PS	**Brenan**
Brenckle, Jacob Frederic(k) (1875–1958) MS	**Brenckle**
Brenckle, Paul L.　A	**P.L.Brenckle**
Brenner, Marten Magnus Widar (1843–1930) MS	**Brenner**
Brenner, Wilhelm (1875–　) S	**W.Brenner**
Brera, Valeriano Luigi (1772–1840) S	**Brera**
Brereton, John Andrew (1787–1839) S	**Brereton**
Bresadola, Giacopo (1847–1929) M	**Bres.**
Bresinsky, Andreas (1935–　) M	**Bresinsky**
Bresler, Moritz (1802–c.1851) S	**Bresler**
Bressan, Guido (1944–　) A	**Bressan**
Breteler, Franciscus Joseph(Jozef) (1932–　) S	**Breteler**
Breton, André (fl. 1964) M	**Breton**
Bretschneider, Emil (Vasilievic) (1833–1901) S	**Bretschn.**
Bretschneider, Ludwig Hermann (1899–1964) A	**L.H.Bretschn.**
Brett, D.W. (fl. 1963–83) F	**D.W.Brett**
Brett, Robert Lindsay Gordon (1898–1975) S	**Brett**
Brettell, R.D.　S	**Brettell**
Bretting, P.K. (fl. 1982) S	**Bretting**
Bretz, Theodore Walter (1908–1968) M	**Bretz**
Breuil (fl. 1915) M	**Breuil**
Breuss, Othmar (fl. 1987) M	**Breuss**
Breutel, Johann Christian (1788–1875) BS	**Breutel**
Brewer, D. (fl. 1974) M	**D.Brewer**
Brewer, James Alexander (1818–1886) S	**Brewer**
Brewer, William Henry (1828–1910) PS	**W.H.Brewer**
Brewster, David (1781–1868) S	**Brewster**

Brewster, M.S. (fl. 1952) M	**M.S.Brewster**
Breyne, Jacob (1637–1697) L	**Breyne**
Brezhnev, Dmitrĭi Danilovich (1905–1982) S	**Brezhnev**
Brezhnev, I.E. (fl. 1964) M	**I.E.Brezhnev**
Briant, Andrew K. (fl. 1929) M	**Briant**
Briaotta, G. (fl. 1989) M	**Briaotta**
Briard, H. (fl. 1888) M	**H.Briard**
Briard, Pierre Alfred (1811–1896) MS	**Briard**
Bricaud, Olivier (fl. 1991) M	**Bricaud**
Briceño–Iragorry, L. (fl. 1938) M	**Bric.-Irag.**
Brichan, James B. (fl. 1832–45) P	**Brichan**
Brick, Carl (1863–1924) M	**Brick**
Brickell, Christopher David (1932–) S	**C.D.Brickell**
Brickell, John (1748–1809) S	**Brickell**
Bridarolli, Albino J. S	**Bridar.**
Bride, A. (fl. 1950) M	**Bride**
Brideaux, Wayne W. A	**Brideaux**
Bridel, Samuel Élisée von (1761–1828) BS	**Brid.**
(Bridel–Brideri, Samuel Élisée von)	
see Bridel, Samuel Élisée von	**Brid.**
Bridge, Paul Dennis (1956–) M	**Bridge**
Bridges, C.H. (fl. 1961) M	**C.H.Bridges**
Bridges, Edwin L. (fl. 1989) S	**E.L.Bridges**
Bridges, J.Robert (fl. 1987) M	**J.R.Bridges**
Bridges, Thomas Charles (1807–1865) S	**Bridges**
Bridson, Diane Mary (1942–) S	**Bridson**
Bridwell, A.W. (fl. 1925) S	**Bridwell**
Brieger, Friedrich Gustav (1900–1985) S	**Brieger**
Briganti, Francesco (1802–1865) M	**F.Brig.**
Briganti, Vincenzo (1766–1836) BMS	**V.Brig.**
Brigger, A.L. A	**Brigger**
Briggs, Barbara Gillian (1934–) S	**B.G.Briggs**
Briggs, F.E. S	**F.E.Briggs**
Briggs, George Edward (1893–) S	**G.E.Briggs**
Briggs, Scott Munro (1889–1917) S	**S.M.Briggs**
Briggs, Thomas Richard Archer (1836–1891) S	**Briggs**
Briggs, Winslow Russell (1928–) S	**W.R.Briggs**
Bright, John (1872–1952) S	**Bright**
Brighton, Christine A. (1945–) S	**Brighton**
Brightwell, Thomas (1787–1868) A	**Brightw.**
Brignoli di Brunnhoff, Giovanni de (1774–1857) S	**Brign.**
Brilli–Cattarini, Aldo Josef Bernhard (1923–) S	**Brilli-Catt.**
Brimont, E. A	**Brimont**
Bringmann, Gottfried (1913–) A	**Bringmann**
Brink, D.E. (fl. 1983) S	**Brink**
Brinker, Robert R. (1905–1970) S	**Brinker**
Brinkman, Alfred Henry (1873–1928) B	**Brinkm.**
Brinkman, S.A. (fl. 1962) M	**S.A.Brinkm.**
Brinkmann, Wilhelm (fl. 1909) M	**Brinkmann**

Brinsley, W. (fl. 1968) S	**Brinsley**
Briosi, Giovanni (1846-1919/1921) CM	**Briosi**
Briot, P. (fl. 1983) M	**P.Briot**
Briot, Pierre Louis (Charles, in error) (1804-1888) S	**Briot**
Briquet, John Isaac (1870-1931) PS	**Briq.**
Brisson, J.D. (1945-) M	**Brisson**
Bristol, B.Muriel A	**Bristol**
Bristow, Henry William (1817-1889) F	**Bristow**
(Brito, Antonio Luiz Vieira Toscano de)	
see Toscano, Antonio Luiz Vieira	**Toscano**
Brito, E.C.Souza S	**E.C.S.Brito**
Brito, Ignacio Machado (fl. 1965-67) A	**I.M.Brito**
Brittan, Norman Henry (1920-) S	**Brittan**
Britten, James (1846-1924) PS	**Britten**
Britten, Lillian Louisa (1886-1952) S	**L.L.Britten**
Brittinger, Christian Casimir (1795-1869) S	**Brittinger**
Brittlebank, Charles Clifton (1862-1945) M	**Brittleb.**
Britto, S.John (1946-) S	**Britto**
Britton, Charles Edward (1872-1944) S	**C.E.Britton**
Britton, Donald Macphail (1923-) PS	**D.M.Britton**
Britton, Elizabeth Gertrude (1858-1934) BPS	**E.Britton**
Britton, M.P. (fl. 1969) M	**M.P.Britton**
Britton, Max Edwin (1912-) A	**M.E.Britton**
Britton, Nathaniel Lord (1859-1934) BMPS	**Britton**
Britzelmayr, Max (1839-1909) M	**Britzelm.**
Brizi, Ugo (1868-1949) BM	**Brizi**
Brizicky, George Konstantin (1901-1968) S	**Brizicky**
Broadhurst, Jean Alice (1873-1954) PS	**Broadh.**
Broadway, Walter Elias (1863-1935) S	**Broadway**
Broady, Paul A. A	**Broady**
Brocchi, Giovanni Battista (1772-1826) S	**Brocchi**
Broch, Hjalmar (1882-1964) A	**Broch**
Brochmann, Christian (1953-) S	**Brochmann**
Brockie, Walter Boa (1897-1972) S	**Brockie**
(Brockman-Jerosch, Heinrich)	
see Brockmann-Jerosch, Henryk	**Brockm.-Jer.**
Brockmann, Christoph (1878-1962) A	**C.Brockmann**
Brockmann, Ingrid (fl. 1976) M	**Brockmann**
Brockmann-Jerosch, Henryk (1879-1939) MPS	**Brockm.-Jer.**
Brockmüller, Hans Joachim Heinrich (1821-1882) BCMPS	**Brockm.**
Broddeson, Otto Edward (1880-1957) S	**Brodd.**
Broderick, Thomas Monteith (1889-1965) A	**Brod.**
Brodie, David Arthur (1868-) S	**Brodie**
Brodie, Harold Johnston (1907-1989) M	**H.J.Brodie**
Brodie, Juliet A	**J.Brodie**
Brodo, Irwin Murray (1935-) M	**Brodo**
Brodskii, Abram Lvovič (1883-) A	**Brodskii**
Broeck, Henri van den (1845-1926) B	**Broeck**

(Broecke, R.Van den)	
see Van den Broecke, R.	**Van den Broecke**
(Broegelmann, Wilhelm)	
see Brögelmann, Wilhelm	**Brög.**
Brögelmann, Wilhelm S	**Brög.**
Broich, S.L. (fl. 1987) S	**Broich**
Bromfield, William Arnold (1801–1851) S	**Bromf.**
Brondeau, Louis de (1794–1859) CM	**Brond.**
Brongniart, Adolphe Théodore (de) (1801–1876) ABFMPS	**Brongn.**
Brongniart, Alexandre (1770–1847) S	**Al.Brongn.**
Bronikovsky, Natalia A	**Bronik.**
Bronn, Heinrich Georg (1800–1862) AS	**Bronn**
Brönnimann, Paul A	**Brönnimann**
Brook, Alan John (1923–) A	**Brook**
Brooke, Jocelyn (1908–1966) S	**Brooke**
Brooke, Robert C. A	**R.C.Brooke**
Brooker, Murray Ian Hill (1934–) S	**Brooker**
Brooks, Albert Nelson (1897–1966) M	**A.N.Brooks**
Brooks, Cecil Joslin (1875–1953) PS	**Brooks**
Brooks, Charles (1872–) M	**C.Brooks**
Brooks, D.R. B	**D.R.Brooks**
Brooks, Frederick Thom(as) (1882–1952) M	**F.T.Brooks**
(Brooks, Marie Elena de)	
see De Paula de Brooks, Marie Elena	**De Paula**
Brooks, Maurice Graham (1900–) S	**M.G.Brooks**
Brooks, R.D. (fl. 1977) M	**R.D.Brooks**
Brooks, R.L. S	**R.L.Brooks**
Brooks, Ralph Edward (1950–) PS	**R.E.Brooks**
Brooks, Roy C. (1947–) S	**R.C.Brooks**
Brooks, Travis E. (fl. 1971) M	**T.E.Brooks**
Broome, Carmen Rose (1939–) S	**C.R.Broome**
Broome, Christopher Edmund (1812–1886) M	**Broome**
Brosius, Marita A	**Brosius**
Brotero, Felix de (Silva) Avellar (1744–1828) BMS	**Brot.**
Brothers, Margaret P. (fl. 1981) M	**Brothers**
Brotherson, Jack D. (fl. 1967) S	**Brotherson**
Brotherus, Viktor Ferdinand (1849–1929) BS	**Broth.**
Brotzen, Fritz (1902–1968) A	**Brotzen**
(Brouard, Arsène Gustave Joseph)	
see Arsène, Gustave (Gerfroy)	**Arsène**
Brouard, Edward Jacques (fl. 1820) S	**Brouard**
Broughton, Arthur (–1796) S	**Broughton**
Brouillet, Luc (1954–) S	**Brouillet**
Broun, Alfred Forbes (1858–) S	**Broun**
Broun, Maurice (1906–) PS	**M.Broun**
Brousset, M. (fl. 1948) M	**Brousset**
Broussonet, Pierre Marie Auguste (1761–1807) S	**Brouss.**
Browicz, Kasimierz (1925–) S	**Browicz**
Brown, Addison (1830–1913) S	**A.Br.**

Brown, Agnes H.S. (née Onions) (fl. 1957) M	A.H.S.Br.
Brown, Alex A	Alex Br.
Brown, C.E. A	C.E.Br.
Brown, Clair A. (1903–) M	C.A.Br.
Brown, Eleanor–Margaret A	E.–M.Br.
Brown, Elizabeth Dorothy (Wuist) (1880–1972) PS	E.D.Br.
Brown, Forest Buffen Harkness (1873–1954) PS	F.Br.
Brown, Gregory K. (1951–) S	G.K.Br.
Brown, Helen Jean (1903–) A	H.J.Br.
Brown, James Greenlief (1880–1957) M	J.G.Br.
Brown, John Ednie (1848–1899) S	J.E.Br.
Brown, John Thomas (1941–) BF	J.T.Br.
Brown, Juliet C. (fl. 1958) BM	J.C.Br.
Brown, Larry E. (1937–) S	L.E.Br.
Brown, Lawrence G. (fl. 1977) M	L.G.Br.
Brown, Margaret E. (fl. 1957) M	M.E.Br.
Brown, Margaret Sibella (1866–1961) B	M.S.Br.
Brown, Nellie Adalesa (1877–1956) M	N.A.Br.
Brown, Nicholas Edward (1849–1934) AS	N.E.Br.
Brown, P.D. B	P.D.Br.
Brown, Robert Neal Rudmose (1879–1957) PS	R.N.R.Br.
Brown, Richard Malcolm (1939–) A	R.M.Br.
Brown, Robert (1773–1858) BMPS	R.Br.
Brown, Robert, of Campster (1842–1895) S	R.Br.ter
Brown, Robert, of NZ (1820–1906) BS	R.Br.bis
Brown, Roland Wilbur (1893–1961) BFS	R.W.Br.
Brown, Roy Curtiss (1947–) S	R.C.Br.
Brown, Spencer Wharton (1918–) S	S.W.Br.
Brown, Stewardson (1867–1921) S	S.Br.
Brown, Virginius Elholm (1902–) A	V.E.Br.
Brown, W. (1888–1975) M	W.Br.
Brown, W.R. (fl. 1943) F	W.R.Br.
Brown, Walter Varian (1913–1977) S	W.V.Br.
Browne, C.A. (fl. 1919) M	C.A.Browne
Browne, Daniel Jay (1804–1867) S	D.J.Browne
Browne, Patrick (1720–1790) MPS	P.Browne
Brownlie, Garth (1923–1986) PS	Brownlie
Brownsey, Patrick J. (1948–) P	Brownsey
(Brshezicki, Michail Vasiljevich)	
see Brzhesitzky, Michail Vasiljevich	Brzhes.
Bruant, Georges (1842–1912) S	Bruant
Brubacher, D.C. (fl. 1984) M	Brub.
Bruce, David (1855–1931) A	D.Bruce
Bruce, Eileen Adelaide (1905–1955) S	E.A.Bruce
Bruce, James (1730–1794) S	Bruce
Bruce, James G. (fl. 1976–1979) P	J.G.Bruce
Bruch, Philipp (1781–1847) BS	Bruch
Brücher, Heinz (1915–) S	Brücher
Bruchet, G. (fl. 1965) M	Bruchet

Brückner, Adam (1862–1933) S   **A.Brückn.**
Brückner, Adolf Friedrich Albrecht (1781–1818) PS   **Brückn.**
Brückner, Gerhard (1902–) S   **G.Brückn.**
Brückner, Gustav Adam (1789–1860) S   **G.A.Brückn.**
Bruderlein, Jean (fl. 1916) M   **Bruderl.**
(Brueckner, Adam)
  see Brückner, Adam   **A.Brückn**
(Brueckner, Adolf Friedrich Albrecht)
  see Brückner, Adolf Friedrich Albrecht   **Brückn.**
(Brueckner, Gerhard)
  see Brückner, Gerhard   **G.Brückn.**
(Brueckner, Gustav Adam)
  see Brückner, Gustav Adam   **G.A.Brückn.**
(Bruegger von Churwalden, Christian Georg)
  see Brügger, Christian Georg   **Brügger**
Bruehl, G.W. (fl. 1965) M   **G.W.Bruehl**
Bruehl, Paul Johannes (1855–1935) B   **Bruehl**
Brug, S.L. A   **Brug**
Bruggeman–Nannenga, Maria Alida (1944–) B   **Brugg.-Nann.**
Bruggen, A.C.van (1929–) S   **A.Bruggen**
Bruggen, Heinrich Wilhelm Eduard van (1927–) S   **H.Bruggen**
Bruggen, Theodore van (1926–) S   **T.Bruggen**
Brügger, Christian Georg (1833–1899) APS   **Brügger**
(Brügger von Churwalden, Christian Georg)
  see Brügger, Christian Georg   **Brügger**
Brugmans, Sebald Justinus (1763–1819) S   **Brugmans**
Bruguière, Jean Baptiste Louis (1838–) S   **J.B.L.Brug.**
Bruguière, Jean Guillaume (1750–1798) S   **Brug.**
Bruhin, Peter Thomas van Aquinas (1835–1896) P   **Bruhin**
Brühl, Paul Johannes (1855–) AS   **Brühl**
Bruhne, Karl (fl. 1894) M   **Bruhne**
Bruhns, Carl (fl. 1928) M   **Bruhns**
(Bruijn, Ary Johannes de)
  see de Bruijn, Ary Johannes   **Bruijn**
Bruinsma, Josephus Johannes (1805–1888) S   **Bruinsma**
Brullo, Salvatore (1947–) S   **Brullo**
Brumhard, Phillipp (1879–) S   **Brumh.**
Brummelen, Johannes van (1932–) M   **Brumm.**
Brummer–Dinger, C.H. (fl. 1955) S   **Brumm.-Ding.**
Brummitt, Richard Kenneth (1937–) S   **Brummitt**
Brumpt, Émile Josef Alexander (1877–1951) AM   **Brumpt**
Brun, Jacques–Joseph (1826–1908) A   **Brun**
Brunaud, Paul (18??–1903) M   **Brunaud**
Brunchorst, Jörgen (1862–1917) MS   **Brunch.**
(Bruneau de Miré, Philippe)
  see Miré, Philippe   **Miré**
Brunel, Jules F. (1905–1986) AS   **Brunel**
Bruner, Stephen Cole (1891–1953) M   **Bruner**
Brunerye, Luc Jean Loup (1939–) S   **Brunerye**

Brunfels, Otto (1488–1534) L	**Brunfels**
Bruni, Achille (1817–1881) S	**Bruni**
Bruni, Alessandro (1944–) A	**A.Bruni**
Brunken, Jere N. (fl. 1977) S	**Brunken**
Brunnbauer, W. (fl. 1989) M	**Brunnb.**
Brunner, Carl (1796–1869) A	**C.Brunner**
Brunner, Ivano L. (fl. 1989) M	**I.L.Brunner**
Brunner, Samuel (1790–1844) S	**S.Brunner**
Brunner von Wattenwyl, Carl (1823–1914) S	**Brunner**
Brunnthaler, Josef (1871–1914) AB	**Brunnth.**
Bruno (fl. 1760) S	**Bruno**
Brunori, A. (fl. 1989) M	**Brunori**
Brunsfeld, Steven J. (1953–) S	**Brunsfeld**
Brunton, Daniel F. (fl. 1989) P	**D.F.Brunt.**
Brunton, William (1775–1806) B	**Brunt.**
Bruschi, Diana (fl. 1912) M	**Bruschi**
Brush, C.Z. (fl. 1955) F	**Brush**
Brusis, Otto A. (fl. 1972) M	**Brusis**
Brusse, Franklin Andrej (1951–) M	**Brusse**
Brussoff, Alexander von (1878–) A	**Brussoff**
(Brussow, Alexander von)	
see Brussoff, Alexander von	**Brussoff**
Brutschy, A. (1885–1955) A	**Brutschy**
Bruttan, Andreas (1829–1893) C	**Bruttan**
Bruun, Helga Gösta (1897–) S	**Bruun**
Bruylants, J. (fl. 1970) M	**Bruyl.**
(Bruyn, Ary Johannes de)	
see de Bruijn, Ary Johannes	**Bruijn**
Bruyn, H.P. de (fl. 1984) M	**Bruyn**
Bruyns, Peter Vincent (1957–) S	**Bruyns**
Bruzelius, Arvid Sture (1799–1865) A	**Bruzelius**
Bryan, Hilah F. A	**H.F.Bryan**
Bryan, Mary Katherine (1877–) M	**Bryan**
Bryan, Virginia Schmitt (1922–) B	**V.S.Bryan**
Bryant, Charles (–1799) S	**Bryant**
Bryant, Truman R. (fl. 1961) P	**T.R.Bryant**
Bryce Derni, Clara A	**Bryce Derni**
Brygoo, E.-R. (fl. 1964) M	**Brygoo**
Bryhn, Niels (1854–1916) BS	**Bryhn**
Bryson, Charles T. (1950–) S	**Bryson**
(Brzezicki, Michail Vasiljevich)	
see Brzhesitzky, Michail Vasiljevich	**Brzhes.**
Brzezinski, József (fl. 1906) M	**Brzez.**
Brzhesitzky, Michail Vasiljevich (1887–) S	**Brzhes.**
(Brzhezitzky, Michail Vasiljevich)	
see Brzhesitzky, Michail Vasiljevich	**Brzhes.**
Bubák, František (1865–1925) MS	**Bubák**
Bubani, Pietro (1806–1888) BPS	**Bubani**
Buben–Zurey, Mary Jo (fl. 1986) M	**Bub.–Zurey**
Bubnova, S.V. (1960–) S	**Bubnova**

Buc'hoz, Pierre Joseph (1731–1807) S	**Buc'hoz**
Buch, Christian Leopold von (1774–1853) BPS	**Buch**
Buch, Hans Robert Viktor (1883–1964) B	**H.Buch**
Buch, T.G. (fl. 1988) S	**T.G.Buch**
Buchanan, A.M. (1944–) BS	**A.M.Buchanan**
Buchanan, D.E. (fl. 1981) M	**D.E.Buchanan**
(Buchanan, Francis)	
see Buchanan–Hamilton, Francis	**Buch.-Ham.**
Buchanan, John (1819–1898) BS	**Buchanan**
Buchanan, Peter K. (1957–) M	**P.K.Buchanan**
Buchanan, Robert Earle (1883–) AM	**R.E.Buchanan**
Buchanan, T.S. (fl. 1940) M	**T.S.Buchanan**
Buchanan–Hamilton, Francis (1762–1829) PS	**Buch.-Ham.**
(Buchanan–White, Francis)	
see White, Francis Buchanan	**F.B.White**
Buchegger, Josef (1886–) S	**Buchegger**
Buchenau, Franz Georg Philipp (1831–1906) S	**Buchenau**
Bucher, Katina E. (1950–) A	**Bucher**
Buchet, Samuel (1875–1956) S	**Buchet**
Buchetet, Théodore (1824–1883) S	**Buchetet**
Buchheim, A. (fl. 1924) M	**A.Buchheim**
Buchheim, Arno Fritz Günther (1924–) S	**Buchheim**
Buchheim, Mark A. A	**M.A.Buchheim**
Buchheister, John C. P	**Buchheist.**
Buchholz, Feodor (Fedor) Vladimirovic (1872–1924) MS	**Buchholz**
Buchholz, John Theodore (1888–1951) PS	**J.Buchholz**
Buchinger, Jean Daniel (1805–1888) BS	**Buchinger**
Buchinger, Maria S	**M.Buchinger**
Buchli, Harro H.R. (fl. 1952) M	**Buchli**
Buchloh, Günther (1923–) BS	**Buchloh**
Buchner, P. (fl. 1912) M	**Buchner**
(Bucholtz, Feodor (Fedor) Vladimirovic)	
see Buchholz, Feodor (Fedor) Vladimirovic	**Buchholz**
(Buchoz, Pierre Joseph)	
see Buc'hoz, Pierre Joseph	**Buc'hoz**
Buchs (fl. 1936) M	**Buchs**
Buchtien, Otto (1859–) S	**Buchtien**
Buchwald, Johannes (1869–1927) S	**Buchw.**
Buchwald, Niels Fabritius (1898–) M	**N.F.Buchw.**
Buck, Hilda Teresa (fl. 1979) S	**H.T.Buck**
Buck, J.D. (fl. 1988) M	**J.D.Buck**
Buck, Reinhardt A	**R.Buck**
Buck, William Russell (1950–) BMPS	**W.R.Buck**
Bucka, Halina A	**Bucka**
Buckland, William (1784–1856) S	**Buckland**
Buckley, Helen R. (fl. 1968) M	**H.R.Buckley**
Buckley, Samuel Botsford (1809–1884) PS	**Buckley**
Buckley, W.D. (fl. 1920) M	**W.D.Buckley**
Buckman, James (1814–1884) BS	**Buckman**

Bucknall, Cedric (1849–1921) MS	**Buckn.**
Bucquoy, Eugène (1837–1904) S	**Bucq.**
Budai, József (1851–1939) S	**Budai**
(Budantsev, A.L.)	
see Budantzev, A.L.	**A.L.Budantzev**
(Budantsev, L.Yu.)	
see Budantzev, L.Yu.	**Budantzev**
Budantzev, A.L. (1957–) S	**A.L.Budantzev**
Budantzev, L.Yu. (fl. 1967) F	**Budantzev**
Budathoki, Usha (fl. 1989) M	**Budathoki**
Budde, Ernst A	**E.Budde**
Budde, Hermann (1890–1954) A	**Budde**
Buddenhagen, Ivan W. (fl. 1957) M	**Buddenh.**
Buddin, Walter (1890–1962) M	**Buddin**
Buddle, Adam (c.1660–1715) L	**Buddle**
Büdel, B. (fl. 1986) M	**Büdel**
Buder, Johannes (1884–1966) A	**Buder**
Budington, A.B. (fl. 1970) M	**Budington**
Budry, David A	**Budry**
Buek, Heinrich Wilhelm (1796–1878) S	**H.Buek**
Buek, Johannes Nicolaus (1779–1856) S	**J.N.Buek bis**
Buek, Johannes Nikolaus (1736–1812) S	**J.N.Buek**
Büel, Hans (fl. 1972) S	**Büel**
Buell, Helen Foot (1902–) A	**Buell**
Buen y del Cos, O. de (1863–1945) S	**Buen**
Buendía, A.G. (fl. 1986) M	**Buendía**
Buenecker, R. (fl. 1977) S	**Buenecker**
Buerger, F. S	**F.Buerger**
Buetschli, Johann Adam Otto (1848–1920) AP	**Buetschli**
(Buettner, Richard)	
see Büttner, Richard	**Büttner**
Buffham, Thomas Hughes (1840–1896) A	**Buffham**
Buffin, N. (fl. 1984) M	**Buffin**
Buffon, Georges Louis Leclerc de (1707–1788) S	**Buffon**
Bugaev, N.N. (fl. 1985) S	**Bugaev**
Bugała, Władysław (1924–) S	**Bugała**
Bugg, S.C. A	**Bugg**
Bùggieri (fl. 1935) M	**Bùggieri**
Bugnicourt, F. (fl. 1939) M	**Bugnic.**
Buhagiar, R.W.M. (fl. 1973) M	**Buhagiar**
Buhr, Herbert (1902–1968) M	**Buhr**
(Buhr, Larry Eugene De)	
see DeBuhr, Larry Eugene	**DeBuhr**
Buhrer, Edna Marie (1898–) M	**Buhrer**
Buhse, Friedrich Alexander (1821–1898) S	**Buhse**
Bui, Ngoc–Sanh (fl. 1964) S	**Bui**
Buia, Alexandra (1911–1964) S	**Buia**
Buijsen, J.R.M. (fl. 1988) S	**Buijsen**
Buining, Albert Frederik Hendrik (1901–1980) S	**Buining**
Buisman, Christine Johanna (1900–1936) M	**Buisman**

Buisson, J.P. (fl. 1779) S	**Buisson**
Buist, George S	**G.Buist**
Buist, Robert (1805–1880) S	**Buist**
Bujak, J.P. A	**J.P.Bujak**
Bujakiewicz, Anna M. (fl. 1968) M	**Bujak.**
Bujorean, George (1893–1971) S	**Bujor.**
Bukasov, Sergej (Sergei) Mikhailovich (1891–1983) S	**Bukasov**
Bukhalo, Asja S. (1932–) M	**Bukhalo**
Bukry, John David (1941–) A	**Bukry**
Bula–Meyer, German A	**Bula-Meyer**
Bulach, E.M. (fl. 1987) M	**Bulach**
Bulanov, P.A. (fl. 1965) M	**Bulanov**
Bulany, Ju.(Yu.) I. (fl. 1987) S	**Bulany**
(Bulanyi, Ju.(Yu.) I.)	
see Bulany, Ju.(Yu.) I.	**Bulany**
Bulgakova, L.L. (1929–) S	**Bulgakova**
Bull, Alec Leonard (fl. 1988) S	**A.L.Bull**
Bull, Jeffrey D. A	**J.D.Bull**
Bull, William (1828–1902) PS	**W.Bull**
Buller, Arthur Henry Reginald (1879–1944) M	**Buller**
Bulley, A.K. (1861?–1942) S	**Bulley**
Bulliard, Jean Baptiste François ('Pierre') (1752–1793) MS	**Bull.**
Bullock, Arthur Allman (1906–1980) S	**Bullock**
Bullock–Webster, George Russell (1858–1934) A	**Bull.-Webst.**
Bulnhein, Carl Otto (1820–1865) A	**Bulnh.**
Bumžaa, D. (fl. 1986) M	**Bumžaa**
Bunbury, Charles James Fox (1809–1886) S	**Bunbury**
Bunbury, L. (fl. 1861) F	**L.Bunbury**
Bunce, Maureen E. (fl. 1961) M	**Bunce**
Bunch, J.L. (fl. 1901) M	**Bunch**
Bundy, William F. (fl. 1883) M	**Bundy**
Bunescu, S. (fl. 1975) M	**Bunescu**
Bunge, Alexander Andrejewitsch (Aleksandr Andreevic, Aleksandrovic) von	
(1803–1890) BS	**Bunge**
Buniva, Michele Francesco (1761–1834) S	**Buniva**
Bunkina, I.A. (fl. 1960) M	**Bunkina**
Bunting, George Sydney (1927–) S	**G.S.Bunting**
Bunting, R.H. (fl. 1923) M	**Bunting**
Burbank, Luther (fl. 1914) S	**Burbank**
Burbidge, Frederick William Thomas (1847–1905) S	**Burb.**
Burbidge, Nancy Tyson (1912–1977) S	**N.T.Burb.**
Burbidge, Robert Brinsley (1943–) S	**R.B.Burb.**
Burbridge, Patricia P. A	**Burbridge**
Burch, Derek George (1933–) S	**D.G.Burch**
Burch, George J. (1852–1914) A	**G.J.Burch**
Burchard, G. (fl. 1929) M	**G.Burchard**
Burchard, Oscar (1863–1949) BS	**Burchard**
Burchell, William John (1781–1863) BPS	**Burch.**
Burchill, R.T. (fl. 1986) M	**Burchill**

Burck, William (1848–1910) PS	**Burck**
Burckle, Lloyd H. A	**Burckle**
Burdet, Hervé Maurice (1939–) PS	**Burdet**
Burdin, H. S	**Burdin**
Burdjukova, Ludmila I. (1948–) M	**Burdjuk.**
Burdsall, Harold H. (1940–) M	**Burds.**
(Burdyukova, Ludmila I.)	
see Burdjukova, Ludmila I.	**Burdjuk.**
Bureau, Louis Édouard (1830–1918) FS	**Bureau**
Büren, G.von (fl. 1922) M	**Büren**
Burenin, V.I. (fl. 1983) S	**Burenin**
Burge, M.N. (fl. 1974) M	**Burge**
Burgeff, Hans Edmund Nicola (1883–1976) BMS	**Burgeff**
Burger, D. A	**D.Burger**
Bürger, Heinrich (1806–1858) S	**Bürger**
Burger, William Carl (1932–) S	**W.C.Burger**
Burger–Wiersma, Tineke A	**Burger-Wiersma**
Burgersdijk, Leendert Alexander Johannes (1828–1900) S	**Burgersd.**
(Burgersdyk, Leendert Alexander Johannes)	
see Burgersdijk, Leendert Alexander Johannes	**Burgersd.**
Burges, Norman Alan (1911–) BF	**Burges**
Burgess, Edward Sandford (1855–1928) PS	**E.S.Burgess**
Burgess, Henry W. (fl. 1827–1833) S	**Burgess**
Burgess, Jack D. (1924–) A	**J.D.Burgess**
Burgess, L.W. (fl. 1982) M	**L.W.Burgess**
Burgh, Johannes van der (1937–) S	**Burgh**
Burghouts, T. (fl. 1989) M	**Burghouts**
Burgman, Mark A. (1956–) S	**Burgman**
Burgos, Julio A. (fl. 1987) M	**Burgos**
Burgsdorff, Friedrich August Ludwig von (1747–1802) S	**Burgsd.**
Burgwitz, George Konstantinovich (1889–) M	**Burgwitz**
Burgyukova, A.I. (fl. 1987) M	**Burgyuk.**
Buriticá, Pablo (1943–) M	**Buriticá**
Burk, William R. (fl. 1991) M	**Burk**
Burkart, Arturo Erhardo (Erardo) (1906–1975) APS	**Burkart**
Burke, G.C. (fl. 1951) M	**Burke**
Burke, John M. A	**J.M.Burke**
Burkhardt, Christian Friedrich (1785–1854) S	**Burkhardt**
Burkholder, JoAnn M. A	**J.M.Burkh.**
Burkholder, Walter Hagemeyer (1891–) MS	**Burkh.**
Burkill, Humphrey Morrison (1914–) S	**H.M.Burkill**
Burkill, Isaac Henry (1870–1965) S	**Burkill**
Burkwood, A. (fl. 1929) S	**Burkwood**
Burle–Marx, Roberto (1909–) S	**Burle-Marx**
Burley, John Stuart (1950–) B	**Burley**
Burlingham, Gertrude Simmons (1872–1952) M	**Burl.**
Burman, Alasdair Graham (1942–) S	**A.G.Burm.**
Burman, Johannes (1707–1779) PS	**Burm.**
Burman, Nicolaas Laurens (Nicolaus Laurent) (1734–1793) APS	**Burm.f.**

Burman, William Alfred (1856–1909) S	W.A.Burm.
Burmann, Gusti A	Burmann
Burmeister, Hermann Carl Conrad (1807–1851) P	Burmeist.
Burnat, Émile (1828–1920) PS	Burnat
Burnett, Gilbert Thomas (1800–1835) PS	Burnett
Burnett, Harry C. (fl. 1965) M	H.C.Burnett
Burnett, John Harrison (1922–) S	J.H.Burnett
Burnham, Stewart Henry (1870–1943) BPS	Burnham
Burnier, R. (fl. 1912) M	Burnier
Burns, A.C. (fl. 1927) M	Burns
Burns, D.A. A	D.A.Burns
Burns–Balogh, Pamela (fl. 1983) S	Burns–Bal.
Burnside, C.E. (fl. 1928) M	Burnside
Burollet, Pierre Andre (1889–) S	Burollet
Burpee, L.L. (fl. 1980) M	Burpee
Burrell, William Holmes (1865–1945) B	Burrell
Burret, (Maximilian) Karl Ewald (1883–1964) S	Burret
Burri, R. (fl. 1921) M	Burri
Burrill, Thomas Jonathan (1839–1916) M	Burrill
Burrows, Colin James (1931–) S	C.J.Burrows
Burrows, Elsie May (1913–1986) A	Burrows
Burrows, John Eric (1950–) PS	J.E.Burrows
Burrows, Sandra Margaret (1959–) P	S.M.Burrows
Bursa, Adam S. (1908–) A	Bursa
Burser, Joachim (1583–1639) L	Burser
Burt, Edward Angus (1859–1939) M	Burt
Burt–Utley, Kathleen (1944–) S	Burt–Utley
Burton, Christine M. (fl. 1988) S	C.M.Burton
Burton, K.A. (fl. 1956) M	K.A.Burton
Burton, Mary Gwendolyn (1917–) M	Burton
Burton, Rodney McGuire (1936–) S	R.M.Burton
Burtt, Bernard Dearman (1902–1938) S	Burtt
Burtt, Brian Laurence ('Bill') (1913–) S	B.L.Burtt
Burtt Davy, Joseph (1870–1940) S	Burtt Davy
Burvenich, Fréderic (1857–1917) S	Burv.
Bury, Priscilla Susan (fl. 1831–1837) S	Bury
Buscalioni, Luigi (1863–1954) S	Buscal.
Busch, Anton (1823–1895) S	Busch
Busch, Elizaveta Aleksandrovna (1886–1960) S	E.A.Busch
Busch, L.V. (fl. 1962) M	L.V.Busch
Busch, Nicolaï Adolfowitsch (Nikolaj Adolfovich) (1869–1941) PS	N.Busch
Busch, Werner (1901–) A	W.Busch
Buschardt, Arthur (1935–) M	Buschardt
Buschmann, Adolphine (1908–) S	Buschm.
Büse, Lodewijk Hendrik (1819–1888) BS	Büse
Busek, J. (fl. 1986) S	Busek
Busen, K.E. A	Busen
Buser, Robert (1857–1931) S	Buser
Büsgen, Moritz (1858–1921) M	Büsgen

Bush, Benjamin Franklin (1858–1937) PS	**Bush**
(Bush, Elizabeth Aleksandrovna)	
see Busch, Elizaveta Aleksandrovna	**E.A.Busch**
Businsky, Roman S	**Businsky**
Busse, Walter Carl Otto (1865–1933) MS	**Busse**
Busson, Félix A	**Busson**
But, Paul Pui–Hay (fl. 1982) S	**But**
Butcher, Roger William (1897–1971) AS	**Butcher**
(Bute, John Stuart, Earl of)	
see Stuart, John, Earl of Bute	**Stuart**
Butin, H.von (fl. 1957) M	**Butin**
Butkov, A.Y.(J.) (1911–1981) S	**Butkov**
Butkus, V.F. (fl. 1972) S	**Butkus**
Butler, Bertram Theodore (1872–) S	**B.T.Butler**
Butler, Edward Eugene (1919–) M	**E.E.Butler**
Butler, Edwin John (1874–1943) MS	**E.J.Butler**
Butler, Ellys Theodora (1906–) M	**E.T.Butler**
Butler, Garl L. A	**G.L.Butler**
Butsch, T.G. (1929–) S	**Butsch**
(Bütschli, Johann Adam Otto)	
see Buetschli, Johann Adam Otto	**Buetschli**
Butters, Frederick King (1878–1945) APS	**Butters**
Buttler, Karl Peter (1942–) S	**Buttler**
Büttner, Julius (1873–) A	**J.Büttner**
Büttner, Richard (fl. 1890) S	**Büttner**
Butts, Charles (1863–1946) A	**Butts**
Butz, George C. (1863–1907) S	**Butz**
Butzin, Friedhelm Reinhold (1936–) AMS	**Butzin**
Buwalda, Pieter (1909–1947) S	**Buwalda**
Buxbaum, Franz (1900–1979) S	**Buxb.**
Buxbaum, Johann Christian (1693–1730) M	**J.C.Buxb.**
Buxton, Bertram Henry (1852–1934) S	**B.H.Buxton**
Buxton, Richard (1786–1865) S	**Buxton**
Buyck, Bart (fl. 1983) M	**Buyck**
Buysson, Robert du (1871–1893) B	**Buyss.**
(Buze, Fedor Aleksandrovich)	
see Buhse, Friedrich Alexander	**Buhse**
Buzunova, I.O. (1946–) S	**Buzunova**
Buzzati–Traverso, Adriano (fl. 1935) M	**Buzz.-Trav.**
Byam–Grounds, J.S. (fl. 1987) S	**Byam-Grounds**
Byatt, Jean Irene (fl. 1970) S	**Byatt**
Bybell, Laurel A	**Bybell**
Byczennikova, N.K. (fl. 1956) S	**Byczenn.**
Bydgosz, W.W. (fl. 1932) M	**Bydgosz**
Bye, R.A. (fl. 1976) S	**Bye**
Byford, W.J. (fl. 1963) M	**Byford**
(Byhouwer, Jan Tijs Pieter)	
see Bijhouwer, Jan Tijs Pieter	**Bijh.**
Bykov, Boris Aleksandrovich (1910–1990) S	**Bykov**

Bykova, E.V. A	**Bykova**
Bykovskaya, S.V. (fl. 1974) M	**Bykovsk.**
Bykstra, J. (fl. 1983) M	**Bykstra**
(Byl, Paul Andries Van der)	
see Van der Byl, Paul Andries	**Van der Byl**
Byles, Ronald Stewart (fl. 1957) S	**Byles**
Bynum, H.H. (fl. 1972) M	**Bynum**
Byrnes, Norman Brice (1922–) S	**Byrnes**
Bystrek, Jan (1934–) M	**Bystrek**
Bystrická, Hedwiga A	**Bystrická**
Bystrický, Ján (1922–) A	**Bystrický**
Bystrov (fl. 1959) M	**Bystrov**
Bywater, Joan (fl. 1959) M	**Bywater**
Bywater, Marie (1951–) S	**M.Bywater**
Byzova, Z.M. (fl. 1960) M	**Byzova**
Caballero, Arturo (1877–1950) AMS	**Caball.**
Caballero Deloya, Miguel (fl. 1969) S	**M.Caball.**
Caballero, Francisca A	**F.Caball.**
(Caballero y Segares, Arturo)	
see Caballero, Arturo	**Caball.**
Cabañes, Javier (fl. 1990) M	**J.Cabañes**
Cabanès, Jean Gustave (1864–1944) S	**Cabanès**
Cabejszekówna, Irena (1910–1972) A	**Cabejsz.**
Cabello, Marta N. (1953–) M	**Cabello**
(Cabezudo Artero, Baltasar)	
see Cabezudo, Baltasar	**Cabezudo**
Cabezudo, Baltasar (1946–) S	**Cabezudo**
Cabioch, Jacqueline A	**Cabioch**
Cabral, Daniel (1946–) M	**Cabral**
Cabral, Elsa Leonor (1951–) S	**E.L.Cabral**
Cabrera, Angel Lulio (1908–) AS	**Cabrera**
Cabrera, Margarita (1925–) S	**M.Cabrera**
Cabrera, R.Leticia S	**R.L.Cabrera**
Cacciato, Alfredo (1907–1986) S	**Cacciato**
Cáceres, Eduardo J. (1943–) A	**Cáceres**
Cachon, Jean A	**Cachon**
(Cachon, Monique)	
see Cachon-Enjumet, Monique	**Cachon-Enj.**
Cachon–Enjumet, Monique (née Cachon, M.) A	**Cachon-Enj.**
Cadet, Thérésian (1937–1987) S	**Cadet**
Cadete, Antonio (fl. 1978) S	**Cadete**
(Cadevall i Diars, Juan)	
see Cadevall y Diars, Juan	**Cadevall**
Cadevall y Diars, Juan (1846–1921) PS	**Cadevall**
Cadman, Mandy–Jane (later Balkwill, M.–J.) (1964–) S	**Cadman**
Čado, Ivan A	**Čado**
Cady, Leo I. (fl. 1964) S	**Cady**
Caels, Theodor Peter (1739–1819) S	**Caels**

Caffin, Jacques François (1778–1854) S   **Caffin**
Caflisch, Jacob Friedrich (1817–1882) S   **Caflisch**
Cai, Shi Xun (fl. 1985) A   **S.X.Cai**
Caillet, Michel (fl. 1980) M   **Caillet**
Cailleux, Roger (1929–) M   **Cailleux**
Cailliaud, Frédéric (1787–1869) S   **Caill.**
Caillon, P. (fl. 1984) M   **Caillon**
Cain, Roy Franklin (1906–) M   **Cain**
Cain, Stanley Adair (1902–) B   **S.A.Cain**
Caine, Thelma Saran (fl. 1980) M   **Caine**
Cajander, Aimo Aarno Antero (after 1935, Kalela, A.A.A.) (1908–) S   **A.Cajander**
Cajander, Aimo Kaarlo (1879–1943) S   **Cajander**
Calandra, François   A   **Calandra**
Calandron, André (fl. 1953) M   **Calandron**
Calaway, Wilson Thayer (1912–) A   **Calaway**
Caldas, Francisco José de (1771–1816) S   **Caldas**
(Caldas y Tenorio, Francisco José de)
  see Caldas, Francisco José de   **Caldas**
Calder, James (Jim) Alexander (1915–1990) S   **Calder**
Calder, Mary G. (fl. 1953) F   **M.G.Calder**
Calderón, Cléofe E. (1940?) S   **C.E.Calderón**
(Calderón de Rzedowski, Graciela)
  see Calderón, Graciela   **Calderón**
Calderón, Graciela (1931–) S   **Calderón**
Caldesi, Ludovico (Luigi) (1821–1884) ABMS   **Caldesi**
Caldis, Panos Demetrius (1896–) M   **Caldis**
(Calduch Almela, Manuel)
  see Calduch, Manuel   **Calduch**
Calduch, Manuel (1901–1981) S   **Calduch**
Caldwell, Ralph Merrill (1903–) M   **Caldwell**
Calestani, Vittorio (1882–) S   **Calest.**
Caley, George (1770–1829) S   **Caley**
Calhoun, Cornelia J. (fl. 1989) M   **Calhoun**
Calkins, Gary Nathan (1869–1943) A   **G.N.Calk.**
Calkins, William Wirt (1842–1914) M   **Calk.**
Callaghan, A.A. (fl. 1989) M   **Callaghan**
Callan, Brenda E. (1960–) M   **Callan**
Callay, Eugène Albert Athenase (1822–1896) S   **Callay**
Callé, Jean (1906–) PS   **Callé**
Callebaut, E. (fl. 1983) M   **Callebaut**
Callejas, Ricardo (fl. 1990) S   **Callejas**
(Callen, D.Lobreau)
  see Lobreau–Callen, D.   **Lobr.-Call.**
Callen, Eric Ottleben (1912–1970) MS   **Callen**
Callier, Alfons S. (1866–1927) S   **Callier**
Calloni, Silvio (1850–1931) S   **Calloni**
Calmette, Léon Charles Albert (1863–1933) M   **Calmette**
Calnegru, Ion (fl. 1961) M   **Calnegru**
Calonge, Francisco de (1938–) M   **Calonge**

Calponzos, L. (fl. 1956) M   **Calp.**
Calvelo, Susana (fl. 1989) M   **Calvelo**
Calvert, Amelia Catherine (1876–   ) S   **A.C.Calvert**
Calvert, Philip Powell (1871–   ) S   **Calvert**
Calviello, Beatriz O. (1934–   ) M   **Calviello**
(Calviello de Roldán, Beatriz O.)
   see Calviello, Beatriz O.   **Calviello**
Calvino, M. (fl. 1938) M   **Calvino**
(Calvo, A.)
   see Calvo Torras, M.A.   **Calvo**
Calvo Torras, M.Ángeles (–1988) M   **Calvo**
Calwer, Carl Gustav (1821–1873) S   **Calwer**
Calzolari, F. (1522–1609) L   **Calzolari**
Camain, R. (fl. 1959) M   **Camain**
Cámara, Fernando (1906–1985) S   **Cámara**
(Cámara Hernández, Julián A.)
   see Cámara, Julián A.   **J.A.Cámara**
Cámara, Julián A. (1932–   ) S   **J.A.Cámara**
(Cámara, Manoel Arruda da)
   see Arruda da Cámara, Manoel   **Arruda**
(Câmara, Manuel Emmanuele)
   see Sousa da Câmara, Manuel Emmanuele de   **Sousa da Câmara**
(Cámara Niño, Fernando)
   see Cámara, Fernando   **Cámara**
Camarda, Ignazio (1946–   ) S   **Camarda**
Camargo, Felisberto Cardoso de (1896–1943) S   **Camargo**
(Camargo G., Luis A.)
   see Camargo, Luis A.   **L.A.Camargo**
Camargo, Luis A. (fl. 1966) S   **L.A.Camargo**
Çamarrone, V. (1923–   ) S   **Camarrone**
Cambage, Richard Hind (1859–1928) S   **Cambage**
Cambessèdes, Jacques (1799–1863) BS   **Cambess.**
Cambridge, M.L. (fl. 1979) S   **Cambridge**
Camburn, Keith E.   A   **Camburn**
(Camel, Georg Joseph)
   see Kamel, Georg Joseph   **Kamel**
(Camelli, Georgius Josephus)
   see Kamel, Georg Joseph   **Kamel**
(Camellus, Georgius Josephus)
   see Kamel, Georg Joseph   **Kamel**
Cameron, Alexander Kenneth (1908–   ) S   **Cameron**
Cameron, Judith V. (fl. 1972) M   **J.V.Cameron**
Camfield, Julius Henry (1852–1916) S   **Camfield**
Camici, Leontina (fl. 1948) M   **Camici**
Caminhoá, Joaquim Monteiro (1835–1896) S   **Caminhoá**
Camino, Mayra (1962–   ) M   **Camino**
Camisola, Giuseppe (1781–1856) S   **Camisola**
Camp, Wendell Holmes (1904–1963) S   **Camp**
Campana, Antonio Francesco (1751–1832) S   **Campana**

Campanile

Campanile, Giulia Rivera (fl. 1922) M — **Campan.**
Campbell, A.K. (fl. 1986) S — **A.K.Campb.**
Campbell, Christopher S. (1946–) S — **C.S.Campb.**
Campbell, Colin K. (fl. 1987) M — **C.K.Campb.**
Campbell, Douglas Houghton (1859–1953) BPS — **Campb.**
Campbell, Elaine Jean Foulds (1959–) S — **E.J.F.Campb.**
Campbell, Ella Orr (1910–) AB — **E.O.Campb.**
Campbell, Gloria Rae (1924–) S — **G.R.Campb.**
Campbell, Guy (fl. 1939) F — **G.Campb.**
Campbell, I. (fl. 1973) M — **I.Campb.**
Campbell, Julian J.N. (1953–) S — **J.J.N.Campb.**
Campbell, Leo (1894–) M — **L.Campb.**
Campbell, R. (fl. 1977) M — **R.Campb.**
Campbell, Susan E.   A — **S.E.Campb.**
Campbell, William Andrew (1906–) M — **W.A.Campb.**
(Campdera Camius, Francisco (François))
  see Campderá, Francisco (François) — **Campd.**
Campderá, Francisco (François) (1793–1862) S — **Campd.**
Camper, Peter (Petrus) (1722–1789) F — **Camper**
Campion–Alsumard, Thérèse Le (fl. 1969) M — **Camp.-Als.**
Campo, Petrio del (fl. 1855) S — **Campo**
(Campo–Duplan, Madeleine Van)
  see Van Campo, Madeleine — **Van Campo**
(Campos Novães, José de)
  see Novães, José de Campos — **Novães**
(Campos Porto, Paulo)
  see Porto, Paulo Campos — **Porto**
Campos, Silvio T.C. (fl. 1956) M — **Campos**
Camposano, Anna (fl. 1951) M — **Camposano**
Camus, Aimée Antoinette (1879–1965) S — **A.Camus**
Camus, Edmond Gustav(e) (1852–1915) S — **E.G.Camus**
Camus, Ferdinand Antonin (1852–1922) B — **F.A.Camus**
Camus, Giulio (1847–1917) S — **Camus**
Camus, Josephine M. (1949–) P — **J.M.Camus**
(Camus, Philippe Xavier)
  see Philippe, Xavier — **Philippe**
Canabaeus, Lotte   A — **Canab.**
Canak, M. (fl. 1979) S — **Canak**
(Canal Feijoo, E.J.)
  see Feijoo, E.J. — **Feijoo**
Canby, Margaret Leslie (1904–) S — **M.L.Canby**
Canby, William Marriott (1831–1904) S — **Canby**
Candargy, Paléologos C. (1870–) S — **P.Candargy**
Candargy, T.   S — **Candargy**
Candeias, Alberto   A — **Candeias**
Candolle, Alphonse Louis Pierre Pyramus de (1806–1893) S — **A.DC.**
Candolle, Anne Casimir Pyramus de (1836–1918) S — **C.DC.**
Candolle, Augustin Pyramus de (1778–1841) BMPS — **DC.**
Candolle, Richard Émile Augustin de (1868–1920) S — **Aug.DC.**

Candoussau, Françoise (1934–) M	**Cand.**
Candusso, Massimo (fl. 1982) M	**Candusso**
(Canessa A., Edwin)	
see Canessa, Edwin	**Canessa**
Canessa, Edwin (fl. 1988) S	**Canessa**
Caneva, Giulia (fl. 1981) M	**Caneva**
Canfield, Elmer R. (fl. 1916) M	**Canf.**
Cañigueral Cid, Juan (1912–) S	**Cañig.**
Cann, J.P. A	**J.P.Cann**
Cann, Sheila F. A	**S.F.Cann**
Cannart d'Hamale, Frédéric (1804–1888) S	**Cannart**
Cannon, John Francis Michael (1930–) S	**Cannon**
Cannon, Margaret Joy (1928–) S	**M.J.Cannon**
Cannon, Paul Francis (1956–) M	**P.F.Cannon**
Cano, J. (fl. 1987) M	**Cano**
Canonaco, A. (fl. 1936) M	**Canonaco**
Canter, Hilda M. (fl. 1949) M	**Canter**
Cantino, E.C. (fl. 1953) M	**Cantino**
Cantino, Philip Douglas (1948–) S	**P.D.Cantino**
Cantley, Nathaniel (–1888) S	**Cantley**
(Cantó Ramos, P.)	
see Cantó, P.	**Cantó**
Cantó, Paloma (1956–) S	**Cantó**
Cantón, Eliseo (fl. 1898) M	**Cantón**
Cantoni, C. (fl. 1979) M	**Cantoni**
Cantor, Theodore Edward (1809–1854) A	**Cantor**
Cantoria, M.C. (fl. 1982) S	**Cantoria**
Cantournet, Jean (fl. 1948) M	**Cant.**
Cantrell, David J. (fl. 1987) F	**Cantrell**
Cao, Jin Zhong (fl. 1987) M	**J.Z.Cao**
Cao, Rui (1958–) S	**R.Cao**
Cao, Tong (1946–) B	**T.Cao**
Cao, Ya Ling (1937–) S	**Y.L.Cao**
Cao, Yu Hui (fl. 1981) S	**Y.H.Cao**
Capanema (fl. 1862) S	**Capan.**
Capdevielle, Paul A	**Capdev.**
Capeder, Giuseppe A	**Capeder**
Capelari, Marina (1958–) M	**Capelari**
Capellano, A. (fl. 1963) M	**Capellano**
Capelli, Carlo Matteo (1763–1831) S	**Capelli**
Capellini, Giovanni (1833–1922) F	**Capellini**
Capitaine, Louis (1883–1923) S	**Capit.**
Capon, Brian (1931–) A	**Capon**
Capot, J. (fl. 1975) S	**Capot**
Capote, R. (fl. 1977) S	**Capote**
Cappelletti, Carlo (1900–) B	**Capp.**
Cappelli, A. (fl. 1982) M	**Cappelli**
Capponi, M. (fl. 1959) M	**Capponi**
Capriotti, Augusto (1920–1970) M	**Capr.**

Capron, E. (fl. 1871) M	**Capron**
Capuron, René Paul Raymond (1921–1971) S	**Capuron**
Capurro, Roberto Horacio (1910–) PS	**Capurro**
Caputo, Giuseppe (1926–) S	**Caputo**
Carabia, José Perez (1910–) S	**Carabia**
Caraccioli (fl. 1934) M	**Caracc.**
Caram, Bernadette A	**Caram**
Carano, Enrico (1877–1943) S	**Carano**
Caratini, Claude A	**Caratini**
Carauta, Jorge Pedro Pereira (1930–) S	**Carauta**
Carbajal, A.J. (fl. 1901) M	**Carbajal**
Carbo, R. (fl. 1988) S	**Carbo**
(Carbonell, José Borja)	
see Borja Carbonell, José	**Borja**
Carbonnier, J. (fl. 1977) S	**Carb.**
(Carbonnière, Louis François Elisabeth Ramond de)	
see Ramond de Carbonnière, Louis François Elisabeth	**Ramond**
Carbono, Eduino (1950–) S	**Carbono**
Card, Hamilton Hye (1877–1953) S	**Card**
Cardenas, Giovanna de Martino S	**G.M.Cardenas**
Cárdenas, Hermosa Martin (1899–1973) S	**Cárdenas**
Cardenas, Lourdes (1933–) S	**L.Cardenas**
Cardeñas Mayoral, Jesus (fl. 1983) S	**J.Cardeñas**
Cárdenas Soriano, M.Angeles (1949–) B	**M.A.Cárdenas**
Cardew, R.M. (fl. 1910) S	**Cardew**
Cardiel, José María (fl. 1990) S	**Cardiel**
Cardinal, André A	**Cardinal**
(Cardona Florit, María de los Angeles)	
see Cardona, María de los Angeles	**Cardona**
Cardona, María de los Angeles (1940–) S	**Cardona**
Cardone (fl. 1905) M	**Cardone**
Cardoso, L. (fl. 1944) M	**Cardoso**
Cardoso, Mary B. A	**M.B.Cardoso**
Cardot, Jules (1860–1934) BS	**Cardot**
Carestia, Antonio (1825–1908) BM	**Carestia**
Caretta, G. (fl. 1960) M	**Caretta**
Carey, John (1797–1880) S	**J.Carey**
Carey, S.W. (fl. 1935–89) F	**S.W.Carey**
Carey, Susan T. (fl. 1976) M	**S.T.Carey**
Carey, William (1761–1834) S	**Carey**
Carhart, Macy (fl. 1916) P	**Carhart**
Carini, A. (fl. 1940) M	**Carini**
Carion, Jules Émile (1796–1863) S	**Carion**
Cariot, Antoine (1820–1883) PS	**Cariot**
Carl, Helmut (fl. 1931) B	**Carl**
Carl, K.P. (fl. 1976) M	**K.P.Carl**
Carleton, Mark Alfred (1866–1925) M	**Carleton**
Carlin, G. (fl. 1982) M	**Carlin**
Carlin–Silväng, U. (fl. 1982) M	**Carl.-Silv.**
Carlquist, Sherwin (1930–) S	**Carlquist**

Carlson, G.W.F.　A	**G.W.F.Carlson**
Carlson, Margery Claire (1892–　) S	**Carlson**
Carlström, Annette L. (1957–　) S	**Carlström**
Carman, P.E. (fl. 1956) M	**Carman**
Carmichael, Dugald (1772–1827) AMPS	**Carmich.**
Carmichael, John W. (fl. 1962) M	**J.W.Carmich.**
Carmignani, Vincenzo (1779–1859) S	**Carmign.**
Carmo Souza, Lidia do (fl. 1957) M	**Carmo Souza**
Carne, Walter Mervyn (1885–1952) M	**Carne**
(Carneiro de Albuquerque, Fernando)	
see Albuquerque, Fernando Carneiro de	**F.C.Albuq.**
Carneiro, L.S. (fl. 1956) M	**Carneiro**
Carnevale, P.　A	**Carnevale**
Carnevale-Ricci, F. (fl. 1926) M	**Carn.-Ricci**
(Carnevali F.C., Germán)	
see Carnevali, Germán	**Carnevali**
Carnevali, Germán (fl. 1987) S	**Carnevali**
Carney, Heath J.　A	**Carney**
Carnham, Susan C. (fl. 1969) M	**Carnham**
Caro, José Aristida (Alfredo) (1919–1985) S	**Caro**
Caro, Yves　A	**Y.Caro**
Carol, W.L.L. (fl. 1928) M	**Carol**
Carolin, Roger Charles (1929–　) PS	**Carolin**
(Caron, Thomas Joseph)	
see Irénée-Marie, Thomas Joseph Caron	**Irénée-Marie**
Caroselli, Nestor E. (fl. 1949) M	**Caros.**
Carozzi, Albert V. (1925–　) A	**Carozzi**
Carpano, M. (fl. 1940) M	**Carpano**
Carpenter, Clarence Willard (1888–1946) M	**C.W.Carp.**
Carpenter, Edward J. (1942–　) A	**E.J.Carp.**
Carpenter, Steven E. (fl. 1974) M	**S.E.Carp.**
Carpenter, William (1797–1874) S	**Carp.**
Carpenter, William Benjamin (1813–1885) S	**W.B.Carp.**
Carpentier, Alfred (1878–1952) F	**Carpentier**
Carr, Cedric Errol (1892–1936) S	**Carr**
Carr, Dennis John (1915–　) BS	**D.J.Carr**
Carr, Geoffrey William (1948–　) S	**G.W.Carr**
Carr, Gerald D. (1945–　) S	**G.D.Carr**
Carr, Lloyd George (1917–　) S	**L.G.Carr**
Carr, Robert Leroy (1940–　) S	**R.L.Carr**
Carr, Stella Grace Maisie (1912–1988) S	**S.G.M.Carr**
Carradori, Giovacchino (1758–1818) C	**Carradori**
Carralves, Matilde Goncalves　A	**Carralves**
Carranza, José Maria (1922–　) M	**Carranza**
Carranza, M.R. (fl. 1983) M	**M.R.Carranza**
Carranza-Morse, J. (fl. 1986) M	**J.Carranza**
Carré, D.　A	**Carré**
Carreiro, M.M. (fl. 1989) M	**Carreiro**
Carrera, César José Mari (1903–　) M	**Carrera**

Carreras, R. (fl. 1986) S	**Carreras**
Carretero, José Luis (1941–) S	**Carretero**
Carrick, John (1914–1978) S	**Carrick**
Carrière, Élie Abel (1818–1896) PS	**Carrière**
(Carrillo Ll., R.)	
see Carrillo, R.	**Carrillo**
Carrillo, R. (fl. 1984) M	**Carrillo**
Carrington, Benjamin (1827–1893) BM	**Carrington**
Carrio, J.S. (fl. 1981–89) S	**Carrio**
Carrion, A.L. (fl. 1952) M	**Carrion**
Carrión, José Sebastian (fl. 1986) BS	**J.S.Carrión**
Carris, Lori M. (fl. 1987) M	**Carris**
Carrisso, Luis Wettnich (1886–1937) S	**Carrisso**
Carroll, George C. (fl. 1986) M	**G.C.Carroll**
Carroll, Isaac (1828–1880) M	**Carroll**
(Carrtero Cervero, José Luis)	
see Carrtero, José Luis	**Carrtero**
Carrtero, José Luis (1941–) S	**Carrtero**
Carruette-Valentin, J. A	**Carr.-Val.**
Carruthers, L. (fl. 1974) S	**L.Carruth.**
Carruthers, William (1830–1922) ABFMPS	**Carruth.**
Carse, Harry (1857–1930) PS	**Carse**
Carsner, Eubanks (1891–) M	**Carsner**
Carson, Joseph (1808–1876) S	**Carson**
Carter, A. (fl. 1982) M	**A.Carter**
Carter, Annetta Mary (1907–1991) S	**A.M.Carter**
Carter, Carlos Newton (fl. 1915) M	**C.N.Carter**
Carter, Charles R. (fl. 1980s) S	**C.R.Carter**
Carter, Herbert James (1858–1940) A	**H.J.Carter**
(Carter Holmes, Susan)	
see Carter, Susan	**S.Carter**
Carter, James Cedric (1905–1981) M	**J.C.Carter**
Carter, John R. A	**J.R.Carter**
Carter, M.V. (fl. 1957) M	**M.V.Carter**
Carter, Nellie (1895–) A	**N.Carter**
Carter, Stanley (fl. 1972) M	**Stan.Carter**
Carter, Susan (1933–) S	**S.Carter**
Carter, William R. (fl. 1921) S	**W.R.Carter**
Cartier, A. (fl. 1980) S	**A.Cartier**
Cartier, Delphine (1935–) S	**Cartier**
Carty, Susan A	**Carty**
Caruana Gatto, Alfredo (1868–1926) S	**Caruana**
Caruel, Théodore (Teodoro) (1830–1898) S	**Caruel**
Carus, Carl Gustav (1789–1869) AM	**Carus**
(Carus Stern, Ernst Ludwig)	
see Krause, Ernst Ludwig	**E.L.Krause**
Caruso, F.L. (fl. 1988) M	**F.L.Caruso**
Caruso, S. (fl. 1926) M	**Caruso**
Carvajal Hernández, Servando (fl. 1981) S	**Carvajal**

Carvalho, A.O. (fl. 1989) M	A.O.Carvalho
(Carvalho e Vasconcellos, João de)	
see Vasconcellos, João de Carvalho e	Vasc.
Carvalho, Lucia d'Avila Freire de (fl. 1969) S	Carvalho
Carvalho, T.de (fl. 1986) M	T.Carvalho
Carver, George Washington (1864–1943) M	Carver
Casadoro, G. A	Casadoro
Casagrande, F. (fl. 1969) M	Casagr.
Casagrandi, O. (fl. 1897) M	Casagrandi
Casares–Gil, Antonio (1871–1929) B	Casares–Gil
Casaretto, Giovanni (1812–1879) S	Casar.
Casas, Creu (1913–) B	Casas
(Casas de Puig, Creu)	
see Casas, Creu	Casas
(Casas, Francisco Javier)	
see Fernández Casas, Francisco Javier	Fern.Casas
Casaseca, Bartolomé (1920–) S	Casaseca
(Casaseca Mena, Bartolomé)	
see Casaseca, Bartolomé	Casaseca
(Casati, Francesco Beltramini de)	
see Beltramini de Casati, Francesco	Beltr.
Casaviella, Juan Ruíz (1835–1897) S	Casav.
Case, Frederick W. (1927–) S	Case
Case, Roberta Burckhardt (fl. 1976) S	R.B.Case
Cash, Edith Katherine (1890–) M	E.K.Cash
Cash, J. (fl. 1904) M	Cash
Cashmore, Alec Brooke (1907–1988) S	Cashmore
Caspary, Johann Xaver Robert (1818–1887) ABFMPS	Casp.
Casper, Siegfried Jost (1929–) AS	Casper
Caspers, R. S	Caspers
Cassebeer, Johann Heinrich (1785–1850) B	Casseb.
Cassel, Franz Peter (1784–1821) S	Cassel
Cassidy, James (1847–1889) S	Cassidy
Cassie, U.Vivienne (1926–) A	Cassie
Cassin, John A	Cassin
Cassini, Alexandre Henri Gabriel de (1781–1832) BS	Cass.
Cassini, R. (fl. 1966) M	R.Cass.
Cassone, Felice (1815–1854) S	Cassone
Castagne, Jean Louis Martin (1785–1858) AMS	Castagne
Castagnola, M. (fl. 1986) M	Castagn.
Castaldo, Rosa A	Castaldo
Castan, P. (fl. 1924) M	Castan
Castañeda, M. (fl. 1941) S	Castañeda
Castañeda Ruíz, Rafael F. (fl. 1986) M	R.F.Castañeda
Castanei, A. (fl. 1927) M	Castanei
Castaner, David (1934–) M	Castaner
Castaño Ramírez, Guillermo (fl. 1984) S	Castaño
Castel, Monique A	Castel
Castellan, A. (fl. 1980) S	Castellan
Castellani, Aldo (1877–1971) M	Castell.

Castellani, Ettore (1909–   ) M   **E.Castell.**
Castellano, Michael A. (fl. 1986) M   **Castellano**
Castellanos, Alberto (1896–1968) S   **A.Cast.**
Castellanos, Salvador Acosta   S   **S.A.Cast.**
Castelli, Laurent A.L. (1914–   ) B   **Castelli**
Castelli, T. (fl. 1938) M   **T.Castelli**
(Castello de Paiva, Antonio)
    see Paiva, Antonio da Costa   **A.Paiva**
Castelnau, François Louis Nompar de Caumat de Laporte (1810–1880) S   **Castelnau**
Castelo–Branco, R. (fl. 1961) M   **Cast.-Branco**
Castetter, Edward Franklin (1896–1978) S   **Castetter**
Castiglioni, Aicardo (fl. 1829) S   **A.Castigl.**
Castiglioni, Julio A. (1920–   ) S   **J.A.Castigl.**
Castiglioni, Luigi Gomes (1757–1832) S   **Castigl.**
Castillo, J. (fl. 1979) M   **J.Castillo**
(Castillo, Jacobo Ruíz del)
    see Ruíz del Castillo, Jacobo   **Ruíz Cast.**
Castillo, Rafael (fl. 1986) S   **Castillo**
Castillo–Campos, Gonzalo (fl. 1985) S   **Cast.-Campos**
Castle, Hempstead (1894–   ) B   **Castle**
Castracane degli Antelminelli, Francesco Saverio (1817–1899) A   **Castrac.**
Castrillon, A.Lopez (fl. 1964) M   **Castr.**
Castro, Duarte de (fl. 1941) S   **Castro**
Castro, María Luisa (fl. 1982) M   **M.L.Castro**
Castro, Vitorina Paiva (fl. 1983) S   **V.P.Castro**
Castro–Mendoza, E. (fl. 1983) M   **Castro-Mend.**
(Castroviejo Bolíbar, Santiago)
    see Castroviejo, Santiago   **Castrov.**
Castroviejo, Santiago (1946–   ) S   **Castrov.**
Casulli, F. (fl. 1981) M   **Casulli**
Catalan, Jordi   A   **J.Catalan**
Catalán, Pilar (1958–   ) P   **Catalán**
(Catalán Rodríguez, Pilar)
    see Catalán, Pilar   **Catalán**
Catalano, Giuseppe (1888–   ) S   **Catal.**
Catanei, A. (fl. 1944) M   **Catanei**
Catasus Guerra, L.J. (fl. 1980) S   **Catasus**
Catcheside, David Guthrie (1907–   ) BS   **Catches.**
Catesby, Mark (1683–1749) L   **Catesby**
Catharino, Eduardo Luis Martins (1960–   ) S   **Cath.**
Cati, Franco   A   **Cati**
Catling, Paul Miles (1947–   ) S   **Catling**
Catling, V.R. (fl. 1988) S   **V.R.Catling**
Catouillard, G.   A   **Catouill.**
Cattaneo, Achille (1839–   ) M   **Catt.**
Cattell, S.Allen   A   **Cattell**
Cauchon, R. (fl. 1989) M   **Cauchon**
(Caulinus, Filippo)
    see Cavolini, Filippo   **Cavolini**

Caullery, Maurice (1868–1958) AM   **Caullery**
Caum, Edward Leonard (1893–1952) MS   **Caum**
(Caumel, Jean Baptiste)
   see Héribaud, Joseph   **Hérib.**
Cauro, R.   A   **Cauro**
Cauvet, D.   S   **Cauvet**
Cauwet, Ann Marie   S   **Cauwet**
(Cauwet–Marc, Ann Marie)
   see Cauwet, Ann Marie   **Cauwet**
Cavaco, Alberto Judice Leote (1916–) S   **Cavaco**
(Cavalcante, W.A.)
   see Cavalcanti, W.A.   **Cavalc.**
(Cavalcante de Lima, Haroldo)
   see Lima, Haroldo Cavalcante de   **H.C.Lima**
Cavalcante, Paulo Bezerra (1922–) S   **Cavalcante**
Cavalcanti, Taciana Barbosa (1961–) S   **T.B.Cavalc.**
Cavalcanti, Wlandemir de Albuquerque (fl. 1967) M   **Cavalc.**
Cavalier, A. (fl. 1835) M   **Cavalier**
Cavalier–Smith, T.   A   **Caval.-Sm.**
Cavaliere, A.R. (fl. 1963) M   **A.R.Caval.**
Cavaliere, Domènico   S   **Caval.**
Cavallero, C. (fl. 1939) M   **Cavall.**
Cavanagh, Lucy Mary (1871–1936) B   **Cavanagh**
Cavanilles, Antonio José (Joseph) (1745–1804) PS   **Cav.**
(Cavanilles Palop, Antonio José (Joseph))
   see Cavanilles, Antonio José (Joseph)   **Cav.**
Cavara, Fridiano (1857–1929) MS   **Cavara**
Cave, George H. (1870–1965) S   **Cave**
Cavender, James C. (1936–) M   **Cavender**
Cavers, Francis (1876–1936) AB   **Cavers**
Cavet, J. (fl. 1992) M   **Cavet**
Cavillier, François Georges (1868–1953) S   **Cavill.**
Cavolini, Filippo (1756–1810) S   **Cavolini**
Cavolo, F.   A   **Cavolo**
Cayeux, F. (fl. 1921) S   **F.Cayeux**
Cayeux, M.H.   S   **Cayeux**
Cayla, Victor   S   **Cayla**
Cayouette, J. (1944–) S   **J.Cay.**
Cayouette, Richard (1914–) S   **Cay.**
Cayrol, Jean–Claude (fl. 1979) M   **Cayrol**
Cayzer, A. (fl. 1922) S   **Cayzer**
Cazalbou, L. (fl. 1914) M   **Cazalbou**
Cazares, Efren (fl. 1990) M   **Cazares**
Cazau, Cecilia (fl. 1990) M   **Cazau**
(Ceballos Fernández de Córdoba, Luis)
   see Ceballos, Luis   **Ceballos**
Ceballos, Luis (1896–1967) S   **Ceballos**
Cebolla, Consuelo (fl. 1988) S   **Cebolla**
(Cebolla Lozano, Consuelo)
   see Cebolla, Consuelo   **Cebolla**

Cedeño, L. (fl. 1987) M	**Cedeño**
Cedercreutz, Carl Wilhelm (1893–1968) AS	**Cedercr.**
Cedergren, (Israel) Gösta Robert (1888–1954) AS	**Cedergr.**
Cederstråhle, Erik Carl Johan (1835–1886) S	**Cederstr.**
Cejp, Karel (1900–1979) M	**Cejp**
Celafu, R. (1900–1979) M	**Celafu**
(Čelakavský, Ladislav Frantisek)	
see Čelakovský, Ladislav Frantisek	**L.F.Čelak.**
Čelakovský, Ladislav Frantisek (1864–1916) MPS ·	**L.F.Čelak.**
Čelakovský, Ladislav Josef (1834–1902) PS	**Čelak.**
Celan, Maria (Marie) S. A	**Celan**
Celarier, Robert P. (1921–1959) S	**Celarier**
Çelebiŏglu, Tülay (fl. 1989) S	**Çeleb.**
Celino, M.S. (fl. 1931) M	**Celino**
Celotti, L. (fl. 1887) M	**Celotti**
Cels, F. S	**F.Cels**
Cels, Jacques Philippe Martin (1743–1806) S	**Cels**
Celsius, Olof (1670–1756) S	**Celsius**
(Cengia Sambo, Maria)	
see Sambo, Maria	**Sambo**
Cépède, C. (fl. 1907) M	**Cépède**
Cepek, Pavel A	**Cepek**
Cerana, Maria Micaela (1949–) S	**Cerana**
Cerati, T.M. (fl. 1987) S	**Cerati**
Cercós, A.P. (fl. 1954) M	**Cercós**
(Černeva, O.V.)	
see Tscherneva, O.V.	**Tscherneva**
(Černhorsky, Zdenek (T.))	
see Černohorsky, Zdenek (T.)	**Čern.**
Černjavski, Paule (1892–1969) S	**Černjavski**
Černohorsky, Zdenek (T.) (1910–) M	**Čern.**
Černý, A. (fl. 1963) M	**Černý**
Cerrate, Emma (1920–) S	**Cerrate**
(Cerrate Valenzuela, Emma)	
see Cerrate, Emma	**Cerrate**
Certes, (Louis Adolphe) Adrien (1835–1903) A	**Certes**
Ceruta, A. (fl. 1960) M	**Ceruta**
Ceruti, Arturo (1908–) M	**Ceruti**
Cervantes, Vicente (Vincente) de (1755–1829) S	**Cerv.**
Cervi, Armando C. (1944–) S	**Cervi**
Cesalpino, Andrea (1519–1603) L	**Cesalpino**
Cesari Rossi, M.Graziella (fl. 1977) M	**Ces.Rossi**
Cesati, Vincenzo de (1806–1883) BMPS	**Ces.**
(Češchmedjiev, Iliya Vasilev)	
see Cheshmedjiev, Iliya Vasilev	**Cheshm.**
(Češmedžiev, Iliya Vasilev)	
see Cheshmedjiev, Iliya Vasilev	**Cheshm.**
Cestoni, Giacinto (1637–1718) L	**Cestoni**
Çetin, Barbaros (1958–) B	**Çetin**

Cetto, B. (fl. 1983) M	**Cetto**
Chabanne, Charles Gabriel (1862–1906) S	**Chabanne**
Chabaud, J.Benjamin (1833–1915) S	**Chabaud**
Chabelska–Frydman, Chaja (fl. 1964) M	**Chab.-Frydm.**
Chabert, Alfred Charles (1836–1916) S	**Chabert**
Chabert, Pierre (1796–1867) S	**P.Chabert**
Chaboisseau, Théodore (1828–1894) S	**Chaboiss.**
Chaborski, Gabriela (fl. 1918) M	**Chaborski**
Chabrey, Dominique (1610–1669) L	**Chabrey**
Chachina, A.G. A	**Chachina**
Chacko, P.M. A	**Chacko**
Chacón Aumente, Rosa (fl. 1987) S	**R.Chacón**
Chacón Roldán, Gloria (1940–) A	**G.Chacón**
Chadefaud, Marius (1900–) AM	**Chadef.**
Chadha, Asha A	**Chadha**
Chadim, V.A. (fl. 1966) S	**Chadim**
Chadwick, Lewis Charles (1902–) S	**Chadwick**
Chaffaut, Simon Amaudric du A	**Chaffaut**
Chagas, Carlos Ribeiro Justiniano (1879–1934) AM	**Chagas**
Chahal, Devinder Singh (fl. 1962) M	**Chahal**
Chai–Anan, Chumsri (1930–) S	**Chai-Anan**
(Chaianan, Chumsri)	
see Chai–Anan, Chumsri	**Chai-Anan**
Chaillet, Jean Frédéric de (1749–1839) BMS	**Chaillet**
Chaix, Dominique (1730–1799) S	**Chaix**
Chakrabarty, T. (fl. 1984) S	**Chakrab.**
Chakraborty, P. (fl. 1979) S	**Chakr.**
Chakravarty, H.L. (fl. 1982) S	**Chakrav.**
(Chalabi–Kab'i, Z.)	
see Chalabi–Kabi, Z.	**Chal.-Kabi**
Chalabi–Kabi, Z. (fl. 1964) S	**Chal.-Kabi**
Chalabuda, T.V. (fl. 1967) M	**Chalab.**
Chalard, A.du S	**Chalard**
Chalaud, Germain (1889–1967) B	**Chalaud**
(Chalilov, E.Ch.)	
see Khalilov, E.Kh.	**Khalilov**
Chalk, Laurence (1895–1979) S	**Chalk**
Challiar S	**Challiar**
Challice, J. (fl. 1981) S	**Challice**
Challinor, Richard Westman (1871–1951) S	**Challinor**
Chalmers, Albert John (1870–1920) AM	**Chalm.**
Chalmers, James (–before 1834) A	**J.Chalm.**
Chalon, Jean Charles Antoine (1846–1921) A	**Chalon**
Chaloner, William Gilbert (1928–) F	**Chaloner**
Chalons S	**Chalons**
(Chalubinski, Titus)	
see Chalubinsky, Titus	**Chal.**
Chalubinsky, Titus (1820–1889) B	**Chal.**
Chamberlain, Charles Joseph (1863–1943) S	**Chamb.**
Chamberlain, David Franklin (1941–) BS	**D.F.Chamb.**

Chamberlain, Donald William (1905–) M	D.W.Chamb.
Chamberlain, Edward Blanchard (1878–1925) B	E.B.Chamb.
Chamberlain, Yvonne Mary (1933–) A	Y.M.Chamb.
Chambers, Kenton Lee (1929–) S	K.L.Chambers
Chambers, Thomas Carrick (1930–) FPS	T.C.Chambers
Chambray, Georges de (1783–1849) S	Chambray
(Chamisseau de Boncourt, Louis Charles Adelaïde)	
see Chamisso, Ludolf Karl Adelbert von	Cham.
Chamisso, Ludolf Karl Adelbert von (1781–1838) ABPS	Cham.
Champion, John George (1815–1854) S	Champ.
Champluvier, Dominique (1953–) S	Champl.
Champy, P. (fl. 1844) S	Champy
Chamuris, George Peter (1954–) M	Chamuris
Chan, Chu Lun (fl. 1990) S	C.L.Chan
Chance, Helena C. (fl. 1921) M	Chance
Chancerel, Lucien (1858–) S	Chanc.
Chand, Ramesh (fl. 1990) M	Chand
Chandal, D.S. (fl. 1986) M	Chandal
Chandhyok, M.S. A	Chandhyok
(Chandian, Nasik S.)	
see Chandjian, Nasik S.	Chandjian
Chandjian, Nasik S. (1946–) S	Chandjian
Chandler, Alfred (1804–1896) S	Chandler
Chandler, Harley Pierce (1875–1918) S	H.P.Chandler
Chandler, Marjorie Elizabeth Jane (1897–) AFS	M.Chandler
Chandra, A.K. A	A.K.Chandra
Chandra, Aindrila (fl. 1974) M	A.Chandra
Chandra, Anil A	Anil Chandra
Chandra, D. (fl. 1979) S	D.Chandra
Chandra, Prakash (1937–) PS	P.Chandra
Chandra, Subhash (1943–) PS	Subh.Chandra
Chandra, Sudhir (fl. 1964) M	S.Chandra
Chandra, Vinod (1953–) B	V.Chandra
Chandrabose, M. (1940–) S	Chandrab.
Chandrasekaran, S. (fl. 1961) M	Chandras.
Chandrasekaran, V. (1940–) S	V.Chandras.
Chandrasekharan, P. (fl. 1957) S	Chandrasekh.
Chandrashekara, K.V. (fl. 1977) M	Chandrash.
(Chandzhian, Nasik S.)	
see Chandjian, Nasik S.	Chandjian
Chaney, Lucian West (1857–1935) A	Chaney
Chaney, Ralph Works (fl. 1944) F	R.W.Chaney
Chang, Ben Neng (fl. 1980s) S	B.N.Chang
Chang, Ch Tseng S	C.T.Chang
Chang, Chao Chien (1900–) S	C.C.Chang
Chang, Chen Shou (fl. 1990) M	Chen S.Chang
Chang, Chen Wan (1924–) S	C.W.Chang
(Chang, Chi Cheng)	
see Chen, Chi Chang	C.C.Chen

Chang, Chi Yung (fl. 1985) S	C.Yung Chang
Chang, Chi(h) Yu (fl. 1980s) S	C.Yu Chang
Chang, Ching En (fl. 1963) S	C.E.Chang
Chang, Chong Shu (c.1883) S	C.S.Chang
Chang, Chun Fu A	C.F.Chang
Chang, Dian Min(e) (fl. 1978) S	D.M.Chang
Chang, F.H. A	F.H.Chang
Chang, Gui Yi (fl. 1987) S	G.Y.Chang
Chang, Hai Yan (fl. 1983) S	H.Y.Chang
Chang, Ho Shii (1938–) M	H.S.Chang
Chang, Ho Tseng (1898–) S	H.T.Chang
Chang, Hung Ta (1914–) S	Hung T.Chang
Chang, J.I. (fl. 1984) P	J.I.Chang
Chang, K. A	K.Chang
Chang, Ki Hong A	K.H.Chang
Chang, Kung (Kuang) Chu (1939–) B	K.C.Chang
Chang, Mam Sian S	M.S.Chang
Chang, Man Hsiang (1934–) B	M.H.Chang
Chang, Mei Chen (1933–) S	M.C.Chang
Chang, Ming Chang (fl. 1980s) M	Ming C.Chang
Chang, Pei Zin S	P.Z.Chang
Chang, Roh Hwei (1931–) S	R.H.Chang
Chang, Shao Yao (fl. 1963) S	S.Y.Chang
Chang, Shao Yun (fl. 1983) S	S.Yun Chang
Chang, Siu Shih (fl. 1984) S	S.S.Chang
Chang, T.T. (fl. 1984) M	T.T.Chang
Chang, Te Jui A	T.J.Chang
Chang, Ting Chien (fl. 1981) S	T.C.Chang
Chang, Tsang Pi A	T.P.Chang
Chang, Tseh Yung (1932–) S	T.Y.Chang
Chang, Wen Jin (fl. 1983) S	W.J.Chang
Chang, Yan Fu (fl. 1983) S	Y.F.Chang
Chang, Yi Bi (fl. 1989) S	Y.B.Chang
Chang, Yong Tian (1936–) S	Y.T.Chang
Chang, Yui Liang (fl. 1958) S	Y.L.Chang
Chang, Yung (fl. 1967) M	Y.Chang
Chang, Zhong Ren (fl. 1983) S	Z.R.Chang
Chang, Zi An A	Z.A.Chang
Chang, Zung Yao (fl. 1981) S	Z.Y.Chang
Changsri, Winit (fl. 1958) M	Changsri
Channamma, K.A.Lucy (fl. 1966) M	Chann.
Channell, Robert Bennie (1924–) S	Channell
Channon, A.G. (fl. 1963) M	Channon
Chantanachat, Srisumon A	Chantan.
(Chantrans, Justin Girod)	
see Girod–Chantrans, Justin	Gir.-Chantr.
Chao, Ai Cheng (fl. 1958) S	A.C.Chao
Chao, Chen Yu (fl. 1979) M	Chen Y.Chao
(Chao, Chi Ding)	
see Zhao, Ji Ding	J.D.Zhao

Chao, Chi Son (1936–) S	C.S.Chao
Chao, Chuan Ying (1920–) S	C.Y.Chao
Chao, Hsiu Chien (1918–) S	H.C.Chao
Chao, Hung Chi (fl. 1981) S	Hung C.Chao
Chao, Jew Ming (fl. 1977) S	J.M.Chao
Chao, Lew Ming (fl. 1973) S	L.M.Chao
Chao, Liang Neng (fl. 1983) S	L.N.Chao
Chao, Neng (1931–) S	N.Chao
Chao, R.L.C. (fl. 1970) M	R.L.C.Chao
Chao, Tien Bung (Bang) (fl. 1981) S	T.B.Chao
Chao, Zi En (fl. 1982) S	Z.E.Chao
Chapelier, Louis Armand (1779–1802) S	Chapel.
Chapman, A.R.O. (1944–) A	A.R.O.Chapm.
Chapman, Alvan (Alvin) Wentworth (1809–1899) PS	Chapm.
Chapman, B. (fl. 1952) M	B.Chapm.
Chapman, C. (fl. 1909) F	C.Chapm.
Chapman, Clara J. (fl. 1941) B	C.J.Chapm.
Chapman, Floyd Barton (1911–) A	F.B.Chapm.
Chapman, Frederick (1864–1943) AF	F.Chapm.
Chapman, Frederick Revans (1849–1936) S	F.R.Chapm.
Chapman, G.B. (fl. 1959) M	G.B.Chapm.
Chapman, S.W. (fl. 1980) M	S.W.Chapm.
Chapman, Valentine Jackson (1910–1980) A	V.J.Chapm.
Chappellier, Paul (1820–1919) S	Chapp.
(Charadze, Anna Lukianovna)	
see Kharadze, Anna Lukianovna	Kharadze
Charbonnel, Jean Baptiste (fl. 1908–1920) S	Charb.
Chardón, Carlos Eugenio (1897–1965) M	Chardón
(Chardón Palacios, Carlos Eugenio)	
see Chardón, Carlos Eugenio	Chardón
Charif (fl. 1952) S	Charif
(Charkevicz, Sigismund Semenovich)	
see Kharkevich, Sigismund Semenovich	Kharkev.
Charles, Vera Katherine (1877–1954) M	Charles
Charoenphol, Charalch (fl. 1973) S	Char.
Charollais, J. A	Charoll.
Charpentié, M.-J. (fl. 1978) M	Charp.
Charpentier, Jean G.F.de (1786–1855) S	Charpent.
Charpin, André (1937–) S	Charpin
Charrel, Louis (fl. 1888) S	Charrel
Charvát, I. (fl. 1961) M	Charvát
Chary, S.J. (fl. 1972) M	Chary
Chase, Corrie Denew (1878–1965) S	C.D.Chase
Chase, Ethel Winifred Bennett (1877–1949) S	E.W.B.Chase
Chase, H.H. A	H.H.Chase
Chase, M.W. (fl. 1981) S	M.W.Chase
Chase, Mary Agnes (née Merrill) (1869–1963) S	Chase
Chassagne, Maurice (1880–1960) S	Chass.
Chassain, Maurice (fl. 1977) M	Chassain
Chastenay, Victorine de (1770–c.1830) S	Chastenay

Châteauneuf, J.J. A	**Chât.**
Châtelain, Jean Jacques (1736–1822) S	**Châtel.**
Chatenier, Constant (1849–1926) PS	**Chatenier**
Chater, Arthur Oliver (1933–) S	**Chater**
Chater, E.H. A	**E.H.Chater**
Chatin, Gaspard Adolphe (1813–1901) MS	**Chatin**
Chatrath, M.S. (fl. 1959) M	**Chatr.**
Chattaway, Mary Margaret (1899–) S	**Chattaway**
Chatterjee, A.A.K. (fl. 1985) M	**A.A.K.Chatterjee**
Chatterjee, Debabarta (1911–1960) S	**Chatterjee**
Chatterjee, G.C. A	**G.C.Chatterjee**
Chatterji, A.K. A	**A.K.Chatterji**
Chatterji, R.N. (fl. 1953) S	**Chatterji**
Chatterley, L.M. (fl. 1985) S	**Chatterley**
Chatton, Édouard (1883–1947) AM	**Chatton**
Chattopadhay, S.B. (fl. 1957) M	**Chattop.**
Chattopadhayay, B.K. (fl. 1991) M	**Chattopadh.**
Chaturvedi, S.K. (fl. 1973) M	**Chaturv.**
Chaturvedi, U.K. A	**U.K.Chaturv.**
Chaubard, Louis Athanase (Anastase) (1785–1854) AMPS	**Chaub.**
Chaube, A.N. A	**Chaube**
Chaudhary, Bansh Raj (1947–) A	**B.R.Chaudhary**
Chaudhary, D.N. (fl. 1983) M	**D.N.Chaudhary**
Chaudhary, Kailash Chandra Basu (1931–) S	**K.C.B.Chaudhary**
Chaudhary, M.C. A	**M.C.Chaudhary**
Chaudhary, Shaukat Ali (1931–) S	**Chaudhary**
Chaudhri, Mohammad Nazeer (1932–) S	**Chaudhri**
Chaudhuri, A.B. (fl. 1972) PS	**A.B.Chaudhuri**
Chaudhuri, H. (fl. 1924–42) AM	**Chaudhuri**
Chaudhury, Renuka (fl. 1964) M	**Chaudhury**
Chaudkhari, R.P. (fl. 1988) S	**Chaudkhari**
Chauhan, A.S. (fl. 1984) S	**A.S.Chauhan**
Chauhan, L.S. (fl. 1965) M	**L.S.Chauhan**
Chauhan, R.K.S. (fl. 1985) M	**R.K.S.Chauhan**
Chauhan, Shashi (fl. 1983) M	**S.Chauhan**
Chauhan, V.D. A	**V.D.Chauhan**
Chaumeton, François Pierre (1775–1819) S	**Chaumeton**
Chautems, A. (fl. 1984) S	**Chautems**
Chauvin, François Joseph (1797–1859) APS	**Chauv.**
Chavaillon, J. A	**Chavaillon**
Chavan, Appasaheb Ramachandaroo (fl. 1937) B	**Chavan**
Chavan, P.B. (fl. 1969) M	**P.B.Chavan**
Chavannes, Édouard Louis (1805–1861) S	**Chav.**
Chaverri, Adelaida (fl. 1988) M	**Chaverri**
(Chávez B., Ma. Luisa)	
see Chávez, Ma. Luisa	**Chávez**
Chávez, Ma. Luisa A	**Chávez**
Chaw, Shu Miaw (1954–) S	**Chaw**
Chawla, D.M. A	**Chawla**

Chawla, G.C. (fl. 1987) M	G.C.Chawla
Chaytor, D.A. (fl. 1937) S	Chaytor
Chazelles de Prizy, Laurent Marie (fl. 1790) S	Chaz.
Chea, Chark Yen (fl. 1979) M	Chea
Cheban, M.E. (fl. 1935) M	Cheban
Chebataroff, J. (fl. 1981) S	Chebat.
Checa, J. (fl. 1984) M	Checa
Chee, K.H. (fl. 1969) M	Chee
Cheek, Martin Roy (1960–) S	Cheek
Cheel, Edwin (1872–1951) MS	Cheel
Cheeseman, Thomas Frederic (1846–1923) PS	Cheeseman
Cheesman, Ernest Entwistle (1898–) S	Cheesman
(Chefranova, Z.V.)	
see Czefranova, Z.V.	Czefr.
Cheissin, E.M. (1907–1968) A	Cheissin
Chemin, Émile (1876–1945) AM	Chemin
Chen, Bao Liang (1944–) S	B.L.Chen
Chen, C.I. (fl. 1934) M	C.I.Chen
Chen, C.Y. (fl. 1976) S	C.Y.Chen
Chen, Chao Lin (fl. 1978) A	C.L.Chen
Chen, Cheih (1928–) S	C.Chen
Chen, Chi Chang (1903–) M	C.C.Chen
Chen, Chia Jui (1935–) S	C.J.Chen
Chen, Ching Hsia S	C.H.Chen
Chen, Ching Tao (fl. 1973) M	C.T.Chen
Chen, Chun Gen (fl. 1981) S	C.G.Chen
Chen, Chung Lien S	Chung L.Chen
Chen, De Mao (fl. 1983) S	D.M.Chen
Chen, Fa Jun (fl. 1991) M	F.J.Chen
Chen, Fei Pong (fl. 1988) S	F.P.Chen
Chen, Fen (fl. 1990) F	Fen Chen
Chen, Feng Hwai (1900–) S	F.H.Chen
Chen, Gui Qing (1939–) M	G.Q.Chen
Chen, Hsiu Ying ('Luetta') S	H.Y.Chen
Chen, J.L. (fl. 1989) M	J.L.Chen
Chen, Ji Di (1923–) M	J.D.Chen
Chen, Jia Kuan (1947–) S	J.K.Chen
(Chen, Jia Rui)	
see Chen, Chia Jui	C.J.Chen
Chen, Jia You (fl. 1983) A	J.Y.Chen
Chen, Jian Bin (fl. 1989) M	J.B.Chen
(Chen, K.T.)	
see Chen, Ching Tao	C.T.Chen
Chen, Ku Lin (fl. 1988) S	K.L.Chen
Chen, M.J. (fl. 1956) M	M.J.Chen
Chen, Min Pen A	M.P.Chen
Chen, Ming Che S	M.C.Chen
Chen, Mo Mei (fl. 1979) M	M.M.Chen
Chen, Pan Chieh (1907–1970) B	P.C.Chen

Chen, Pang Yu (1936–) S	P.Y.Chen
Chen, Pei Shan (fl. 1979) S	P.S.Chen
Chen, Qian Hai (fl. 1987) S	Q.H.Chen
Chen, Qiao A	Q.Chen
Chen, Qing Lian (fl. 1982) S	Q.L.Chen
Chen, Qing Tao (1932–) M	Q.T.Chen
Chen, Sen Jen (1933–) S	S.J.Chen
Chen, Shan (fl. 1985) S	Shan Chen
Chen, Shao Qin (fl. 1991) M	S.Q.Chen
Chen, Shao Yun (fl. 1983) S	S.Y.Chen
Chen, Shi (fl. 1982) S	Shi Chen
Chen, Shou Chan (fl. 1984) M	Shou C.Chen
Chen, Shou Liang (1921–) S	S.L.Chen
Chen, Shu Kun (1936–) S	S.K.Chen
Chen, Sing Chi (1931–) S	S.C.Chen
Chen, T. (fl. 1985) M	T.Chen
Chen, T.P. (fl. 1975) S	T.P.Chen
Chen, Tao (fl. 1986) S	Tao Chen
Chen, Tê Chao (1926–) S	T.C.Chen
Chen, Tzu Tsung (fl. 1983) S	T.T.Chen
Chen, Wei Chiu (1934–) S	W.C.Chen
Chen, Wei Lie (fl. 1986) S	W.L.Chen
Chen, Wei Qun (fl. 1982) A	W.Q.Chen
Chen, Xi Dian (fl. 1989) S	X.D.Chen
Chen, Xi Mu (fl. 1984) S	X.M.Chen
Chen, Xiao Ya (1955–) S	X.Y.Chen
Chen, Xin Lu (fl. 1989) S	X.L.Chen
Chen, Xing Qiu (fl. 1988) A	X.Q.Chen
Chen, Xiu Xiang (fl. 1988) S	X.X.Chen
Chen, Xun (fl. 1987) S	X.Chen
Chen, Yao Dong (fl. 1981) S	Y.D.Chen
Chen, Yi Ling (1930–) S	Y.L.Chen
Chen, Yow Yuh A	Y.Y.Chen
Chen, Yu An (fl. 1986) S	Y.A.Chen
Chen, Yung S	Y.Chen
Chen, Ze Ying (fl. 1988) S	Ze Y.Chen
Chen, Ze Yu (fl. 1989) M	Ze Yu Chen
Chen, Zhao Xuan (fl. 1982) M	Z.X.Chen
Chen, Zhen Feng (fl. 1988) S	Z.F.Chen
Chen, Zheng Hai (fl. 1989) S	Z.H.Chen
Chen, Zhi Rong (fl. 1983) S	Z.R.Chen
Chen, Zhi Xiu (fl. 1989) S	Zhi X.Chen
Chen, Zhong Yi (fl. 1989) S	Z.Y.Chen
Chen, Zhu An (fl. 1990) M	Z.A.Chen
Chen, Zhuo Hua A	Zhuo H.Chen
Chen, Zi Bai (fl. 1988) S	Z.B.Chen
Chen, Zong Lian (fl. 1985) S	Z.L.Chen
Chen, Zui (Zuei) Ching (1930–) M	Z.C.Chen
Chenantais, Jules E. (1854–1942) M	Chenant.
Chenault, Léon (1853–1930) S	Chenault

Chenevard, Paul (1839–1919) PS	**Chenevard**
Chenevière, E. A	**Chenev.**
Cheney, Donald P. (1945–) A	**D.P.Cheney**
Cheney, G.M. (fl. 1932) M	**G.M.Cheney**
Cheney, Lellen Stirling (1858–1938) B	**Cheney**
Cheney, Ralph Holt (1896–) S	**R.H.Cheney**
Cheng, Chao Tsung S	**C.T.Cheng**
Cheng, Ching Yung (J.) (1918–) S	**C.Y.Cheng**
Cheng, J.K. S	**J.K.Cheng**
Cheng, Jian (Jing) Fu (fl. 1987) PS	**J.F.Cheng**
Cheng, K.J. (fl. 1989) M	**K.J.Cheng**
Cheng, K.K. (fl. 1985) M	**K.K.Cheng**
Cheng, Mien S	**M.Cheng**
Cheng, P.L. A	**P.L.Cheng**
Cheng, Shi Jun (fl. 1982) S	**S.J.Cheng**
Cheng, Shu Lan (fl. 1967) M	**S.L.Cheng**
Cheng, Shu Zhi (fl. 1981) S	**S.Z.Cheng**
Cheng, Sze Hsu (fl. 1959) S	**S.H.Cheng**
Cheng, Wan Chun (1904–1983) PS	**W.C.Cheng**
Cheng, Wu Tsang (1940–) S	**W.T.Cheng**
Cheng, X.Y. (fl. 1985) M	**X.Y.Cheng**
Cheng, Xiao (1957–) P	**X.Cheng**
Cheng, Zhao Di A	**Z.D.Cheng**
Chennaveeraiah, M.S. (1924–) S	**Chennav.**
Chenon, L.J. (fl. 1751) S	**Chenon**
Cheo, C.C. (1911–) M	**C.C.Cheo**
Cheo, Shuh Yuen S	**S.Y.Cheo**
Cheo, Tai Yien (fl. 1981) S	**T.Y.Cheo**
Cheremisinov, N.A. (fl. 1936) M	**Cherem.**
Cheremisinova, E.A. A	**Cheremis.**
(Cherepanova, N.P.)	
see Czerepanova, N.P.	**Czerepan.**
Cherfils, Herbert (fl. 1937) S	**Cherfils**
Cherler, Johann Heinrich (1570–1610) S	**Cherler**
Chermezon, Henri (1885–1939) S	**Cherm.**
Chermsirivathana, C. (fl. 1982) S	**Chermsir.**
Chernetskaya, Z.S. (fl. 1929) M	**Chernetsk.**
(Cherneva, O.V.)	
see Tscherneva, O.V.	**Tscherneva**
Chernjavskaja, M.A. A	**Chernjavsk.**
Chernov, I.Yu. (fl. 1988) M	**Chernov**
Chernyakovska, E.G. (fl. 1983) S	**Chernyak.**
Cherubini, A. (fl. 1989) M	**Cherubini**
Cheshmedjiev, Iliya Vasilev (1930–) S	**Cheshm.**
(Cheshmedzhiev, Iliya Vasilev)	
see Cheshmedjiev, Iliya Vasilev	**Cheshm.**
Chesney, Francis Rawdon (1789–1872) S	**Chesney**
Chester, Frederick Dixon (1861–1943) M	**Chester**
Chesters, Charles G.C. (fl. 1964) M	**Chesters**

Chevalier, Auguste Jean Baptiste (1873–1956) BPS	**A.Chev.**
Chevalier, Charles A	**C.Chev.**
Chevalier, E. (1826–1914) S	**Chev.**
Chevallier, François Fulgis (1796–1840) AMPS	**Chevall.**
Chevallier, Louis Pierre Désiré (1852–1938) S	**L.Chevall.**
Chevassut, Georges (fl. 1965) MS	**Chevassut**
Chevaugeon, J. (fl. 1950) M	**Chevaug.**
Chevtaeva, V.A. (fl. 1981) S	**Chevtaeva**
Chew, Wee–Lek (1932–) S	**Chew**
Cheype, J.–L. (fl. 1983) M	**Cheype**
(Chhabi Ghora)	
see Ghora, S.C.	**Ghora**
Chi, C.C. (fl. 1950) M	**C.C.Chi**
Chi, Chin Wen (1915–) S	**C.W.Chi**
Chi, Pei Kun (fl. 1988) M	**P.K.Chi**
Chi, Yuh Tzao A	**Y.T.Chi**
Chia, Liang Chi (1921–) S	**L.C.Chia**
Chia, Tsu Chang S	**T.C.Chia**
Chiaje, Stephano delle (1794–1860) A	**Chiaje**
Chian, Chen Yu (fl. 1981) A	**C.Y.Chian**
(Chiang Cabrera, Fernando)	
see Chiang, Fernando	**F.Chiang**
Chiang, Fernando (1943–) S	**F.Chiang**
Chiang, Mou Yen (fl. 1989) M	**M.Y.Chiang**
Chiang, Tzen Yuh (1960–) B	**T.Y.Chiang**
Chiang, Young Meng A	**Y.M.Chiang**
Chiao, Chi Yuen (1901–) S	**C.Y.Chiao**
Chiarugi, Alberto (1901–1960) AS	**Chiarugi**
Chickering, John White (1831–1913) S	**Chick.**
Chiddarwar, P.P. (fl. 1955) M	**Chidd.**
(Chien, Chian Chü)	
see Chien, Jia Ju	**J.J.Chien**
Chien, Chiu Yuan (fl. 1971) M	**C.Y.Chien**
Chien, Jia Ju (fl. 1957–84) PS	**J.J.Chien**
Chien, Sung Shu (1885–) PS	**S.S.Chien**
Chifflot, Julien B.J. S	**Chifflot**
Chihara, Mitsuo (1927–) A	**Chihara**
(Chikhachef, Petr Aleksandrovich)	
see Tchichatscheff, Petr Aleksandrovich	**Tchich.**
Child, A. (fl. 1979) S	**A.Child**
Child, John (1922–1984) CM	**J.Child**
Child, Marion (fl. 1932) M	**Child**
Child, P. (fl. 1986) M	**P.Child**
Childs, J.L. S	**J.L.Childs**
Childs, Le Roy (1888–) M	**Childs**
Chilton, Charles (1860–1929) A	**Chilton**
Chilton, John E. (fl. 1954) M	**J.E.Chilton**
Chilton, St.John Poindexter (1909–) M	**S.P.Chilton**
Chin, T.G. A	**T.G.Chin**
Chin, Tsen Li (fl. 1981) S	**T.L.Chin**

Ching, Ren Chang (1898–1986) PS   **Ching**
Chinnappa, B. (fl. 1968) M   **Chinnappa**
Chinnock, Robert James (1943–) PS   **Chinnock**
Chintamani, Ahalya A   **Chintam.**
Chintchuk, A.G. (1897–) S   **Chintchuk**
Chinthibidze, Leonida S. (1923–) S   **Chinth.**
Chiovenda, Emilio (1871–1941) ABPS   **Chiov.**
Chiplonkar, A. (fl. 1969) M   **Chipl.**
Chipp, Thomas Ford (1886–1931) S   **Chipp**
Chippendale, George McCartney (1921–) S   **Chippend.**
Chippindall, Lucy Katherine Armitage (1913–1992) S   **Chippind.**
Chiritescu-Arva, Marin (1889–1935) A   **Chirit.-Arva**
Chisholm, G.G. S   **Chisholm**
Chistyakov, L.D. A   **Chistyakov**
Chisumpa, Sylvester Mudenda (1948–) S   **Chisumpa**
Chitaley, S.D. (fl. 1973) FMS   **Chitaley**
Chithra, V. (1947–) S   **Chithra**
Chitoku, T. A   **Chitoku**
Chitrovo, Anatolii (fl. 1904) S   **A.Chitr.**
Chitrovo, Vladimir Nikolaevich (1879–1949) S   **Chitr.**
(Chitrowo, Anatolii)
   see Chitrovo, Anatolii   **A.Chitr.**
(Chitrowo, Vladimir Nikolajewic)
   see Chitrovo, Vladimir Nikolaevich   **Chitr.**
Chittenden, Frederick James (1873–1950) S   **Chitt.**
Chitzanidis, A. (fl. 1956) M   **Chitzan.**
Chiu, Lee Ching (fl. 1979) S   **L.C.Chiu**
Chiu, Lien Ching S   **Lien C.Chiu**
Chiu, P.C. (1919–) PS   **P.C.Chiu**
Chiu, Pao Ling (fl. 1974) PS   **P.L.Chiu**
Chiu, Pei Shi (fl. 1974) PS   **P.S.Chiu**
Chiu, Wei Fan (fl. 1949) M   **W.F.Chiu**
Chivers, Arthur Houston (1880–) M   **Chivers**
Chlebicki, Andrzej (1949–) M   **Chleb.**
Chmel, L. (fl. 1966) M   **Chmel**
Chmielevsky, Vincent A   **V.Chmiel.**
Chmielewski, J.G. (fl. 1987) S   **Chmiel.**
(Chochriakov, Michael Kuzmich)
   see Khokhriakov, Michael Kuzmich   **Khokhr.**
(Chochrjakov, Michael Kuzmich)
   see Khokhriakov, Michael Kuzmich   **Khokhr.**
(Chochryakov, Michael Kuzmich)
   see Khokhriakov, Michael Kuzmich   **Khokhr.**
Chodat, Fernand François Louis (1900–1974) A   **F.Chodat**
Chodat, Robert Hippolyte (1865–1934) AMS   **Chodat**
Choe, Dumun A   **Choe**
Choi, Byoung Hee (1956–) S   **B.H.Choi**
Choi, Hong Kun (Keun) (1952–) P   **H.K.Choi**
Choisy, Jacques Denys (Denis) (1799–1859) MS   **Choisy**

Choisy, Maurice Gustave Benoit (1897–1966) M	M.Choisy
Cholewa, Anita F. (1953–) S	Cholewa
Cholil, Abdul (fl. 1989) M	Cholil
Cholnoky, Béla Jenö (1899–1972) A	Cholnoky
Cholnoky-Pfannkuche, Käthe A	Choln.-Pfannk.
Cholodnyj, Nikolay Grigorievich (1882–1953) A	Cholodnyj
Chona, Behari Lall (1906–) M	Chona
Choo, T.S. (fl. 1984) S	T.S.Choo
Chopik, V.I. (1929–) S	Chopik
Chopinet, R.G. (1914–) S	Chopinet
Chopra, Bashambar Nath (1898–) B	B.N.Chopra
Chopra, Goverdhan Lal (1903–) M	G.L.Chopra
Chopra, R.K. (fl. 1990) M	R.K.Chopra
Chopra, Ram Saran (1904–) B	R.S.Chopra
Choquette, George B. A	Choquette
Chorin, Mathilda (fl. 1961) M	Chorin
Choroschailov, N.G. (1904–) S	Chorosch.
Choroschkov, A.A. (fl. 1907) S	Chor.
(Choroshkov, A.A.)	
see Choroschkov, A.A.	Chor.
Chou, H.C. (fl. 1954) M	H.C.Chou
Chou, Pan Kai (fl. 1980) S	P.K.Chou
Chou, Ruth Chen-Ying (fl. 1947) APS	R.C.Y.Chou
Chou, Tai Yen S	T.Y.Chou
Chou, Y.H. (fl. 1980) S	Y.H.Chou
Chou, Yi Liang (1922–) AS	Y.L.Chou
Chouard, Pierre (1903–1983) S	Chouard
Choubert, B. A	Choubert
Choudhary, Madhav C. A	Choudhary
Choudhury, A. A	A.Choudhury
Choudhury, B.P. (fl. 1986) S	B.P.Choudhury
Choudhury, R.B.R. (fl. 1958) S	R.B.R.Choudhury
Choudhury, Ratan A	R.Choudhury
Chouhan, J.S. (fl. 1980) M	Chouhan
Choulette, Sebastian (1803–) S	Choul.
Choux, Pierre (1890–1983) S	Choux
Chovrenko (fl. 1925) M	Chovrenko
Chow, Chung Hwang (fl. 1935) M	C.H.Chow
Chow, Hang Fan (fl. 1934) S	H.F.Chow
Chow, Liang Dong (fl. 1975) S	L.D.Chow
Chow, S. (fl. 1962) S	S.Chow
Chowdary, Y.B.K. A	Chowdary
Chowdhery, H.J. (1949–) MP	H.J.Chowdhery
Chowdhery, K. (fl. 1984) M	K.Chowdhery
Chowdhry, P.N. (fl. 1974) M	Chowdhry
Chowdhuri, Pramatha Kanta (1923–) S	Chowdhuri
Chowdhury, Nira Pad (1911–) PS	N.P.Chowdhury
Chowdhury, S. (fl. 1969–1972) MP	S.Chowdhury
Chowdhury, S.R. (fl. 1970) M	S.R.Chowdhury

Chretiennot–Dinet, Marie–Josèphe A	**Chret.–Dinet**
Christ, D.H. S	**D.H.Christ**
Christ, Konrad Hermann Heinrich (1833–1933) FPS	**H.Christ**
Christen, H.R. A	**Christen**
Christenberry, George A. (fl. 1940) M	**Christenb.**
Christener, Christian (1810–1872) S	**Christener**
Christensen, Carl Frederik Albert (1872–1942) PS	**C.Chr.**
Christensen, Clyde Martin (1905–) M	**C.M.Chr.**
Christensen, Knud Ib (1955–) S	**K.I.Chr.**
Christensen, Martha (1932–) M	**M.Chr.**
Christensen, Tyge Ahrengot (1918–) AB	**T.A.Chr.**
Christenson, E.A. (fl. 1985) S	**Christenson**
Christian, Hugh Basil (1871–1950) S	**Christian**
Christian, J. (fl. 1991) M	**J.Christian**
Christiansen, A. (fl. 1910) P	**A.Christ.**
Christiansen, D.N. (1879–1952) S	**D.N.Christ.**
Christiansen, Elizabeth B. A	**E.B.Christ.**
Christiansen, Mads Peter (1889–1975) MS	**M.P.Christ.**
Christiansen, Mogens Skytte (1918–) M	**M.S.Christ.**
Christiansen, Willi Friedrich (1885–1966) S	**W.F.Christ.**
Christie, Jesse Roy (1889–) M	**Christie**
Christmann, Carl Friedrich (1752–1836) S	**C.Christm.**
Christmann, Gottlieb Friedrich (1752–1836) PS	**Christm.**
Christodoulakis, D. (fl. 1982) S	**Christod.**
Christoff, A. (fl. 1939) M	**Christoff**
Christophel, David Charles (1947–) F	**Christophel**
Christopher, Raymond A. A	**R.A.Christopher**
Christopher, Warren Neil (1895–) M	**Christopher**
Christophersen, Erling (1898–) BPS	**Christoph.**
Christopherson, J.B. (fl. 1914) M	**Christopherson**
Christova, E. (fl. 1939) M	**Christova**
Chrshanovski, Vladimir Gennadievich (1912–1985) S	**Chrshan.**
(Chrshanovsky, Vladimir Gennadievich)	
see Chrshanovski, Vladimir Gennadievich	**Chrshan.**
Chrtek, Jindřich (1930–) S	**Chrtek**
Chrtková, Anna (formerly Žertova, Anna) (1930–) S	**Chrtková**
(Chrtková–Žertová, Anna)	
see Chrtková, Anna	**Chrtková**
Chrysler, Mintin Asbury (1871–1963) FMPS	**Chrysler**
(Chrzanovskij, Vladimir Gennadievich)	
see Chrshanovski, Vladimir Gennadievich	**Chrshan.**
Chu, Chao Yi S	**C.Y.Chu**
Chu, Cheng De (fl. 1970s) S	**C.D.Chu**
Chu, Chia Yen A	**Chia Y.Chu**
Chu, Chung Hsiang S	**C.H.Chu**
Chu, Ge Lin(g) (1934–) S	**G.L.Chu**
Chu, Hao Jan A	**H.J.Chu**
Chu, Shu Ping A	**S.P.Chu**
Chu, Wei Ming (1930–) PS	**W.M.Chu**

Chu, Zong Yuan (fl. 1978) S	Z.Y.Chu
Chuang, Ching Chang (1931–) B	C.C.Chuang
Chuang, Hsuan (1938–) S	H.Chuang
Chuang, Tsan Iang (1933–) S	T.I.Chuang
Chudeau, René (1864–1921) B ˙	Chud.
Chudyba, Henryk A	Chudyba
Chudybowa, Danuta A	Chudybowa
(Chukavina, Anna Prokofevna)	
see Czukavina, Anna Prokofevna	Czukav.
Chukhrukidze, A.N. (fl. 1976) S	Chukhr.
Chun, Faith (Shu Chen) S	F.Chun
Chun, Woon Young (1889–) PS	Chun
Chun, Yung Ho (1924–) P	Y.H.Chun
Chung, Chi Hsing S	C.H.Chung
Chung, Hsin Hsuan S	H.H.Chung
Chung, I S	I Chung
Chung, In Cho (1918–) S	I.C.Chung
Chung, Myong Gi (fl. 1989) S	M.G.Chung
Chung, Nian June (fl. 1977) S	N.J.Chung
Chung, Yung Ho (1924–) P	Y.H.Chung
Chupp, Charles David (1886–1967) M	Chupp
Church, George Lyle (1903–) S	G.L.Church
Church, Margaret Brooks (1889–1949?) M	Church
Churchill, David Maughan (1933–) AF	Churchill
Churchill, Steven Paul (1948–) B	S.P.Churchill
Churchland, L.M. (fl. 1969) M	Churchl.
Chuvaschov, B.I. A	Chuv.
Cialdella, A.M. (fl. 1984) S	Ciald.
Ciarrocchi, L. (fl. 1933) M	Ciarr.
Cibrián, D. (fl. 1986) M	Cibrián
Cibula, William G. (fl. 1979) M	Cibula
Ciccarone, Antonio S. (1909–1982) M	Ciccar.
Ciccarone, Claudio (fl. 1985) M	C.Ciccar.
(Cicin, Nikolai Vasiljevich)	
see Tsitsin, Nikolai Vasiljevich	Tsitsin
Cienkowski, Leo de (1822–1887) AM	Cienk.
Ciesielski, Paul F. A	Cies.
Ciferri, Orio A	O.Cif.
Ciferri, Raffaele (1897–1964) MS	Cif.
(Cifuentes B., Joaquin)	
see Cifuentes, Joaquin	Cifuentes
Cifuentes, Joaquin (fl. 1981) M	Cifuentes
Cimerman, A. (fl. 1984) M	Cimerman
Cincović, T. (fl. 1972) S	Cincović
Cinovskis, Raĭmond Ekabovich (1930–) S	Cinovskis
Cinq–Mars, Lionel (1919–1973) PS	Cinq-Mars
Ciobanu, I.R. (1929–) S	Ciobanu
Ciocan, Constantin (1904–1987) M	Ciocan
Ciocîrlan, Vasile (1926–) S	Ciocîrlan

Cirik, Semra A	Cirik
Cirik–Altindag, S. A	Cirik-Alt.
Cirillo, Domenico Maria Leone (1739–1799) S	Cirillo
Cîrţu, D. (fl. 1972) S	D.Cîrţu
Cîrţu, Mariana (fl. 1972) S	M.Cîrţu
Cirujano, Santos (1950–) S	Cirujano
Citerne, Paul Émile Charles (1857–) S	Citerne
Ciurchea, Maria (1931–) M	Ciurchea
Claassen, M.Isabella A	Claassen
Claessens, B. (fl. 1934) M	Claess.
Claire, Charles (?1867–1931) S	Claire
Clairville, Joseph Philippe de (1742–1830) BPS	Clairv.
Claisse, M.L. (fl. 1987) M	Claisse
Clancy, K.J. (fl. 1979) M	Clancy
Claparède, Jean Louis René Antoine Éuard (1830–1871) A	Clap.
Clapham, Abraham (fl. 1860–1880) P	Clapham
Clapham, Arthur Roy (1904–1990) S	A.R.Clapham
Clapp, Grace Lucretia (1881–) M	Clapp
Clapperton, Bain Hugh (1788–1827) S	Clapperton
Claqué (fl. 1923) M	Claqué
Clara, Feliciano Mercado (1896–) M	Clara
Clarion, Jacques (1776–1844) S	Clarion
Clarion, Jean (1780–1856) S	J.Clarion
Clark, Betty M. (fl. 1974) M	B.M.Clark
Clark, Carolyn A. (fl. 1979) S	C.A.Clark
Clark, Howard Walton (1870–) A	H.W.Clark
Clark, J.Y. (fl. 1983) S	J.Y.Clark
Clark, James Curtis (1951–) S	C.Clark
Clark, John Joseph (1898–) S	Clark
Clark, Judson Freeman (1890–) M	J.F.Clark
Clark, Lois (1884–1967) B	L.Clark
Clark, Lynn G. (1956–) S	L.G.Clark
Clark, Robert Brown (1914–) S	R.B.Clark
Clark, Ross C. (1940–) S	R.C.Clark
Clark, W.Dennis (1948–) S	W.D.Clark
Clark, William Andrew (1911–1983) S	W.A.Clark
Clarke, Benjamin (1813–1890) S	Clarke
Clarke, Brian A	B.Clarke
Clarke, Charles Baron (1832–1906) PS	C.B.Clarke
Clarke, Edward Daniel (1769–1822) S	E.D.Clarke
Clarke, Giles C.S. (1944–) S	G.C.S.Clarke
Clarke, John Mason (1857–1925) S	J.M.Clarke
Clarke, K.J. A	K.J.Clarke
Clarke, R.T. (fl. 1965) M	R.T.Clarke
Clarke, Robin F.A. A	R.F.A.Clarke
Clarke, Walter Bosworth (1879–) S	W.B.Clarke
Clarkson, Edward Hale (1866–1934) PS	Clarkson
Clarkson, John Richard (1950–) S	J.R.Clarkson
Clarkson, Quentin Deane (1925–) S	Q.D.Clarkson

Claudia, P. (fl. 1986) M	**Claudia**
Claus, George Gyorgy A	**G.G.Claus**
Claus, Karl Ernst (1796–1864) S	**Claus**
Clausen, Andrea Martina (1949–) S	**A.M.Clausen**
Clausen, Jens Christian (1891–1969) S	**J.C.Clausen**
Clausen, Pedro Cláudio Dinamarquez (Peter) (1855–) S	**Clausen**
Clausen, Robert Theodore (1911–1981) PS	**R.T.Clausen**
Clausen, Roy Elwood (1891–1956) MS	**R.E.Clausen**
Clauson, Th. (1817–1860) S	**Clauson**
Claussen, N.H. (fl. 1940) M	**N.H.Claussen**
Claussen, Peter (1877–1959) M	**Claussen**
Clauzade, F.J.Georges A. (1914–) M	**Clauzade**
Clauzet, J.-P. (fl. 1984) M	**Clauzet**
Clavaud, Armand (1828–1890) S	**Clavaud**
Claverie, Pascal S	**Claverie**
Clavino, Mario (1875–) S	**Clavino**
Clay, Keith (fl. 1989) M	**Clay**
Clayberg, Carl Dudley (1931–) S	**Clayberg**
Claypole, Edward Waller (1835–1901) S	**Claypole**
Clayton, Deborah A	**D.Clayton**
Clayton, John (1686–1773) S	**J.Clayton**
Clayton, Margaret N. (1943–) A	**M.N.Clayton**
Clayton, T. (fl. 1981) S	**T.Clayton**
Clayton, William Derek (1926–) S	**Clayton**
Clayton, Y.M. (fl. 1985) M	**Y.M.Clayton**
Clear, I.D. (–c.1967) B	**Clear**
Cleef, A.M. (1951–) B	**Cleef**
Clegg, M.T. (fl. 1907) M	**Clegg**
Cleland, John Burton (1878–1971) M	**Cleland**
Cleland, Ralph Erskine (1892–1971) AS	**R.E.Cleland**
Clemen, R.E. S	**Clemen**
Clemenceau (fl. 1866–1872) S	**Clemenc.**
Clémencet, L. (fl. 1932) M	**Clémencet**
Clémençon, Heinz (1935–) AM	**Clémençon**
Clement, Ian Duncan (1917–) S	**Clement**
Clement–Westerhoff F	**Clem.-West.**
(Clemente, M.)	
see Clemente-Muñoz, M.	**Clem.-Muñoz**
Clemente y Rubio, Simón de Roxas (Rojas) (1777–1827) ABMPS	**Clemente**
Clemente–Muñoz, M. (fl. 1986) S	**Clem.-Muñoz**
Clementi, Giuseppe C. (1812–1873) S	**Clementi**
Clements, Edith Gertrude (Schwartz) (1877–) S	**E.G.Clem.**
Clements, Frederick Edward (1874–1945) AMS	**Clem.**
Clements, Mark Alwin (1949–) S	**M.A.Clem.**
Clemesha, Stephen Chapman (1942–) PS	**Clemesha**
Clerc, George Onésime (1845–1920) S	**Clerc**
Clerc, Philippe (1955–) M	**P.Clerc**
Clericuzio, Marco (fl. 1989) M	**Cleric.**
Clerk, Leon le S	**Clerk**

Cleve, Astrid (later Cleve-Euler, A.) (1875–1968) A	**A.Cleve**
Cleve, Per Theodor (1840–1905) A	**Cleve**
Cleve–Euler, Astrid	
see Cleve, Astrid	**A.Cleve**
Cleveland, Daniel (–1929) A	**Clevel.**
Cleveland, Lemuel R. (1892–1969) A	**L.R.Clevel.**
Clevenger, Sarah (1926–) S	**Clev.**
Clewell, Andre Francis (1934–) S	**Clewell**
Cleyer, André (–1967/68) S	**Cleyer**
Clifford, George (1685–1760) S	**G.Clifford**
Clifford, Harold Trevor (1927–) BFS	**Clifford**
Clifton, George (1823–1913) A	**Clifton**
Cline, Molly Niedbalski (fl. 1983) M	**M.N.Cline**
Cline, S.D. (fl. 1983) M	**S.D.Cline**
Clint, Katherine Lamberton (1904–1982) S	**Clint**
Clinton, George Perkins (1867–1937) M	**G.P.Clinton**
Clinton, George William (1807–1885) M	**Clinton**
Clokey, Ira Waddell (1878–1950) S	**Clokey**
Clos, Dominique (1821–1908) S	**Clos**
Clos, Enrique C. (1894–1977) S	**E.C.Clos**
Closon, Jules (fl. 1897) S	**Closon**
Clover, Elzada Urseba (1896–1980) S	**Clover**
Clowes, Christopher D. A	**C.D.Clowes**
Clowes, Frederic (fl. 1850–1860) P	**Clowes**
Clum, Floyd M. (fl. 1955) M	**Clum**
Clusius, Carolus (1526–1609) LM	**Clus.**
Clute, Willard Nelson (1869–1950) PS	**Clute**
Cmelitschek, H. (fl. 1976) S	**Cmel.**
Cmiech, Helena A. A	**Cmiech**
Co, L.L. (fl. 1980) P	**Co**
Coaz, Johann Wilhelm Fortunat (1822–1918) S	**Coaz**
Cobb, Fields W. (fl. 1987) M	**F.W.Cobb**
Cobb, L. (fl. 1983) S	**L.Cobb**
Cobb, Nathan Augustus (1859–1932) M	**Cobb**
Cobbold, Arthur (fl. 1903) S	**Cobbold**
Cobres, Joseph Paul von (1747–1823) S	**Cobres**
Coccia, Michele (fl. 1991) M	**Coccia**
Cocconi, Girolamo (1822–1904) MS	**Cocc.**
Cochet, G. (fl. 1957) M	**G.Cochet**
Cochet, Pierre Charles Marie (1866–1936) S	**Cochet**
Cochran, L.C. (fl. 1937) M	**Cochran**
Cochrane, Theodore Stuart (1942–) S	**Cochrane**
Cociu, V. (1923–) S	**Cociu**
(Cock, A.W.A.M.De)	
see De Cock, A.W.A.M.	**De Cock**
Cockayne, Leonard C. (1855–1934) FPS	**Cockayne**
Cockerell, Theodore Dru Alison (1866–1948) ABMPS	**Cockerell**
Cockrell, Robert Alexander (1909–) S	**Cockrell**
Cocks, John (1787–1861) A	**Cocks**

Cocquand, M. (fl. 1978) M — Cocquand
Cocucci, Alfredo Elio (1926–) S — Cocucci
Codd, Leslie Edward Wastell (1908–) S — Codd
Codogno, Michele (1951–) M — Codogno
Codomier, Louis A — Codomier
Codoreanu, Vasile (1917–) M — Codor.
Codreanu, Margareta A — M.Codreanu
Codreanu, Radu (1904–1987) AM — Codreanu
Codreanu–Balcescu, D. (fl. 1981) M — Codr.-Balc.
Cody, William James (1922–) PS — Cody
Coe, Ernest F. (1867–1951) S — Coe
Coe, Howard Sheldon (1888–1918) S — H.S.Coe
Coe–Teixeira, Beulah (1925–) S — Coe-Teix.
Coelho, R.P. (fl. 1959) M — Coelho
Coello Marín, Pilar (1955–) S — Coello
Coemans, Henri Eugène Lucien Gaëtan (1825–1871) MS — Coem.
Coesel, Peter F.M. (1941–) A — Coesel
Coetzee, Hester (1945–) S — Coetzee
Coffin, R.L. (1940–) PS — Coffin
Cogniaux, Célestin Alfred (1841–1916) BS — Cogn.
Cohen, C.L.D. A — Cohen
Cohen, Y. (fl. 1966) M — Y.Cohen
Cohen–Bazire, G. A — Cohen-Baz.
Cohen–Stuart, Combertus Pieter (1889–1945) S — Cohen-Stuart
Cohn, E. (fl. 1904) M — E.Cohn
Cohn, Ferdinand Julius (1828–1898) ABMS — Cohn
Cohrs, Albert (fl. 1963) S — Cohrs
Coile, N.C. (fl. 1981) S — Coile
Coincy, Auguste Henri Cornut de (1837–1903) S — Coincy
Coker, Dorothy (1894–) B — D.Coker
Coker, William Chambers (1872–1953) AMS — Coker
Colasante, Maria Antonietta (Maretta) (1944–) S — Colas.
Colbath, G.Kent A — Colbath
Colbert, Jean Baptiste S — Colbert
Colby, Arthur Samuel (1887–) M — Colby
Coldea, Gheorghe (1939–) S — Coldea
Colden, Jane (1724–1766) S — Colden
Colditz, Friedrich Volkmar A — Colditz
Cole, A.L.J. (fl. 1964) M — A.L.J.Cole
Cole, Desmond Thorne (1922–) S — D.T.Cole
Cole, Donald (1925–) S — D.Cole
Cole, Gary T. (fl. 1987) M — G.T.Cole
Cole, J.B. (fl. 1932) M — J.B.Cole
Cole, John Rufus (1900–) M — Cole
Cole, Kathleen M. (1924–) A — K.M.Cole
Colebrooke, Henry Thomas (1765–1837) S — Colebr.
Coleman, Edith (–1951) S — E.Coleman
Coleman, James Robert (1934–) S — J.R.Coleman
Coleman, Leslie Charles (1887–) M — L.C.Coleman
Coleman, Nathan (1825–1887) S — N.Coleman

Coleman, William Higgins (?1816–1863) S	**Coleman**
Colenso, (John) William (1811–1899) BMPS	**Colenso**
Coles, R.B. (fl. 1988) M	**R.B.Coles**
Coles, Susan Margaret (fl. 1971) S	**S.M.Coles**
Colinvaux, Llewellyn Hillis A	**Colinv.**
Coll, J. A	**Coll**
Colla, Luigi(Aloysius) (1766–1848) PS	**Colla**
Colla, Silvia (fl. 1932) M	**S.Colla**
Colladon, Louis Théodore Frederic (1792–1862) S	**Collad.**
Collard, S.B. (fl. 1966) M	**Collard**
Collenette, Iris Sheila (1927–) S	**Collen.**
Collett, Henry (1836–1901) S	**Collett**
Colley, R.H. (fl. 1927) M	**Colley**
Collie, Alexander (1793–1835) S	**Collie**
Collier, Albert Walker (1910–) A	**Collier**
Collier, Jane A	**J.Collier**
Collin, Bernard (1881–1915) AM	**Collin**
Collin, P. (fl. 1987) M	**P.Collin**
Collinder, Erik (1848–1920) S	**Collinder**
Collins, E.J. (fl. 1933) S	**E.J.Collins**
Collins, Frank Shipley (1848–1920) A	**Collins**
Collins, Gary B. (1940–) A	**G.B.Collins**
Collins, Guy N. (1872–1938) S	**G.N.Collins**
Collins, Lawrence Turner (1937–) S	**L.T.Collins**
Collins, W.B. (fl. 1935) M	**W.B.Collins**
Collins, Zacchaeus (1764–1831) S	**Z.Collins**
Collinson, Peter (1694–1768) S	**Collinson**
Colmeiro, Miguel (1816–1901) MPS	**Colmeiro**
(Colmeiro y Penido, Miguel)	
see Colmeiro, Miguel	**Colmeiro**
(Colom Casanovas, Guillermo)	
see Colom, Guillermo	**Colom**
Colom, Guillermo (1900–) A	**Colom**
Colomb, Marie Louis Georges (1856–) PS	**Colomb**
Colozza, A. (fl. 1908) S	**Colozza**
Colsmann, Johannes (1771–1830) S	**Colsm.**
Coltman–Rogers, Charles (1854–1929) S	**Coltm.-Rog.**
Colton, Harold Sellers (1881–1970) A	**Colton**
Coluzzi, M. (fl. 1959) M	**Coluzzi**
Coman, A. (1881–1972) S	**Coman**
Coman, Nicolae (1937–) M	**N.Coman**
Comas, Augusto (1949–) A	**Comas**
(Comas González, Augusto)	
see Comas, Augusto	**Comas**
Comber, Harold Frederick (1897–1969) S	**H.F.Comber**
Comber, James Boughtwood (1929–) S	**J.B.Comber**
Comber, Thomas (1837–1902) A	**Comber**
Combescot, Charles A	**Combescot**
Combs, Robert (1872–1899) S	**Combs**

Comby, L. (fl. 1942) M	**Comby**
Comeaux, B.L. (fl. 1987) S	**Comeaux**
Comelli, Francesco (1793–1852) A	**Comelli**
Comère, Joseph (1854–1932) A	**Comère**
Comes, Orazio (1848–1923) MS	**Comes**
Comes, Salvatore (1880–) A	**S.Comes**
Commelijn, Caspar(us) (1667/8–1731) M	**C.Commelijn**
Commelijn, Jan (1629–1692) L	**J.Commelijn**
(Commelin, Caspar(us))	
see Commelijn, Caspar(us)	**C.Commelijn**
(Commelin, Jan)	
see Commelijn, Jan	**J.Commelijn**
Commerson, Philibert (1727–1773) PS	**Comm.**
Commons, A. (1829–1919) M	**Commons**
Comolli, Giuseppe (1780–1849) PS	**Comolli**
Comparetti, Andrea (1745–1801) S	**Compar.**
Compère, Pierre (1934–) AS	**Compère**
Compton, Gladys Ruth (later Hutchison) (1909–) S	**G.R.Compton**
Compton, Robert Harold (1886–1979) APS	**Compton**
Conant, David S. (1949–) P	**D.S.Conant**
Conant, Norman Francis (1908–) M	**Conant**
Conard, A. A	**A.Conard**
Conard, Henry Shoemaker (1874–1971) BS	**Conard**
Condamine, Charles Marie de la (1701–1774) S	**Cond.**
Condit, C. (fl. 1944) F	**Condit**
Condra, George Evert (1869–1958) A	**Condra**
Conert, Hans Joachim (1929–) S	**Conert**
Conforti, V. A	**Conforti**
Congdon, Joseph Whipple (1834–1910) S	**Congdon**
Conger, Paul Sidney (1897–1979) A	**Conger**
Conil, Raphaël A	**Conil**
Conill, Joseph Léon Emile (1872–1944) S	**Conill**
Conklin, George Hall (1866–1940) B	**Conkl.**
Conn, Barry John (1948–) S	**B.J.Conn**
Conn, Harold Joel (1886–1952) M	**H.J.Conn**
Conn, Herbert William (1859–1917) A	**Conn**
Connant, N.F. (fl. 1941) M	**Connant**
Connell, Frank Herman (1905–) A	**Connell**
Conners, Ibra Lockwood (1894–) M	**Conners**
Connole, M.D. (fl. 1968) M	**Connole**
Connor, Henry Eamonn (1922–) S	**Connor**
Conomos, T.John (1938–) A	**Conomos**
Conover, John T. (1918–) AM	**Conover**
Conrad, J. S	**J.Conrad**
Conrad, Marc A. A	**M.A.Conrad**
Conrad, Marcelle (1897–1990) S	**M.Conrad**
Conrad, Solomon White (1779–1831) S	**Conrad**
Conrad, Timothy Abbot (1803–1877) AF	**T.A.Conrad**
Conrad, Walter (1888–1943) A	**W.Conrad**
Conradi, Frans Edvard (1890–1907) B	**Conradi**

Conran, John Godfrey (1960–) S	**Conran**
Conrard, Louis P.H. S	**Conrard**
Conrath, P. (1892–) S	**P.Conrath**
Conrath, Paul (1861–1931) S	**Conrath**
Console, Michelangelo (1812–1897) S	**Console**
Constance, Lincoln (1909–) S	**Constance**
Constantineanu, Ion (Ioan, Joan) C. (1860–1931) MS	**Const.**
(Constantineau, Ion (Ioan, Joan) C.)	
see Constantineanu, Ion (Ioan, Joan) C.	**Const.**
Constantinescu, Ovidiu (1933–) M	**Constant.**
Constantinou, P.T. (fl. 1960) M	**Constantinou**
Contandriopoulos, Juliette (1922–) S	**Contandr.**
Contejean, Charles Louis (1824–1907) S	**Contejean**
Conti, Pasquale (Pascal) (1874–1898) S	**Conti**
Conti, Sergio F. A	**S.F.Conti**
Contorni, Mauro (1959–) S	**Contorni**
Contré, Emile (1916–1981) S	**Contré**
Contreras Jiménez, José Luis (1952–) S	**J.L.Contr.**
Contreras Jiménez, Luz Maria (fl. 1981) S	**L.M.Contr.**
Contu, Marco (E.) (fl. 1984) M	**Contu**
Conway, Elsie A	**E.Conway**
Conway, Kenneth E. (fl. 1976) M	**Conway**
Conwentz, Hugo Wilhelm (1855–1922) F	**Conw.**
Conzatti, Cassiano (1862–1951) PS	**Conz.**
Coode, Mark James Elgar (1937–) S	**Coode**
Coogan, A.H. (1929–) A	**Coogan**
Cook, Christopher David Kentish (1933–) S	**C.D.K.Cook**
Cook, James (1728–1779) S	**Cook**
Cook, Laurence J. (–?1959) S	**L.J.Cook**
Cook, Melville Thurston (1869–1952) M	**M.T.Cook**
Cook, Orator Fuller (1867–1949) MPS	**O.F.Cook**
Cook, Philip William (1936–) AM	**P.W.Cook**
Cook, R.James (fl. 1967) M	**R.J.Cook**
Cook, R.T.A. (1940–) M	**R.T.A.Cook**
Cook, Varner James (1904–) S	**V.J.Cook**
Cook, Walter Robert Ivimey	
see Ivimey Cook, Walter Robert	**Ivimey Cook**
Cooke, Charles Montague (1874–1948) B	**C.M.Cooke**
Cooke, David Alan (1949–) S	**D.A.Cooke**
Cooke, John C. (fl. 1973) M	**J.C.Cooke**
Cooke, Mordecai Cubitt (1825–1914) ABM	**Cooke**
Cooke, R.C. (fl. 1966) M	**R.C.Cooke**
Cooke, R.James (fl. 1967) M	**R.J.Cooke**
Cooke, Theodore (1836–1910) S	**T.Cooke**
Cooke, William Bridge (1908–) AM	**W.B.Cooke**
Cookson, Isabel Clifton (1893–1973) ABFMS	**Cookson**
Cookson, J. (fl. 1926) M	**J.Cookson**
Cool, Cath. (fl. 1913) M	**Cool**
Coombe, David Edwin (1927–) S	**Coombe**

Coombs, Frank Andrew (1877–1964) S	**Coombs**
Cooney, Donald G. (fl. 1964) M	**Cooney**
Coons, George Herbert (1885–) M	**Coons**
Cooper, A.W. (fl. 1898) M	**A.W.Cooper**
Cooper, D.C. (fl. 1928) MS	**D.C.Cooper**
Cooper, Daniel (1817?–1842) S	**Cooper**
Cooper, Dorothy (1941–) S	**D.Cooper**
Cooper, E.C. S	**E.C.Cooper**
Cooper, Ellwood (1829–1918) S	**E.Cooper**
Cooper, I.C.G. (–1957) A	**I.C.G.Cooper**
Cooper, James Graham (1830–1902) S	**J.G.Cooper**
Cooper, Robert Cecil (1917–) S	**R.C.Cooper**
Cooper, Rowland Edgar (1890–1962) S	**R.E.Cooper**
Cooper–Driver, Gillian (1936–) P	**Cooper–Driver**
Cooperrider, Tom Smith (1927–) S	**Cooperr.**
Cope, Thomas Arthur (1949–) S	**Cope**
Copeland, Edwin Bingham (1873–1964) MP	**Copel.**
Copeland, Herbert Faulkner (1902–1968) MS	**H.F.Copel.**
Copeland, Joseph J. (1907–) A	**J.J.Copel.**
Copley, Peter Bruce (1956–) S	**Copley**
Coppejans, Eric (1948–) A	**Coppejans**
Coppey, Amedee (1874–1913) B	**Copp.**
Coppins, Brian John (1949–) M	**Coppins**
(Coquebert de Montbret, Gustave)	
see Montbret, Gustave Coquebert de	**Montbret**
Coquerel, Charles (1822–1867) A	**Coquerel**
Corbasson, M. (fl. 1968) S	**Corbasson**
Corbaz, Roger (fl. 1955) M	**Corbaz**
Corbetta, G. (fl. 1963) M	**Corbetta**
Corbière, François Marie Louis (1850–1941) BS	**Corb.**
Corbière, L. (fl. 1929) M	**L.Corb.**
Corbin, James B. (fl. 1979) M	**Corbin**
Corbishley, Amy Gertrude (–1977) S	**Corbishley**
Corda, August Karl Joseph (1809–1872) ABFMPS	**Corda**
Cordas, David I. (fl. 1976) M	**Cordas**
Cordeiro, Inês (1958–) S	**Cordeiro**
Cordeiro–Marino, Marilza (1939–) A	**Cord.–Mar.**
Cordemoy, C. (1840–1909) S	**C.Cordem.**
Cordemoy, Eugène Jacob de (1835–1911) PS	**Cordem.**
Cordemoy, F. S	**F.Cordem.**
Cordemoy, Hubert Louis Philippe Eugène Jacob de (1866–1927) S	**H.L.Cordem.**
Corden, Malcolm (fl. 1979) M	**Corden**
Cordero, Paciente A. (1943–) A	**Cordero**
Cordes, W.A. (fl. 1920) M	**Cordes**
Cordier, François Simon (1797–1874) M	**Cordier**
Cordley, Arthur Burton (1864–1936) M	**Cordley**
Cordroch, M. (fl. 1936) M	**Cordroch**
Cordus, Euricius (1486–1535) L	**E.Cordus**
Cordus, Valerius (1514–1544) L	**V.Cordus**

Core, Earl Lemley (1902–1984) S	**Core**
Corer, W.C. (1872–1953) M	**Corer**
Corfixen, Peer (fl. 1990) M	**Corfixen**
Corillion, Robert J. (1908–) AS	**Corill.**
Corinaldi, Giacomo (1782–1847) A	**Corin.**
Corlett, Michael P. (1937–) M	**Corlett**
Corley, Hugh Vanner (1914–) P	**Corley**
Corley, Martin Francis Vanner (1944–) B	**M.F.V.Corley**
Corliss, J.O. (1922–) A	**Corliss**
Cormaci, Mario (1944–) A	**Cormaci**
Cornaz, Charles Auguste Édouard (1825–1911) S	**Cornaz**
Corneju, A. (fl. 1934) M	**Corneju**
Cornelissen, Égide Norbert (1769–1849) S	**Cornel.**
Cornelisson, C.H. S	**Cornelisson**
Cornelius, Ruth Vera (1908–) S	**Cornelius**
Corner, Edred John Henry (1906–) MS	**Corner**
Cornford, C.E. (fl. 1960) M	**Cornford**
Cornman, John Farnsworth (1913–) S	**Cornman**
Cornu, Marie Maxime (1843–1901) ACM	**Cornu**
Cornut, Jacques-Philippe (1606?–1651) L	**Cornut**
Corradi, (Bartolomeo Giacomo) Rinaldo (1897–1976) P	**Corradi**
Corradini, D. A	**Corradini**
Corrêa da Serra, José Francisco (1751–1823) S	**Corrêa**
(Corrêa de Méllo, Joachim)	
see Correia de Méllo, Joachim	**Corr.Méllo**
Correa, Juan A(lberto) (1955–) A	**J.A.Correa**
Correa, Maevia Noemi (1914–) S	**M.N.Correa**
Corrêa, Manoel Pio (1874–1934) S	**M.P.Corrêa**
Correia, Ana I.V.D. (1953–) S	**A.I.V.Correia**
Correia de Méllo, Joachim (1816–1877) S	**Corr.Méllo**
Correia, R.I.de S. (fl. 1967) S	**R.I.S.Correia**
Correll, Donovan Stewart (1908–1983) PS	**Correll**
Correll, Helen Butts (1907–) S	**H.B.Correll**
Correns, Carl Franz Joseph Erich (1864–1933) ABMS	**Correns**
Correvon, Louis Henry (1854–1939) S	**Correvon**
Corrias, Bruno (1939–) S	**Corrias**
Corrick, Margaret Georgina (1922–) S	**Corrick**
Corsin, Paul M. (1880–) AF	**Corsin**
Corte, Aurora Montemartini (fl. 1920–1957) AM	**A.M.Corte**
(Corte, C.de)	
see de Corte, C.	**de Corte**
Cortés Latorre, Cayetano (1896–1966) AB	**C.Cortés**
Cortés, Santiago (1854–1924) S	**Cortés**
Cortesi, Fabrizio (1879–1949) S	**Cortesi**
Corti, Bonaventura (1729–1813) M	**Corti**
Corti, Roberto (1909–1986) S	**R.Corti**
Cortini, J.C. (fl. 1921) M	**Cortini**
Cortini-Pedrotti, Carmela (1931–) B	**Cort.-Pedr.**
Cory, Victor Louis (1880–1964) S	**Cory**

Cosandey, Florian (1897–) A	**Cosandey**
Cosentini, Ferdinando (1769–1840) PS	**Cosent.**
Cosson, Ernest Staint–Charles (1819–1889) ABMPS	**Coss.**
Costa, Alvaro do Santos (1912–) M	**A.S.Costa**
Costa, Carlos Alberto Amaral (fl. 1958) M	**C.A.A.Costa**
Costa, C.P. (fl. 1980) M	**C.P.Costa**
Costa, Cecilia Gonçalves (1928–) S	**C.G.Costa**
Costa, José Gonçalves de (1899–1967) S	**J.G.Costa**
Costa, José Porfirio (1910–) M	**J.P.Costa**
Costa, Lucy I. A	**L.I.Costa**
Costa, Manuel Jacinto (1938–) S	**M.J.Costa**
Costa, Maria Eugénia Amorim Pereira da (fl. 1952) M	**M.E.A.Costa**
(Costa, Nara Leane Moreira da)	
see Leane Moreira da Costa, Nara	**Leane**
(Costa, Nuno Maria Sousa)	
see Sousa Costa, N.M.	**Sousa Costa**
Costa, P. (fl. 1976) M	**P.Costa**
(Costa Talens, Manuel Jacinto)	
see Costa, Manuel Jacinto	**M.J.Costa**
Costa y Cuxart, Antonio Cipriano (1817–1886) S	**Costa**
Costantin, Julien Noël (1857–1936) AMS	**Costantin**
Costas, Martha Alcira (1938–) S	**Costas**
Coste, F. (fl. 1970) M	**F.Coste**
Coste, Hippolyte Jacques (1858–1924) S	**H.J.Coste**
Coste, Jean François (1741–1819) S	**Coste**
Coste, Michel A	**M.Coste**
Costin, J. A	**Costin**
Cota, J.Hugo (1956–) S	**Cota**
(Cota S., J.Hugo)	
see Cota, J.Hugo	**Cota**
Cothenius, Christian Andreas von (1708–1789) S	**Cothen.**
Cotta, Carl Bernhard von (1808–1879) F	**Cotta**
Cottam, Walter Pace (1894–) S	**Cottam**
Cotter, H.Van T. (fl. 1989) M	**Cotter**
Cottet, Michel (1825–1896) S	**Cottet**
Cottini, G.B. (fl. 1939) M	**Cottini**
Cotton, Arthur Disbrowe (1879–1962) AMS	**Cotton**
Cotton, John Storrs (1875–) S	**J.S.Cotton**
Cotton, Raymond (1948–) S	**R.Cotton**
Couch, John Nathaniel (1896–1986) M	**Couch**
Coudenberg, Peter (fl. 1566) L	**Coudenb.**
Couderc, George (1850–1928) M	**Couderc**
Couderc, J.M. (fl. 1917) B	**J.M.Couderc**
Coudert, J. (fl. 1955) M	**Coudert**
Coulter, John Merle (1851–1928) MS	**J.M.Coult.**
Coulter, Thomas (1793–1843) S	**Coult.**
Couper, Robert A. (1923–) B	**Couper**
Coupin, Henri Eugène Victor (1868–) CM	**Coupin**
Cour, P. (fl. 1966) M	**Cour**
Courbai, A. S	**Courbai**

Courchet, Lucien Désiré Joseph (1851–1924) S   **Courchet**
(Courset, George(s) Louis Marie Dumont de)
    see Dumont de Courset, George(s) Louis Marie   **Dum.Cours.**
(Coursey, De)
    see De Coursey   **De Coursey**
Court, Arthur Bertram (1927–) MS   **Court**
Courtecuisse, Régis (1956–) M   **Courtec.**
Courteville, H.   A   **Courtev.**
Courtin, Albert (fl. 1850–1858) S   **Courtin**
Courtinat, Bernard   A   **Courtinat**
Courtois, Richard Joseph (1806–1835) S   **Courtois**
Cousturier, Paul (–1921) S   **Coustur.**
Couté, Alain (1939–) A   **Couté**
Coutière, H. (fl. 1911) AM   **Coutière**
Coutinho, António Xavier Pereira (1851–1939) BMPS   **Cout.**
Coutts, John (1872–1952) S   **Coutts**
Covas, Guillermo (1915–) S   **Covas**
Coventry, Bernard Oakes (–1920) S   **Coventry**
Coville, Frederick Vernon (1867–1937) BMPS   **Coville**
Coville, James (fl. c.1820) S   **J.Coville**
Cowan, C.P. (fl. 1986) S   **C.P.Cowan**
Cowan, John Macqueen (1891–1960) S   **Cowan**
Cowan, Richard Sumner (1921–) S   **R.S.Cowan**
Cowell, John Francis (1852–1915) S   **J.F.Cowell**
Cowell, M.H. (fl. 1839) S   **Cowell**
Cowley, Elizabeth Jill (1940–) S   **Cowley**
Cowper, Denis (–1974) S   **Cowper**
Cox, B.A. (fl. 1974) M   **B.A.Cox**
Cox, Billy Joe (1941–) S   **B.J.Cox**
Cox, Darrell E. (fl. 1939) M   **D.E.Cox**
Cox, Don K. (fl. 1958) S   **D.K.Cox**
Cox, Edmond R. (1932–) A   **Ed.R.Cox**
Cox, Eileen J.   A   **E.J.Cox**
Cox, Eleanor R.   A   **El.R.Cox**
Cox, Euan Hillhouse Methven (1893–1977) S   **Cox**
Cox, H.T. (fl. 1939) M   **H.T.Cox**
Cox, J.   A   **J.Cox**
Cox, P.A. (fl. 1985) S   **P.A.Cox**
Cox, P.B.   S   **P.B.Cox**
Cox, Vivienne J. (fl. 1957) M   **V.J.Cox**
Coyuette   S   **Coyuette**
Crabbe, James Albert (1914–) P   **Crabbe**
Crabtree, D.R. (fl. 1983–90) F   **Crabtree**
(Craene, Albert De)
    see De Craene, Albert   **De Craene**
Cragin, Francis Whittemore (1858–1937) M   **Cragin**
Craib, William Grant (1882–1933) S   **Craib**
Craig, Ethelda Jane (fl. 1939) B   **E.J.Craig**
Craig, Robert T. (1847–1927) S   **R.T.Craig**

Craig, Thomas (1839–1916) S	**Craig**
Craig, Thomas Theodore (1907–) S	**T.T.Craig**
Craig, William Nicol (fl. 1960s) S	**W.N.Craig**
Craigie, J.H. (fl. 1929) M	**Craigie**
Craik, Robert (fl. 1921) M	**Craik**
Cralley, Elza Monroe (1905–) M	**Cralley**
(Cramer, Carl)	
see Kramer, Carl	**C.Kramer**
Cramer, Carl Eduard (1831–1901) AMS	**C.E.Cramer**
Cramer, Fritz H. A	**F.H.Cramer**
Cramer, Johann Christian (fl. 1803) S	**Cramer**
Cramer, L.H. (fl. 1991) S	**L.H.Cramer**
Cramer, Pieter Johannes Samuel (1879–1952) S	**P.J.S.Cramer**
Crampton, Beecher (1918–) S	**Crampton**
Crandall, Bowen Sinclair (1909–) M	**Crand.**
Crane, Fern Ward (1906–) PS	**Crane**
Crane, J.Leland (1935–) M	**J.L.Crane**
Cranfill, Raimond (fl. 1981) P	**Cranfill**
Crantz, Heinrich Johann Nepomuk von (1722–1799) BS	**Crantz**
Cranwell, Lucy May (1907–) AS	**Cranwell**
Cranz, David (1723–1777) B	**Cranz**
Cratty, Robert Irvin (1853–1940) S	**Cratty**
Craven, Lyndley Alan (1945–) S	**Craven**
Crawford, D.A. (fl. 1954) M	**D.A.Crawford**
Crawford, Daniel J. (1942–) S	**D.J.Crawford**
Crawford, Lloyd C. (fl. 1951) PS	**Crawford**
Crawford, Richard M. A	**R.M.Crawford**
Crawshay, Richard (fl. 1930) M	**Crawshay**
Creager, Don B. (fl. 1963) M	**Creager**
Cree, Gavin S	**Cree**
Creelman, D.W. (fl. 1951) M	**Creelman**
Cremades, Javier (fl. 1990) A	**Cremades**
Cremer, Leo (1866–) FS	**Cremer**
Cremers, Georges (1936–) PS	**Cremers**
Crepet, W.L. (fl. 1990) F	**Crepet**
Crépin, François (1830–1903) F	**Crép.**
Crescenti, U. A	**Crescenti**
Crespo, Ana (fl. 1989) MS	**A.Crespo**
Crespo, Manuel B. (fl. 1988) S	**M.B.Crespo**
Crespo, Susana (1928–) S	**Crespo**
(Crespo–Killalba, Manuel B.)	
see Crespo, Manuel B.	**M.B.Crespo**
Creţiou, I. (f. 1983) M	**Creţiou**
Cretzoiu, G. (fl. 1983) M	**G.Cretz.**
Cretzoiu, Paul (1909–1946) BMPS	**Cretz.**
Creveld, M. (fl. 1981) M	**Creveld**
Crevost, Charles (1858–) S	**Crevost**
Crewz, David W. (1947–) S	**Crewz**
Cribb, Alan Bridson (1925–) AM	**Cribb**
Cribb, Joan Winifred (1949–) M	**J.W.Cribb**

Cribb, Phillip James (1946–) S	P.J.Cribb
Cribbs, J.E. (fl. 1939) F	Cribbs
Crichton, George A. (fl. 1960s–1980s) M	Crichton
Cridland, Arthur A. (1936–) AF	Cridland
Crié, Louis Auguste (1850–1912) FM	Crié
(Criekinge, Louis van)	
see Van Criekinge, Louis	Van Criek.
Crins, William J. (1955–) S	Crins
Cripps (fl. 1876–1880) S	Cripps
Crişan, Aurelia (1930–) M	Crişan
Crisci, Jorge Victor (1945–) S	Crisci
Crisp, Michael Douglas (1950–) S	Crisp
Cristinzio, M. (fl. 1938) M	Cristinzio
Cristóbal, Carmen Lelia (1932–) S	Cristóbal
Cristofolini, Giovanni (1939–) S	Cristof.
Cristurean, Ioan (1935–) S	Cristur.
Critchfield, R.L. (fl. 1988) M	R.L.Critchf.
Critchfield, William Burke (1923–1989) S	Critchf.
Crivelli, Paolo Giuseppe (fl. 1981) M	Crivelli
Croall, Alexander (1809–1885) S	Croall
Croasdale, Hannah T. (1905–) A	Croasdale
Croat, Thomas Bernard (1938–) S	Croat
Crocals, A. (1861–1932) M	Crocals
Crockett, Lawrence J. A	Crockett
Croft, B.J. (fl. 1989) M	B.J.Croft
Croft, James R. (1951–) P	J.R.Croft
Croft, William Noble (1915–1953) A	Croft
(Croix, Isobyl Florence la)	
see la Croix, Isobyl Florence	la Croix
Croizat, Léon Camille Marius (1894–1982) S	Croizat
Croll, Donald A. A	Croll
Crombie, James Mascall Morrison (1830–1906) M	Cromb.
Crome, Georg Ernst Wilhelm (1781–1813) BS	Crome
Crompton, Clifford William (fl. 1978) S	Crompton
Crompton, Jeremy G. (fl. 1984) M	J.G.Crompton
Cronberg, Gertrud A	Cronberg
Cronk, Quentin C.B. (fl. 1980) P	Cronk
Cronquist, Arthur John (1919–1992) APS	Cronquist
Crook, H.Clifford (1882–1974) S	Crook
Crookall, Robert (fl. 1930–1932) F	Crookall
Crookes, Marguerite E. (fl. 1951) P	Crookes
Crooks, Kathleen M. (fl. 1935) M	Crooks
Crookshank, Edgar March (1858–1928) A	Crooksh.
Croom, A.B. S	Croom
Croom, Hardy Bryan (1797–1837) S	H.B.Croom
Croome, Roger L. A	Croome
(Cros i Matas, Rosa M.)	
see Cros, Rosa M.	Cros
Cros, Pierre A	P.Cros

Cros, Rosa M. (1945–) B	**Cros**
Crosa, Orfeo (fl. 1975) S	**Crosa**
Crosby, Marshall Robert (1943–) BP	**Crosby**
Crosby–Smith, Joseph (1853–1930) S	**Crosby–Sm.**
Crosier, W.V. (fl. 1965) M	**Crosier**
Cross, David O. S	**D.O.Cross**
Cross, Joy Barnes (1897–) A	**Cross**
Crossland, Charles (1844–1916) M	**Crossl.**
Crosswhite, Frank Samuel (1940–) S	**Crosswh.**
Crouan, Hippolyte Marie (1802–1871) AM	**H.Crouan**
Crouan, Pierre Louis (1798–1871) AM	**P.Crouan**
Crouch, Herbert Branch (1906–) A	**Crouch**
Croucher, George (1833–1905) S	**Croucher**
Crous, P.W. (fl. 1989) M	**Crous**
(Crovetto, Raul Martínez)	
see Martinez Crovetto, Raul	**Mart.Crov.**
Crow, Garrett E. (1942–) S	**G.E.Crow**
Crow, William Bernard (1895–) A	**Crow**
Crowden, R.K. (fl. 1986) S	**Crowden**
Crowell, Ivan Herrett (1904–) M	**Crowell**
Croxall, John P. (fl. 1975–1982) P	**Croxall**
Crozals, André de (1861–1932) BM	**Croz.**
Cruchet, Denis (1847–1926) MS	**Cruchet**
Cruchet, Paul (1875–) M	**P.Cruchet**
Cruckshanks, Alexander (fl. 1831) S	**Crucksh.**
Cruden, Robert William (1936–) S	**Cruden**
Crueger, Hermann (1818–1864) S	**Crueg.**
(Crüger, Hermann)	
see Crueger, Hermann	**Crueg.**
(Cruickshanks, Alexander)	
see Cruckshanks, Alexander	**Crucksh.**
Cruise, James E. (1925–) S	**Cruise**
Crum, Ethel Katherine (1886–1943) S	**Crum**
Crum, Howard Alvin (1922–) B	**H.A.Crum**
Crump, D.H. (fl. 1980) M	**D.H.Crump**
Crump, William Bunting (1868–1950) S	**Crump**
Crundwell, Alan Cyril (1923–) BS	**Crundw.**
Cruse, Carl Friedrich Wilhelm (1803–1873) S	**Cruse**
Crusio, W. (fl. 1979) S	**Crusio**
Cruz, Neusa Diniz da (fl. 1965) S	**Cruz**
(Cruz, R.d')	
see D'Cruz, R.	**D'Cruz**
(Csakó, Kalman)	
see Czakó, Kalman	**Czakó**
Csapody, I. (fl. 1985) S	**I.Csapody**
Csapody, Vera (1890–1985) S	**Csapody**
Csató, J. (1833–1913) S	**Csató**
Csongor, Gy. (1915–) S	**Csongor**
Csürös, I. (1914–) S	**Csürös**

Cuatrecasas, José (1903–) MS	**Cuatrec.**
(Cuatrecasas y Arvoni, José)	
see Cuatrecasas, José	**Cuatrec.**
(Cubas Domínguez, Paloma J.A.)	
see Cubas, Paloma J.A.	**Cubas**
Cubas, Paloma J.A. (fl. 1986) PS	**Cubas**
Cuboni, Giuseppe (1852–1920) M	**Cub.**
Cuccuini, Piero (1948–) S	**Cuccuini**
Cueto Romero, Miguel (1955–) S	**Cueto**
Cuevas, Ramón (fl. 1989) S	**Cuevas**
Cufodontis, Georg (1896–1974) PS	**Cufod.**
Cugini, Gino (1852–1907) MS	**Cugini**
Cugnac, Antoine de (1898–) S	**Cugnac**
Cui, Cun Qi (fl. 1989) M	**C.Q.Cui**
Cui, Da Fang (fl. 1987) S	**D.F.Cui**
Cui, Nai Ran (1933–) S	**N.R.Cui**
Cui, Shun Chang (fl. 1989) S	**S.C.Cui**
Cui, Tie Cheng (fl. 1987) S	**T.C.Cui**
Cui, Xian Ju (fl. 1987) S	**X.J.Cui**
Cui, Xiang Dong (fl. 1981) S	**X.D.Cui**
Culberson, Nan Chicita Frances (1931–) M	**C.F.Culb.**
Culberson, William Louis (1929–) M	**W.L.Culb.**
Cullen, James (1936–) S	**Cullen**
Cullinger, B.D. (fl. 1927) M	**Cull.**
Cullman, Willy (1905–) S	**Cullman**
Cullum, Thomas Gery (1741–1831) S	**Cullum**
Culmann, Paul Frederic (1860–1936) B	**Culm.**
Cuming, Hugh (1791–1865) S	**Cuming**
Cumino, Paolo (fl. 1900s) MS	**Cumino**
Cumino, U. (fl. 1900s) M	**U.Cumino**
Cumming, William Archibald (1911–) S	**Cumming**
Cummings, Carlos Emmons (1878–) M	**C.E.Cumm.**
Cummings, Clara Eaton (1855–1906) M	**Cumm.**
Cummings, Robert H. A	**R.H.Cumm.**
Cummins, George Baker (1904–) M	**Cummins**
Cummins, H.A. (fl. 1932) M	**H.A.Cummins**
Cúneo, Rubén (1956–) F	**Cúneo**
Cunha, A. Gonçalves da A	**A.G.Cunha**
Cunha, Aristides Marques da (fl. 1923) AM	**A.M.Cunha**
Cunha, Maria Christina da Silva (fl. 1982) S	**M.C.S.Cunha**
(Cunha, Narciso Soares da)	
see Soares da Cunha, Narciso	**Soares da Cunha**
Cunnell, G.J. (fl. 1956) M	**Cunnell**
Cunningham, Alida Mabel (1868–) S	**A.M.Cunn.**
Cunningham, Allan (1791–1839) APS	**A.Cunn.**
Cunningham, David Douglas (1843–1914) AM	**D.D.Cunn.**
Cunningham, Gordon Herriott (Heriot) (1892–1962) M	**G.Cunn.**
Cunningham, John L. (fl. 1964) M	**J.L.Cunn.**
Cunningham, Richard (1793–1835) S	**R.Cunn.**

Cunningham, Stewart (fl. 1990) S	**S.Cunn.**
Cuomo, V. (fl. 1983) M	**Cuomo**
Cupp, Easter E. (1904–) A	**Cupp**
Curl, Herbert Charles (1928–) A	**Curl**
Curnow, William (1809–1887) B	**Curn.**
Currah, Randolph S. (1954–) M	**Currah**
Curran, Hugh McCullum (1875–1920) S	**H.M.Curran**
Curran, Mary Katherine (later Brandegee, M.K.) (1844–1920) S	**Curran**
Curreli, Luigi (fl. 1988) M	**Curreli**
Currey, Frederick (1819–1881) AMP	**Curr.**
Currie, James Nimrod (1883–) M	**Currie**
Curtis, Carlton Clarence (1864–1945) S	**C.C.Curtis**
Curtis, Charles Henry (1869–1958) S	**C.H.Curtis**
Curtis, John Thomas (1913–1961) MS	**J.T.Curtis**
Curtis, K.M. (fl. 1926) M	**K.M.Curtis**
Curtis, Moses Ashley (1808–1872) MS	**M.A.Curtis**
Curtis, Samuel (1779–1860) S	**S.Curtis**
Curtis, William (1746–1799) BMS	**Curtis**
Curtis, Winifred Mary (1905–) S	**W.M.Curtis**
Curtiss, A.H. S	**Curtiss**
Curzi, Mario (1898–1944) M	**Curzi**
Cushman, Joseph Augustine (1881–1949) A	**Cushman**
Cusick, Alison W. (1941–) P	**Cusick**
Cusin (fl. 1958) S	**Cusin**
Cusma–Velari, Tiziana (1947–) S	**Cusma–Vel.**
Cusset, Colette (1944–) S	**C.Cusset**
Cusset, Gérard Henri Jean (1936–) S	**G.Cusset**
Cusson, Pierre (1727–1783) S	**Cusson**
Custer, Jakob Laurenz (1755–1828) S	**Custer**
Custers, M.T.J. (fl. 1940) M	**Custers**
Cutak, Ladislaus (1908–1973) S	**Cutak**
Cutanda, Vicente (1804–1866) S	**Cutanda**
(Cutanda y Jarauta, Vicente)	
see Cutanda, Vicente	**Cutanda**
Cutler, David Frederick (1939–) S	**D.F.Cutler**
Cutler, Donald Ward (1890–) A	**D.W.Cutler**
Cutler, E.C. (fl. 1916) M	**Cutler**
Cutler, Hugh Carson (1912–) S	**H.C.Cutler**
Cutter, Victor Macomber (1917–) M	**Cutter**
Cuvier, Georges Léopold Chrétien Frédéric Dagobert (1769–1832) F	**Cuvier**
Cuvier, Georges–Frédéric (Frédéric) (1773–1838) S	**F.Cuvier**
Cypers von Landrecy, Viktor (1857–1930) BS	**Cypers**
(Cyrillo, Domenico Maria Leone)	
see Cirillo, Domenico Maria Leone	**Cirillo**
Czabator, F.J. (fl. 1976) M	**Czabator**
Czakó, Kalman (1843–1895) S	**Czakó**
Czarnecki, David B. A	**Czarn.**
Czech, Gerald (1930–) S	**Czech**
Czechov, V.P. S	**Czechov**

Czeczott, Hanna (1888–) S	**Czeczott**
Czefranova, Z.V. (1923–) S	**Czefr.**
Czempyrek, Hanna A	**Czemp.**
Czerdantseva, V.Ja. B	**Czerd.**
Czerepanov, Sergei Kirillovich (1921–) PS	**Czerep.**
Czerepanova, N.P. (fl. 1952) M	**Czerepan.**
Czerepnin, L.M. S	**Czerepnin**
Czernajew, Vassiliĭ Matveievitch (1796–1871) MS	**Czern.**
(Czerneva, O.B.)	
see Tscherneva, O.V.	**Tscherneva**
Czerni, B. (1842–) S	**Czerni**
(Czerniaiev, Basil Matveievich)	
see Czernajew, Vassiliĭ Matveievitch	**Czern.**
(Czerniakovska, Ekaterina Georgiewna)	
see Czerniakowska, Ekaterina Georgiewna	**Czerniak.**
Czerniakowska, Ekaterina Georgiewna (1892–1942) S	**Czerniak.**
(Czernjaëv, Basil Matveĭvich)	
see Czernajew, Vassiliĭ Matveievitch	**Czern.**
(Czernjaëw, Basil Matveĭvich)	
see Czernajew, Vassiliĭ Matveievitch	**Czern.**
Czernov, E.G. (1908–) S	**Czernov**
Czernova, N.M. (1901–) S	**Czernova**
Czetz, Anton (1801–1865) S	**Czetz**
Czevrenidi, S.Kh. (1911–) S	**Czevr.**
Czevtaeva, V.A. S	**Czevtaeva**
Czihak, J. (1800–1888) S	**Czihak**
Czopanov, P.Cz. (1927–) S	**Czopanov**
Czosnowski, Jerzy (1918?–1990) A	**Czosn.**
Czukavina, Anna Prokofevna (1929–1985) S	**Czukav.**
Czurda, Viktor (1897–) A	**Czurda**
Czygan, Franz–Christian A	**Czygan**
Czyzewska, Sabina (fl. 1968) M	**Czyz.**
(d'Albequerque, I.L.)	
see Albequerque, I.L.d'	**Albeq.**
(d'Aleizette, Aymar Charles)	
see Alleizette, Aymar Charles d'	**Alleiz.**
(d'Almeida, J.F.R.)	
see Almeida, J.F.R.d'	**J.F.R.Almeida**
D'Amatao, R.F. (fl. 1977) M	**D'Amatao**
D'Amato, Giovanni Frederico (1941–) S	**D'Amato**
D'Archiac, E.J.A.D.de Saint Simon (1802–1868) A	**D'Archiac**
D'Arcy, William Gerald (1931–) S	**D'Arcy**
D'Cruz, R. (fl. 1973) S	**D'Cruz**
D'Eliscu, P.N. (fl. 1977) M	**D'Eliscu**
D'Emerico, Saverio (1950–) S	**D'Emerico**
d'Hérelle, F.H. (fl. 1909) M	**d'Hérelle**
D'hose, R. (fl. 1976) S	**D'hose**

(d'Incarville, Pierre Nicolas le Chéron)	
see Incarville, Pierre Nicolas le Chéron	**Incarv.**
D'Lacoste, L.G. A	**D'Lacoste**
(d'Oliveira, António Lopes Branquino)	
see Oliveira, António Lopes Branquino d'	**B.Oliveira**
(D'Ombrain, Henry Honywood)	
see Dombrain, Henry Honywood	**Dombrain**
(d'Orbigny, (Alcide) Charles Victor Dessalines)	
see Orbigny, (Alcide) Charles Victor Dessalines d'	**Orb.**
(d'Orbigny, Alcide Dessalines)	
see Orbigny, Alcide Dessalines d'	**A.D.Orb.**
D'Orey, J.D.S. (fl. 1982) S	**D'Orey**
d'Ukedem, J. (fl. 1857) M	**d'Ukedem**
d'Urville, Jules Sébastian César Dumont (1790–1842) S	**d'Urv.**
(da Albuquerque, Byron Wilson Pereira)	
see Albuquerque, Byron Wilson Pereira da	**Albuq.**
(da Câmara, Manuel Emmanuele)	
see Sousa da Câmara, Manuel Emmanuele de	**Sousa da Câmara**
(da Costa, José Porfirio)	
see Costa, José Porfirio	**J.P.Costa**
(da Costa, Maria Eugénia Amorim Pereira)	
see Costa, Maria Eugénia Amorim Pereira da	**M.E.A.Costa**
(da Costa Neto, José Porfirio)	
see Costa, José Porfirio	**J.P.Costa**
(da Costa Paiva, Antonio)	
see Paiva, Antonio da Costa	**A.Paiva**
(da Fonseca, O.)	
see Fonseca, Olympio Oliveira Ribeiro da	**Fonseca**
(da Fonseca–Vaz, A.M.S.)	
see Vaz, A.M.S.da Fonseca	**Vaz**
(da Gama, José de Saldanha)	
see Saldanha da Gama, José de	**Saldanha**
(da Luz, Carlos Gomes)	
see Luz, Carlos Gomes da	**Luz**
(da Maia, M.)	
see Maia, M.da	**Maia**
Da Matta, A. (fl. 1928) M	**Da Matta**
Da Ponte, J.Júlio (fl. 1966) M	**Da Ponte**
(da Silva Cunha, Maria Christina)	
see Cunha, Maria Christina da Silva	**M.C.Cunha**
(da Silva Lacaz, C.)	
see Silva Lacaz, C. de	**Silva Lacaz**
(da Silva Maia, Heraldo)	
see Maia, Heraldo da Silva	**H.Maia**
(da Silva Manso, António Luiz Patricio)	
see Silva Manso, António Luiz Patricio da	**Silva Manso**
(da Silva Manso, José)	
see Silva Manso, José da	**J.Silva Manso**

(da Silva, Manuel Augusto Pirajá)
   see Pirajá da Silva, Manuel Augusto   **Pirajá**
(da Silva, Marlene Freitas)
   see Silva, Marlene Freitas da   **M.F.Silva**
(da Silva, Nilda Marquete Ferreira)
   see Marquete Ferreira da Silva, Nilda   **Marquete**
da Silva, Pirajá
   see Pirajá da Silva, Manuel Augusto   **Pirajá**
(da Silva, Rudolfo Albino Dias)
   see Dias da Silva, Rodolfo Albino   **Dias da Silva**
(da Silva Vidal, R.)
   see Silva Vidal, R.da   **Silva Vidal**
(da Silveira Azevedo, Nearch)
   see Azevedo, Nearch da Silveira   **N.S.Azevedo**
(da Torre, Antonio Rocha)
   see Torre, Antonio Rocha da   **Torre**
Da Veiga, A. (fl. 1929) M   **Da Veiga**
(da Vinha, Sérgio Guimarães)
   see Vinha, Sérgio Guimarães da   **Vinha**
Daber, Rudolf (1929–) A   **Daber**
Dabrowska, J. (fl. 1982) S   **Dabrowska**
Dabrowski, Z. (fl. 1928) M   **Dabrowski**
Dackman, C. (fl. 1988) M   **Dackman**
Dadalauri, T.G. (fl. 1978) M   **Dadal.**
Daday, Eugen von (Jenő–től) A   **Daday**
Dade, Harry Arthur (1895–1978) M   **Dade**
Dadwal, V.S. (fl. 1984) M   **Dadwal**
(Daeniker, Albert Ulrich)
   see Däniker, Albert Ulrich   **Däniker**
Dafni, A. (fl. 1971) S   **Dafni**
Dafteri, L.N. (fl. 1966) M   **Dafteri**
Dagaeva, V.K. (1896–1947) S   **Dagaeva**
Daghlian, Charles P. (1950–) FM   **Daghlian**
Daglio, C.A. (fl. 1947) M   **Daglio**
Dahl, Andreas (Anders) (1751–1789) S   **Dahl**
Dahl, Å. Eilif (1916–) MS   **Å.E.Dahl**
Dahl, Ove Christian (1862–1940) S   **O.C.Dahl**
Dahlberg, Carl Gustav (fl. 1753) S   **Dahlb.**
Dahlgren, Bror Eric (1877–1961) S   **Dahlgren**
Dahlgren, Rolf Martin Theodor (1932–1987) S   **R.Dahlgren**
Dahlstedt, Gustav Adolf Hugo (1856–1934) S   **Dahlst.**
Dai, Chan Din (fl. 1985) S   **C.D.Dai**
Dai, Kiyoshi (fl. 1990) M   **K.Dai**
Dai, Lun Kai (fl. 1963) S   **L.K.Dai**
Dai, Qi Hui (fl. 1986) S   **Q.H.Dai**
Dai, S.L. (fl. 1986) S   **S.L.Dai**
Dai, Tian (Tien) Lun (1923–) S   **T.L.Dai**
Daiber, J. (fl. 1866) S   **Daiber**
Daily, Eva Fay Kenoyer (1911–) A   **E.F.K.Daily**

Daily, William Allen (1912–) A   **W.A.Daily**
Dainat, Hanne Lore (fl. 1970) M   **H.L.Dainat**
Dainat, J. (fl. 1973) M   **J.Dainat**
Dakin, J.C. (fl. 1966) M   **Dakin**
Dakshini, K.M.M. (fl. 1970) M   **Dakshini**
Dal Bello, G.M. (fl. 1988) M   **Dal Bello**
Dal Vesco, Giovanna (fl. 1955) M   **Dal Vesco**
Dalbey, Nora Elizabeth (1888–1932) M   **Dalbey**
Dalby, Dunkery Hugh (1930–) S   **Dalby**
Dalchow, J. (fl. 1975) M   **Dalchow**
Dale, Barrie   A   **B.Dale**
Dale, Elizabeth (fl. 1926) MP   **E.Dale**
Dale, Ivan Robert (1904–1963) S   **Dale**
Dale, Samuel (1659–1739) L   **S.Dale**
Dale, William Thomas (1918–) M   **W.T.Dale**
(Daléchamp, Jacques)
    see Daléchamps, Jacques   **Daléchamps**
Daléchamps, Jacques (1513–1588) L   **Daléchamps**
Daley, Brian   A   **Daley**
Dalgaard, Vilhelm (1946–) S   **Dalgaard**
Dalla Torre, Karl (Carl) Wilhelm von (1850–1928) BMPS   **Dalla Torre**
(Dalla Torre von Thurnberg-Sternhoff, Karl Wilhelm von)
    see Dalla Torre, Karl (Carl) Wilhelm von   **Dalla Torre**
Dallière, Alexis (1823–1901) S   **Dallière**
Dallimore, William (1871–1959) S   **Dallim.**
Dallman, Arthur Augustine (1883–1963) S   **Dallman**
Dallwitz, M.J. (fl. 1985) S   **Dallwitz**
Dalmau, Luz M.   A   **Dalmau**
Dalpé, Yolande (1948–) M   **Dalpé**
Dalton, James (1764–1843) S   **Js.Dalton**
Dalton, John (1766–1844) S   **Jn.Dalton**
Daly, Douglas C. (1953–) S   **Daly**
Dalzell, Nicol (Nicolas) Alexander (1817–1878) PS   **Dalzell**
Dalziel, John McEwan (1872–1948) S   **Dalziel**
Dam, Herman van (1947–) A   **Dam**
Damanhuri, Ahmad   B   **Damanhuri**
Damanti, Paolo (1858–post 1899) S   **Damanti**
Damassa, Sarah Pierce   A   **Damassa**
Damazio, Léonidas Botelho (1854–1922) PS   **Damazio**
Damblon, J. (fl. 1956) M   **Damblon**
Damboldt, Jürgen (1937–1978) PS   **Damboldt**
Dames e Silva, Júlia (fl. 1970) S   **Dames e Silva**
Damle, Kamala (fl. 1955) M   **Damle**
Dammann, Hildegard   S   **Dammann**
Dammer, Carl Lebrecht Udo (1860–1920) S   **Dammer**
Dammerman, Karel Willem (1885–1951) S   **Damm.**
Damon, Samuel C. (fl. 1950) M   **Damon**
Dampier, William (1652–1715) L   **Dampier**
Dams, Erich (fl. 1905) S   **Dams**

Damsholt, Kell (1938–) B   **Damsh.**
Dana, Bliss F. (1891–) M   **B.F.Dana**
Dana, James Dwight (1813–1895) F   **Dana**
Dancik, B.P. (fl. 1985) S   **Dancik**
Dandy, James Edgar (1903–1976) S   **Dandy**
Danebekov, A.E. (fl. 1973) M   **Daneb.**
Danert, Siegfried (1926–1973) S   **Danert**
Danesch, Edeltraud (1922–) S   **E.Danesch**
Danesch, Othmar (1919–) S   **O.Danesch**
Danesi, L. (1851–1915) S   **Danesi**
Dang, J.K. (fl. 1983) M   **J.K.Dang**
Dange, S.R.S. (fl. 1980) M   **Dange**
Dangeard, Pierre Clement Augustin (1862–1947) AMPS   **P.A.Dang.**
Dangeard, Pierre Jean Louis (1895–1970) A   **P.J.L.Dang.**
Danguy, Paul Auguste (1862–1942) S   **Danguy**
Dangwal, A.K. A   **Dangwal**
Daniel (Hermano) (fl. 1933–1949) S   **Daniel**
Daniel, J.T. (fl. 1973) M   **J.T.Daniel**
Daniel, Lucien Louis (1856–1940) S   **L.L.Daniel**
Daniel, P. (1943–) S   **P.Daniel**
Daniel, Thomas Franklin (1954–) S   **T.F.Daniel**
Daniell, William Freeman (1818–1865) S   **Daniell**
Daniels, Eve Ypin (fl. 1927) M   **E.Y.Daniels**
Daniels, Francis Potter (1869–1947) S   **Daniels**
Daniels, G.S. (fl. 1979) S   **G.S.Daniels**
Daniels, Joan (fl. 1961) M   **J.Daniels**
Daniels, Kenneth A   **K.Daniels**
Daniels, Roger Edward (1943–) B   **R.E.Daniels**
Danielsen, Anders (1919–) S   **Danielsen**
Däniker, Albert Ulrich (1894–1957) PS   **Däniker**
Danilewsky, B. A   **Danil.**
Danilov, A.L. (fl. 1922) M   **A.L.Danilov**
Danilov, Afanasij Nikolaevich (1903–1942) AS   **A.N.Danilov**
Danilov, Alexander Danilovich (1903–) S   **A.D.Danilov**
Danilova, T.D. (fl. 1957) M   **Danilova**
Danin, Avinoam (1939–) S   **Danin**
Danley, D.M. (fl. 1985) S   **Danley**
Dann, Valerie (fl. 1968) M   **Dann**
(Dannenberg, Markgraf Ingeborg)
  see Markgraf–Dannenberg, Ingeborg   **Markgr.–Dann.**
Danquah, O.-A. (fl. 1975) M   **Danquah**
Danser, Benedictus Hubertus (1891–1943) S   **Danser**
Dansereau, Pierre Mackay (1911–) S   **Dans.**
Danthoine, D. (Etienne) (fl. 1788) S   **Danthoine**
Dar, G.H. (1950–) S   **Dar**
Darakov, O.B. (fl. 1986) M   **Darakov**
Darasse, H. (fl. 1959) M   **Darasse**
Darbishire, Otto Vernon (1870–1934) AM   **Darb.**
Darby, John (1804–1877) S   **Darby**

Darbyshire, S.J. (1953–   ) M   **Darbysh.**
(Dardano de Andrade Lima, Arturo)
  see de Andrade–Lima, Arturo Dardano   **Andrade-Lima**
Darevskaja, E.M. (fl. 1950) S   **Darevsk.**
Dargan, J.S. (fl. 1977) M   **Dargan**
Dariev, A.S. (fl. 1982) S   **Dariev**
Darijma   S   **Darijma**
Darimont, F. (fl. 1956) M   **Darimont**
(Darjono–Kramadibrata, K.)
  see Kramadibrata, K.   **Kramad.**
Darker, Grant Dooks (1898–1979) M   **Darker**
Darley, W.Marshall   A   **Darley**
Darling, Louise (fl. 1940) M   **L.Da·ling**
Darling, Richard B.   A   **R.B.Darling**
Darling, Samuel Taylor (1872–   ) AM   **Darling**
Darlington, Cyril Dean (1903–   ) S   **C.D.Darl.**
Darlington, Henry Townsend (1875–1964) B   **H.T.Darl.**
Darlington, Josephine (1905–   ) S   **J.Darl.**
Darlington, William (1782–1863) S   **Darl.**
Darluc, Michel (1707–1783) S   **Darluc**
Darnaedi, Deddy (1952–   ) P   **Darnaedi**
Darnell, Anthony William (1880–   ) S   **Darnell**
Darnell–Smith, George Percy (1868–1942) M   **Darnell-Sm.**
Darracq, Ulysse (–1872) S   **Darracq**
Darwin, Charles Robert (1809–1882) S   **Darwin**
Darwin, Robert Waring (1724–1816) S   **R.W.Darwin**
Darwin, Steven P. (1949–   ) S   **S.P.Darwin**
Das, A.C. (fl. 1962) M   **A.C.Das**
Das, Anjali (fl. 1978) PS   **A.Das**
Das, Ashok Kumar (fl. 1990) AM   **A.K.Das**
Das, C.M. (fl. 1984) M   **C.M.Das**
Das, C.R. (1932–   ) AS   **C.R.Das**
Das, Debika (1938–   ) S   **D.Das**
Das, Doli (fl. 1973) S   **Doli Das**
Das, G.C. (fl. 1982) S   **G.C.Das**
(das Gracas, M.)
  see Gracas, M.das   **Gracas**
Das Gupta, B.M.   A   **B.M.Das Gupta**
Das Gupta, C. (fl. 1957) M   **C.Das Gupta**
Das, N.   A   **N.Das**
Das, S.K. (fl. 1980) M   **S.K.Das**
Das, Sitanath (fl. 1958) M   **S.Das**
Das, Sitesh (fl. 1965) S   **Sitesh Das**
Das, Veena   A   **V.Das**
Das–Gupta, Hem Chandra C. (1878–1933) A   **Das-Gupta**
Das–Gupta, Matiranjan   A   **M.Das-Gupta**
Dasgupta, M.K. (fl. 1978) M   **M.K.Dasgupta**
Dasgupta, S. (fl. 1976) S   **S.Dasgupta**
Dasgupta, S.N. (fl. 1953) M   **S.N.Dasgupta**

Dassanayake, Meliyasena Dhammadasa (1921–) S   **Dassan.**
Dassier de la Chassagne, Henri Gabirel Benoit (1748–1816) M   **Dass.**
Dassonville, C. (fl. 1899) M   **Dassonv.**
Daston, J.S.   S   **Daston**
Dastugue, C.   A   **Dastugue**
Dastur, Jehangir Fardunji (1886–1952) M   **Dastur**
Daszewska, Wanda (fl. 1912) M   **Dasz.**
Date, K.G. (fl. 1972) M   **Date**
Datta, S. (fl. 1932) M   **S.Datta**
Datta, T.K. (fl. 1932) AM   **Datta**
Dattilo–Rubbo, S. (fl. 1938) M   **Datt.-Rubbo**
Daubeny, Charles Giles Bridle (1795–1867) S   **Daubeny**
Daubs, Edwin Horace (fl. 1965) S   **Daubs**
Dauge, J. (fl. 1980) S   **Dauge**
Daugherty, Lyman H. (fl. 1941) FM   **Daugherty**
Dauncey, Elizabeth Anne (1965–) S   **Dauncey**
Dauphin, J. (fl. 1904) M   **Dauphin**
Dautez, Gustave (fl. 1889) S   **Dautez**
Dauvesse (fl. 1880) S   **Dauvesse**
Davaine, C.   A   **Davaine**
Davall, Edmund (1763–1798) S   **Davall**
Davazamc, S. (fl. 1965) S   **Davazamc**
Daveau, Jules Alexandre (1852–1929) S   **Daveau**
Davenport, George Edward (1833–1907) PS   **Davenp.**
Davenport, L.J. (fl. 1988) S   **L.J.Davenp.**
Davey, Frederick Hamilton (1868–1915) S   **Davey**
Davey, R.A. (fl. 1977) M   **R.A.Davey**
Davey, Roger Jack   A   **R.J.Davey**
David, Alix (D.) (1927–) M   **A.David**
David, Armand (1826–1900) S   **David**
David, Irenée (1791–1862) S   **I.David**
David, John Charles (1964–) M   **J.C.David**
Davidian, Hagop Haroutune (1907–) S   **Davidian**
(Davidoff, Bozimir)
    see Davidov, Bozimir   **Davidov**
Davidov, Bozimir (1870–1927) S   **Davidov**
Davids, M. (fl. 1984) S   **Davids**
Davidse, Gerrit (1942–) S   **Davidse**
Davidson, Anstruther (1860–1932) S   **Davidson**
Davidson, Carol (1944–) S   **C.Davidson**
Davidson, D.E. (fl. 1976) M   **D.E.Davidson**
Davidson, John Fraser (1911–) S   **J.F.Davidson**
Davidson, John G.N. (fl. 1972) M   **J.G.N.Davidson**
Davidson, Lynette Elizabeth (1916–) S   **L.E.Davidson**
Davidson, Robert A. (1927–) S   **R.A.Davidson**
Davidson, Ross Wallace (1902–) M   **R.W.Davidson**
Davies, A. (fl. 1985) M   **A.Davies**
Davies, E.H.   A   **E.H.Davies**
Davies, Elizabeth Winsome (1924–) S   **E.W.Davies**
Davies, Frances G. (1944–) S   **F.G.Davies**

Davies, George (1834–1892) B	**G.Davies**
Davies, Hugh (1739–1821) AMS	**Davies**
Davies, J.B. (fl. 1985) MS	**J.B.Davies**
Davies, Olive B. (fl. 1917) S	**O.B.Davies**
Davies, R.R. (fl. 1955) AM	**R.R.Davies**
Davies, Una Violet (fl. 1949) B	**U.V.Davies**
Davis, Alva Raymond (1887–) M	**A.R.Davis**
Davis, Benjamin Harold (1905–) M	**B.H.Davis**
Davis, Bradley Moore (1871–1957) A	**B.M.Davis**
Davis, Charles Carroll (1911–) A	**C.C.Davis**
Davis, Diana Helen (1945–) S	**D.H.Davis**
Davis, E.E. (fl. 1975) M	**E.E.Davis**
Davis, E.Wade (fl. 1983) S	**E.W.Davis**
Davis, Gwenda Louise Rodway (1911–) S	**G.L.R.Davis**
Davis, Hannibal Albert (1899–) S	**H.A.Davis**
Davis, Herbert Spencer (1875–) A	**H.S.Davis**
Davis, J.S. (fl. 1974) S	**J.S.Davis**
Davis, Jerrold I. (fl. 1990) S	**J.I.Davis**
Davis, John Jefferson (1852–1937) M	**Davis**
Davis, Kary Cadmus (1867–1936) S	**K.C.Davis**
Davis, L.D. (fl. 1878) S	**L.D.Davis**
Davis, Lily H. (fl. 1950) M	**L.H.Davis**
Davis, Louie Irby (1897–) S	**L.I.Davis**
Davis, M.R. (fl. 1969) F	**M.R.Davis**
Davis, Marguerite Carolyn (1903–) M	**M.C.Davis**
Davis, Paul G. A	**P.G.Davis**
Davis, Peter Hadland (1918–1992) S	**P.H.Davis**
Davis, Ray Joseph (1895–1984) S	**R.J.Davis**
Davis, Thomas Arthur Warren (1899–1980) S	**T.A.W.Davis**
Davis, Tyreeca S	**T.Davis**
Davis, William Harold (1876–1948) M	**W.H.Davis**
Davis, William Stanley (1930–) S	**W.S.Davis**
Davis, William Thompson (1862–1945) S	**W.T.Davis**
Davison, E.M. (fl. 1983) M	**E.M.Davison**
Davison, J.D. S	**Davison**
Davlianidze, M.T. (1934–) S	**Davlian.**
(Davy, Joseph Burtt)	
see Burtt Davy, Joseph	**Burtt Davy**
Davydkina, T.A. (1938–) M	**Davydkina**
Dawe, Morley Thomas (1880–1943) S	**Dawe**
Dawes, Clinton J. (1935–) AM	**Dawes**
Dawit Abebe (fl. 1989) S	**Dawit**
Dawson, Christine O. (fl. 1961) M	**C.O.Dawson**
Dawson, Elmer Yale (1918–1966) AS	**E.Y.Dawson**
Dawson, Genoveva (Geneviève) (1918–) S	**G.Dawson**
Dawson, John William (1820–1899) AFS	**Dawson**
Dawson, John Wyndham (1928–) S	**J.W.Dawson**
Dawson, Marion Lucile (1909–) M	**M.L.Dawson**
Dawson, Penelope A. A	**P.A.Dawson**
Dawson, Principal A	**P.Dawson**

Day, Alva George (later Grant, A.D.) (1920–) S	A.G.Day
Day, John (1824–1888) S	J.Day
Day, Mary Anna (1852–1924) S	Day
Day, Maxwell Frank Cooper (1915–) A	M.F.C.Day
Day, R.T. (fl. 1983) S	R.T.Day
Dayal, Ram (1931–) M	Dayal
(Daydon Jackson, Benjamin)	
see Jackson, Benjamin Daydon	B.D.Jacks.
Dayton, R.S. (fl. 1985) S	R.S.Dayton
Dayton, William Adams (1885–1958) S	Dayton
(de A.Andrade, M.H.)	
see Andrade, M.H.de A.	M.H.A.Andrade
De, A.B. (fl. 1983) MS	A.B.De
(de Abreu, Cordélia Luiza)	
see Abreu, Cordélia Luiza	Abreu
(de Aguiar, Joaquim Macedo)	
see Aguiar, Joaquim Macedo de	Aguiar
(de Albuquerque, B.W.P.)	
see Albuquerque, B.W.P de	Albuq.
(de Albuquerque, Fernando Carneiro)	
see Albuquerque, Fernando Carneiro de	F.C.Albuq.
(de Almeida, Floriano Paulo)	
see Almeida, Floriano Paulo de	F.P.Almeida
(de Almeida, José Veríssimo)	
see Almeida, José Veríssimo de	J.V.Almeida
(de Almeida, Maria Teresa)	
see Almeida, Maria Teresa de	M.T.Almeida
(de Almeida Pinto, Joaquin)	
see Almeida, Joaquin de Almeida Pinto	Almeida
(de Ana Magán, F.J.)	
see Fernández de Ana Magán, F.J.	Fern.Magán
(de Andrade, Aydil)	
see Andrade, Aydil Grave de	A.Andrade
de Andrade Chiappeta, Alda (fl. 1971) S	de Andrade
(de Andrade, J.C.)	
see Andrade, J.C.de	J.C.Andrade
(de Andrade-Lima, Arturo Dardano)	
see Andrade-Lima, Acturo Dardano de	Andrade-Lima
(de Araujo Schwarz, E.)	
see Schwarz, E.de Araujo	E.A.Schwarz
(de Arêa Leão, A.E.)	
see Arêa Leão, A.E.de	Arêa Leão
(de Arrábida, D.Francisco Antonio)	
see Arrábida, D.Francisco Antonio de	Arráb.
(de Asso y del Rio, Ignacio Jordán)	
see Asso y del Rio, Ignacio Jordán de	Asso
(de Azambuja, David)	
see Azambuja, David de	Azamb.

(de Azevedo, A.M.G.)	
see Azevedo, A.M.G.de	**A.M.G.Azevedo**
(de Azevedo, P.C.)	
see Azevedo, P.C.de	**P.C.Azevedo**
(de Azevedo Penna, Ivo)	
see Azevedo Penna, I.de	**I.A.Penna**
(de Azevedo Penna, Leonam de)	
see Azevedo Penna, Leonam de	**L.A.Penna**
(de Bacelar, J.J.A.H.)	
see Bacelar, J.J.A.H.de	**Bacelar**
De Barbieri, Esther (fl. 1932) M	**De Barb.**
De Bardessi, E. (fl. 1932) M	**De Bard.**
(de Barnola, Joaquín Maria)	
see Barnola, Joaquín Maria de	**Barnola**
(de Barros, Fábio)	
see Barros, Fábio de	**F.Barros**
(de Barros, Maud L.)	
see Barros, Maud L.de	**M.Barros**
de Bary, Heinrich Anton (1831–1888) AMS	**de Bary**
(de Beauvois, Ambroise Marie François Joseph)	
see Palisot de Beauvois, Ambroise Marie François Joseph	**P.Beauv.**
(de Belleval, Pierre Richer)	
see Richer de Belleval, Pierre	**Rich.Bell.**
(de Berenger, Giuseppe Adolpho)	
see Berenger, Giuseppe Adolpho de	**Berenger**
De Bertoldi, Marco (fl. 1976) M	**De Bert.**
(de Beurmann, Charles Lucien)	
see Beurmann, Charles Lucien de	**Beurmann**
(de Beuzeville, Wilfred Alexander Watt)	
see Beuzeville, Wilfred Alexander Watt de	**Beuzev.**
(de Bivona–Bernardi, Antonius)	
see Bivona–Bernardi, Antonius de	**Biv.**
(de Blainville, Henri Marie Ducrotay)	
see Blainville, Henri Marie Ducrotay de	**Blainv.**
de Boer, Hendrik Wijbrand (1885–1970) S	**de Boer**
(de Boisduval, Jean Baptiste Alphonse Déchauffour(e))	
see Boisduval, Jean Baptiste Alphonse Déchauffour(e) de	**Boisduval**
(de Boissieu, Henri)	
see Boissieu, Henri de	**H.Boissieu**
(de Bolòs, Oriol)	
see Bolòs, Oriol de	**O.Bolòs**
(de Bolòs y Vayreda, Antonio)	
see Bolòs y Vaireda, Antonio de	**A.Bolòs**
(de Boncourt, Louis Charles Adelaïde Chamisseau)	
see Chamisso, Ludolf Karl Adelbert von	**Cham.**
(de Brébisson, Louis Alphonse)	
see Brébisson, Louis Alphonse de	**Bréb.**
(de Brignoli di Brunnhoff, Giovanni)	
see Brignoli di Brunnhoff, Giovanni de	**Brign.**

(de Brondeau, Louis)
  see Brondeau, Louis de   **Brond.**
de Bruijn, Ary Johannes (1811–1896) MS   **Bruijn**
(de Bruyn, Ary Johannes)
  see de Bruijn, Ary Johannes   **Bruijn**
(de Bruyn, H.P.)
  see Bruyn, H.P. de   **Bruyn**
de Bruyne, C. (fl. 1890) M   **de Bruyne**
(de Buffon, Georges Louis Leclerc)
  see Buffon, Georges Louis Leclerc de   **Buffon**
(De Buhr, Larry Eugene)
  see DeBuhr, Larry Eugene   **DeBuhr**
(de Caldas, Francisco José)
  see Caldas, Francisco José de   **Caldas**
(de Calonge, Francisco)
  see Calonge, Francisco de   **Calonge**
(de Camargo, Felisberto Cardoso)
  see Camargo, Felisberto Cardoso de   **Camargo**
(de Campos Novães, José)
  [see Novães, José de Campos   **Novães**
(de Candolle, Alphonse Louis Pierre Pyramus)
  see Candolle, Alphonse Louis Pierre Pyramus de   **A.DC.**
(de Candolle, Anne Casimir Pyramus)
  see Candolle, Anne Casimir Pyramus de   **C.DC.**
(de Candolle, Augustin Pyramus)
  see Candolle, Augustin Pyramus de   **DC.**
(de Candolle, Richard Émile Augustin)
  see Candolle, Richard Émile Augustin de   **Aug.DC.**
(de Carbonnière, Louis François Elisabeth Ramond)
  see Ramond de Carbonnière, Louis François Elisabeth   **Ramond**
(de Carvalho e Vasconcellos, João)
  see Vasconcellos, João de Carvalho e   **Vasc.**
(de Carvalho, T.)
  see Carvalho, T.de   **T.Carvalho**
(de Cassini, Alexandre Henri Gabriel)
  see Cassini, Alexandre Henri Gabriel de   **Cass.**
(de Castro, Duarte)
  see Castro, Duarte de   **Castro**
(de Cervantes, Vicente (Vincente))
  see Cervantes, Vicente (Vincente) de   **Cerv.**
(de Cesati, Vincenzo)
  see Cesati, Vincenzo de   **Ces.**
(de Chaillet, Jean Frédéric)
  see Chaillet, Jean Frédéric de   **Chaillet**
(de Chambray, Georges)
  see Chambray, Georges de   **Chambray**
(de Charpentier, Jean G.F.)
  see Charpentier, Jean G.F.de   **Charpent.**
(de Chastenay, Victorine)
  see Chastenay, Victorine de   **Chastenay**

(de Clairville, Joseph Philippe)
    see Clairville, Joseph Philippe de   **Clairv.**
De Cock, A.W.A.M. (fl. 1986) M   **De Cock**
(de Coincy, Auguste Henri Cornut)
    see Coincy, Auguste Henri Cornut de   **Coincy**
de Corte, C. (fl. 1909) S   **de Corte**
(de Costa, José Gonçalves)
    see Gonçalves de Costa, José   **J.Gonç.Costa**
(de Courset, George(s) Louis Marie)
    see Dumont de Courset, George(s) Louis Marie   **Dum.Cours.**
De Coursey (fl. 1927) M   **De Coursey**
De Craene, Albert (1905–) S   **De Craene**
(de Crozals, André)
    see Crozals, André de   **Croz.**
(de Cugnac, Antoine)
    see Cugnac, Antoine de   **Cugnac**
(de Díaz, Elsa Nélida Lacoste)
    see Lacoste de Díaz, Elsa Nélida   **E.N.Lacoste**
(de Ezcurdia, Luis)
    see Ezcurdia, Luis de   **Ezcurdia**
(de Faria, J.Gomes)
    see Gomes de Faria, J.   **Gomes Faria**
(de Ferré, Yvette)
    see Ferré, Yvette de   **Ferré**
(de Férussac, André Étienne Justin Pascal François d'Audibert)
    see Férussac, André Étienne Justin Pascal François
d'Audibert de   **Férussac**
(de Fonvert, Alexandre Jean Baptiste Amédée Reinaud)
    see Reinaud de Fonvert, Alexandre Jean Baptiste   **Rein.Fonv.**
(de Forestier)
    see Forestier, de   **Forest.**
(de Foucault, Emmanuel)
    see Foucault, Emmanuel de   **Foucault**
De France, Jesse Allison (1899–) S   **De France**
(de Franqueville, Albert)
    see Franqueville, Albert de   **Franquev.**
(de Freitas, Gilberto)
    see Freitas, Gilberto de   **Freitas**
(de Freycinet, Henri Louis Claude de Saulces)
    see Freycinet, Henri Louis Claude de Saulces de   **Freyc.**
(de Fromentel, Louis Édouard Gourdan)
    see Fromentel, Louis Édouard Gourdan de   **Fromentel**
(De Gaetano, L.)
    see Gaetano, L. De   **Gaetano**
(de Garsault, François Alexandre Pierre)
    see Garsault, François Alexandre Pierre de   **Garsault**
(de Geer van Jutphaas, Jan Lodewijk Willem)
    see Geer van Jutphaas, Jan Lodewijk Willem de   **Geer**
(de Girard, Frédéric)
    see Girard, Frédéric de   **Girard**

(de Gorter, David)	
see Gorter, David de	**Gorter**
de Graaf, A. (fl. 1986) S	**A.de Graaf**
de Graaf, Frederik (1913–1971) A	**de Graaf**
(de Grandjot, Gertrude F.)	
see Grandjot, Gertrud F.de	**G.F.Grandjot**
(de Granville, Jean–Jacques)	
see Granville, Jean–Jacques de	**Granv.**
(de Grateloup, Jean Pierre A.Sylvestre)	
see Grateloup, Jean Pierre A.Sylvestre de	**Gratel.**
(de Grosourdy, René)	
see Grosourdy, René de	**Grosourdy**
(de Guzmán, Enriquito D.)	
see Guzmán, Enriquito D.de	**E.D.Guzmán**
de Haas, A.J.P. (fl. 1983) S	**A.J.P.de Haas**
de Haas, Th. (1888–) S	**de Haas**
(de Halperín, Delia R.)	
see Halperín, Delia R.de	**Halperín**
de Hoog, G.S. (1948–) M	**de Hoog**
(de Jersey, Noel J.)	
see Jersey, Noel J.de	**Jersey**
de Joncheere, Gerardus J. (1909–1989) PS	**de Jonch.**
De Jong, C.B. (fl. 1968) M	**C.B.De Jong**
(De Jong, Diederik Cornelius Dignus)	
see DeJong, Diederik Cornelius Dignus	**DeJong**
(de Jong, P.C.)	
see DeJong, P.C.	**P.C.DeJong**
De Jonge, Alide E.van Hall (1871–1951) M	**De Jonge**
De Jonghe, Jean (1804–1876) S	**De Jonghe**
(de Jussieu, Adrien Henri Laurent)	
see Jussieu, Adrien Henri Laurent de	**A.Juss.**
(de Jussieu, Antoine)	
see Jussieu, Antoine de	**Ant.Juss.**
(de Jussieu, Antoine Laurent)	
see Jussieu, Antoine Laurent de	**Juss.**
(de Jussieu, Bernard)	
see Jussieu, Bernard de	**B.Juss.**
(de Jussieu, Christophe)	
see Jussieu, Christophe de	**C.Juss.**
(de Jussieu, Joseph)	
see Jussieu, Joseph de	**J.Juss.**
De Kam, M. (fl. 1969) M	**De Kam**
de Koning, R. (fl. 1983–) S	**de Koning**
de Kort, I. (fl. 1984) S	**de Kort**
de Korte, W.E. (fl. 1918) M	**de Korte**
de Kruif, A.P.M. (fl. 1984) S	**de Kruif**
De Kruyff, E. (fl. 1908) M	**De Kruyff**
(De l'Arbre, Antoine)	
see Delarbre, Antoine	**Delarbre**

(de l'Escluse, Charles)
   see Clusius, Carolus   **Clus.**
(de la Condamine, Charles Marie)
   see Condamine, Charles Marie de la   **Cond.**
(de la Devansaye, Alphonse)
   see Devansaye, Alphonse de la   **Devansaye**
(de la Fontaine de Coincy, Auguste Henri Cornut)
   see Coincy, Auguste Henri Cornut de   **Coincy**
(de la Fuente García, Vicente)
   see Fuente García, Vicente de la   **Fuente García**
(de la Llave, Pablo)
   see La Llave, Pablo de   **La Llave**
de la Luz de Armas, Matilde (fl. 1985) S   **de la Luz**
(de la Paz Graells, Mariano)
   see Graells, Mariano de la Paz   **Graells**
(de la Peirouse, Philippe Picot)
   see Lapeyrouse, Philippe Picot de   **Lapeyr.**
(de la Pylaie, Auguste Jean Marie Bachelot)
   see Bachelot de la Pylaie, Auguste Jean Marie   **Bach.Pyl.**
(De la Roche, Daniel)
   see Delaroche, Daniel   **D.Delaroche**
(De la Roche, François)
   see Delaroche, François   **F.Delaroche**
(de la Sagra, Ramón)
   see Sagra, Ramón de la   **Sagra**
De la Soie, Gaspard Abdon (1818–1877)   **De la Soie**
de la Sota, Elías Ramón (1932–   ) FPS   **de la Sota**
De la Torre, Antonio (fl. 1990) S   **De la Torre**
(de la Torre, Margarita)
   see Torre, Margarita de la   **M.Torre**
(de la Torre, Wolfredo Wildpret)
   see Wildpret de la Torre, Wolfredo   **Wildpret**
(De la Vigne, Gislain François)
   see Delavigne, Gislain François   **Delavigne**
(de Labillardière, Jacques Julien Houtton)
   see Labillardière, Jacques Julien Houtton de   **Labill.**
(de Lacerda, F.S.)
   see Lacerda, F.S.de   **Lacerda**
(de Lachenal, Werner)
   see Lachenal, Werner de   **Lachen.**
(de Lacoizqueta, José Maria)
   see Lacoizqueta, José Maria de   **Lacoizq.**
(de Lacroix, Edouard Georges)
   see Lacroix, Edouard Georges de   **E.G.Lacroix**
(de Lacroix, Louis–Sosthène Veyron)
   see Lacroix, Louis–Sosthène Veyron de   **Lacroix**
(de Lacroix, S.)
   see Lacroix, S.de   **S.Lacroix**

(de Laharpe, Jean Jacques Charles)
   see Laharpe, Jean Jacques Charles de   **Laharpe**
(de Lamarck, Jean Baptiste Antoine Pierre de Monnet)
   see Lamarck, Jean Baptiste Antoine Pierre de Monnet de   **Lam.**
de Lamare, Elizabeth Herrera (fl. 1990) S   **de Lamare**
(de Lamare Occhioni, E.M.)
   see Occhioni, E.M.de Lamare   **Occhioni f.**
(de Lambertye, Léonce)
   see Lambertye, Léonce de   **Lambertye**
(de Lanessan, Jean Marie Antoine)
   see Lanessan, Jean Marie Antoine de   **Laness.**
De Langhe, Charles   S   **C.De Langhe**
De Langhe, Joseph Edgard (1907-) PS   **De Langhe**
de Lannoy (fl. 1863) S   **de Lannoy**
(de Lapeyrouse, Philippe Picot)
   see Lapeyrouse, Philippe Picot de   **Lapeyr.**
(de Laplanche, Maurice Counjard)
   see Laplanche, Maurice Counjard de   **Lapl.**
De Larambergue   S   **De Laramb.**
(de Lasteyrie-Dusaillant, Charles Philippe)
   see Lasteyrie-Dusaillant, Charles Philippe de   **Last.-Dus.**
de Laubenfels, David John (1925-) S   **de Laub.**
(de Launay)
   see Launay, de   **Laun.**
(de Layens, Georges)
   see Layens, Georges de   **Layens**
(de Lazzarini Peckollt, Waldemar)
   see Peckolt, Waldemar de Lazzarini   **W.Peckolt**
(de Lazzarini Peckolt, Oswaldo de Lazzarini)
   see Peckolt, Oswaldo de Lazzarini   **O.Peckolt**
(de Lazzarini Peckolt, Waldemar)
   see Peckolt, Waldemar de Lazzarini   **W.Peckolt**
(de Lemos Pereira, Alice)
   see Lemos Pereira, Alice de   **Lemos**
(de Lengyel, B.)
   see Lengyel, B.de   **B.Lengyel**
(de Lens, Adrien Jacques)
   see Lens, Adrien Jacques de   **Lens**
de Lesdain, Maurice Bouly (1869-1965) M   **de Lesd.**
(de Lexarza, Juan José Martinez)
   see Lexarza, Juan José Martinez de   **Lex.**
(de Lima, Haroldo Cavalcante)
   see Lima, Haroldo Cavalcante de   **H.C.Lima**
(de Lima, José Américo)
   see Lima, José Américo de   **J.A.Lima**
(de Litardière, René Verriet)
   see Litardière, René Verriet de   **Litard.**
(de Loja, Berta Herrera Alarcón)
   see Herrera, Berta   **B.Herrera**
de los Rios, Miguel (fl. 1931) M   **de los Rios**

(de Loureiro, João)	
see Loureiro, João de	**Lour.**
De Luca, Paolo (1944–) A	**De Luca**
(de Macedo, Manuel Antonio)	
see Macedo, Manuel Antonio de	**M.A.Macedo**
(de Magalhães Gomes, Alberto Augusto)	
see Magalhães Gomes, Alberto Augusto de	**A.A.Magalh.**
(de Magalhães, O.)	
see Magalhães, O.de	**O.Magalh.**
(de Maisonneuve, Michel Charles Durieu)	
see Durieu de Maisonneuve, Michel Charles	**Durieu**
(de Malzine, Omer de Pierre Antoine Hyacinth Recq)	
see Malzine, Omer de Pierre Antoine Hyacinth Recq de	**Malzine**
De Man, Johannes Govertus (1850–1930) M	**De Man**
De Marbaix, J. (fl. 1976) M	**De Marb.**
De Marco, Giovanni (1939–) S	**De Marco**
(de Mariz, Joaquim)	
see Mariz, Joaquim de	**Mariz**
(de Marsilly, Louis Joseph Auguste)	
see Marsilly, Louis Joseph Auguste de Commines de	**Marsilly**
(de Martrin-Donos, Julien Victor)	
see Martrin-Donos, Julien Victor de	**Martrin-Donos**
(de Matos Araujo, Paulo Agostinho)	
see Araujo, Paulo Agostinho de Matos	**Araujo**
(de Mattos Filho, Armando)	
see Mattos, Armando de	**A.Mattos**
(de Mattos, João Rodrigues)	
see Mattos, Joáo Rodrigues de	**Mattos**
(de Mattos, Nilza Fischer)	
see Mattos, Nilza Fischer de	**N.F.Mattos**
(de Medeiros, Arnaldo Gomes)	
see Medeiros, Arnaldo Gomes de	**A.G.Medeiros**
(de Medeiros–Costa, J.T.)	
see Medeiros–Costa, J.T.de	**Med.–Costa**
de Meijer, André A.R. (fl. 1990) M	**de Meijer**
(de Mello, Alfredo Froilano Bachmann)	
see Mello, Alfredo Froilano Bachmann de	**Mello**
(de Mello Barreto, Henrique Lamahyer)	
see Barreto, Henrique Lamahyer de Mello	**Barreto**
(de Mello, Froilano)	
see Mello, Alfredo Froilano Bachmann	**Mello**
(de Méllo, Joachim Correia)	
see Correia de Méllo, Joachim	**Corr.Méllo**
(de Mello, Luiz Emygdio)	
see Emygdio de Mello, Luiz	**Emygdio**
(de Mello–Silva, Renato)	
see Mello–Silva, Renato de	**Mello–Silva**
(de Meneses, O.J.Azancot)	
see Meneses, O.J. Azancot de	**Meneses**

(de Menezes, Carlos Azevedo)
   see Menezes, Carlos Azevedo de   **Menezes**
(de Menezes, Nanusa L.)
   see Menezes, Nanusa Luiza de   **N.L.Menezes**
(de Mier y Terán, Manuel)
   see Terán, Manuel de Mier y   **Terán**
(de Miklouho-Maclay, Nikolaj Nikolajewitsch)
   see Miklouho-Maclay, Nikolaj Nikolajewitsch   **Mikl.-Maclay**
(de Miranda, Francisco E.L.)
   see Miranda, Francisco E.L.de   **F.E.L.Miranda**
(de Miranda, Lennart Rodrigues)
   see Rodrigues de Miranda, Lennart   **Rodr.Mir.**
(de Miranda, Veja Velloso)
   see Miranda, Veja Velloso de   **V.V.Miranda**
(de Mirbel, Charles François Brisseau)
   see Mirbel, Charles François Brisseau de   **Mirb.**
(de Miré, Philippe)
   see Miré, Philippe   **Miré**
(de Montemayor, Lorenzo)
   see Montemayor, Lorenzo de   **Montemayor**
De Moor, V.P.G. (1827-1895) S   **De Moor**
(de Morais, Antonio A.Taborda)
   see Taborda de Morais, Antonio A.   **Tab.Morais**
(de Morais, J.O.Falcão)
   see Morais, J.O.Falcão de   **Morais**
(de Morales, Sebastiàn Alfredo)
   see Morales, Sebastiàn Alfredo de   **Morales**
(de Moura, Carlos Alberto Ferreira)
   see Moura, Carlos Alberto Ferreira de   **Moura**
De Nardi, Ján Christina (1937-) S   **De Nardi**
(de Necker, Noel Martin Joseph)
   see Necker, Noel Martin Joseph de   **Neck.**
(de Nevers, G.C.)
   see Nevers, G.C.de   **Nevers**
(de Nobele, L.)
   see Nobele, L.de   **Nobele**
(de Noë, Friedrich Wilhelm)
   see Noë, Friedrich Wilhelm   **Noë**
De Notaris, Giuseppe (Josephus) (1805-1877) ABMPS   **De Not.**
(de Noter, Raphaël)
   see Noter, Raphaël de   **Noter**
(de Oliveira, Eurico Cabral)
   see Oliveira, Eurico Cabral de   **E.C.Oliveira**
(de P.Azevedo, Maria Teresa)
   see Azevedo, Maria Teresa de P.   **M.T.P.Azevedo**
(de Palezieux, Philippe)
   see Palezieux, Philippe de   **Palez.**
De Paula de Brooks, Marie Elena (1940-) AS   **De Paula**
(de Paula, José Elsio)
   see Paula, José Elsio de   **Paula**

(de Pereda Sáez, José María)
  see Pereda, José María   **Pereda**
(de Peyerimhoff de Fontenelle, Paul)
  see Peyerimhoff de Fontenelle, Paul de   **Peyerimh.**
(de Poisy, M.B.)
  see Poisy, M.B.de   **Poisy**
(de Pouques, M.L.)
  see Poucques, M.L.de   **Poucques**
(de Pouzolz, Pierre Marie Casimir)
  see Pouzolz, Pierre Marie Casimir de   **Pouzolz**
(de Préfontaine, M.Bruletout)
  see Préfontaine, M.Bruletout de   **Préf.**
(de Pronville, Auguste)
  see Pronville, Auguste de   **Pronville**
(de Puget, Louis)
  see Puget, Louis de   **L.Puget**
De Puydt, Emile (1810–1891) S   **De Puydt**
(de Puyfol, Jordan)
  see Jordan de Puyfol   **Jord.Puyf.**
(de Puymaly, André Henri Laurent)
  see Puymaly, André Henri Laurent de   **Puym.**
De Quatre, B.D. (fl. 1982) M   **De Quatre**
(de Queiroz, Lusinete A.)
  see Queiroz, Lusinete A.de   **L.A.Queiroz**
(de Queiroz, Luciano Paganucci)
  see Queiroz, Luciano Paganucci de   **L.P.Queiroz**
De, R.N. S   **R.N.De**
(de Ramatuelle, Thomas Albin Joseph d'Audibert)
  see Ramatuelle, Thomas Albin Joseph d'Audibert de   **Ramat.**
(de Reboul, Eugène)
  see Reboul, Eugène de   **Reboul**
(de Resende, Flávio P.)
  see Resende, Flávio P.de   **Resende**
De Retz, Bernard G.G. (1910–) S   **De Retz**
(de Rey-Pailhade, Constantin)
  see Rey-Pailhade, Constantin de   **Rey-Pailh.**
(de Rochebrune, Alphonse Trémeau)
  see Rochebrune, Alphonse Trémeau de   **Rochebr.**
(de Roemer, R.)
  see Roemer, R.de   **R.Roem.**
(de Romero, Maria E.Múlgura)
  see Múlgura de Romero, Maria E.   **Múlgura**
de Roon, Adrianus Cornelis (1928–) S   **de Roon**
(de Rosny, Louis Léon Lucien Prunol)
  see Rosny, Louis Léon Lucien Prunol de   **Rosny**
de Rossi, Giuseppe (fl. 1855–1865) S   **de Rossi**
(de Roussel, Henri François Anne)
  see Roussel, Henri François Anne de   **Roussel**
(de Rouville, Paul Gervaise)
  see Rouville, Paul Gervais de   **Rouville**

de Rulamort

(de Rulamort, M.)
  see Rulamort, M.de   **Rulamort**
(de S. Secco, R.)
  see Secco, R.de S.   **Secco**
(de Saint–Amans, Jean Florimond Boudon)
  see Saint–Amans, Jean Florimond Boudon de   **St.–Amans**
(de Saint–Germain, J.J.)
  see Saint–Germain, J.J.de   **St.–Germ.**
(de Saint–Hilaire, Auguste François César Prouvençal)
  see Saint–Hilaire, Auguste François César Prouvençal de   **A.St.–Hil.**
(de Saint–Moulin, Vincentius Josephus)
  see Saint–Moulin, Vincentius Josephus de   **St.–Moul.**
(de Saint–Pierre, Jacques Nicolas Ernest Germain de)
  see Germain de Saint–Pierre, Jacques Nicolas Ernest   **Germ.**
(de Saint–Simon, Maximilien Henri)
  see Saint–Simon, Maximilien Henri de   **St.–Simon**
(de Saldanha da Gama, José)
  see Saldanha da Gama, José de   **Saldanha**
(de Sampaio, Alberto José)
  see Sampaio, Alberto José de   **A.Samp.**
(de Saporta, Louis Charles Joseph Gaston)
  see Saporta, Louis Charles Joseph Gaston de   **Saporta**
(de Sarmiento, M.N.R.)
  see Sarmiento, M.N.R.de   **Sarmiento**
(de Saussure, Horace Bénédict)
  see Saussure, Horace Bénédict de   **Sauss.**
(de Saussure, Nicolas Théodore)
  see Saussure, Nicolas Théodore de   **N.T.Sauss.**
(de Savigny, Marie Jules César Lélorgne)
  see Savigny, Marie Jules César Lélorgne de   **Savigny**
(de Schoenefeld, Wladimir)
  see Schoenefeld, Wladimir de   **Schoenef.**
(de Secondat, Jean Baptiste)
  see Secondat, Jean Baptiste de   **Secondat**
(de Seizilles de Mazancourt, J.)
  see Mazancourt, J.de Seizilles de   **Mazanc.**
(de Sélys–Longchamps, Michel Edmond)
  see Sélys–Longchamps, Michel Edmond de   **Sélys–Longch.**
(de Serres, (Pierre) Marcel (Toussaint))
  see Serres, (Pierre) Marcel (Toussaint) de   **M.Serres**
(de Servais, Gaspard Joseph)
  see Servais, Gaspard Joseph de   **Servais**
De Seynes, Jules (1833–1912) M   **De Seynes**
(de Sigaldi,)
  see Sigaldi, de   **Sigaldi**
(de Silvestri, Antonio)
  see Silvestri, Antonio de   **A.Silvestri**
(de Siqueira, Josafá Carlos)
  see Siqueira, Josafá Carlos de   **J.C.Siqueira**

(de Siqueira, Mauro Wanderley)
  see Siqueira, Mauro Wanderley de   **Siqueira**
De Sloover, J.R. (fl. 1986) M   **J.R.De Sloover**
De Sloover, Jean Louis J.A. (1936–) BS   **De Sloover**
De Smet, Louis (1813–1887) S   **De Smet**
De Smet, S. (fl. 1952) M   **S.De Smet**
De Smidt, F.P.G. (fl. 1927) M   **De Smidt**
(de Smyttère, Philippe Joseph Emmanuel)
  see Smyttère, Philippe Joseph Emmanuel de   **Smyttère**
(de Sornay, Pierre)
  see Sornay, Pierre de   **Sornay**
(de Sousa e Silva, Estela)
  see Silva, Estela de Sousa e   **E.S.Silva**
(de Sousa, Magdalena Peña)
  see Peña de Sousa, Magdalena   **Peña de Sousa**
(de Souza Barreiros, Humberto)
  see Barreiros, Humberto de Souza   **Barreiros**
(de Souza, Simone)
  see Souza, Simone de   **Souza**
(de St.–Amans, Jean Florimond Boudon)
  see Saint–Amans, Jean Florimond Boudon de   **St.-Amans**
(De Stefani, Carlo)
  see Stefani, Carlo de   **Stefani**
(de Talou, A.)
  see Talou, A.de   **Talou**
(de Théis, Alexandre (Étienne Guillaume))
  see Théis, Alexandre (Étienne Guillaume) de   **Théis**
(de Toledo, Joaquim Franco)
  see Toledo, Joaquim Franco de   **Toledo**
(de Toledo, Laura S. Domínguez)
  see Domínguez de Toledo, Laura S.   **L.S.Domínguez**
De Toni, Ettore (Hector) (1858–) S   **E.De Toni**
De Toni, Giovanni Batista (1864–1924) AMP   **De Toni**
De Toni, Giuseppe (fl. 1900) A   **G.De Toni**
(de Tristan, Jules Marie Claude)
  see Tristan, Jules Marie Claude de   **Tristan**
(de Tussac, François Richard)
  see Tussac, François Richard de   **Tussac**
(de Urries de Azara, M.J.)
  see Urries de Azara, M.J.de   **Urries**
de Valéra, Máirín (1912–1984) A   **de Valéra**
(de Vasconcelos, Augusto Teixeira)
  see Vasconcelos, Augusto Teixeira de   **A.T.Vasconc.**
(de Vasconcelos, C.T.)
  see Vasconcelos, C.T.de   **C.T.Vasconc.**
(de Vattimo, Ítalo)
  see Vattimo, Ítalo de   **Vattimo**
(de Vattimo–Gil, Ida)
  see Vattimo–Gil, Ida de   **Vattimo-Gil**

(de Vicq, (Léon–Bonaventure) Éloy)
   see Vicq, (Léon–Bonaventure) Éloy de   **Vicq**
(de Villafañe Lastra, T.)
   see Villafañe Lastra, T.de   **Villafañe**
De Ville, Jean Baptiste (fl. 1689) L   **De Ville**
De Ville, Nicolas (fl. 1707) L   **N.De Ville**
De Villiers, J.J.R. (fl. 1989) M   **De Villiers**
(de Vilmorin, (Auguste Louis) Maurice (Lévêque))
   see Vilmorin, (Auguste Louis) Maurice (Lévêque) de   **M.Vilm.**
(de Vilmorin, (Charles Philippe) Henry (Lévêque))
   see Vilmorin, (Charles Philippe) Henry (Lévêque) de   **H.Vilm.**
(de Vilmorin, (Joseph Marie) Philippe (Lévêque))
   see Vilmorin, (Joseph Marie) Philippe (Lévêque) de   **P.Vilm.**
(de Vilmorin, (Philippe) Victoire (Lévêque))
   see Vilmorin, (Philippe) Victoire (Lévêque) de   **V.Vilm.**
(de Vilmorin, (Pierre Philippe) André (Lévêque))
   see Vilmorin, (Pierre Philippe) André (Lévêque) de   **A.Vilm.**
(de Vilmorin, (Pierre) Louis (François Lévêque))
   see Vilmorin, (Pierre) Louis (François Lévêque) de   **Vilm.**
(de Vilmorin, Jacques Lévêque)
   see Vilmorin, Jacques Lévêque de   **J.Vilm.**
(de Vilmorin, Roger Philippe Vincent)
   see Vilmorin, Roger (Marie Vincent Philippe Lévêque) de   **R.Vilm.**
De Vis, Charles Walter (1829–1915) FS   **De Vis**
(de Visiani, Roberto)
   see Visiani, Roberto de   **Vis.**
de Vogel, Eduard Ferdinand (1942–   ) S   **de Vogel**
De Vol, Charles Edward (1903–1989) PS   **De Vol**
de Voogd, W.B. (fl. 1977) S   **de Voogd**
de Vos, André (Pascal Alexandre) (1834–   ) S   **A.de Vos**
de Vos, Anna Petronella Cornelia (1893–1958) S   **A.P.C.de Vos**
de Vos, Cornelis (1806–1895) S   **de Vos**
de Vos, Miriam Phoebe (1912–   ) S   **M.P.de Vos**
de Vries, Bernhard W.L. (1941–   ) M   **B.de Vries**
de Vries, Gerardus Albertus (fl. 1953) M   **G.A.de Vries**
de Vries, Hugo (1848–1935) S   **de Vries**
de Vries, N.F. (fl. 1938) M   **N.F.de Vries**
de Vriese, Willem Hendrik (1806–1862) PS   **de Vriese**
De Vroey, C. (fl. 1977–) M   **De Vroey**
de Wet, Johannes Martenis Jacob (1927–   ) S   **de Wet**
de Wilde, Jan Jacobus Friedrich Egmond (1932–   ) S   **J.J.de Wilde**
de Wilde, Willem Jan Jacobus Oswald (1936–   ) S   **W.J.de Wilde**
(de Wilde–Duyfjes, Brigitta Emma Elisabeth)
   see Duyfjes, Brigitta Emma Elisabeth   **Duyfjes**
De Wildeman, Émile Auguste(e) Joseph (1866–1947) MPS   **De Wild.**
De Winter, Bernard (1924–   ) S   **De Winter**
de Wit, Hendrik Cornelius Dirk (1909–   ) S   **de Wit**
de Wit, R.   A   **R.de Wit**
(De Wolf, Gordon Parker Jr.)
   see DeWolf, Gordon Parker Jr.   **DeWolf**

(de Zigno, Achille)
  see Zigno, Achille de   **Zigno**
Deacon, J.W. (fl. 1974) M   **Deacon**
Deák, M.H.   A   **Deák**
Deakin, Richard (1808–1873) MPS   **Deakin**
Deam, Charles Clemon (1865–1953) S   **Deam**
Dean, Richard (1830–1905) S   **Dean**
Deane, Henry (1847–1924) AFS   **H.Deane**
Deane, Walter (1848–1930) S   **W.Deane**
(Dear, Christine Melanie Wilmot-)
  see Wilmot-Dear, Christine Melanie   **Wilmot-Dear**
Dearden, E.Ruth (fl. 1949) M   **Dearden**
Dearness, John (1852–1954) M   **Dearn.**
Deasey, M.C. (fl. 1982) M   **Deasey**
Deason, Temd R. (1931–) A   **Deason**
Deb, Debendra Bijoy (1924–) S   **Deb**
Debaecke, D. (fl. 1988) S   **Debaecke**
Debaisieux, P. (fl. 1919) M   **Debais.**
Debat, Louis (1822–1906) BM   **Debat**
Debay (fl. 1859) F   **Debay**
Debbarman, Peary Mohon (1887–1925) S   **Debb.**
Debbert, P. (fl. 1987) S   **Debbert**
Debeaux, Jean Odon (1826–1910) MS   **Debeaux**
Debey, Matthias Dominikus Hubert Maria (1817–1884) AFM   **Debey**
Debouck, D.G. (fl. 1986) S   **Debouck**
Debras, Édouard Henri Alfred (1889–) S   **Debras**
Debray, Marcel M. (fl. 1964) S   **Debray**
Debroczajeva, D.N. (1916–) S   **Debrocz.**
DeBuhr, Larry Eugene (1948–) S   **DeBuhr**
Deby, Julien Marc (1826–1895) A   **Deby**
Decaisne, Joseph (1807–1882) APS   **Decne.**
Decaisne, M.J. (fl. 1844) M   **M.J.Decne.**
Decary, Raymond (c.1890–1973) S   **Decary**
Dechatres, R. (fl. 1974) M   **Dechatres**
Déchy, Moritz von (1851–) S   **Déchy**
Decken, Carl Claus von der (1833–1865) PS   **Decken**
Deckenbach, K.N. (fl. 1924) M   **K.N.Deckenb.**
Deckenbach, V. (fl. 1896) M   **Deckenb.**
Decker, A.   S   **A.Decker**
Decker, D.S. (fl. 1988) S   **D.S.Decker**
Decker, Henry Fleming (1930–) S   **H.F.Decker**
Decker, Paul (1867–) S   **Decker**
Decraene, Louis-Philippe Ronse
  see Ronse Decraene, Louis-Philippe   **Ronse Decr.**
Decuille, Charles (fl. 1893) M   **Decuille**
Dedecca, D.M. (fl. 1956) S   **Dedecca**
Dedecek, Josef (1843–1915) B   **Dedecek**
Dedusenko, Nina Timofeevna (–1962) A   **Dedus.**
(Dedusenko-Shchegoleva, Nina Timofeevna)
  see Dedusenko, Nina Timofeevna   **Dedus.**

Deecke, Johannes Ernst Wilhelm (1862–1934) F   **Deecke**
Défago, Geneviève (fl. 1968) M   **Défago**
Deferrari, Amelia M. (fl. 1988) P   **Deferrari**
DeFilipps, Robert Anthony (1939–   ) S   **DeFilipps**
Deflandre, Georges (–Victor) (1897–1973) AF   **Deflandre**
Deflandre, Marthe   A   **M.Deflandre**
(Deflandre–Rigaud, Marthe)
   see Deflandre, Marthe   **M.Deflandre**
Deflers, M.A. (fl. 1894) S   **Deflers**
Defleurs, Albert (1841–1921) S   **Defleurs**
Defrance, Jacques Louis Marin (1758–1850) AF   **Defrance**
Degelius, Gunnar Bror Fritiof (né Nilsson, G.B.F.) (1903–   ) M   **Degel.**
Degen, Árpád von (1866–1934) PS   **Degen**
Degener, Isa (Irmgard) (née Hansen) (1924–   ) PS   **I.Deg.**
Degener, Otto (1899–1988) PS   **O.Deg.**
Deginani, Norma B. (1950–   ) S   **Deginani**
Degland, Jean Vincent Yves (1773–1841) S   **Degl.**
Deguchi, Hironori (1948–   ) B   **Deguchi**
Dehgan, B. (fl. 1988) S   **Dehgan**
Dehnhardt, Friedrich (1787–1870) S   **Dehnh.**
(Deichmann Branth, Jakob Severin)
   see Branth, Jakob Severin Deichmann   **Branth**
Deighton, Frederick Claude (1903–1992) M   **Deighton**
Deinboll, Peter Vogelius (Wegelius) (1783–1874) S   **Deinboll**
Deinema, M.H. (fl. 1961) M   **Deinema**
(Dejean de Saint–Marcel, Gaspard)
   see Saint–Marcel, Gaspard Dejean de   **St.-Marcel**
Dejeva, N.G. (fl. 1967) M   **Dejeva**
DeJong, Diederik Cornelius Dignus (1931–   ) S   **DeJong**
DeJong, P.C. (fl. 1976) S   **P.C.DeJong**
Deka, G.K. (fl. 1965) S   **G.K.Deka**
Deka, Hareswan (fl. 1970) S   **H.Deka**
(Dekaprelevič, Leonard Leonardovič)
   see Dekaprelevicz, Leonard Leonardovicz   **Dekapr.**
(Dekaprelevich, Leonard Leonardovich)
   see Dekaprelevicz, Leonard Leonardovicz   **Dekapr.**
Dekaprelevicz, Leonard Leonardovicz (1886–   ) S   **Dekapr.**
Dekenbakh, Konstantin von Deckenbach   A   **Dekenbakh**
Dekhkan–Khodzhaeva, N.A. (fl. 1982) M   **Dekhk.-Khodzh.**
Dekker, A.J.F.M. (fl. 1983) S   **A.Dekker**
Dekker, Ernst   A   **E.Dekker**
Dekker, Johannes (1879–   ) M   **Dekker**
Deku, Vasile   A   **Deku**
del Arco Aguilar, M. (fl. 1982) S   **del Arco**
Del Corral, P. (fl. 1934) M   **Del Corral**
Del Prete, Carlo (1949–   ) S   **Del Prete**
Del Puerto, Osvaldo (1931–   ) S   **Del Puerto**
Del Rosario, Romualdo M (1939–   ) B   **Del Ros.**
(Del–Amo y Mora, Mariano)
   see Amo y Mora, Mariano del   **Amo**

Delacroix, Edouard Georges (1858–1907) M	**Delacr.**
Delamain, Jean (1902–1989) S	**Delamain**
Delamare, G. (fl. 1929) M	**Delamare**
Delaméterie, Jean Claud S	**Delamét.**
Delaney, Kris R. S	**Delaney**
(Delannoy)	
see de Lannoy	**de Lannoy**
Delarbre, Antoine (1724–1813) BS	**Delarbre**
Delaroche, Daniel (1743–1813) S	**D.Delaroche**
Delaroche, François (1780–1813) S	**F.Delaroche**
(Delascio Chitty, Francisco)	
see Delascio, Francisco	**Delascio**
Delascio, Francisco (fl. 1987) S	**Delascio**
(Delasoie, Gaspard Abden)	
see Soie, Gaspard Abden	**Soie**
Delastre, Charles Jean Louis (1792–1859) MS	**Delastre**
Delavay, Pierre Jean Marie (1834–1895) S	**Delavay**
Delay, J. (fl. 1960–1981) S	**Delay**
Delavigne, Gislain François (–1805) S	**Delavigne**
Delendick, T.J. (fl. 1978) S	**Delendick**
Delépine, René A	**Delépine**
Delessert, Jules Paul Bejamin (1773–1847) S	**Deless.**
Delétang, Luis F. A	**Delétang**
Deleuil, J.B.A. (fl. 1874) S	**Deleuil**
Deleuze, Joseph Philippe François (1753–1835) S	**Deleuze**
Delevoryas, Theodore (1929–) FS	**Delev.**
Delf, Ellen Marion (1883–1980) A	**Delf**
Delforge, C. (fl. 1984) S	**C.Delforge**
Delforge, Pierre (fl. 1981) S	**P.Delforge**
Delgadillo, Claudio (1945–) B	**Delgad.**
(Delgadillo Moya, Claudio)	
see Delgadillo, Claudio	**Delgad.**
Delgado, Joachim Filippe Nery (1835–1908) F	**Delgado**
Delgado Salinas, Alfonso (1950–) S	**A.Delgado**
Delhaye, R.J. (fl. 1952) S	**Delhaye**
Delile, Alire Raffeneau (1778–1850) ABMPS	**Delile**
Delipavlov, Dimitàr Danailov (1919–) S	**Delip.**
Delise, Dominique François (1780–1841) CM	**Delise**
Delitsch, Heinrich (fl. 1943) M	**Delitsch**
Delivopoulos, Stylianos G. A	**Delivop.**
Dell, Bernard D. (1947–) S	**Dell**
Della Lucia, Terezhina M.C. (fl. 1990) M	**Della Lucia**
Della Torre, B. (fl. 1962) M	**Della Torre**
(Delle Chiaie, Stefano)	
see Delle Chiaje, Stefano	**Delle Chiaie**
Delle Chiaje, Stefano (1794–1860) MS	**Delle Chiaje**
Dellow, Vivienne A	**Dellow**
Delogne, Charles Henri (1834–1901) AB	**Delogne**
Delon, R. (fl. 1983) M	**Delon**

Deloney, L.N. S   **Deloney**
Delphy, Jean A   **Delphy**
Delpino, Giacomo Giuseppe Federico (1833–1905) S   **Delpino**
Delponte, Giovanni Battista (1812–1884) AS   **Delponte**
Demanche (fl. 1907) M   **Demanche**
Demandt, Ernst (1883–) S   **Demandt**
Demaree, Juan Brewer (1885–) M   **Demaree**
Demares, M. (fl. 1988) S   **Demares**
Demaret, Fernand M.H. (1911–) BPS   **Demaret**
Demarsin, J.-P. S   **Demarsin**
Dematra (1742–1824) S   **Dematra**
Demelius, P. (fl. 1923) M   **Demelius**
Demeshkina, Zh.V. A   **Demeshk.**
Demeter, Karoly (Charles) (1852–1890) B   **Demet.**
Demetrescu, Emanuel A   **Demetrescu**
Demetriades, S.D. (fl. 1951) M   **Demetr.**
Demetrio, Charles Hermann (1845–1936) M   **Demetrio**
(Demidoff, Anatole)
    see Demidov, Anatoli Nikolajewitsch   **A.N.Demidov**
Demidov, Anatoli Nikolajewitsch (1812–1870) MS   **A.N.Demidov**
Demidov, Prokofii Akinfievitch (1710–1786) S   **P.A.Demidov**
Demidova, L.I. (1922–) S   **L.I.Demidova**
Demidova, Zinaida Afanasevna (fl. 1926) M   **Z.A.Demidova**
(Demidow, Anatoli Nikolajewitsch)
    see Demidov, Anatoli Nikolajewitsch   **A.N.Demidov**
Demiriz, Hüsnü (1920–) P   **Demiriz**
(Demissew, Sebsebe)
    see Sebsebe Demissew   **Sebsebe**
Demjanenko, A.N. S   **A.N.Demjan.**
Demjanenko, O.N. (1894–) S   **Demjan.**
Deml, Günther (1947–) M   **G.Deml**
Deml, Irmingard (1944–) S   **I.Deml**
Demme, R. (fl. 1890) M   **Demme**
Demoly, Jean-Pierre (fl. 1985) S   **Demoly**
(DeMoor, V.P.G.)
    see De Moor, V.P.G.   **De Moor**
DeMott, K. (fl. 1983) S   **DeMott**
Demoulin, Vincent (1946–) M   **Demoulin**
Dempsey, Graeme P. A   **Dempsey**
Dempster, Lauramay Tinsley (1905–) S   **Dempster**
Demurova, R.A. (1909–) S   **Demurova**
Den Dooren de Jong, L.E. (fl. 1926) M   **Den Dooren**
(den Hartog, Cornelius)
    see Hartog, Cornelis den   **Hartog**
(den Hollander, G.)
    see Hollander, G.den   **Hollander**
Den Nijs, J.C.M. (1946–)S   **Den Nijs**
Denford, Keith Eugene (1946–) S   **Denford**
Deng, Chao Yi (fl. 1987) S   **C.Y.Deng**

Deng, Jia Qi (fl. 1987) S	J.Q.Deng
Deng, Mao Bin (fl. 1987) S	M.B.Deng
Dengubenko, A.V. (fl. 1980) S	Dengub.
Denham, Dale Lee (1922–) S	D.L.Denham
Denham, Dixon (1786–1828) S	Denham
Denholm, Ian M. S	Denholm
Deniak, V. (fl. 1938) M	Deniak
Deninger, Karl (1878–1919) A	Deninger
Denis, Marcel (1897–1929) AS	Denis
Denis, Thomas (1830–) S	T.Denis
Denison, William C. (1928–) M	Denison
Denisse, Étienne (fl. 1843–1846) S	Denisse
Denissova, G.A. (1923–) S	Denissova
Denizot, Michel George (1931–) A	Denizot
Dennes, George Edgar (1817–post 1859) S	Dennes
Dennis, C. (fl. 1973) M	C.Dennis
Dennis, Richard William George (1910–) M	Dennis
Dennis, Robert L. (fl. 1970) FM	R.L.Dennis
Dennis, W.M. (fl. 1979) S	W.M.Dennis
(Dennstaedt, August Wilhelm)	
see Dennstedt, August Wilhelm	Dennst.
Dennstedt, August Wilhelm (1776–1826) S	Dennst.
Denny, Patrick (fl. 1973) AS	Denny
Denton, Melinda Fay (1944–) S	Denton
Deppe, Ferdinand (1794–1861) S	Deppe
Deppenaar, B.J. (fl. 1930) M	Deppenaar
Dequatre, B.D. (fl. 1982) M	Dequatre
Derbès, August Alphonse (1818–1894) A	Derbès
Derbsch, Helmut (fl. 1977) M	Derbsch
Deremiah, J.W. (fl. 1931) M	Deremiah
Derenberg, Julius(?) (1873–1928) S	Derenb.
Derera, N.F. (fl. 1964) S	Derera
Derganc, Leodegar (fl. 1897–1903) S	Derganc
Dermek, A. (1925–1989) M	Dermek
Dermen, Haig (1895–) S	Dermen
Derrick, E.M. (fl. 1926) S	Derrick
Derrick, Lewis N. (fl. 1987) P	L.N.Derrick
Derrieu A	Derrieu
(Dersal, William Richard Van)	
see Van Dersal, William Richard	Van Dersal
Derville, P.Henry A	Derville
Derviz–Sokolova, T.G. (1928–) S	Derv.-Sok.
Derx, H.G. (fl. 1930) M	Derx
Des Étangs, Nicolas Stanislas Chaâles (1801–1876) S	Des Étangs
Des Moulins, Charles Robert Alexandre (1798–1875) S	Des Moul.
Des Pomeys, Marie (fl. 1961) M	Des Pomeys
Desai, B.G. (fl. 1970) M	B.G.Desai
Desai, M.V. (fl. 1955) M	M.V.Desai
Desai, Manherlal Cnhotalal (1910–) S	Desai

Désaulniers, N.L. (fl. 1987) M   **Désauln.**
Desbois, François (1827–1902) S   **Desbois**
Descals, Enrique (fl. 1987) M   **Descals**
Desch, H.E. (fl. 1931–1940) S   **Desch**
Deschamps, J.R. (fl. 1972) M   **J.R.Deschamps**
Deschamps, Louis Auguste (1765–1842) S   **Deschamps**
Descoings, Bernard M. (1931–   ) S   **Desc.**
Descole, Horacio Raul (1910–   ) S   **Descole**
Descourtilz, Michel Étienne (1775–1836) S   **Descourt.**
Déséglise, Pierre Alfred (1823–1883) S   **Déségl.**
Desfontaines, René Louiche (1750–1833) ABMPS   **Desf.**
Deshmukh, P.G. (fl. 1975) M   **Deshmukh**
Deshpande, K.B. (fl. 1964) M   **K.B.Deshp.**
Deshpande, K.S. (fl. 1965) M   **K.S.Deshp.**
Deshpande, Sandhya D. (fl. 1979) M   **S.D.Deshp.**
Deshpande, U.R. (1930–   ) S   **Deshp.**
(Desiatova–Shostenko, Nathalie A.)
   see Desjatova–Shostenko, Nathalie A.   **Des.-Shost.**
Desikachary, Thamarapu Vedanta (1919–   ) A   **Desikachary**
Desjardin, Dennis E. (1950–   ) M   **Desjardin**
Desjatova–Shostenko, Nathalie A. (later Roussine, N.A.) (1889–1969) S   **Des.-Shost.**
Desmarais, Yves (1918–   ) S   **Desmarais**
Desmazières, John Baptiste Henri Joseph (1786–1862) ABMPS   **Desm.**
(Desmoulins, Charles Robert Alexandre)
   see Des Moulins, Charles Robert Alexandre   **Des Moul.**
Desole, Luigi (1904–1979) S   **Desole**
Despaty, Marcel (fl. 1920) S   **Despaty**
Desplanques, A. (fl. 1972) B   **Despl.**
Desportes, F. (fl. 1807) M   **F.Desp.**
Desportes, Jean Baptiste Réné Pouppé (1704–1748) S   **Desp.**
Desportes, Narcisse Henri François (1776–1856) S   **N.H.F.Desp.**
Despreaux, J.M. (1794–1843) BM   **Despr.**
Desrousseaux, Louis Auguste Joseph (1753–1838) PS   **Desr.**
Dessai, Vanctexa   A   **Dessai**
Dessureault, M. (fl. 1988) M   **Dessur.**
Destombes, P. (fl. 1961) M   **Destombes**
Desvaux, Étienne–Émile (1830–1854) S   **E.Desv.**
Desvaux, Nicaise Auguste (1784–1856) ABMPS   **Desv.**
Determann, R.O. (fl. 1981) S   **Determann**
Déterville (fl. 1800) S   **Déterv.**
Detharding, Georg Gustav (1765–1838) S   **Dethard.**
Dethier, D. (fl. 1982) S   **Dethier**
Detling, LeRoy Ellsworth (1909–1967) S   **Detling**
Dettmann, Mary Elizabeth (1935–   ) F   **M.E.Dettmann**
Dettmann, Ute (1933–   ) S   **U.Dettmann**
Deunff, Jean   A   **J.Deunff**
Deunff, M.J.   A   **M.J.Deunff**
Dev, Sukh (fl. 1957) M   **Dev**
Deva, S. (fl. 1986) S   **Deva**
Devansaye, Alphonse de la (1845–1900) S   **Devansaye**

DeVay, J.E. (fl. 1968) M	**DeVay**
(Devesa Alcaraz, Juan Antonio)	
see Devesa, Juan Antonio	**Devesa**
Devesa, Juan Antonio (1955–) S	**Devesa**
Devi, Ambika A	**A.Devi**
Devi, Kamla (fl. 1969) P	**K.Devi**
Devi, S.Uma (fl. 1979) M	**S.U.Devi**
Deville, L. (fl. 1859) S	**Deville**
(Devis, Charles Walter)	
see De Vis, Charles Walter	**De Vis**
Devlin, E.G. A	**Devlin**
Devred, René (1921–) S	**Devred**
(DeVroey, C.)	
see De Vroey, C.	**De Vroey**
Devyatov, A.G. (fl. 1987) S	**Devyatov**
Dewald, C.L. (fl. 1985) S	**Dewald**
Dewel, Ruth Ann (fl. 1985) M	**Dewel**
Dewer, D. (fl. 1891) S	**Dewer**
(deWet, Johannes Martenis Jacob)	
see de Wet, Johannes Martenis Jacob	**de Wet**
Dewèvre, Alfred (1866–1897) MS	**Dewèvre**
Dewey, Chester (1784–1867) S	**Dewey**
Dewey, Douglas R. (1929–) S	**D.R.Dewey**
Dewey, Lyster Hoxie (1865–1944) S	**L.H.Dewey**
Dewit, Jeanine (1929–) S	**Dewit**
DeWolf, Gordon Parker Jr. (1927–) S	**DeWolf**
Dexter, Raymond (fl. 1935) S	**Dexter**
Dey, Jonathan Paul (1943–) M	**J.P.Dey**
Dey, P.K. (fl. 1919) M	**Dey**
Deyl, Miloš (1906–1985) S	**Deyl**
Deyrolle, E. S	**Deyrolle**
Dhande, G.W. (fl. 1949) M	**G.W.Dhande**
Dhande, R.S. (fl. 1975) M	**R.S.Dhande**
Dhar, A.K. (fl. 1977) M	**A.K.Dhar**
Dhar, U. (fl. 1983) S	**U.Dhar**
Dhar, Vishwa (fl. 1981–) M	**V.Dhar**
Dhargalkar, V.K. A	**Dharg.**
Dharne, C.G. (fl. 1965) M	**Dharne**
Dhawan, Saroj (fl. 1968) M	**Dhawan**
Dhaware, A.S. (fl. 1980) M	**Dhaware**
Dhillon, S.S. (fl. 1978) M	**Dhillon**
Dhingra, G.S. (fl. 1979) M	**Dhingra**
Dhir, K.K. (fl. 1981–89) S	**Dhir**
Di Capua, Ernesta (fl. 1904) S	**Di Capua**
Di Fulvio, Teresa Emil (1930–) S	**Di Fulvio**
Di Martino, Andrea (1926–) S	**Di Martino**
Di Menna, Margaret E. (fl. 1954) M	**Di Menna**
Di Negro, Giovanni Carlo (1769–1857) S	**Di Negro**
Di, Wei Zhong (1936–) S	**W.Z.Di**

Dia, Giovanna (1952–) S	**Dia**
Diamandis, S. (fl. 1978) M	**Diam.**
Diana, Silvana (1942–) S	**Diana**
Diao, Zheng Su (fl. 1988) S	**Z.S.Diao**
Diard, Pierre (1784–1849) S	**Diard**
Dias da Silva, Rodolfo Albino (1889–1931) S	**Dias da Silva**
Dias, Maria Rośalia de Sousa (fl. 1952) M	**Dias**
Díaz, Consuelo (1952–) S	**C.Díaz**
(Díaz de la Guardia Guerrero, Consuelo)	
see Díaz, Consuelo	**C.Díaz**
Díaz Dumas, Marta Aleida (1950–) S	**M.A.Díaz**
(Díaz, Elsa Nélida Lacoste de)	
see Lacoste de Díaz, Elsa Nélida	**E.N.Lacoste**
Díaz, G. (fl. 1987) M	**G.Díaz**
(Díaz González, Tomás Emilio)	
see Díaz, Tomás Emilio	**T.E.Díaz**
Díaz Lifante, Zoila (1962–) S	**Z.Díaz**
(Díaz, Marta Aleida)	
see Díaz Dumas, Marta Aleida	**M.A.Díaz**
Díaz Piedrahíta, Santiago (1944–) S	**S.Díaz**
Díaz, Tomás Emilio (1949–) P	**T.E.Díaz**
Díaz–Miranda, David (1946–) S	**Díaz-Mir.**
Díaz–Piferrer, Manuel (1914–) A	**Díaz-Pif.**
Dibben, Martyn James (1943–) M	**Dibben**
Dichtl, Alois (1841–) S	**Dichtl**
Dick, Esther Amelia (1909–) M	**E.A.Dick**
Dick, James (–1775) A	**Dick**
Dick, Margaret (fl. 1983) M	**M.Dick**
Dick, Michael W. (fl. 1960) M	**M.W.Dick**
Dickens, J.S.W. (fl. 1989) M	**Dickens**
Dickie, George (1812–1882) AB	**Dickie**
Dickinson, C.H. (fl. 1964) M	**C.H.Dickinson**
Dickinson, Carola I. (1900–1970) A	**C.I.Dickinson**
Dickinson, Joseph (1805?–1865) S	**Dickinson**
Dickson, Bertram Thomas (fl. 1925) M	**B.T.Dicks.**
Dickson, Edward Dalzell (–1900) S	**E.D.Dicks.**
Dickson, George Frederick (fl. 1839) S	**G.F.Dicks.**
Dickson, J. (1784–1856) M	**J.Dicks.**
Dickson, James H. (1937–) S	**J.H.Dicks.**
Dickson, James (Jacobus) J. (1738–1822) BMPS	**Dicks.**
DiCosmo, Frank (fl. 1977) M	**DiCosmo**
Diddell, Mary Blain (fl. 1953) PS	**Diddell**
Diddens, Harmanna Antonia (fl. 1939) M	**Diddens**
(Diderich, Paul)	
see Diederich, Paul	**Dieder.**
Didier, Alexandre (fl. 1956) S	**Didier**
Didier, J. (fl. 1956) S	**J.Didier**
Didrichsen, Didrik Ferdinand (1814–1887) S	**Didr.**

Didrichsen, P.   S | **P.Didr.**
Dieck, Georg (1847–1925) S | **Dieck**
Diederich, Paul (1959–   ) M | **Diederich**
Diederichs, Christoph (fl. 1989) M | **Dieder.**
Diedicke, Hermann (1865–1940) M | **Died.**
(Diego Calonge, Francisco de)
  see Calonge, Francisco de | **Calonge**
Diehl, William Webster (1891–   ) M | **Diehl**
Diels, Friedrich Ludwig Emil (1874–1945) BPS | **Diels**
Diem, Hoang G. (fl. 1974) M | **H.G.Diem**
Diem, José (1899–1986) PS | **Diem**
Dierbach, Johann Heinrich (1788–1845) S | **Dierb.**
Dierckx, R.P. (fl. 1901) M | **Dierckx**
Diesing, Karl(Carl) Moritz (1800–1867) AS | **Diesing**
Dietel, Paul (1860–1947) M | **Dietel**
Dieterich, Carl Friedrich (1734–1805) S | **Dieter.**
Dieterle, Jennie van Ackeren (1909–   ) S | **Dieterle**
Dietrich, Albert Gottfried (1795–1856) MS | **A.Dietr.**
Dietrich, David Nathaniel Friedrich (1799–1888) BMS | **D.Dietr.**
Dietrich, Friedrich Gottlieb (1768–1850) S | **F.Dietr.**
Dietrich, Heinrich August (fl. 1852) M | **H.A.Dietr.**
Dietrich, Helga (fl. 1982) S | **H.Dietr.**
Dietrich, Werner (1938–   ) S | **W.Dietr.**
Dietrich, Wilhelm Otto (1881–1964) A | **W.O.Dietrich**
Dietrichson, E. (fl. 1954) M | **Dietrichson**
Dietz, Samuel M. (fl. 1970) M | **Dietz**
Dietzow, W.Ludwig von (fl. 1914) B | **Dietzow**
Díez de Cramer, M.del   A | **Díez Cramer**
Díez, María del Carmen R.   A | **Díez**
Digby, Stephanie (fl. 1987) M | **Digby**
Diggs, G.M. (fl. 1982) S | **Diggs**
Digilio, A.P.L. (fl. 1954) M | **Digilio**
Diguet, Léon (fl. 1925) S | **Diguet**
Dihoru, Gheorghe C. (1933–   ) S | **Dihoru**
Dijk, D.Eduard van   B | **Dijk**
Dijkstra, S.J. (fl. 1949) FM | **Dijkstra**
Dikii, S.P. (fl. 1984) S | **Dikii**
Diklić, Nikola (1925–   ) S | **Diklić**
Dikshit, B.K. (fl. 1985) S | **Dikshit**
Dikshit, K.L.   A | **K.L.Dikshit**
Dikshit, R.P.   A | **R.P.Dikshit**
Dilcher, David L. (1936–   ) AFM | **Dilcher**
Dill, I. (fl. 1984) M | **I.Dill**
Dill, O.   A | **O.Dill**
Dillard, Gary E. (1938–   ) A | **Dillard**
Dillenius, Johann Jacob (1684–1747) BLMP | **Dill.**
Dillon, Lawrence S. (1910–   ) A | **Dillon**
Dillon, Michael O. (1947–   ) S | **M.O.Dillon**
(Dillon, Richard Quartin)
  see Quartin-Dillon, Richard | **Quart.-Dill.**

Dillwyn, Lewis Weston (1778–1855) AMPS	**Dillwyn**
Dilmy, Anwari (1915–1979) S	**Dilmy**
DiMichelle, William A. (fl. 1976) B	**DiMich.**
Dimitri, Milan Jorge S	**Dimitri**
Dimitrieva, A.R. (1905–) S	**Dimitr.**
Dimitrov, Stojan Genther (1918–) S	**Dimitrov**
Dimitrova, Evtimija G. (fl. 1980) M	**E.G.Dimitrova**
Dimitrova, V.N. (1923–) S	**Dimitrova**
Dimitrova-Konaklieva, S.D. A	**Dim.-Konakl.**
Dimmitt, M.A. (fl. 1984) S	**Dimmitt**
Dimock, A.W. (fl. 1949) M	**Dimock**
Din, Laily bin (fl. 1991) M	**Din**
(Dinegro, Giovanni Carlo)	
see Di Negro, Giovanni Carlo	**Di Negro**
Dinelli, Angela (1943–) S	**Dinelli**
Ding, Bing Yang (1953–) S	**B.Y.Ding**
Ding, Chen Sen (1933–) S	**C.S.Ding**
Ding Hou (1921–) S	**Ding Hou**
Ding, Su Qin (fl. 1984) S	**S.Q.Ding**
Ding, Wen Jiang (fl. 1985) S	**W.J.Ding**
Ding, Zuo Chao (fl. 1988) PS	**Z.C.Ding**
Dinghra, G.S. (fl. 1987) M	**Dinghra**
Dingler, Hermann (1846–1935) S	**Dingler**
Dingley, Joan M. (fl. 1951) M	**Dingley**
Dingwall, Ingrid (1945–) S	**Dingwall**
Dininno, V. A	**Dininno**
Diniz, F. (fl. 1967) F	**F.Diniz**
Diniz, Manuel (de) Assunção (fl. 1957) S	**Diniz**
Diniz, Maria Adélia (1941–) S	**M.A.Diniz**
Dinklage, Max Julius (1864–1935) S	**Dinkl.**
Dinsmore, John Edward (1862–1951) S	**Dinsm.**
Dinter, Moritz Kurt (1868–1945) S	**Dinter**
Diogo, Julius Cesar (1876–1936) S	**Diogo**
Dippel, Leopold (1827–1914) AS	**Dippel**
Dippenaar, B. (fl. 1930) M	**Dippen.**
Diratzouyan, Nersès (1875–) S	**Diratz.**
Diskus, Alfred A	**Diskus**
Dismier, Gabriel (1856–1942) B	**Dism.**
Dissing, Henry (1931–) M	**Dissing**
Dissmann, E. (fl. 1931) M	**Dissmann**
Distefano, Concetto (fl. 1956) S	**Distefano**
Ditlevsen, E. (fl. 1964) M	**Ditlevsen**
Ditmar, L.P.Fr. (fl. 1806–1817) M	**Ditmar**
Ditmer, E.E. S	**Ditmer**
DiTomaso, J.M. (fl. 1984) S	**DiTomaso**
Ditrichsen, P. S	**Ditrichsen**
Dittrich, Manfred (1934–) MS	**Dittrich**
Diţu, Ion (fl. 1963) M	**Diţu**
(Ditzu, Ion)	
see Diţu, Ion	**Diţu**

Divakaran, K. (fl. 1959) S   **Divak.**
Dive, Daniel G.   A   **Dive**
Diwald, Karl   A   **Diwald**
Dix, William Leroy (1875–   ) MP   **Dix**
(Dixit, B.K.)
    see Dikshit, B.K.   **Dikshit**
Dixit, Ram Das (1942–   ) PS   **R.D.Dixit**
Dixit, S.C.   A   **S.C.Dixit**
Dixit, S.N. (fl. 1974) S   **S.N.Dixit**
Dixon, Dennis M. (fl. 1986) M   **D.M.Dixon**
Dixon, Hugh Neville (1861–1944) B   **Dixon**
Dixon, John R. (fl. 1975) M   **J.R.Dixon**
Dixon, Marjorie (fl. 1968) M   **M.Dixon**
Dixon, Peter Stanley (1928–   ) A   **P.S.Dixon**
Dixon, R.V.   A   **R.V.Dixon**
Dixon, Sylvia M. (fl. 1952) M   **S.M.Dixon**
Dixon, William E. (fl. 1888) S   **W.E.Dixon**
Dixon–Stewart, Dorothy (fl. 1932) M   **Dixon-Stew.**
Djavanchir–Khoie, Karim (fl. 1963) S   **Djav.-Khoie**
Djugaeva, T.M.   S   **Djugaeva**
Dmitriev, J.J. (fl. 1964) S   **J.J.Dmitriev**
Dmitriev, V.D.   S   **V.D.Dmitriev**
Dmitrieva, Alexandra A. (1905–   ) S   **Dmitrieva**
(Dmitriew, V.D.)
    see Dmitriev, V.D.   **V.D.Dmitriev**
(do Carmo Souza, Lidia)
    see Carmo Souza, Lidia do   **Carmo Souza**
Doan, T.   S   **Doan**
Doassans, Jacques Emile (1852–1908) M   **Doass.**
Döbbeler, Peter (1946–   ) BM   **Döbbeler**
Dobbie, H.B. (fl. c.1915–1918) P   **Dobbie**
Dobbs, C.G. (fl. 1954) M   **Dobbs**
Dobell, C.Clifford (1886–1949) A   **Dobell**
Dobell, Patricia   A   **P.Dobell**
Dobignard, A. (fl. 1987) S   **Dobignard**
Dobrescu, Constantin (1912–   ) MS   **Dobrescu**
(Dobrochaeva, D.N.)
    see Dobroczajeva, D.N.   **Dobrocz.**
(Dobrochayeva, D.N.)
    see Dobroczajeva, D.N.   **Dobrocz.**
Dobrochotova, K.V. (1898–   ) S   **Dobrochot.**
Dobroczajeva, D.N. (1916–   ) S   **Dobrocz.**
(Dobroczajeva–Kovalczuk, D.N.)
    see Dobroczajeva, D.N.   **Dobrocz.**
Dobrovolksy, M.E. (fl. 1914) M   **Dobrov.**
Dobrozrakova, Taisiia Leonidovna (fl. 1927) M   **Dobrozr.**
Dobson, F.H. (fl. 1977) S   **Dobson**
Doby–Dubois, M. (fl. 1957) M   **Doby-Dub.**
Docea, Eugen (1920–   ) M   **Docea**

Dochnahl, Friedrich Jakob (1820–1904) S   **Dochnahl**
Dockrill, Alick William (1915–) S   **Dockrill**
(Docturovsky, Vladimir Semenovic)
    see Dokturovsky, Vladimir Semenovic   **Doktur.**
Dod, Donald D. (1912–) S   **Dod**
Dodds, Lionel G. (fl. 1937) S   **Dodds**
Dode, Louis-Albert (1875–1943) S   **Dode**
Dodékova, L.   A   **Dodékova**
Dodge, Bernard Ogilvie (1872–1960) M   **B.O.Dodge**
Dodge, Carroll William (1895–1988) M   **C.W.Dodge**
Dodge, H.R. (fl. 1947) M   **H.R.Dodge**
Dodge, John D.   A   **J.D.Dodge**
Dodge, Raynal (1844–1918) P   **R.Dodge**
Dodin, A. (fl. 1964) M   **Dodin**
Dodoens, Rembert (1518–1585) L   **Dodoens**
(Dodonaeus, Rembertus)
    see Dodoens, Rembert   **Dodoens**
Dodson, Calaway H. (1928–) S   **Dodson**
Doebley, John F. (fl. 1980) S   **Doebley**
(Doell, Johann(es) Christoph (Christian))
    see Döll, Johann(es) Christoph (Christian)   **Döll**
Doellinger, Ignatz (1770–1841) S   **Doell.**
(Doelz, Bruno)
    see Dölz, Bruno   **Dölz**
(Doerfler, Ignaz)
    see Dörfler, Ignaz   **Dörfl.**
Doerrien, Catharina Helena (1717–1795) S   **Doerr.**
Doflein, Franz (1873–1925) A   **Doflein**
Dogadina, T.V.   A   **Dogadina**
Dogan, M. (fl. 1982) S   **Dogan**
Dogiel, Valentin Alexandrovitch (1882?–1955) A   **Dogiel**
Dogma, I.J. (fl. 1966) M   **Dogma**
Doguet, Gaston (fl. 1952) M   **Doguet**
Dohi, K. (fl. 1921) M   **Dohi**
Doi, Tôhei (1882–1946) S   **Doi**
Doi, Yoshimichi (1939–) M   **Yoshim.Doi**
Doi, Yoshio (fl. 1926) S   **Y.Doi**
Doidge, Ethel Mary (1887–1965) M   **Doidge**
Doignon, Pierre (1912–) B   **Doign.**
Dôke, G. (fl. 1947) M   **Dôke**
Dokosi, O.B. (1930–) S   **Dokosi**
(Dokturovskiĭ, Vladimir Semenovic)
    see Dokturovsky, Vladimir Semenovic   **Doktur.**
Dokturovsky, Vladimir Semenovic (1884–1935) BS   **Doktur.**
Dolan, Desmond (fl. 1947) M   **Dolan**
Dolcher, Tullis (1921–) S   **Dolcher**
Dold, H. (fl. 1910) M   **Dold**
Dole, Eleazer Johnson (1888–) S   **Dole**
Dolejš, K. (fl. 1973) M   **Dolejš**

Dolgoff, G.I. A	**Dolgoff**
Dolianiti, E. (fl. 1953) F	**Dolianiti**
Döll, Johann(es) Christoph (Christian) (1808–1885) PS	**Döll**
Doll, Reinhard (1941–) MS	**R.Doll**
Dollfus, Gustav Frédéric (1850–1931) A	**Dollfus**
Dolliner, George (1794–1872) S	**Dolliner**
(Döllinger, Ignaz)	
see Doellinger, Ignatz	**Doell.**
Doluchanov, A.G. (1900–) S	**Doluch.**
Dölz, Bruno (–1945) S	**Dölz**
Domac, Radovan (1918–) S	**Domac**
Domański, Stanisław (1916–) M	**Domański**
Domashova, A.A. (fl. 1957) M	**Domashova**
Dombey, Joseph (1742–1796) S	**Dombey**
Dombrain, Henry Honywood (1818–1905) S	**Dombrain**
(Dombrovskaia, Anna V.)	
see Dombrovskaja, Anna V.	**Dombr.**
Dombrovskaja, Anna V. (1926–) M	**Dombr.**
(Dombrowskaja, Anna W.)	
see Dombrovskaja, Anna V.	**Dombr.**
Domercq, S. (fl. 1956) M	**Domercq**
Domin, Karel (1882–1953) PS	**Domin**
Domínguez de Toledo, Laura S. (fl. 1985) M	**L.S.Domínguez**
Domínguez, Eugenio (c.1945) S	**E.Domínguez**
Domínguez, Juan A. (fl. 1933) M	**Domínguez**
(Domínguez Vilches, Eugenio)	
see Domínguez, Eugenio	**E.Domínguez**
Dominik, Tadeus (fl. 1934) M	**Dominik**
Domjan, Anna (fl. 1936) M	**Domjan**
Domke, Friedrich Walter (1899–1988) S	**Domke**
Domokos, János (1904–) S	**Domokos**
Domsch, K.H. (fl. 1969) M	**Domsch**
(Don Bresadola, Giacomo)	
see Bresadola, Giacopo	**Bres.**
Don, David (1799–1841) MPS	**D.Don**
Don, George (1798–1856) BPS	**G.Don**
Don, George (1764–1814) BS	**Don**
Don, P.N. (fl. 1845) S	**P.N.Don**
Donadille, P. (1936–) S	**Donad.**
Donadini, Jean-Claud (1939–1987) M	**Donadini**
Donald, John Donald (1923–) S	**Donald**
Donat, Artur A	**Donat**
Donati, Vitaliano (1713–1762) A	**Donati**
Donckelaar, André (1783–1858) S	**Donckel.**
Done, C.C. (fl. 1986) S	**Done**
Dong, Bui Xuan (fl. 1973) M	**B.X.Dong**
Dong, Mei Ling A	**M.L.Dong**
Dong, Quan Zhong (fl. 1983) S	**Q.Z.Dong**

Doniţă, N. (1929–) S	**Doniţă**
Donk, Marinus Anton (1908–1972) MP	**Donk**
(Donkelaar, André)	
see Donckelaar, André	**Donckel.**
Donkin, Arthur Scott (fl. 1858–1873) A	**Donkin**
Donn, James (1758–1813) S	**Donn**
Donné, A. A	**Donné**
Donneaux, A. (fl. 1981) S	**Donneaux**
Donnell Smith, F. PS	**F.Donn.Sm.**
Donnell Smith, John (1829–1928) PS	**Donn.Sm.**
Donnersmark, Leo Victor Felix Henkel von (1785–1861) S	**Donnersm.**
Donoso, J. (fl. 1986) M	**Donoso**
Donovan, B. (fl. 1988) M	**B.Donovan**
Donovan, C. A	**Donovan**
Dons, J.A. A	**Dons**
Donselaar, Johannes van (1928–) PS	**Donsel.**
Dony, John George (1899–1991) S	**Dony**
Dönz, O.Christ. A	**Dönz**
Donze, Pierre A	**Donze**
Dooms, L. (fl. 1971) M	**Dooms**
Doorenbos, Jan (1921–) S	**J.Door.**
Doorenbos, Simon Gottfried Albert (1891–1980) S	**Door.**
Doorn–Hoekman, H.van (fl. 1970) S	**Doorn–Hoekm.**
Dop, Alexis Johannes A	**A.J.Dop**
Dop, Paul Louis Amans (1876–1954) S	**Dop**
Döpp, Walter (fl. 1939–1963) P	**Döpp**
Doppelbaur, Hans Walter (1927–1970) M	**Doppelb.**
Dor, Inka A	**Dor**
Dorado, O. (fl. 1987) S	**Dorado**
Dorai, M. (fl. 1986) M	**Dorai**
Doran, L.J. (fl. 1976) S	**L.J.Doran**
Doran, William Leonard (1893–) S	**Doran**
Dorda Alcaraz, Elena (Helena) (fl. 1987) S	**Dorda**
(Dordević, Petor)	
see Georgevitch, Peter	**Georgev.**
Doré, F. A	**F.Doré**
Dore, William George (1912–) S	**Dore**
Dorenbosch, Maria M.J. (fl. 1968) M	**Dorenb.**
Dorf, Erling (fl. 1930–1938) F	**Dorf**
Dörfelt, Heinrich (fl. 1973) M	**Dörfelt**
Dorff, Paul A	**Dorff**
Dörfler, Ignaz (1866–1950) PS	**Dörfl.**
Dörhöfer, Gunther A	**Dörhöfer**
Döring, Harry A	**Döring**
Dorman (fl. 1927) M	**Dorman**
Dorman, Keith William (1910–) S	**K.W.Dorman**
Dorn, Robert D. (fl. 1970) S	**Dorn**
Dorner, József (1808–1873) S	**Dorner**
Dorofeev, P.I. (fl. 1957) FS	**P.I.Dorof.**

Dorofeev, V.F. (1919–1988) S	**Dorof.**
Dorofeev, V.I. (1956–) S	**V.I.Dorof.**
Dorogin, Georgi Nikolaevich (fl. 1926) M	**Dorogin**
Dorogostaïsky, V. A	**Dorogost.**
Doroguine, Georges (fl. 1911) M	**Dorog.**
Doronina, Ju.(Yu.) A. (1938–) S	**Doronina**
Doronkin, V.M. (1950–) S	**Doronkin**
Doroszenko, A. (fl. 1980) S	**Doroszenko**
Doroszevska, Alina (fl. 1974) S	**Dorosz.**
(Doroszewska, Alina)	
see Doroszevska, Alina	**Dorosz.**
Dorr, Laurence J. (1953–) S	**Dorr**
Dorrien Smith, A.A. (1876–1955) S	**Dorr.Sm.**
Dorschner, Janos S	**Dorschner**
Dorsett, Palemon Howard (1862–1943) M	**Dorsett**
Dorthes, Jacob (Jacques) Anselme (1759–1794) S	**Dorthes**
Dorworth, C.E. (fl. 1988) M	**Dorworth**
(dos Santos, Aniceta C.)	
see Santos, Aniceta C.dos	**A.C.Santos**
dos Santos, J.U.Moreira (fl. 1980) S	**J.U.M.Santos**
Dosch, Ludwig (fl. 1873–1888) P	**Dosch**
Doshi, Y.A. A	**Doshi**
Døssing, L. (fl. 1961) M	**Døssing**
Dostál, Josef (1903–) PS	**Dostál**
Dostalek, J. (fl. 1979) S	**Dostalek**
Dottori, Nilda M. (1944–) S	**Dottori**
Doty, Maxwell Stanford (1916–) AMS	**Doty**
Doubinger, Jeanne (1951–) AFMS	**Doub.**
Doucette, Gregory J. A	**Doucette**
Dougherty, Ellsworth C. (–1965) A	**Dougherty**
Douglas, David (1798–1834) S	**Douglas**
Douglas, G.W. (fl. 1986) S	**G.W.Douglas**
Douglas, K.H. (fl. 1968) S	**K.H.Douglas**
Douglas, Mackenzie (fl. 1921) M	**M.Douglas**
Douin, Charles Isidore (1858–1944) B	**Douin**
Douin, Robert Charles Victor (1892–) B	**R.C.V.Douin**
Doumergue, François (?1858–1938) S	**Doum.**
Douvillé, Joseph Henri Ferdinand (1846–1937) A	**Douvillé**
Dovaston, H.F. (fl. 1948) M	**Dovaston**
(Dover, Cindy Lee Van)	
see Van Dover, Cindy Lee	**Van Dover**
Dowding, Eleanor Silver (fl. 1944) M	**Dowding**
Dowell, Philip (1864–1936) PS	**Dowell**
Dowell, Ruth Isabel (1897–) M	**R.I.Dowell**
Downar, N.V. (fl. 1855–1862) S	**Downar**
Downie, Charles A	**C.Downie**
Downie, Dorothy G. (–1960) S	**Downie**
Downie, S.R. (fl. 1986) S	**S.R.Downie**
Downs, Robert Jack (1923–) S	**Downs**

Dowsett, J.A. (fl. 1981) M	**Dowsett**
Dowson, Walter John (1887–1963) M	**Dowson**
Doyer, Catharina M. (fl. 1925) M	**Doyer**
Doyle, Conrad Bartling (1884–1973) S	**Doyle**
Doyle, William T. (1929–) B	**W.T.Doyle**
Dozy, Frans (François) (1807–1856) BM	**Dozy**
Drabble, Eric Frederic (1887–1933) S	**Drabble**
Drábek, Karel A	**Drábek**
Draganov, St.Iord. A	**Draganov**
Dragastan, Ovidiu A	**Dragastan**
Dragendorff, O. S	**Dragend.**
Dragesco, Jean A	**Dragesco**
Dragoni, I. (fl. 1979) M	**Dragoni**
Drake del Castillo, Emmanuel (1855–1904) PS	**Drake**
Drake, William Fitt M. (1786–1874) B	**W.F.M.Drake**
Dransfield, John (1945–) S	**J.Dransf.**
Dransfield, Soejatmi (née Soenarko, S.) (1939–) S	**S.Dransf.**
Drapalik, Donald Joseph (1934–) S	**Drapalik**
Draparnaud, Jacques Philippe Raymond (1772–1804) AS	**Drap.**
Drapiez, Pierre Auguste Joseph (1778–1856) S	**Drapiez**
Drar, Mohammed (1894–1964) S	**Drar**
Drayton, Frank Lisle (1892–) M	**Drayton**
Drebes, Gerhard (fl. 1968) AM	**Drebes**
Drechsler, Charles (E.) (1892–1986) M	**Drechsler**
Dreer, H.A. (1818–1873) S	**Dreer**
Drees, Friedrich Wilhelm S	**Drees**
Drège, Carl Friedrich (Charles Frederick) (1791–1867) S	**C.F.Drège**
Drège, Jean François (Johann Franz) (1794–1881) ABS	**Drège**
Drejer, Solomon (Salomon) Thomas Nicolai (1813–1842) S	**Drejer**
Drenkovski, Rade (1932–) S	**Drenk.**
Drenovsky, A.K. (1879–1967) S	**Dren.**
(Drenowski, A.K.)	
see Drenovsky, A.K.	**Dren.**
Drenth, Engberg (1945–) S	**Drenth**
Drescher, Aubrey A. (1910–) S	**Drescher**
Dress, William John (1918–) S	**Dress**
Dressler, Robert Louis (1927–) S	**Dressler**
Dreves, Johann Friedrich Peter (1772–1816) S	**Dreves**
Drew, Elmer Reginald (1865–1930) S	**Drew**
Drew, Kathleen M. (1901–1957) A	**K.M.Drew**
Drew, William Brooks (1908–) BS	**W.B.Drew**
Drewes, J.P.F. S	**Drewes**
Drewes, Susanna I. (fl. 1991) S	**S.I.Drewes**
Drews, G. A	**Drews**
Dreyfuss, Michael (fl. 1976) M	**Dreyfuss**
Drezepolski, Roman A	**Drezep.**
Drid, F. S	**Drid**
(Drimpelman, Ernst Wilhelm)	
see Drümpelmann, Ernst Wilhelm	**Drümp.**
Dring, Donald M. (1932–1978) M	**Dring**

Dring, Vivienne (1932–   ) M   **V.Dring**
Drinnan, A.N. (fl. 1986) F   **Drinnan**
Drobnick, Ruby (fl. 1961) S   **Drobnick**
(Drobov, Vasiliĭ Petrovich)
   see Drobow, Vasiliĭ Petrovich   **Drobow**
Drobow, Vasiliĭ Petrovich (1885–1956) S   **Drobow**
Dromer, J. (fl. 1984) M   **Dromer**
Dromgoole, F.I.   A   **Dromgoole**
Droop, M.R.   A   **Droop**
Droop, Stephen (fl. 1987) S   **S.Droop**
Drosdov, N.A. (1897–   ) S   **Drosdov**
Drosdova, N.A.   A   **Drosdova**
Drouet, Francis Elliott (1907–1982) A   **F.E.Drouet**
Drouet, Henri (1829–   ) S   **Drouet**
Drouhot, E. (fl. 1966) M   **Drouhot**
Drouillon, R. (fl. 1951) M   **Drouillon**
Druce, Anthony Peter (1920–   ) S   **A.P.Druce**
Druce, George Claridge (1850–1932) PS   **Druce**
Drude, Carl Georg Oscar (1852–1933) S   **Drude**
Druery, Charles Thomas (1843–1917) P   **Druery**
Drugg, Warren S. (1929–   ) A   **Drugg**
Drummond, James (1784–1863) MS   **J.Drumm.**
Drummond, James Lawson (1783–1853) AS   **J.L.Drumm.**
Drummond, James Ramsay (1851–1921) S   **J.R.Drumm.**
Drummond, Octavio de Almeida (1912–   ) M   **O.A.Drumm.**
Drummond, Robert Bailey (1924–   ) S   **R.B.Drumm.**
Drummond, Thomas (1780–1835) B   **Drumm.**
Drümpelmann, Ernst Wilhelm (1760–1830) S   **Drümp.**
Drury, David Geoffrey (1942–   ) S   **D.G.Drury**
Drury, Heber (1819–1872) S   **Drury**
Drury, William Holland (1921–   ) S   **W.H.Drury**
Druten, Denise van (1930–   ) S   **Druten**
Dryander, Jonas Carlsson (1748–1810) PS   **Dryand.**
Drygalski, Erich Dagobert von (1865–1949) B   **Dryg.**
Dshanaëva, V.M.   S   **Dshanaëva**
Du Bois, C. (fl. 1910) M   **Du Bois**
(du Chaffaut, Simon Amaudric)
   see Chaffaut, Simon Amaudric du   **Chaffaut**
(du Chalard, A.)
   see Chalard, A.du   **Chalard**
Du, F. (fl. 1976) M   **Du**
Du Manoir, J. (fl. 1981) M   **Du Manoir**
(du Petit–Thouars, Abel Aubert)
   see Thouars, Abel Aubert du Petit–   **A.Thouars**
(du Petit–Thouars, Louis Marie Aubert)
   see Thouars, Louis Marie Aubert du Petit   **Thouars**
du Plessis, S.J. (fl. 1933) M   **du Plessis**
du Pont, Pierre S.   S   **du Pont**
Du Puy, David J. (1958–   ) S   **Du Puy**

(du Quesnay, M.C.)	
see duQuesnay, M.C.	duQuesnay
Du Rietz, Gustaf Einar (1895–1967) MS	Du Rietz
Du Roi, Johann Philipp (1741–1785) S	Du Roi
du Toit, Alex L. F	A.L.du Toit
du Toit, C.A. A	C.A.du Toit
du Toit, J.W. (fl. 1967) M	J.W.du Toit
Du Toit, Petrus Cornelis Vermuelen (1945–) S	Du Toit
Du Tour de Salvert (fl. 1803–1815) S	Du Tour
Duan, R.L. (fl. 1985) M	R.L.Duan
Duan, Xian Zhen (fl. 1987) S	X.Z.Duan
Duanmu, Shing (fl. 1963) S	Duanmu
Duarte, Aparício Pereira (1910–1984) S	Duarte
Duarte Bello, P.P. (1922–) B	Duarte Bello
Duarte, Maria de Lourdes Reis (1942–) M	M.L.Duarte
Dubard, Marcel Marie Maurice (1873–1914) S	Dubard
Düben, Magnus Wilhelm von (1814–1845) S	Düben
Dubey, A.K. (fl. 1983) S	A.K.Dubey
Dubey, P.K. (fl. 1983) M	P.K.Dubey
Dubey, R.K. (fl. 1990) M	R.K.Dubey
Dubin, Jesse (fl. 1969) M	Dubin
Dubitskiĭ, A.M. (fl. 1973) M	Dubitskiĭ
Dubjansky, Vladimir Andrejevich (1877–1962) S	Dubj.
Dublin, Mary Virginia (fl. 1979) M	Dublin
Dublish, P.K. (fl. 1977) M	Dublish
Dubois, François Noel Alexandre (1752–1824) MS	Dubois
Dubois, Louis (1819–1864) S	L.Dubois
Dubois, Saint Séverin (fl. 1898) M	S.S.Dubois
Dubois–Tylski, Th. A	Dub.–Tylski
Dubos, Bernadette (fl. 1986) M	Dubos
Duboscq, Octave Joseph (1868–) M	Duboscq
Dubovik, Oljga N. (1935–) S	Dubovik
Dubuis, A. (fl. 1959) S	Dubuis
DuBuysson, R. (fl. 1888) P	DuBuysson
Duby, Jean Étienne (1798–1885) ABMPS	Duby
Ducellier, François A	F.Ducell.
Ducellier, Léon Octave (1878–1937) S	Ducell.
Duchartre, Henri S	H.Duch.
Duchartre, Pierre Étienne Simon (1811–1894) S	Duch.
Duchassaing de Fontbressin, Édouard Placide (1818–1873) S	Duchass.
Duché, Jacques (fl. 1952) M	Duché
Duchemin, T. (fl. 1975) M	Duchemin
Duchesne, Antoine Nicolas (1747–1827) S	Duchesne
Ducke, Walter Adolpho (1876–1959) S	Ducke
Ducker, Sophie Charlotte (1909–) AM	Ducker
Duckering, Mae Webster (1894–) B	Duck.
Duclaux, E.P. (fl. 1864) M	Duclaux
Duclaux, Geneviève A	G.Duclaux
Ducluzeau, J.A.P. (fl. 1805) AS	Ducluz.

Ducomet, V. (fl. 1907) M	**Ducomet**
Ducommun, Jules César (1829–1892) S	**Ducommun**
Ducros de St. Germain (fl. ?1828) S	**Ducros**
Duda, Josef (1925–) B	**Duda**
Duddington, C.L. (fl. 1950) M	**Dudd.**
Dudeja, S.K. (fl. 1984) M	**Dudeja**
Dudey, R.K. (fl. 1990) M	**Dudey**
Dudka, Irina A. (1934–) M	**Dudka**
Dudley, Margaret Gertrude (1888–) S	**M.G.Dudley**
Dudley, Theodore ('Ted') Robert (1936–) S	**T.R.Dudley**
Dudley, William Russel (1849–1911) MS	**Dudley**
Duek, Jacobo Jack (1936–) PS	**Duek**
(Duell–Hermanns, I.)	
see Düll–Hermanns, I.	**Düll–Herm.**
Dueñas, H.J. (fl. 1979) M	**Dueñas**
Dueñas, Margarita (1959–) M	**M.Dueñas**
Düesberg, Walter (fl. 1890) S	**Düesberg**
Duffort, Louis (1846–1923) S	**Duffort**
Dufft, Adolf (1803–1875) M	**Dufft**
Dufour, Jean–Marie Léon (1780–1865) BMS	**Dufour**
Dufour, Léon (Marie) (1862–1942) M	**L.M.Dufour**
Dufour, Luigi (1830–1901) A	**L.Dufour**
Dufrenoy, Jean (1894–) M	**Dufrenoy**
Dufresne, Pierre (1786–1836) S	**Dufr.**
Duftschmid, Johann Baptiste (1804–1866) S	**Duftschmid**
Dugand, Armando (1906–1971) S	**Dugand**
Dugas, Marguerite (fl. 1928) B	**Dugas**
Dugès, Alfred Auguste Dalsescantz (1826–1910) S	**Dugès**
Duggar, Benjamin Minge (1872–1956) M	**Duggar**
Dughi, Raymond (1932–) M	**Dughi**
Duhamel du Monceau, Henri Louis (1700–1782) S	**Duhamel**
Duharte Gongora, María Elvira (fl. 1985) S	**Duharte**
Duhem, Bernard (fl. 1989) M	**Duhem**
Duin, W.van (fl. 1991) M	**Duin**
Duistermaat, H. (1962–) S	**Duist.**
Dujardin, Félix (1801–1860) A	**Dujard.**
Duke, James A. (1929–) S	**J.A.Duke**
Duke, Maude M. (fl. 1926) M	**Duke**
Duke, Norman C. (1952–) S	**N.C.Duke**
Dulac, Joseph (1827–1897) PS	**Dulac**
Dular, Ram (fl. 1966) M	**Dular**
Dulfer, Hans (1900–1975) S	**Dulfer**
Düll, Ruprecht Peter George (1939–) BS	**Düll**
Düll–Hermanns, I. B	**Düll–Herm.**
Duman, Hayri (fl. 1990) S	**H.Duman**
Duman, Maximilian George (1906–1990) S	**Duman**
(Dumaz–le Grand, Noëlle)	
see Dumaz–le–Grand, Noëlle	**Dumaz–le–Grand**
Dumaz–le–Grand, Noëlle (fl. 1953) S	**Dumaz–le–Grand**
Dumbadze, Thamar A. (1902–) S	**Dumbadze**

Dumée, Paul (1849–1930) M	**Dumée**
Dumitraş, Lucreţia (1928–) M	**Dumitraş**
Dumitrica, Paulian A	**Dumitrica**
Dumitriu–Tătăranu, I. (1927–) S	**Dum.-Tăt.**
Dummer, Richard Arnold (1887–1922) S	**Dummer**
Dümmer, Richard Arnold	
see Dummer, Richard Arnold	**Dummer**
(Dumont Courset, George(s) Louis Marie)	
see Dumont de Courset, George(s) Louis Marie	**Dum.Cours.**
(Dumont d'Urville, Jules Sébastian César)	
see d'Urville, Jules Sébastian César Dumont	**d'Urv.**
Dumont de Courset, George(s) Louis Marie (1746–1824) S	**Dum.Cours.**
Dumont, Kent P. (fl. 1971) M	**Dumont**
Dumont, Michael P. A	**M.P.Dumont**
Dumortier, Barthélemy Charles Joseph (1797–1878) ABMS	**Dumort.**
Dumoulin (fl. 1782) S	**Dumoulin**
Dun, W.S. (fl. 1890–1930) F	**Dun**
Dunaev, A.S. (fl. 1986) M	**Dunaev**
Dunal, Michel Félix (1789–1856) AMPS	**Dunal**
Dunant de Salatin, Philippe (1797–1866) S	**Dunant**
Dunbar, Henry Fowler (1889–) S	**Dunbar**
Duncan, Andrew (1773–1828) S	**A.Duncan**
Duncan, D.D. (fl. 1982) S	**D.D.Duncan**
Duncan, F.R.S. (fl. 1876) FM	**F.R.S.Duncan**
Duncan, James (1802–1876) S	**Duncan**
Duncan, Peter Martin (1824–1891) A	**P.Duncan**
Duncan, Thomas (1948–) S	**T.Duncan**
Duncan, Wilbur Howard (1910–) S	**W.H.Duncan**
Dundas, Frederic Winn (1911–) S	**Dundas**
Dunegan, John Clymer (1898–) M	**Dunegan**
Dungs, Fritz (1915–1977) S	**Dungs**
Duniway, J.M. (fl. 1985) M	**Duniway**
Dunk, Klaus von der (1943–) B	**K.Dunk**
Dunk, Kurt von der (1897–1985) B	**Dunk**
Dunker, Wilhelm Bernhard Rudolph Hadrian (1809–1885) F	**Dunker**
Dunkerly, J.S. A	**Dunkerly**
Dunkle, Meryle Byron (1888–) S	**Dunkle**
Dunkley, Harvey Lawrence (1910–) S	**Dunkley**
Dunlap, Albert Atkinson (1902–) M	**Dunlap**
Dunlop, Clyde Robert (1946–) S	**Dunlop**
Dunn, David Baxter (1917–) S	**D.B.Dunn**
Dunn, Edward John (1844–1937) S	**E.J.Dunn**
Dunn, M.T. (fl. 1982) M	**M.T.Dunn**
Dunn, Paul H. (fl. 1957) AM	**P.H.Dunn**
Dunn, Stephen Troyte (1868–1938) PS	**Dunn**
Dunsterville, Galfrid Clement Keyworth (1905–1988) S	**Dunst.**
Dupain, M.V. (fl. 1913) M	**Dupain**
Duperrey, Louis Isodore (1786–1865) PS	**Duperrey**
Dupias, George (fl. 1951) M	**Dupias**
Dupin, Françoise A	**Dupin**

Duplessey, F.S. (fl. 1800) S   **Dupl.**
Dupont, A.E.   S   **A.E.Dupont**
Dupont, J.B. (fl. 1825) S   **Dupont**
Dupont, Suzanne (1922–   ) S   **S.Dupont**
Duprat, E. (fl. 1970) M   **Duprat**
Dupray, L.   A   **Dupray**
Dupret, François Hippolyte (1853–1932) B   **Dupr.**
Dupuy (fl. 1882) S   **Dupuy**
Dupuy–Jamain, F.N. (1817–1888) S   **Dup.-Jam.**
Duque, Jaramillo Jesus Maria (1785–1862) S   **Duque**
duQuesnay, M.C. (fl. 1971) S   **duQuesnay**
Durairaj, P. (fl. 1971) M   **Durairaj**
Durairatnam, M.   A   **Durair.**
Durán, Miguel   A   **M.Durán**
Durán Oliva, F. (fl. 1991) S   **Durán Oliva**
Durán, Rubén (fl. 1961) M   **Durán**
Durand, Bernard Michel (1928–   ) S   **B.M.Durand**
Durand, Elias Judah (1870–1922) MS   **E.J.Durand**
Durand, Elias (Elie) Magloire (1794–1873) BS   **Durand**
Durand, Ernest Armand (1872–1910) S   **E.A.Durand**
Durand, H. (fl. 1990) M   **H.Durand bis**
Durand, Hélène (1883–1934) S   **H.Durand**
Durand, J.E.   S   **J.E.Durand**
Durand, L.   S   **L.Durand**
Durand, Philippe   S   **P.Durand**
Durand, Théophile Alexis (1855–1912) BPS   **T.Durand**
Durand–Duquesney, Jean Victor (1785–1862) S   **Dur.-Duq.**
Durande, Jean François (1732–1794) PS   **Durande**
Durando, Gaetano Leone (1811–1892) S   **Durando**
Durazzini, Antonio (fl. 1772) S   **Durazz.**
Durazzo, Ippolito (1750–1818) S   **Durazzo**
Durdević, L. (fl. 1982) S   **Durdević**
Durham, John Wyatt (1907–   ) A   **Durham**
Durie, E.Beatrix (fl. 1957) M   **Durie**
Durieu de Maisonneuve, Michel Charles (1796–1878) ABMPS   **Durieu**
During, Hein Johannes (1947–   ) B   **During**
Düringer, Ingeborg   A   **Düringer**
Durkee, L.H. (1927–   ) S   **Durkee**
(Duroi, Johann Philipp)
   see Du Roi, Johann Philipp   **Du Roi**
Durrell, L.W. (fl. 1966) M   **Durrell**
Durrieu, Guy (1931–   ) M   **Durrieu**
Durrieu, Louis   A   **L.Durrieu**
Dürrnberger, A. (fl. 1898) S   **Dürrnb.**
Dürrschmidt, Monika   A   **Dürrschm.**
Dury, Marie Noelle (1933–   ) B   **Dury**
Dusén, Per Karl Hjalmar (1855–1926) BFPS   **Dusén**
Dusi, H.   A   **Dusi**
Dusn, P. (fl. 1907) F   **Dusn**
Duss, Antoine (1840–1924) MS   **Duss**

Dustan, A.G. (fl. 1924) M	**Dustan**
Duthie, Augusta Vera (1881–1963) BPS	**A.V.Duthie**
Duthie, H.C. (1938–) A	**H.C.Duthie**
Duthie, John Firminger (1845–1922) S	**Duthie**
Dutilly, Arthème (1896–) P	**Dutilly**
Dutra, João (1862–1939) PS	**Dutra**
Dutrie, Louis (fl. 1939) S	**Dutrie**
Dutrochet, René Joachim Henri (1776–1847) S	**Dutr.**
Dutt, A.K. (fl. 1981) AS	**Dutt**
Dutta, B.G. (fl. 1962) M	**B.G.Dutta**
Dutta, R.M. (fl. 1963) S	**R.M.Dutta**
Dutta, S. (fl. 1951) AM	**S.Dutta**
Dutton, J.Everett A	**Dutton**
Duval, Charles–Jeunet (Carl–Jeanet) (1751–1828) MS	**C.-J.Duval**
Duval, E. (fl. 1974) M	**E.Duval**
Duval, Henri August (1777–1814) S	**Duval**
Duval–Jouve, Joseph (1810–1883) PS	**Duval-Jouve**
Duveen, Dennis (fl. 1974) S	**Duveen**
Duvel, Albert Walter S	**Duvel**
Duvernoy, A. (fl. 1920) M	**Duvernoy**
Duvigneaud, Jacques (1920–) S	**J.Duvign.**
Duvigneaud, Paul Auguste (1913–) AMS	**P.A.Duvign.**
Duxbury, S. A	**Duxbury**
Duyfjes, Brigitta Emma Elisabeth (1936–) S	**Duyfjes**
Duyfjes, Hendrik Gerard Pieter (1908–1943) M	**H.G.P.Duyfjes**
(Dvigubsky, Ivan Alexievich)	
see Dwigubskij, Iwan Alexievič	**Dwig.**
Dvoïnos, L.M. (fl. 1982) M	**Dvoïnos**
Dvořák, František (1921–) S	**F.Dvořák**
Dvořák, R. (fl. 1930) M	**Dvořák**
Dvořáková, Marie (1940–) S	**Dvořáková**
Dwidjoseputro, Dakimah (fl. 1970) M	**Dwidjos.**
Dwigubskij, Iwan Alexievič (1771/2–1839) S	**Dwig.**
Dwivedi, J.N. (fl. 1959) AM	**J.N.Dwivedi**
Dwivedi, R.S. (fl. 1961) M	**R.S.Dwivedi**
Dwyer, John Duncan (1915–) S	**Dwyer**
Dyal, Sarah Creecie (1907–) S	**Dyal**
Dybowski, Jean (1855–1928) S	**J.Dyb.**
Dybowski, Wladyslaw Benedikt (1833–1930) S	**Dyb.**
(Dybowsky, Jean)	
see Dybowski, Jean	**J.Dyb.**
Dyce, James Wood (1905–) P	**Dyce**
(Dyck, Joseph Franz Maria Anton Hubert Ignatz Fürst zu Salm–Reifferscheid)	
see Salm–Reifferscheid–Dyck, Joseph Franz Maria Anton Hubert Ignatz	
Fürst zu	**Salm-Dyck**
Dyer, C.B. A	**C.B.Dyer**
Dyer, Robert Allen (1900–1987) S	**R.A.Dyer**
Dyer, William Turner Thiselton (Thistleton) (1843–1928) PS	**Dyer**
(Dyke, C.Gerald)	
see Van Dyke, C.Gerald	**Van Dyke**

Dykes, William Rickatson (1877–1925) S	**Dykes**
Dyko, Barbara J. (fl. 1977) M	**Dyko**
Dykstra, Michael J. (fl. 1975) M	**Dykstra**
Dykstra, Richard F. A	**R.F.Dykstra**
Dylis, Nikolaĭ Vladislavovich (1915–1985) S	**Dylis**
Dyr, J. (fl. 1939) M	**Dyr**
Dyring, Johan Peder Michael (1849–1930) S	**Dyring**
Dyurinski (fl. 1934) M	**Dyur.**
Dzagania, A. (fl. 1986) M	**Dzagania**
Dzekov, S. (fl. 1970) S	**Dzekov**
Dzerzhinskiĭ, V.A. (fl. 1973) M	**Dzerzh.**
Dzhafarov, S.A. (fl. 1959) M	**Dzhaf.**
Dzhalagoniya, K.T. (fl. 1962) M	**Dzhalag.**
Dzhangurasov, F.Ch. (1918–) S	**Dzhang.**
Dzhanuzokov, A. (fl. 1964) M	**Dzhanuz.**
Dzhuraeva, Z. (fl. 1974) M	**Dzhur.**
(e Silva, Júlia Dames)	
see Dames e Silva, Júlia	**Dames e Silva**
Eade, George William (1905–) S	**Eade**
Eakle, Thomas W. A	**Eakle**
Eames, Arthur Johnson (1881–1969) S	**A.J.Eames**
Eames, Edward Ashley (1872–) S	**E.A.Eames**
Eames, Edwin Hubert (1865–1948) PS	**Eames**
Eardley, Constance Margaret (1910–1978) S	**Eardley**
Earle, Franklin Sumner (1856–1929) MS	**Earle**
Earle, Sylvia Alice (1935–) A	**S.A.Earle**
Earle, W.Hubert (1906–1984) S	**W.H.Earle**
Early, M.P. (fl. 1973) M	**Early**
Easterly, Nathan William (1927–) S	**Easterly**
Eastman, C.R. S	**C.R.Eastman**
Eastman, Helen (1863–) P	**Eastman**
Eastwood, Alice (1859–1953) S	**Eastw.**
Eaton, Alvah Augustus (1865–1908) PS	**A.A.Eaton**
Eaton, Amos (1776–1842) BMPS	**Eaton**
Eaton, Daniel Cady (1834–1895) ABPS	**D.C.Eaton**
Eaton, Geoffrey Leonard A	**G.L.Eaton**
Eaton, Hezekiah Hulbert (1809–1832) S	**H.H.Eaton**
Eaton, James R. S	**J.R.Eaton**
Eaton, R.A. (fl. 1971) M	**R.A.Eaton**
Eaton, Richard Jefferson (1890–1976) P	**R.J.Eaton**
Ebben, Marion H. (fl. 1961) M	**Ebben**
Ebbesen, G. (fl. 1937) M	**Ebbesen**
Ebel, F. (fl. 1988) S	**F.Ebel**
Ebel, Paul Wilhelm Sosistheus Eugen (1815–1884) S	**Ebel**
Éber, Zoltán A	**Éber**
Eberhardt, Philippe Albert (1874–1942) B	**Eberh.**
Eberle, E. (fl. 1957) S	**Eberle**
Eberly, William Robert (1926–) A	**Eberly**

Ebermaier, Carl Heinrich (1802–1870) S — C.H.Eberm.
Ebermaier, Johann Erdwin Christopher (1769–1825) S — Eberm.
Ebinger, John Edwin (1933–) S — Ebinger
Eboh, Dan O. (fl. 1986) M — Eboh
Echandi, E. (fl. 1957) M — Echandi
Eck-Boorsboom, M.H.J.van S — Eck-Boorsb.
Eckblad, Finn–Egil (1923–) M — Eckblad
Eckel, Patricia (1950–) B — Eckel
Eckenwalder, James E. (1949–) S — Eckenw.
Eckfeldt, John Wiegand (1851–1933) M — Eckfeldt
Ecklon, Christian Friedrich (Frederik) (1795–1868) S — Eckl.
Économidou, Eva (fl. 1971) S — Écon.
Economou-Amilli, Athena (1947–) A — Econ.-Amilli
Eddelbütel, H. (fl. 1912) M — Eddelb.
Eddy, Alan (1937–) B — A.Eddy
Eddy, Caspar Wistar (1790–1828) S — Eddy
Eddy, E.D. (1893–) M — E.D.Eddy
Eddy, Samuel (1897–) A — S.Eddy
Edees, Eric Smoothby (fl. 1970) S — Edees
Edel, J. (fl. 1853) S — Edel
Edelberg, Lennart (1915–1981) S — Edelb.
Edelstein, Tivkah (1926–1977) A — Edelst.
Edgar, Elizabeth (1929–) S — Edgar
Edgecombe, Walter Brian (1945–) S — Edgecombe
Edgell, H.S. A — Edgell
Edgerton, Claude Wilbur (1880–1965) M — Edgerton
Edgeworth, Michael Pakenham (1812–1881) PS — Edgew.
Ediger, Robert Ike (1937–) S — Ediger
Ediger, V.S. (fl. 1989) M — V.S.Ediger
Edison, Joseph (1904–1987) S — Edison
Edlich, Fritz A — Edlich
Edlin, Herbert Leeson (1913–1977) S — Edlin
Edmonds, Jennifer M. (fl. 1971–86) S — Edmonds
Edmondson, Charles Howard (1914–) A — Edm.
Edmondson, John Richard (1948–) S — J.R.Edm.
Edmondston, Thomas (1825–1846) ABS — Edmondston
Edoh, D.O. (fl. 1968) M — Edoh
Edsbagge, Hans A — Edsbagge
Edson, Howard Austin (1875–1960) M — Edson
Edwall, Gustavo (Gustaf) (1862–1946) S — Edwall
Edward, J.C. (fl. 1958) M — Edward
Edwards, Alexander (1904–) S — A.Edwards
Edwards, Arthur Mead (1836–1914) A — A.M.Edwards
Edwards, David S. (fl. 1986) P — D.S.Edwards
Edwards, Eric Thomas (1905–) M — E.T.Edwards
Edwards, John (fl. 1763–1818) S — J.Edwards
Edwards, Lucy E. A — L.E.Edwards
Edwards, Peter A — P.Edwards

Edwards, Peter John (1947–) P	**P.J.Edwards**
Edwards, R.L. (fl. 1987) M	**R.L.Edwards**
Edwards, Sean Rowan (1943–) B	**S.R.Edwards**
Edwards, Steven W. (fl. 1983) S	**S.W.Edwards**
Edwards, Sydenham Teast (1769–1819) S	**Edwards**
Edwards, Trevor J. (fl. 1989) S	**T.J.Edwards**
Edwards, Wilfred Norman (1890–1956) AM	**W.N.Edwards**
Edwin, Gabriel (1926–) S	**Edwin**
Eeden, Frederik Willem van (1829–1901) S	**Eeden**
Eek, Th.van (fl. 1937) M	**Eek**
(Eel, Ludo Van)	
see Van Eel, Ludo	**Van Eel**
Een, Gillis (1924–) B	**Een**
Eerkens (fl. 1983) S	**Eerkens**
Eftimie, Elena (1929–) M	**Eftimie**
Egami, T. (fl. 1935) M	**Egami**
Egan, Robert Shaw (1945–) M	**Egan**
(Egea Fernández, José Maria)	
see Egea, José Maria	**Egea**
Egea, José Maria (1956–) M	**Egea**
Egede, Hans Poulsen (1686–1758) L	**Egede**
Egeland, I. (fl. 1969) M	**I.Egeland**
Egeland, John (1871–1928) M	**Egeland**
Egeling, Gustav (1858–1922) M	**Egeling**
Egemen S	**Egemen**
Egerod, Lois Eubank (1919–) A	**Egerod**
Egger, Keith N. (fl. 1982) M	**Egger**
Eggers, Henrik Franz Alexander von (1844–1903) S	**Eggers**
Eggers Ware, D.M. (fl. 1983) S	**Egg.Ware**
Eggert, D.A. (fl. 1967) F	**D.A.Eggert**
Eggert, Heinrich Karl Daniel (1841–1904) PS	**Eggert**
Eggerth, Karl (Carl) (1860–1888) M	**Eggerth**
Eggins, H.O.W. (fl. 1966) M	**Eggins**
Eggleston, Willard Webster (1863–1935) PS	**Eggl.**
Eggli, Urs (1959–) S	**Eggli**
Egler, Frank Edwin (1911–) S	**Egler**
Egler, Walter Alberto (1924–1961) S	**W.A.Egler**
Egli, Bernhard (fl. 1990) S	**Egli**
Egorova, Elena Markelovna (fl. 1964) S	**E.M.Egorova**
Egorova, Tatiana V. (1930–) S	**T.V.Egorova**
Egües, A. (fl. 1930) M	**Egües**
Egunyomi, Adeyemi (1944–) B	**Egunyomi**
Egusa, S. (fl. 1977) M	**Egusa**
Ehlers, R. (fl. 1982) S	**Ehlers**
Ehrenberg, Carl August (1801–1849) S	**C.Ehrenb.**
Ehrenberg, Christian Gottfried (1795–1876) AMS	**Ehrenb.**
Ehrendorfer, Friedrich (1927–) S	**Ehrend.**
Ehresmann, Douglas W. A	**Ehresm.**
Ehret, Georg Dionysius (1708–1770) S	**Ehret**

Ehrhart, Jakob Friedrich (1742–1795) BMPS	**Ehrh.**
Ehrhart, Johann Balthasar (1700–1756) S	**J.B.Ehrh.**
Ehrler, A. (fl. 1950) FP	**Ehrler**
Ehrlich, Aline A	**A.Ehrlich**
Ehrlich, John (1907–) M	**Ehrlich**
Eiben, C.E. (fl. 1870) C	**Eiben**
Eichelbaum, F. (fl. 1906) M	**Eichelb.**
Eichelberg, Johann Friedrich Andreas (1808–1887) S	**Eichelberg**
Eichhorn, Eugen (1878–1963) M	**Eichhorn**
Eichlam, Friedrich (Federico) (–1911) S	**Eichlam**
Eichler, August Wilhelm (1839–1887) S	**Eichler**
Eichler, Bogumił (1843–1905) A	**B.Eichler**
Eichler, Hansjörg (1916–) S	**H.Eichler**
Eichler, W. (fl. 1952) M	**W.Eichler**
Eichwald, E.V. (fl. 1843) M	**E.V.Eichw.**
Eichwald, Karl (–1976) S	**K.Eichw.**
Eichwald, Karl Eduard von (1794–1876) AFS	**Eichw.**
Eicker, Albert (1935–) M	**Eicker**
Eidam, Michael Emil Howard (1845–1901) M	**Eidam**
Eifert, Imre János (1934–) S	**Eifert**
Eifrig, H. (fl. 1937) B	**Eifrig**
Eig, Alexander (1894–1938) S	**Eig**
Eiger, J. (fl. 1956) P	**Eiger**
Eijk, G.W.van (1932–) M	**Eijk**
Einhellinger, A.E. (fl. 1973) M	**Einhell.**
Eiseman, Nathaniel J. A	**Eiseman**
Eisenach, Paul Heinrich Otto (1847–1917) S	**Eisenach**
Eisenack, Alfred (1891–1982) M	**Eisenack**
Eisenbeis, Gerhard A	**Eisenb.**
Eisengrein, Georg Adolf (1799–1857) S	**Eisengr.**
Eisfelder, Irmgard (fl. 1962) M	**Eisf.**
Eiten, George (1923–) S	**G.Eiten**
Eiten, Liene Teixeira (1925–1979) S	**L.T.Eiten**
Eitner, Eugen (1839–1921) M	**Eitner**
Ekart, Tobias Philipp (1799–1877) BC	**Ekart**
Ekberg, Lars (1940–) S	**Ekberg**
Ekim, T. (fl. 1980) S	**Ekim**
Ekimov, V.P. S	**Ekimov**
Eklund (fl. 1883) M	**Eklund**
Eklund, Ole Arthur (1899–1946) MS	**O.A.Eklund**
Ekman, Erik Leonard (1883–1931) PS	**Ekman**
Ekman, Hedda Maria Emerence Adelaïde Elisabeth (1862–1936) S	**E.Ekman**
Ekman, Stefan (1965–) M	**S.Ekman**
Ekstam, Otto Josef Agaton (1870–1931) BS	**Ekstam**
Ekstrand, Emil Viktor (1841–1884) B	**Ekstr.**
Ekstrand, H. (fl. 1937) M	**H.Ekstr.**
Ekutimischvili, M. S	**Ekutim.**
(Ekvtimischvili, M.)	
see Ekutimischvili, M.	**Ekutim.**

El Gazzar, Adel Ibrahim Hamed (1942–) S   **El Gazzar**
(El Hadidi, Mohammed Nabil)
  see Hadidi, Mohammed Nabil   **Hadidi**
El Karemy, Zeinab A.R. (fl. 1989) S   **El Karemy**
El Naggar, Salah M. (fl. 1989) S   **El Naggar**
El Shafie, A.E. (fl. 1980) M   **El Shafie**
El-Abyed, M.Samy (fl. 1964) M   **El-Abyed**
El-Ani, Arif S. (fl. 1966) M   **El-Ani**
El-Buni, A.M. (fl. 1976) M   **El-Buni**
(El-din, Ahmed Gamal)
  see Gamal-Eldin, Ahmed   **A.Gamal-Eldin**
El-Gholl, Nabih E. (fl. 1978) M   **El-Gholl**
El-Helaly, A.F. (fl. 1962) M   **El-Helaly**
El-Masry, H.G. (fl. 1966) M   **El-Masry**
El-Oqlah, Ahmad Ali (1947–) S   **El-Oqlah**
El-Sayed, Sayed Z. A   **El-Sayed**
Elarosi, H. (fl. 1962) M   **Elarosi**
Elberskirch, W. A   **Elbersk.**
Elborne, S.A. (fl. 1991) M   **Elborne**
Elbrächter, Malte A   **Elbr.**
Elders, M.C.C. (fl. 1986) M   **Elders**
Elenev, P. (fl. 1923) M   **Elenev**
(Elenevskii, Andrej G.)
  see Elenevsky, Andrej G.   **Elenevsky**
Elenevsky, Andrej G. (1928–) S   **Elenevsky**
Elenkin, Alexander (Aleksander) Alexandrovich (1873–1942) ABMS   **Elenkin**
Elffers, Joan (1928–) S   **Elffers**
Elfstrand, Marten (1859–1927) S   **Elfstr.**
Elfving, Frederik Emil Volmar (1854–1942) A   **Elfving**
Eliade, Eugenia (1930–) M   **Eliade**
Elias, H. (fl. 1907–1944) S   **Elias**
Elias, Maxim Konradovich (1873–1942) AF   **M.K.Elias**
Elias, Thomas Sam (1942–) S   **T.S.Elias**
Eliasson, Albin Gottfrid (1860–) M   **A.G.Eliasson**
Eliasson, Uno H. (1939–) MS   **Eliasson**
Eliçin, Gökhan (1935–) S   **Eliçin**
Elisarjeva, M.F. S   **Elis.**
(Eliscu, P.N.d')
  see D'Eliscu, P.N.   **D'Eliscu**
Elisei, F.G. (fl. 1938) M   **Elisei**
Elisens, Wayne J. (1948–) S   **Elisens**
Elix, John Alan (1941–) M   **Elix**
Elizondo, Jorge Elizondo (–1989) S   **Elizondo**
Elkan, Louis (Ludvig) (1815–1851) S   **Elkan**
Elladi, E.V. (1889–) S   **Elladi**
Ellett, C.Wayne (fl. 1974) M   **Ellett**
(Elliot, George Francis Scott)
  see Scott-Elliot, George Francis   **Scott-Elliot**
Elliot, J.E. (fl. 1916) M   **J.E.Elliot**

Elliot, Walter (1803–1887) S	**Elliot**
Elliott, Charlotte (1883–) M	**C.Elliott**
Elliott, Edward S. (fl. 1956) M	**E.S.Elliott**
Elliott, Graham Francis A	**G.F.Elliott**
(Elliott, Jessie Sproat)	
see Bayliss Elliott, Jessie Sproat	**Bayl.Ell.**
Elliott, John Asbury (1887–1923) M	**J.A.Elliott**
Elliott, Mary E. (fl. 1961) M	**M.E.Elliott**
Elliott, R.F. (fl. 1964) M	**R.F.Elliott**
Elliott, Stephen (1771–1830) AS	**Elliott**
Ellis, Charles Howard (1929–) A	**C.H.Ellis**
Ellis, David (1874–1937) AM	**D.Ellis**
Ellis, David H. (fl. 1978) M	**D.H.Ellis**
Ellis, Don Edwin (1908–) M	**D.E.Ellis**
Ellis, Edward A. (fl. 1956) M	**E.A.Ellis**
Ellis, Jamuel Leopold (1934–) S	**J.L.Ellis**
Ellis, Janet Pamela (fl. 1974) M	**J.P.Ellis**
Ellis, Job Bicknell (1829–1905) M	**Ellis**
Ellis, John (1710–1776) S	**J.Ellis**
Ellis, John J. (fl. 1962) M	**J.J.Ellis**
Ellis, John William (1857–1916) M	**J.W.Ellis**
Ellis, Len Thomas (1955–) B	**L.T.Ellis**
Ellis, M. (fl. 1940) M	**M.Ellis**
Ellis, Martin Beazor (1911–) M	**M.B.Ellis**
Ellis, R.P. (fl. 1984) S	**R.P.Ellis**
Ellis, Robert Gwynn (1942–) S	**R.G.Ellis**
Ellis, T.L. (fl. 1981) BM	**T.L.Ellis**
Ellis, W.Neale A	**W.N.Ellis**
Ellis–Adam, A.C. (1937–) A	**Ellis–Adam**
Ellison, B.R. (fl. 1945) M	**Ellison**
Ellison, William Lee (1923–) S	**W.L.Ellison**
Ellrodt, Theodor Christian (1767–1804) MS	**Ellr.**
Ellwanger, George (1816–1906) S	**Ellw.**
Ellwein, Hermann (fl. 1926) B	**Ellwein**
Elmer, Adolph Daniel Edward (1870–1942) PS	**Elmer**
Elovskij, V.A. A	**Elovskij**
Elphick, J.J. (fl. 1968) M	**Elphick**
Elsik, William Clinton (1935–) AM	**Elsik**
Elvander, Patrick E. (1950–) S	**Elvander**
Elven, Reidar (1947–) S	**Elven**
Elwert, Johann Caspar Philipp (1760–1827) S	**Elwert**
Elwes, Henry John (1846–1922) S	**Elwes**
Emberger, Jacques A	**J.Emb.**
Emberger, Marie Louis (1897–1969) S	**Emb.**
Embree, Robert W. (1932–) M	**Embree**
Emden, J.H.van (fl. 1968) M	**Emden**
Emeric, D. (fl. c.1828) S	**Emeric**
Emerson, J.K. (fl. 1984) P	**J.K.Emers.**
Emerson, Julia Titus (1877–1962) B	**Emers.**

Emerson, Ralph (1912-1979) M	**R.Emers.**
Emery, William H.P. (1924–) S	**Emery**
Emig, William Harrison (1888-1967) B	**Emig**
Émile-Weil, P. (fl. 1919) M	**Émile-Weil**
Emmerich, Margarete (1933–) S	**Emmerich**
Emmert, Friedrich (–1868) S	**Emmert**
Emmons, Chester Wilson (1900–) M	**C.W.Emmons**
Emmons, Ebenezer (1799-1863) B	**Emmons**
Emmott, Janet I. (fl. 1964) P	**Emmott**
Emory, William Hemsley (1811-1887) S	**Emory**
Emoto, Yoshikadzu (1872–) AM	**Emoto**
Emura, Kazuko M.K. (fl. 1972) S	**Emura**
(Emygdio, L.)	
see Emygdio de Mello Filho, Luiz	**Emygdio**
Emygdio de Mello Filho, Luiz (1913–) S	**Emygdio**
Emys, J.D. (1837-1912) PS	**Emys**
Enander, Sven Johan (1847-1928) S	**Enander**
Encarnación, Filomeno (fl. 1984) S	**Encarn.**
Encke, Fritz (1861-1931) S	**Encke**
Ende, Willem Pieter van den (fl. 1823) P	**Ende**
Ender, Ernst Eduard (1837-1893) S	**Ender**
Enderle, Manfred (1947–) M	**Enderle**
Enderlein, Günther A	**Enderlein**
Endert, Frederik Hendrik (1891-1953) S	**Endert**
Endlicher, Stephan Friedrich Ladislaus (1804-1849) ABFMPS	**Endl.**
Endo, Akira (fl. 1987) M	**A.Endo**
Endo, Riuji (1891-1969) A	**R.Endo**
Endō, S. (fl. 1927) M	**S.Endō**
Endo, Y. (fl. 1986) S	**Y.Endo**
Endre (fl. 1981-89) S	**Endre**
Endres, A.R. (–1877) S	**Endres**
Endress, Peter Karl (1942–) S	**P.K.Endress**
Endress, Philipp Anton Christoph (1806-1831) S	**Endress**
Endroth, Johannes (1956–) B	**Endroth**
Endtmann, Jürgen Klaus (1938–) S	**Endtm.**
Ener, Stephen A. (fl. 198?) M	**Ener**
Engel, Franz (fl. 1865) S	**Engel**
Engel, Heinz (fl. 1977) M	**H.Engel**
Engel, John Jay (1941–) B	**J.J.Engel**
Engel, K. (fl. 1976) S	**K.Engel**
Engel, L. (fl. 1877) M	**L.Engel**
Engelhard, A.W. (fl. 1986) M	**Engelhard**
Engelhardt, Hermann (1839-1918) AF	**Engelh.**
Engelhardt-Zasada, C. (fl. 1974) M	**Eng.-Zas.**
Engelhorn, Tamra (later Raven, T.E.) (1945–) S	**Engelhorn**
Engelke, J. (fl. 1912) M	**Engelke**
Engelmann, Georg (George) (1809-1884) MPS	**Engelm.**
Engelmark, Fred Thor-Bjorn (1947–) B	**Engelmark**
Engesser, Carl (fl. 1852) S	**Engesser**

Engin, A. (fl. 1983) S	**Engin**
England, J.L. (fl. 1982) S	**England**
Engler, Heinrich Gustav Adolf (1844–1930) ABMPS	**Engl.**
Engler, Viktor (1885–1917) S	**V.Engl.**
Englerth, George Henry (1907–) M	**Englerth**
English, Carl Schurz (1904–) S	**English**
English, Edith (fl. 1948) S	**E.English**
English, Mary P. (fl. 1954) M	**M.P.English**
English, William Harley (1911–) M	**W.H.English**
Englmaier, P. (fl. 1984) S	**Englmaier**
Englund, Bengt (1908–) S	**Englund**
Engstrand, Lennart (1942–) S	**Engstrand**
Enjoji, Sazao (fl. 1931) M	**Enjoji**
Enjumet, Monique A	**Enjumet**
Enken, V.B. (1900–) S	**Enken**
Enkina, T.V. (fl. 1969) M	**Enkina**
Enlows, Ella Morgan (Austin) (1884–) M	**Enlows**
Ensermu Kelbessa (1952–) S	**Ensermu**
Ensign, Margaret Ruth (later Lewis, M.R.E.) (1919–) S	**Ensign**
Enslen, Aloysius (–1818) S	**Enslen**
Entwisle, Timothy John (1960–) A	**Entwisle**
Entz, Géza (1875–1943) AM	**Entz**
Enwald, Kurt H. A	**Enwald**
Epis, R.C. A	**Epis**
Epling, Carl Clawson (1894–1968) MS	**Epling**
Epple, Paul (fl. 1951) S	**Epple**
(Eppling, Carl Clawson)	
see Epling, Carl Clawson	**Epling**
Erady, N.A. (fl. 1953) AM	**Erady**
Erb, R. (fl. 1971) M	**Erb**
Erbe, Lawrence Wayne (1924–) S	**Erbe**
Erben, M. S	**Erben**
Ercegović, Ante (1895–) A	**Erceg.**
Ercsi, J. (1792–1868) S	**Ercsi**
Erdman, Kimball Stewart (1937–) S	**Erdman**
Erdmann, Carl Gottfried (1774–1835) S	**Erdmann**
Erdner, Eugen (1869–1927) S	**Erdner**
Erdos, Gregory W. (fl. 1971) M	**Erdos**
Erdtman, Otto Gunnar Elias (1897–1973) BS	**Erdtman**
(Erdtmann, Otto Gunnar Elias)	
see Erdtman, Otto Gunnar Elias	**Erdtman**
Eremejewa, A.M. (fl. 1939) M	**Erem.**
Eremin, G.V. (fl. 1979) S	**Eremin**
Erfurth, Ch.B. (–1893) S	**Erfurth**
Ergolskaia, Z.V. A	**Ergolsk.**
Erhardt, W. (fl. 1988) S	**Erhardt**
Erichsen, Christian Friedo Eckhard (1867–1945) M	**Erichsen**
Erickson, Frederica ('Rica') Lucy (1908–) S	**F.L.Erickson**
Erickson, Ralph Orlando (1914–) S	**R.O.Erickson**

Ericsson, Stefan (1954–) S   **Ericsson**
Erik, Sadik (1948–) S   **Erik**
Eriksen, Bente (1960–) S   **B.Eriksen**
Eriksson, Birgitta (fl. 1970) M   **B.Erikss.**
Eriksson, Jakob (1848–1931) M   **Erikss.**
Eriksson, John (1921–) M   **J.Erikss.**
Eriksson, K.-E. (fl. 1985) M   **K.-E.Erikss.**
Eriksson, Ove E. (fl. 1987) M   **O.E.Erikss.**
Eriksson, Roger (1958–) S   **R.Erikss.**
Eristavi, M.I. (fl. 1977) S   **Eristavi**
Erkamo, Viljo U.K. (1912–) S   **Erkamo**
Erkmen, Ugur   A   **Erkmen**
Erlandsson, Stellan (1902–) A   **Erl.**
Erlanson, Eileen Jessie (1899–) S   **Erlanson**
Erlee, M.P.M. (fl. 1957) S   **Erlee**
Erman, Georg Adolf (1806–1877) S   **Erman**
Ermolaeva, L.M.   A   **Ermol.**
(Ermolajeva, L.M.)
   see Ermolaeva, L.M.   **Ermol.**
Ern, Hartmut (1935–) S   **Ern**
Ernst, Adolf (Adolfo) (1832–1899) S   **Ernst**
Ernst, Alfons (Alfred) (1875–1968) AS   **A.Ernst**
Ernst, Julius   A   **J.Ernst**
Ernst, Wallace Roy (1928–1971) S   **W.R.Ernst**
Ernst, Wilf   S   **W.Ernst**
Ernsting, Arthur Conrad (1709–1768) S   **Ernsting**
Errera, Leo (–Abram) (1858–1905) S   **Errera**
Ershad, Djafar (1936–) M   **Ershad**
Erskine, David S. (fl. 1960) S   **Erskine**
Ertter, Barbara Jean (1953–) S   **Ertter**
(Erve, A.W.Van)
   see Van Erve, A.W.   **Van Erve**
Ervin, Marion D. (fl. 1956) M   **Ervin**
Erwin, Donald C. (fl. 1958) M   **Erwin**
Escalante, Manuel G.   S   **Escal.**
(Eschbaugh, William Hardy)
   see Eshbaugh, William Hardy   **Eshbaugh**
Eschenbach, Johann Friedrich (1757–) S   **Eschenb.**
Eschscholtz, Johann Friedrich Gustav von (1793–1831) S   **Eschsch.**
Eschweiler, Franz Gerhard (Franciscus Gerardus) (1796–1831) MPS   **Eschw.**
(Escobar A., M.)
   see Escobar, M.   **M.Escobar**
Escobar, Dennis E. (fl. 1976) M   **D.E.Escobar**
Escobar, Gustavo Adolfo (fl. 1976) M   **G.A.Escobar**
Escobar, Linda K. (fl. 1989) S   **L.K.Escobar**
Escobar, M.   B   **M.Escobar**
(Escobar R., Rodrigo)
   see Escobar, Rodrigo   **R.Escobar**
Escobar, Rodrigo (fl. 1978) S   **R.Escobar**

Escomel, Edmundo (fl. 1924) AM — **Escomel**
Escriche Esteban, Manuel (1913–) S — **Escriche**
Escudeiro, Alexandra C.S. (1949–) S — **Escud.**
(Esenbeck, Christian Gottfried Daniel Nees)
   see Nees von Esenbeck, Christian Gottfried Daniel — **Nees**
(Esenbeck, Theodor Friedrich Ludwig Nees von)
   see Nees von Esenbeck, Theodor Friedrich Ludwig — **T.Nees**
Esenova, Kh.E. (fl. 1976) S — **Esenova**
Esfandiari, Esfandiar (von) (1909–) MS — **Esfand.**
Eshbaugh, William Hardy (1936–) S — **Eshbaugh**
Eshonkolov, N. (fl. 1974) M — **Eshonk.**
Eskuche, Ulrich G. (1926–) S — **Eskuche**
Eslyn, Wallace E. (fl. 1961) M — **Eslyn**
Esmarch, F. (fl. 1927) M — **F.Esm.**
Esmarch, Heinrich Peter Christian (1745–1830) S — **Esm.**
Esnault, Joël (fl. 1987) M — **Esnault**
Espadaler, Xavier (fl. 1987) M — **Espad.**
Espejo Serna, Adolfo (fl. 1987) S — **Espejo**
Esper, Eugen Johann Christoph (1742–1810) S — **Esper**
Espinal, Luis Sigifredo (1929–) S — **Espinal**
(Espinal–T., Luis Sigifredo)
   see Espinal, Luis Sigifredo — **Espinal**
Espinar, M.C. (fl. 1988) S — **Espinar**
Espinosa Bustos, Marcial Ramón (1874–1959) MPS — **Espinosa**
Espinosa, Francisco Javier (fl. 1984) S — **F.J.Espinosa**
(Espinosa Garduño, Judith)
   see Espinosa, Judith — **J.Espinosa**
Espinosa, Judith (1935–) S — **J.Espinosa**
(Espinosa–Garcia, F.J.)
   see Espinosa, F.J. — **F.J.Espinosa**
Espuelas, I. (fl. 1987) S — **Espuelas**
Esquivel, M.A. (fl. 1985) S — **Esquivel**
(Ess El–Din, E.K.)
   see Ezz–Eldin, E.K. — **Ezz–Eldin**
Essary, Samuel Henry (1870–1935) M — **Essary**
Essed, E. (fl. 1911) M — **Essed**
Essenova, Ch.E. (1936–) S — **Essenova**
Essette, Henri (fl. 1965) M — **Essette**
Essig, Frederick Burt (1947–) S — **Essig**
Esslinger, Theodore Lee (1944–) M — **Essl.**
Estadès, Alain (fl. 1988) M — **Estadès**
Estee, L.M. (fl. 1913) M — **Estee**
Estep, Kenneth W. A — **Estep**
Esterhuysen, Elsie Elizabeth (1912–) S — **Esterh.**
Estes, Frederick Earle (1902–) S — **Estes**
Estes, James R. (1937–) S — **J.R.Estes**
(Esteve Chueca, Fernando)
   see Esteve, Fernando — **Esteve**
Esteve, Fernando (1919–1988?) S — **Esteve**
Esteve–Raventós, Fernando (fl. 1985) M — **Estéve–Rav.**

Esteves Pereira, Eddie (fl. 1979) S	**Esteves**
Estey, R.H. (fl. 1977) M	**Estey**
(Étangs, Nicolas Stanislas Chaâles des)	
see Des Étangs, Nicolas Stanislas Chaâles	**Des Étangs**
Etayo, J. (fl. 1991) M	**Etayo**
Etchells, J.E. (fl. 1950) M	**Etchells**
Etheridge, D.E. (fl. 1961) M	**D.E.Ether.**
Etheridge, Robert (1847–1920) AF	**Ether.**
Étienne, Georges (–1910) B	**Etienne**
Etlinger, Andreas Ernst (fl. 1777) S	**Etl.**
Etting S	**Etting**
Ettingshausen, Constantin (Konstantin) von (1826–1897) ABFMPS	**Ettingsh.**
Ettl, Hanuš A	**H.Ettl**
Ettl, O. A	**O.Ettl**
(Ettlinger, D.M. Turner)	
see Turner Ettlinger, D.M.	**Turner Ettl.**
Ettlinger, Leopold (fl. 1943) M	**Ettl.**
Eubank, Lois L. A	**Eubank**
Eudes–Deslongchamps, M. (fl. 1824) M	**Eudes–Desl.**
Eugène, R.P. (fl. 1868) S	**Eugène**
Eulenstein, Theodor (–1875) A	**Eulenst.**
see Philippi, Federico	**F.Phil.**
Euphrasén, Bengt Anders (1756–1796) S	**Euphrasén**
Eurola, Seppo (1930–) A	**Eurola**
Evans, A.Murray (1932–) P	**A.M.Evans**
Evans, Alexander William (1868–1959) BM	**A.Evans**
Evans, Arthur Humble (1855–1943) S	**A.H.Evans**
Evans, Dan K. (1938–) S	**D.K.Evans**
Evans, E.H. (fl. 1971) M	**E.H.Evans**
Evans, Harry Charles (fl. 1971) M	**H.C.Evans**
Evans, J.H. A	**J.H.Evans**
Evans, J.Whitman S	**J.W.Evans**
Evans, L.V. A	**L.V.Evans**
Evans, Maurice Smethurst (1854–1920) S	**M.S.Evans**
Evans, Obed David (1889–1975) PS	**O.D.Evans**
Evans, Walter Harrison (1863–1941) S	**W.H.Evans**
Evans, William Edgar (1882–1963) S	**W.E.Evans**
Evelegh, M. A	**Evelegh**
Eveleigh, D.E. (fl. 1961) M	**Eveleigh**
Eveleigh, E.S. (fl. 1990) M	**E.S.Eveleigh**
Evenari, Michael (fl. 1940) S	**Evenari**
Evensen, Dale L. A	**Evensen**
Evenson, V.S. (1933–) M	**V.S.Evenson**
Evenson, Virginia (fl. 1967) S	**Evenson**
Everett, Joy (1953–) S	**J.Everett**
Everett, Thomas Henry (1903–) S	**Everett**
Everhart, Benjamin Matlack (1818–1904) M	**Everh.**
Everist, Selwyn Lawrence (1913–1981) S	**Everist**
Everken, V. (–1881?) B	**Everk.**
Everly, Mary Louise (1922–) S	**Everly**

Evers, Georg (1837–1916) S   **Evers**
Eversmann, Eduard Friedrich von (1794–1860) MS   **Eversm.**
Evert, Erwin F. (1940–) S   **Evert**
Eves, Donald Smith (1911–) S   **Eves**
Evitt, William Robert (1923–) A   **Evitt**
Evlachova, A.A. (fl. 1939) M   **Evlachova**
(Evlakhova, A.A.)
   see Evlachova, A.A.   **Evlachova**
Evolceanu, Radu (fl. 1959) M   **Evolc.**
Evrard, Charles Marie (1926–) S   **C.M.Evrard**
Evrard, François (1885–1957) S   **Evrard**
Ewald, E. (fl. 1983) S   **Ewald**
Ewan, Joseph Andorfer (1909–) PS   **Ewan**
Ewart, Alfred James (1872–1937) S   **Ewart**
Ewen, A.B. (fl. 1985) M   **Ewen**
Exell, Arthur Wallis (1901–) S   **Exell**
Exell, Mildred A. S   **M.A.Exell**
Exner, Beatrice (fl. 1953) M   **Exner**
Eyde, Richard H. (1928–1990) F   **Eyde**
Eyer, Jerome A. (1934–) A   **Eyer**
Eyles, Frederick (1864–1937) S   **Eyles**
Eyma, Pierre Joseph (1903–1945) S   **Eyma**
Eyndhoven, G.L.van (fl. 1942) M   **Eyndh.**
Eyre, William L.W. (fl. 1903) M   **Eyre**
Eysenhardt, Carl Wilhelm (1794–1825) AS   **Eysenh.**
Eyster, William Henry (1889–) S   **Eyster**
(Ez–El–din, E.K.)
   see Ezz–Eldin, E.K.   **Ezz–Eldin**
Ezcurdia, Luis de (fl. 1899) P   **Ezcurdia**
Ezcurra, Cecilia (1954–) S   **C.Ezcurra**
Ezcurra, E. (fl. 1986) S   **E.Ezcurra**
Ezekiel, Walter Naphtali (1901–) M   **Ezekiel**
Ezuka, Akinori (fl. 1959) M   **Ezuka**
Ezz–Eldin, E.K. (fl. 1989) M   **Ezz–Eldin**

Faber, Friedrich Carl von (1880–1954) AM   **Faber**
Fabian, F.W. (fl. 1933) M   **Fabian**
Fabijan, D.M. (fl. 1987) S   **Fabijan**
Fabre, Esprit (fl. c.1850) S   **E.Fabre**
Fabre, G. (fl. 1987) S   **G.Fabre**
Fabre, Jean Henri Casimir (1823–1915) M   **Fabre**
Fabre–Domergue A   **Fabre–Dom.**
Fabricatore, Jolande A. (fl. 1951) M   **Fabric.**
Fabricius, Johan Christian (1745–1808) S   **J.Fabr.**
Fabricius, Philipp Conrad (1714–1774) S   **Fabr.**
Fabris, Humberto Antonio (1924–1976) S   **Fabris**
Fábry, H. (fl. 1971) M   **Fábry**
Facchini, Francesco (1788–1852) S   **Facchini**
Faden, Robert B. (1942–) PS   **Faden**

Faegri, Knut (1909-) S   **Faegri**
Fage, Louis (1883–1964) A   **Fage**
Fagerland, E. A   **Fagerland**
Fagerlind, Folke (1907-) S   **Fagerl.**
Fagersten, Reino (1931-) B   **Fagersten**
Fagerström, Lars Fajalar (1914-) S   **Fagerstr.**
Fagon, Gui–Crescent (1638–1718) L   **Fagon**
Fahlberg, Samuel (1758–1834) S   **Fahlberg**
(Fahmy, Ahmed Gamal–Eldin Mohammed)
   see Gamal–Eldin, Ahmed   **A.Gamal-Eldin**
Fahmy, Tewfik (fl. 1927) M   **Fahmy**
Fahn, Abraham (1916-) S   **Fahn**
Fahrendorff, Ernst (fl. 1941) M   **Fahrend.**
Fairall, A.R. S   **Fairall**
Fairchild, D. S   **D.Fairchild**
Fairchild, Thomas R. A   **Fairchild**
Fairman, Charles Edward (1856–1934) M   **Fairm.**
Faizieva, F.Kh. (fl. 1986) M   **Faizieva**
Faizova, S.S. (fl. 1983) M   **Faizova**
Fakhr, M.S. S   **Fakhr**
Fakhrieva, G.D. (fl. 1975) S   **Fakhrieva**
Fakirova, Violeta (fl. 1973) M   **Fakirova**
Fakutsu, R. (fl. 1962) M   **Fakutsu**
Falanruw, Marjorie V.C. (1943-) S   **Falanruw**
(Falcão de Morais, J.O.)
   see Morais, J.O.Falcão de   **Morais**
Falcão, Garnier S. (fl. 1967) M   **G.S.Falcão**
(Falcão Ichaso, Carmen Lúcia)
   see Ichaso, Carmen Lúcia Falcão   **Ichaso**
Falcão, Joaquim Ignácio de Almeida (fl. 1951) S   **J.I.Falcão**
Falck, Olga (fl. 1947) M   **O.Falck**
Falck, Richard (1868–1955) M   **Falck**
Falconer, Hugh (1808–1865) S   **Falc.**
Falconer, William (1774–1824) S   **W.Falc.**
Falk, H.G. (fl. 1874) M   **H.G.Falk**
Falk, Johan Peter (Pehr) (1733–1774) S   **Falk**
Falkenberg, Paul (fl. 1900) A   **Falkenb.**
Fall, Joan (fl. 1951) M   **Fall**
Fallahyan, F. (fl. 1973) M   **Fallahyan**
Fallen, Mary E. (fl. 1983) S   **Fallen**
Falqui, Giuseppe S   **Falqui**
Fambach, D. (fl. 1925) M   **Fambach**
Familler, Ignaz (1863–1923) B   **Fam.**
Famin, M. A   **Famin**
Fan, Guang Jin (fl. 1970) S   **G.J.Fan**
Fan, Guo Sheng (fl. 1988) S   **G.S.Fan**
Fan, Kung Chu (1928-) A   **K.C.Fan**
Fan, Li (fl. 1988) M   **L.Fan**
Fan, Mei Zhen (fl. 1990) M   **M.Z.Fan**

Fan, Shan Gren (fl. 1981) M	S.G.Fan
Fan, Wen Pei (fl. 1966) S	W.P.Fan
Fan, Y.P. A	Y.P.Fan
Fang, C.T. (fl. 1948) M	C.T.Fang
Fang, Cheng Fu (fl. 1979) S	C.F.Fang
Fang, Ding (1920–) S	D.Fang
Fang, Min Man (fl. 1987) S	M.M.Fang
Fang, Ming Yuan (fl. 1984) S	M.Y.Fang
Fang, Q.X. (fl. 1983) M	Q.X.Fang
Fang, Rhui Cheng (fl. 1970s) S	R.C.Fang
(Fang, Rui Zheng)	
see Fang, Rhui Cheng	R.C.Fang
Fang, Si Tzen A	S.T.Fang
Fang, Sin Fang (fl. 1966) M	S.F.Fang
Fang, Wen Pei (1899–1983) S	W.P.Fang
Fang, Wen Zhe (fl. 1984) S	W.Z.Fang
Fang, Xin Ping (fl. 1987) S	X.P.Fang
Fang, Yun Yi (1920–) S	Y.Y.Fang
Fanlo, Rosario (1947–) S	Fanlo
Fantham, Harold Benjamin (1876–1937) AM	Fantham
Fantini, A.A. (fl. 1961) M	Fantini
Fantz, Paul R. (1941–) S	Fantz
Farden, Carl Alexander (fl. 1935) S	C.A.Farden
Farden, Richard S. (–1950) S	R.S.Farden
Farenholtz, Hermann (1884–) S	Farenholtz
Farghaly, M.S. A	Farghaly
(Faria, J.Gomes de)	
see Gomes de Faria, J.	Gomes Faria
Faridi, Mohammad Aslam Farooq (1927–1985) A	Faridi
Farille, Michel A. (1945–) S	Farille
Farinacci, A. A	Farin.
Faris, James Abraham (1890–1933) M	Faris
Farjon, A.K. (1946–) S	Farjon
Farkas, Edit (1959–) M	Farkas
(Farkaš–Vukotinovic, Ljudevit (Ludwig von))	
see Vukotinović, Ljudevit Farkaš	Vuk.
Farlow, William Gilson (1844–1919) ACM	Farl.
Farmar, Leo (1875–1907) S	Farmar
Farmelo, M.S. (fl. 1984) M	Farmelo
Farmer, J. (fl. 1985) S	J.Farmer
Farmer, Mark A. A	Farmer
Farneti, Rodolfo (1859–1919) BM	Farneti
Farooq, Shahid (1953–) A	Farooq
Farooqui, Phool Begum A	Farooqui
Farquhar, John Keith Marshall Lang (1858–1921) S	Farquhar
Farr, David Frederick (1941–) M	D.F.Farr
Farr, Edith May (1864–1956) S	Farr
Farr, Marie L. (fl. 1962) M	M.L.Farr
Farr, R.C. (fl. 1985) M	R.C.Farr

Farrant, C.A. (fl. 1986) M	**C.A.Farrant**
Farrant, P.A. (fl. 1991) P	**P.A.Farrant**
Farrar, Donald R. (1941–) P	**Farrar**
Farrell, L. (fl. 1930) M	**Farrell**
Farrell, P.G. (fl. 1979) S	**P.G.Farrell**
Farrer, Reginald John (1880–1920) S	**Farrer**
Farrington, Edward Irving (1876–1958) S	**Farrington**
Farron, Claude (1935–) S	**Farron**
Farrow, W.M. (fl. 1955) M	**Farrow**
Faruqi, Shamin Ahmad (1933–) S	**Faruqi**
Farwell, Oliver Atkins (1867–1944) PS	**Farw.**
Farwig, S. (fl. 1987) S	**Farwig**
Fasano, Angelo (fl. 1787) S	**Fasano**
Fasola, A. A	**Fasola**
Fasolo-Bonfante, Paolo (fl. 1971) M	**Fas.-Bonf.**
Fassatiová, Olga (fl. 1953) M	**Fassat.**
Fassett, Norman Carter (1900–1954) S	**Fassett**
Fathima Khaleel, Tasneem (fl. 1972) S	**Fathima**
Fatichenti, F. (fl. 1969) M	**Fatich.**
Fattohy, Zohair Ibraheem A	**Fattohy**
Fauché, M. (fl. 1832) S	**Fauché**
Fauconnet, Charles Isaac (1811–1876) S	**Fauc.**
Faull, Joseph Horace (1876–1961) MPS	**Faull**
Faure, Alfred S	**A.Faure**
Faure, Alphonse (fl. 1923) S	**Faure**
Fauré-Fremiet, Emmanuel (1883–1971) A	**Fauré-Frem.**
Faure-Reaynaud, M. (fl. 1979) M	**Faure-Reayn.**
Faurel, Louis (1907–1973) MS	**Faurel**
Faurie, Urbain Jean (1847–1915) MS	**Faurie**
Faus, J. (fl. 1984) M	**Faus**
Faust, Ernst Carroll (fl. 1937) AM	**E.C.Faust**
Faust, Frances K. (fl. 1937) M	**F.K.Faust**
Fautrey, François (fl. 1892) M	**Fautrey**
Favarger, Claude P.E. (1913–) S	**Favarger**
Favrat, Louis (1827–1893) S	**Favrat**
Favre, A. (fl. 1983) M	**A.Favre**
Favre, Jules (fl. 1937) M	**J.Favre**
Favre, Louis (1822–1904) S	**Favre**
Fawcett, Howard Samuel (1877–1948) M	**H.S.Fawc.**
Fawcett, S.G.M. (fl. 1940) M	**S.G.M.Fawc.**
Fawcett, William (1851–1926) S	**Fawč.**
Faxon, Charles Edward (1846–1918) P	**Faxon**
Fay, Alice D.A. (fl. 1973) P	**A.Fay**
Fay, Dolores J. (fl. 1947) M	**Fay**
Fay, John J. (fl. 1970) S	**J.J.Fay**
Fay, John Michael (fl. 1986) S	**J.M.Fay**
Faye, Léon (–1855) S	**Faye**
Fayed, A.-A. (fl. 1979) S	**Fayed**
Fayet (fl. 1926) M	**Fayet**

Fayod, Victor (1860–1900) M	**Fayod**
Fazalnoor, K. (fl. 1977) M	**Fazalnoor**
Fearn, Brian (1937–) S	**B.Fearn**
Fearn, P. (fl. 1963) S	**P.Fearn**
Featherly, Henry Ira (1893–) S	**Feath.**
Featherman, Americus (1822–c.1880) BPS	**Featherm.**
Fechner, G.A. (fl. 1849) S	**Fechner**
Fechser, H. (fl. 1960) S	**Fechser**
Fedak, G. (fl. 1985) S	**Fedak**
(Fedchenko, Aleksei Pavlovich (Alexei Pawlowitsch))	
see Fedtschenko, Aleksei Pavlovich (Alexei Pawlowitsch)	**A.Fedtsch.**
(Fedchenko, Boris Alexsevič (Alexeevich))	
see Fedtschenko, Boris Alexjewitsch (Alexeevich)	**B.Fedtsch.**
(Fedchenko, Olga Aleksandrovna)	
see Fedtschenko, Olga Alexandrowna	**O.Fedtsch.**
Fedde, Friedrich Karl Georg (1873–1942) S	**Fedde**
Feddema, Charles (1920–) S	**Feddema**
Federer, Z. (fl. 1931) S	**Federer**
Federici, E. (fl. 1928) M	**Federici**
Federle, E. A	**Federle**
Fedoronczuk, Nikolaj M. (1948–) S	**Fedor.**
Fedorov, Alexander Alexandrovich (1906–1982) S	**Al.Fed.**
Fedorov, Andrej Aleksandrovich (1909–1987) S	**Fed.**
Fedorov, V.G. A	**V.G.Fed.**
Fedossejew, M. S	**M.Fedoss.**
Fedossejew, S. S	**S.Fedoss.**
Fedtschenko, Aleksei Pavlovich (Alexei Pawlowitsch) (1844–1873) S	**A.Fedtsch.**
Fedtschenko, Boris Alexjewitsch (Alexeevich) (1872–1947) BPS	**B.Fedtsch.**
Fedtschenko, Olga Alexandrowna (1845–1921) S	**O.Fedtsch.**
Feduchy, E. (fl. 1943) M	**Feduchy**
Fée, Antoine Laurent Apollinaire (1789–1874) MPS	**Fée**
Feer, Heinrich (1857–1892) S	**Feer**
Feest, A. (fl. 1982) M	**Feest**
Fefilova, Likiia Aleksandrovna (fl. 1978) B	**Fefil.**
Fehlner, Karl (1859–1884) B	**Fehln.**
Fei, Yong (fl. 1991) M	**Y.Fei**
Feichtinger, Sandor (Alexander) (1817–1907) S	**Feicht.**
Feige, Guido Benno (1937–) M	**Feige**
Feijoo, E.J. (fl. 1930) M	**Feijoo**
Feika, Abdul M.B. (fl. 1978) S	**Feika**
Feilberg, J. (fl. 1984) P	**Feilberg**
Feinbrun, Naomi (1900–) S	**Feinbrun**
(Feinbrun–Dothan, Naomi)	
see Feinbrun, Naomi	**Feinbrun**
Feist-Castel, Monique A	**Feist-Castel**
Feistmantel, Ottokar (fl. 1881–86) F	**Feistm.**
Fekete, L. (1837–1916) S	**Fekete**
Feldman, M. (fl. 1978) S	**Feldman**
Feldman-Muhsam, B. (fl. 1963) M	**Feldm.-Muhs.**

(Feldmann, Geneviève)	
see Feldmann–Mazoyer, Geneviève	**Feldm.-Maz.**
Feldmann, Jean (1905–1978) AM	**Feldmann**
Feldmann–Mazoyer, Geneviève (fl. 1955) AM	**Feldm.-Maz.**
Felger, Richard S. (fl. 1968) S	**Felger**
Felippone, Florentino (1849–1939) B	**Felipp.**
Félix, Armand (fl. 1912) S	**A.Félix**
Felix, Charles J. A	**C.J.Felix**
(Félix, Henri Jacques)	
see Jacques–Félix, Henri	**Jacq.-Fél.**
Félix, Johannes Paul (1858–1941) FMS	**Félix**
Fell, Jack W. (fl. 1970) M	**Fell**
Fellenberg, Emanuel S	**Fellenb.**
Fellmann, Jacob (1795–1875) S	**J.Fellm.**
Fellmann, Nils Isak (1841–1919) S	**N.I.Fellm.**
Fellner, Maximilian Johann Nepomuk (fl. 1775) M	**Fellner**
Fellner, R. (fl. 1986) M	**R.Fellner**
Feltgen, Johann (1833–1904) M	**Feltgen**
Fenaroli, Luigi (1899–1980) PS	**Fen.**
Fenchel, Tom A	**Fenchel**
Fendler, August (1813–1883) P	**Fendler**
Feng, Chang Xue (fl. 1987) S	**C.X.Feng**
Feng, Gui Hua (fl. 1989) S	**G.H.Feng**
Feng, Kuo Mei (1917–) S	**K.M.Feng**
Feng, Xue Zhou (fl. 1988) S	**X.Z.Feng**
Feng, Yung Hua B	**Y.H.Feng**
Fenley, Kittie Lucille (later Parker, K.L.) (1910–) S	**Fenley**
Fennell, Dorothy Irene (1916–) M	**Fennell**
Fennell, James H. S	**J.H.Fennell**
Fenner, Ellen Aline (1889–) M	**Fenner**
Fenner, Juliane A	**J.Fenner**
Fenninger, Alois A	**Fenninger**
Fensholt, Dorothy Eunice (1911–) A	**Fensholt**
Fensome, Robert A. A	**Fensome**
Fenton, Carroll Lane (1900–) A	**C.L.Fenton**
Fenton, J.P.G. A	**J.P.G.Fenton**
Fenton, Mildred Adams (1899–) A	**M.A.Fenton**
(Fenzi, Emanuele Orazio)	
see Franceschi, Francesco	**Franceschi**
Fenzl, Eduard (1808–1879) S	**Fenzl**
Feo, Mildred (fl. 1967) M	**Feo**
(Feodorov, Alexander Alexandrovich)	
see Fedorov, Alexander Alexandrovich	**Al.Fed.**
(Feodorov, Andrej Aleksandrovich)	
see Fedorov, Andrej Aleksandrovich	**Fed.**
Feráková, Viera (1938–) S	**Feráková**
Ferdinandsen, Carl Christian Frederic(k) (1879–1944) MS	**Ferd.**
Fergus, C.L. (fl. 1964) M	**Fergus**
Ferguson, Alexander McGowan (1874–1955) S	**A.M.Ferguson**

Ferguson, Allan Ross (1943–) S	**A.R.Ferguson**
Ferguson, David J. (fl. 1989) S	**D.J.Ferguson**
Ferguson, David Kay (1942–) S	**D.K.Ferguson**
Ferguson, Ian Keith (1938–) S	**I.K.Ferguson**
Ferguson, Margaret Clay (1863–1951) S	**M.C.Ferguson**
Ferguson, William (1820–1887) PS	**Ferguson**
Fergusson, John (1834–1907) B	**Fergusson**
Ferlan, Leo (1928–1961) S	**Ferlan**
Ferlatte, W.J. (fl. 1988) S	**Ferlatte**
Ferle, F. (fl. 1912) M	**Ferle**
Fermi, C. (fl. 1895) M	**Fermi**
Fernald, Merritt Lyndon (1873–1950) PS	**Fernald**
Fernandes, Abílio (1906–) S	**A.Fern.**
Fernandes, Afrânio (fl. 1984) S	**Afr.Fern.**
Fernandes, J.F.St.Antonio (fl. 1919) M	**J.F.Fern.**
Fernandes, L.G. (fl. 1918) M	**L.G.Fern.**
Fernandes, Robert R. (1931–) S	**R.R.Fern.**
Fernandes, Rosette Mercedes Saraiva Batarda (1916–) PS	**R.Fern.**
(Fernandes Vital, A.)	
see Vital, A.Fernandes	**A.F.Vital**
Fernandes–Baquero, G. (fl. 1959) M	**Fern.-Baq.**
Fernández–Alonso, José Luis (fl. 1986) S	**Fern.Alonso**
Fernández Areces, Maria Pilar (fl. 1990) S	**Fern.Areces**
Fernández, C.F. (fl. 1987) S	**C.F.Fernández**
(Fernández Carvajal, Maria del Carmen)	
see Fernández–Carvajal, Maria del Carmen	**Fern.-Carv.**
Fernández Casado, Maria de los Angeles (1952–) S	**Fern.Casado**
Fernández Casas, Francisco Javier (1945–) S	**Fern.Casas**
Fernández de Ana Magán, F.J. (fl. 1983) M	**Fern.Magán**
Fernández González, Federico (1956–) S	**Fern.Gonz.**
(Fernández Magán, F.J.)	
see Fernández de Ana Magán, F.J.	**Fern.Magán**
Fernández, Manuel (fl. 1983) S	**M.Fernández**
(Fernández, María Luisa López)	
see López Fernández, María Luisa	**López Fern.**
(Fernández Nava, Rafael)	
see Fernández, Rafael	**R.Fernández**
Fernández Prieto, José Antonio (1950–) S	**Fern.Prieto**
Fernández, Rafael (1956–) S	**R.Fernández**
Fernández, Z.Mayra (1948–) S	**Z.M.Fernández**
Fernández Zequeira, M. (fl. 1981) S	**M. Fernández Zeq.**
Fernández–Arias Gonzalez, M.I. (fl. 1990) S	**Fern.-Arias**
Fernández–Carvajal, Maria del Carmen (1951–) S	**Fern.-Carv.**
Fernández–García, Elena (fl. 1990) M	**Fern.-García**
(Fernández–Meldonado Villalobos, Francisco)	
see Villalobos, Francisco	**Villalobos**
Fernández–Pérez, Alvaro (1920–) S	**A.Fernández**
Fernández–Villar, Celestino (1838–1907) S	**Fern.-Vill.**
Fernando, Edwino S. (1953–) S	**Fernando**
Fernier, H. (fl. 1954) M	**Fernier**

Ferrada, Luiz V.Ferrada Urzúa (fl. 1949) M	**Ferrada**
Ferrari, E. (fl. 1984) M	**E.Ferrari**
Ferrari, Enrico (1945–) S	**Ferrari**
Ferrari, G. (fl. 1982) S	**G.Ferrari**
Ferrari, J.M. (fl. 1977) S	**J.M.Ferrari**
Ferrari, O. (fl. 1985) S	**O.Ferrari**
Ferrarini, Erminio (1919–) S	**Ferrarini**
Ferraris, Teodoro (1874–1943) M	**Ferraris**
Ferraro, Lidia Itati (1951–) BM	**L.I.Ferraro**
Ferraro, Matilde (fl. 1955) S	**M.Ferraro**
Ferraz, S. (fl. 1976) M	**Ferraz**
Ferré, Yvette de S	**Ferré**
Ferreira, Alexandre (Alejandro) Rodrigues (1756–1815) S	**Ferreira**
Ferreira, Francisco A. (fl. 1986) M	**F.A.Ferreira**
Ferreira, Hilda Manhâ (fl. 1978) S	**H.M.Ferreira**
Ferreira, M.B. (fl. 1977) S	**M.B.Ferreira**
Ferreira, N.P. (fl. 1987) M	**N.P.Ferreira**
Ferreira, V.Flechtmann (fl. 1977) S	**V.F.Ferreira**
Ferren, Wayne R. (1948–) S	**Ferren**
Ferrer, R.L. (fl. 1980) M	**Ferrer**
Ferrer–Ziogas, C. (fl. 1969) M	**Ferr.-Ziogas**
Ferrero (fl. c.1821) S	**Ferrero**
Ferreyra, Ramón Alejandro (1912–) S	**Ferreyra**
Ferris, Bernard J. (1922–) A	**B.J.Ferris**
Ferris, Roxana Judkins (1895–1978) S	**Ferris**
Ferriss, James Henry (1849–1926) S	**Ferriss**
Ferro, Giovanni (fl. 1907) M	**Ferro**
Ferrucci, María Silvia (1950–) S	**Ferrucci**
Ferry, René Joseph Justin (1845–1924) M	**Ferry**
Fert, Charles A	**Fert**
Fertig, Irene Gruenberg (fl. 1978)	**Fertig**
Feruglio, Egidio (fl. 1933) F	**Feruglio**
Ferukawa, Hisahiko (fl. 1963) M	**Feruk.**
Férussac, André Étienne Justin Pascal François d'Audibert de (1786–1836) S	**Férussac**
Fessenden, Anna Parker A	**Fess.**
Fetherston, Edith S	**Feth.**
Feuer, S. (fl. 1982) S	**Feuer**
Feuerer, Tassilo (1949–) M	**Feuerer**
Feuillebois, P.V.A. (1840–1899) M	**Feuilleb.**
Feuillée, Louis Éconches (1660–1732) L	**Feuillée**
Feuillet, Christian (1948–) S	**Feuillet**
Feurich, G. (fl. 1915) M	**Feurich**
Feurstein, Pankratia A	**Feurstein**
Fiala, Franz (1861–1898) S	**Fiala**
Fiala, J.L. (fl. 1988) S	**J.L.Fiala**
Fiard, Jean–Pierre (fl. 1976) MS	**Fiard**
Fiasson, J.–L.F. (fl. 1977) M	**Fiasson**
Fiatte, M.C. A	**Fiatte**
Ficalho, Franciso Manoel Carlos de Mello de (1837–1903) S	**Ficalho**

Ficinus, Heinrich David August (1782–1857) CMS	**Ficinus**
Fidalgo, Maria Eneydna Pacheco Kauffman (1928–1970) M	**M.Fidalgo**
Fidalgo, Oswaldo (1928–) M	**O.Fidalgo**
Fieber, Franz Xaver (1807–1872) S	**Fieber**
Fiebrig, Karl August Gustav (1879–1951) S	**Fiebrig**
(Fiebrig–Gertz, Karl August Gustav)	
see Fiebrig, Karl August Gustav	**Fiebrig**
Fiedler, Karl Friedrich Bernhard (1807–1869) BM	**Fiedl.**
Fiek, Emil (1840–1897) PS	**Fiek**
Field, Barron (1786–1846) S	**B.Field**
Field, David Vincent (1937–) S	**D.V.Field**
Field, Ethel Content (fl. 1915) M	**E.C.Field**
Field, Henry Claylands (1825–1912) PS	**Field**
Field, John Horace (1933–) B	**J.H.Field**
Fielding, Cecil Henry (1848–1918) S	**C.H.Fielding**
Fielding, Henry Barron (1805–1851) PS	**Fielding**
Fields, William G. (fl. 1963) M	**Fields**
Fieschi, V. (c.1910) S	**Fieschi**
Fife, Allan James (1951–) B	**Fife**
Figari, Antonio Bey (1804–1870) AS	**Fig.**
Figert, Ernst (1848–1925) S	**Figert**
Figueirado, Mário Barreto (fl. 1967) M	**Figueirado**
(Figueiras, Manuel)	
see López–Figueiras, Manuel	**López–Fig.**
Figueras, M.J. (fl. 1986) M	**Figueras**
(Figuerola Lamata, Ramón)	
see Figuerola, Ramón	**Figuerola**
Figuerola, Ramón (1953–) S	**Figuerola**
Figureau, C. (fl. 1987) S	**Figureau**
Fijałkowski, Dominik (1922–) S	**Fijałk.**
Fijten, F. (fl. 1975) S	**Fijten**
Filarszky, Nándor (Ferdinand) (1858–1941) APS	**Fil.**
Filatova, N.S. (1930–) S	**Filatova**
Filer, T.H. (fl. 1969) M	**Filer**
Filet, G.J. (1825–1891) S	**Filet**
Filewicz, M.V. A	**Filewicz**
Filgueiras, Tarciso S. (1950–) S	**Filg.**
Filhol, Henri (1843–1902) S	**Filhol**
Filigheddu, Rossella (1959–) S	**Filigh.**
Filimonova, N.M. (fl. 1970) M	**N.M.Filim.**
Filimonova, Z.N. (1929–) S	**Filim.**
Filipchenko, Ju.A. S	**Filipch.**
Filipello Marchisio, Valeria (fl. 1977) M	**Fil.March.**
Filipello, Sebastiano (1892–1936) B	**Filipello**
Filipescu, Miltiade G. A	**Filipescu**
Filipev, Ivan Nikolaevich (1889–1937?) M	**Filipev**
Filippi, Ilario (fl. 1979) M	**Filippi**
Filippopulos, Giorgio K. (fl. 1927) M	**Filippop.**

Filippov, G.S. (fl. 1934) M	**Filippov**
Filla, F. (fl. 1926) S	**Filla**
Fillion, Roger (fl. 1989) M	**Fillion**
Filov, A.I. (1903–) S	**Filov**
Filson, Rex Bertram (1930–) BM	**Filson**
Filtenborg, Ole (fl. 1987) M	**Filt.**
Finch, R.A. (1939–) S	**Finch**
Finck, K.E. (fl. 1988) M	**Finck**
Findlay, Judy N. (fl. 1970) S	**Findlay**
Findley, John L. (fl. 1973) S	**Findley**
Fine, Karen E. A	**Fine**
Fineran, B.A. (fl.1969) M	**Fineran**
Finet, Achille Eugène (1863–1913) S	**Finet**
Finger, J. (fl. 1987) M	**Finger**
Fingerhuth, Carl Anton (1802–1876) MS	**Fingerh.**
Fink, Bruce (1861–1927) M	**Fink**
Finkel, A. (1917–) S	**Finkel**
Finlayson, George (1790–1823) S	**Finl.**
Finley, David E. (fl. 1965) M	**Finley**
Finocchio, Alfred F. (1930–) M	**Finocchio**
Finschow, Günter (1926–) S	**Finschow**
Fintelmann, Gustav Adolf (1803–1871) S	**Fintelm.**
Fiol, J.B. (fl. 1976) M	**Fiol**
Fiore, James A	**J.Fiore**
Fiore, Maria (fl. 1932) M	**Fiore**
Fiori, Adriano (1865–1950) MPS	**Fiori**
Fiori, Alfonso S	**Alf.Fiori**
Fiori, Andrea (1854–) S	**Andr.Fiori**
Fiori, Attilio S	**Att.Fiori**
Fiorini–Mazzanti, Elisabetta (1799–1879) AB	**Fior.-Mazz.**
Firc, Alberto V. (1882–1944) S	**Firc**
Firdousi, S.A. (fl. 1990) M	**Firdousi**
Firtion, F. A	**Firtion**
Fisch, Carl (1859–19??) M	**C.Fisch**
Fischer, (Emanuel Friedrich) Ludvig (1828–1907) S	**L.Fisch.**
Fischer, Alfred (1858–1913) M	**A.Fisch.**
Fischer, Cecil Ernest Claude (1874–1950) S	**C.E.C.Fisch.**
Fischer, D.P. (fl. 1976) M	**D.P.Fisch.**
Fischer, Eberhard (fl. 1989) S	**Eb.Fisch.**
Fischer, Eduard (1861–1939) M	**E.Fisch.**
Fischer, F.B. (fl. 1853) S	**F.B.Fisch.**
Fischer, F.E. (fl. 1950) M	**F.E.Fisch.**
Fischer, F.G.L. (fl. 1852) S	**F.G.L.Fisch.**
Fischer, Friedrich Ernst Ludwig von (Fedor Bogdanovic) (1782–1854) PS	**Fisch.**
Fischer, Georg (–1941) S	**G.Fisch.**
Fischer, George William (1906–) M	**G.W.Fisch.**
Fischer, Gustav (1889–) S	**Gust.Fisch.**
Fischer, H.P. (fl. 1962) S	**H.P.Fisch.**
Fischer, Hugo (fl. 1904) P	**H.Fisch.**

Fischer, Ida (fl. 1940) M	I.Fisch.
Fischer, Jacob Benjamin (1730–1793) S	J.Fisch.
Fischer, Jean–Claude A	J.-C.Fisch.
Fischer, Johann Carl (1804–1885) M	J.C.Fisch.
Fischer, M. (fl. 1986) M	M.Fisch.
Fischer, Manfred A. (1942–) S	M.A.Fisch.
Fischer, Pierre C. (fl. 1980) S	P.C.Fisch.
Fischer, Robert (fl. 1930) AM	R.Fisch.
Fischer von Waldheim, (Johann) Gotthelf (Friedrich) (1771–1853) F	G.Fisch.Waldh.
Fischer von Waldheim, Alexander Gregorevich (1803–1884) S	Fisch.Waldh.
Fischer von Waldheim, Alexandr Alexandrovich (Alexander Alexandrowitz)	
(1839–1920) BMS	A.A.Fisch.Waldh.
Fischer, W. (fl. 1928) M	W.Fisch.
Fischer–Benzon, Rudolf J.D.von (1839–1911) S	Fisch.-Benz.
Fischer–Ooster, Carl von (1807–1875) A	Fisch.-Oost.
Fischman, O. (fl. 1967) M	Fischman
Fiscus, S.A. (fl. 1986) M	Fiscus
Fiserová, Daniela (fl. 1970) S	Fiserová
Fisher, Beryl Stranack (1907–1951) S	B.S.Fisher
Fisher, Eileen E. (fl. 1940) M	E.E.Fisher
Fisher, Elmon McLean (1861–1938) S	Fisher
Fisher, Francine E. (fl. 1950) M	F.E.Fisher
Fisher, Francis John Fulton (1926–) S	F.J.F.Fisher
Fisher, J.H. A	J.H.Fisher
Fisher, M.J. A	M.J.Fisher
Fisher, P.J. (fl. 1987) M	P.J.Fisher
Fisher, Tharl Richard (1921–) S	T.R.Fisher
Fisjun, V.V. (1922–) S	Fisjun
Fitch, Walter Hood (1817–1892) S	Fitch
Fitschen, Jost (1869–1947) AS	Fitschen
Fitter, Richard Sydney Richmond (1913–) S	Fitter
Fittkau, H.W. (fl. 1971) S	Fittkau
Fitze, K. (fl. 1954) S	Fitze
Fitzgerald, Laurence R. (fl. 1943) M	L.R.Fitzg.
Fitzgerald, Robert Desmond (David) (1830–1892) S	Fitzg.
Fitzgerald, William Vincent (1867–1929) S	W.Fitzg.
Fitzherbert, Wyndham (–1916) S	Fitzh.
Fitzpatrick, Harry Morton (1886–1949) MS	Fitzp.
Fitzpatrick, Mary Alida S	M.A.Fitzp.
Fize, Antoinette A	Fize
Fjerdingstad, E. A	Fjerd.
(Fjodorov, Andrej Aleksandrovich)	
see Fedorov, Andrej Aleksandrovich	Fed.
Flageolet, J. (fl. 1891) M	Flageolet
Flagey, Camille (1837–1898) M	Flagey
Flahault, Charles Henri Marie (1852–1935) AMS	Flahault
Flaksberger, Constantin Andreevich (1880–1942) S	Flaksb.
Flamant, L. (fl. 1922) M	Flamant
Flaster, Bernardo (1934–) S	Flaster

Flatau, L. (fl. 1987) M	**Flatau**
Flatberg, Kjeld Ivar (1943–) BS	**Flatberg**
Flatt von Alföld, Károly (1853–1906) S	**Flatt**
(Flechtmann Ferreira, V.)	
see Ferreira, V.F.	**V.F.Ferreira**
Fleischer, Bohumil (1847–1913) S	**B.Fleisch.**
Fleischer, Franz von (1801–1878) S	**F.Fleisch.**
Fleischer, Johann Gottlieb (Theophilus) (1797–1838) S	**J.Fleisch.**
Fleischer, Max (1861–1930) B	**M.Fleisch.**
Fleischhack (fl. 1865) M	**Fleischh.**
Fleischmann, Andreas (1805–1867) S	**Fleischm.**
Fleischmann, H. S	**H.Fleischm.**
Fleming, John (1785–1857) AS	**Fleming**
Fleming, Mary (fl. 1966) M	**M.Fleming**
Fleming, R.F. A	**R.F.Fleming**
Flenley, J.R. (fl. 1978) P	**Flenley**
Flensburg, Tom A	**Flensburg**
Flentje, Noel Thomas (1921–) M	**Flentje**
(Flerov, Alexandr Fedorovich)	
see Flerow, Alexandr Fedorovich	**Flerow**
Flerow, Alexandr Fedorovich (1872–1960) S	**Flerow**
Fletcher, Anthony (1944–) M	**A.Fletcher**
Fletcher, Harold Roy (1907–1978) S	**H.R.Fletcher**
Fletcher, James (1852–1908) S	**Fletcher**
Fletcher, R.A. (fl. 1984) S	**R.A.Fletcher**
Fleury, J.F. (fl. 1819) S	**Fleury**
Fliche, Paul Henri Maria Thérèse André (1836–1908) AFP	**Fliche**
Flinck, Karl Evert (1915–) S	**Flinck**
Flinders, Matthew (1774–1814) S	**Flinders**
Fling, Eva Myrtelle (later Roush, E.M.) (1886–) S	**Fling**
Flint, Elizabeth Alice (1909–) A	**E.A.Flint**
Flint, Lewis Herrick (1893–) A	**Flint**
Flipphi, R.C.H. (fl. 1987) S	**Flipphi**
Floderus, Björn Gustaf Oscar (1867–1941) S	**Flod.**
(Floerke, Heinrich Gustav)	
see Flörke, Heinrich Gustav	**Flörke**
Flögel, J.H.L. A	**Flögel**
Florence, E.Jacques M. (1951–) M	**Florence**
Florenzano, Gino (fl. 1974) M	**G.Florenz.**
Florenzano, J. (fl. 1948) M	**Florenz.**
Flores, José (fl. 1987) S	**Flores**
Floret, J.J. (1939–) S	**Floret**
Flórián, E. (fl. 1964) M	**Flórián**
Florin, Carl Rudolf (1894–1965) AFS	**Florin**
Florin, Maj–Britt (1915–) A	**M.Florin**
Flörke, Heinrich Gustav (1764–1835) BMS	**Flörke**
Florovskaya, Elena Fedorowna (1902–) M	**Florovsk.**
Florschutz, Peter Arnold (1923–197?) B	**Florsch.**
Florschutz–de Waard, Jeanne (1924–) B	**J.Florsch.**
Florström, Bruno Leonard (Leonhard) (1879–1914) S	**Florstr.**

Floruţiu, A. (fl. 1967) M	**Floruţiu**
Flory, Walter S. S	**Flory**
Floto, Ernst Vilhelm (1902–1983) S	**Floto**
Flotow, Julius Christian Gottlieb Ulrich Gustav Georg Adam Ernst Adam Friedrich von (1788–1856) AMS	**Flot.**
Flous, Fernande (1908–) S	**Flous**
Flower, R.J. A	**Flower**
Flowers, Seville (1900–1968) BP	**Flowers**
Floyd, Frad Gillan (fl. 1902) P	**Floyd**
Floyd, Gary L. (1940–) A	**G.L.Floyd**
Flu, P.C. A	**Flu**
Flueckiger, Friedrich August (1828–1894) S	**Flueck.**
(Flueggé, Johannes)	
see Flüggé, Johannes	**Flüggé**
Flügel, Erik A	**E.Flügel**
Flügel, Helmut A	**H.Flügel**
Flüggé, Johannes (1775–1816) S	**Flüggé**
Flyr, Lowell David (1937–1971) S	**Flyr**
Foà, Anna A	**Foà**
Fobe, Fr. (fl. 1897–1900) S	**Fobe**
Focke, Gustav Woldemar A	**G.W.Focke**
Focke, Hendrik Charles (1802–1856) S	**H.Focke**
Focke, Rosmarie A	**R.Focke**
Focke, Wilhelm Olbers (1834–1922) S	**Focke**
Foerste, August Frederick (1862–1936) F	**Foerste**
Foerster, Arnold (1810–1884) S	**Foerster**
Foessner, Ilse A	**Foessner**
Foëx, Étienne Edmond (1876–1944) M	**E.E.Foëx**
Foëx, Gustave Louis Émile (1844–1906) S	**Foëx**
Foged, Niels (1906–1988) A	**Foged**
Fogel, Robert (1947–) M	**Fogel**
Foggi, Bruno (1955–) S	**Foggi**
Foggitt, William (1835–1917) S	**Foggitt**
Foissner, Ilse (fl. 1986) M	**I.Foissner**
Foissner, Wilhelm (fl. 1986) M	**W.Foissner**
Foister, Charles Edward (fl. 1931) M	**Foister**
Fokin, A.D. (fl. 1926) M	**Fokin**
Foko, J. (fl. 1971) M	**Foko**
(Folch i Guillèn, Ramón)	
see Folch, Ramón	**Folch**
Folch, Ramón (1946–) S	**Folch**
Foldago, R. (fl. 1984) M	**Foldago**
Foldats, Ernesto (1916–) S	**Foldats**
Foldvik, Ninja A	**Foldvik**
Foley, H. (fl. 1949) M	**Foley**
Folin, Th. S	**Folin**
Folk, Robert L. (1925–) A	**Folk**
Follin, J.-C. (fl. 1966) M	**Follin**
Follman, S. (fl. 1968) M	**Follman**

Follmann, Gerhard (1930–) M	**Follmann**
Folsom, James P. (1950–) S	**Folsom**
Fomin, Aleksandr Vasiljevich (1869–1935) PS	**Fomin**
Fonnegra G., Ramiro (fl. 1985) S	**Fonnegra**
Fonseca, Alvaro (fl. 1991) M	**A.Fonseca**
Fonseca, Luis da A	**L.Fonseca**
Fonseca, O.M. (fl. 1965) M	**O.M.Fonseca**
Fonseca, Olympio Oliveira Ribeiro da (fl. 1922) AM	**Fonseca**
(Fonseca Vaz, A.M.S.da)	
see Vaz, A.M.S. da Fonseca	**Vaz**
(Font i Quer, Pio)	
see Font Quer, Pio (Pius)	**Font Quer**
Font Quer, Pio (Pius) (1888–1964) MPS	**Font Quer**
Fontaine, William Morris (1835–1913) F	**Fontaine**
Fontana, Anna (fl. 1971) M	**A.Fontana**
Fontana, Felice (1730–1805) S	**Fontana**
Fontana, Pietro (1876–1948) S	**P.Fontana**
Fontell, Carl Wilhelm (1873–) A	**Fontell**
Fontella, Jorge (1936–) S	**Fontella**
(Fontella Pereira, Jorge)	
see Fontella, Jorge	**Fontella**
Fontes, Fernando Carvalho (1915–) S	**Fontes**
Fontoynont, M. (fl. 1923) M	**Fontoyn.**
Fonty, Gérard (fl. 1989) M	**Fonty**
(Fonvert, Amédée)	
see Reinaud de Fonvert, Alexandre Jean Baptiste	**Rein.Fonv.**
Foord, N.John A	**Foord**
Forbes, Charles Noyes (1883–1920) S	**C.N.Forbes**
Forbes, Edward (1815–1854) S	**E.Forbes**
Forbes, Evelyn J. (fl. 1935) M	**E.J.Forbes**
Forbes, F.Malcolm (fl. 1980s) S	**F.M.Forbes**
Forbes, Fayette Frederick (1851–1935) S	**F.F.Forbes**
Forbes, Francis Blackwell (1839–1908) S	**F.B.Forbes**
Forbes, Helena M.L. (1900–1959) S	**H.M.L.Forbes**
Forbes, Henry Ogg (1851–1932) PS	**H.O.Forbes**
Forbes, James (1773–1861) S	**J.Forbes**
Forbes, John (1798–1823) S	**Forbes**
Forbes, P.L. (fl. 1969) S	**P.L.Forbes**
Forchhammer, Johan Georg (1794–1865) AF	**Forchh.**
Forchheimer, Sylvia A	**Forchheimer**
Forchiassin, F. (fl. 1988) M	**Forch.**
(Forckhammer, Johan Georg)	
see Forchhammer, Johan Georg	**Forchh.**
Ford, E.J. (fl. 1967) M	**E.J.Ford**
Ford, Neridah Clifton (1926–) S	**Ford**
Ford, Trevor David (1932–) A	**T.D.Ford**
Ford–Lloyd, Brian V. S	**Ford-Lloyd**
Ford–Robertson, F.S. (fl. 1924) M	**Ford-Rob.**
Forde, Neville (fl. 1958) S	**Forde**
Fordinandsen, C.C.F. (1879–1944) M	**Fordin.**

Foreau, Georges (1882–1967) B	**Foreau**
Foreman, Donald Bruce (1945–) S	**Foreman**
Forero, Enrique (1942–) S	**Forero**
(Forero–Gonzalez, Enrique)	
see Forero, Enrique	**Forero**
Forest, Herman Silva (1921–) A	**H.S.Forest**
Forestier, de (fl. 1853) S	**Forest.**
Foret, James Aloysius (1921–) S	**Foret**
Fóriss, Ferenc. (1892–) M	**Fóriss**
Forman, Lewis Leonard (1929–) S	**Forman**
Formánek, Eduard (1845–1900) PS	**Formánek**
Formiggini, Leone S	**Formigg.**
Fornachon, J.C.M. (fl. 1964) M	**Forn.**
Fornaciari, Giovanni (fl. 1949–1973) S	**Fornac.**
Forneris, Guiliana (1946–) S	**Forneris**
Forquignon, L. (fl. 1880s) M	**Forq.**
Forrest, George (1873–1932) S	**Forrest**
Fors, N.S. (fl. before 1828) S	**Fors**
Forselles, Jacob Henrik af (1785–1855) S	**Forselles**
(Forskaol, Pehr (Peter))	
see Forsskål, Pehr (Peter)	**Forssk.**
Forssell, Karl Bror Jakob (1856–1898) M	**Forssell**
Forsskål, Pehr (Peter) (1732–1763) APS	**Forssk.**
Forsström, Johan Eric (1775–1824) S	**Forsstr.**
Forsteneichner, Franz (fl. 1931) M	**Forsten.**
Förster, A. (fl. 1878) S	**A.Först.**
Forster, Benjamin Meggot (1764–1829) M	**B.M.Forst.**
Förster, Carl Friedrich (fl. 1846) S	**C.F.Först.**
Forster, Edward (1765–1849) S	**E.Forst.**
Forster, Jakob S	**J.Forst.**
Forster, Johann Georg Adam (1754–1794) PS	**G.Forst.**
Forster, Johann Reinhold (1729–1798) S	**J.R.Forst.**
Förster, Karl (1874–) S	**K.Först.**
Förster, Kurt A	**Kurt Först.**
Forster, Paul Irwin (1961–) S	**P.I.Forst.**
Forster, Thomas Furley (1761–1825) S	**T.F.Forst.**
Forstner, S. (fl. 1975) M	**Forstner**
Forsyth, William (1737–1804) S	**Forsyth**
Forsyth–Major, Charles Immanuel (1843–1923) S	**Fors.–Major**
Fort, F. (fl. 1984) M	**Fort**
Forti, Achille Italo (1878–1937) A	**Forti**
Fortis, F. S	**Fortis**
Fortoul, R. (fl. 1976) M	**Fortoul**
Fortuin (fl. 1979) S	**Fortuin**
Fortunato, Renée H. (1957–) S	**Fortunato**
Fortune, Robert (1812–1880) S	**Fortune**
Fosberg, Francis Raymond (1908–) MPS	**Fosberg**
Foschi, S. (fl. 1955) M	**Foschi**
Foslie, Michael (Mikaľ) Heggelund (1855–1909) B	**Foslie**
Foster, A.A. (fl. 1958) M	**A.A.Foster**

Foster, Adriance Sherwood (1901–1973) S   **A.S.Foster**
Foster, Arthur Crawford (1893–) M   **A.C.Foster**
Foster, B.T. (fl. 1982) S   **B.T.Foster**
Foster, Clinton B. (fl. 1981) F   **C.B.Foster**
Foster, H.Lincoln (fl. 1954) P   **H.L.Foster**
Foster, John H.   A   **J.H.Foster**
Foster, Michael (1836–1907) S   **Foster**
Foster, Michael S. (1942–) AS   **M.S.Foster**
Foster, Mulford Bateman (1888–1978) S   **M.B.Foster**
Foster, Robert Crichton (1904–1986) S   **R.C.Foster**
Foster, William Robert (1905–) M   **W.R.Foster**
Fosteris, Stelian (fl. 1943) M   **Fosteris**
Fothergill, John (1712–1780) S   **Foth.**
Fotjanova, L.I. (fl. 1963) F   **Fotjanova**
Fotsch, Karl Albert (–1940?) S   **Fotsch**
Fott, Bohuslav (1908–1976) AM   **Fott**
Foucaud, Julien (1847–1904) S   **Foucaud**
Foucault, Emmanuel de (fl. 1813–1850) S   **Foucault**
Foucher, Jean–Claude   A   **Foucher**
Fougeroux de Bondaroy, Auguste Denis (1732–1789) S   **Foug.**
Fouillade, Amédée (1870–) S   **Fouill.**
Fouilloy, R. (fl. 1963) S   **Fouilloy**
Foulerton, Alexander G.R. (fl. 1900) M   **Foul.**
Fourcade, Georges Henri (1866–1948) S   **Fourc.**
Fourmont, R. (fl. 1938) M   **Fourmont**
Fournet, J. (fl. 1973) M   **Fournet**
Fournier, Eugène Pierre Nicolas (1834–1884) PS   **E.Fourn.**
Fournier, Paul Victor (1877–1964) PS   **P.Fourn.**
Fournier, Robert O.   A   **R.O.Fourn.**
Fourreau, Jules Pierre (1844–1871) S   **Fourr.**
Fowler, Robert Lawrence (1910–) PS   **Fowler**
Fowlie, Jack A. (fl. 1965) S   **Fowlie**
Fox, Denis L. (1901–) A   **D.L.Fox**
Fox, Margaret   A   **M.Fox**
Fox, T.Colcott (fl. 1896) M   **Fox**
Fox Talbot, William Henry (1800–1877) S   **Fox Talbot**
Fox, W.T. (fl. 1973) M   **W.T.Fox**
Fox–Strangways, William Thomas Horner (1795–1865) FS   **Fox-Strangw.**
Foxworthy, Frederick William (1877–1950) S   **Foxw.**
Föyn, Björn (1898–) A   **Föyn**
Fraas, Carl Nicolaus (1810–1875) S   **Fraas**
Fragman, Ori (1965–) S   **Fragman**
Frágner, Peter (fl. 1956) M   **Frágner**
(Fragoso, Romualdo González)
    see González Fragoso, Romualdo   **Gonz.Frag.**
Frahm, G. (fl. 1898) S   **Frahm**
Frahm, Jahn–Peter   S   **J.-P.Frahm**
Franc, A. (fl. 1969) M   **A.Franc**
Franc van Berkhey, Jan (Johannes) le (1729–1812) S   **Franc**
França, Carlos   A   **França**

(France, Jesse Allison De)
  see De France, Jesse Allison   **De France**
Francé, Raoul Heinrich (1874–1943) AP   **Francé**
Frances, S.P. (fl. 1987) M   **Frances**
Franceschi, Francesco (1843–1924) S   **Franceschi**
Francey, Pierre (fl. 1933) S   **Francey**
Franchet, Adrien René (1834–1900) PS   **Franch.**
Franchetti, Giovanna (1930–) S   **Franchetti**
Franchini, G. A   **Franchini**
Francis, George William (1800–1865) PS   **Francis**
Francis, Sheila M. (fl. 1975) M   **S.M.Francis**
Francis, William Douglas (1889–1959) S   **W.D.Francis**
Francisco-Ortega, Javier (1958–) S   **Franc.-Ort.**
Francke-Grosman, Helene (fl. 1961) M   **Francke-Grosm.**
(Franco de Toledo, Joaquim)
  see Toledo, Joaquim Franco de   **Toledo**
Franco, João Manuel Antonio do Amaral (1921–) PS   **Franco**
François, J. S   **J.François**
François, Louis Ernst Eugène (1856–) S   **François**
Frandsen, N.O. (fl. 1943) M   **Frandsen**
Frank, Albert Bernhard (1839–1900) AMP   **A.B.Frank**
Frank, E. (fl. 1982) S   **E.Frank**
Frank, Gerhard R.W. (fl. 1963–1978) S   **G.Frank**
Frank, Joseph C. (1782–1835) S   **Frank**
Frank, R. (fl. 1982) S   **R.Frank**
Franke, Max A   **Franke**
Franken, N.A.P. (fl. 1978) S   **Franken**
Franklin, Dudly Arthur (1936–) S   **D.A.Franklin**
Franklin, E.F. (fl. 1983) S   **E.F.Franklin**
Franklin, George C.H. (fl. 1940) M   **G.C.Franklin**
Franklin, John (1786–1847) PS   **Franklin**
(Franklin–Hennessy, Esmée)
  see Hennessy, Esmée Franklin   **Hennessy**
Franks, Millicent ? (1886–1961) S   **Franks**
Frankton, Clarence (1906–) S   **Frankton**
Franquet, Robert Fernand (1897–1930) S   **Franquet**
Franqueville, Albert de (–1891) S   **Franquev.**
Fransen, Sven Goran (1949–) B   **Fransen**
Frantskevich, N.A. (fl. 1983) S   **Frantsk.**
Franz, C. S   **C.Franz**
(Franz de Laet-Contich, von)
  see Laet-Contich, Franz von   **Laet-Cont.**
Franz, Erich (fl. 1908) S   **E.Franz**
Franz, G. (fl. 1973) M   **G.Franz**
Franz, R. (fl. 1981–89) S   **R.Franz**
(Franzé, Reszö (Resso))
  see Francé, Raoul Heinrich   **Francé**
Franzén, Roy (1957–) S   **Franzén**
Franzoni, A. (fl. 1861) M   **Franzoni**

Frappier, Alphonse (fl. 1853–1883) S	A.Frapp.
Frappier, Charles (fl. 1853–1895) S	Frapp.
Fraser, Charles (c.1788–1831) S	C.Fraser
Fraser, Elizabeth M. (fl. 1968) M	E.M.Fraser
Fraser, George (1854–) S	G.Fraser
Fraser, Hugh (1834–1904) S	H.Fraser
Fraser, James (1854–1935) S	J.Fraser
Fraser, John (1750–1811) S	Fraser
Fraser, Lilian Ross (1908–) MS	L.R.Fraser
Fraser, Patrick Neill (1830–1905) S	P.N.Fraser
Fraser, Samuel Victorian (1890–1972) S	S.V.Fraser
Fraser, William Pollock (1867–1943) M	W.P.Fraser
Fraser-Jenkins, Christopher Roy (1948–) P	Fraser-Jenk.
Frauenfeld, Georg von (1807–1873) AM	Frauenf.
Fraymouth, Joan Elise (1924–) M	Fraym.
Freckmann, Robert W. (1939–) S	Freckmann
Frederick, Lafayette (fl. 1969) MS	Frederick
Fredericq, Suzanne A	Fredericq
Frederiksen, Signe (1942–) S	Fred.
Fredrickson, Anders Theodor (1868–1905) S	Fredr.
Freeman, Alice E.H. (fl. 1979) M	A.E.Freeman
Freeman, C.C. (fl. 1983) S	C.C.Freeman
Freeman, Edward Monroe (1875–) M	E.M.Freeman
Freeman, Florence Lucinda (later Jones, F.L.) (1912–) S	F.L.Freeman
Freeman, George Fouché (1876–1930) S	G.F.Freeman
Freeman, Glenn W. (fl. 1979) M	G.W.Freeman
Freeman, J.R. (fl. 1985) S	J.R.Freeman
Freeman, John Daniel (1941–) S	J.D.Freeman
Freeman, Oliver Myles (1891–) S	O.M.Freeman
(Freeman–Mitford, Algernon Bertram)	
see Mitford, Algernon Bertram Freeman	Mitford
Freese, Leonard Roy (1906–) A	Freese
Frege, Christian August (1758–1834) M	Frege
Frei, W. (fl. 1921) M	Frei
Freiberg, Wilhelm (fl. 1911) PS	Freiberg
Freijsen, Arnoldus Henricus Joseph (1934–) S	Freijsen
Freïndling, M.V. (fl. 1949) M	Freïndl.
Freine, Susana E. S	Freine
Freire, Domingos (fl. 1886) M	Freire
(Freire Allemão, Manoel)	
see Allemão, Manoel	M.Allemão
Freire, Carlos Vianna S	C.V.Freire
Freire, F.O. (fl. 1979) M	F.O.Freire
Freire, Luis (fl. 1977) M	L.Freire
Freire, R. (fl. 1963) M	R.Freire
Freire, Susan E. (fl. 1986) S	S.E.Freire
Freitag, Helmut E.(1932–) S	Freitag
(Freitas da Silva, Marlene)	
see Silva, Marlene Freitas da	M.F.Silva

Freitas, Gilberto de A	**Freitas**
Frémont, John Charles (1813–1890) S	**Frém.**
Frémy, Pierre (1880–1944) A	**Frémy**
French, Charles (1840–1890) PS	**French**
French, James C. (1947–) S	**J.C.French**
Frenguelli, Joaquím (1883–1958) AF	**Freng.**
Frentrop, J.G.J. (fl. 1981) M	**Frentrop**
Frentzen, Kurt (fl. 1922) F	**Frentzen**
Frenzel, Johannes A	**Frenzel**
Fresenius, Johann Baptist Georg (George) Wolfgang (1808–1866) AMS	**Fresen.**
Fresneda, José A. (fl. 1989) M	**Fresneda**
Fresnel, Jacqueline A	**Fresnel**
Freudenthal, Hugo D. A	**Freud.**
Frew, Barbara P. (fl. 1962) M	**Frew**
Frey, Dorothea (fl. 1957) M	**D.Frey**
Frey, Eduard (–Stauffer) (1888–1974) M	**Frey**
Frey, Ruedi S	**R.Frey**
Frey, Wolfgang (1942–) B	**W.Frey**
Freycinet, Henri Louis Claude de Saulces de (1779–1842) PS	**Freyc.**
Freyer, Heinrich (1802–1866) S	**Freyer**
Freyer, Johann Gottfried S	**J.G.Freyer**
Freyer, K. (fl. 1975) M	**K.Freyer**
Freyn, Josef Franz (1845–1903) S	**Freyn**
Freyre, Rafael A	**Freyre**
(Freyreis, Georg Wilhelm)	
see Freyreiss, Georg Wilhelm	**Freyr.**
Freyreiss, Georg Wilhelm (1789–1825) S	**Freyr.**
Freytag, George F. S	**Freytag**
Frezier, Amédée François (1682–1773) S	**Frez.**
Frezzi, Mariano J. (fl. 1968) M	**Frezzi**
Frić, Alberto Vojtech (1882–1944) S	**Frić**
Friche–Joset, François (1799–1856?) S	**Friche-Joset**
Frick, Gustav Adolph (1878–) S	**Frick**
Frick, Thomas Adam (1901–) S	**T.A.Frick**
Frickhinger, Christian Albrecht (1818–1907) S	**Frickh.**
Frickhinger, Hermann (fl. 1911) S	**H.Frickh.**
Friderichsen, Peter Kristian Nicolaj (1853–1932) S	**Frid.**
Friedland, Solomon (1912–) S	**Friedl.**
Friedman, L. (fl. 1962) M	**Friedman**
Friedmann, Emerich Imre (1921–) M	**Friedmann**
Friedmann, F. (1941–) S	**F.Friedmann**
Friedrich, Hans Christian (1925–) S	**Friedrich**
Friedrich, Karl (fl. 1974) S	**K.Friedrich**
Friedrich, Ludwig A	**L.Friedrich**
(Friedrich, Martha)	
see Friedrich–Holzhammer, Martha	**Friedr.-Holzh.**
Friedrich–Holzhammer, Martha (1928–) S	**Friedr.-Holzh.**
Friedrichs, G. (fl. 1929) M	**Friedrichs**
Friedrichsthal, Emanuel von (1809–1842) S	**Friedr.**

Fries, (Klas) Robert Elias (1876–1966) MPS	R.E.Fr.
Fries, Elias Arne (1887–) M	E.A.Fr.
Fries, Elias Magnus (1794–1878) AMPS	Fr.
Fries, Elias Petrus (1834–1858) M	E.P.Fr.
Fries, Harald (1896–1963) S	H.Fr.
Fries, Oscar Robert (1840–1908) M	O.R.Fr.
Fries, Theodor (Thore) Magnus (1832–1913) AMPS	Th.Fr.
Fries, Thore Christian Elias (1886–1930) MS	T.C.E.Fr.
Friese, C. (fl. 1986) M	Friese
Friesen, C.von (fl. 1933) S	Friesen
Friesen, N.V. (1957–) S	N.V.Friesen
Friesner, Ray Clarence (1894–1952) PS	Friesner
Friis, Ib (1945–) S	Friis
Frikart, Carl Ludwig (1879–1964) S	Frikart
(Friquenon, Marie Émile)	
see Bianor, (Frère)	Bianor
Frisbie, Leonard F. (fl. 1959) S	Frisbie
Fristedt, Robert Fredric (Fredrik) (1832–1893) S	Fristedt
Frisvad, Jens C. (fl. 1987) M	Frisvad
Frisvoll, Anna Arnfinn B	Frisvoll
Fritel, Paul Honoré (1867–) FM	Fritel
Fritsch, C. (fl. 1898) S	C.Fritsch
Fritsch, Felix Eugen (1832–1893) A	F.E.Fritsch
Fritsch, Karl (1864–1934) MPS	Fritsch
Fritze, Richard (1841–1903) S	Fritze
Fritzen, Johannes (1891–) S	Fritzen
(Frivaldsky, Emerich (Imre))	
see Frivaldszky von Frivald, Emerich (Imre)	Friv.
Frivaldszky von Frivald, Emerich (Imre) (1799–1870) PS	Friv.
Frizen, N.V. (fl. 1985) S	Frizen
Frizzi, Giuliano (1948–) S	Frizzi
Fröberg, Lars (1957–) M	Fröberg
Fröderström, Harald August (1876–1944) S	Fröd.
Frodin, David Gamman (1940–) S	Frodin
Froebel, Karl Otto (1844–1906) S	Froebel
(Froederstroem, Harald August)	
see Fröderström, Harald August	Fröd.
(Froehlich, Antonin)	
see Fröhlich, Antonin	A.Fröhl.
(Froehlich, Friedrich Heinrich Wilhelm)	
see Frohl, Friedrich Heinrich Wilhelm	Frohl
Froehlich, Josef (1891–1966) B	J.Froehl.
Froehner, Albrecht (fl. 1897) S	A.Froehner
Froelich, Joseph Aloys von (1766–1841) BS	Froel.
Fróes, Ricardo de Lemos (1891–1960) S	Fróes
Fröes, Heitor Prayuer (fl. 1931) M	H.P.Fröes
Frohl, Friedrich Heinrich Wilhelm (1769–1845) B	Frohl
Fröhlich, Antonin (1882–1969) S	A.Fröhl.
Fröhlich, E. S	E.Fröhl.

Fröhlich, W. (fl. 1939) M	**W.Fröhl.**
(Fröhner, Albrecht)	
see Froehner, Albrecht	**A.Froehner**
Fröhner, Eugene (1858–1940) S	**E.Fröhner**
Fröhner, Sigurd E. (1941–) S	**S.E.Fröhner**
Froidevaux, L. (fl. 1973) M	**Froid.**
(Froilano Bachmann de Mello, Alfredo)	
see Mello, Alfredo Froilano Bachmann de	**Mello**
(Froilano de Mello)	
see Mello, Alfredo Froilano Bachmann de	**Mello**
(Frölich, Joseph Aloys von)	
see Froelich, Joseph Aloys von	**Froel.**
Frollo, M.M. A	**Frollo**
Frolov, I.P. (fl. 1967) M	**Frolov**
Frolov, Yu.M. (fl. 1985) S	**Yu.M.Frolov**
Fromentel, Louis Édouard Gourdan de (1824–) A	**From.**
Fromentin, Huguette (fl. 1964) M	**Fromentin**
Fromm Trinta, Elza (1934–) S	**Fromm**
(Fromm–Trinta, Elza)	
see Fromm Trinta, Elza	**Fromm**
Fromme, Fred Denton (1886–1966) M	**Fromme**
Fron, Georges (1870–1957) M	**Fron**
Froriep, Ludwig Friedrich von S	**Froriep**
Frost, Charles Christopher (1805–1880) MS	**Frost**
Frost, John (1803–1840) S	**J.Frost**
Frouin, H. (fl. 1986) M	**Frouin**
Fruwirth, Karl (1862–1930) S	**Fruwirth**
Fry, Wayne L. (1922–) A	**Fry**
Frye, Theodore Christian (1896–1962) APS	**Frye**
Fryer, Alfred (1826–1912) S	**Fryer**
Fryxell, Greta A. (1926–) A	**G.A.Fryxell**
Fryxell, Joan E. (fl. 1983) S	**J.E.Fryxell**
Fryxell, Paul Arnold (1927–) S	**Fryxell**
Fu, Cheng Xin (fl. 1986) S	**C.X.Fu**
Fu, De Zhi (1952–) S	**D.Z.Fu**
Fu, Guo Ai (fl. 1982) S	**G.A.Fu**
Fu, Guo Xun (1932–) S	**G.X.Fu**
Fu, Hiang Chian (1926–) S	**H.C.Fu**
Fu, Hua Long (fl. 1982) AS	**H.L.Fu**
Fu, Jing Qiou (fl. 1985) S	**J.Q.Fu**
Fu, Kun Tsun (1912–) S	**K.T.Fu**
Fu, Li Kuo (1934–) S	**L.K.Fu**
Fu, Pei Yun (1926–) S	**P.Y.Fu**
Fu, Ping Du (fl. 1981) S	**P.D.Fu**
Fu, Shu Hsia (1916–) PS	**S.H.Fu**
Fu, Xiu Hui (1950–) M	**X.H.Fu**
Fu, Yu Chin (fl. 1970) S	**Y.C.Fu**
Fu, Zhen Sheng (fl. 1987) M	**Z.S.Fu**
Fuchs, Alfred (1872–1927) S	**A.Fuchs**
Fuchs, Hans Peter (1928–) PS	**H.P.Fuchs**

Fuchs, Leonard (1501–1566) LP	**L.Fuchs**
Fuchs, Theodor (1842–1925) A	**Fuchs**
Fuchs, W.H. (fl. 1960) M	**W.H.Fuchs**
(Fuchs-Eckert, Hans Peter)	
see Fuchs, Hans Peter	**H.P.Fuchs**
Fuchsig, H. (1926–) B	**Fuchsig**
Fucini, Alberto (1864–1940) A	**Fucini**
Fuckel, Karl Wilhelm Gottlieb Leopold (1821–1876) M	**Fuckel**
Fuente García, Vicente de la (1950–) S	**Fuente**
Fuentes, César A. (fl. 1956) M	**C.A.Fuentes**
Fuentes Fiallo, Victor R. (fl. 1986) S	**V.R.Fuentes**
Fuentes, Maturana Francisco (1876–1934) S	**Fuentes**
Fuentes, S.F. (fl. 1964) M	**S.F.Fuentes**
(Fuernrohr, August Emanuel)	
see Fürnrohr, August Emanuel	**Fürnr.**
Fuertes, Javier S	**Fuertes**
Fuge, Dingley P. A	**Fuge**
Fuhlrott, Johann Carl (1804–1877) S	**Fuhlrott**
Füisting, Wilhelm (fl. 1860) M	**Füisting**
Fuji, Norio A	**Fuji**
Fujii, H. (fl. 1931) M	**H.Fujii**
Fujii, K. (fl. 1909) F	**Fujii**
Fujii, Mutue T. A	**M.T.Fujii**
Fujikuro, Yasaburô (fl. 1965) M	**Fujik.**
Fujishima, Hirosumi (fl. 1990) S	**Fujish.**
Fujita, Daisuke (1958–) A	**Fujita**
Fujiwara, R. (fl. 1987) S	**Fujiw.**
Fukarek, P. (fl. 1974) S	**Fukarek**
Fukazawa, Y. (fl. 1967) M	**Fukaz.**
Fukui, Takeji (fl. 1916) M	**Fukui**
Fukuoka, Nobuyuki (1904–) S	**Fukuoka**
Fukushi, Teikichi (fl. 1921) M	**Fukushi**
Fukushima, Hiroshi (1924–) A	**Fukush.**
Fukuyama, Noriaki (1912–1946) S	**Fukuy.**
Fukuyo, Yasuwo (1948–) A	**Fukuyo**
Fulford, Margaret Hannah (1904–) S	**Fulford**
Fuller, Albert Morse (1899–1981) S	**A.M.Fuller**
Fuller, George Damon (1869–1961) S	**Fuller**
Fullerton, R.A. (fl. 1977) M	**Full.**
(Fulvio, Teresa Emil di)	
see Di Fulvio, Teresa Emil	**Di Fulvio**
Funahashi, Stesno (1924–) A	**Funah.**
Funck, Heinrich Christian (1771–1839) BMS	**Funck**
(Funcke, Heinrich Christian)	
see Funck, Heinrich Christian	**Funck**
Fünfstück, Leberecht Moritz (1856–1925) S	**Fünfstück**
Fung, Hok Lam (fl. 1988) S	**H.L.Fung**
Funk, Alvin (fl. 1965) M	**A.Funk**
Funk, Georg (–1958) A	**Funk**

Funk, Victoria Ann (1947–) BS	**V.A.Funk**
Furlow, John J. (1942–) S	**Furlow**
Furnari, Francesco (1933–) S	**Furnari**
Furnari, Giovanni (1947–) A	**G.Furnari**
Fürnkranz, Dietrich (1936–) S	**Fürnkranz**
Fürnrohr, August Emanuel (1804–1861) MS	**Fürnr.**
Fürnrohr, Heinrich (1841/2–1918) S	**H.Fürnr.**
Furrazola Gomez, Gustavo (fl. 1987) S	**Furrazola**
Furrer–Ziogas, C. (fl. 1952) M	**Furrer–Ziogas**
Fursa, T.B (fl. 1972) S	**Fursa**
Furst, P. (fl. 1924) B	**Furst**
Furtado, Caetano Xavier (1897–1980) S	**Furtado**
Furtado, João S. (fl. 1965) M	**J.S.Furtado**
Furtado, Rocilda P. A	**R.P.Furtado**
Furukawa, Hisahiki (fl. 1966) M	**H.Furuk.**
Furukawa, T. (fl. 1977) M	**T.Furuk.**
Furuki, Tatsuwo (fl. 1989) B	**Furuki**
Furumi, Masatomi (1888–1930) S	**Furumi**
Furuse, M. (fl. 1988) S	**M.Furuse**
Furuya, Kouhei (F.) (fl. 1973) M	**Furuya**
Fusey, Pierre A	**Fusey**
Fuson, G.B. (fl. 1980) M	**Fuson**
Fuss, Johann Mihály (Michael) (1814–1883) MS	**Fuss**
Futák, Ján (1914–1980) PS	**Futák**
Futó, Mihaly (1882–1929) PS	**Futó**
Futrell, M.C. (fl. 1963) M	**Futrell**
Fyles, Faith (fl. 1915) M	**Fyles**
Fyson, Philip Furley (1877–) S	**Fyson**
Gaarder, Karen Ringdal (fl. 1954) A	**Gaarder**
Gabaev, S.G. S	**Gabaev**
Gabel, James Russel (1918–) A	**Gabel**
Gäbel, Max A	**M.Gäbel**
Gabelli, Lucio (1872–1918) S	**Gabelli**
Gable, Joseph Benson (1886–1972) S	**Gable**
Gabriel, Madeleine (fl. 1965) M	**Gabriel**
(Gabrielian, Elenora Tzolakovna)	
see Gabrieljan, Elenora Tzolakovna	**Gabrieljan**
Gabrieljan, Elenora Tzolakovna (1929–) S	**Gabrieljan**
Gabrielson, Ira Noel (1889–) S	**Gabrielson**
Gabrielson, Paul W. A	**P.W.Gabrielson**
Gabrielsson, Johan August (1860–1888) S	**Gabr.**
(Gabrielyan, Elenora Tsolakovna)	
see Gabrieljan, Elenora Tzolakovna	**Gabrieljan**
(Gachechiladze, K.A.)	
see Gatschetschiladze, K.A.	**Gatsch.**
Gachet, Antoine Hippolyte (1798–1842) M	**Gachet**
Gack, C. (fl. 1986) S	**Gack**

Gadd, Caleb Herbert (fl. 1947) M — **Gadd**
Gaddy, L.L. (fl. 1986) S — **Gaddy**
Gadgil, P.D. (fl. 1983) M — **Gadgil**
Gadzhieva, G.G. (1940–   ) S — **Gadzh.**
Gadzhieva, M.A.   A — **M.A.Gadzh.**
(Gadzieva, G.G.)
   see Gadzhieva, G.G. — **Gadzhieva**
Gaede (fl. 1828) S — **Gaede**
Gaertner, Alwin (fl. 1954) M — **A.Gaertn.**
Gaertner, Carl (Karl) Friedrich von (1772–1850) S — **C.F.Gaertn.**
Gaertner, Joseph (1732–1791) MS — **Gaertn.**
Gaertner, Philipp Gottfried (1754–1825) BMS — **P.Gaertn.**
Gaetano, L.De (fl. 1897) M — **Gaetano**
Gaggianese, Ettore (fl. 1990) M — **Gaggian.**
Gagliardi, Giuseppe (1812–1881) S — **Gagliardi**
Gagnebin, Abraham (1707–1800) S — **Gagnebin**
(Gagnebin de la Ferrière, Abraham)
   see Gagnebin, Abraham — **Gagnebin**
Gagnepain, François (1866–1952) S — **Gagnep.**
Gagnepain, L.François (1833–1911) S — **L.F.Gagnep.**
Gagnidze, Revas Ivanovich (1932–   ) S — **Gagnidze**
Gahn, Henric (1747–1816) S — **Gahn**
Gaidukov, Nikolai Mikhailovich (1874–1928) A — **Gaidukov**
Gaikwad, Y.B. (fl. 1973) M — **Gaikwad**
Gail, Floyd Whitney (1884–   ) S — **Gail**
Gaillard, Albert (1858–1903) M — **Gaillard**
Gaillard, Brigitte (fl. 1989) M — **B.Gaillard**
Gaillard, Jeanne   A — **J.Gaillard**
Gaillardot, Charles (1814–1883) S — **Gaill.**
Gaillat, A. (fl. 1925) M — **Gaillat**
Gaillon, François Benjamin (1782–1839) AM — **Gaillon**
Gaimard, Joseph–Paul (1790–1858) A — **Gaimard**
Gain, Luis (Louis)   A — **Gain**
Gaiser, Lulu Odell (1896–1965) S — **Gaiser**
Gaither, Thomas Walter (fl. 1979) M — **Gaither**
Gaja, L. (fl. 1911) M — **Gaja**
Gajewski, Wacław (1911–   ) S — **Gajewski**
Gajić, Milovan (fl. 1954) S — **Gajić**
Galán Cela, Pablo   S — **Galán Cela**
Galán de Mera, Antonio (fl. 1988) — **A.Galán**
Galán, Ricardo (fl. 1983) M — **R.Galán**
Galang, Marilyn M. (fl. 1965) S — **Galang**
Galanin, A.V. (fl. 1985) S — **Galanin**
Galdeano, Helvio L.   S — **Galdeano**
Galdieri, Agostino (1870–   ) A — **Galdieri**
(Galé, Julián Acuña)
   see Acuña Galé, Julián Baldomero — **Acuña**
Gale, Shirley (1915–   ) S — **Gale**
Galea, V.J. (fl. 1986) M — **Galea**
Galeano, Gloria A. (1958–   ) S — **Galeano**

(Galeano–Garcés, Gloria A.)
   see Galeano, Gloria A.   **Galeano**
Galeotti, Henri Guillaume (1814–1858) PS   **Galeotti**
Galgoczy, J. (fl. 1962) M   **Galgoczy**
Galiano, Emilio Fernández (1921–) S   **Galiano**
Galiano, Maria Elena (fl. 1974) S   **M.E.Galiano**
Galimova, L.M. (fl. 1989) M   **Galimova**
Galindo, Jorge (1931–) M   **Galindo**
(Galindo–A., Jorge)
   see Galindo, Jorge   **Galindo**
Galippe, V. (fl. 1883) M   **Galippe**
Galiulina (fl. 1951) M   **Galiulina**
Galkin, M.A. (fl. 1970) S   **Galkin**
Gallagher, Susan B. A   **Gallagher**
(Galland, Isodore)
   see Gallaud, Ernest–Isidore   **Gallaud**
Galland, Marie Cécile (fl. 1971) M   **Galland**
Galland, Nicole (1955–) S   **N.Galland**
Gallaud, Ernest–Isidore (fl. 1907) S   **Gallaud**
Gallé, László (1908–1980) M   **Gallé**
Gallego Cidoncha, María Jesús (1952–) S **Gallego**
Gallego, E. (fl. 1986) M   **E.Gallego**
Gallenmüller, Joseph (fl. 1876) S   **Gallenm.**
Gallerand, R. (fl. 1904) S   **Gall.**
Gallesio, Georgio (1772–1839) S   **Gallesio**
Galli, Roberto (fl. 1983) M   **Galli**
Galli–Vallerio, B. (fl. 1898) M   **Galli–Vall.**
Gallik, Oszvald A   **Gallik**
Gallo, Alfonso (1890–1952) S   **Gallo**
Galløe, Olaf (1881–1965) M   **Galløe**
Galloway, Beverly Thomas (1863–1938) M **Galloway**
Galloway, David John (1942–) M   **D.J.Galloway**
Galloway, David L. (fl. 1968) M   **D.L.Galloway**
Galloway, Leo A. (1921–) S   **L.A.Galloway**
Galloway, Leslie Douglas (fl. 1933) M   **L.D.Galloway**
Gallucci–Rangone, M.M. (fl. 1954) M   **Gall.–Rang.**
(Gallucci–Rangoni, M.M.)
   see Gallucci–Rangone, M.M.   **Gall.–Rang.**
Galpin, Ernest Edward (1858–1941) S   **Galpin**
Galstyan–Avanesyan, S.Kh. (fl. 1986) S **Galst.–Avan.**
Galtier, Jean (fl. 1970–77) F   **Galtier**
(Galučko, Anatol I.)
   see Galushko, Anatol I.   **Galushko**
Galun, Margalith (1927–) M   **Galun**
Galushko, Anatol I. (1926–) S   **Galushko**
Galuzo, I.G. A   **Galuzo**
Galway, Desma Hall (1917–) S   **Galway**
Galzin, Amédée (1853–1925) M   **Galzin**
(Gama, José de Saldanha da)
   see Saldanha da Gama, José de   **Saldanha**

Gamajunova, Aleksandra Pavlovna (1904–   ) S   **Gamajun.**
Gamal-Eldin, Ahmed (c.1964–) S   **A.Gamal-Eldin**
Gamal-Eldin, Elsayeda (1943–   ) S   **E.Gamal-Eldin**
(Gamalitzkaya, N.A.)
  see Gamalizkaya, N.A.   **Gamalizk.**
Gamalizkaya, N.A. (fl. 1958) M   **Gamalizk.**
Gambardella, Raffaele   A   **Gambard.**
Gambhir, S.P. (fl. 1979) M   **Gambhir**
Gamble, James Sykes (1847–1925) S   **Gamble**
Gambogi, P. (fl. 1957) M   **Gambogi**
Gambold, Gerhard (fl. 198?) M   **Gambold**
Gamerro, Juan .C. (1923–   ) F   **Gamerro**
Gamisans, Jacques (1944–   ) S   **Gamisans**
Gammel, John Anthony Stephan (1894–   ) M   **Gammel**
Gammie, George Alexander (1864–1935) PS   **Gammie**
Gams, Helmut (1893–1976) ABMS   **Gams**
Gams, K.Walter (1934–   ) M   **W.Gams**
Gamundí de Amos, Irma Josefa (1925–   ) M   **Gamundí**
Gan, C.F. (fl. 1983) S   **C.F.Gan**
Ganapathi, Alhagananthan (fl. 1978) M   **Ganap.**
Gančev, Ivan   S   **Gančev**
Gančev, Slavco Petrov (1920–   ) S   **S.P.Gančev**
Gandara, Guillermo (1879–1939) M   **Gandara**
Gander, Hieronymus (1832–1902) B   **Gander**
Gandhe, R.V. (fl. 1973) M   **Gandhe**
Gandhi, H.P.   A   **H.P.Gandhi**
Gandhi, Kancheepuram N. (1948–   ) S   **Gandhi**
Gandhi, S.K. (fl. 1979) M   **S.K.Gandhi**
Gandilyan, P.A. (1929–   ) S   **Gandilyan**
Gandoger, Michel (1850–1926) MPS   **Gand.**
Ganem, R.S.   A   **Ganem**
Ganencar, Narana   A   **Ganencar**
Ganesan, E.K. (1938–   ) A   **Ganesan**
Ganeschin, Sergej Sergejewitsch (1879–1930) S   **Ganesch.**
Ganesh, P.N. (fl. 1988) M   **Ganesh**
Gangawane, L.V. (fl. 1974) M   **Gang.**
Gangla, Kusum S.   A   **Gangla**
Gangopadhyay, Mohan G. (fl. 1988–) S   **M.G.Gangop.**
Gangopadhyay, N. (fl. 1989) S   **N.Gangop.**
Gangulee, Hirendra Chandra (1914–   ) B   **Gangulee**
Ganguly, D. (fl. 1963) M   **Ganguly**
Gankin, Roman W. (fl. 1971) S   **Gankin**
Ganz, Carol S.   A   **Ganz**
Ganzon-Fortes, Edna T. (1952–   ) A   **Ganz.-Fort.**
(Gao, Bao Chun)
  see Kao, Pao Chun   **P.C.Kao**
Gao, Cheng Zhi (fl. 1981) S   **C.Z.Gao**
Gao, Chien H. (1928–   ) B   **C.H.Gao**
Gao, Dong Fan (fl. 1980) S   **D.F.Gao**

Gao, Han Liang (fl. 1991) M   **H.L.Gao**
Gao, Hong Tu (fl. 1979) M   **H.T.Gao**
Gao, L.Z. (fl. 1987) S   **L.Z.Gao**
Gao, Lian Da   A   **L.D.Gao**
Gao, R.X. (fl. 1982) M   **R.X.Gao**
Gao, Run Qing (fl. 1984) S   **R.Q.Gao**
Gao, Wei Heng (fl. 1987) S   **W.H.Gao**
Gao, Wua (fl. 1979) S   **W.Gao**
Gao, Xiang Qun (1954–   ) M   **X.Q.Gao**
Gao, Y. (fl. 1982) M   **Y.Gao**
Gao, Zeng Yi (fl. 1985) S   **Z.Y.Gao**
Gao, Zhi Hai (fl. 1989) S   **Z.H.Gao**
Garabedjan, M.K. (fl. 1970) M   **M.K.Garab.**
Garabedjan, Star (1895–1978) S   **Garab.**
Garassini, Luiz A. (fl. 1972) M   **Garass.**
(Garavaglio, Santo)
    see Garovaglio, Santo   **Garov.**
Garaventa, H.Agustin (1911–1981) S   **Garaventa**
Garay, Leslie Andrew (1924–   ) S   **Garay**
Garbari, Fabio (1937–   ) S   **Garbari**
Garbary, David J. (fl. 1980) A   **Garbary**
Garber, Abram Paschal (1838–1881) P   **Garber**
Garber, Edward David (1918–   ) S   **E.D.Garber**
Garbini, Adriano (1857–1940) A   **Garbini**
Garbowski, I. (fl. 1923) M   **Garb.**
(Garcés Orejuela, Carlos)
    see Orejuela, Carlos Garcés   **Orejuela**
García Arévalo, Abel (fl. 1990) S   **García Arév.**
García Aser, Concepción (fl. 1972) M   **C.García**
García Caluff, Manuel (fl. 1991) P   **García Caluff**
Garcia de Orta (fl. 1490–1570) LS   **Garcia de Orta**
García, Donato (1782–1855) BPS   **D.García**
García Gonzalez, M.E. (fl. 1989) S   **M.E.García**
García Jacas, Núria (fl. 1988) S   **N.García**
García, Jesus (fl. 1981) M   **J.García**
García, José Gonçalves (1904–1971) S   **J.G.García**
García, L. (fl. 1988) M   **L.García**
(García Martín, Felipe)
    see García–Martín, Felipe   **García–Martín**
García Martínez, Xosé Ramón (fl. 1986) S   **García Mart.**
(García Murillo, Pablo)
    see García–Murillo, Pablo   **García–Mur.**
(García Nogueruela, Donato)
    see García, Donato   **D.García**
García, Rodolfo (1873–   ) S   **R.García**
García Vallejo, Isabel (fl. 1992) S   **García Vall.**
García Zúñiga, A. (fl. 1984) S   **A.García**
García–Barriga, Hernando (1913–   ) S   **García–Barr.**
García–Granados (fl. 1984) S   **García–Gran.**
García–Martín, Felipe (1954–   ) S   **García–Martín**

García–Mendoza, A. (fl. 1986) S	**García-Mend.**
García–Murillo, Pablo (1962–) S	**García-Mur.**
García–Rada, German (1902–) M	**García-Rada**
García–Zorrón, Noemi (fl. 1977) M	**García-Zorrón**
Garcias Font, Llorenç (1885–1975) S	**Garcias Font**
(Garcias y Font, Lorenzo (Lorens))	
see Garcias Font, Llorenç	**Garcias Font**
Garcin, R. (fl. 1984) M	**Garcin**
Garcke, Christian August Friedrich (1819–1904) S	**Garcke**
Gard, Médéric A	**Gard**
Garden, Alexander (1730–1792) S	**Garden**
Garden, Joy (after 1956, Thompson, Joy) (1923–) S	**J.Garden**
(Gardés, Anisochor)	
see Orejuela, Carlos Garcés	**Orejuela**
Gardet, G. (of Nancy) (fl. 1931) AB	**Gardet**
Gardet, Monique A	**M.Gardet**
Gardiner, William (1808–1852) S	**Gardiner**
Gardner, C.S. (fl. 1984) S	**C.S.Gardner**
Gardner, Charles Austin (1896–1970) S	**C.A.Gardner**
Gardner, Donald E. (fl. 1985) M	**D.E.Gardner**
Gardner, George (1812–1849) PS	**Gardner**
Gardner, John Sharkie (fl. 1886) F	**J.S.Gardner**
Gardner, Max William (1890–1979) M	**M.W.Gardner**
Gardner, Nathaniel Lyon (1864–1937) AM	**N.L.Gardner**
Gardner, Rhys Owen (1949–) S	**R.O.Gardner**
Gardoni, Luigi (1819–1880) S	**Gardoni**
Gardou, Christiane S	**Gardou**
(Gareth Jones, E.B.)	
see Jones, E.B. Gareth	**E.B.G.Jones**
Garg, A.K. (fl. 1966) M	**Garg**
Garg, Sunita (née Agrawal, S.) (fl. 1987) S	**S.Garg**
Gargiulo, Gaetano M. (1955–) A	**Gargiulo**
Garibaldi, Eleni Accati (1938–) M	**Garib.**
Gariod, Charles Henri (1836–1892) S	**Gariod**
Garitty, J. S	**Garitty**
Garkovenko, V.M. (fl. 1989) S	**Garkov.**
Garman, Philip (1891–) M	**Garman**
Garnet, John Roslyn (1906–) S	**Garnet**
Garnier, G. (fl. 1991) M	**G.Garnier**
Garnier, Max (fl. 1895–1918) S	**Garnier**
Garnier, R. (fl. 1961) M	**R.Garnier**
Garnock–Jones, Philip John (1950–) S	**Garn.-Jones**
Garofano, L. (fl. 1973) M	**Garofano**
Garoglio, P.G. (fl. 1953) M	**Garoglio**
Garovaglio, Santo (1805–1882) BCMS	**Garov.**
Garre, M. (fl. 1984) S	**Garre**
Garretson, J.D. S	**Garretson**
Garrett, Albert Osbun (1870–1948) M	**Garrett**

(Garrido G., Norberto)	
see Garrido, Norberto	**Garrido**
Garrido, Norberto (1952–) M	**Garrido**
Garsault, François Alexandre Pierre de (1691–1778) PS	**Garsault**
Garside, Sidney (1889–1961) B	**Garside**
Gärtner, Georg (1946–) A	**G.Gärtner**
Gartner, Hans (fl. 1939) S	**Gartner**
Gartner, Stefan, jr. (1937–) A	**S.Gartner**
(Gartwiss, Nicolai Anders von)	
see Hartwiss, Nicolai Anders von	**Hartwiss**
Garud, A.B. (fl. 1968) M	**Garud**
Garwood, Edmund Johnston (1864–1949) A	**Garwood**
Garzia (fl. 1838) S	**Garzia**
(Gasa, Eric la)	
see la Gasa, Eric	**la Gasa**
Gasc, C. (fl. 1972) M	**Gasc**
(Gashimov, D.K.)	
see Hashimov, D.K.	**Hash.**
(Gasilien, Frère)	
see Parrique, Géraud	**Parrique**
Gaspar, Frida C. S	**Gaspar**
Gaspariková, Viera A	**Gasparik.**
Gasparrini, Guglielmo (1804–1866) MPS	**Gasp.**
Gasperini, G. (fl. 1886) M	**Gasperini**
Gassner, Johann Gustav (1881–1955) M	**Gassner**
Gasson, Peter Eric (1957–) S	**Gasson**
Gastaldo, Paola (1927–) S	**Gastaldo**
Gastony, Gerald Joseph (1940–) P	**Gastony**
Gaterau (fl. 1789) S	**Gaterau**
Gates, Bronwen (1945–) S	**B.Gates**
Gates, David M. (1921–) S	**D.M.Gates**
Gates, Frank Caleb (1887–1955) S	**F.C.Gates**
Gates, Howard Elliott (1895–1957) S	**H.E.Gates**
Gates, Reginald Ruggles (1882–1962) S	**R.R.Gates**
Gathoye, Jean–Louis (1963–) S	**Gathoye**
Gatschetschiladze, K.A. (fl. 1949) S	**Gatsch.**
Gattefossé, Jean (1899–) S	**Gattef.**
Gattenhof, Georg Matthias (1722–1788) S	**Gattenhof**
Gatterer, Christoph Wilhelm Jacob (1759–1838) B	**Gatterer**
Gatti, C. (fl. 1929) M	**Gatti**
Gattinger, Augustin (1825–1903) S	**Gatt.**
Gatty, Margaret Scott (1809–1873) A	**Gatty**
Gauba, Erwin (1891–1964) S	**Gauba**
Gauckler, Konrad (1898–1983) S	**Gauckler**
(Gaudichaud, Charles)	
see Gaudichaud–Beaupré, Charles	**Gaudich.**
Gaudichaud–Beaupré, Charles (1789–1854) ABMPS	**Gaudich.**
(Gaudichaud–Beaupres, Charles)	
see Gaudichaud–Beaupré, Charles	**Gaudich.**

Gaudin, Jean François Aimée Gottlieb Philippe (1766–1833) F   **Gaudin**
Gaudin, L. (fl. 1919) M   **L.Gaudin**
Gaufin, Arden Rupert (1936–   ) A   **Gaufin**
Gaugué, G. (fl. 1973) M   **Gaugué**
Gäumann, Ernest (Ernst) Albert (1893–1963) M   **Gäum.**
Gaume, Marie Leon Camille Raymond (1885–1964) B   **Gaume**
Gaur, R.C. (1933–   ) S   **R.C.Gaur**
Gaur, R.D. (1945–   ) MS   **R.D.Gaur**
Gaussen, Henri Marcel (1891–1981) S   **Gaussen**
Gautam, S.K. (fl. 1982) M   **S.K.Gautam**
Gautam, S.P. (fl. 1983) M   **S.P.Gautam**
Gauthier, M. (fl. 1961) AM   **M.Gauthier**
Gauthier, Robert (1941–   ) S   **R.Gauthier**
Gauthier–Lièvre, Lucienne Emilienne (1897–   ) AB   **Gauth.-Lièvre**
Gautier, Marie Clément Gaston (1841–1911) S   **Gaut.**
Gavaudan, Pierre   A   **Gavaudan**
Gavioli, Orazio (1871–1944) S   **Gavioli**
Gaviria, J. (fl. 1987) S   **Gaviria**
Gavrilenko, B.D.   S   **Gavr.**
Gavrilenko, I.G. (1956–   ) S   **I.G.Gavr.**
(Gawler, John Bellenden Ker)
    see Ker Gawler, John Bellenden   **Ker Gawl.**
Gawłowska, Maria J. (1910–   ) S   **Gawł.**
Gay, (Henri Félix) François (1858–1898) AS   **F.Gay**
Gay, Claude (1800–1873) ABMPS   **Gay**
Gay, Jacques Étienne (1786–1864) PS   **J.Gay**
Gayão, Teresa de Jesus (fl. 1951) M   **Gayão**
Gáyer, Gyula (Julius) (1883–1932) S   **Gáyer**
Gaynor, B.D. (fl. 1987) S   **Gaynor**
Gayoso, A.M.   A   **Gayoso**
Gayral, Paulette (1921–   ) A   **Gayral**
(Gayral–Engerbaud, Paulette)
    see Gayral, Paulette   **Gayral**
(Gazzar, Adel Ibrahim Hamed El)
    see El Gazzar, Adel Ibrahim Hamed   **El Gazzar**
Ge, Qi Xin (1911–   ) M   **Q.X.Ge**
Gea, F.J. (fl. 1991) M   **Gea**
Gebhard, Johann Nepomuk (1774–1827) S   **Gebh.**
Geddes, E.T.   S   **E.T.Geddes**
Geddes, Patrick (1854–1932) A   **Geddes**
Gedoelst, L.M. (fl. 1902–1911) M   **Gedoelst**
(Geel, Pierre Corneille (Petrus Cornelius) Van)
    see Van Geel, Pierre Corneille (Petrus Cornelius)   **Van Geel**
Geer van Jutphaas, Jan Lodewijk Willem de (1784–1857) S   **Geer**
Geerinck, Daniel (1945–   ) S   **Geerinck**
(Geert, August(e) van)
    see Van Geert, August(e)   **Van Geert**
Geesink, J. (fl. 1976) MS   **J.Geesink**

Geesink, Robert (1945–) S	**R.Geesink**
Geeson, J.D. (fl. 1975) M	**Geeson**
Geevarghese, K.K. (fl. 1981–89) P	**Geev.**
Geheeb, Adalbert (1842–1909) B	**Geh.**
Geholt, C.S. (fl. 1972) M	**Geholt**
Gehrmann, Karl (1885–1925) MS	**Gehrm.**
Géhu, Jean–Marie (fl. 1976–1983) S	**Géhu**
Gehu-Franck, J. (fl. 1981) S	**Gehu-Franck**
Geiger, Arthur (fl. 1910) M	**A.Geiger**
Geiger, Philipp Lorenz (1785–1836) S	**Geiger**
Geilinger, Gottlieb (1881–1955) S	**Geil.**
Geiman, Quentin Monroe (1904–) A	**Geiman**
Geinitz, Franz Eugen (1854–1925) F	**F.E.Geinitz**
Geinitz, Hans Bruno (1814–1900) FM	**Geinitz**
Geise, M.J. S	**Geise**
Geiseler, Eduard Ferdinand (1781–1827) S	**Geiseler**
Geisenheyner, Franz Adolf Ludwig (1841–1926) PS	**Geisenh.**
Geissbühler, Jakob (1896–) A	**Geissb.**
Geissert, Fritz (fl. 1980) PS	**Geissert**
Geissler, Ursula (1931–) A	**Geissler**
Geisy, R.M. (fl. 1962) B	**Geisy**
Geitler, Lothar (1899–1990) AM	**Geitler**
Geitzenauer, Kurt R. (1940–) A	**Geitzen.**
Gelbrecht, J. (fl. 1987) S	**Gelbr.**
Geldikhanov, A.M. (1953–) S	**Geld.**
(Geldykhanov, A.M.)	
see Geldikhanov, A.M.	**Geld.**
Gelert, Otto Christian Leonor (Kristian Laurits) (1862–1899) PS	**Gelert**
(Gelmersen, G.P.)	
see Helmersen, G.P.	**Helmersen**
Gelmi, Enrico (1855–) S	**Gelmi**
Gelonesi, G. (fl. 1927) M	**Gelonesi**
Gelting, Paul Emil Elliot (1905–1964) M	**Gelting**
Geltman, Dmitri Victorovich (1957–) S	**Geltman**
Gelyuta, V.(W.)P. (fl. 1984) M	**V.P.Gelyuta**
Gembardt, C. (fl. 1987) S	**Gembardt**
Gemeinhardt, Konrad (1883–) A	**Gemeinh.**
Gemici, Yusuf (1957–) S	**Gemici**
Gemma, Jane N. (1951–) M	**Gemma**
Gemsch, Norbert A	**Gemsch**
Gena, C.B. (fl. 1987) PS	**Gena**
Gené, J. (fl. 1990) M	**Gené**
Géneau de Lamarlière, L. (fl. 1894) M	**Géneau**
Genersich, Samuel (1768–1844) S	**Geners.**
Genestar Serra, Rafael (fl. 1956) M	**Genestar**
Genevier, Léon Gaston (1830–1880) MS	**Genev.**
Genini, A. (fl. 1985) F	**Genini**

(Genissel–Homolle (Madame Le))	
see Homolle, A.–M.	**Homolle**
Genkal, S.I. A	**Genkal**
Genkel, A. (fl. 1913) M	**Genkel**
Gennadius, Panagiotes G. S	**Gennad.**
Gennari, Patrizio (1820–1897) PS	**Gennari**
Genoud, E.G. (fl. 1913) M	**Genoud**
(Genoud, J.Toilliez)	
see Toilliez–Genoud, J.	**Toill.-Gen.**
Genova, M.K. (fl. 1970) M	**Genova**
Genth, Carl Friedrich Ferdinand (1810–1837) BM	**Genth**
Gentil, Ambroise (1842–1929) S	**Gentil**
Gentil, Théodore August Louis François (1874–1949) S	**L.Gentil**
Gentile, Richard (1929–) A	**Gentile**
Gentili, J. (fl. 1953) M	**Gentili**
Gentles, J.C. (fl. 1961) M	**Gentles**
Gentner, Georg (1877–1940) M	**Gentner**
Gentner, Walter Andrew (1922–) S	**W.A.Gentner**
Gentry, Alwyn Howard (1945–) S	**A.H.Gentry**
Gentry, Howard Scott (1903–) S	**Gentry**
Gentry, Johnnie Lee (1939–) S	**J.L.Gentry**
Genty, Paul André (1861–1955) MS	**P.A.Genty**
Genty, T.G. (1843–) S	**Genty**
Geoffrey, A. (fl. 1978) B	**Geoffrey**
(Georg, Alexander Black)	
see George, Alexander Black	**A.B.George**
Georg, Lucille K. (fl. 1952) M	**Georg**
George, Alexander Black (1916–1957) S	**A.B.George**
George, Alexander Segger (1939–) S	**A.S.George**
George, Edward (1830–1900) AB	**George**
George, Eric Alan (1920–) A	**E.A.George**
George, K.V. (fl. 1960) M	**K.V.George**
Georges, G. A	**Georges**
Georgescu, C.Constantin (1898–1968) MS	**Georgescu**
Georgevitch, Jivoin A	**J.Georgev.**
Georgevitch, Peter (Pierre) (1874–1947) AM	**Georgev.**
Georgi, Johann Gottlieb (1729–1802) PS	**Georgi**
Georgia, Ada E. (–1921) S	**Georgia**
Georgiadis, G. (fl. 1981) S	**G.Georgiadis**
Georgiadis, T. (fl. 1977) S	**T.Georgiadis**
Georgiadou, E. (fl. 1978) S	**Georgiadou**
(Georgieff, Stephan)	
see Georgiev, Stephan	**Georgiev**
(Georgieff, Toma Coev)	
see Georgiev, Toma Coev	**T.Georgiev**
Georgiev, Stephan (1859–1900) S	**Georgiev**
Georgiev, Toma Coev (1883–) S	**T.Georgiev**
Gepp, Anthony (Antony) (1862–1955) ABPS	**A.Gepp**
Gepp, Ethel Sarel Barton (née Barton, E.S.) (1864–1922) A	**E.Gepp**

Gérard, F. (fl. 1885) S F.Gérard
Gerard, John (1545–1612) L J.Gerard
Gérard, Louis (1733–1819) S Gérard
Gerard, René Constant Joseph (1853–1935) S R.C.J.Gerard
Gerard, William Ruggles (1841–1914) M W.R.Gerard
(Gerarde, John)
   see Gerard, John J.Gerard
(Gérardin de Mirécourt, Sébastien)
   see Gérardin, Sébastien Gérardin
Gérardin, Sébastien (1751–1816) PS Gérardin
Gerasimenko, I.I. (1939–) S Geras.
Gerbeaux S Gerbeaux
Gerber, G. (of Bavaria) (fl. 1860) B Gerber
Gerber, Kurt (1904–) M K.Gerber
Gerbino, Xaverio (Saverio) (1814–1898) S Gerbino
Gerdemann, James W. (1921–) M Gerd.
Gereau, Roy E. (1947–) S Gereau
Geremicca, M. S Geremicca
Gergely, Ivan János (Ioan) (1928–1990) S Gergely
Gerhardt, Elizabeth (1931–) S El.Gerhardt
Gerhardt, Ewald (1947–) M Ew.Gerhardt
Gerhardt, Julius (1827–) C Gerhardt
Gerhold, N. (fl. 1980) M Gerhold
Gerlach, Ellen A E.Gerlach
Gerlach, G. (fl. 1987) S G.Gerlach
Gerlach, W. (fl. 1959) M Gerlach
Gerlagh, M. (fl. 1968) M Gerlagh
Gerloff, Johannes Hermann (1915–) A Gerloff
Gerloff, N. (fl. 1990) S N.Gerloff
Germain de Saint-Pierre, Jacques Nicolas Ernest (1815–1882) AMPS Germ.
(Germain, Ernest)
   see Germain de Saint-Pierre, Jacques Nicolas Ernest Germ.
Germain, Henry (1903–1989) A H.Germ.
Germain, René Gerard Antoine (1914–1982) S R.Germ.
German, M.T. (fl. 1985) S German
(German R., M.T.)
   see German, M.T. German
Germano, G. (fl. 1977) M Germano
Germar, Ernst Freidrich (1786–1853) S Germar
Germishuizen, Gerrit (1950–) S Germish.
Gerneck, Rudolf A Gerneck
Gérôme, Joseph (1863–1928) S Gérôme
Gerrard, William Tyrer (–1866) PS Gerrard
Gerrath, Joseph F. (1936–) A Gerrath
Gerrettson-Cornell, L. (fl. 1980) M Gerr.-Corn.
Gerstberger, Pedro (1951–) S Gerstb.
Gerstlauer, Lorenz (1863–1949) S Gerstl.
Gerstner, Jacob (1888–1948) S Gerstner
Gerth, H. (fl. 1932) F H.Gerth

Gerth van Wijk, Hugo Leonardus (1849–1921) S — **Gerth**
Gervais, Camille (1933–) S — **Gervais**
Geschele, E.E. (fl. 1927) M — **Geschele**
Gesner, Conrad von (1516–1565) S — **Gesner**
Gessner, Fritz (1905–1972) A — **F.Gessner**
Gessner, Johann (1709–1790) S — **Gessner**
Gessner, Robert V. (1948–) M — **R.V.Gessner**
Gestro, Raffaele (Raffaelo) (1845–1936) S — **Gestro**
Getachew Aweke (1937–) S — **Getachew**
Getliffe, Fiona Mary (1941–) S — **Getliffe**
(Getliffe Norris, Fiona Mary)
    see Getliffe, Fiona Mary — **Getliffe**
Geuns, Matthias (1735–1817) S — **M.Geuns**
Geuns, Steven Jan van (1767–1795) S — **S.Geuns**
Gevers Deynoot, Pieter Marie Eduard (1816–1860) S — **Gev.Deyn.**
Geyer, Carl (Charles) Andreas (1809–1853) S — **Geyer**
Geyer, Mechthild (fl. 1987) M — **M.Geyer**
Geyler, Hermann Theodor (1834–1889) AFM — **Geyl.**
Ghaffer, A. (fl. 1972) M — **Ghaffer**
Ghafoor, Abdul (1938–) S — **Ghafoor**
Ghahreman, Ahmed (1928–) S — **Ghahr.**
Ghandhi, H.P.   A — **Ghandhi**
Ghardi   S — **Ghardi**
Gharse, P.S. (fl. 1964) M — **Gharse**
Ghatak, Jagadananda (1928–) P — **J.Ghatak**
Ghatak, Prafulla Nath (1902–1939) M — **Ghatak**
Ghate, N. (fl. 1985) M — **Ghate**
Ghazanfar, Shahina Agha (1949–) S — **Ghaz.**
Gheldoff, Constantin (fl. 1845) S — **Gheldoff**
(Gheorghieff, Stephan)
    see Georgiev, Stephan — **Georgiev**
Gherardi, Bartolomeo (1783–1857) S — **Gherardi**
Ghesquière, Jean H.P.A. (1888–) MS — **Ghesq.**
Ghiesbreght, Auguste Boniface (1810–1893) S — **Ghiesbr.**
Ghika, N.D.   S — **Ghika**
Ghildyal, N. (fl. 1986) S — **Ghildyal**
Ghini, Luca (1490–1556) S — **Ghini**
Ghişa, Eugen Victor (1909–1984) S — **Ghişa**
(Ghobi, Khristophos (Christophos) Jacoblewitsch)
    see Gobi, Christophos Jacoblewitsch (Khristofor Yakovlevich) — **Gobi**
Ghora, S.Chhabi (fl. 1985) S — **Ghora**
(Ghoryaninov, Paul (Paulus) Fedorowitsch)
    see Horaninow, Paul (Paulus) Fedorowitsch — **Horan.**
Ghose, Birenda Nath (1885–1983) S — **Ghose**
Ghose, R.L.M.   S — **R.L.M.Ghose**
Ghose, Surenda Lal (1893–1945) A — **S.L.Ghose**
Ghosh, B. (fl. 1980) PS — **B.Ghosh**
Ghosh, Gouri Rani (fl. 1985) M — **G.R.Ghosh**
Ghosh, R.K. (fl. 1981–89) S — **R.K.Ghosh**

Ghosh, Rathindra Nath A	**R.N.Ghosh**
Ghosh, S.N. (fl. 1986) M	**S.N.Ghosh**
Ghosh, S.R. (1942–) PS	**S.R.Ghosh**
Giaccone, Giuseppe (1936–) A	**Giaccone**
Giacobbe, Andrea (1891–1981) S	**Giacobbe**
Giacomini, Valerio (1914–1981) BS	**Giacom.**
Giaiotti, Andreina (L.) (fl. 1977) M	**Giaiotti**
Giangualani, R.N. (fl. 1982) S	**Giang.**
Giard, Alfred–Matheiu (1846–1908) AM	**Giard**
Giardelli, María Luisa S	**Giardelli**
(Giardiniere, Colombano)	
see Nocca, Domenico	**Nocca**
Giatgong, P. (fl. 1971) M	**Giatgong**
Gibbes, Lewis Reeve (1810–1894) S	**Gibbes**
Gibbs, Lilian Suzette (1870–1925) ABMPS	**Gibbs**
Gibbs, Peter Edward (1938–) S	**P.E.Gibbs**
Gibbs, Vicary (1853–1932) S	**V.Gibbs**
Gibbs–Russell, Garland Elizabeth (1945–) S	**Gibbs–Russ.**
Gibby, Mary (1949–) P	**Gibby**
Gibelli, Giuseppe (1831–1898) MPS	**Gibelli**
Gibert, José Ernesto (1818–1886) S	**Gibert**
Gibson, Alexander (1800–1867) S	**A.Gibson**
Gibson, Arthur Charles (1947–) S	**A.C.Gibson**
Gibson, Dorothy L.Nash (née Nash, D.L.) (1921–) S	**D.N.Gibson**
Gibson, Frederick (1892–1953) S	**F.Gibson**
Gibson, George Stacey (1818–1883) S	**Gibson**
Gibson, Ian A.S. (fl. 1972) M	**I.A.S.Gibson**
Gibson, Jack L. (–1990) MS	**J.L.Gibson**
Gibson, Robert Adams (1946–) A	**R.A.Gibson**
Gibson, Samuel (1790–1849) S	**S.Gibson**
Gibson, William Hamilton (1850–1896) M	**W.H.Gibson**
Gibulescu, R. (fl. 1968) F	**Gibulescu**
Gickelhorn, Joseph (Josef) (1891–1957) AM	**Gickelh.**
Giddens, J.E. (fl. 1958) M	**Giddens**
Giddings, Nahum James (1883–1966) M	**Giddings**
Gideon, O. (fl. 1983) S	**Gideon**
Giebel, Christoph Gottfried Andreas (1820–1881) F	**Giebel**
Gier, Leland Jacob (1904–1983) B	**Gier**
Gierczak, Maria (fl. 1963) M	**Gierczak**
Giese, Monika (1959–) B	**Giese**
Giesen, Hans (fl. 1938) S	**Giesen**
Giesenhagen, Karl Friedrich Georg (1860–1928) BMPS	**Giesenh.**
Giess, Johan Wilhelm Heinrich (1910–) S	**Giess**
Giesy, Robert Marshall (1922–) B	**Giesy**
Giffen, Malcolm Hutchison (1902–) A	**Giffen**
Giglioli, Italo (1852–1920) S	**Giglioli**
Gijać, M. S	**Gijać**
(Gikachvili, K.G.)	
see Gikaschvili, K.G.	**Gikaschvili**

Gikaschvili, K.G. (fl. 1948) M	**Gikaschvili**
(Gikashvili, K.G.)	
see Gikaschvili, K.G.	**Gikaschvili**
Gil Llano, J.R. (fl. 1990) S	**Gil Llano**
Gilbert, Benjamin Davis (1835–1907) PS	**Gilbert**
Gilbert, Edward Martinius (1875–1956) M	**E.M.Gilbert**
Gilbert, Elizabeth F. (1929–) PS	**E.F.Gilbert**
Gilbert, Georges Charles Clément (1908–1983) S	**G.C.C.Gilbert**
Gilbert, Henry Clark (1891–) M	**H.C.Gilbert**
Gilbert, Jean–Edouard (Edward) (fl. 1918) M	**J.-E.Gilbert**
Gilbert, John Lester (1920–1985) S	**J.L.Gilbert**
Gilbert, M.E. (fl. 1926) M	**M.E.Gilbert**
Gilbert, Michael George (1943–) S	**M.G.Gilbert**
Gilbert, Oliver Lathe (1936–) BM	**O.L.Gilbert**
Gilbert, P.A. (fl. 1942) S	**P.A.Gilbert**
Gilbert, William James (1916–) A	**W.J.Gilbert**
Gilbert, William Williams (1880–1940) M	**W.W.Gilbert**
Gilbertson, G.L. (fl. 1986) M	**G.L.Gilb.**
Gilbertson, Robert Lee (1925–) M	**Gilb.**
Gilchrist, T.Caspar (fl. 1896) M	**Gilchrist**
Gilenstam, G. (fl. 1969) M	**Gilenstam**
Gilert, Elizabeth (1947–) M	**Gilert**
Giles, B.E. (fl. 1988) S	**Giles**
Gilg, E. (fl. 1934) M	**E.Gilg**
Gilg, Ernest Friedrich (1867–1933) S	**Gilg**
Gilg–Benedict, Charlotte S	**Gilg-Ben.**
Gilibert, Jean–Emmanuel (1741–1814) BMPS	**Gilib.**
Gilkey, Helen Margaret (1886–1972) MS	**Gilkey**
Gilkinet, A. (fl. 1880) M	**Gilkinet**
Gill, Denzell Leigh (1909–) M	**D.L.Gill**
Gill, H.S. (fl. 1961) M	**H.S.Gill**
Gill, Jiri (1936–) S	**J.Gill**
Gill, Lake Shore (1900–) MS	**L.S.Gill**
Gillekens, Leopold Guillaume (1833–1905) S	**Gillek.**
Gilles, Gérard (fl. 1969) M	**Gilles**
Gillespie, James P. (fl. 1962) P	**J.P.Gillespie**
Gillespie, John Wynn (–1932) S	**Gillespie**
Gillet, Abel (1857–1927) M	**A.Gillet**
Gillet, Claude–Casimir (1806–1896) MS	**Gillet**
Gillet, Hubert (1924–) B	**H.Gillet**
Gillet, Justin (1866–1943) S	**J.Gillet**
Gillet–Lefebvre, Jeanne (fl. 1965) B	**Gillet-Lef.**
Gillett, George Wilson (1917–1976) S	**G.W.Gillett**
Gillett, Jan Bevington (1911–) S	**J.B.Gillett**
Gillett, John Montague (1918–) S	**J.M.Gillett**
Gillett, Margaret Clark (1878–1962) S	**Gillett**
Gilli, Alexander (1904–) PS	**Gilli**
Gilliam, Martina S. (fl. 1975) M	**Gilliam**
Gillies, Concettinam A	**C.Gillies**
Gillies, John (1792–1834) B	**Gillies**

Gilliland, Hamish Boyd (1911–1965) S	**Gilliland**
Gilliland, R.B. (fl. 1962) M	**R.B.Gilliland**
Gillis, William Thomas (1933–1979) S	**Gillis**
Gillman, Henry (1833–1915) S	**Gillman**
Gillman, Linnea Stewart (fl. 1977) M	**L.S.Gillman**
Gillot, François Xavier (1842–1910) MS	**Gillot**
Gilly, Charles Louis (1911–1970) S	**Gilly**
Gilman, Joseph Charles (1890–1966) M	**J.C.Gilman**
Gilman, Marshall French (1871–1944) S	**Gilman**
Gilmartin, Amy Jean (1932–1989) S	**Gilmartin**
Gilmore, J.W. (fl. 1966) M	**Gilmore**
Gilmour, John Scott Lennox (1906–1986) S	**Gilmour**
Gilmour, P.M. (fl. 1987) S	**P.M.Gilmour**
Gilomen, Hans (1886–1940) S	**Gilomen**
Gilpin, R.H. (fl. 1954) M	**Gilpin**
Giménez-Jurado, G. (fl. 1990) M	**Gim.-Jurado**
Gimesi, M. (fl. 1924) M	**M.Gimesi**
Gimesi, Nándor István (1892–1953) A	**Gimesi**
Ginanni, Francesco (1716–1766) S	**F.Ginanni**
Ginanni, Giuseppe (1692–1753) L	**Ginanni**
Gindrat, D. (fl. 1967) M	**Gindrat**
Gineste, Ch. A	**Gineste**
Gingins de la Sarraz, Frédéric Charles Jean (1790–1863) S	**Ging.**
(Gingins, Frédéric Charles Jean)	
see Gingins de la Sarraz, Frédéric Charles Jean	**Ging.**
Ginns, James Herbert (1938–) M	**Ginns**
Ginori Conti, Piero (1865–1939) S	**Gin.Conti**
Gintis, B.O. (fl. 1984) M	**Gintis**
Gintl, W. (fl. 1856) M	**Gintl**
Gintovt, E.A. (fl. 1959) M	**Gintovt**
Ginzberger, August (1873–1940) B	**Ginzb.**
Gioelli, Felice (1901–1970) S	**Gioelli**
Giordano, Antonio (fl. 1939) M	**A.Giord.**
Giordano, Ferdinando (fl. early 19th Cent.) S	**Giord.**
Giordano, Guiseppe Camillo (1841–1901) B	**G.C.Giord.**
Giordano, M. (fl. 1918) M	**M.Giord.**
Giovannetti, Manuela (fl. 1991) M	**Giovann.**
Giovannozzi, M. (fl. 1948) M	**Giov.**
Giralt, Mireia (fl. 1990) M	**Giralt**
Girard, Frédéric de (1810–1851) S	**Girard**
Girard, Henri S	**H.Girard**
Girard, R. (fl. 1956) M	**R.Girard**
Girard, V. (fl. 1987) S `	**V.Girard**
Girardet, A. (fl. c.1891) P	**Girardet**
Girardi, Annamaria M. (fl. 1970) S	**Girardi**
Girardin, Jean Pierre Louis (1803–1884) S	**Girardin**
Giraud, G. A	**Giraud**
Giraudias, Ludovic (1848–1922) S	**Giraudias**
Girbal, J. (fl. 1982) M	**Girbal**

Girenko, M.M. (1930–) S	**Girenko**
Girgensohn, Gustav Karl (1786–1872) B	**Girg.**
Giri, B.K. (fl. 1966) M	**B.K.Giri**
Giri, G.S. (1950–) S	**G.S.Giri**
Girod–Chantrans, Justin (1750–1841) AS	**Gir.-Chantr.**
Giroux, Mathilde (fl. 1933) S	**Giroux**
Girzitska, Z. (fl. 1929) M	**Girz.**
Giseke, Paul Dietrich (1741–1796) S	**Giseke**
Gisler, Roland A	**Gisler**
Gistl, Rudolf (1891–) A	**Gistl**
Gitmez, Gulden Usman A	**Gitmez**
Giudici de Nicola, Marina (1923–) A	**Giudici**
Giulietti, Anna Maria (1945–) S	**Giul.**
Giuma, A.Y. (fl. 1972) M	**Giuma**
Giusti, Lionel (fl. 1984) S	**Giusti**
Given, David Roger (1943–) PS	**Given**
Givulescu, Răzvan (1920–) BM	**Givul.**
Gizhitskaya, Z.K. (fl. 1929) M	**Gizhitsk.**
Gjaerevoll, Olav (1916–) BS	**Gjaerev.**
Gjaerum, Halvor B. (fl. 1961) M	**Gjaerum**
Gjurašin, Stjepan (1867–1936) MS	**Gjurašin**
Glaab, Ludwig Jakob (1858–1928) S	**Glaab**
Gladders, P. (fl. 1988) M	**Gladders**
Gladkova, V.N. (1936–) S	**Gladkova**
Gladstones, J.S. (fl.1970) S	**Gladst.**
Glaetzle, W. (fl. 1983) S	**Glaetzle**
Glare, T.R. (fl. 1987) M	**Glare**
Glas, W.P. (fl. 1959) B	**Glas**
Glasau, Fritz (fl. 1961) S	**Glasau**
Glaser, T. (fl. 1953) M	**Glaser**
(Glaser, Z.I.)	
see Glezer, Z.I.	**Glezer**
Glass, Charles (fl. 1975) S	**Glass**
Glassman, Sidney Frederick (1919–) PS	**Glassman**
Glatfelter, Noah Miller (1837–1911) S	**Glatf.**
Glawe, Dean A. (fl. 1987) M	**Glawe**
Glazebrook, Thomas Kirkland (1780–1885) S	**Glazebr.**
Glaziou, Auguste François Marie (1828–1906) PS	**Glaz.**
Gleason, Henry Allan (1882–1975) PS	**Gleason**
Gleba, Y.Y. S	**Gleba**
Gledhill, David (1929–) S	**Gledhill**
Gleditsch, Johann Gottlieb (1714–1786) MPS	**Gled.**
Gleeson, P. (fl. 1976) M	**Gleeson**
Glehn, Peter von (1835–1876) S	**Glehn**
Gleichen, Wilhelm Friedrich von (1717–1783) S	**Gleichen**
Gleisberg, Walther (1891–) A	**Gleisberg**
Glen Bott, Janet I. (fl. 1951) M	**Glen Bott**
Glen, Hugh Francis (1950–) S	**Glen**
Glendinning, Robert (1805–1862) S	**Glend.**

Glenk, Hans–Otto A	**Glenk**
(Gleser, S.I.)	
see Glezer, Z.I.	**Glezer**
Glezer, Z.I. (fl. 1959) AM	**Glezer**
Glick, Perry Aaron (1895–) M	**Glick**
Glime, Janice Mildred (1941–) B	**Glime**
(Glinka, Eduard Janczewski von)	
see Janczewski, Eduard von Glinka	**Jancz.**
Glinsukon, T. (fl. 1977) M	**Glins.**
Glintzboeckel, Ch. A	**Glintzb.**
Glocker, Ernst Friedrich ('Constantino') von (1793–1858) AF	**Glocker**
Glogau, Arthur (1874–) S	**Glogau**
Glover, James (1844–1925) S	**Glover**
Glover, Phillip Earle (1912–) S	**P.E.Glover**
(Glowacki, Julius)	
see Głowacki, Julius	**Głow.**
Głowacki, Julius (1846–1915) BM	**Głow.**
Glowinski, H. (fl. 1982) M	**Glowinski**
Gloxin, Benjamin Peter (1765–1794) S	**Gloxin**
Gloyer, Walter Oscar (1886–) M	**Gloyer**
Glück, Christian Maximilian Hugo (1868–1940) AMPS	**Glück**
Gluge, G. (fl. 1857) M	**Gluge**
Gmelin, Carl (Karl) Christian (1762–1837) S	**C.C.Gmel.**
Gmelin, Johann Friedrich (1748–1804) ABMPS	**J.F.Gmel.**
Gmelin, Johann Georg (1709–1755) S	**J.G.Gmel.**
Gmelin, Philipp Friedrich (1721–1768) S	**P.F.Gmel.**
Gmelin, Samuel Gottlieb (1743/45–1774) APS	**S.G.Gmel.**
Gnilovskaja, M.B. A	**Gnil.**
Goadby, Bede Theodoric (1863–1944) S	**Goadby**
Goaty, Etienne Louis Henri (1830–1890) S	**Goaty**
Göbelez, M. (fl. 1960) M	**Göbelez**
Gobi, Christophos Jacoblewitsch (Khristofor Yakovlevich) (1847–1919) AM	**Gobi**
Gochenaur, S.E. (fl. 1963) M	**Goch.**
Gochnat, Frédéric Charles (–1816) S	**Gochnat**
Gocht, Hans A	**Gocht**
Góczán, F. A	**Góczán**
Goddard, H.N. (fl. 1913) M	**Goddard**
Goddijn, Wouter Adriaan (1884–1960) PS	**Goddijn**
Godeas, Alicia M. (fl. 1971) M	**Godeas**
Godeau, M. (fl. 1987) S	**Godeau**
Godefroy, F.F. S	**Godefroy**
Godefroy–Lebeuf, Alexandre (1852–1903) S	**God.-Leb.**
Godet, Charles Henri (1797–1879) PS	**Godet**
Godey (fl. 1874) AM	**Godey**
Godfery, Masters John (1856–1945) S	**Godfery**
Godfrey, Charles Cartlidge (1855–1927) S	**Godfrey**
Godfrey, George Harold (1888–) M	**G.H.Godfrey**
Godfrey, Judith Dean (1947–) B	**J.D.Godfrey**
Godfrey, Robert Kenneth (1911–) S	**R.K.Godfrey**

Godkin, James (1891–   ) M   **Godkin**
Godman, Frederick Du Cane (1834–1919) B   **Godm.**
Godron, Dominique Alexandre (1807–1880) PS   **Godr.**
Godward, Maud Beatrice Ethel (1910–   ) A   **Godward**
Godwinski, M.I. (1915–   ) B   **Godw.**
(Godwinsky, M.I.)
   see Godwinski, M.I.   **Godw.**
Goebel, Karl Christian Traugott Friedemann (1794–1851) S   **Goebel**
Goebel, Karl Immanuel Eberhard (1855–1932) ABP   **K.I.Goebel**
Goel, A.K. (fl. 1983) S   **Goel**
Goeldi, Emil August (1859–1917) S   **Goeldi**
(Goeppert, Johann Heinrich Robert)
   see Göppert, Johann Heinrich Robert   **Göpp.**
Goeppinger, Robert   S   **R.Goepp.**
Goering, Philip Friedrich Wilhelm (1809–1876) S   **Goering**
Goerz, Rudolf (1879–1935) S   **Goerz**
Goeschke, Franz (1844–1912) S   **Goeschke**
Goetghebeur, Paul (1952–   ) S   **Goetgh.**
Goethart, Jan Willem Christiaan (1866–1938) S   **Goethart**
Goethe, Johann Wolfgang von (1749–1832) S   **Goethe**
Goetz, S.G. (fl. 1984) M   **Goetz**
Goeze, Edmund (1838–1929) S   **E.Goeze**
Goeze, Gunther (fl. 1885) S   **G.Goeze**
Goffart, Jules (1864–1954) S   **Goffart**
Gogina, E.E. (1928–1985) S   **Gogina**
Goh, Teik Khiang (fl. 1987) M   **Goh**
Goheen, Donald J. (fl. 1978) M   **Goheen**
Gohrbandt, Klaus   A   **Gohrb.**
Goidànich, Gabriel (1912–   ) M   **Goid.**
Goiran, Agostino (Augustin) (1835–1909) BPS   **Goiran**
Gojdics, Mary (1900–   ) A   **Gojdics**
(Gokale, V.P.)
   see Gokhale, V.P.   **Gokhale**
Gokhale, V.P. (fl. 1951) M   **Gokhale**
Gokujiv, A.N.   A   **Gokujiv**
Gola, Giuseppe (1877–1956) BM   **Gola**
Golan, Y. (fl. 1987) M   **Golan**
Golatkar, V.V. (fl. 1976) M   **Golatkar**
Gold, Harvey S. (fl. 1957) M   **H.S.Gold**
Gold, J.J. (fl. 1988) M   **J.J.Gold**
Gold, Kenneth (1932–   ) A   **Gold**
Goldbach, Carl (Karl) Ludwig (1793–1824) S   **Goldb.**
Goldberg, Aaron (1917–   ) S   **Goldberg**
Goldblatt, Peter (1943–   ) S   **Goldblatt**
Golden, Larry   A   **Golden**
Goldenberg, Carl Friedrich (1798–1881) F   **Goldenb.**
Goldfuss, Georg August (1782–1848) S   **Goldfuss**
Goldie, Hugh   S   **H.Goldie**
Goldie, John (1793–1886) PS   **Goldie**

Goldie–Smith, E.Kathleen (fl. 1951) M   **Goldie–Sm.**
Golding, Jack (1918–) S   **Golding**
Goldman, Edward Alphonso (1873–1946) S   **Goldman**
Goldman, Marcus Isaac (1881–) A   **M.I.Goldman**
Goldmann, J.G. (1810–1848) PS   **Goldm.**
Goldring, William (1854–1919) S   **Goldring**
Goldring, Winifred (fl. 1924) F   **W.Goldring**
Goldschmidt, M. (fl. 1920) S   **M.Goldschm.**
Goldschmidt, Richard (1878–1958) A   **Goldschm.**
Goldstein, Melvin E. (1936–) A   **M.E.Goldst.**
Goldstein, Solomon (1929–) M   **S.Goldst.**
Golenia, A. (fl. 1962) M   **Golenia**
Golenkin, Michail Iljitsch (1864–1941) S   **Golenkin**
(Golicin, Sergey Vladimirovich (Sergius V.))
    see Golitsin, Sergey Vladimirovich (Sergius V.)   **Golitsin**
Golitsin, Sergey Vladimirovich (Sergius V.) (1897–1968) PS   **Golitsin**
(Golitzyn, Sergey Vladimirovich (Sergius V.))
    see Golitsin, Sergey Vladimirovich (Sergius V.)   **Golitsin**
Goller, Andreas (1840–1912) S   **Goller**
(Gollerbach, Maksimillian Maksimillianovitch)
    see Hollerbach, Maximilian Maximilianovich   **Hollerb.**
(Gollerbakh, Maximilian Maximilianovich)
    see Hollerbach, Maximilian Maximilianovich   **Hollerb.**
Göllner, J. (fl. 1931) M   **Göllner**
Golneva, I.   S   **Golneva**
Golonkowa, J. (fl. 1970) M   **Golonk.**
Goloskokov, Vitaliĭ Petrovich (1913–) S   **Golosk.**
Golovin, Petr Nikolaevich (1897–1968) M   **Golovin**
Golovina, N.P. (fl. 1955) M   **N.P.Golovina**
Golovina, W.P. (fl. 1960) M   **W.P.Golovina**
Golovleva, L.A. (fl. 1974) M   **Golovleva**
Golub, V.A. (fl. 1936) S   **Golub**
Golubev, V.(W.)I. (fl. 1977) M   **Golubev**
Golubeva, O.G. (fl. 1988) M   **Golubeva**
Golubić, Stjepko   A   **Golubić**
Golubkova, E.I. (fl. 1987) S   **E.I.Golubk.**
Golubkova, Nina Siergeovna (1932–) M   **N.S.Golubk.**
Golubkova, V.F. (1923–) S   **Golubk.**
Gölz, Peter (1935–) S   **Gölz**
Gomankov, A.V.   B   **Gomankov**
Gombault, René (1871–) S   **Gomb.**
Gombócz, Endre (1882–1945) S   **Gombócz**
Gombos, Andrew Michael (1948–) A   **Gombos**
(Gomes, Alberto Augusto de Magalhães)
    see Magalhães Gomes, Alberto Augusto de   **A.A.Magalh.**
Gomes, Antonio Ildefonso (1816–1843) S   **A.I.Gomes**
Gomes, Bernardino António (1806–1877) S   **B.A.Gomes**
Gomes, Bernardino António (1769–1823) S   **Gomes**

(Gomes, Carlos Thomaz Magalhães)	
see Magalhães Gomes, Carlos Thomaz	**Magalh.**
Gomes da Silva, Ary (fl. 1985) S	**Gomes da Silva**
Gomes de Faria, J. A	**Gomes de Faria**
(Gomes de Medeiros, Arnaldo)	
see Medeiros, Arnaldo Gomes de	**A.G.Medeiros**
Gomes, José Corrêa, jr. (1919–1965) S	**J.C.Gomes**
Gomes, M.R.M. (fl. 1952) M	**M.R.M.Gomes**
Gomes Machado, Carlos Maria (fl. 1868) S	**Gomes Mach.**
Gómez, Calixto León (fl. 1976) M	**C.L.Gómez**
(Gómez Campo, César)	
see Gomez–Campo, César	**Gomez-Campo**
Gómez de la Maza y Jiménez, Manuel (1867–1916) S	**M.Gómez**
(Gómez de Ortega, Casimiro)	
see Ortega, Casimiro Gómez de	**Ortega**
(Gomez, Gustavo Furrazola)	
see Furrazola Gomez, Gustavo	**Furrazola**
Gómez García, José Daniel (1958–) S	**D.Gómez**
Gómez, H.D. (fl. 1983) M	**H.D.Gómez**
(Gómez, José Daniel)	
see Gómez García, José Daniel	**D.Gómez**
Gómez, Luis Diego (1944–) MPS	**L.D.Gómez**
Gómez Moreno, Manuel (fl. 1946) S	**Gomez Mor.**
(Gómez Pignataro, Luis Diego)	
see Gómez, Luis Diego	**L.D.Gómez**
Gómez Pompa, Arturo (1934–) S	**Gómez Pompa**
Gómez, S.A. (fl. 1952/3) S	**S.A.Gómez**
Gómez–Bolea, A. (fl. 1990) M	**Gómez-Bolea**
Gomez–Campo, César (1933–) S	**Gomez-Campo**
Gómez–Laurito, Jorge (fl. 1989) P	**Gómez-Laur.**
Gómez–Mercado, F. (fl. 1988) S	**Gómez-Merc.**
Gómez–Sosa, Edith V. (1942–) S	**Gómez-Sosa**
Gomont, Maurice Augustin (1839–1909) A	**Gomont**
Gon, Samuel M. (fl. 1987) S	**S.M.Gon**
Gonçalves, Antonio Estevan (Esteves) (1939–) S	**A.E.Gonç.**
(Gonçalves Costa, Cecilia)	
see Costa, Cecilia Gonçalves	**C.G.Costa**
(Gonçalves da Cunha, A.)	
see Cunha, A. Gonçalves de	**A.G.Cunha**
(Gonçalves de Costa, José)	
see Costa, José Gonçalves de	**J.G.Costa**
Gonçalves, Maria Leonor (1934–) S	**Gonç.**
Gonçalves–Carralves, Matilde F. (1943–) B	**Gonç.-Carr.**
Gönczöl, J. (1942–) M	**J.Gönczöl**
Gönczöl, M. (fl. 1985) M	**M.Gönczöl**
Gong, Gu Tang (1965–) S	**G.T.Gong**
Gong, Yao Ming (1958–) S	**Y.M.Gong**
Gonnermann, Gabriel (1806–1884) M	**Gonn.**
Gönnert, R. A	**Gönnert**
Gonsoulin, Gene J. (fl. 1970) S	**Gonsoulin**

(Gontcharov, Nikolai Fedorovich)
  see Gontscharow, Nikolai Fedorovich   **Gontsch.**
(Gontscharov, Nikolai Fedorovich)
  see Gontscharow, Nikolai Fedorovich   **Gontsch.**
Gontscharow, Nikolai Fedorovich (1900–1942) S   **Gontsch.**
Gonzáles, L.G. (fl. 1975) S   **L.G.Gonzáles**
Gonzáles, M.S. (fl. 1984) M   **M.S.Gonzáles**
González Albo Campillo, José (1913–1990) S   **Gonz.Albo**
González, Aldo E. (fl. 1984) M   **A.E.González**
González, Carmen C. (fl. 1984) M   **C.C.González**
González Canalejo, Antonino (fl. 1984) S   **A.González**
González Elizondo, M.Socorro (fl. 1985) S   **S.González**
González Fragoso, Romualdo (1862–1928) M   **Gonz.Frag.**
González, Francisco (Franciscous) (fl. 1877) S   **González**
González Géigel, Lutgarda (fl. 1985) S   **Gonz.Géigel**
González Guerrero, Pedro (1901–1984) A   **P.González**
(González, Julian)
  see Daniel, (Hermano)   **Daniel**
(González, Luz Maria)
  see González Villarreal, Luz Maria   **L.M.González**
(González, M.I.Fernández–Arias)
  see Fernández–Arias González, M.I.   **Fern.–Arias**
(González Medrano, Francisco)
  see Medrano, Francisco González   **Medrano**
(González Ortega, Jesús)
  see Ortega, Jesús González   **J.G.Ortega**
González Raposo, José (Padre) (fl. 1991) S   **J.González**
González, San Martín Felipe (fl. 1989) M   **S.F.González**
González Tamayo, Roberto (1945–) S   **R.González**
González Villarreal, Luz Maria (1954–) S   **L.M.González**
(Gonzalez y Azaola, Inigo)
  see Azaola, Inigo Gonzalez y   **Azaola**
(González–Albo Campillo, José)
  see González Albo Campillo, José   **Gonz.Albo**
(Gonzalez–Albo, José)
  see González Albo Campillo, José   **Gonz.Albo**
(González–Medrano, Francisco)
  see Medrano, Francisco González   **Medrano**
Gonzalo, (Hermano)   S   **Gonzalo**
Gonzalves, Ella Anne (1906–) A   **Gonzalves**
Gooch, F.S. (fl. 1938) M   **Gooch**
Good, D.A. (fl. 1984) S   **D.A.Good**
Good, Peter (–1803) S   **Good**
Good, Ronald D'Oyley (1896–) S   **R.D.Good**
Goodding, Charlotte Olive (later Reeder, C.O.) (1916–) S   **C.O.Goodd.**
Goodding, Leslie Newton (1880–1967) MPS   **Goodd.**
Goode, D.L. (f. 1988) S   **D.L.Goode**
Goode, John B. (fl. 1881) P   **Goode**
Goodenough, Samuel (1743–1827) A   **Gooden.**
Goodey, J.Basil (fl. 1951) M   **Goodey**

Goodman, David K. A	D.K.Goodman
Goodman, George Jones (1904–) S	Goodman
Goodspeed, Thomas Harper (1887–1966) S	Goodsp.
Goodwyn, Lyle E. S	Goodwyn
Goor, Andreas Cornelis Joseph van (1881–1925) A	Goor
Goos, Roger Delmon (1924–) M	Goos
(Goossens, (J.A.A.)M.(H.))	
see Goossens–Fontana, (J.A.A.)M.(H.)	Gooss.-Font.
Goossens, Antonie Petrus Gerhardy (1896–1972) S	Gooss.
Goossens–Fontana, (J.A.A.)M.(H.) (fl. 1951) M	Gooss.-Font.
Gopal, Brij (1944–) P	Gopal
Gopal, K. (fl. 1983) M	K.Gopal
Gopalakrishnan, P. A	Gopalakr.
Gopalan, R. (1947–) S	Gopalan
Gopalkrishnan, K.S. (fl. 1950) M	Gopalkr.
Gopinathan (fl. 1987) S	Gopin.
Göppert, Johann Heinrich Robert (1800–1884) BFMPS	Göpp.
Gorbunova, N.V. (1936–) S	Gorbunova
Gordenko, V.I. (fl. 1975) M	Gordenko
Gordienko, M.O. A	Gordienko
Gordon, C.C. (fl. 1961) M	C.C.Gordon
(Gordon, Elizabeth M.)	
see Gordon–Mills, Elizabeth M.	Gordon-Mills
Gordon, George (1806–1879) S	Gordon
Gordon, Gregory D. A	G.D.Gordon
Gordon, Hugh Douglas (1912–1978) M	H.D.Gordon
Gordon, M. (fl. 1920) S	M.Gordon
Gordon, Morris A. (fl. 1951) M	M.A.Gordon
Gordon, William Laurence (1901–1963) M	W.L.Gordon
Gordon, William Thomas (1884–1950) A	W.T.Gordon
Gordon–Gray, Kathleen Dixon Huntley (1918–) BS	Gordon-Gray
Gordon–Mills, Elizabeth M. (1943–) A	Gordon-Mills
Goreau, Thomas F. (1924–1970) A	Goreau
Goree, Harold (fl. 1974) M	Goree
Gorenflot, Robert (fl. 1968) S	Gorenflot
Gorenz, August Mark (1920–) M	Gorenz
Gorgidze, A.D. (fl. 1985) S	Gorgidze
Gorgulevskaya, E.I. A	Gorgul.
(Gorianinov, Pavel Fedorovich)	
see Horaninow, Paul (Paulus) Fedorowitsch	Horan.
Gorin, P.A.J. (fl. 1970) M	Gorin
(Göring, Philip Friedrich Wilhelm)	
see Goering, Philip Friedrich Wilhelm	Goering
Gorini, Constantino (1865–1950) S	Gorini
Górka, Hanna A	Górka
Gorlacheva, Z.S. (fl. 1989) S	Gorl.
(Gorlaczeva, Z.S.)	
see Gorlacheva, Z.S.	Gorl.
(Gorlakeva, Z.S.)	
see Gorlacheva, Z.S.	Gorl.

Gorlenko, Mikhail Vladimirovich (1908–) M	**Gorlenko**
Gorlenko, R.V. (fl. 1969) M	**R.V.Gorlenko**
Gornall, Richard John (1951–) S	**Gornall**
Gornostaĭ, V.I. (fl. 1972) M	**Gornostaĭ**
Gorodkov, Boris Nikolaevich (1890–1953) BPS	**Gorodkov**
(Gorodkow, Boris Nikolaevič)	
see Gorodkov, Boris Nikolaevich	**Gorodkov**
(Goroschankin, Johann N.)	
see Gorozhankin, Ivan Nikolaevich	**Gorozh.**
(Gorovii, L.F.)	
see Gorovij, L.F.	**Gorovij**
Gorovij, L.F. (fl. 1977) M	**Gorovij**
(Gorovoi, Petr G.)	
see Gorovoj, Petr G.	**Gorovoj**
Gorovoj, Petr G. (1936–) S	**Gorovoj**
(Gorovoy, Petr G.)	
see Gorovoj, Petr G.	**Gorovoj**
Gorozhankin, Ivan Nikolaevich (1848–1904) A	**Gorozh.**
Gorschkova, Sofia Gennadievna (1889–1972) S	**Gorschk.**
(Gorshkova, Sofia Gennadievna)	
see Gorschkova, Sofia Gennadievna	**Gorschk.**
Gorski, Stanislaw Batys (1802–1864) S	**Gorski**
Gortani, Luigi (1850–1908) S	**Gortani**
Gortani, Michele (1883–1966) S	**M.Gortani**
Gorter, David de (1717–1783) S	**Gorter**
Gorter, Gerhard Jacobus Marinus Anne (1913–) M	**G.J.M.Gorter**
Gorter, Klaas (fl. 1912) S	**K.Gorter**
(Görts van Rijn, Anne Renée Ariette)	
see Görts–van Rijn, Anne Renée Ariette	**Görts**
Görts–van Rijn, Anne Renée Ariette (1940–) S	**Görts**
Goryacheva, G.I. A	**Goryacheva**
(Görz, Rudolph)	
see Goerz, Rudolf	**Goerz**
Gosden, R. (1893–) A	**Gosden**
Gosklags, James H. (fl. 1957) M	**Gosklags**
Gosse, Philip Henry (1810–1888) S	**Gosse**
Gosselin, R.Roland (fl. 1905) S	**Gosselin**
Gossen, B.D. (fl. 1986) M	**Gossen**
Gossot, P. (1900–1971) S	**Gossot**
Gossweiler, John (1873–1952) S	**Gossw.**
Goswami, R.N. (fl. 1979) M	**R.N.Goswami**
Goswami, S.K. (1850–1951) B	**Goswami**
Goswamy, Hit Kishore (1942–) PS	**Goswamy**
Gothan, Walther Ulrich Eduard Friedrich (1879–1954) F	**Gothan**
Goto, Kazvo (fl. 1962) M	**K.Goto**
Goto, Shoji (fl. 1987) M	**S.Goto**
(Gotō, Shōji)	
see Goto, Shoji	**Goto**

Gotoh, Toshikazu A	**Gotoh**
Gott, Cora L. (fl. 1956) M	**Gott**
Göttinger (fl. 1884) M	**Göttinger**
Gottlieb, L.D. S	**Gottlieb**
Gottlieb-Tannenhain, Paul von (1879–) S	**Gottl.-Tann.**
Gottsberger, Gerhard Karl (1940–) M	**Gottsb.**
Gottsche, Carl Moritz (1808–1892) B	**Gottsche**
Gottwald, Tim R. (fl. 1979) M	**Gottwald**
Götz, Erich (1940–) S	**Götz**
Götz, Hans A	**H.Götz**
Gouan, Antoine (1733–1821) APS	**Gouan**
Gouas, Léon (fl. 1858) S	**Gouas**
Goubel, K.C. (fl. 1884) M	**Goubel**
Gouda, E.J. (fl. 1986) S	**Gouda**
Goudot, Justin (1822–1845) S	**Goudot**
Gouet, Philippe (fl. 1989) M	**Gouet**
Gougerot, Henri (1881–) M	**Gougerot**
Goujaud, Aimé Jacques Alexandre (later Bonpland, A.J.A.) (1773–1858) S	**Goujaud**
Goujet, D. (fl. 1979) M	**Goujet**
Goujon, Joseph (1858–) S	**Goujon**
Gould, Charles Jay (1912–) M	**C.J.Gould**
Gould, D. (fl. 1985) M	**D.Gould**
Gould, Frank Walton (1913–1981) S	**Gould**
Gould, R.W. A	**R.W.Gould**
Gould, Rodney Edward (1944–) F	**R.E.Gould**
Gould, S. (1942–) S	**S.Gould**
(Goulimis, C.N.)	
see Goulimy, C.N.	**Goulimy**
Goulimy, C.N. (1886–1963) S	**Goulimy**
Goupil, Clément Jacques (1784–1858) S	**Goupil**
Gour, H.N. (fl. 1979) M	**Gour**
Gourbiere, F. (fl. 1979) M	**Gourb.**
Gourlay, Henry William (1881–1956) AS	**Gourlay**
Gourlie, William (1815–1856) F	**Gourlie**
Gourret, Paul A	**Gourret**
Goutaland, P. S	**Goutaland**
Gouwentak, Cornelia (fl. 1924) M	**Gouw.**
Gouws, Jozef Benjamin (1909–) S	**Gouws**
Govindarajalu, Ethirajalu (1925–) S	**Govind.**
Govindarajan, V.S. (fl. 1962) M	**Govindar.**
Govindaswami, S. (fl. 1958) S	**Govindasw.**
Govindu, Heirehalli Chenniah (1918–) M	**Govindu**
Govorov, Leonid Ipatevich (1885–1941) S	**Govorov**
Govoruchin, Vasilii Sergeevich (1903–1970) S	**Govor.**
(Govorukhin, Vasilii Sergeevich)	
see Govoruchin, Vasilii Sergeevich	**Govor.**
Gowan, Sharon P. (fl. 1988) M	**Gowan**
Gowans, Charles Shields (1923–) A	**Gowans**
Goward, Trevor (fl. 1984) M	**Goward**

Gowda, M.H.Mari (fl. 1951) S   **Gowda**
Gowda, S.Sambe (1925–1981) A   **S.S.Gowda**
Gowen, James Robert (–1862) S   **Gowen**
Gower, William Hugh (1835–1894) S   **Gower**
Gowthaman, S.   A   **Gowth.**
Goy, Doris Alma (later Smith, D.A.) (1912–) PS   **Goy**
Goyal, J.P. (fl. 1974) M   **Goyal**
Goyal, S.K.   A   **S.K.Goyal**
Goyder, David John (1959–) S   **Goyder**
(Goyena, Miguel Ramírez)
   see Ramírez Goyena, Miguel   **Ram.Goyena**
(Graaf, A.de)
   see de Graaf, A.   **A.de Graaf**
(Graaf, Frederik de)
   see de Graaf, Frederik   **de Graaf**
Grabherr, Walter (fl. 1936–1942) S   **Grabherr**
Gräbner, K.-E. (fl. 1954) M   **Gräbner**
Grabovetzkaya, A.N.   S   **Grabov.**
Grabovskaja, A.E. (née Borodina, A.E.) (1953–) S   **Grabovsk.**
Grabowski, Heinrich Emanuel (1792–1842) S   **Grab.**
Gracas, M.das (fl. 1987) S   **Gracas**
Gràcia, E. (1952–) M   **Gràcia**
Gradaille, J.L. (fl. 1990) P   **Gradaille**
Graddon, William Douglas (1896–1989) M   **Graddon**
Gradstein, Stephan Robbert (1943–) B   **Gradst.**
Graebener, Leopold (1849–) S   **Graebener**
Graebner, Karl Otto Robert Peter Paul (1871–1933) PS   **Graebn.**
Graebner, Paul (1900–1978) S   **P.Graebn.**
Graeffer, John   S   **Graeffer**
Graells, Mariano de la Paz (1809–1898) S   **Graells**
(Graells y de la Agüera, Mariano de la Paz)
   see Graells, Mariano de la Paz   **Graells**
(Graeve, Pehr Henrik Fredrik)
   see Graewe, Per Henrik Fredrik   **Graewe**
Graewe, Per Henrik Fredrik (1819–1866) M   **Graewe**
Graf, Alfred Byrd (1901–) S   **A.B.Graf**
Graf, Ferdinand (1833–1877) S   **F.Graf**
Graf, Rainer (1811–1872) S   **R.Graf**
Graf, Siegmund (Sigismund) (1801–1838) S   **Graf**
Graff, Eberhard Gottlieb (1780–1841) S   **Graff**
Graff, Paul Weidemeyer (1880–) M   **P.W.Graff**
Grafl, Ina (1915–1970) S   **Grafl**
Graham, Edward Harrison (1902–1966) S   **E.H.Graham**
Graham, Eileen A.   A   **E.A.Graham**
Graham, George Gordon (1917–) S   **G.G.Graham**
Graham, Herbert William (1905–) A   **H.W.Graham**
Graham, J.H. (fl. 1953) M   **J.H.Graham**
Graham, John (1805–1839) S   **J.Graham**
Graham, Linda Edwards (1946–) A   **L.E.Graham**

Graham, Rex Alan (1915–1958) S	**R.A.Graham**
Graham, Robert C. (1786–1845) S	**Graham**
Graham, Robert James Douglas (1884–1950) S	**R.J.D.Graham**
Graham, Shirley Ann Tousch (1935–) S	**S.A.Graham**
Graham, Victoria Anne Wassell (1950–) S	**V.A.W.Graham**
Grall, J.-R. A	**Grall**
Gram, Kai Jørgen Arthur (1897–1961) S	**Gram**
Grambast, Louis J. (1927–1977) A	**Grambast**
Grambast, Nicole A	**N.Grambast**
Gramberg, Eugen (fl. 1920) M	**Gramberg**
Grampini, Ottavio (1845–) S	**Grampini**
Gran, Haaken Hasberg (1870–1955) AM	**Gran**
Granata, L. (fl. 1909) MP	**Granata**
Granby, Rut (1938–) S	**Granby**
Grand, Larry F. (1940–) M	**Grand**
(Grand, O.Le)	
see Le Grand, O.	**O.Le Grand**
Grand'Eury, François Cyrille (1839–1917) F	**Grand'Eury**
Granda, Manuel M. (fl. 1986) S	**Granda**
Grande, Loreto (1878–1965) S	**Grande**
Grandidier, Alfred (1836–1921) S	**Grandid.**
Grandjean, M. (fl. 1912) M	**Grandjean**
Grandjot, Gertrud F.de (fl. 1936) S	**G.F.Grandjot**
Grandjot, Karl (Carlos) (1900–1979) S	**K.Grandjot**
(Grandvaux Barbosa, Luis Agosto)	
see Barbosa, Luis Agosto Grandvaux	**Barbosa**
Granick, Elsa Backman S	**Granick**
(Granié, Étienne Marcellin)	
see Sennen, E.C.	**Sennen**
(Granié–Blanc, Étienne Marcellin)	
see Sennen, E.C.	**Sennen**
Graniti, Antonio (1926–) M	**Graniti**
Granmo, A. (fl. 1978) M	**Granmo**
Grant, Adele Gerard (Lewis) (1881–1969) S	**A.L.Grant**
Grant, Alva Day (née Day, A.G (1920–) S	**A.D.Grant**
Grant, Elizabeth (fl. 1943) S	**E.Grant**
Grant, Freeman Augustus (1898–) M	**F.A.Grant**
Grant, George Bernard (1849–1917) S	**G.B.Grant**
Grant, James Augustus (1827–1892) S	**Grant**
Grant, Karen S.Alt (née Alt, K.S.) (1936–) PS	**K.A.Grant**
Grant, Martin Lawrence (1907–1968) S	**M.L.Grant**
Grant, Verne Edwin (1917–) PS	**V.E.Grant**
Grant, William Frederick (1924–) S	**W.F.Grant**
Grant, William M. A	**W.M.Grant**
Grantzow, Carl (Karl) (–1894) S	**Grantzow**
Granville, Jean–Jacques de (1943–) S	**Granv.**
Granzow de la Cerda, Inigo B	**Granzow**
Grapengiesser, Sten (1868–1955) S	**Grapeng.**
Gras, Auguste (1819–1874) S	**Gras**

Gräser, Robert (1893–1977) S	**Gräser**
Grashoff, J.L. S	**Grashoff**
Grassé, Pierre–Paul (1895–1985) A	**Grassé**
Grassi, Giovanni Battista (1854–1925) A	**Grassi**
Grassl, Carl Otto (1908–) S	**Grassl**
Grasso, M.P. (fl. 1984–) S	**M.P.Grasso**
Grasso, Vincenzo (1914–) M	**Grasso**
Grateloup, Jean Pierre A.Sylvestre de (1782–1861) AS	**Gratel.**
Gratz, Levi Otto (1894–1968) M	**Gratz**
Grau, A. (fl. 1983) S	**A.Grau**
Grau, Hans Rudolph Jürke (1937–) S	**Grau**
Grauer, Sebastian (1758–1820) S	**Grauer**
(Graumuller, Johann Christian Friedrich)	
see Graumüller, Johann Christian Friedrich	**Graum.**
Graumüller, Johann Christian Friedrich (1770–1824) S	**Graum.**
Grauwinkel, B. (fl. 1987) M	**Grauw.**
Gravely, Frederic Henry (1885–) S	**Gravely**
Graves, Arthur Harmount (1879–1962) M	**A.H.Graves**
Graves, Charles Burr (1860–1936) S	**C.Graves**
Graves, Edward Willis (1882–1936) P	**E.W.Graves**
Graves, George (1784–1839) S	**Graves**
Graves, James Ansel (1828–1909) P	**J.A.Graves**
Graves, Louis (1791–1857) S	**L.Graves**
Gravet, Pierre Joseph Frédéric (1827–1907) B	**Grav.**
Gravina, Pasquale (fl. 1812–1815) S	**Gravina**
Gravis, A. A	**A.Gravis**
Gravis, Jean Joseph Auguste (1857–1937) S	**Gravis**
Grawitz, P. (fl. 1877) M	**Grawitz**
Gray, Alan Maurice (1943–) PS	**A.M.Gray**
Gray, Asa (1810–1888) ABPS	**A.Gray**
Gray, Bruce (1939–) PS	**B.Gray**
Gray, Dennis J. (fl. 1979) M	**D.J.Gray**
Gray, Elizabeth G. (fl. 1954) M	**E.G.Gray**
Gray, Frederick William (1878–) MPS	**F.W.Gray**
Gray, John Edward (1800–1875) AB	**J.E.Gray**
Gray, Louis Harold (1905–1955) S	**L.H.Gray**
Gray, Max (1929–) S	**M.Gray**
Gray, Netta Elizabeth (1913–1970) S	**N.E.Gray**
Gray, Samuel Frederick (1766–1828) ABMPS	**Gray**
Gray, Samuel Octavus (1828–1902) A	**S.O.Gray**
Grayum, Michael Howard (1949–) P	**Grayum**
Graziani, Antoine (fl. 1890) M	**Graziani**
Grear, John Wesley (1937–) MS	**Grear**
Greathead, S.K. (fl. 1961) M	**Greath.**
Greatrex, F.C. (fl. 1965) S	**Greatrex**
Grebe, Carl (1852–1922) B	**Grebe**
Grebenjuk, I.N. (fl. 1971) M	**Grebenjuk**
Grebenščikov, Igor Sergeevich	
see Grebenshchikov, Igor Sergeevich	**Greb.**

Grebenshchikov, Igor Sergeevich (1912–1986) S	**Greb.**
Grecescu, Dimitrie (Demetrius) (1841–1910) S	**Grecescu**
Grech Delicata, Giovanni Carlo (Johann Carl) (1811–1882) S	**Grech**
Greco, N.V. (fl. 1916) M	**Greco**
(Gredilla, Apolinar Federico y Gauna)	
see Gredilla y Gauna, Apolinar Federico	**Gredilla**
Gredilla y Gauna, Apolinar Federico (1859–1919) S	**Gredilla**
Greef, Richard (1829–1892) M	**Greef**
Green, John Christopher (1939–) A	**J.C.Green**
Green, John William (1930–) AS	**J.W.Green**
Green, Mary Letitia (later Sprague, M.L.) (1886–1978) S	**M.L.Green**
Green, Peter Shaw (1920–) PS	**P.S.Green**
Green, Roderick B. A	**R.B.Green**
Green, Ted S	**T.Green**
Green, Thomas (fl. 1820) B	**Green**
Greenall, Judith M. (fl. 1963) M	**Greenall**
Greene, Benjamin Daniel (1793–1862) S	**B.D.Greene**
Greene, Dorothy M. B	**D.M.Greene**
Greene, Edward Lee (1843–1915) MPS	**Greene**
Greene, Frank Cook (1886–) PS	**F.C.Greene**
Greene, Henry Campbell (1904–1967) M	**H.C.Greene**
Greene, Stanley Wilson (1928–1989) B	**S.W.Greene**
Greenhalgh, Geoffry N. (fl. 1972) M	**Greenh.**
Greenleaf, M.A. (fl. 1980) M	**M.A.Greenleaf**
Greenleaf, R.C. (fl. 1866) A	**Greenleaf**
Greenman, Jesse More (1867–1951) PS	**Greenm.**
Greenway, Percy James ('Peter') (1897–1980) S	**Greenway**
Greenwell, Amy Beatrice Holdsworth (1921–1974) S	**Greenwell**
Greenwood, Edward Warren (1918–) S	**E.W.Greenw.**
Greenwood, William Frederick Neville (fl. 1917–31) B	**Greenw.**
Gregor, Mary J.F. (fl. 1935) M	**Gregor**
Gregory, Charles Truman (1887–) M	**C.T.Greg.**
Gregory, David Palache (1930–) S	**D.P.Greg.**
Gregory, Eliza Standerwick (1840–1932) S	**Greg.**
Gregory, Luis E. S	**L.E.Greg.**
Gregory, N.M. (fl. 1986) M	**N.M.Greg.**
Gregory, Philip Herries (1907–) M	**P.H.Greg.**
Gregory, Walton Carlyle (1910–) S	**W.C.Greg.**
Gregory, William (1803–1858) A	**W.Greg.**
Greguss, Pál (Paul) (1889–) AB	**Greguss**
Greig-Smith, Peter (1922–) B	**Greig-Sm.**
Greis, Hans (1912–1947) M	**Greis**
Greis-Dengler, Ida (fl. 1940) M	**Greis-Dengler**
Grélet, Louis-Joseph (1870–1945) M	**Grélet**
Gremblich, Pater Julius (1851–1905) S	**Grembl.**
Gremli, August(e) (1833–1899) PS	**Gremli**
Gremmen, J. (fl. 1953) M	**Gremmen**
Grenier, Jean Charles Marie (1808–1875) PS	**Gren.**

Grenier–Blanc, Etienne Marcelin	
see Sennen, frère	**Sennen**
Grenzebach, Myrle Eunice (fl. 1926) S	**Grenzeb.**
Gres, Yu.A. (fl. 1982) M	**Gres**
Greschik, Victor (1862–1946) B	**Greschik**
Greshoff, Maurits (1862–1909) S	**Greshoff**
Gresino, Giacomo (1859–1946) S	**Gresino**
Greter, P.Fintan (1899–) B	**Greter**
Grether, David Frank (1920–) PS	**Grether**
Grether, Rosaura (1950–) S	**R.Grether**
Gretz, Michael R. (1955–) A	**Gretz**
Greuet, Claude A	**Greuet**
Greuter, Werner Rodolfo (1938–) PS	**Greuter**
Greves, S. (fl. 1927) S	**Greves**
Greville, Robert Kaye (1794–1866) ABMPS	**Grev.**
Grewel, J.S. (fl. 1970) M	**Grewel**
Grey, Charles Hervey (1875–1955) S	**Grey**
Grey–Wilson, Christopher (1944–) S	**Grey–Wilson**
Grgurinovic, Cheryl A. (fl. 1982) M	**Grgur.**
Grichenko, A.H.Hertha (fl. 1964) M	**Grich.**
Grierson, Andrew John Charles (1929–1990) S	**Grierson**
Grierson, James Douglas (1931–) F	**J.D.Grierson**
Griesselich, Ludwig (1804–1848) S	**Griess.**
Griesser, B. (fl. 1987) M	**Griesser**
Griessmann, Karl A	**Griessm.**
Griffin, Dana Gove, III (1938–) AB	**D.G.Griffin**
Griffin, H.D. (fl. 1968) M	**H.D.Griffin**
Griffin, Michael James (fl. 1979) M	**M.J.Griffin**
Griffini, Luigi (fl. 1873) M	**Griffini**
Griffioen, K. S	**Griffioen**
Griffith, John Edward (1843–1933) S	**J.E.Griff.**
Griffith, John William (1819–1910) A	**J.W.Griff.**
Griffith, William (1810–1845) BPS	**Griff.**
Griffiths, Benjamin Millard (1886–1942) A	**B.M.Griffiths**
Griffiths, D.A. (fl. 1974) M	**D.A.Griffiths**
Griffiths, David (1867–1935) MS	**Griffiths**
Griffon, Édouard (1869–1912) M	**Griffon**
Grifo, F.T. (fl. 1988) S	**Grifo**
Griggs, Robert Fiske (1881–1962) AMS	**Griggs**
Grignan, G.T. (fl. 1900) S	**Grignan**
Grignani, Dario A	**Grignani**
Grigoraki, Léon (fl. 1921) M	**Grigoraki**
(Grigoriev, Yuri Sergeevich)	
see Grigorjev, Juri Sergeevich	**Grig.**
Grigorieva–Manoilova, O.C. (fl. 1915) M	**Grig.-Man.**
Grigoriu, A.C. (fl. 1975) M	**Grigoriu**
Grigorjev, Juri Sergeevich (1905–1975) S	**Grig.**
Grigorovitch, A.S. A	**Grigorov.**
Grilli, Edmondo (fl. 1987) M	**Grilli**

Grimaldi, Clemente S	**Grimaldi**
Grimes, James Walter (1953–) PS	**J.W.Grimes**
Grimes, M. (fl. 1930) M	**Grimes**
Grimm, Johann Friedrich Carl (1737–1821) S	**Grimm**
Grimme (fl. 1911) S	**Grimme**
Grimme, Arnold T. (fl. 1936) B	**A.T.Grimme**
Grinbergs, J. (fl. 1967) M	**Grinb.**
Grindel, David Hieronymus (1776–1836) B	**Grindel**
Grindon, Leopold Hartley (1818–1904) S	**Grindon**
(Grinj, F.A.)	
see Grynj, F.A.	**Grynj**
Grinling, Kenneth (fl. 1967) M	**Grinling**
Grinsven, A.M.van (fl. 1984) M	**Grinsven**
Grinţescu, George P. (1870–1956) S	**Grinţ.**
Grinţescu, Ioan (1874–1963) APS	**I.Grinţ.**
(Grintzescu, Jean)	
see Grinţescu, Ioan	**I.Grinţ.**
Gris, Jean Antoine Arthur (1829–1872) S	**Gris**
Grischina, L.V. (fl. 1974) M	**Grischina**
Griscom, Ludlow (1890–1959) S	**Griscom**
Grisebach, August Heinrich Rudolf (1814–1879) PS	**Griseb.**
Griselini, Francesco (1717–1783) S	**Grisel.**
(Grisina, L.V.)	
see Grischina, L.V.	**Grisch.**
(Gritsenko, P.P.)	
see Gritzenko, P.P.	**P.P.Gritz.**
Gritzenko, N.V. (1927–) S	**N.V.Gritz.**
Gritzenko, P.P. (fl. 1979) S	**P.P.Gritz.**
Gritzenko, R.I. (1922–) S	**R.I.Gritz.**
Grizel, Henri (fl. 1971) AM	**Grizel**
Grjebine, A. (fl. 1985) M	**Grjebine**
Grobéty, A.-E. A	**Grobéty**
(Grodsinsky, Léon)	
see Grodzinsky, Léon	**Grodz.**
Grodzińska, Krystyna (1934–) S	**Grodzińska**
Grodzinsky, Leon (1908–) M	**Grodz.**
Groen, L.E. (1946–) S	**L.E.Groen**
Groenewald, B.H. S	**Groenew.**
Groenewege, J. (fl. 1921) M	**Groen.**
Groenhart, Pieter (1894–1965) M	**Groenh.**
Groenland, Johannes (1824–1891) S	**Groenl.**
Gröger, F. (fl. 1967) M	**Gröger**
Grognier, Louis Furcy (1776–1837) S	**Grognier**
Grognot, Camille (1792–1869) MS	**Grognot**
Grohmann, Fritz (fl. 1978) S	**Grohmann**
Grolle, Riclef (1934–) B	**Grolle**
Grom, Sezana, Sr (fl. 1960) B	**Grom**
Gromov (fl. 1817) S	**Gromov**
Gromov, B.V. A	**B.V.Gromov**

(Gromow)	
see Gromov	**Gromov**
Gronbach, E. (fl. 1984) S	**Gronbach**
Grönblad, Rolf Leo (1895–1962) A	**Grönblad**
Gronchi (fl. 1931) M	**Gronchi**
Gronde, Keympe van der (fl. 1980) B	**Gronde**
Grondona, Eduardo M. (1911–) S	**Grondona**
Grønlund, Carl Christian Howitz (1825–1901) M	**Grønlund**
Gronovius, Johan Frederik (Jan Fredrik) (1686–1762) S	**Gronov.**
Grøntved, Johannes (1882–1956) S	**Grøntved**
Gröntved, Julius (1899–1967) A	**J.Gröntved**
Gronvall, Troed Axel Ludwig (1838–1892) B	**Gronvall**
(Groonewald, B.H.)	
see Groenewald, B.H.	**Groenew.**
(Groot, W.Sandino)	
see Sandino–Groot, W.	**Sand.-Groot**
Grootendorst, Herman Johannes (1911–) S	**Groot.**
Groover, Robert D. A	**Groover**
Gros, G. A	**Gros**
Grosbüsch, J. (fl. 1914) M	**Grosb.**
Grosdemange, Charles (fl. 1893–1897) S	**Grosdem.**
Grose, Elizabeth S. (fl. 1968) M	**Grose**
Grosmann, Helene (fl. 1932) M	**Grosmann**
Grosourdy, René de (fl. 1864) S	**Grosourdy**
Gross, E. (fl. 1985) S	**E.Gross**
Gross, Gerhard (fl. 1968) M	**G.Gross**
Gross, Henry Lawrence (1930–) M	**H.L.Gross**
Gross, Hugo (1888–) S	**H.Gross**
Gross, Ludwig (1860–) S	**L.Gross**
Gross, Roland (1890–1945) S	**R.Gross**
Gross, Rudolf (1872–) S	**Gross**
Grossenbacher, John Gasser (1875–) M	**Grossenb.**
Grosser, Wilhelm Carl Heinrich (1869–) S	**Grosser**
Grosset, Hugo Edgarovich (1903–1981) S	**Grosset**
(Grossgeim, Alexander Alfonsovich)	
see Grossheim, Alexander Alfonsovich	**Grossh.**
Grossheim, Alexander Alfonsovich (1888–1948) PS	**Grossh.**
Grote, Paul J. F	**Grote**
Grotenfelt, Karl Gustaf Johannes (1855–) M	**Grotenf.**
ʳGrothe, E.H.M. (fl. 1984) S	**Grothe**
Grout, Abel Joel (1867–1947) BMP	**Grout**
Grout, Frank Fitch (1880–) A	**F.F.Grout**
Grove, Arthur (1865–1942) S	**A.Grove**
Grove, Edmund (fl. 1886) A	**E.Grove**
Grove, William Bywater (1848–1938) AM	**Grove**
Grover, Frederick Orville (1868–1964) S	**Grover**
Groves, Eric William (1923–) S	**E.W.Groves**
Groves, Henry (1855–1912) AS	**H.Groves**
Groves, James (1858–1933) AFS	**J.Groves**

Groves, James Walton (1906–1970) M	**J.W.Groves**
Gruas–Cavagnetto, C. A	**Gruas–Cav.**
Grube, M. (fl. 1989) M	**Grube**
Gruber, August A	**A.Gruber**
Gruber, Calvin Luther (1864–) P	**˙Gruber**
Gruber, Eduard A	**E.Gruber**
Gruber, Ilse (fl. 1970) M	**I.Gruber**
Gruber, Michel (1943–) S	**M.Gruber**
Gruber, Rudolf (fl. 1982) S	**R.Gruber**
Grubov, Valery Ivanovich (1917–) PS	**Grubov**
Gruby, David (1810–1898) AM	**Gruby**
Grudzinskaja, Irina Aleksandrovna (1920–) S	**Grudz.**
(Grudzinskaya, Irina Aleksandrovna)	
see Grudzinskaja, Irina Aleksandrovna	**Grudz.**
(Gruenberg–Fertig, Irene)	
see Fertig, Irene Gruenberg	**Fertig**
Gruezo, William Sm. (1951–) M	**Gruezo**
Grufberg, Isaac Olof (1736–1764) S	**Grufberg**
Grufberg, J.V. S	**J.V.Grufberg**
Gruff, Susan Carol (1953–) M	**Gruff**
Gruhn, U. (fl. 1991) M	**Gruhn**
Gruia, Lucian (fl. 1964) AM	**Gruia**
Gruithuisen, Franz von Paula (Franciscus de Paula) (1774–1852) AM	**Gruith.**
Grulich, Vit (1956–) S	**Grulich**
(Grum–Grzhimailo, Grigor Efimowitsch)	
see Grumm–Grzhimailo, Grigor Efimowitsch	**Grumm–Grzhim.**
Grumm–Grzhimailo, Grigor Efimowitsch (fl. 1889–90) S	**Grumm–Grzhim.**
(Grumman, Vitus Johannes)	
see Grummann, Vitus Johannes	**Grummann**
Grummann, Vitus Johannes (1899–1967) M	**Grummann**
Grün, Walter A	**Grün**
Grund, Darryl W. (1938–) M	**Grund**
Gruner, Leopold F. (1839–) MS	**Gruner**
Gruner, S.A. S	**S.A.Gruner**
Grüning, G.R. (1862–1926) S	**Grüning**
Grunov, Albert (1826–1914) A	**Grunov**
(Grunow, Albert)	
see Grunov, Albert	**Grunov**
Grupe, Donald A. S	**Grupe**
(Grushvitskii, Igor Vladimirovich)	
see Grushvitzky, Igor Vladimirovich	**Grushv.**
Grushvitzky, Igor Vladimirovich (1916–) S	**Grushv.**
Grüss, G. (fl. 1927) M	**G.Grüss**
Grüss, Johannes (1860–) AFMS	**Grüss**
Grutterink, L.H. S	**Grutt.**
Grütz, O. (fl. 1925) M	**Grütz**
Gruzov, E.N. A	**Gruzov**
Grynj, F.A. (1902–) S	**Grynj**
Gsell, Rudolf (1892–1953) S	**Gsell**

Gu, An Gen (fl. 1988) S   **A.G.Gu**
Gu, De Xing (fl. 1983) S   **D.X.Gu**
Guadagno, Michele (1878–1930) P   **Guadagno**
Guaglianone, Encarnación Rosa (1932–   ) S   **Guagl.**
Gualtieri (fl. 1790) S   **Gualt.**
Guan, Shao Fei (fl. 1987) S   **S.F.Guan**
Guaranha, Jane Maria Rollo (fl. 1981) S   **Guaranha**
Guard, Arthur Thomas (1897–   ) S   **Guard**
Guareschi, Icilio   S   **Guareschi**
Guarim, Germano (fl. 1979) S   **Guarim**
(Guarim Neto, Germano)
    see Guarim, Germano   **Guarim**
Guarrera, Sebastián A.   A   **Guarrera**
Guarrigues, O. (fl. 1979) S   **Guarr.**
Guarro Artigas, Josep (fl. 1982) M   **Guarro**
(Guarro, Josep)
    see Guarro Artigas, Josep   **Guarro**
Guatam, S.P. (fl. 1978) M   **Guatam**
Guatteri, Giovanni Battista (1743–1793) S   **Guatteri**
Guba, Emil Frederick (1897–   ) M   **Guba**
Gubanov, Ivan A. (1933–   ) S   **Gubanov**
Gubellini, Leonardo (1954–   ) S   **Gubellini**
Gucevič, S.A. (fl. 1952) M   **Gucevič**
(Gucevicz, S.A.)
    see Gucevič, S.A.   **Gucevič**
Gudjónsson, Gudni (1913–1948) S   **Gudj.**
Gudkova, E.P. (1933–   ) S   **Gudkova**
Gudoschnikov, Sergei Vasilevich (1916–   ) S   **Gudoschn.**
Gudymovich, S.S.   A   **Gudym.**
Guebhard, Ch. (fl. 1842–1848) S   **Guebhard**
Guécho, E. (fl. 1984) M   **Guécho**
Guédès, Manoel (fl. 1977) B   **Man.Guédès**
Guédès, Michel (1942–1985) BPS   **Guédès**
Guéguen, Fernand Pierre Joseph (1872–1915) M   **Guég.**
Guého, E.L.Joseph (1937–   ) S   **J.Guého**
Guého, Eveline (fl. 1986) M   **E.Guého**
Gueldenstaedt, Anton Johann von (1745–1781) PS   **Gueldenst.**
Güemes Heras, Jaime (fl. 1990) S   **Güemes**
(Guenther, Johann Christian Carl)
    see Günther, Johann Christian Carl   **Günther**
Guépin, Jean Pierre (1779–1858) MS   **Guépin**
Guérin, Henry P.   S   **H.P.Guérin**
Guérin, J.X.B. (1775–1850) PS   **Guérin**
(Guerke, Robert Louis August Maximilian)
    see Gürke, Robert Louis August Maximilian   **Gürke**
Guerke, Wayne R. (fl. 1978) B   **W.R.Guerke**
Guerlesquin, Micheline Y. (1928–   ) A   **Guerl.**
Guermeur, P.   A   **Guermeur**
Guerra, A.G. (fl. 1983) M   **A.G.Guerra**

(Guerra Montes, Juan)
  see Guerra, Juan   **J.Guerra**
Guerra, Juan (1952–) B   **J.Guerra**
Guerra, P. (fl. 1938) M   **Guerra**
Guerrero, Rosa T. (fl. 1971) M   **Guerrero**
Guersent, Louis Ben (1776–1848) BMS   **Guers.**
Guest, Evan Rhuvon (1902–1992) MS   **Guest**
Guétrot, M. (1873–1941) PS   **Guétrot**
Guettard, Jean Étienne (1715–1786) S   **Guett.**
Guevara–Fefer, F. (fl. 1980) S   **Guevara**
Guffroy, Charles–Émile (fl. 1938) S   **Guffroy**
Gugelberg von Moos, Maria Barbara Flandrina von (1836–1918) B   **Gugelb.**
Gugerli, Karl (fl. 1939) S   **Gugerli**
Gugler, Wilhelm (1874–1909) S   **Gugler**
Guglielmetti, Giovanni (1876–1915) A   **Guglielm.**
Gugnacka, Wanda (1945–) S   **Gugn.**
(Gugnacka–Fiedor, Wanda)
  see Gugnacka, Wanda   **Gugn.**
Guha Bakshi, D.N. (1931–) S   **Guha Bakshi**
Guiart, Jules (1870–) M   **Guiart**
Guibert, Victor (1826–1866) S   **Guib.**
Guicciardi, Giacinto (fl. 1855) S   **Guicc.**
Guidi, G. (fl. 1896) M   **Guidi**
Guidotti, Rolando S   **Guid.**
Guignard, G. (fl. 1986) S   **G.Guignard**
Guignard, Jean–Louis–Léon (1852–1928) A   **Guignard**
Guilding, Lansdown (1798–1831) S   **Guilding**
Guiler, Eric R. A   **Guiler**
Guilfoyle, William Robert (1840–1912) S   **Guilf.**
Guillard, Robert Russell Louis (1921–) A   **Guillard**
(Guillarmod, Amy Frances May Gordon Jacot)
  see Jacot Guillarmod, Amy Frances May Gordon   **Jacot Guill.**
Guillaumet, Jean L. (1934–) S   **Guillaumet**
Guillaumin, André (1885–1974) PS   **Guillaumin**
Guillaumot, Marius Georges (1884–1970) BM   **Guillaumot**
Guillemette, M.K. (fl. 1988) M   **Guillem.**
Guillemin, Jean Baptiste Antoine (1796–1842) APS   **Guill.**
(Guillén, Julio E.López)
  see López Guillén, Julio E.   **López Guillén**
Guilliermond, Marie Antoine Alexandre (1876–1945) M   **Guillierm.**
Guillon, E.le (1906–) S   **E.Guillon**
Guillon, Pierre Anatole (1819–1908) S   **Guillon**
Guillot, Jean (fl. 1989) S   **Guillot**
Guilmot, Gustave (1818–1885) S   **Guilmot**
Guimarães, António (Antoine) Luis Machado (1883–1969) B   **Guim.**
(Guimarães da Vinha, Sérgio)
  see Vinha, Sérgio Guimarães da   **Vinha**
Guimarães, Elsie Franklin (1935–) S   **E.F.Guim.**
Guimarães, José d'Ascensão (1862–1922) S   **J.A.Guim.**

Guimarães, Silva Maria Pita de Beauclair (1944–   ) A   **S.M.Guim.**
Guimpel, Friedrich (1774–1839) S   **Guimpel**
Guinberteau, J. (fl. 1978) M   **Guinb.**
Guinea, Emilio (1907–1985) MPS   **Guinea**
(Guinea López, Emilio)
   see Guinea, Emilio   **Guinea**
Guinet, Jean Étienne Auguste (1846–1928) B   **Guinet**
Guinet, Philippe (fl. 1951) S   **P.Guinet**
Guinier, Philibert (1876–1962) S   **Guinier**
Guinochet, Marcel (1909–   ) S   **Guin.**
Guirão y Navarro, Angel (–1890) S   **Guirão**
Guiry, Michael Dominic R. (1949–   ) A   **Guiry**
Guittonneau, Guy Georges (1934–   ) S   **Guitt.**
Gujarati, S. (fl. 1965) M   **Gujarati**
Gulden, Gro Sissel (1939–   ) M   **Gulden**
(Güldenstädt, Anton Johann von)
   see Gueldenstaedt, Anton Johann von   **Gueldenst.**
(Güldenstaedt, Anton Johann von)
   see Gueldenstaedt, Anton Johann von   **Gueldenst.**
Gulia, Gavino (1835–1889) S   **Gulia**
Gullino, G. (fl. 1971) M   **Gullino**
Gümbel, Carl Wilhelm von (1823–1898) AF   **C.Gümbel**
Gümbel, Wilhelm Theodor (1812–1858) B   **W.Gümbel**
Gumbinger, M. (fl. 1982) M   **Gumb.**
Gumbleton, William Edward (1840–1911) S   **Gumbl.**
Gumprecht, Reinhart (1900–   ) S   **Gumpr.**
Gunckel, Luer Hugo (1901–   ) S   **Gunckel**
Güner, Adil (1950–   ) S   **Güner**
Gunn, Charles R. (1927–   ) S   **C.R.Gunn**
Gunn, Ronald Campbell (1808–1881) S   **Gunn**
Gunnarsson, Johan Gottfrid (1866–1944) S   **Gunnarsson**
Gunnell, Jean (fl. 1961) M   **Gunnell**
Gunnell, Pamela S. (fl. 1987) M   **P.S.Gunnell**
Gunnerbeck, Erik (1942–   ) M   **Gunnerb.**
Gunnerus, Johan Ernst (1718–1773) ABMS   **Gunnerus**
Gunter, L.S. (fl. 1927) M   **Gunter**
Günther, Ernst Karl Franz   A   **E.K.F.Günther**
Günther, Johann Christian Carl (1769–1833) S   **Günther**
Günther, Karl–Friedrich (1941–   ) S   **K.-F.Günther**
Guo, Chao (fl. 1990) M   **C.Guo**
Guo, Cheng Ze (fl. 1988) S   **C.Z.Guo**
Guo, H.L. (fl. 1986) M   **H.L.Guo**
Guo, Hua Shan (fl. 1988) A   **H.S.Guo**
Guo, Jian Hua (fl. 1988) S   **J.H.Guo**
Guo, Ke (fl. 1986) S   **K.Guo**
Guo, Lan Bin (fl. 1989) S   **L.B.Guo**
Guo, Lin (1949–   ) M   **L.Guo**
Guo, Qing Fing (fl. 1987) S   **Q.F.Guo**
Guo, Rong Lin (fl. 1980) S   **R.L.Guo**
Guo, Xin Hu (fl. 1988) S   **X.H.Guo**

Guo, Xin Qing (fl. 1979) S	**X.Q.Guo**
Guo, Ying Lan (1942–) M	**Y.L.Guo**
Guo, You Hao (1952–) S	**Y.H.Guo**
(Guo, Yujie)	
see Kuo, Yu Chieh	**Y.C.Kuo**
Guo, Zheng Tang (fl. 1987) MS	**Z.T.Guo**
Gupta, A. (fl. 1985) M	**A.Gupta**
Gupta, A.K. (fl. 1991) M	**A.K.Gupta**
Gupta, Arya Bhushan (1919–) A	**A.B.Gupta**
Gupta, B.K. (fl. 1987) MS	**B.K.Gupta**
Gupta, B.L. S	**B.L.Gupta**
Gupta, Chaturbhuji (fl. 1987) M	**C.Gupta**
Gupta, Durga (fl. 1981) MS	**D.Gupta**
Gupta, Dwijendra S	**Dwij.Gupta**
Gupta, Jatis Chandra Sen (fl. 1956) M	**J.C.S.Gupta**
Gupta, K.K. (fl. 1961) A	**K.K.Gupta**
Gupta, Kedarmal M. (1908–1987) BFPS	**K.M.Gupta**
Gupta, Meera A	**M.Gupta**
Gupta, N.N. (fl. 1979) M	**N.N.Gupta**
Gupta, Neelima A	**N.Gupta**
Gupta, P.C. (1938–) M	**P.C.Gupta**
Gupta, R. (fl. 1983) M	**R.Gupta**
Gupta, R.C. (fl. 1988) M	**R.C.Gupta**
Gupta, R.S. A	**R.S.Gupta**
Gupta, S.K. (fl. 1980) M	**S.K.Gupta**
Gupta, Satish Chandra (1937–) M	**S.C.Gupta**
Gupta, Shikha (fl. 1988) M	**S.Gupta**
Gupte, S.Y. A	**Gupte**
Gurgenidze, M.Z. (fl. 1965) S	**Gurgen.**
Gürich, Georg Julius Ernst (1859–1938) AF	**Gürich**
Guridi–Gómez, Lydia I. (fl. 1988) S	**Guridi-Gómez**
Gürke, Robert Louis August Maximilian (1854–1911) S	**Gürke**
Gursky, Anatolii Valeryanovich (1906–1967) S	**Gursky**
Gürtler, C. (1879–196?) S	**Gürtler**
Gurung, Vidya Laxmi (fl. 1989) P	**Gurung**
Gururaja, M.N. A	**Gururaja**
Gurzitska (fl. 1950) M	**Gurz.**
Gusarova, I.S. A	**Gusarova**
(Guseinov, E.S.)	
see Husseinov, E.S.	**E.S.Husseinov**
(Guseinova, B.F.)	
see Husseinova, B.F.	**Husseinova**
Gusev, Yu.D. (1922–1986) S	**Gusev**
Guseva, K.A. A	**Guseva**
Gushczin, G.G. (1896–1939) S	**Gushczin**
Gusmus, Hermann (1843–) S	**Gusmus**
(Gussarova, I.S.)	
see Gusarova, I.S.	**Gusarova**
(Gussejnov, Sch.A.)	
see Husseinov, Sch.A.	**Husseinov**

Gussone, Giovanni (1787–1866) PS   **Guss.**
Güssow, Hans Theodor (1879–) M   **Güssow**
Gustafson, B.A. (fl. 1955) M   **Gustafson**
Gustafsson, Åke (1908–) S   **Å.Gust.**
Gustafsson, Carl Emil (1868–1939) S   **Gust.**
(Gustafsson, Lars–Åke)
  see Gustavsson, Lars–Åke   **Gustavsson**
Gustafsson, Mats A. (1941–) MS   **M.A.Gust.**
Gustavsson, Arne (fl. 1959) M   **A.Gustavsson**
Gustavsson, Lars–Åke (1946–) S   **Gustavsson**
Gustops, A. S   **Gustops**
Guşuleac, Mihail (Michail) (1887–1960) S   **Guşul.**
Gutbier, Christian August von (1798–1866) FS   **Gutbier**
Guterman, Carl Edward Frederick (1903–1957) M   **Guterman**
Gutermann, Walter Eckard (1935–) S   **Gutermann**
Guthnick, Heinrich Joseph (1800–1880) S   **Guthnick**
Guthnik, K. S   **K.Guthnick**
Guthrie, Ernest John (fl. 1959) M   **E.J.Guthrie**
Guthrie, Francis (1831–1899) S   **Guthrie**
Guthrie, Louise (1879–1966) S   **L.Guthrie**
Gutiérrez Amaro, Jorge E. (fl. 1981) S   **J.E.Gut**
Gutiérrez, Carmen (fl. 1981) M   **C.Gut.**
Gutiérrez de Sanguinetti, Maria M. (fl. 1986) S   **M.M.Gut.**
Gutiérrez, Gil (1917–) S   **Gut.**
Gutiérrez, Hermes G. (1933–) PS   **H.G.Gut.**
Gutmann, H. S   **Gutmann**
Gutner, L.S. (fl. 1933) M   **Gutner**
(Gutsevich, S.A.)
  see Gucevič, S.A.   **Gucev.**
Gutte, Peter (1939–) S   **Gutte**
Gutteridge, R.J. (fl. 1977) M   **Gutter.**
Gutwinski, Roman (1860–1932) A   **Gutw.**
Güvenc, Tuncer A   **Güvenc**
Guyénot, V.G. (fl. 1922) M   **Guyénot**
Guyer, Oskar A   **Guyer**
Guyétant, Sebastien (1777–1865) S   **Guyét.**
Guymer, Gordon Paul (1953–) S   **Guymer**
Guyot, Alain Lucien (fl. 1932) M   **A.L.Guyot**
Guyot, H. (fl. 1917) M   **Guyot**
(Guzmán Dávalos, Laura)
  see Guzmán–Dávalos, Laura   **Guzm.-Dáv.**
Guzmán, Enriquito D.de (fl. 1984) M   **E.D.Guzmán**
Guzmán, Gastón (1938–) M   **Guzmán**
(Guzmán Grimaldi, Gaston)
  see Guzman, Gastón   **Guzmán**
Guzmán Mejía, Rafael (1950–) S   **R.Guzmám**
(Guzmán, Rafael)
  see Guzmán Mejía, Rafael   **R.Guzmán**
Guzmán–Dávalos, Laura (1961–) M   **Guzm.-Dáv.**
Gvinianidze, Z.I. (1931–) S   **Gvinian.**

Gviniaschvili, Ts.N. (1939–) S	**Gvin.**
(Gviniashviliz, Ts.N.)	
see Gviniaschvili, Ts.N.	**Gvin.**
Gvritischvili, M.N. (1950–) M	**Gvrit.**
(Gvritishvili, M.N.)	
see Gvritischvili, M.N.	**Gvrit.**
Gwynne–Vaughan, David Thomas (fl. 1907) F	**Gwynne–Vaughan**
Gyelnik, Vilmos Köfaragó (1906–1945) M	**Gyeln.**
Gyola, F. (fl. 1909) S	**Gyola**
Györffy, Irma S	**I.Györffy**
Györffy, Istvan (1880–1959) AB	**Györffy**
Ha, Thi Dung (fl. 1970) S	**Ha**
Haage, Ferdinand (1859–1930) S	**F.Haage**
Haage, Friedrich Adolph (1796–1866) S	**Haage**
Haage, Johann Nicolaus (1826–1872) S	**J.N.Haage**
Haage, Walther (1899–) S	**W.Haage**
Haager, J.R. (fl. 1984) S	**Haager**
Haan, Willem de (1801–1855) S	**Haan**
Haarmeyer, J. S	**Haarm.**
Haartman, Johan Gustav (1777–1799) S	**Haartm.**
(Haas, A.J.P.de)	
see de Haas, A.J.P.	**A.J.P.de Haas**
Haas, Adolf Peter S.J. (1914–1982) S	**Haas**
Haas, Hans (1904–) M	**H.Haas**
Haas, Judith E. (fl. 1975) S	**J.E.Haas**
Haas, Robert (fl. 1988) S	**R.Haas**
(Haas, Th.de)	
see de Haas, Th.	**de Haas**
Haasis, Ferdinard Wead (1889–) M	**Haasis**
Haasis, Frank Arling (1908–1965) M	**F.A.Haasis**
Habeeb, Herbert (1917–) AB	**Habeeb**
Haber, Erich (1943–) S	**Haber**
Haberer, Joseph Valentine (1855–1925) P	**Haberer**
Haberfield, W. (fl. 1919) M	**Haberf.**
Haberle, Carl Constantin Christian (1764–1832) S	**Haberle**
Habermann, V. S	**Haberm.**
Habib, Daniel (1936–) A	**Habib**
Habirshaw, Frederick (fl. 1877) A	**Habirshaw**
Hablitz, Carl Ludwig von (1752–1821) S	**Hablitz**
(Hablizl, Carl Ludwig von)	
see Hablitz, Carl Ludwig von	**Hablitz**
Hackel, Eduard (1850–1926) S	**Hack.**
Hackel, Josef (1783–1869) S	**J.Hack.**
Hackel, M. S	**M.Hack.**
Hacket, K. (fl. 1979) M	**Hacket**
Hackett, Harold E. A	**Hackett**
Hacquaert, Armand L. (1906–) A	**Hacquaert**
Hacquet, Balsazar (Balthasar) A. (1739–1815) MS	**Hacq.**

Hada, Yoshine A	**Hada**
Hadač, Emil Franziskov Lazne (1914–) B	**Hadač**
Hadfield, W. (fl. 1960) M	**Hadfield**
Hadidi, Mohammed Nabil El (1934–)S	**Hadidi**
Hadiwidjaja, T. (fl. 1950) M	**Hadiw.**
Hadj–Moustapha, Moustapha (fl. 1957) S	**Hadj–Moust.**
Hadjisterkoti, E. (fl. 1984) M	**Hadjist.**
Hadlock, R. (fl. 1969) M	**Hadlock**
Hadwen, S. A	**Hadwen**
Haeckel, Ernst Heinrich Philipp August (1834–1919) A	**Haeckel**
Haecker, Gottfried Renatus (1789–1864) S	**Haecker**
Haegi, Laurence Arnold Robert (1952–) S	**Haegi**
Haehn, H. (fl. 1922) M	**Haehn**
Haenke, Thaddäus (Tadeáš) Peregrinus Xaverius (1761–1817) S	**Haenke**
Haenseler, Conrad Martin (1888–) S	**C.Haens.**
Haenseler, Felix (1766–1841) S	**Haens.**
(Haesendonck, (Gérard) Constant Van)	
see Van Haesendonck, (Gérard) Constant	**Van Haes.**
Hafellner, Joseph (Josef) (1951–) M	**Hafellner**
Häffner, Jürgen (fl. 1983) M	**Häffner**
Hafiz Khan, Mohammad Abdul (fl. 1955) M	**Hafiz Khan**
(Hafiz, Mohammad Abdul)	
see Hafiz Khan, Mohammad Abdul	**Hafiz Khan**
Häfliger, Ernst S	**Häfl.**
Hagara, Ladislav (fl. 1990) M	**Hagara**
Hagberg, Mats (1955–) S	**Hagberg**
Hagborg, Walter Arnold Ferdinand (1908–) M	**Hagborg**
Hagelstein, Robert (1870–1945) AM	**Hagelst.**
Hagem, Otto Christian (1885–) AM	**Hagem**
Hagemann, Isolde (1944–) S	**I.Hagemann**
Hagemann, Wilhelm (fl. 1908) M	**Hagemann**
Hagemann, Wolfgang (1929–) P	**W.Hagemann**
Hagen, Ingebrigt Severin (1852–1917) B	**I.Hagen**
Hagen, Johann Heinrich (1738–1775) S	**Hagen**
Hagen, Johnny A	**J.Hagen**
Hagen, Karl (Carl) Gottfried (1749–1829) MS	**K.G.Hagen**
Hagen, Stanley Harlan (1913–) S	**S.H.Hagen**
Hagena, Karl (1806–1882) S	**Hagena**
Hagenah, Dale J. (fl. 1954) P	**Hagenah**
Hagenah, Ethelda (fl. 1969) P	**E.Hagenah**
Hagenbach, Carl (Karl) Friedrich (1771–1849) S	**Hagenb.**
Hagendijk, A. (1942–) S	**Hagend.**
Hagerup, Olaf (1889–1961) S	**Hagerup**
Hagiwara, Hiromitsu (1945–) M	**H.Hagiw.**
Hagiwara, Tokio (1896–) S	**Hagiw.**
Hagler, A.N. (fl. 1985) M	**Hagler**
Haglund, Erik Emil (1877–1938) M	**Haglund**
Haglund, Gustaf Emanuel (1900–1955) S	**G.E.Haglund**
Hágsater, Eric (1945–) S	**Hágsater**

Hagström, Elisabeth (fl. 1977) M	E.Hagstr.
Hagström, Johan Oskar (1860–1922) S	Hagstr.
(Hahlov, V.A.)	
see Khakhlov, V.A.	Khakhlov
Hahn, Glen Gardner (1889–1968) M	G.G.Hahn
Hahn, Gotthold (fl. 1875–1911) CM	G.Hahn
Hahn, Otto (1828–1904) S	Hahn
Hahn, W.J. (fl. 1988) S	W.J.Hahn
Hahne, August Hermann (1873–1942) PS	Hahne
Hahnel, G. (fl. 1977) M	Hahnel
Haicour, R. (fl. 1988) S	Haicour
Haiduk, E. (fl. 1989) M	Haiduk
Haillant, Nicolas (fl. 1886) S	Haill.
Haines, Adelbert Lee (1915–) S	A.L.Haines
Haines, Henry Haselfoot (1867–1945) PS	Haines
Haines, John H. (1938–) M	J.H.Haines
Haines, Richard Wheeler (1906–1982) S	R.W.Haines
Hains, John J. A	Hains
Hainz, Richard Wheeler (fl. 1955) S	Hainz
Hair, John Bruce (1909–1979) S	Hair
Hajdu, L. A	Hajdu
Hajós, Márta (1916–) A	Hajós
Hajra, P.K. (1940–) S	Hajra
Hajsig, M. (fl. 1958) M	Hajsig
Håkansson, Hannalore A	Håk.
Hakelier, Nils Georg (fl. 1967) BM	Hakelier
Hakki, Madjit Ismail (1934–) S	Hakki
Hakulinen, Rainer Alarik (1918–) M	Hakul.
Halácsy, Eugen von (Eugène de) (1842–1913) BP	Halácsy
Halász, Márta (1905–1971) A	Halász
Halbinger, Federico (1925–) S	Halb.
Halda, Josef J. (fl. 1971) S	Halda
Hale, Mason Ellsworth (1928–1990) M	Hale
Halfin, L.L. S	Halfin
Halim, Youssef A	Halim
Hall, ?Derek (fl. 1972) M	D.Hall
Hall, Alfred Daniel (1864–1942) S	A.D.Hall
Hall, Anthony Vincent (1936–) S	A.V.Hall
Hall, Carlotta Case (née Case) (1880–1949) PS	C.C.Hall
Hall, Charles Albert (1872–1965) S	C.A.Hall
Hall, Constant Johann Jakob van (1875–) M	C.J.J.Hall
Hall, David Walter (1940–) S	D.W.Hall
Hall, Edwin Cuthbert (1874–1953) S	E.C.Hall
Hall, Elihu (1822–1882) S	E.Hall
Hall, Franklin Wilson (1852–1888) S	F.W.Hall
Hall, Harlow H. (fl. 1961) M	H.H.Hall
Hall, Harry (1906–1986) S	Harry Hall
Hall, Harvey Monroe (1874–1932) S	H.M.Hall
Hall, Herman van (1830–1890) S	H.Hall

Hall, Herman (Hermanus) Christiaan van (1801–1874) PS	H.C.Hall
Hall, I.R. (fl. 1976) M	I.R.Hall
Hall, Irvin M. (fl. 1957) M	I.M.Hall
Hall, Ivan Victor (1924–) S	I.V.Hall
Hall, James (1811–1898) AFP	J.Hall
Hall, John Bartholomew (1932–1984) PS	J.B.Hall
Hall, John Galentine (1870–) M	J.G.Hall
Hall, John W. (1918–) FP	J.W.Hall
Hall, Joyce A. (fl. 1972) M	J.A.Hall
Hall, Leslie Beeching (1878–1945) S	L.B.Hall
Hall, Lisabel Irene (fl. 1984) S	L.I.Hall
Hall, Marion Trufant (1920–) S	M.T.Hall
Hall, Patrick Martin (1894–1941) S	P.M.Hall
Hall, R.A. (fl. 1983) M	R.A.Hall
Hall, Richard P. (1900–1969) A	R.P.Hall
Hall, Sam Rutherford (1905–) A	S.R.Hall
Hall, Thomas Batt (1814–1886) S	T.B.Hall
Hall, William (1743–1800) S	Hall
Hall, William T. A	W.T.Hall
Hallas, Emma Dorothea Kathinka Helene (1849–1926) A	Hallas
Hallbauer, D.K. (fl. 1977) M	Hallbauer
Hallberg, F. (–1924) S	Hallb.
Hallberg, Mildred S	M.Hallb.
Halldal, Per Haakon Haller (1922–) A	Halldal
Hallé, Francis (1938–) S	F.Hallé
Halle, Johann Samuel (1727–1810) S	Halle
Hallé, Nicolas (1927–) S	N.Hallé
Halle, Thore Gustav (Gustafson) (1884–1964) F	T.Halle
Hallegraeff, Gustaaf M. A	Hallegr.
Hallenberg, Nils (1947–) M	Hallenb.
Haller, Albrecht von (1758–1823) S	Haller f.
Haller, B. (fl. 1969) M	B.Haller
Haller, Gottlieb Emmanuel von (1735–1786) S	G.Haller
Haller, John Robert (1930–) S	J.R.Haller
Haller, R., of Aarau (fl. 1950) M	R.Haller Aar.
Haller, R., of Suhr (fl. 1950) M	R.Haller Suhr
Haller, Victor Albrecht von (1708–1777) BMPS	Haller
Hallett, A.F. A	Hallett
Hallett, I.C. (fl. 1983) M	I.C.Hallett
Halley, S. S	Halley
Hällfors, Guy A	G.Hällfors
Hällfors, Seija A	S.Hällfors
Hallidad, Per A	Hallidad
Halliday, Geoffrey (1933–) S	G.Halliday
Halliday, Patricia (1930–) S	P.Halliday
Hallier, Ernst Hans (1831–1904) MS	Hallier
Hallier, Hans (Johannes) Gottfried (1868–1932) S	Hallier f.
Halling, Roy Edward (1950–) M	Halling
Halloy, Stephan (1953–) P	Halloy

Halos, M.Th. A	**Halos**
Halperín, Delia R.de A	**Halperín**
Halse, Richard Ray (fl. 1981) S	**Halse**
Halsey, Abraham (1790–1857) M	**Halsey**
Halsey, F. (fl. 1936) M	**F.Halsey**
Halsted, Byron David (1852–1918) AM	**Halst.**
Haluwyn, C.van (fl. 1983) M	**Haluwyn**
Halvorsen, T. (fl. 1986) S	**Halvorsen**
Halvorson, W.L. (fl. 1990) M	**Halvorson**
Ham, R.W.J.M.van der (1951–) S	**R.W.Ham**
Hama, Eisuke (fl. 1987) S	**E.Hama**
Hama, Taketo (fl. 1958) AM	**Hama**
Hamamoto, Makiko (fl. 1988) M	**Hamam.**
Hamann, Ole Jorgen (1944–) PS	**O.J.Hamann**
Hamann, Ulrich (1931–) S	**U.Hamann**
Hamant, Claude (fl. 1955) M	**Hamant**
Hamaoka, T. (fl. 1986) M	**Hamaoka**
Hamar, J. A	**Hamar**
Hamaya, Toshio (1928–) S	**Hamaya**
Hamberg, Knut Hermann Rudolf (1873–1920) S	**Hamb.**
Hambler, David John (1930–) S	**Hambler**
Hamburger, Clara A	**Hamburger**
Hameed, S.F. (fl. 1983) M	**Hameed**
Hamel, A. A	**A.Hamel**
Hamel, Gontran Georges Henri (1883–1944) AS	**Hamel**
Hamer, Fritz (fl. 1970) S	**Hamer**
Hamers, Maria E.C. (fl. 1987) M	**Hamers**
Hamerton, A.E. A	**Hamerton**
(Hamet, Raymond)	
see Raymond–Hamet	**Raym.-Hamet**
Hämet–Ahti, Raija–Lena (1931–) PS	**Hämet–Ahti**
Hamid, Abdul (fl. 1942) M	**Hamid**
Hamidullah (fl. 1987) S	**Hamidullah**
Hamilton, Alexandra Greenlaw (1852–1941) S	**A.G.Ham.**
Hamilton, Anthony Parke (1939–) S	**A.P.Ham.**
Hamilton, Arthur (fl. 1832) S	**A.Ham.**
Hamilton, Arthur Andrew (1855–1929) S	**A.A.Ham.**
Hamilton, Clement W. (1954–) S	**C.W.Ham.**
(Hamilton, Francis Buchanan)	
see Buchanan–Hamilton, Francis	**Buch.-Ham.**
Hamilton, W.S. (fl. 1885) S	**W.S.Ham.**
Hamilton, William (1783–1856) S	**Ham.**
Hamlin, Bruce Gordon (1929–1976) BS	**Hamlin**
Hamm, Phil B. (fl. 1982) M	**Hamm**
(Hammack, Mary Louise Everly)	
see Everly, Mary Louise	**Everly**
Hammar, Olaf Niklas (1821–1875) BS	**Hammar**
Hammarlund, C. (fl. 1925) M	**Hammarl.**
Hammel, Barry E. (1946–) S	**Hammel**

Hammelev

Hammelev, D. (fl. 1988) M	**Hammelev**
Hammen, T.van der (1924–) FM	**Hammen**
Hammer, B.W. (fl. 1920) M	**B.W.Hammer**
Hammer, Christopher (1720–1804) S	**Hammer**
Hammer, K. (1944–) S	**K.Hammer**
Hammer, Samuel (fl. 1989) M	**S.Hammer**
Hämmerle, Juan Andreas (1876–1930) S	**Hämmerle**
Hammerschmid, Anton (1851–1933) B	**Hamm.**
Hammill, Terrence Michael (fl. 1970) M	**Hammill**
Hammond, Datus M. A	**Hammond**
Hammond, P.C. (fl. 1985) S	**P.C.Hammond**
Hampe, Georg Ernst Ludwig (1795–1880) BMS	**Hampe**
Hamsen, Louis (fl. 1953) M	**Hamsen**
Hamy, Théodor Jules Ernest (1842–1908) S	**Hamy**
Hamzah, Amis S	**Hamzah**
Han, Fu Shan (fl. 1982) A	**F.S.Han**
Han, J. (fl. 1988) S	**J.Han**
Han, Quan Zhong (fl. 1980) S	**Q.Z.Han**
Han, Shu Jin (fl. 1978) M	**S.J.Han**
Han, Y.L. (fl. 1984) S	**Y.L.Han**
Hanan, E.B. (fl. 1938) M	**Hanan**
Hanawa, S. (fl. 1920) M	**Hanawa**
Hanbury, Daniel (1825–1875) S	**D.Hanb.**
Hanbury, Frederick Janson (1851–1938) S	**F.Hanb.**
Hanbury, Thomas (1832–1907) S	**T.Hanb.**
Hance, Henry Fletcher (1827–1886) PS	**Hance**
Hancock, A. (fl. 1969) FM	**A.Hancock**
Hancock, Thomas (1783–1849) S	**Hancock**
Hancock, William (1847–1914) S	**W.Hancock**
Handa, M.R. A	**Handa**
Handel–Mazzetti, Heinrich R.E. (1882–1940) BMPS	**Hand.-Mazz.**
Handlos, Wayne L. S	**Handlos**
Handro, Osvaldo (1908–1986) PS	**Handro**
Handschuch, Carl Friedrich Gottfried Albert (fl. 1832) S	**Handsch.**
Hänel, Klaus A	**Hänel**
Hanelt, Peter H. (1930–) S	**Hanelt**
Hanes, Clarence Robert (1876–) S	**Hanes**
Hanganu, Elisabeta A	**Hanganu**
Hanid, M.Afzel (fl. 1970) S	**Hanid**
Hanin, L. (fl. 1800) S	**Hanin**
Hankó, Belá (fl. 1983) AM	**Hankó**
Hanks, Lena (Lenda) Tracy (1879–1944) AS	**Hanks**
Hänlein, Friedrich Hermann (1851–) S	**Hänlein**
Hanlin, Richard Thomas (1931–) M	**Hanlin**
Hanna, G Dallas (1887–1970) AF	**Hanna**
Hanna, William Fielding (fl. 1929) M	**W.F.Hanna**
Hannaford, Samuel (1828–1874) S	**Hannaford**
Hannibal, L.H. S	**Hannibal**
Hannibal, Lester Stuart (1906–) S	**L.S.Hannibal**

Hannon, Chancellor I. (fl. 1955) M	**C.I.Hannon**
Hannon, Joseph Desiré (1822–1870) S	**Hannon**
Hanry, Hippolyte (1807–1893) S	**Hanry**
Hansbrough, John Raymond (1903–) M	**Hansbr.**
Hansen, Alfred (1925–) S	**A.Hansen**
Hansen, Anders Jørgen (1857–1911) B	**A.J.Hansen**
Hansen, Bertel (1932–) B	**B.Hansen**
Hansen, Bruce Frederick (1944–) S	**B.F.Hansen**
Hansen, Carl (1848–1903) S	**Carl Hansen**
Hansen, Carlo (1932–1991) S	**C.Hansen**
Hansen, Emil Christian (1842–1909) M	**E.C.Hansen**
Hansen, Eric Steen (1943–) M	**E.S.Hansen**
Hansen, Everett M. (1946–) M	**E.M.Hansen**
Hansen, Gayle I. A	**G.I.Hansen**
Hansen, George (1863–1908) S	**G.Hansen**
Hansen, Hans Marius Mølholm (1899–1960) S	**H.M.M.Hansen**
Hansen, Hans Nicholas (1891–) M	**H.N.Hansen**
Hansen, Hans Vilhelm (1951–) S	**H.V.Hansen**
Hansen, Irmgard (Degener) (1924–) S	**I.Hansen**
Hansen, Jens Morten A	**J.M.Hansen**
Hansen, Jørgen Benth (1917–) AS	**J.B.Hansen**
Hansen, Lars (1788–1876) S	**Hansen**
Hansen, Lise (1928–) M	**L.Hansen**
Hansen, Niels Ebbesen (1866–1950) S	**N.E.Hansen**
Hansen, Ove Juel (1945–) S	**O.J.Hansen**
Hansford, Clifford Gerald (1900–1966) M	**Hansf.**
Hansgirg, Anton (Antonin) (1854–1917) A	**Hansg.**
Hänskä, H. (fl. 1986) M	**Hänskä**
Hanson, A.M. (fl. 1944) M	**A.M.Hanson**
Hanson, Craig Alfred (fl. 1962) S	**C.A.Hanson**
Hanson, Earl D. (1927–) A	**E.D.Hanson**
Hanson, Peter (1824–1887) S	**Hanson**
Hanson, William L. (1931–) A	**W.L.Hanson**
Hanstein, Johannes Ludwig Emil Robert von (1822–1880) APS	**Hanst.**
Hantke, Rene (fl. 1974) B	**Hantke**
Hantken, Miksa von (1821–1893) A	**Hantken**
Hantschel, Franz (1884–) S	**Hantsch.**
Hantschke, D. (fl. 1969) M	**Hantschke**
Hantzsch, Carl August (1825–1886) A	**Hantzsch**
Hanzawa, Jun (fl. 1914) M	**Hanzawa**
Hao, Kin Shen (1903–) S	**K.S.Hao**
Happ, George Bippus (1893–) S	**Happ**
Happe, Andreas Friederich (1733–1802) S	**Happe**
Haq, Bilal Ul A	**B.U.Haq**
Haq, Imtiazul A	**I.Haq**
Hara, Hiroshi (1911–1986) BPS	**H.Hara**
Hara, Kanesuke (1885–1962) MS	**Hara**
Hara, Mikio (1929–) B	**M.Hara**
Hara, Shigemitsu A	**S.Hara**

Hara, Yoshiaki (1944–   ) A   Y.Hara
Haračić, Ambrosio (1855–1916) PS   Haračić
Harada, Hiroshi (fl. 1987) M   H.Harada
Harada, Kenichi   A   K.Harada
Harada, Yukio (fl. 1977) M   Y.Harada
Haraguchi, Kazuo   A   Harag.
Harbison, Thomas Grant (1862–1936) S   Harb.
Harbour, D.S.   A   Harbour
Harcombe, P.A.   S   Harc.
Hard, Miron Elisha (1849–1914) M   Hard
Harden, H. (fl. 1936) S   Harden
Harder, Richard (1888–   ) M   Harder
Hardham, Clare B. (fl. 1964) S   Hardham
Hardin, James Walker (1929–   ) S   Hardin
Harding, P.R. (fl. 1953) M   Harding
Hardison, John Robert (1918–   ) M   Hardison
Hardman, R.   S   Hardman
Hardouin, L. (1800–1858) S   Hardouin
Hardtke, H.-J. (fl. 1987) M   Hardtke
Härdtl, Heinrich   A   Härdtl
Hardwicke, Thomas (1757–1835) S   Hardw.
Hardy, Alfred Douglas (1870–1958) A   A.D.Hardy
Hardy, Apollon Joseph (1846–1929) S   A.J.Hardy
Hardy, David Spencer (1931–   ) S   D.S.Hardy
Hardy, James (fl. 1849) S   J.Hardy
Hardy, John (1817–1884) B   Hardy
Hardy, John T.   A   J.T.Hardy
Hare, Raleigh Frederick (1870–1934) S   Hare
Harger, Edgar Burton (1867–1946) S   Harger
Hargitai, Z. (1921–1945) S   Hargitai
Hargraves, Paul E. (1941–   ) A   Hargraves
Hargreaves, Bruce James (1942–   ) S   Hargr.
Haridasan, K. (fl. 1985) S   Harid.
Hariot, Paul Auguste (1854–1917) ABM   Har.
Harkness, Bernard Emerson (1907–   ) S   B.E.Harkn.
Harkness, Harvey Wilson (1821–1901) M   Harkn.
Harkness, Robert (1816–1878) A   R.Harkn.
Härkönen, Marja (1939–   ) M   Härk.
Harl (fl. 1838) F   Harl
Harlan, Jack Rodney (1917–   ) S   J.R.Harlan
Harlan, Richard (1796–1843) A   Harlan
Harland, Rex   A   Harland
Härle, Albert (1905–   ) S   Härle
Harley, Raymond Mervyn (1936–   ) S   Harley
Harley, Winifred J. (1895–   ) P   W.J.Harley
Harling, Gunnar Wilhelm (1920–   ) S   Harling
Harlton, Bruce H.   A   Harlton
Harmaja, Harri (1944–   ) MS   Harmaja
Harmand, A. (fl. 1930) M   A.Harm.

Harmand, Julien Herbert Auguste Jules (1844–1915) M · Harm.
Harmon, William E. (1941–) S · Harmon
Harms, Hermann August Theodor (1870–1942) S · Harms
Harms, Larry J. S · L.J.Harms
Harms, Vernon Lee (1930–) S · V.L.Harms
Harmsen, L. (fl. 1960) M · Harmsen
Harned, Joseph Edward (1870–) S · Harned
Haroon, A.K.Y. A · Haroon
Harper, Francis (1886–) S · F.Harper
Harper, Robert Almer (1862–1946) S · R.A.Harper
Harper, Roland McMillan (1878–1966) PS · R.M.Harper
Harr, Jost (fl. 1971) M · Harr
Harries, Heinrich (fl. 1963) S · Harries
Harriman, Neil A. (1938–) S · N.A.Harriman
Harriman, Philip Ainslie S · Harriman
Harrington, Alan John (1942–) B · A.J.Harr.
Harrington, Francis A. (fl. 1990) M · F.A.Harr.
Harrington, Harold David (1903–1981) S · H.D.Harr.
Harrington, Mark Walrod (1848–1926) PS · Harr.
Harrington, Thomas C. (1952–) M · T.C.Harr.
Harris, Carolyn Wilson (1849–1910) M · C.W.Harris
Harris, David Charles (fl. 1982) M · D.C.Harris
Harris, Denny Olan (1937–) A · D.O.Harris
Harris, Elizabeth (fl. 1960) M · E.Harris
Harris, Hubert Andrew (1909–) M · H.A.Harris
Harris, John R. A · J.R.Harris
Harris, Katharine A · K.Harris
Harris, Richard C. (1939–) M · R.C.Harris
Harris, Stuart Kimball (1906–) PS · S.K.Harris
Harris, Thomas Maxwell (1903–1983) ABFS · T.M.Harris
Harris, Warwick (1940–) S · W.Harris
Harris, Wayne K. (fl. 1981) AF · W.K.Harris
Harris, William H. (1860–1920) S · Harris
Harris, Wilson P. (1869–1929) S · W.P.Harris
Harrison, Arthur Leslie (1905–) M · A.L.Harrison
Harrison, Errol Rhodes (fl. 1972) S · E.R.Harrison
Harrison, Francis Charles St.Barbe (fl. 1928) M · F.C.Harrison
(Harrison, John Heslop)
see Heslop–Harrison, John · Hesl.-Harr.f.
(Harrison, John William Heslop)
see Heslop–Harrison, John William · Hesl.-Harr.
Harrison, Kenneth Archibald (1901–) M · K.A.Harrison
Harrison, Martin Bernard A · M.B.Harrison
Harrison, Sydney Gerald (1924–1988) S · S.G.Harrison
Harrison, Travis Henry John (1901–) M · T.H.J.Harrison
Harrow, Robert Lewis (1867–1954) S · R.L.Harrow
Harrow, William (1861–1945) S · W.Harrow
Harsch, N.S.K. (fl. 1991) M · Harsch
Harshberger, John William (1869–1929) S · Harshb.

(Hart, E.'t)
  see 't Hart, E.   E.'t Hart
(Hart, Henk 't)
  see 't Hart, Henk   't Hart
Hart, Henry Chichester (1847–1908) S   Hart
Hart, J.A. (fl. 1985) S   J.A.Hart
Harter, Leonard Lee (1875–1952) M   Harter
Hartge, Lena Medora (fl. 1923) M   Hartge
Hartig, Heinrich Julius Adolph Robert (1839–1901) MS   R.Hartig
Hartig, Theodor (1805–1880) M   Hartig
Harting, Pieter (1812–1885) AMP   Harting
Hartinger, Anton (1806–1890) S   Hartinger
Hartl, Dimitri (1926–   ) S   Hartl
Hartland, William Baylor (fl. 1903) S   Hartland
Hartley, Bernard   A   B.Hartley
Hartley, Carl Pierce (1887–   ) M   C.P.Hartley
Hartley, Thomas Gordon (1931–   ) S   T.G.Hartley
Hartley, William (1829–1907) S   Hartley
Hartman, Carl (1824–1884) BS   C.Hartm.
Hartman, Carl Johan(n) (1790–1849) ABPS   Hartm.
Hartman, Carl Vilhelm (1862–1941) S   C.V.Hartm.
Hartman, E.L. (fl. 1958) P   E.L.Hartm.
Hartman, G.L. (fl. 1988) M   G.L.Hartm.
Hartman, Robert Wilhelm (1827–1891) S   R.W.Hartm.
Hartman, Ronald Lee (1945–   ) S   R.L.Hartm.
Hartmann, Emmanuel (fl. 1926) M   E.Hartmann
Hartmann, Franz Xaver von (1737–1791) S   Hartmann
Hartmann, Hans   S   H.Hartmann
Hartmann, Heidrun Elsbeth Klara Osterwald (1942–   ) S   H.E.K.Hartmann
Hartmann, Max (1876–   ) AM   M.Hartmann
Hartmann, Wilhelm (fl. 1794) S   W.Hartmann
Hartog, Cornelis den (1931–   ) AS   Hartog
Hartog–Adams, I.den   A   Hartog-Adams
Hartung, C.A.F.A.Heinrich (fl. 1812) S   Hartung
Hartung, William John (1882–   ) S   W.J.Hartung
Hartvig, Per (1941–   ) S   Hartvig
Hartweg, Andreas (–1831) S   A.Hartw.
Hartweg, Karl Theodor (1812–1871) S   Hartw.
Hartwich, Carl (fl. 1902) S   Hartwich
Hartwig, August Karl Julius (1823–1913) S   Hartwig
Hartwiss, Nicolai Anders von (1791–1860) S   Hartwiss
Harvey, Francis Leroy   A   F.L.Harv.
Harvey, I.C. (fl. 1975) M   I.C.Harv.
Harvey, James Vernon (fl. 1925) M   J.V.Harv.
Harvey, LeRoy Hatfield (1911–   ) S   L.H.Harv.
Harvey, Margaret (1919–   ) S   M.Harv.
Harvey, William Henry (1811–1866) ABMS   Harv.
Harvey, Yvette Berenice (1966–   ) S   Y.B.Harv.
Harvey–Gibson, Robert John (1860–1929) A   Harv.-Gibs.
Harvill, Alton McCaleb,Jr (1916–   ) B   Harvill

Harway, Carl J. S	**Harway**
Harz, Carl (Karl) Otto (1842–1906) MS	**Harz**
Harz, Kurt (fl. 1914–1935) S	**K.Harz**
Harzer, Carl August Friedrich (1784–1846) S	**Harzer**
Hasanain, S.Z. (fl. 1960) M	**Hasanain**
Hasegawa, A. (fl. 1975) M	**A.Haseg.**
Hasegawa, Jiro (fl. 1979) B	**J.Haseg.**
Hasegawa, Masahisa (fl. 1927) M	**Haseg.**
Hasegawa, Takezi (fl. 1960) M	**T.Haseg.**
Hasegawa, Yoshio (1919–) A	**Y.Haseg.**
Hásek, J. (fl. 1959) M	**Hásek**
Hasenbusch, V.L. (1897–) S	**Hasenb.**
Hashimoto, Goro (1913–) S	**Hashim.**
Hashimoto, Tamotsu (1933–) MS	**T.Hashim.**
Hashimov, D.K. (fl. 1970) S	**Hashimov**
Hashioka, Yoshio (fl. 1941) M	**Hashioka**
Hashmi, M.H. (fl. 1972) M	**M.H.Hashmi**
Hashmi, S.H. (fl. 1971) S	**S.H.Hashmi**
Hasija, S.K. (fl. 1962) M	**Hasija**
Haskins, E.F. (fl. 1968) M	**E.F.Haskins**
Haskins, Reginald Hinton (fl. 1950) M	**Haskins**
Hasle, Grethe Rytter (1920–) A	**Hasle**
Hasler, Alfred von (fl. 1922) M	**Hasler**
Hassall, Arthur Hill (1817–1894) A	**Hassall**
Hassall, David C. (fl. 1977) S	**D.C.Hassall**
Hassan, Samy K.M. (fl. 1982) M	**Hassan**
Hasse, Clara H. (1880–) M	**C.H.Hasse**
Hasse, Hermann Edward (1836–1915) BM	**Hasse**
Hassebrauk, Kurt (fl. 1965) M	**Hassebr.**
(Hassee, Hermann Edward)	
see Hasse, Hermann Edward	**Hasse**
Hässel, Gabriel G. S	**Hässel**
Hasselbring, Heinrich (1875–1932) M	**Hasselbr.**
Hasselquist, Fredric (1722–1752) BS	**Hasselq.**
Hasselrot, Torsten Edvard (Edverd) (1903–1970) M	**Hasselrot**
Hasselt, Arend Ludolf van (1848–1909) S	**A.Hasselt**
Hasselt, Johan Coenraad van (1797–1823) MS	**Hasselt**
Hasskarl, Justus Carl (1811–1894) BPS	**Hassk.**
Hässler, Arne (1904–1952) S	**A.Hässl.**
Hassler, Émile (1861–1937) PS	**Hassl.**
Hasslow, Olof Johnsson (1871–1952) A	**Hasslow**
Hata, Tutomu (fl. 1953) M	**Hata**
Hatai, Kishio (fl. 1977) M	**Hatai**
Hatakeyama, S. (fl. 1984) S	**Hatak.**
Hatanaka, Shin-ichi (fl. 1984–1988) MPS	**Hatan.**
Hatch, Edwin Daniel (fl. 1946) S	**Hatch**
Hatch, Stephan LaVor (1945–) S	**S.L.Hatch**
Hatcher, Raymond Edward (1930–1967) B	**Hatcher**
Hatmanu, Mircea (1927–) M	**Hatmanu**
Hatta, Hiroaki (fl. 1986) S	**H.Hatta**

Hattink, T.A. (fl. 1970) S	**Hattink**
Hattori, Akihiko A	**A.Hatt.**
Hattori, Hirotarô (1875–1965) S	**Hatt.**
Hattori, Shizuo (1902–) S	**Shiz.Hatt.**
Hattori, Sinske (1915–) B	**S.Hatt.**
Hatusima, Sumihiko (1906–) PS	**Hatus.**
Haub, Peter Ferdinand (1804–1867) S	**Haub**
Haubold F	**Haubold**
Hauck, Ferdinand (1845–1889) A	**Hauck**
Hauduroy, Paul A	**Haud.**
Hauerslev, K. (fl. 1969) M	**Hauerslev**
Haufler, Christopher H. (1950–) P	**Haufler**
Haug, Emil A	**Haug**
Hauge, Halvor Vegard (1914–) A	**Hauge**
Haughton, Samuel (1821–1897) F	**Haught.**
Hauke, Richard L. (1930–) P	**Hauke**
Hauman, Lucien Leon (1880–1965) BMS	**Hauman**
(Hauman–Merck, Lucien Leon)	
see Hauman, Lucien Leon	**Hauman**
Haupt, Arthur Wing (1874–) B	**Haupt**
Hauptfleisch, Paul (1861–1906) AMS	**Hauptfl.**
Hauser, Edward J.P. (fl. 1981) S	**E.J.P.Hauser**
Hauser, Margit Luise S	**Hauser**
Hausknecht, Anton (fl. 1988) M	**Hauskn.**
(Hausman, Franz von)	
see Hausmann, Franz von	**Hausm.**
Hausmann, Franz von (1810–1878) PS	**Hausm.**
Hausmann, W. (fl. 1861) M	**W.Hausm.**
(Hausmann zu Stetten, Franz von)	
see Hausmann, Franz von	**Hausm.**
Haussknecht, Heinrich Carl (1838–1903) PS	**Hausskn.**
Hautzinger, Leo.K.K.H. (c.1920) S	**Hautz.**
Havaas, Johan Jonson (1864–1956) M	**Hav.**
Havard, Valéry (1846–1927) S	**Havard**
(Havås, Johan Jonsen)	
see Havaas, Johan Jonson	**Hav.**
Haviland, George Darby (1857–1901) S	**Havil.**
Havivi, Y. (fl. 1963) M	**Havivi**
Havlicek, R. (fl. 1981) S	**Havlicek**
Haw, S.G. (fl. 1980s) S	**S.G.Haw**
Haware, M.P. (fl. 1971) M	**Haware**
Hawker, Lilian Edith (fl. 1951) M	**Hawker**
Hawkes, Alex Drum (1927–1977) PS	**A.D.Hawkes**
Hawkes, John Gregory (1915–) PS	**Hawkes**
Hawkes, Michael W. A	**M.W.Hawkes**
Hawkeswood, Trevor J. (1956–) S	**Hawkeswood**
Hawksworth, David Leslie (1946–) M	**D.Hawksw.**
Hawksworth, Frank Goode (1926–) S	**Hawksw.**
Hawley, Henry Cusack Wingfield (1876–1923) M	**Hawley**

Hawlitschka, Eva   A — **Hawl.**
Haworth, Adrian Hardy (1768–1833) S — **Haw.**
Haworth, Christopher C. (1934–1989) S — **C.C.Haw.**
Haworth, Elizabeth Y.   A — **E.Y.Haw.**
Haworth–Booth, Michael (fl. 1950) S — **Haw.-Booth**
Hawthorne, P.L. (fl. 1954) M — **Hawth.**
Hay, A. (fl. 1988) S — **A.Hay**
Hay, Cameron H.   A — **C.H.Hay**
Hay, George Upham (1844–1913) S — **G.Hay**
Hay, Judith Ann (fl. 1961) S — **J.A.Hay**
Hay, Thomas (1875–1953) S — **T.Hay**
Hay, William Delisle (fl. 1887) S — **W.D.Hay**
Hay, William Perry (1872–1947) S — **Hay**
Hay, William Winn (1934–   ) A — **W.W.Hay**
Hayakawa, Shiroo (fl. 1978) M — **Hayak.**
Hayakawa, Tetsu   A — **T.Hayak.**
Hayashi, Yasaka (1911–   ) S — **Hayashi**
Hayashi, Yasuo (fl. 1974) M — **Y.Hayashi**
Hayata, Bunzô (1874–1934) PS — **Hayata**
Hayden, Ferdinand Vanderveer (1829–1887) S — **Hayden**
Hayden, M.Victoria   S — **M.V.Hayden**
Hayden, Walter John (1951–   ) S — **W.J.Hayden**
Hayduck, F. (fl. 1922) M — **Hayduck**
Haye, Anna M.   S — **Haye**
Hayek, August von (1871–1928) PS — **Hayek**
Haynald, Cardinal Stephan Franz Ludwig (Lajos) (1816–1891) S — **Haynald**
Hayne, Friedrich Gottlob (1763–1832) S — **Hayne**
Haynes, Caroline Coventry (1858–1951) B — **Haynes**
Haynes, Robert Ralph (1945–   ) S — **R.R.Haynes**
Häyrén, Ernst Fredrik (1878–1957) AMS — **Häyrén**
Haythorn, J.M. (fl. 1980) M — **Haythorn**
Hayward, Glenys C. (fl. 1977) M — **G.C.Hayw.**
Hayward, Ida Margaret (1872–1949) S — **Hayw.**
Hayward, W.R. (fl. 1868–1895) S — **W.R.Hayw.**
Hayward, Wyndham (1903–   ) S — **W.Hayw.**
Hazen, Tracy Elliot (1874–1943) AS — **Hazen**
Hazit, Y. (fl. 1982) S — **Hazit**
Hazlinszky von Hazslin, Friedrich August (Frigyes Ágost)
   (1818–1896) BMPS — **Hazsl.**
(Hazslin, Friedrich August (Frigyes Ágost))
   see Hazlinszky von Hazslin, Friedrich August (Frigyes Ágost) — **Hazsl.**
(Hazslinszky, Friedrich August (Frigyes Ágost))
   see Hazlinszky von Hazslin, Friedrich August (Frigyes Ágost) — **Hazsl.**
He, Bing Zhang (1937–   ) M — **B.Z.He**
He, Jia Qing (fl. 1989) S — **J.Q.He**
He, Jing Biao (fl. 1988) S — **J.B.He**
He, Ming You (fl. 1985) S — **M.Y.He**
He, Pi Xu (fl. 1984) S — **P.X.He**
He; Re Ru   A — **R.R.He**

He, S.Y. (fl. 1984) S   **S.Y.He**
He, Shao Chang (fl. 1985) M   **S.C.He**
He, Si (1959–   ) B   **S.He**
He, Ting Nong (fl. 1980) S   **T.N.He**
He, Xian (fl. 1989) M   **X.He**
He, Yue (fl. 1989) S   **Y.He**
He, Yun He (fl. 1988) S   **Y.H.He**
He, Zong Zhi (1933–   ) M   **Z.Z.He**
Heads, Michael J. (1957–   ) S   **Heads**
Heald, Frederick De Forest (1872–1954) M   **Heald**
Heard, Albert John (fl. 1927) BF   **Heard**
Heard, S.B. (fl. 1988) S   **S.B.Heard**
Heath, F.H.Rodier   S   **F.H.R.Heath**
Heath, Fannie Mahood (–1931) S   **Heath**
Heath, I.Brent (fl. 1983) M   **I.B.Heath**
Heath, Paul V. (fl. 1983) S   **P.V.Heath**
Hebenstreit, Johann Christian (1720–1795) S   **Hebenstr.**
Hebert, L.P. (fl. 1983) S   **L.P.Hebert**
Hebert, Teddy T. (fl. 1971) M   **T.T.Hebert**
Heblack, Russell K. (fl. 1979) M   **Heblack**
Hebrard, Jean-Pierre (1943–   ) B   **Hebr.**
Hécart, Gabriel Antoine Joseph (1755–1838) S   **Hécart**
Hechel, Wilhelm (–1905) S   **Hechel**
Hechler, Jürgen (fl. 1992) M   **Hechler**
Hecht, Julius Gottfried Conrad (1771–1837) S   **Hecht**
Heckard, Lawrence Ray (1923–1991) S   **Heckard**
Heckel, Édouard Marie (1843–1916) S   **Heckel**
Hecker, Johann Julius (1707–1768) S   **Hecker**
Hector, James (1834–1907) S   **Hector**
Hectot, Jean Alexandre (1764–   ) P   **Hectot**
Hedayetullah, S. (fl. 1964) M   **Hedayet.**
Hedberg, Inga (1927–   ) S   **I.Hedberg**
Hedberg, Karl Olov (1923–   ) S   **Hedberg**
Hedderson, Terry Albert John (1962–   ) B   **Hedd.**
Hedenas, Lars (1955–   ) B   **Hedenas**
Hedgcock, George Grant (1863–1946) M   **Hedgc.**
Hedge, Ian Charleson (1928–   ) S   **Hedge**
Hedgecock, D. (fl. 1983) M   **Hedgec.**
Hedger, J.N. (fl. 1970) M   **Hedger**
Hedges, Florence (1878–1956) M   **Hedges**
Hedin, Sven Anders von (1865–1952) B   **Hedin**
Hedjaroude, Ghorban-Ali (fl. 1969) M   **Hedjar.**
Hedley, George Ward (1871–1941) S   **Hedley**
Hedlund, Johan Teodor (1861–1953) MS   **Hedl.**
Hedlung, Theodor J.   S   **Hedlung**
Hedrén, Bo Christer Mikael (1956–   ) S   **Hedrén**
Hedrick, Joyce (1897–   ) M   **J.Hedrick**
Hedrick, Ulysses Prentiss (1870–1951) S   **Hedrick**
Hedwig, Johann (1730–1799) BMPS   **Hedw.**
Hedwig, Romanus (Romanes) Adolf (1772–1806) BMPS   **R.Hedw.**

Heed, W.B. (fl. 1976) M	**Heed**
Heede, Adolphe van den (1841–1928) S	**Heede**
Heeg, Moritz (–1902) B	**Heeg**
(Heek, Werner van)	
see Van Heek, Werner	**Van Heek**
Heer, Oswald von (1809–1883) ABFS	**Heer**
Heerdt, P.François van S	**Heerdt**
Heerebout, G.R. A	**Heereb.**
Heering, Wilhelm Christian August (1876–1916) A	**Heering**
Heese, Emil (1862–1914) S	**Heese**
Hegde, G.R. A	**G.R.Hegde**
Hegde, R.K. (fl. 1976) M	**Hegde**
Hegde, S.N. (fl. 1981) S	**S.N.Hegde**
Hegelmaier, Christoph Friedrich (1833–1906) S	**Hegelm.**
Hegetschweiler, Johannes Jacob (1789–1839) MS	**Hegetschw.**
(Hegetschweiler–Bodmer, Johannes Jacob)	
see Hegetschweiler, Johannes Jacob	**Hegetschw.**
Hegewald, Eberhard Heiz (1942–) AB	**E.H.Hegew.**
Hegewald, Pirkko Dagmar (1934–) B	**P.D.Hegew.**
Hegi, Gustav (1876–1932) PS	**Hegi**
Hegner, Robert W. (1880–1942) A	**Hegner**
Hegyi, Dezsö (fl. 1911) M	**Hegyi**
Heiberg, Peder Andreas Christian (1837–1875) A	**Heib.**
Heideman, M.E. S	**Heideman**
Heiden, Heinrich A	**Heiden**
Heidinger, Wilhelm A	**Heidinger**
(Heijden, E.van der)	
see Heyden, E.van der	**Heyden**
Heijný, Slavomil (1924–) S	**Heijný**
Heikinheimo, V. (fl. 1932) M	**Heikinh.**
Heil, Hans Albrecht (1899–) S	**Heil**
Heil, Kenneth D. (1941–) S	**K.D.Heil**
Heilborn, Otto (1892–1943) S	**Heilborn**
Heilbronn, Alfred (1885–1961) S	**Heilbr.**
Heim, Frédéric Louis (1869–) S	**F.Heim**
Heim, Georg Christoph (1743–1807) S	**Heim**
Heim, Roger (Jean) (1900–1979) M	**R.Heim**
Heimans, Eli (1861–1914) S	**E.Heimans**
Heimans, Jacobus (1889–1978) BS	**J.Heimans**
Heimdal, Berit Riddervold A	**Heimdal**
Heimen, G. (fl. 1980) S	**Heimen**
Heimerl, Anton (1857–1942) AP	**Heimerl**
Hein, Burghard (1944–) M	**B.Hein**
Hein, Heinrich (fl. 1877) S	**Hein**
Heine, Hermann Heino (1922–) PS	**Heine**
Heine, J.W. (fl. 1965) M	**J.W.Heine**
Heine, Otto Rudolph (1920–) S	**O.R.Heine**
Heineck, Otto (1860–) S	**Heineck**
Heineke, T.E. (fl. 1982) S	**Heineke**
Heinemann, Paul (1916–) M	**Heinem.**

(Heinhold, Gustav)
   see Heynhold, Gustav — Heynh.
Heiniger, Helen (fl. 1953) S — Heiniger
Heinrich, F. (fl. 1916) M — F.Heinrich
Heinrich, Walter (fl. 1925) S — Heinrich
Heinricher, Emil Johann Lambert (1856–1934) A — Heinr.
Heinrichson–Normet, T. (fl. 1969) M — Heinr.-Norm.
Heinsen, Ernst (fl. 1894) MP — Heinsen
Heintsch, Lucja A — Heintsch
Heiny, Dana Kelly (fl. 1990) M — Heiny
Heinze, B. (fl. 1904) M — Heinze
Heinzel, Philipp Gustav (1816–) S — Heinzel
Heinzerling, Otto A — Heinzerl.
Heisecke, Ana Maria A — Heisecke
Heiser, Charles Bixler (1920–) S — Heiser
Heiser, Dorothy Gaebler S — D.G.Heiser
Heiser, Dorothy R. (fl. 1945) M — D.R.Heiser
Heister, Lorenz (1683–1758) S — Heist.
(Hejný, Slavomil)
   see Heijný, Slavomil — Heijný
Hejtmanek, M. (fl. 1964) M — Hejtm.
Hekking, William Henri Alphonse Maria (1930–) BS — Hekking
Hektoen, Ludwig (1863–1954) M — Hektoen
Heldmann, C. (fl. 1837) S — Heldmann
Heldreich, I.von S — I.Heldr.
Heldreich, Theodor Heinrich Hermann von (1822–1902) S — Heldr.
Helenes, Javier A — Helenes
Hell, Kurt Gunther (fl. 1969) B — K.G.Hell
Hellbom, Pehr (Per) Johan (1827–1903) BM — Hellb.
(Hellén, Carl Niclas)
   see Hellenius, Carl Niclas — Hell.
Hellenius, Carl Niclas (1745–1820) S — Hell.
Hellenius, Johannes S — J.Hell.
Heller, Alfonse Henry (–1973) S — A.H.Heller
Heller, Amos Arthur (1867–1944) PS — A.Heller
Heller, Carl Bartholomäus (1824–1880) S — C.Heller
Heller, David (1936–) S — D.Heller
Heller, Franz Xaver (1775–1840) S — F.Heller
Hellerman, Joan (1929–) A — Hellerman
Hellinger, Esther (fl. 1947) M — Hellinger
Hellmayr, Carl Eduard (1877–1944) S — Hellm.
Hellmayr, K.J. S — K.Hellm.
Hellquist, C.Baare (1940–) S — Hellq.
Hellström, Fredrik (1824–1889) S — Hellstr.
Hellweg, Frank (1958–) S — Hellweg
Hellwig, Franz Carl (1861–1889) PS — Hellw.
Helm, Friedrich Gustav (fl. 1809–1828) S — Helm
Helm, Hermann Wilhelm Johannes (1906–) S — H.W.J.Helm
Helm, Julius (–1844) S — J.Helm

Helmcke, J.-G. (1908–) A	**Helmcke**
Helmersen, G.P. S	**Helmersen**
Helmrich, Carl (1833–1868) S	**Helmr.**
Helms, Richard (1842–1914) S	**Helms**
Helson, G.A.H. A	**Helson**
Heluta, V.P. (fl. 1990) M	**Heluta**
Helwig, Burghard S	**Helwig**
Helwing, Georg Andreas (1668–1748) L	**Helwing**
Hema, A. A	**Hema**
Hemadri, Koppula (fl. 1970) S	**Hemadri**
Hembree, S.C. (fl. 1985) M	**Hembree**
Hemer, Darwin O. A	**Hemer**
Hemmendorf, Ernst (1866–1928) S	**Hemmend.**
Hemmi, Takewo (1890–) M	**Hemmi**
Hempel, Adolph A	**A.Hempel**
Hempel, Gustav (1842–1904) S	**Hempel**
Hempel, Werner (fl. 1980) S	**W.Hempel**
Hempfling, W.P. (fl. 1979) M	**Hempfling**
Hemsley, James Hatton (1923–) S	**J.H.Hemsl.**
Hemsley, William Botting (1843–1924) APS	**Hemsl.**
Henckel von Donnersmarck, Leo Victor Felix (1785–1861) S	**Henckel**
Hendel, Johann Christian (1742–1823) S	**Hendel**
Henderson, Andrew Augustus (1816–1876) S	**A.A.Hend.**
Henderson, Archibald (1921–) S	**A.Hend.**
Henderson, Douglas Mackay (1927–) BMS	**D.M.Hend.**
Henderson, Edward George (1782–1876) S	**Hend.**
Henderson, George (1836–1929) S	**G.Hend.**
Henderson, Louis Forniquet (1853–1942) PS	**L.F.Hend.**
Henderson, Mayda Doris (1928–) S	**M.D.Hend.**
Henderson, Murray Ross (1899–1982) S	**M.R.Hend.**
Henderson, Nellie Frater (1885–1952) B	**N.F.Hend.**
Henderson, Norlan C. (fl. 1962) S	**N.C.Hend.**
Henderson, Rodney John Francis (1938–) S	**R.J.F.Hend.**
Hendey, Norman Ingram (1903–) A	**Hendey**
Hendrick, L.R. (fl. 1951) M	**Hendrick**
Hendrickx, Fred L. (fl. 1948) M	**Hendr.**
Hendrix, Floyd Fuller (fl. 1968) M	**F.F.Hendrix**
Hendrix, James W. (fl. 1967) M	**J.W.Hendrix**
Hendrych, Radovan (1926–) S	**Hendrych**
Héneau, Alphonse (fl. 1889) S	**Héneau**
Henfrey, Arthur (1819–1859) A	**Henfr.**
Hengstmengel, J. (fl. 1982) M	**Hengstm.**
Henkel, A.von (fl. 1923) M	**A.Henkel**
Henkel, Heinrich (fl. 1897–1914) S	**H.Henkel**
Henkel, Johann Baptist (1815–1871) S	**Henkel**
Henneberg, Wilhelm (fl. 1926) M	**Henneberg**
Hennebert, Grégoire L. (1929–) M	**Hennebert**
Hennedy, Roger (1809–1877) PS	**Hennedy**
Henneguy, Louis–Félix (1850–1928) AB	**Henneg.**

Hennen, Joe Fleetwood (1928–) M	J.F.Hennen
Hennen, Joseph (1852–1927) B	Hennen
Hennert, Carl Wilhelm (1739–1800) S	Hennert
Hennessy, Esmée Franklin (fl. 1986) S	Hennessy
Henney, Mary R. (fl. 1980) M	Henney
Henning, Ernst Johan (1857–1929) M	Henning
Hennings, Paul Christoph (1841–1908) ABMS	Henn.
Hennipman, Elbert (1937–) MPS	Hennipman
Hénon, Jacques Louis (1802–1872) S	Hénon
Henrard, Johannes (Jan) Theodoor (1881–1974) MS	Henrard
Henrich, James Emil (fl. 1987) S	Henrich
Henrichsen, Martin Vahl	
see Vahl, Martin (Henrichsen)	Vahl
Henrici, Arthur Trautwein (fl. 1930) M	Henrici
Henrickson, James Solberg (1940–) S	Henr.
Henriques, Julio Augusto (1838–1928) ABMPS	Henriq.
Henry, Aimé Constant Fidèle (1801–1875) S	Henry
Henry, Ambrose Nathaniel (1936–) S	A.N.Henry
Henry, Arthur Wellesley (1896–) M	A.W.Henry
Henry, Augustine (1857–1930) S	A.Henry
Henry, Berch Waldo (1915–) M	B.W.Henry
Henry, Eric C (1949–) A	E.C.Henry
Henry, J.de N. (fl. 1980) S	J.N.Henry
Henry, Jean–Luis A	J.-L.Henry
Henry, Joseph (1816–1887) S	J.Henry
Henry, Joseph Kaye (1866–1930) S	J.K.Henry
Henry, LeRoy Kershaw (1905–) S	L.K.Henry
Henry, Louis (1853–1903) S	L.Henry
Henry, Mary K. (1884–1967) S	M.K.Henry
Henry, Max S	M.Henry
Henry, Rene (1884–1960) B	R.Henry
Henry, Robert (fl. 1944) M	Rob.Henry
Henschel, August Wilhelm Eduard Theodor (1790–1856) S	Hensch.
Henschen, Salomon Eberhard (1847–1930) S	Henschen
Hensen, Karel S.W. (1918–) S	K.S.W.Hensen
Hensen, Victor (–1924) A	Hensen
Henshaw, Julia Willmothe (fl. 1906) S	Henshaw
Henslow, George (1835–1925) S	G.Hensl.
Henslow, John Stevens (1796–1861) S	Hensl.
Hensold, N. (fl. 1988) S	Hensold
Henssen, Aino (1925–) M	Henssen
Hentig, H. (fl. 1881) S	Hentig
Hentze, Wilhelm (1793–1874) S	Hentze
Hepp, Ernst (1878–1968) M	E.Hepp
Hepp, Johann Adam Philipp (1797–1867) M	Hepp
Hepper, Frank Nigel (1929–) S	Hepper
Hepting, George Henry (1907–) M	Hepting
Herak, Milan (1917–) A	Herak
(Herasimenko, I.I.)	
see Gerasimenko, I.I.	Geras.

Herbert, Desmond Andrew (1898–1976) MS	**D.A.Herb.**
Herbert, Joan W. (fl. 1954) M	**J.W.Herb.**
Herbert, William (1778–1847) S	**Herb.**
Herbich, Franz (1791–1865) S	**Herbich**
Herborg, J. (fl. 1987) S	**Herborg**
Herbst, Derral Raymon (1934–) S	**D.R.Herbst**
Herbst, Nora A	**N.Herbst**
Herbst, Rafael (fl. 1964–77) F	**R.Herbst**
Herbst, William (1833–1907) M	**Herbst**
Herd, G.W. (fl. 1965) M	**Herd**
Herder, Ferdinand Gottfried Theobald Maximilian von (1828–1896) S	**Herder**
Herdman, E.Catherine A	**Herdman**
(Hérelle, F.H.d')	
see d'Hérelle, F.H.	**d'Hérelle**
Herendeen, Patrick S. (fl. 1990) F	**Herend.**
Hergt, Bernhard (fl. c.1912) P	**B.Hergt**
Hergt, Johann Ludwig (fl. 1822) S	**Hergt**
Héribaud, Joseph (1841–1918) PS	**Hérib.**
(Héribaud–Joseph, (Frère))	
see Héribaud, Joseph	**Hérib.**
(Heribert–Nilsson, Nils)	
see Nilsson, Nils Heribert	**N.H.Nilsson**
Hérincq, François (1820–1891) S	**Hérincq**
Hering, Constantijn J. (1800–1880) S	**Hering**
Hering, Karl (1896–1943) A	**K.Hering**
Heringer, Ezechias Paulo (1905–) S	**Heringer**
Herink, Josef (fl. 1949) M	**Herink**
(Héritier, Charles Louis l')	
see L'Héritier de Brutelle, Charles Louis	**L'Hér.**
Herman, Alberta I. (fl. 1971) M	**Herman**
Herman, Eliot Mark A	**E.M.Herman**
Hermanides–Nijhof, E.J. (fl. 1977) M	**Herm.-Nijhof**
Hermann, Frederick Joseph (1906–1987) BS	**F.J.Herm.**
Hermann, Friedrich (1873–1967) S	**F.Herm.**
Hermann, Johannes Gotthelf (1741–) A	**J.Herm.**
Hermann, Paul (1646–1695) L	**Herm.**
Hermann, Paul (1876–) M	**P.Herm.**
Hermann, Wilhelm W. (fl. 1936) M	**W.W.Herm.**
Hermes, Heather A	**Hermes**
Hermjakob, G. (fl. 1970) S	**Hermj.**
Hermosilla, Jorge A	**Hermos.**
(Hermosilla S., Jorge)	
see Hermosilla, Jorge	**Hermos.**
(Hernández Bermejo, J.Esteban)	
see Hernández-Bermejo, J. Esteban	**Hern.-Berm.**
Hernández Camacho, Jorge (fl. 1958) S	**Hern.Cam.**
Hernández Cardona, A.M. (fl. 1976) S	**A.M.Hern.**
(Hernández, Effraim Ildefonso)	
see Hernández–Xolocotzi Guzman, E.I.	**Hern.-Xol.**

Hernández, Héctor Manuel (1954–) S	**H.M.Hern.**
Hernández, Julián Camara S	**J.Hern.**
(Hernández Macías, Héctor)	
see Hernández, Héctor Manuel	**H.M.Hern.**
(Hernández, Mateo Rodríguez)	
see Rodríguez Hernández, Mateo	**Rodr.Hern.**
Hernández Torres, I. (fl. 1987) S	**Hern.Torres**
Hernández–Bermejo, J.Esteban (1949–) S	**Hern.-Berm.**
Hernández–Padrón, C. (1953–) M	**Hern.-Padr.**
Hernández–Xolocotzi Guzman, Effraim Ildefonso (1913–1991) S	**Hern.-Xol.**
Herndon, A. (fl. 1981) S	**A.Herndon**
Herndon, Walter Roger (1926–) A	**Herndon**
Herold, Johann Moritz David (1790–1862) S	**Herold**
Heron, D.A. (fl. 1929) M	**Heron**
Herpell, Gustav Jacob (1828–1912) M	**Herp.**
Herre, Adolar Gottlieb Julius (Hans) (1895–1979) S	**A.G.J.Herre**
Herre, Albert William Christian Theodore (1868–1962) M	**Herre**
(Herrera Alarcón de Loja, Berta)	
see Herrera, Berta	**B.Herrera**
Herrera Arrieta, Yolanda (1954–) S	**Y.Herrera**
Herrera, Berta (1930–) S	**B.Herrera**
Herrera Figueroa, Sara (fl. 1982) M	**S.Herrera**
Herrera Oliver, Pedro (fl. 1981) S	**P.Herrera**
Herrera, R.A. (fl. 1980) M	**R.A.Herrera**
Herrera, Teófilo (fl. 1953) M	**T.Herrera**
Herrera y Garmendia, Fortunado Luciano (1875–1945) S	**Herrera**
Herrich–Schaeffer, Gottlieb August (1799–1874) S	**Herr.-Schaeff.**
Herrmann, Eugen Adolf (fl. 1920) M	**E.A.Herrm.**
Herrmann, Johann (Jean) (1738–1800) S	**Herrm.**
Herrmann, Rudolf Albert Wolfgang (1885–) S	**R.A.W.Herrm.**
(Herrmann–Erlee, M.P.M.)	
see Erlee, M.P.M.	**Erlee**
Herrnstadt, Ilana (1940–) BS	**Herrnst.**
Hershenov, B. A	**Hershenov**
Hertel, Hannes (1939–) M	**Hertel**
Herter, Wilhelm (Guillermo) Gustav(o) Franz (Francis) (1884–1958) MPS	**Herter**
Hertlein, Leo George (1898–) A	**Hertlein**
Hertrich, William (1878–1966) S	**Hertrich**
Hertsch, Hermann (1819–1856) S	**Hertsch**
Hertwig, Richard (1850–1937) AM	**Hertwig**
Hervey, Alpheus Baker (1839–1931) AB	**Herv.**
Hervey, Eliphalet Williams (1834–1925) S	**E.W.Herv.**
(Hervier, Gabriel Marie Joseph)	
see Hervier–Basson, Gabriel Marie Joseph	**Hervier**
Hervier–Basson, Gabriel Marie Joseph (1846–1900) PS	**Hervier**
Herzberg, Paul (1865–) M	**Herzberg**
Herzer, Herman(n) (1833–1912) AFMS	**Herzer**
Herzfeld, Stephanie (1868–1930) S	**Herzfeld**
Herzig, W.N. A	**Herzig**

Herzog, E. (fl. 1986) S	**E.Herzog**
Herzog, Michael S	**M.Herzog**
Herzog, Theodor Carl (Karl) Julius (1880–1961) BPS	**Herzog**
Hesler, Lexemuel Ray (1888–1977) M	**Hesler**
Heslop–Harrison, Helena (–1984) S	**H.Hesl.-Harr.**
Heslop–Harrison, John (1920–) S	**Hesl.-Harr.f.**
Heslop–Harrison, John William (1881–1967) S	**Hesl.-Harr.**
Heslot, Henri (fl. 1953) S	**Heslot**
Hespenheide, Henry August (fl. 1968) S	**Hespenh.**
Hess, G. (fl. 1944) M	**G.Hess**
Hess, Hans E. (1920–) MS	**H.E.Hess**
Hess, Johann Jakob (1844–1883) S	**Hess**
Hess, Wilfred Moser (fl. 1951) M	**W.M.Hess**
Hess, William John (1934–) S	**W.J.Hess**
Hesse, Ed. (fl. 1905) AM	**E.Hesse**
Hesse, Hermann Albrecht (1852–1937) S	**Hesse**
Hesse, Julius Oswald (1835–1917) S	**J.Hesse**
Hesse, Michael (1943–) S	**M.Hesse**
Hesse, Rudolph (Rudolf) (1844–1912) MS	**R.Hesse**
Hesselbo, Christen August (1874–1952) B	**Hesselbo**
Hesseltine, Clifford William (fl. 1954) M	**Hesselt.**
Hessler, Karl (fl. 1822) B	**Hessl.**
(Hest, J.J.Van)	
see Van Hest, J.J.	**Van Hest**
Hester, J.Pinckney (fl. 1943) S	**Hester**
Hestmark, G. (fl. 1990) M	**Hestmark**
Hetrick, B.A.D. (fl. 1991) M	**Hetrick**
Hetterscheid, W.L.A. (fl. 1981–89) P	**Hett.**
Hettige, Gloria (fl. 1983) M	**Hettige**
Heucher, Johann Heinrich von (1677–1747) L	**Heuch.**
(Heuer, Ilse R.)	
see Mendoza–Heuer, Ilse R.	**Mend.-Heuer**
Heuffel, János (Johann) A. (1800–1857) S	**Heuff.**
(Heufler, Ludwig Samuel Joseph David Alexander)	
see Heufler zu Rasen und Perdonneg, Ludwig Samuel Joseph David Alexander	
	Heufl.
Heufler zu Rasen und Perdonneg, Ludwig Samuel Joseph David Alexander	
(1817–1885) MPS	**Heufl.**
Heugel, Carl August (1802–1876) BM	**Heugel**
Heukels, Hendrik (1854–1936) PS	**Heukels**
Heukels, P. (fl. 1973) B	**P.Heukels**
(Heurck, Henri Ferdinand Van)	
see Van Heurck, Henri Ferdinand	**Van Heurck**
Heurn, Willem Cornelis van (1887–1972) S	**Heurn**
Heusden, E.C.H.van (fl. 1988) S	**Heusden**
(Heusi, William McKinley)	
see Hiesey, William McKinley	**Hiesey**
Hevly, Richard Holmes (1934–) PS	**Hevly**
Heward, Robert (1791–1877) PS	**Heward**

Hewings, A.D. (fl. 1981) M	**Hewings**
Hewitt, Florence Ellen (1910–1979) A	**Hewitt**
Hewson, Helen Joan (1938–) BS	**Hewson**
Heybroek, Hans M. S	**Heybroek**
Heyden, E.van der (fl. 1990) S	**Heyden**
Heydenreich, Karl (fl. 1933) S	**Heydenr.**
Heyder, Edward (1808–1884) S	**Heyder**
Heydrich, Franz (–1911) A	**Heydr.**
Heydt, Adam (fl. 1932) S	**Heydt**
Heyer, Carl (1797–1856) S	**Heyer**
Heykoop, Michel (fl. 1987) M	**Heykoop**
Heyn, Chaia Clara (1924–) S	**Heyn**
Heyne, Benjamin (1770–1819) S	**B.Heyne**
Heyne, Friedrich Adolf (1760–1826) S	**F.Heyne**
Heyne, Karel (1877–1947) S	**K.Heyne**
Heynhold, Gustav (1800–1860) S	**Heynh.**
Heynig, Hermann A	**Heynig**
Heywood, Vernon Hilton (1927–) PS	**Heywood**
Hibberd, David John A	**D.J.Hibberd**
Hibberd, James Shirley (1825–1890) S	**Hibberd**
Hibbits, J. (fl. 1981) M	**Hibbits**
Hick, Thomas (1840–1896) S	**Hick**
Hickel, Barbara A	**B.Hickel**
Hickel, Paul Robert (1865–1935) S	**Hickel**
Hicken, Cristóbal Mariá (1875–1933) PS	**Hicken**
Hickey, Ralph James (1950–) PS	**Hickey**
Hickman, C.J. (fl. 1940) M	**Hickman**
Hickman, James Craig (1941–) S	**J.C.Hickman**
Hickok, Leslie G. (1946–) P	**Hickok**
Hicks, Gilbert Henry (1861–1899) M	**Hicks**
Hicks, John Braxton (1823–1897) A	**J.B.Hicks**
Hidaka, Zyun (fl. 1934) M	**Hidaka**
(Hidén, Henrik Ilmari Augustus)	
see Hiitonen, Henrik Ilmari Augustus	**Hiitonen**
Hiep, N.T. (fl. 1980) S	**Hiep**
Hiepko, Paul Hubertus (1932–) S	**Hiepko**
Hiern, William Philip (1839–1925) BS	**Hiern**
Hieronymus, Georg Hans Emmo (Emo) Wolfgang (1846–1921) APS	**Hieron.**
Hiesey, William McKinley (1903–) S	**Hiesey**
Higa, S. (fl. 1988) S	**Higa**
Higgins, Bascombe Britt (1887–) M	**B.B.Higgins**
Higgins, Henry Hugh (1814–1893) B	**Higgins**
Higgins, Larry C. (1936–) S	**L.C.Higgins**
Higgins, Vera (1892–1968) S	**V.Higgins**
Higinbotham, Betty Louise (1910–) B	**B.L.Higinb.**
Higinbotham, Noe Levi (1913–) AB	**N.L.Higinb.**
Higley, William Kerr (1860–1908) S	**Higley**
Higuchi, Masanobu (1955–) B	**Higuchi**
Hiitonen, Henrik Ilmari Augustus (1898–) PS	**Hiitonen**

Hijman, Maria E.E. (fl. 1977) S — **Hijman**
Hijwegen, T. (fl. 1979) M — **Hijwegen**
Hikino, Hiroshi (1931–) S — **Hikino**
Hilber, Oswald (1942–) M — **O.Hilber**
Hilber, Růžena (fl. 1983) M — **R.Hilber**
Hildebrand, Alexander Anderson (1896–) M — **A.A.Hildebr.**
Hildebrand, Earl Martin (1902–) AM — **E.M.Hildebr.**
Hildebrand, Frederik Hendrik (1900–1975) S — **F.H.Hildebr.**
Hildebrand, Friedrich Hermann Gustav (1835–1915) MS — **Hildebr.**
Hildebrandt, Johann Maria (1847–1881) S — **Hildebrandt**
Hildmann, H. (–1895) S — **Hildm.**
Hildreth, K.C. A — **Hildreth**
Hildt, Johann Adolph (–1805) S — **Hildt**
Hilend, Martha Luella (1902–) S — **Hilend**
Hilferty, Frank Joseph (1920–) B — **Hilf.**
Hilgard, Theodore Charles (1828–1875) MS — **Hilg.**
Hilger, Harmut H. (1948–) S — **Hilger**
Hilitzer, Alfred C. (1899–1940) M — **Hilitzer**
Hill, Adrian Edward (fl. 1979) F — **A.E.Hill**
Hill, Albert Frederick (1889–1977) S — **A.F.Hill**
Hill, Albert Joseph (1940–) S — **A.J.Hill**
Hill, Arthur William (1875–1941) S — **A.W.Hill**
Hill, C.F. (fl. 1990) M — **C.F.Hill**
Hill, David A. (fl. 1990) S — **D.A.Hill**
Hill, David R.A. (1962–) A — **D.R.A.Hill**
Hill, Ellsworth Jerome (1833–1917) BS — **E.J.Hill**
Hill, Grace Alma A — **G.A.Hill**
Hill, Hibbert A — **H.Hill**
Hill, John (1716–1775) AMPS — **Hill**
Hill, Kenneth D. (1948–) S — **K.D.Hill**
Hill, Mark Oliver (1945–) B — **M.O.Hill**
Hill, Paul John A — **P.J.Hill**
Hill, Robert Southey (1954–) F — **R.S.Hill**
Hill, Steven Richard (1950–) S — **S.R.Hill**
Hill, Walter (1820–1904) PS — **W.Hill**
Hillcoat, Jean Olive Dorothy (1904–1990) S — **Hillc.**
Hillebrand, Wilhelm (William) B. (1821–1886) PS — **Hillebr.**
Hillebrand, William Francis (1853–1925) S — **W.F.Hillebr.**
Hillebrandt, Franz (1805–1860) S — **Hillebrandt**
Hillegas, Arthur Burdette (1907–) M — **Hillegas**
Hillhouse, William (1850–1910) S — **Hillh.**
Hilliard, Douglas K. A — **D.K.Hilliard**
Hilliard, Olive Mary (1926–) S — **Hilliard**
Hillier, Louis E.J. (1869–1962) B — **Hillier**
(Hillis, Llewellyn W.)
  see Hillis–Colinvaux, Llewellyn W. — **Hillis–Col.**
Hillis–Colinvaux, Llewellyn W. (1930–) A — **Hillis–Col.**
Hillmann, Johannes (1881–1943) M — **Hillmann**
Hills, Len V. A — **Hills**

Hilpert, Friedrich Wilhelm (1907–   ) B   **Hilp.**
Hilse, Friedrich Wilhelm (1820–1871) A   **Hilse**
Hilsenbeck, Richard A. (1952–   ) S   **Hilsenb.**
Hilsenberg, Karl Theodor (1802–1824) S   **Hils.**
Hiltebrandt, F.   S   **F.Hiltebr.**
Hiltebrandt, V.M. (1898–1941) S   **Hiltebr.**
Hilton, Richard L. (1933–   ) AS   **R.L.Hilton**
Hilton, Roger Norman (1927–   ) M   **R.N.Hilton**
Hilu, Khidir W. (fl. 1981) S   **Hilu**
Himeno, T.   B   **Himeno**
Himmelbauer, Wolfgang (1886–1937) S   **Himmelb.**
Himpel, J.Stephan (fl. 1891) S   **Himpel**
Hind, David John Nicholas (1957–   ) S   **D.J.N.Hind**
Hind, F.J.   S   **F.J.Hind**
Hind, William Marsden (1815–1894) S   **Hind**
Hindák, Frantisek (1937–   ) A   **Hindák**
Hindle, Edward   A   **Hindle**
Hindmarsh, Mary MacLean   S   **Hindm.**
Hinds, Richard Brinsley (1812–1847) S   **Hinds**
Hinds, Thomas E. (1923–   ) M   **T.E.Hinds**
Hindson, W.R. (fl. 1958) M   **Hindson**
Hine, A.E.   A   **Hine**
Hingorani, M.K. (fl. 1957) M   **Hing.**
Hinkova, T.S. (fl. 1961) M   **Hinkova**
Hinneri, Sakari Lauri (1940–   ) B   **Hinn.**
Hino, Iwao (fl. 1929) M   **I.Hino**
Hino, Takayuki (fl. 1931) M   **T.Hino**
Hinode, Taketoshi   A   **Hinode**
Hinterhuber, Julius (1810–1880) S   **J.Hinterh.**
Hinterhuber, Rudolph (1802–1892) S   **R.Hinterh.**
Hintikka, Toivo Juho (fl. 1924) M   **Hintikka**
Hintze, Fl. (fl. 1893–1932) B   **Hintze**
Hinz, P.–A. (fl. 1987) S   **Hinz**
Hinze, Gustav (1879–   ) A   **Hinze**
Hinzelin, F. (fl. 1991) M   **Hinzelin**
(Hipólito Unanue, José)
  see Unanue, José Hipólito   **Unanue**
Hippe, Ernst (fl. 1878) S   **Hippe**
Hirahara, S. (fl. 1970) M   **Hirahara**
Hirahata, T (fl. 1962) S   **Hirahata**
Hirane, Seiichi (fl. 1960) M   **Hirane**
Hirano, Kasuzura (fl. 1962) M   **K.Hirano**
Hirano, Minoru (1910–   ) A   **Hirano**
Hirao (fl. 1981) S   **Hirao**
Hirata, Koji (fl. 1952) M   **Hirata**
Hirata, S. (fl. 1979) M   **S.Hirata**
Hiratsuka, Naoharu (1873–1946) M   **Hirats.**
Hiratsuka, Naohide (1903–   ) M   **Hirats.f.**
Hiratsuka, Toshiko (1931–   ) M   **T.Hirats.**

Hiratsuka, Yasuyuki (1933–) M	**Y.Hirats.**
Hirayama, Shigekatsu (fl. 1931) M	**Hiray.**
Hirc, Dragutin (1852?–1921) S	**Hirc**
Hiremath, R.V. (fl. 1977) M	**Hiremath**
Hiriart, Patricia (fl. 1981) S	**Hiriart**
(Hiriart Valencia, Patricia)	
see Hiriart, Patricia	**Hiriart**
Hirmer, Max (1893–1981) FPS	**Hirmer**
Hirn, Karl Engelbrecht (1872–1907) A	**Hirn**
Hiroë, Isamu Matsuura (1905–) M	**Hiroë**
Hiroe, Minosuke (1914–) S	**M.Hiroe**
Hirose, Hiroyuki (1912–1985) A	**Hirose**
Hirota, Naonori S	**Hirota**
Hirsch, Allan (1929–) A	**Hirsch**
Hirsch, Gerald (1953–) M	**G.Hirsch**
Hirschhorn, Elisa (1914–) M	**Hirschh.**
Hirscht, Karl S	**Hirscht**
Hirth, Alfred (fl. 1908) P	**Hirth**
Hirtz, A.C. (fl. 1987) S	**Hirtz**
Hirzel, Willy (1913–) S	**Hirzel**
(Hisauchi, Kiyotaka)	
see Hisauti, Kiyotaka	**Hisauti**
Hisauti, Kiyotaka (1884–1981) S	**Hisauti**
Hisinger, Wilhelm (1766–1852) F	**Hising.**
Hitchcock, Albert Spear (1865–1935) PS	**Hitchc.**
Hitchcock, Charles H. (fl. 1862) F	**C.H.Hitchc.**
Hitchcock, Charles Leo (1902–1986) S	**C.L.Hitchc.**
Hitchcock, Edward (1793–1864) MPS	**E.Hitchc.**
Hitomi, Tsuyoshi (fl. 1931) M	**Hitomi**
Hitwegen, T. (fl. 1991) M	**Hitwegen**
Hiura, Makato (fl. 1929) M	**Hiura**
Hiyama, Kôzô (1905–) PS	**Hiyama**
Hjaltalín, Oddur Jönsson (1782–1840) S	**Hjalt.**
Hjelmquist, Karl Jesper Hakon (1905–) S	**Hjelmq.**
(Hjelmqvist, Karl Jesper Hakon)	
see Hjelmquist, Karl Jesper Hakon	**Hjelmq.**
Hjelt, Albert Hjalmar (1851–1925) S	**Hjelt**
Hjerting, Jan Peter Knudsen (1917–) S	**Hjert.**
Hjort, Aarse (fl. 1954) M	**A.Hjort**
Hjort, Johan A	**J.Hjort**
Hjortstam, Kurt (1933–) M	**Hjortstam**
Hladnik, Franz (1773–1844) S	**Hladnik**
Hladun, Néstor Luis (1951–) M	**Hladun**
Hlava, J. (fl. 1887) M	**Hlava**
Hlaváček, Jiří (fl. 1958) M	**Hlaváček**
Hnatiuk, Roger James (1946–) S	**Hnatiuk**
Ho, Ching S	**C.Ho**
(Ho, Ching Shêng)	
see Hao, Kin Shen	**K.S.Hao**

Ho, Chun Nien (fl. 1955) S	C.N.Ho
Ho, Feng Chi (fl. 1982) S	F.C.Ho
Ho, H.H. (fl. 1990) M	H.H.Ho
Ho, Huei Tin (fl. 1932) S	H.T.Ho
Hô, Pham-Hoàng (1931–) AS	P.H.Hô
Ho, Shan Bao (1930–) S	S.B.Ho
Ho, Ting Nung (1938–) S	T.N.Ho
Ho, Y.W. (fl. 1990) M	Y.W.Ho
Ho, Ye Chi (fl. 1989) S	Y.C.Ho
Hoare, Cecil A. A	Hoare
Hoban, Michael A. A	Hoban
Hobbs, Thomas W. (fl. 1985) M	Hobbs
Hobdy, R.W. (fl. 1984) S	Hobdy
Hobkirk, Charles Codrington Pressick (1837–1902) B	Hobk.
Hobson, Edward (1782–1830) B	Hobson
Hobson, Julian A	J.Hobson
Hochapfel, Heinz (H.) (fl. 1931) M	Hochapfel
Hochberzanka, E. (fl. 1930) M	Hochb.
Hochgesand, E. (1959–) M	Hochg.
Hochreutiner, Bénédict Pierre Georges (1873–1959) S	Hochr.
Hochstetter, Christian Ferdinand Friedrich (1787–1860) AMPS	Hochst.
Hochstetter, Christian Gottlob Ferdinand von (1829–1884) S	C.G.F.Hochst.
Hochstetter, Karl Christian Friedrich (1818–1880) S	K.Hochst.
Hochstetter, Wilhelm Christian (1825–1881) S	W.Hochst.
Höck, Fernando (1858–1915) S	Höck
Hockey, John Frederick (1895–) M	Hockey
Hocking, Ailsa D. (fl. 1977) M	A.D.Hocking
Hocking, George Macdonald (1908–) S	Hocking
Hocquart, Léopold François Joseph (1760–1818) S	Hocq.
Hodel, Donald R. (fl. 1985) S	Hodel
Hodgdon, Albion Reed (1909–1976) S	Hodgdon
Hodge, Walter Hendricks (1912–) PS	Hodge
Hodges, Charles S. (fl. 1965) M	Hodges
Hodgetts, William J. A	Hodgetts
Hodgson, E.A. (fl. 1962) B	E.A.Hodgs.
Hodgson, H.J. (fl. 1956) S	H.J.Hodgs.
Hodgson, William (1824–1901) S	Hodgs.
Hodişan, Ioan I. (1928–) MS	Hodişan
Hodson, Elmer Reed (1875–) M	Hodson
Hoe, William Joseph (1941–) B	Hoe
(Hoeck, Fernando)	
see Höck, Fernando	Höck
Hoefer, Jean Chrétien Ferdinand (1811–1878) S	Hoef.
(Hoefft, Franz M.S.V.)	
see Höfft, Franz M.S.V.	Höfft
Hoefker, Heinrich (1859–1945) S	Hoefker
(Höeg, Ove Arbo)	
see Høeg, Ove Arbo	Høeg
Høeg, Ove Arbo (1898–) AFM	Høeg

Hoehne, Frederico Carlos (1882–1959) S   **Hoehne**
Hoehne, Wilson (1908–　) S   **W.Hoehne**
(Hoehnel, Franz Xaver Rudolf von)
  see Höhnel, Franz Xaver Rudolf von   **Höhn.**
Hoejer, Johan Christian (fl. 1753) S   **Hoejer**
Hoek, Christiaan van den (1933–　) A   **C.Hoek**
Hoek, Julie L. (fl. 1901) S   **J.L.Hoek**
(Hoek, L.Van)
  see Van Hoek, L.   **Van Hoek**
Hoekstra, E.S. (fl. 1982) M   **Hoekstra**
Hoerich, O. (fl. 1921) F   **Hoerich**
Hoerl, Ruth A. (fl. 1939) M   **Hoerl**
Hoerner, Godfrey Richard (1893–　) M   **Hoerner**
Hoerold, Rudolf　S   **Hoerold**
Hoes, J.A. (fl. 1965) M   **Hoes**
(Hoess, Franz)
  see Höss, Franz   **Höss**
Hoëtte, Shirley (fl. 1935) M   **Hoëtte**
Hoeven, E.P.van der (fl. 1977) M   **E.P.Hoeven**
Hoeven, Jan van der (1801–1868) S   **Hoeven**
Hoever, J. (1801–　) S   **Hoever**
Hofberg, Kohan Herrman (1823–1880) S   **Hofberg**
Hofeneder, Heinrich　A   **Hofen.**
Hofer, Johannes (Jean) (1697–1781) S   **Hofer**
Hoffberg, Carl Fredrick (1729–1790) S   **Hoffb.**
Hoffman, Emily J.　A   **E.J.Hoffman**
Hoffman, Larry Ronald (1936–　) A   **L.R.Hoffman**
Hoffman, M.J. (fl. 1984) M   **M.J.Hoffman**
Hoffman, Ralph (1870–1932) S   **Hoffman**
Hoffmann, Carl (Karl) (1802–1883) S   **C.Hoffm.**
Hoffmann, Ferdinand (1860–1914) S   **F.Hoffm.**
Hoffmann, Fred Walter (1897–　) S   **F.W.Hoffm.**
Hoffmann, George Franz (1761–1826) BMPS   **Hoffm.**
Hoffmann, Gerhard (fl. 1935) B   **G.Hoffm.**
Hoffmann, Heinrich Karl (Carl) Hermann (1819–1891) MS   **H.Hoffm.**
Hoffmann, Johan Frederik (1823–1841) S   **J.F.Hoffm.**
Hoffmann, Johann Joseph (1805–1878) S   **J.J.Hoffm.**
Hoffmann, Karl August Otto (1853–1909) S   **O.Hoffm.**
Hoffmann, Käthe (Kaethe) (1883–c.1931) S   **K.Hoffm.**
Hoffmann, Lucien (1961–　) A   **L.Hoffm.**
Hoffmann, Norbert　A   **N.Hoffm.**
Hoffmann, Philipp (fl. 1868) S   **P.Hoffm.**
Hoffmann, Ralph (1870–1932) P   **Ralph Hoffm.**
Hoffmann, Reinhold (1885–　) S   **R.Hoffm.**
Hoffmann, Wilhelm (fl. 1856) S   **W.Hoffm.**
Hoffmann, William Edwin (1896–　) A   **W.E.Hoffm.**
Hoffmann–Grobéty, Amelie Ennenda (1884–　) S   **Hoffm.-Grob.**
Hoffmannsegg, Johann Centurius von (1766–1849) PS   **Hoffmanns.**
Hoffmeister, Werner (1819–1845) S   **Hoffmeister**

Hoffstad, Olaf Alfred (1865–1943) S	**Hoffstad**
Höfft, Franz M.S.V. (fl. 1826) S	**Höfft**
Hofker, H. (1859–) S	**Hofker**
Höfle, Marc Aurel (–1855) S .	**Höfle**
Höfler, Karl (1893–) AP	**Höfler**
(Hofman Bang, Niels)	
see Bang, Niels	**Bang**
(Hofman, Niels)	
see Bang, Niels	**Bang**
(Hofman–Bang, Niels)	
see Bang, Niels	**Bang**
Hofmann, Adolf (1853–1913) F	**A.Hofm.**
Hofmann, Eduard (1802–1875) S	**Hofm.**
Hofmann, Elise (fl. 1929–1952) F	**E.Hofm.**
(Hofmann, Heinrich Karl (Carl) Hermann)	
see Hoffmann, Heinrich Karl (Carl) Hermann	**H.Hoffm.**
Hofmann, Herman J. (–1918) A	**H.J.Hofm.**
Hofmann, Hermann (–1923) P	**H.Hofm.**
Hofmann, Joseph (1822–1901) S	**J.Hofm.**
Hofmann, Joseph Vincenz (1800–1863) S	**J.V.Hofm.**
Hofmeister, Wilhelm Friedrich Benedict (1824–1877) S	**Hofmeist.**
Hofsten, Angelica v. (fl. 1968) M	**Hofsten**
Hofstetter, Adrian Marie A	**Hofst.**
Hofstra, J. S	**Hofstra**
Högberg, Johan Daniel (1823–1843) S	**Högb.**
Hogg, John (1800–1869) S	**J.Hogg**
Hogg, Robert (1818–1897) S	**R.Hogg**
Hogg, Thomas (1820–1892) S	**T.Hogg**
Hogg, Thomas (1777–1855) S	**Hogg**
Hogholen, E. (fl. 1977) M	**Hogholen**
Hoham, Ronald William (1942–) A	**Hoham**
Hohenacker, Rudolph Friedrich (1798–1874) PS	**Hohen.**
(Hohenbühel Heufler, Ludwig Samuel Joseph David Alexander von)	
see Heufler zu Rasen und Perdonneg, Ludwig Samuel Joseph David	
Alexander	**Heufl.**
Hohenwarth, Sigismund (1745–1825) S	**Hohenw.**
Hohl, Hans Rudolph (fl. 1984) M	**H.R.Hohl**
Hohl, Marianne (fl. 1979) M	**M.Hohl**
Hohmeyer, Helmuth H. (fl. 1984) M	**Hohmeyer**
Hohn, Matthew Henry (1920–) A	**M.H.Hohn**
Höhnel, Franz Xaver Rudolf von (1852–1920) ABMS	**Höhn.**
Höhnel, Ludwig von (1857–1942) S	**L.Höhn.**
Höhnk, Johann Willy Georg von (fl. 1932) M	**Höhnk**
Høiland, Klaus (1948–) M	**Høil.**
(Höjer, Johan Christian)	
see Hoejer, Johan Christian	**Hoejer**
Hök, C.T. (fl. 1836) M	**Hök**
Holanda, M.W. (fl. 1963) M	**Holanda**
Holandre, Jean Joseph Jacques (1778–1857) S	**Holandre**

Holcomb, Gordon E. (fl. 1976) M — Holcomb
Holden, Henry Smith (fl. 1955) F — H.S.Holden
Holden, Isaac (1832–1903) A — Holden
Hole, Robert Selby (1875–1938) S — Hole
Holferty, George Mellinger (1859–) B — Holf.
Holien, H. (fl. 1986) M — Holien
Holkema, Franciscus (1840–1870) S — Holkema
Holl, Friedrich (fl. 1820–1850) PS — Holl
Holl, Friedrich C. (1815–) M — F.C.Holl
Holland, Alma A. (1925–) M — A.A.Holland
Holland, Dorothy Fitzgerald (1897–) M — D.F.Holland
Holland, John Henry (1869–1950) S — Holland
Holland, Peter Charles (1927–) M — P.C.Holland
Hollande, A.-Ch. (fl. 1925) M — Hollande
Hollande, André A — A.Hollande
Hollander, G.den (fl. 1983) S — Hollander
Hollands, R. (fl. 1990) M — Hollands
Holle, Georg Karl Hans Dietrich (1825–1893) M — Holle
Hollenberg, George Jacob (1897–1988) A — Hollenb.
Holler, August (1835–1904) B — Holler
Hollerbach, Maximilian Maximilianovich (1907–) AC — Hollerb.
(Hollerbakh, Maximilian Maximilianovich)
  see Hollerbach, Maximilian Maximilianovich — Hollerb.
Holley, R.A. (fl. 1984) M — Holley
Hollick, Charles Arthur (1857–1933) ABFMS — Hollick
Holliday, Paul Cyrus (1923–) M — Holliday
Hollós, Ladislaus (Lászlo) (1859–1940) M — Hollós
Holloway, John Ernest (1881–1945) PS — Holloway
Hollrung, Max Udo (1858–1937) AMP — Hollrung
Holm, Herman Theodor (1854–1932) S — Holm
Holm, Jørgen Tyche (Georg Tycho) (1726–1759) S — J.T.Holm
Holm, Kerstin (1924–) M — K.Holm
Holm, Lennart (1921–) M — L.Holm
Holm, Richard William (1925–) S — R.W.Holm
Holm, Sv. Nørgaard (fl. 1989) M — S.N.Holm
Holm, Theodor (1880–1943) S — T.Holm
(Holm, Theodor)
  see Holmskjold, Theodor — Holmsk.
Holm-Nielsen, Lauritz B. (1946–) BS — Holm-Niels.
Holman, R.T. (fl. 1984) S — Holman
Holmberg, Eduardo Ladislao (1852–1937) S — E.Holmb.
Holmberg, Otto Rudolf (1874–1930) BP — Holmb.
Holmboe, Jens (1880–1943) AS — Holmboe
Holmen, Kield Áxel (1921–1974) B — Holmen
(Holmen, Kjeld Áxel)
  see Holmen, Kield Áxel — Holmen
Holmes, Edward Morell (1843–1930) AS — Holmes
Holmes, Francis Oliver (1899–) M — F.O.Holmes

Holmes, Robert William (1925–) A	R.W.Holmes
Holmes, W.B.K. (fl. 1977) F	W.B.K.Holmes
Holmes, Walter C. (1944–) S	W.C.Holmes
Holmgren, Arthur Hermann (1912–) S	A.H.Holmgren
Holmgren, Bjorn Frithiofsson (1872–1946) S	B.F.Holmgren
Holmgren, Hjalmar Josef (1822–1885) B	Holmgren
Holmgren, Noel Herman (1937–) S	N.H.Holmgren
Holmgren, Patricia K. (née Kern, P.K.) (1940–) S	P.K.Holmgren
Holmlund, P.-E. (fl. 1980) S	Holmlund
(Holmskiod, Theodor)	
see Holmskjold, Theodor	Holmsk.
Holmskjold, Theodor (1732–1794) MS	Holmsk.
Holøs, Sverre (fl. 1990) M	Holøs
Holsinger, Edward C.T. A	Hols.
Holsinger, Kent Eugene (1956–) S	K.E.Hols.
Holst, Bruce (fl. 1986) S	B.Holst
Holst, E.C. (fl. 1936) M	Holst
Holt, Erhard Moller (fl. 1985) S	E.M.Holt
Holt, George Alfred (1852–1921) B	Holt
Holtan-Hartwig, Jon (fl. 1988) M	Holt.-Hartw.
(Holterman, Carl)	
see Holtermann, Carl	Holterm.
Holtermann, Carl (1866–1923) M	Holterm.
Holthuis, L.B. (fl. 1942) S	Holthuis
Holtmann, Theodor A	Holtm.
Holton, Charles Stewart (fl. 1957) M	Holton
Holttum, Richard Eric (1895–1990) PS	Holttum
Holtz, Johann Friedrich Ludwig (1824–1907) AC	Holtz
Holub, Josef Ludwig (1930–) PS	Holub
Holubová-Jechová, Věra (fl. 1963–1983) M	Hol.-Jech.
(Holubová-Klasková, Anna)	
see Klásková, Anna	Klásk.
Holuby, Josef (Jószef) Ludovit (1836–1923) PS	Holuby
Holway, Edward Willet Dorland (1853–1923) M	Holw.
Holý, František (1935–1984) S	Holý
Holzapfel, Christina Marie (1942–) S	Holzapfel
Hölzer (fl. 1939–1943) P	Hölzer
Holzfuss, Ernst (1868–) S	Holzfuss
Holzhammer, Elisabeth (fl. 1955) S	E.Holzh.
(Holzhammer, Martha)	
see Friedrich-Holzhammer, Martha	Friedr.-Holzh.
Holzinger, John Michael (1853–1929) BS	Holz.
Holzinger, Josef Bonaventura (1835–1912) M	J.B.Holz.
Holzmann, Timoleon (1843–) S	Holzm.
Holzschu, D.L. (fl. 1981) M	Holzschu
Homann, Georg Gothilf Jacob (fl. 1828–1835) PS	· Homann
Hombron, Jacques Bernard (1800–1852) PS	Hombr.
Homès, P. S	Homès
Homfeld, H. A	Homfeld

Homma, Yasu (1892–) M　Homma
Hommersand, Max Hoyt (1930–) A　Hommers.
Homola, Richard L. (1934–) M　Homola
Homolle, Anne–Marie (fl. 1937–1950) S　Homolle
Homrich, Marí H. (fl. 1973) M　Homrich
Honckeny, Gerhard August (1724–1805) S　Honck.
Honda, Masaji (Masazi) (1897–1984) PS　Honda
Honda, Sachiko　A　S.Honda
Honey, Edwin Earle (1891–1956) M　Honey
Hong, De Yuan(g) (1936–) S　D.Y.Hong
Hong, Jian Yuan (fl. 1987) S　J.Y.Hong
Hong, Soon Woo (fl. 1986) M　S.W.Hong
Hong, Suk Pyo (fl. 1989) S　S.P.Hong
Hong, Tao (fl. 1963) S　T.Hong
Hong, Won Shic (1919–) B　W.S.Hong
Hongo, Tsuguo (1923–) M　Hongo
Honigberg, Bronislaw　A　Honigb.
Honigmann, Hans Leo　A　Honigm.
Honjo, Susumu　A　Honjo
(Honrubia García, Mario)
　see Honrubia, Mario　Honrubia
Honrubia, Mario (1956–) M　Honrubia
Hoo, Gin (fl. 1951) S　G.Hoo
Hood, I.A. (fl. 1988) M　Hood
Hoof, H.A.van (fl. 1950) M　Hoof
(Hoog, G.S.de)
　see de Hoog, G.S.　de Hoog
Hoog, Johannes Marius Cornelis (John) (1865–1950) S　Hoog
Hoog, Thomas M. (1899–) S　T.M.Hoog
Hoogkamer–Te Niet, M.C. (fl. 1984) M　Hoogk.Niet
Hoogland, Ruurd Dirk (1922–) S　Hoogland
Hooibrenk, Danield (fl. 1848–1861) S　Hooibr.
Hook, James Mon van (1870–1935) M　J.M.Hook
Hooker, Joseph Dalton (1817–1911) ABMPS　Hook.f.
Hooker, William (1779–1832) S　W.Hook.
Hooker, William Jackson (1785–1865) ABMPS　Hook.
Hoola van Nooten, Bertha (fl. 1914) S　Hoola van Nooten
Hooper, John (1802–1869) A　Hooper
Hooper, Robert G. (1946–) A　R.G.Hooper
Hooper, Sheila Spenser (1925–) S　S.S.Hooper
Hoopes, Josiah (1832–1904) S　Hoopes
(Hoorebeke, Charles Joseph Van)
　see Van Hoorebeke, Charles Joseph　Van Hooreb.
Hoover, Robert Francis (1913–1970) S　Hoover
Hoozemans, A.C.M. (fl. 1970) M　Hooz.
Hope, Brigt　A　B.Hope
Hope, Charles William Webley (1832–1904) PS　C.Hope
Hope, John (1725–1786) S　Hope
Hopffer, Carl (1810–) S　Hopffer

Höpfner, C. (–1904) S	Höpfner
Hopkin, A.A. (fl. 1988) M	Hopkin
Hopkins, E. (fl. 1869) S	E.Hopkins
Hopkins, Edwin Fraser (1891–) M	E.F.Hopkins
Hopkins, Helen Collingwood (1953–) S	H.C.Hopkins
Hopkins, John Collier Frederick (fl. 1931) M	J.C.F.Hopkins
Hopkins, Lewis Sylvester (1872–1945) PS	Hopkins
Hopkins, Milton (1906–) S	M.Hopkins
Hopkinson, S.J. (fl. 1984) M	Hopk.
Hopkirk, Thomas (1785–1841) S	Hopkirk
Hoppauch, Katherine Wells A	Hoppauch
Hoppe, David Heinrich (1760–1846) BMPS	Hoppe
Hoppe, Tobias Konrad Johann (fl. 1750) S	T.Hoppe
Hopper, Stephen Donald (1951–) S	Hopper
Höppner, Hans (1873–1946) S	Höppner
Hopsu–Havu, V.K. (fl. 1978) M	Hopsu–Havu
Hora, Frederich Bayard (1908–1984) M	Hora
Horák, Bonslaw (1877–1942) S	Horák
Horak, Egon (1937–) M	E.Horak
Horak, Karl E. (fl. 1978) S	K.E.Horak
(Horaninov, Paul (Paulus) Fedorowitsch)	
see Horaninow, Paul (Paulus) Fedorowitsch	Horan.
Horaninow, Paul (Paulus) Fedorowitsch (1796–1865) MPS	Horan.
Horánszky, Andreas (1928–) S	Horánszky
Horecká, Mária A	Horecká
Horenstein, Evelyn A. (fl. 1961) M	Horenst.
Hori, Shôtarô (1865–1945) M	Hori
Hori, Terumitsu (1938–) A	T.Hori
Horich, Clarence Kl. (fl. 1982) S	Horich
Horie, H. (fl. 1979) M	H.Horie
Horie, Yoshikazu (fl. 1978) M	Y.Horie
Horiguchi, Mankichi A	M.Horig.
Horiguchi, Takeo A	T.Horig.
Horikawa, Tomiya (1920–1956) S	T.Horik.
Horikawa, Yoshiwo (1902–1976) BM	Horik.
Horino, S. (fl. 1981) S	Horino
Horkel, Johann (1769–1846) S	Horkel
(Horlacheva, Z.S.)	
see Gorlacheva, Z.S.	Gorl.
Hormann, Hans (1902–1981) B	Hormann
Hormuzaki, Constantin (von) (1863–1937) S	Hormuz.
Horn af Rantzien, Henning (1922–1960) AS	Horn
Horn, B.W. (fl. 1987) M	B.W.Horn
Horn, C.N. (fl. 1986) S	C.N.Horn
(Horn, Gene Stanley Van)	
see Van Horn, Gene Stanley	Van Horn
Horn, Norman L. (fl. 1954) M	N.L.Horn
Hornby, A.J.Ward (1893–) A	Hornby
Hornby, David (fl. 1977) M	D.Hornby

Horne, Arthur Samuel (fl. 1920) M	**A.S.Horne**
Horne, David Bertram (1940–) S	**D.B.Horne**
Horne, John (1835–1905) S	**Horne**
Horne, William Titus (1872–1944) M	**W.T.Horne**
Hornemann, Jens Wilken (1770–1841) ABMPS	**Hornem.**
Horner, Harry Theodore (fl. 1968) M	**Horner**
Hörnhammer, L. (fl. 1933) F	**Hörnh.**
Hornibrook, Murray (1874–1949) S	**Hornibr.**
Horniček, Emil (fl. 1958) M	**Horniček**
Hornschuch, Christian Friedrich (1793–1850) BP	**Hornsch.**
Hornstedt, Claës (Claudius) Fredric (1758–1809) S	**Hornst.**
Hornung, Ernst Gottfried (1795–1862) S	**Hornung**
Horný, Radvan (fl. 1974) S	**Horný**
Horo, F.B. (fl. 1978) M	**Horo**
Horobin, John F. (fl. 1990) S	**Horobin**
(Hörold, Rudolf)	
see Hoerold, Rudolf	**Hoerold**
Horowitz, Aharon A	**Horowitz**
Horr, Asa (Hor) (1817–1896) S	**Horr**
Horrell, Ernst Charles (1870–1937) B	**Horrell**
Horsey, Richard Edgar (1883–1972) S	**Horsey**
Horsfall, James Gordon (1905–) M	**Horsfall**
Horsfield, Thomas (1773–1859) PS	**Horsf.**
Horst, Ulrich A	**Horst**
Hort, Fenton John Anthony (1828–1892) S	**Hort**
Horta, Paul Parreiras (fl. 1924) M	**Horta**
Hortobágyi, Tibor (1912–) AM	**Hortob.**
Horton, Diana Gail (1949–) B	**D.G.Horton**
Horton, James Heathman (1931–) S	**Horton**
Horton, P. (fl. 1981) S	**P.Horton**
Horvát, Adolf Oliver (1907–) S	**A.O.Horvát**
Horvat, F. (fl. 1983) S	**F.Horvat**
Horvat, Ivo (1897–1963) S	**Horvat**
Horvat, Marija Dvořák (1909–) S	**M.D.Horvat**
Horvatić, Stjepan(Stefan) (1899–1975) BS	**Horvatić**
Horvátovszky, Tsigmond (Zsigmond) (fl. 1774) S	**Horv.**
Horwood, Arthur Reginald (1879–1937) FMS	**Horw.**
Hosack, David (1769–1835) S	**Hosack**
Hosagoudar, Virupakshagouda Bhimanagouda (fl. 1984) M	**Hosag.**
Hosaka (fl. 1935) S	**Hosaka**
Hosé, Johan Albert (1769–1800) S	**Hosé**
Hosford, David R. (fl. 1975) M	**Hosford**
Hoshaw, Robert William (1921–) A	**Hoshaw**
Hoshi, Hiroshi (1959–) S	**Hoshi**
Hoshino, Kazuo (fl. 1969) M	**Hoshino**
Hoshizaki, Barbara Joe (1928–) PS	**Hoshiz.**
Hosius, August (1825–1896) AF	**Hosius**
Hosmani, S.P. A	**Hosmani**
Hosokawa, Takahide (1909–) PS	**Hosok.**

Höss, Franz (1756–1840) S   **Höss**
Hossain, A.B.M.Enayet (1945–) S   **E.Hossain**
Hossain, Mosharraf (1928–) S   **M.Hossain**
Hosseus, Carl Curt (1878–1950) MPS   **Hosseus**
Host, Nicolaus Thomas (1761–1834) BS   **Host**
Höstermann, Gustav (1872–) M   **Hösterm.**
Hotchkiss, Arland Tillotson (1918–) AS   **Hotchk.**
Hotson, Hugh Howison (1916–) M   **H.H.Hotson**
Hotson, John Williams (1870–) M   **Hotson**
Hotta, Mitsuru (1935–) S   **M.Hotta**
Hotta, Teikichi (1899–) S   **Hotta**
Hottes, Alfred Carl (1891–1955) S   **Hottes**
Hotyat, Lucienne (fl. 1982) S   **Hotyat**
Hötzl, Heinz   A   **Hötzl**
(Hou, Ding)
    see Ding Hou   **Ding Hou**
Hou, Tian Jue (fl. 1981) M   **T.J.Hou**
Hou, Wen Hu (fl. 1986) S   **W.H.Hou**
Hou, Yue Shin (fl. 1986) S   **Y.S.Hou**
Houba, Julien (1843–) S   **Houba**
Houby, J. (1882–1964) M   **Houby**
Houdard, Jules (fl. 1911) S   **Houdard**
Hough, Romeyn Beck (1857–1924) S   **Hough**
Houghton, Arthur Duvernoix (1870–1938) S   **Houghton**
Houlbert, Constant Vincent (1857–1947) CF   **Houlbert**
Houllet, R.J.B. (1811/15–1890) S   **Houllet**
Houlston, John (fl. 1848–52) PS   **Houlston**
Houmeau, Jean-Michel (fl. 1982) M   **Houmeau**
House, H.H.   S   **H.H.House**
House, Homer Doliver (1878–1949) MPS   **House**
Houston, Byron Robinson (1914–) M   **Houston**
Houstoun, William (1695–1733) LS   **Houst.**
(Houtte, Louis Benoît Van)
    see Van Houtte, Louis Benoît   **Van Houtte**
(Houtte, Louis Van)
    see Van Houtte, Louis   **L.Van Houtte**
Houttuyn, Maarten (Martin) (1720–1798) BP   **Houtt.**
Houtzagers, Gÿsbertus (1888–1957) S   **Houtz.**
Houzeau de Lehaie, J. (1867–1959) S   **J.Houz.**
Houzeau de Lehaie, Jean Charles (1820–1888) S   **Houz.**
Hovasse, Raymond (1895–) AM   **Hovasse**
Hovelacque, Maurice Jean Alexandre (1858–1898) FMS   **Hovel.**
Hoven, Frederick Johan Jacob Slingsbij van (1815–1879) S   **Hoven**
Hovenkamp, Peter Hans (1953–) PS   **Hovenkamp**
Hovens, J. (fl. 1984) S   **Hovens**
Hovey, Charles Mason (1810–1887) S   **Hovey**
How, Foon Chew (fl. 1956) S   **F.C.How**
How, K.C.   S   **K.C.How**
Howard, Grace Elizabeth (1886–1978) M   **G.E.Howard**

Howard, Harold W. (1913–) S	**H.W.Howard**
Howard, Henry J. (–1957) M	**H.J.Howard**
Howard, John Eliot (1807–1883) S	**Howard**
Howard, R.Vince A	**R.V.Howard**
Howard, Richard Alden (1917–) S	**R.A.Howard**
Howard, Thaddeus Monroe (1929–) S	**T.M.Howard**
Howarth, Willis Openshaw (1890–1964) S	**Howarth**
Howcroft, N.H.S. (fl. 1981) S	**Howcroft**
Howe, Clifton Durant (1874–) S	**C.Howe**
Howe, Elliot Calvin (1828–1899) MPS	**Howe**
Howe, Marshall Avery (1867–1936) ABM	**M.Howe**
Howe, Reginald Heber (1875–1932) CM	**R.Howe**
Howe, W.E. (–1891) P	**W.Howe**
Höweler, Louise H. (fl. 1967) M	**Höweler**
Howeler, R.H. (fl. 1984) M	**R.H.Howeler**
Howell, Benjamin Franklin (1890–) A	**B.F.Howell**
Howell, John F. A	**J.F.Howell**
Howell, John Thomas (1903–) S	**J.T.Howell**
Howell, S.R. (fl. 1934) S	**S.R.Howell**
Howell, Thomas Jefferson (1842–1912) MS	**Howell**
Howie, Charles (1811–1899) S	**Howie**
Howitt, Alfred William (1830–1908) S	**A.W.Howitt**
Howitt, Beatrice F. (1891–1981) S	**B.F.Howitt**
Howitt, Godfrey (1800–1873) M	**Howitt**
Howse, Richard (1821–1901) S	**Howse**
Hoyer, Carl (Karl) August Heinrich (fl. 1838) S	**Hoyer**
Hoyle, Arthur Clague (1904–1986) S	**Hoyle**
Hoyle, Mitchell D. (1942–) A	**M.D.Hoyle**
Hoyt, William Dana (1880–) A	**Hoyt**
Hranova, Ana (fl. 1925) M	**Hranova**
Hrabětova–Uhrová, Anezka (1900–1981) S	**Hrabětova**
Hromadnik, H. (fl. 1983) S	**H.Hrom.**
Hromadnik, L. (fl. 1987) S	**L.Hrom.**
Hruby, Johan (1882–1964) MS	**Hruby**
Hrubý, Karel (1910–1962) S	**K.Hrubý**
Hryniewiecki, Boleslaw B. (1875–1963) S	**Hryn.**
Hsai, Wei Yi S	**W.Y.Hsai**
Hseu, R.S. (fl. 1989) M	**R.S.Hseu**
Hsia, Kuang Cheng (fl. 1982) S	**K.C.Hsia**
Hsiao, Ju Ying S	**J.Y.Hsaio**
Hsiao, Pei Ken (fl. 1964) S	**P.K.Hsiao**
Hsieh, A.Tsai (fl. 1955) S	**A.T.Hsieh**
Hsieh, Chang Fu (1947–) S	**C.F.Hsieh**
Hsieh, Chen Ko (fl. 1987) S	**C.K.Hsieh**
Hsieh, Huei Mei (fl. 1987) M	**H.M.Hsieh**
Hsieh, L.S. (fl. 1990) M	**L.S.Hsieh**
Hsieh, S.Y. (fl. 1990) M	**S.Y.Hsieh**
Hsieh, Shen (1904–) M	**S.Hsieh**
Hsieh, Wen Hsui (fl. 1986) M	**W.H.Hsieh**

Hsieh, Y. (fl. 1987) P	**Y.Hsieh**
Hsieh, Yin Tang (fl. 1974–1988) PS	**Y.T.Hsieh**
Hsing, C.H. (fl. 1986) S	**C.H.Hsing**
Hsing, K.H. (fl. 1979–1986) P	**K.H.Hsing**
Hsiung, Ta Shih A	**T.S.Hsiung**
Hsiung, Wen Yue (fl. 1980) S	**W.Y.Hsiung**
Hsu, Chien Chang (1932–) S	**C.C.Hsu**
Hsu, Kuo Shih S	**K.S.Hsu**
Hsu, Lian Wang (fl. 1974) M	**L.W.Hsu**
Hsu, Ping Sheng (1924–) S	**P.S.Hsu**
Hsu, Shi Chang (fl. 1966) M	**S.C.Hsu**
Hsu, Ting Zhi (1941–) S	**T.Z.Hsu**
Hsu, Wei Ying S	**W.Y.Hsu**
Hsu, Wen Hau (fl. 1989) M	**W.H.Hsu**
Hsu, Y.T. (fl. 1955) S	**Y.T.Hsu**
Hsu, Yang Pong S	**Yang P.Hsu**
Hsu, Yin S	**Y.Hsu**
Hsu, Ying Peng ('Ben') (1933–) PS	**Y.P.Hsu**
Hsu, Yung Chun (fl. 1983) S	**Y.C.Hsu**
Hsuan, Shive Jye (fl. 1965) S	**S.J.Hsuan**
Hsue, Hsiang Hao (fl. 1963) S	**H.H.Hsue**
Hsue, Qing Song (fl. 1978) S	**Q.S.Hsue**
Hsue, Ruey Shyang (fl. 198?) M	**R.S.Hsue**
(Hsueh, Chi Ju)	
see Xue, Ji Ru	**J.R.Xue**
(Hsueh, Ji Ru)	
see Xue, Ji Ru	**J.R.Xue**
(Hsueh, Jia Rong)	
see Xue, Jia Rong	**Jia R.Xue**
Hu, Cheng Hua (1933–) S	**C.H.Hu**
Hu, Chi Ming (1935–) S	**C.M.Hu**
Hu, Chia Chi S	**C.C.Hu**
Hu, Fu Mei (fl. 1988) M	**F.M.Hu**
Hu, Hai (fl. 1989) M	**H.Hu**
Hu, Hong Jun A	**H.J.Hu**
Hu, Hsen Hsu (1894–1968) APS	**Hu**
Hu, Hung Chuen A	**H.C.Hu**
Hu, Hung Tao (fl. 1988) M	**H.T.Hu**
Hu, Lin Cheng (fl. 1984) S	**L.C.Hu**
Hu, Mei Rong (fl. 1991) M	**M.R.Hu**
Hu, R. (fl. 1982) M	**R.Hu**
Hu, Ren Liang (1932–) B	**R.L.Hu**
Hu, Shiu Ying (1910–) S	**S.Y.Hu**
Hu, Wen Kwang (Kuang) (1922–) S	**W.K.Hu**
(Hu, Xian Su)	
see Hu, Hsen Hsu	**Hu**
Hu, Xiao Hong (fl. 1984) S	**X.H.Hu**
Hu, Xin Wen (fl. 1990) M	**X.W.Hu**
Hu, Yan Xing (fl. 1989) M	**Y.X.Hu**

Hu, Yu Cheng (fl. 1991) S   **Y.C.Hu**
Hu, Zhi Bi (fl. 1982) S   **Z.B.Hu**
Hu, Zhi Hao (1937–) S   **Z.H.Hu**
Hu, Zhing Xian (fl. 1988) M   **Z.X.Hu**
Hua, Henri (1861–1919) S   **Hua**
Hua, Jack J. (fl. 1989) M   **J.J.Hua**
Hua, Mao Sen   A   **M.S.Hua**
Huang, Cai Fen (fl. 1984) S   **C.F.Huang**
Huang, Cheng Chiu (1922–) S   **C.C.Huang**
(Huang, Cheng Jiu)
   see Huang, Cheng Chiu   **C.C.Huang**
Huang, Gui Qin (fl. 1987) S   **G.Q.Huang**
Huang, J.F. (fl. 1985) S   **J.F.Huang**
Huang, Jin Xiang (fl. 1989) S   **J.X.Huang**
Huang, Ke Fu (fl. 1984) S   **K.F.Huang**
Huang, Keng Tang (fl. 1956) M   **K.T.Huang**
Huang, L.H. (fl. 1973) M   **L.H.Huang**
Huang, M.S. (fl. 1976) S   **M.S.Huang**
Huang, Peng Cheng (1932–) S   **P.C.Huang**
Huang, Pu Hwa (fl. 1978) S   **P.H.Huang**
Huang, Quan (fl. 1985) S   **Q.Huang**
Huang, R.   A   **R.Huang**
Huang, Ren Huang (fl. 1982) S   **R.H.Huang**
Huang, Rong Fu (fl. 1981) S   **R.F.Huang**
Huang, S.M. (fl. 1977) S   **S.M.Huang**
Huang, Se Zei (fl. 1980s) S   **S.Z.Huang**
(Huang, Shan Bin)
   see Huang, Tian Zhang   **T.Z.Huang**
Huang, Shing Fan (fl. 1987) S   **S.F.Huang**
Huang, Shu Chung (1921–) S   **S.C.Huang**
Huang, Shu Hua (fl. 1980s) S   **S.H.Huang**
Huang, Tian Zhang (1934–) M   **T.Z.Huang**
Huang, Tseng Chieng (1931–) MPS   **T.C.Huang**
Huang, Wei Lian (fl. 1984) S   **W.L.Huang**
Huang, Wen Da   S   **W.D.Huang**
Huang, Xie Cai (fl. 1984) S   **X.C.Huang**
Huang, Xiu Lan (1932–) S   **X.L.Huang**
Huang, Yun Fei (1934–)S   **Y.F.Huang**
Huang, Yong Chin   S   **Y.C.Huang**
Huang, Yun Huei   S   **Y. Huei Huang**
Huang, Yun Hui (fl. 1983) S   **Y. Hui Huang**
Huard, Jean (fl. 1966) B   **Huard**
Hubálek, Z. (fl. 1973) M   **Hubálek**
Hubart, J.–M. (fl. 1987) M   **Hubart**
Hubbard, Charles Edward (1900–1980) S   **C.E.Hubb.**
Hubbard, Frederic Tracy (1875–1962) S   **F.T.Hubb.**
Hubbard, J.W.   S   **J.W.Hubb.**
(Hübener, Johann Wilhelm Peter)
   see Huebener, Johann Wilhelm Peter   **Huebener**

Hubeny, Joseph (fl. 1830–1843) S — **Hubeny**
Huber, C. (fl. 1874) S — **C.Huber**
Huber, Glenn Anthony (1899–   ) M — **G.A.Huber**
(Huber, Gottfried Eduard)
  see Huber–Pestalozzi, Gottfried Eduard — **Hub.-Pest.**
Huber, Hans (1919–   ) BS — **Hans Huber**
(Huber, Hans) (Johan Christoph)
  see Huber, Johann Christoph(er) — **J.C.Huber**
Huber, Harvey Evert (1884–   ) S — **H.E.Huber**
Huber, Heribert Franz Josef (1931–   ) S — **H.Huber**
Huber, Jakob ('Jacques') E. (1867–1914) AS — **Huber**
Huber, Johann Christoph(er) (1830–1913) MS — **J.C.Huber**
Huber, Josef Anton (1899–   ) S — **J.A.Huber**
Huber, M.J.   S — **M.J.Huber**
Huber, Walter (1958–   ) S — **W.Huber**
Huber–Morath, Arthur (1901–1990) S — **Hub.-Mor.**
Huber–Pestalozzi, Gottfried Eduard (1877–1966) A — **Hub.-Pest.**
Hubert, Ernest Everett (1887–   ) M — **Hubert**
Hübl, Paul   S — **Hübl**
Hübner, Wolfgang (fl. 1933) S — **Hübner**
Huck, Robin Bovard (1935–   ) S — **Huck**
Huddlestun, Paul   A — **Huddl.**
Hudelo, L. (fl. 1920) M — **Hudelo**
Hudson, James (1846–   ) S — **J.Huds.**
Hudson, H.J. ('Harry') (fl. 1960) M — **H.J.Huds.**
Hudson, William (1730–1793) ABMPS — **Huds.**
Hudziok, Georg W. (1929–   ) S — **Hudziok**
Hue, Auguste–Marie (1840–1917) M — **Hue**
Huebener, Johann Wilhelm Peter (1807–1847) BS — **Huebener**
Hueber, Francis Maurice (1929–   ) BF — **Hueber**
Huebner, G.   S — **Huebner**
Hueck, Kurt (1897–1965) S — **Hueck**
(Huegel, Carl (Karl) Alexander Anselm von)
  see Hügel, Carl (Karl) Alexander Anselm von — **Hügel**
(Huelphers, K.A.)
  see Hülphers, K.A. — **Hülph.**
Huerta, G.Guzman (fl. 1958) M — **G.G.Huerta**
Huerta, Laura (1913–   ) A — **L.Huerta**
(Huerta M., Laura)
  see Huerta, Laura — **L.Huerta**
Huet, Augustin Louis Pierre (1814–1888) S — **A.L.P.Huet**
Huet du Pavillon, Alfred (1829–1907) S — **A.Huet**
Huet du Pavillon, Édouard (1819–1908) S — **É.Huet**
Huet, L. (fl. 1928) M — **L.Huet**
Huett, John William (1832/3) S — **Huett**
Hufford, Terry Lee (1935–   ) A — **Hufford**
Hufschmitt, G. (fl. 1930s) M — **Hufschm.**
Huft, Michael J. (1949–   ) S — **Huft**
Hügel, Carl (Karl) Alexander Anselm von (1794–1870) S — **Hügel**

Huggert, L. (fl. 1973) M — **Huggert**
Hughes, Colin Edward (1957–) S — **C.E.Hughes**
Hughes, Dorothy Kate (later Popenoe, D.K.) (1899–1932) S — **Hughes**
Hughes, Elwyn Owen (1916–) A — **E.O.Hughes**
Hughes, Gilbert C. (fl. 1969) M — **G.C.Hughes**
Hughes, Norman F. (1918–) A — **N.F.Hughes**
Hughes, Ralph H. (fl. 1985–1987) P — **R.H.Hughes**
Hughes, Stanley John (1918–) M — **S.Hughes**
Hughes, William Elfyn (1948–) S — **W.E.Hughes**
Hugin, G. (fl. 1987) S — **Hugin**
Hugo, Loretta (1942–) S — **Hugo**
Hugueney, R. (fl. 1988) M — **Hugueney**
Huguenin, Auguste (1780–1860) S — **Huguenin**
Huguenin, B. (fl. 1967) M — **B.Huguenin**
(Huguet del Villar, Emile (Emilio))
   see Villar, Emile (Emilio) Huguet del — **Villar**
Huguet, P. S — **Huguet**
Huhndorf, S.M. (fl. 1990) M — **Huhndorf**
Huhtinen, Seppo (1956–) M — **Huhtinen**
Huijsman, H.S.C. (1900–) M — **Huijsman**
Huisman, John M. (fl. 1980s) A — **Huisman**
Huitfeldt-Kaas, H. A — **Huitf.-Kaas**
Hulbary, Robert Louis (1916–1981) M — **Hulbary**
Hulburt, Edward M. A — **Hulburt**
Huldén, Larry (fl. 1983) M — **Huldén**
Hulea, Ana (1915–) M — **Hulea**
Hulják, János (1883–1942) S — **Hulják**
Hull, John (1761–1843) B — **Hull**
Hülphers, K.A. (1882–1948) S — **Hülph.**
Hülsen, Rudolf (1837–1912) S — **Hülsen**
Hult, Ragnar (1857–1899) B — **Hult**
Hultén, Oskar Eric Gunnar (1894–1981) BPS — **Hultén**
Hulting, Johan (1842–1929) M — **Hulting**
Humber, Richard A. (1947–) M — **Humber**
Humbert, Jean-Henri (1887–1967) PS — **Humbert**
Humblot, Léon (fl. 1848) S — **Humblot**
Humboldt, Friedrich Wilhelm Heinrich Alexander von (1769–1859) BMPS — **Humb.**
Hume, Allan Octavian (1829–1912) S — **Hume**
Hume, Edward P. S — **E.P.Hume**
Hume, Hardrada Harold (1875–1965) MS — **H.H.Hume**
Humm, Harold Judson (1912–) AM — **Humm**
Hummel, Edward C. (1903–) S — **Hummel**
Hummelinck, Pieter Wagenaar (1907–) S — **Hummelinck**
Humnicki, Valentin (1815–) S — **Humn.**
Humpert, Friedrich (fl. 1887) S — **Hump.**
Humphrey, Clarence John (1882–1970) M — **C.J.Humphrey**
Humphrey, James Ellis (1861–1897) M — **Humphrey**
Humphreys, Edwin William (1883–) A — **Humphreys**
Humphries, Christopher John (1947–) S — **Humphries**

(Hunde, Asfaw)
  see Asfaw Hunde   **Asfaw**
Hundeshagen, Johann Christian (1785–1834) S   **Hundesh.**
Huneck, Siegfried (1928–) M   **Huneck**
Huneycutt, M.B. (fl. 1952) M   **Huneycutt**
Hungerford, E.W. (fl. 1923) M   **Hungerf.**
Hunt, David Richard (1938–) S   **D.R.Hunt**
Hunt, George Edward (1841–1873) B   **Hunt**
Hunt, John (fl. 1956) M   **J.Hunt**
Hunt, Nicholas Rex (1885–1963) M   **N.R.Hunt**
Hunt, Peter Francis (1936–) S   **P.F.Hunt**
Hunt, Richard Stanley (1944–) M   **R.S.Hunt**
Hunt, Trevor Edgar (1913–) S   **T.E.Hunt**
Hunt, Willis Roberts (1893–) M   **W.R.Hunt**
Hunter, Alexander (1729–1809) S   **Hunter**
Hunter, Gordon Eugene (1930–) S   **G.E.Hunter**
Hunter, Ingrid L. (fl. 1977) M   **I.L.Hunter**
Hunter, Joan (fl. 1971) M   **J.Hunter**
Hunter, Larry J. (fl. 1977) M   **L.J.Hunter**
Hunter, Lydia Lillian Mary (1892–) M   **L.L.M.Hunter**
Hunter, William (1755–1812) PS   **W.Hunter**
Hunter, William E. (fl. 1975) M   **W.E.Hunter**
Huntington, John Warren (1853–) P   **Huntington**
Hunziker, Armando Theodoro (1919–) S   **Hunz.**
Hunziker, Jakob S   **J.Hunz.**
Hunziker, Juan Héctor (1925–) S   **J.H.Hunz.**
Huq, Ahmed Mozaharul (1947–) S   **Huq**
Huq, Molla Fazlul A   **M.F.Huq**
Hurcombe, Ruth S   **Hurcombe**
Hurka, Herbert A   **Hurka**
Hurley, M.A. A   **Hurley**
Hürlimann, Hans (1921–) BS   **Hürl.**
Hurst, Charles Chamberlain (1870–1947) S   **Hurst**
Hurst, Evelyn Inez S   **E.I.Hurst**
Hurter, E. A   **Hurter**
Hurusawa, Isao (1916–) S   **Hurus.**
Hus, Henri Theodore Antoine de Leng (1876–) AS   **Hus**
Husain, Akhtar (fl. 1966) M   **A.Husain**
Husain, S.S. (fl. 1963) M   **S.S.Husain**
Husain, Syed Afaq (fl. 1970) S   **S.A.Husain**
Husain, T. (fl. 1986) S   **T.Husain**
Husnot, Pierre Tranquille (1840–1929) BPS   **Husn.**
Huss, Martin J. (fl. 1991) M   **Huss**
Hussain, Farrukh A   **Hussain**
Hussain, Syed Murtaza (1912–) M   **S.M.Hussain**
Husseinov, E.S. (fl. 1989) M   **E.S.Husseinov**
Husseinov, Sch.A. (1939–) S   **Husseinov**
Husseinova, B.F. (fl. 1961) M   **Husseinova**
Hussenot, Louis Cincinnatus Sévérin Léon (1809–1845) S   **Hussenot**

Hussey, A.M. (fl. 1847) M   **Hussey**
Husson, A.M. (fl. 1952) S   **Husson**
Hustedt, Friedrich (1886–1968) A   **Hust.**
Husz, Béla (1892–1954) M   **Husz**
Hutchins, Charles Robert (1928–) S   **C.R.Hutchins**
Hutchins, Lee Milo (1888–) M   **Hutchins**
Hutchinson, George Evelyn (1903–c.1968) A   **G.E.Hutch.**
Hutchinson, John (1884–1972) S   **Hutch.**
Hutchinson, Joseph Burtt (1902–1988) S   **J.B.Hutch.**
Hutchinson, S.A. (fl. 1955) M   **S.A.Hutch.**
Hutchinson, Wesley Gillis (1903–) M   **W.G.Hutch.**
Hutchison, James A. (fl. 1962) M   **J.A.Hutchison**
Hutchison, Leonard J. (fl. 1988) M   **L.J.Hutchison**
Hutchison, Paul Clifford (1924–) S   **Hutchison**
Huteau, H. (fl. 1894) S   **Huteau**
Huter, Rupert (1834–1919) PS   **Huter**
Huth, Ernst (1845–1897) PS   **Huth**
Huth, Karl (1902–) M   **K.Huth**
Hutner, Seymour Herbert (1911–) A   **Hutner**
Hutoh, M. (fl. 1962) P   **Hutoh**
Hütter, Ralf (fl. 1962) M .   **Hütter**
Hüttig, Heinrich (fl. 1872–1890) S   **Hüttig**
Huttleston, Donald Grunert (1920–) S   **Huttl.**
Hutton, William (1797–1860) BFM   **Hutton**
Huuskonen, Avi Johannes (1902–) M   **Huusk.**
Huvé, Hélène   A   **H.Huvé**
Huvé, P.   A   **P.Huvé**
Huxley, Anthony Julian (1920–) S   **Huxley**
Huxley, Camilla Rose (1952–) S   **C.R.Huxley**
Huynh, Kim–Lang (1935–) S   **Huynh**
Huyssteen, Dorothea Christina van (fl. 1937) S   **Huysst.**
Huzioka, Kazno (Kazmo) (fl. 1960) BF   **Huzioka**
Hwang, F.Y. (fl. 1936) M   **F.Y.Hwang**
Hwang, Liang (1906–) M   **L.Hwang**
Hwang, Shu Mei (1933–) S   **S.M.Hwang**
Hwang, T.S. (fl. 1982) S   **T.S.Hwang**
Hy, Félix Charles (1853–1918) ABMPS   **Hy**
Hyatt, Mildred Travis (1915–1971) M   **Hyatt**
Hyde, Harold Augustus (1892–1973) S   **Hyde**
Hyde, Kevin D. (fl. 1986) M   **K.D.Hyde**
Hygen, Georg (1908–) A   **Hygen**
Hyland, Bernard Patrick Matthew (1937–) S   **B.Hyland**
Hyland, Fay (1900–) S   **Hyland**
Hylander, Clarence John (1897–) S   **C.Hyl.**
Hylander, Hjalmar (1877–1965) S   **H.Hyl.**
Hylander, Nils (1904–1970) MPS   **Hyl.**
Hylmö, Bertil (1915–) S   **B.Hylmö**
Hylmö, David Einar (1883–1940) A   **Hylmö**
Hyvönen, Jaakko (1959–) B   **Hyvönen**

Iacob, Viorica C. (1940–) M	**Iacob**
Iacobescu, Nicolae (1863–1931) M	**Iacobescu**
Iaconis, Celina L. (fl. 1953) M	**Iaconis**
(Ianischevsky, Dmitrij E.)	
see Janischewsky, Dmitrij E.	**Janisch.**
Iatroú, Gregory (1949–) S	**Iatroú**
Ibarra, Florinda E. (fl. 1947) S	**Ibarra**
Ibbotson, Henry (1814–1886) S	**Ibbots.**
(Ibragimov, G.R.)	
see Ibrahimov, H.R.	**Ibrah.**
Ibrahimov, H.R. (fl. 1955) M	**Ibrah.**
Ichaso, Carmen Lúcia Falcão (1940–) S	**Ichaso**
Ichikawa, Wataru (1902–) A	**Ichik.**
Ichinoe, Masakatsu (fl. 1968) M	**Ichinoe**
Ichitani, Takio (fl. 1986) M	**Ichit.**
Ide, Kiyoharu (fl. 1981) S	**Ide**
Idei, Masahito (1953–) A	**M.Idei**
Idei, Yoshihiko A	**Y.Idei**
Ideta, Arata (1870–) M	**Ideta**
Idrobo, Jesús Medardo (1917–) S	**Idrobo**
(Idrobo–Muñoz, Jesús Medardo)	
see Idrobo, Jesús Medardo	**Idrobo**
Ietswaart, J.H. (1940–) S	**Ietsw.**
Igmándy, J. (1897–1950) S	**Igmándy**
Igmándy, Z. (fl. 1966) M	**Z.Igmándy**
Ignataviciute, M. (fl. 1962) M	**Ignat.**
Ignatov, Mikhail (Misha) S. (1956–) BS	**Ignatov**
Igolkin, G.I. (1883–1942) S	**Igolkin**
Igoschina, Kapitolina Nikolaevna (1894–1975) S	**Igoschina**
(Igoshina, Kapitolina Nikolaevna)	
see Igoschina, Kapitolina Nikolaevna	**Igoschina**
Iguchi, Takashi (fl. 1978) M	**Iguchi**
Ihlenfeldt, Hans–Dieter (1932–) S	**Ihlenf.**
Ihnatowicz, A. (fl. 1975) M	**Ihnat.**
Iida, K. (fl. 1964) M	**Iida**
Iijima, T. (fl. 1985) M	**Iijima**
Iinuma, Yokusai (1782–1865) S	**Iinuma**
Iisiba, Eikichi (1873–1936) B	**Iisiba**
Iizuka, Hiroshi (fl. 1953) M	**Iizuka**
Ikari, Jirô A	**Ikari**
Ikata, Suehiko (1894–) M	**Ikata**
Ikebe, Chikako (fl. 1977–1978) P	**Ikebe**
Ikeda, K. (fl. 1986) M	**Ikeda**
Ikeno, Seiichirô (Seiitirô) (1867–1944) MS	**Ikeno**
Ikonnikov, Sergei Sergeevich (1931–) S	**Ikonn.**
Ikonnikov–Galitzky, Nikolai Petrovic (1892–1942) S	**Ikonn.–Gal.**
Ikramov, M.I. (1923–) S	**Ikramov**
Ilgaard, Benjamin (1943–) S	**Ilgaard**
Iliescu, Constantin Horia Teodor (1940–) M	**Iliescu**

(Ilin, Modest Mikhaïlovich)	
see Iljin, Modest Mikhaïlovich	**Iljin**
Iljin, Modest Mikhaïlovich (1889–1967) PS	**Iljin**
Iljinskaja, J.A. (1921–) FS	**Iljinsk.**
(Iljinskaya, Irina Alekseevna)	
see Iljinskaja, J.A.	**Iljinsk.**
Iljinski, Alexei Porfirievich (1888–1945) S	**Iljinski**
(Iljinskij, Alexei Porfirievic)	
see Iljinski, Alexei Porfirievich	**Iljinski**
(Iljinsky, Alexei Porfirievich)	
see Iljinski, Alexei Porfirievich	**Iljinski**
Illana, Carlos (fl. 1986) M	**Illana**
Illario, Teresina S	**Illario**
Illarionova, I.A. S	**I.A.Illar.**
Illarionova, N.B. (fl. 1957) S	**N.B.Illar.**
Illick, Joseph Simon (1884–) S	**Illick**
Illitschevsky, S. S	**Illitsch.**
(Illitschewski, S.)	
see Illitschevsky, S.	**Illitsch.**
Illman, William Irwin (1921–) M	**Illman**
Ilse, Hugo (1835–1900) S	**Ilse**
Iltis, A. A	**A.Iltis**
Iltis, Hugh Hellmut (1925–) S	**H.H.Iltis**
Iltis, Hugo (1882–1952) A	**Iltis**
(Ilyin, Modest Mikhaïlovich)	
see Iljin, Modest Mikhaïlovich	**Iljin**
(Im Thurn, E.F.)	
see Thurn, Everard Ferdinand im	**Thurn**
Imada, O. (fl. 1972) M	**Imada**
Imahori, Kôzô (1917–) A	**Imahori**
Imai, Morieko (Moriko) (1942–) M	**M.Imai**
Imai, Sanshi (1900–) M	**S.Imai**
Imaizumi, Rikizo A	**Imaiz.**
Imamura, Saun–ichirô (Shun–ichirô) (1903–) S	**Imamura**
Imazeki, Rokuya (1904–) M	**Imazeki**
Imazu, Michio (fl. 1989) M	**Imazu**
Imbach, Emil J. (1897–1970) MS	**Imbach**
(Imchanitskaya, N.N.)	
see Imkhanitskaya, N.N.	**Imkhan.**
(Imchanitzkaja, N.N.)	
see Imkhanitskaya, N.N.	**Imkhan.**
Imhäuser, Ludwig A	**Imhäuser**
Imhof, Franz Jacob (fl. 1874) S	**Imhof**
Imhof, Othmar Emil (1855–1936) A	**O.E.Imhof**
Imkhanitskaya, N.N. (1935–) S	**Imkhan.**
Imlay, George (Gilbert) (1754–) S	**Imlay**
Imlay, Joan B. (fl. 1939) S	**J.B.Imlay**
Imle, Ernest Paul (1910–) M	**Imle**
Imler, Louis (fl. 1955) M	**Imler**
Immelman, Kathleen Leonore (1955–) S	**Immelman**

Imms, Augustus Daniel (1880–) A	**Imms**
Imšenecki, A.A. (fl. 1934) M	**Imšen.**
Imshaug, Henry Andrew (1925–) M	**Imshaug**
Imshenetskii, Aleksander Aleksandrovich (1905–) M	**Imshen.**
Inaba, Tadaaki A	**Inaba**
Inagaki, Kan–ichi (1908–) AS	**Inagaki**
Inagaki, Naoki (1929–) M	**N.Inagaki**
Inaschvili, Tsimi N. (fl. 1964) M	**Inaschv.**
Incarville, Pierre Nicolas le Chéron (1706–1757) S	**Incarv.**
Indelicato, Stephen R. A	**Indel.**
Indoh, Hiroharu (fl. 1939) M	**Indoh**
Infantes Vera, Juana G. (fl. 1962) S	**Infantes**
Ing, Bruce (1937–) M	**Ing**
Ingegnatti, Annibale (1838–) S	**Ingegn.**
Ingerslev, Leif (1903–1968) S	**Ingerslev**
Ingham, Norman D. (fl. 1908) S	**N.D.Ingham**
Ingham, William (1854–1923) B	**Ingham**
Inglis, John A	**Inglis**
Ingold, Cecil Terence (1905–) M	**Ingold**
Ingram, A. (fl. 1921) M	**A.Ingram**
Ingram, Collingwood (1880–1981) S	**Ingram**
Ingram, Edwin G. (fl. 1976) M	**E.G.Ingram**
Ingram, John William (1924–) S	**J.W.Ingram**
Ingrouille, Martin John (1955–) S	**Ingr.**
Ingwersen, Walter Edward Theodore (1885–1960) S	**Ingw.**
Inman, A.J. (fl. 1991) M	**A.J.Inman**
Inman, Ondess Lamar (1890–1942) S	**Inman**
Innes, Clive Frederick (1909–) S	**Innes**
Inobe, T. S	**Inobe**
Inokuma, Taizô (1904–1973) S	**Inokuma**
Inoue, Hiroshi (1932–1989) BS	**Inoue**
(Inoue, Masakane)	
see Inoue, Mayumi	**May.Inoue**
Inoue, Mayumi (fl. 1982) M	**May.Inoue**
Inoue, Mizoshi (fl. 1958) B	**M.Inoue**
Inoue, Satoru (1919–) B	**S.Inoue**
(Inoué, Tsutomu)	
see Inouye, Tsutomu	**Inouye**
Inouye, Isao A	**I.Inouye**
Inouye, Tsutomu (fl. 1951) B	**Inouye**
Insam, J. A	**Insam**
Inui, Tamaki (fl. 1965) M	**Inui**
Inumaru, Sunao (1899–) M	**Inumaru**
Inzenga, Giuseppe (1816–1887) MS	**Inzenga**
Ioannides, Nicos S. A	**Ioann.**
Ionescu, M.A. (1900–) S	**Ionescu**
Ioriya, Teru (1941–) A	**Ioriya**
Iqbal, M.M. A	**M.M.Iqbal**
Iqbal, S.H. (fl. 1967) M	**S.H.Iqbal**

Iqbaluddin   A   **Iqbaluddin**
Iranshahr, Mousa (1923–   ) S   **Iranshahr**
Iranzo Reig, Julio (1942–   ) S   **Iranzo**
Ireland, Robert Root (1932–   ) B   **Ireland**
Irénée-Marie, Thomas Joseph Caron (1898–1960) AS   **Irénée-Marie**
Irgang, Bruno Edgar (1941–   ) S   **Irgang**
Irigoyen, Luis H. (fl. 1917) S   **Irigoyen**
Irish, Henry Clay (1868–1960) S   **Irish**
Irizawa, T. (fl. 1930) M   **Irizawa**
Irlet, B. (fl. 1984) M   **Irlet**
Irmisch, Johann Friedrich Thilo (1816–1879) S   **Irmisch**
Irmscher, Edgar (1887–1968) BPS   **Irmsch.**
Irvine, Alexander (1793–1873) S   **Irvine**
Irvine, Anthony Kyle (1937–   ) S   **A.K.Irvine**
Irvine, David Edward Guthrie (1924–   ) A   **D.E.G.Irvine**
Irvine, Frederick Robert (1898–1962) S   **F.Irvine**
Irvine, Linda Mary (1928–   ) A   **L.M.Irvine**
Irving, E.G. (1816–1855) S   **Irving**
Irving, Robert Stewart (1942–   ) S   **R.S.Irving**
Irving, Walter (1867–1934) S   **W.Irving**
Irwin, Howard Samuel (1928–   ) S   **H.S.Irwin**
Irwin, J.A.G. (fl. 1987) M   **J.A.G.Irwin**
Irwin, James Bruce (1921–   ) S   **Irwin**
Isaac, Frances Margaret (Leighton) (née Leighton, F.M.) (1909–   ) S   **Isaac**
Isaac, Ivor (1914–   ) M   **I.Isaac**
Isaac, William Edwyn (1905–   ) A   **W.E.Isaac**
Isabella, P.K. (George) (1556–1633) AL   **Isabella**
Isabelle, Arsène (1795–1879) S   **Isab.**
Isagi, Yuji   A   **Isagi**
Isagulova, E.Z.   A   **Isagul.**
Isaka, M. (fl. 1979) M   **Isaka**
Isakov, K. (fl. 1962) S   **Isakov**
Isanbaeva, G.S. (fl. 1973) M   **Isanb.**
Isely, Duane (1918–   ) S   **Isely**
Isépy, I. (fl. 1965) S   **Isépy**
Isert, Paul Erdmann (1756–1789) M   **Isert**
Ishiba, C. (fl. 1982) M   **Ishiba**
Ishibashi, T. (fl. 1927) M   **Ishib.**
Ishidoya, Tsutomu (1891–1958) S   **Ishid.**
Ishijima, Wataru (–1980) A   **Ishij.**
Ishikawa, Mitsuhara (1887–   ) M   **Ishik.**
Ishitani, Takasuke (fl. 1987) M   **Ishit.**
Ishiyama, Shinichi (fl. 1934) M   **Ishiy.**
Ising, Ernest Horace (1884–1973) S   **Ising**
Iskanderova, Dz.G.   S   **Iskand.**
Iskenderov, A.T. (1923–   ) S   **Iskend.**
Isla, Maria Lourdes de la (fl. 1964) M   **Isla**
Islam, Abul Khayer Mohammed Nurul   A   **A.K.Islam**
Islam, M.A.   A   **M.A.Islam**

Islamov, E.M. (fl. 1981) M	**Islamov**
Isley, P.T. (fl. 1984) S	**Isley**
Ismail, A.L.S. (fl. 1977) M	**Ismail**
Işmen, Hikmet (fl. 1947) M	**Işmen**
Isnard, Antoine-Tristan Danty d' (1663–1743) L	**Isnard**
Isoviita, Pekka (1931–) B	**Isov.**
Israelson, Gunnar (1910–) A	**Israelson**
Issaiev, J. S	**Issaiev**
Issatschenko, Boris Laurentiewicz (1871–) M	**Issatsch.**
Issel, Raffaele A	**Issel**
Issler, Émile (Emil) (1872–1952) S	**Issler**
(Istvánffi de Csík Madéfalva, Gyula von)	
see Istvanfy de Csík Madéfalva, Gyula von	**Istv.**
Istvanfy de Csík Madéfalva, Gyula von (né Schaarschmidt, J.) (1860–1930)	
AMS	**Istv.**
(Istvanfy, Gyula von)	
see Istvanfy de Csík Madéfalva, Gyula von	**Istv.**
Isvet, Paul Erdmann S	**Isvet**
Itô, Hiroshi (1909–) PS	**H.Itô**
Itô, Kazuo (1915–) M	**Kaz. Itô**
Ito, Keisuke (1803–1901) S	**Ito**
Ito, Koji (1931–) S	**Koji Ito**
Itô, Mituo (fl. 1934) S	**Mit.Itô**
Ito, Motomi (1956–) S	**Mot.Ito**
Ito, Seiya (1883–1962) M	**S.Ito**
Ito, Tadayoshi (fl. 1983) M	**Tad.Ito**
Itô, Takeo (1911–) A	**Tak.Itô**
Itô, Tokutarô (1868–1941) MS	**T.Itô**
Itô, Yoshi (1907–) S	**Y.Itô**
Itoh, Mutsumi (fl. 1987) M	**Itoh**
Itono, Hiroshi (1943–1987) A	**Itono**
Itria, Carlos D. S	**Itria**
(Iturriaga, (M.) Teresita)	
see Iturriaga de Capiello, (M.) Teresita	**Iturr.**
Iturriaga de Capiello, (M.) Teresita (fl. 1987) M	**Iturr.**
Itzerott, Heinz (1912–1983) M	**Itzerott**
Itzigsohn, Ernst Friedrich Hermann (1814–1879) ABMS	**Itzigs.**
Ivanenko, Yu. (Ju.) A. (1962–) P	**Ivanenko**
Ivanić, M. (fl. 1937) M	**Ivanić**
Ivanina, L.I. (1917–) S	**Ivanina**
Ivanischvili, M.A. (1937–) S	**Ivan.**
(Ivanishvili, M.A.)	
see Ivanischvili, M.A.	**Ivan.**
Ivanitzkaja, A. S	**Ivanitzk.**
(Ivanitzkaya, A.)	
see Ivanitzkaja, A.	**Ivanitzk.**
Ivanitzky, Nikolaj (Nikolai) Aleksandrovich (1845–1899) S	**Ivanitzky**
Ivanjukovich, L.K. (1937–) S	**Ivanjuk.**
Ivanov, A.I. (1934–) M	**A.I.Ivanov**

Ivanov, A.P. (1903–) S   **A.P.Ivanov**
Ivanov, Nikolai Rodionovich (1902–1978) S   **N.R.Ivanov**
Ivanova, K.V. (1903–) S   **K.V.Ivanova**
Ivanova, L.G. (fl. 1970) M   **L.G.Ivanova**
Ivanova, M.M. (fl. 1965) S   **M.M.Ivanova**
Ivanova, N.A. (1893–1942) S   **N.A.Ivanova**
(Ivanyukovich, L.K.)
  see Ivanjukovich, L.K.   **Ivanjuk.**
Ivaschin, D.S. (1912–) S   **Ivaschin**
Ivens, A.J. (fl. 1953) S   **Ivens**
Iversen, Johannes (1904–1971) PS   **Iversen**
Ives, Eli (1779–1861) S   **E.Ives**
Ives, Joseph Christmas (1828–1868) S   **Ives**
Ivimey Cook, Walter Robert (1902–1952) M   **Ivimey Cook**
Ivo (fl. 1981) S   **Ivo**
Ivory, M.H. (fl. 1983) M   **M.H.Ivory**
Ivory, T.H. (fl. 1967) M   **Ivory**
Iwadare, Satoru (fl. 1934) M   **Iwadare**
Iwahashi, Yasumi A   **Iwah.**
Iwamasa, Sadazi (1911–) B   **Iwamasa**
Iwanoff, Leonidas A   **Iwanoff**
Iwarsson, M. (1948–) S   **Iwarsson**
Iwasaki, Hiroyoshi (fl. 1987) M   **H.Iwasaki**
Iwasaki, Nizo (fl. 1941) B   **Iwasaki**
Iwata, Jirô (1909–) S   **Iwata**
Iwatake, H. (fl. 1927) M   **Iwatake**
Iwatsu, Tokio (fl. 1984) M   **Iwatsu**
Iwatsuki, Kunio (1934–) PS   **K.Iwats.**
Iwatsuki, Zennoske (1929–) B   **Z.Iwats.**
Iyengar, A.V.V. (fl. 1935) M   **A.V.V.Iyengar**
Iyengar, Mandeyam Osuri Parthasarathy (1886–1963) AM   **M.O.P.Iyengar**
Iyer, Rohini (fl. 1972) M   **Iyer**
Izaguirre de Arturio, Primavera (1930–) S   **Izag.**
Izco, Jesús (1940–) S   **Izco**
(Izco Sevillano, Jesús)
  see Izco, Jesús   **Izco**
Izhboldina, L.A. A   **Izhb.**

Jaag, Otto (1900–1978) ACM   **Jaag**
Jaap, Otto (1864–1922) BM   **Jaap**
(Jaarsveld, Ernst Jacobus van)
  see Van Jaarsveld, Ernst Jacobus   **Van Jaarsv.**
Jaaska, Vello S   **Jaaska**
Jaasund, Erik (1918–1986) A   **Jaasund**
(Jablonski, Eugene)
  see Jablonszky, Eugene   **Jabl.**
Jablonszky, Eugene (1892–1975) S   **Jabl.**
Jabrova–Kolakovskaja, V.S. (fl. 1953) S   **Jabr.-Kolak.**
Jaccard, Henri (1844–1922) S   **Jaccard**

Jack, James Robertson (1863–1955) A   **J.R.Jack**
Jack, John George (1861–1949) S   **J.G.Jack**
Jack, Joseph Bernard (Josef Bernhard) (1818–1901) BC   **J.B.Jack**
Jack, Michele A. (fl. 1975) M   **M.A.Jack**
Jack, William (1795–1822) S   **Jack**
Jackes, Betsy Rivers (née Paterson, B.R.) (1935–) S   **Jackes**
Jackman, George (1837–1887) S   **Jackman**
Jackson, Albert Bruce (1876–1947) BMS   **A.B.Jacks.**
Jackson, Arthur Keith (1914–) S   **A.K.Jacks.⁻**
Jackson, Benjamin Daydon (1846–1927) S   **B.D.Jacks.**
Jackson, C.H.N. S   **C.H.N.Jacks.**
Jackson, Curtis Rukes (1927–) M   **C.R.Jacks.**
Jackson, George (1790–1811) S   **Jacks.**
Jackson, Henry Alexander Carmichael (1877–1961) M   **H.A.C.Jacks.**
Jackson, Herbert Spencer (1883–1951) M   **H.S.Jacks.**
Jackson, Joseph (1847–1924) BC   **J.Jacks.**
Jackson, Lyle Wendell Redverse (1900–) M   **L.W.R.Jacks.**
(Jackson, Michael B. Wyse)
   see Wyse Jackson, Michael B.   **Wyse Jacks.**
Jackson, N. (fl. 1989) M   **N.Jacks.**
Jackson, P.J. (fl. 1981–89) P   **P.J.Jacks.**
Jackson, Raymond Carl (1928–) S   **R.C.Jacks.**
Jackson, Vincent William (1876–) S   **V.W.Jacks.**
Jackson, William D. (fl. 1958) S   **W.D.Jacks.**
Jacky, Ernst (Ernest) (1874–) M   **Jacky**
(Jacob de Cordemoy, Eugène)
   see Cordemoy, Eugène Jacob de   **Cordem.**
(Jacob de Cordemoy, Hubert Louis Philippe Eugène)
   see Cordemoy, Hubert Louis Philippe Eugène Jacob de   **H.L.Cordem.**
Jacob, Edward (c.1710–1788) S   **Jacob**
Jacob, F.H. (fl. 1979) M   **F.H.Jacob**
Jacob, J.L. (fl. 1962) M   **J.L.Jacob**
Jacob, K. (fl. 1937–53) F   **K.Jacob**
Jacob, Kurumthottical Cherian (1890–1972) S   **K.C.Jacob**
Jacob-Makoy, Lambert (1790–1873) S   **Jacob-Makoy**
Jacobesco, Nicolas (fl. 1906) M   **Jacobeso**
Jacobi, Georg Albano von (1805–1874) S   **Jacobi**
Jacobovics, Anton (fl. 1835) S   **Jacobov.**
Jacobs, Donald Leroy (1919–) AB   **D.L.Jacobs**
Jacobs, Homer L. (1899–1981) M   **H.L.Jacobs**
Jacobs, Marius (1929–1983) S   **M.Jacobs**
Jacobs, Maxwell Ralph (1905–1979) S   **Jacobs**
Jacobs, Surrey Wilfrid Laurance (1946–) S   **S.W.L.Jacobs**
Jacobsen, Barry J. (1947–) M   **B.J.Jacobsen**
Jacobsen, Bodil Aavad A   **B.A.Jacobsen**
Jacobsen, H.C. A   **H.C.Jacobsen**
Jacobsen, Hans (1815–1891) S   **Jacobsen**
Jacobsen, Hermann Johannes Heinrich (1898–1978) AS   **H.Jacobsen**
Jacobsen, Jens Peter (1847–1885) A   **J.Jacobsen**

Jacobsen, Niels Henning Günther (1941–) PS	N.Jacobsen
Jacobsen, Terry Dale (1950–) S	T.D.Jacobsen
Jacobsen, Werner Bahne Georg (1909–) PS	W.Jacobsen
Jacobsson, Stig (1938–) M	Jacobsson
Jacono, I. (fl. 1933) M	Jacono
Jacot Guillarmod, Amy Frances May Gordon (1911–1992) MS	Jacot Guill.
Jacquemart, Albert (1808–1875) S	Jacquemart
Jacquemont, Venceslas Victor (1801–1832) S	Jacquem.
Jacquemoud, Fernand (1946–) S	Jacquemoud
Jacques, G. A	G.Jacques
Jacques, Henri Antoine (1782–1866) S	Jacques
Jacques, Joseph Emile (1908–) M	J.E.Jacques
Jacques–Félix, Henri (1907–) S	Jacq.-Fél.
Jacques–Félix, M. (fl. 1983) M	M.Jacq.-Fél.
Jacques–Vuarambon, Roger S	Jacq.-Vuar.
Jacquetant, E. (fl. 1984) M	Jacquet.
Jacquier, H. (fl. 1984) M	Jacquier
Jacquin, Hector (fl. 1832) S	H.Jacq.
Jacquin, Joseph Franz von (1766–1839) S	J.Jacq.
Jacquin, Nicolaus (Nicolaas) Joseph von (1727–1817) ABMPS	Jacq.
Jacquinot, Honoré (1814–1887) PS	Jacquinot
Jaczewski, Arthur Louis Arthurovič (1863–1932) M	Jacz.
Jaczewski, P.A. (fl. 1931) M	P.A.Jacz.
(Jaczewsky, Arthur Louis Arthurovič de)	
see Jaczewski, Arthur Louis Arthurovič	Jacz.
(Jaczewsky, P.A.)	
see Jaczewski, P.A.	P.A.Jacz.
Jaderholm, Axel Elof (1868–1927) B	Jaderh.
Jadhav, V.K. (fl. 1972) M	Jadhav
Jadin, Fernand (1862–) A	Jadin
Jaeger, August (1842–1877) B	A.Jaeger
Jaeger, Edmund Carroll (1887–1983) S	E.Jaeger
Jaeger, Georg Friedrich (von) (1785–1866) A	Jaeger
Jaeger, Hermann (1815–1890) S	H.Jaeger
Jaennicke, Johann Friedrich (1831–1907) S	Jaennicke
Jafar, Syed Abbas A	Jafar
Jaffrezo, Michel A	Jaffrezo
Jafri, Saiyad Masudal (Saiyid Masudul) Hasan (1927–1986) S	Jafri
Jagadeeswar, P. (fl. 1991) M	Jagad.
Jaganathan, T. (fl. 1972) M	Jagan.
Jäger, F.J. (fl. 1985) S	Jäger
(Jäger, Edmund Carroll)	
see Jaeger, Edmund Carroll	E.Jaeger
Jagger, Ivan Claude (1889–1939) M	Jagger
Jäggi, Jakob (Jacob) (1829–1894) S	Jäggi
Jäggli, Mario (1880–1959) B	Jäggli
Jagiełło, Małgorzata (1956–) S	Jagiełło
Jagtap, A.P. (fl. 1991) S	A.P.Jagtap
Jagtap, T.G. A	Jagtap

Jahandiez, Émile (1876–1938) BP	**Jahand.**
Jahn, A. (fl. 1970) S	**A.Jahn**
Jahn, Alfredo (1867–1940) S	**Al.Jahn**
Jahn, August Friedrich William Ernst (fl. 1774) S	**Jahn**
Jahn, Eduard Adolf Wilhelm (1871–1942) M	**E.Jahn**
Jahn, Hermann (1911–1987) M	**H.Jahn**
Jahn, Theodore Louis (1905–1979) A	**T.L.Jahn**
Jahnke, Erna A	**Jahnke**
Jahns, H.Martin (fl. 1977) M	**Jahns**
Jahoda, Rosa A	**Jahoda**
Jain, A.C. (fl. 1960) M	**A.C.Jain**
Jain, B.L. (fl. 1966) M	**B.L.Jain**
Jain, G.L. (fl. 1962) M	**G.L.Jain**
Jain, Kanti P. (fl. 1964) AFM	**K.P.Jain**
Jain, Prakash Chandra (fl. 1979) M	**P.C.Jain**
Jain, Raj K. (fl. 1968) MS	**R.K.Jain**
Jain, Rana R. (fl. 1981) M	**R.R.Jain**
Jain, S. (fl. 1990) M	**S.Jain**
Jain, S.C. A	**S.C.Jain**
Jain, S.S. (1952–) S	**S.S.Jain**
Jain, Sudhanshu Kumar (1926–) S	**S.K.Jain**
Jain, V.P. (fl. 1970) M	**V.P.Jain**
Jairajpuri, Durdana S. A	**Jairajp.**
Jajó, Bedrich (fl. 1947) S	**Jajó**
Jakober, K.D. (fl. 1989) M	**Jakober**
Jakobsen, I. (fl. 1989) M	**I.Jakobsen**
Jakobsen, K. (1929–) S	**Jakobsen**
Jakowatz, Anton (1872–) S	**Jakow.**
Jakubiec, Hanna A	**Jakubiec**
Jakubziner, Moisej Markovič (Markovich) (1898–) S	**Jakubz.**
Jákucs, Pál (Paul) (1928–) S	**Jákucs**
Jakuschevskii, Efram Sergeevich (1902–) S	**Jakusch.**
Jakuschkina, O.V. S	**Jakuschk.**
Jalaludin, S. (fl. 1990) M	**Jalaludin**
Jalan, S. (fl. 1965) M	**Jalan**
Jalas, Arvo Jaakko Juhanni (1920–) S	**Jalas**
Jamalainen, E.A. (fl. 1943) M	**Jamal.**
Jamaluddin (fl. 1975) M	**Jamaluddin**
Jaman, R. (fl. 1981–1988) P	**Jaman**
James, Charles William (1929–) S	**C.W.James**
James, E.J. A	**E.J.James**
James, Edwin (1797–1861) S	**E.James**
James, Joseph Francis (1857–1897) AM	**J.James**
James, Lois Elsie (1914–) S	**L.E.James**
James, Peter Wilfred (1930–) M	**P.James**
James, Sydney Herbert (1933–) S	**S.H.James**
James, T.A. (fl. 1980s) S	**T.A.James**
James, Thomas Potts (1803–1882) B	**James**
Jameson, Alexander Pringle (1886–) A	**A.P.Jameson**

Jameson, Hampden Gurney (1852–1936) B	H.Jameson
Jameson, William (Guilielmo) (1796–1873) S	Jameson
(Jameson (of Quito), William (Guilielmo))	
see Jameson, William (Guilielmo)	Jameson
Jamieson, Clara Octavia (1879–) M	Jamieson
Jamieson, David W. (1943–) B	D.W.Jamieson
Jamir, N.S. (fl. 1981–89) P	Jamir
Jamoni, P.Giovanni (fl. 1985) M	Jamoni
Jamzad, Ziba (1951–) S	Jamzad
Jan du Chêne, Roger E. A	Jan du Chêne
Jan, Georg (Giorgio) (1791–1866) PS	Jan
Janaki Ammal, Edavaleth Kakkath (1897–) S	Jan.Ammal
Janaki, V.C. A	V.C.Janaki
Janakidevi, Kilambi A	Janakid.
Janardhanan, K.K. (fl. 1964) M	Janardh.
Janardhanan, K.P. (1933–) S	K.P.Janardh.
Janarthanam, M.K. (fl. 1988) S	Janarth.
Jancey, R.C. S	Jancey
Janchen, Erwin Emil Alfred (1882–1970) APS	Janch.
(Janchen–Michel von Westland, Erwin Emil Alfred)	
see Janchen, Erwin Emil Alfred	Janch.
Jančić, R. (fl. 1988) S	Jančić
Janczewski, Eduard von Glinka (1846–1918) AMS	Jancz.
(Janczewski von Glinka, Eduard)	
see Janczewski, Eduard von Glinka	Jancz.
Jandaik, C.L. (fl. 1982) M	Jandaik
Jane, Frank William (1901–1963) AM	Jane
Janečkova, V. (fl. 1977) M	Janečkova
Janet, Charles (1849–1932) A	Janet
Janet, Mercia (fl. 1941) A	M.Janet
Janev S	Janev
Jang, J.C. (fl. 1985) M	J.C.Jang
Jang, Yong Suk (fl. 1986) M	Y.S.Jang
Janicki, C. A	Janicki
Janisch, Carl (1825–1900) A	C.Janisch
Janischewsky, Dmitrij E. (1875–1944) S	Janisch.
Janka, Victor von (1837–1900) S	Janka
(Janka von Bulcs, Victor)	
see Janka, Victor von	Janka
Jankauskas, T.V. A	Jankauskas
Janke, Alexander (1887–) M	Janke
Jankó, Johan (fl. 1890) S	Jankó
Janković, Milorad M. (1924–) S	Janković
Jankowska, Krystyna (fl. 1929) M	Jank.
Jannin, L. (fl. 1913) M	Jannin
Janos, D.P. (fl. 1982) M	Janos
Janowski, Margaretha S	Janowski
Janse, J.D. (fl. 1981) M	J.D.Janse
Janse, Johannes Albertus (1911–1977) S	Janse

Jansen, A.E. (fl. 1978) M	A.E.Jansen
Jansen, Gerrit M. (fl. 1982) M	G.M.Jansen
Jansen, Johannes Theodorus (1890–1948) PS	J.T.Jansen
Jansen, M.E. (fl. 1979) S	M.E.Jansen
Jansen, P.C.M. (1943–) S	P.C.M.Jansen
Jansen, Pieter (1882–1955) BPS	Jansen
Jansen, Robert K. (1954–) S	R.K.Jansen
Jansen–Jacobs, Marion Josephine (1944–) S	Jans.-Jac.
Jansonius, Jan (1928–) AM	Janson.
Janssens, Joannes Arnoldus P. (1952–) B	Janssens
Jansson, Carl-Axel (1925–) S	C.-A.Jansson
Jansson, H.-B. (fl. 1981) M	H.-B.Jansson
Janzen, Daniel H. S	D.H.Janzen
Janzen, Peter (1851–1922) B	Janzen
Jao, Chin Chih (1900–) A	C.C.Jao
Jaquet, Firmin (1858–1933) S	Jaquet
Jaquotot Villalonga, Maria Concepción (1932–) P	Jaquotot
Jarai-Komlodi, M. (fl. 1983) S	Jarai-Koml.
(Jaramillo Azanza, Jaime)	
see Jaramillo, Jaime	J.Jaram.
Jaramillo, Jaime (1944–) S	J.Jaram.
(Jaramillo L., Victor)	
see Jaramillo, Victor	V.Jaram.
Jaramillo, Victor (fl. 1984–) S	V.Jaram.
(Jarceva, M.V.)	
see Jartseva, M.V.	Jartseva
Jardin, Désiré Édélestan Stanislas Aimé (1822–1896) A	Jard.
Jardine, Nicholas (1943–) S	N.Jardine
Jardiné, Serge M. (fl. 1965) AM	S.Jardiné
Jarman, S.Jean S	Jarman
Jarmolenko, A.V. (1905–1944) S	Jarm.
Järnefelt, Heikki Arvid (1891–1963) A	Järnefelt
Jaroschenko, Pavel Dionisievich (1906–) S	Jarosch.
Jaroscz, Franz Eduard Felix (1799–) S	Jaroscz
Jarowaja, Nelly (fl. 1968) M	Jarow.
Jarrett, Frances Mary (1931–) P	F.M.Jarrett
Jarrett, Phyllis Heather (fl. 1929) S	P.H.Jarrett
Jarrett, V.H.C. (fl. 1936) S	V.H.C.Jarrett
Jarry, Denise T. (1936–) M	Jarry
Jartseva, M.V. A	Jartseva
Jarvie, J.K. (fl. 1987) S	Jarvie
Järvinen, Irma (1947–) B	Järvinen
Jarvis, Charles Edward (1954–) S	C.E.Jarvis
Jarvis, William Robert (1927–) M	Jarvis
Jasevoli, G. (fl. 1924) M	Jasevoli
Jasiewicz, Adam (1928–) S	Jasiewicz
Jasnitsky, W. A	Jasn.
Jastrzebowski, Wojciech (1799–1882) S	Jastrz.
Játiva, Carlos D. (fl. 1963) S	Játiva

Jatta, Antonio (1852–1912) M .... **Jatta**
Jaubert, Hippolyte François (1798–1874) PS ...... **Jaub.**
Jauch, Clotilde (1910–   ) M .... **Jauch**
Jauffret, J. (fl. 1955) M .... **Jauffret**
Jauhar, Prem Prakash (1937–   ) S .... **Jauhar**
(Jaume Saint-Hilaire, Jean Henri)
    see Saint-Hilaire, Jean Henri Jaume .... **J.St.-Hil.**
Jausar, A.G. (fl. 1961) M .... **Jausar**
Jauvy, Fr.P. (1760–1822) S .... **Jauvy**
Javeid, G.N. (fl. 1973) S .... **Javeid**
Jávorka, G.V.   S .... **G.V.Jáv.**
Jávorka, Sándor (Alexander) (1883–1961) S ..... **Jáv.**
Javornický, Pavel (1932–   ) A .... **Javorn.**
Jayagopal, K.   A .... **Jayag.**
Jayasuriya, Anthony H.M. (1944–   ) S .... **Jayas.**
Jayaweera, Don Martin Arthur (1912–1982) S ... **Jayaw.**
Jazkulieva, V.E.   A .... **Jazk.**
(Jazkulijeva, V.E.)
    see Jazkulieva, V.E. .... **Jazk.**
Jeanbernat, Ernest–Jules–Marie (1835–1888) S .. **Jeanb.**
Jeanjean, Alexis Félix (1867–1941) S .... **Jeanj.**
Jeanmonod, D. (fl. 1953–) S .... **Jeanm.**
Jeannenay, A.   S .... **Jeann.**
Jeanpert, Henri Édouard (1861–1921) PS .... **Jeanp.**
Jeanplong, Josef (József) (1919–   ) S .... **Jeanpl.**
Jeaume, G. (fl. 1923) M .... **Jeaume**
Jebb, Matthew (1958–   ) S .... **Jebb**
(Jechová, Vera)
    see Holubová-Jechová, Vera .... **Hol.-Jech.**
Jedlička, Josef (1912–1959) B .... **Jedl.**
Jedwabnick, Elizabeth (Elisabeth) (fl. 1924) S ... **Jedwabn.**
Jeeji-bai, N.   A .... **Jeeji-bai**
Jeekel, C.A.W. (fl. 1959) M .... **Jeekel**
Jefferies, R.L. (fl. 1987) S .... **Jefferies**
Jeffers, Walter Fulton (1915–1989) M .... **Jeffers**
Jeffrey, Charles (1934–   ) S .... **C.Jeffrey**
Jeffrey, Edward Charles (1866–1952) FM .... **E.Jeffrey**
Jeffrey, John (1826–1854) S .... **J.Jeffrey**
Jeffrey, John Frederick (1866–1943) S .... **Jeffrey**
Jeffrey, S.W.   A .... **S.W.Jeffrey**
Jeffs, Royal Edgar (1879–1933) S .... **Jeffs**
Jeggle, Walter (fl. 1973) S .... **Jeggle**
(Jegun de Marans, Antoine Louis Georges)
    see Brondeau, Louis de .... **Brond.**
Jehle, Robert Andrew (1882–   ) M .... **Jehle**
Jehlik, V. (fl. 1979) S .... **V.Jehlik**
Jekhowsky, B.de   A .... **Jekh.**
Jekyll, Gertrude (1843–1932) S .... **Jekyll**
Jelenc, Féodor (1911–   ) B .... **Jelenc**

Jelenevsky, A.G. (1930–) S	**Jelen.**
Jelić, M.(B.) (fl. 1975) M	**Jelić**
Jelinek, Anton (fl. 1857–1859) S	**Jelin.**
Jelliffe, Smith Ely (1866–1945) S	**Jell.**
Jellis, G.J. (fl. 1991) M	**G.J.Jellis**
Jellis, Sally (fl. 1984) S	**Jellis**
Jellison, William Livingston (1906–) M	**Jellison**
Jen, Bu Jun (fl. 1982) S	**B.J.Jen**
Jên, Hsien Wang S	**H.Wang Jên**
Jen, Hsien Wei (fl. 1984) S	**H.Wei Jen**
Jen, Xiang Wei (fl. 1983) S	**X.W.Jen**
Jeng, R.S. (fl. 1977) M	**Jeng**
Jenik, Jan (1929–) S	**Jenik**
Jenkina, T.V. (fl. 1966) M	**Jenkina**
Jenkins, Anna Eliza (1886–1973) M	**Jenkins**
Jenkins, D.J. (fl. 1982) M	**D.J.Jenkins**
Jenkins, D.T. (fl. 1933) M	**D.T.Jenkins**
Jenkins, Edmund Howard (1856–1921) S	**E.H.Jenkins**
Jenkins, James Angus (1904–1965) S	**J.A.Jenkins**
Jenkins, S.F. (fl. 1964) M	**S.F.Jenkins**
Jenkins, Wilbert Armonde (1905–1956) M	**W.A.Jenkins**
Jenkinson, James (1739?–1808) S	**Jenk.**
Jenman, George Samuel (1845–1902) PS	**Jenman**
Jenner, Charles (1810–1893) S	**C.Jenner**
Jenner, Edward (1803–1872) A	**Jenner**
Jennings, A.J. (fl. 1913) S	**A.J.Jenn.**
Jennings, Alfred Vaughan (1864–1903) S	**A.Jenn.**
Jennings, Herbert Spencer (1868–1947) M	**H.S.Jenn.**
Jennings, Otto Emery (1877–1964) PS	**Jenn.**
Jennison, Harry Milliken (1885–1940) S	**Jennison**
Jenny, R. (fl. 1985) S	**Jenny**
Jensen, Christian Erasmus Otterstrøm (Otterström) (1859–1941) B	**C.E.O.Jensen**
Jensen, Christian Nephi (1880–) M	**C.N.Jensen**
Jensen, Emron Alfred (1925–) A	**E.A.Jensen**
Jensen, Hans Laurits (1898–) M	**H.L.Jensen**
Jensen, J.D. (fl. 1981) M	**J.D.Jensen**
Jensen, J.P. (fl. 1945) M	**J.P.Jensen**
Jensen, James Bernard (1936–) A	**J.B.Jensen**
Jensen, Jens Ludwig (1836–1904) M	**J.L.Jensen**
Jensen, Johan Georg Keller (1818–1886) S	**Jensen**
Jensen, S.N. (fl. 1912) M	**S.N.Jensen**
Jensen, Thomas (1824–1877) B	**T.Jensen**
Jensen, V. (fl. 1967–1980) M	**V.Jensen**
Jensen, Vilhelm Peter Herlof (1870–) M	**V.P.H.Jensen**
Jenssen, G.M. (fl. 1985) M	**G.M.Jenssen**
Jenssen, K.M. (fl. 1986) M	**K.M.Jenssen**
Jentys–Szaferowa, Janina (1895–) S	**Jent.-Szaf.**
Jeppesen, Stig (1943–) BS	**Jeppesen**
Jepps, Margaret W. A	**Jepps**

Jepson, Willis Linn (1867–1946) PS   **Jeps.**
Jérémie, J. (1944–　) S   **Jérémie**
Jerković, Lazar　A   **Jerković**
Jermalaviczjute, S.I. (fl. 1966) M   **Jermal.**
Jermy, Anthony Clive (1932–　) MPS   **Jermy**
Jersey, Noel J.de (1923–　) B   **Jersey**
Jervis, ?W.R. (fl. 1960) S   **Jervis**
Jervis, Roy Newell (1913–　) S   **R.N.Jervis**
Jerzer (fl. 1987) M   **Jerzer**
Jessen, Karl Friedrich Wilhelm (1821–1889) A   **Jess.**
Jessen, Knud (1884–1971) S   **K.Jess.**
Jesson, E.M. (fl. 1915–1916) S   **Jesson**
Jessop, John Peter (1939–　) PS   **Jessop**
Jessup, L.W. (1947–　) S   **Jessup**
Jesup, H.Phillips (fl. 1964) S   **H.P.Jesup**
Jesup, Henry Griswold (1826–1903) S   **Jesup**
Jeswiet, Jacob (1879–1966) S   **Jeswiet**
Jeuken, M. (fl. 1952) S   **Jeuken**
Jewell, Herbert Winship (1872–　) P   **Jewell**
Jezek, Vojtech (1892–1960) B   **Jez.**
Jha, D.　A   **D.Jha**
Jha, J. (fl. 1988) P   **J.Jha**
Ji, Jin Xiang (fl. 1989) S   **J.X.Ji**
Jia, Ju Sheng (fl. 1989) M   **J.S.Jia**
Jiang, C.L. (fl. 1985) M   **C.L.Jiang**
Jiang, De Hu (fl. 1987) S   **D.H.Jiang**
Jiang, Guan(g) Zheng (1922–　) M   **G.Z.Jiang**
Jiang, Hua Ren (fl. 1986) S   **H.R.Jiang**
Jiang, Huei Ming (fl. 1989) S   **H.M.Jiang**
Jiang, M.J.　A   **M.J.Jiang**
Jiang, Shou Zhong (1928–　) M   **S.Z.Jiang**
Jiang, X.P. (fl. 1984) S   **X.P.Jiang**
Jiang, Xing Lin (1927–　) S   **X.L.Jiang**
Jiang, You Chuan (fl. 1985) S   **Y.C.Jiang**
Jiang, Yu Mei (1940–　) M   **Y.M.Jiang**
Jiang, Zi De (fl. 1989) M   **Z.D.Jiang**
Jien, Zhuo Po (fl. 1963) S   **Z.P.Jien**
Jílek, Bohumil (1905–1972) S   **Jílek**
Jiménez Albarran, Maria Josefa (fl. 1980) S   **Jiménez Alb.**
Jiménez Almonte, José de Jesús (1905–1982) BS   **J.Jiménez Alm.**
Jiménez M., Quírico (fl. 1989) S   **Q.Jiménez**
Jiménez Machorro, Rolando (1961–　) S   **R.Jiménez**
Jiménez Munuera, Francisco de Paula (fl. 1899–1913) S   **Jiménez Mun.**
Jiménez, Oton (1895–　) PS   **Jiménez**
Jiménez Ramírez, Jaime (fl. 1989) S   **J.Jiménez Ram.**
Jin, H. (fl. 1985) M   **H.Jin**
Jin, Shu Ying (1935–　) S   **S.Y.Jin**
Jin, Yue Xing (1934–　) PS   **Y.X.Jin**
Jing, Yi Xin (fl. 1980) M   **Y.X.Jing**

(Jing, Yue Xing)
  see Jin, Yue Xing   **Y.X.Jin**
Jirasek, Jon (Johann) (1754–1797) S   **Jirasek**
Jirásek, Václav (1906–) S   **V.Jirásek**
Jírovec, O. (fl. 1939) M   **Jírovec**
Joachim, Leon (1873–1945) M   **Joachim**
Joaquin de Barnola, R.P. (fl. 1912) P   **Joaquin**
Job de Francis, Maria Manuela   S   **Job**
Jochems, Sarah Cornelius Johannes (1891–) M   **Jochems**
(Joenckema, Rembert van)
  see Dodoens, Rembert   **Dodoens**
Joerger, A.P.   A   **Joerger**
Joffe, Abraham Z. (fl. 1974) M   **Joffe**
Joguet, Raymond (1893–1958) M   **Joguet**
Johan–Olsen, Olav (later Sopp, O.J.) (1860–1931) M   **Johan–Olsen**
Johann Baptist, Fabian Sebastian (1782–1859) S   **Joh.Baptist**
Johannes, Heinrich (fl. 1950) M   **Johannes**
Johannes, Robert Earl (1936–) A   **R.E.Johannes**
Johannesen, E.W. (fl. 1980) M   **Johannesen**
Johannsen, Elzbieta (Ella) W. (fl. 1973) M   **Johannsen**
Johansen, Donald Alexander (1901–) S   **D.A.Johans.**
Johansen, Frits (1882–1957) S   **Johans.**
Johansen, G. (fl. 1949) M   **G.Johans.**
Johansen, Hans William (1932–) A   **H.W.Johans.**
Johansen, Inger (1951–) M   **I.Johans.**
Johansen, Jeffrey R.   A   **J.R.Johans.**
Johanson, A.E. (fl. 1943) M   **A.E.Johanson**
Johanson, Carl Johan (1858–1888) MS   **Johanson**
Johansson, Dick (fl. 1974) S   **D.Johanss.**
Johansson, Jan Thomas (fl. 1988) S   **J.T.Johanss.**
Johansson, Karl (1856–1928) S   **Johanss.**
Johansson, Nils Thure (1893–1939) F   **N.T.Johanss.**
Johar, D.S. (fl. 1955) M   **Johar**
John, Albin   S   **John**
John, David Michael (1942–) A   **D.M.John**
John, Jacob   A   **J.John**
John, R.P.   A   **R.P.John**
John, Rachel (fl. 1955) M   **R.John**
John, Volker (1952–) S   **V.John**
Johns, Robert James (1944–) S   **R.J.Johns**
Johns, Robert Marvin (1928–1963) M   **Johns**
Johnson, Aaron Guy (1880–) M   **Aar.G.Johnson**
Johnson, Albert G. (1912–1977) S   **Alb.G.Johnson**
Johnson, Anne (1928–) B   **A.Johnson**
Johnson, Arthur Monrad (1878–1943) S   **A.M.Johnson**
Johnson, Asa Emery (1825–1906) M   **A.E.Johnson**
Johnson, Bertil Lennart (1909–) S   **B.L.Johnson**
Johnson, C.D.   A   **C.D.Johnson**
Johnson, C.J. (fl. 1980) M   **C.T.Johnson**

Johnson, Charles (1791–1880) S	**Johnson**
Johnson, Charles Pierpont (–1893) S	**C.P.Johnson**
Johnson, David M. (1955–) PS	**D.M.Johnson**
Johnson, Dennis A. (fl. 1991) M	**D.A.Johnson**
Johnson, Dorothy L. S	**D.L.Johnson**
Johnson, Duncan Starr (1867–1937) B	**D.S.Johnson**
Johnson, Edward Marshall (1896–) M	**E.M.Johnson**
Johnson, Eileen Ruth Laithlain (1896–1972) S	**E.R.L.Johnson**
Johnson, Eric A. (fl. 1978) M	**E.A.Johnson**
Johnson, G.A.L. A	**G.A.L.Johnson**
Johnson, Howard Wilfred (1901–) M	**H.W.Johnson**
Johnson, James (1886–1952) M	**J.Johnson**
Johnson, James Yates (1820–1900) PS	**J.Y.Johnson**
Johnson, Jesse Harlan (1892–1969) A	**J.H.Johnson**
Johnson, Joseph Elias (Ellis) (1817–1882) S	**J.E.Johnson**
Johnson, Joseph Harry (1894–) S	**H.Johnson**
Johnson, Lawrence Alexander Sidney (1925–) S	**L.A.S.Johnson**
Johnson, Leland Parrish (1910–) A	**L.P.Johnson**
Johnson, Lorenz(o) Nickerson (1862–1897) A	**L.N.Johnson**
Johnson, Miles F. (1936–) S	**M.F.Johnson**
Johnson, Paul W. A	**P.W.Johnson**
Johnson, Peter Neville (1946–) S	**P.N.Johnson**
Johnson, R.G. (fl. 1980) M	**R.G.Johnson**
Johnson, Raymond Roy (1932–) S	**R.R.Johnson**
Johnson, Robert William (1930–) S	**R.W.Johnson**
Johnson, Terry Walter (1923–) M	**T.W.Johnson**
Johnson, Thomas (1863–1954) AFM	**T.Johnson**
Johnson, Thorvaldur (1897–) M	**Thorv.Johnson**
Johnson, William (1844–1919) S	**W.Johnson**
Johnston, B.C. (fl. 1985) S	**B.C.Johnst.**
Johnston, Christopher (fl. 1860) A	**C.Johnst.**
Johnston, George (1797–1855) AMS	**Johnst.**
Johnston, Harry Hamilton (1858–1927) P	**H.H.Johnst.**
Johnston, Ivan Murray (1898–1960) PS	**I.M.Johnst.**
Johnston, Jen (1954–) M	**J.Johnst.**
Johnston, John Robert (1880–1953) MS	**J.R.Johnst.**
Johnston, Laverne Albert (1930–) S	**L.A.Johnst.**
Johnston, Marshall Conring (1930–) S	**M.C.Johnst.**
Johnston, P.R. (1952–) M	**P.R.Johnst.**
Johnston, Robert Mackenzie (1844–1918) F	**R.M.Johnst.**
Johnston, Thomas Harvey (1881–1951) S	**T.H.Johnst.**
Johnstone, George Henry (1881–1960) S	**G.H.Johnstone**
Johnstone, Ian M. A	**I.M.Johnstone**
Johnstone, William Grosart (–c.1860) A	**Johnstone**
Johow, Friedrich (Federico) Richard Adelbert (Adelbart) (1859–1933) MPS	**Johow**
Johri, Bhavdish Narain (1945–) M	**B.N.Johri**
Johri, Brij Mohan (1909–) BM	**Johri**
Johri, S.C. (fl. 1984) S	**S.C.Johri**
Jokela, Paavo Sigfrid (1896–1971) S	**Jokela**

Jokerst, James D. (fl. 1990) S	Jokerst
Jokl, M. (fl. 1918) M	Jokl
Jolis, Auguste François (1823–1904) BM	Jolis
Jolivet, Pierre (fl. 1959) M	Jolivet
Jolley, D.W. A	Jolley
Jollos, Victor A	Jollos
Joly, Aylthon Brandão (1924–1975) A	A.B.Joly
Joly, Nicolas (1812–1885) A	Joly
Joly, Patrick (fl. 1966) M	P.Joly
Joly, S. (fl. 1956) M	S.Joly
Jolyclerc, Nicolas Marie Thérèse (1746–1817) B	Jolycl.
Jonathan, Carl (1799–1837) S	Jonathan
(Joncheere, Gerardus J.de)	
see de Joncheere, Gerardus J.	de Jonch.
Jones, Alan Philip Dalby (1918–1946) S	A.P.D.Jones
Jones, Almut Gitter (1923–) S	A.G.Jones
Jones, Arthur Mowbray (1820–1889) P	A.M.Jones
Jones, Brian Michael Glyn (1933–) S	B.M.G.Jones
Jones, D. (fl. 1968) M	D.Jones
Jones, Daniel Angell (1861–1936) B	D.A.Jones
Jones, David Lloyd (1944–) PS	D.L.Jones
Jones, David T. (1900–) A	D.T.Jones
Jones, E.B.Gareth (fl. 1962) M	E.B.G.Jones
Jones, Eustace Wilkinson (1909–) B	E.W.Jones
Jones, F.R. (fl. 1964) M	F.R.Jones bis
Jones, Florence Lucinda (née Freeman, F.L.) (1912–) S	F.L.Jones
Jones, Fred Reuel (1884–1956) M	F.R.Jones
Jones, Frederick Butler (1909–) S	F.B.Jones
Jones, George Howard (fl. 1924) M	G.H.Jones
Jones, George Neville (1903–1970) BPS	G.N.Jones
Jones, Henry G. (fl. 1961) S	H.G.Jones
Jones, Herbert Lyon (1866–1898) S	H.L.Jones
Jones, John F. (fl. 1931) B	J.F.Jones
Jones, John Pike (1790–1857) A	J.P.Jones
Jones, Keith (1926–) S	K.Jones
Jones, Leon Kilby (1895–1966) M	L.K.Jones
Jones, Lewis Ralph (1864–1945) MS	L.R.Jones
Jones, M.A. S	M.A.Jones
Jones, Marcus Eugene (1852–1934) MPS	M.E.Jones
Jones, O.A. (fl. 1947) F	O.A.Jones
Jones, Paul L. A	P.L.Jones
Jones, Philip Malory (1892–) M	P.M.Jones
Jones, Quentin (1920–) S	Q.Jones
Jones, Rena T. (fl. 1971) M	R.T.Jones
Jones, Rod A	R.Jones
Jones, S. (fl. 1861) P	S.Jones
Jones, Samuel Boscom (1933–) S	S.B.Jones
Jones, Theobald (R.) (1790–1868) A	T.R.Jones
Jones, W.W. (fl. 1905) S	W.W.Jones

Jones, Walter (fl. 1929) M	W.Jones
Jones, William (1746–1794) S	Jones
(Jong, C.B.De)	
see De Jong, C.B.	De Jong
(Jong, Diederik Cornelius Dignus De)	
see DeJong, Diederik Cornelius Dignus	D.C.D.DeJong
(Jong, P.C.de)	
see DeJong, P.C.	P.C.DeJong
Jong, Shung Chang (1936–) M	S.C.Jong
(Jonge, Alide E.van Hall De)	
see De Jonge, Alide E.van Hall	De Jonge
(Jonghe, Adriaen)	
see Junius, Hadrianus	Junius
(Jonghe, Jean De)	
see De Jonghe, Jean	De Jonghe
Jongmans, Willem Josephus (1878–1957) BFS	Jongm.
Jonker, Anni Margriet Emma (1920–) S	A.M.E.Jonker
Jonker, Fredrik Pieter (1912–) S	Jonker
(Jonker-Verhoef, Anni Margriet Emma)	
see Jonker, Anni Margriet Emma	A.M.E.Jonker
Jonkman, Hendricus Franciscus (fl. 1879) S	Jonkman
Jonsell, Bengt Edvard (1936–) S	Jonsell
Jónsson, Helgi (1867–1925) AS	Jónss.
Jónsson, Sigurdur A	S.Jónss.
Jonston, Johannes (1603–1675) LS	Jonst.
Joó, I. (1806–1881) S	Joó
Jooste, Wonter Johannes (1933–) M	Jooste
Jordal, Louis Henrik (1919–1951) S	Jordal
Jordan, A.J. (1873–1906) S	A.J.Jord.
Jordan, Claude Thomas Alexis (1814–1897) S	Jord.
Jordan, David Star (1851–1931) S	D.S.Jord.
Jordan de Puyfol (1819–1891) S	Jord.Puyf.
Jordan, M.M. (fl. 1986) M	M.M.Jord.
Jordan, William Paul (fl. 1973) M	W.P.Jord.
(Jordanoff, Daki)	
see Jordanov, Daki	Jordanov
Jordanov, Daki (1893–1978) S	Jordanov
Jordi, Ernst (1899–1933) S	E.Jordi
Jordi, Ernst A. (1877–1933) M	Jordi
Jörgensen, A. (fl. 1898) M	A.Jörg.
Jørgensen, Carl Adolph (1899–)M	C.A.Jørg.
Jørgensen, Erik G. (1921–)A	E.G. Jørg.
Jörgensen, Eugen Honoratius (Honoratus) (1862–1938) AB	Jörg.
Jørgensen, O. A	C.Jørg.
Jørgensen, Per Magnus (1944–) BMS	M.Jørg.
Jørgensen, R.B. (fl. 1985) S	R.B.Jørg.
Jörgensen, Sigurd S	S.Jörg.
Jorissenne, G. (fl. 1882) S	Joriss.

Jörlin, Engelbert (1733–1810) S	**Jörl.**
Jørstad, Ivar (1887–1967) M	**Jørst.**
Josch, Eduard von (1799–1874) S	**Josch**
Joseph, J. (fl. 1964–1979) PS	**J.Joseph**
Joseph, K.T. (fl. 1979) S	**K.T.Joseph**
Joshi, Amar Chaud (1908–1971) S	**Joshi**
Joshi, B.D. A	**B.D.Joshi**
Joshi, G.T. (fl. 1971) M	**G.T.Joshi**
Joshi, H.V. A	**H.V.Joshi**
Joshi, I.J. (fl. 1981) M	**I.J.Joshi**
Joshi, M.C. (fl. 1955) MS	**M.C.Joshi**
Joshi, Mamta (fl. 1979) M	**M.Joshi**
Joshi, S.P. (fl. 1986) M	**S.P.Joshi**
Joshua, William (1828–1898) AM	**Joshua**
Josserand, Marcel (1900–) M	**Joss.**
Josst, Franz (1815–1862) S	**Josst**
Jost, Ludwig (1865–1947) A	**Jost**
Josten, K.H. (fl. 1962) F	**Josten**
Josue, A.R. (fl. 1976) M	**Josue**
Jôtani, Yukio (1904–) PS	**Jôtani**
Jotter, Mary Lois (1914–1942) S	**Jotter**
Joubert, Andor M. (1952–) S	**Joubert**
Jouin, V. S	**Jouin**
(Joukowsky, O.)	
see Zhukovsky, O.	**O.Zhuk.**
(Joukowsky, Peter Mikhailovich)	
see Zhukovsky, Peter Mikhailovich	**Zhuk.**
Jourdan, (Claude) Pascal (1835–1881) S	**Jourd.**
Jousé, Anastasia P. (1905–) A	**Jousé**
Jouvenaz, D.P. (fl. 1991) M	**Jouvenaz**
Jovanović, Branislav (fl. 1972) S	**Jovan.**
Jovanović–Dunjić, Rajna S	**Jovan.-Dunj.**
Jovet, Paul Albert (1896–1991) S	**Jovet**
(Jovet-Ast, Suzanne (née Ast, S.))	
see Ast, Suzanne	**Ast**
Jowett, Rosemary A	**Jowett**
Jowitt, John F. S	**Jowitt**
Joyal, Elaine (fl. 1986) S	**Joyal**
Joyeux, C. (fl. 1941) M	**Joyeux**
Joyon, L. (fl. 1966) AM	**Joyon**
Ju, Y.M. (fl. 1985) M	**Y.M.Ju**
Juarez, F. (fl. 1986) S	**Juarez**
Juch, Karl (Carl) Wilhelm (1774–1821) S	**Juch**
Judd, Walter Stephen (1951–) S	**Judd**
Judziewicz, Emmet J. (1953–) S	**Judz.**
Juel, Hans Oscar (1863–1931) BMS	**Juel**
(Juergens, Georg Heinrich Bernhard)	
see Jürgens, Georg Heinrich Bernhard	**Jürg.**
Juge de Saint Martin, Jacques Joseph (1743–1824) S	**Juge St.Mart**
Juhlin-Dannfelt, Hermann Julius Brorson (1852–1937) AS	**Juhl.-Dannf.**

Juillard–Hartmann, G. (fl. 1919) M	**Juill.-Hartm.**
Jukhananov, D.Ch. S	**Jukhan.**
(Juksip, Albert Jakovlevič)	
see Juxip, Albert Jakovlevič	**Juxip**
Jülich, Walter (1942–) M	**Jülich**
Julien, Alfred Cyprien (–1902) S	**Julien**
Julin, E. (fl. 1986) S	**Julin**
Jullien (fl. 1789) M	**Jullien**
Jumelle, Henri Lucien (1866–1935) S	**Jum.**
Juminer, B. (fl. 1965) M	**Juminer**
(Jundsill, Józef)	
see Jundzill, Józef	**J.Jundz.**
(Jundsill, Stanislaw Bonifacy)	
see Jundzill, Stanislaw Bonifacy	**S.B.Jundz.**
Jundzill, Józef (1794–1877) S	**J.Jundz.**
Jundzill, Stanislaw Bonifacy (1761–1847) S	**S.B.Jundz.**
Junell, Lena (fl. 1965) M	**L.Junell**
Junell, Sven Albert Brynolt (1901–) AS	**Junell**
Jung, H.S. (fl. 1987) M	**H.S.Jung**
Jung, Joachim (1587–1657) L	**Jung**
Jung, S.L. (fl. 1979) S	**S.L.Jung**
Jung, W. (fl. 1832) S	**W.Jung**
Jung–Mendaçolli, Sigrid Luiza (1952–) S	**Jung-Mend.**
Jungck, Max (1849–) S	**Jungck**
Junge, Alexander S	**A.Junge**
Junge, Paul (1881–1919) PS	**Junge**
Junger, Ernst (fl. 1891) S	**Junger**
Junghans, Philipp Kaspar (Caspar) (1738–1797) S	**Junghans**
Jungherr, E. (fl. 1934) M	**Jungherr**
Junghuhn, (Friedrich) Franz Wilhelm (1809–1864) BMPS	**Jungh.**
Jungk, Christian Ludwig (fl. 1807) S	**Jungk**
Jungner, Johan Richard (1858–1929) S	**Jungner**
Jungnickel, Fritz (fl. 1984) S	**Jungn.**
Jüngst, Ludwig Volrad (1804–1880) S	**Jüngst**
Juniper, A.J. (fl. 1953) M	**Juniper**
Junius, Hadrianus (1511–1575) M	**Junius**
Junussov, S.Ju. (1934–) S	**Junussov**
Jurair, A.M.M. (fl. 1960) M	**Jurair**
Jurányi, Lajos (Ludwig) (1837–1897) S	**Jurányi**
Juratzka, Jakob (Jacob) (1821–1878) BS	**Jur.**
(Jurcevič, Ivan Danilovich)	
see Jurkevich, Ivan Danilovich	**Jurk.**
Jürgens, Georg Heinrich Bernhard (1771–1846) AMS	**Jürg.**
Jurilj, Anto A	**Jurilj**
Juriš, Štefan (1928–) A	**Juriš**
Jurišić, Zivojin J. (1863–1921) S	**Jurišić**
Jurkevich, Ivan Danilovich (1902–) S	**Jurk.**
Jurtzev, B.A. (1932–) S	**Jurtzev**
Jury, Stephen Leonard (1949–) S	**Jury**

Jurzitza, G. (fl. 1970) M	**Jurzitza**
Juschev, A.A. S	**Juschev**
Juse, A. A	**Juse**
Juslenius, Abrahamus Danielis (1732–1803) S	**Jusl.**
Jussieu, Adrien Henri Laurent de (1797–1853) S	**A.Juss.**
Jussieu, Antoine de (1686–1758) S	**Ant.Juss.**
Jussieu, Antoine Laurent de (1748–1836) BMPS	**Juss.**
Jussieu, Bernard de (1699–1777) S	**B.Juss.**
Jussieu, Christophe de (1685–1758) S	**C.Juss.**
Jussieu, Joseph de (1704–1779) S	**J.Juss.**
Just, Johann Leopold (1841–1891) BS	**Just**
Just, Theodor Karl (1904–1960) S	**K.Just**
Justin, Rajko (1865–1938) S	**Justin**
Jux, Ulrich A	**Jux**
Juxip, Albert Jakovlevič (1886–1966) S	**Juxip**
Juzepczuk, Sergei Vasilievich (1893–1959) S	**Juz.**
Kaalaas, Baard Bastian Larsen (1851–1918) B	**Kaal.**
Käarik, A. (fl. 1991) M	**Käarik**
Kaas, Hanne A	**Kaas**
Kaastra, Roelof Cornelis (1942–) S	**Kaastra**
Kabanov, N.M. A	**N.M.Kabanov**
Kabanov, Nikolai Evgenievich (1905–) S	**Kabanov**
Kabát, Josef Emanuel (1849–1925) M	**Kabát**
Kabath, Hermann (1816–1888) S	**Kabath**
Kabiersch, Waldefried (fl. 1936) B	**Kabiersch**
Kablíková, Josephine (Josefina) (1787–1925) MS	**Kablík.**
Kabsch, Wilhelm August Walther (1835–1864) S	**Kabsch**
Kabulov, Dzhabbar Tilljaevich (1911–) S	**Kabulov**
Kabuye, Christine H Sophie (1938–) S	**Kabuye**
Kácha, A. A	**Kácha**
Kache, Paul (1882–1945) S	**Kache**
Kachler, Johann (1782–c.1863) S	**Kachl.**
Kachroo, J.V. (fl. 1966) M	**J.V.Kachroo**
Kachroo, Prem Nath (1924–) BPS	**Kachroo**
Kaczmarek, Regidius Marion Clemens (1888–) S	**Kaczm.**
Kaczmarek, Sławomir (fl. 1987) M	**S.Kaczm.**
Kaczmarska, Irena A	**Kaczmarska**
Kadereit, Joachim W. (1956–) S	**Kadereit**
Kadiri, M. (fl. 1984) M	**Kadiri**
Kadlubowska, Joanna Zofia A	**Kadlub.**
Kadono, Y. (fl. 1983) S	**Kadono**
Kadota, Yuichi (1949–) S	**Kadota**
Kaempfer, Engelbert (1651–1716) LS	**Kaempf.**
Kaercher, W. (1933–) S	**Kaercher**
Kaever, Matthias A	**Kaever**
Kafi, Ab. (fl. 1955) M	**Kafi**
Kagami, M. (fl. 1975) M	**Kagami**
Kagan, J.S. (fl. 1986) S	**Kagan**
Kahl, Alfred (1877–1946) A	**Kahl**

Kahn, B. (fl. 1986) S	**B.Kahn**
Kahn, F. (fl. 1988) S	**F.Kahn**
Kain, Charles Henri (–1913) A	**Kain**
Kain, Joanna M. (1930–) A	**J.M.Kain**
(Kairamo, Alfred Oswald)	
see Kihlman, Alfred Oswald	**Kihlm.**
Kaiser, Paul Ernst Ewald (1857–1935) AS	**Kaiser**
Kaiser, W.J. (fl. 1979) M	**W.J.Kaiser**
Kajan, E. (fl. 1987) M	**Kajan**
Kajanus, Birger (also Nilson, B.) (1882–1931) M	**Kajanus**
Kajimura, Mitsuo (1935–) A	**Kajim.**
Kajiwara, T. (fl. 1976) M	**Kajiw.**
Kakishima, Makoto (fl. 1985) M	**Kakish.**
Kakkar, R.K. (fl. 1964) M	**Kakkar**
Kaku (fl. 1981) S	**Kaku**
Kakudidi, Esezah K.Z. (1955–) S	**Kakudidi**
Kalaméés, Kuulo A. (1934–) M	**Kalaméés**
Kalanda, Kankenza (1947–) S	**Kalanda**
Kalani, I.K. (fl. 1961) M	**Kalani**
Kalb, Klaus (1942–) M	**Kalb**
Kalbe, Lothar A	**Kalbe**
Kalchbrenner, Károly (Karl) (1807–1886) M	**Kalchbr.**
Kale, J.C. (fl. 1971) M	**J.C.Kale**
Kale, S.B. (fl. 1965) M	**S.B.Kale**
Kale, Sou V.S. (fl. 1971) M	**S.V.S.Kale**
Kale, Sudha R. A	**S.R.Kale**
Kalela, Aimo Aarno Antero (before 1935, Cajander, A.A.A.) (1808–1977) S	**Kalela**
(Kalenichenko, M.G.)	
see Kaleniczenko, M.G.	**M.G.Kalen.**
Kaleniczenko, Ivan Osipovich (1805–1876) S	**Kalen.**
Kaleniczenko, M.G. (1931–) S	**M.G.Kalen.**
Kaliaperumal, N. A	**Kaliap.**
Kalina, Tomás (1935–) A	**Kalina**
Kalinsky, Robert George (1945–) A	**Kalinsky**
Kalkat, R.S. (fl. 1984) M	**Kalkat**
Kalkman, Cornelis (1928–) S	**Kalkman**
Kallénbach, Franz Joseph (1893–1944) M	**Kallenb.**
Källersjö, Mari (1954–) S	**Källersjö**
Kallio, Paavo Pauli (1914–) AS	**Kallio**
Kallunki, Jacquelyn Ann (1948–) S	**Kallunki**
Kalm, Matthias (1973–1833) S	**M.Kalm**
Kalm, Pehr (1716–1779) S	**Kalm**
Kalmbacher, George Anthony (1897–1977) S	**Kalmb.**
Kalmuss, Friedrich (1843–1910) B	**Kalmuss**
Kalopissis, J. (fl. 1980) S	**Kalop.**
Kalteisen, M. (fl. 1987) S	**Kalteisen**
Kaltenbach, Johann Heinrich (1807–1876) S	**Kaltenb.**
Kalugina, Alexandra Arkhipovna (1929–) A	**Kalugina**
Kalymbetov, B.K. (fl. 1952) M	**Kalymb.**

Kalymbetov, V. S	V.Kalymb.
(Kam, M.De)	
see De Kam, M.	De Kam
Kam, Yee Kiew (–1981) S	Y.K.Kam
Kamal (fl. 1963) M	Kamal
Kamari, Georgia (1943–) S	Kamari
Kamat, M.N. (1897–1980) M	Kamat
Kamat, N.D. A	N.D.Kamat
Kambayashi, Toyoaki (–1939) M	Kambay.
Kamble, S.Y. (1940–) S	Kamble
Kambly, Paul E. (fl. 1936) M	Kambly
Kamei, Senji (1893–) M	Kamei
Kamel, Georg Joseph (1661–1706) L	Kamel
Kamelin, R.V. (1938–) S	Kamelin
Kamenski, F.M. (fl. 1899) M	Kamenski
Kamerling, Zeno (1872–) S	Kamerling
Kamibayashi, Keijiro (fl. 1915) S	Kamib.
Kamieński, Franciszek (Frans) Michailow von (1851–1912) S	Kamieński
Kamieński, T. (fl. 1899) M	T.Kamieński
(Kamieńsky, Franciszek (Frans) Michailow von)	
see Kamieński, Franciszek (Frans) Michailow von	Kamieński
(Kamieńsky, T.)	
see Kamieński, T.	T.Kamieński
Kamikoti, Sizuka (1910–) S	Kamik.
Kamilov, R.D. (fl. 1972) M	Kamilov
Kamimura, Minoru (1909–) B	Kamim.
Kaminski, Elfriede A	Kaminski
Kamjaipai, W. (fl. 1971) M	Kamj.
Kammathy, R.V. (1932–) S	Kammathy
Kämmer, Franco (von) (1945–) S	Kämmer
Kammerer, Gertraud A	Kammerer
Kampe, E. (fl. 1888) S	Kampe
Kämpf, A.N. (fl. 1975) S	Kämpf
Kämpfer, Engelbert	
see Kaempfer, Engelbert	Kaempf.
Kampmann, Frédéric-Edouard (1797–1814) S	Kampm.
Kampmann, Frédéric-Edouard ('Fritz') (1830–1914) S	Kampm.f.
Kamptner, Erwin (1889–) A	Kamptner
Kamÿschko, O.P. (fl. 1960) M	Kamÿschko
(Kamyschtko, O.P.)	
see Kamÿschko, O.P.	Kamyschko
(Kamyshko, O.P.)	
see Kamÿschko, O.P.	Kamyschko
Kanai, Hiroo (1930–) S	Kanai
Kanaya, Taro A	Kanaya
(Kancaveli, Zaiharias A.)	
see Kantschaweli, Zaiharias A.	Kantsch.
(Kanchaveli, L.(A.))	
see Kantschaweli, L.(A.)	L.A.Kantsch.

(Kanczaveli, Zaiharias A.)
see Kantschaweli, Zaiharias A.   **Kantsch.**
Kanda, Hiroshi (1946–) B   **Kanda**
Kandaswamy, M. (fl. 1955) M   **M.Kandasw.**
Kandaswamy, T.K. (fl. 1977) M   **T.K.Kandasw.**
Kandinskaya, L.I. (fl. 1971) M   **Kandinsk.**
Kane, D.F. (fl. 1976) M   **D.F.Kane**
Kane, J. (fl. 1977) M   **J.Kane**
Kane, Katharine Sophia Bailey (1811–1886) S   **Kane**
Kanehira, Ryôzô (1882–1948) PS   **Kaneh.**
Kaneko, Shigeru (fl. 1977) M   **S.Kaneko**
Kaneko, Takashi A   **T.Kaneko**
Kanér, Oskar Richard (1878–) S   **Kanér**
Kanes, William H. (1934–) A   **Kanes**
Kang, M.S. (fl. 1968) M   **Kang**
Kanis, Andrias (Andries, Andrew) (1934–1986) S   **Kanis**
Kanitkar, U.K. (fl. 1975) M   **Kanitkar**
Kanitz, August (Agoston, Agost) (1843–1896) AMPS   **Kanitz**
Kanjilal, Praphulla Chandra (1886–1972) S   **P.C.Kanjilal**
Kanjilal, Upendranath N. (1859–1928) S   **Kanjilal**
Kann, Edith (1907–1987) A   **Kann**
Kano, K. (fl. 1937) M   **Kano**
Kanouse, Bessie Bernice (1889–1969) M   **Kanouse**
Kanthamma, S. A   **Kanth.**
(Kantschaveli, L.(A.))
see Kantschaweli, L.(A.)   **L.A.Kantsch.**
Kantschaweli, L.(A.) (fl. 1928) M   **L.A.Kantsch.**
Kantschaweli, Zaiharias A. (1894–1932) S   **Kantsch.**
Kantvilas, Ginteras (1956–) M   **Kantvilas**
Kanuma, Mosaburô A   **Kanuma**
Kanwal, H.S. (fl. 1979) B   **Kanwal**
Kanzawa, S. (fl. 1970) M   **Kanzawa**
Kao, Muh Tsuen (1925–) S   **M.T.Kao**
Kao, Pao Chun (1935–) S   **P.C.Kao**
Kao, Tso Ching (1926–) S   **T.C.Kao**
Kapadia, Zarir Jamasji (née Patel, V.) (1935–) S   **Kapadia**
Kapanadze, D.A. (fl. 1985) S   **D.A.Kapan.**
Kapanadze, I.S. (fl. 1985) S   **I.S.Kapan.**
Kapanadze, M.B. (1952–) S   **Kapan.**
Kapeller, Olga Antonovna (1892–1975) S   **Kapeller**
Kapellos, Christos A   **Kapellos**
Kapétanidis, I. S   **Kapét.**
Kaplan, William (fl. 1969) M   **Kaplan**
Kapoor, Brij (Brig) Mohan (1936–) S   **B.M.Kapoor**
Kapoor, Jagmohan Nath (1930–) M   **J.N.Kapoor**
Kapoor, S.L. (1931–) S   **S.L.Kapoor**
Kapoor, Saroj (fl. 1970) M   **S.Kapoor**
Kappert, Hans (1890–) S   **Kappert**
Kappler, August (1815–1887) S   **Kappl.**

Kappus, A. S	**Kappus**
Kapraun, Donald Frederick (1945–) A	**Kapraun**
(Kapšanaki Gotsi, Evangelia)	
see Kapšanaki–Gotsi, Evangelia	**Kapš.-Gotsi**
Kapšanaki–Gotsi, Evangelia (1950–) M	**Kapš.-Gotsi**
Kar, A.K. (fl. 1968) M	**A.K.Kar**
Kar, R.A. (fl. 1969) FM	**R.A.Kar**
Kar, Ranajit Kumar (1936–) AM	**R.K.Kar**
Karaca, Ibrahim (1926–) M	**Karaca**
Karaeva, N.I. A	**Karaeva**
Karakulin, Boris Palladiyerrich (1888–1942) M	**Karak.**
Karamboloff (fl. 1931) M	**Karamb.**
Karan, D. (fl. 1964) M	**Karan**
Karandikar, K.R. A	**Karand.**
Karasawa, Kôtarô (fl. 1982) S	**Karas.**
Karasjuk S	**Karasjuk**
Karatygin, I.V. (fl. 1989) M	**Karatygin**
Karavaev, Mikhail Nikolaevich (1903–) S	**Karav.**
Kärcher, Reinhold (fl. 1987) M	**Kärcher**
Karczewska, H. (fl. 1969) M	**Karcz.**
Karczewska, Jadwiga A	**J.Karcz.**
Karczmarz, Kazimierz (1933–) B	**Karczm.**
Karegeannes, C. (fl. 1975) S	**Kareg.**
Karel, Güngör (fl. 1947) M	**Karel**
Karelin, Grigorij Silyč (Gregor Silič (Silitsch, Siliovitsch)) (1801–1872) S	**Kar.**
(Karemy, Zeinab A.R. El)	
see El Karemy, Zeinab A.R.	**El Karemy**
Kargupta, Amarendra Nath A	**Kargupta**
Karhu, Niilo S	**Karhu**
Karim, Abdel G.A. A	**Karim**
Karimov, Mir–Kadyr Absalovich (1906–) M	**Karimov**
Karimova, V.V. (fl. 1978) S	**Karimova**
Karis, H.(Kh.) (fl. 1980) M	**H.Karis**
Karis, Per Ola (1955–) S	**P.O.Karis**
Karjagin, Ivan Ivanovič (1894–1966) S	**Karjagin**
Karl, János (1842–1882) S	**Karl**
Karlén, Thomas (fl. 1983) S	**Karlén**
Karling, John Sidney (1898–) AM	**Karling**
Karlsson, Thomas (1945–) S	**Karlsson**
Karlström, Per–Olof (fl. 1975) S	**Karlström**
Karmyscheva, N.H. (1913–) S	**Karmysch.**
Kärnefelt, Ingvar (1944–) M	**Kärnefelt**
Kárpáti, Zoltan E. (1909–1972) S	**Kárpáti**
(Karpinskii, Aleksandr Petrovič)	
see Karpinsky, Aleksandr Petrovich	**Karpinsky**
Karpinsky, Aleksandr Petrovich (1846–1936) A	**Karpinsky**
(Karpisonova, Rimma A.)	
see Karpissonova, Rimma A.	**Karpiss.**
Karpissonova, Rimma A. (1931–) S	**Karpiss.**

Karpova–Benua, E.I. (fl. 1973) M	**Karp.-Benua**
Karr, G.W. (fl. 1976) M	**Karr**
Karrenberg, C.L. (fl. 1928) M	**Karrenb.**
Karrer, Sigmund (1881–1954) S	**Karrer**
Karsakoff, N. A	**Karsakoff**
Karsch, Anton (1822–1892) S	**Karsch**
Karsten, (Maria) Caroline (1902–) S	**M.C.Karst.**
Karsten, Gustav Karl Wilhelm Hermann (1817–1908) AMPS	**H.Karst.**
Karsten, George Henry Hermann (1863–1937) A	**G.Karst.**
Karsten, Petter (Peter) Adolf (1834–1917) M	**P.Karst.**
Kartesz, John T. (fl. 1990) S	**Kartesz**
Karthikeyan, Satavanam (1940–) S	**Karth.**
Karttunen, Krister (1960–) B	**Kartt.**
Karwacki, Léon (fl. 1911) M	**Karwacki**
Karwinsky von Karwin, Wilhelm Friedrich von (1780–1855) S	**Karw.**
Karyagin, Ivan Ivanovich (1894–1966) S	**Karyagin**
Kasach, A.E. (fl. 1970) S	**Kasach**
Kasai, Mikio (–1944) M	**Kasai**
Kasaki, Hideo (né Morioka, H.) (1917–) A	**Kasaki**
Kasanowsky, Viktor Ivanovich (–1920/21) A	**Kasan.**
Kasapligil, Baki (1918–) S	**Kasapligil**
Kaschina, L.I. (1929–) S	**Kaschina**
Kaschmensky, B.F. (–1907) S	**Kaschm.**
Käser, Friedrich (1853–1915/45) S	**Käser**
Kashina, L.I. (fl. 1986) S	**Kashina**
Kashiwadani, Hiroyuki (1944–) M	**Kashiw.**
Kashtanova, A.E. A	**Kasht.**
Kashyap, Shiv Ram (1882–1934) B	**Kashyap**
Kashyapa, Kamleshwar (fl. 1965) S	**Kashyapa**
Kaska, Harold Victor (1926–) A	**Kaska**
Käsler, Robert A	**Käsler**
Kasper, Andrew Edward (1942–) F	**Kasper**
Kass, R.J. (fl. 1985) S	**Kass**
Kassacz, A.E. S	**Kassacz**
Kassau, Erich (1903–) S	**Kassau**
Kassumov, F.Ju. (1939–) S	**Kassumov**
Kassumova, T.A. (fl. 1985) S	**Kassumova**
Kästner, A. (1936–) S	**Kästner**
(Kaszakewich, L.J.)	
see Kazakevicz, L.I.	**Kazak.**
Katagiri, H. (fl. 1950) M	**Katag.**
Katajev, I.A. (fl. 1952) M	**Katajev**
Katenin, A.E. (1935–) S	**Katenin**
Kater, John McAllister (1901–) A	**Kater**
Katić, Danilo Ljubissawić (1873–) B	**Katić**
Katina, Z.F. (fl. 1952) S	**Katina**
Kato, Masahiro (1946–) PS	**M.Kato**
Kato, Sueo A	**S.Kato**
Kato, Tadayuki (fl. 1987) S	**T.Kato**

Katoă, H. (fl. 1929) M	H.Katoă
Katoă, Y. (fl. 1926) M	Y.Katoă
Katsuki, Shigetaka (fl. 1952) M	Katsuki
Katsura, Kiichi (fl. 1971) M	Katsura
Kattermann, F. (fl. 1983) S	Katt.
Katumoto, Ken (1927–) M	Katum.
Katz, B. (fl. 1974) M	Katz
Katznelson, J. (fl. 1958–72) S	Katzn.
Kauffman, Calvin Henry (1869–1931) MS	Kauffman
Kauffmann, Gary (fl. 1988) M	G.Kauffm.
Kauffmann, Nikolai Nikolajevich (1834–1870) S	Kauffm.
Kaufman, L. (fl. 1987) M	Kaufman
(Kaufman, Nikolai Nikolajevich)	
see Kauffmann, Nikolai Nikolajevich	Kauffm.
Kaufmann, M.J. (fl. 1967) M	Kaufm.
Kaul, J.L. (fl. 1988) M	J.L.Kaul
Kaul, Kailash Nath (1905–1983) S	Kaul
Kaul, M.K. (fl. 1984) S	M.K.Kaul
Kaul, Veriendranath P. (1934–) M	V.P.Kaul
Kaulfuss, Georg Friedrich (1786–1830) BP	Kaulf.
Kaulfuss, Johannes (Johann) Simon (–1947) BM	J.S.Kaulf.
Kaulich, L. (fl. 1948) M	Kaulich
Kaur, Surjit (1936–) P	S.Kaur
Kaur, Swaru Jeet (fl. 1975) MPS	S.J.Kaur
Kaurin, Christian (1831–1898) B	Kaurin
Kausel, Eberhard Max Leopold (1910–1972) S	Kausel
Kaushal, R. (fl. 1982) M	R.Kaushal
Kaushal, S.C. (fl. 1978) M	S.C.Kaushal
Kava, I.B. S	Kava
Kavanagh, James Aloysius (1937–) M	Kavanagh
Kavanagh, K.P. (fl. 1982) P	K.P.Kavanagh
Kaveriappa, K.M. (fl. 1991) M	Kaver.
Kavina, Karel (1890–1948) BMS	Kavina
Kavka, Bohumil (1901–1977) S	Kavka
Kavkasidze, D.K. (fl. 1953) M	Kavkas.
Kawabata, Seisaku (1906–1985) A	Kawab.
Kawabe, K. (fl. 1980) M	Kawabe
Kawagoe, Sh. (fl. 1916) M	Kawagoe
Kawai, Katsumi (fl. 1932) M	Kawai
Kawakami, Hiroshi (1955–) AM	H.Kawak.
Kawakami, Noboru (fl. 1955) M	N.Kawak.
Kawakami, Takiya (1871–1915) M	Kawak.
Kawamoto, Isao (fl. 1985) M	Kawamoto
Kawamura, A. (fl. 1958) M	A.Kawam.
Kawamura, Seiichi (–1946) M	Kawam.
Kawano, Shoichi (1936–) S	Kawano
Kawasaki, Hiroko (fl. 1988) M	H.Kawas.
Kawasaki, Tetsuya (fl. 1959) S	Kawas.
Kawase, Y. (fl. 1954) M	Kawase

Kawatani, Toyohiko (fl. 1976) M	**Kawat.**
Kawatsure, S. (fl. 1933) M	**Kawats.**
Kawchuk, L.M. (fl. 1988) M	**Kawchuk**
Kawecki, Zbigniew S	**Kawecki**
Kay, Quentin Oliver Newton (1939–) S	**Kay**
Kayser, E. (fl. 1892) M	**Kayser**
Kayser, Konrad (fl. 1932) S	**K.Kayser**
Kazakevicz, I.I. (1893–) S	**Kazak.**
Kazakova, A.A. (1917–) S	**Kazakova**
(Kazámierczak, J.)	
see Kaźmierczak, J.	**Kaźm.**
Kazarjan, Enok S. S	**Kazarjan**
Kazenas, L.D. (fl. 1959) M	**Kazenas**
Kazimierski, T. (1924–) S	**Kazim.**
Kazmi, Syed Muhammad Anwar (1926–) S	**Kazmi**
Kaźmierczak, J. A	**Kaźm.**
Kaznowski, L. (fl. 1925) M	**Kazn.**
Ke, Ping (1925–) S	**P.Ke**
Keane, P.J. (fl. 1971) M	**Keane**
Kearney, Thomas Henry (1874–1956) S	**Kearney**
Keating, Richard Clark (1937–) S	**R.C.Keating**
Keating, William Hippolitus (Hypolitus) (1799–1840) S	**Keating**
Keay, Ronald William John (1920–) S	**Keay**
Keble Martin, William (1877–1969) S	**Keble Martin**
Kechekmadze, L.A. (fl. 1965) M	**Kechekm.**
Keck, David Daniels (1903–) S	**D.D.Keck**
Keck, Karl (1825–1894) S	**Keck**
Kedves, M. A	**Kedves**
Keefe, Anselm Maynard (1895–) A	**Keefe**
Keeler, Charles A. A	**Keeler**
Keeler–Wolf, Tod (fl. 1983) S	**Keeler–Wolf**
Keeley, Frank James (1868–1949) A	**Keeley**
Keeley, Sterling C. (1948–) S	**S.C.Keeley**
Keenan, James (1924–1983) S	**Keenan**
Keener, Carl Samuel (1931–) S	**Keener**
Keeping, Walter (1854–1888) A	**Keeping**
Kegel, Hermann Aribert Heinrich (1819–1856) S	**Kegel**
Kegel, W. (fl. 1906) M	**W.Kegel**
Keighery, Gregory John (1950–) S	**Keighery**
Keijzer, Frans Gaspard A	**Keijzer**
Keil, David John (1946–) S	**D.J.Keil**
Keil, Franz R. (1822–1876) B	**Keil**
Keil, Rudolf A	**R.Keil**
Keilin, D. (fl. 1921) M	**Keilin**
Keissler, Karl (Carl) von (1872–1965) ABMS	**Keissl.**
Keith, James (1825–1905) M	**Keith**
Kekhlibarova, L. (fl. 1980) M	**Kekhlib.**
Kelaart, Edward Frederick (Eduard Frederik) (1818?–1860) A	**Kelaart**
(Kelbessa, Ensermu)	
see Ensermu Kelbessa	**Ensermu**

319

Keld, Edvard (1867–1945) S	**Keld**
Kelhofer, Ernst (1877–1917) S	**Kelh.**
Kelkar, P.V. (fl. 1963) M	**Kelkar**
Kelkar, S.S. (fl. 1986) M	**S.S.Kelkar**
Keller, Alfred (1849–1925) S	**A.Keller**
Keller, Allan Charles (1914–) S	**A.C.Keller**
Keller, Barbara Thomas (1946–) S	**B.T.Keller**
Keller, Boris Aleksandrovich (Alexandrovic) (1874–1945) S	**B.Keller**
Keller, Gerwin (fl. 1982) M	**Gerw.Keller**
Keller, Gottfried (1873–1945) S	**G.Keller**
Keller, Harold Willard (1937–) M	**H.W.Keller**
Keller, Ida Augusta (1866–1932) S	**I.Keller**
Keller, J. (1917–1945) S	**J.Keller**
Keller, Jean (1941–) M	**Jean Keller**
Keller, Jenó B.von (1841–1897) S	**J.B.Keller**
Keller, Johann Christoph (1737–1796) S	**Keller**
Keller, Louis (1850–1915) PS	**L.Keller**
Keller, Robert (1854–1939) BPS	**R.Keller**
Keller, S. (fl. 1980) M	**S.Keller**
Keller-Schierlein, W. (fl. 1957) M	**Kell.-Schierl.**
Kellerer, Johann (1859–) S	**Kellerer**
Kellerman, Maude (1888–) S	**M.Kellerm.**
Kellerman, William Ashbrook (1850–1908) MS	**Kellerm.**
Kelley, W.A. (fl. 1986) S	**Kelley**
Kellner, Gisela A	**Kellner**
Kellogg, Albert (1813–1887) PS	**Kellogg**
Kellogg, Elizabeth Anne ('Toby') (1951–) S	**E.A.Kellogg**
Kellogg, Royal Shaw (1874–) S	**R.S.Kellogg**
Kellogg, Vernon Myman Lyman (1867–1937) S	**V.M.L.Kellogg**
Kelly, Clara J. (1909–) B	**C.J.Kelly**
Kelly, Howard Atwood (1858–1943) M	**Kelly**
Kelly, James Peter (1885–1955) S	**J.Kelly**
Kelsey, Beth Low S	**B.L.Kelsey**
Kelsey, Francis Duncan (1849–1905) AM	**Kelsey**
Kelsey, Harlan Page (1872–1959) S	**H.P.Kelsey**
Kelso, E.H. S	**E.H.Kelso**
Kelso, Leon Hugh (1907–) S	**Kelso**
Kelso, Sylvia (1953–) S	**S.Kelso**
Kelway, James (1815–1899) S	**Kelway**
Kemmler, Carl (Karl) Albert (1813–1888) S	**Kemmler**
Kemp, Elizabeth M. (fl. 1960) FM	**Kemp**
Kemp, Klaus–D. A	**K.-D.Kemp**
Kemperman, T.C.M. (fl. 1984) AM	**Kemperman**
Kempton, Phyllis E. (fl. 1968) M	**Kempton**
(Kemularia Natadze, Liubov Manucharovna)	
see Kemularia–Nathadze, Liubov Manucharovna	**Kem.-Nath.**
Kemularia–Nathadze, Liubov Manucharovna (1891–1985) S	**Kem.-Nath.**
Kendrick, James Blair (1893–1962) M	**J.B.Kendr.**
Kendrick, William Bryce (1933–) M	**W.B.Kendr.**

Kenfield, Dougl (fl. 1991) M   **Kenfield**
Keng, Hsüan (1923–) S   **H.Keng**
(Keng, Kwan Hou)
    see Keng, Pai Chieh   **Keng f.**
Keng, Pai Chieh (1917–) S   **Keng f.**
Keng, Yi Li (1897–1975) S   **Keng**
Kenneally, Kevin Francis (1945–) S   **Kenneally**
Kennedy, George C. (fl. 1976) S   **G.C.Kenn.**
Kennedy, George Golding (1841–1918) B   **Kenn.**
Kennedy, Helen Alberta (1944–) S   **H.A.Kenn.**
Kennedy, James Domoné (1898–) S   **J.D.Kenn.**
Kennedy, John (1759–1842) S   **J.Kenn.**
Kennedy, Lorene L. (fl. 1959) M   **L.L.Kenn.**
Kennedy, Patrick Beveridge (1874–1930) S   **P.B.Kenn.**
Kennedy–O'Byrne, John Kevin Patrick (1927–) S   **Kenn.-O'Byrne**
Kennelly, Violet C.E. (fl. 1930) M   **Kennelly**
Kenneth, Archibald Graham (1915–1989) S   **Kenneth**
Kenneth, Robert G. (fl. 1977) M   **R.G.Kenneth**
Kensit, Harriet Margaret Louisa (later Bolus, H.M.L.) (1877–1970) S   **Kensit**
Kent, Adolphus Henry (1828–1913) S   **A.H.Kent**
Kent, Douglas Henry (1920–) S   **D.H.Kent**
Kent, Leslie E. A   **L.E.Kent**
Kent, William Saville (1845–1908) A   **Kent**
Kenyon, William (fl. 1847) S   **Kenyon**
Keppel, Johannes Cornelius van (1922–1982) S   **Keppel**
Ker, Charles Henry Bellenden (1785–1871) S   **Ker**
(Ker, John Bellenden)
    see Ker Gawler, John Bellenden   **Ker Gawl.**
Ker Gawler, John Bellenden (1764–1842)   **Ker Gawl.**
Keraudren, Monique (1928–1981) S   **Keraudren**
(Keraudren–Aymonin, Monique)
    see Keraudren, Monique   **Keraudren**
Kerber, Edmund (fl. 1882–83) S   **Kerber**
Kerchove de Denterghem, Oswald Charles Eugène Marie Ghislain de
    (1844–1906) S   **Kerch.**
Kereszty, Zoltán (1937–) S   **Kereszty**
Kerguélen, Michel François–Jacques (1928–) S   **Kerguélen**
Kerimova, R.S. S   **Kerimova**
Kerken, Amelia E.van (fl. 1960) M   **Kerken**
Kern, Frank Dunn (1883–1973) M   **F.Kern**
Kern, Friedrich (1850–1925) B   **Kern**
Kern, Hartmut (1929–) M   **H.Kern**
Kern, Johannes Hendrikus (1903–1974) S   **J.Kern**
Kern, Patricia M. (later Holmgren, P.K.) (1940–) S   **P.M.Kern**
Kernan, M.J. (fl. 1983) M   **Kernan**
Kerner, Anton Joseph (1831–1898) BPS   **A.Kern.**
Kerner, Johann Simon von (1755–1830) S   **J.Kern.**
Kerner, Josef (1829–1906) S   **Jos.Kern.**
(Kerner von Marilaun, Anton Joseph)
    see Kerner, Anton Joseph   **A.Kern.**

Kerner von Marilaun, Friedrich (1866–) S	**F.Kern.**
Kernstock, Ernst (1852–1900) M	**Kernst.**
Kerp, Hans F	**Kerp**
Kerpel, D.A. S	**Kerpel**
Kerr, Allen D. (fl. 1973) S	**A.D.Kerr**
Kerr, Arthur Francis G. (1877–1942) S	**Kerr**
Kerr, K.M. (fl. 1985) S	**K.M.Kerr**
Kerr, Lesley Ruth (1900–1927) S	**L.R.Kerr**
Kerrigan, Richard W. (fl. 1985) M	**Kerrigan**
Kerry, B.R. (fl. 1980) M	**Kerry**
Kers, Lars Erik (1931–) MS	**Kers**
Kerschen, P. (fl. 1977) M	**Kerschen**
Kershaw, E.M. (fl. 1910) F	**Kershaw**
Kersten, Otto (1839–1900) P	**Kerst.**
Keshava Murthy, R.K. (fl. 1987) S	**Kesh.Murthy**
Keskin, B. (fl. 1964) M	**Keskin**
Kesler, Willy (fl. 1955) S	**Kesler**
(Kessel, Stephen Lackey)	
see Kessell, Stephen Lackey	**Kessell**
Kessell, Stephen Lackey (1897–1979) S	**Kessell**
Kesselring, F.W. (1876–1966) S	**Kesselr.**
Kesselring, Jakob (1835–1909) S	**J.Kesselr.**
Kessler, Dietrich (fl. 1977) M	**D.Kessler**
Kessler, Erich Eduard (1927–) A	**E.E.Kessler**
Kessler, J.W. (fl. 1987) S	**J.W.Kessler**
Kessler, Kenneth J. (fl. 1984) M	**K.J.Kessler**
Kessler, Patricia J.A. (1948–) S	**P.J.A.Kessler**
Kesteren, H.A.van (fl. 1972) M	**Kesteren**
Kesteven, Hereward Leighton (fl. 1920s–1940s) M	**Kesteven**
Ketchledge, Edwin Herbert (1924–) B	**Ketchl.**
Ketskoveli, Nikolay (Nikolai) Nikolaevich (1897–1982) S	**Ketsk.**
(Ketzchoveli, Nikolaya Nikolaevič)	
see Ketskoveli, Nikolay (Nikolai) Nikolaevich	**Ketsk.**
Keupp, H. A	**Keupp**
Kevorkian, Arthur George (1905–) M	**Kevorkian**
Keys, Isaiah Waterloo Nicholson (1818–1890) S	**Keys**
Keyserling, (Andreëvich) Alexander Friedrich Michael Leberecht Arthur von (1815–1891) PS	**Keyserl.**
Keysselitz, G. A	**Keyss.**
Khakhina, A.G. A	**Khakhina**
Khakhlov, V.A. A	**Khakhlov**
Khakimova, A.G. (fl. 1982) S	**Khak.**
(Khaleel, Fathima Tasneem)	
see Fathima Khaleel, Tasneem	**Fathima**
Khalfina, N.A. A	**Khalfina**
Khalilov, E.Kh. (1913–) S	**Khalilov**
Khalis, N. (fl. 1982) M	**Khalis**
Khalkuziev, P. (fl. 1970) S	**Khalk.**
Khamidchodzhaev, S.A. S	**Khamidch.**

Khan, A.Z.M.Nowsher A. (fl. 1976) M	N.Khan
Khan, Azmatullah (fl. 1944) M	A.Khan
Khan, Inam Ullah (fl. 1962) M	I.U.Khan
Khan, K.R. (fl. 1979) M	K.R.Khan
Khan, M. A	M.Khan
Khan, M.A. A	M.A.Khan
Khan, M.K. (fl. 1989) M	M.K.Khan
Khan, Mohammad (Mohan) Salar (1924–) S	M.S.Khan
(Khan, Mohammad Abdul Hafiz)	
see Hafiz Khan, Mohammad Abdul	Hafiz Khan
Khan, Mohammed L.S. (1924–) S	M.L.S.Khan
Khan, R.S. (1938–) M	R.S.Khan
Khan, S.M. (fl. 1979) M	S.M.Khan
Khan, S.N. (fl. 1987) M	S.N.Khan
Khan, S.R. (fl. 1986) M	S.R.Khan
Khan, Shakil Ahmad (fl. 1962) M	S.A.Khan
Khan, Shama–ul–Islam A	S.Khan
Khan, Sultan Ahmad (1928–) B	Sultan Khan
Khan, Taj Malook A	T.M.Khan
Khan, Y.S.A. A	Y.S.A.Khan
Khan, Z.R. (fl. 1983) M	Z.R.Khan
Khandelwal, Sharda (1949–) P	Khand.
(Khandzhyan, Nasik S.)	
see Chandjian, Nasik S.	Chandjian
Khangura, Ravjit K. (fl. 1990) M	Khangura
Khánh, Trân Công (1936–) S	Khánh
Khani, K. (fl. 1987) S	Khani
Khanminchun, V.M. (1945–) S	Khanm.
Khanna, Ashok K. (1951?–1984) A	A.K.Khanna
Khanna, Lalit Prasad (–1946) B	Khanna
Khanna, M. (fl. 1980) M	M.Khanna
Khanna, P.K. (fl. 1963) M	P.K.Khanna
Khara, H.S. (fl. 1988) M	Khara
Kharadze, Anna Lukianovna (1905–1971) S	Kharadze
Kharbjueova, E.D. (1895–1943) S	Kharb.
Khare, K.B. (fl. 1972) M	K.B.Khare
Khare, P.K. S	P.K.Khare
Kharkevich, Sigismund Semenovich (1921–) S	Kharkev.
Khassanov, O.Ch. (1928–) S	Khass.
Khatamsaz, Maboubeh (1949–) S	Khat.
Khatoon, Surayya (1957–) MS	Khatoon
Khatri, K.S. (fl. 1986) S	Khatri
Khattab, Ahmed (1907–) S	Khattab
Khawkine, W. A	Khawkine
Khazanoff, Amram (1890–) M	Khaz.
Khek, Eugen Johan (1861–1927) S	Khek
Khep, N.T. (fl. 1981) S	Khep
Kheswalla, Kavasji Framaji (fl. 1941) M	Khesw.
Khetarpal, R.K. (fl. 1984) M	Khetarpal

(Khinchuk, A.G.)
  see Chintchuk, A.G.   **Chintchuk**
(Khinthibze, Leonida S.)
  see Chinthibidze, Leonida S.   **Chinth.**
(Khinthjbidze, Leonida S.)
  see Chinthibidze, Leonida S.   **Chinth.**
(Khintibidze, Leonida S.)
  see Chinthibidze, Leonida S.   **Chinth.**
Khmeleva, N.N. A   **Khmeleva**
Khôi, Nguyên Dang (fl. 1963) S   **N.D.Khôi**
Khoi, Nguyen Khac (fl. 1980) S   **N.K.Khoi**
(Khoie, Karim Djavanchir)
  see Djavanchir–Khoie, Karim   **Djav.-Khoie**
Khokhrjakov, A.P. (1933–) PS   **A.P.Khokhr.**
Khokhrjakov, Michael Kuzmich (1905–) M   **Khokhr.**
(Khokhryakoff, Michael Kuzmich)
  see Khokhrjakov, Michael Kuzmich   **Khokhr.**
(Khokhryakov, A.P.)
  see Khokhrjakov, A.P.   **A.P.Khokhr.**
Kholia, B.S. (fl. 1988) P   **Kholia**
Khoon, Meng Wong (fl. 1982) P   **Khoon**
(Khouri, J.)
  see Khoury, J.   **Khoury**
Khoury, J. (fl. 1902) M   **Khoury**
Khristyuk, P.M. A   **Khristyuk**
(Khrzhanovsky, Vladimir Gennadievich)
  see Chrshanovski, Vladimir Gennadievich   **Chrshan.**
(Khrzhanowski, Vladimir Gennadievich)
  see Chrshanovski, Vladimir Gennadievich   **Chrshan.**
Khulbe, R.D. (fl. 1977) M   **Khulbe**
Khullar, Surinder Pal (fl. 1976) PS   **Khullar**
Khune, N.N. (fl. 1978) M   **Khune**
Khurana, I.P.S. (fl. 1977) M   **Khurana**
Khvorova, Irina Vasilevna A   **Khvorova**
Kiaer, Frantz Caspar (1835–1893) B   **Kiaer**
(Kiaerskou, Hjalmar Frederik Christian)
  see Kiaerskov, Hjalmar Frederik Christian   **Kiaersk.**
Kiaerskov, Hjalmar Frederik Christian (1835–1900) PS   **Kiaersk.**
Kickx, Jean (1775–1831) S   **J.Kickx**
Kickx, Jean (1803–1864) ABMPS   **J.Kickx f.**
Kickx, Jean Jacques (1842–1887) M   **J.J.Kickx**
Kidd, David Eugene (1930–) A   **D.E.Kidd**
(Kidd, Franklin)
  see Kidd, Mary Nest   **Kidd**
Kidd, Mary Nest (née Owen, M.N.) (1890–1974) M   **Kidd**
Kidson, R. (fl. 1921) M   **Kidson**
Kidston, Robert (1852–1924) AFS   **Kidst.**
Kieffer, Frédéric (1827–1927) S   **F.Kieff.**
Kieffer, Jean-Jacques (1857–1925) M   **Kieff.**

Kiehn, M. (fl. 1986) S	**Kiehn**
(Kiellberg, Gunnar Konstantin)	
see Kjellberg, Gunnar Konstantin	**Kjellb.**
Kielmeyer, Carl Friedrich von (1765–1844) S	**Kielm.**
Kiely, Temple B. (fl. 1948) M	**Kiely**
Kiem, J. (fl. 1981) S	**Kiem**
Kienholz, Jesse Reuben (1904–) M	**Kienholz**
Kienitz–Gerloff, Johann Heinrich Emil Felix (1851–1914) S	**Kien.-Gerl.**
Kierulff (fl. 1852) S	**Kierulff**
Kies, Ludwig A	**L.Kies**
Kies, Pauline (later Bohnen) (1918–) S	**Kies**
Kieser, Dietrich Georg von (1779–1862) S	**Kieser**
Kiesling, Christian Gotthilf (1724–1754) S	**Kiesling**
Kiesling, Roberto (1941–) S	**R.Kiesling**
Kiew, Ruth (fl. 1989) S	**Kiew**
Kiffer, E. (fl. 1970) M	**Kiffer**
Kiffmann, R. (fl. 1952) S	**Kiffmann**
Kiger, Robert William (1940–) S	**Kiger**
Kiggelaer, François (Franciscus) (1648–1722) LS	**Kiggel.**
Kihlman, Alfred Oswald (1858–1938) MS	**Kihlm.**
Kikuchi, Akio (1883–1951) S	**Kikuchi**
Kikuchi, Masao (1908–1969) S	**M.Kikuchi**
Kilani, A. (fl. 1962) M	**Kilani**
Kilbertus, Gerard (fl. 1974) B	**Kilb.**
Kile, G.A. (fl. 1978) M	**Kile**
Kilgus, G. S	**Kilgus**
Kilian, Günter (1960–) S	**Kilian**
Kilian, N. (fl. 1987) S	**N.Kilian**
Kilias, Harald (1949–) M	**H.Kilias**
Kilias, R. (fl. 1981) M	**R.Kilias**
Killermann, Sebastian (1870–1956) M	**Killerm.**
Killian, Charles (1887–1957) M	**Kill.**
Killias, Eduard (1829–1891) BM	**Killias**
Killick, Donald Joseph Boomer (1926–) S	**Killick**
Killip, Ellsworth Paine (1890–1968) PS	**Killip**
(Kilpady, Sripadrao)	
see Rao, K.Sripada	**K.Sr.Rao**
Kilpatrick, R.A. (fl. 1967) M	**Kilp.**
Kim, Dong Ho (1940–) A	**D.H.Kim**
Kim, Jong Won (fl. 1991) S	**J.W.Kim**
Kim, Sang J. (fl. 1974) M	**S.J.Kim**
Kim, Won Hyung A	**W.H.Kim**
Kim, Yun Shik (1934–) S	**Y.S.Kim**
Kimati, H. (fl. 1980) M	**Kimati**
Kimball, James Putnam (1836–1913) F	**Kimball**
Kimbrough, James William (1934–) M	**Kimbr.**
Kimchi, M. (fl. 1987) M	**Kimchi**
Kimeridze, Kukuri Romanovich (1927–) S	**Kimer.**
Kimimura, M. B	**Kimim.**

Kimnach, Myron William (1922–) S	**Kimnach**
Kimura, Arika (1900–) S	**Kimura**
Kimura, G.G. A	**G.G.Kimura**
Kimura, Kojiro (fl. 1940) MS	**K.Kimura**
Kimura, Yojiro (1912–) AS	**Y.Kimura**
Kimyai, Abbas A	**Kimyai**
Kinahan, John Robert (1828–1863) P	**Kinahan**
Kindberg, Nils Conrad (1832–1910) BS	**Kindb.**
Kindra, G.S. A	**Kindra**
Kindt, Christian Sommer (1816–1903) M	**Kindt**
King, Bruce L. (1943–) S	**B.L.King**
King, C.C. (fl. 1991) M	**C.C.King**
King, Chalmers Jackson (1893–1945) M	**C.J.King**
King, Charlotte Maria (1864–1937) M	**C.M.King**
King, D.S. (fl. 1975) M	**D.S.King**
King, George (1840–1909) S	**King**
King, James E. (1940–) M	**J.E.King**
King, Joe Mack (1944–) A	**J.M.King**
King, John W. A	**J.W.King**
King, Robert J. (1945–) A	**R.J.King**
King, Robert Merrill (1930–) S	**R.M.King**
King, Rosemary Anne (1952–) S	**R.A.King**
Kingdon-Ward, Francis (1885–1958) S	**Kingdon-Ward**
(Kingma, J.F.H.van Beyma Thoe)	
see Beyma, J.F.H.van	**J.F.H.Beyma**
(Kingma, T.H.van Beyma Thoe)	
see Beyma, T.H.van	**T.H.Beyma**
Kingman, Chester Cole (1873–1913) B	**Kingm.**
Kingston, John F. (18 –) A	**Kingston**
Kinh, Luong-Cong A	**Kinh**
Kinkar, V.N. A	**Kinkar**
Kinkelin, F. (fl. 1884–1908) F	**Kink.**
Kinney, Abbot (1850–1920) PS	**Kinney**
Kinoshita, K. (fl. 1931) M	**Kinosh.**
Kinoshita-Gouvêa, Luíza Sumiko (1947–) S	**Kin.-Gouv.**
Kinscher, Heinrich S	**Kinscher**
(Kinzicaeva, G.K.)	
see Kinzikaëva, G.K.	**Kinzik.**
Kinzikaëva, G.K. (1931–) S	**Kinzik.**
Kippist, Richard (1812–1882) S	**Kippist**
Kiran, Usha (fl. 1991) M	**Kiran**
Kirby, Harold (1900–1952) A	**H.Kirby**
Kirby, Mary (later Gregg) (1817–1893) S	**Kirby**
Kirby, Robert Stearns (1892–1962) M	**R.S.Kirby**
Kirchheimer, Franz (1911–) A	**Kirchh.**
Kirchhoff, Christina Hedwig (1947–) S	**Kirchhoff**
Kirchner, B. (fl. 1981) S	**B.Kirchn.**
Kirchner, Emil Otto Oskar von (1851–1925) M	**Kirchn.**
Kirchner, Georg (1837–1885) S	**G.Kirchn.**
Kirchner, Leopold Anton (–1879) M	**L.A.Kirchn.**

Kirchner, Walter Charles George (1875–) B	**W.C.G.Kirchn.**
(Kirgizbaeva, Kh.M.	
see Kirgizbajeva, H.M.	**Kirgizb.**
Kirgizbajeva, H.M. (fl. 1986) M	**Kirgizb.**
Kiričkova, A.I. S	**Kiričk.**
Kirikova, N.N. (fl. 1974) M	**Kirikova**
Kirilenko, T.S. (fl. 1967) M	**Kiril.**
(Kirillow, Peter)	
see Kirilov, Peter	**P.Kir.**
Kirilov, Ivan Petrovich ('Johann') (1821–1842) S	**Kir.**
Kirilov, Peter (fl. 1849) S	**P.Kir.**
(Kirilow, Iwan Petrovič ('Johann'))	
see Kirilov, Ivan Petrovich ('Johann')	**Kir.**
Kirjakov, Ivan A	**Kirjakov**
Kirjalova, D.N. (fl. 1935) M	**Kirjalova**
Kirjanov, V.V. A	**Kirjanov**
Kirjanova, E.S. (fl. 1965) S	**Kirjanova**
Kirk, Catherine Mary (1953–) M	**C.M.Kirk**
Kirk, John (1832–1922) S	**J.Kirk**
Kirk, John William Carnegie (1878–1962) S	**J.W.C.Kirk**
Kirk, Paul Michael (1952–) M	**P.M.Kirk**
Kirk, Paul W. (fl. 1966) M	**P.W.Kirk**
Kirk, Thomas (1828–1898) PS	**Kirk**
Kirkbride, Joseph Harold (1943–) S	**J.H.Kirkbr.**
Kirkbride, M.C.G. (fl. 1982) S	**M.C.G.Kirkbr.**
Kirkman, Katherine (fl. 1981) S	**K.Kirkman**
Kirkman, W.B. (fl. 1990) S	**W.B.Kirkman**
Kirkpatrick, James Barrie (1946–) S	**J.B.Kirkp.**
Kirkpatrick, R.S. (fl. 1986) S	**R.S.Kirkp.**
Kirkpatrick, Randolph (1863–1950) A	**Kirkp.**
Kirkup, Donald William (1958–) S	**Kirkup**
Kirkwood, Joseph Edward (1872–1928) S	**Kirkw.**
Kirmis, Max (–1926) S	**Kirmis**
(Kirouac, Joseph Louis Conrad)	
see Marie-Victorin, Joseph Louis Conrad	**Vict.**
(Kirouak, Joseph Louis Conrad)	
see Marie-Victorin, Joseph Louis Conrad	**Vict.**
(Kirpichnikov, Moiseï Elevich)	
see Kirpicznikov, Moisey Elevich	**Kirp.**
Kirpicznikov, Moisey Elevich (1913–) S	**Kirp.**
Kirscher, Jan S	**Kirscher**
Kirschleger, Frédéric R. (1804–1869) PS	**Kirschl.**
Kirschner, J. (fl. 1979) S	**Kirschner**
Kirschstein, Wilhelm (1863–1946) M	**Kirschst.**
Kirtikar, Kanhoba (Kanoba) Ranchoddâs (1850–1917) S	**Kirt.**
Ķirulis, A. (fl. 1942) M	**Ķirulis**
(Kiselev, Ivan Aleksandrovich)	
see Kisselev, Ivan Aleksandrovich	**Kisselev**
Kislik, Serge (fl. 1991) M	**Kislik**

Kiss, Á. (1889–1968) PS	**Kiss**
Kiss, István A	**I.Kiss**
Kiss von Zilah, Endre (1873–) P	**E.Kiss**
Kisselev, Ivan Aleksandrovich (1888–) A	**Kisselev**
Kisseleva, E.I. A	**Kisseleva**
Kit Tan (1953–) BS	**Kit Tan**
Kita, G. (fl. 1913) M	**Kita**
Kitagawa, Masao (1909–) PS	**Kitag.**
Kitagawa, Naofumi (1935–) B	**N.Kitag.**
Kitaibel, Pál (Paul) (1757–1817) PS	**Kit.**
Kitajima, H. (fl. 1970) M	**Kitaj.**
Kitami, Takehiko (1935–) A	**Kitami**
Kitamura, Siro (1906–) S	**Kitam.**
(Kitanoff, Boris Pavlov	
see Kitanov, Boris Pavlov	**Kitanov**
Kitanov, Boris Pavlov (1912–) AS	**Kitan.**
Kitasima, K. (fl. 1922) M	**Kitas.**
Kitazawa, K. (fl. 1989) M	**Kitaz.**
Kits van Waveren, E. (1906–) M	**Kits van Wav.**
Kittel, Baldwin (Balduin) Martin (1798–1885) BS	**Kitt.**
(Kittell, (Sister))	
see Kittell, Mary Teresita	**Kittell**
Kittell, Mary Teresita (1892–) S	**Kittell**
Kittler, Christian (–1919) S	**Kittler**
Kittlitz, Friedrich Heinrich von (1799–1871) S	**Kittlitz**
Kitton, Frederic (1827–1895) A	**Kitton**
Kittredge, Elsie May (1870–1954) P	**Kittr.**
Kittredge, Walter (1953–) S	**W.Kittr.**
Kitz, Dennis J. (fl. 1979) M	**Kitz**
Kivirikko, Kaalo Eemeli (né Stenroos, K.E.) (1870–1947) S	**Kivir.**
Kiwak, Annie (fl. 1963) S	**Kiwak**
Kiyohara, Tisato A	**Kiyoh.**
Kizikelashvili, O.G. (fl. 1985) M	**Kizik.**
(Kjaerskou, Hjalmar Frederik Christian)	
see Kiaerskov, Hjalmar Frederik Christian	**Kiaersk.**
(Kjaerskov, Hjalmar Frederik Christian)	
see Kiaerskov, Hjalmar Frederik Christian	**Kiaersk.**
Kjansep–Romasckina, N.P. A	**Kjansep-Rom.**
Kjeldsen, Chris K. (1939–) A	**Kjeldsen**
Kjellberg, Gunnar Konstantin (1885–1943) PS	**Kjellb.**
Kjellman, Frans Reinhold (1846–1907) A	**Kjellm.**
Kjellqvist, Ebbe (fl. 1972) PS	**Kjellq.**
Kjøller, Annelise (fl. 1960) M	**Kjøller**
Klackenberg, Jens (1951–) S	**Klack.**
Kladiwa, Leo (1920–) S	**Kladiwa**
Klán, J. (fl. 1979) M	**Klán**
Klas, Zora A	**Klas**
Klashorst, G.van de (fl. 1981) M	**Klashorst**
Klásková, Anna (later Skalická, A.) (1932–) S	**Klásk.**

Klášterský, Ivan (1901–1979) S	**Klášt.**
Klatt, Friedrich Wilhelm (1825–1897) A	**Klatt**
(Klaus, Karl Ernst)	
see Claus, Karl Ernst	**Claus**
Klebahn, Heinrich (1859–1942) AM	**Kleb.**
Klebs, Georg Albrecht (1857–1918) ABS	**G.A.Klebs**
Klebs, Richard (1850–1911) M	**R.Klebs**
Kleeberger, George Reinhard (1849–) S	**Kleeb.**
Kleef, C.van (fl. 1877) S	**Kleef**
Kleen, Emil Andreas Gabriel (1847–1923) A	**Kleen**
Klein, Edmond Joseph (1866–1942) S	**E.J.Klein**
Klein, Erich (fl. 1989) S	**E.Klein**
Klein, J. (fl. 1870) M	**J.Klein**
Klein, Jacob Theodor (1685–1759) L	**Klein**
Klein, Ludwig (1857–1928) M	**L.Klein**
Klein, Roberto Miguel (1926–) S	**R.M.Klein**
Klein, William E. (fl. 1965) S	**W.E.Klein**
Klein, William McKinley (1933–) S	**W.M.Klein**
Kleine-Natrop, H.E. (fl. 1957) M	**Kleine-Natrop**
Kleinhans, Rodolphe (1828–) BS	**Kleinh.**
Kleinholz, J.R. (fl. 1937) M	**Kleinholz**
Kleinhoonte, Anthonia (1887–1960) S	**Kleinhoonte**
Kleinig, David Arthur (1947–) S	**Kleinig**
Klement, Karl Walter (1931–1982) A	**K.W.Klem.**
Klement, Oscar (1897–1980) M	**Klem.**
Klenert, Wilhelm (fl. 1912) S	**Klenert**
Kleopow, Jurij Dmitrievič (1902–1942) S	**Kleopow**
Klercker, John Echard Fredrik af (1866–1930) A	**Klercker**
Kletshetov, A.N. (fl. 1929) M	**Kletsh.**
Klett, Gustav Theodor (–1827) S	**Klett**
Klett, W. S	**W.Klett**
Klift, W.C.van der (fl. 1971) M	**Klift**
Kligman, A.M. (fl. 1945) M	**Kligman**
Klika, Jaromír (1888–1957) MS	**Klika**
Kling, Hedi J. A	**H.J.Kling**
Kling, Stanley A. A	**Kling**
Klinge, A.B. (fl. 1944) M	**A.B.Klinge**
Klinge, Johannes Christoph (1851–1902) PS	**Klinge**
Klinggräff, Carl Julius Meyer von (1809–1879) S	**C.Klinggr.**
Klinggräff, Hugo Erich Meyer von (1820–1902) BP	**H.Klinggr.**
Klingman, A.M. (fl. 1945) M	**Klingman**
Klingmuller, Walter (fl. 1958) B	**Klingm.**
Klingstedt, Frederik Woldemar (1881–1964) S	**Klingst.**
Klingström, A. (fl. 1965) M	**Klingström**
Klinkowski, Maximilian (1904–1971) S	**Klink.**
(Klinsman, Ernst Ferdinand)	
see Klinsmann, Ernst Ferdinand	**Klinsm.**
Klinsmann, Ernst Ferdinand (1794–1865) PS	**Klinsm.**
Kliphuis, Edske (1924–) S	**Kliphuis**

Klisiewicz, John M. (fl. 1970) M   **Klis.**
Klittich, C.J.R. (fl. 1989) M   **Klittich**
Kljashtorin, L.B.   A   **Kljasht.**
Kljuykov, E.V. (1950–   ) S   **Kljuykov**
(Klobukova, Eugenija Nikolaevna Alissova)
   see Alissova–Klobukova, Eugenija Nikolaevna   **Aliss.**
Klöcker, Albert (1862–1923) M   **Klöcker**
(Kloczkova Akulova, N.G.)
   see Kloczkova, N.G.   **Kloczkova**
Kloczkova, N.G. (1952–   ) A   **Kloczkova**
Klokov, Michail Vasiljevich (Mikhail Vasilevich) (1896–1981) S   **Klokov**
Klokov, V.M. (fl. 1972) S   **Klokov f.**
Kloos, Abraham Willem (1880–1952) S   **Kloos**
Klopotek, Agnes von (fl. 1967) M   **Klopotek**
Kloppenburg, Dale (1922–   ) S   **Kloppenb.**
Kloppstech, K.   A   **Kloppst.**
Klotter, Hans–Erich   A   **Klotter**
Klotz, Gerhard (1928–   ) S   **G.Klotz**
Klotz, Leo Joseph (1895–   ) M   **Klotz**
Klotzsch, Johann Friedrich (1805–1860) MPS   **Klotzsch**
Klug, Guillermo (fl. 1930) A   **Klug**
Klugh, Alfred Brooker (1882–1932) A   **Klugh**
Kluk, Krzysztof (Christoph) (1739–1796) S   **Kluk**
Klumpp, Barbara   A   **Klumpp**
(Kluykov, E.V.)
   see Kljuykov, E.V.   **Kljuykov**
Kluyver, Albert Jan (1888–1956) M   **Kluyver**
Klyncheva, M.V. (fl. 1965) M   **Klyncheva**
(Klyuikov, E.V.)
   see Kljuykov, E.V.   **Kljuykov**
Klyver, Frederick Detlev (fl. 1929) A   **Klyver**
Kmet, Andrej (Andras(s)) (1841–1908) S   **Kmet**
(Knaap–van Meeuwen, M.S.)
   see Meeuwen, M.S.Knaap–van   **Meeuwen**
Knaben, Gunvor S. (1911–   ) S   **Knaben**
Knaf, Joseph (Josef) Friedrich (1801–1865) BS   **Knaf**
Knaf, Karl (Karel) (1852–1878) S   **K.Knaf**
Knagg, Mary M.B.   S   **Knagg**
Knapp, A. (fl. 1921) M   **A.Knapp**
Knapp, Edgar Wolfram (1906–   ) AB   **E.W.Knapp**
Knapp, Elisa P. (fl. 1956) M   **E.P.Knapp**
Knapp, F.H. (fl. 1846) S   **F.Knapp**
Knapp, J.H.   S   **J.H.Knapp**
Knapp, John Leonard (1767–1845) S   **Knapp**
Knapp, Josef Armin (1843–1899) M   **J.A.Knapp**
Knapp, Rüdiger (1917–   ) S   **R.Knapp**
Knapp, Sandy (fl. 1985) S   **S.Knapp**
Knapp, Susan D.   A   **S.D.Knapp**
Knauss, J.F. (fl. 1975) M   **Knauss**

Knaut, Christian (1654–1716) L — **Knaut**
Knauth, Bernard (fl. 1926) M — **B.Knauth**
Knauth, Christoph(orus) (1638–1694) L — **Knauth**
Knebel, Gottfried (1908–) A — **Knebel**
Kneiff, Friedrich Gotthard (1785–1832) S — **Kneiff**
Knerr, Ellsworth Brownell (1861–1942) S — **Knerr**
Kneucker, Johann Andreas ('Andrees') (1862–1946) ABS — **Kneuck.**
Kniep, Karl Johannes Hans (1881–1930) M — **Kniep**
Knieskern, Peter D. (1798–1871) S — **Kniesk.**
Knight, Charles (1818–1895) BM — **C.Knight**
Knight, G.M. (fl. 1890) S — **G.M.Knight**
Knight, Henry (1834–1896) S — **H.Knight**
Knight, Joseph (1777?–1855) S — **Knight**
Knight, Ora Willis (1874–1913) S — **O.W.Knight**
Knight, Thomas Andrew (1759–1838) S — **T.Knight**
Knight, Walter (fl. 1983) S — **W.Knight**
Kniphof, Johann(es) Hieronymus (1704–1763) AS — **Kniph.**
Knize (fl. 1969) S — **Knize**
Knoblauch, Emil Friedrich (1864–1936) S — **Knobl.**
Knobloch, Irwin William (1907–) FPS — **Knobloch**
Knoche, (Edward Louis) Herman (1870–1945) S — **Knoche**
Knoell, Hilde (fl. 1935) F — **Knoell**
Knoepffler-Péguy, Michèle A — **Knoepffler-Péguy**
Knoester, J. (fl. 1986) S — **Knoester**
Knokha, P. (fl. 1968) M — **Knokha**
Knoll, Andrew Herbert (1951–) A — **A.H.Knoll**
Knoll, Fritz (1883–1981) FPS — **Knoll**
Knoop, Johann Hermann (c.1700–1769) S — **Knoop**
Knoph, Johannes-G. (fl. 1984) M — **Knoph**
Knorr, Georg Wolfgang (1705–1761) S — **Knorr**
Knorring, O.E. (1896–1979) S — **Knorring**
(Knorring-Neustrujeva, O.E.)
   see Knorring, O.E. — **Knorring**
Knowles, George Beauchamp (fl. 1820–1852) S — **Knowles**
Knowles, Matilda Cullen (1864–1933) BM — **M.Knowles**
Knowles, R. A — **R.Knowles**
Knowlton, Clarence Hinckley (1875–) S — **C.H.Knowlt.**
Knowlton, Frank Hall (1860–1926) ABF — **Knowlt.**
Knox, David (fl. 1986) M — **D.Knox**
Knox, Elizabeth M. (fl. 1950) P — **E.M.Knox**
Knox, John S. (fl. 1973) M — **J.S.Knox**
Knox, M.D.E. (fl. 1983) M — **M.D.E.Knox**
Knox, R.Bruce A — **R.B.Knox**
Knox–Davies, Peter Sidney (1929–) M — **Knox–Dav.**
Knoyle, J.Mary (fl. 1961) M — **Knoyle**
Knudsen, Henning (1948–) M — **Knudsen**
Knudsen, Martin A — **M.Knudsen**
Knudson, Brenda M. A — **B.M.Knudson**
Knudtzon, S.H. (fl. 1986) S — **Knudtzon**

Knuth, Frederik Marcus (1904–1970) S   F.M.Knuth
Knuth, P. (fl. 1916) M   P.Knuth
Knuth, Paul Erich Otto Wilhelm (1854–1899) S   Knuth
Knuth, Reinhard Gustav Paul (1874–1957) PS   R.Knuth
(Knuth-Knuthenborg, Frederik Marcus)
    see Knuth, Frederik Marcus   F.M.Knuth
Kny, Carl Ignaz Leopold (1841–1916) BM   Kny
Knyazheva, L.A. (fl. 1974) M   Knyazheva
Ko, W.H. (fl. 1978) M   ·   W.H.Ko
Ko, Wan Chang (1916–   ) S   W.C.Ko
(Ko-Bayashi, Tsuyako)
    see Kobayashi, Tsuyako   Ts.Kobay.
Ko-no, Gakuichi (fl. 1906) B   Ko-no
Koach, J. (fl. 1986) S   Koach
Kobara, Takaaki   A   Kobara
Kobayashi, Hideaki   A   H.Kobay.
Kobayashi, K. (fl. 1991) M   K.Kobay.
Kobayashi, Sumiko (1922–   ) S   S.Kobay.
Kobayashi, Takao (Takeo) (1929–   ) M   Tak.Kobay.
Kobayashi, Tsuyako (1928–   ) A   Ts.Kobay.
(Kobayashi, Yoshio)
    see Kobayasi, Yosio   Kobayasi
Kobayasi, Hiromu (1926–   ) A   H.Kobayasi
Kobayasi, Yosio (1907–   ) AM   Kobayasi
Kobel, Fritz (1896–   ) M   Kobel
Kobelev, V.K. (1894–1940) S   Kobelev
Kobendza, Roman (1886–   ) S   Kobendza
Köberlin, Christoph Ludwig (c.1830–1862) S   Köb.
Kobus, Jan Derk (1858–1910) S   Kobus
Kobuski, Clarence Emmeren (1900–1963) S   Kobuski
Kobyljanskij, V.D. (1928–   ) S   Kobyl.
(Kobylyanskii, V.D.)
    see Kobyljanskij, V.D.   Kobyl.
Kocev, H. (fl. 1968) S   Kocev
Koch, Claus   A   C.Koch
Koch, Erwin (fl. 1895) S   E.Koch
Koch, Georg Friedrich (1809–1874) S   G.Koch
Koch, Grünberg Christian Theodor (1872–   ) S   G.C.T.Koch
Koch, H.A. (fl. 1969) M   H.A.Koch
Koch, Hans Peter Gyllembourg (1807–1883) S   H.P.G.Koch
Koch, Heinrich (1805–1887) S   H.Koch
Koch, Joachim (1908–1981) M   J.Koch
Koch, Johann Friedrich Wilhelm (1759–1831) BMS   Koch
Koch, Karl (fl. 1934) S   K.Koch bis
Koch, Karl (Carl) Heinrich Emil (Ludwig) (1809–1879) PS   K.Koch
Koch, Leo Francis (1916–   ) AB   L.F.Koch
Koch, Ludwig Konrad Albert (1850–1938) S   L.K.A.Koch
Koch, Peter (fl. 1888–1889) A   P.Koch
Koch, Stephen D. (1940–   ) S   S.D.Koch

Koch, Walo (1896–1956) PS	**W.Koch**
Koch, Wilhelm Daniel Joseph (1771–1849) ABPS	**W.D.J.Koch**
Koch, William Julian (1924–) M	**W.J.Koch**
Kochansky-Devidé, Vanda A	**Koch.-Dev.**
Kochhar, P.L. (fl. 1935) M	**Kochhar**
Kochkareva, T.F. (fl. 1975) S	**Kochk.**
Kochman, József (fl. 1934) M	**Kochman**
Kochs, Julius (1900–) S	**Kochs**
Kochummen, Kizhakkedathu Mathai (1931–) S	**Kochummen**
Kociolek, John Patrick A	**Kociolek**
Kocková-Kratochvílova, Anna (fl. 1958) M	**Kock.-Krat.**
Koczwara, Marian (Maryan) (1893–1970) AS	**Koczw.**
Kodama, K. (fl. 1962) M	**K.Kodama**
Kodama, S. (fl. 1962) M	**S.Kodama**
Kodama, Shinsuke (1884–) PS	**Kodama**
Kodama, Takashi (fl. 1986) M	**Tak.Kodama**
Kodama, Tsutomu (fl. 1956) B	**T.Kodama**
Koechlin, Jean (1926–) S	**Koechlin**
(Koehler, Johann Christian Gottlieb)	
see Köhler, Johann Christian Gottlieb	**Köhler**
Koehler, Max (fl. 1936) B	**M.Koehler**
Koehne, Bernhard Adalbert Emil (1848–1918) S	**Koehne**
Koelderer, Johann Georg (fl. 1747) S	**Koeld.**
Koeler, Georg Ludwig (1765–1807) S	**Koeler**
Koelle, Johann Ludwig Christian (1763–1797) S	**Koelle**
Koelliker, Rudolf Albert von (1817–1905) S	**Koell.**
Koelpin, Alexander Bernhard (1839–1801) S	**Koelp.**
Koelreuter, I.T. A	**I.T.Koelr.**
(Koelreuter, Joseph Gottlieb)	
see Kölreuter, Joseph Gottlieb	**Kölr.**
Koeman, R.P.T. A	**Koeman**
Koenig, Adolph (1855–1932) F	**A.Koenig**
Koenig, Charles (fl. 1925) F	**C.Koenig**
Koenig, D. (1909–) S	**D.Koenig**
(Koenig, Johann Gerhard)	
see König, Johann Gerhard	**J.König**
Koenig, Karl (fl. 1841) S	**K.Koenig**
Koenig, Karl Dietrich Eberhard (1774–1851) S	**K.D.Koenig**
Koeniguer, J.-C. (fl. 1979) M	**Koeniguer**
Koeppen, Nicolas A	**Koeppen**
(Koerber, Gustav Wilhelm)	
see Körber, Gustav Wilhelm	**Körb.**
(Koernicke, Friedrich August)	
see Körnicke, Friedrich August	**Körn.**
(Koerte, Heinrich Friedrich Franz (Ernst))	
see Körte, Heinrich Friedrich Franz (Ernst)	**Körte**
Koeva, Jordanka (1935–) M	**Koeva**
(Köfaragó–Gyelnik, Vilmos)	
see Gyelnik, Vilmos Köfaragó	**Gyeln.**

Kofler, Lucie (1910–) M	**Kofler**
Kofoid, Charles Atwood (1865–1947) A	**Kof.**
Kogan, Sholom Iosifovich (1911–) A	**Kogan**
Kohl, Friedrich Georg (1855–1910) S	**Kohl**
Kohl, Heinrich (1787–1821) S	**H.Kohl**
Kohlbrugge, J.H.F. (fl. 1912) M	**Kohlbr.**
Kohler, C. (fl. 1987) S	**C.Kohler**
Köhler, Egon (1932–) S	**Eg.Köhler**
Köhler, Erich (fl. 1920–1924) MP	**Er.Köhler**
Köhler, F.E. (1889–) S	**F.E.Köhler**
Kohler, Hermann Adolph (1834–1879) S	**H.A.Kohler**
Kohler, J. (fl. 1921) S	**J.Kohler**
Köhler, Johann Christian Gottlieb (1759–1833) S	**Köhler**
Köhler, Udo (1911–1983) S	**U.Köler**
Kohlmeyer, Erika Ottilie (1930–1979) M	**E.Kohlm.**
Kohlmeyer, Jan W. (fl. 1960) M	**Kohlm.**
Kohlmuller, R. (fl. 1988) S	**Kohlmuller**
Kohmoto, K. (fl. 1981) M	**Kohmoto**
Kohn, Linda M. (1950–) M	**L.M.Kohn**
Kohn, Maximilian (fl. 1846) S	**Kohn**
Kohts, Fritz (1853–c.1872) S	**Kohts**
Koidzumi, Gen'ichi (Geniti, Gen–Iti) (1883–1953) APS	**Koidz.**
Koidzumi, Hideo (1886–1945) S	**H.Koidz.**
Koidzumi, Kadzuo S	**K.Koidz.**
Koidzumi, Makoto A	**M.Koidz.**
Köie, Mogens Engell (1911–) S	**Köie**
Koike, Tuneo S	**Koike**
Koizuma, M. (fl. 1986) M	**Koizuma**
Kojić, M. S	**Kojić**
Kok, Peter Daniel François (1944–) S	**Kok**
Kokaya, Ts.D. (fl. 1983) S	**Kokaya**
Kokkini, Stella (1953–) S	**Kokkini**
Kokko, E.G. (fl. 1974) M	**Kokko**
Kokolija, T.G. (fl. 1971) M	**Kokol.**
Kokwaro, John Ongayo (1940–) S	**Kokwaro**
Kol, Erszébet (Elizabet) (1897–1980) AM	**Kol**
(Kolakovskiï, Alfred Alekseevich)	
see Kolakovsky, Alfred Alekseevich	**Kolak.**
Kolakovsky, Alfred Alekseevich (1906–) FS	**Kolak.**
(Kolakowsky, Alfred Alekseevich)	
see Kolakovsky, Alfred Alekseevich	**Kolak.**
Kolandavelu, K. (fl. 1985) M	**Koland.**
Kolarik, J. (fl. 1986) S	**Kolarik**
Kolb, Max (1829–1915) S	**Kolb**
Kolbáni, Pál (Paul) (1757–1816) S	**Kolbáni**
Kolbe, Robert Wilhelm (1882–1960) A	**Kolbe**
Kölbing, F.W. (–1840) S	**Kölbing**
Kolenati, Friedrich August (Anton) Rudolf (1813–1864) BMS	**Kolen.**
Kolesnikov, Boris Pavlovich (1909–1980) S	**Kolesn.**

Kolfschoten, G.A. (fl. 1970) M — **Kolfsch.**
Kolhatkar, G.G. (1902–) PS — **Kolh.**
Kolipinski, M.C. (fl. 1962) M — **Kolip.**
Kolkwitz, Richard (1873–1956) A — **Kolkw.**
(Kölliker, Rudolf Albert von)
  see Koelliker, Rudolf Albert von — **Koell.**
Kollmann, Fania Weissmann- (1916–) S — **Kollmann**
Kołodziejćzyk, January (1889–1949) A — **Kołodz.**
(Koloknikov, L.B.)
  see Kolokolnikov, L.B. — **Kolok.**
Kolokolnikov, L.B. (fl. 1956) S — **Kolok.**
Kolosov, P.N. A — **Kolosov**
Kolpakova, T. A — **Kolp.**
Kölreuter, Joseph Gottlieb (1733–1806) S — **Kölr.**
Koltin, Y. (fl. 1966) M — **Koltin**
Koltz, Jean Pierre Joseph (1827–1907) BM — **Koltz**
Komagata, Kazuo (fl. 1987) M — **Komag.**
Komárek, Jiří (1931–) A — **Komárek**
Komarenko, L.E. A — **Komar.**
Komárková–Legnerová, Jaroslava A — **Komárk.-Legn.**
Komarov, B.M. S — **B.M.Kom.**
Komarov, Nikolay Fedorovich (Nikolai Fedorovič) (1901–1942) S — **N.F.Kom.**
Komarov, Vladimir Leontjevich (Leontevich) (1869–1945) AMPS — **Kom.**
Komarova, Emma Petrovna (1929–) M — **Komarova**
Komatsu, A. (fl. 1956) M — **A.Komatsu**
Komatsu, Shunzô (1879–1932) S — **Komatsu**
Komaya, Ginji (fl. 1924) M — **Komaya**
Kominami, Kiyoshi (1883–) M — **Komin.**
(Komirnaia, Olga Nikolaevna)
  see Komirnaya, Olga Nikolaevna — **Komirn.**
Komirnaya, Olga Nikolaevna (1907–) M — **Komirn.**
Komiya, Sadashi (1932–) S — **Komiya**
Komlódi, Magda (1931–) S — **Komlódi**
Komzha, A.L. (1954–) S — **Komzha**
Kondo, Katsuhiko (fl. 1984) S — **K.Kondo**
Kondô, Kingo (fl. 1984) S — **Kondô**
Kondrateva, A.A. (fl. 1952) M — **Kondrat.**
Kondrateva, Nadezhda Vasilevna (1925–) A — **N.V.Kondrat.**
(Kondratjeva, Nadezhda Vasilevna)
  see Kondrateva, Nadezhda Vasilevna — **N.V.Kondrat.**
Kondratjuk, Evgeniï Nikolaevich (1914–) S — **Kondr.**
(Kondratyuk, Evgeniï Nikolaevich)
  see Kondratjuk, Evgeniï Nikolaevich — **Kondr.**
Kondratyuk, Ye.M. (fl. 1985) S — **Ye.M.Kondr.**
(Konechnaja, G.Ju.)
  see Konechnaya, G.Yu. — **Konechn.**
Konechnaya, G.Yu. (1951–) S — **Konechn.**
Kong, Hua Zhong (1940–) M — **H.Z.Kong**
Kong, Wen Yan (fl. 1988) S — **W.Y.Kong**

Kong, Xian Xu (fl. 1983) PS — X.X.Kong
Kongar, E.T. (1937–) S — Kongar
Kongisser, Rudolph A. A — Kong.
(König, Carl Dietrich Eberhard)
  see Koenig, Karl Dietrich Eberhard — K.D.Koenig
König, Carl Wilhelm (1808–1874) S — C.W.König
König, Johann Gerhard (1728–1785) BMPS — J.König
König, Peter (fl. 1988) S — P.König
Königer, Willibald (fl. 1982) S — Koniger
(Koning, R.de)
  see de Koning, R. — de Koning
Konishi, Kenji (1929–) A — Konishi
Konno, E. S — E.Konno
Konno, Kaoru (fl. 1969) AM — K.Konno
Konno, Toshinori (1939–) A — T.Konno
Konokotina, A.G. (fl. 1913) M — Konok.
Konop, R. (fl. 1981) S — Konop
Konopacka, W. (fl. 1926) M — Konopacka
Konopova, K. (fl. 1984) S — Konopova
Konrad, Paul (1877–1948) M — Konrad
Kooijman, Havik Nicolaas (1893–) A — Kooijman
(Kooiman, Havik Nicolaas)
  see Kooijman, Havik Nicolaas — Kooijman
Koopmann, Karl (fl. 1879–1900) S — Koopmann
Koopowitz, H. (fl. 1986) S — Koop.
Koorders, Sijfert Hendrik (1863–1919) AMS — Koord.
Koorders–Schumacher, Anna (1870–1934) S — Koord.-Schum.
Kooy, F. (fl. 1982) S — Kooy
(Kooyman, H.)
  see Kooijman, H.N. — Kooijman
(Kopachevskaia, E.G.)
  see Kopachevskaya, E.G. — Kopach.
Kopachevskaya, E.G. (fl. 1972) M — Kopach.
Köpff, Friedrich (fl. 1892) S — Köpff
Koponen, Aune Kyllikki (1938–) B — A.K.Kop.
Koponen, Timo Juhani (1939–) B — T.J.Kop.
Kopp, Julius (1880–) PS — Kopp
Kopp, Lucille E. (fl. 1964) S — L.E.Kopp
Koppe, Carl (Karl) Friedrich August (1803–1874) S — Koppe
Koppe, Karl (1890–1979) B — K.Koppe
Köppen, Friedrich Theodor (Petrowitsch) von (1833–1908) S — Köppen
Koppen, John D. A — J.D.Koppen
Kops, Jan (1765–1849) S — Kops
Kopsch, August (1913–) B — Kopsch
Korab, J.J. (fl. 1930) M — Korab
Korac, M. (fl. 1974) S — Korac
Korb, E. (1873–) S — Korb
Körber, Gustav Wilhelm (1817–1885) M — Körb.
(Korbonshaja, Ya.I.)
  see Korbonskaja, Ja.I. — Korbonsk.

Korbonskaja, Ja.I. (fl. 1956) M	**Korbonsk.**
(Korbonskaya, Ya.I.)	
see Korbonskaja, Ja.I.	**Korbonsk.**
Korczagin, Aleksandr Aleksandrovich (Alexander Alexandrovich)	
(1900–1987) BS	**Korcz.**
Korde, Kira Borisovna (1912–) A	**Korde**
Kordon, K.F. (1884–1962) S	**Kordon**
Kores, Paul J. (1950–) S	**Kores**
Korf, Richard Paul (1925–) M	**Korf**
Korhonen, K. (1943–) M	**Korhonen**
Kôriba, Kwan (1882–1957) S	**Kôriba**
Korica, Bogdan (1918–) S	**Korica**
Korkina, V. (1924–) S	**Korkina**
Korkishko, R.I. (fl. 1986) S	**Kork.**
Korkonossova, L.J. S	**Korkon.**
(Korn–Huber, (Georg) Andreas von)	
see Kornhuber, (Georg) Andreas von	**Kornh.**
Kornaś, Jan Kazimierz (1923–) PS	**Kornaś**
Körner, Heinz A	**Körner**
Korneva, E.I. S	**Korneva**
Kornhuber, (Georg) Andreas von (1824–1905) S	**Kornh.**
Körnicke, Friedrich August (1828–1908) MPS	**Körn.**
Kornilova, Valentina Stepanovna (1908–) S	**Kornil.**
Kornmann, Peter (1907–) A	**Kornmann**
Kornuch–Trotzky, Petrus (Peter) (1803–1877) S	**Korn.-Trotzky**
Korobkov, A.A. (1940–) S	**Korobkov**
Koroleva, A.S. (fl. 1971) S	**A.S.Korol.**
Koroleva, G.S. A	**G.S.Korol.**
Koroleva, K.M. S	**K.M.Korol.**
Koroleva, Valentina Alekseevna (1898–) S	**Korol.**
(Koroleva–Pavlova, Valentina Alekseevna)	
see Koroleva, Valentina Alekseevna	**Korol.**
Korotkevich, O.S. A	**Korotk.**
(Korotkij, M.F.)	
see Korotky, M.F.	**Korotky**
(Korotkiy, M.F.)	
see Korotky, M.F.	**Korotky**
Korotkova, Elena Evgenevna (1906–1977) S	**Korotkova**
Korotky, M.F. (–1915) S	**Korotky**
Korovin, Evgenii (Yevgeni, Eugeny) Petrovich (1891–1963) S	**Korovin**
Korovin, S.P. (fl. 1962) S	**S.P.Korovin**
Korshikov, Aleksandr Arkadievich (1889–1942) AM	**Korshikov**
Korshinsky, Sergei Ivanovitsch (1861–1900) S	**Korsh.**
Kort, Antoine Joseph (1874–1957) S	**Kort**
(Kort, I.de)	
see de Kort, I.	**de Kort**
Körte, Heinrich Friedrich Franz (Ernst) (1782–1845) S	**Körte**
(Korte, W.E.de)	
see de Korte, W.E.	**de Korte**

Korthals, Pieter Willem (1807–1892) S	**Korth.**
(Korzhinsky, Sergei Ivanovich)	
see Korshinsky, Sergei Ivanovitsch	**Korsh.**
(Koržinskij, Sergei Iwanowitsch)	
see Korshinsky, Sergei Ivanovitsch	**Korsh.**
Košanin, Nedelyko (1874–1934) S	**Košanin**
Kosanke, Robert M. (fl. 1955) F	**Kosanke**
(Koschelova, E.N.)	
see Koshkelova, Je.N.	**Koshk.**
Kościelny, S. (fl. 1935) M	**Kościelny**
Kosenko, Z.A. A	**Kosenko**
Koshewnikow, Dmitrij Alexandrowitsch (1858–1882) S	**Kosh.**
Koshkelova, Je.N. (fl. 1961) M	**Koshk.**
Kosina, Cyril (fl. 1984) M	**Kosina**
(Kosinskaja, Ekaterina Konstantinovna)	
see Kossinskaja, Ekaterina Konstantinovna	**Kossinsk.**
Koske, Richard E. (fl. 1972) M	**Koske**
Koskela, P. (fl. 1983) M	**Koskela**
Koskinen, Arvo Adiel (1898–) M	**Koskinen**
Koslovsky, V.L. (fl. 1878) S	**Kosl.**
(Koso–Poliansky, Boris Mikhailovic)	
see Koso–Poljansky, Boris Mikhailovic	**Koso–Pol.**
Koso–Poljansky, Boris Mikhailovic (1890–1957) S	**Koso–Pol.**
(Koss, G.)	
see Koss, Jurij Ivanovich	**Koss**
Koss, Jurij Ivanovich (1889–1961) S	**Koss**
Kossenko, Ivan Sergeevič (1896–) S	**Kossenko**
Kossinskaja, Ekaterina Konstantinovna (1900–) A	**Kossinsk.**
(Kossinski, K.K.)	
see Kossinsky, C.C.	**Kossinsky**
Kossinsky, C.C. (1874–1928) PS	**Kossinsky**
Kossko, I.N. (1924–1956) S	**Kossko**
Kossobudzka, Hanna (fl. 1936) M	**Kossob.**
Kossych, V.M. (1931–) S	**Kossych**
Kost, G. (fl. 1979) M	**G.Kost**
Köst, H.P. A	**H.P.Köst**
(Kostelcky, Vincenc František)	
see Kosteletzky, Vincenz Franz	**Kostel.**
Kosteletzky, Vincenz Franz (1801–1887) S	**Kostel.**
Koster, Henry (1793–1820) S	**H.Kost.**
Koster, Joséphine Thérèse (1902–1986) A	**J.Kost.**
Kosterman, Y. A	**Kosterman**
Kostermans, André Joseph Guillaume Henri (1907–) S	**Kosterm.**
Kostichka, C.J. (fl. 1979) M	**Kostichka**
Kostin, V.A. (fl. 1981) M	**Kostin**
Kostina, Klaudia Fedorovna (1900–) S	**Kostina**
Kostka, G. (fl. 1927) M	**Kostka**
(Kostoff, Dontcho)	
see Kostov, Dontcho	**Kostov**

Kostov, Dontcho (1897–1949) S   **Kostov**
Kostov, V. (fl. 1971) M   **V.Kostov**
Kostrun, Gertrud   A   **Kostrun**
Kosuge, K. (fl. 1989) S   **K.Kosuge**
Kotejowa, Elizbieta (1933–   ) S   **Kotejowa**
Kothari, K.L. (fl. 1966) M   **K.L.Kothari**
Kothari, M.J. (1943–   ) S   **Kothari**
Kothari, N.M.   A   **N.M.Kothari**
Kotila, John Earnest (1893–1951) M   **Kotila**
Kotilainen, Mauno Johannes (1895–1961) S   **Kotil.**
Kotiranta, Heikki (1954–   ) M   **Kotir.**
Kotlaba, František (1927–   ) M   **Kotl.**
Kotlán, A.   A   **Kotlán**
Kotov, Mikhail Ivanovich (1896–1978) S   **Kotov**
Kotschy, Carl (Karl) Georg Theodor (1813–1866) BS   **Kotschy**
Kott, L.S. (fl. 1981–89) P   **Kott**
Kotte, Walter (1893–   ) M   **Kotte**
Kotthoff, Peter (1883–   ) M   **Kotthoff**
(Kotuchov, Yu.A.)
    see Kotukhov, Yu.A.   **Kotukhov**
Kotukhov, Yu.A. (fl. 1966) P   **Kotukhov**
Kotula, Andrzej (Andreas) (1822–1891) S   **Kotula**
Kotula, Boleslaw (1849–1898) S   **B.Kotula**
Kotulowa, W. (fl. 1970) M   **Kotulowa**
Kotze, Janetta D.S. (fl. 1990) S   **Kotze**
Koul, K.K. (fl. 1984) S   **Koul**
Kounu, K. (fl. 1972) M   **Kounu**
Koutnik, Daryl Lee (1951–   ) S   **Koutnik**
Kouwets, Frans A.C. (1954–   ) A   **Kouwets**
Kouyeas, H. (fl. 1963) M   **H.Kouyeas**
Kouyeas, V. (fl. 1963) M   **V.Kouyeas**
(Kovacevski, Eduard Friedrich)
    see Eversmann, Eduard Friedrich von   **Eversm.**
(Kovačevski, Ivan Christow)
    see Kovatschevski, Ivan Christow   **Kovatsch.**
(Kovachev, Ivan Galabov)
    see Kovatschev, Ivan Galabov   **Kov.**
Kovachich, W.G. (fl. 1954) M   **Kovachich**
Kovácik, Lubomír   A   **Kovácik**
(Kovács, Gyula)
    see Kováts von Szent–Lélek, Julius   **Kováts**
Kovács–Lang, Edit (1938–   ) S   **Kovács–Lang**
Koval, Eleonora Zakharovna (1930–   ) M   **Koval**
Kovalenko, A.E. (fl. 1988) M   **Kovalenko**
Kovalev, Nikolai Vasilevich (1888–   ) S   **Kovalev**
Kovalevskaja, S.S. (1929–   ) S   **Kovalevsk.**
(Kovalevskaya, S.S.)
    see Kovalevskaja, S.S.   **Kovalevsk.**

Kovanda, Miloslav (1936–) S	**Kovanda**
Kovár, Napsal Filip (1863–1925) M	**Kovár**
Kováts, Ferenc (1873–1956) S	**F.Kováts**
(Kováts, Julius)	
see Kováts von Szent-Lélek, Julius	**Kováts**
Kováts von Szent-Lélek, Julius (1815–1873) FS	**Kováts**
(Kováts von Szentélek, Julius)	
see Kováts von Szent-Lélek, Julius	**Kováts**
Kovatschev, Ivan Galabov (1927–) S	**Kov.**
Kovatschevski, Ivan Christow (1904–) M	**Kovatsch.**
(Kovtonjuk, N.K.)	
see Kovtonyuk, N.K.	**Kovt.**
Kovtonyuk, N.K. (fl. 1987) S	**Kovt.**
Kowal, Tadeusz (1924–1979) S	**Kowal**
Kowalczyk, Stanley A. A	**Kowalczyk**
Kowalski, Donald T. (1938–) M	**Kowalski**
Koyama, Hiroshige (1937–) S	**H.Koyama**
Koyama, Mitsuo (1885–1935) S	**Koyama**
Koyama, Tetsuo Michael (1933–) PS	**T.Koyama**
Koyapillil, Mathai Matthew Michael (1930–) S	**Koyap.**
Koyuncu, Mehmet (1944–) S	**M.Koyuncu**
Kozakiewicz, Zofia (later Lawrence, Z.) (1947–) M	**Kozak.**
Kozenko, E.P. A	**Kozenko**
Kozhanchikov, V.I. (fl. 1971) S	**Kozhanch.**
Kozhevnikov, A.E. (1955–) S	**A.E.Kozhevn.**
Kozhevnikov, Yu.P (1941–) S	**Kozhevn.**
(Kozhuchov, I.V.)	
see Kozhukhov, I.V.	**Kozhukhov**
(Kozhuharov, Stefan Ivanov)	
see Kožuharov, Stefan Ivanov	**Kožuharov**
Kozhukhov, I.V. (1899–1952) S	**Kozhukhov**
Kozik, E. (fl. 1983) S	**Kozik**
Kozina, C. (fl. 1987) M	**Kozina**
Kozlova, Olga Georgievna A	**O.G.Kozlova**
Kozlova, T.M. (fl. 1981) M	**Kozlova**
(Kozlovskaja, Natalia Vitalievna)	
see Kozlowskaja, Natalija Vitalevna	**Kozlowsk.**
(Kozlovskij, V.L.)	
see Koslovsky, V.L.	**Koslovsky**
Kozlowska, Aniela (1898–) S	**Kozlowska**
Kozlowskaja, Natalija Vitalevna (1923–) S	**Kozlowsk.**
(Kozlowski, Roman)	
see Kozlowsky, Roman	**R.Kozlowsky**
Kozlowsky, Roman (1899–) B	**R.Kozlowsky**
(Kozo-Poliansky, Boris Mikhailovic)	
see Koso-Poljansky, Boris Mikhailovic	**Koso-Pol.**
Kožuharov, Stefan Ivanov (1933–) S	**Kožuharov**
Kozur, H. A	**Kozur**
Kozyrenko, T.F. A	**Kozyr.**
Krabbe, Gustav (Heinrich) (1855–1895) S	**Krabbe**

Krach, B. (fl. 1988) S	**Krach**
Kraehenbuhl, Felix S	**Kraehenb.**
Kraenzlin, Friedrich (Fritz) Wilhelm Ludwig (1847–1934) S	**Kraenzl.**
Kraepelin, G. (fl. 1982) M	**G.Kraep.**
Kraepelin, Karl (Carl) (Matthias Friedrich) (1848–1915) S	**Kraep.**
Kraetzl, Franz (1852–) S	**Kraetzl**
Kraft, Gerald Thompson (1939–) A	**Kraft**
Krahmer, B. (fl. 1909) B	**Krahmer**
Krahulcova, A. (fl. 1982) S	**Krahulc.**
Kraicz, Isa A	**Kraicz**
Krainsky, A. (fl. 1913) M	**Krainsky**
Krainz, Hans (1906–1980) S	**Krainz**
Krajina, Vladimir Joseph (1905–) BPS	**Krajina**
Král, Miloš (1932–) S	**M.Král**
Kral, Robert (1926–) S	**Kral**
Kralik, Jean–Louis (1813–1892) S	**Kralik**
Kramadibrata, K. (1952–) M	**Kramad.**
Kramer, Carl (1843–1882) S	**C.Kramer**
Kramer, Charles Lawrence (1928–) M	**C.L.Kramer**
Kramer, Franz August (fl. 1875) S	**F.Kramer**
Kramer, J.P. (fl. 1981) M	**J.P.Kramer**
Kramer, Karl Ulrich (1928–) PS	**K.U.Kramer**
Kramer, Wilhelm Heinrich (–1765) S	**Kramer**
Kramer, Wolfgang Anton (1947–) B	**W.A.Kramer**
Krammer, Kurt A	**Krammer**
Krampe, Oskar (fl. 1926) M	**Krampe**
Kranz, Cajetan Anton (1839–1886) S	**Kranz**
Kranz, Jürgen von (fl. 1966) M	**J.Kranz**
Kränzlin, Fred (fl. 1984) M	**F.Kränzl.**
(Kränzlin, Friedrich (Fritz) Wilhelm Ludwig)	
see Kraenzlin, Friedrich (Fritz) Wilhelm Ludwig	**Kraenzl.**
Krapf, Karl J.von (1782–) S	**Krapf**
Krapivina, I.G. (fl. 1975) M	**Krapiv.**
Krapovickas, Antonio (1921–) S	**Krapov.**
Krašan, Franz (1840–1907) S	**Krašan**
Krascheninnikov, Ippolit (Hippolit) Mikhailovich (1884–1947) S	**Krasch.**
Krascheninnikov, Stephan (Stepan) Petrovich (1713–1755) S	**S.Krasch.**
(Krasilnikov, N.A.)	
see Krassilnikov, N.A.	**Krassiln.**
(Krasilov, Valentin Abramovich)	
see Krassilov, Valentin Abramovich	**Krassilov**
(Krasilskhik, Isaak Matvyeevich)	
see Krassilstschik, Isaak Matvyeevich	**Krass.**
(Krasilstschik, Isaak Matvyeevich)	
see Krassilstschik, Isaak Matvyeevich	**Krass.**
Krasnikov, A.A. (fl. 1984) S	**Krasnikov**
Krasnoborov, Ivan M. (1931–) S	**Krasnob.**
Krasnopeeva, P.S. A	**Krasnop.**
Krasnoperova, L.A. A	**Krasnoper.**
Krasnov, Andrej Nikovaevich (1862–1914) S	**Krasn.**

Krasnova, A.N. (1938–) S	**Krasnova**
(Krasovskaja, L.S.)	
see Krassovskaja, L.S.	**Krassovsk.**
(Krasovskaya, L.S.)	
see Krassovskaya, L.S.	**Krassovsk.**
Krassavina, L.K. A	**Krassav.**
Krasser, Fridolin (1863–1923) FPS	**Krasser**
Krassilnikov, Nikolay Aleksandrovich (1896–) AM	**Krassiln.**
Krassilov, Valentin Abramovich (fl. 1967) BFM	**Krassilov**
Krassilstschik, Isaak Matvyeevich (1857–) AM	**Krass.**
Krasske, Georg (1889–1951) A	**Krasske**
(Krassnov, Andrej Nikovaevich)	
see Krasnov, Andrej Nikovaevich	**Krasn.**
(Krassnow, Andrej Nikovaevič)	
see Krasnov, Andrej Nikovaevich	**Krasn.**
Krassochkin, V.T. (1904–) S	**Krassochkin**
Krassovskaja, L.S. (1948–) S	**Krassovsk.**
Krasucka, Werner (fl. 1968) S	**Krasucka**
Kratz, Joseph (fl. 1861) S	**Kratz**
(Krätzl, Franz)	
see Kraetzl, Franz	**Kraetzl**
Krauer, Johann George (1794–1845) S	**Krauer**
Kraus, Gregor Conrad Michael (1841–1915) S	**Kraus**
Krause, Christian Ludwig (–1774) S	**C.Krause**
Krause, David L. (fl. 1972) S	**D.L.Krause**
Krause, Ernst (–1858) S	**E.Krause**
Krause, Ernst Hans Ludwig (1859–1942) BMS	**E.H.L.Krause**
Krause, Ernst Ludwig (1839–1903) S	**E.L.Krause**
Krause, Johann Wilhelm (1764–1842) S	**Krause**
Krause, Johannes S	**J.Krause**
Krause, Karin (1927–) S	**Kar.Krause**
Krause, Kurt (1883–1963) S	**K.Krause**
Krause, M. S	**M.Krause**
Krause, R.A. (fl. 1972) M	**R.A.Krause**
Kräusel, Richard (1890–1966) ABFM	**Kräusel**
Krauskopf, Englebert (1820–1881) S	**Krauskopf**
Krauss, Christian Ferdinand Friedrich von (1812–1890) ABS	**C.Krauss**
Krauss, G. (fl. 1812) S	**G.Krauss**
Krauss, Helen K. (fl. 1947) S	**H.K.Krauss**
Krauss, Johan Carl (1759–1826) S	**Krauss**
Krauss, Otto (–1935) S	**O.Krauss**
Krauss, Robert Wallfar (1921–) S	**R.W.Krauss**
Krautter, Louis (1880–1909) S	**Krautter**
Kravtzev, B.I. (fl. 1961) M	**Kravtzev**
(Krawtzew, B.I.)	
see Kravtzev, B.I.	**Kravtzev**
Krebs, (Georg) Ludwig Engelhard (1792–1844) S	**Krebs**
Krebs, F.L. (fl. 1827–1835) S	**F.Krebs**
Krebs, William N. (1948–) A	**W.N.Krebs**

(Kreczetovicz, Lev Melkhisedekovich)	
see Kreczetowicz, Lev Melkhisedekovich	**Krecz.**
(Kreczetovicz, V.I.)	
see Kreczetowicz, V.I.	**V.I.Krecz.**
Kreczetowicz, Lev Melkhisedekovich (1878–) MS	**Krecz.**
Kreczetowicz, V.I. (1901–1942) PS	**V.I.Krecz.**
Kreger van Rij, Nelly Jenne Wilhelmina (fl. 1964) M	**Kreger**
(Kreger–van Rij, Nelly Jenne Wilhelmina)	
see Kreger van Rij, Nelly Jenne Wilhelmina	**Kreger**
Kreh, Wilhelm (1884–1954) M	**Kreh**
Kreisel, Hans (1931–) M	**Kreisel**
Krejčí, Johann (Jan) (1825–1887) AF	**Krejčí**
Krejzová, Rőžena (fl. 1976) M	**Krejzová**
Kremer, Bruno P. A	**B.P.Kremer**
Kremer, Jean Pierre (1812–1867) BS	**Kremer**
Krempelhuber, August von (1813–1882) AM	**Kremp.**
Krendl, Franz Xaver (1926–) S	**Krendl**
Krenner, Joseph Andor (fl. 1944) AM	**Krenner**
Kress, (Walter) John (1951–) S	**W.J.Kress**
Kress, Alarich Alban Herwig Ludwig (1932–) S	**Kress**
(Kress–Deml, Irmingard)	
see Deml, Irmingard	**I.Deml**
Krestovskaja, Tatyana Valerievna (1953–) S	**Krestovsk.**
(Kretschetovitsch, Lev Melkhisedekovitsch)	
see Kreczetowicz, Lev Melkhisedekovich	**Krecz.**
(Kretschetovitsch, V.I.)	
see Kreczetowicz, V.I.	**V.I.Krecz.**
Kreutz, C.A.J. S	**Kreutz**
Kreutzer, Karl (Carl) Joseph (1809–1866) S	**Kreutzer**
Kreutzer, William R. S	**W.R.Kreutzer**
(Kreutzinger, Kurt G.)	
see Kreuzinger, Kurt G.	**Kreuz.**
Kreuzinger, Kurt G. (1905–1989) S	**Kreuz.**
Kreyer, Georgij Kalowic (Karlovič) (1887–1942) MS	**Kreyer**
Kreysig (fl. 1829) S	**Kreysig**
Krhovský, J. A	**Krhovský**
Kriechbaum, Wilhelm (fl. 1925) S	**Kriechb.**
Krieger, Hans A	**H.Krieg.**
Krieger, Karl Wilhelm (1848–1921) BM	**K.Krieg.**
Krieger, Louis Charles Christopher (1873–1940) M	**L.Krieg.**
Krieger, Walther (1880–) BP	**W.Krieg.**
Krieger, Willi (1886–1954) AM	**Willi Krieg.**
Krieglsteiner, German J. (fl. 1982) M	**Krieglst.**
Krieglsteiner, L.G. (fl. 1991) M	**L.G.Krieglst.**
Krienitz, Lothar A	**Krienitz**
Křísa, Bohdan (1936–) S	**Křísa**
Krisai, I. (fl. 1989) M	**Krisai**
(Krischtofowitsch, African Nikolaevich)	
see Kryshtofowicz, African Nikolaevich	**Krysht.**

Krishna, A. (fl. 1980) M	A.Krishna
Krishna, Bal (fl. 1990) F	Bal Krishna
Krishna, Bijay (1937–) S	B.Krishna
Krishnamurthy, C.S. (fl. 1947) M	C.S.Krishnam.
Krishnamurthy, K.H. (fl. 1958) S	K.H.Krishnam.
Krishnamurthy, R. A	R.Krishnam.
Krishnamurthy, Vasudeva B. (fl. 1965) A	V.Krishnam.
Krishnaswamy, N. (fl. 1957) S	Krishnasw.
Kriss, A.E. (fl. 1933) M	Kriss
Krist, Vladimír (1905–1942) M	Krist
Kristiansen, Jørgen (1931–) A	Kristiansen
Kristiansen, Roy (fl. 1987) M	R.Kristiansen
Kristinsson, Hörður (1937–) M	Kristinsson
(Krištofovič, African Nikolaevich)	
see Kryshtofowicz, African Nikolaevich	Krysht.
(Kritska, Ljubor I.)	
see Krytzka, Ljubor I.	Krytzka
Kriván–Hutter, Erika A	Kriván–Hutter
Křivanec, K. (fl. 1977) M	Křivanec
Krivotulenko, U.F. (fl. 1955) S	Krivot.
Kröber, H. (fl. 1981) M	Kröber
Krocker, Anton Johann (1744–1823) S	Krock.
Kroeger, Paul (fl. 1987) M	Kroeger
Kroemer, Karl Maximilian Wilhelm (1871–1956) M	Kroemer
Krog, Hildur (1922–) M	Krog
Krok, Thorgny Ossian Bolivar Napoleon (1834–1921) PS	Krok
Krolikowski, L. (fl. 1928) M	Krolik.
Krombach, Johann Heinrich Wilhelm (1791–1881) S	Kromb.
Krombholz, Julius Vincenz von (1782–1843) M	Krombh.
Kroner, G. (fl. 1980) S	Kroner
Kronfeld, Ernst Moriz (Mauriz) (1865–1942) MS	Kronf.
Kronfeldt, G. S	Kronfeldt
(Kropákova, A.Metzelova)	
see Metzelova–Kropácova, A.	Metz.-Krop.
Kropp, Bradley R. (fl. 1984) M	Kropp
Krösche, Ernst (fl. 1909–28) S	Krösche
Krotov, A.I. A	A.I.Krotov
Krotov, A.S. (1907–) S	Krotov
Krousheva, R. (fl. 1967) M	Krousheva
Krstinic, Ante (fl. 1991) S	Krstinic
Kruckeberg, Arthur Rice (1920–) S	Kruckeb.
(Krueger, Wilhelm)	
see Krüger, Wilhelm	W.Krüger
Krug, Carl (Karl) Wilhelm Leopold (1833–1898) PS	Krug
Krug, John Christian (1938–) M	J.C.Krug
Kruganova, Evfrosina Akimova (1914–) S	Kruganova
Krugel, T. (fl. 1983) S	Krugel
Krüger, Ernst August (1860–1942) S	E.Krüger
Krüger, Friedrich (1864–1914) S	F.Krüger

Krüger, Marcus Salomonides (fl. 1841) S	**Krüger**
Krüger, Wilhelm (1857–) AM	**W.Krüger**
(Kruif, A.P.M.de)	
see de Kruif, A.P.M.	**de Kruif**
Kruis (fl. 1891) M	**Kruis**
Krukoff, Boris Alexander (1898–1983) S	**Krukoff**
Krupinsky, J.M.(K.) (fl. 1982) M	**Krup.**
Kruschewa, R. (fl. 1959) M	**Krusch.**
Kruschke, Emil Paul (1907–1976) S	**Kruschke**
Kruse, Friedrich (–1890) S	**Kruse**
Kruse, Paul (fl. 1896) M	**P.Kruse**
(Kruseman, Maria Johanna)	
see Steenis-Kruseman, Maria Johanna van	**Steen.-Krus.**
Krusenstjerna, Axel Henrik Edvard Filip von (1908–) B	**Krus.**
Krüssmann, Johann Gerd (1910–1980) S	**Krüssm.**
Krutzsch, Wilfried (fl. 1970) AM	**Krutzsch**
(Kruyff, E.De)	
see De Kruyff, E.	**De Kruyff**
Krylov, Georgii Vasilevich (1910–) S	**G.V.Krylov**
Krylov, Porphyry Nikitic (1850–1931) S	**Krylov**
(Krylow, Porphyry Nikitic)	
see Krylov, Porphyry Nikitic	**Krylov**
(Kryshtofovich, African Nikolaevich)	
see Kryshtofowicz, African Nikolaevich	**Krysht.**
Kryshtofowicz, African Nikolaevich (1885–1953) BFS	**Krysht.**
Krytzka, Ljubov I. (1941–) S	**Krytzka**
(Krytzkaja, Ljubov I.)	
see Krytzka, Ljubov I.	**Krytzka**
Krzemecki, A. (fl. 1913) M	**Krzemecki**
Krzemieniewska, Helena (1878–1966) M	**Krzemien.**
Ku, Tsue Chih (1931–) S	**T.C.Ku**
Kuan, Chung Tian (fl. 1983) S	**C.T.Kuan**
Kuan, Ke Chien (1913–) S	**K.C.Kuan**
Kuang, Ko Rjên (fl. 1941) S	**Kuang**
(Kuang, Ko Zen)	
see Kuang, Ko Rjên	**Kuang**
Kubanskaja, Zinaida Victorovna (1908–) S	**Kubansk.**
Kubát, Karel (1941–) S	**Kubát**
Kuber, G. (fl. 1964) S	**Kuber**
Kubicek, Q.B. (fl. 1984) M	**Kubicek**
Kubička, Jiřího (Jiri) (1913–) M	**Kubička**
Kubitzki, Klaus (1933–) S	**Kubitzki**
Kubo, Hitoshi (fl. 1986) M	**Kubo**
Kubono, T. (fl. 1989) M	**Kubono**
Kubota, Hideo S	**H.Kubota**
Kubota, I.-I. S	**Kubota**
Kuc, Marian (1932–) B	**Kuc**
Kučera, J. (fl. 1927) M	**Kučera**
Kuchel, Rex Harold (1917–1985/6) S	**Kuchel**
Küchenmeister, Gottlob Friedrich Heinrich (1821–1890) M	**Küchenm.**

Kuckuck, Ernst Hermann Paul (1866–1918) A   **Kuck.**
Kucowa, Irena (1912–) S   **Kucowa**
Küçüködük, M. (fl. 1983) S   **Küçük.**
Kucyniak, James (1919–1962) B   **Kucyn.**
Kudabaeva, G.M. (1951–) S   **Kudab.**
Kudo, H. (fl. 1989) M   **H.Kudo**
Kudo, Richard Roksabro (1886–1967) A   **R.R.Kudo**
Kudô, Yûshun (1887–1932) PS   **Kudô**
(Kudriavzev, W.I.)
    see Kudrjanzev, V.I.   **Kudrjanzev**
Kudrjanzev, V.I. (fl. 1932) M   **Kudrjanzev**
(Kudrjanzew, W.I.)
    see Kudrjanzev, V.I.   **Kudrjanzev**
Kudrjaschev, S.N. (1907–1943) S   **Kudr.**
Kudrjaschova, G.L. (1940–) S   **Kudrjasch.**
Kudřna, K. (fl. 1928) M   **Kudřna**
(Kudryaschova, G.L.)
    see Kudrjaschova, G.L.   **Kudrjasch.**
(Kudryashova, G.L.)
    see Kudrjaschova, G.L.   **Kudrjasch.**
Kuehn, Harold H. (fl. 1956) M   **Kuehn**
Kuehne, Paul E. A   **Kuehne**
(Kuekenthal, Georg)
    see Kükenthal, Georg   **Kük.**
(Kuetzing, Friedrich Traugott)
    see Kützing, Friedrich Traugott   **Kütz.**
Kufferath, Hubert (1882–1957) AM   **Kuff.**
Kugler, E. (fl. 1874–1883) P   **Kugler**
Kugrens, Paul (1942–) A   **Kugrens**
Kuhbier, Manfred Heinrich (1834–) S   **Kuhbier**
Kuhl, Heinrich (1796–1821) MS   **Kuhl**
Kuhlenkamp, Ralph A   **Kuhlenk.**
Kühlewein, Paul Eduard (1798–1870) PS   **Kühlew.**
Kuhlman, E.G. (fl. 1969) M   **Kuhlman**
Kuhlmann, João Geraldo (1882–1958) S   **Kuhlm.**
Kuhlmann, Moysés (1906–) S   **M.Kuhlm.**
Kühlwein, Hans (1911–) M   **Kühlw.**
Kuhn, E. (fl. 1979) S   **E.Kuhn**
Kuhn, Friedrich Adalbert Maximilian ('Max') (1842–1894) MPS   **Kuhn**
Kühn, Julius Gotthelf (1825–1910) AM   **J.G.Kühn**
Kuhn, Jürg A. A   **J.A.Kuhn**
Kuhn, Oskar (fl. 1955) F   **O.Kuhn**
Kuhn, Richard (1863–) A   **R.Kuhn**
Kuhnemann, Oscar (fl. 1937) AB   **Kuhnem.**
Kühner, Robert (1904–) M   **Kühner**
Kuhnholtz–Lordat, G. (fl. 1949) M   **Kuhnh.-Lord.**
Kuhnt, Otto S   **Kuhnt**
Kuijt, Job (1930–) MS   **Kuijt**
Kuiper, J. (1953–) A   **Kuiper**

Kujala, Viljo (Vilho) (1891–) M	**Kujala**
Kükenthal, Georg (1864–1955) S	**Kük.**
Kukk, Erich G. (1828–) A	**Kukk**
Kukkonen, Ilkka Toiva Kalervo (1926–) MPS	**Kukkonen**
Kukreti, B.C. A	**Kukreti**
Kulczyński, Stanislaw (1895–1975) S	**Kulcz.**
Kulescha, Michael Jossiphowitcz (1878–) S	**Kulescha**
Kuleshov, Nikolai Nikolayevich (1890–1968) S	**Kuleshov**
Kulesza, Witold (1891–1938) S	**Kulesza**
Kulieva, Kh. (Ch.) G. (1927–) S	**Kulieva**
Kulik, E.I. A	**E.I.Kulik**
Kulik, Martin Michael (1932–) M	**Kulik**
Kulkar, S.S. (fl. 1986) M	**Kulkar**
Kulkarni, A.R. (fl. 1969) MS	**A.R.Kulk.**
Kulkarni, B.G.Patil (fl. 1971) MS	**B.G.P.Kulk.**
Kulkarni, C.R. (fl. 1977) M	**C.R.Kulk.**
Kulkarni, Chandrakant Ganapatrao (1900–) S	**C.G.Kulk.**
Kulkarni, Chintamani (fl. 1977) M	**C.Kulk.**
Kulkarni, D.K. (1950–) M	**D.K.Kulk.**
Kulkarni, Gopal Subrao (1884–) M	**Kulk.**
Kulkarni, N.B. (fl. 1951) M	**N.B.Kulk.**
Kulkarni, R.L. (fl. 1974) M	**R.L.Kulk.**
Kulkarni, S.M. (1955–) M	**S.M.Kulk.**
Kulkarni, U.K. (fl. 1958) M	**U.K.Kulk.**
Kulkarni, U.V. (fl. 1975) M	**U.V.Kulk.**
Kulkarni, Y.S. (fl. 1949) M	**Y.S.Kulk.**
Kull, Ulrich Otto (1938–) S	**Kull**
Kullhem, Henrik August (1839–1877) M	**Kullh.**
Kullman, Bellis B. (1947–) M	**Kullman**
Kulshrestha, D.D. (fl. 1966) M	**Kulshr.**
Kult, Karel (fl. 1956) M	**K.Kult**
Kultenko, E.S. (fl. 1981–89) S	**Kultenko**
Kultiasow, Michail Vasilevich (1891–1968) S	**Kult.**
(Kultiassow, Michail Vasilevich)	
see Kultiasow, Michail Vasilevich	**Kult.**
Kulumbaeva, A.A. A	**Kulumb.**
Kumamoto, Toshihiko (fl. 1987) M	**Kumam.**
Kumano, Shigeru (1930–) A	**Kumano**
Kumar, Anand (c. 1955–) S	**An.Kumar**
Kumar, Arun AS	**A.Kumar**
Kumar, Arvind (fl. 1988) S	**Arv.Kumar**
Kumar, B.K.Vijay (fl. 1983) S	**B.K.V.Kumar**
Kumar, C.Sathish (fl. 1989) S	**C.S.Kumar**
Kumar, D.Ramesh (1963–) M	**D.R.Kumar**
Kumar Das, Asok A	**Kumar Das**
Kumar, Dinesh (fl. 1968) BM	**D.Kumar**
Kumar, Har Darshan (1934–) A	**H.D.Kumar**
Kumar, P. (fl. 1979) M	**P.Kumar**
Kumar, P.K.Ratheesh (fl. 1990) S	**P.K.R.Kumar**

Kumar, Rhagu (fl. 1987) M	**R.Kumar**
Kumar, S.M. (fl. 1969) M	**S.M.Kumar**
Kumar, S.S. (1936–) B	**S.S.Kumar**
Kumar, Surendra A	**S.Kumar**
Kumari, Gorti Raghawa (Raghava) (1929–) S	**Kumari**
Kumazawa, Masao (1904–) MS	**Kumaz.**
Kumbhojkar, Mohan Shamrao (1942–) M	**Kumbh.**
Kuminova, Aleksandra Vladimirovna (1911–) S	**Kuminova**
Kummer, Ferdinand (1820–1870) S	**Kumm.**
Kummer, Georg (1885–1954) S	**G.Kumm.**
Kummer, Paul (1834–1912) BM	**P.Kumm.**
Kümmerle, Jenö Béla (1876–1931) PS	**Kümmerle**
Kümmerling, H. (fl. 1991) M	**Kümmerl.**
Kümpel, Horst S	**Kümpel**
Kumsare, Antonia Jakobovna (1902–) A	**Kumsare**
Kunc, K. (fl. 1962) M	**Kunc**
Kundalkar, B.D. (fl. 1983) M	**Kund.**
Kündig, Jakob (1863–1933) S	**Kündig**
Kundu, Balsi Chand (1905–) AS	**Kundu**
Kunert, Friedrich (1863–) S	**Kunert**
Kunert, J. (fl. 1968) M	**J.Kunert**
Kung, Hsian Shiu (1932–) P	**H.S.Kung**
Kung, Hsien Wu (1897–) S	**H.W.Kung**
Kunkel, Günther W.H. (1928–) PS	**G.Kunkel**
Kunkel, Louis Otto (1884–1960) M	**Kunkel**
Künkele, Siegfried (1931–) S	**Künkele**
Kunow, Gustav (1847–1912) S	**Kunow**
Künstler, J. A	**Künstler**
Kunth, Karl (Carl) Sigismund (1788–1850) ABMPS	**Kunth**
Kuntze, Carl (Karl) Ernst (Eduard) Otto (1843–1907) ABFMPS	**Kuntze**
Kuntze, J.E. (fl. 1979) M	**J.E.Kuntze**
Kunwar, I.K. (fl. 1986) M	**Kunwar**
Kunz, Hans (1904–) S	**Kunz**
Kunze, Gustav (1793–1851) AMPS	**Kunze**
Kunze, Johannes (–1881) M	**J.Kunze**
Kunze, Karl Sebastian Heinrich (1774–1820) S	**K.Kunze**
Kunze, Richard Ernest (1838–1919) S	**R.E.Kunze**
Kunzmann, H. (fl. 1985) S	**Kunzmann**
Kuo, Chen Meng (1948–) BP	**C.M.Kuo**
Kuo, Chin Chen S	**Chin C.Kuo**
Kuo, Chiu Chen (1933–) S	**Chiu C.Kuo**
Kuo, John (1938–) S	**J.Kuo**
Kuo, Pung (Pen) Chao (fl. 1980) S	**P.C.Kuo**
Kuo, Yu Chieh A	**Y.C.Kuo**
Kuono, K. (fl. 1972) M	**Kuono**
Kupatadze, G.A. (fl. 1968) S	**Kupat.**
(Kupčok, Samuel)	
see Kupcsok, Samuel	**Kupcsok**
Kupcsok, Samuel (1850–1914) S	**Kupcsok**

Kupfer, Elsie M. (fl. 1902) M — **Kupfer**
Küpfer, Philippe (1942–) S — **P.Küpfer**
Kupffer, Karl Reinhold (1872–1935) S — **Kupffer**
Kupicha, Frances Kristina (1947–) S — **Kupicha**
Kupper, E.M. S — **E.M.Kupper**
Kupper, Walter (1874–1953) S — **Kupper**
Kuprevich, Vasiliĭ Feofilovich (1897–1969) M — **Kuprev.**
(Kuprevič, Vasiliĭ Feofilovich)
   see Kuprevich, Vasiliĭ Feofilovich — **Kuprev.**
Kuprianov, A.N. (fl. 1972) S — **Kupr.**
Kuprianova, E.S. (fl. 1981) M — **E.S.Kuprian.**
Kuprianova, Lyudmila Andreyeva (Andreevna) (1914–1987) S — **Kuprian.**
(Kupriyanova, Lyudmila Andreyeva (Andreevna))
   see Kuprianova, Lyudmila Andreyeva (Andreevna) — **Kuprian.**
Kuraishi, H. (fl. 1958) M — **Kuraishi**
Kurata, S. (fl. 1931) M — **Kurata**
Kurata, Satoru (1922–1978) PS — **Sa.Kurata**
Kurata, Shigeo (fl. 1965–72) S — **Sh.Kurata**
Kurbanbekov, Z.K. (1935–) S — **Kurbanb.**
Kurbanov, D.K. (1946–) S — **Kurbanov**
Kurbansachatov, A.K. (fl. 1968) M — **Kurbans.**
(Kurbansakhatov, A.K.)
   see Kurbansachatov, A.K. — **Kurbans.**
Kurbatski, Vladimir Ivanovich (1941–) S — **Kurbatski**
(Kurbatskii, V.I.)
   see Kurbatski, V.I. — **Kurbatski**
Kuribayashi, Kazue (–1954) M — **Kurib.**
Kurita, Siro (fl. 1969–77) P — **Kurita**
Kuriyal, S.K. (fl.1990) M — **Kuriyal**
Kuriyama, Nobuhiro A — **Kuriy.**
Kurogi, Munengo (1921–1988) AS — **Kurogi**
Kurokawa, Syo (1926–) M — **Kurok.**
Kurosaki, N. (fl. 1989) S — **N.Kurosaki**
Kurosaki, S. (fl. 1983) S — **S.Kurosaki**
Kurosawa, Eiiti (fl. 1926) M — **Kuros.**
Kurosawa, Gompei (fl. 1908) M — **G.Kuros.**
Kurosawa, Sachiko (1927–) S — **S.Kuros.**
(Kurozawa, Gompei)
   see Kurosawa, Gompei — **G.Kuros.**
Kurr, Johann Gottlieb von (1798–1870) F — **Kurr**
Kurschner, Harald (1950–) B — **Kurshcner**
Kurth, H. A — **Kurth**
Kurtto, Arto (1951–) S — **Kurtto**
Kurtz, Fritz (Federico) (1854–1920) FS — **Kurtz**
Kurtz, Karl Marie Max (1846–1910) S — **K.Kurtz**
Kurtzman, Cletus P. (1938–) M — **Kurtzman**
Kuru, M. (fl. 1932) M — **Kuru**
(Kurybayashi, Kazue)
   see Kuribayashi, Kazue — **Kurib.**

Kurz, Albert (1886–) A	**A.Kurz**
Kurz, Hermann (1886–) B	**H.Kurz**
Kurz, Wilhelm Sulpiz (see also Amann, J.) (1834–1878) PS	**Kurz**
Kurzynski, T.A. (fl. 1978) M	**Kurzynski**
Kusaka, Masao (1915–) S	**Kusaka**
Kusakabe, I. (fl. 1983) S	**Kusak.**
Kušan, Fran (1902–) M	**Kušan**
Kusano, Shunsuke (1874–1962) M	**Kusano**
Kuschel, G. (fl. 1963) S	**Kuschel**
(Kuschke, G.)	
see Kushke, G.	**Kuschke**
(Kuseneva, Olga Iakinfovna)	
see Kuzeneva, Olga Iakinfovna	**Kuzen.**
Kushke, G. (fl. 1913) M	**Kuschke**
Kushwaha, R.K.S. (fl. 1976) M	**Kushwaha**
(Kusnetzou, Nicolai Ivanovitch)	
see Kusnezow, Nicolai Ivanowicz	**Kusn.**
(Kusnezov, Ivan V.)	
see Kusnezow, Ivan V.	**I.V.Kusn.**
(Kusnezov, Nicolai Ivanovitch)	
see Kusnezow, Nicolai Ivanovitch	**Kusn.**
(Kusnezov, S.)	
see Kusnezow, S.	**S.Kusn.**
(Kusnezov, Vladimir Alexandrovic)	
see Kusnezow, Vladimir Alexandrovich	**V.A.Kusn.**
Kusnezow, Ivan V. (1881–1945) S	**I.V.Kusn.**
Kusnezow, N.M. S	**N.M.Kusn.**
Kusnezow, Nicolai Ivanovitch (1864–1932) PS	**Kusn.**
Kusnezow, S. (fl. 1971) S	**S.Kusn.**
Kusnezow, Vladimir Alexandrovich (1887–1940) S	**V.A.Kusn.**
Kusnezowa, T.T. (fl. 1952) M	**Kusnezowa**
Kušta, Jan (1900–) A	**Kušta**
Küster, Ernst (1874–1953) A	**Küster**
Kuthan, Ing.Jan (fl. 1976) M	**Kuthan**
Kuthatheladze, A.I. (1912–) S	**A.I.Kuth.**
Kuthatheladze, D.Sh. (1937–) S	**D.S.Kuth.**
Kuthatheladze, Schushana Ilyinichna (1905–) S	**Kuth.**
Kuthubutheen, Ahmed Jalaludin (fl. 1979) M	**Kuthub.**
Kuttin, E.S. (fl. 1981) M	**Kuttin**
Kützing, Friedrich Traugott (1807–1893) AMS	**Kütz.**
Kuusk, Vilma (1931–) S	**Kuusk**
Kuvaev, V.B. (1918–) S	**Kuvaev**
Kuwabara, Yoshiko (fl. 1969) M	**Kuwab.**
Kuwahara, Yukinobu (1927–) B	**Kuwah.**
Kuyper, Thomas W. (fl. 1986) M	**Kuyper**
Kuyumdzhieva, A. (fl. 1989) M	**Kuyumdzh.**
Kuzaha, S. (fl. 1973) M	**Kuzaha**
Kuzeneva, Olga Iakinfovna (1887–1978) PS	**Kuzen.**
(Kuzeneva–Prochorova, Olga Iakinfovna)	
see Kuzeneva, Olga Iakinfovna	**Kuzen.**

Kuzinsky, P.A.von (fl. 1889) S   **Kuzinsky**
Kuzmanov, Bogdan Antonov (1934–1991) S   **Kuzmanov**
Kuzmin, G.V.   A   **G.V.Kuzmin**
Kuzmin, J.A. (1925–) S   **Kuzmin**
Kuzmina, Angelina Ivanovna (1922–) A   **A.I.Kuzmina**
Kuzmina, L.V. (1932–) S   **Kuzmina**
(Kuznetszov, Ivan V.)
   see Kusnezow, Ivan V.   **I.V.Kusn.**
(Kuznetzov, N.M.)
   see Kusnezow, N.M.   **N.M.Kusn.**
(Kuznetzov, Nicolai Ivanovitch)
   see Kusnezow, Nicolai Ivanovitch   **Kusn.**
(Kuznetzov, S.)
   see Kusnezow, S.   **S.Kusn.**
(Kuznetzov, Vladimir Alexandrovic)
   see Kusnezow, Vladimir Alexandrovich   **V.A.Kusn.**
(Kuznetzova, T.T.)
   see Kusnezowa, T.T.   **Kusnezowa**
(Kuznezov, Nicolai Ivanovitch)
   see Kusnezow, Nicolai Ivanowicz   **Kusn.**
Kvaček, Zlatko (1937–) S   **Kvaček**
Kvasnikov, E.I. (fl. 1979) M   **Kvasn.**
Kvist, Gustav (1909–) S   **G.Kvist**
Kvist, L.P. (fl. 1988) S   **L.P.Kvist**
Kwashnina, E.S. (fl. 1928) M   **Kwashn.**
Kwok, Kin Fong (fl. 1963) S   **K.F.Kwok**
(Kwon, Kyung Joo)
   see Kwon-Chung, Kyung Joo   **Kwon-Chung**
Kwon-Chung, Kyung Joo (fl. 1965) M   **Kwon-Chung**
Kyde, M.M. (fl. 1986) M   **Kyde**
Kyhos, Donald William (1929–) S   **Kyhos**
Kylin, Johan Harald (1879–1949) A   **Kylin**
Kyono, T. (fl. 1962) M   **Kyono**
Kytövuori, Ilkka (1941–) M   **Kytöv.**

L'Hardy-Halos, Marie-Thérèse   A   **L'Hardy-Halos**
L'Héritier de Brutelle, Charles Louis (1746–1800) S   **L'Hér.**
L'Herminier, Ferdinand (1802–1866) S   **L'Herm.**
(L'Obel, Mathias de)
   see Lobel, Mathias de   **Lobel**
(La Billardière, Jacques Julien Houtton de)
   see Labillardière, Jacques Julien Houtton de   **Labill.**
(la Condamine, Charles Marie de)
   see Condamine, Charles Marie de la   **Cond.**
la Croix, Isobyl Florence (1933–) S   **la Croix**
La Duke, John C. (1950–) S   **La Duke**
La Favre, J.S. (fl. 1981) M   **La Favre**
La Fons, Alexandre de Mélicocq (1802–1867) S   **La Fons**
la Gasa, Eric (fl. 1970) S   **la Gasa**

(La Gasca, Mariano)	
see Lagasca y Segura, Mariano	**Lag.**
La Llave, Pablo de (1773–1833) S	**La Llave**
(la Peirouse, Philippe Picot de)	
see Lapeyrouse, Philippe Picot de	**Lapeyr.**
La Porte, Juan S	**La Porte**
(la Pylaie, Auguste Jean Marie Bachelot de)	
see Bachelot de la Pylaie, Auguste Jean Marie	**Bach.Pyl.**
La Rivers, Ira A	**La Rivers**
(la Roche, Daniel de)	
see Delaroche, Daniel	**D.Delaroche**
(la Roche, François de)	
see Delaroche, François	**F.Delaroche**
La Serna Ramos, Irene E. (fl. 1981) S	**La Serna**
La Touche, C.J. (fl. 1950) M	**La Touche**
la Valva, Vincenzo (1947–) S	**la Valva**
La, Yong Joon (fl. 1988) M	**Y.J.La**
Laan, F.M.van der (fl. 1986) S	**Laan**
Laban, Friedrich Christian (fl. 1866) S	**Laban**
Labbé, Alphonse A	**Labbé**
Labh, L. A	**Labh**
Labillardière, Jacques Julien Houtton de (1755–1834) AMPS	**Labill.**
Laborde, J. (fl. 1897) M	**Laborde**
Laborde, L. S	**L.Laborde**
Laborel, Jacques A	**Laborel**
Labouret, J. (fl. 1853–58) S	**Labour.**
Laboureur, Pierre (fl. 1952) M	**Laboureur**
Labram, Jonas David (1785–1852) S	**Labram**
Labrie, Jean Joseph (–1927) S	**Labrie**
Labrousse, Francis Jean (fl. 1931) M	**Labr.**
Labroy (fl. 1903–1924) S	**Labroy**
Lacaita, Charles Carmichael (1853–1933) PS	**Lacaita**
Lacassagne, Marcel S	**Lacass.**
Lace, John Henry (1857–1918) S	**Lace**
Lacerda, F.S.de A	**Lacerda**
Lacey, John (1937–) M	**J.Lacey**
Lacey, William Springthorpe (1917–) BF	**Lacey**
Lachance, Marc–André (fl. 1981) M	**Lachance**
(Lachashvili, I.Ja.)	
see Latschaschvili, I.Ja.	**Latsch.**
Lachenal, Werner de (1736–1800) S	**Lachen.**
Lachenaud, Georges (fl. 1901) B	**Lachenaud**
Lachmann, Alphonse (1917–1961) B	**A.Lachm.**
Lachmann, Heinrich Wilhelm Ludolf (1801–1861) S	**Lachm.**
Lachmann, Johannes (1832–1860) A	**J.Lachm.**
Lachmann, P. (1851–1907) PS	**P.Lachm.**
Lachot, Henri (1850–) S	**Lachot**
Lack, Hans Walter (1949–) S	**Lack**
Lackey, Elsie Wattie (1901–) A	**E.W.Lackey**
Lackey, J.A. (fl. 1978) S	**J.A.Lackey**

Lackey, James Bridges (1893–) A — **Lackey**
Lackowitz, August Wilhelm (1836–1916) S — **Lackow.**
(Lackschewitz, Paul)
   see Lakschewitz, Paul — **Laksch.**
Lackström, Emil Frithiof (–1883) P — **Lackström**
Lacoizqueta, José Maria de (1831–1891) S — **Lacoizq.**
Lacoste de Díaz, Elsa Nélida (1923–) A — **E.N.Lacoste**
Lacoste, Louis (1927–) M — **Lacoste**
Lacouture, Charles (1832–1908) B — **Lacout.**
Lacroix, Edouard Georges de (1858–1907) M — **E.G.Lacroix**
Lacroix, Louis–Sosthène Veyron de (1818–1864) BMS — **Lacroix**
Lacroix, S.de (fl. 1859) S — **S.Lacroix**
Lacy, R.C. (fl. 1949) M — **Lacy**
Ladau, Johan Fredrik (fl. 1799) S — **Ladau**
Ladbrook, James (1870–) S — **Ladbr.**
Ladd, Douglas (fl. 1987) M — **Ladd**
Lademann, Johann Matthias Friedrich (1760–1810) S — **Lademann**
(Ladero Alvarez, Miguel)
   see Ladero, Miguel — **Ladero**
Ladero, Miguel (1939–) S — **Ladero**
Ladizinsky, Gideon (1936–) S — **Ladiz.**
Ladó, Carlos (1955–) M — **Ladó**
(LaDuke, John C.)
   see La Duke, John C. — **La Duke**
Ladygina, G.M. (1929–1989) S — **Ladygina**
Ladyzenskaya, Claudia Ivanovna (1900–) B — **Ladyz.**
Laegaard, Simon (1933–) S — **Laegaard**
Laencina, José (fl. 1987) S — **Laencina**
(Laer, H.Van)
   see Van Laer, H. — **Van Laer**
Laessøe, Thomas (1958–) M — **Laessøe**
Laestadius, Carl Petter (1835–1920) M — **C.Laest.**
Laestadius, Lars Levi (1800–1861) PS — **Laest.**
Laet–Contich, Franz von (fl. 1904) S — **Laet-Cont.**
Lafargue, Francoise A — **Lafargue**
Lafaurie, Danielle A — **Lafaurie**
Laferrière, Joseph E. (fl. 1990) MS — **Laferr.**
Lafferty, Henry Aloysius (1891–1954) M — **Laff.**
Laffitte, Charles (1811–1895) S — **Laffitte**
Laffon, J.C. (fl. 1848) S — **Laffon**
Laflamme, (Par) Gaston (fl. 1976) M — **Lafl.**
(Lafons de Mélicocq, Alexandre de)
   see La Fons, Alexandre de Mélicocq — **La Fons**
Lafont, A. A — **Lafont**
LaFrankie, J.V. (fl. 1986) S — **LaFrankie**
Lag, Jul (1915–) B — **J.Lag**
Lagarde, Joannes Joseph (1866–) M — **Lagarde**
Lagasca y Segura, Mariano (1776–1839) BPS — **Lag.**
Lagerberg, Karl Erik Torsten (1882–1970) MS — **Lagerb.**

Lagerheim, Nils Gustaf (von, de) (1860–1926) ACMS	**Lagerh.**
Lagerstedt, Nils Gerhard Wilhelm (1847–1925) A	**Lagerst.**
Lagger, Franz Joseph (1802–1870) MS	**Lagger**
Lagière, R. (fl. 1946) M	**Lagière**
Lago, Elena (fl. 1989) S	**Lago**
(Lagrèse–Fossat, Adrian Rose Arnaud)	
see Lagrèze–Fossat, Adrian Rose Arnaud	**Lagr.-Foss.**
Lagrèze–Fossat, Adrian Rose Arnaud (1818–1874) S	**Lagr.-Foss.**
Laguna y Villanueva, Máximo (1826–1902) PS	**Laguna**
Lahaie, D.G. (fl. 1978) M	**Lahaie**
Laharpe, Jean Jacques Charles de (1802–1877) S	**Laharpe**
Lahm, Johann Gottlieb Franz–Xaver (1811–1888) M	**J.Lahm**
Lahm, Wilhelm (1856–) S	**W.Lahm**
Lahman, Bertha Marion (1872–) S	**Lahman**
Lahmeyer, R.H. S	**Lahmeyer**
Lahondere, C. (fl. 1988) S	**Lahondere**
Lai, Ming Jou (1949–) BMPS	**M.J.Lai**
Lai, Qi Rui (fl. 1989) S	**Q.R.Lai**
Lai, Xing Hua (fl. 1991) M	**X.H.Lai**
Lai, Yi Qi (fl. 1979) M	**Y.Q.Lai**
Laibach, Friedrich (1885–) M	**Laib.**
Laicharding, Johann Nepomuk von (1754–1797) S	**Laichard.**
(Laicharting, Johann Nepomuk von)	
see Laicharding, Johann Nepomuk von	**Laichard.**
Laine, Unto Olavi (1930–) S	**Laine**
Laing, Robert Malcolm (1865–1941) APS	**Laing**
Laing, S.A.K. (fl. 1991) M	**S.A.K.Laing**
(Laínz Gallo, P.Manuel)	
see Laínz, P.Manuel	**M.Laínz**
Laínz, J.M. (1900–) S	**Laínz**
Laínz, P.Manuel (1923–) PS	**M.Laínz**
Laio, Yin Zhang (fl. 1980) M	**Y.Z.Laio**
Laird, Marshall (fl. 1956) AM	**Laird**
Lakela, Olga Korhoven (1890–) S	**Lakela**
Lakhanpal, R.N. (fl. 1955) F	**R.N.Lakh.**
Lakhanpal, T.N. (fl. 1968) M	**T.N.Lakh.**
Lakon (fl. 1907) S	**Lakon**
Lakon, Georg (fl. 1935) M	**G.Lakon**
Lakowitz, Conrad Waldemar (1859–1945) AF	**Lakow.**
Lakschewitz, Paul (1865–1936) S	**Laksch.**
Lakshmanan, K.K. (fl. 1988) M	**Lakshm.**
Lakshminarasimhan, P. (fl. 1986) S	**Lakshmin.**
Lakshnakara, Mom Chao S	**Lakshn.**
Lakusić, Radomir S	**Lakusić**
Lal, Akshaibar (fl. 1951) M	**Lal**
Lal, B. (fl. 1981) M	**B.Lal**
Lal, Jagdish A	**J.Lal**
Lal, S.B. (fl. 1973) M	**S.B.Lal**
Lal, Shashi P. (fl. 1974) M	**S.P.Lal**
Lall, G. (fl. 1962) M	**Lall**

Lallana, A.M. (fl. 1981) S	**Lallana**
(Lallemant, Julius Léopold Edouard Ave)	
see Avé–Lallemant, Julius Léopold Edouard	**Avé-Lall.**
Lalli, Giorgio (1952–) M	**Lalli**
(Lalonde, Louis-Marie)	
see Louis-Marie, (Père)	**Louis-Marie**
Laloraya, Vinod K. A	**Laloraya**
Lalung–Bonnaire A	**Lal.-Bonn.**
Lam, Annie (fl. 1986) M	**A.Lam**
Lam, Herman Johannes (1892–1977) BPS	**H.J.Lam**
Lamarck, Jean Baptiste Antoine Pierre de Monnet de (1744–1829) AMPS	**Lam.**
(Lamare Occhioni, E.M.de)	
see Occhioni, Elêna Maria de Lamare	**Occhioni f.**
Lamb, Anthony L. (fl. 1982) S	**A.L.Lamb**
Lamb, Brian M. (fl. 1958) S	**B.M.Lamb**
Lamb, Edgar (–1980) S	**E.Lamb**
Lamb, Ivan Mackenzie (1911–1990) AMS	**I.M.Lamb**
Lambert, Aylmer Bourke (1761–1842) S	**Lamb.**
Lambert, Edmund Bryand (1897–) M	**E.B.Lamb.**
Lambert, François (1859–1940) M	**F.Lamb.**
Lambert, Fred Dayton (1871–) A	**F.D.Lamb.**
Lambert, J.G. (fl. 1985) S	**J.G.Lamb.**
Lambert, Léon Célestin (1867–) S	**L.C.Lamb.**
Lambert, Wilhelm (1827–1860) S	**W.Lamb.**
Lambertye, Léonce de (1810–1877) S	**Lambertye**
Lambinon, Jacques (Ernest Joseph) (1936–) BMS	**Lambinon**
Lambotte, Jean Baptiste Émil (Ernest) (1832–1905) M	**Lambotte**
Lamboy, W.F. (fl. 1988) S	**Lamboy**
Lame, Elton M.C. S	**Lame**
Lami, Robert (fl. 1935) AS	**Lami**
Lamkey, Ernest Michael Rudolph (1890–) M	**Lamkey**
Lammers, Thomas G. (1955–) S	**Lammers**
Lammersdorff, Johann Anton (fl. 1781) S	**Lammersd.**
Lammes, Tapio (1934–) B	**Lammes**
Lamond, Jenifer M. (1936–) S	**Lamond**
Lamont, Byron Bernard (1945–) S	**Lamont**
Lamore, Bette J. (fl. 1977) M	**Lamore**
Lamotte, Martial (1820–1883) S	**Lamotte**
Lamotte, Robert Smith (fl. 1952) ABFM	**R.S.Lamotte**
(LaMotte, Robert Smith)	
see Lamotte, Robert Smith	**R.S.Lamotte**
Lamoure, D. (fl. 1965) M	**Lamoure**
Lamoureux, Charles H. (1933–) S	**Lamoureux**
Lamouroux, Jean Vincent Félix (1779–1825) AS	**J.V.Lamour.**
Lamouroux, Jean (Jeanin) Pierre Péthion (1797–1866) S	**J.P.Lamour.**
(Lamson–Scribner, Frank)	
see Scribner, Frank Lamson-	**Scribn.**
Lamy de la Chapelle, Pierre Marie Édouard (1804–1886) CM	**Lamy**
(Lamy, Pierre Marie Édouard de la Chapelle)	
see Lamy de la Chapelle, Pierre Marie Édouard	**Lamy**

Lan, Kai Min (fl. 1983) S	K.M.Lan
Lan, Y.C. (fl. 1977) P	Y.C.Lan
Lan, Yong Zhen (1933–) S	Y.Z.Lan
Lancaster, Charles Roy (1937–) S	Lancaster
Lańcucka–Środoniowa, Maria (1913–) S	Lańc.-Środ.
Landa, J. (fl. 1986) M	Landa
Lande, Max (fl. 1907) B	Lande
Lander, Nicholas Sean (1948–) S	Lander
(Landingham, Sam L. Van)	
see VanLandingham, Sam L.	VanLand.
Landolt, Elias (1926–) S	Landolt
Landon, John W. (fl. 1975) S	J.W.Landon
Landon, K.C. (fl. 1978) S	K.C.Landon
Landoz, J. (1793–1866) S	Landoz
(Landrecy, Viktor Cypers von)	
see Cypers von Landrecy, Viktor	Cypers
Landrieu, M. (fl. 1912) M	Landrieu
Landrum, Leslie Roger (1946–) BS	Landrum
Landry, Garrie Paul (1951–) PS	G.P.Landry
Landry, Pierre (1951–) S	P.Landry
Landsborough, David (1779–1854) A	Landsb.
Landschoot, P.J. (fl. 1989) M	Landsch.
Landwehr, J. (fl. 1969) S	Landwehr
Lane, Irwin E. (1926–) PS	Lane
Lane, L.C. (fl. 1984) M	L.C.Lane
Lane, Meredith A. (1951–) S	M.A.Lane
Lane, T.M. (fl. 1988) S	T.M.Lane
Lane-Poole, Charles Edward (1885–1970) S	Lane-Poole
Lanessan, Jean Marie Antoine de (1843–1919) S	Laness.
Lanfranco, E. (fl. 1982) S	Lanfr.
Lang, A.G. (fl. 1975) F	A.G.Lang
Láng, Adolph (Adolf) Franz (1795–1863) PS	Láng
Lang, Frank Alexander (1937–) PS	F.A.Lang
Lång, K.Gösta W. (1875–1912) M	K.G.W.Lång
Lang, Kai Yung (1936–) S	K.Y.Lang
Lang, Karl Heinrich (1800–1843) S	K.H.Lang
Lang, Norma Jean (1931–) A	N.J.Lang
Lang, Otto Friedrich (1817–1847) S	O.Lang
Lang, Qing (fl. 1987) S	Q.Lang
Lang, Thomas (fl. c.1853) S	T.Lang
Lang, William Henry (1874–) AFM	W.H.Lang
Langdon, Raymond Forbes Newton (1916–) M	Langdon
Lange, Axel Edward (1871–1941) S	A.E.Lange
Lange, Bodil (1918–) B	B.Lange
Lange, F.W. A	F.W.Lange
Lange, Jakob Emanuel (1864–1941) M	J.E.Lange
Lange, Johan Martin Christian (1818–1898) ABMPS	Lange
Lange, Lene (fl. 1977) M	L.Lange
Lange, Morten (1919–) M	M.Lange

Lange, Morten Thomsen (1824–1875) B	M.T.Lange
Lange, Otto Ludwig (1927–) M	O.L.Lange
Lange, Robert Terence (1934–) M	R.T.Lange
Lange, Thorvald Arthur (1872–1957) S	T.A.Lange
Lange-Bertalot, Horst A	Lange-Bert.
Langenbach, Gustav (1831–1873) S	Langenb.
Langenfeld, Voldemar Theodorovich (1923–) S	Langenf.
Langenhan (fl. 1905) F	Langenhan
Langenheim, Jean Harmon (1925–) S	Langenh.
Langer, Ewald (fl. 1990) M	Langer
Langer, Sándor (Alexander) A	S.Langer
Langeron, Maurice Charles Pierre (1874–1950) BFM	Langeron
Langethal, Christian Eduard (1806–1878) S	Langeth.
Langford, G. (fl. 1982) S	G.Langford
Langford, Martha F. A	M.F.Langford
(Langhe, Charles De)	
see De Langhe, Charles	C.De Langhe
(Langhe, Joseph Edgard De)	
see De Langhe, Joseph Edgard	De Langhe
Langkavel, Bernhard August (1825–1902) S	Langk.
Langlois, Auguste Barthélemy (1832–1900) BM	Langl.
Langman, Johann Friedrich (fl. 1841) S	Langm.
Langridge, D.F. A	Langr.
Langsam, D.M. (fl. 1986) M	Langsam
Langsdorff, Georg Heinrich von (1774–1852) PS	Langsd.
Langstedt, Friedrich Ludwig (1750–1804) S	Langst.
Langvad, F. (fl. 1989) M	Langvad
Lanier, L. (fl. 1978) M	Lanier
Lanjouw, Joseph (1902–1984) S	Lanj.
Lankester, Edwin Ray (1814–1874) A	Lank.
Lanne, C. (fl. 1979) M	Lanne
Lanneau, Christiane (fl. 1967) M	Lanneau
Lanner, Ronald M. (fl. 1974) S	Lanner
(Lannoy, de)	
see de Lannoy	de Lannoy
Lannoy, Gilbert (fl. 1991) M	Lannoy
Lanquetin, Paule (fl. 1973) M	Lanq.
Lantzius-Béninga, Georg Boyung Scato (1815–1871) S	Lantz.-Bén.
Lantzsch, Kurt A	Lantzsch
Lányi, Béla (–1917) S	Lányi
Lanza, Domenico (1868–1940) S	Lanza
Lanzi, Matteo (1824–1908) AM	Lanzi
Lanzoni, Gianbattista (fl. 1985) M	Lanzoni
Lapage, G. A	Lapage
(Lapeirouse, Philippe Picot de)	
see Lapeyrouse, Philippe Picot de	Lapeyr.
Lapeyrouse, Philippe Picot de (1744–1818) PS	Lapeyr.
Lapham, Increase Allen (1811–1875) S	Lapham
Lapie, Georges (fl. 1914) S	Lapie

Lapierre, Jean Marie (1754–1834) S	**Lapierre**
Lapinpuro, L. (fl. 1986) S	**Lapinpuro**
Laplanche, Maurice Counjard de (1843–1904) M	**Lapl.**
Laporte, Louis Jacques (fl. 1930) A	**Laporte**
Laporte, M. A	**M.Laporte**
Laptev, Yu.P. (1933–) S	**Laptev**
Lapukhova, S. (fl. 1971) M	**Lapukhova**
(Lara, José Maria Pérez)	
see Pérez Lara, J.M.	**Pérez Lara**
Lara, Raúl (1934–) S	**R.Lara**
(Lara Rico, Raúl)	
see Lara, Raúl	**R.Lara**
(Larambergue, De)	
see De Larambergue	**De Laramb.**
Larbalestier, Charles du Bois (1838–1911) M	**Larbal.**
Larber, Giovanni (1785–1845) S	**Larber**
Larch, J. (fl. 1967) F	**Larch**
Largent, David Lee (1937–) M	**Largent**
Larisey, Mary Maxine (1909–) S	**Larisey**
Laristschev, A.A. A	**Laristschev**
Laroche, G. (fl. 1909) M	**Laroche**
Laroche, Rose Claire (fl. 1973) S	**R.C.Laroche**
Laroche–Collet, S. (fl. 1983) M	**Lar.-Coll.**
Larrañaga, Dámaso Antonio (1771–1848) PS	**Larrañaga**
Larreategui, Joseph (José) Dionisio (fl. 1795–c.1805) S	**Larreat.**
Larrondo, J.V. (fl. 1990) M	**Larrondo**
Larsen, B.B. (fl. 1986) S	**B.B.Larsen**
Larsen, Egon (1928–1969) S	**E.Larsen**
Larsen, Esther Louise (1901–) S	**Larsen**
Larsen, H.J. (fl. 1975) M	**H.J.Larsen**
Larsen, J.E. Bregnhøj (fl. 1970) M	**J.E.B.Larsen**
Larsen, Jacob A	**J.Larsen**
Larsen, Kai (1926–) S	**K.Larsen**
Larsen, Karin (fl. 1971) M	**Kar.Larsen**
Larsen, Michael J. (1938–) M	**M.J.Larsen**
Larsen, Poul (1864–1938) M	**P.Larsen**
Larsen, Supee Saksuwan (1939–) S	**S.S.Larsen**
Larson, Russell Harold (1904–1961) M	**Larson**
Larsson, Karl–Henrik (1948–) M	**K.H.Larss.**
Larsson, Lars Magnus (1822–1884) S	**Larss.**
Larter, Clara Ethelinda (1847–1936) B	**Larter**
Lasch, Wilhelm Gottfried (1787–1863) MPS	**Lasch**
Lasché, A. (fl. 1882) M	**Lasché**
Lasebna, A.M. (1922–) S	**Lasebna**
Lasègue, Antoine (1793–1873) S	**Lasègue**
Laskaris, Thomas (fl. 1950) M	**Laskaris**
Laskowski, Chester Walter (1941–) S	**Lask.**
Lasseigne, Alex A. (1944–) S	**Lass.**
Lassen, Per (1942–) S	**Lassen**
Lasser, Tobias (1911–) S	**Lasser**

Lasseur, Ph. (fl. 1922) M — **Lasseur**
Lassimonne, Simon Etienne (fl. 1932) S — **Lassim.**
Lasteyrie–Dusaillant, Charles Philippe de (1759–1849) S — **Last.-Dus.**
Lata, Kanchan (fl. 1979) M — **Lata**
Latapie, François de Paule A. (1839–1823) S — **Latap.**
Latch, G.C.M. (fl. 1965) M — **Latch**
(Latchaschvili, I.Ja.)
   see Latschaschvili, I.Ja. — **Latsch.**
Laterrade, Jean François (1784–1858) S — **Laterr.**
Latgé, Jean–Paul (fl. 1984) M — **Latgé**
Latha, Y. A — **Latha**
Latham, Dennis Harold (fl. 1934) M — **Latham**
Lathrop, Earl W. (1924–) M — **Lathrop**
Latiff, A. (fl. 1982) S — **Latiff**
Latour–Marliac, Joseph (Bory) (1830–1911) S — **Lat.-Marl.**
Latourrette, Marc Antoine Louis Claret de (Fleurieu de) (1729–1793) MS — **Latourr.**
(Latrow, Nikolai Nicolaevich)
   see Lavrov, Nikolai Nicolaevich — **Lavrov**
Latschaschvili, I.Ja. (1918–) S — **Latsch.**
Latterell, Frances M. (fl. 1984) M — **Latterell**
Latz, Peter K. (1941–) S — **Latz**
Latzel, Albert (1858–1946) BS — **Latzel**
Latzel, G. S — **G.Latzel**
(Latzel–Ragusa, A.)
   see Latzel, A. — **Latzel**
Lau, Alfred B. (fl. 1940–1980) S — **A.B.Lau**
Lau, Lan Fong (fl. 1963) S — **L.F.Lau**
Lau, S. S — **S.Lau**
Laubenburg, K.E. (fl. 1899) P — **Laubenb.**
(Laubenfels, David John de)
   see de Laubenfels, David John — **de Laub.**
Laubert, Karl Richard (1870–) M — **Laubert**
Lauby, Antoine A — **Lauby**
Lauche, (Friedrich) Wilhelm (Georg) (1827–1883) PS — **Lauche**
Lauder, Henry Scott A — **Lauder**
Lauener, Lucien André (Andrew) (1918–1991) S — **Lauener**
Laugaste, R.A. A — **Laugaste**
Laughlin, Kendall S — **Laughlin**
Laughton, Elaine M. (fl. 1948) M — **Laughton**
Lauman, George Neuman (1874–1944) S — **Lauman**
Launay, de (fl. 1793) S — **Laun.**
Laundon, Geoffrey Frank (Gillian) (1938–1984) M — **G.F.Laundon**
Laundon, Jack Rodney (1934–) AMS — **J.R.Laundon**
Launert, Georg Oskar Edmund (1926–) PS — **Launert**
Laurent, Émile (1861–1904) S — **Laurent**
Laurent, Jules (1860–1918) S — **J.Laurent**
Laurent, Louis Aimé Alexandre (1873–1947) AF — **L.Laurent**
Laurent, Marcel Désiré Joseph (1879–1924) S — **M.Laurent**
(Laurent–Täckholm, Vivi)
   see Täckholm, Vivi — **Täckh.**

Laurer, Johann Friedrich (1798–1873) BMS   **Laurer**
Lauret, Michel   A   **Lauret**
Laurila, Matti (1915–1942) M   **Laurila**
Laursen, Gary A. (fl. 1973) M   **Laursen**
Lausi, Diulio (1923–) S   **Lausi**
Lautenschlager, E. (fl. 1982) S   **Lautenschl.**
Lauterbach, Carl (Karl) Adolf Georg (1864–1937) BPS   **Lauterb.**
Lauterborn, Robert (1869–1952) A   **Lauterborn**
Lauterer, Joseph (1848–1911) S   **Lauterer**
Lauth, G. (1793–1817) S   **G.Lauth**
Lauth, Thomas (1758–1826) S   **Lauth**
Lauvergne, Hubert (fl. 1829) S   **Lauv.**
Lauzac-Marchal, Marguerite (fl. 1974) S   **Lauz.-March.**
Laval, Edouard (1871–) M   **Laval**
Laval, Michèle   A   **M.Laval**
Lavalle, Jean (1820–) S   **Lavalle**
Lavallée, Pierre Alphonse Martin (1836–1884) S   **Lavallée**
Lavarack, Peter S. ('Bill') (fl. 1975) S   **Lavarack**
Laven, Ludwig (1881–) M   **Laven**
Laveran, A. (1845–1922) AM   **Laveran**
Lavette, Andrée   A   **Lavette**
Lavie, Pierre (fl. 1954) MS   **Lavie**
Lavier, George L.   S   **G.L.Lavier**
Lavier, Georges (fl. 1935) AM   **Lavier**
Lavin, Matt (1956–) S   **Lavin**
Lavioli, C. (fl. 1961) M   **Lavioli**
Lavis, Mary Gwendolene (1902–) S   **Lavis**
Lavoie, G. (fl. 1984) S   **Lavoie**
Lavorato, Carmine (fl. 1986) M   **Lavorato**
Lavranos, John Jacob (1926–) S   **Lavranos**
Lavrenko, Eugeny Mikhailovič (1900–1987) S   **Lavrenko**
Lavrentiades, Georgios (1920–) S   **Lavrent.**
Lavrov, Nikolai Nicolaevich (fl. 1926) M   **Lavrov**
Lavrova, T.V. (1949–) S   **Lavrova**
Lavy, Jean (1775–1851) S   **Lavy**
Law, Yuh Wu (1917–) S   **Y.W.Law**
Lawalrée, André Gilles Célestin (1921–) PS   **Lawalrée**
Lawesson, Jonas Erik (1959–) S   **Lawesson**
Lawhavinit, Ong-ard (fl. 1986) M   **Lawhav.**
Lawler, Adrian Russell (1940–) A   **Lawler**
Lawrance, Mary (fl. 1790–1831) S   **Lawrance**
Lawrence, George (1827–1895) S   **Lawr.**
Lawrence, George Hill Mathewson (1910–1978) S   **G.H.M.Lawr.**
Lawrence, J.   S   **J.Lawr.**
Lawrence, Margaret Elizabeth (1953–) S   **M.E.Lawr.**
Lawrence, W.H. (fl. 1912) M   **W.H.Lawr.**
Lawrence, William Evans (1883–1950) S   **W.E.Lawr.**
Lawrence, William John Cooper (1899–1985) S   **W.J.C.Lawr.**
Lawrence, Zofia (née Kozakiewicz, Z.) (1947–) M   **Z.Lawr.**

Ławrynowicz, Maria (1943–) M	**Ławryn.**
Lawson, Charles (1794–1873) S	**C.Lawson**
Lawson, George (1827–1895) FPS	**G.Lawson**
Lawson, George W. A	**G.W.Lawson**
Lawson, Marmaduke Alexander (1840–1896) S	**M.A.Lawson**
Lawson, Peter (–1820) S	**Lawson**
Lawton, Elva (1896–) B	**E.Lawton**
Lawton, H.W. S	**Lawton**
Laxa, Otakar (fl. 1930) M	**Laxa**
Laxmann, Erich (Erik) G. (1737–1796) S	**Laxm.**
Lay, Ko Ko S	**Lay**
Layens, Georges de (1834–1897) S	**Layens**
Layser, Earle F. (fl. 1971) S	**Layser**
Laza Palacios, Modesto (1901–1981) S	**Laza**
Lazăr, Alexandru (1927–1990) M	**A.Lazăr**
Lazar, Jože (1903–1975) AS	**Lazar**
Lazarenko, Andrei Sazontovich (1901–1979) B	**Laz.**
Lazareva, E.P. A	**Lazareva**
Lazarides, Michael (1928–) S	**Lazarides**
Lázaro é Ibiza, Blas (1858–1921) AMS	**Lázaro Ibiza**
(Lázaro Ibiza, Blas)	
see Lázaro é Ibiza, Blas	**Lázaro Ibiza**
Lazo, Waldo (fl. 1972) M	**Lazo**
Lazzari, Giacomo (fl. 1983) M	**Lazzari**
(Lazzarini Peckollt, Waldemar de)	
see Peckolt, Waldemar de Lazzarini	**W.Peckolt**
(Le Campion–Alsumard, Thérèse)	
see Campion–Alsumard, Thérèse Le	**Camp.-Als.**
Le Clerc, Léon A	**Le Clerc**
(le Clerk, Leon)	
see Clerk, Leon le	**Clerk**
Le Cohu, R. A	**Le Cohu**
Le Cointe, Paul (1870–) S	**Le Cointe**
Le Conte, John (1818–1891) S	**J.Le Conte**
(Le Conte, John (Eatton))	
see Leconte, John (Eatton)	**Leconte**
Le Gal, Marcelle Louise Fernande (1895–1979) M	**Le Gal**
Le Gall de Kerlinou, Nicholas Joseph Marie (1787–1860) S	**Le Gall**
(Le Gall, Nicholas Joseph Marie)	
see Le Gall de Kerlinou, Nicholas Joseph Marie	**Le Gall**
(Le Grand, Antoine)	
see Legrand, Antoine	**Legrand**
Le Grand, O. (fl. 1895) S	**O.Le Grand**
(le Guillon, E.)	
see Guillon, E.le	**E.Guillon**
Le Héricher, Édouard (1812–1890) F	**Le Héricher**
Le Houérou, Henri–Noël (1928–) S	**Le Houér.**
Le Jolis, Auguste François (1823–1904) AS	**Le Jol.**
Le Lièvre, F. (fl. 1838) S	**Le Lièvre**

Le Maout, Jean Emmanuel Maurice (1799–1877) S — Le Maout
Le Monnier, (Alexandre Alexis) George (1843–   ) M — G.Le Monn.
Le Monnier, Louis Guillaume (1717–1799) M — Le Monn.
Le Peletier de Saint-Fargeau, Amédée Louis Michel (1770–1845) S — Le Pelet.
Le Prévost, Auguste (1787–1860) M — Le Prévost
(Le Provost, Auguste)
  see Le Prévost, Auguste — Le Prévost
Le Roux, S.F. (fl. 1963) F — Le Roux
Le Thomas, Annick (1936–   ) S — Le Thomas
(Le Thomas-Hommay, Annick)
  see Le Thomas, Annick — Le Thomas
Le Turquier de Longchamp, Joseph Alexandre (1748–1829) S — Le Turq.
Lea, Thomas Gibson (1785–1844) MS — Lea
Leach, C.M. (fl. 1972) M — C.M.Leach
Leach, Gregory John (1952–   ) S — G.J.Leach
Leach, Julian Gilbert (1894–   ) M — J.G.Leach
Leach, Leslie ('Larry') Charles (1909–   ) S — L.C.Leach
Leach, Robert (1893–   ) M — R.Leach
Leadbeater, Barry S.C.   A — B.Leadb.
Leadbeater, G. (fl. 1957) M — Leadb.
Leadlay, Etelka A. (fl. 1990) S — Leadlay
Leahy, R.M. (fl. 1989) M — Leahy
Leakey, Colin L.A. (fl. 1964) M — C.L.Leakey
Leakey, D.G.B. (fl. 1932) S — Leakey
Leal, Adrian Ruíz   S — A.R.Leal
Leal, Carlos G. (fl. 1951) S — Leal
Leal, F.B. (fl. 1969) M — F.B.Leal
Leander, P.A. (1872–1935) S — Leander
Leandri, Jacques Désiré (1903–1982) MS — Leandri
Leandro do Sacramento, P. (1778–1829) S — Leandro
Leane Moreira da Costa, Nara (fl. 1982) S — Leane
Leão, A.E.de Arêa (fl. 1927) M — Leão
(Leão, Antonio Pacheco)
  see Pacheco Leão, Antonio — Pach.Leão
Leary, Richard L.   A — Leary
Leask, Barbara G.S. (fl. 1976) M — Leask
Leatherdale, Donald (1920–   ) M — Leath.
Leathers, Chester Ray (1929–   ) M — Leathers
Leavenworth, Melines Conklin (1796–1862) S — Leavenw.
Leavitt, Robert Greenleaf (1865–1942) S — Leav.
Lebas, E. (1800s) S — Lebas
Lebasque, J. (fl. 1933) M — Lebasque
Lebeau, Jean (fl. 1972) S — Lebeau
Lebedev, E.L. (fl. 1968) F — Lebedev
Lebedeva, Lydia Alexsandrovna (fl. 1922) M — Lebedeva
Lebednik, Phillip A.   A — Lebednik
Lebel, Jacques Eugène (1801–1878) AS — Lebel
Leberle, H. (fl. 1909) M — Leberle

Lebert, Hermann (1813–1878) MS	**Lebert**
Lebeskva (fl. 1933) M	**Lebeskva**
Lebeurier (fl. 1988) M	**Lebeurier**
Lebezhinskaja, L.D. (fl. 1959) M	**Lebezh.**
Leblebici, Erkuter (1939–) S	**Leblebici**
Leboime, René (1885–1956) A	**Leboime**
Lebouché, Marie–Claire A	**Lebouché**
Lebour, George Alexander Louis (1847–1918) F	**G.Lebour**
Lebour, Marie Victoria (fl. 1925) A	**M.Lebour**
Lebreton, F. (fl. 1787) M	**Lebreton**
Lebreton, P. (fl. 1981) S	**P.Lebreton**
Lebron–Luteyn, María L. (fl. 1983) S	**Lebron–Luteyn**
Lebrun, Jean–Paul Antoine (1906–1985) S	**Lebrun**
Lebrun, Jean–Pierre (1932–) S	**J.-P.Lebrun**
(Lecal, Juliette)	
see Lecal–Schlauder, Juliette	**Lec.-Schlaud.**
Lecal–Schlauder, Juliette A	**Lec.-Schlaud.**
Lechevalier, H. (fl. 1957) M	**H.Lechev.**
Lechevalier, Mary P. (fl. 1957) M	**M.P.Lechev.**
Lechler, Wilibald (1814–1856) S	**Lechl.**
Lechmere, Arthur Eckley (1885–1919) M	**Lechmere**
Lechner, A.A.van Pelt S	**Lechner**
Lechnovich, V.S. (1902–) S	**Lechn.**
Lechtova–Trnka, M. (fl. 1931) M	**Lecht.-Trnka**
Leckenby, John (1814–1877) BF	**Leck.**
Leclair, A. (fl. 1932) M	**Leclair**
LeClair, P.M. (fl. 1975) M	**P.M.LeClair**
Leclercq, Suzanne Céline Marie (1901–) AF	**Leclercq**
Leclercq, Suzanne (1939–1975) F	**S.Leclercq**
Lecompte, Marius (1902–1973) A	**Lecompte**
Lecompte, O. (fl. 1965) S	**O.Lecompte**
Lecomte, M.H. S	**M.H.Lecomte**
Lecomte, Paul Henri (1856–1934) PS	**Lecomte**
Leconte, John (Eatton) (1784–1860) S	**Leconte**
Lecoq, Henri (1802–1871) S	**Lecoq**
Lecoufle S	**Lecoufle**
Lecoyer, Joseph Cyprien (1835–1899) S	**Lecoy.**
Lecron, J.M. (fl. 1989) S	**Lecron**
Ledebour, Carl (Karl) Friedrich von (1785–1851) PS	**Ledeb.**
Ledermann, Carl Ludwig (1875–1958) S	**Ledermann**
Ledermüller, Martin Frobenius (1719–1769) S	**Lederm.**
Ledingham, George Aleck (1903–) M	**Ledingham**
Ledochowski, J.S.M. (fl. 1953) M	**Ledoch.**
Ledoux, E.P. (1898–) S	**Ledoux**
Lee, Alma Theodora (née Melvaine, A.T.) (1912–1990) S	**A.T.Lee**
Lee, B.J.S. (fl. 1958) S	**B.J.S.Lee**
Lee, Bing Kwe (fl. 1981) S	**B.K.Lee**
Lee, C.C. A	**C.C.Lee**
Lee, Chang Shook (fl. 1986) P	**C.S.Lee**
Lee, Fwu Ling (fl. 1986) M	**F.L.Lee**

Lee

Lee, Gabriel Wharton (1880–1928) A	G.W.Lee
Lee, Henry Atherton (1894–) MS	H.A.Lee
Lee, In Kyu (1936–) A	I.K.Lee
Lee, J.D. (fl. 1976) M	J.D.Lee
Lee, James (1715–1795) S	J.Lee
Lee, Ji Yul (fl. 1972) M	J.Y.Lee
Lee, John J. (1933–) A	J.J.Lee
Lee, John Ramsay (1868–1959) S	J.R.Lee
Lee, Kong Teh A	K.T.Lee
Lee, Kwok Wah A	K.W.Lee
Lee, Robert Edward A	R.Ed.Lee
Lee, Robert Edwin (1911–) S	R.E.Lee
Lee, Robert K.S. (1931–) A	R.K.S.Lee
Lee, Shin Chiang (fl. 1964) B	Shin C.Lee
Lee, Shu Kang (1915–) S	S.K.Lee
Lee, Shun Ching (1892–) S	S.C.Lee
Lee, Wei Siang (fl. 1957) M	W.S.Lee
Lee, William George (1950–) S	W.G.Lee
Lee, Yao Yin A	Y.Y.Lee
Lee, Yin Tse (fl. 1973) S	Y.T.Lee
Lee, Yong Bo (fl. 1986) M	Y.B.Lee
Lee, Yong No (1920–) S	Y.N.Lee
Lee, Yong Pil A	Y.P.Lee
Leechmann, Alleyhe S	Leechm.
Leedale, Gordon Frank (1932–) A	Leedale
Leeds, Arthur Newlin (1870–) P	Leeds
Leefe, John Ewbank (1824–1889) S	Leefe
Leeke, Georg Gustav Paul (1883–1933) S	Leeke
Leelavathy, K.M. (fl. 1966) M	Leelav.
Leenhouts, Pieter Willem (1926–) S	Leenh.
Leers, Johann Georg Daniel (1727–1774) ABMS	Leers
Lees, Edwin (1800–1887) S	Lees
Lees, Frederick Arnold (1847–1921) B	F.Lees
Leeuwen, Betsy Louise Jacoba van (1946–) S	Leeuwen
Leeuwenberg, Anthonius Josephus Maria (1930–) S	Leeuwenb.
Leewis, R.J. A	Leewis
Lefébure, Louis F.Henri (1754–1839) S	Leféb.
Lefébure, P. A	P.Leféb.
Lefebvre, Camille Léon (1905–) M	Lefebvrc
Lefebvre, Jeanne (fl. 1960) B	J.Lefebvre
Lefèvre, Édouard (1839–1894) S	E.Lefèvre
Lefèvre, Louis Victor (1810–) S	Lefèvre
Lefèvre, Marcel (1897–) A	M.Lefèvre
Lefevre, Roger A	R.Lefevre
Lefor, Michael William (fl. 1975) S	Lefor
Lefroy, John Henry (1817–1890) S	Lefroy
Legakis, P.A. (fl. 1961) M	Legakis
Legand, C.Diego (1946–) S	Legand
Legendre, Charles Valentin Alexandre (1841–1935) S	Legendre

Legendre, J. (fl. 1911) M	**J.Legendre**
Léger, André A	**A.Léger**
Léger, J.C. (fl. 1968) M	**J.C.Léger**
Léger, L. (fl. 1922) M	**L.Léger**
Léger, Louis Raoul Urbain Théophile Maurice (1866–1901) A	**Léger**
Léger, Marcel A	**M.Léger**
Leggett, M. (fl. 1981) M	**M.Legg.**
Leggett, William Henry (1816–1882) S	**Legg.**
Legler, Fritz A	**Legler**
Legname, Pablo Paul (1930–1989) S	**Legname**
Legnerová, Jaroslava A	**Legnerová**
Legrain, E. (fl. 1900) M	**Legrain**
Legrand, Antoine (1839–1905) PS	**Legrand**
Legrand, Carlos Maria Diego Enrique (1901–) PS	**D.Legrand**
Legrand, Ed. S	**E.Legrand**
Legré, Ludovic (1838–1904) S	**Legré**
Legué, Alphonse Marie Léon (1841–1920) S	**Legué**
Leguizamón, Raul Rene (fl. 1971) F	**Leguiz.**
(Lehaie, J.Houzeau de)	
see Houzeau de Lehaie, J.	**J.Houz.**
(Lehaie, Jean Charles Houzeau de)	
see Houzeau de Lehaie, Jean Charles	**Houz.**
Lehman, Samuel George (1887–) M	**Lehman**
Lehmann, Alexander (1814–1842) S	**Al.Lehm.**
Lehmann, August (fl. 1869) S	**Aug.Lehm.**
Lehmann, Carl Berhard (1811–1875) S	**C.B.Lehm.**
Lehmann, Christian O. (1926–) S	**C.O.Lehm.**
Lehmann, Donald L. A	**D.L.Lehm.**
Lehmann, Eduard (1841–1902) S	**E.Lehm.**
Lehmann, Ernst B.Johann (1880–1957) AS	**E.B.J.Lehm.**
Lehmann, Friedrich Carl (1850–1903) MS	**F.Lehm.**
Lehmann, Johann Friedrich (fl. 1809) S	**J.F.Lehm.**
Lehmann, Johann Georg Christian (1792–1860) ABPS	**Lehm.**
Lehoczky, J. (fl. 1959) M	**Lehoczky**
Lehodey, Y. (fl. 1965) M	**Lehodey**
Lehr, Georg Philipp (1756–1807) S	**Lehr**
Lehtola, V.B. (fl. 1940) M	**Lehtola**
Leiberg, John Bernhard (1853–1913) BS	**Leiberg**
Leiblein, Valerius (Valentin) (1799–1869) AS	**Leiblein**
Leibling, Otto (fl. 1884) S	**Leibling**
Leibold, Friedrich Ernst (1804–1864) S	**Leibold**
Leichhardt, Friedrich Wilhelm Ludwig (1813–1848) S	**Leichh.**
Leichtlin, Maximilian (1831–1910) S	**Leichtlin**
Leidy, Joseph (1823–1891) AFM	**Leidy**
Leighton, Frances Margaret (later Isaac, F.M.L.) (1909–) S	**F.M.Leight.**
Leighton, William Allport (1805–1889) MS	**Leight.**
Leimer, Franz (fl. 1854) S	**Leimer**
Leinati, F. (fl. 1928) M	**Leinati**

Leiner, Ludwig (1830–1901) BC   **Leiner**
Leinig, M. (fl. 1967) S   **Leinig**
Leins, Peter (1937–) S   **Leins**
Leisman, Gilbert Arthur (1924–) FMP   **Leisman**
Leister, Geoffrey L. A   **Leister**
Leistikow, Klaus V. (fl. 1962–1979) F   **Leistikow**
Leistner, Otto Albrecht (1931–) S   **Leistner**
(Leitão Filho, Hermógenes de Freitas)
    see Leitão, Hermógenes de Freitas   **Leitão**
Leitão, Hermógenes de Freitas (1944–) S   **Leitão**
Leite, Célia Romano (1949–) A   **C.R.Leite**
Leite, Clarice Loguercio (1955–) M   **C.L.Leite**
Leite, José Eugenio (1907–1980) S   **Leite**
Leitgeb, Hubert (1835–1888) BM   **Leitg.**
Leithe, Friedrich A   **Leithe**
Leitner, Edward Frederick (Friedrich August Ludwig) (1812–1838) S   **Leitn.**
Leitner, Jane (fl. 1938) M   **J.Leitn.**
Leiva Sanchez, A.T. (fl. 1987) S   **Leiva**
Leizerson, N.V. (fl. 1979) S   **Leizerson**
Lejeune, Alexandre Louis (Alexander Ludwig) Simon (1779–1858) PS   **Lej.**
Lejeune-Carpentier, Maria (–1951) A   **Lej.-Carp.**
Lejoly, Jean (1945–) S   **Lejoly**
(Lekhnovich, V.S.)
    see Lechnovich, V.S.   **Lechn.**
Lele, Keshav Makund (1931–1981) AF   **Lele**
(Lelièvre, J.F.)
    see Le Lièvre, F.   **Le Lièvre**
Lellep, Elli (fl. 1973) S   **Lellep**
Lellinger, David Bruce (1937–) PS   **Lellinger**
Lelong, Michel G. (1932–) S   **Lelong**
Lely, Hugh Vandervaes (1891–) S   **Lely**
Lemaigne, Yves (fl. 1968) F   **Lemaigne**
Lemaire, (Antoine) Charles (1801–1871) PS   **Lem.**
Lemaire, Ad. A   **A.Lem.**
(Lemaire, M.)
    see Neveu-Lemaire, M.   **Neveu-Lem.**
Léman, Dominique Sébastien (1781–1829) ACMPS   **Léman**
(Leman, Eduard (Eduardovich))
    see Lehmann, Eduard   **E.Lehm.**
Lembcke, M. (fl. 1959) S   **Lembcke**
Lemcke, Alfred Max Bernhard (1864–) S   **Lemcke**
Leme, Elton M.C. (1960–) S   **Leme**
Lemée, Albert Marie Victor (1872–1900) S   **Lemée**
Lemke, David E. (1953–) S   **D.E.Lemke**
Lemke, Paul Arenz (fl. 1964) M   **P.A.Lemke**
Lemke, Willi (1893–1973) S   **Lemke**
Lemmermann, Ernst Johann (1867–1915) AB   **Lemmerm.**
Lemmon, Betty Ann Elberson (1925–) B   **B.A.E.Lemmon**
Lemmon, John Gill (1832–1908) PS   **Lemmon**

Lemmon, W.P. (fl. 1938) S	**W.P.Lemmon**
Lemoine, (Pierre Louis) Victor (1823–1911) S	**Lemoine**
Lemoine, Émile (1862–1943) S	**É.Lemoine**
Lemoine, Marcel A	**M.Lemoine**
Lemoine, Marie (1887–1984) A	**Me.Lemoine**
(Lemonnier, Louis Guillaume)	
see Le Monnier, Louis Guillaume	**Le Monn.**
(Lemos Pereira, Alice de)	
see Pereira, Alice de Lemos	**A.L.Pereira**
Lems, Kornelius (1931–1968) S	**Lems**
Lemus, A.J. A	**Lemus**
Lendner, Alfred (1873–1948) M	**Lendn.**
Lengyel, B.de (fl. 1936) S	**B.Lengyel**
Lengyel, Géza Emlékezete (1884–1965) S	**Lengyel**
Lenné, Peter Joseph (1789–1866) S	**Lenné**
Lennox, Joanne Williams (fl. 1979) M	**Lennox**
Lenoble, André A	**A.Lenoble**
Lenoble, Félix (1867–1948) S	**Lenoble**
Lenoir, A. (fl. 1986) M	**Lenoir**
Lenormand, (Sébastien) René (1796–1871) AS	**Lenorm.**
Lens, Adrien Jacques de (fl. 1828) S	**Lens**
Lentin, J.K. A	**Lentin**
Lentz, Paul Lewis (1918–) M	**Lentz**
Lenz, Harald (Harold) Othmar (1798–1870) M	**Lenz**
Lenz, Lee Wayne (1915–) MS	**L.W.Lenz**
Lenz, M.von (fl. 1969) M	**M.Lenz**
Lenzenweger, Rupert A	**Lenzenw.**
Léon, (Hermano) (1871–1955) MPS	**Léon**
León Arencibia, M.C. (1951–) S	**M.C.León**
León de la Luz, José Luis (fl. 1989) S	**León de la Luz**
(León Gallegos, Hector M.)	
see León-Gallegos, Hector M.	**León-Gall.**
Léon, Jules (fl. 1949) S	**J.Léon**
León-Gallegos, Hector M. (fl. 1972) M	**León-Gall.**
Leonard, Emery Clarence (1892–1968) S	**Leonard**
Léonard, Guy (1927–) S	**G.Léonard**
Léonard, Jean Joseph Gustave (1920–) S	**J.Léonard**
Leonard, Kurt J. (fl. 1974) M	**K.J.Leonard**
Leong, W.F. (fl. 1990) M	**W.F.Leong**
Leonhardi, Peter Carl Pius Gustav Hermann von (1809–1875) A	**Leonh.**
Leonhardt, Carl (1902–) S	**Leonhardt**
Leonian, Leon Hatchig (1888–1945) M	**Leonian**
Leonova, T.G. (1930–) S	**Leonova**
Leopold, G. (fl. 1900) M	**Leopold**
Leou, Chong Sheng S	**C.S.Leou**
Lepage, Ernest (1905–1981) BPS	**Lepage**
Lepailleur, Henri A	**Lepaill.**
Lepechin, Ivan Ivanovich (1737–1802) AS	**Lepech.**

Lepelletier, Gabr. (fl. 1822) MS   **Lepell.**
Lepeschkin, S.N.   S   **Lepeschk.**
Lepesme, P. (fl. 1942) M   **Lepesme**
(Lepik, Elmar Emil)
   see Leppik, Elmar Emil   **Leppik**
Lepper, Lothar (1932–   ) S   **Lepper**
Leppik, Elmar Emil (fl. 1932) M   **Leppik**
Leprieur, F.M.R. (1799–1869) MPS   **Lepr.**
Lepşi, I. (1895–1966) A   **Lepşi**
Léránth, J. (fl. 1974) M   **Léránth**
Lerche, Johann Jakob (Jacob) (1703–1780) S   **Lerche**
Lerche, Witta   A   **W.Lerche**
Lerchenfeld, Josef Radnitzky von (1753–1812) S   **Lerchenf.**
Leredde, Claude (fl. 1953) S   **Leredde**
Leresche, Louis François Jules Rodolphe (1808–1885) PS   **Leresche**
Leroy, André (1801–1875) S   **Leroy**
Leroy, Jean–F. (1915–   ) S   **J.-F.Leroy**
Leroy, Louis (1808–1887) S   **L.Leroy**
Leroy, Victor (1836–1908) B   **V.Leroy**
Les, Donald H. (1954–   ) S   **Les**
Lesacher, Eugène (1824–   ) S   **Lesacher**
Leschenault de la Tour, Jean Baptiste Louis (Claude) Théodore
   (1773–1826) S   **Lesch.**
Lescot, Michèle (1939–   ) S   **Lescot**
Lescuyer, O.H. (fl. 1855–72) S   **Lesc.**
Lesieur, Ch. (fl. 1912) M   **Lesieur**
Lesins, I. (fl. 1979) S   **I.Lesins**
Lesins, K.A. (fl. 1979) S   **K.A.Lesins**
Leske, Nathanael Gottfried (1751–1786) S   **Leske**
Leskov, Alexandr Ivanovich (1902–1942) S   **Leskov**
Lesley, J.P. (fl. 1889) F   **Lesley**
Leslie, Alan Christopher (1950–   ) S   **A.C.Leslie**
Leslie, John F. (fl. 1989) M   **J.F.Leslie**
Lespinasse, Jean Martial Gustave (1807–1876) S   **Lesp.**
Lesquereux, Charles Léo (1806–1889) ABFS   **Lesq.**
Lesser, Edmund (fl. 1885) M   **Lesser**
Lessing, Christian Friedrich (1809–1862) AMS   **Less.**
Lesson, Adolphe Pierre Primivère (1805–1888) S   **A.Lesson**
Lesson, Réné Primivère (1794–1849) S   **R.Lesson**
Lester–Garland, Lester Vallis (1860–1944) S   **Lest.-Garl.**
Lestiboudois, François Joseph (–1815) S   **F.Lestib.**
Lestiboudois, Jean Baptiste (1715–1804) S   **J.Lestib.**
Lestiboudois, Thémistocle Gaspard (1797–1876) APS   **T.Lestib.**
Leszczyc–Sumiński, Michael Hieronim (1820–1898) S   **Leszcz.-Sum.**
Letacq, Arthur Louis (1855–1923) S   **Letacq**
Letellier, A.   A   **A.Letell.**
Letellier, Jean Baptiste Louis (1817–1898) MS   **Letell.**
Letendre, Jean Baptiste Pierre (1828–1886) MS   **Letendre**
Letgé, J.–P. (fl. 1988) M   **Letgé**

Letourneux, Aristide-Horace (1820–1890) MPS — **Letourn.**
Letouzey, Réné (1918–1989) S — **Letouzey**
Letov, Aleksandr Sergeevich (1904–   ) M — **Letov**
(Letow, Aleksandr Sergeevich)
  see Letov, Aleksandr Sergeevich — **Letov**
Letrouit-Galinou, Marie-Agnes (1931–   ) M — **Letr.-Gal.**
Lett, Henry William (1836–1920) B — **Lett**
Lettau, Georg (1878–1951) M — **Lettau**
Lettsom, John Coakley (1744–1815) S — **Lettsom**
Letty, Cythna Lindenberg (Mrs Forssman) (1895–1985) S — **Letty**
Leu, S.Y. (fl. 1975) S — **S.Y.Leu**
Leuba, Fritz (1848–1910) M — **Leuba**
Leuchtmann, Adrian (1956–   ) M — **Leuchtm.**
Leuckart, (Karl George Friedrich) Rudolph (1823–1898) A — **Leuck.**
Leuckert, Christian (1930–   ) M — **Leuckert**
Leuduger-Fortmorel, Georges (1830–1902) A — **Leud.-Fortm.**
Leuenberger, Beat Ernst (1946–   ) S — **Leuenb.**
Leunis, Johannis (1802–1873) P — **Leunis**
Leupold, W. A — **Leupold**
(Leus, Simeona)
  see Leus-Palo, Simeona — **Leus-Palo**
Leus-Palo, Simeona (fl. 1931) M — **Leus-Palo**
Leute, Gerfried Horand (1941–   ) S — **Leute**
Levadnav, G.D. A — **Levadnav**
Levander, K.M. (1867–   ) A — **Levander**
Léveillé, Augustin Abel Hector (1863–1918) PS — **H.Lév.**
Léveillé, Joseph-Henri (1796–1870) MS — **Lév.**
Levi, A. (fl. 1888) M — **Levi**
(Levi, David)
  see Levi-Morenos, David — **Levi-Morenos**
Levi-Morenos, David (1863–   ) A — **Levi-Morenos**
Levichev, I.G. (1945–   ) S — **Levichev**
Levier, Emile (Emilio) (1839–1911) BMPS — **Levier**
Levin, Emmanuel Grigorjevich (1915–1944) S — **E.G.Levin**
Levin, Ernst Ivar (1868–   ) S — **Levin**
Levin, G.M. (fl. 1980) S — **G.M.Levin**
Levin, Harold Leonard (1929–   ) A — **H.L.Levin**
Levina, Rosa Efimovna (1908–1987) S — **Levina**
Levine, Norman Dion (1912–   ) A — **Levine**
Levittan, Edwin D. (fl. 1948) F — **Levittan**
Levring, Carl Tore Christian (1913–1980) A — **Levring**
Lévy, J. A — **Lévy**
Levyns, Margaret Rutherford Bryan (1890–1975) S — **Levyns**
Lewalle, J. (fl. 1977) S — **Lewalle**
Lewejohann, Klaus (1937–   ) S — **Lewej.**
Lewin, Joyce (M.) Chismore (1926–   ) A — **J.C.Lewin**
Lewin, Maria A — **M.Lewin**
Lewin, Ralph Arnold (1921–   ) A — **Lewin**
Lewinsky, Jette (1948–   ) B — **Lewinsky**

Lewis, Ann L. (fl. 1968) M	**A.L.Lewis**
Lewis, Beverley Ann (1966–) S	**B.A.Lewis**
Lewis, C.E. (fl. 1912) M	**C.E.Lewis**
Lewis, David P. (1946–) M	**D.P.Lewis**
Lewis, Francis West (1825–1902) A	**F.W.Lewis**
Lewis, Frank Harlan (1919–) S	**F.H.Lewis**
Lewis, Frederick (1857–1930) S	**F.Lewis**
Lewis, Gwendoline Joyce (1909–1967) S	**G.J.Lewis**
Lewis, Gwilym Peter (1952–) S	**G.P.Lewis**
Lewis, Harrison Flint (1893–1974) S	**H.F.Lewis**
Lewis, Herbert Price (1895?–1947) A	**H.P.Lewis**
Lewis, Isaac McKinney (1878–1943) M	**I.M.Lewis**
Lewis, Ivey Foreman (1882–1964) A	**I.F.Lewis**
Lewis, John (1921–) S	**J.Lewis**
Lewis, Margaret Ruth Ensign (née Ensign, M.R.) (1919–) S	**M.R.Lewis**
Lewis, Marko Alexander (1947–) B	**M.A.Lewis**
Lewis, Meriwether (1774–1809) S	**Lewis**
Lewis, Patricia (1924–) S	**P.Lewis**
Lewis, R.C. A	**R.C.Lewis**
Lewis, Sara M. A	**S.M.Lewis**
Lewis, Walter Hepworth (1930–) S	**W.H.Lewis**
Lewis, Walter W. A	**W.W.Lewis**
Lewis, Y.S. (fl. 1955) M	**Y.S.Lewis**
Lewton, Frederick Lewis (1874–1959) S	**Lewton**
Lexarza, Juan José Martinez de (1785–1824) S	**Lex.**
Ley, Augustin (1842–1911) S	**Ley**
Ley, Frances Arlene (1919–) S	**F.A.Ley**
(Ley, Shang Hao)	
see Li, Shang Hao	**S.H.Li**
Ley, W. S	**W.Ley**
Leybold, Friedrich (1827–1879) PS	**Leyb.**
Leybold, Friedrich Ernst (1804–1864) S	**F.E.Leyb.**
(Leyboldt, Friedrich Ernst)	
see Leybold, Friedrich Ernst	**F.E.Leyb.**
Leydolt, Franz (1810–1859) S	**Leyd.**
(Leyser, Friedrich Wilhelm von)	
see Leysser, Friedrich Wilhelm von	**Leyss.**
Leysser, Friedrich Wilhelm von (1731–1815) ABMS	**Leyss.**
Lezaud, Lucien A	**Lezaud**
Lhotsky, Johann (Jan) (1800–1860s) S	**Lhotsky**
(Lhotzky, Johann (Jan))	
see Lhotsky, Johann (Jan)	**Lhotsky**
Li, A.D. (1922–) S	**A.D.Li**
Li, An Jen (fl. 1981) S	**A.J.Li**
Li, Ben Liang (fl. 1989) S	**B.L.Li**
Li, Bin (1952–) M	**B.Li**
Li, Bo Sheng (fl. 1986) S	**B.S.Li**
Li, Chao Da (fl. 1990) M	**C.D.Li**
Li, Chao Lan (fl. 1988) M	**C.Lan Li**

Li, Chao Luan(g) (1938–   ) S   C.L.Li
Li, Cheng Sen (fl. 1982) F   C.S.Li
Li, Chia Wei   A   C.W.Li
Li, Chien (fl. 1987) S   C.Li
Li, De Zhu (1963–   ) S   D.Z.Li
Li, Deng Ke (1937–   ) B   D.K.Li
Li, Fa Zeng (fl. 1982–1988) PS   F.Z.Li
Li, Feng (fl. 1988) P   F.Li
Li, Guang Zhao (1940–   ) S   G.Z.Li
Li, Hen (1929–   ) S   H.Li
Li, Hsi Wen (1902–   ) S   H.W.Li
Li, Hui Lin (1911–   ) PS   H.L.Li
Li, Hui Zhong (fl. 1986) M   H.Z.Li
Li, Jia Ying (fl. 1984) AS   J.Y.Li
Li, Jian Qiang (fl. 1988) S   J.Q.Li
Li, Jian Xiu (fl. 1983–1988) PS   J.X.Li
Li, Jian Yi (1916–   ) M   J.Yi Li
Li, Jin Liang (fl. 1990) M   J.L.Li
Li, Jing Hua (fl. 1988) S   J.H.Li
Li, L.J. (fl. 1985) M   L.J.Li
Li, Li Tzu (fl. 1990) M   L.T.Li
Li, Liang Ching   A   L.C.Li
Li, Liang Qian (1952–   ) S   L.Q.Li
Li, Lin Chu (fl. 1982) S   L.Chu Li
Li, Ming Xia (1932–   ) M   M.X.Li
Li, Ming Yuan (fl. 1977) M   M.Y.Li
Li, Pei Chun (Qiong) (fl. 1981) S   P.C.Li
Li, Pei Yuan (fl. 1987) S   P.Y.Li
Li, Ping Tao (1936–   ) S   P.T.Li
Li, Ren He (fl. 1979) M   R.H.Li
Li, Rong Xi (fl. 1990) M   R.X.Li
Li, S.Q. (fl. 1983) M   S.Q.Li
Li, Shang Hao (1917–   ) A   S.H.Li
Li, Shi You (fl. 1988) S   S.Y.Li
Li, Shu Chun (fl. 1982) S   S.C.Li
Li, Shu Jiu (fl. 1988) S   S.J.Li
Li, Shu Xin (Shu Hsin) (1926–   ) S   S.X.Li
(Li, Shun Ching)
   see Lee, Shun Ching   S.C.Lee
Li, T.B. (fl. 1984) M   T.B.Li
Li, T.Q. (fl. 1987) M   T.Q.Li
Li, Tai Hui (fl. 1987) MS   T.H.Li
Li, Ti Zhi (fl. 1989) S   T.Z.Li
Li, Tion Juan (fl. 1981) M   T.J.Li
Li, Wan Cheng (fl. 1981) S   W.C.Li
Li, Wei Hsin   A   W.H.Li
Li, Wen De (fl. 1985) S   W.D.Li
Li, Wen Zheng (fl. 1987) S   W.Z.Li
Li, Xing Jiang (1932–   ) B   X.J.Li

Li, Xing Wen (fl. 1986) S	X.W.Li
Li, Xiu Jiu (fl. 1989) S	Xui J.Li
Li, Xue Yu (fl. 1983) S	X.Y.Li
Li, Y.M. (fl. 1988) M	Y.M.Li
Li, Ya Ru (fl. 1979) S	Y.R.Li
Li, Yan Hui (1930–) S	Y.H.Li
Li, Yao Ying (1930–) A	Yao Y.Li
Li, Yi Jan (fl. 1985) A	Y.J.Li
Li, Yin (fl. 1970) S	Yin Li
Li, Yong Kang (fl. 1986) S	Y.K.Li
Li, Yu (1932–) M	Yu Li
Li, Yu Shan (fl. 1988) S	Y.S.Li
Li, Yue Ying (fl. 1979) M	Yue Y.Li
Li, Z.Y. (fl. 1976) M	Z.Y.Li
Li, Ze Xian (fl. 1990) S	X.Z.Li
Li, Zeng Zhi (fl. 1984) M	Z.Z.Li
Li, Zhen Qing (fl. 1989) S	Zhen Q.Li
Li, Zheng Yu (fl. 1987) S	Z.Yu Li
Li, Zhi Hua (1935–) B	Z.H.Li
Li, Zhong Ming (fl. 1983) F	Z.M.Li
Li, Zhong Qing (fl. 1983) M	Zhong Q.Li
Li, Zi Ping (fl. 1991) M	Z.P.Li
Li, Zi Quan (fl. 1979) M	Zi Q.Li
Li, Zong Xiu (1937–) S	Z.X.Li
Liais, E. S	Liais
Lian, Yong (Yung) Shan (1939–) S	Y.S.Lian
Liang, Chou Fen(g) (1921–) S	C.F.Liang
Liang, Jia Ji (fl. 1979) A	J.J.Liang
Liang, Jian(g) Ying (1943–) S	J.Y.Liang
Liang, Min Qing (fl. 1984) S	M.Q.Liang
Liang, Pao Han S	P.H.Liang
Liang, Sheng Ye(h) (fl. 1980) S	S.Ye Liang
Liang, Shu Bin (fl. 1988) S	S.B.Liang
Liang, Sung (Song) Yun (1935–) S	S.Yun Liang
Liang, W. (fl. 1975) M	W.Liang
Liang, Wei Jian (fl. 1988) S	W.J.Liang
Liang, Zong Qi (fl. 1981) M	Z.Q.Liang
Liao, Jih Ching (fl. 1971) S	J.C.Liao
Liao, Lawrence M. (1959–) A	L.M.Liao
Liao, Rong Gui (fl. 1988) S	R.G.Liao
Liao, W.Q. (fl. 1983) M	W.Q.Liao
Liao, Y.R. (fl. 1984) M	Y.R.Liao
Liao, Yin Zhang (fl. 1980) M	Y.Z.Liao
Liben, Louis (1926–) S	Liben
Libert, Marie–Anne (1782–1865) ABMS	Lib.
Liberta, Anthony E. (1933–) M	Liberta
Libon, Joseph (1821–1861) S	Libon
Libonati–Barnes, Susan D. (fl. 1979) M	Lib.–Barnes
Liboschitz, Joseph (1783–1824) MS	Libosch.
Licata, Giovanni Battista (1850–1886) S	Licata

(Licea Duran, Sergio)
   see Licea, Sergio   **Licea**
Licea, Sergio A   **Licea**
Licent, E. (fl. 1920) M   **Licent**
Lichtenstein, August Gerhard Gottfried (1780–1851) S   **A.Licht.**
Lichtenstein, Georg Rudolph (1745–1807) S   **G.Licht.**
Lichtenstein, J.L. (fl. 1916) M   **J.L.Licht.**
Lichtenstein, Juana S.de (1902–) PS   **J.S.Licht.**
Lichtenstein, Martin Heinrich Karl von (1780–1857) S   **Licht.**
Lichti–Federovich, Sigrid A   **Lichti-Fed.**
Lichtwardt, Robert W. (fl. 1954) M   **Lichtw.**
Lichvar, R.W. (fl. 1981) S   **Lichvar**
Lickleder, Max (1826–1893) B   **Lickl.**
Licopoli, Gaetano (1833–1897) M   **Licop.**
Lid, Johannes (1886–1971) BPS   **Lid**
Lidbeck, Anders (1772–1829) S   **A.Lidb.**
Lidbeck, Eric Gustav (1724–1803) S   **Lidb.**
Lidén, Magnus (1951–) S   **Lidén**
Lidforss, Bengt (1868–1913) S   **Lidf.**
Lie, B. (fl. 1982) M   **B.Lie**
Liebe, Karl Leopold Theodor (1828–1894) S   **K.L.T.Liebe**
Liebe, Theodor (fl. 1862) S   **T.Liebe**
Liebetanz, B. A   **B.Liebet.**
Liebetanz, Erwin A   **E.Liebet.**
Lieblein, Franz Kaspar (Caspar) (1744–1810) S   **Liebl.**
Liebmann, Frederik Michael (1813–1856) ABPS   **Liebm.**
Liebner, C. (fl. 1895) S   **Liebner**
Liede, Siegrid (1957–) S   **Liede**
Liégard, Auguste (1801?–1892) S   **Liég.**
Liegel, Georg (1777–1861) S   **Liegel**
Lien, J.C. (fl. 1980) M   **J.C.Lien**
Lien, Wen Yen S   **W.Y.Lien**
Lienau, W. (fl. 1863) S   **Lienau**
Lieneman, Catherine Mary (1899–) M   **Lieneman**
Liengjarern, M. A   **Liengjar.**
Liesner, Ron L. (1944–) S   **Liesner**
Lieteranz, E. (fl. 1910) M   **Lieter.**
Lieth, Helmut (fl. 1961) S   **Lieth**
(Lièvre, F.Le)
   see Le Lièvre, F.   **Le Lièvre**
Lifshitz, R. (fl. 1984) M   **Lifsh.**
Light, S.F. (1886–1947) A   **Light**
Lightfoot, John (1735–1788) ABMS   **Lightf.**
Lignier, Elie Antoine Octave (1855–1916) ABFS   **Lign.**
Lignières, J. (fl. 1903) M   **Lignières**
Lihnell, D. (fl. 1939) M   **Lihnell**
Likhonos, F.D. (1897–) S   **Likhonos**
Lilienfeld Toal, O.A.von (fl. 1927) M   **Lil.Toal**
Lilja, Nils (1808–1870) S   **Lilja**

Liljeblad, Samuel (1761–1815) ABMPS	**Lilj.**
Liljefors, Alf W. (1904–) S	**Liljef.**
Lillick, Lois Carol (1913–) A	**Lillick**
Lillieroth, C.G. S	**C.G.Lill.**
Lillieroth, Sigvard A	**S.Lill.**
Lillo, Miguel (1862–1931) S	**Lillo**
Lim, Gloria (fl. 1968) M	**Lim**
Lim, Siew Ngo (fl. 1972) S	**S.N.Lim**
(Lima, Américo Pires de)	
see Pires de Lima, Américo	**Pires de Lima**
Lima, Carlos E. (fl. 1988) M	**C.E.Lima**
(Lima, Dardano de Andrade)	
see Andrade–Lima, Arturo Dardano de	**Andrade–Lima**
Lima, I.Hollanda (fl. 1958) M	**I.H.Lima**
Lima, Haroldo Cavalcante de (1955–) S	**H.C.Lima**
Lima, José Américo de (fl. 1960) M	**J.A.Lima**
Lima, Marli Pires Morim de (1952–) S	**M.P.Lima**
Limber, Donald Philips (1894–) M	**Limber**
Limminghe, Alfred Marie Antoine (1834–1861) C	**Limm.**
Limpricht, Hans Wolfgang (1877–) S	**H.Limpr.**
Limpricht, Karl Gustav (1834–1902) B	**Limpr.**
Lin, B. (fl. 1981) M	**B.Lin**
Lin, Bi Qin (fl. 1985) A	**B.Q.Lin**
Lin, C.F. (fl. 1975) M	**C.F.Lin**
Lin, Jia Yi (fl. 1988) S	**J.Y.Lin**
Lin, Lang Ying A	**L.Y.Lin**
Lin, Mu Mu (fl. 1987) S	**M.M.Lin**
Lin, Pan(g) Juan (1936–) B	**P.J.Lin**
Lin, Qin(g) Zhong (1950–) S	**Q.Z.Lin**
Lin, Quan (fl. 1990) S	**Q.Lin**
Lin, S.F. (fl. 1977) M	**S.F.Lin**
Lin, Sang Hsiung (fl. 1979) M	**Sang H.Lin**
Lin, Shan Hsiung (1942–) B	**S.H.Lin**
Lin, Tsan Piao (fl. 1975–1987) S	**T.P.Lin**
Lin, Tsung Yau (fl. 1971) M	**T.Y.Lin**
Lin, Wan Tao (fl. 1980s) S	**W.T.Lin**
Lin, Wei Chih (fl. 1970) S	**W.C.Lin**
Lin, Ying Ren (fl. 1988) M	**Y.R.Lin**
(Lin, You Xing)	
see Ling, You Xin	**Y.X.Ling**
Linardić, Josip (1914–1941) A	**Linardić**
Linares, E. (fl. 1987) S	**Linares**
Lincke, Johann Rudolph (fl. 1840) S	**Lincke**
Linczevski, Igorj Alexandrovich (1908–) S	**Lincz.**
Linczevski, O.A. (1910–) S	**O.A.Lincz.**
Lind, Edna M. A	**E.M.Lind**
Lind, Jens Wilhelm August (1874–1939) M	**Lind**
Lindahl, Per-Olof (1922–) M	**Lindahl**
Lindau, Gustav (1866–1923) MPS	**Lindau**

Lindau, J. (fl. 1900) M	**J.Lindau**
Lindauer, Victor W. (1888–1964) A	**Lindauer**
Lindberg, Gustaf Anders (1832–1900) S	**G.Lindb.**
Lindberg, Harald (1871–1963) BS	**H.Lindb.**
Lindberg, J.B.W. S	**J.Lindb.**
Lindberg, Sextus Otto (1835–1889) BS	**Lindb.**
Lindblad, Matts Adolf (1821–1899) MS	**Lindblad**
Lindblom, Alexis Edvard (Eduard) (1807–1853) BS	**Lindblom**
(Linde, E.J.van der)	
see Van der Linde, E.J.	**Van der Linde**
Lindeberg, Brita (fl. 1959) M	**B.Lindeb.**
Lindeberg, Carl Johan (1815–1900) S	**Lindeb.**
Lindegg, Giovanna (fl. 1935) M	**Lindegg**
Lindeman, Jan Christiaan (1921–) S	**Lindeman**
Lindemann, Eduard Emanuilovitch von (1825–1900) S	**Lindem.**
Lindemann, Emanuel von (1795–1845) S	**Em.Lindem.**
Lindemann, Erich (1892–) A	**Er.Lindem.**
Linden, B.L.van der (fl. 1959) S	**B.L.Linden**
Linden, Jean Jules (1817–1898) PS	**Linden**
Linden, Lucien (1851–1940) S	**L.Linden**
Lindenberg, Johann Bernhard Wilhelm (1781–1851) ABS	**Lindenb.**
Linder, David Hunt (1899–1946) M	**Linder**
Linder, Hans Peter (1954–) S	**H.P.Linder**
Linder, Theodor (fl. 1909) B	**T.Linder**
Linderman, Robert G. (fl. 1976) M	**Linderman**
Lindfors, (Karl Magnus) Theodor (fl. 1920) M	**Lindf.**
Lindgren, Sven Johan (1810–1849) BMS	**Lindgr.**
Lindheimer, Ferdinand Jacob (1801–1879) S	**Lindh.**
Lindig, Alejandro (Alexander) (fl. 1861) P	**Lindig**
Lindig, D. (fl. 1987) S	**D.Lindig**
Lindinger, Karl Hermann Leonhard (1879–) S	**Linding.**
Lindley, John (1799–1865) ABFMPS	**Lindl.**
Lindman, Carl Axel Magnus (1856–1928) PS	**Lindm.**
Lindner, Paul (1861–1945) M	**Lindner**
Lindquist, Juan Carlos (1899–) M	**J.C.Lindq.**
Lindquist, Sven Bertil Gunvald (1904–1963) S	**Lindq.**
Lindroth, Johan Ivar (later Liro, J.I.) (1872–1943) MS	**Lindr.**
Lindsay, Archibald K. (–1781) S	**A.K.Linds.**
Lindsay, David C. (fl. 1971) M	**D.C.Linds.**
Lindsay, George Edmund (1916–) S	**G.E.Linds.**
Lindsay, John (1785–1803) S	**J.Linds.**
Lindsay, Robert (1846–1913) S	**R.Linds.**
Lindsay, William Lauder (1829–1880) ABMPS	**Linds.**
Lindsey, J.Page (1948–) M	**Lindsey**
Lindstedt, Alf A	**A.Lindst.**
Lindstedt, Karl (1846–) M	**Lindst.**
Lindström, Axel Albert (1864–1946) S	**Lindstr.**
Lindström, Håkan (fl. 1989) M	**H.Lindstr.**
Lindstrom, Sandra C. (1948–) A	**S.C.Lindstr.**

Lindt, W. (fl. 1889) M	**Lindt**
Lindtner, Vojtech H. (1904–1965) MS	**Lindtner**
Linford, Maurice Blood (1901–1960) M	**Linford**
Ling, Chüan (fl. 1970) S	**C.Ling**
Ling, H.U. A	**H.U.Ling**
Ling, H.Y. A	**H.Y.Ling**
Ling, Lai Kuan S	**L.K.Ling**
Ling, Lee (1911–) M	**L.Ling**
Ling, Ping Ping (1934–) S	**P.P.Ling**
Ling, Yeou Ruenn (1937–) S	**Y.R.Ling**
Ling, Yong (fl. 1930) M	**Y.Ling**
Ling, Yong Yuan (1903–1981) S	**Ling**
Ling, You Xin (1934–) P	**Y.X.Ling**
Ling, Yuan Jie (fl. 1980) A	**Y.J.Ling**
(Ling–Yong)	
see Ling, Yong	**Y.Ling**
(Ling–Young)	
see Ling, Yong	**Y.Ling**
Lingappa, B.T. (fl. 1953) M	**Lingappa**
Lingelsheim, Alexander von (1874–1937) MS	**Lingelsh.**
Linhart, György (Georg, George) (1844–1925) M	**Linh.**
Link, George Konrad Karl (1888–) M	**G.K.Link**
Link, Johann Heinrich Friedrich (1767–1851) ABMPS	**Link**
Linke, August (fl. 1853–57) S	**Linke**
Linkola, Kaarlo (1888–1942) MS	**Linkola**
Linn, Manson Bruce (1908–) M	**Linn**
Linnaeus, Carl von (1707–1778) ABMPS	**L.**
Linnaeus, Carl von (1741–1783) ABMPS	**L.f.**
(Linné, Carl)	
see Linnaeus, Carl von	**L.f.**
(Linné, Carl von)	
see Linnaeus, Carl von	**L.**
Linnemann, Germaine (fl. 1936) M	**Linnem.**
Linney, William Marcus (1835–1887) S	**Linney**
Linsbauer, Karl (1872–1934) A	**Linsbauer**
Lint, Harold LeRoy (1917–) S	**Lint**
Linton, Edward Francis (1848–1928) S	**E.F.Linton**
Linton, William James (1812–1898) P	**Linton**
Linton, William Richardson (1850–1908) S	**W.R.Linton**
(Liogier Allut, Enrique Eugenio)	
see Alain, (Brother)	**Alain**
(Liogier, Enrique Eugenio)	
see Alain, (Brother)	**Alain**
(Liogier, Henri Alain)	
see Alain, (Brother)	**Alain**
(Liou, Ho)	
see Liu, Hou	**H.Liu**
Liou, Lian(g) (1933–) S	**L.Liou**
Liou, Ming Yuan (fl. 1987) S	**M.Y.Liou**

Liou, S.C. (fl. 1977) M	**S.C.Liou**
Liou, Shou Lu (1933–) S	**S.L.Liou**
Liou, Shu Run (fl. 1985) S	**S.R.Liou**
Liou, Tchen Ngo (fl. 1929–1958) MPS	**Liou**
Liou, Ying Xing (fl. 1985) S	**Y.X.Liou**
Lipa, Jerzy J. A	**Lipa**
Lipatova, V.V. (fl. 1963) S	**Lipatova**
Lipinski, W. (fl. 1924) M	**Lipinski**
Lipkin, Yaakov (fl. 1974) M	**Lipkin**
(Lipman, Teodor)	
see Lippmaa, Teodor	**Lippmaa**
Lipnik, E.S. A	**Lipnik**
Lipp, Franz Joseph (1731–1775) S	**Lipp**
Lippert, Chr. (fl. 1894) M	**C.Lippert**
Lippert, Wolfgang (1937–) S	**W.Lippert**
Lippert, Xaver Joseph (fl. 1786) S	**Lippert**
Lippmaa, Teodor (1892–1943) S	**Lippmaa**
Lippold, Hans (1932–1980) S	**Lippold**
Lipps, Jere Henry (1939–) A	**J.H.Lipps**
Lipps, P.E. (fl. 1980) M	**P.E.Lipps**
Lipschitz, Sergej Julievitsch (1905–1983) S	**Lipsch.**
Lipscomb, Barney L. (1950–) S	**Lipscomb**
(Lipshits, Sergej Julievitsch)	
see Lipschitz, Sergej Julievitsch	**Lipsch.**
(Lipsic, Sergej Julievitsch)	
see Lipschitz, Sergej Julievitsch	**Lipsch.**
(Lipskij, Wladimir Hippolitowitsch)	
see Lipsky, Vladimir Ippolitovich	**Lipsky**
Lipsky, Vladimir Ippolitovich (1863–1937) S	**Lipsky**
Liptovszky, Josef (fl. 1970) M	**Lipt.**
Lira Saade, Rafael (fl. 1991) S	**Lira**
Liro, Johan Ivar (né Lindroth, J.I.) (1872–1943) MS	**Liro**
Lisa, Domenico (1801–1867) BS	**Lisa**
Lisal, K. (fl. 1986) S	**Lisal**
Lisavenko, M.A. (fl. 1970) S	**Lisav.**
Lisboa, H.M. S	**H.M.Lisboa**
Lisboa, José Camilla (1822–1897) S	**Lisboa**
Lisboa, Moacyr do Amaral (fl. 1974) S	**M.Lisboa**
Liskun, I.G. A	**Liskun**
Lisowski, Stanisław (1924–) BS	**Lisowski**
Lissone, Sebastiano S	**Lissone**
List, Friedrich Ludwig (fl. 1828–1837) S	**List**
Listeman, H. (fl. 1973) M	**Listeman**
Lister, Arthur (1830–1908) M	**Lister**
Lister, Guilielma (1860–1949) M	**G.Lister**
Lister, Joseph Jackson (1857–1927) S	**J.L.Lister**
Lister, T.R. A	**T.R.Lister**
Liston, A. (fl. 1986) S	**Liston**
Litardière, René Verriet de (1888–1957) PS	**Litard.**
Litschauer, Viktor (1879–1939) M	**Litsch.**

Litten, Walter (1915–) M	**Litten**
Littini, G. (fl. 1984) M	**Littini**
Little, Elbert Luther (1907–) BS	**Little**
Little Flower, (Sister) (fl. 1983) M	**Little Flower**
Littlejohn, Lewis H. S	**Littlej.**
Littler, Mark Masterson (1939–) A	**Littler**
Littlewood, Ray Charles (1924–1967) S	**Littlew.**
Littrell, R.H. (fl. 1970) M	**Littrell**
Litvinenko, R.M. A	**Litvin.**
Litvinov, Dimitri Ivanovich (1854–1929) MS	**Litv.**
Litvinov, M.A. (fl. 1967) M	**M.A.Litv.**
(Litwinow, Dmitrij Ivanovitsch)	
see Litvinov, Dimitri Ivanovich	**Litv.**
(Litwinow, M.A.)	
see Litvinov, M.A.	**M.A.Litv.**
Liu, A.T. (fl. 1982) S	**A.T.Liu**
Liu, Ai Ying (fl. 1991) M	**A.Y.Liu**
Liu, B.W. (fl. 1981) S	**B.W.Liu**
Liu, Bo (fl. 1976) M	**B.Liu**
Liu, D. (fl. 1982) M	**D.Liu**
Liu, Da Ji (1940–) S	**D.J.Liu**
Liu, Dai Ming (fl. 1986) S	**D.M.Liu**
Liu, Dao Qing (fl. 1984) S	**D.Q.Liu**
Liu, Deng Yi (fl. 1988) S	**D.Y.Liu**
Liu, Fang Yuan (fl. 1988) S	**F.Y.Liu**
Liu, Gui Sen (fl. 1989) S	**G.S.Liu**
Liu, Guo Hou (fl. 1989) S	**G.H.Liu**
Liu, Guo Jun (1933–) S	**G.J.Liu**
Liu, H.G. (fl. 1986) M	**H.G.Liu**
Liu, H.H. (fl. 1981) S	**H.H.Liu**
Liu, Hou (fl. 1932) S	**H.Liu**
Liu, Huo Lin (fl. 1982) S	**H.L.Liu**
Liu, J.K. (fl. 1981) S	**J.K.Liu**
Liu, Jian Hua (fl. 1984) S	**J.H.Liu**
Liu, Jian Lin (fl. 1989) S	**J.L.Liu**
Liu, Jian Sheng (1955–) S	**J.S.Liu**
Liu, Ju Chang (1895–) S	**J.C.Liu**
Liu, Jun Zhe (fl. 1989) S	**J.Z.Liu**
Liu, Ke Wang (1940–) S	**K.W.Liu**
(Liu, Lian(g))	
see Liou, Lian(g)	**L.Liou**
Liu, Lii Jang (fl. 1971) M	**L.J.Liu**
Liu, Mei Hua (fl. 1987) M	**M.H.Liu**
Liu, Ming Gang (fl. 1987) S	**M.G.Liu**
Liu, Ming Ting (fl. 1988) S	**M.T.Liu**
Liu, Nian (fl. 1987) S	**N.Liu**
Liu, Pei Song (fl. 1984) S	**P.S.Liu**
Liu, Qi Hong (fl. 1983) S	**Q.H.Liu**
Liu, Qi Xing (fl. 1989) S	**Q.X.Liu**
Liu, S.Q. (fl. 1987) M	**S.Q.Liu**

Liu, Shang Wu (1934–   ) S   **S.W.Liu**
(Liu, Shên O)
  see Liou, Tchen Ngo   **Liou**
(Liu, Shou Lu)
  see Liou, Shou Lu   **S.L.Liou**
Liu, Shou Yang (fl. 1987) S   **S.Y.Liu**
Liu, T.R. (fl. 1985) M   **T.R.Liu**
Liu, Tang Shui (fl. 1975) S   **Tang S.Liu**
Liu, Tung Shui (1911–   ) S   **T.S.Liu**
Liu, W.D. (fl. 1984) S   **W.D.Liu**
Liu, Wo Peng (fl. 1991) M   **W.P.Liu**
Liu, X.C. (fl. 1986) S   **X.C.Liu**
Liu, Xi Jin (1911–   ) M   **X.J.Liu**
Liu, Xian Zhang (fl. 1991) S   **X.Z.Liu**
Liu, Xiang Jun (fl. 1989) S   **X.Jun Liu**
Liu, Xiao Long (fl. 1988) S   **X.L.Liu**
Liu, Yeh Ching (fl. 1976) S   **Y.C.Liu**
Liu, Yin Hua (fl. 1984) M   **Yin H.Liu**
Liu, Ying Hsin (fl. 1984) M   **Ying H.Liu**
Liu, Yong Min (fl. 1984) S   **Y.M.Liu**
(Liu, Yu Hu)
  see Law, Yuh Wu   **Y.W.Law**
Liu, Yu Lan (1931–   ) S   **Y.L.Liu**
Liu, Z.K. (fl. 1985) M   **Z.K.Liu**
Liu, Zhao Guang (fl. 1984) S   **Z.G.Liu**
Liu, Zhen Hau (fl. 1988) S   **Z.H.Liu**
Liu, Zhen Qiu (fl. 1985) AS   **Z.Q.Liu**
Liu, Zheng Yu (fl. 1983) PS   **Z.Y.Liu**
Liu, Zhong Ling (fl. 1983) S   **Z.L.Liu**
Liuti, Alfredo (1957–   ) S   **Liuti**
Livera, E.J.   S   **Livera**
Livingston, W.H. (fl. 1987) M   **Livingston**
Livsey, Susan M.E. (fl. 1987) M   **Livsey**
Lizgunova, Tatyana Vasilyevna (1901–1984) S   **Lizg.**
Ljava, J.L.   S   **Ljava**
Ljubarsky, L.V. (fl. 1961–1975) M   **Ljub.**
Ljubitskaya, Lydia Ivanovna (1886–   ) B   **Ljubitsk.**
(Ljubitz, Lydia Ivanovna)
  see Savicz–Lubitskaya, Lydia Ivanovna   **L.I.Savicz**
Ljungström, Ernst Leopold (1854–1943) S   **Ljungstr.**
Llamas, Felix (1952–   ) S   **Llamas**
Llambías, J. (fl. 1930) M   **Llambías**
Llaña, Alfredo H.   A   **Llaña**
Llano, George Albert (1911–   ) M   **Llano**
Llanos, Antonio (1806–1881) S   **Llanos**
(Llave, Pablo de la)
  see La Llave, Pablo de   **La Llave**
Llenas y Fernández, Manuel (fl. 1900) M   **Llenas**
Lleras, Eduardo (1944–   ) S   **Lleras**

(Llimona Pages, X.)
   see Llimona, X.   **Llimona**
Llimona, X. (fl. 1973) M   **Llimona**
Llorens de Ros, Antonio (1917–) S   **Llorens**
(Llorens García, Leonardo)
   see Llorens, Leonardo   **L.Llorens**
Llorens, Leonardo (Lleonard) (1946–) S   **L.Llorens**
Lloyd, Curtis Gates (also McGinty, C.G.) (1859–1926) M   **Lloyd**
Lloyd, David G. (1937–) S   **D.G.Lloyd**
Lloyd, Francis Ernest (1868–1947) PS   **F.E.Lloyd**
Lloyd, George N. (1804–1889) S   **G.N.Lloyd**
Lloyd, H.L. (fl. 1988) M   **H.L.Lloyd**
Lloyd, James (1810–1896) PS   **J.Lloyd**
Lloyd, John (1791–1870) S   **Jn.Lloyd**
Lloyd, Robert Michael (1938–) S   **R.M.Lloyd**
Lo Bianco, Salvatore A   **Lo Bianco**
Lo, Hang Chiang (fl. 1977) S   **H.C.Lo**
Lo, Hsien Shui (1927–) S   **H.S.Lo**
Lo, T.C. (fl. 1953) M   **T.C.Lo**
Lo, T.T. (fl. 1961) M   **T.T.Lo**
Lo, Tien Yu S   **T.Y.Lo**
Lob, U.W.A. (fl. 1985) S   **Lob**
Lobachev, A.Ya. (fl. 1977) S   **Lobachev**
Lobarzewski, Hyacinth Strzemù von (1816–1862) AB   **Lobarz.**
Lobata de Faria, C. (fl. 1938) M   **Lob.Faria**
Lobato, Rosa Corrêa (fl. 1973) S   **Lobato**
Lobb, Thomas (1820–1894) S   **T.Lobb**
Lobb, William (1809–1863) S   **W.Lobb**
Lobban, Christopher S. (1950–) A   **Lobban**
Lobel, Mathias de (1538–1616) S   **Lobel**
(Lobelius, Mathias de)
   see Lobel, Mathias de   **Lobel**
Lobik, Alexis Iulianovich (fl. 1928) M   **Lobik**
Lobin, Wolfram (1951–) S   **Lobin**
Lôbo, Jorge (fl. 1940) AM   **Lôbo**
Lobreau–Callen, D. (1940–) S   **Lobr.-Callen**
Lochhead, A.G. (fl. 1940) M   **A.G.Lochhead**
Lochhead, R. A   **R.Lochhead**
Lock, John Michael (1942–) S   **Lock**
Locke, John (1792–1856) S   **Locke**
Locker, Sigurd A   **Locker**
Lockhart, David (–1846) S   **Lockh.**
Lockwood, Tommie Earl (1941–1975) S   **Lockwood**
Locquin, Marcel V. (fl. 1943) M   **Locq.**
Locquin–Linard, Monique (fl. 1990) M   **Locq.-Lin.**
Lodder, Jacomina (1905–1987) M   **Lodder**
Loddiges, Conrad (L.) (1738–1826) S   **Lodd.**
Loddiges, George (1784–1846) S   **G.Lodd.**
Lodge, D.Jean (fl. 1988) M   **Lodge**

Lodh, A.R. (fl. 1977) M	**Lodh**
Lodha, B.C. (fl. 1971) M	**Lodha**
Lodhi, S.A. (fl. 1962) M	**Lodhi**
Loe, A. (fl. 1981–89) S	**Loe**
Loeblich, Alfred R. (1914– †A	**Loebl.**
Loeblich, Alfred R.(III) (1941–) AM	**A.R.Loebl.**
Loeblich, Laurel A. A	**L.A.Loebl.**
Loeffler, Wolfgang (fl. 1957) M	**Loeffler**
Loefgren, (Johan) Albert(o) (Constantin) (1854–1918) S	**Loefgr.**
Loefling, Pehr (1729–1756) S	**Loefl.**
Loehr, Egid von (fl. 1848) S	**E.Leohr**
Loehr, Matthias Joseph (1800–1882) S	**M.Loehr**
(Loennroth, Erik Johannes)	
see Lönnroth, Erik Johannes	**E.J.Lönnr.**
Loerakker, W.M. (fl. 1975) M	**Loer.**
Loesch, Alfred (1865–1945) S	**Loesch**
Loescher, Eduard (fl. 1852) S	**Loescher**
Loesel, Johannes (1607–1655) L	**Loesel**
(Loeselius, Johannes)	
see Loesel, Johannes	**Loesel**
Loesener, Ludwig Eduard Theodor (1865–1941) S	**Loes.**
Loesener, Otto (fl. c.1826) S	**O.Loes.**
Loeske, Leopold (1865–1935) B	**Loeske**
Loew, Ernst (1843–1908) S	**Loew**
Loewenbaum, M.E. (fl. 1973) M	**Loewenb.**
(Loewenthal, W.)	
see Löwenthal, W.	**Löwenthal**
(Löfgren, (Johan) Albert(o) (Constantin))	
see Loefgren, (Johan) Albert(o) (Constantin)	**Loefgr.**
(Löfling, Pehr)	
see Loefling, Pehr	**Loefl.**
Logemann, H. (fl. 1987) M	**Logem.**
Loh, T.C. (fl. 1982) M	**Loh**
Lohammar, G. (fl. 1953) M	**Lohammar**
Lohar, L.W. (fl. 1955) M	**Lohar**
Lohde, Georg (fl. 1874) M	**Lohde**
Loher, August (1874–1930) S	**Loher**
Lohman, Kenneth E. (1897–) A	**Lohman**
Lohman, Marion Lee (1903–) M	**M.L.Lohman**
Lohman, William H. A	**W.H.Lohman**
Lohmann, Hans (1863–1934) A	**Lohmann**
Lohmeyer, T.R. (fl. 1979) M	**Lohmeyer**
(Löhr, Matthias Joseph)	
see Loehr, Matthias Joseph	**M.Loehr**
Lohsomboon, Pongvipa (fl. 1986) M	**Lohsomb.**
Lohwag, Heinrich (1884–1945) M	**Lohwag**
Lohwag, Irmgard (fl. 1972) M	**I.Lohwag**
Lohwag, Kurt (1913–1970) M	**K.Lohwag**
(Loidi Aguirre, Javier José)	
see Loidi, Javier José	**Loidi**

Loidi, Javier José (1953–) S	**Loidi**
Loiseaux, Susan A	**Loiseaux**
Loisel, R.J. (1938–) S	**R.J.Loisel**
Loiseleur-Deslongchamps, Jean Louis August(e) (1774–1849) APS	**Loisel.**
Loitlesberger, Karl (1857–1943) B	**Loitl.**
Lojacono, Michele (1853–1919) BPS	**Lojac.**
(Lojacono Pojero, Michele)	
see Lojacono, Michele	**Lojac.**
Lojka, (W.) Hugo (1844–1887) M	**Lojka**
Løjtnant, Bernt (1946–) S	**Løjtnant**
Lokhorst, Gijsbert M. (1943–) A	**Lokhorst**
Lomakin, Aleksandr Aleksandrovich (1863–1930) S	**Lomakin**
Lomakin, E.N. S	**E.N.Lomakin**
Lomakina, N.V. S	**Lomakina**
Lomax, Alban Edward (1861–1894) S	**Lomax**
Lombard, Eugene H. A	**E.H.Lombard**
Lombard, Frances Faust (1915–) M	**Lombard**
Lombardo, Atilio (fl. 1958) P	**Lombardo**
Lomelí Gonzalez, Irma Rosalina (fl. 1981) S	**Lomelí**
Lomnicki, A.M. (fl. 1886) FM	**Lomnicki**
Lomonosova, Maria N. (1949–) S	**Lomon.**
Lon, Lian Li (fl. 1980) S	**L.L.Lon**
Lona, Fausto (fl. 1949) S	**Lona**
Lonaczewski, Alexandrovic (1885–1938) S	**Lonacz.**
Lonard, Robert I. (1942–) S	**Lonard**
Lonati, Giuliano (fl. 1985) M	**Lonati**
Lonay, Hyacinthe (1871–1934) S	**Lonay**
Londes, Friedrich Wilhelm (1780–1807) S	**Londes**
Londoño, Ximena (fl. 1987) S	**Londoño**
Long, Bayard Henry (1885–1969?) S	**B.H.Long**
Long, David Geoffrey (1948–) BS	**D.G.Long**
Long, G. (1928–) S	**G.Long**
Long, John A. A	**J.A.Long**
Long, Robert William (1927–1976) S	**R.W.Long**
Long, Tung Lun (fl. 1989) S	**T.L.Long**
Long, William Henry (1867–1947) M	**Long**
Long, Yao Hwa (fl. 1984) S	**Y.H.Long**
Longa, Massimo (fl. 1915–26) P	**Longa**
Longcore, Joyce E. (fl. 1989) M	**Longcore**
(Longhi, Hilda Maria)	
see Longhi-Wagner, Hilda Maria	**Longhi-Wagner**
Longhi-Wagner, Hilda Maria (fl. 1977) S	**Longhi-Wagner**
Longis, Danièle (fl. 1961) M	**Longis**
Longo, Biagio (1872–1950) S	**Longo**
Longpre, Edwin Keith (fl. 1970) S	**Longpre**
Longyear, Burton Orange (1868–1969) M	**Longyear**
(Lonicerus, Adam)	
see Lonitzer, Adam	**Lonitzer**
Lonitzer, Adam (1528–1586) L	**Lonitzer**

Lönnbohm, Oskar Anders (1856–1927) S	**Lönnb.**
Lönnegren, August Valfrid (1842–1904) M	**Lönn.**
Lönnrot, Elias (1802–1884) S	**Lönnrot**
Lönnroth, Erik Johannes (1883–) S	**E.J.Lönnr.**
Lönnroth, Knut Johan (1826–1885) MS	**Lönnr.**
Lonsing, A. (fl. 1939) S	**Lonsing**
Loock, E.E.M. (1905–1973) S	**Loock**
Looken, H.van (fl. 1987) S	**Looken**
Looman, Jan (1919–) M	**Looman**
Loomis, Nina Hosler (1883–1963) A	**Loomis**
Loos, C.A. (fl. 1950) M	**Loos**
Looser, Gualterio (1898–1982) PS	**Looser**
Loosjes, Adriaan (1761–1818) S	**Loosjes**
(Lopes, José Pinto)	
see Pinto Lopes, José	**Pinto Lopes**
Lopes, M.Helena Ramos (fl. 1970)	**R.Lopes**
López, A. (fl. 1980) M	**A.López**
(López Castrillon, A.)	
see Castrillon, A. López	**Castr.**
López, E.A. (1951–) S	**E.A.López**
López, F. (fl. 1979) M	**F.López**
(López F., L.C.)	
see López, L.C.	**L.C.López**
López Fernández, María Luisa (1940–) S	**López Fern.**
(López Figueiras, Manuel)	
see López–Figueiras, Manuel	**López–Fig.**
López González, Ginés Alejandro (1950–) S	**G.López**
López Guillén, Julio E. (1914–) S	**López Guillén**
López Jaramillo, Luis E. (fl. 1983) S	**L.E.López**
López, L.C. (fl. 1965) M	**L.C.López**
López Pacheco, M.J. (fl. 1985) S	**López Pach.**
López, Silvia E. (1949–) M	**S.E.López**
López–Figueiras, Manuel (1915–) MS	**López–Fig.**
López–Franco, Rosa María (fl. 1990) M	**López–Franco**
(López–González, Ginés Alejandro)	
see López González, Ginés Alejandro	**G.López**
López–Lastra, Claudia Cristina (fl. 1990) M	**López–Lastra**
López–Palacios, Santiago (1918–) S	**López–Pal.**
Lopriore, Giuseppe (1865–1928) S	**Lopr.**
Lopuchin, Alexander S.D. A	**Lopuchin**
Lorbeer, Gerhard (–1945) B	**Lorb.**
Lorch, Jacob W. (1924–) AS	**J.W.Lorch**
Lörch, Ph.J. (fl. 1890) S	**P.J.Lörch**
Lorch, Wilhelm (1867–1954) BS	**Lorch**
Lord, Ernest E. (1899–1970) S	**Lord**
Lordkipanidze, A.D. (fl. 1947) S	**Lordkip.**
Lorea–Hernández, Francisco G. (1956–) PS	**Lorea–Hern.**
Lorek, Christian Gottlieb (1788–1871) S	**Lorek**
Lorence, David H. (1946–) PS	**Lorence**
Lorens, L. (fl. 1983) S	**Lorens**

Lorent, J.August von (1812–1884) S	**Lorent**
Lorente y Asensi, Vincente Alfonso (1758–1813) S	**Lorente**
Lorentz, Hendrik Anton (1853–1928) S	**H.Lorentz**
Lorentz, Paul (Pablo) Günther (1835–1881) ABMS	**Lorentz**
Lorenz, Annie (1879–1927) S	**A.Lorenz**
Lorenz, B. (fl. 1894) S	**B.Lorenz**
Lorenz, Josef Roman (1825–1911) A	**Lorenz**
Lorenz, R. (fl. 1987) S	**R.Lorenz**
Lorenz, Rolland Carl (1904–) M	**R.C.Lorenz**
Lorenz, Theodor (1875–1909) A	**T.Lorenz**
Lorenzen, Marcus (1847–1928) S	**Lorenzen**
Lorenzo, Laura (fl. 1989) M	**Lorenzo**
Lorenzoni, Giovanni Giorgio (1938–) S	**Lorenzoni**
Loret, Henri (1811–1888) S	**Loret**
Loret, Victor (1859–) S	**V.Loret**
Lorey, A. (fl. 1851) S	**A.Lorey**
Lorey, Félix-Nicolas (–1841) S	**Lorey**
Loria, M.L. (fl. 1967) S	**Loria**
Loriente, E. (fl. 1982) S	**Loriente**
Lörinczi, Francisc (1924–) M	**Lörinczi**
Lorinser, Friedrich Wilhelm (1817–1895) M	**F.Lorinser**
Lorinser, Gustav (1811–1863) S	**G.Lorinser**
Lorougnon, J.Guédé (fl. 1964) S	**Lorougnon**
Lortet, Clémence (1772–1835) S	**Lortet**
Lortet, Louis Charles (1836–1909) S	**L.Lortet**
Los Rios, P.M.U. (fl. 1931) M	**Los Rios**
(Losa España, Taurino Mariano)	
see Losa, Taurino Mariano	**Losa**
Losa, Taurino Mariano (1893–1965) MS	**Losa**
Losa-Quintana, José Maria (fl. 1969) M	**J.M.Losa**
Losana, Mathaeo A	**Losana**
(Lösch, Alfred)	
see Loesch, Alfred	**Loesch**
Losch, Hermann (1888–) S	**Losch**
Losch, Ilse (1920–) PS	**I.Losch**
Loscos, C. S	**C.Loscos**
Loscos y Bernál, Francisco (1823–1886) S	**Loscos**
Lösecke, August Georg von (1837–1912) M	**Lösecke**
Loseva, E.I. A	**Loseva**
(Losina–Losinskaja, A.S.)	
see Losinskaja, A.S.Losina–	**Losinsk.**
Losinskaja, A.S.Losina– (1903–1958) S	**Losinsk.**
Loss, Guiseppe (Josef) (1831–1880) P	**Loss**
Lothian, Thomas Robert Noel (1915–) S	**Lothian**
Lotsy, Johannes Paulus (1867–1931) MPS	**Lotsy**
Lott, Emily J. (fl. 1982) S	**E.J.Lott**
Lott, Henry J. (fl. 1938) S	**Lott**
Lotte, Gabr.(fl. 1928) M	**Lotte**
Lotus, S.S. (fl. 1936) M	**Lotus**

Lotze, Maurizio (Maritius) A	**Lotze**
Lou, Jian Shing (1935–) B	**J.S.Lou**
Lou, Lian Li (fl. 1980) S	**L.L.Lou**
Lou, Zhi Cen (fl. 1988) S	**Z.C.Lou**
Loubière, M.Auguste (fl. 1923) M	**Loubière**
Loudon, Jane Wells (1807–1858) S	**J.W.Loudon**
Loudon, John Claudius (1783–1843) PS	**Loudon**
(Louis, (Frère))	
see Rolland–Germain	**Roll.-Germ.**
Louis, A. A	**A.Louis**
Louis, Adriaan M. (1944–) S	**A.M.Louis**
Louis, Jean Laurent Prosper (1903–1947) S	**Louis**
Louis–Marie, (Père) (1896–1978) S	**Louis-Marie**
Loundó, Dattá A	**Loundó**
Lounsberry, Alice (1872–1949) S	**Lounsb.**
Loureiro, João de (1717–1791) ABPS	**Lour.**
Lourteig, Alicia (1913–) S	**Lourteig**
Lousley, Job Edward (1907–1976) S	**Lousley**
Louw, A.J. (fl. 1941) M	**Louw**
Louw, C.D. (fl. 1976) M	**C.D.Louw**
Löve, Áskell (1916–) BPS	**Á.Löve**
Löve, Doris Benta Maria (1918–) BMPS	**D.Löve**
Love, Robert Merton (1909–) S	**Love**
Lovejoy, Ruth Ellen Harrison (1882–) M	**Lovejoy**
Loveless, A. Ray (fl. 1965) M	**Loveless**
Lovelius, Olga L. (fl. 1970) S	**Lovelius**
Lovén, Fredrik August (1847–1929) S	**Lovén**
Lovis, John Donald (1930–) PS	**Lovis**
Lövkvist, Börje (1919–) S	**Lövkvist**
Lovrić, Andrija–Zelimir (1943–) S	**Lovrić**
Low, G.C. (fl. 1913) M	**G.C.Low**
Low, Hugh (1824–1905) S	**H.Low**
Löw, Immanuel (1854–1944) S	**I.Löw**
Löw, Irmentraut A	**Irm.Löw**
Low, Stuart Henry (1826–1890) S	**S.H.Low**
Löw, Ulrich (1922–) S	**U.Löw**
Lowater, W.R. (fl. 1925) M	**Lowater**
Lowden, Richard M. (1943–) S	**Lowden**
Lowe, Charles H. S	**C.H.Lowe**
Lowe, Charles William (1885–1969) AS	**C.Lowe**
Lowe, Edward Joseph (1825–1900) PS	**E.J.Lowe**
Lowe, Ephriam Noble (1864–1933) S	**E.N.Lowe**
Lowe, Josiah Lincoln (1905–) M	**J.Lowe**
Lowe, Michael R. (fl. 1990) S	**M.R.Lowe**
Lowe, Rex Loren (1943–) A	**R.L.Lowe**
Lowe, Richard Thomas (1802–1874) PS	**Lowe**
Lowen, Rosalind (fl. 1986) M	**Lowen**
Löwenthal, W. (fl. 1904) M	**Löwenthal**
Lowrey, Timothy K. (1953–) S	**Lowrey**

Lowrie, Allen S	**Lowrie**
Lowry, Porter Peter (1956–) S	**Lowry**
Lowther, J.Stewart (fl. 1955) F	**Lowther**
Lowy, Bernard (1916–) M	**Lowy**
Loyal, D.S. (fl. 1969) P	**Loyal**
(Lozano, Gustavo)	
see Lozano–Contreras, Gustavo	**Lozano**
Lozano, Matteo S	**M.Lozano**
Lozano–Contreras, Gustavo (1938–) S	**Lozano**
Lozeron, Henri A	**Lozeron**
(Lozina–Lozinscaia, A.S.)	
see Losinskaja, A.S.Losina–	**Losinsk.**
Lu, An Min(g) (1939–) S	**A.M.Lu**
Lu, Bao Ren A	**B.Ren Lu**
Lu, Bao Rong (fl. 1988) S	**B.Rong Lu**
Lu, D.A. (fl. 1900s) M	**D.A.Lu**
Lu, De Quan (1936–) S	**D.Q.Lu**
Lu, Fu Yuan (fl. 1976) S	**F.Y.Lu**
Lu, Jia Yun (fl. 1987) M	**J.Y.Lu**
Lu, Karen L. (fl. 1983) S	**K.L.Lu**
Lu, Ling Ti (Di) (1930–) S	**L.T.Lu**
Lu, Rui Lin (fl. 1989) S	**R.L.Lu**
Lu, Sheng You (1946–) S	**S.Y.Lu**
Lu, Xiang Huai (fl. 1989) M	**X.H.Lu**
Lu, Yen Hao A	**Y.H.Lu**
Lu, Yi Xin (1945–) S	**Y.X.Lu**
Lu, Zhu (fl. 1988) S	**Z.Lu**
Lubbers, Louis (1832–1905) S	**Lubbers**
Lübeck, H.G. (1809–1900) S	**Lübeck**
Lüben, August Heinrich Philipp (1804–1874) S	**Lüben**
Lubenetz, P.A. S	**Lubenetz**
Lubinsky, G. (fl. 1955) M	**Lubinsky**
Luc, H. B	**H.Luc**
Luc, M. (fl. 1951) M	**Luc**
(Luca, Paolo De)	
see De Luca, Paolo	**De Luca**
Lucae, August Friedrich Theodor (1800–1848) S	**Lucae**
Lucand, Jean Louis (1821–1896) MS	**Lucand**
Lucas, A. (fl. 1966) M	**A.Lucas**
Lucas, Arthur Henry Shakespeare (1853–1936) A	**A.H.S.Lucas**
Lucas, Grenville Llewellyn (1935–) S	**G.Ll.Lucas**
Lucas, I.A.N. A	**I.A.N.Lucas**
Lucas, J.A.W. A	**J.A.W.Lucas**
Lucas, Leon T. (fl. 1973) M	**L.T.Lucas**
Lucas, Maria Teresa (fl. 1952) M	**M.T.Lucas**
Lucas Rodríguez, Rafael (1915–1981) S	**Lucas Rodr.**
Lucas–Clark, Joyce A	**Lucas–Clark**
Lucchini, Gianfelice (1942–) M	**Lucchini**
Lucé, Johann Wilhelm Ludwig von (–1862) S	**Lucé**
Luceño, Modesto (fl. 1987) S	**Luceño**

Luces de Febres, Zoraida (1922–   ) S   **Luces**
Lucet, A. (fl. 1897) M   **Lucet**
Luchetti, G. (fl. 1936) M   **Luchetti**
Luchnik, Z.I. (1909–   ) S   **Luchnik**
Lucien, D. (fl. 1923) M   **Lucien**
Luck, H.D. (fl. 1987) S   **Luck**
Luck–Allen, E.Robena (fl. 1960) M   **Luck-Allen**
Lückel, Emil (fl. 1978) S   **Lückel**
Lückhoff, Carl August (1914–1960) S   **C.A.Lückh.**
Lückhoff, Hilmar Albert (1916–   ) S   **H.A.Lückh.**
Lücking, Robert (fl. 1991) M   **Lücking**
Lucksch, Ina   A   **Lucksch**
Luckwill, L.C. (1914–   ) S   **Luckwill**
(Lucznik, Z.I.)
    see Luchnik, Z.I.   **Luchnik**
Lüders, Friedrich Wilhelm Anton (1751–1810) S   **Lüders**
Lüderwaldt, Albert (1861–1917) S   **Lüderw.**
Lüdi, Werner (1888–1968) MS   **Lüdi**
Ludlow, Frank (1885–1972) S   **Ludlow**
Ludvík, Jirí   A   **Ludvík**
Ludwig, Alfred (1879–1964) S   **A.Ludw.**
Ludwig, Carl (fl. 1801) B   **C.Ludw.**
Ludwig, Christian Friedrich (1757–1823) BS   **C.F.Ludw.**
Ludwig, Christian Gottlieb (1709–1773) ABPS   **Ludw.**
Ludwig, Clinton Albert (1886–1941) M   **C.A.Ludw.**
Ludwig, E. (fl. 1988) M   **E.Ludw.**
Ludwig, Francis W. (1904–   ) A   **F.W.Ludw.**
Ludwig, Friedrich (1851–1918) AM   **F.Ludw.**
Ludwig, Rudolph August Birminhold Sebastien (1812–1880) MS   **R.Ludw.**
Ludwig, Wolfgang (1923–   ) S   **W.Ludw.**
Luebke, Niel T. (1950–   ) PS   **Luebke**
(Lueckel, Emil)
    see Lückel, Emil   **Lückel**
Lueder, Franz Hermann Heinrich (–1791) S   **Lueder**
Luedersdorff, Friedrich Wilhelm (1801–   ) S   **Luedersd.**
Luehmann, Johann George W. (1843–1904) S   **Luehm.**
Luer, Carlyle A. (1922–   ) S   **Luer**
Luer, G.M. (fl. 1969) S   **G.M.Luer**
Luerssen, Christian (1843–1916) APS   **Luerss.**
Luetzelburg, Philipp von (1880–1948) S   **Luetzelb.**
Luferov, A.N. (fl. 1989) S   **Luferov**
Lühne, Vincenz   A   **Lühne**
Lühnemann, G.H. (fl. 1809) AM   **Lühnem.**
Luijk, Abraham van (1874–   ) M   **Luijk**
Luisier, Alphonse (1872–1957) BS   **Luisier**
Luiten, Bernada (fl. 1975) M   **Luiten**
Luizet, Marie Dominique (1851–1930) S   **Luizet**
Lukas, Karen Jeanne (1941–   ) A   **Lukas**
Lukasheva, L.M. (fl. 1979) M   **Lukasheva**

Lukavský, Jaromír A	**Lukavský**
Luke, Padmabai (fl. 1972) M	**Luke**
Lukin, V. (fl. 1974) M	**Lukin**
(Lukischeva, A.N.)	
see Lukitscheva, A.N.	**Lukitsch.**
Lukitscheva, A.N. S	**Lukitsch.**
Lukmanoff, Athanase de (fl. 1889) S	**Lukman.**
Lum, P.T.M. (fl. 1985) M	**Lum**
Lumb, Dennis (1871–1951) S	**Lumb**
Lumbsch, Helge Thorsten (1964–) M	**Lumbsch**
Lumley, Peter Francis (1938–) S	**Lumley**
Lumnitzer, István (Stephan) (1750–1806) MS	**Lumn.**
Lunan, John (fl. 1814) S	**Lunan**
Lund, A. (fl. 1930) M	**A.Lund**
Lund, Anders Axel Wilhelm (1839–1925) S	**A.A.W.Lund**
Lund, Everett Eugene (1907–) A	**E.E.Lund**
Lund, Ingelise D. (1957–) S	**I.D.Lund**
Lund, John Walter Guerrier (1913–) A	**J.W.G.Lund**
(Lund, Nicolai)	
see Lund, Nils	**N.Lund**
Lund, Niels Tønder (1749–1809) S	**Lund**
Lund, Nils (Nicolai) (1814–1847) MS	**N.Lund**
Lund, Peter Jorgensen (1870–1938) B	**P.J.Lund**
Lund, Peter Wilhelm (1801–1880) S	**P.W.Lund**
Lund, Søren (1905–1974) A	**S.Lund**
Lundberg, Folke Reginald (1890–) A	**Lundb.**
Lundblad, Anna Birgitta (1920–) B	**Lundbl.**
Lundblad, Britton (fl. 1950) F	**B.Lundbl.**
Lundell, Cyrus Longworth (1907–) AS	**Lundell**
Lundell, Peter Magnus (Manne) (1841–1930) AB	**P.Lundell**
Lundell, Seth (1892–1966) M	**S.Lundell**
Lundequist, Nils Wilhelm (1804–1863) S	**Lundeq.**
Lundevall, Carl–Fredrik (1921–) S	**Lundev.**
Lundgren, Jan (1941–) S	**Lundgren**
Lundh–Almestrand, Asta A	**Lundh–Alm.**
Lundin, Roger (1955–) S	**Lundin**
Lundmark, Johan Daniel (1755–1792) S	**Lundmark**
Lundqvist, (Adolf) Gösta (1894–1967) F	**Lundq.**
Lundqvist, Nils G. (1930–) M	**N.Lundq.**
Lundström, Axel Nicolaus (1847–1905) S	**Lundstr.**
Lundström, Carl Erik (1882–) S	**C.E.Lundstr.**
Lunell, Joël (1851–1920) PS	**Lunell**
Lunemann, Georg Heinrich (1780–1830) B	**Lun.**
Luneva, L.S. (1937–) S	**Luneva**
Luneva, N.N. (1949–) S	**N.N.Luneva**
Lungescu, Elena (1929–) M	**Lungescu**
Lunghini, Dario L. (1946–) M	**Lunghini**
Lunn, Joyce E. (fl. 1986) M	**Lunn**
Lunsford, J.N. (fl. 1976) M	**Lunsford**

Luo, Guang Yu (1938–) MS	**G.Y.Luo**
Luo, Jian Xin (fl. 1989) B	**J.X.Luo**
Luo, Jin Yu (1938–) S	**J.Y.Luo**
Luo, Ping (fl. 1987) S	**P.Luo**
Luo, Run Liang (fl. 1987) S	**R.L.Luo**
Luo, S.B. (fl. 1982) M	**S.B.Luo**
Luo, W. (fl. 1982) M	**W.Luo**
Luo, Xi Gao (fl. 1984) S	**X.G.Luo**
Luond, R. (fl. 1980) S	**Luond**
Luong, Ngoc–Toan (fl. 1965) S	**Luong**
Luparini, F. (fl. 1948) M	**Luparini**
Lupikina, E.G. A	**Lupikina**
Luppi Mosca, Anna Maria (1929–) M	**Luppi Mosca**
(Luque Palomo, Teresa)	
see Luque, Teresa	**Luque**
Luque, Teresa (c.1950) S	**Luque**
Lusby, David H. S	**D.H.Lusby**
Lusby, Phillip S. (1953–) S	**P.S.Lusby**
Lüscher, Hermann (1859–1920) S	**Lüscher**
Lushington, Alfred Wyndham (c.1860–1920) S	**Lush.**
Lusina, Giuseppe (1893–1963) PS	**Lusina**
Lusk, Demaris E. (fl. 1987) M	**Lusk**
Luss, A.I. (1899–1942) S	**Luss**
Lüstner, G. (fl. 1935) M	**Lüstner**
Lustrati, Gina (fl. 1980) M	**Lustrati**
(Lutati, Ferdinando Vignolo)	
see Vignolo–Lutati, Ferdinando	**Vignolo**
Luteraan, Philippe–Jacques (fl. 1954) M	**Luteraan**
Luteyn, James Leonard (1948–) S	**Luteyn**
Luther, Alexander Ferdinand (1877–1955) A	**Luther**
Luther, Hans Edmund (1915–1982) A	**H.E.Luther**
Luther, Harry E. (1952–) S	**H.Luther**
Luthi, R. (fl. 1967) M	**Luthi**
Luthmer, Oton Jimenez (1896?–1988) S	**Luthmer**
Luthy, J. (fl. 1987) S	**Luthy**
Lütjeharms, Wilhelm Jan (1907–1983) MS	**Lütjeh.**
Lütkemüller, Johannes (1850–1913) A	**Lütkem.**
Lutken, Christian Frederik (1827–1901) B	**Lutken**
Luttrell, Everett Stanley (1916–1988) M	**Luttr.**
Lutz, Bertha Maria Julia (1894–) S	**Lutz**
Lutz, Karl (fl. 1920–1935) S	**K.Lutz**
Lutze, Günther (1840–1930) S	**Lutze**
Lützow, G. (fl. 1895) B	**Lützow**
Luxford, George (1807–1854) S	**Luxf.**
Luyken, Johann (Joannes) Albert (1785–1867) M	**Luyk.**
Luykx, M.H.F. (fl. 1986) M	**Luykx**
Luz, Carlos Gomes da (1871–1952) M	**Luz**
(Luz de Armas, Matilde de la)	
see de la Luz de Armas, Matilde	**de la Luz**

Luz, R. (fl. 1981) S	R.Luz
Luzina, Z.A. (1901–) S	Luzina
Luzzani, Filiberto (1909–1943) S	Luzzani
Luzzatto, Gina (fl. 1937–38) S	Luzzatto
Lwoff, André (1902–) A	A.Lwoff
Lwoff, Marguerite A	M.Lwoff
Lý, Trân Ðinh (1939–) S	Lý
Lyakavichyus, A.A. (fl. 1972) S	Lyak.
Lyakh, S.P. (fl. 1981) M	Lyakh
Lyall, David (1817–1895) S	Lyall
Lye, Kaare Arnstein (1940–) S	Lye
Lyell, Charles (1767–1849) BM	Lyell
Lyell, Katherine Murray (1817–1915) P	K.Lyell
Lyka, Károly (1869–1965) S	Lyka
Lyle, Lilian (1864–) A	Lyle
Lyman, George Richard (1871–1926) M	Lyman
Lynch, Richard Irwin (1850–1924) S	Lynch
Lyngbye, Hans(en) Christian (1782–1837) AS	Lyngb.
Lynge, Bernt Arne (1884–1942) MS	Lynge
Lyon, Alexander Geoffrey (fl. 1950–1960) A	A.G.Lyon
Lyon, George Francis (1795–1832) S	G.F.Lyon
Lyon, Harold Lloyd (1879–1957) AMPS	Lyon
Lyons, Albert Brown (1841–1926) S	A.Lyons
Lyons, Israel (1739–1775) S	Lyons
Lyons, John Charles (1792–1874) S	J.Lyons
Lys, Maurice A	Lys
Lyssov, V.N. (1897–) S	Lyssov
Lythgoe, J.N. (fl. 1958) M	Lythgoe
Lyttkens, August (1845–1925) S	Lyttkens
(Lyubarskiĭ, L.V.)	
see Ljubarsky, L.V.	Ljub.
Lyulbeva, S.A. A	Lyulbeva
(M'Calla, William)	
see McCalla, William	McCalla
(M'Clelland, John)	
see McClelland, John	McClell.
(M'Coy, Frederick)	
see McCoy, Frederick	McCoy
(M'Intosh, Charles)	
see McIntosh, Charles	McIntosh
M'Keever, F.L. A	M'Keever
(M'Ken, Mark Johnston)	
see McKen, Mark Johnston	McKen
M'Mahon, Bernard (c.1775–1816) S	M'Mahon
Ma, Cheng Gong (1936–) S	C.G.Ma
Ma, Chi S	C.Ma
Ma, Chi Yun (fl. 1982) S	C.Y.Ma
Ma, En Wei (fl. 1980) S	E.W.Ma

Ma, Guo Zhong (1956–) M	**G.Z.Ma**
Ma, Jie (fl. 1988) S	**J.Ma**
Ma, Jin Shuang (1955–) S	**J.S.Ma**
Ma, Nai Xun (fl. 1985) S	**N.X.Ma**
Ma, Ping (fl. 1989) S	**P.Ma**
Ma, Shu Tai (fl. 1983) PS	**S.T.Ma**
Ma, T.G. (fl. 1988) S	**T.G.Ma**
Ma, Wei Liang (1936–) S	**W.L.Ma**
Ma, Yi Lun (fl. 1985) P	**Y.L.Ma**
Ma, Yi Zhong (fl. 1985) S	**Y.Z.Ma**
Ma, Yu Chuan (1916–) S	**Ma**
(Ma, Yuquan)	
see Ma, Yu Chuan	**Ma**
Maack, G.August S	**G.Maack**
Maack, Richard Karlovich (1825–1886) BPS	**Maack**
(Maak, Richard Karlovich)	
see Maack, Richard Karlovich	**Maack**
Maas Geesteranus, Rudolph Arnold (1911–) MS	**Maas Geest.**
Maas, Hillegonda (1941–) S	**H.Maas**
Maas, John L. (fl. 1971) M	**J.L.Maas**
Maas, Paulus Johannes Maria (1939–) S	**Maas**
(Maas–van de Kamer, Hillegonda)	
see Maas, Hillegonda	**H.Maas**
Maasen, A. (fl. 1913) M	**Maasen**
Maass, C.A. (1859–1929) S	**C.A.Maass**
Maass, H.I. (fl. 1986) S	**H.I.Maass**
Maass, Wolfgang Siegfried Gunther (1929–) BM	**Maass**
Maassoumi, Ali Asghan Ramak (1948–) S	**Maassoumi**
Maatjis, Martin S	**Maatjis**
Maatsch, Richard Franz Theodor (1904–) S	**Maatsch**
Mabberley, David John (1948–) PS	**Mabb.**
Mabille, Jean A	**J.Mabille**
Mabille, Jules Paul (1835–1923) S	**Mabille**
(for prefix Mac, see also Mc)	
MacAdam, Raymond B. A	**MacAdam**
MacAllister, Frederick (1874–1949) B	**MacAllister**
Macarthur, William (1800–1882) S	**Macarthur**
Macbride, James (1784–1817) S	**J.Macbr.**
Macbride, James Francis (1892–1976) PS	**J.F.Macbr.**
Macbride, Thomas Huston (1848–1934) M	**T.Macbr.**
MacBryde, Bruce (1941–) S '	**MacBryde**
Maccari, C. (fl. 1931) M	**Maccari**
MacCarthy, Lee (fl. 1922) M	**MacCarthy**
MacCaughey, Vaughan (1887–1954) S	**MacCaughey**
Macchiati, Luigi (1852–1921) AMS	**Macch.**
MacDaniels, Laurence Howland (1888–) S	**MacDan.**
MacDonald, J.D. (fl. 1981) M	**J.D.MacDon.**
MacDonald, James A. (1908–) M	**MacDon.**
MacDonald, Mary B. A	**M.B.MacDon.**

Macdonald, Norman (fl. 1915) M · N.Macdon.
MacDougal, Daniel Trembly (1865–1958) S · MacDougal
MacDougal, John M. S · J.M.MacDougal
MacDougall, T.M. (–1973) S · T.M.MacDoug.
MacDougall, Thomas Baillie (1895–1973) S · T.MacDoug.
Macedo, Amaro (1914–) S · A.Macedo
(Macedo de Aguiar, Joaquim)
   see Aguíar, Joaquim Macedo de · Aguíar
Macedo, Manuel Antonio de (–1867) S · Macedo
MacElwee, Alexander (1869–1923) S · MacElwee
MacEntee, Frank J. A · MacEntee
Macfadyen, James (1798–1850) S · Macfad.
Macfall, J.S. (fl. 1986) M · Macfall
MacFarlane, I. (fl. 1968) M · I.MacFarl.
Macfarlane, John Muirhead (1855–1943) S · Macfarl.
Macfarlane, Terry Desmond (1953–) S · T.D.Macfarl.
Macfie, J.W.S. (fl. 1921) M · Macfie
MacGarvie, Quentin D. (fl. 1965) M · MacGarvie
(MacGibbon, James)
   see McGibbon, James · McGibb.
MacGillivray, John (1822–1867) S · J.MacGill.
MacGillivray, Paul Howard (1834–1895) S · P.MacGill.
MacGillivray, William (1796–1852) S · W.MacGill.
MacGinitie, Henry Dunlap (fl. 1936) F · MacGinitie
MacGregor, Ernest Alexander (1880–) S · MacGregor
Machacek, John Emil (1902–) M · Machacek
Machado, A.A. (fl. 1956) M · A.A.Machado
Machado Cazorla, Edgardo (1938–) S · E.Machado
(Machado Guimarães, António (Antoine) Luis)
   see Guimarães, António (Antoine) Luis Machado · A.L.M.Guim.
Machado, Othon Xavier de Brito (1896–1951) S · Machado
Machatschki–Laurich, B. (fl. 1926) S · Mach.-Laur.
Machida, H. (fl. 1973) M · Machida
(Machmedov, A.M.)
   see Makhmedov, A.M. · Makhm.
Machmudova, M.A. (1957–) S · Machm.
Machule, Martin (1899–) S · Machule
Maciejowska, Zofia (fl. 1963) M · Maciej.
(MacIntosh, Charles)
   see McIntosh, Charles · McIntosh
Mack, Bruno A · B.Mack
Mack, Denis A · D.Mack
Mackay, Alexander Howard (1848–1929) MP · A.Mackay
Mackay, James Townsend (1775–1862) BMS · J.Mackay
MacKee, Hugh S. (1912–) MS · MacKee
MacKee, Margaret (–1990) S · M.MacKee
(MacKen, Mark Johnston)
   see McKen, Mark Johnston · McKen
Mackensen, Bernard (1863–1914) S · Mackensen
Mackenzie, D.W.R. (fl. 1961) M · D.W.R.Mack.

Mackenzie, Kenneth Kent (1877–1934) PS	**Mack.**
Mackey, J. (fl. 1954) S	**Mackey**
MacKie, F.P. A	**F.P.MacKie**
Mackie, William Wylie (1873–) M	**Mackie**
Mackin, J.G. (fl. 1966) M	**Mackin**
Mackinder, Barbara Ann (1958–) S	**Mackinder**
MacKinnon, Doris L. A	**D.L.MacKinnon**
Mackinnon, Ewen (fl. 1912) M	**E.Mackinnon**
Mackinnon, J.Andrew (fl. 1983) M	**J.A.Mackinnon**
Mackinnon, Juan E. (fl. 1929) M	**J.E.Mackinnon**
Mackinnon, K.A. (fl. 1989) M	**K.A.Mackinnon**
Mackintosh, M.Edmond S	**Mackintosh**
Macklin, E.D. (fl. 1925) S	**Macklin**
Macklot, Heinrich Christian (1799–1832) S	**Macklot**
MacLeish, Nanda F.Fleming (1953–) S	**MacLeish**
Macleod, D.M. (fl. 1970) M	**D.M.Macleod**
MacLeod, Julius (1857–1919) S	**MacLeod**
Macloskie, George (1834–1919) PS	**Macloskie**
Maclure, William (1763–1840) S	**Maclure**
Macmillan, Bryony Hope (1933–) S	**B.H.Macmill.**
MacMillan, Conway (1867–1929) MS	**MacMill.**
Macnab, James (1810–1878) S	**Macnab**
(MacNab, William Ramsay)	
see McNab, William Ramsay	**W.R.McNab**
MacNeal, Ward J. (fl. 1945) M	**MacNeal**
Maconochie, John Richard (1941–1984) S	**Maconochie**
Macoun, James Melville (1862–1920) MS	**J.M.Macoun**
Macoun, John (1831–1920) BMS	**Macoun**
MacOwan, Peter (1830–1909) MS	**MacOwan**
Macpherson, C.R. (fl. 1971) M	**Macph.**
Macquart, Justin (Jean) (Pierre Marie) (1778–1855) S	**Macquart**
Macrae, Ruth (fl. 1955) M	**Macrae**
MacRaild, G.N. A	**MacRaild**
Macreight, Daniel Chambers (1799–1857) S	**Macreight**
Macvicar, John Gibson (1801–1884) S	**J.Macvicar**
Macvicar, Symers Macdonald (1857–1932) B	**Macvicar**
Macy, J.M. (fl. 1971) M	**Macy**
Madaan, R.L. (fl. 1975) M	**Madaan**
Madala, Ravi Kiran (fl. 1982) S	**Madala**
Madalski, József (1902–) S	**Madalski**
Madar, Z. (fl. 1987) M	**Madar**
Madden, Edward (1805–1856) S	**Madden**
Madej, T. (fl. 1965) M	**Madej**
Mädel–Angeliewa, Erika (fl. 1969) F	**Mädel–Ang.**
Madelin, Michael Francis (fl. 1960) M	**Madelin**
Madenis, Claude Benoit (1798–1863) S	**Madenis**
Mader, U. S	**Mader**
Madhuravani, Agnes (fl. 1972) M	**Madhur.**
Madhusoodanam, P.V. (fl. 1984–1986) PS	**Madhus.**

(Madhusoodanan, P.V.)
  see Madhusoodanam, P.V.   **Madhus.**
Madiot, Paul (1780–1832) S   **Madiot**
Madison, Michael (fl. 1977) S   **Madison**
Mädler, Karl (1907–) AF   **Mädler**
Madrazo–Garibay, M. (fl. 1985) M   **Madr.-Gar.**
Madre, V.E.   A   **Madre**
Madrigal–Sanchez, X. (fl. 1969) S   **Madrigal**
Madulid, Domingo A. (1946–) S   **Madulid**
Maecke (fl. 1941) M   **Maecke**
Maeda, M. (fl. 1954) M   **Maeda**
(Maehara, Kanjiro)
  see Mayebara, Kanjiro   **Mayeb.**
Maekawa, Fumio (1908–1984) APS   **F.Maek.**
Maekawa, Nitaro (1954–) M   **N.Maek.**
Maekawa, Tokujirô (Tokijiro) (1886–) S   **Maek.**
Maekawa, Y. (fl. 1988) S   **Y.Maek.**
Mäemets, A.A.   S   **Mäemets**
Maerklin, Georg Friedrich (1761–1823) S   **Maerkl.**
Maerter, Franz Joseph (1753–1827) S   **Maerter**
Maesen, L.J.G.van der (1944–) S   **Maesen**
Maessen, Käthe (fl. 1955) M   **Maessen**
(Maevskii, Petr Feliksovich)
  see Majevski, Petr Feliksovich   **Majevski**
Maffei, Siro Luigi (1879–) M   **Maffei**
(Maffeucci, A.)
  see Maffucci, A.   **Maffucci**
Maffucci, A. (fl. 1895) M   **Maffucci**
Magalaschvili, T.D. (fl. 1981) S   **Magal.**
Magalhães, Geraldo Mendes (1906–) S   **G.M.Magalh.**
Magalhães Gomes, Alberto Augusto de (1871–1934) S   **A.A.Magalh.**
Magalhães Gomes, Carlos Thomaz (1855–1944) S   **Magalh.**
Magalhães, O.de (fl. 1927) M   **O.Magalh.**
Magalon, Maurius (fl. 1929) S   **Magalon**
(Magán, F.J.Fernández de Ana)
  see Fernández de Ana Magán, F.J.   **Fern.Magán**
Magana, P. (fl. 1986) S   **Magana**
Magasi, Laszlo P. (1935–) M   **Magasi**
Magdeburg, Paul   A   **Magdeb.**
Mägdefrau, Karl (1907–) BM   **Mägd.**
Magdun, S.G. (fl. 1972) M   **Magdun**
Maggi, Oriana (1951–) M   **Maggi**
Maggs, Christine A. (1956–) A   **Maggs**
Maggs, Donald H.   A   **D.H.Maggs**
Maghraby, A.M.   A   **Maghraby**
Magill, Robert Earle (1947–) BS   **Magill**
Magloire, L. (fl. 1965) AM   **Magloire**
Magnaghi, Angelo (fl. 1904) M   **Magnaghi**
Magnarelli, L.A. (fl. 1979) M   **Magnar.**
Magne, Francis (1924–) A   **F.Magne**

Magne, Jean Henri (1804–1885) S	**Magne**
Magne, Marie-France A	**M.-F.Magne**
Magnes, Martin (fl. 1991) M	**Magnes**
Magnier, Charles (fl. 1882) PS	**Magnier**
Magnin, Antoine (1848–1926) A	**Magnin**
Magnol, Pierre (1638–1715) L	**Magnol**
Magnotta, Angelina A	**Magnotta**
Magnus, Paul Wilhelm (1844–1914) AMS	**Magnus**
Magnusson, Adolf Hugo (1885–1964) M	**H.Magn.**
Magnusson, Axel Alfred (1847–1909) S	**Magn.**
Mágocsy-Dietz, Sándor (Alexander) (1855–1945) S	**Mágocsy**
Magrath, Lawrence K. (1943–) S	**Magrath**
Magrou, Joseph (1883–) M	**Magrou**
Maguire, Bassett (1904–1991) PS	**Maguire**
Magulaev, A.Yu. (fl. 1987) S	**Magulaev**
Magyar, J. (1884–1945) S	**Magyar**
Magyar, Pál (fl. 1930) S	**P.Magyar**
Mahabalé, Tryambak Shankur (1909–1983) BMP	**Mahab.**
Mahadevan, T.M. A	**Mahad.**
Mahdihassan, S. (fl. 1929) M	**Mahdih.**
Maheshwari, C.L. A	**C.L.Maheshw.**
Maheshwari, H.K. (fl. 1967) F	**H.K.Maheshw.**
Maheshwari, Jai Kisahn (1931–) BS	**Maheshw.**
Maheu, Jacques Marie Albert (1873–1937) BM	**Maheu**
Mahler, William F. (1930–) S	**Mahler**
Mahmood, M. (fl. 1983) M	**M.Mahmood**
Mahmood, Tariq (fl. 1968) M	**T.Mahmood**
Mahmud, K.A. (fl. 1950) M	**Mahmud**
Mahoney, Daniel P. (fl. 1977) M	**Mahoney**
Mahoney, M.K. (fl. 1990) M	**M.K.Mahoney**
Mahoney, Ron K. (1951–) A	**R.K.Mahoney**
Mahood, Albert D. A	**Mahood**
Mahu, Manuel (1930–) BS	**Mahu**
Mahunnah, R.L.A. (fl. 1980) S	**Mahunnah**
Mahyar, U.W. (1956–) S	**Mahyar**
Maia, Heraldo da Silva (fl. 1955) M	**H.Maia**
Maia, M.da (fl. 1920) M	**Maia**
Maia, Maria Helena Dália (fl. 1959) M	**M.H.Maia**
Maidana, Nora I. (1953–) A	**Maidana**
Maiden, Joseph Henry (1859–1925) PS	**Maiden**
Maier, Dorothea A	**Maier**
Mailho, Jean-Baptiste (fl. 1889) S	**Mailho**
Maillard, Pierre Néhémie (1813–1883) P	**Maillard**
Maillard, Roger A	**R.Maillard**
Maille, Alphonse (1813–1865) S	**Maille**
Maillefer, Arthur (1880–1960) AS	**Maill.**
(Mailun, Z.A.)	
see Majlun, Z.A.	**Majlun**
Maime, G.D.A. (1864–1937) M	**Maime**
Maingay, Alexander Carroll (1836–1869) S	**Maingay**

Maino, Evelyn (1911–1958) S	**Maino**
Mains, Edwin Butterworth (1890–1968) M	**Mains**
Mainx, Felix A	**Mainx**
Maire, Louis (1885–) M	**L.Maire**
Maire, Réné Charles Joseph Ernest (1878–1949) ABMPS	**Maire**
Mairet, Ethel M. S	**Mairet**
Mairh, Om Prakash A	**Mairh**
Maironi da Ponte, Giovanni (1748–1833) S	**Mair.Ponte**
Maisaya, I.I. (fl. 1985) S	**Maisaya**
Maithy, P.K. (fl. 1974) AF	**Maithy**
Maiti, G.G. (fl. 1980) S	**Maiti**
Maitland, Thomas Douglas (1885–1976) S	**Maitland**
Maitulina, Yu.K. (fl. 1982) S	**Maitul.**
Maity, M.K. (fl. 1970) M	**Maity**
Maiwald, Vincenz Fridolin (1862–) S	**Maiwald**
Majchenberg, Mario (fl. 1980s) M	**Majchenb.**
Majchrowicz, T. (fl. 1964) M	**Majchr.**
Majeed Kak, A. (fl. 1982) S	**Majeed Kak**
Majevski, Petr Feliksovich (1851–1892) S	**Majevski**
(Majewski, Petr Feliksovich)	
see Majevski, Petr Feliksovich	**Majevski**
Majewski, Tomasz (1940–) M	**T.Majewski**
Majlun, Z.A. (fl. 1962) S	**Majlun**
Majoli, Cesare (1746–1823) S	**Majoli**
(Major, Charles Immanuel Forsyth)	
see Forsyth–Major, Charles Immanuel	**Fors.–Major**
Majorov, A.A. (1924–) S	**Majorov**
Májovský, Jozef (1920–) S	**Májovský**
Majumdar, Nimai Chandra (1932–) S	**Majumdar**
Majumdar, R.B. (fl. 1971) S	**R.B.Majumdar**
Majumdar, S.C. (1949–) S	**S.C.Majumdar**
(Makaravicz, M.F.)	
see Makarevich, M.F.	**Makar.**
Makarevich, M.F. (1906–) M	**Makar.**
Makarov, Vladilen V. (1935–) S	**Makarov**
Makarova, G.I. (fl. 1982) M	**Makarova**
Makarova, Iraida Viktorovna A	**I.V.Makarova**
Makaschvili, A.K. S	**Makaschv.**
Makasheva, R.Kh. (1917–) S	**Makasheva**
Makhaev, V.N. A	**Makhaev**
Makhija, Urmila V. (1950–) M	**Makhija**
Makhmedov, A.M. (1951–) S	**Makhm.**
Maki, T. (fl. 1932) M	**Maki**
Maki, Y. (fl. 1940) M	**Y.Maki**
Makienko, V.F. A	**Makienko**
Makiguchi, N. (fl. 1976) M	**Makig.**
Makin, J.G. (fl. 1974) M	**Makin**
Mäkinen, A.I. (1935–) S	**A.Mäkinen**
Mäkinen, Liisa (1936–) S	**L.Mäkinen**

Mäkinen, Yrjö Lauri Antero (1931–) MS   **Y.Mäkinen**
Makino, Tomitarô (1862–1957) PS   **Makino**
Makowsky, Alexander (1833–1908) S   **Makowsky**
(Makoy, Lambert Jacob)
    see Jacob–Makoy, Lambert   **Jacob-Makoy**
Makryï, T.V. (fl. 1984) M   **Makryï**
(Maksimova, M.I.)
    see Maximova, M.I.   **M.I.Maximova**
(Maksimova, M.M.)
    see Maximova, M.M.   **M.M.Maximova**
Makushina, E.N. S   **Makush.**
Malabaila von Canal, Emanuel Joseph S   **Malab.**
Malacarne, Claro Guiseppe (fl. 1810) S   **Malac.**
(Malagarriga Heras, Teodoro Luis Ramón de Peñaflor)
    see Malagarriga, Teodoro Luis Ramón Peñaflor   **Malag.**
Malagarriga, Teodoro Luis Ramón Peñaflor (1904–) S   **Malag.**
Malaisse, François (1934–) S   **Malaisse**
Malajczuk, N. (fl. 1985) M   **Malajczuk**
Malama, A.A. (fl. 1965) M   **Malama**
Malan, C.E. (fl. 1947) M   **Malan**
Malathi, C.P. (fl. 1988) S   **Malathi**
Malathrakis, N.E. (fl. 1972) M   **Malathr.**
Malato–Beliz, João Vicente Cordeiro (1920–) S   **Malato-Beliz**
Malaviya, N. (fl. 1983) M   **Malaviya**
Malbranche, Alexandre François (1818–1888) MS   **Malbr.**
Malchevskaya, N.N. (fl. 1939) M   **Malchevsk.**
Malcolm, William McLagan (1936–) S   **Malcolm**
Malcolmson, Jean F. (fl. 1958) M   **Malc.**
Małecka, J. (1926–) S   **Małecka**
Maleev, Vladimir Petrovic (1894–1941) S   **Maleev**
Maleki, Zeynol-Abedin (Zeinolabedin) (1913–) S   **Maleki**
Malençon, Georges Jean Louis (1898–1984) M   **Malençon**
Malevich, O.A. (fl. 1936) M   **Malevich**
Malguth, R. (fl. 1928) M   **Malguth**
Malherbe, Alfred (1804–1866) S   **Malherbe**
Malhotra, C.L. (1936–) S   **Malhotra**
Malhotra, G. (fl. 1974) M   **G.Malhotra**
Malick, K.C. (1938–) S   **Malick**
Malik, K.A. (fl. 1970) M   **Malik**
Malinowski, Edmund (1885–1979) S   **Malin.**
Malinvaud, Louis Jules Ernst (1836–1913) PS   **Malinv.**
Malinverni, Alessio (1830–1887) S   **Malinverni**
Malkovský, K.M. (fl. 1932) M   **Malk.**
Malla, Devindra S. (fl. 1971) M   **Malla**
Malla, Samar Bahadur (1933–) S   **S.B.Malla**
Malladra, Alessandro (1865–1944) S   **Mall.**
Mallet, J. (fl. 1985) S   **Mallet**
Mallett, George B. (fl. 1902) S   **Mallett**
Mallik, F. (fl. 1976) M   **Mallik**

Malloch, David Warren (1940–) M	**Malloch**
Malm, Jacob von (1901–) S	**Malm**
Malmberg, August Johan (later Mela, A.J.) (1846–1904) S	**Malmberg**
Malmborg, Sten von (fl. 1933) B	**Malmb.**
Malme, Gustaf Oskar Andersson (né Andersson, G.O.) (1864–1937) BMS	**Malme**
Malmgren, Anders Johan (1834–1897) S	**Malmgren**
Malmio, Bruno Julius (1883–1966) S	**Malmio**
Malo, B. (fl. 1978) S	**Malo**
Maloch, Frantisek (1862–1940) S	**Maloch**
Malos, Constantin (1933–) S	**Malos**
Malta, Nicolajs (1890–1944) B	**Malta**
Malte, Malte Oscar (1880–1933) S	**Malte**
Malthé, Gottlieb Friedrich (fl. 1787) S	**Malthé**
(Maltsev, Al.Ivanovich)	
see Malzev, Al.Ivanovich	**Malzev**
Malvesin–Fabre, Georges S	**Malv.-Fabre**
Malý, Franz de Paula (1823–1891) S	**F.Malý**
Maly, Joseph Karl (Carl) (1797–1866) S	**Maly**
Malý, Karl Franz Josef (1874–1951) PS	**K.Malý**
Malyavkina, V.S. A	**Malyavk.**
Malyer, Hulusi (1954–) S	**Malyer**
Malyschev, Leonid I. (1931–) S	**Malyschev**
(Malysev, Leonid I.)	
see Malyschev, Leonid I.	**Malyschev**
Malyutin, K.G. (fl. 1972) S	**Malyutin**
Malyutin, N.I. (fl. 1987) S	**N.I.Malyutin**
Malyutina, E.T. (1930–) S	**Malyutina**
Malzev, Al.Ivanovich (1879–1948) S	**Malzev**
Malzev, I.I. (1948–) S	**I.I.Malzev**
Malzeva, A.T. S	**A.T.Malzeva**
Malzeva, I.I. (1943–) S	**Malzeva**
(Malzew, Al.Ivanovich)	
see Malzev, Al.Ivanovich	**Malzev**
Malzine, Omer de Pierre Antoine Hyacinth Recq de (1820–1881) S	**Malzine**
Mamaev, Stanislav A. (1928–) S	**Mamaev**
Mamay, Sergius Henry (Harry) (1920–) AF	**Mamay**
Mambetalieva, Sonun (1934–1963) A	**Mambet.**
Mamede, Maria Candida Henrique (1956–) S	**Mamede**
Mameli Calvino, Giuliana Eva (1886–1978) S	**Mameli**
Mamet, Bernard L. A	**Mamet**
Mamgain, S.K. (fl. 1985) S	**Mamgain**
Mamkaeva, K.A. A	**Mamkaeva**
Mamukaschvili, T.S. (fl. 1977) M	**Mamuk.**
(Man, Johannes Govertus De)	
see De Man, Johannes Govertus	**De Man**
(Managetta, Günther Beck von)	
see Beck, Günther von Mannagetta und Lërchenau	**Beck**
Manandhar, J.B. (fl. 1986) M	**J.B.Manandhar**
Manandhar, V. (fl. 1986) M	**V.Manandhar**

Manciet, Jean   A   **Manciet**
Mancini, Vincenzo (1853–   ) M   **Mancini**
Manda, Robert F.   S   **R.F.Manda**
Manda, W.Albert (fl. 1892) S   **Manda**
Mandal, M. (fl. 1978) M   **M.Mandal**
Mandal, N.C. (fl. 1983) M   **N.C.Mandal**
Mandenova, I.P. (1907–   ) S   **Manden.**
Mandl, K.   S   **Mandl**
Mandon, E. (fl. 1898) S   **E.Mandon**
Mandon, Gilbert (Gustav) (1799–1866) S   **Mandon**
Mandra, York T. (1922–   ) A   **Mandra**
Manetti, Giuseppe (fl. 1831–1858) S   **G.Manetti**
Manetti, Xaverio (1723–1785) S   **Manetti**
Maneval, Willis Edgar (1877–1956) M   **Maneval**
Mangaly, Jose K. (fl. 1989) S   **Mangaly**
Mangan, Aedine (fl. 1965) M   **A.Mangan**
Mangano, V. (fl. 1983) M   **Mangano**
Manganotti, Antonio (1810–1892) S   **Mangan.**
Mangelsdorf, Paul Christoph (1899–   ) S   **Mangelsd.**
Mangen, J.M. (fl. 1986) S   **Mangen**
Mangenot, C. (fl. 1883) A   **C.Mangenot**
Mangenot, François (fl. 1953) M   **F.Mangenot**
Mangenot, George Marie (1899–1985) S   **Mangenot**
Mangenot, Georges (1913–   ) A   **G.Mang.**
Mangin, Arthur (1824–1887) S   **A.Mangin**
Mangin, Louis Alexandre (1852–1937) ACM   **L.Mangin**
Manguin, Émile Etienne (1893–1966) AS   **Manguin**
Manhart, J.R. (fl. 1987) S   **Manhart**
Mani, K.J. (fl. 1985) S   **Mani**
Manian, S. (fl. 1989) M   **Manian**
Manickam, V.S. (1941–   ) P   **Manickam**
Manier, Jehanne-Françoise (fl. 1955) AM   **Manier**
Manilal, K.S. (1938–   ) S   **Manilal**
Manimohan, P. (fl. 1986) M   **Manim.**
Maniotis, J. (fl. 1963) M   **Maniotis**
Manitz, Hermann (1941–   ) S   **Manitz**
Manivit, Hélène   A   **Manivit**
Manjón, José L. (fl. 1984) M   **Manjón**
Manjula, B. (fl. 1981) M   **Manjula**
Mańka, K. (fl. 1958) M   **Mańka**
Mankau, R. (fl. 1978) M   **Mankau**
Mankin, C.J. (fl. 1963) M   **Mankin**
Mankou, R. (fl. 1978) M   **Mankou**
Mann, Albert (1853–1935) A   **A.Mann**
Mann, Benjamin Pickmann (1848–1926) S   **B.Mann**
Mann, David G. (1953–   ) A   **D.G.Mann**
Mann, Gustav (1836–1916) S   **G.Mann**
Mann, Hans (fl. 1938) M   **Hans Mann**
Mann, Horace (1844–1868) MPS   **H.Mann**

Mann, Johann Gottlieb (fl. 1829) S	**J.G.Mann**
Mann, Robert James (1817–1886) S	**R.Mann**
Mann, Wenzeslaus (Wenzel) Blasius (1799–1839) BMS	**W.Mann**
Manners, J.G. (fl. 1953) M	**Manners**
(Mannetje, Len 't)	
see 't Mannetje, Len	**'t Mannetje**
Mannina, F.M. (fl. 1984) M	**Mannina**
Manning, J.C. (fl. 1985) S	**J.C.Manning**
Manning, Jacob Warren (1826–1904) S	**Manning**
Manning, Wayne Eyer (1899–) S	**W.E.Manning**
Mannoni, Octave (1899–) S	**Mannoni**
Manns, Thomas Franklin (1876–1954) M	**Manns**
Manoch, Leka (fl. 1981) M	**Manoch**
Manocha, M.S. (fl. 1964) M	**Manocha**
Manoharachary, C. (fl. 1971) M	**Manohar.**
Manoliu, Alexandru (fl. 1976) M	**Manoliu**
Manoury, Charles Ambroise (fl. 1869) A	**Manoury**
Mansanet, José (1915–1990) S	**Mansanet**
Mansel, John Clavell (1817–1902) S	**Mansel**
(Mansel–Pleydell, John Clavell)	
see Mansel, John Clavell	**Mansel**
Mansfeld, Rudolf (1901–1960) S	**Mansf.**
Mansion, Arthur (1863–1905) BS	**Mans.**
(Manso, António Luiz Patricio da Silva)	
see Silva Manso, António Luiz Patricio da	**Silva Manso**
(Manso, José da Silva)	
see Silva Manso, José da	**J.Silva Manso**
Mantell, Gideon Algernon (1790–1852) AF	**Mantell**
Manten, Jacob (Jack) Manten (1898–1958) S	**Manten**
Manter, Harold Winfred (1898–) M	**Manter**
Mantin, Georges Antoine (1850–1910) S	**Mantin**
Manton, Irene (1904–1988) APS	**Manton**
Mantri, J.M. (fl. 1965) M	**Mantri**
Manuel, Monte Gregg (1947–1981) B	**Manuel**
Manum, Svein Bendik (1926–) A	**Manum**
Manus (fl. 1932) M	**Manus**
Manza, Artemio Valderrama (1896–1964) A	**Manza**
Mao, Pin I (fl. 1964) S	**P.I Mao**
(Mao, Ping Yi)	
see Mao, Pin I	**P.I Mao**
Mao, Xue Ying (fl. 1985) S	**X.Y.Mao**
Mao, Zu Mei (1933–) S	**Z.M.Mao**
Mapletoft, Heather A	**Mapletoft**
Mappus, Marcus (1666–1736) L	**Mappus**
Maquet, Philippe (fl. 1985) S	**Maquet**
Marabini, J. (fl. 1983) S	**Marabini**
Marais, Elizabeth M. (1945–) S	**E.M.Marais**
Marais, Wessel (1929–) S	**Marais**
Marak, Joe H. A	**Marak**

Maranta, Bartolomeo (c.1500–1571) S	**Maranta**
Marasas, Walter Friedrick Otto (1941–) MS	**Marasas**
Maratti, Giovanni Francesco (1723–1777) AMS	**Maratti**
Maratti, J.F. (fl. 1822) M	**J.F.Maratti**
(Marbaix, J.De)	
see De Marbaix, J.	**De Marb.**
(Marc des Clercs du Saint Viateur, (Frère))	
see Marc, François	**Marc**
Marc, François (1862–1912) BM	**Marc**
Marcailhou d'Ayméric, Alexandre Lucien Marie (1839–1897) S	**Marcailhou**
Marcano, C. (fl. 1973) M	**Marcano**
Marcano, V. (fl. 1989) P	**V.Marcano**
Marcano–Berti, Luis (fl. 1967) S	**Marc.-Berti**
Marcelli, M.P. (fl. 1986) M	**Marcelli**
Marcello, Alessandro (1894–1980) M	**Marcello**
(Marčenko, P.D.)	
see Marchenko, P.D.	**P.D.Marchenko**
Marcenò, Cosimo (1939–) S	**Marcenò**
Marcet y Poal, Adeodato Francisco (1875–1964) S	**Marcet**
Marcgilhou d'Aylméric, Hippolyte (1855–1909) S	**Marcgilhou**
(Marcgraf, Friedrich)	
see Markgraf, Friedrich	**Markgr.**
(Marcgraf, Georg)	
see Marcgrave, Georg	**Marcgr.**
Marcgrave, Georg (1610–1644) L	**Marcgr.**
(Marcgravius, Georg)	
see Marcgrave, Georg	**Marcgr.**
Marchal, Élie (1839–1923) MS	**Marchal**
Marchal, Émile Jules Joseph (1871–1954) BM	**É.J.Marchal**
(Marchal, Marguerite Lauzac)	
see Lauzac–Marchal, Marguerite	**Lauz.-March.**
Marchand, André (fl. 1971) M	**A.Marchand**
Marchand, Louis (1807–1843) M	**L.Marchand**
Marchand, Nestor Léon (1833–1911) MS	**Marchand**
Marchand, S. (fl. 1976) M	**S.Marchand**
Marchant, C.J. (fl. 1973) S	**C.J.Marchant**
Marchant, Harvey J. A	**H.J.Marchant**
Marchant, Neville Graeme (1939–) S	**N.G.Marchant**
Marchant, William James (1886–1952) S	**Marchant**
Marchenko, A.M. (fl. 1985) M	**Marchenko**
Marchenko, P.D. (fl. 1987) M	**P.D.Marchenko**
Marchesetti, Carlo de (Carl von) (1850–1926) PS	**Marches.**
Marchesi, Eduardo (1943–) S	**Marchesi**
Marchesoni, Vittorio (1912–1963) A	**Marchesoni**
Marchewianka, Marja A	**Marchew.**
Marchi, Palmer (1931–) S	**Marchi**
Marchionatto, Juan Bautista (1898–) M	**Marchion.**
Marchisotto, John A	**Marchis.**

Marchoux, Émil (fl. 1897) M	**Marchoux**
Marcialis, Efisio (fl. 1889) S	**Marcialis**
Marcich, Z.G. (fl. 1963) M	**Marcich**
Marcilla, J. (fl. 1970) M	**Marcilla**
Marcinkiewicz, T. (fl. 1979) M	**Marcink.**
Marck, Wilhelm von der (1814–1900) S	**Marck**
Marcks, Brian (fl. 1974) S	**Marcks**
(Marco, Giovanni De)	
see De Marco, Giovanni	**De Marco**
Marcos Pascual, Antonio (1900–) S	**Marcos**
Marcos Samaniego, Nieves (fl. 1987) S	**N.Marcos**
Marcot–Coqueugniot, Jacqueline A	**Marcot-Coq.**
(Marcovicz, Simon)	
see Marcovitch, Simon	**S.Marcov.**
(Marcovicz, Vasil Vasilevicz)	
see Marcowicz, Vasil Vasilevicz	**Marcow.**
Marcovitch, Simon (1890–) S	**S.Marcov.**
Marcowicz, Vasil Vasilevicz (1865–1942) S	**Marcow.**
Marcucci, Emilio (1837–1890) P	**Marcucci**
Marcy, Randolph Barnes (1812–1887) S	**Marcy**
(Mardaleischvili, T.K.)	
see Mardalejschvili, T.K.	**Mardal.**
(Mardaleishvili, T.K.)	
see Mardalejschvili, T.K.	**Mardal.**
Mardalejschvili, T.K. (1940–) S	**Mardal.**
Maréchal, Robert Joseph Jean–Marie (1926–) S	**Maréchal**
Marès, Paul (1826–1900) S	**Marès**
Maresch, E. A	**E.Maresch**
Maresch, Otto A	**O.Maresch**
Maresquelle, Henri J. (fl. 1924) M	**H.J.Maresq.**
Maresquelle, M. (fl. 1924) M	**M.Maresq.**
Margadant, Willem Daniel (1916–) B	**Margad.**
Margain–Hernández, Romerto M. (1954–) A	**Marg.-Hern.**
Margalef, Ramón A	**Margalef**
(Marggraf, Georg)	
see Marcgrave, Georg	**Marcgr.**
Margheri, Maria C. (fl. 1971) M	**Margheri**
Marginson, J.C. (fl. 1988) S	**Marginson**
Margittai, Antal (1880–1939) S	**Margittai**
Margot, Henri (1807–1894) S	**Margot**
Margulis, Lynn (1938–) A	**Margulis**
Marhold, Karol (1959–) S	**Marhold**
Mariani, Gindetta (fl. 1911) M	**Mariani**
Mariappan, V. (fl. 1977) M	**Mariappan**
Marie, Edouard Auguste (1835–1888) B	**Marie**
(Marie–Victorin, Joseph Louis Conrad)	
see Victorin, Joseph Louis Conrad Marie–	**Vict.**
Marignoni, Giuseppi Br. (fl. 1909) M	**Marignoni**

(Marilaun, Anton Joseph Kerner von)	
see Kerner, Anton Joseph	A.Kern.
Marimpietri, Luigi S	Marimp.
Marinkella, C.J. (fl. 1968) M	Marink.
Marino, C. (fl. 1979) M	Marino
Marino, Donato A	D.Marino
Marion, Antoine–Fortuné (1846–1900) AFS	Marion
Mariotti, Mauro Giorgio (1954–) S	Mariotti
Marissal, Félix Victor (1824–1881) S	Marissal
Mariti, Giovanne (1736–1806) S	Mariti
Mariz, Geraldo (1923–) S	G.Mariz
Mariz, Joaquim de (1847–1916) S	Mariz
Marjollet, Josef Marie (1823–1894) S	Marj.
Markali, Joar A	Markali
Markey, Donald R. A	Markey
Markgraf, Friedrich (1897–1987) PS	Markgr.
Markgraf–Dannenberg, Ingeborg (1911–) S	Markgr.-Dann.
Markham, Clements Robert (1830–1916) S	Markham
Marklund, Erik (1893–) S	E.Markl.
Marklund, George Gunnar (1892–1964) S	Markl.
Marklund, Hans (fl. 1991) M	H.Markl.
Markötter, Erike Irene (1905/06–1983) S	Markötter
Marková, J. (fl. 1987) M	J.Marková
Markova, M. (fl. 1973) S	Markova
Markova–Letova, Marie Federovna (1901–) M	Mark.-Let.
(Marlier–Spirlet, Marie–Louise)	
see Spirlet, Marie–Louise	Spirlet
Marloth, Hermann Wilhelm Rudolf (1855–1931) S	Marloth
Marmey, Françoise (fl. 1958) S	Marmey
Marmolejo, J.G. (fl. 1990) M	Marm.
Marnac, Émile (1853–1929) S	Marnac
Marneffe, H. (fl. 1946) M	Marneffe
Marner, Serena K. (fl. 1989) S	Marner
Marnier–Lapostolle, Julien (1902–1976) S	Marn.-Lap.
Marnock, Robert (1800–1889) S	Marnock
Maron, Charles (1851–1926) S	Maron
Marples, Mary J. (fl. 1954) M	Marples
Marpmann, G. (fl. 1886) M	Marpmann
Marquand, Cecil Victor Boley (1897–1943) S	C. Marquand
Marquand, Ernest David (1848–1918) BS	Marquand
Marquart, Clamor Ludwig (1804–1881) S	Marquart
Marquart, Friedrich (fl. 1842) M	F.Marquart
(Marques da Cunha, Aristides)	
see Cunha, Aristides Marques da	A.M.Cunha
Marques, Maria do Carmo Mendes (1934–) S	Marques
Marquete Ferreira da Silva, Nilda (fl. 1972–1987) S	Marquete
Marquis, Alexandre Louis (1777–1828) S	Marquis
Marr, C.D. (fl. 1973) M	Marr
Marras, F. (fl. 1962) M	Marras

Marrat, Frederick Price (1820–1904) BF   **Marrat**
Marrero Rodriguez, A. (fl. 1988) S   **Marrero Rodr.**
Marret, Léon (fl. 1900–1929) BF   **Marret**
Marroquín, Jorge S. (1935–) S   **Marroq.**
Marsais, P. (fl. 1927) M   **Marsais**
Marschall von Bieberstein, Friedrich August (1768–1826) PS   **M.Bieb.**
(Marschlins, Carl Ulysses Adalbert von Salis)
    see Salis–Marschlins, Carl Ulysses Adalbert von   **Salis**
Marsden, Craig R. (fl. 1976) P   **C.R.Marsden**
Marsden, William (1754–1836) S   **Marsden**
Marsden-Jones, Eric (1887–1960) S   **Marsden-Jones**
Marsh, Alfred H. (fl. 1973) M   **A.H.Marsh**
Marsh, Charles Dwight (1855–1932) S   **Marsh**
Marsh, Cythia (1956–) B   **C.Marsh**
Marsh, Judith Anne (1951–) S   **J.A.Marsh**
Marsh, Ray Stanley (1894–) S   **R.S.Marsh**
Marshall, (Caroline) Nina Lovering (1861–1921) M   **N.L.Marshall**
Marshall, A. (fl. 1914) M   **A.Marshall**
Marshall, D.L. (fl. 1983) S   **D.L.Marshall**
Marshall, Edward Shearburn (1858–1919) S   **E.S.Marshall**
Marshall, Humphry (1722–1801) S   **Marshall**
Marshall, John Braybrooke (1913–) S   **J.B.Marshall**
Marshall, Joseph Jewison (1860–1934) S   **J.Marshall**
Marshall, K.L. A   **K.L.Marshall**
Marshall, Reginald Charles (1893–) S   **R.C.Marshall**
Marshall, Rush Porter (1891–) M   **R.P.Marshall**
Marshall, Sheina M. A   **S.M.Marshall**
Marshall, W.Taylor (fl. 1952) S   **W.T.Marshall**
Marshall, William (1815–1890) S   **W.Marshall**
Marshall, William Emerson (1872–1937) S   **W.E.Marshall**
Marsili, Giovanni M. (1727–1794) S   **Marsili**
Marsili, Luigi (1656–1730) S   **L.Marsili**
Marsilly, Louis Joseph Auguste de Commines de (1811–1890) S   **Marsilly**
Marson, Guy (1953–) M   **G.Marson**
Marson, Paul (fl. 1980s) M   **P.Marson**
Marsson, Karl Maximilian (1845–1909) ACS   **M.Marsson**
Marsson, Theodor Friedrich (1816–1892) S   **T.Marsson**
Martel, Eduardo (1846–1929) A   **Martel**
Martelli, G.P. (fl. 1961) M   **G.P.Martelli**
Martelli, Niccolo (–1829) S   **N.Martelli**
Martelli, Ugolino (1860–1934) BMS   **Martelli**
(Martellius, Nicolaus)
    see Martelli, Niccolo   **N.Martelli**
Martens, Brunhilde A   **B.Martens**
Martens, Eduard (Karl) von (1831–1904) S   **E.Martens**
Martens, George Matthias von (1788–1872) AB   **G.Martens**
Martens, Martin (1797–1863) PS   **M.Martens**
Martensen, Hans Oluf (1928–) S   **Martensen**

Martensson, Olle (Olof) (1915–   ) B — **Martensson**
Martersteck, Johann Clemens (fl. 1792) S — **Marterst.**
Marthe, François (fl. 1801) P — **Marthe**
Marti, L. (fl. 1972) M — **Marti**
Marticorena, Clodomiro (1929–   ) S — **Martic.**
Martin, Adolf (fl. 1851) S — **A.Martin**
Martin, Bernardin-Antoine (1813–1897) S — **B.-A.Martin**
Martín Bolaños, Manuel (1897–c.1970) S — **Martín Bol.**
Martin, C.H.   A — **C.H.Martin**
Martin, Charles-Édouard (1847–1937) M — **C.Martin**
Martin de Argenta, Vicente (fl. 1862) S — **V.Martin**
Martin, Émile (Pierre) (1810–1895) S — **E.Martin**
Martin, F.N. (fl. 1987) M — **F.N.Martin**
Martin, Floyd Leonard (1909–   ) S — **F.L.Martin**
Martin, Francine   A — **F.Martin**
Martin, George (1827–1886) M — **G.Martin**
Martin, George Hamilton (1887–   ) M — **G.H.Martin**
Martin, George Willard (1886–1971) AM — **G.W.Martin**
Martin, James Stillman (1914–   ) S — **J.S.Martin**
Martin, Kenneth J. (1942–   ) M — **K.J.Martin**
Martin, Lucille (1925–   ) S — **L.Martin**
Martin, Margaret Trevena (1905–   ) A — **M.T.Martin**
Martin, Paul (1923–1982) AS — **P.Martin**
Martin, Philip Michael Dunlop (fl. 1969) M — **P.M.D.Martin**
Martin, Robert Franklin (1910–   ) S — **R.F.Martin**
Martin, W.Wallace (fl. 1972) M — **W.W.Martin**
Martin, Weston Joseph (1917–   ) M — **W.J.Martin**
Martin, William (1886–1975) BMS — **W.Martin**
Martin, William C. (fl. 1980) S — **W.C.Martin**
(Martin (of New Zealand), William)
    see Martin, William — **W.Martin**
Martin-Sans, E. (fl. 1929) M — **Martin-Sans**
Martindale, Issac Comly (1842–1893) M — **I.C.Martind.**
Martindale, Joseph Anthony (1837–1914) FM — **J.A.Martind.**
Martinelli, Gustavo (1954–   ) S — **Martinelli**
Martinet, Jean Baptiste Henri (1840–   ) S — **Martinet**
Martínez, Angel T. (fl. 1957) M — **A.T.Martínez**
Martinez Crovetto, Raul (1921–1988) S — **Mart.Crov.**
Martínez, D. (fl. 1987) M — **D.Martínez**
(Martinez de Lexarza, Juan José)
    see Lexarza, Juan José Martinez de — **Lex.**
Martínez, Esteban   S — **E.Martínez**
(Martínez García, Julieta)
    see Martínez, Julieta — **J.Martínez**
Martínez, J.J.   B — **J.J.Martínez**
Martínez, José Benito (fl. 1931) M — **J.B.Martínez**
Martínez, Julieta (fl. 1989) S — **J.Martínez**
Martínez Macchiavello, J.Carlos (1931–   ) A — **J.C. Martínez**

Martínez Martínez, Alfonso (fl. 1982) S   **A.Mart.Mart.**
Martínez Martínez, M. (1907–1936) S   **Mart.Mart.**
Martínez, Maximino (1888–1964) S   **Martínez**
Martínez, Milagrosa R.   A   **M.R.Martínez**
Martinez Parras, José María (1953–   ) S   **Mart.Parras**
(Martínez–Crovetto, Raul)
   see Martinez Crovetto, Raul   **Mart.Crov.**
Martinez–Laborde, Juan B. (fl. 1983) S   **Mart.-Laborde**
Martini, Alessandro (1934–   ) S   **Martini**
Martini, Erlend   A   **E.Martini**
Martini, Fabrizio (1949–   ) S   **F.Martini**
Martinis, Z. (fl. 1973)   **Martinis**
Martinoli, Giuseppe (1911–1970) S   **Martinoli**
Martinov, Ivan Ivanovič (fl. 1826) S   **Martinov**
Martinovský, Jan Otakar (1903–1980) S   **Martinovský**
Martins, Angela B. (1945–   ) S   **A.B.Martins**
Martins, C. (fl. 1928) M   **C.Martins**
Martins, Charles Frédéric (1806–1889) S   **Martins**
Martins, Elena Maria Occioni (1948–   ) S   **E.M.O.Martins**
Martins, Enrico Sampaio (1944–   ) S   **E.S.Martins**
Martins, Henrique Ferreira (fl. 1988) S   **H.F.Martins**
Martins, Jordelina L. (fl. 1971) S   **J.L.Martins**
Martins, Maria Adélia
   see Diniz, Maria Adélia   **M.A.Diniz**
Martinsson, Anders   A   **Martinsson**
Martis, B. de (fl. 1991) M   **Martis**
Martius, Carl Friedrich Phil.Sigm. (1829–1899) MS   **C.Mart.**
Martius, Carl (Karl) Friedrich Philipp von (1794–1868) AMPS   **Mart.**
Martius, Ernst Wilhelm (1756–1849) S   **E.Mart.**
Martius, Heinrich von (1781–1831) MS   **H.Mart.**
Martius, Theodor Wilhelm Christian (1796–1863) S   **T.Mart.**
Marton, Kela (fl. 1970) M   **Marton**
Martrin–Donos, Julien Victor de (1800–1870) S   **Martrin-Donos**
Martsikh, Zh.G. (fl. 1965) M   **Martsikh**
Marty, Francis   A   **F.Marty**
Marty, Pierre (1868–   ) BF   **Marty**
Martyn, Eldred Bridgeman (1903–   ) M   **E.B.Martyn**
Martyn, John (1699–1768) S   **J.Martyn**
Martyn, Thomas M. (1736–1825) S   **Martyn**
Martynyuk, T.D. (fl. 1986) M   **Martynyuk**
Marullaz, M.   A   **Marullaz**
Marum, Martin (Martinus) van (1750–1837) S   **Marum**
Maruyama, I. (fl. 1979) S   **Maruy.**
Marvan, Peter (Petr) (1929–   ) AM   **Marvan**
Marvanová, Ludmila (fl. 1963) M   **Marvanová**
Marx, Gerhard (1956–   ) S   **Marx**
(Marxmüller, Hermann)
   see Merxmüller, Hermann   **Merxm.**
Marzari–Pencati, Guiseppe (1779–1836) S   **Marz.-Penc.**

Marzell, Heinrich (1885–1970) S	**Marzell**
Marziano, F. (fl. 1987) M	**Marziano**
Marzina, L.A. (fl. 1965) M	**Marzina**
Masaki, H. (fl. 1980) S	**Masaki**
Masaki, Tomitarô A	**T.Masaki**
Masalles, R.M. (1948–) S	**Masalles**
Masamune, Genkei (1899–) PS	**Masam.**
Masamura, Satoshi S	**Masamura**
Masart, F. (fl. 1975) M	**Masart**
Mascherpa, J.M. (fl. 1978) S	**Mascherpa**
Maschke, Joachim (fl. 1976) B	**Maschke**
Masclans, Francesco (1905–) S	**Masclans**
(Masclans i Girvès, Francesco)	
see Masclans, Francesco	**Masclans**
Masclef, Amédée (1858–) S	**Masclef**
Masferrer y Arquimbau, Ramón (1850–1884) S	**Masf.**
Mashiba, S. (fl. 1987) S	**Mashiba**
Mashkina, E.S. (fl. 1986) M	**Mashkina**
Masilamoney, P. (fl. 1982) S	**Masil.**
Masing, Victor (1925–) S	**Masing**
Maslen F	**Maslen**
Maslennikova, T.I. (fl. 1957) S	**Maslenn.**
Maslin, Bruce R. (1946–) S	**Maslin**
Maslov, V.P. A	**Maslov**
Masner, Josef (fl. 1956) S	**Masner**
Mason, Charles Thomas (1918–) S	**C.T.Mason**
Mason, Edmund William (1890–1975) M	**E.W.Mason**
Mason, Francis (1799–1874) B	**Mason**
Mason, Francis Archibald (1878–1936) M	**F.A.Mason**
Mason, Herbert Louis (1896–) S	**H.Mason**
Mason, Lucile Roush (1896–1986) A	**L.R.Mason**
Mason, M.H. (fl. 1956) S	**M.H.Mason**
Mason, Ruth (1913–1990) AS	**R.Mason**
Mason, Silas Cheever (1857–1935) S	**S.C.Mason**
Massa, C. (fl. 1912) M	**Massa**
Massales, R.M. (fl. 1987) S	**Massales**
Massalongo, Abramo Bartolommeo (1824–1860) ABFM	**A.Massal.**
Massalongo, Caro Benigno (1852–1928) BMP	**C.Massal.**
Massalski, V.J. (fl. 1885) S	**Massalski**
(Massalskij, V.J.)	
see Massalski, V.J.	**Massalski**
Massanell i Mira, Maria–Antònia A	**Massanell**
Massara, Giuseppe Filippo (1792–1839) S	**Massara**
Massari, Michele (fl. 1897) B	**Massari**
Massart, Francis (fl. 1963) M	**F.Massart**
Massart, Jean (1865–1925) AS	**Massart**
Massé, Jean–Pierre (fl. 1805) AM	**Massé**
Massee, George Edward (1850–1917) M	**Massee**
Massenot, M. (fl. 1948) M	**Massenot**

Massey, Louis Melville (1890–) M	**Massey**
Massey, M. (fl. 1816) S	**M.Massey**
Massey, R.L. S	**R.L.Massey**
Massia, G. (fl. 1924) M	**Massia**
Massieux, Michèle A	**Massieux**
(Masslennikova, T.I.)	
see Maslennikova, T.I.	**Maslenn.**
Massner, W. (fl. 1987) M	**Massner**
Masson, Daniel S	**D.Masson**
Masson, Francis (1741–1805) S	**Masson**
Master, I.M. (fl. 1975) M	**Master**
Masters, Bruce Allen (1936–) A	**B.A.Mast.**
Masters, John William (c.1792–1873) S	**J.W.Mast.**
Masters, Margaret J. (fl. 1971) M	**M.J.Mast.**
Masters, Maxwell Tylden (1833–1907) PS	**Mast.**
Masters, William (1796–1874) S	**W.Mast.**
Mastrorilli, V.I. A	**Mastr.**
Masuda, Michio (1943–) A	**Masuda**
Masyuk, N.P. A	**Masyuk**
Matekaitis, P.A. (fl. 1981) S	**Matek.**
Mateo, Gonzalo (1953–) S	**Mateo**
(Mateo Sanz, Gonzalo)	
see Mateo, Gonzalo	**Mateo**
Materassi, R. (fl. 1971) M	**Mater.**
Mateu, Isabel (1952–) S	**Mateu**
Máthé, Imre (1912–) S	**Máthé**
Matheis, W. (fl. 1974) M	**Matheis**
Mathew, Brian Frederick (1936–) S	**B.Mathew**
Mathew, Kunjamma (fl. 1962) S	**K.Mathew**
Mathew, Philip (fl. 1985) S	**P.Mathew**
Mathew, Sam P (fl. 1990) S	**S.P.Mathew**
Mathews, Ferdinand Schuyler (1854–1938) S	**F.S.Mathews**
Mathews, Joseph ('Jimmy') William (1871–1949) S	**J.W.Mathews**
Mathews, William (1828–1901) S	**Mathews**
Mathey, Annick (fl. 1971) M	**Mathey**
Mathez, Joël (fl. 1969) S	**Mathez**
Mathias, Mildred Esther (1906–) S	**Mathias**
Mathiassen, Geir (1954–) M	**Math.**
(Mathiesen, Aino)	
see Mathiesen–Kaarik, Aino	**Math.–Kaarik**
Mathiesen–Kaarik, Aino (fl. 1954) M	**Math.–Kaarik**
Mathieson, Arthur C. (1937–) A	**A.C.Mathieson**
Mathieson, Jean (fl. 1949) M	**Mathieson**
Mathieu, Antoine Auguste (1814–1890) S	**A.Mathieu**
Mathieu, Charles Louis Guillaume (1828–1904) S	**C.L.G.Mathieu**
Mathieu, Charles Marie Joseph (1791–1873) BM	**Mathieu**
Mathieu, Louis (1793–1867) S	**L.Mathieu**
Mathis, C. A	**Mathis**
Mathon, Claude Charles (1924–) S	**Mathon**

Mathou, Thérèse (1900–) S — **Mathou**
Mathre, Judith H. (fl. 1962) M — **Mathre**
Mathsson, Albert (–1898) S — **Mathsson**
Mathur, B.L. (fl. 1959) M — **B.L.Mathur**
Mathur, B.N. A — **B.N.Mathur**
Mathur, P.N. (fl. 1969) M — **P.N.Mathur**
Mathur, R. (1948–) S — **R.Mathur**
Mathur, R.L. (fl. 1959) M — **R.L.Mathur**
Mathur, R.S. (–1981) M — **R.S.Mathur**
Mathur, S.B. (fl. 1977) M — **S.B.Mathur**
Mathur, S.M. A — **S.M.Mathur**
(Matikachvili, V.)
  see Matikaschvili, V. — **Matik.**
Matikaschvili, V. S — **Matik.**
Matin, Faridenh (fl. 1989) S — **Matin**
Matinyan, A.B. (fl. 1960) S — **Matinyan**
Maton, William George (1774–1835) S — **Maton**
(Matos Araujo, Paulo Agostinho de)
  see Araujo, Paulo Agostinho de Matos — **Araujo**
Matouschek, Franz (1871–) B — **Matousch.**
Matruchot, (Alphonse) Louis (Paul) (1863–1921) M — **Matr.**
Matsson, Lars Peter Reinhold (1870–1938) S — **Matsson**
Matsuda, Ichiro (fl. 1959) M — **I.Matsuda**
Matsuda, Sadahisa (1857–1921) PS — **Matsuda**
Matsui, Tohru (1963–) B — **Matsui**
Matsumae, A. (fl. 1964) M — **Matsumae**
Matsumoto, Hiroyoshi (fl. 1934) M — **H.Matsumoto**
Matsumoto, K. (fl. 1957) M — **K.Matsumoto**
Matsumoto, Sadamu (fl. 1990) P — **S.Matsumoto**
Matsumoto, Takashi (1891–) M — **Tak.Matsumoto**
Matsumoto, Teikichi (1898–1968) M — **T.Matsumoto**
Matsumura, Jinzô (1856–1928) BPS — **Matsum.**
Matsumura, Shorien (1872–1960) S — **S.Matsum.**
Matsuno, Jútaró (1868–1946) S — **Matsuno**
Matsuoka, Kazumi (fl. 1991) AM — **Matsuoka**
Matsushima, Takashi (fl. 1971) M — **Matsush.**
Matsuura, Isamu (fl. 1928) M — **Matsuura**
(Matta, A.da)
  see Da Matta, A. — **Da Matta**
Matta, Alberto (fl. 1973) M — **A.Matta**
Matta, E.A.F.da (fl. 1955) M — **Matta**
Mattauch, F. A — **Mattauch**
Mattei, Giovanni Ettore (1865–1943) S — **Mattei**
Mattei, Luigi (fl. 1908) M — **L.Mattei**
Matten, L.C. (fl. 1968) F — **Matten**
Matteri, Celina Maria (1943–) B — **Matteri**
Mattfeld, Johannes (1895–1951) S — **Mattf.**
Matthäs, Ursula (1949–) S — **Matthäs**
Matthei, Oscar R. (1935–) S — **Matthei**

Matthew, Charles Geekie (1862–1936) PS	C.G.Matthew
Matthew, George Frederick (1837–1923) AB	Matthew
Matthew, Koyapillil Mathai (1930–) S	K.M.Matthew
Matthews, Dawn (fl. 1977) M	D.Matthews
Matthews, Henry Bleneowe (1861–1934) S	H.B.Matthews
Matthews, Henry John (1859–1909) S	Matthews
Matthews, James F. (1935–) P	J.F.Matthews
Matthews, Velma Dare (1904–1958) M	V.D.Matthews
Matthews, Victoria Ann (1941–) S	V.A.Matthews
Matthias, I.G. (fl. 1985) S	Matthias
Matthiesen, Franz (1878–1914) S	Matthiesen
Matthieu, C. (fl. 1853) S	Matthieu
(Matthiolus, Petrus Andreas)	
see Mattioli, Pietro (Pier) Andrea Gregorio	Mattioli
Mattick, Wilhelm Fritz (1901–1984) MS	Mattick
Mattioli, Pietro (Pier) Andrea Gregorio (1500?–1577) L	Mattioli
Mattirolo, Oreste (1856–1947) MPS	Mattir.
Mattlet, G. (fl. 1926) M	Mattlet
Mattos, Armando de (fl. 1968) S	A.Mattos
(Mattos Filho, Armando de)	
see Mattos, Armando de	A.Mattos
Mattos, Joáo Rodrigues de (1926–) S	Mattos
Mattos, Nilza Fischer de (1931–) S	N.F.Mattos
Mattox, Karl R. (1936–) A	Mattox
Mattson, J. (fl. 1987) M	Mattson
Mattuschka, Heinrich Gottfried von (1734–1779) MS	Matt.
Matuda, Eizi (1894–1978) PS	Matuda
Matuo, Takken (fl. 1954) M	Matuo
Matuszkiewicz, Wladyslaw (1921–) S	Matuszk.
Matvienko, O.M. A	Matv.
Mátyás, V. (fl. 1971) S	Mátyás
Matz, Julius (1886–) M	Matz
Matzenauer, Lothar A	Matzen.
Matzenbàcher, N.I. (fl. 1978) S	Matzenb.
Matzenko, A.E. (1930–) S	Matzenko
Matzer, Mario (fl. 1990) M	Matzer
Maublanc, André (1880–1958) M	Maubl.
Maude, R.B. (fl. 1986) M	Maude
Mauer, Fedor Mihajlovič (1897–1963) S	Mauer
Mauksch, Johann Daniel Thomas (1748–1831) S	Mauksch
Maul, Richard (fl. 1894) M	Maul
Maulny, Louis J.C. (1758–1815) S	Maulny
Maumené, Albert (fl. 1894–1902) S	Maumené
Maund, Benjamin (1790–1863) S	Maund
Maurand, J. (fl. 1970) M	Maurand
Maurer, Heinrich (Ludwig) (1818–1885) S	Maurer
Maurer, Willibald (1926–) S	W.Maurer
Mauri, Ernesto (1791–1836) MS	Mauri
Mauricio, H. (fl. 1930–1933) S	Mauricio

Maurin, A.M.   S   **Maurin**
Maurizio, Adam M. (1862–1941) MS   **Maurizio**
Maury, Paul Jean Baptiste (1858–1893) PS   **Maury**
Maw, George (1832–1912) S   **Maw**
(Maximilian, Alexander Philipp zu Wied–Neuwied)
   see Wied–Neuwied, Maximilian Alexander Philipp zu   **Wied–Neuw.**
Maximova, M.I. (1938–   ) S   **M.I.Maximova**
Maximova, M.L. (1939–   ) S   **Maximova**
Maximova, M.M. (fl. 1981–89) S   **M.M.Maximova**
Maximowicz, Carl Johann (Ivanovič) (1827–1891) BPS   **Maxim.**
Maximowicz–Ambodik, Nestor (1740–1812) S   **Maxim.–Amb.**
Maxon, William Ralph (1877–1948) PS   **Maxon**
Maxted, Nigel (1954–   ) S   **Maxted**
Maxwell, J.F. (fl. 1983) S   **J.F.Maxwell**
Maxwell, Richard Howard (1926–   ) S   **R.H.Maxwell**
Maxwell, T.C. (1822–1908) S   **Maxwell**
May (fl. 1870–1887) S   **May**
May, Curtis (1898–   ) M   **C.May**
May, Fred E.   A   **F.E.May**
May, Luiza Cardos (fl. 1972) M   **L.C.May**
May, R. (fl. 1988) S   **R.May**
May, Valerie (1916–   ) A   **V.May**
Mayama, Shigeki   A   **Mayama**
Maycock, James Dottin (–1837) S   **Maycock**
Mayebara, Kanjiro (1890–   ) BP   **Mayeb.**
Mayer, A.F.I.Carolus   A   **C.Mayer**
Mayer, Adolf (1843–   ) S   **Ad.Mayer**
Mayer, Adolf Theodor (1871–1952) S   **A.Mayer**
Mayer, Anton (1867–1951) A   **Mayer**
Mayer, Ernest (1920–   ) S   **E.Mayer**
Mayer, F.K. (fl. 1978) S   **F.K.Mayer**
Mayer, Johann (1754–1807) S   **J.Mayer**
Mayer, Johann Christoph Andreas (1747–1801) S   **J.C.Mayer**
Mayer, Johann Prokop (1737–1804) S   **J.P.Mayer**
Mayo, Simon Joseph (1949–   ) S   **Mayo**
Mayol, Maria (fl.1992) S   **Mayol**
Mayor, Eugéne (1877–1976) MS   **Mayor**
Mayr, Heinrich (1856–1911) MS   **Mayr**
Mayr, Johann (fl. 1797) S   **J.Mayr**
Mayrhofer, Helmut (1953–   ) M   **H.Mayrhofer**
(Mayrhofer, I.N.)
   see Mayrhoffer, Johann Nepomuk   **Mayrh.**
Mayrhofer, Karl (–1853) S   **Mayrhofer**
Mayrhofer, Michaela (1955–   ) M   **M.Mayrhofer**
Mayrhofer, P.Joseph (1831–1881) S   **J.Mayrhofer**
Mayrhoffer, Johann Nepomuk (1764–1832) S   **Mayrh.**
Mayuranathan, P.V.   S   **Mayur.**
(Maza y Jiménez, Manuel Gómez de la)
   see Gómez de la Maza y Jiménez, Manuel   **M.Gómez**

Mazade, M. (fl. 1978) S	**Mazade**
Mazaffarian, V. S	**Mazaff.**
Mazancourt, J.de Seizilles de A	**Mazanc.**
Mazé, Hippolyte Pierre (1818–1892) A	**Mazé**
Mazé, P. (fl. 1910) M	**P.Mazé**
Mazel, Eugène (fl. 1981) S	**Mazel**
Mazen, M.B. (fl. 1979) M	**Mazen**
Maziero, Rosana (fl. 1988) M	**Maziero**
Mazkevich, V.I. (1896–1943) S	**Mazk.**
(Mazoyer, Geneviève)	
see Feldmann–Mazoyer, Geneviève	**Feldm.-Maz.**
Mazuc, E. (fl. 1854) S	**Mazuc**
(Mazuk, E.)	
see Mazuc, E.	**Mazuc**
Mazza, Angelo (1844–1929) A	**Mazza**
Mazza, Giorgio (fl. 1987) M	**G.Mazza**
Mazza, R. (fl. 1987) M	**R.Mazza**
Mazza, S. (fl. 1930) M	**S.Mazza**
Mazzer, Samuel J. (1934–) M	**Mazzer**
Mazziari, Alessandro Domenico (–1857) S	**Mazziari**
Mazzola, Pietro (1945–) S	**Mazzola**
Mazzoni, Antonio (1958–) S	**Mazzoni**
Mazzucato, Giovanni (1787–1814) S	**Mazzuc.**
(Mazzuccato, Giovanni)	
see Mazzucato, Giovanni	**Mazzuc.**
Mazzucchelli, Vittorio (1859–1941) S	**Mazzucch.**
Mazzuchetti, G. (fl. 1965) M	**Mazzuch.**
McAdam, S.V. (fl. 1981) B	**McAdam**
McAfee, B.J. (fl. 1982) M	**McAfee**
McAleer, J.H. (fl. 1959) M	**McAleer**
McAlpin, Bruce (fl. 1986) P	**McAlpin**
McAlpine, Daniel (1849–1932) M	**McAlpine**
McAndrew, James (1836–1917) BS	**McAndr.**
McArdle, David (1849–1934) BM	**McArdle**
McArthur, E.Durant (1941–) S	**McArthur**
McAtee, Waldo Lee (1883–1962) S	**McAtee**
McAulay, Alexander Leicester (1895–1969) S	**McAulay**
McBride, Douglas L. A	**McBride**
McBurney, Jean (1909–) S	**McBurney**
McCabe, Dennis E. (fl. 1979) M	**McCabe**
McCain, A.H. (fl. 1967) M	**McCain**
McCain, John W. (fl. 1982) M	**J.W.McCain**
McCall, David A	**McCall**
McCalla, William (1814–1849) A	**McCalla**
McCalla, William Copeland (1872–1962) S	**W.C.McCalla**
McCandless, Esther Lieb (1923–1983) A	**McCandless**
McCann, Yale Mervin Charles (1899–1980) S	**McCann**
McCarthy, G.J.P. (fl. 1976) M	**G.J.P.McCarthy**

McCarthy, Michael Gerard (1858–1915) S   **McCarthy**
McCarthy, Patrick Martin (1955–) M   **P.M.McCarthy**
McCartney, Robert B. (fl. 1991) S   **McCartney**
McCauley, V.J.E. (fl. 1971) M   **McCauley**
McClain, R.L. (fl. 1925) M   **McClain**
McClatchie, Alfred James (1861–1906) BS   **McClatchie**
McClean, Alan Percy Douglas (1902–) S   **McClean**
McClelland, John (1805–1883) FS   **McClell.**
McClellen, Wilbur Dwight (1914–) M   **McClellen**
McClintock, David Charles (1913–) S   **D.C.McClint.**
McClintock, Elizabeth May (1912–) S   **E.M.McClint.**
McClintock, James Albertine (1889?) M   **McClint.**
McClung, L.S. (fl. 1940) M   **McClung**
McClure, Floyd Alonzo (1897–1970) S   **McClure**
McClure, T.T. (fl. 1951) M   **T.T.McClure**
McColl, W.R. (1855–1933) P   **McColl**
McColloch, Lacy Porter (1907–) M   **McColloch**
McCollum, D.W. A   **McCollum**
McComb, Jennifer Anne (1943–) S   **McComb**
McCord, David Ross (1844–) P   **McCord**
McCormick, J.Frank (fl. 1971) S   **J.F.McCormick**
McCormick, Robert (1800–1890) S   **McCormick**
McCoy, C.W. (fl. 1984) M   **C.W.McCoy**
McCoy, Frederick (1817–1899) AF   **McCoy**
McCoy, John J. A   **J.J.McCoy**
McCoy, Scott (1897–) S   **S.McCoy**
McCoy, Thomas Nevil (1905–) PS   **T.N.McCoy**
McCracken, Derek Albert (1943–) A   **McCracken**
McCue, G. (fl. 1960) S   **McCue**
McCulloch, Irene (1886–) A   **McCulloch**
McCulloch, Janet S. (fl. 1977) M   **J.S.McCulloch**
McCulloch, Lucia (1873–1955) M   **L.McCulloch**
McCullough, Herbert Alfred (1914–) M   **McCull.**
McDade, Lucinda A. (1953–) S   **McDade**
McDaniel, Sidney T. (1940–) S   **McDaniel**
McDearman, W. (fl. 1985) S   **McDearman**
McDermid, Karla J. A   **McDermid**
McDermott, Laura Frances (1882–1923) S   **McDermott**
McDonald, Donald B. A   **D.B.McDonald**
McDonald, J.Andrew (fl. 1989) S   **J.A.McDonald**
McDonald, James C. (fl. 1965) M   **J.C.McDonald**
McDonald, William H. (1837–1902) P   **McDonald**
McDonough, E.S. (fl. 1961) M   **McDonough**
McDougal, Walter Byron (1883–) M   **McDougal**
McDougall, Patricia J. (fl. 1968) M   **P.J.McDougall**
McDougall, Walter Byron (1883–) M   **W.McDougall**
McDowell, J.A. (fl. 1896) S   **McDowell**
McEnery, Marie E. A   **McEnery**
McFadden, G.I. (1958–) A   **McFadden**

McFadden, Mabel Effie A	**M.E.McFadden**
McFarlin, James B. (fl. 1932) S	**McFarlin**
McGee, P.A. (fl. 1986) M	**McGee**
McGhee, R.Barclay (1918–) A	**McGhee**
McGibbon, James (fl. 1848–1864) S	**McGibb.**
McGillivray, Donald John (1935–) S	**McGill.**
McGinnis, Michael R. (fl. 1974) M	**McGinnis**
McGinty (also Lloyd, C.G.) (1859–1926) M	**McGinty**
McGivney, Mary Vincent de Paul (fl. 1938) S	**McGivney**
McGlynn, William Henry (1876–1956) S	**McGlynn**
McGregor, Ronald Leighton (1919–) BS	**McGregor**
McGuinness, M.D. (fl. 1983) M	**McGuinn.**
McGuire, J.M. (fl. 1941) M	**McGuire**
McGuire, P.E. (fl. 1983) S	**P.E.McGuire**
McHenry, Jerry (fl. 1968) M	**McHenry**
McHugh, R. (fl. 1986) M	**McHugh**
McIlvaine, Charles (1840–1909) M	**McIlv.**
McInteer, Berthus Boston (1887–1978) A	**McInteer**
McIntire, C.David (1932–) A	**McIntire**
McIntosh, Charles (1794–1864) S	**McIntosh**
McIntosh, Terry T. (1948–) B	**T.T.McIntosh**
McIntyre, Andrew (1931–) A	**McIntyre**
McIntyre, D.J. A	**D.J.McIntyre**
McIvor, William Graham (–1876) S	**McIvor**
McKay, Hazel H. (fl. 1959) M	**H.H.McKay**
McKay, J.A. (fl. 1938) S	**McKay**
McKean, Douglas Ronald (1948–) S	**McKean**
(McKee, Hugh S.)	
see MacKee, Hugh S.	**MacKee**
McKeen, Colin D. (fl. 1957) M	**McKeen**
McKeen, Wilber E. (fl. 1964) M	**W.E.McKeen**
McKelvey, Susan Adams (1888–1964) S	**McKelvey**
McKemy, John M. (fl. 1991) M	**McKemy**
McKen, Mark Johnston (1823–1872) PS	**McKen**
McKenzie, Eric H.C. (1946–) M	**McKenzie**
McKeown, B.M. (fl. 1985) M	**McKeown**
McKibben, Wendell R. A	**McKibben**
McKibbin, Dale L. A	**D.L.McKibbin**
McKibbin, J.N. (fl. 1883) S	**McKibbin**
McKie, Ernest Norman (1882–1948) S	**McKie**
McKinlay, Alexander (fl. 1869) B	**McKinlay**
McKinney, Harold Hall (1889–) M	**McKinney**
McKnight, Kent H. (fl. 1968) M	**McKnight**
McLachlan, Jack L. (1930–) A	**McLachlan**
(McLaren, Henry Duncan)	
see Aberconway, Henry Duncan McLaren, Lord	**Aberc.**
McLarty, H.R. (fl. 1941) M	**McLarty**
McLaughlin, David Jordan (1940–) M	**D.J.McLaughlin**
McLaughlin, Esther G. (fl. 1978) M	**E.G.McLaughlin**

McLaughlin, J.Harvey (1915–   ) M   **J.H.McLaughlin**
McLaughlin, John J.Anthony (1924–   ) A   **J.A.McLaughlin**
McLaughlin, Robert Penfield (1898–   ) S   **McLaughlin**
McLaughlin, Steven Paul (1948–   ) S   **S.P.McLaughlin**
McLean, Dewey M.   A   **D.M.McLean**
McLean, R.A.   S   **R.A.McLean**
McLean, Robert Colquhoun (1890–1981) M   **McLean**
McLean, Robert J. (1940–   ) A   **R.J.McLean**
McLennan, Ethel Irene (1891–1983) M   **McLennan**
McLure, Ken   A   **McLure**
(McMahon, Bernard)
  see M'Mahon, Bernard   **M'Mahon**
McMillan, A.J.S. (fl. 1990) S   **McMillan**
McMinn, Howard Earnest (1891–1963) S   **McMinn**
McMunn, R.L. (fl. 1925) M   **McMunn**
McMurran, Stockton Mosby (1887–1920) M   **McMurran**
McMurtry, D. (fl. 1984) S   **McMurtry**
McNab, William (1780–1848) S   **McNab**
McNab, William Ramsay (1844–1889) APS   **W.R.McNab**
McNabb, R. (fl. 1989) M   **R.M̧cNabb**
McNabb, R.F.R. (1934–1972) M   **McNabb**
McNair, James Birtley (1889–   ) S   **McNair**
McNamara, Kenneth J. (fl. 1983) F   **McNamara**
McNeal, Dale W. (1939–   ) S   **McNeal**
McNeil, Patrick Gordon (1908–1986) S   **McNeil**
McNeil, R. (fl. 1983) M   **R.McNeil**
McNeill, Ellis Meade (1901–   ) A   **E.M.McNeill**
McNeill, John (1933–   ) S   **McNeill**
McNichols, Elaine L. (fl. 1974) M   **McNichols**
McPartland, John M. (fl. 1983) M   **McPartl.**
McPherson, Gordon (1947–   ) S   **McPherson**
McQueen, Cyrus B. (1951–   ) B   **C.B.McQueen**
McQueen, D.R. (fl. 1955) F   **McQueen**
McRae, William (1878–1952) M   **McRae**
McRitchie, J.J. (fl. 1978) M   **McRitchie**
McVaugh, Rogers (1909–   ) S   **McVaugh**
McVickar, David L. (fl. 1942) M   **McVickar**
McWeeney, Edm.J. (fl. 1895) M   **McWeeney**
McWhorter, Frank Paden (1896–   ) M   **McWhorter**
Mead, Samuel Barnum (1798–1880) S   **Mead**
Meakin, Samuel H. (1867–1955) A   **Meakin**
Mears, James Austin (1944–   ) S   **Mears**
Meavahd, Khasan (fl. 1967) M   **Meavahd**
Mecenović, Karl   S   **Mecen.**
(Mechtieva, N.A.)
  see Mekhtijeva, N.A.   **Mekht.**
Medagli, Pietro (1951–   ) S   **Medagli**
Medd, A.W.   A   **Medd**
Medeiros, A.C. (fl. 1988) S   **A.C.Medeiros**
Medeiros, Arnaldo Gomes de (fl. 1962) M   **A.G.Medeiros**

Medeiros–Costa, Judas Tadeu de (fl. 1987) S	**Med.-Costa**
Medelius, Sigrid Olof (1878–1930) B	**Medelius**
(Medicus, Friedrich Casimir)	
see Medikus, Friedrich Kasimir	**Medik.**
Medicus, Ludwig Wallrad (1771–1850) S	**L.W.Medicus**
Medicus, W. (c.1892) S	**W.Medicus**
Medikus, Friedrich Kasimir (1736–1808) S	**Medik.**
(Medina Cota, José Miguel)	
see Medina–Cota, José Miguel	**Medina–Cota**
(Medina, José Maria Muñoz)	
see Muñoz Medina, José Maria	**Muñoz Med.**
Medina Lemos, Rosalinda (fl. 1986) S	**Medina**
Medina–Cota, José Miguel (1952–) S	**Medina–Cota**
Medlar, Edgar Mathias (1887–) M	**Medlar**
Medley, Max E. S	**Medley**
Medlicott, Henry Benedict (1829–1905) S	**Medlicott**
Medlin, Linda K. A	**Medlin**
Medrano, Francisco González (1939–) S	**Medrano**
(Medvedev, Jakob Sergejewitsch)	
see Medwedew, Jakob Sergejevitsch	**Medw.**
(Medvedev, Pavel Michaelovich)	
see Medwedew, Pavel Michaelovich	**P.Medw.**
Medwedew, Jakob Sergejevitsch (1847–1923) S	**Medw.**
Medwedew, Pavel Michaelovich (1900–1968) S	**P.Medw.**
Medwell, Lorna M. (fl. 1954) BF	**Medwell**
Meehan, Frances (fl. 1946) M	**F.Meehan**
Meehan, Thomas (1826–1901) S	**Meehan**
Meeker, Joseph Andrew (fl. 1971) M	**Meeker**
(Meer, John P.Van der)	
see Vandermeer, John P.	**Vanderm.**
Meerburgh, Nicolaas (1734–1814) S	**Meerb.**
Meerendonk, J.P.M.van de (fl. 1984) S	**Meerend.**
Meerow, Alan W. (1952–) S	**Meerow**
Meerssche, J.van de (fl. 1986) M	**Meerssche**
Meerstadt, A. (fl. 1987) S	**Meerst.**
Meese, David (1723–1770) S	**Meese**
Meeuse, Adrianus Dirk Jacob (1914–) S	**A.Meeuse**
Meeuse, Bastiaan Jacob D. (1916–) S	**B.Meeuse**
Meeuwen, M.S.Knaap-van (1936–) S	**Meeuwen**
Mefferd, R.B. (fl. 1955) M	**Mefferd**
Meffert, Abramovic V.V. S	**Meffert**
Mégnin, P. (fl. 1881) M	**Mégnin**
Mehlenbacher, Lyle E. S	**Mehlenb.**
Mehlisch, K. (fl. 1935) M	**Mehlisch**
Mehr, J.L. (fl. 1943) M	**Mehr**
Mehra, Bharati A	**B.Mehra**
Mehra, K.L. (fl. 1962) S	**K.L.Mehra**
Mehra, Kamini R. A	**K.R.Mehra**
Mehra, Pran Nath (1907–) BPS	**Mehra**

Mehrlich, Ferninand Paul (1905–   ) M   **Mehrl.**
Mehrotra, B.M. (fl. 1978) M   **B.M.Mehrotra**
Mehrotra, Brahma Swarup (fl. 1962) M   **B.S.Mehrotra**
Mehrotra, Bishan N. (1936–   ) PS   **Mehrotra**
Mehrotra, Brij Rani (fl. 1970) M   **B.R.Mehrotra**
Mehrotra, M.D. (fl. 1962) M   **M.D.Mehrotra**
Mehrotra, Naresh C.   A   **N.C.Mehrotra**
Mehrotra, Nishi (fl. 1985) M   **N.Mehrotra**
Mehrotra, R.K.   A   **R.K.Mehrotra**
Mehta, Karm Chand (1892–1950) B   **Mehta**
Mehta, P.R. (fl. 1951) M   **P.R.Mehta**
Méhu, Marie Antoine Adolphe (1840–1881) S   **Méhu**
Meier, Fred Campbell (1893–1938) M   **Meier**
Meigen, Friedrich Carl (1864–   ) S   **F.Meigen**
Meigen, Johann Wilhelm (1764–1845) S   **Meigen**
Meigen, Wilhelm (1826/7) S   **W.Meigen**
Meijden, R.van (1945–   ) S   **Meijden**
(Meijer, André A.R.de)
   see de Meijer, André A.R.   **de Meijer**
Meijer, Frans Nicholaas (1875–1918) S   **F.N.Meijer**
Meijer, Willem (1923–   ) BS   **Meijer**
Meikle, Robert Desmond (1923–   ) S   **Meikle**
Meinecke, Emilio Pepe Michael (1869–1957) M   **E.Meinecke**
Meinecke, Johann Ludwig Georg (1721–1823) S   **Meinecke**
Meinesz, A.   A   **Meinesz**
Meinhardt, Uta (fl. 1973) S   **Meinh.**
Meinke, R.J. (fl. 1983) S   **Meinke**
Meins, Claus (1806–1873) S   **Meins**
Meinshausen, Karl Friedrich (1819–1899) S   **Meinsh.**
Meisel, Max (1892–1969) S   **Meisel**
Meisner, Carl Daniel Friedrich (1800–1874) S   **Meisn.**
(Meissner, Carl Daniel Friedrich)
   see Meisner, Carl Daniel Friedrich   **Meisn.**
Meissner, Carl Friedrich Wilhelm (1792–1853) M   **C.F.W.Meissn.**
Meister, Friedrich (fl. 1893–1919) A   **F.Meister**
Meister, Jakob (1850–1927) S   **J.Meister**
Mejen, S.V.   B   **Mejen**
Mejer, Ludwig (1825–1895) S   **Mejer**
Mejia, K. (fl. 1988) S   **K.Mejia**
Mejia, M.M. (fl. 1986) S   **M.M.Mejia**
(Mejia P., M.M.)
   see Mejia, M.M.   **M.M.Mejia**
Mekhtijeva, N.A. (fl. 1957) M   **Mekht.**
Mela, Aukusti Juhana (né Malmberg, A.J.) (1846–1904) S   **Mela**
Melan, M.A. (fl. 1984) M   **Melan**
Melander, C. (fl. 1887) S   **Melander**
Melander, Leonard William (1893–   ) S   **L.W.Melander**
Mélard, L. (fl. 1910) M   **Mélard**
Melchert, Thomas E. (1936–   ) S   **Melchert**

Melchior, Hans (1894–1984) S	**Melch.**
Melderis, Aleksandre (1909–1986) S	**Melderis**
Melén, Eric Gustaf (1801–1828) S	**Melén**
Meléndez–Howell, Leda–María (fl. 1965) M	**Mel.-Howell**
(Meleščenko, V.S.)	
see Meleshchenko, V.S.	**Meleshch.**
Meleshchenko, V.S. S	**Meleshch.**
Meletti, Paolo (1927–) S	**Meletti**
Melhus, Irving E. (1881–1969) MP	**Melhus**
Melia, M.S. (fl. 1975) M	**Melia**
Melibaev, S. (fl. 1982) S	**Melibaev**
Melik–Khachatryan, D.G. (fl. 1959) M	**Melik-Khach.**
Melikyan, A.P. (1935–) S	**Melikyan**
Melin, Johannes Botwid Elias (1889–) BM	**Melin**
Meling, E.V. (fl. 1982) S	**Meling**
Melis, P. (fl. 1988) S	**Melis**
Melkonian, M. A	**Melkonian**
Mell, Clayton Dissinger (1875–1945) S	**Mell**
Melle, Peter Jacobus van (1891–1953) S	**Melle**
Mellichamp, T.Lawrence (1935–) S	**Mellich.**
Melliss, John Charles (fl. 1863–1875) BS	**Melliss**
Mello, Alfredo Froilano Bachmann de (fl. 1918) AM	**Mello**
Mello, Arthur Ferreira de (fl. 1977) S	**A.F.Mello**
(Mello Barreto, Henrique Lamahyer de)	
see Barreto, Henrique Lamahyer de Mello	**Barreto**
(Mello de Ficalho, Franciso Manoel Carlos de)	
see Ficalho, Franciso Manoel Carlos de Mello de	**Ficalho**
(Mello Filho, Luiz Emygdio de)	
see Emygdio de Mello, Luiz	**Emygdio**
(Méllo, Joachim Corrêa)	
see Correia de Méllo, Joachim	**Corr.Méllo**
(Méllo, Joachim Correia de)	
see Correia de Méllo, Joachim	**Corr.Méllo**
(Mello, Luiz Emygdio de)	
see Emygdio de Mello, Luiz	**Emygdio**
(Mello Netto, Ladislau de Souza)	
see Netto, Ladislau de Souza Mello	**Netto**
Mello–Silva, Renato de (1961–) S	**Mello-Silva**
Melnichuk, Vsevolod Maximovich (fl. 1960) B	**Meln.**
(Melničuk, Vsevolod Maksimovič)	
see Melnichuk, Vsevolod Maximovich	**Meln.**
Melnik, V.A. (1937–) M	**Melnik**
Melnikova, V.V. A	**Melnikova**
Melo, Ireneia (1947–) M	**Melo**
Melot, Jacques (fl. 1987) M	**Melot**
Melsheimer, Marcellus (1827–1920) S	**Melsh.**
Melvaine, Alma Theodora (later Lee, A.T.) (1912–) PS	**Melvaine**
Melvill, James Cosmo (1845–1929) S	**Melvill**

Melville, Ronald (1903–1985) S	**Melville**
Melvin, Lionel (fl. 1956) S	**Melvin**
Melzer, Helmut (1922–) MPS	**H.Melzer**
Melzer, Vaclav (fl. 1925) M	**Melzer**
Melzheimer, Volker (1939–) S	**Melzh.**
Memminger, Edward Read (1856–) S	**Memm.**
Men, Fan Son (fl. 1976) B	**F.S.Men**
Mena, A.J. (fl. 1970) M	**Mena**
Mena Portales, Julio (fl. 1987) M	**J.Mena**
Menabde, A.M. S	**A.M.Menabde**
Menabde, Vladimir Levanovich (1898–) S	**Menabde**
Menadue, Y. (fl. 1986) S	**Menadue**
Menapace, F.J. (fl. 1986) S	**Menapace**
Mencher, E. (fl. 1968) F	**Mencher**
Mendel, Gregor Johann (1822–1884) S	**Mendel**
Mendes, Eduardo José Santos Moreira (1924–) APS	**Mendes**
(Mendes Magalhães, Geraldo)	
see Magalhães, Geraldo Mendes	**G.M.Magalh.**
Mendes, O. (fl. 1953) M	**O.Mendes**
Mendiola, Blanca Rojas E.de A	**B.R.Mendiola**
Mendiola, Victoria B. (fl. 1931) M	**Mendiola**
Mendonça, Carlos Victor (1965–) S	**C.V.Mendonça**
(Mendonça Filho, Carlos Victor)	
see Mendonça, Carlos Victor	**C.V.Mendonça**
Mendonça, Francisco de Ascencão (1889–1982) PS	**Mendonça**
Mendonça–Hagler, L.C. (fl. 1985) M	**Mend.–Hagler**
Mendoza, D.R. (fl. 1925) M	**Mend.**
Mendoza, José Miguel (fl. 1925) M	**J.M.Mend.**
Mendoza, L. (fl. 1987) M	**L.Mend.**
Mendoza, María Laura A	**M.L.Mendoza**
Mendoza–Heuer, Ilse R. (1919–) S	**Mend.–Heuer**
Meneghini, Guiseppe Giovanni Antonio (1811–1889) AF	**Menegh.**
Menéndez Amor, Josefa (1916–) S	**Men.Amor**
Menéndez, Carlos Alberto (1921–1976) AF	**C.A.Menéndez**
(Menendez, Josefa)	
see Menédez Amor, Josefa	**Men.Amor**
Menès, J.C. (fl. 1989) M	**Menès**
Meneses, Isabel A	**I.Meneses**
Meneses, Oscar J.Azancot de (fl. 1956) S	**Meneses**
Menet, André S	**Menet**
Meñez, Ernani G. (1931–) A	**Meñez**
Menezes, Carlos Azevedo de (1863–1928) PS	**Menezes**
Menezes, L.C. (fl. 1987) S	**L.C.Menezes**
Menezes, Nanusa Luiza de (1934–) S	**N.L.Menezes**
Meng, Ren Xian (fl. 1986) S	**R.X.Meng**
Meng, You Ru (fl. 1989) M	**Y.R.Meng**

Mengascini, Adriana S. (fl. 1987) M	**Mengasc.**
Mengaud, L. A	**Mengaud**
Menge, Franz Anton (1808–1880) BFS	**Menge**
Menge, John A. (fl. 1984) M	**J.A.Menge**
Menges, Rainer (fl. 1980) S	**Menges**
Menier, Charles (1913–) M	**Menier**
(Menitski, Yu.L.)	
see Menitsky, Ju.L.	**Menitsky**
Menitsky, Ju.L. (G.) (1937–) S	**Menitsky**
Mennega, Alberta Maria Wilhelmina (1912–) S	**Mennega**
Mennema, Jacob (1930–) S	**Mennema**
Menon, Radha (fl. 1954) M	**R.Menon**
Menon, S.K. (fl. 1959) M	**S.K.Menon**
Menozzi, Angelo S	**Menozzi**
Menten, J.O.M. (fl. 1980) M	**Menten**
Mentz, August (1867–1944) S	**Mentz**
Mentzer, Vera P. (fl. 1936) M	**Mentzer**
Menyhárt, Lászlò (Ladislav) (1849–1897) S	**Menyh.**
(Ményharth, László)	
see Menyhárt, Lászlò (Ladislav)	**Menyh.**
Menzel, Margaret Y. (fl. 1983) S	**M.Y.Menzel**
Menzel, Mario (1958–) B	**M.Menzel**
Menzel, Paul Julius (1864–1927) F	**Menzel**
Menzies, Archibald (1754–1842) BPS	**Menzies**
Mer, Émile (fl. 1881) P	**Mer**
Mer, G.S. (fl. 1984) M	**G.S.Mer**
Mérat de Vaumartoise, François Victor (1780–1851) ABMPS	**Mérat**
Mercado, Ángel (1937–) M	**Mercado**
(Mercado Sierra, Ángel)	
see Mercado, Ángel	**Mercado**
Mercer, C. (fl. 1957) M	**C.Mercer**
Mercer, W.B. (fl. 1913) M	**Mercer**
Merchant, Y. (fl. 1963) S	**Merchant**
Mercier, (Marie) Philippe (1781–1831) S	**P.Mercier**
(Mercier de Coppet, Elysée)	
see Mercier, Elysée	**Mercier**
Mercier, Elysée (1802–1863) S	**Mercier**
Mercier, J. A	**J.Mercier**
Merck, E. S	**Merck**
Mercklin, Karl Eugen (Eugeniewitsch) von (1821–1904) S	**Merckl.**
Mercuri, O.A. (fl. 1976) M	**Mercuri**
Meredith, Louisa Anne (1812–1895) S	**Meredith**
(Mérejkovski, Constantin Sergeevič)	
see Mereschkowski, Konstantin Sergejewicz	**Mereschk.**
Mereles, Fátima (1953–) S	**Mereles**
(Mereschkovsky, Konstantin Sergejewicz)	
see Mereschkowski, Konstantin Sergejewicz	**Mereschk.**

Mereschkowski, Konstantin Sergejewicz (1854–1921) AM	**Mereschk.**
(Mereschkowsky, Konstantin Sergejewicz)	
see Mereschkowski, Konstantin Sergejewicz	**Mereschk.**
Merezhko, Tatjana A.(O.) (1942–) M	**Merezhko**
Meriat (fl. 1981–89) S	**Meriat**
Merino, B.Baltasar (1845–1917) PS	**Merino**
(Merino y Román, P.Baltasar)	
see Merino, B.Baltasar	**Merino**
Merker, Gustav (1871–) S	**Merker**
Merklein, Friedrich (fl. 1861) S	**Merklein**
Merlet de la Boulaye, Gabriel-Eléonor (1736–1807) S	**Merlet**
Merlin, J.A. (fl. 1929) M	**Merlin**
Mermiér, F. (fl. 1963) M	**Mermiér**
Merny, G. (fl. 1962) M	**Merny**
Merola, Aldo (1924–1980) S	**Merola**
Merrem, Blasius (1761–1824) S	**Merrem**
Merriam, Clinton Hart (1855–1942) S	**Merriam**
Merrill, Elmer Drew (1876–1956) MPS	**Merr.**
Merrill, George Knox (1864–1927) M	**G.Merr.**
Merrill, J.A. A	**J.A.Merr.**
Merriman, Mabel L. A	**M.L.Merriman**
Merriman, Paul Rossiter (1882–) S	**Merriman**
Mertens, Franz Karl (Carl) (1764–1831) PS	**Mert.**
Mertens, Karl (Carl) Heinrich (1796–1830) S	**K.Mert.**
Mertens, Thomas Robert (1930–) S	**T.R.Mert.**
Mertz, Dieter A	**Mertz**
Merucci, E.L. (fl. 1942) M	**Merucci**
(Merwe, Frederick Ziervogel van der)	
see Van der Merwe, Frederick Ziervogel	**Van der Merwe**
(Merwe, Jacoba Johanna Maria(?) van der)	
see Van der Merwe, Jacoba Johanna Maria(?)	**J.J.M.van der Merwe**
(Merwe, W.J.J.van der)	
see Van der Merwe, W.J.J.	**W.J.J.van der Merwe**
Merxmüller, Hermann (1920–1988) MS	**Merxm.**
Merzbacher, Gottfried (1843–1926) S	**Merzb.**
Mesa, Aldo (fl. 1971) P	**Mesa**
Meschinelli, Aloysius (Luigi) (1865–) FM	**Mesch.**
Mesczerjakov, A.A. (1916–) S	**Mesczer.**
Mesfin Tadesse (1951–) S	**Mesfin**
Mesícek, Joseph (fl. 1970) S	**Mesícek**
Meslin, Roger E. A	**Meslin**
Mesnil, Felix (1868–1938) AM	**Mesnil**
Mesplède, V.H. (fl. 1980) M	**Mesplède**
Mesquita, A.L. (fl. 1986) S	**Mesquita**
(Mesquita Rodrigues, José E.de)	
see Rodrigues, José E.de Mesquita	**Rodrigues**
Messel (fl. 1933) S	**Messel**
Messeri, Albina (1904–1972) S	**Messeri**
Messiaen, C.M. (fl. 1959) M	**Messiaen**

Messikommer, Edwin A	**Messik.**
Messmer, Pearl R. S	**Messmer**
Messner, Kurt (fl. 1971) MS	**Messner**
Metcalf, Franklin Post (1892–1955) S	**F.P.Metcalf**
Metcalf, Haven (1875–1940) M	**Metcalf**
Metcalfe, Charles Russell (1904–1991) S	**C.R.Metcalfe**
Metcalfe, G. (fl. 1957) M	**G.Metcalfe**
Metcalfe, Orrick Baylor (1879–1936) S	**Metcalfe**
Metcalfe, S.E. A	**S.E.Metcalfe**
Methven, Andrew S. (fl. 1985) M	**Methven**
Metianu, T. (fl. 1966) M	**Metianu**
Metlesics, H. (fl. 1976) S	**Metlesics**
Métrod, Georges (fl. 1938) M	**Métrod**
Metsch, Johann Christian (1796–1856) S	**Metsch**
Metschnikoff, Ilya (1845–1916) M	**Metschn.**
Mettenheimer, C. A	**Mettenh.**
Mettenius, Georg Heinrich (1823–1866) PS	**Mett.**
Metting, Blaine A	**Metting**
Metz, Mary Clare (1907–) S	**Metz**
Metzelova-Kropácova, A. (1922–) S	**Metz.-Krop.**
Metzger, Jean (fl. 1990) S	**J.Metzg.**
Metzger, Johann (1789–1852) S	**Metzg.**
Metzler, Berthold (fl. 1989) M	**B.Metzler**
Metzler, Jakob Adolf ('H') (1812–1883) CM	**Metzler**
Metzner, Jerome (1911–) A	**J.Metzner**
Metzner, Paul (1893–) M	**P.Metzner**
Metzven, John (fl. 1900) S	**Metzven**
Meulenhoff, J.S. (1867–1936) M	**Meulenh.**
Meunier, Alphonse F. (1857–1918) A	**Meunier**
Meunissier, A.A. (1876–1947) S	**Meun.**
Meurer, Franz Ferdinand (1809–1882) S	**F.Meurer**
Meurer, Pfarrer (fl. 1848) S	**Meurer**
Meurling, Petrus (fl. 1820) S	**Meurling**
Meurs, Abraham (1904–) M	**Meurs**
Meusel, Hermann (1909–) S	**Meusel**
Mewborn, A.D. (fl. 1902) M	**Mewborn**
Mexia, Ynes Enriquetta Julietta (1870–1938) S	**Mexia**
(Meyden, R.van)	
see Meijden, R.van	**Meijden**
Meyen, Franz Julius Ferdinand (1804–1840) MPS	**Meyen**
Meyen, S.V. (fl. 1963–1969) F	**S.V.Meyen**
Meyer, Adolf (1868–) S	**Ad.Mey.**
Meyer, Albert (fl. 1884) S	**Alb.Mey.**
Meyer, Arthur (1850–1922) A	**Art.Mey.**
Meyer, August (fl. 1872) S	**Aug.Mey.**
Meyer, Bernhard (1767–1836) BS	**B.Mey.**
Meyer, Carl Anton (Andreevič) von (1795–1855) PS	**C.A.Mey.**
Meyer, Dieter Erich (1926–1982) PS	**D.E.Mey.**
Meyer, Ernst Heinrich Friedrich (1791–1858) BS	**E.Mey.**

Meyer, F.C. (fl. 1922) M   F.C.Mey.
(Meyer, Frank Nicholas)
   see Meijer, Frans Nicholaas   F.N.Meijer
Meyer, Frederick Gustav (1917–) S   F.G.Mey.
Meyer, Friedrich Karl (1926–) S   F.K.Mey.
Meyer, G.L. (fl. 1881) S   G.L.Mey.
Meyer, Georg Friedrich Wilhelm (1782–1856) MPS   G.Mey.
Meyer, German Bula   A   G.B.Mey.
Meyer, Hans (Johannes August Theodor) (1885–1935) B   H.J.A.Mey.
Meyer, Herbert   A   H.Mey.
Meyer, J. (fl. 1932) M   J.Mey.
Meyer, J.A. (fl. 1957) M   J.A.Mey.
Meyer, J.C. (fl. 1854) S   J.C.Mey.
Meyer, J.G.Fr. (fl. 1809) P   J.G.F.Mey.
Meyer, Konstantin Ignatevich (1881–) A   K.I.Mey.
Meyer, M. (fl. 1988) P   M.Mey.
Meyer, Marianne (fl. 1990) M   Mar.Mey.
Meyer, Paul Gerhard (1934–) S   P.G.Mey.
Meyer, Richard Lee (1931–) A   R.L.Mey.
Meyer, Robert (1881–) S   Rob.Mey.
Meyer, Robert J. (1954–) M   R.J.Mey.
Meyer, Rudolph (fl. 1896–1914) S   Rud.Mey.
Meyer, S.R.   A   S.R.Mey.
Meyer, Sally A. (fl. 1967) M   S.A.Mey.
Meyer, Susan Lynn Fricke (fl. 1943) M   S.L.F.Mey.
Meyer, Teodore (1910–) S   T.Mey.
Meyer–Abich, Adolf (1893–) S   Mey.-Abich
Meyers, Samuel P. (fl. 1960) M   Meyers
Meylan, Charles (1868–1941) BCM   Meyl.
Meyrán, Jorge (1918–) S   J.Meyrán
Meyran, Octave (1858–1944) S   Meyran
Meyrat, A. (fl. 1984) S   Meyrat
Mez, Carl Christian (1866–1944) S   Mez
Mezzatesta   S   Mezzat.
Mhaiskar, V.G. (fl. 1968) M   Mhaiskar
Mhaskar, D.N. (fl. 1974) M   Mhaskar
Mhatre, J.R. (fl. 1945) M   Mhatre
Mhoro, Boniface (fl. 1980) S   Mhoro
Miall, Louis Compton (1842–1921) F   Miall
Miao, Bo Mao (fl. 1985) S   B.M.Miao
Miao, Ru Huai (1943–) S   R.H.Miao
(Miau, Ru Huai)
   see Miao, Ru Huai   R.H.Miao
Micales, Jessica A. (fl. 1987) M   Micales
Miceli, P. (1947–) S   Miceli
Micevski, Kiril (1926–) S   Micevski
(Micevsky, Kiril)
   see Micevski, Kiril   Micevski
Michael, Edmund (1849–1920) M   Michael

Michael, Paul Oskar (1861–) S   **P.Michael**
Michael, Peter William (fl. 1978) S   **P.W.Michael**
Michaelides, J. (fl. 1979) M   **Michaelides**
(Michailovski, S.I.)
   see Michajlovski, S.I.   **Michajl.**
Michajlov, V.I. A   **V.I.Michajlov**
Michajlova, M.A. (fl 1950) S   **Michajlova**
Michajlovski, L.V. (fl. 1974) M   **L.V.Michajl.**
Michajlovski, S.I. S   **Michajl.**
Michajłow, Włodzimierz A   **Michajłow**
Michalet, (Louis) Eugène (1829–1862) S   **Michalet**
Michaleva, V.M. (1929–) S   **Michaleva**
Michard, André A   **Michard**
Michaux, André (1746–1803) BMPS   **Michx.**
Michaux, François André (1770–1855) S   **F.Michx.**
Micheels, Henri (1862–1922) S   **Micheels**
Michel, Étienne (fl. 1816) S   **Michel**
Michel, Mathieu Joseph (1825–1890) S   **M.Michel**
Michel, Rudolf Wilhelm S   **R.W.Michel**
Micheletti, Luigi (1844–1912) BMS   **Micheletti**
Micheli, Marc (1844–1902) S   **Micheli**
Micheli, Pier (Pietro) Antonio (1679–1737) BLMS   **P.Micheli**
Michelin, Jean–Louis Hardouin (1786–1867) AF   **Michelin**
Michelis, Friedrich Bernard Ferdinand (1818–1886) S   **Michelis**
Michell, Margaret R. A   **Michell**
Michelozzi Clavarino, Guidetta Roti
   see Roti Michelozzi, Guidetta   **Roti Mich.**
Michelson, A.J. S   **Michelson**
Michener, Ezra (1794–1887) S   **E.Michener**
Michener, Josephine Rigden A   **J.R.Michener**
Michimi, R. (fl. 1977) M   **Michimi**
Michon (fl. 1854) S   **Michon**
Michot, Norbert Louis (1803–1887) S   **Michot**
Michotte, D. (1907–) S   **Michotte**
Miciurin, I.V. (1855–1935) S   **Miciurin**
Mickel, John Thomas (1934–) PS   **Mickel**
Mickovski, M. (fl. 1962) M   **Mick.**
Middelhoek, A. (fl. 1952) M   **Middelh.**
Middelhoven, W.J. (fl. 1984) M   **Middelhoven**
(Middendorf, Aleksandr Fedorovic)
   see Middendorff, Alexander Theodorowitsch   **Middend.**
Middendorff, Alexander Theodorowitsch (1815–1894) BS   **Middend.**
Middleborg, J. (fl. 1987) M   **Middleb.**
Middleton, A.T. (fl. 1984) S   **A.T.Middleton**
Middleton, David John (1963–) S   **D.J.Middleton**
Middleton, John Tylor (1912–) M   **Middleton**
Mieckley, W. S   **Mieckley**
Mieg, Achilles (1731–1799) S   **Mieg**
Miège, Émile (fl. 1910) S   **É.Miège**

Miège, Jacques (1914–) S	**J.Miège**
Miégeville, Joseph (1819–1901) S	**Miégev.**
Miehe, Hugo (1875–1932) AM	**Miehe**
Mielcarek, R. (fl. 1982) S	**Mielcarek**
Mielichhofer, Mathias (1772–1847) BS	**Miel.**
(Mier y Terán, Manuel de)	
see Terán, Manuel de Mier y	**Terán**
Miers, John (1789–1879) PS	**Miers**
Migliardi, V. (fl. 1911) M	**Migliardi**
Migliorato-Garavini, Erminio S	**Migl.-Gar.**
Migliozzi, Vincenzo (fl. 1985) M	**Migl.**
Mignot, Jean-Pierre A	**Mignot**
Migo, Hisao (1900–) S	**Migo**
Migout, Abel (1830–) S	**Migout**
Miguel, J.R. (fl. 1987) S	**Miguel**
Migula, Emil Friedrich August Walther (1863–1938) M	**Mig.**
Migushova, E.F. (1923–) S	**Migush.**
Mik, Josef (1839–1900) S	**Mik**
Mikami, Hideo (1918–) A	**Mikami**
Mikan, Johann Christian (1769–1844) S	**J.C.Mikan**
Mikan, Josef Gottfried (1743–1814) S	**J.G.Mikan**
Mikawa, Takashi (fl. 1975) M	**Mikawa**
Mikelsson, R.M. A	**Mikelsson**
Mikeschin, Georgij Vladimirovič (1911–1965) S	**Mikeschin**
(Mikhailova, M.A.)	
see Michajlova, M.A.	**Michajlova**
(Mikhailovski, S.I.)	
see Michajlovski, S.I.	**Michajl.**
(Mikhailovskii, L.V.)	
see Michajlovski, L.V.	**L.V.Michajl.**
Mikheev, A.D. (1933–) S	**Mikheev**
Mikheladze, I.A. PS	**Mikhel.**
Miki, Shigeru (1901–1974) FS	**Miki**
Miklouho-Maclay, Nikolaj Nikolajewitsch (1846–1888) S	**Mikl.-Maclay**
Mikutowicz, Johann Mattias (Johannes Mathias) (1872–1951) B	**Mikut.**
Milanez, Adauto Ivo (1937–) M	**A.I.Milanez**
Milanez, Fernando Romano (1905–) S	**Milanez**
Milani, Giovanni Battista (1858–) S	**Milani**
Milano, V.A. (1921–) S	**Milano**
Milanović, M. A	**Milanović**
Milbraith, David Gallus (1880–) M	**Milbr.**
Milbrath, J.A. (fl. 1942) M	**Milbrath**
Mildbraed, Gottfried Wilhelm Johannes (1879–1954) BPS	**Mildbr.**
Milde, Carl August Julius (1824–1871) BMPS	**Milde**
Milde, Julius (1824–1902) S	**J.Milde**
Mildner, R. (fl. 1971) S	**Mildner**
Miles, Beverley Alan (1937–1970) S	**B.A.Miles**
Miles, Lee Ellis (1890–1941) M	**Miles**
Milesi, Marco (fl. 1904) M	**Milesi**

Miliarakis, Spyridon (–1919) S	**Miliar.**
Militzer, Max (1894–1971) S	**Militzer**
Milko, A.A. (fl. 1964) M	**Milko**
Milkovits, I. (fl. 1987) S	**Milk.**
Milkuhn, F.G. (fl. 1949) S	**Milkuhn**
Mill, Robert Reid (1950–) S	**R.R.Mill**
Mill, S.W. (fl. 1987) S	**S.W.Mill**
Mill, Walter (1820–1904) S	**W.Mill**
Millais, John Guille (1865–1931) S	**Millais**
Millán, Aníbal Roberto (1892–) S	**Millán**
Millan, José Henrique (fl. 1967–1977) F	**J.H.Millan**
Millar, Alan J.K. (1957–) A	**A.Millar**
Millar, C.S. (fl. 1974) M	**Millar**
Millar, K.R. (fl. 1990) M	**K.R.Millar**
Millard, Wilfrid Arthur (1880–) M	**Millard**
Millardet, Pierre Marie Alexis (1838–1902) CM	**Millardet**
Mille, Luis (Louis, Aloysius) (1873–1953) PS	**Mille**
Millepied, P. A	**Millepied**
Miller, A.G. (1951–) S	**A.G.Mill.**
Miller, C.E. (fl. 1955) M	**C.E.Mill.**
Miller, Charles (1739–1817) S	**C.Mill.**
Miller, Charles N. (1938–) F	**C.N.Mill.**
Miller, D.D. (fl. 1986) M	**D.D.Mill.**
Miller, Elihu Sanford (1848–1940) P	**E.S.Mill.**
Miller, Gerrit Smith (1869–1956) S	**G.S.Mill.**
Miller, Gertrude Nevada (1919–) S	**G.N.Mill.**
Miller, Harvey Alfred (1928–) BP	**H.A.Mill.**
Miller, J.D. (fl. 1985) M	**J.D.Mill.**
Miller, J.N. (fl. 1934) M	**J.N.Mill.**
Miller, John (1849–1915) S	**J.Mill.**
Miller, John Frederick (1715–1794) S	**J.F.Mill.**
Miller, John M. (–1796) S	**J.M.Mill.**
Miller, John S. (before 1744, Mueller, J.S.) (1715–1780) S	**J.S.Mill.**
Miller, Julian Howell (1890–1961) M	**J.H.Mill.**
Miller, Kathy Ann (1953–) A	**K.A.Mill.**
Miller, Kim Irving (1936–) S	**K.I.Mill.**
Miller, Lee Wallace (1904–1970) M	**L.W.Mill.**
Miller, Lillian Wood (1937–) S	**Lill.W.Mill.**
Miller, M.A. (fl. 1976) M	**M.A.Mill.**
Miller, M.W. (fl. 1960) M	**M.W.Mill.**
Miller, Norton George (1942–) B	**N.G.Mill.**
Miller, Oliphant Bell (1882–1966) S	**O.B.Mill.**
Miller, Orson K. (1930–) M	**O.K.Mill.**
Miller, Paul William (1901–) MS	**P.W.Mill.**
Miller, Peter L. A	**P.L.Mill.**
Miller, Philip (1691–1771) S	**Mill.**
Miller, Robert Barclay (1875–) S	**R.B.Mill.**
Miller, Samuel A. (1837–1897) S	**S.A.Mill.**

Miller, Vera A.Mentzer (fl. 1941) M	**V.A.M.Mill.**
Miller, Viktor V. (W.W.) (fl. 1934) AM	**V.V.Mill.**
Miller, Wilhelm (William Tyler) (1869–1938) S	**W.T.Mill.**
Miller, William (c.1831–1898) S	**W.Mill.**
Milliken, Jessie (1877–1951) S	**Milliken**
Millioud, Marcel E. A	**Millioud**
Millner, Patricia D. (fl. 1975) M	**Millner**
Milloy, M.A. (fl. 1990) F	**Milloy**
Mills, Alan K. (fl. 1989) M	**A.K.Mills**
Mills, Frederick William (1868–1949) A	**Mills**
Mills, Jimmy T. A	**J.T.Mills**
Mills, John Norton (1914–1977) S	**J.N.Mills**
Mills, William Hobson (1873–1959) S	**W.H.Mills**
Millspaugh, Charles Frederick (1854–1923) BMPS	**Millsp.**
Milne, Colin (1743–1815) S	**Milne**
Milne-Edwards, Alphonse (1835–1900) S	**Milne-Edw.**
Milne-Redhead, Edgar Wolston Bertram Handsley (1906–) S	**Milne-Redh.**
Milner, R.J. (fl. 1986) M	**Milner**
Milochevich, S. (fl. 1931) M	**Miloch.**
Milovanova, I.V. A	**Milovan.**
Milovtzova, M.A. (fl. 1937) M	**Milovtz.**
Milstead, Wayne Lavine (1932–) S	**Milstead**
Miltitz, Fiiedrich Joseph Franz Xaver von (–1840) S	**Miltitz**
Mimeur, Geneviève S	**Mimeur**
Mimura, Akio (fl. 1978) M	**Mimura**
Min, Cheng Lin (fl. 1986) S	**C.L.Min**
Min–Thein, U. A	**Min-Thein**
Minà–Palumbo, Francesco (1814–1899) S	**Minà–Pal.**
Minakata, K. (fl. 1925) M	**Minakata**
Minami, K. (fl. 1978) M	**Minami**
(Minatullaev, N.A.)	
see Minatullayev, N.A.	**Minat.**
Minatullayev, N.A. (fl. 1965) S	**Minat.**
Minchin, A. (fl. 1905) M	**Minchin**
Minden, Max D.von (1871–) M	**Minden**
Minderhoud, M.E. (fl. 1986) S	**Minderh.**
Minderova, E.V. (fl. 1957) S	**Minderova**
Miner, Ernest Lavon (1900–1972) BF	**E.L.Miner**
Miner, Harriet Stewart (fl. 1885) S	**Miner**
Mineta, M. (fl. 1984) M	**Mineta**
Ming, Tien Lu (1937–) S	**T.L.Ming**
Ming, Yow Nung (fl. 1964) M	**Y.N.Ming**
Mingbaeva, Sh.N. (fl. 1982) M	**Mingb.**
Mingrone (fl. 1969) S	**Mingrone**
Miniaev, Nikolai Aleksandrovich (1909–) PS	**Miniaev**
(Miniati, Daniela)	
see Miniati–Radin, Daniela	**Miniati-Radin**
Miniati–Radin, Daniela A	**Miniati-Radin**

(Miniayev, Nikolai Aleksandrovich)
   see Miniaev, Nikolai Aleksandrovich   **Miniaev**
Minio, Michelangelo (1872–1960) C   **Minio**
Minks, Arthur (1846–1908) M   **Minks**
Minkwitz, Zenaida Alexandrovna (1878–1918/19) S   **Minkw.**
Minne, Achille J. (fl. 1907) M   **Minne**
Minod, Marcel Maurice (1887–1939) S   **Minod**
Minoura, Kyûbei (fl. 1954) M   **Minoura**
Minoura, N. A   **N.Minoura**
Minter, David William (1950–) M   **Minter**
Minuart, Juan (1693–1768) S   **Minuart**
Minutillo, Francesco (fl. 1985) S   **Minut.**
(Minyaev, Nikolai Aleksandrovich)
   see Miniaev, Nikolai Aleksandrovich   **Miniaev**
Miou (fl. 1980's) P   **Miou**
Miquel, Friedrich Anton Wilhelm (1811–1871) ABMPS   **Miq.**
Mirakhur, R.K. (fl. 1977) M   **Mirakhur**
Miranda, Faustino (1905–1964) AS   **Miranda**
Miranda, Francisco E.L.de (fl. 1941) S   **F.E.L.Miranda**
(Miranda González, Faustino)
   see Miranda, Faustino   **Miranda**
(Miranda, Lennart Rodrigues de)
   see Rodrigues de Miranda, Lennart   **Rodr.Mir.**
Miranda, Mary (fl. 1976) M   **M.Miranda**
Miranda, Veja Velloso de S   **V.V.Miranda**
Mirande, Marcel (1864–1930) M   **Mirande**
Mirbel, Charles François Brisseau de (1776–1854) BPS   **Mirb.**
Miré, Philippe (fl. 1956) S   **Miré**
Mirek, Zbigniew (1951–) S   **Mirek**
Mironov, B.A. (1907–1944) S   **Mironov**
Mironova, N.V. A   **Mironova**
Mirov, Nicholas Tihomitovich (1893–1980) S   **Mirov**
Mirza, F. (fl. 1962) M   **Mirza**
Mirza, J.H. (fl. 1979) M   **J.H.Mirza**
Mirzoeva, Nina Vasilevna (1908–) S   **Mirzoeva**
(Mirzojeva, Nina Vasilevna)
   see Mirzoeva, Nina Vasilevna   **Mirzoeva**
Misas, Guillermo (fl. 1978) S   **Misas**
(Misas Urreta, Guillermo)
   see Misas, Guillermo   **Misas**
(Mischenko, Pavel Ivanovich)
   see Misczenko, Pavel Ivanovich   **Miscz.**
Mischkin, B.A. (1914–1950) S   **Mischkin**
Mischler, T.C. (fl. 1983) S   **Mischler**
Misczenko, Pavel Ivanovich (1869–1938) S   **Miscz.**
Mishra, J.N. (fl. 1953) M   **Mishra**
Mishra, R.P. (fl. 1979) M   **R.P.Mishra**
Mishra, R.R. (fl. 1962) M   **R.R.Mishra**
Mishra, S.C. (fl. 1984) S   **S.C.Mishra**

Mišík, Milan (fl. 1983) AM	**Mišík**
Misirdali, Hüseyin (1946–) S	**H.Misirdali**
Miski, M. (fl. 1985) S	**Miski**
Misonne, Th. (fl. 1893) S	**Misonne**
Misra, A.P. (fl. 1976) M	**A.P.Misra**
Misra, B.M. (fl. 1987) M	**B.M.Misra**
Misra, J.K. (fl. 1981) M	**J.K.Misra**
Misra, Jogendra Nath A	**J.N.Misra**
Misra, K.K. A	**K.K.Misra**
Misra, Kailash Chandra (1915–) B	**Misra**
Misra, L. S	**L.Misra**
Misra, P.C. (fl. 1975) M	**P.C.Misra**
Misra, P.L. A	**P.L.Misra**
Misra, R.C. A	**R.C.Misra**
Misra, S. (fl. 1981) S	**S.Misra**
Missbach, Ernst Robert (1864–1938) S	**Missbach**
Mitchel, D.H. (1917–) M	**Mitchel**
Mitchell, Andrew Stewart (1952–) S	**A.S.Mitch.**
Mitchell, Anthony R. (1938–) S	**A.R.Mitch.**
Mitchell, D.J. (fl. 1987) M	**D.J.Mitch.**
Mitchell, D.W. (fl. 1976) M	**D.W.Mitch.**
Mitchell, David Searle (1935–) PS	**D.S.Mitch.**
Mitchell, J.E. (fl. 1973) M	**J.E.Mitch.**
Mitchell, John (1711–1768) S	**Mitch.**
Mitchell, John (1762–) S	**J.Mitch.**
Mitchell, Margaret O. A	**M.O.Mitch.**
Mitchell, Michael Edward (1934–) M	**M.E.Mitch.**
Mitchell, P.J. (fl. 1983) S	**P.J.Mitch.**
Mitchell, Robert J. (1936–) S	**R.J.Mitch.**
Mitchell, T.L. (fl. 1927) S	**T.L.Mitch.**
Mitchell, Thomas Livingstone (1792–1855) S	**T.Mitch.**
Mitchell, William Warren (1923–) S	**W.W.Mitch.**
(Mitchell (of Stanstead), John)	
see Mitchell, John	**J.Mitch.**
(Mitchourine, I.V.)	
see Miciurin, I.V.	**Miciurin**
Mitford, Algernon Bertram Freeman (1837–1916) S	**Mitford**
Mítítíuc, Mihai Şt. (1937–) M	**Mítítíuc**
Mitra, A.K. A	**A.K.Mitra**
Mitra, A.N. A	**A.N.Mitra**
Mitra, Asim (fl. 1984) M	**A.Mitra**
Mitra, Bimal (1947–) S	**B.Mitra**
Mitra, Debabrata (1938–) S	**D.Mitra**
Mitra, J.N. (fl. 1965) S	**J.N.Mitra**
Mitra, Manoranjan (1895–1942) M	**Mitra**
Mitra, Tapati A	**T.Mitra**
Mitrofanov, P. A	**Mitrofanov**
Mitrofanova, O.V. (fl. 1969) M	**Mitrof.**
Mitroiu, Natalia A	**Mitroiu**

(Mitrophanova, O.V.)	
see Mitrofanova, O.V.	**Mitrof.**
(Mitrophanow, P.)	
see Mitrofanov, P.	**Mitrofanov**
Mitroshina, E.I. (fl. 1949) M	**Mitrosh.**
Mitscherlich, Gustav Alfred (1832–) S	**Mitscherl.**
(Mitsevski, Kiril)	
see Micevski, Kiril	**Micevski**
Mitsuta, Shigeyuki (fl. 1979–1986) P	**Mitsuta**
(Mittelbacher von Mitterburg, Ludwig)	
see Mitterpacher von Mitterburg, Ludwig	**Mitterp.**
Mitten, William (1819–1906) BM	**Mitt.**
Mitter, Julian Herron (1881–) M	**Mitter**
(Mitterberg, Ludwig Mitterpacher von)	
see Mitterpacher von Mitterburg, Ludwig	**Mitterp.**
Mitterpacher von Mitterburg, Ludwig (1734–1818) S	**Mitterp.**
Mitui, Kunio (1940–1988) P	**Mitui**
Miura, Akio (1928–) A	**A.Miura**
Miura, Koichiro (fl. 1971) M	**K.Miura**
Miura, Michiya (fl. 1914) M	**Miura**
Miwa, H. (fl. 1930) M	**Miwa**
Mix, Arthur Jackson (1888–1956) M	**Mix**
Mix, Marianne A	**M.Mix**
Miyabe, Kingo (1860–1951) AMPS	**Miyabe**
Miyabe, Tsutome (1880–1921) MP	**T.Miyabe**
Miyaji, Makoto (fl. 1976) M	**Miyaji**
Miyake, Cbuichi (fl. 1926) M	**C.Miyake**
Miyake, Ichiro (fl. 1910) M	**I.Miyake**
Miyake, Kiichi (1876–) BP	**Miyake**
Miyake, Tsutomu (1880–) MS	**T.Miyake**
Miyamoto, Futoshi (fl. 1983–1988) P	**Miyam.**
Miyawaki, K. (fl. 1940) M	**Miyaw.**
Miyazawa, Yoichi (fl. 1969) M	**Miyaz.**
Miyoshi, Manabu (1861–1939) AMS	**Miyoshi**
Mizgireva, O.F. (1908–) S	**Mizg.**
Mizuno, Makoto (1950–) AS	**M.Mizuno**
Mizushima, Masami (1925–1972) S	**M.Mizush.**
Mizushima, Urara (1927–) B	**Mizush.**
Mizutani, Masami (1930–) B	**Mizut.**
Mlady, Franciscus (fl. 1838) M	**Mlady**
(Mlokosevich, J.L.)	
see Mlokossewicz, J.L.	**Mlok.**
Mlokossewicz, J.L. S	**Mlok.**
Mo, Sin Li (fl. 1980–1987) PS	**S.L.Mo**
(Mo, X.L.)	
see Mo, Sin Li	**S.L.Mo**
Mo, Ze Qian (fl. 1989) S	**Z.Q.Mo**
(Mobaven, Sadegh)	
see Mobayen, Sadegh	**Mobayen**

Mobayen, Sadegh (1919–) S	**Mobayen**
Mobberley, David George (1921–) S	**Mobb.**
Moberg, J.Roland (1939–) M	**Moberg**
Möbius, Karl (1825–1908) A	**K.Möbius**
Mobley, R.L. (fl. 1948) M	**Mobley**
Mochizuka, Hideyasu (fl. 1979) M	**Mochiz.**
Mochizuki, Rikuo (fl. 1969) S	**Mochizuki**
Moçiño, José Mariano (1757–1820) PS	**Moç.**
Moczydłowska, Małgorzata A	**Mocz.**
Modak, C.D. (fl. 1973) M	**Modak**
Moe, Asche (1867–1937) S	**A.Moe**
Moe, Nils Green (1812–1892) S	**Moe**
Moe, Richard Lee (1946–) A	**R.L.Moe**
Moebius, Martin August Johannes (1859–1946) S	**Moebius**
(Moehring, Paul Heinrich Gerhard)	
see Möhring, Paul Heinrich Gerhard	**Möhring**
Moeliono, B. (fl. 1960) S	**Moeliono**
Moeller, Arturo F. (fl. 1922–1930) S	**A.F.Moeller**
Moeller, Henry (fl. 1927) S	**H.Moeller**
Moench, Conrad (1744–1805) BPS	**Moench**
Moënne-Loccoz, Pierre (fl. 1989) M	**Moënne-Locc.**
(Moens, Johann Bernalot)	
see Bernelot Moens, Jacob Carel	**Bern.Moens**
Moerch, Otto Josias Nicolai (1799–1842) S	**Moerch**
(Moeschl, Wilhelm)	
see Möschl, Wilhelm	**Möschl**
Moeser, Walter S	**Moeser**
(Moessler, Johann Christoph)	
see Mössler, Johann Christoph	**Mössler**
Moestrup, Øjvind A	**Moestrup**
Moesz, Gusztáv (Gustáv) von (1873–1946) MS	**Moesz**
Moewus, Franz (1856–1937) A	**F.Moewus**
Moewus, Liselotte A	**L.Moewus**
Moezel, P.G. van der (fl. 1987) S	**Moezel**
Moffatt, Will (William) Sayer (fl. 1909) M	**Moffatt**
Moffett, Rodney Oliver (1937–) S	**Moffett**
Moffler, M.D. (fl. 1978) S	**Moffler**
Mogarkar, C.D. (fl. 1973) M	**C.D.Mogarkar**
Mogarkar, K.M. (fl. 1979) M	**K.M.Mogarkar**
Mogea, J.P. (1947–) S	**Mogea**
Mogensen, Gert Steen (1944–) B	**Mogensen**
Moggi, Guido (1927–) S	**Moggi**
Moggridge, Johann Traherne (1842–1874) S	**Moggr.**
Mogi, M. (fl. 1939) M	**Mogi**
Mohamed, M.A.Haji (1948–) B	**Mohamed**
Mohammed, M.S. (fl. 1987) M	**Mohammed**
Mohanan, C.N. (fl. 1985) MS	**C.N.Mohanan**
Mohanan, M. (fl. 1985) S	**M.Mohanan**
Mohanty, N.N. (fl. 1958) M	**Mohanty**

(Mohinder Nath, Dewan Nair)
  see Nath Nair, Dewan Mohinder   **Nath Nair**
Mohl, Hugo von (1805–1872) BS   **Mohl**
Mohlenbrock, Robert H. (1931–) APS   **Mohlenbr.**
Mohler, Hanspeter A   **H.Mohler**
Mohler, J.C. (fl. 1928) S   **Mohler**
Mohlo, D. (fl. 1981) M   **Mohlo**
Mohnike, Otto Gottlieb Johan (1813–1887) S   **Mohn.**
Mohr, Charles Theodore (Karl Theodor) (1824–1901) BMS   **C.Mohr**
Mohr, Daniel Matthias Heinrich (1780–1808) ABMPS   **D.Mohr**
Mohr, Hartmut (fl. 1984) S   **H.Mohr**
Mohr, Nicolai (i Pedersen) (1742–1790) S   **Mohr**
Möhring, Paul Heinrich Gerhard (1710–1792) S   **Möhring**
Moir, William Whitmore Goodale (1896–1985) S   **Moir**
Moisan, Charles Auguste (fl. 1839) S   **Moisan**
Moissejeva, A.I. A   **Moiss.**
Moiz, A.A. A   **Moiz**
Mokry, Franz (fl. 1986) P   **Mokry**
Molau, Ulf (1949–) S   **Molau**
(Moldenhauer, Johann Jacob Paul)
  see Moldenhawer, Johann Jacob Paul   **Moldenh.**
Moldenhawer, Johann Jacob Paul (1766–1827) S   **Moldenh.**
Moldenke, Charles Edward (1860–1935) S   **C.Moldenke**
Moldenke, Harold Norman (1909–) S   **Moldenke**
Mölder, Karl (1899–1975) A   **Mölder**
Moldovan, Ioan (1934–) M   **Moldovan**
Molendo, Ludwig (1833–1902) B   **Molendo**
(Molero Briones, Julián)
  see Molero, Julián   **Molero**
Molero, Julián (1946–) S   **Molero**
Molero Mesa, Joaquín (1952–) S   **Molero Mesa**
(Molero–Briones, Julián)
  see Molero, Julián   **Molero**
Molesworth Allen, Betty Eleanor Gosset
  see Allen, B.E.G. Molesworth   **B.M.Allen**
Molfino, José Furtado (–1964) S   **Molfino**
Molina, Ana María (1947–) S   **A.M.Molina**
Molina, Antonio R. (1926–) S   **A.R.Molina**
(Molina de Riera, Ana María)
  see Molina, Ana María   **A.M.Molina**
Molina, Giovanni Ignazio (Juan Ignacio) (1737–1829) PS   **Molina**
Molina, José Andrés (1956–1991) S   **A.Molina**
(Molina Maruenda, José Andrés)
  see Molina, José Andrés   **A.Molina**
Molina–Valero, L.A. (fl. 1980) M   **Mol.-Val.**
Molinet, Amorós Eugenio S   **Molinet**
Molinier, René (1899–1975) S   **Molin.**
Molkenboer, Julian(us) Hendrik (1816–1854) BMS   **Molk.**
Moll, Jan Willem (1851–1933) S   **J.Moll**

Moll, Karl Maria Ehrenbert von (1760–1838) S	**K.Moll**
Møller, Birgitte Christina (1857–1934) S	**B.Møller**
Møller, F.H. (1887–) M	**F.H.Møller**
Møller, Friedrich Alfred Gustav Jobst (1860–1922) MS	**Møller**
Moller, G.N. S	**G.N.Moller**
Möller, Hjalmar August (1866–1941) BS	**H.Möller**
Möller, Johann Dietrich (Diedrich) (1844–1907) S	**J.D.Möller**
Möller, Ludwig (1847–1910) S	**L.Möller**
Möller, Ludwig Heinrich Ferdinand (1820–1877) S	**L.H.F.Möller**
Møller, Max E.K. (1909–) A	**M.Møller**
Moller, Sophie Maren (1840–1920) B	**S.M.Moller**
Möller, Valerian von A	**V.Möller**
Molliard, Marin (1866–1944) M	**Molliard**
Molloy, Brian Peter John (1930–) S	**Molloy**
Molly, M.J. (fl. 1985) P	**Molly**
Molnar, A.C. (fl. 1965) M	**Molnar**
Molon, Francesco (1820–1885) F	**Molon**
Moloney, Cornelius Alfred (1848–1913) S	**Moloney**
Molseed, Elwood Wendell (1938–1967) S	**Molseed**
Molyneux, William Mitchell (1935–) S	**Molyneux**
Momeu, Laura A	**Momeu**
Momiyama, Yasuichi (1904–) S	**Momiy.**
Mommaerts, J.P. A	**Momm.**
Momose, Sizuo (1906–1968) PS	**Momose**
Momotani, Yoshihide (1928–) S	**Momot.**
Monachino, Joseph Vincent (1911–1962) S	**Monach.**
Moncada Ferrera, Milagros (1937–) S	**Moncada**
Monchenko, W.I. A	**Monchenko**
Monckton, Horace Woollaston (1857–1931) S	**Monckton**
Mondal, D.C. (fl. 1982) S	**Mondal**
Monguillon, Eugène Louis Honoré (1865–1940) BM	**Mong.**
Moniéro, Jordano (fl. 1951) F	**Moniéro**
(Moniushko, Vladimire A.)	
see Monjuschko, Vladimire A.	**Monjuschko**
Moniz, L. (fl. 1964) M	**Moniz**
Monjuschko, Vladimire A. (1903–1935) S	**Monjuschko**
(Monjuško, Vladimire A.)	
see Monjuschko, Vladimire A.	**Monjuschko**
Mönkemeyer, Wilhelm (1862–1938) B	**Mönk.**
Monkman, Charles (fl. 1860s) P	**Monkman**
Monnard, Jean Pierre (1791–) S	**Monnard**
Monnet, Paul–Louis (–1915) S	**Monnet**
Monnier, Alfred (1874–1917) S	**Alf.Monnier**
Monnier, Auguste (1800–) S	**Monnier**
Monnier, Denys (1834–1898) S	**D.Monnier**
Monnier, Françoise (fl. 1959) M	**F.Monnier**
Monnier, Paul Camille Jacques (1922–) PS	**P.Monnier**
Monod de Froideville, Charles (1896–) S	**C.Monod**
Monod, M. (fl. 1984) M	**M.Monod**

Monod, Théodore (1902–) S	**Monod**
Monoson, H.L. (fl. 1969) M	**Monoson**
Monoyer, A. (fl. 1937) M	**Monoyer**
Monro, Claude Frederick Hugh (1863–1918) S	**Monro**
Monroe, Charles Edwin (1857–1931) S	**Monroe**
Montagne, Jean Pierre François Camille (1784–1866) AMS	**Mont.**
Montalva, Soledad A	**Montalva**
Montalvo, A.M. (fl. 1986) S	**Montalvo**
Montandon, F.Jules (fl. 1856) S	**Montandon**
Montbret, Gustave Coquebert de (1805–1837) S	**Montbret**
Monte, Michela (fl. 1967) M	**Monte**
(Montealegre A., Jaime R.)	
see Montealegre, Jaime R.	**Monteal.**
Montealegre, Jaime R. (fl. 1987) M	**Monteal.**
Montecchi, Amer (fl. 1987) M	**Montecchi**
Montecucchi, P. (fl. 1973) M	**Montec.**
Montegut, J. (fl. 1948) M	**Montegut**
(Monteiro Filho, Honorio da Costa)	
see Monteiro, Honorio da Costa	**Monteiro**
Monteiro, Honória da Costa (1923–) S	**H.C.Monteiro**
Monteiro, Honorio da Costa (1900–1978) S	**Monteiro**
(Monteiro Neto, Honória da Costa)	
see Monteiro, Honória da Costa	**H.C.Monteiro**
Montel, R. A	**Montel**
Montele, L. S	**Montele**
Montell, Justus Elias (1869–1954) S	**Montell**
Montelucci, Giuliano (1889–1983) S	**Montel.**
(Montemartini, Aurora)	
see Corte, Aurora Montemartini	**A.M.Corte**
(Montemartini Corte, Aurora)	
see Corte, Aurora Montemartini	**A.M.Corte**
Montemartini, Luigi (1869–1952) M	**Montemart.**
Montemayor, Lorenzo de (fl. 1949) M	**Montem.**
Montgomery, Frederick Howard (1902–1978) S	**Montgom.**
Montgomery, Margaret A	**M.Montgom.**
Monthoux, O. (1932–) M	**Monthoux**
Monti, Gaetano Lorenzo (1712–1797) S	**Monti**
Montiel, O.M. (fl. 1988) S	**Montiel**
Montin, Lars Jonasson (1723–1785) BS	**Montin**
Montlaur, H. (fl. 1920) M	**Montl.**
Montoya–Bello, Leticia (1963–) M	**Montoya**
Montpellier, J. (fl. 1929) M	**Montpell.**
Montresor, Marina A	**Montresor**
Montrocher, R. (fl. 1967) M	**Montrocher**
Montrouzier, Xavier (1820–1897) S	**Montrouz.**
Montserrat Marti, Gabriel María (1956–) S	**G.Monts.**
Montserrat Marti, Josep Maria (1955–) S	**J.M.Monts.**
Montserrat Recoder, Pedro (1918–) PS	**P.Monts.**
Monville, M.Chevalier (de) (fl. 1838) S	**Monv.**

Moo, Chuen Tau S	**C.T.Moo**
Moody, Sophy (fl. 1864) S	**Moody**
Moon, Alexander (–1825) S	**Moon**
Moor (fl. 1927) M	**Moor**
(Moor, V.P.G.De)	
see De Moor, V.P.G.	**De Moor**
Moore (fl. 1927) M	**Moore bis**
Moore, Albert Hanford (1883–) PS	**A.H.Moore**
Moore, Barbara Jo (1938–) M	**B.J.Moore**
Moore, Charles (1820–1905) S	**C.Moore**
Moore, Daniel David Tompkins (1820–1892) S	**D.D.T.Moore**
Moore, David (1808–1879) BPS	**Moore**
Moore, David Moresby (1933–) S	**D.M.Moore**
Moore, Dwight Munson (1891–1985) S	**Dw.Moore**
Moore, Frederick William (1857–1949) S	**F.W.Moore**
Moore, George Thomas (1871–1956) A	**G.Moore**
Moore, Harold Emery (1917–1980) S	**H.E.Moore**
Moore, John William (1901–) PS	**J.W.Moore**
Moore, Justin Payson (1841–1923) M	**J.P.Moore**
Moore, L.R. (fl. 1963) FM	**L.R.Moore**
Moore, Lucy Beatrice (1906–1987) S	**L.B.Moore**
Moore, M. (fl. 1933) M	**M.Moore**
Moore, R.M. S	**R.M.Moore**
Moore, Raymond Cecil (1892–) A	**R.C.Moore**
Moore, Raymond John (1918–) S	**R.J.Moore**
Moore, Royall T. (fl. 1954) M	**R.T.Moore**
Moore, Spencer Le Marchant (1850–1931) S	**S.Moore**
Moore, Thomas (1821–1887) PS	**T.Moore**
Moore, Thomas Bather (1850–1919) S	**T.B.Moore**
Moore, Thomas Verner (1877–1969) S	**T.V.Moore**
Moore, William Dewey (1897–) M	**W.D.Moore**
Moore, Winifred Olivia (1904–) S	**W.O.Moore**
Moorman, Mary A	**Moorman**
Moorthy, S. (fl. 1981) S	**Moorthy**
Moquin-Tandon, Christian Horace Bénédict Alfred (1804–1863) S	**Moq.**
Mora, Alberto M.Brenes (1870–1945) S	**Mora**
(Mora, Luis Eduardo)	
see Mora–Osejo, Luis Eduardo	**L.E.Mora**
Mora, V. (fl. 1984) M	**V.Mora**
Mora, W.K. (fl. 1985) M	**W.K.Mora**
Mora–Osejo, Luis Eduardo (1931–) APS	**L.E.Mora**
Moraes, M. (fl. 1950) M	**Moraes**
(Morais, Antonio Arthur Taborda de)	
see Taborda de Morais, Antonio Arthur	**Tab.Morais**
Morais, J.O.Falcão de (fl. 1957) M	**Morais**
Moraldo, Benito (1938–) S	**Moraldo**
Morales, Concepción (1944–) S	**C.Morales**
Morales, E. A	**E.Morales**
Morales, G. (fl. 1982) S	**G.Morales**
Morales, Gustavo A. A	**G.A.Morales**

(Morales L., G.)	
see Morales, G.	**G.Morales**
Morales, María I. (fl. 1973) M	**M.I.Morales**
Morales, Sebastiàn Alfredo de (1823–1900) S	**Morales**
(Morales Torres, Concepción)	
see Morales, Concepción	**C.Morales**
Morales Valverde, Ramón (fl. 1984) S	**R.Morales**
Moran, Reid Venable (1916–) S	**Moran**
Moran, Robbin C. (fl. 1986) P	**R.C.Moran**
Morandi, Giambattista (Giovanni Battista) (fl. 1744) S	**Morandi**
Morante Serrano, Gregorio (fl. 1984) S	**Morante**
Morariu, Iuliu (1905–1989) MS	**Morariu**
Morat, Philippe (1937–) S	**Morat**
Moravec, Jaroslav (1929–) S	**Moravec**
Moravec, Jiri (fl. 1965) M	**J.Moravec**
Moravec, Zdeněk (fl 1954) M	**Z.Moravec**
Morawetz, Wilfried (1951–) S	**Morawetz**
Morbelli, M.A. (fl. 1973) F	**Morbelli**
Morbey, Sidney Jack A	**Morbey**
Morch, Axel Moller (1797–1876) B	**Morch**
(Morcowicz, Vasil Vasilevicz)	
see Marcowicz, Vasil Vasilevicz	**Marcow.**
Mordak, Elena V. (1934–) S	**Mordak**
Mordant de Launay, Jean Claude Mien (c.1750–1816) S	**Mord.Laun.**
Morden, Clifford W. (fl. 1986) S	**Morden**
Mordue, Janet Elizabeth Mary (1936–) M	**Mordue**
Mordvinkina, A.I. (1894–) S	**Mordv.**
More, Alexander Goodman (1830–1895) S	**More**
More, W.D. (fl. 1964) M	**W.D.More**
Moreau, Claude (fl. 1952) M	**C.Moreau**
Moreau, Fernand (1886–1980) M	**Moreau**
Moreau, Mireille (fl. 1949) M	**M.Moreau**
Moreau, Richard (fl. 1959) M	**R.Moreau**
(Moreau, Román A.Pérez)	
see Pérez–Moreau, Román A.	**Pérez-Mor.**
(Moreau, Román L.Pérez)	
see Peréz–Moreau, Román L.	**R.L.Pérez-Mor.**
Moreau, Valentine (fl. 1952) M	**V.Moreau**
Moreau–Benoît, Arlette A	**Moreau-Ben.**
Morefield, James D. (fl. 1988) S	**Morefield**
Moreira, Alvaro Xavier (1920–) S	**A.X.Moreira**
(Moreira da Costa, Nara Leane)	
see Leane Moreira da Costa, Nara	**Leane**
(Moreira dos Santos, J.U.)	
see Santos, J.U.Moreira dos	**J.U.Santos**
Moreira, Nicolau Joaquim (1824–1894) S	**Moreira**
Morel, Charles S	**C.Morel**
Morel, Francisque (c.1849–1925) S	**Morel**
Morel, Georges M. (1916–1973) S	**G.M.Morel**
Morel, J.F.Nicolas (fl. 1805) S	**J.F.N.Morel**

Morel, Jean-Marie (1728–1810) S	**J.M.Morel**
Morel, L.F. (fl. 1865) M	**L.Morel**
Morelet, Michel (fl. 1965) M	**M.Morelet**
Morelet, Pierre Marie Arthur (1809–1892) S	**Morelet**
Morellet, Jean (1882–1945) A	**J.Morellet**
Morellet, Lucien (1882–1945) A	**L.Morellet**
Moreno, Gabriel (fl. 1977) MS	**G.Moreno**
Moreno Sanz, Margarita (1948–) S	**Moreno**
Morenz, J. (fl. 1964) M	**Morenz**
Moret, Jacques (fl. 1987) S	**Moret**
Moretti, Aldo (1948–) S	**A.Moretti**
Moretti, Giuseppe L. (1782–1853) MS	**Moretti**
Morgan, Andrew Price (1836–1907) BM	**Morgan**
Morgan, David R. (fl. 1990) S	**D.R.Morgan**
Morgan, Jeanne (fl. 1952) F	**J.Morgan**
Morgan, Randall S	**Rand.Morgan**
Morgan, Robert (1863–1900) S	**R.Morgan**
Morgan-Jones, Gareth (1940–) M	**Morgan-Jones**
Morgenroth, Eduard (1861–) F	**Morgenr.**
Morgenroth, Peter A	**P.Morgenr.**
Mori, Antonio (1847–1902) S	**Mori**
Mori, H. (fl. 1968) M	**H.Mori**
Mori, Kunihiko (1905–) S	**K.Mori**
Mori, Michiyasu (1909–) A	**M.Mori**
Mori, Scott A. (1941–) S	**S.A.Mori**
Mori, Tamezô (1884–1962) PS	**T.Mori**
Moriarty, Henrietta Maria (fl. 1806–1843?) S	**Moriarty**
Moricand, Moïse Étienne (Stefano) (1779–1854) S	**Moric.**
Morière, Pierre Gilles (Jules) (1817–1888) S	**Morière**
Morikawa, Kinichi (1898–1936) S	**Morik.**
Morikawa, T. (fl. 1939) M	**T.Morik.**
Morillo, Gilberto N. (1944–) S	**Morillo**
(Morim de Lima, Marli)	
see Lima, Marli Pires Morim de	**M.P.Lima**
Morimoto, Y. (fl. 1953) M	**Morim.**
Morin, François Malat Marie (1856–1900?) B	**F.Morin**
Morin, Nancy Ruth (1948–) S	**Morin**
Morinaga, Tsutomu (fl. 1977) M	**Morinaga**
Morini, Fausto (1858–1927) M	**Morini**
Morioka, Hideo (later Kasaki, H.) (1917–) A	**Morioka**
Moriondo, Francesco (fl. 1961) M	**Moriondo**
Moris, Giuseppe Giacinto (Joseph Hyacinthe) (1796–1869) ABS	**Moris**
Morison, Mary O. A	**M.O.Morison**
Morison, Robert (1620–1683) L	**Morison**
(Moritz, Alexandre)	
see Moritzi, Alexandre	**Moritzi**
Moritz, Johann Wilhelm Karl (1797–1866) P	**Moritz**
Moritz, Otto (1904–) S	**O.Moritz**
Moritzi, Alexandre (1807–1850) BMPS	**Moritzi**
Moriyon, I. (fl. 1974) M	**Moriyon**

437

Morla Juaristi, Carlos (fl. 1985) S	**Morla**
Morley, Brian D. (1943–) S	**B.D.Morley**
Morley, F.H.W. (fl. 1983) S	**F.H.W.Morley**
Morley, R.J. (fl. 1978) P	**R.J.Morley**
Morley, Thomas (1917–) S	**Morley**
Morling, Greta A	**Morling**
Mornand, J. (fl. 1984) M	**Mornand**
Mörner, Carl Thore (1864–1940) S	**Mörner**
(Morochkovskii, S.F.)	
see Morochkovsky, S.F.	**Morochk.**
Morochkovsky, S.F. (fl. 1933) M	**Morochk.**
(Moroczkovskii, S.F.)	
see Morochkovsky, S.F.	**Morochk.**
Moroff, Theodor A	**Moroff**
Morong, Thomas (1827–1894) S	**Morong**
Morot, Louis René Marie François (1854–1915) S	**Morot**
(Morotschkowsky, S.F.)	
see Morochkovsky, S.F.	**Morochk.**
Morozova, A.A. S	**Morozova**
Morozova–Vodyanitskaya, N. A	**Moroz.-Vod.**
Morquer, R. (fl. 1933) M	**Morquer**
Morrall, R.A.A. (fl. 1968) M	**Morrall**
Morren, Auguste (1804–1870) A	**Morren**
Morren, Charles François Antoine (1807–1858) S	**C.Morren**
Morren, Charles Jacques Édouard (1833–1886) AS	**E.Morren**
Morrill, Joy A	**J.Morrill**
Morrill, L.C. A	**L.C.Morrill**
Morris, Brian (1936–) S	**B.Morris**
Morris, Daniel (1844–1933) S	**D.Morris**
Morris, Dennis Ivor (1924–) S	**D.I.Morris**
Morris, Edward Lyman (1870–1913) S	**E.Morris**
Morris, Everett F. (fl. 1956) M	**E.F.Morris**
Morris, F.John A. (1869–) S	**F.J.A.Morris**
Morris, George Edward (1853–1916) M	**G.Morris**
Morris, J.E. (fl. 1960) F	**J.E.Morris**
Morris, John (1810–1886) F	**Morris**
Morris, Michael I. (fl. 1970) S	**M.I.Morris**
Morris, Muriel (fl. 1943) M	**M.Morris**
Morris, Patrick Francis (1896–1974) S	**P.Morris**
Morris, Richard (fl. 1820–1830) S	**R.Morris**
Morris, Samuel (1896–) A	**S.Morris**
Morrison, A.W. (fl. 1938) B	**A.W.Morrison**
Morrison, Alexander (1849–1913) S	**Morrison**
Morrison, David A. (fl. 1989) S	**D.A.Morrison**
Morrison, John Laurence (1911–) S	**J.L.Morrison**
Morrow, Angela C. A	**Morrow**
Morse, Elizabeth Eaton (1864–1955) MS	**Morse**
Morse, Warner Jackson (1872–1931) P	**W.J.Morse**
Mortensen, Hans (1825–1908) S	**H.Mort.**

Mortensen, Morten Larsen (1881–1911) S	**M.Mort.**
Morthier, Paul (1823–1886) M	**Morthier**
Mortola, W.R. (fl. 1985) S	**Mortola**
Morton, Conrad Vernon (1905–1972) P	**C.V.Morton**
Morton, F.J. (fl. 1963) M	**F.J.Morton**
Morton, Friedrich (1890–1969) P	**F.Morton**
Morton, Gary H. (fl. 1974) S	**G.H.Morton**
Morton, John Kenneth (1928–) S	**J.K.Morton**
Morton, Joseph B. (fl. 1986) M	**J.B.Morton**
Morton, Julius Sterling (1832–1902) S	**Morton**
Moruzi, Constanţa (Constance) (1899–) M	**Moruzi**
Mory, Birgit (1944–) S	**Mory**
Mosca, Anna Maria L. (fl. 1954) M	**Mosca**
Mosca, Luigi S	**L.Mosca**
Möschl, Wilhelm (1906–1980) S	**Möschl**
(Moscoso Puello, Rafael M.)	
see Moscoso, Rafael M.	**Moscoso**
Moscoso, Rafael M. (1874–1951) S	**Moscoso**
Moseley, Edwin Lincoln (1865–1948) S	**E.Moseley**
Moseley, Henry Nottidge (1844–1890) S	**H.Moseley**
Mosén, Carl Wilhelm Hjalmar (1841–1887) BS	**Mosén**
Moser, G. (fl. 1971) S	**G.Moser**
Moser, Heinrich Christoph (fl. 1794) S	**Moser**
Moser, Jean Jacques (1846–1934) S	**J.J.Moser**
Moser, Meinhard M. (1924–) M	**M.M.Moser**
Moser, Walter (1910–) S	**W.Moser**
Moses, A.S. (fl. 1990) M	**A.S.Moses**
Moses, Arthur (fl. 1913) M	**Moses**
Mosher, Edna (1878–1972) S	**Mosher**
Moshkin, V.A. (fl. 1980) S	**Moshkin**
Moskovetz, Simon (fl. 1933) M	**Mosk.**
Moskowitz, Norman (1922–) A	**Moskowitz**
Mosquin, Theodore (1932–) S	**Mosquin**
Moss, Charles Edward (1870–1930) S	**Moss**
Moss, Ezra Henry (1892–1963) S	**E.Moss**
Moss, Margaret (1885–1953) S	**M.Moss**
Moss, Marion Beatrice (1903–) S	**M.B.Moss**
Moss, Stephen T. (fl. 1975) M	**S.T.Moss**
Mossa, Luigi (1938–) S	**Mossa**
Mosseray, Raoul (1908–1940) MS	**Mosseray**
Mössler, Johann Christoph (fl. 1805–1835) S	**Mössler**
Mostert, S.C. (fl. 1987) S	**Mostert**
Mosto, Patricia A	**Mosto**
Mota, J. (fl. 1988) S	**Mota**
Motelay, Léonce (1830–1917) PS	**Motelay**
Motley, James (–1859) S	**Motley**
Mott, Frederick Thompson (1825–1908) S	**Mott**
Motte, Jean (1897–) M	**Motte**
Mottet, Séraphin Joseph (1861–1930) S	**Mottet**
Mottini, Pietro S	**Mottini**

Mottram, R. (fl. 1980) S	**Mottram**
Motyka, József (1900–) M	**Motyka**
(Mou, Chuan Jing)	
see Mu, Chuan Jing	**C.J.Mu**
Moubasher, A.H. (fl. 1973) M	**Moub.**
Mouchacca, Jean (fl. 1971) M	**Mouch.**
Mougeot, (Joseph) Antoine (1815–1889) FS	**A.Moug.**
Mougeot, Jean Baptiste (1776–1858) BMPS	**Moug.**
Mouillefert, Pierre (1845–1903) S	**Mouill.**
Moul, Edwin Theodore (1929–) B	**Moul**
(Moulins, Charles Robert Alexandre des)	
see Des Moulins, Charles Robert Alexandre	**Des Moul.**
Mounce, Irene (Stewart) (1894–) M	**Mounce**
Moura, Armando Reis A	**A.R.Moura**
Moura, Carlos Alberto Ferreira de (1929–) S	**Moura**
Moura, Nilsa Ramos de (fl. 1966) M	**N.R.Moura**
(Mourashkinsky, K.E.)	
see Murashkinsky, K.E.	**Murashk.**
Moureau, J. (fl. 1949) M	**Moureau**
Mouret, Marcellin (1881–1915) S	**Mouret**
Mousnier, Jules (fl. 1873) S	**Mousnier**
(Moustafa, A.F.)	
see Mustafa, A.F.	**Mustafa**
Mouterde, Paul (1892–1972) S	**Mouterde**
Moutinho, J.L.de A. (fl. 1980) S	**Moutinho**
(Moutinho Neto, J.L.de A.)	
see Moutinho, J.L.de A.	**Moutinho**
Mouton, Victor (1875–1901) M	**Mouton**
Mouton-Fontenille de la Clotte, Marie Jacques Phillipe (1769–1837) S	**Mouton-Font.**
Mouzouras, R. (fl. 1988) M	**Mouzouras**
Movssesjan, L.I. (fl. 1967) M	**Movss.**
Moxley, George Loucks (1871–) P	**Moxley**
Moyeen, M. (fl. 1989) S	**Moyeen**
Moyen, J. (fl. 1889) S	**Moyen**
Moyer, Isaac Shoemaker (1838–1898) S	**Moyer**
Moyne, Gilbert (fl. 1980) M	**Moyne**
Mozaffarian, V. (1953–) S	**Mozaff.**
(Moziño Suarez de Figueroa, José Mariano)	
see Moçiño, José Mariano	**Moç.**
Mozzetti, Ferdinando (1786–1850) S	**Mozz.**
Mrak, E.M. (fl. 1952) M	**Mrak**
Mrkos, O. (fl. 1926) M	**Mrkos**
Mrozińska, Teresa (1931–) A	**Mrozińska**
Mshigeni, Keto E. (fl. 1960) A	**Mshigeni**
Mtschvetadze, D.I. (1937–) S	**Mtskhvet.**
(Mtskhvetadze, D.I.)	
see Mtschvetadze, D.I.	**Mtskhvet.**
(Mtzchvetadze, D.I.)	
see Mtschvetadze, D.I.	**Mtskhvet.**
Mu, Chuan Jing (fl. 1979) M	**C.J.Mu**

Mu, X. (fl. 1977) M	**X.Mu**
Muchina, V.V. A	**Muchina**
Muchovej, J.J. (fl. 1987) M	**J.J.Muchovej**
Muchovej, R.M.C. (fl. 1991) M	**R.M.C.Muchovej**
Mücke, Deitriche (fl. 1959) M	**Mücke**
Mudaliar, C.R. (fl. 1957) S	**C.R.Mudaliar**
Mudaliar, S.K. (1933–) S	**S.K.Mudaliar**
Mudaliar, S.V. A	**S.V.Mudaliar**
Mudd, William A. (1830–1879) M	**Mudd**
Mudie, Robert (1777–1842) S	**Mudie**
Müdler, Karl (fl. 1939) F	**Müdler**
Muehldorf, Anton (1890–) S	**Muehld.**
Muehlenbeck, Heinrich Gustav (1798–1845) S	**Muehlenbeck**
(Muehlenberg, Gotthilf Heinrich Ernest)	
see Muhlenberg, Gotthilf Heinrich Ernest	**Muhl.**
(Muehlenberg, Henry Ludwig)	
see Mühlenberg, Henry Ludwig	**H.L.Mühl.**
Muehlenpfordt, F. (fl. 1849–1891) S	**Muehlenpf.**
Muehlstein, Lisa K. (fl. 1991) M	**Muehlst.**
Mueller, Anton (1798–1864) S	**A.Muell.**
Mueller Argoviensis, Johannes (Jean)	
see Müller Argoviensis, Johannes (Jean)	**Müll.Arg.**
(Mueller, Cornelius Herman)	
see Müller, Cornelius Herman	**C.H.Müll.**
Mueller, Ferdinand Jacob Heinrich von (1825–1896) ABFMPS	**F.Muell.**
(Mueller, Franz August ('Friedrich'))	
see Müller, Franz August ('Friedrich')	**F.A.Müll.**
(Mueller, Fritz (Johan Friedrich Theodor))	
see Müller, Fritz (Johan Friedrich Theodor)	**F.J.Müll.**
Mueller, Gregory M. (1953–) M	**G.M.Muell.**
(Mueller, Jean Baptiste (Baptista))	
see Müller, Jean Baptiste (Baptista)	**J.B.Müll.**
(Mueller, Johann Karl (Carl) August (Friedrich Wilhelm))	
see Müller, Johann Karl (Carl) August (Friedrich Wilhelm)	**Müll.Hal.**
Mueller, Johann Sebastian (after 1744, Miller, J.S.) (1715–1780) S	**J.S.Muell.**
(Mueller, Johannes (Jean))	
see Müller Argoviensis, Johannes (Jean)	**Müll.Arg.**
Mueller, Joseph M. (1802–1870) S	**J.M.Muell.**
(Mueller, Karl (Carl))	
see Müller, Karl (Carl)	**Müll.Berol.**
Mueller, Nicolaus Jacob Carl (1842–1901) S	**N.J.C.Muell.**
(Mueller, Otto Friedrich (Friderich, Fridrich, Frederik))	
see Müller, Otto Friedrich (Friderich, Fridrich, Frederik)	**O.F.Müll.**
(Mueller, Philipp Jakob)	
see Müller, Philipp Jakob	**P.J.Müll.**
Mueller, Wilhelm (fl. 1898) MS	**W.Muell.**
(Muench, Ernst)	
see Münch, Ernst	**Münch**
(Muenchhausen, Otto von)	
see Münchhausen, Otto von	**Münchh.**

Muende, I. (fl. 1940) M	**Muende**
Muenscher, Walter Leopold Conrad (1891–1963) A	**Muenscher**
Muftic, Mahmoud K.S. (fl. 1957) M	**Muftic**
Mühlberg, Friedrich ('Fritz') (1840–1915) S	**Mühlberg**
Mühlenbach, Viktor (1898–) S	**Mühlenb.**
Muhlenberg, Gotthilf Heinrich Ernest (1753–1815) MPS	**Muhl.**
Mühlenberg, Henry Ludwig (1756–1817) PS	**H.L.Mühl.**
(Mühlenpfordt, F.)	
see Muehlenpfordt, F.	**Muehlenpf.**
Muhr, Lars–Erik (fl. 1987) M	**Muhr**
Muhsin, T.M. (fl. 1981) M	**Muhsin**
Muid, Sepia (fl. 1985) M	**Muid**
Muijs, D. (fl. 1921) M	**Muijs**
(Muir, David)	
see Moore, David	**Moore**
Muir, John (1838–1914) S	**J.Muir**
Muir, John A. (fl. 1973) M	**J.A.Muir**
Muir, Marjorie D. (fl. 1970) AM	**M.D.Muir**
Muirhead, Clara Winsome (1915–1985) S	**Muirhead**
Mukelar, A. (fl. 1976) M	**Mukelar**
Mukerjee, Paromita (fl. 1979) M	**P.Mukerjee**
Mukerjee, Sushie Murmar (1896–1934) S	**Mukerjee**
Mukerji, Krisha Gopal (1934–) M	**Mukerji**
Mukherjee, A.K. (1932–) S	**A.K.Mukh.**
Mukherjee, Debapriya A	**D.Mukh.**
Mukherjee, Kalyan Kumar (fl. 1975) S	**K.K.Mukh.**
Mukherjee, N. S	**N.Mukh.**
Mukherjee, Pronob Kumar (1934–) S	**P.K.Mukh.**
Mukherjee, Subhra A	**S.Mukh.**
Mukherjee, Sunil Kumar (1914–) S	**Mukh.**
Mukhina, Valentina V. A	**Mukhina**
Mukhopadhyay, A.N. (fl. 1965) M	**Mukhop.**
Mukunya, Daniel M. (fl. 1973) M	**Mukunya**
Mulas, Bonaria (1945–) S	**Mulas**
(Muldaschev, A.A.)	
see Muldashev, A.A.	**Muldashev**
Muldashev, A.A. (1954–) S	**Muldashev**
Mulder, A.S. A	**A.S.Mulder**
Mulder, D. (fl. 1981) M	**D.Mulder**
Mulder, Jack L. (fl. 1973) M	**J.L.Mulder**
Mulder, Nicolaas ('Claas') (1796–1867) S	**Mulder**
Mulertt, Hugo S	**Mulertt**
Mulford, A.Isabel (fl. 1896) S	**Mulford**
Múlgura de Romero, Maria E. (1943–) S	**Múlgura**
Mulkidjanian, Yakov Ivanovich (1914–) S	**Mulk.**
(Mulkidzhanyan, Yakov Ivanovich)	
see Mulkidjanian, Yakov Ivanovich	**Mulk.**
Mulleavy, Perry (fl. 1977) M	**Mulleavy**
Mullenders, W. S	**Mullend.**

(for Müller, see also Mueller)

Muller, Albert Stanley (1901–) M	**A.S.Mull.**
Müller Argoviensis, Johannes (Jean) (1828–1896) ABMPS	**Müll.Arg.**
(Müller, August Binz)	
see Binz, August	**Binz**
Muller, Carl Alfred (1855–) S	**C.A.Mull.**
Müller, Carl (Karl) (1820–1889) MS	**Müll.Stuttg.**
Müller, Carla A	**C.Müll.**
Müller, Cornelius Herman (1909–) S	**C.H.Müll.**
Müller, Dieter G. A	**D.G.Müll.**
Müller, E.G.Otto (1857–) S	**E.G.O.Müll.**
Müller, Emil (von) (1920–) M	**E.Müll.**
Muller, Félix (1818–1896) S	**F.Mull.**
(Müller, Ferdinand Jacob Heinrich von)	
see Mueller, Ferdinand Jacob Heinrich von	**F.Muell.**
Müller, Franz August ('Friedrich') (1798–1871) S	**F.A.Müll.**
Muller, Franz Sebastian S	**F.S.Mull.**
Muller, Friederich Alfred Gustav Jobst (1860–) S	**F.A.G.Mull.**
Muller, Frits Mari (1907–) S	**F.M.Mull.**
Müller, Fritz (Johan Friedrich Theodor) (1822–1897) S	**F.J.Müll.**
Müller, (Georg Ferdinand) Otto (1837–1917) A	**O.Müll.**
Müller, Gerd K. (1929–) S	**G.K.Müller**
Muller, Gino (1948–) S	**G.Mull.**
(Müller (Halle), Johann Karl (Carl) August (Friedrich Wilhelm))	
see Müller, Johann Karl (Carl) August (Friedrich Wilhelm)	**Müll.Hal.**
Müller, Heinrich Ludwig Hermann (1829–1883) BS	**H.Müll.**
(Müller, Herman)	
see Müller Thurgau, Herman	**Müll.-Thurg.**
Müller, I. (fl. 1983) S	**I.Müll.**
Muller, I.H. (fl. 1990) S	**I.H.Muller**
Müller, Jiří (fl. 1981) M	**J.Müll.**
Müller, J.P. (fl. 1878) S	**J.P.Müll.**
Müller, Jean Baptiste (Baptista) (1806–1894) M	**J.B.Müll.**
Müller, Johann Karl (Carl) August (Friedrich Wilhelm) (1818–1899) BMPS	**Müll.Hal.**
Müller, Johannes (1861–1920) S	**Joh.Müll.**
(Müller, Johannes (Jean))	
see Müller Argoviensis, Johannes (Jean)	**Müll.Arg.**
Müller, Justus A	**Justus Müll.**
Müller, Karl (1881–1955) B	**Müll.Frib.**
Müller, Karl Alfred Ernst (1855–1907) S	**K.A.E.Müll.**
Müller, Karl (Carl) (1817–1870) MPS	**Müll.Berol.**
Müller, Konrad (1857–) S	**K.Müll.**
Muller, Mary Taylor? (1907–) S	**M.T.Mull.**
Müller Melchers, F.C. A	**Müll.Melch.**
(Müller of Freiburg, Karl)	
see Müller, Karl	**Müll.Frib.**
Müller, Otto Friedrich (Friderich, Fridrich, Frederik) (1730–1784) ABMPS	**O.F.Müll.**
(Müller, Otto)	
see Müller (Georg Ferdinand) Otto	**O.Müll.**
Muller, P.J. (fl. 1973) M	**P.J.Mull.bis**

Müller, Philipp Jakob (1832–1889) S	**P.J.Müll.**
(Muller, (Pierre) Felix)	
see Muller, Felix	**F.Mull.**
Müller, Theodor (1894–1969) M	**T.Müll.**
Müller Thurgau, Herman (1850–1927) MS	**Müll.-Thurg.**
Müller, Walther Otto (1833–1887) C	**W.O.Müll.**
(Müller, Wilhelm)	
see Mueller, Wilhelm	**W.Muell.**
(Müller (of Stuttgart), Carl (Karl))	
see Müller, Carl (Karl)	**Müll.Stuttg.**
Müller–Doblies, Dietrich (1938–) S	**D.Müll.-Doblies**
Müller–Doblies, Ute (1938–) S	**U.Müll.-Doblies**
Müller–Knatz (fl. 1912) P	**Müll.-Knatz**
Müller–Kögler, E. (fl. 1970) M	**Müll.-Kög.**
Müller–Stoll, W.R. (fl. 1969) F	**Müll.-Stoll**
(Müller–Thurgau, Herman)	
see Müller Thurgau, Herman	**Müll.-Thurg.**
Müller–Uri, C. (fl. 1986) M	**Müll.-Uri**
Mullet, John E. A	**Mullet**
Mulligan, Brian Orson (1907–) S	**Mulligan**
Mulligan, Gerald Alfred (1928–) PS	**G.A.Mulligan**
Mullins, J.T. (fl. 1961) M	**Mullins**
Mulsow, K. (fl. 1911) M	**Mulsow**
Muma, M.H. (fl. 1966) M	**Muma**
Mumford, Thomas Frenzel (1944–) A	**Mumford**
Munasinghe, H.L. (fl. 1955) M	**Munas.**
Munby, Giles (1813–1876) S	**Munby**
Münch, Ernst (1876–1946) MS	**Münch**
Münchhausen, Otto von (1716–1774) S	**Münchh.**
Munda, Ivka M. (1927–) A	**Munda**
Munday, J. (1928–) S	**Munday**
Münderlein, Pfarrer (fl. 1898) P	**Münderl.**
Mundkur, Bhalchendra Bhavanishankar (1896–1952) M	**Mundk.**
Mundt, H. S	**H.Mundt**
Mundt, Walter (1853–1927) S	**Mundt**
Munier–Chalmas, Ernst Charles (1843–1903) A	**Mun.-Chalm.**
Munir, Ahmad Abid (1936–) S	**Munir**
Muniyamma, M. (fl. 1983) S	**Muniy.**
Muniz, Julio (fl. 1923) AM	**Muniz**
Muñiz, O. (1937–) S	**O.Muñiz**
Munjal, R.L. (fl. 1950) M	**Munjal**
Munjala, B. (fl. 1986) M	**Munjala**
Munk, Anders (fl. 1950) M	**Munk**
Munn, E.A. (fl. 1981) M	**E.A.Munn**
Munn, Mancel Thornton (1887–1956) M	**Munn**
Muñoz Garmendia, José Félix (1949–) PS	**Muñoz Garm.**
Muñoz, Jesús M. (1955–) S	**J.M.Muñoz**
Muñoz Medina, José Maria (1895–) S	**Muñoz Med.**
Muñoz–Pizarro, Carlos (1913–1976) S	**Muñoz**

Munro, Derek B. (fl. 1990) S   **D.B.Munro**
Munro, William (1818–1889) S   **Munro**
Munshi, A.H. (fl. 1986) S   **Munshi**
Munson, Thomas Volney (1843–1913) S   **Munson**
Münster, Georg (1776–1844) S   **Münster**
Munster, R. (fl. 1990) S   **R.Munster**
(Muntañola, Maria)
    see Muntañola–Cvetković, Maria   **Munt.-Cvetk.**
Muntañola–Cvetković, Maria (fl. 1952) M   **Munt.-Cvetk.**
Münter, Johann Andreas Heinrich August Julius (1815–1885) S   **Münter**
Munting, Abraham (1626–1683) L   **Munting**
Müntzing, A. (fl. 1936) S   **Müntzing**
Munz, Philip Alexander (1892–1974) PS   **Munz**
Mur, Luuc R.   A   **Mur**
Muradyan, L.G. (1938–) S   **Muradyan**
Murakami, H. (fl. 1966) M   **Murak.**
Murakami, Noriaki (1959–) PS   **N.Murak.**
Muralt, Johann von (1645–1733) LS   **Muralt**
(Murashkinski, K.E.)
    see Murashkinsky, K.E.   **Murashk.**
Murashkinsky, K.E. (fl. 1924) M   **Murashk.**
Murata, Gen (1927–) S   **Murata**
Murata, Jin (1952–) S   **J.Murata**
Muratova, V.S. (1890–1948) S   **Muratova**
Muravjeva, O.A. (1900–) S   **Murav.**
(Muravjova, O.A.)
    see Muravjeva, O.A.   **Murav.**
Murayama, D. (fl. 1943) M   **Muray.**
Murbeck, Svante Samuel (1859–1946) PS   **Murb.**
Murchison, Roderick Impey (1792–1871) AP   **Murch.**
Murdy, William H. (1928–) P   **Murdy**
Muret, Edouard   S   **E.Muret**
Muret, J. (1799–1877) S   **Muret**
Murillo, Adolfo (1840–1899) S   **Murillo**
(Murillo, Giberto)
    see Morillo, Gilberto N.   **Morillo**
Murillo, María Teresz (1929–) P   **M.T.Murillo**
Murillo–Pulido, Maria Teresz
    see Murillo, Maria Teresz   **M.T.Murillo**
Murith, Laurent Joseph (1742–1816) S   **Murith**
Muroi, Hiroshi (1914–) S   **Muroi**
Muroi, Tetsuo (fl. 1977) M   **T.Muroi**
Murphy, Alice   A   **A.Murphy**
Murphy, Edmund (1828–1866) S   **Murphy**
Murphy, Hickman Charles (1902–1968) MS   **H.C.Murphy**
Murphy, Margaret K. (fl. 1989) M   **M.K.Murphy**
Murr, Josef (1864–1932) S   **Murr**
Murray, Albert Edward (1935–) S   **A.E.Murray**
Murray, Alexander (1798–1838) S   **Al.Murray**
Murray, Andrew (180?–1850) S   **A.Murray**

Murray, Andrew (1812–1878) S	**A.Murray bis**
Murray, B.J. (fl. 1926) M	**B.J.Murray**
Murray, Barbara Mitchell (1938–) BM	**B.M.Murray**
Murray, D.I.L. (fl. 1982) M	**D.I.Murray**
Murray, Edward (fl. 1969) S	**E.Murray**
Murray, George Robert Milne (1858–1911) AC	**G.Murray**
Murray, J.S. (fl. 1960) M	**J.S.Murray**
Murray, James A. (1923–1961) BM	**Js.Murray**
Murray, Johan Andreas (1740–1791) APS	**Murray**
Murray, John (1841–1914) AB	**J. Murray**
Murray, Richard Paget (1842–1908) S	**R.P.Murray**
Murray, Stewart (1789?–1858) S	**S.Murray**
Murray, Thomas Jefferson (1891–) M	**T.J.Murray**
Murrill, William Alphonso (1869–1957) MS	**Murrill**
Murti, S.K. (1943–) S	**Murti**
Murton, R.K. (fl. 1977) M	**Murton**
Murvanischvili, I.K. (fl. 1991) M	**Murvan.**
Mus, Maurici (fl. 1987) S	**Mus**
Musacchio, Aldo A	**A.Musacchio**
Musacchio, Eduardo A. A	**E.A.Musacchio**
(Musaev, S.G.)	
see Mussajev, S.G.	**Mussajev**
Muschegjan, A.M. (fl. 1958) S	**Musch.**
Muschler, Reinhold (Reno) Conrad (1883–1957) AS	**Muschl.**
Musgrave, W.E. (fl. 1907) M	**Musgrave**
Musil, Albina Frances (1894–) S	**Musil**
(Musin–Puschkin, Apollos Apollosovich)	
see Mussin–Puschkin, Apollos Apollossowitsch	**Muss.Puschk.**
Muskat, J. (fl. 1958) M	**Muskat**
Musper, Fritz A	**Musper**
Muspratt, J. (fl. 1985) M	**Muspratt**
Mussa, Enrico (1865–1942) S	**Mussa**
Mussajev, S.G. (1937–) S	**Mussajev**
Mussche, Jean Henri (1765–1834) P	**Mussche**
Musselman, Lytton J. (1943–) S	**Musselman**
Musset, Charles Raymond (1826–1892) S	**Musset**
Mussin–Puschkin, Apollos Apollossowitsch (1760–1805) S	**Muss.Puschk.**
Mustafa, A.F. (fl. 1973) M	**Mustafa**
Mustafaev, I.D. (fl. 1981) S	**Mustafaev**
Mustafee, T.P. (fl. 1966) M	**Mustafee**
Mustapha, H. (fl. 1975–1977) F	**Mustapha**
Mutel, Pierre Auguste Victor (1795–1847) PS	**Mutel**
Muthaiyan, M.C. (fl. 1984) M	**Muthaiyan**
Muthappa, B.N. (fl. 1966) M	**Muthappa**
Muthumary, J. (fl. 1986) M	**Muthumary**
Mutis, José Celestino Bruno (1732–1808) S	**Mutis**
Mutis, Sinforoso (1773–1822) S	**S.Mutis**
(Mutis y Bosio, José Celestino Bruno)	
see Mutis, José Celestino Bruno	**Mutis**

(Mutis y Consuegra, Sinforoso)	
see Mutis, Sinforoso	S.Mutis
Muzafarov, A.M. A	Muzafarov
Mvukiyumwami, J. (fl. 1982) S	Mvukiy.
Mycock, D.J. (fl. 1988) M	Mycock
Mygind, Franz (1710–1789) S	Mygind
Myint, Tin (1936–) S	Myint
Myre, Mario (1908–) S	Myre
Myrin, Claus (Claës) Gustaf (1803–1835) BS	Myrin
Myrzakulov, P. S	Myrz.
Mziray, William R. (fl. 1980) S	Mziray
(N.Santa Maria, Juan)	
see Santa María, Juan	Santa María
(N.Santos Guerra, Arnoldo)	
see Santos Guerra, Arnoldo	A.Santos
Na–Thalang, Obchant (fl. 1980) B	Na–Thalang
Nabel, Kurt (fl. 1939) M	Nabel
Nábělek, Frantisek (1884–1965) PS	Nábělek
Nabiev, M.M. (1926–) S	Nabiev
Naccari, Fortunato Luigi (1793–1860) A	Naccari
Nachmony, Shoshana (fl. 1955) B	Nachm.
Nachtigal, Gustav Hermann (1834–1885) S	Nacht.
Nackejima, Cunio (fl. 1971) S	Nackej.
Naczi, Robert Francis Cox (1963–) S	Naczi
Nadakavukaren, Mathew J. A	Nadak.
Nadeaud, Jean (1834–1898) BPS	Nadeaud
Nadji, Abd–Ur Bahman (fl. 1892) S	Nadji
Nadson, Georgii Adamovich (1867–) AM	Nadson
Nádvorník, Joseph (1906–1977) M	Nádv.
Naef, Jacques A	Naef
(Naegeli, Carl Wilhelm von)	
see Nägeli, Carl Wilhelm von	Nägeli
(Naegeli, Otto)	
see Nägeli, Otto	O.Nägeli
Naezén, Daniel Eric (1752–1808) S	Naezén
Nag, Kalpana (1937–) P	Nag
Nag Raj, Tumkur R. (fl. 1960) M	Nag Raj
Nagaeva, G.A. A	Nagaeva
Nagai, Masaji (1905–1966) AM	Nagai
Nagamasu, H. (fl. 1986) S	Nagam.
Nagami, Shuzo (1911–) S	Nagami
Naganishi, H. (fl. 1968) M	H.Nagan.
Naganishi, Nirosuke (fl. 1915) M	Nagan.
Nagano, Iwao (1930–) B	I.Nagano
Nagano, Ken (fl. 1928) M	Nagano
Nagarkar, M.B. (1951–) M	Nagarkar
Nagasawa, Eiji (1948–) M	Nagas.
Nagase, H. (fl. 1988) S	Nagase

Nagata, Kenneth M. (1945–) S	**Nagata**
Nagatomo, Isamu (fl. 1937) M	**Nagat.**
Nägeli, Carl Wilhelm von (1817–1891) AMS	**Nägeli**
Nägeli, Otto (1871–1938) MS	**O.Nägeli**
Nagendran, C.R. (fl. 1975) S	**Nagendran**
Nageswara Rao, A. (fl. 1983) S	**Nag.Rao**
(Naggar, Salah M. El)	
see El Naggar, Salah M.	**El Naggar**
Naggi, A. (fl. 1905) S	**Naggi**
Nägler, Kurt A	**Nägler**
Nagornaya, S.S. (fl. 1979) M	**Nagornaya**
Nagorny, P.J. (fl. 1926) M	**Nagorny**
Nagumo, Tamotsu (1950–) A	**Nagumo**
Nagy, Eszter (Esther) AB	**E.Nagy**
Nagy, F. (fl. 1970) M	**Nagy**
Nagy, Lois Anne A	**L.A.Nagy**
Nagy–Petri, Zoltan (fl. 1970) M	**Nagy-Petri**
Nagy–Tóth, F. A	**Nagy-Tóth**
Naidu, B.Appala (fl. 1953) S	**Naidu**
Naik, C.D. (fl. 1986) M	**C.D.Naik**
Naik, V.N. (fl. 1966) S	**Naik**
Naim, Z. (fl. 1958) M	**Naim**
(Nair, Dewan Mohinder (Nath))	
see Nath Nair, Dewan Mohinder	**Nath Nair**
Nair, G.Bhadran (1942–) P	**G.B.Nair**
Nair, G.U. A	**G.U.Nair**
Nair, K.K.N. (1948–) S	**K.K.N.Nair**
Nair, K.R.Gopinathan (fl. 1963) M	**K.R.G.Nair**
Nair, K.Vasudevan (1943–) S	**K.V.Nair**
Nair, L.N. (fl. 1983) M	**L.N.Nair**
Nair, M.K. (fl. 1988) S	**M.K.Nair**
Nair, N.Balakrishnan (fl. 1967) M	**N.B.Nair**
Nair, N.Chandrasekharan (1927–) MPS	**N.C.Nair**
Nair, N.G. (1948–) MS	**N.G.Nair**
Nair, R.Vasudevan (fl. 1975) S	**R.V.Nair**
Nair, V.J. (1940–) S	**V.J.Nair**
Nair, V.M.G. (fl. 1979) M	**V.M.G.Nair**
Nair, V.S. (fl. 1983) M	**V.S.Nair**
Nairne, Alexander Kyd (fl. 1880s) S	**Nairne**
Naithani, B.D. (1938–) S	**B.D.Naithani**
Naithani, H.B. (1944–) S	**H.B.Naithani**
Naito, Toshiyuki (1939–) M	**Naito**
Najarajan, K. (fl. 1980) M	**Najar.**
Najera, Marta (fl. 1972) S	**Najera**
Najim, L. (fl. 1984) M	**Najim**
Nakagawa, Y. (fl. 1988) M	**Nakagawa**
Nakagiri, Akira (1957–) M	**Nakagiri**
Nakahi, Takashi (fl. 1980s) M	**Nakahi**
Nakai, Kei M. (1954–) S	**K.M.Nakai**

Nakai, Takenoshin (Takenosin) (1882–1952) ABPS	**Nakai**
Nakaike, Toshiyuki (1943–) PS	**Nakaike**
Nakajima, Kunio (fl. 1951) S	**K.Nakaj.**
Nakajima, Sadao S	**S.Nakaj.**
Nakajima, Tokuichiro (1910–) B	**Nakaj.**
Nakamoto, Lynda A	**Nakamoto**
Nakamura, M. (fl. 1936) S	**M.Nakam.**
Nakamura, T. (fl. 1932) M	**Nakam.**
Nakamura, Takehisa (1932–) PS	**T.Nakam.**
Nakamura, Takeshi A	**Tak.Nakam.**
Nakamura, Yositeru (1910–) A	**Y.Nakam.**
Nakanishi, H. (fl. 1982) S	**H.Nakan.**
Nakanishi, Minoru (fl. 1966) M	**M.Nakan.**
Nakanishi, Satoshi (1928–) M	**Nakan.**
Nakano, Harufusa (1883–1972) AS	**Nakano**
Nakano, Taketo (1943–) AS	**Tak.Nakano**
Nakano, Toru (fl. 1978) M	**T.Nakano**
Nakase, Takashi (fl. 1987) M	**Nakase**
Nakashima, Kazuo (1904–) S	**Nakash.**
Nakasima, ?S. (fl. 1935) P	**Nakasima**
Nakasone, Karen K. (1953–) M	**Nakasone**
Nakata, Kakugoro (1886–1939) M	**Nakata**
Nakata, N. (fl. 1928) M	**N.Nakata**
Nakato, Narumi (fl. 1981) P	**Nakato**
Nakazawa, R. (fl. 1913) M	**Nakaz.**
Nakinishiki, Kôji (fl. 1906) B	**Nakin.**
Nalepina, L.N. (fl. 1964) M	**Nalepina**
Nam, E.A. (fl. 1977) M	**E.A.Nam**
Nam, G.A. (fl. 1977) M	**G.A.Nam**
Nam, Ki Wan A	**K.W.Nam**
Nam, Vincenzo (1855–) S	**Nam**
Namalina (fl. 1929) M	**Namalina**
Namba, Tsuneo (fl. 1959) S	**Namba**
Nambu, N. (fl. 1909) M	**Nambu**
Namegata, Tomitaro (fl. 1952–1961) P	**Nameg.**
Namekata, T. (fl. 1967) M	**Namek.**
Namyslowski, Boleslaw (1882–1929) AM	**Namysl.**
Nanakorn, Weerachai (1948–) S	**Nanakorn**
Nand, Krishna (fl. 1966) M	**Nand**
Nanir, S.P. (fl. 1973) M	**Nanir**
(Nannenga–Bremekamp, Neeltje Elizabeth)	
see Bremekamp, Neeltje Elizabeth	**Nann.-Bremek.**
Nannfeldt, John (Johan) Axel Frithiof (1904–1985) MS	**Nannf.**
Nannizzi, Arturo (1877–1961) M	**Nann.**
Nanta, A. (fl. 1927) M	**Nanta**
Nantiyal, D.D. (fl. 1984) F	**Nantiyal**
(Naoumoff, M.N.)	
see Naumov, M.N.	**M.N.Naumov**

(Naoumoff, Nikolai Alexsandrovič)
   see Naumov, Nikolai Alexsandrovich   **Naumov**
Naphade, S.R. (fl. 1971) M   **Naphade**
Napolitano, Joseph J. (1935–) A   **Napolit.**
Napper, Diana Margaret (1930–1972) S   **Napper**
Naqshi, Abdul Rashid (1944–) S   **Naqshi**
Narain, Narsingh A   **N.Narain**
Narain, Udit (fl. 1971) M   **Narain**
Narasimhan, D. (fl. 1988) S   **D.Naras.**
Narasimhan, M.J. (fl. 1922) M   **Naras.**
Narasimhan, Tyagarajan A   ·   **T.Naras.**
Narayan, P. (fl. 1986) M   **Narayan**
Narayana, B.M. (fl. 1981) S   **Narayana**
Narayanan, S.Anant (fl. 1962) M   **Narayanan**
Narayanasamy, P. (fl. 1968) M   **Narayanas.**
Narayanaswami, S. (fl. 1957) M   **S.Naray.**
Narayanaswami, V. (fl. 1949) S   **V.Naray.**
Nardi, Enio (1942–) P   **E.Nardi**
Nardi, G. (fl. 1976) M   **G.Nardi**
(Nardi, Jan Christina de)
   see De Nardi, Ján Christina   **De Nardi**
Nardina, N.S. (fl. 1965) S   **Nardina**
Nardo, Giovanni Domenico (Giandomenico) (1802–1877) A   **Nardo**
Narendra, D.V. (fl. 1972) M   **Narendra**
Narita, Seiichi A   **S.Narita**
Narita, Takeshi (fl. 1959) M   **Narita**
Naruhashi, Naohito (1941–) S   **Naruh.**
Narula, A.M. (fl. 1984) M   **Narula**
(Nasarov, M.I.)
   see Nasarow, M.I.   **Nasarow**
Nasarow, M.I. (1882–1942) S   **Nasarow**
Nascimento, Maria L. (fl. 1956) M   **Nascim.**
Nash, Dorothy L. (later Gibson, D.L.) (fl. 1970) S   **D.L.Nash**
Nash, George Valentine (1864–1921) S   **Nash**
Nash, R.C. S   **R.C.Nash**
Nash, Thomas H. (III) (1945–) M   **T.H.Nash**
Nasimova, T. (1946–) S   **Nasimova**
Nasir, Eugene (1908–) S   **Nasir**
Nasir, Yasin J. (1943–) S   **Y.J.Nasir**
Nasr, Abdul Halim (1908–) A   **Nasr**
Nassar, N.M.A. (fl. 1986) S   **Nassar**
Nasser, M. (fl. 1973) M   **Nasser**
(Nassimova, T.)
   see Nasimova, T.   **Nasimova**
Nast, Charlotte G. (fl. 1943) S   **Nast**
Nasution, R.E. (fl. 1977) S   **Nasution**
Nasyrov, O.N. (fl. 1972) M   **Nasyrov**
Natali, A. (fl. 1988) S   **Natali**
Natalyina, O.B. (fl. 1931) M   **Natalyina**

Natarajan, K.V. (fl. 1980) AM	**Natarajan**
Nates, Pedro Pablo S	**Nates**
Nath Nair, Dewan Mohinder (1908–1971) S	**Nath Nair**
Nath, R. (fl. 1970) M	**R.Nath**
Nath, V. (1949–) M	**V.Nath**
Nathhorst, Theophilus Erdmann (fl. 1756) S	**Nathh.**
Natho, Günther (1930–) S	**Natho**
Nathorst, Alfréd Gabriel (1850–1921) FS	**Nath.**
Nathorst–Windahl, T. (fl. 1961) M	**Nath.-Wind.**
Natour, Rashad M. A	**Natour**
Nattras, Roland Marshall (1895–) M	**Nattras**
(Nattrass, Roland Marshall)	
see Nattras, Roland Marshall	**Nattras**
Naudin, Charles Victor (1815–1899) S	**Naudin**
Nauenburg, Johannes D. (1951–) S	**Nauenb.**
Nauman, Clifton E. (fl. 1979–1982) P	**Nauman**
Naumann, Arno (1862–1932) M	**A.Naumann**
Naumann, Einar Christian Leonard (1891–1934) A	**Naumann**
Naumann, W. (fl. 1970) M	**W.Naumann**
Naumburg, Samuel Johann (1768–1799) S	**Naumb.**
Naumov, G.I. (fl. 1979) M	**G.I.Naumov**
Naumov, M.N. (fl. 1935) M	**M.N.Naumov**
Naumov, Nikolai Alexsandrovich (1888–1959) M	**Naumov**
Naumova, N.N. (fl. 1955) M	**Naumova**
Naumova, S.N. A	**S.N.Naumova**
Nauss, R.N. (fl. 1949) M	**Nauss**
Nauta, M.M. (1959–) M	**Nauta**
Nautiyal, Avinash Ch. A	**Nautiyal**
Nauwerck, Arnold A	**Nauwerck**
Nava Fernández, Herminio Severiano (1956–) S	**Nava**
Navale, G.K.B. (fl. 1963) F	**Navale**
Navarathnam, E.Susan A	**Navarath.**
Navarro Andres, Florentino (fl. 1984) S	**F.Navarro**
Navarro Aranda, Carmen (1949–) S	**C.Navarro**
(Navarro de Andrade, Edmundo)	
see Navarro, Edmundo	**Navarro**
Navarro, Edmundo (fl. 1914–1941) S	**Navarro**
Navarro, Gonzalo (fl. 1991) S	**G.Navarro**
Navarro, J.Nelson A	**J.N.Navarro**
Navarro, R. (fl. 1986) M	**R.Navarro**
Navarro, T. (fl. 1988) S	**T.Navarro**
Navarro–Rosinés, Père (fl. 1986) M	**Nav.-Ros.**
Navas, Antonio J. (fl. 1978) M	**A.J.Navas**
Navas Bustamante, Luisa Eugenia (fl. 1973) S	**L.E.Navas**
Navas, Longinos (1858–1938) BM	**Navas**
Navashin, Sergei Gavrilovich (1857–1930) BM	**Navashin**
Nave, Johann (1831–1864) A	**Nave**
Naveau, (Georges) Raymond (Léonard) (1889–1932) BM	**Naveau**

Náves, Andrés (1839–1910) S	**Náves**
Naville, A. (fl. 1922) M	**Naville**
(Nawaschin, Sergius Gawrilowitsch)	
see Navashin, Sergei Gavrilovich	**Navashin**
Nawawi, A. (fl. 1974) M	**Nawawi**
Nayal, A.A. A	**Nayal**
Nayar, Bala Krishnan (1927–) PS	**B.K.Nayar**
Nayar, Madhavan Parameswarau (1932–) S	**M.P.Nayar**
Nayar, S.K. (fl. 1972) M	**S.K.Nayar**
Naylor, Frederick (1811–1882) S	**Naylor**
Naylor, Gladys L. A	**G.L.Naylor**
Nazarova, Estella A. (1936–) S	**Nazarova**
Nazarova, M.M. (fl. 1991) M	**M.M.Nazarova**
Nazimuddin, S. (fl. 1981) S	**Nazim.**
Nazor, Vladimir (fl. 1903–1904) PS	**Nazor**
Ndabaneze, Pontien (1952–) S	**Ndab.**
Ndimande, B.N. (fl. 1979) M	**Ndimande**
Neagu–Tîrcovnicu, Marina (fl. 1961) M	**Neagu–Tîrc.**
(Neal, Adam)	
see Neale, Adam	**Neale**
Neal, David Carleton (1890–) M	**Neal**
Neal, Marie Catherine (1889–1965) S	**M.Neal**
Neale, Adam (fl. 1770) S	**Neale**
Neale, J.W. A	**J.W.Neale**
Neale, Walter T. S	**W.T.Neale**
Nealley, Greenleaf Cilley (1846–1896) S	**Nealley**
Nearing, G.C. (fl. 1947) M	**Nearing**
Necchi, Orlando A	**Necchi**
Nechaev, A.A. (fl. 1979) S	**Nechaev**
Nechaeva, T.I. (1938–) S	**Nechaeva**
Nechitsche, A. (fl. 1904) M	**Nechitsche**
Necker, Noel Martin Joseph de (1730–1793) ABMPS	**Neck.**
Necski, Sergei Arsejevic S	**Necski**
Nectoux, Hippolyte (fl. 1808) S	**Nectoux**
Neczajev, A.A. (1945–) S	**Neczajev**
Neda, Hitoshi (1957–) M	**Neda**
Nedjalkov, S. (fl. 1956) S	**Nedjalkov**
Nedolushko, V.A. (1953–) S	**Nedol.**
(Nedoluzhko, V.A.)	
see Nedolushko, V.A.	**Nedol.**
Nee, Albert H. S	**A.H.Nee**
Née, Luis (fl. 1789–1794) S	**Née**
Nee, Michael (1947–) S	**M.Nee**
Neergaard, Pierre Paul Ferdinand Mourier (1907–) M	**Neerg.**
(Nees, Christian Gottfried Daniel Nees von)	
see Nees von Esenbeck, Christian Gottfried Daniel	**Nees**
(Nees, Theodor Friedrich Ludwig)	
see Nees von Esenbeck, Theodor Friedrich Ludwig	**T.Nees**
Nees von Esenbeck, Christian Gottfried Daniel (1776–1858) ABMPS	**Nees**

Nees von Esenbeck, Theodor Friedrich Ludwig (1787–1837) MS	**T.Nees**
Neese, Elizabeth C. (1934–) S	**Neese**
Neger, Franz (Friedrich) Wilhelm (1868–1923) MS	**Neger**
Neger, Johannes (fl. 1871) S	**J.Neger**
Negishi, H. (fl. 1991) M	**Negishi**
Negodi, Giorgio Carlo (1900–1974) MS	**Negodi**
Negoro, Ken–Ichiro (Ken–Itirô) (1910–) A	**Negoro**
Nègre, Robert (1924–) S	**Nègre**
Negrean, Gavril A. (1932–) MS	**Negrean**
Negri, Francesco (1841–1924) S	**Negri**
Negri, Giovanni (1877–1960) BS	**G.Negri**
Negrillo Galindo, A.Ma. (1937–) S	**Negrillo**
Negrín–Sosa, M.L. (fl. 1986) S	**Negrín**
Negroni de Bonvehi, Marta Beartriz (fl. 1963) M	**Negr.Bonv.**
Negroni, Pablo (fl. 1929) AM	**Negroni**
Negru, Alexander (1914–) M	**Negru**
Negrul, Alexander Mikhailovič (1900–) S	**Negrul**
Negrutzky, S.F. (fl. 1966) M	**Negr.**
Nehira, Takeo (fl. 1951) M	**Nehira**
Nehring, A. S	**Nehring**
Nehrling, Henry (1853–1929) S	**Nehrl.**
Neil, Cornelius Bernardus (1897–) M	**Neil**
Neill, David A. (1953–) S	**D.A.Neill**
Neill, Patrick (1776–1851) S	**Neill**
Neilreich, August (1803–1871) PS	**Neilr.**
Neish, G.A. (fl. 1981) M	**Neish**
Nejburg, M.F. S	**Nejburg**
Nejceff, I. (1870–1913) S	**Nejceff**
(Nekrasova, Vera Leontievna)	
see Nekrassova, Vera Leontievna	**Nekr.**
Nekrassova, Vera Leontievna (1881–1979) S	**Nekr.**
(Nekrassowa, Vera Leontievna)	
see Nekrassova, Vera Leontievna	**Nekr.**
Nel, D.C. (fl. 1935) S	**D.Nel**
Nel, Ellen E. (fl. 1963) M	**E.E.Nel**
Nel, Gert Cornelius (1885–1950) S	**Nel**
Nelen, E.S. (fl. 1959) M	**Nelen**
Nelmes, Ernest (1895–1959) S	**Nelmes**
Nelson, Aven (1859–1952) MPS	**A.Nelson**
Nelson, Cirilo H. (1938–) S	**C.Nelson**
Nelson, David (–1789) S	**Nelson**
Nelson, Edward Milles (1851–1938) A	**E.M.Nelson**
Nelson, Edward William (1855–1934) S	**E.W.Nelson**
Nelson, Elias Emanuel (1876–1949) PS	**E.E.Nelson**
Nelson, Erich (1897–1980) S	**E.Nelson**
Nelson, Ernest Charles (1951–) S	**E.C.Nelson**
Nelson, Guy L. S	**G.L.Nelson**
Nelson, Ira Schreiber (1911–1965) S	**I.S.Nelson**
Nelson, J.P. (fl. 1980) S	**J.P.Nelson**

Nelson, James Carlton (1867–1944) S	J.C.Nelson
Nelson, John (fl. 1860) S	J.Nelson
Nelson, John Gudgeon (1818–1882) S	J.G.Nelson
Nelson, Paul E. (fl. 1956) M	P.E.Nelson
Nelson, R.J. A	R.J.Nelson
Nelson, R.R. (fl. 1959) M	R.R.Nelson
Nelson, Ray (1893–) M	R.Nelson
(Nelson S., Cirilo H.)	
see Nelson, Cirilo H.	C.H.Nelson
Nelson, T.W. (fl. 1981) S	T.W.Nelson
Nelson, Wendy Alison (1954–) A	W.A.Nelson
Nema, K.G. (fl. 1969) M	Nema
Němec, Bohumil Řehoř (1873–1966) M	Němec
Němejc, František (1901–1976) AFS	Němejc
Németh, József A	Németh
Nemirova, F.S. (1946–) S	Nemirova
Nemlich, Hanna (fl. 1974) M	Nemlich
Nemoto, Kwanji (1860–1936) PS	Nemoto
Nemoto, Takahisa A	Tak.Nemoto
Nemoto, Tomoyuki (1958–) S	T.Nemoto
Nendtvich, Carl Maximilian (Károli Miksa) (1811–1892) S	Nendtv.
Nendvich, Tamas (1782–1858) S	Nendvich
Nene, K.M. (fl. 1968) M	Nene
(Nenjukov, Fedor Stepanovich)	
see Nenukow, Fedor Stepanowitsch	Nenukow
(Nenjukow, Fedor Stepanowitsch)	
see Nenukow, Fedor Stepanowitsch	Nenukow
Nenukow, Fedor Stepanowitsch (1906–1942) S	Nenukow
Neophytova, V.K. (fl. 1955) M	Neophyt.
Nepli, G.N. (1910–) S	Nepli
Nepomnjasczaja, O.A. (fl. 1984) S	Nepomn.
(Nepomnyashchaya, O.A.)	
see Nepomnjasczaja, O.A.	Nepomn.
Néraud, Jules (fl. 1826) S	Néraud
Neresheimer, Eugen A	Neresh.
Nervander, Johan Hugo Emmerik (1827–1909) S	Nerv.
(Nery da Silva, José)	
see Silva, José Nery da	J.Silva
Neshataeva, G.Yu. (fl. 1982) S	Neshat.
Nesio, Maria L. Ribeiro (fl. 1987) M	Nesio
Nesis, K.N. A	Nesis
Nesom, Guy L. (1945–) S	G.L.Nesom
Nesom, Margaret (fl. 1972) M	M.Nesom
Nespiak, A. (fl. 1960) M	Nespiak
Ness, Helge (1861–1928) S	Ness
Nessel, Hermann (1877–1949) PS	Nessel
Nesselhauf, Rudolf S	Nesselh.
Nestler, A. (fl. 1812) S	A.Nestl.
Nestler, Chrétien Géofroy (Christian Gottfried) (1778–1832) BMS	Nestl.

Netto, Ladislau de Souza Mello e (1837–1893) S	**Netto**
Netzer, D. (fl. 1987) M	**Netzer**
Neuberg, Mariya Fedorovna (1894–1962) BF	**Neuberg**
Neuberger, Joseph (1854–1924) S	**Neuberger**
Neubert, Hermann (1935–) M	**H.Neubert**
Neubert, Wilhelm (1808–1905) S	**Neubert**
Neubner, Jacob Eduard (1855–) M	**Neubner**
Neuendorf, Magnus (1942–) M	**Neuendorf**
Neuenhahn, Carl August S	**Neuenh.**
Neuhoff, Walther (fl. 1926) M	**Neuhoff**
Neuman, Julius John (fl. 1914) M	**J.Neuman**
Neuman, Leopold Martin (1852–1922) S	**Neuman**
Neumann, Alfred (1916–1973) S	**A.Neumann**
Neumann, Ferdinand (fl. 1844) S	**F.Neumann**
Neumann, Joseph Henri François (1800–1858) S	**Neumann**
Neumann, K.G. (1774–1850) S	**K.G.Neumann**
Neumann, Louis (1827–1903) S	**L.Neumann**
Neumann, Richard (1884–1910) S	**R.Neumann**
Neumayer, Georg Balthasar von (1862–1909) B	**Neumayer**
Neumayer, Hans (1887–1945) S	**H.Neumayer**
Neumayer, Josef (1791–1840) S	**J.Neumayer**
Neureuter, Franz (fl. 1910) S	**Neureuter**
Neushul, Michael (1933–) A	**Neushul**
Neusser, B. S	**Neusser**
Neutelings, T.M.W. (fl. 1986) S	**Neutel.**
Neuweiler, Ernst (1875–) BF	**Neuweiler**
(Neuwied, Maximilian Alexander Philipp zu Wied)	
see Wied–Neuwied, Maximilian Alexander Philipp zu	**Wied-Neuw.**
Neuwirth, F. (fl. 1949) M	**F.Neuwirth**
Neuwirth, J. (–1944) S	**Neuwirth**
Nevers, G.C.de (fl. 1987) S	**Nevers**
Neves, J.Aroeira (fl. 1927) M	**J.A.Neves**
Neves, José de Barros (1914–1982) S	**J.B.Neves**
Neves, R. A	**R.Neves**
Neves–Armond, Amaro Ferreira das (1854–1944) S	**Neves**
Neveu–Lemaire, M. (fl. 1912) M	**Neveu-Lem.**
Neville, P. (fl. 1988) M	**Neville**
Nevling, Lorin Ives (1930–) S	**Nevling**
Nevodovsky, G.S. (fl. 1950) M	**Nevod.**
Nevski, Sergei Arsenjevic (1908–1938) S	**Nevski**
(Nevskiĭ, Sergei Arsenjevič)	
see Nevski, Sergei Arsenjevic	**Nevski**
Newberry, John Strong (1822–1892) FS	**Newb.**
Newbold, Patty Thum S	**Newbold**
Newbould, William Williamson (1819–1886) S	**Newbould**
Newcombe, Frederick Charles (1858–1927) S	**Newc.**
Newell, Steven Y. (fl. 1970) M	**S.Y.Newell**
Newell, Thomas Kenneth (1939–) S	**Newell**

Newhall, Allan Goodrich (1894–) M	**A.G.Newhall**
Newhall, Charles Stedman (1842–1935) S	**Newhall**
Newhouse, Jan A	**Newhouse**
Newman, Edward (1801–1876) PS	**Newman**
Newman, Francis William (c.1796–1859) S	**F.Newman**
Newman, Karl Robert (1931–) A	**K.R.Newman**
Newnham, M.R. (fl. 1986) S	**Newnham**
Newroth, P.R. A	**Newroth**
Newsom, Vesta Marie (1902–1958) S	**Newsom**
Newton, Alan (1927–) S	**A.Newton**
Newton, Angela (fl. 1990) S	**Ang.Newton**
Newton, Edwin Tully (1840–1930) A	**E.T.Newton**
Newton, George Albert (1879–) M	**G.A.Newton**
Newton, Isaac (1840–1906) B	**Newton**
Newton, James (1639–1718) L	**J.Newton**
Newton, Leonard Eric (1936–) S	**L.E.Newton**
Newton, Lily (–1981) A	**L.Newton**
Newton, Linda Mary (1928–) A	**L.M.Newton**
Newton, Martha Elizabeth (1941–) B	**M.E.Newton**
Neygenfind, Friedrich Wilhelm (fl. 1821) S	**Neygenf.**
Neyraut (fl. 1904) S	**Neyraut**
(Nezdoĭminogo, E.L.)	
see Nezdojminogo, E.L.	**Nezdojm.**
Nezdojminogo, E.L. (fl. 1970) M	**Nezdojm.**
Ng, Francis S.P. (1940–) S	**Ng**
Ng, N.Q. (fl. 1981) S	**N.Q.Ng**
Ngachan, S.V. (fl. 1979) M	**Ngachan**
Ngan, Phung Trung (fl. 1965) S	**Ngan**
Nguen, N.Thin (fl. 1983) S	**Nguen**
(Nguen, To Quyen)	
see Nguyen, To Quyen	**T.Q.Nguyen**
Nguen, Van Tkhan (fl. 1977) M	**V.T.Nguen**
Nguyen, Huu Dinh A	**H.D.Nguyen**
Nguyen, Huu Hien (1939–) S	**H.H.Nguyen**
Nguyen, Tha Lau (fl. 1966) M	**T.L.Nguyen**
Nguyen, To Quyen (fl. 1965) S	**T.Q.Nguyen**
Ni, Chen Kai (fl. 1982) PS	**C.K.Ni**
Ni, Chi Cheng (fl. 1980) S	**C.C.Ni**
Nicholas, Ashley (1954–) S	**Nicholas**
Nicholls, K.W. (fl. 1977) S	**K.W.Nicholls**
Nicholls, Kenneth Howard (1946–) A	**K.H.Nicholls**
Nicholls, M.S. (fl. 1986) S	**M.S.Nicholls**
Nicholls, William Henry (1885–1951) S	**Nicholls**
Nichols, George Elwood (1882–1939) B	**Nichols**
Nichols, Herbert Wayne (1934–) A	**H.W.Nichols**
Nichols, Maurice Barstow A	**M.B.Nichols**
Nicholson, George (1847–1908) BPS	**G.Nicholson**
Nicholson, Henry Alleyne (1844–1899) AF	**H.Nicholson**
Nicholson, William Alexander (1858–1935) S	**W.A.Nicholson**

Nicholson, William Edward (1866–1945) B   **W.E.Nicholson**
Ničić, Djoerje (1856–1920) S   **Ničić**
Nickerson, Norton Hart (1926–　) S   **N.H.Nick.**
Nickerson, Walter J. (fl. 1944) M   **W.J.Nick.**
Nicklès, Napoléon (　–1878) S   **Nicklès**
Nickrent, Daniel L. (1956–　) S   **Nickrent**
Nicolai, Marie Françoise Emilie (1900–1961) A   **M.F.E.Nicolai**
Nicolai, Ernst August (1800–1874) S   **Nicolai**
Nicolaj, Pietro (fl. 1975) M   **Nicolaj**
(Nicolajaeva, M.I.)
    see Nikolajeva, M.I.   **M.I.Nikol.**
Nicolari S   **Nicolari**
Nicolas, (Léon Marie Joseph) Gustave (1879–1955) BM   **Nicolas**
Nicolau, Stefan (fl. 1912) M   **Nicolau**
Nicoli, R.M. (fl. 1974) M   **Nicoli**
Nicolson, Alice C. (fl. 1990) S   **A.C.Nicolson**
Nicolson, Dan Henry (1933–　) PS   **Nicolson**
Nicolson, T.H. (fl. 1968) M   **T.H.Nicolson**
Nicolucci, Giustiniano (1818–1904) S   **Nicolucci**
(Nicora de Panza, Elisa G.)
    see Nicora, Elisa G.   **Nicora**
Nicora, Elisa G. (1912–　) S   **Nicora**
Nicot, Jacqueline (fl. 1953) M   **Nicot**
Nicot–Toulouse, J. (fl. 1947) M   **Nicot-Toul.**
Nicotra, Leopoldo (1846–1940) BPS   **Nicotra**
Nie, Da Shu　A   **D.S.Nie**
Nie, Min Xiang (fl. 1989) S   **M.X.Nie**
Nie, Shao Quan (1932–　) S   **S.Q.Nie**
Nie, Shou Chuan (fl. 1982) S   **S.C.Nie**
Niebuhr, Carsten (1733–1815) S   **Niebuhr**
Niedenzu, Franz Josef (1857–1937) S   **Nied.**
Niederlein, Gustavo (1858–1924) S   **Niederl.**
Niehaus, C.J.G. (fl. 1932) M   **Niehaus**
Niel, C.B.van (fl. 1929) M   **C.B.Niel**
Niel, Pierre Eugène (1836–1905) S   **Niel**
Nielsen, Etlar Lester (1905–　) S   **E.L.Nielsen**
Nielsen, H.S. (fl. 1968) M   **H.S.Nielsen**
Nielsen, Ivan Christian (1946–　) S   **I.C.Nielsen**
Nielsen, L.T. (fl. 1985) M   **L.T.Nielsen**
Nielsen, L.W. (fl. 1957) M   **L.W.Nielsen**
Nielsen, Peter (1829–1897) MS   **Nielsen**
Nielsen, Ruth (1944–　) A   **R.Nielsen**
Niemalä, Tuomo (1940–　) M   **Niemalä**
Niemetz, W.F. (fl. 1898–1922) S   **Niemetz**
Niemi, Ake　A   **Niemi**
Nienburg, Karl Ortgies Wilhelm (1882–1932) S   **Nienb.**
Nieschalk, Albert (1904–1985) PS   **A.Niesch.**
Nieschalk, Charlotte (1913–　) S   **C.Niesch.**
Nieschulz, Otto　A   **Nieschulz**

(Niessel von Meyendorf, Gustav)
  see Niessl von Mayendorf, Gustav   **Niessl**
Niessen, Joseph (1864–) M   **Niessen**
Niessl von Mayendorf, Gustav (1839–1919) CMP   **Niessl**
Niethammer, Anneliese (1901–) M   **Nieth.**
Nieto Feliner, Gonzalo (1958–) S   **Nieto Fel.**
Nieuwdorp, P.J. (fl. 1969) M   **Nieuwdorp**
Nieuwenhuis, A.W. (fl. 1908) M   **Nieuwenh.**
Nieuwland, Julius (Aloysius) Arthur (1878–1936) BMPS   **Nieuwl.**
Niezabitowski, Edward Lubicz (fl. 1913) AM   **Niezab.**
Niezgoda, Christine J. (1950–) S   **Niezgoda**
Nigmatullin, F.A. (1929–) S   **Nigmat.**
Niino, Hiroshi A   **Niino**
Nijalingappa, B.H.M. (fl. 1979) S   **Nijal.**
Nikhra, K.M. (fl. 1981) M   **Nikhra**
Nikiforova, N.B. (1912–) S   **Nikif.**
Nikiforova, O.D. (1950–) S   **O.D.Nikif.**
Nikitin, S.A. (fl. 1937) S   **S.A.Nikitin**
Nikitin, Sergei Nikolaevic (1850–1909) S   **Nikitin**
Nikitin, V.A. (1906–1974) S   **V.A.Nikitin**
Nikitin, Vasilii Vasilevich (Vasily Vasilyevich) (1906–1988) S   **V.V.Nikitin**
Nikitina, E.V. (1893–) S   **Nikitina**
Nikitina, V.N. A   **V.N.Nikitina**
Niklas, Karl Joseph (1948–) A   **Niklas**
Nikler, Leon A   **Nikler**
(Nikogosian, V.G.)
  see Nikogossjan, V.G.   **Nikog.**
Nikogossjan, V.G. (fl. 1965–) BM   **Nikog.**
(Nikolaev, V.N.)
  see Nikolajev, V.N.   **V.N.Nikolajev**
(Nikolaeva, M.I.)
  see Nikolajeva, M.I.   **M.I.Nikol.**
Nikolajev, E.V. (1951–) S   **Nikolajev**
Nikolajev, V.A. A   **V.A.Nikolajev**
Nikolajev, V.J. (1960–) S   **V.J.Nikolajev**
Nikolajev, V.N. (1930–) S   **V.N.Nikolajev**
Nikolajeva, M.I. (fl. 1970) M   **M.I.Nikol.**
Nikolajeva, T.L. (fl. 1933) M   **Nikol.**
Nikolić, E. (fl. 1904) P   **Nikolić**
Nikolić, Vojislav (1925–1989) S   **V.Nikolić**
Nikolskaya, V.D. A   **Nikolsk.**
Niles, Cornelia D. (fl. 1962) S   **Niles**
Nilson, Birger (also Kajanus, B.) (1882–1931) M   **Nilson**
Nilsson, Arvid (1897–) S   **A.Nilsson**
Nilsson, Gunnar Bror Fritiof (later Degelius, G.B.F.) (1903–) M   **G.B.F.Nilsson**
Nilsson, Lars Albert (1860–1906) S   **L.A.Nilsson**
Nilsson, Nils Heribert (né Nilsson) (1883–1955) S   **N.H.Nilsson**
(Nilsson, Nils Herman)
  see Nilsson–Ehle, Nils Herman   **Nilsson–Ehle**

Nilsson, Nils Hjalmar (1856–1925) S	**Nilsson**
Nilsson, Orjan Eric Gustaf (1933–) S	**O.Nilsson**
Nilsson, Siwert (1933–) S	**S.Nilsson**
Nilsson, Sven (1929–) M	**Sv.Nilsson**
Nilsson, T. (fl. 1977) M	**T.Nilsson**
(Nilsson–Degelius, Gunnar Bror Fritiof)	
see Nilsson, Gunnar Bror Fritiof	**G.B.F.Nilsson**
Nilsson–Ehle, Nils Herman (1873–1949) S	**Nilsson-Ehle**
Nimis, Pier Luigi (1953–) M	**Nimis**
Nimmo, Joseph (–1854) S	**Nimmo**
Nimura, Hiroyoshi (fl. 1962) M	**Nimura**
Ninbu (fl. 1981) S	**Ninbu**
Ninck, André (1872–1950) BF	**Ninck**
Ning, Zhu Hua (fl. 1988) S	**Z.H.Ning**
Ninh, Tran (fl. 1981) B	**Ninh**
Niño, F.L. (fl. 1930) M	**Niño**
Niolle, P. (fl. 1944) M	**Niolle**
Nipkow, Fritz (1886–1963) AM	**Nipkow**
Nirenberg, Helgard I. (fl. 1976) M	**Nirenberg**
Nisbet, Gladys T. (1895–) S	**G.T.Nisbet**
Nisbet, John (1853–1914) S	**Nisbet**
Nishida, Florence H. (fl. 1988) M	**F.H.Nishida**
Nishida, Harofumi (fl. 1982–) FP	**H.Nishida**
Nishida, Makoto (1927–) FPS	**M.Nishida**
Nishida, Shiro A	**S.Nishida**
Nishida, Toji (1874–1927) M	**Nishida**
Nishihara, Natsuki (1920–1990) M	**Nishih.**
Nishikado, Yosikazu (1892–) M	**Nishik.**
Nishikawa, Tsunehiko A	**Nishikawa**
Nishimura, Kazuko (fl. 1976) M	**Nishim.**
Nishimura, Naoki (1951–) B	**N.Nishim.**
Nishimura, S. (fl. 1981) M	**S.Nishim.**
Nishio, K. (fl. 1957) M	**Nishio**
Nishiwaki, Y. (fl. 1929) M	**Nishiw.**
Nishizawa, Tadahiro (fl. 1955) M	**Nishiz.**
Nisikado, Y. (fl. 1936) M	**Y.Nisik.**
Nisman, Carmen (fl. 1971) P	**Nisman**
(Nisman S., Carmen)	
see Nisman, Carmen	**Nisman**
Nissole, Guillaume (1647–1735) L	**Nissole**
(Nissolle, Guillaume)	
see Nissole, Guillaume	**Nissole**
Nitardy, E. A	**Nitardy**
Nitare, J. (fl. 1982) M	**Nitare**
Nitecki, Matthew H. (1925–) A	**Nitecki**
Nitsche, P. A	**P.Nitsche**
Nitsche, Walter (1883–) S	**Nitsche**
Nitschke, Theodor Rudolph Joseph (1834–1883) BM	**Nitschke**
Nitzelius, Tor (1914–) S	**Nitz.**

Nitzsch, Christian Ludwig (1782–1837) A	**Nitzsch**
Nival, Paul A	**Nival**
Niven, (David) James (1774–1826) S	**Niven**
Niviani, Nino (fl. 1833) F	**Niviani**
Nixon, Kevin C. (1953–) S	**Nixon**
Niyomdham, Chawalit (1949–) S	**Niyomdham**
Nizamuddin, Mohammed (1930–) A	**Nizam.**
Nkongmeneck, B.-A. (fl. 1986) S	**Nkongm.**
Noack, Fritz (1863–) M	**F.Noack**
Noack, Martin (1888–1927) S	**Noack**
Nobele, L.de (fl. 1893) S	**Nobele**
Noble, Elmer Ray (1909–) A	**E.R.Noble**
Noble, Glenn Arthur (1909–) A	**G.A.Noble**
Noble, Joanne M. A	**J.M.Noble**
Noble, Mary (fl. 1954) M	**Noble**
Noble, W.J. (fl. 1980–) M	**W.J.Noble**
Nobles, Mildred Katherine (1903–) M	**Nobles**
Nobre, Augusto (1865–) S	**Nobre**
Nobs, Malcolm A. (1916–) S	**Nobs**
Nocca, Domenico (1758–1841) BMS	**Nocca**
Noda, Mitsuzo (1909–) A	**Noda**
Noday, Olivier du (fl. 1877) B	**Noday**
Noë, Friedrich Wilhelm (18 –1858) S	**Noë**
Noeggerath, Johann Jacob (1788–1877) FS	**Noegg.**
Noehden, Heinrich Adolph (August) (1775–1804) MS	**Noehd.**
Noel, Charles William Francis (1850–1926) S	**Noel**
Noël, Denise A	**D.Noël**
Noël, Paul (fl. 1910) M	**P.Noël**
Noeldeke, Johann Ludwig Carl (Karl) (1815–1898) S	**Noeld.**
Noelli, Alberto (–1927) S	**Noelli**
Noerdlinger, Hermann (1818–1897) S	**Noerdl.**
Nogellehner, Dieter (fl. 1965) F	**Nogell.**
(Nöggerath, Johann Jacob)	
see Noeggerath, Johann Jacob	**Noegg.**
Nograsek, Andrea (fl. 1990) M	**Nograsek**
Noguchi, Akira (1907–) B	**Nog.**
Nogueira, Isabel Mariana Simões (1935–) S	**I.Nogueira**
Nogueira, P.J.C. (fl. 1922) M	**Nogueira**
Nöhden, H.A. (1775–1804) M	**Nöhden**
Noher de Halac, Rita I. (fl. 1973) S	**Noher**
Nohýnková, Eva A	**Nohýnk.**
Noicszewski, K. (fl. 1890) M	**Noicsz.**
Noisette, Louis Claude (1772–1849) S	**Nois.**
Nojima, T. (fl. 1927) M	**Nojima**
Nolan, Richard A. (fl. 1975) M	**Nolan**
Noll, Fritz (1858–1908) S	**Noll**
Noll, Henry R. (fl. 1852) S	**H.Noll**
Nolla, José Antonio Bernabé (1902–) M	**Nolla**
Nolte, Ernst Ferdinand (1791–1875) BS	**Nolte**

Noltie, Henry John (1957–) S	**Noltie**
Nomi, Ryosaku (fl. 1960) M	**Nomi**
Nomura, H. (1897–) M	**Nomura**
Nomura, Yukihiko (fl. 1960) M	**Y.Nomura**
Nongonierma, A. (fl. 1977) S	**Nongon.**
Nonis, Umberto (fl. 1977) M	**Nonis**
Nonne, Johann Philipp (1729–1772) S	**Nonne**
Nonomura, Arthur Michio A	**A.M.Nonom.**
Nonomura, H. (fl. 1954) M	**Nonom.**
Noor, M.N. A	**Noor**
Noordeloos, Machiel Evert (1949–) M	**Noordel.**
Nooteboom, Hans Peter (1934–) S	**Noot.**
(Nooten, Bertha Hoola van)	
see Hoola van Nooten, Bertha	**Hoola van Nooten**
Norberg, Ann M. (fl. 1979) M	**Norberg**
Nordal, Inger (previously Bjørnstad, I.N.) (1942–) S	**Nordal**
Nordborg, Gertrud (1931–) S	**Nordborg**
Nordenhed, A.-C. (fl. 1980) S	**Nordenhed**
Nordenskiöld, (Nils) Adolph Erik (von) (1832–1901) S	**Nordensk.**
Nordenskiöld, Hedda (fl. 1969) S	**H.Nordensk.**
Nordenstam, Rune Bertil (1936–) S	**B.Nord.**
Nordenstam, Sten Roland (1892–) S	**Nord.**
Nordgaard, O. A	**Nordg.**
Nordhagen, Rolf (1894–1979) S	**Nordh.**
Nordin, Ingvar (fl. 1964) M	**Nordin**
Nordli, Ottar A	**Nordli**
Nördlinger, Herman von (1818–1897) S	**Nördl.**
Nordmann, Alexander (Davidovič) von (1803–1866) S	**Nordm.**
Nordmann, V. A	**V.Nordm.**
Nordnes, J. (fl. 1982) M	**Nordnes**
Nordstedt, Carl Fredrik Otto (1838–1924) ABS	**Nordst.**
Nordstein, Stein (fl. 1990) M	**Nordstein**
Norem, W.L. (1910–) A	**Norem**
Norén, Carl Otto Gustav (1876–1914) S	**Norén**
Norkett, Alan Henry (1915–) B	**Nork.**
Norkrans, Birgitte (fl. 1968) M	**Norkrans**
Norlind, Josef Yngve Valentin (1887–) S	**Norlind**
Norlindh, Nils Tycho (1906–) S	**Norl.**
Norman, Cecil (1872–1947) S	**C.Norman**
Norman, Eliane Meyer (1931–) S	**E.M.Norman**
Norman, George (1824–1882) A	**G.Norman**
Norman, Johannes Musaeus (1823–1903) MS	**Norman**
Norman, Marion S	**M.Norman**
Normand, Didier (1908–) S	**Normand**
Normandin, Robert F. (1927–) A	**Normandin**
Normatova, R. (fl. 1981) S	**Normatova**
Noro, Shun–ichi (fl. 1988) M	**Noro**
Noroña, Francisco (c.1748–1787) PS	**Noronha**
Noronha, E.de A. (fl. 1952) M	**E.A.Noronha**

(Noronha, François (Fernando))
  see Noroña, Francisco — **Noronha**
Norris, Daniel Howard (1933–) B — **D.H.Norris**
Norris, Geoffrey (1937–) A — **G.Norris**
Norris, J.C. (fl. 1931) M — **Norris**
Norris, James Newcomb (1942–) A — **J.N.Norris**
Norris, Richard Earl (1926–) A — **R.E.Norris**
Norrlin, Johan Petter (Peter) (1842–1917) BMS — **Norrl.**
Northington, David K. (1944–) S — **North.**
Northrop, Alice Belle (Rich) (1864–1922) S — **Northr.**
Northstrom, T.Edward (fl. 1974) S — **Northstrom**
Norton, Jessie Baker (1877–1938) S — **J.B.Norton**
Norton, John Bitting Smith (1872–1966) MS — **Norton**
Norum, Elisaeus Pettersen (1868–1908) A — **Norum**
Noruzov, V.S. (fl. 1974) M — **Noruzov**
Norvick, M.S. A — **Norvick**
Nosaka, Shirô (1933–) S — **Nosaka**
Nose, H. (fl. 1933) M — **Nose**
Noshiro, S. (fl. 1985) S — **Noshiro**
(Notaris, Giuseppe (Josephus) De)
  see De Notaris, Giuseppe (Josephus) — **De Not.**
Notcutt, William Lowndes (1819–1868) S — **Notcutt**
Noter, Raphaël de (1857–1936) S — **Noter**
Notø, Andreas (1865–1948) S — **Notø**
Nott, Charles Palmer (fl. 1896–1900) A — **Nott**
Nötzold S — **Nötzold**
Noulet, Jean Baptiste (1802–1890) MS — **Noulet**
Nour, M.A. (fl. 1957) M — **Nour**
Nouvel, H. A — **Nouvel**
Novães, José de Campos (–1932) S — **Novães**
Novák, E.K. (fl. 1964) M — **E.K.Novák**
Novák, František Antonín (1892–1964) PS — **Novák**
Novák, Josef (1846–1917) S — **J.Novák**
Nováková, Marcela A — **Nováková**
Novelli, N. S — **Novelli**
Növgaard, V.A. (fl. 1901) M — **Növgaard**
Novichkova, L.N. A — **Novichk.**
Noviello, Carmine (fl. 1962) M — **Noviello**
Novik, E.O. (fl. 1968) F — **Novik**
Novikov, V.S. (1940–) S — **Novikov**
Novikova, L.N. (fl. 1989) S — **L.N.Novikova**
Novikova, M.V. (fl. 1989) S — **M.V.Novikova**
Novikova, N.G. S — **Novikova**
Novinskaya, V.F. A — **Novinsk.**
Novobranova, T.I. (fl. 1972) M — **Novobr.**
Novopokrovsky, Ivan Vassiljevich (1880–1951) S — **Novopokr.**
Novoselova, E.D. (fl. 1936) M — **Novos.**
Novotelnova, N.S. (fl. 1950) M — **Novot.**
Novouspenskiy, S.P. (fl. 1927) M — **Novousp.**

Novozhilov, Ju.K. (fl. 1985) M	**Novozh.**
(Novožilov, Yu.K.)	
see Novozhilov, Ju.K.	**Novozh.**
Novruzov, V.S. (fl. 1972) M	**Novruzov**
Nowacki, Edmund K. (1930–) S	**Nowacki**
Nowak, Janusz (1931–) M	**Nowak**
Nowak, K.A. (fl. 1982) M	**K.A.Nowak**
Nowakowska–Waszczuk, A. (fl. 1983) M	**Nowak.-Waszcz.**
Nowakowski, Leon (1847–19??) AM	**Nowak.**
Nowell, John (1802–1867) B	**Nowell**
Nowell, William (1880–1968) MS	**W.Nowell**
Nowicke, Joan W. (1938–) S	**Nowicke**
Nowotny, Wolfgang (fl. 1989) M	**Nowotny**
Nozaki, Hisayoshi A	**Nozaki**
Nozawa, Koji (1925–) A	**K.Nozawa**
Nozawa, Yuriko (1924–) A	**Nozawa**
Nozeran, René (1920–1989) S	**Nozeran**
Ntepe-Nyame, C. (fl. 1984) S	**Ntepe-Nyame**
Nubling (1876–1953) S	**Nubling**
Nüesch, Emil (1877–1959) M	**Nüesch**
Nüesch, Jakob (fl. 1960) M	**J.Nüesch**
Nunes, Valter Fraga (fl. 1991) S	**Nunes**
Nuno, Mariko (1932–) M	**Nuno**
Nurse, Frances R. A	**Nurse**
Nusrath, M. (fl. 1960) M	**Nusrath**
Nuss, Ingo (1941–) M	**Nuss**
Nutman, F.J. (fl. 1952) M	**Nutman**
Nuttall, Lawrence William (1857–1933) MS	**L.Nutt.**
Nuttall, Thomas (1786–1859) MPS	**Nutt.**
Nyananyo, B.L. (fl. 1986–) S	**Nyananyo**
Nyárády, Anton (1920–) S	**A.Nyár.**
Nyárády, Erasmus Julius (1881–1966) S	**Nyár.**
Nyberg, W. (fl. 1945) M	**Nyberg**
Nydegger, M. (fl. 1981) S	**Nydegger**
Nygaard, Gunnar (1903–) A	**Nygaard**
Nygreen, Paul W. (1925–) A	**Nygreen**
Nygren, Axel (1912–) S	**Nygren**
Nyholm, Elsa Tufvessen (1911–) B	**Nyholm**
Nyland, George (fl. 1952) M	**Nyland**
Nylander, (Wilhelm) William (1822–1899) ABMS	**Nyl.**
Nylander, Andreas Edvinus (1831–1890) M	**A.Nyl.**
Nylander, Fredrik (Frederick) (1820–1880) MPS	**F.Nyl.**
Nylander, Olof O. (1864–1943) S	**O.Nyl.**
Nyman, Carl Frederik (1820–1893) PS	**Nyman**
Nyman, Erik Olof August (1866–1900) BMS	**E.Nyman**
Nyman, Per Olof (fl. 1950) B	**P.Nyman**
Nyst, (Henry Joseph) Pierre (1780–1846) S	**Nyst**
O'Brien, Charlotte Grace (1845–1909) S	**C.O'Brien**
O'Brien, James (1842–1930) S	**O'Brien**

O'Brien, Muriel J. (fl. 1967) M	**M.J.O'Brien**
O'Connor, F.W. A	**F.W.O'Connor**
O'Connor, M. (fl. 1932) M	**M.O'Connor**
O'Connor, Patrick (1889–1969) S	**O'Connor**
O'Donell, Carlos Alberto (1912–1954) S	**O'Donell**
O'Donnell, Kerry L. (fl. 1973) M	**O'Donnell**
O'Gara, Patrick Joseph (1872–1927) M	**O'Gara**
O'Hara, Colonel A	**O'Hara**
O'Kane, S.L. (fl. 1987) S	**O'Kane**
O'Kelly, Charles J. (1953–) A	**O'Kelly**
O'Meara, Eugene (c.1815–1880) A	**O'Meara**
O'Neill, Hugh Thomas (1894–1969) S	**O'Neill**
O'Reilly, T. (fl. 1983) S	**O'Reilly**
O'Rourke, C.J. (fl. 1969) M	**O'Rourke**
O'Shanahan, J. (fl. 1968) S	**O'Shan.**
O'Shaugnessy (fl. 1842) S	**O'Shaugn.**
Oakes, William (1799–1848) MPS	**Oakes**
Oates, Kenneth A	**Oates**
Obenhaus, Renate (fl. 1959) M	**Obenhaus**
Oberdorfer, Erich (1905–) S	**Oberd.**
Oberhollenzer, H. (fl. 1984) M	**Oberholl.**
Oberholzer, Ernst (fl. 1886–1950) P	**Oberh.**
Obermayer, Walter (1960–) M	**Obermayer**
(Obermeijer, Anna Amelia)	
see Obermeyer, Anna Amelia	**Oberm.**
(Obermejer, Anna Amelia)	
see Obermeyer, Anna Amelia	**Oberm.**
Obermeyer, Anna Amelia (1907–) PS	**Oberm.**
Obermeyer, Wilhelm (fl. 1908) M	**W.Oberm.**
(Obermeyer–Mauve, Anna Amelia)	
see Obermeyer, Anna Amelia	**Oberm.**
Oberste–Brink, K. S	**Ob.-Brink**
Oberwinkler, Franz (1939–) M	**Oberw.**
Obón de Castro, C. (fl. 1990) S	**Obón**
Oborný, Adolf (1840–1924) S	**Oborný**
Obradović, M. (fl. 1974) S	**Obrad.**
Obregón–Botero, Rafael (1909–) M	**Obreg.-Bot.**
Obrhel, Jiří A	**Obrhel**
Obrist, Johann (fl. 1879) S	**Obrist**
Obrist, Walter (fl. 1959) M	**W.Obrist**
Obukhova, V.M. A	**Obukhova**
Ocakverdi, H. (fl. 1986) S	**Ocakv.**
Ocampo–Paus, Roseli (1937–) A	**Ocampo-Paus**
Occhioni, Elêna Maria de Lamare (fl. 1985) S	**Occhioni f.**
Occhioni, Paul (1915–) S	**Occhioni**
Ochi, Harumi (1920–) B	**Ochi**
Ochiauri, D.A. (fl. 1981) S	**Ochiauri**
Ochoa, C.M. (fl. 1952) S	**Ochoa**
Ochoterena, E.L. (fl. 1924) M	**E.L.Ochot.**

Ochoterena, Isaac (1885–1950) S	**Ochot.**
Ochse, Jacob Jonas (1891–1970) S	**Ochse**
Ochsner–Christen, Fritz (1899–) S	**Ochsner**
Ochyra, Ryszard (1949–) B	**Ochyra**
Ockendon, David Jeffery (1940–) S	**Ockendon**
Oda, M. (fl. 1932) M	**Oda**
Odashima, Kojirô S	**Odash.**
Oddo, Nicola (fl. 1967) M	**Oddo**
(Odinzova, E.N.)	
see Odinzowa, E.N.	**Odinzowa**
Odinzowa, E.N. (fl. 1947) M	**Odinzowa**
Oeder, George Christian Edler von Oldenburg (1728–1791) BMPS	**Oeder**
Oefelein, Hans (1905–1970) S	**Oefelein**
Oehler, D.Z. A	**D.Z.Oehler**
Oehler, J.H. A	**J.H.Oehler**
Oehlkers, F. (fl. 1922) M	**Oehlkers**
Oehme, Hans (fl. 1940) S	**Oehme**
Oehninger, Carl Johannes (fl. 1908) S	**Oehn.**
Oehrens B., Edgardo (fl. 1987) M	**Oehrens**
Oelhafen von Schoellenbach, Carl Christoph (1709–1783) S	**Oelhafen**
(Oellgaard, Benjamin)	
see Øllgaard, Benjamin	**B.Øllg.**
(Oellgaard, Hans)	
see Øllgaard, Hans	**H.Øllg.**
(Oers, F.Van)	
see Van Oers, F.	**Van Oers**
Oersted, Anders Sandoe (1816–1872) AMS	**Oerst.**
Oertel, C.Gustav (1833–1908) BM	**Oertel**
Oescu, C.V. (1891–1973) M	**Oescu**
Oeser, R. (fl. 1976) S	**Oeser**
Oesterreich, Gernard S	**Oesterr.**
Oettel, Carl Christian (1742–1819) S	**Oettel**
Oettingen, Heinrich von (fl. 1890–1925) S	**Oett.**
Oettli, Max A	**Oettli**
Offner, Jules (1873–1957) M	**Offner**
Ofosu–Asiedu, A. (fl. 1975) M	**Ofosu–As.**
Oftedahl, Orrin A	**Ofted.**
Oganesian, Marina E. (1954–) S	**Ogan.**
(Oganesyan, Marina E.)	
see Oganesian, Marina E.	**Ogan.**
Oganova, É.A. (fl. 1960) M	**Oganova**
Ogarkov, B.N. (fl. 1979) M	**Ogarkov**
Ogata, Masasuke (1883–1944) PS	**Ogata**
Ogata, S. (fl. 1929) M	**S.Ogata**
Ogawa, J.M. (fl. 1990) M	**J.M.Ogawa**
Ogawa, T. (fl. 1952) M	**Ogawa**
Ogden, Eugene Cecil (1905–) S	**Ogden**
Ogihara, T. (fl. 1981) M	**Ogihara**

Ogilvie, Barbara M.L. (fl. 1930s) S	**Ogilvie**
Ogimi, C. (fl. 1972) M	**Ogimi**
Ogle, C.C. (fl. 1987) P	**Ogle**
Ogoshi, A. (fl. 1986) M	**Ogoshi**
Oguchi, T. (fl. 1980) M	**Oguchi**
Ogura, Yudzuru (Yuzuro) (1895–1981) BFPS	**Ogura**
Oguri, H. (fl. 1983) M	**Oguri**
Oh, Byoung Un (1954–) S	**B.U.Oh**
Oh, S.Y. (fl. 1978) S	**S.Y.Oh**
Oh, Yong Cha (fl. 1986) P	**Y.C.Oh**
Ohara, J. (fl. 1985) S	**J.Ohara**
Ohara, Y. (fl. 1954) M	**Y.Ohara**
Ohashi, Hiro (1882–) AS	**Ohashi**
Ohashi, Hiroyoshi (1936–) S	**H.Ohashi**
Ohba, Hideaki (1943–) PS	**H.Ohba**
Ohba, Tatsuyuki (1936–) S	**Ohba**
Ohinata, Z. (fl. 1912) B	**Ohinata**
Ohira, Ikuo (1947–) M	**Ohira**
Ohishi, T. (fl. 1980) S	**Ohishi**
Ohki, Kiichi (1882–) S	**Ohki**
Ohki, Masao (1930–) S	**M.Ohki**
Ohl, J.A. (fl. 1922) M	**Ohl**
Ohle, H. (1937–) S	**Ohle**
Ohlendorff, J.H. (–1857) S	**Ohlend.**
Ohlert, Otto Ludwig Arnold (1816–1875) M	**Ohlert**
Ohlsén, Adolf Ragnar (1875–1957) S	**Ohl.én**
Ohlsen, R. S	**R.Ohlsen**
(Ohlson, Karl E.)	
see Ohlsson, Karl E.	**Ohlsson**
Ohlsson, Karl E. (1940–) M	**Ohlsson**
Ohmi, Hikoei (1914–) A	**Ohmi**
Ohmura, T. (fl. 1960s) P	**Ohmura**
Ohno, N. A	**Ohno**
Oho, O. (fl. 1919) M	**Oho**
Ohr, Howard D. (fl. 1989) M	**Ohr**
Ohsaki, Manji (1958–) A	**Ohsaki**
Ohsawa, Takeshi (Takashi) (fl. 1989) M	**Tak.Ohsawa**
Ohsawa, Tatsuro (fl. 1983) S	**T.Ohsawa**
Ohta, Masataka (1950–) A	**Ohta**
Ohta, Tatsuo A	**T.Ohta**
Ohtani, Shigeru (fl. 1961) PS	**Ohtani**
Ohtsuki, Torao (fl. 1962) M	**Ohtsuki**
Ohwi, Jisaburo (1905–1977) PS	**Ohwi**
Oishi, C. (fl. 1979) M	**C.Oishi**
Oishi, Saburo (fl. 1931–49) F	**Oishi**
Oka, K. (1918–) PS	**Oka**
Oka, T. (fl. 1980) P	**T.Oka**
Okada, Gen (fl. 1984) M	**G.Okada**
Okada, Hiroshi (fl. 1984) S	**H.Okada**

Okada, Hisatake A	**His.Okada**
Okada, K.A. (fl. 1985) S	**K.A.Okada**
Okada, M. (fl. 1990) S	**M.Okada**
Okada, Yoshikazu (1902–1984) A	**Okada**
Okafor, Jonathan Chukuemeka (1934–) S	**Okafor**
Okami, Yoshio (fl. 1954) M	**Okami**
Okamoto, Motoharu (1947–) S	**Okamoto**
Okamura, H. (fl. 1979) S	**H.Okamura**
Okamura, Kintarô (1867–1935) AS	**Okamura**
Okamura, Shûtai (1877–1947) B	**S.Okamura**
Okeden, Fitzmaurice A	**Okeden**
Oken, Lorenz (1779–1851) MS	**Oken**
(Okenfuss, Lorenz)	
see Oken, Lorenz	**Oken**
Oki, T. (fl. 1972) M	**Oki**
Okón, Lázaro A	**Okón**
Okpala, E.U. (fl. 1979) M	**Okpala**
Oksner, Alfred Nikolaevich (1898–1973) MS	**Oksner**
Ôkubo, Mariko (fl. 1953) M	**M.Ôkubo**
Ôkubo, Saburô (1857–1914) S	**Ôkubo**
Okuda, T. (fl. 1977) M	**Okuda**
Okuma, Yutaka (fl. 1987) M	**Okuma**
Okuno, Haruo A	**Okuno**
Okunuki, K. (fl. 1931) M	**Okunuki**
Okuyama, Shunki (1909–) PS	**Okuyama**
Ola'h, G.–M. (fl. 1937) M	**Ola'h**
Olafsen, Claus S	**Olafsen**
(Olah, G.M.)	
see Ola'h, G.–M.	**Ola'h**
Olaio, Amélia Almeida Ribeiro (fl. 1990) S	**Olaio**
Olarinmoye, S.O. (fl. 1978) B	**Olar.**
Olchowecki, Alexander (fl. 1974) M	**Olchow.**
Oldeman, Roelof Arend Albert (1937–) S	**Oldeman**
Olden, E.van S	**Olden**
Oldenland, Henrik Bernard (–1699) L	**Oldenl.**
Oldham, Thomas (1816–1878) FS	**Oldham**
Olech, Maria (1941–) M	**Olech**
Olexia, M. (fl. 1969) M	**Olexia**
Olexia, P.D. (fl. 1986) M	**P.D.Olexia**
Olin, Emil Hjalmar Frederik (1869–1915) S	**E.H.F.Olin**
Olin, Johan Henrick (1769–1824) S	**Olin**
Oliva, Alberti (1879–1953) S	**A.Oliva**
Oliva, Leonardo (1805–1873) S	**Oliva**
Oliva, Pedro (fl. 1989) M	**P.Oliva**
Oliva, R. (fl. 1982) B	**R.Oliva**
Olive, Edgar William (1870–1971) AMS	**Olive**
Olive, Lindsay Shepherd (1917–) M	**L.S.Olive**
Oliveira, António Lopes Branquino d' (fl. 1940) M	**Oliveira**
Oliveira, Eurico Cabral de (1940–) A	**E.C.Oliveira**
Oliveira, Fernando (fl. 1973) S	**F.Oliveira**

(Oliveira Filho, Eurico Cabral de)
  see Oliveira, Eurico Cabral de   **E.C.Oliveira**
Oliveira, Julieta Celestino   A   **J.C.Oliveira**
Oliveira, Sérgio A.A. de (fl. 1990) S   **S.Oliveira**
Oliver, Daniel (1830–1916) PS   **Oliv.**
Oliver, Edward George Hudson (1938–) S   **E.G.H.Oliv.**
Oliver, Francis Wall (1864–1951) S   **F.Oliv.**
Oliver, Royce Ladell (1929–) S   **R.L.Oliv.**
Oliver, Walter Reginald Brook (1883–1957) PS   **W.R.B.Oliv.**
Oliveros, M.B. (fl. 1982) S   **Oliveros**
Olivi, Giuseppe (1769–1795) A   **Olivi**
Olivier, (Joseph) Ernst (1844–1914) S   **E.Olivier**
Olivier, Dorothea L. (fl. 1978) M   **D.L.Olivier**
Olivier, G. (fl. 1926) M   **G.Olivier**
Olivier, Guillaume Antoine (1756–1814) S   **Olivier**
Olivier, Henri Jacques François (1849–1923) M   **H.Olivier**
Øllgard, Benjamin (1943–) PS   **B.Øllg.**
Øllgaard, Hans (1943–) S   **H.Øllg.**
Olliver, Mamie (fl. 1933) M   **Olliver**
Ollivier, Gaston–Maurice (1890–1929) AM   **Ollivier**
Ollivier, P.O. (fl. 1930) M   **P.O.Ollivier**
Olmsted, Frederick Law (1822–1903) S   **Olmsted**
Olney, Stephen Thayer (1812–1878) AS   **Olney**
Olofsson, Paul (1896–) S   **Olofsson**
Olorode, Onotoye (fl. 1970) S   **Olorode**
Olowokudejo, J.D. (fl. 1986) S   **Olow.**
Olsen, Jeanine Louise (1952–) A   **J.L.Olsen**
Olsen, John S. (1950–) S   **Olsen**
Olsen, Sigurd   A   **S.Olsen**
Olsen–Sopp, O.J. (fl. 1912) M   **Olsen–Sopp**
Olson, Alver J. (fl. 1941) M   **A.J.Olson**
Olson, H.C. (fl. 1937) M   **H.C.Olson**
Olson, Mary E. (fl. 1897) M   **Olson**
(Olsson, Pehr Hjalmar)
  see Olsson–Seffer, Pehr Hjalmar   **Olss.–Seff.**
Olsson, Peter (1833–1906) P   **P.Olsson**
Olsson, Peter (Petter) (1838–1923) S   **P.Olsson bis**
Olsson–Seffer, Pehr Hjalmar (1873–1911) S   **Olss.–Seff.**
Oltean, Mircea   A   **Oltean**
Oltmanns, Friedrich (1860–1945) A   **Oltm.**
Omang, Simon Oskar Fredrik (1867–1953) S   **Omang**
Omarov, D.S. (fl. 1978) S   **Omarov**
(Omelchuk–Myakushko, Taisija Ja.)
  see Omelczuk, Taisija Ja.   **Omelczuk**
Omelczuk, Taisija Ja. (1933–) S   **Omelczuk**
Omer, Saood (1957–) S   **Omer**
Ommering, G.van (fl. 1977) S   **Ommering**
Omvik, Aasa (fl. 1955) M   **Omvik**
Ondratschek, Karl   A   **Ondr.**

Ondřej, Michal (fl. 1969) M   **Ondřej**
Ondrushova, Dagmar (fl. 1971) M   **Ondrush.**
Ondrušová, Dagmar
  see Ondrushová, Dagmar   **Ondrush.**
Oniki, M. (fl. 1986) M   **Oniki**
Onions, Agnes H.S. (later Brown, Agnes H.S.) (fl. 1966) M   **Onions**
Onishi, H. (fl. 1957) M   **Onishi**
Onnis, Antonio (1929–) S   **Onnis**
Onno, Max (1903–) S   **Onno**
Ono, Chitari   A   **C.Ono**
Ono, Motoyoshi (1837–1890) S   **Ono**
Ono, Yoshitaka (fl. 1978) M   **Y.Ono**
Onochie, Charles Francis Akado (1914–) S   **Onochie**
Onofri, Silvano (1951–) M   **Onofri**
Onoyama, N. (fl. 1964) M   **Onoyama**
Onraedt, Maurice (1904–) B   **Onr.**
Onsberg, Per (fl. 1972) M   **Onsberg**
Onuki, Masatoshi (fl. 1990) M   **Onuki**
Onuma, Fusaji (fl. 1930) M   **Onuma**
(Ookubo, Mariko)
  see Ôkubo, Mariko   **M.Ôkubo**
Oolbekkink, G.T. (fl. 1991) M   **Oolbekk.**
Oomen, H.A.P.C. (fl. 1935) M   **Oomen**
Oorschot, Connie A.N.van (fl. 1977) M   **Oorschot**
Oort, Arend Joan Petrus (1903–) MS   **Oort**
Ooshima, Kaiichi (1948–) A   **Ooshima**
Oosten, M.W.B van (fl. 1840) S   **Oosten**
Oosterveld, P. (fl. 1978) S   **Oosterv.**
Ooststroom, Simon Jan van (1906–1982) PS   **Ooststr.**
Opatowski, Heinrich (1812–) S   **H.Opat.**
Opatowski, Wilhelm (Guilelmus) (1810–1838) M   **Opat.**
Ophel, I.L.   A   **Ophel**
Opiz, Philipp (Filip) Maximilian (1787–1858) BMPS   **Opiz**
Oppenheim, Leo Paul (1863–1934) A   **Oppenheim**
Oppenheimer, Hillel (Heinz) Reinhard (1899–) S   **Oppenh.**
Opperman, A. (fl. 1986) M   **A.Opperman**
Opperman, P.A. (–1942) S   **Opperman**
Orange, Alan (1955–) M   **Orange**
Orazmuchommedov, A. (1924–) S   **Orazm.**
Orazova, A.O. (1928–) S   **Orazova**
Orban, Sandor (1947–) B   **Orban**
Orbigny, (Alcide) Charles Victor Dessalines d' (1806–1876) CMP   **Orb.**
Orbigny, Alcide Dessalines d' (1802–1857) BFS   **A.D.Orb.**
(Orčenášek, M.)
  see Orchenashek, M.   **Orchen.**
Orchard, Anthony Edward (1946–) S   **Orchard**
Orchenashek, M. (fl. 1977) M   **Orchen.**
Orcutt, Charles Russell (1864–1929) ABS   **Orcutt**
Ordoyno, Thomas (fl. 1790s–1810s) S   **Ordoyno**

Ordway, John Morse (1823–1909) S — **Ordway**
Oredsson, Alf (1938–) S — **Oredsson**
Orejuela, Carlos Garcés (1915–) M — **Orejuela**
(Orell Casasnovas, Jeroni)
  see Orell, Jeroni — **Orell**
Orell, Jeroni (1924–) P — **Orell**
Orellana, R.G. (fl. 1959) M — **Orellana**
(Orey, J.D.S.d')
  see D'Orey, J.D.S. — **D'Orey**
Orishimo, Yoshinobu (1881–) M — **Orish.**
Orlicz, Anna (1941–) M — **Orlicz**
Orlós, H. (fl. 1967) M — **Orlós**
Orlov, A.A. (1888–) S — **Orlov**
Orlov, J.A. S — **J.A.Orlov**
Orlova, A. (fl. 1925) M — **Orlova**
Orlova, N.I. (1921–) S — **N.I.Orlova**
Ormeno–Nuněz, J. (fl. 1988) M — **Ormeno–Nuñez**
Ormezzano, Quentin Jean Baptiste (1854–1912) S — **Ormezz.**
Ormières, René (fl. 1961) M — **Ormières**
Ormonde, José Eduardo Martins (1943–) PS — **Ormonde**
Ornduff, Robert (1932–) S — **Ornduff**
Örösi–Pál, Z. (fl. 1936) M — **Örösi–Pál**
Orozco, Clara Inés (1952–) S — **C.I.Orozco**
Orozco, E. (fl. 1921) M — **Orozco**
Orphanides, Theodhoros Georgios (1817–1886) S — **Orph.**
Orpin, C.G. (fl. 1981) M — **Orpin**
Orpurt, Philip A. (fl. 1955) M — **Orpurt**
Orr, G.F. (fl. 1963) M — **G.F.Orr**
Orr, Leslie Wayne (1902–) M — **L.W.Orr**
Orr, Matthew Young (?1883–1953) S — **Orr**
Orshanskaya, V. (fl. 1915) M — **Orshansk.**
Orsi, Maria Cristina (1947–) S — **Orsi**
Orssich (fl. 1985) S — **Orssich**
Örstadius, L. (fl. 1984) M — **Örstadius**
(Örsted, Anders Sandö)
  see Oersted, Anders Sandoe — **Oerst.**
(Örsted, Anders Sandö)
  see Oersted, Anders Sandoe — **Oerst.**
Orszag–Sperber, F. A — **Orsz.–Sperb.**
(Orta, Garcia de)
  see Garcia de Orta — **Garcia de Orta**
Ortega, A. (1943–) M — **A.Ortega**
Ortega, Casimiro Gómez de (1740–1818) PS — **Ortega**
Ortega, Jesús González (1876–1936) S — **J.G.Ortega**
Ortega, L.M. (fl. 1987) S — **L.M.Ortega**
Ortega, María Dolores (fl. 1985) S — **M.D.Ortega**
(Ortega Menes, María Dolores)
  see Ortega, María Dolores — **M.D.Ortega**
Ortega, Olivencia A. S — **O.A.Ortega**

(Ortega Torres, L.M.)	
see Ortega, L.M.	**L.M.Ortega**
Ortgies, Karl Eduard (1829–1916) S	**Ortgies**
Ortíz, Evangelina S	**E.Ortíz**
Ortíz, J.Javier (1957–) S	**J.J.Ortíz**
(Ortíz Núñez, Santiago)	
see Ortíz, Santiago	**S.Ortíz**
Ortíz, Santiago (1957–) S	**S.Ortíz**
Ortíz Valdivieso, Pedro (fl. 1976–) S	**P.Ortíz**
Ortloff, Fr. (–1896) B	**Ortloff**
Ortmann, Anton (1801–1861) S	**Ortmann**
Ortmann, Arnold Edward (1863–1927) S	**A.E.Ortmann**
Ortmann, Johann (1814–1890) S	**J.Ortmann**
Orton, Clayton Roberts (1885–1955) M	**Orton**
Orton, Peter D. (1916–) M	**P.D.Orton**
Orton, William Allen (1877–1930) M	**W.Orton**
Ortuño Medina, Francisco (1919–) S	**Ortuño**
Orzell, Steve L. (fl. 1989) S	**Orzell**
Osada, Keigo (1956–) A	**K.Osada**
Osada, Takemasa (1912–) B	**Osada**
Osbeck, Pehr (1723–1805) MS	**Osbeck**
Osborn, Arthur (1878–1964) S	**Osborn**
Osborn, Henry Stafford (1823–1894) S	**H.Osborn**
Osborn, Jeffrey M. (fl. 1989) M	**J.M.Osborn**
Osborn, Theodore George Bentley (1887–1973) MS	**T.Osborn**
(Osborne, Theodore George Bentley)	
see Osborn, Theodore George Bentley	**T.Osborn**
Osbornova–Kosinova, J. (fl. 1981) S	**Osb.–Kos.**
Oshio, Masayoshi (1937–) M	**Oshio**
Oshite, Kei (1919–) A	**Oshite**
Osipian, Lia Levonevna (1930–) M	**Osipian**
(Osipyan, Lia Levonevna)	
see Osipian, Lia Levonevna	**Osipian**
Oskamp, Dirk Leonard (1768–) S	**Oskamp**
Óskarsson, Ingimar (1892–1981) S	**Ósk.**
Osmaston, Arthur Edward (–1972) S	**Osmaston**
Osment, W. S	**Osment**
Osmolovskaya, M.B. (fl. 1969) M	**Osmol.**
Osner, George Adin (1888–) M	**Osner**
Osorio, Hector S. (1928–) M	**Osorio**
Osorio, N. S	**N.Osorio**
Osorio Tafall, B.F. A	**B.F.Osorio**
Ospina, H.Mariano (fl. 1973) S	**Ospina**
Ossa, José Antonio de la (–1829) S	**Ossa**
Ossycznjuk, V.V. (1918–) S	**Ossyczn.**
Ostapko, V.M. (1950–) S	**Ostapko**
Ostazeski, Stanley A. (fl. 1968) M	**Ostaz.**
Osten, Cornelius (1863–1936) PS	**Osten**
Osten–Sacken, Friedrich P.von der (Fedor Romanovič) (1832–) S	**Ost.–Sack.**

(Osten–Saken, Friedrich P.von der (Fedor Romanovič))
   see Osten–Sacken, Friedrich P.von der (Fedor Romanovič)   **Ost.-Sack.**
Ostenfeld, Carl Emil Hansen (1873–1931) AMS   **Ostenf.**
Osterhout, George Everett (1858–1937) S   **Osterh.**
Ostermeyer, R. S   **Osterm.**
Osterwald, Karl (1853–1923) B   **Osterwald**
Osterwalder, Adolph (Adolf) (1872–1948) M   **Osterw.**
Østhagen, Haavard (fl. 1976) M   **Østh.**
Östman, Magnus (1852–1927) S   **Östman**
Ostolaza, C. (fl. 1983) S   **Ostolaza**
(Ostolaza N., C.)
   see Ostolaza, C.   **Ostolaza**
Ostrejko, S.A. S   **Ostrejko**
Østrup, Ernst Vilhelm (1845–1917) A   **Østrup**
Osvacil (fl. 1950) S   **Osvacil**
Osvačilová, Vera (1924–) S   **Osvač.**
Oswald, Fred W. (fl. 1971) S   **Oswald**
Oswald, T. (fl. 1979) M   **T.Oswald**
Osychnyuk, V.V. (fl. 1976) S   **Osychnyuk**
Ota, M. (fl. 1927) M   **M.Ota**
Ota, Noboru (fl. 1923) M   **N.Ota**
Otani, Hironao (fl. 1932) M   **Otani**
Otani, Yoshio (fl. 1958) M   **Y.Otani**
Otedoh, M.O. (fl. 1982) S   **Otedoh**
Oteng–Yeboah, A.A. (fl. 1970) S   **Oteng–Yeb.**
Otero Jañez, José Idilio (1893–) S   **Otero**
Oti, Kazuo (1908–) B   **Oti**
Otieno, N.C. (fl. 1969) M   **Otieno**
Otis, Charles Herbert (1886–1976) S   **Otis**
Otruba, Josef (1889–1952) S   **Otruba**
Otschiauri, D.A. (1917–) S   **Otsch.**
Otsuka, K. (fl. 1976) P   **Otsuka**
Ott, Donald W. A   **D.W.Ott**
Ott, Ernst A   **E.Ott**
Ott, Johann (fl. 1851) S   **Ott**
Ott, Jonathan (fl. 1977) M   **J.Ott**
Ottaviano, Vincenzo (1790–1853) S   **Ottav.**
Otth, Carl (Karl) Adolph (1803–1839) S   **Otth**
Otth, Gustav Heinrich (1806–1874) M   **G.H.Otth**
Ottley, Alice Maria (1882–1971) S   **Ottley**
Otto, Bernhard Christian (1745–1835) S   **B.Otto**
Otto, Carlos Frederico Eduardo (1812–1885) S   **Ed.Otto**
Otto, Christoph Friedrich (1783–1856) S   **Otto**
Otto, Ernst von (1799–1863) S   **E.Otto**
Otto, Johann Gottfried (1761–1832) MS   **J.Otto**
Ottolander, Kornelius Johannes Willem (1822–1887) S   **Ottol.**
Ottone, E.G. (fl. 1983) F   **Ottone**
Otu, M. (fl. 1957) M   **Otu**
Ou, Chern Hsiung (fl. 1976) S   **C.H.Ou**

Ou, S.H. (fl. 1936) M   **S.H.Ou**
Ou, Shan Hua (fl. 1985) S   **Shan Hua Ou**
Oudejans, Robertus Cornelis Hilarius Maria (fl. 1990) S   **Oudejans**
Oudemans, Cornelius Anton Jan Abraham (Corneille Antoine Jean Abram)
    (1825–1906) MS   **Oudem.**
Oudney, Walter (1790–1824) S   **Oudney**
Ouellette, G.B. (fl. 1966) M   **Ouell.**
Ouyahya, A. (fl. 1982) S   **Ouyahya**
(Ovadiahu–Yavin, Ziva)
    see Yavin, Z.   **Yavin**
(Ovchinnikov, Pavel Nikolaevich)
    see Ovczinnikov, Pavel Nikolaevich   **Ovcz.**
Ovczinnikov, Pavel Nikolaevich (1903–1979) S   **Ovcz.**
Oven, E.van (fl. 1905) M   **Oven**
Over, William Henry (1866–) S   **Over**
Overeem, Casper van (1893–1927) M   **Overeem**
Overeem, D.van (fl. 1922) M   **D.Overeem**
(Overeem de Haas, Casper van)
    see Overeem, Casper van   **Overeem**
Overholts, Lee Oras (1890–1946) M   **Overh.**
(Overin, Alexander Pavlovic)
    see Owerin, Alexander Pavlovic   **Owerin**
Overkott, Ortrud (1914–) S   **Overkott**
Ovrebo, Clark L. (fl. 1983) M   **Ovrebo**
Øvstedal, Dag Olav (1944–) M   **Øvstedal**
Ŏwaki, K. (fl. 1966) M   **Ŏwaki**
Owen, John H. (fl. 1956) M   **J.H.Owen**
Owen, Maria Louisa (1825–1913) S   **Owen**
Owen, Mary Nest (later Kidd, M.N.) (fl. 1919) M   **M.N.Owen**
Owen, Peter Edward (1966–) S   **P.E.Owen**
Owen, W.L. (fl. 1948) M   **W.L.Owen**
Owerin, Alexander Pavlovic S   **Owerin**
Ownbey, Francis Marion (1910–1974) S   **Ownbey**
Ownbey, Gerald Bruce (1916–) S   **G.B.Ownbey**
Oxley, Frederick A   **Oxley**
Oxner, A.M. (fl. 1986) M   **Oxner**
(Oxner, Alfred Nikolaevich)
    see Oksner, Alfred Nikolaevich   **Oksner**
Oyasu, N. (fl. 1954) M   **Oyasu**
Oye, Paul Hermann Gustav von (1886–1969) A   **Oye**
Oyewole, S.O. (fl. 1973) S   **Oyewole**
Oyster, John Houck (1849–after 1904) S   **Oyster**
Ozanin, C. (1835–1909) S   **Ozanin**
Ozanon, H.Charles (1835–1909) S   **Ozanon**
Ožegović, J.W. (fl. 1973) M   **Ožeg.**
Ozenda, Paul (1920–) MS   **Ozenda**
Özhatay, Neriman (1947–) S   **Özhatay**
(Özhatey, Neriman)
    see Özhatay, Neriman   **Özhatay**

Özkan, Hamdi (fl. 1947) M   Özkan
Özkan, Mediha (fl. 1947) M   M.Özkan
Öztürk, Avni (fl. 1978) S   Öztürk

P'ei, Chien (1903–) S   C.P'ei
(P.de Resende, Flávio)
  see Resende, Flávio P.de   Resende
Paasche, Eystein (1932–) A   Paasche
Pabot, Henri A. (fl. 1966) S   Pabot
Pabrez, A. S   A.Pabrez
Pabrez, J. S   J.Pabrez
Pabst, Carl (1825/6–1863) S   C.Pabst
Pabst, Guido Frederico João (1914–1980) S   Pabst
Pabst, Gustav (–1911) BS   G.Pabst
Paccard, Ernesto (fl. 1905) S   Paccard
Pace, Loretta (1961–) S   Pace
Pacey, Jane M. B   Pacey
Pacha–Aue, R. (fl. 1968) M   Pacha–Aue
Pacheco Leão, Antonio (1872–1931) S   Pach.Leão
Pacher, David (1816–1902) PS   Pacher
Pachkhede, A.U. (fl. 1985) M   Pachkhede
Pachomova, M.G. (1925–) S   Pachom.
Pacioni, Giovanni (1948–) M   Pacioni
Packard, Alpheus Spring (1839–1905) A   Packard
Packard, Charles E. (–1970) A   C.E.Packard
Packard, Patricia L. (1927–) S   P.L.Packard
Packe, Charles (1826–1896) S   Packe
Packer, John G. (1929–) S   Packer
Paclt, Jiři(Jiré) (1925–) MS   Paclt
Pacottet, P. (fl. 1904) M   Pacottet
Pacyna, Anna (1940–) S   Pacyna
Paczoski, Jozef Konradovich (Joseph Conradovich) (1864–1942) S   Pacz.
Padberg, M. B   Padberg
Paddock, T.B.B. (1942–) A   Paddock
Paden, John W. (1933–1990) M   Paden
Padhi, B. (fl. 1954) M   Padhi
Padhye, Arvind A. (fl. 1962) M   A.A.Padhye
Padhye, Y.A. (fl. 1954) M   Y.A.Padhye
Padma, S.D. (fl. 1985) M   Padma
Padmanabhan, S.Y. (fl. 1965) M   Padman.
Padmore, Patricia Anne (1929–) S   Padmore
Paădure, Elena (fl. 1972) M   Paădure
Padwardhan, P.G. (fl. 1962) M   Padwardhan
Padwick, Geoffrey Watts (1909–) M   Padwick
Paechnatz, E. (fl. 1987) M   Paechn.
Paegle, B. (fl. 1927) S   Paegle
Paes, Antonio S.Ana (fl. 1918–) M   Paes
Paes, Luiz Emundo (fl. 1971) S   L.E.Paes
Paéz de Badillo, I. (fl. 1982) M   Paéz

Pagan, Francisco Mariano (1896–1942) B	**Pagan**
Page, Christopher Nigel (1942–) PS	**C.N.Page**
Page, Jennifer S. (1949–) S	**J.S.Page**
Page, Joanna R.Ziegler (1938–) A	**J.R.Z.Page**
Page, V. (fl. 1967) F	**V.Page**
Page, William Bridgewater (1790–1871) S	**Page**
Page, Winifred Mary (1887–1965) M	**W.M.Page**
Pages, Xavier (fl. 1975) M	**Pages**
Paget, James (1814–1899) S	**Paget**
Paglia, Enrico (1834–1889) S	**Paglia**
Pahnsch, Gerhard (–1880) S	**Pahnsch**
Pai, Chin Kai (fl. 1957) M	**C.K.Pai**
Pai, K.M. A	**K.M.Pai**
Pai, Pei Yu (1938–) S	**P.Y.Pai**
Pai, Yin Yüan (fl. 1933) S	**Y.Y.Pai**
Paiche, Philippe (1842–1911) S	**Paiche**
Paige, Edward Winslow (1844–1918) S	**Paige**
Pailleux, (Nicolas) Auguste (1812–1898) S	**Pailleux**
Paillot, Justin (1829–1891) S	**Paill.**
Paine, Frederick Sylvanus (1883–) M	**F.S.Paine**
Paine, John Alsop (1840–1912) S	**Paine**
Paine, M.J.H. S	**M.J.H.Paine**
Paine, Sydney Gross (1881–1937) M	**S.G.Paine**
Painter, Joseph Hannum (1879–1908) S	**J.H.Painter**
Painter, William Hunt (1835–1910) S	**Painter**
(Pais, Antonio S.Ana)	
see Paes, Antonio S.Ana	**Paes**
Paiva, Antonio da Costa (1806–1879) S	**A.Paiva**
Paiva, Jorge Américo Rodrigues (1933–) S	**Paiva**
Pajarón, Santiago (1954–) S	**Pajarón**
Pakaln, D.A. (fl. 1971) S	**Pakaln**
Pákh, Erzsébet H. A	**Pákh**
(Pakhomova, M.G.)	
see Pachomova, M.G.	**Pachom.**
Pal, Ajit Kumar A	**A.K.Pal**
Pal, Benjamin Peary (1906–1989) A	**Pal**
Pal, D.C. (fl. 1975) S	**D.C.Pal**
Pal, G.D. (fl. 1986) S	**G.D.Pal**
Pal, J. (fl. 1986) M	**J.Pal**
Pal, K.P. (fl. 1968) M	**K.P.Pal**
Pal, Mahendra (fl. 1970) M	**M.Pal**
Pal, Niranjan (1927–) P	**N.Pal**
Pal, Sunanda (1933–) P	**S.Pal**
(Palacios Chavez, Rodolfo)	
see Palacios–Chavez, Rodolfo	**R.Palacios**
(Palacios, Modesto Laza)	
see Laza Palacios, Modesto	**Laza**
Palacios, Pedro S	**P.Palacios**
Palacios, Ramón A. (1941–) S	**R.A.Palacios**

Palacios–Chavez, Rodolfo (1929–) S	**R.Palacios**
Palacký, Johann Baptiste (Jan) (1830–1908) S	**Palacký**
Palagetti, Mario (fl. 1902) M	**Palag.**
Palamar–Mordvintseva, G.M. A	**Pal.-Mordv.**
Palamarev, Em. A	**Palam.**
Palamedi, B. (fl. 1932) M	**Palamedi**
Palaniswami, A. (fl. 1972) M	**Palan.**
Palanov, A.V. (fl. 1988) S	**Palanov**
Palanza, Alfonso (1851–1899) S	**Palanza**
(Palasso, Pierre Bernard)	
see Palassou, Pierre Bernard	**Palassou**
Palassou, Pierre Bernard (1745–1830) S	**Palassou**
Palassow, M.L. (fl. 1953) S	**Palassow**
Palau, Pedro (1881–1956) S	**P.Palau**
Palau y Verdera, Antonio (–1793) S	**Palau**
(Palav Ferrer, Pedro)	
see Palau, Pedro	**P.Palau**
Paley, Frederick Apthorp (1815–1888) S	**Paley**
Palezieux, Philippe de (1871–1957) S	**Palez.**
Palhinha, Ruy Telles (1871–1957) S	**Palhinha**
Palibin, Ivan Vladimirovich (1872–1949) PS	**Palib.**
Palik, Piroska (1895–1966) A	**Palik**
Palisot de Beauvois, Ambroise Marie François Joseph (1752–1820) ABMPS	**P.Beauv.**
Palitz, R. (fl. 1935) S	**Palitz**
Palla, Eduard (1864–1922) AMS	**Palla**
Palla, Piero A	**P.Palla**
Pallari, Eino (1901–1959) S	**Pallari**
Pallas, Johann Dietrich (fl. 1758) S	**J.Pall.**
Pallas, Peter (Pyotr) Simon von (1741–1811) AMPS	**Pall.**
Pallis, M. (fl. 1938) S	**Pallis**
Palm, Björn Torvald (1887–1956) AM	**Palm**
Palm, Mary E. (1954–) M	**M.E.Palm**
Palma, M. (fl. 1982) S	**Palma**
Palmberg, Johannes Olai (1640–1691) L	**Palmberg**
Palmer, Charles Mervin (1900–) A	**C.Palmer**
Palmer, Edward (1831–1911) S	**Palmer**
Palmer, Elmore (1839–1909) S	**E.Palmer**
Palmer, Ernest Jesse (1875–1962) S	**E.J.Palmer**
Palmer, Frederic T. (fl. 1867) S	**F.Palmer**
Palmer, James Terence (1923–) M	**J.T.Palmer**
Palmér, Johan Ernst (1863–1946) S	**J.E.Palmér**
Palmer, Johann Ludwig (1784–1836) S	**J.L.Palmer**
Palmer, Julius Auboineau (1840–1899) M	**J.A.Palmer**
Palmer, Theodore Sherman (1860–1962) S	**T.S.Palmer**
Palmer, Thomas Chalkley (1860–1935) AP	**T.C.Palmer**
Palmer, William (1856–1921) PS	**W.Palmer**
Palmgren, Alvar (1880–1960) S	**Palmgr.**
Palmova, E.F. (1882–1954) S	**Palmova**
Palmstruch, Johan Wilhelm (1770–1811) S	**Palmstr.**
Palouzier, Émile (fl. 1891) S	**Palouz.**

Palun, Maurice (1777–1860) S   **Palun**
Palustre (fl. 1840) S   **Palustre**
Pammel, Louis Hermann (1862–1931) MS   **Pammel**
Pampaloni, Luigi (1897–   ) S   **Pampal.**
Pampanini, Renato (1875–1949) MPS   **Pamp.**
Pamplin, William (1806–1899) S   **Pamplin**
Pampuch, Albert (fl. 1840) S   **Pampuch**
Pamukçuoğlu, Adil (1929–   ) S   **Pamukç.**
Pan, Jin Tang (1935–   ) S   **J.T.Pan**
Pan, Kai Yu (1937–   ) S   **K.Y.Pan**
Pan, Kuo Ying   A   **Kuo Y.Pan**
Pan, Sheng Li (fl. 1986) S   **S.L.Pan**
Pan, Ti Chang (fl. 1988) S   **T.C.Pan**
Pan, Ze Hui (1938–   ) S   **Z.H.Pan**
Panagopoulos, C.G. (fl. 1960) M   **Panag.**
Panasenko, V.T. (fl. 1929) M   **Panas.**
Panayotatou, A. (fl. 1927) M   **Panay.**
Pancher, Jean Armand Isidore (1814–1877) S   **Pancher**
(Panchic, Josif)
  see Pančić, Joseph   **Pančić**
Pančić, Joseph (1814–1888) S   **Pančić**
(Pancio, Giuseppe)
  see Pančić, Joseph   **Pančić**
Panda, P.C. (fl. 1986) S   **Panda**
Pande, A.K. (fl. 1968) M   **A.K.Pande**
Pande, Alaka (fl. 1973) M   **A.Pande**
Pande, C.B. (fl. 1953) M   **C.B.Pande**
Pandé, S.K. (1899–1960) B   **Pandé**
Pander, C. (1794–1865) S   **Pander**
Pandey, A.K. (fl. 1989) M   **A.K.Pandey**
Pandey, B.N. (fl. 1990) M   **B.N.Pandey**
Pandey, Devesh Chandra (–1988) A   **D.C.Pandey**
Pandey, P.C. (fl. 1971) M   **P.C.Pandey**
Pandey, R.P. (1952–   ) S   **R.P.Pandey**
Pandey, R.S.   A   **R.S.Pandey**
Pandey, U.C.   A   **U.C.Pandey**
Pandiani, Arturo   S   **Pandiani**
Pando, Francisco (1962–   ) M   **Pando**
Pandotra, V.R. (fl. 1964) M   **Pandotra**
Panero, José L. (fl. 1988–) S   **Panero**
Panfilova, T.S. (fl. 1963) M   **Panf.**
Pang, Jin Hu (fl. 1983) S   **Pang**
Pangalo, Konstantin Ivanovič (1883–1965) S   **Pangalo**
Pangtey, Y.P.S. (fl. 1990) P   **Pangtey**
Pangua, Emilia (fl. 1988) P   **Pangua**
Panigrahi, Gopinath (1924–   ) PS   **Panigrahi**
Panigrahi, Sarojini G. (fl. 1979) S   **S.G.Panigrahi**
Panizzi, Antonio   S   **A.Panizzi**
Panizzi, Francesco (1817–1893) MS   **Panizzi**

Panizzi–Savio, Francesco
  see Panizzi, Francesco   **Panizzi**
Panja, G.(J.) (fl. 1927) M   **Panja**
Panjutin, P.S. S   **Panjut**
Pankow, Helmut (1929–) A   **Pankow**
Pannell, Caroline M. (fl. 1982) S   **Pannell**
Panov, P.P. (1932–) S   **Panov**
Pansart, J. A   **Pansart**
Pansche, Adolf (1841–1887) S   **Pansche**
Pant, Daya Nand (fl. 1951) M   **D.N.Pant**
Pant, Devesh Chandra (1941–) M   **D.C.Pant**
Pant, Divya Darshan (1919–) ABFP   **D.D.Pant**
Pant, Garima (fl. 1989) M   **G.Pant**
Pantanelli, Dante (1844–1913) S   **Pantan.**
Pantanelli, Enrico Francesco (1881–1951) M   **E.Pantan.**
Pantastico, Julia Baldia A   **Pantast.**
Pantazidou, Andriana (1955–) A   **Pantaz.**
Pantel, C. (fl. 1885) S   **Pantel**
Panter, C. (fl. 1987) M   **C.Panter**
Panter, Jacqueline Anne (1945–) S   **Panter**
Panthaki, Dhun Phiroze (1933–) S   **Panthaki**
Pantić, S. A   **Pantić**
Pantidou, Maria E. (fl. 1963) M   **Pantidou**
Pantling, Robert (1856–1910) S   **Pantl.**
Pantocsek, Jószef (Joseph) (1846–1916) AS   **Pant.**
Panţu, Zacharia C. (1866–1934) S   **Panţu**
(Pantzu, Zacharia C.)
  see Panţu, Zacharia C.   **Panţu**
Panwar, A.B. (fl. 1987) M   **A.B.Panwar**
Panwar, K.S. (fl. 1972) M   **Panwar**
Panzer, Georg Wolfgang Franz (1755–1829) PS **Panz.**
(Pao, Shi Ying)
  see Bao, Shi Ying   **S.Y.Bao**
Paoletti, Giulio (1865–1941) AMPS   **Paol.**
Paoli, Guido (1881–1947) MS   **Paoli**
Paoli, Paolo (1938–1988) S   **P.Paoli**
Paolucci, Luigi (1849–1935) S   **Paolucci**
Pap, S. (1911–1969) S   **Pap**
Papafava, D. (fl. 1847) S   **Papaf.**
Papanicolaou, Konstantinos (1947–) S   **Papan.**
Paparisto, K. (1914–1980) S   **Papar.**
Papatsou, S. (fl. 1975) S   **Papatsou**
Papava, V. (fl. 1952) S   **Papava**
Papavizas, G.C. (fl. 1961) M   **Papav.**
Pape, Georg Carl (1834–1868) S   **Pape**
Pape, Heinrich (1891–) M   **H.Pape**
Papegaaig, J. (fl. 1925) M   **Papeg.**
Papendorf, M.C. (fl. 1967) M   **Papendorf**
Papenfuss, George Frederick (Frederik) (1903–1981) A   **Papenf.**

Papetti, C. (fl. 1985) M	**Papetti**
Papierok, B.P. (fl. 1982) M	**Papierok**
Papp, Adolf A	**A.Papp**
Papp, Constantin (1896–1972) BS	**Papp**
Papp, J. (fl. 1944) S	**J.Papp**
Pappe, Karl (Carl) Wilhelm Ludwig (1803–1862) PS	**Pappe**
Pâque, Égide (1850–1918) S	**Pâque**
Paracer, Chetan Swarup (fl. 1962) M	**Paracer**
Parada, V.M. (fl. 1984) S	**Parada**
Paradkar, S.A. (fl. 1971) FM	**Paradkar**
Parameswaran, N. (fl. 1972) M	**Param.**
Paramonova, N.V. A	**Paramon.**
Paravicini, E. (fl. 1918) M	**Parav.**
Paray, Ladislao (fl. 1954) S	**Paray**
Parbery, D.G. (fl. 1966) M	**Parbery**
Parbery, Ian H. (–1990; fl.posth.1991) M	**I.H.Parbery**
Pardé, Léon Gabriel Charles (1865–1943) S	**Pardé**
Pardeshi, V.N. (fl. 1982) S	**Pardeshi**
Pardo, C. (fl. 1982) S	**C.Pardo**
Pardo de Tavera, Trinidad Herménégilde José (1857–1925) S	**Pardo**
Pardo, José Heeren (fl. 1925–1932) S	**J.H.Pardo**
Pardo y Sastrón, José (1822–1909) S	**J.Pardo**
Paréjas, Edouard (1890–1961) A	**Paréjas**
Parekh, Kalpana S. A	**Parekh**
Parfenov, Victor Ivanovich (1934–) S	**Parfenov**
Parfenova, M.D. S	**Parfenova**
Parfitt, B.D. (fl. 1978–) S	**B.D.Parfitt**
Parfitt, Edward (1820–1893) S	**Parfitt**
Parguey–Leduc, Agnès (fl. 1967) M	**Parg.-Leduc**
Parihar, N.S. (1923–) B	**Parihar**
Parijs, J.P. S	**Parijs**
Paris, Jean Édouard Gabriel Narcisse (1827–1911) B	**Paris**
Parish, Charles Samuel Pollock (1822–1897) S	**C.S.P.Parish**
Parish, Samuel Bonsall (1838–1928) PS	**Parish**
Parisi, Bruno A	**B.Parisi**
Parisi, Rosa (fl. 1921) M	**Parisi**
Parisot, Charles Louis (1820–1890) S	**Parisot**
Park, C.W. (fl. 1986) S	**C.W.Park**
Park, David (fl. 1974) M	**D.Park**
Park, Mungo (1771–1805) S	**Park**
Park, R.F. (fl. 1982) M	**R.F.Park**
Park, Yun Sil (fl. 1989) M	**Y.S.Park**
Parke, David Lewis A	**D.L.Parke**
Parke, Mary (1908–1989) A	**Parke**
Parker, Arthur K. (fl. 1957) M	**A.K.Parker**
Parker, Bruce C. A	**B.C.Parker**
Parker, Charles Sandbach (–1869) P	**C.Parker**
Parker, Charles Stewart (1882–1950) M	**C.S.Parker**
Parker, Dorothy Inez (1910–) B	**D.I.Parker**

Parker, H.M. (fl. 1975) S	**H.M.Parker**
Parker, J. S	**J.Parker**
Parker, John Bernard (1870–) M	**J.B.Parker**
Parker, Kitty Lucille (née Fenley, K.L.) (1910–) S	**K.L.Parker**
Parker, Lee R. (fl. 1990) AF	**L.R.Parker**
Parker, P.F. (fl. 1972) S	**P.F.Parker**
Parker, Richard Neville (1884–1958) S	**R.Parker**
Parker, W.K. (1823–1890) A	**Parker**
Parker–Rhodes, A.Frederick (1914–) M	**Park.-Rhodes**
Parkes, Hilda M. A	**Parkes**
Parkhurst, Howard Elmore (1848–1916) S	**Parkhurst**
Parkin, John (1873–1964) S	**Parkin**
Parkinson, Charles Edward (1890–1945) S	**C.E.Parkinson**
Parkinson, James (1755–1824) S	**J.Parkinson**
Parkinson, John (1567–1650) L	**John Parkinson**
Parkinson, P.G. A	**P.G.Parkinson**
Parkinson, Sydney C. (1745–1771) S	**Parkinson**
Parkinson, Verona O. (fl. 1981) M	**V.O.Parkinson**
Parkman (fl. 1982) S	**Parkman**
Parks, Clifford R. (fl. 1963) S	**C.R.Parks**
Parks, Harold Ernest (1880–1967) M	**Parks**
Parks, Harris Bradley (1879–) S	**H.B.Parks**
Parks, James C. (1942–) S	**J.C.Parks**
Parlatore, Filippo (1816–1877) S	**Parl.**
Parmar, P.J. (1950–) S	**Parmar**
Parmasto, Erast (1928–) M	**Parmasto**
Parmasto, Ilmi (1935–) M	**I.Parmasto**
Parmelee, John A. (1924–) M	**Parmelee**
Parmentier, Antoine Auguste (1737–1813) PS	**Parm.**
Parmentier, Jean Louis Jacques Henri (1775/77–1852) S	**J.L.Parm.**
Parmentier, Joseph Julien Ghislain (1755–1852) S	**J.Parm.**
Parmentier, Paul Evariste (1860–1946/47) PS	**P.Parm.**
Parnell, Dennis Richard (1939–) S	**D.R.Parn.**
Parnell, John Adrian Naicker (1954–) S	**J.Parn.**
Parnell, Richard (1810–1882) S	**Parn.**
Parodi, Domingo (1823–1890) S	**D.Parodi**
Parodi, Lorenzo Raimundo (1895–1966) FS	**Parodi**
Parodi, S.E. (fl. 1930) M	**S.E.Parodi**
Parona, Corrado A	**Parona**
Paroz, Robert (fl. 1971) S	**Paroz**
Parr, Cyril A.E. (1902–1977) S	**Parr**
Parra, J. (fl. 1987) M	**Parra**
Parrettini, Gian Luigi (fl. 1990) M	**Parrett.**
Parriat, Henri (fl. 1951) B	**Parriat**
Parriaud, A. A	**A.Parriaud**
Parriaud, Henri A	**H.Parriaud**
Parriche, M. (fl. 1990) M	**Parriche**
Parrique, Géraud (1851–1907) BMS	**Parrique**
Parris, Barbara Sydney (1945–) PS	**Parris**

Parrot, Aimé G. (fl. 1966) M	**A.G.Parrot**
Parrot, Friedrich E. (fl. 1811–1830) S	**F.E.Parrot**
Parrot, Johann Jacob Friedrich Wilhelm (1792–1841) S	**Parrot**
Parry, Charles Christopher (1823–1890) S	**Parry**
Parry, William Edward (1790–1855) BS	**W.Parry**
Parsa, Ahmed (Ahmad) (1907–) S	**Parsa**
Parsons, Frances Theodora (1861–1952) S	**F.Parsons**
Parsons, Mary Elizabeth (1859–1947) S	**M.Parsons**
Parsons, Murray J. (1941–) A	**M.J.Parsons**
Parsons, Samuel Browne (1819–1906) S	**Parsons**
Parsons, Thomas Henry (fl. 1926) S	**T.Parsons**
Parsons, William Edward S	**W.E.Parsons**
Parukutty, P.R. A	**Paruk.**
Parvela, August Armas (1885–1953) S	**Parvela**
Pascal, Diego Baldassare (1768–1812) S	**Pascal**
Pascalet, M. (fl. 1934) M	**Pascalet**
Pascher, Adolf (Adolph) A. (1881–1945) AMPS	**Pascher**
Pascoe, Ian G. (1949–) M	**Pascoe**
Pascovschi, Serghie (1905–) S	**Pasc.**
(Pascual, Antonio Marcos)	
see Marcos Pascual, Antonio	**Marcos**
Pascual, L. (fl. 1985) S	**Pascual**
Pascual, R. (fl. 1987) M	**R.Pascual**
Pasha, M.K. (fl. 1989) S	**Pasha**
Pashev, I. (fl. 1971) M	**Pashev**
Pashkov, G.D. A	**Pashkov**
Pasinetti, Lauro (fl. 1935) M	**Pasin.**
Pasini, A. (fl. 1911) M	**Pasini**
Pasquale, Fortunato (1856–1917) S	**F.Pasq.**
Pasquale, Guiseppe Antonio (1820–1893) BMS	**Pasq.**
Pasricha, R. (fl. 1987) M	**Pasricha**
Passalacqua, T. (fl. 1926) M	**Passal.**
Passarge, H. (fl. 1981) S	**H.Passarge**
Passarge, Siegfried (1867–1958) S	**Passarge**
Passauer, Uwe (1942–) M	**Passauer**
Passecker, F. (fl. 1932) M	**Passecker**
Passerini, Giovanni (1816–1893) MPS	**Pass.**
Passerini, Napoleone (1862–1951) S	**N.Pass.**
Passini, M.-F. (fl. 1987) S	**Passini**
Passy, Antoine François (1792–1873) S	**Passy**
Pastiels, André (1919–1970) A	**Pastiels**
(Pastor Díaz, Julio Enrique)	
see Pastor, Julio Enrique	**Pastor**
Pastor, Julio Enrique (1954–) S	**Pastor**
Pastore, Ada I. (fl. 1939) P	**Pastore**
Pastour, L. (1822–1895) M	**Pastour**
Patay (fl. 1935) M	**Patay**
Patel, M.K. (fl. 1949) M	**Patel**
Patel, R.J. A	**R.J.Patel**

Patel, Villoo (later Kapadia, Z.J.) (1935–) S	**V.Patel**
Páter, B. (1860–1938) S	**Páter**
Paterson, Betsy Rivers (later Jackes, B.R.) (1935–) S	**B.R.Paterson**
Paterson, R.H. (1814–1889) PS	**R.H.Paterson**
Paterson, R.R.M. (1952–) M	**R.R.M.Paterson**
Paterson, Robert (fl. 1844) F	**R.Paterson**
Paterson, Robert Andrew (1926–) M	**R.A.Paterson**
Paterson, William (1755–1810) S	**Paterson**
Pathak, G.P. (fl. 1978) M	**G.P.Pathak**
Pathak, V.N. (fl. 1969) M	**V.N.Pathak**
Patil, A.S. (fl. 1972) M	**A.S.Patil**
Patil, B.V. (fl. 1975) M	**B.V.Patil**
Patil, C.R. (fl. 1986) M	**C.R.Patil**
Patil, G.V. (fl. 1972) FM	**G.V.Patil**
Patil, M.S. (fl. 1985) M	**M.S.Patil**
Patil, R.B. (fl. 1973) S	**R.B.Patil**
Patil, R.S. (fl. 1985) M	**R.S.Patil**
Patil, S.D. (fl. 1966) M	**S.D.Patil**
Patil, U.R. (fl. 1975) M	**U.R.Patil**
Patnaik, S.N. (fl. 1964) P	**Patnaik**
Paton, Alan James (1963–) S	**A.J.Paton**
Paton, Jean Annette (1929–) B	**Paton**
Patouillard, Narcisse Théophile (1854–1926) BM	**Pat.**
Patout, Marie Rose (fl. 1864) S	**Patout**
Patraw, Pauline Mead (fl. 1936) S	**Patraw**
Patrick, Ruth Myrtle (1907–) A	**R.M.Patrick**
Patrick, T.S. (fl. 1984) S	**T.S.Patrick**
Patrick, W.W. (fl. 1979) M	**W.W.Patrick**
Patrick, William (fl. 1831) S	**Patrick**
Patrin, Eugène Louis Melchior (1742–1815) S	**Patrin**
Patschke, Wilhelm (1888–) S	**Patschke**
(Patschosky, Jozef Konradovich (Joseph Conradovich)) see Paczoski, Jozef Konradovich (Joseph Conradovich)	**Pacz.**
(Patschotsky, Jozef Konradovich (Joseph Conradovich)) see Paczoski, Jozef Konradovich (Joseph Conradovich)	**Pacz.**
Patterson, David J. A	**D.J.Patt.**
Patterson, Flora Wambaugh (1847–1928) M	**F.Patt.**
Patterson, Harry Norton (1853–1919) S	**Patt.**
Patterson, Paul Morrison (1902–) BM	**P.Patt.**
Patterson, Robert (1947–) S	**R.Patt.**
Pattison, Samuel Rowles (1809–1901) F	**Pattison**
Patton, W.S. (fl. 1910) A	**Patton**
Patunkar, B.W. (fl. 1980) S	**Patunkar**
Paturson, B. (fl. 1954) M	**Paturson**
Patwardhan, Parashuram Gangadhar (1935–) M	**Patw.**
Patzak, Alois (1930–) S	**Patzak**
Patze, Carl August (1808–1892) S	**Patze**
Patzelt, Joseph Eduard (fl. 1842) S	**Patzelt**
Patzke, Erwin (1929–) S	**Patzke**

Pau, Carlos (1857–1937) PS   **Pau**
(Pau y Español, Carlos)
   see Pau, Carlos   **Pau**
Paucă, Ana M. (1907–1963) S   **Paucă**
Paul, A.P. (fl. 1986) M   **A.P.Paul**
Paul, A.R. (fl. 1966) M   **A.R.Paul**
Paul, Alison M. (fl. 1987) P   **A.M.Paul**
Paul, Bernard (fl. 1987) M   **B.Paul**
Paul, D. (fl. 1969) M   **D.Paul**
Paul, Henry A   **H.Paul**
Paul, Hermann Karl Gustav (1876–1964) BMS   **H.K.G.Paul**
Paul, J. (fl. 1834) S   **J.Paul**
Paul, R.J. (fl. 1982–) S   **R.J.Paul**
Paul, R.N. (fl. 1979) S   **R.N.Paul**
Paul, S.R. (1933–) S   **S.R.Paul**
Paul, T.K. (fl. 1983) S   **T.K.Paul**
Paul, William (1822–1905) S   **Paul**
Paul, Y.S. (fl. 1983) M   **Y.S.Paul**
(Paula de Brooks, Marie Elena de)
   see De Paula de Brooks, Marie Elena   **De Paula**
Paula, José Elsio de (1934–) S   **Paula**
Paulet, Jean Jacques (1740–1826) MS   **Paulet**
(Pauli, Johan)
   see Paulli, Johan   **Paulli**
Paulin, Alphons (1853–1942) PS   **Paulin**
Paull, Charles Leslie Fairbanks (fl. 1904) S   **Paull**
Paulli, Johan (1732–1804) S   **Paulli**
Paulsen, M.D. (fl. 1976) M   **M.D.Paulsen**
Paulsen, Ove Wilhelm (1874–1947) AS   **Paulsen**
Paulson, Robert (1857–1935) M   **Paulson**
Paulus, H.F. (fl. 1986) S   **Paulus**
Paun, Marin (1924–) S   **Paun**
Paunero, Elena (1906–) MS   **Paunero**
(Paunero Ruíz, Elena)
   see Paunero, Elena   **Paunero**
Pauquy, Charles Louis Constant (1800–1854) S   **Pauquy**
Paust, Susan (1949–) S   **Paust**
Pauwels, Jan Lodewijk Haibrecht (1899–) S   **Pauwels**
Pauwels, L. (fl. 1985) S   **L.Pauwels**
Pauzé, J.F. (fl. 1975) M   **Pauzé**
Pávai, V. (1820–1874) S   **Pávai**
Pavarino, Giovanni Luigi (1867–1937) MS   **Pavar.**
Pavgi, M.S. (fl. 1951) M   **Pavgi**
Pavillard, Jules (1868–1961) A   **Pavill.**
Pavlenko, V.F. (fl. 1965) M   **Pavlenko**
Pavletic, Zlatko (1920–1981) B   **Pavletic**
Pavlick, Leon E. (fl. 1983) S   **Pavlick**
Pavlov, Nikolai Vasilievich (1893–1971) PS   **Pavlov**
Pavlov, V.N. (1929–) S   **V.N.Pavlov**

Pavlova, A.M. (1901–) S	A.M.Pavlova
Pavlova, N.M. (1897–1973) S	Pavlova
Pavlova, N.S. (1938–) S	N.S.Pavlova
Pavlovic, S. (fl. 1981) S	Pavlovic
Pavlyus, M. (fl. 1988) S	Pavlyus
Pavolini, Angiolo Ferdinando S	Pavol.
Pavón, José Antonio (1754–1844) S	Pav.
(Pavón y Jiménez, José Antonio)	
see Pavón, José Antonio	Pav.
Pavone, Pietro (1948–) S	Pavone
Pawar, A.B. (fl. 1987) M	A.B.Pawar
Pawar, D.R. (fl. 1989) M	D.R.Pawar
Pawar, V.H. (fl. 1967) M	V.H.Pawar
Pawłowska, Stanisława (1905–) S	Pawłowska
Pawłowski, Bogumił (1898–1971) S	Pawł.
Pawlus, Maria (1954–) S	Pawlus
Pax, Ferdinand Albin (1858–1942) S	Pax
Paxton, Glen Ernest (1896–) M	G.E.Paxton
Paxton, Joseph (1803–1865) S	Paxton
Payak, M.M. (fl. 1951) M	Payak
Payen, Anselme (1795–1871) S	Payen
Payens, J.P.D.W. (Leyden) (1928–) S	Payens
Payer, Jean–Baptiste (1818–1860) BS	Payer
Payne, Frederick William (1852–1927) A	Payne
Payne, Marie (fl. 1980) M	M.Payne
Payne, R.W. (fl. 1974) M	R.W.Payne
Payne, Willard William (1937–) S	W.W.Payne
Payot, Vénance Marie (1826–1902) BPS	Payot
Payrau, Vincent (fl. 1900) S	Payrau
Payson, Edwin Blake (1893–1927) S	Payson
(Paz Graells, Mariano de la)	
see Graells, Mariano de la Paz	Graells
Pazij, V.K. (fl. 1961) S	Pazij
Pázmány, Dénes (Dionisie) (1931–) MS	Pázmány
Pazout, Frantisek S	Pazout
Pazschke, Franz Otto (1843–1922) M	Pazschke
Pchelkin, A.V. (fl. 1987) M	Pchelkin
Peabody, F.J. (fl. 1979) S	Peabody
Peach, Mary (fl. 1950) M	Peach
(Peacock, Edward Adrian)	
see Woodruffe–Peacock, Edward Adrian	Woodr.-Peac.
Peacock, John T. (fl. 1868) S	Peacock
Peagle, Berta S	Peagle
Pearce, R.B. (fl. 1981) M	R.B.Pearce
Pearce, Sydney Albert (1906–) S	Pearce
Pearcy, L. (fl. 1981) S	Pearcy
Pearl, Raymond (1879–1940) S	Pearl
Pearsall, William Harold (1891–1964) S	Pearsall f.
Pearsall, William Harrison (1860–1936) S	Pearsall

Pearse, Arthur Sperry (1877–1956) A	**Pearse**
Pearson, Arthur Anselm (1874–1954) M	**A.Pearson**
Pearson, Barbara E. A	**B.E.Pearson**
Pearson, Barbara R. A	**B.R.Pearson**
Pearson, Gilbert S	**G.Pearson**
Pearson, Henry Harold Welch (1870–1916) S	**H.Pearson**
Pearson, Robert Hooper (1866–1918) S	**R.H.Pearson**
Pearson, Roger C. (fl. 1986) M	**R.C.Pearson**
Pearson, Thomas Gilbert (1873–1943) S	**T.G.Pearson**
Pearson, William Henry (1849–1923) B	**Pearson**
Pease, Arthur Stanley (1881–1964) PS	**Pease**
Pease, Vinnie A. A	**V.A.Pease**
Peattie, Donald Culross (1898–1964) S	**Peattie**
Pechanek, Jan (fl. 1971) S	**Pechanek**
Peck, Charles Horton (1833–1917) BMPS	**Peck**
Peck, Franz Gustav Magnus (1817–1892) S	**F.Peck**
Peck, Morton Eaton (1871–1959) S	**M.Peck**
Peck, Raymond Elliot (1904–) A	**R.E.Peck**
Peck, William Dandridge (1763–1822) S	**W.Peck**
Peckoit, Gustav (1861–1923) S	**Peckoit**
Peckolt, Oswaldo de Lazzarini S	**O.Peckolt**
Peckolt, Theodor(o) (1822–1912) S	**Peckolt**
Peckolt, Waldemar de Lazzarini S	**W.Peckolt**
Pedersen, Anfred (1920–) S	**A.Pedersen**
Pedersen, O.A. (fl. 1989) M	**O.A.Pedersen**
Pedersen, Poul Mcøller A	**P.M.Pedersen**
Pedersen, Steffen Mariager A	**S.M.Pedersen**
Pedersen, T.A. (fl. 1958) M	**T.A.Pedersen**
Pedersen, Throels Myndel (1916–) S	**Pedersen**
Pedersoli, José Luis (fl. 1970) S	**Pedersoli**
Pedicino, Nicola Antonio (1839–1883) S	**Pedicino**
Pedley, Leslie (1930–) S	**Pedley**
Pedralli, G. S	**Pedralli**
Pedrini, A.G. A	**Pedrini**
Pedro, José Gomes (1915–) S	**Pedro**
(Pedrol i Solanes, Joan)	
see Pedrol, Joan	**Pedrol**
Pedrol, Joan (1959–) S	**Pedrol**
Pedrosa, M.Carlota (fl. 1965) M	**Pedrosa**
Pedrotti, Giovanni (1867–1938) S	**Pedrotti**
Pée–Laby, E. (fl. 1891) A	**Pée–Laby**
Peebles, Robert Hibbs (1900–1955) S	**Peebles**
Peek, Chester A. (fl. 1959) M	**Peek**
Peekel, Gerhard (1876–1949) S	**Peekel**
Peerally, M.A. (fl. 1972) M	**Peerally**
Peeters, F. AS	**Peeters**
Peev, D.R. (fl. 1972) S	**Peev**
Pegg, K.G. (fl. 1982) M	**Pegg**
Pegler, David Norman (1938–) M	**Pegler**

Peglion, Vittorio (1873–) M	**Peglion**
Pehersdorfer, Anna (1849–1925) S	**Pehersd.**
Pei, Ming Hao (fl. 1986) M	**M.H.Pei**
Pei, Sheng (Seng) Ji (1938–) S	**S.J.Pei**
Peinado Lorca, Manuel (1953–) S	**Peinado**
Peirce, George James (1868–1954) S	**Peirce**
Peirson, Frank Warrington (1865–1951) S	**Peirson**
Peitscher, Alfred S	**Peitscher**
Peitz, Eduard (–1984) S	**Peitz**
Peixoto, Ariane Luna (1947–) S	**Peixoto**
Péju, G. (fl. 1921) M	**Péju**
Pekelharing, C.A. (fl. 1885) M	**Pekelh.**
Pekkola, Wäinö A	**Pekkola**
Peláez Campomanes, F. (fl. 1954) M	**Peláez**
Peláez, D. A	**D.Peláez**
Pellanda, Guiseppe (fl. 1904) S	**Pellanda**
Pellandini, Wanda (fl. 1989) M	**Pelland.**
Pellegrin, François (1881–1965) S	**Pellegr.**
Pellegrini, Gaetano (1824–1883) S	**Pell.**
Pelletan, Jules (1833–1892) AS	**Pelletan**
Pelletier, Pierre Joseph (1788–1842) S	**Pellet.**
Pelloe, Emily Harriet (1878–1941) S	**Pelloe**
Pelourde, Fernand (1884–1916) F	**Pelourde**
Peltereau, E. (fl. 1931) M	**Pelt.**
Peltier, A.G. (fl. 1959) S	**A.G.Peltier**
Peltier, M. (fl. 1965) S	**M.Peltier**
Peña de Sousa, Magdalena (fl. 1972) S	**Peña de Sousa**
Pena, Pierre (fl. 1520–1600) L	**Pena**
Peña, Rafael (fl. 1901) S	**R.Peña**
Penard, Eugène (1855–1954) A	**Penard**
Penas, Angel (1948–) S	**Penas**
(Penas Merino, Angel)	
see Penas, Angel	**Penas**
Pendergrass, William R. (fl. 1948) M	**Pend.**
Pendse, G.S. (fl. 1975) M	**G.S.Pendse**
Pendse, M.A. (fl. 1976) M	**M.A.Pendse**
Penecke, Karl Alphons (1858–1944) A	**Penecke**
Penev, Ivan Nikolov (1915–) S	**Penev**
Penfold, Arthur Raymond (1890–) S	**A.R.Penfold**
Penfold, Jane Wallas (fl. 1820–1850) S	**Penfold**
Peng, Ching I (fl. 1984) S	**C.I Peng**
Peng, Chun Liang (fl. 1989) S	**C.L.Peng**
Peng, J.C. (fl. 1990) M	**J.C.Peng**
Peng, J.H. (fl. 1989) M	**J.H.Peng**
Peng, Long Jin (fl. 1989) S	**L.J.Peng**
Peng, Wei Dong (fl. 1986) S	**W.D.Peng**
Peng, Yin Bin (1922–) M	**Y.B.Peng**
Peng, Ze (Tse) Xiang (1924–) S	**Z.X.Peng**

Penhallow, Davis Pearce (1854–1910) AF	**Penh.**
Peniašteková, Magdaléna (1949–) S	**Peniašt.**
Peniguel, G. A	**Peniguel**
Penland, Charles William Theodore (1899–1982) S	**Penland**
Penna, Ivo de Azevedo (fl. 1983) S	**I.A.Penna**
Penna, Leonam de Azevedo (1903–1979) S	**L.A.Penna**
Pennant, Thomas (1726–1798) S	**Pennant**
Pennell, Francis Whittier (1886–1952) S	**Pennell**
Pennick, Nigel C. A	**Pennick**
Pennier de Longchamp, Pierre Barthelemy (fl. 1766) M	**Pennier**
Pennington, Leigh Humboldt (1877–1929) M	**Penn.**
Pennington, Richard Toby (1968–) S	**R.T.Penn.**
Pennington, Terence Dale (1938–) S	**T.D.Penn.**
Pennington, Winifred A	**W.Penn.**
Penny, George (–1838) PS	**Penny**
Pennycook, S.R. (fl. 1982) M	**Pennycook**
Penrose, Deborah A	**Penrose**
Pensiero, José F. (1957–) S	**Pensiero**
Penso, Giuseppe A	**Penso**
Pénzes, Antal (1895–) MS	**Pénzes**
Penzig, Albert Julius Otto (Albertus Giulio Ottone) (1856–1929) MS	**Penz.**
Peola, Paolo (1869–) FS	**Peola**
Pepere, Alberto (fl. 1914) M	**Pepere**
Pépin, Pierre Denis (c.1802–1876) S	**Pépin**
Pepoon, Herman Silos (1860–1941) S	**Pepoon**
Peppers, R.A. (fl. 1970) FM	**Peppers**
Peragallo, Hypollyte (1951–) A	**H.Perag.**
Peragallo, Maurice (1853–) A	**Perag.**
Pérard, Alexandre Jules César (1834–1887) PS	**Pérard**
Pérard, Charles A	**C.Pérard**
Perch–Nielsen, Katharina A	**Perch-Nielsen**
Percival, John (1863–1949) MS	**Percival**
Percival, Stephen F. A	**S.F.Percival**
Perco, Bruno (fl. 1982) M	**Perco**
Percy, J.G. (fl. 1973) M	**Percy**
Percy-Lancaster, Sydney (1886–1972) S	**Percy-Lanc.**
Perdeck, A.C. (fl. 1950) M	**Perdeck**
Perdue, Robert Edward (1924–) S	**Perdue**
Pereboom, Cornelis (fl. 1788) S	**C.Pereb.**
Pereboom, Nicolaas Ewoud (fl. 1787) S	**Pereb.**
Pereda Sáez, José María de (1909–1972) S	**Pereda**
Peredo, H.L. (fl. 1986) M	**Peredo**
Peregrine, W.T.H. (fl. 1972) M	**Peregrine**
Pereira, Alice de Lemos (fl. 1942) S	**A.L.Pereira**
Pereira, C.R. (fl. 1970) M	**C.R.Pereira**
Pereira, Cezio S	**C.Pereira**
(Pereira Coutinho, António Xavier)	
see Coutinho, António Xavier Pereira	**Cout.**

(Pereira da Albuquerque, Byron Wilson)
  see Albuquerque, Byron Wilson Pereira da   **Albuq.**
Pereira de Sousa, Ester Concucão
  see Sousa, Ester Concucão   **E.C.Sousa**
(Pereira, Eddie Esteves)
  see Esteves Pereira, Eddie   **Esteves**
Pereira, Edmundo (1914–1986) S   **E.Pereira**
(Pereira Filho, M.J.)
  see Pereira, M.J.   **M.J.Pereira**
Pereira, Huascar (–1926) S   **Pereira**
(Pereira, Jorge Fontella)
  see Fontella, Jorge   **Fontella**
Pereira, M.J. (fl. 1927) M   **M.J.Pereira**
Pereira, Maria Verônica Leite (fl. 1986) S   **V.Pereira**
Pereiro Miguens, M. (fl. 1968) M   **Pereiro**
Perera, M.L. (fl. 1965) M   **Perera**
Peres, Generosa E.P. (fl. 1959) M   **Peres**
Peresipkin, U.P. (fl. 1979) M   **Peresipkin**
Péresse, Madeleine (fl. 1971) M   **Péresse**
Perestenko, Luiza Pavlovna (1937–) A   **Perest.**
Pérez Arbeláez, Enrique (1896–1972) PS   **Pérez Arbel.**
Pérez, C. (fl. 1907) M   **Pérez**
Pérez Carro, F.Javier (fl. 1990) P   **Pérez Carro**
(Pérez Cueto, Eva)
  see Pérez, Eva   **E.Pérez**
Pérez de la Rosa, Jorge A. (1955–) S   **Pérez de la Rosa**
Pérez de Paz, Pedro Luis (1949–) S   **P.Pérez**
Pérez, Eva (fl. 1988) S   **E.Pérez**
Pérez Lara, José Mariá (1841–1918) S   **Pérez Lara**
Perez Morales, C. (fl. 1989) S   **Perez Morales**
Pérez Raya, Francisco (1956–) S   **Pérez Raya**
Pérez Reyes, Rodolfo (fl. 1961) AM   **Pérez Reyes**
Pérez–Chiscano, J.L. (1930–) S   **Pérez–Chisc.**
Pérez–Cirera, José Luis A   **Pérez–Cirera**
Pérez–Moreau, Román A. (1905–1971) S   **Pérez–Mor.**
Peréz–Moreau, Román L. (1931–) S   **R.L.Pérez–Mor.**
Pérez–Silva, Evangelina (1931–) M   **Pérez–Silva**
Pérez–Vera, F. (fl. 1975) S   **Pérez–Vera**
(Perfiliev, Ivan Aleksandrovich)
  see Perfiljev, Ivan Aleksandrovich   **Perfil.**
Perfiljev, B.V. A   **B.V.Perfil.**
Perfiljev, Ivan Aleksandrovich (1882–1942) S   **Perfil.**
Perfiljeva, V.I. (1928–) S   **Perfiljeva**
Pericás, Joan J. (fl. 1987) S   **Pericás**
Perini, Carlo (1817–1883) S   **Perini**
Perini, Claudia (1951–) M   **C.Perini**
Peris, Juan B. (1948–) S   **Peris**
Perkins, C.F. (fl. 1900) M   **C.F.Perkins**

Perkins, D.D. (fl. 1986) M	**D.D.Perkins**
Perkins, E.E. (fl. 1830–1840) S	**E.Perkins**
Perkins, Frank O. (fl. 1973) M	**F.O.Perkins**
Perkins, George Henry (1844–1933) S	**G.Perkins**
Perkins, Janet Russell (1853–1933) P	**Perkins**
Perkins, John Russell (1868–) S	**J.R.Perkins**
Perktold, Josef Anton (1804–1870) BM	**Perktold**
Perleb, Karl (Carl) Julius (1794–1845) S	**Perleb**
Perlova, G.Ja. (1904–) S	**Perlova**
Perman, Jindřich A	**Perman**
Pernhoffer, Gustav von (1831–1899) S	**Pernh.**
Pernitzsch, Heinrich (fl. 1825) S	**Pern.**
Pero, Paolo A	**Pero**
Perold, Sarie Magdalena (1928–) B	**Perold**
Perotti, Renato (1879–1953) M	**Perotti**
Perpenti, Helena (Candida Lena) (1764–1846) S	**Perp.**
Perrault, Claude (1613–1688) LS	**Perrault**
Perreau–Bertrand, Jacqueline (fl. 1963) M	**Perr.-Bertr.**
Perret, Horace (1853–) S	**Perret**
Perret, Michèle A	**M.Perret**
Perreymond, Jean Honoré (1794–1843) S	**Perreym.**
Perrier, Alfred (1809–1866) S	**Perrier**
Perrier de la Bâthie, Jean Octave Edmond (1843–1916) S	**J.O.E.Perrier**
Perrier de la Bâthie, Joseph Marie Henry Alfred (1873–1958) S	**H.Perrier**
Perrier de la Bâthie, Eugène Pierre (1825–1916) S	**E.P.Perrier**
Perrier, E. (1844–1921) S	**E.Perrier**
Perrin du Lac, François Marie S	**Perrin du Lac**
Perrin, Ida Southwell (1860–) S	**Perrin**
Perrin, Peter W. (fl. 1972) M	**P.W.Perrin**
Perrine, Henry (1797–1840) S	**Perrine**
Perring, Franklyn Hugh (1927–) S	**F.H.Perring**
Perring, Wilhem (1838–1906) S	**Perring**
Perroncito, Eduardo (fl. 1888) M	**Perronc.**
Perrone, V. (fl. 1983) S	**Perrone**
Perrot, Émile Constant (1867–1951) S	**Perrot**
Perrott, P.Elizabeth Thomas (fl. 1955) M	**Perrott**
Perrottet, George (Georges Guerrard) Samuel (1793–1870) S	**Perr.**
Perroud, Louis François (1833–1889) S	**Perroud**
Perry, Gillian (1943–) S	**G.Perry**
Perry, J.P. (fl. 1982) S	**J.P.Perry**
Perry, James Depew (1938–) S	**J.D.Perry**
Perry, Lily May (1895–1992) PS	**L.M.Perry**
Perry, Matthew Calbraith (1794–1858) B	**Perry**
Perry, P.L. (fl. 1984) S	**P.L.Perry**
Perry, Reginald H. (1903–) S	**R.H.Perry**
Perry, Thelma J. (fl. 1987) M	**T.J.Perry**
Perry, Thomas A. (fl. 1850) S	**T.A.Perry**
Perry, William Groves (1796–1863) S	**W.G.Perry**
Persiani, Anna Maria (1953–) M	**Persiani**

Persiel, Ingetraud (fl. 1959) M	**Persiel**
Person, C.P. S	**C.P.Person**
Person, Lee Homer (1904–) M	**Person**
Personnat, Victor (fl. 1854–1870) S	**Personnat**
Persoon, Christiaan Hendrik (1761–1836) AMS	**Pers.**
Perssoa, F.P. (fl. 1963) M	**Perssoa**
Persson, Dagmar (Dagnyz) (1930–) S	**D.Perss.**
Persson, Jimmy (1942–) S	**Jim.Perss.**
Persson, John (1854–1930) B	**J.Perss.**
Persson, Karin M. (1938–) S	**K.M.Perss.**
Persson, Nathan Petter Herman (1893–1978) BS	**Perss.**
Perty, Joseph Anton Maximillian (1804–1884) A	**Perty**
Peruchena, J.G. (fl. 1929) M	**Peruch.**
Perveen, Bushra (fl. 1980) M	**Perveen**
Pesante, Aldo (1910–) M	**Pesante**
Peschier, Jean (1744–1831) S	**Pesch.**
Peschkova, Galina A. (1930–) S	**Peschkova**
Pescott, Edward Edgar (1872–1954) S	**Pescott**
Pesez, G. A	**Pesez**
(Peshkova, Galina A.)	
see Peschkova, Galina A.	**Peschkova**
Peşmen, Hasan (1939–1980) S	**Peşmen**
Pesson, Paul A	**Pesson**
Pestalozzi, J.Anton (1871–1937) S	**Pestal.**
(Pestalozzi–Bürkli, J.Anton)	
see Pestalozzi, J.Anton	**Pestal.**
Pestinskaja, T.V. (fl. 1951) M	**Pestinsk.**
Petagna, Luigi (1779–1832) S	**L.Petagna**
Petagna, Vincenzo (1734–1810) S	**Petagna**
Petauer, T. (fl. 1981) S	**Petauer**
Petch, Thomas (1870–1948) M	**Petch**
Péteaux, Jules Charles Joseph (1840–1896) S	**Péteaux**
Petelin, Dmitry A. (1957–) S	**Petelin**
Pételot, (Paul) Alfred (1885–after 1940) S	**Pételot**
Peter, (Gustav) Albert (1853–1937) PS	**Peter**
Peter, A. S	**A.Peter**
Peter, Elfriede (1905–) S	**E.Peter**
Peter, Robert (1805–1894) S	**R.Peter**
(Peter–Stibal, Elfriede)	
see Peter, Elfriede	**E.Peter**
Péterfi, Leontin Stefan (1906–) A	**L.S.Péterfi**
Péterfi, Márton (Martin) (1875–1922) BS	**Péterfi**
Peterhans, Émile (1899–) A	**Peterhans**
Petermann, Wilhelm Ludwig (1806–1855) S	**Peterm.**
Peters, Harriet Anne (1938–) M	**H.A.Peters**
Peters, Karl (1865–1925) S	**K.Peters**
Peters, Mark David (fl. 1978) F	**M.D.Peters**
Peters, Nicolaus A	**N.Peters**
Peters, Wilhelm Carl Hartwig (1815–1883) S	**Peters**

Petersen, (Lorents Christian) Severin (1840–1929) M	**S.Petersen**
Petersen, Erik Johan (1894–) A	**E.J.Petersen**
Petersen, Hans (1836–1927) S	**H.Petersen**
Petersen, Henning Eiler (Ejler) (1877–1946) AMS	**H.E.Petersen**
Petersen, J.V. A	**J.V.Petersen**
Petersen, Johannes Boye (1887–1961) A	**J.B.Petersen**
Petersen, Karl (fl. 1929) S	**K.Petersen**
Petersen, L.J. (fl. 1972) M	**L.J.Petersen**
Petersen, Niels Frederick (1877–1940) S	**N.Petersen**
Petersen, Otto George (1847–1937) S	**Petersen**
Petersen, Ronald H. (1934–) M	**R.H.Petersen**
Peterson, Alvah (1888–) M	**Peterson**
Peterson, Bo–Hagard (1918–1990) S	**B.Peterson**
Peterson, J.L. (fl. 1986) M	**J.L.Peterson**
Peterson, Kathleen M. (1948–) S	**K.M.Peterson**
Peterson, Paul M. (1954–) S	**P.M.Peterson**
Peterson, R.A. (fl. 1955) M	**R.A.Peterson**
Peterson, Roger S. (fl. 1963) M	**R.S.Peterson**
Peterson, Stephen W. (fl. 1981) M	**S.W.Peterson**
Peterson, Wilbur Louis (1944–) B	**W.L.Peterson**
Petgès, G. (fl. 1923) M	**Petgès**
Pethybridge, George Herbert (1871–1948) M	**Pethybr.**
Petif, C. (fl. 1830) S	**C.Petif**
(Petif de la Gautrois, Johann Friedrich Carl Ludwig Corentin)	
see Petif, Johann Friedrich Carl Ludwig Corentin	**Petif**
Petif, Johann Friedrich Carl Ludwig Corentin (1764–1845) S	**Petif**
Petit, Antoine (–1843) S	**Petit**
Petit, Daniel Pierre (fl. 1987) S	**D.P.Petit**
Petit, Emil Charles Nicolai (1817–1893) S	**E.Petit**
Petit, Ernest M.A. (1927–) S	**E.M.A.Petit**
Petit, Felix (fl. 1824) S	**F.Petit**
Petit, Paul Charles Mirbel (1834–1913) A	**P.Petit**
Petit–Radel, Philippe (1749–1815) S	**Petit-Radel**
(Petit–Thouars, Abel Aubert du)	
see Thouars, Abel Aubert du Petit–	**A.Thouars**
(Petit–Thouars, Louis Marie Aubert)	
see Thouars, Louis Marie Aubert du Petit	**Thouars**
Petitberghien, A. (fl. 1966) M	**Petitb.**
Petitmengin, Marcel Georges Charles (1881–1908) PS	**Petitm.**
Petiver, James (1658–1718) S	**Petiver**
Petkoff, Stefan Pavlikianoff (1866–1956) A	**Petkoff**
(Petkov, Stefan Pavlikianov)	
see Petkoff, Stefan Pavlikianoff	**Petkoff**
Petrak, Franz (1886–1973) MS	**Petr.**
Petrescu, Constantin C. (1879–1936) S	**Petrescu**
Petrescu, Mihai (1925–) M	**M.Petrescu**
Petri, Friedrich (1837–1896) S	**F.Petri**
Petri, Lionello (1875–1946) M	**Petri**
Petrie, A.H.K. (fl. 1926) S	**A.H.K.Petrie**

Petrie, Donald (1846–1925) S   **Petrie**
Petriella, Bruno (fl. 1969–82) F   **Petriella**
Petrini, Liliane E. (fl. 1989) M   **L.E.Petrini**
Petrini, Orlando (1952–   ) M   **Petrini**
Petrochenko, G.N. (fl. 1973) S   **Petroch.**
Petropavlovsky, M.F. (1893–   ) S   **Petrop.**
Petrosvn, N.M.   A   **Petrosvn**
Petrov, A.N. (fl. 1984) M   **A.N.Petrov**
Petrov, K.M.   A   **K.M.Petrov**
Petrov, Marija (fl. 1927) M   **M.Petrov**
Petrov, Vsevolod Alexeevič (1896–1955) PS   **Petrov**
Petrov, Yu.(Ju.) E. (1934–   ) A   **Yu.E.Petrov**
Petrova, A.V. (fl. 1973) S   **Petrova**
Petrová, J.   A   **J.Petrová**
Petrova, K.A.   S   **K.A.Petrova**
Petrovič, Sava (1839–1889) S   **Petrovič**
(Petrovskii, V.V.)
   see Petrovsky, V.V.   **V.V.Petrovsky**
Petrovsky, Andrei (Andreas) Stanislavovič (1832–1882) A   **Petrovsky**
Petrovsky, V.V. (1930–   ) S   **V.V.Petrovsky**
(Petrowski, Andrei (Andreas) Stanislavovič)
   see Petrovsky, Andrei (Andreas) Stanislavovič   **Petrovsky**
(Petrowsky, Andrei (Andreas) Stanislavovič)
   see Petrovsky, Andrei (Andreas) Stanislovič   **Petrovsky**
Petruschky, J. (fl. 1898) M   **Petruschky**
Petruscu, M. (fl. 1958) M   **Petruscu**
Petry, Arthur (1858–   ) S   **Petry**
Petrý, Karin   A   **K.Petrý**
Petryk, A.A.   A   **Petryk**
Petschenko (fl. 1908) M   **Petsch.**
Petter, Franz (1798–1853) S   **Petter**
Petters, Franz (1784–1866) S   **Petters**
(Petterson, Bror Johan (John))
   see Pettersson, Bror Johan (John)   **Pett.**
Pettersson, Bengt (1915–   ) S   **B.Pett.**
Pettersson, Börge (fl. 1984–) S   **Börge Pett.**
Pettersson, Bror Johan (John) (1895–   ) S   **Pett.**
Pettet, Antony (fl. 1964) B   **Pettet**
Pettinari, Carla (fl. 1951) M   **Pettinari**
Pettit, A.   A   **Pettit**
(Petunnikov, Alexej Nikolaievicz)
   see Petunnikow, Alexej Nikolaievič   **Petunn.**
Petunnikow, Alexej Nikolaievič (1842–1919) S   **Petunn.**
Petzhold, Georg Paul Alexander (1810–1889) S   **Petzh.**
Petzi, Franz von Sales (1851–1928) S   **Petzi**
Petzold, (Karl) Wilhelm (1848–1897) S   **W.Petz.**
Petzold, Carl Edward Adolph (1815–1891) S   **Petz.**
Pevalek, Ivo (1893–1967) AS   **Pevalek**
Peybernès, Bernard   A   **Peybernès**

Peyer (fl. 1829) S	**Peyer**
Peyerimhoff de Fontenelle, Paul de S	**Peyerimh.**
Peyl, Josef (fl. 1863) M	**Peyl**
Peynaud, E. (fl. 1956) M	**Peynaud**
Peyre, B.L. (fl. 1823) S	**Peyre**
Peyritsch, Johann Joseph (1835–1889) MS	**Peyr.**
Peyronel, Beniamino (1890–1975) M	**Peyronel**
Peyronel, Bruno (1919–1982) S	**B.Peyronel**
(Peyrouse, Philippe Picot de la)	
see Lapeyrouse, Philippe Picot de	**Lapeyr.**
Pezzolato, A. (fl. 1899) M	**Pezz.**
Pfeffer, Wilhelm Friedrich Philipp (1845–1920) BS	**Pfeff.**
Pfeifer, Howard William (1928–) S	**Pfeifer**
Pfeiffer, C.M. (fl. 1986) M	**C.M.Pfeiff.**
Pfeiffer, Hans Heinrich (1890–) S	**H.Pfeiff.**
Pfeiffer, Johan Philip (1888–1947) S	**J.Pfeiff.**
Pfeiffer, Louis (Ludwig) Karl Georg (1805–1877) BS	**Pfeiff.**
Pfeiffer, Norma Etta (1889–) PS	**N.Pfeiff.**
Pfeil, Friedrich Wilhelm Leopold (1783–1859) S	**Pfeil**
Pfender, Juliette A	**Pfender**
Pfennig, H. (fl. 1977) S	**Pfennig**
Pfiester, Lois A. (1936–) A	**Pfiester**
Pfister, Donald H. (1945–) M	**Pfister**
Pfisterer, R. (fl. 1986) S	**Pfisterer**
Pfitzer, Ernst Hugo Heinrich (1846–1906) AMPS	**Pfitzer**
Pflaum, Fritz (fl. 1897) S	**Pflaum**
Pflug, August S	**A.Pflug**
Pflug, H.D. (fl. 1978) AM	**H.D.Pflug**
Pflug, Hans (1925–) B	**Pflug**
Pfuhl, Fritz C.A. (1853–1913) S	**Pfuhl**
Pfund, Johannes Daniel Christian (1813–1876) S	**Pfund**
Phadke, C.H. (1835–1889) M	**Phadke**
Phaff, Herman J. (fl. 1952) M	**Phaff**
Phang, Siew Moi (1952–) A	**Phang**
(Phanichaphol, Dhanee)	
see Phanichapol, Dhanee	**Phanich.**
Phanichapol, Dhanee (1936–) M	**Phanich.**
Phares, David Lewis (1817–1892) S	**Phares**
Phelps, Almira (formerly Lincoln, née Hart) (1793–1884) S	**A.Phelps**
Phelps, William (1776–1856) S	**Phelps**
Phelsum, Murk (Murck, Mark) van (–c.1780) S	**Phelsum**
Phengklai, Chamlong (1934–) S	**Phengklai**
Philbrick, Ralph N. (1934–) S	**Philbrick**
Philcox, David (1926–) S	**Philcox**
Philemon, Elijah (fl. 1991) M	**Philemon**
Philibert, Henri (1822–1901) B	**H.Philib.**
Philibert, J.C. (fl. 1800) S	**Philib.**
Philip, Robert Harris (1852–1912) A	**Philip**
Philipose, M.Y. A	**M.Y.Philipose**

Philipose

Philipose, Manchayil Thomas (1914–   ) A ... **Philipose**
Philippar, François Haken (Aken) (1802–1849) S ... **Philippar**
Philippe, Xavier (1802–1866) S ... **Philippe**
Philippi, Federico (1838–1910) S ... **F.Phil.**
Philippi, Georg (1936–   ) B ... **G.Phil.**
Philippi, Rudolph Amandus (Rodolfo (Rudolf) Amando) (1808–1904) APS ... **Phil.**
(Philippov, G.S.)
  see Philippow, G.S. ... **Philippow**
(Philippov, J.)
  see Philippow, J. ... **J.Philippow**
Philippow, G.S. (fl. 1926) M ... **Philippow**
Philippow, J.   S ... **J.Philippow**
Philipson, Melva Noeline (1925–   ) S ... **M.N.Philipson**
Philipson, William Raymond (1911–   ) PS ... **Philipson**
Phillipps, A. (fl. 1982) S ... **Phillipps**
Phillips, Edwin Percy (1884–1967) S ... **E.Phillips**
Phillips, F.W.   A ... **F.W.Phillips**
Phillips, Henry (1779–1840) S ... **Phillips**
Phillips, John (1800–1874) ABF ... **J.Phillips**
Phillips, Lyle Llewellyn (1923–   ) S ... **L.Ll.Phillips**
Phillips, Reginald William (1854–1926) A ... **R.W.Phillips**
Phillips, Robert Albert (1866–1945) S ... **R.A.Phillips**
Phillips, Ronald C. (1932–   ) A ... **R.C.Phillips**
Phillips, Sylvia Mabel (1945–   ) S ... **S.M.Phillips**
Phillips, Tom L. (1931–   ) B ... **T.L.Phillips**
Phillips, Walter Sargeant (1905–1975) S ... **W.S.Phillips**
Phillips, William (1822–1905) M ... **W.Phillips**
Phillipson, Peter B. (1957–   ) S ... **Phillipson**
Philp, Eric George (1930–   ) S ... **Philp**
Philson, Paul J.   A ... **Philson**
Phinney, Arthur John (1850–1942) S ... **Phinney**
Phinney, Harry Kenyon (1918–   ) A ... **H.K.Phinney**
Phipps, Constantine John (1744–1792) S ... **Phipps**
Phipps, James Bird (1934–   ) S ... **J.B.Phipps**
Phitos, Demetrius (1930–   ) S ... **Phitos**
Phoebus, Philipp (1804–1880) M ... **Phoebus**
Phon, P.Dy (fl. 1981) S ... **Phon**
Phuong, V.X. (fl. 1982) S ... **Phuong**
Phuphathanaphong, Leena (1936–   ) S ... **Phuph.**
Pia, Julius von (1887–1943) AFMP ... **Pia**
Piane, V. (fl. 1975) M ... **Piane**
Piasecki, Stefan   A ... **Piasecki**
Piazza, Michele Antonio   S ... **Piazza**
Picado, C. (fl. 1932) M ... **Picado**
Picard, Casimir (1806–1841) S ... **Picard**
Picard, François (1879–1939) M ... **F.Picard**
Picbauer, Richard (1886–1955) M ... **Picb.**
Picci, G. (fl. 1956) M ... **Picci**
Picci, Vincenzo (1929–   ) S ... **V.Picci**

494

Piccioli, Antonio (1741–1842) S	**Piccioli**
Piccioli, Elvira (fl. 1932) B	**E.Piccioli**
Piccioli, Giuseppe (fl. 1783–1818) S	**G.Piccioli**
Piccioli, Ludovico (1867–) S	**L.Piccioli**
(Piccivoli, Giuseppe)	
see Piccioli, Giuseppe	**G.Piccioli**
Picco, Vittoria (fl. 1788) S	**Picco**
Piccone, Antonio (1844–1901) ABMS	**Picc.**
Piccone, Giovanni Maria (1772–1832) S	**G.M.Picc.**
Pichi Sermolli, Rodolfo Emilio Giuseppe (1912–) MPS	**Pic.Serm.**
(Pichi-Sermolli, Rodolfo Emilio Giuseppe)	
see Pichi Sermolli, Rodolfo Emilio Giuseppe	**Pic.Serm.**
Pichler, R.Alfred (fl. 1928) B	**Pichl.**
Pichon, Marcel (1921–1954) AS	**Pichon**
Pichon, Thomas (fl. 1811) S	**T.Pichon**
Pickard, Joseph Fry (1876–1943) S	**Pickard**
Pickel, Bento José (1906–1963) S	**Pickel**
Pickens, Andrew Lee (1890–) S	**Pickens**
Pickering, Charles (1805–1878) S	**Pickering**
Pickering, J. (fl. 1990) M	**J.Pickering**
Pickersgill, Barbara (1940–) S	**Pickersgill**
Pickett, Fermen Layton (1881–1940) S	**Pickett**
Pickford, Grace E. (1902–) A	**Pickford**
Pickhardt, W. (fl. 1957) M	**Pickh.**
Piçkoś, Halina (1939–) S	**Piçkoś**
Pico (fl. 1823) M	**Pico**
(Picot de Lapeyrouse, Philippe)	
see Lapeyrouse, Philippe Picot de	**Lapeyr.**
Picquenard, Charles–Armand (1872–1940) AMS	**Picq.**
Piddington, Henry (1797–1858) S	**Pidd.**
Pidoplichko, N.M. (fl. 1938) M	**Pidopl.**
Pidoplitschka, M.M. (fl. 1930) M	**Pidoplitschka**
Piearce, Graham D. (fl. 1980) M	**Piearce**
Piebauer, Richard (1886–) M	**Pieb.**
(Piedrahíta, Santiago Díaz)	
see Díaz Piedrahíta, Santiago	**S.Díaz**
Piehl, Martin Abraham (1932–) S	**Piehl**
Piękoś-Mirkowa, Halina (1939–) S	**Pięk.-Mirk.**
Piel, Kenneth M. A	**Piel**
Piemeisel, Frank Joseph (1891–1925) M	**Piem.**
Pienaar, Barendina Jacoba (1926–) S	**B.J.Pienaar**
Pienaar, Richard N. (1942–) A	**Pienaar**
Piepenbring, Georg Heinrich (1763–1806) S	**Piepenbr.**
Pieper, Gustav Robert (fl. 1908) S	**G.Piep.**
Pieper, Philipp Anton (1798–1851) B	**Piep.**
Pieplow, Ulrich (fl. 1938) M	**Pieplow**
Piérart, Pierre J.A.G. (1927–) S	**Piérart**
Pierce, Gary J. (fl. 1978) S	**G.J.Pierce**
Pierce, John Hwett (1912–) S	**Pierce**

Pierce, Newton Barris (1856–1916) M	N.Pierce
Pierce, Prince (fl. 1975) S	P.Pierce
Pierce, Wright (fl. 1933) S	W.Pierce
Pieri, Michele Trivoli (–1834) S	Pieri
Pierot, Jacques (1812–1841) S	Pierot
Pierrat, Dominique (1820–1893) S	Pierrat
Pierrat, M.D. (fl. 1880) S	M.D.Pierrat
Pierre, Jean Baptiste Louis (1833–1905) S	Pierre
Pierre, Jean-François A	J.-F.Pierre
Pierre, Joachim Isidore (1812–1881) S	J.I.Pierre
Pierre, L. S	L.Pierre
Pierrot, Philogène (1835–1896) S	Pierrot
Pierrot, Raymond Bernard (1915–) B	R.B.Pierrot
Pieters, Adrian John (1866–1940) M	Pieters
Pietka, M. (fl. 1983) M	Pietka
Pietsch, Friedrich Maximillian (1856–) S	Pietsch
Pietschmann, M. (1959–) M	Pietschm.
Piggin, C.M. S	Piggin
Pignal, M.C. (fl. 1962) M	Pignal
Pignatti, Sandro (Alessandro) (1930–) PS	Pignatti
Pijl, Leendert van der (1903–1990) S	Pijl
Pijper, A. (fl. 1918) M	Pijper
Pike, A.V. (fl. 1946) S	A.V.Pike
Pike, L.H. (fl. 1982) M	L.H.Pike
Pike, Nicolas (1815–1905) A	Pike
Pike, Radcliffe Barnes (1903–) S	R.B.Pike
Pikul, F.J. (fl. 1974) M	Pikul
Pilar, Georg (Gjuro) (1847–1893) S	Pilar
Pilát, Albert (1903–1974) FM	Pilát
Pilát, Ignatz Anton (1820–1870) S	I.A.Pilát
Pilbeam, J. (fl. 1981) S	Pilbeam
Pilger, Robert Knud Friedrich (1876–1953) AS	Pilg.
Pilipenko, F.S. (1913–1980) S	Pilip.
Pillai, J.S. (fl. 1968) M	Pillai
Pillans, Neville Stuart (1884–1964) S	Pillans
Pillay, D.T.N. (fl. 1968) M	Pillay
Piller, Mathias (1733–1788) S	Piller
Pilling, Friedrich Oscar (1824–1897) S	Pilling
Pilous, Zdněk (Zdeněk) (1912–) B	Pilous
Piltz, J. (fl. 1980) S	Piltz
Pilz, George E. (1942–) S	Pilz
Pim, Greenwood (fl. 1883) M	Pim
Pimenov, Michael Georgievich (1937–) S	Pimenov
Pimenova, L.V. S	L.V.Pimenova
Pimenova, N.N. S	N.N.Pimenova
Pinatzis, Leonidas (1891–1964) S	Pinatzis
Pinchot, Giffard (1865–1946) S	Pinchot
Pinheiro, F.C. A	Pinheiro
Pinkava, Donald John (1933–) S	Pinkava

Pinkerton, M.E. (fl. 1936) M	**Pinkerton**
Pinkwart, Hermann S	**Pinkw.**
Pinnock, D.E. (fl. 1988) M	**Pinnock**
Pinoy, Pierre Ernest (1873–1948) AM	**Pinoy**
Pinto, A.V. (fl. 1981) S	**A.V.Pinto**
(Pinto Barcellos, Ana Maria)	
see Barcellos, Ana Maria Pinto	**Barcellos**
Pinto, Cesar A	**C.Pinto**
(Pinto da Silva, António Rodrigo)	
see Silva, António Rodrigo Pinto da	**P.Silva**
(Pinto da Silva, Quitéria de Jesus G.)	
see Silva, Quitéria Jesus G. Pinto da	**Q.J.P.Silva**
(Pinto Escobar, Polidoro)	
see Pinto–Escobar, Polidoro	**Pinto-Esc.**
(Pinto, Joaquin de Almeida)	
see Almeida, Joaquin de Almeida Pinto	**Almeida**
(Pinto Lopes, José)	
see Pinto Lopes, José	**Pinto-Lopes**
Pinto–Escobar, Polidoro (1926–) S	**Pinto**
Pinto–Lopes, José (fl. 1952) M	**Pinto-Lopes**
Pinzger, Paul (fl. 1868) S	**Pinzger**
Pio, Giovanni Battista (fl. 1813) S	**Pio**
Pionnat, J.C. (fl. 1963) M	**Pionnat**
Piontelli, Eduardo L. (fl. 1977) M	**Piont.**
Piotrowski, Kazimierz (1876–1897) S	**Piotr.**
Piovano, Giovanni (fl. 1953) B	**Piovano**
Piper, Charles Vancouver (1867–1926) MS	**Piper**
Piper, Richard Upton (1818–1897) S	**R.Piper**
Pipoly, John J., III (1955–) S	**Pipoly**
Pippen, Richard W. (1935–) S	**Pippen**
Piquet, John (1825–1912) S	**Piquet**
Pirajá da Silva, Manuel Augusto (1873–1961) MS	**Pirajá**
Pirani, José Rubens (1958–) S	**Pirani**
Piré, Louis Alexandre Henri Joseph (1827–1887) BS	**Piré**
Pires de Lima, Américo (1886–1966) S	**Pires de Lima**
Pires, João Murça (1916–) S	**Pires**
(Pires Morim de Lima, Marli)	
see Lima, Marli Pires Morim de	**M.P.Lima**
Pires–O'Brien, Maria Joaquina (fl. 1992) S	**Pires-O'Brien**
Piringer, W. (fl. 1939) M	**Piringer**
Pirini, Camilla A	**Pirini**
Piriz, M.L. A	**Piriz**
Pirona, Giulio Andrea (1822–1895) S	**Pirona**
Pirone, Pascal Pompey (1907–) M	**Pirone**
Pirotta, Pietro Romualdo (1853–1936) MPS	**Pirotta**
Pirovano, Alberto (1884?–1973) S	**Pirovano**
Pirozhkova, N.M. (fl. 1987) S	**Pirozhk.**
Pirozynski, Kris. A. (fl. 1962) M	**Piroz.**

Pisano, E. (fl. 1979) S	**Pisano**
(Pisano V., E.)	
see Pisano, E.	**Pisano**
Pisareva, N.F. (fl. 1961) M	**Pisareva**
(Pisareva Pirie, N.F.)	
see Pisareva, N.F.	**Pisareva**
Piscicelli, Maurizio (1871–) S	**Pisc.**
Piso, Willem (c.1611–1678) L	**Piso**
Pišpek, P.A. (fl. 1929) M	**Pišpek**
Pissarev, V.E. (1882–) S	**Pissarev**
Pissjaukova, V.V. (1906–) S	**Pissjauk.**
Pišút, Ivan (1935–) M	**Pišút**
Pitard, Charles–Joseph Marie (1873–1927) BMS	**Pit.**
(Pitard–Briau, Charles–Joseph Marie)	
see Pitard, Charles–Joseph Marie	**Pit.**
Pitcher, Zina (1797–1872) S	**Pitcher**
Piton, Françoise (fl. 1974) M	**Piton**
Pitot, Albert Auguste Louis (1905–) P	**Pitot**
Pitschmann, Hans H. (1922–) A	**Pitschm.**
Pitsyk, G.K. A	**Pitsyk**
Pitt, John Ingram (fl. 1966) M	**Pitt**
Pittendrigh, Colin Stephenson (1918–) S	**Pittendr.**
(Pitter de Fábrega, Henri François)	
see Pittier, Henri François	**Pittier**
Pittier, Henri François (1857–1950) PS	**Pittier**
Pittman, Albert B. S	**Pittman**
Pittoni, Josef Claudius (1797–1878) S	**Pittoni**
(Pitzik, G.K.)	
see Pitsyk, G.K.	**Pitsyk**
(Pitzyk, G.K.)	
see Pitsyk, G.K.	**Pitsyk**
Pivovarova, N.S. (fl. 1985) S	**Pivov.**
Pizarro, H. A	**H.Pizarro**
Pizarro, João Joachim (fl. 1872–1887) S	**Pizarro**
Pjataeva, A.D. S	**Pjataeva**
Plaäts–Niterink, A.J.van der (fl. 1972) M	**Plaäts–Nit.**
Plakidas, Antonios George (1895–1987) M	**Plakidas**
Plamada, Emanuel (1926–) B	**Plam.**
Planchon, (François) Gustav (1833–1900) S	**G.Planch.**
Planchon, Jules Émile (1823–1888) PS	**Planch.**
Planchon, Louis David (1858–1915) S	**L.Planch.**
Planchuelo, Ana Maria (1945–) S	**Planchuelo**
Plancke, Jacqueline (1937–) S	**Plancke**
Planderová, Eva A	**Pland.**
Planellas Giralt, José (1850–1886) S	**Planellas**
Planer, Johann Jacob (1743–1789) S	**Planer**
Plank, S. (fl. 1980) M	**Plank**
Plante, Raphaël A	**Plante**
Plas, F.van der (fl. 1970) S	**Plas**

Plate, H.-P. (fl. 1983) M   **Plate**
Plate, L.   A   **L.Plate**
Platen, Paul Louis (1876–) F   **Platen**
Plats, M.Sh. (fl. 1982) M   **Plats**
Platschek, A.   S   **A.Platschek**
Platschek, E.M. (1878–1955) S   **Platschek**
Platzmann, Julius (1832–1902) S   **Platzm.**
Plaut, Hugo Carl (fl. 1887) M   **Plaut**
Plawski, Alexander (fl. 1830) S   **Plawski**
Playfair, George Israel (1871–1922) A   **Playfair**
Playne (fl. 1856) S   **Playne**
Plaz, Anton Wilhelm (1708–1784) S   **Plaz**
Plée, Auguste (1787–1825) MS   **Plée**
Plée, François (fl. 1827–64) S   **F.Plée**
Pleger, R. (fl. 1983) S   **Pleger**
Plehn, Marianne (fl. 1912) AM   **Plehn**
Pleijel, Carl Gerhard Wilhelm (1866–1937) S   **Pleijel**
Plenck, Joseph Jacob von (1738–1807) S   **Plenck**
(Plenk, Joseph Jacob von)
   see Plenck, Joseph Jacob von   **Plenck**
Plesník, F. (fl. 1964) S   **Plesník**
(Plessis, S.J.du)
   see du Plessis, S.J.   **du Plessis**
Plessl, Annemarie   A   **Plessl**
Plevako, E.A. (fl. 1935) M   **Plevako**
Plitmann, Uzi (1936–) S   **Plitmann**
Plitt, Charles Christian (1869–1933) BM   **Plitt**
Plitzka, Alfred (1861–) S   **Plitzka**
Plocek, Alexander (fl. 1973) S   **Plocek**
Ploeg, D.T.E.van der (fl. 1981) S   **D.T.E.Ploeg**
Ploeg, J.van der (fl. 1983) S   **J.Ploeg**
Plomb, G. (fl. 1968) M   **Plomb**
Plonka, F. (fl. 1988) S   **Plonka**
Plotnikov, I.A.   S   **Plotn.**
Plotnikov, N.   S   **N.Plotn.**
Plöttner, Traugott (1853–1923) M   **Plöttn.**
Plowes, Darrel C.H. (fl. 1986) S   **Plowes**
Plowman, Timothy Charles (1944–1989) S   **Plowman**
Plowright, Charles Bagge (1849–1910) M   **Plowr.**
Plues, Margaret (c.1840–c.1903) S   **Plues**
Plukenet, Leonard (1642–1706) LP   **Pluk.**
Plumel, Marcel–Marie (fl. 1991) S   **Plumel**
Plumier, Charles (1646–1704) LP   **Plum.**
Plummer, Sarah Allen (1836–1923) S   **Plummer**
Plumstead, Edna Pauline (1903–1989) F   **Plumst.**
Plunkett, Orda Allen (1897–) M   **Plunkett**
Pluskal, Franišček Sal. (1811–1901) S   **Pluskal**
(Pluskal–Moravičanský, Franišček Sal.)
   see Pluskal, Franišček Sal.   **Pluskal**
Pluszczewski, Emile (1855–) S   **Pluszcz.**

Pluvinage, D. (fl. 1991) M	**Pluvinage**
Po, Gilda L. (fl. 1979) M	**Po**
Poaletti, G. (fl. 1880) M	**Poaletti**
Pobedimova, Evgeniia (Eugenia) Georgievna (1898–1973) S	**Pobed.**
Pobéguin, Charles Henri Oliver (1856–1951) S	**Pobég.**
Poche, Franz (fl. 1913) AM	**Poche**
Pochmann, Alfred A	**Pochm.**
Pocock, Mary Agard (1886–1977) A	**Pocock**
Pocock, Stanley Albert John (1928–) A	**S.A.J.Pocock**
Pócs, Tamás (1933–) BS	**Pócs**
Počta, Filip (1931–) A	**Počta**
Pöder, R. (fl. 1983) M	**Pöder**
Podger, Francis Denis (1933–) S	**Podger**
Podlahová-Růžena, (fl. 1970) M	**Podl.-Růž.**
Podlech, Dietrich (1931–) S	**Podlech**
Podpěra, Josef (1878–1954) BPS	**Podp.**
Podzimek, Jan (fl. 1927) M	**J.Podzimek**
Podzimek, K. (fl. 1926) M	**K.Podzimek**
Podzorski, Andrew C. A	**Podz.**
Poe, Ione (1901–) S	**Poe**
Poech, Josef (Alois) (1816–1846) BS	**Poech**
Poederlé, Eugène Josef Charles Gilain Hubert d'Olmen (1742–1813) S	**Poederlé**
Poelchau, Harald S. A	**Poelchau**
Poellnitz, Karl von (1896–1945) S	**Poelln.**
Poelt, Josef (1924–) BMS	**Poelt**
Poeppig, Eduard Friedrich (1798–1868) MPS	**Poepp.**
Poetsch, Ignaz Sigismund (1823–1884) BM	**Poetsch**
Poeverlein, Hermann (1874–1957) MS	**Poeverl.**
Poggenburg, Justus Ferdinand (1840–1893) PS	**Poggenb.**
(Poggenburgh, Justus Ferdinand)	
see Poggenburg, Justus Ferdinand	**Poggenb.**
Poggioli, Michelangelo (1775–1850) S	**Poggioli**
Pogorevitsch BF	**Pogor.**
Pohl, Johann Baptist Emanuel (1782–1834) PS	**Pohl**
Pohl, Johann Ehrenfried (1746–1800) S	**J.E.Pohl**
Pohl, Julius (1861–1939) S	**J.Pohl**
Pohl, Richard Walter (1916–) S	**R.W.Pohl**
Pohlad, B.R. (fl. 1978) M	**Pohlad**
Pohle, Richard Richardowitsch (1869–1926) BS	**Pohle**
Pohnert, Helmut (1930–) S	**Pohnert**
Pohribniak, I.I. A	**Pohr.**
Poignant, Alain-François A	**Poignant**
Poinar, George O. (fl. 1990) M	**Poinar**
Poindexter, Robert W. (–1943) S	**Poind.**
Poirault, (Marie Henri) Georges (1858–1936) MPS	**G.Poirault**
Poirault, Jules Pierre François (1830–1907) S	**Poirault**
Poiret, Jean Louis Marie (1755–1834) ABMPS	**Poir.**
Poirier, Jacques (fl. 1990) M	**Poirier**
Poirion, L.P. (1901–) S	**Poirion**

Poisson, Henri Louis (1877–1963) S	**Poiss.**
Poisson, Jules (1833–1919) S	**J.Poiss.**
Poisson, R. (fl. 1929) M	**R.Poiss.**
Poisson, Raymond Alfred (1895–) A	**R.A.Poiss.**
Poisy, M.B.de (fl. 1984) M	**Poisy**
Poiteau, Alexandre (1776–1850) S	**A.Poit.**
Poiteau, Pierre Antoine (1766–1854) MS	**Poit.**
Poitras, Adrian W. (fl. 1950) M	**Poitras**
Poitrasson (fl. 1873–1878) S	**Poitr.**
Poivre, Pierre (1719–1786) S	**Poivre**
Pojarkov, B.V. A	**Pojarkov**
Pojarkova, Antonina Ivanovna (1897–1980) PS	**Pojark.**
Pojarkova, H.N. (1923–) S	**H.N.Pojark.**
Pokle, D.S. (fl. 1986) S	**Pokle**
Pokorny, Alois (Aloys) (1826–1886) BCP	**Pokorny**
Pokorný, Vladimír A	**V.Pokorný**
Polák, Karl (1847–1900) S	**Polák**
Polakowski, Hellmuth (1847–1917) BS	**Pol.**
(Polakowsky, Hellmuth)	
see Polakowski, Hellmuth	**Pol.**
Polatschek, Adolf (1932–) S	**Polatschek**
Poldini, Livio (1930–) S	**Poldini**
Põldmaa, P. (fl. 1971) M	**Põldmaa**
(Pole Evans, Illtyd (Iltyd) Buller)	
see Pole–Evans, Illtyd (Iltyd) Buller	**Pole-Evans**
Pole–Evans, Illtyd (Iltyd) Buller (1879–1968) MS	**Pole-Evans**
Polgár, Sándor (1876–1944) BMS	**Polg.**
Polhill, Roger Marcus (1937–) S	**Polhill**
Politis, Demetrios J. (fl. 1972) M	**D.J.Politis**
Politis, Jean A	**J.Politis**
Politis, John (1886–) M	**Politis**
Politova, O.E. S	**Politova**
Polívka, Frantisek (1860–1923) S	**Polívka**
Poljakov, I.D. (fl. 1983) M	**I.D.Poljakov**
Poljakov, Petr Petrovich (1902–1974) S	**Poljakov**
Poljakova, L.A. (fl. 1975) M	**L.A.Poljak.**
Poljakova, O.M. (fl. 1931) S	**Poljak.**
(Poljanskij, G.I.)	
see Poljansky, G.I.	**Poljansky**
Poljansky, G.I. (fl. 1923) AM	**Poljansky**
Poljansky, Vladimir I. (1907–1959) A	**V.I.Poljansky**
Pôll, Josef (1874–1940) S	**Pôll**
Pollacci, Gino (1872–1963) M	**Pollacci**
Pollack, Flora Green (fl. 1946) M	**Pollack**
Pollanetz, E. S	**Pollanetz**
Pollard, Charles Louis (1872–1945) PS	**Pollard**
Pollard, Glenn E. (1901–1976) S	**G.E.Pollard**
Pollexfen, John Hutton (1813–1899) A	**Pollexf.**
Pollich, Johann Adam (1740–1780) BMS	**Pollich**

Pollingher, Utsa   A   **Pollingher**
Pollini, Ciro (Cyrus) (1782–1833) ABMPS   **Pollini**
Pollock, James Barklay (1863–1934) M   **Pollock**
Pollock, S.H. (fl. 1979) M   **S.H.Pollock**
Polonelli, Luciano (fl. 1978) M   **Polon.**
Polovinko, A.E.   S   **Polov.**
(Polozhii, Antonina Vasilievna)
     see Polozhij, Antonina Vasilievna   **Polozhij**
Polozhij, Antonina Vasilievna (1917–   ) S   **Polozhij**
(Polozii, Antonina Vasilievna)
     see Polozhij, Antonina Vasilievna   **Polozhij**
Polozova, N.L. (fl. 1968) M   **Polozova**
Polscher, W. (1831–1861) S   **Polscher**
Polsinelli, Mario (fl. 1987) M   **Polsin.**
Polunin, Nicholas Vladimir (1909–   ) BS   **Polunin**
Poluzzi, C. (fl. 1965) M   **Poluzzi**
(Polyakov, I.D.)
     see Poljakov, I.D.   **I.D.Poljakov**
(Polyakova, L.A.)
     see Poljakova, L.A.   **L.A.Poljak.**
(Polyanskij, Vladimir I.)
     see Poljansky, Vladimir I.   **V.I.Poljansky**
Pomata y Gisbert, Eladio (fl. 1880) S   **Pomata**
Pomel, Auguste Nicolas (1821–1898) AFS   **Pomel**
Pomerleau, René (1904–   ) M   **Pomerl.**
Pomina, C. (fl. 1963) M   **Pomina**
(Ponce de León, Antonio)
     see Ponce de León y Aimé, Antonio   **A.Ponce de León**
Ponce de León, José (fl. 1814) S   **Ponce de León**
(Ponce de León, Patricio)
     see Ponce de León y Carrillo, Patricio   **P.Ponce de León**
Ponce de León y Aimé, Antonio (1887–1961) S   **A.Ponce de León**
Ponce de León y Carrillo, Patricio (1915–   ) MS   **P.Ponce de León**
Ponce, Marta Monica (1954–   ) PS   **Ponce**
Poncet, Jacques   A   **Poncet**
Ponchet, J. (fl. 1977) M   **Ponchet**
Poncy, O. (fl. 1978) S   **Poncy**
Poneropoulos, Eustathios (fl. 1880) S   **Ponerop.**
Ponert, Jirí (1937–   ) S   **Ponert**
Pong, S.M. (fl. 1932) PS   **Pong**
Pongratz, E. (fl. 1966) M   **Pongratz**
Ponnappa, K.M. (fl. 1967) M   **Ponnappa**
Ponomarchuk, G.I. (1937–   ) S   **Ponom.**
Ponomarenko, V.V. (1938–   ) S   **Ponomar.**
Pons, Alexandre (1838–1893) S   **Pons**
Pons, C.   A   **C.Pons**
Pons, Denise (fl. 1965) FMP   **D.Pons**
Pons, Ninosca (Ninoska) (fl. 1986) M   **N.Pons**
Pons, Simon (1861–   ) S   **S.Pons**

(Pont, Pierre S.du)
  see du Pont, Pierre S.   **du Pont**
Pontarlier, Nicolas Charles (1812–1889) S   **Pontarl.**
(Ponte, J.Júlio da)
  see Da Ponte, J.Júlio   **Da Ponte**
Pontedera, Giulio (1688–1757) S   **Ponted.**
Pontén, Johan (Jon) Peter (1776–1857) S   **Pontén**
Pontiroli, Aida (1919–　) S   **Pontiroli**
Pontual, Duarte (fl. 1948) M   **Pontual**
Ponzo, Antonio (1876–1944) S   **Ponzo**
Pool, Raymond John (1882–1967) S   **Pool**
Pool, William (fl. 1874) M   **W.Pool**
Poole, Alick Lindsay (1908–　) S   **Poole**
Poole, Madeleine Margaret (1945–　) S   **M.M.Poole**
Poot, G.A. (fl. 1989) M   **Poot**
Pop, Emil (1897–1974) M   **Pop**
Pop, Ioan (1922–　) S   **I.Pop**
Pope, Clara Maria (fl. 1760s–1838) S   **C.Pope**
Pope, Gerald Vernon (1941–　) S   **G.V.Pope**
Pope, Seth Alison (1911–　) M   **S.A.Pope**
Pope, Willis Thomas (1873–1961) S   **Pope**
Popek, Ryszard (1937–　) S   **Popek**
Popenoe, Dorothy Kate (née Hughes, D.K.) (1899–1932) S   **D.Popenoe**
Popenoe, Frederick Wilson (1892–1975) S   **Popenoe**
Popescu, A. (1933–　) S   **Popescu**
Poplavskaja, Henrietta Ippolitovna (1885–1956) S   **Popl.**
(Poplavskaja–Sukaczeva, G.I.)
  see Poplavskaja, Henrietta Ippolitovna   **Popl.**
Poplu, M.C. (fl. 1873) S   **Poplu**
Poporskav, G.I.　A   **Poporskav**
Popov, G.B. (fl. 1957) S   **G.B.Popov**
Popov, K.P. (fl. 1959) S   **K.P.Popov**
Popov, Mikhail Grigoríevič (Grigoríevich) (1893–1955) S   **Popov**
Popov, N.P.　S   **N.P.Popov**
Popov, T.I. (fl. 1927) S   **T.I.Popov**
Popova, Galina Mikailovna (1895–　) S   **Popova**
Popova, T.G.　A   **T.G.Popova**
Popova, T.N. (1940–　) S   **T.N.Popova**
Popovici, Alexandru P. (1866–　) S   **Popovici**
Popovskaja, G.I.　A   **Popovsk.**
Popovský, Jiří　A   **Popovský**
Popp, Bonifaz (　–1892) S   **Popp**
Poppelwell, Dugald Louis (1863–1939) S   **Poppelw.**
Poppendieck, Hans–Helmut (1948–　) S   **Poppend.**
Popuschoj, J.S. (fl. 1963) M   **Popuschoj**
Poradielova, N. (fl. 1915) M   **Porad.**
Porcher, Felix (fl. 1848–1879) S   **F.Porcher**
Porcher, Francis Peyre (1825–1895) S   **Porcher**
Porcius, Florian (1816–1907) S   **Porcius**

Pore, R.S. (fl. 1965) M	**Pore**
Poretzky, Artemij Sergejevicz (1901–) S	**Poretzky**
Poretzky, Vadima Sergeevich (1893–1953?) A	**V.S.Poretzky**
Poroca, Danuza José Muniz (1942–) M	**Poroca**
Porsch, Otto (1875–1959) S	**Porsch**
Porsild, Alf Erling (1901–1977) PS	**A.E.Porsild**
Porsild, Morton Pedersen (1872–1956) BPS	**Porsild**
Porta, Pietro (1832–1923) S	**Porta**
(Porte, Paulo Campos)	
see Porto, Paulo Campos	**Porto**
Porte, William Solomon (1891–) M	**Porte**
Portenschlag–Ledermayer, Franz von (1772–1822) PS	**Port.**
Porter, A.F. (fl. 1982) S	**A.F.Porter**
Porter, Annie (1910–) A	**A.Porter**
Porter, Carlos Emilio (1868–1942) S	**C.E.Porter**
Porter, Cedric Lambert (1905–) S	**Ced.Porter**
Porter, Charles Lyman (1889–) MS	**Ch.Porter**
Porter, David (fl. 1974) M	**D.Porter**
Porter, Duncan MacNair (1937–) S	**D.M.Porter**
Porter, J.P. (fl. 1942) M	**J.P.Porter**
Porter, Lilian E. (1885–1973) M	**L.Porter**
Porter, Michael (fl. 1990) S	**M.Porter**
Porter, Robert Ker (1779–1842) S	**R.K.Porter**
Porter, Thomas Conrad (1822–1901) S	**Porter**
Portères, Ronald (1906–1974) S	**Portères**
Porterfield, Willard Merrill (1893–1966) S	**Porterf.**
Portilla, J.J. (fl. 1989) M	**J.J.Portilla**
Portilla, Ma.T. (fl. 1989) M	**M.T.Portilla**
Porto, Maria Luiza (fl. 1974) S	**M.L.Porto**
Porto, Paulo Campos (1889–) S	**Porto**
Posada–Arango, Andres (1859–) S	**Posada-Ar.**
Poscharsky, Gustav Adolf (1832–1915) S	**Posch.**
Poschkurlat, A.P. (1912–) S	**Poschk.**
Poselger, Heinrich (1818–1883) S	**Poselg.**
Posey, Gilbert Bradley (1891–) M	**Posey**
Posluszny, Usher (fl. 1976) S	**Posl.**
Posnova, A.N. A	**Posnova**
Pospelov, A.G. (fl. 1950) M	**Pospelov**
Pospichal, Eduard (1838–1905) PS	**Posp.**
Pospischal, Alfred S	**Pospischal**
Pospíšil, Valentin (1912–) B	**Pospíšil**
Post, Bertram van Dyke (1871–1960) S	**B.D.Post**
Post, Douglas Manners (1920–) S	**D.M.Post**
Post, Erika A	**E.Post**
Post, Ernst Jacob Lennart von (1884–1951) S	**L.Post**
Post, George Edward (1838–1909) S	**Post**
Post, Hampus Adolf von (1822–1911) MS	**H.Post**
Post, Tom (Tomas) Erik von (1858–1912) S	**T.Post**
Postel, Emil A.W. (fl. 1856) S	**Postel**

Postels, Alexander Philipou (1801–1871) A   **Postels**
Posthumus, Cilia J.M. (fl. 1963) M   **C.Posth.**
Posthumus, Oene (1898–1945) FP   **Posth.**
(Postrigan, Sawwa A.)
   see Postriganj, Sawwa A.   **Postr.**
Postriganj, Sawwa A. (1891–　) S   **Postr.**
Potanin, Grigorii Nikolajevic (1835–1920) S   **Potanin**
Potapov, G.M. (1896–1967) S   **Potapov**
Potatosova, E.G. (fl. 1960) M   **Potat.**
(Potebnja, Andrej A.)
   see Potebnia, Andrei A.   **Potebnia**
Potebnia, Andrei A. (1870–1919) M   **Potebnia**
Pothe de Baldis, Elba Diana　A   **Pothe de Baldis**
Potier de la Varde, Robert André Léopold (1878–1961) B   **P.de la Varde**
Potier, E. (fl. 1970) M   **Potier**
Potlaïchuk, V.I. (fl. 1952) M   **Potl.**
Potokina, S.A. (fl. 1988) S   **Potokina**
Potonié, Henry (1857–1913) AF   **Potonié**
Potonié, Robert Henri Hermann Ernst (1889–1974) BFM   **R.Potonié**
Potron, M. (fl. 1911) M   **Potron**
(Pötsch, Ignaz Sigismund)
   see Poetsch, Ignaz Sigismund   **Poetsch**
Pott, Johann Friedrich (1738–1805) S   **Pott**
Pott, Reino (1869–1965) S   **R.Pott**
Potter, Alden Archibald (1884–　) M   **A.A.Potter**
Potter, Michael Cressé (1859–1948) AMS   **Potter**
Pottier, Jacques Georges (1892–　) S   **Pottier**
Pottier–Alapetite, Germaine (1894–　) S   **Pott.-Alap.**
Potts, F.H. (1824–1888) PS   **Potts**
Potůček, Oldrich (1929–　) S   **Potůček**
Potzger, John Ernest (1886–1955) S   **Potzger**
Potztal, Eva Hedwig Ingeborg (1924–　) S   **Potztal**
Pouchet, Albert Maxime (1880–1965) M   **A.Pouchet**
Pouchet, Charles–Henri–Georges (1833–1894) A   **C.H.G.Pouchet**
Pouchet, Félix Archimède (1800–1872) S   **Pouchet**
Poucques, M.L.de　A   **Poucques**
Poulain, M. (fl. 1990) M   **Poulain**
Poulet (fl. 1808) M   **Poulet**
Poulin, Michel (1954–　) A   **Poulin**
Poulsen, Viggo Albert (1855–1919) S   **Poulsen**
Poulter, Barbara A. (fl. 1954) S   **Poulter**
Poulton, Ethel Maud　A   **Poulton**
Pound, F.J.　S   **F.J.Pound**
Pound, Roscoe (1870–1964) M   **Pound**
Pount, Helen (fl. 1974) S   **Pount**
Poupion, J.　S   **Poup.**
(Pourret de Figeac, Pierre André)
   see Pourret, Pierre André   **Pourr.**
Pourret, Pierre André (1754–1818) S   **Pourr.**

(Pourret–Figeac, Pierre André)
  see Pourret, Pierre André   **Pourr.**
Pouschet, Felix Archimede S   **Pouschet**
Pouzar, Zdenek (1932–) MPS   **Pouzar**
Pouzolz, Pierre Marie Casimir de (1785–1858) S   **Pouzolz**
Povah, Alfred Hubert William (1889–1975) M   **Povah**
Poveda, Luis J. (fl. 1978) S   **Poveda**
(Poveda V., Luis J.)
  see Poveda, Luis J.   **Poveda**
Póvoa dos Reis, Cónego M. A   **C.M.Póv.Reis**
Póvoa dos Reis, Manuel A   **M.Póv.Reis**
Povrkov, B.V. A   **Povrkov**
Powell, Albert Michael (1937–) S   **A.M.Powell**
Powell, Charles W. (1854–1927) S   **C.W.Powell**
Powell, Dulcie (fl. 1967) M   **D.Powell**
Powell, Henry Thomas (1925–) A   **H.T.Powell**
Powell, Jocelyn Marie (1939–) S   **J.M.Powell**
Powell, John Wesley (1834–1902) S   **J.W.Powell**
Powell, Martha J. A   **M.J.Powell**
Powell, Paul E. (fl. 1974) M   **P.E.Powell**
Powell, Thomas (1809–1887) S   **Powell**
Powell, William Nottingham (1904–) A   **W.N.Powell**
Power, Thomas (fl. 1845) S   **Power**
Powers, Joseph Horace (1866–) A   **Powers**
Powrie, Elizabeth (1925–1977) S   **Powrie**
Poyser, William Aldworth (1882–1928) PS   **Poyser**
Poyton, R.O. (fl. 1970) AM   **Poyton**
Pozdeeva, N.G. (1913–) S   **Pozdeeva**
Prabhu, A.S. (fl. 1963) M   **Prabhu**
Prabhu, A.V. (fl. 1977) M   **A.V.Prabhu**
Prada, María del Carmen Isabel (1953–) PS   **Prada**
(Prada Moral, María del Carmen Isabel)
  see Prada, María del Carmen Isabel   **Prada**
Pradal, Émile (1795–1874) S   **Pradal**
Pradeep, A.K. (fl. 1990) S   **Pradeep**
Pradhan, S.G. (fl. 1986) M   **S.G.Pradhan**
Pradhan, Udai Ch (fl. 1974) S   **Pradhan**
Prado (fl. 1982) P   **Prado**
Praeger, Robert Lloyd (1865–1953) PS   **Praeger**
Praetorius, Ignaz (1836–1908) S   **Praet.**
Prager, Ernst (Ernest) (1866–1913) BP   **Prag.**
Prager, Jan C. (1934–) A   **J.C.Prag.**
Prahl, Johann Friedrich (fl. 1837) S   **J.Prahl**
Prahl, Peter (1843–1911) B   **Prahl**
Prahn, Hermann (fl. 1887) S   **Prahn**
Prain, David (1857–1944) APS   **Prain**
Prajer, Zbigniew (fl. 1969) S   **Prajer**
Prakasa Rao, Chellapilla Surya (1917–) MS   **Prak.Rao**
Prakash, Gyan (fl. 1960) F   **Prakash**

Prakash, O. (fl. 1976) M	**O.Prakash**
Prakash, P. (fl. 1983) M	**P.Prakash**
Prakash, Uttam (fl. 1968) F	**U.Prakash**
Prakash, Ved (1957–) S	**V.Prakash**
(Pramanick, B.B.)	
see Pramanik, B.B.	**Pramanik**
Pramanik, A. (fl. 1986) S	**A.Pramanik**
Pramanik, B.B. (1933–) S	**Pramanik**
Pramer, D. (fl. 1968) M	**Pramer**
Prance, Ghillean ('Iain') Tolmie (1937–) S	**Prance**
Prantl, Karl Anton Eugen (1849–1893) ABMPS	**Prantl**
Prasad, Akshinthala K.Sai Krishna (1949–) A	**A.K.S.Prasad**
Prasad, Braj Nandan A	**B.N.Prasad**
Prasad, M.N.V. (fl. 1988) S	**M.N.V.Prasad**
Prasad, N. (fl. 1955) M	**Prasad**
Prasad, R. (fl. 1966) M	**R.Prasad**
Prasad, R.N. A	**R.N.Prasad**
Prasad, S.S. (fl. 1968) M	**S.S.Prasad**
Prasada, R. (fl. 1963) M	**Prasada**
Prasannakumar, K.S. (fl. 1985) S	**Prasann.**
Praschil, Wenceslaus Wilhelm (fl. 1840) S	**Praschil**
Prasertphon, S. (fl. 1963) M	**Pras.**
Prashar, I.B. (fl. 1986) M	**Prashar**
Prassler, Maria (1938–) S	**Prassler**
Prat, Henri (1902–) S	**H.Prat**
Prat, Silvestr (1895–) A	**Prat**
Pratesi, Pietro (fl. 1800) S	**Pratesi**
Pratje, O. (fl. 1922) FM	**Pratje**
Pratov, N.P. (1934–) S	**N.P.Pratov**
Pratov, Uktam Pratovich (1934–) S	**U.P.Pratov**
Pratt, Anne (1806–1893) S	**Pratt**
Pratt, O.A. (fl. 1918) M	**O.A.Pratt**
Pratt, Robert G. (fl. 1973) M	**R.G.Pratt**
Praturlon, Antonio A	**Praturlon**
Prauser, Helmut A	**Prauser**
(Pravazek, Stanislaus von)	
see Prowazek, Stanislaus von	**Prowazek**
Pray, Thomas Richard (1923–) PS	**Pray**
Préaubert, Ernest (1852–1933) S	**Préaub.**
Preble, Edward Alexander (1871–1957) S	**Preble**
Preda, Agilulfo (1870–1941) S	**Preda**
Préfontaine, M.Bruletout de (fl. 1763) S	**Préf.**
(Preis, Balthazar)	
see Preiss, Balthazar	**Preiss**
Preisig, Hans Rudolf (1949–) A	**Preisig**
Preiss, Balthazar (1765–1850) S	**Preiss**
Preiss, Johann August Ludwig (1811–1883) S	**L.Preiss**
Preiss, W.V. A	**W.V.Preiss**
Preissler, J. S	**Preissler**

Preissman, Ernst (1844–) P   **Preissm.**
Prelli, Rémy (1947–) P   **Prelli**
Prema, P.   A   **Prema**
Premanath, R.K. (1952–) S   **Premanath**
Prenger, Alfred Gerhard (1860–) S   **Prenger**
Prentice, Charles Brightly (1820–1894) PS   **Prent.**
Prentice, Heather (fl. 1981) S   **H.C.Prent.**
Prentiss, Albert Nelson (1836–1896) S   **Prentiss**
(Preobrajensky, Grigory A.)
   see Preobraschensky, Grigory A.   **Preobr.**
Preobraschensky, Grigory A. (1892–1919) S   **Preobr.**
(Preobrazhensky, Grigory A.)
   see Preobraschensky, Grigory A.   **Preobr.**
Prescott, A.M. (fl. 1983) S   **A.M.Prescott**
Prescott, Gerald Webber (1899–1988) A   **Prescott**
Prescott, John D. (–1837) S   **J.D.Prescott**
Presl, Carl (Karl, Carel, Carolus) Bořivoj (Boriwog, Boriwag)
   (1794–1852) BFMPS   **C.Presl**
Presl, Jan Svatopluk (Swatopluk) (1791–1849) CPS   **J.Presl**
Presley, John Thomas (1906–) M   **Presley**
Press, J.Robert (fl. 1982) S   **Press**
Prestle, K.H. (fl. 1985) S   **Prestle**
Prestoe, Henry (1842–1923) S   **Prestoe**
Preston, Christopher David (1955–) S   **C.D.Preston**
Preston, Isabella (1881–1965) S   **I.Preston**
Preston, Norman C. (fl. 1948) M   **N.C.Preston**
Preston, Thomas Arthur (1838–1905) S   **Preston**
Preti, G. (fl. 1936) M   **Preti**
Preuss, Carl Gottlieb Traugott (1795–1855) AM   **Preuss**
Preuss, Hans (1879–1935) S   **H.Preuss**
Preuss, Paul Rudolph (1861–) S   **P.Preuss**
(Prévost, Auguste le)
   see Le Prévost, Auguste   **Le Prévost**
Prévost, Honoré Albert (1822–1883) S   **H.A.Prévost**
Prévost, Jean Louis (fl. 1760–1810) S   **Prévost**
Preyer, Axel (fl. 1901) M   **Preyer**
Price, C.W. (fl. 1979) M   **C.W.Price**
Price, F.A.E.   S   **F.A.E.Price**
Price, Ian Russell (1940–) A   **I.R.Price**
Price, Isobel P. (fl. 1963) M   **I.P.Price**
Price, James Henry (1932–) A   **J.H.Price**
Price, Michael Greene (1941–) PS   **M.G.Price**
Price, Morgan Phillips (1885–1973) S   **M.P.Price**
Price, R.A. (fl. 1988) S   **R.A.Price**
Price, S.Reginald   A   **S.R.Price**
Price, Sarah (fl. 1864) MS   **S.Price**
Price, Sarah (Sadie) Frances (1849–1903) S   **S.F.Price**
Price, T.V. (fl. 1970) M   **T.V.Price**
Price, W.C. (fl. 1986) M   **W.C.Price**

Price, William Robert (1886–1975) S	W.R.Price
Priemer, Franz (fl. 1893) S	Priemer
Priest, M.J. (1953–) M	Priest
Priestley, Henry (fl. 1914) M	Priestley
(Prieto, F.)	
see Fernández Prieto, J.A.	Fern.Prieto
Prigge, Barry A. (1947–) S	Prigge
Příhoda, Antonín (fl. 1954) M	Příhoda
Priimachenko, G.D. A	Priim.
Prijanto (fl. 1965) S	Prijanto
Prilipko, Leonid Ivanovich (1907–1983) S	Prilipko
Prill, W. (fl. 1913) F	W.Prill
Prillieux, Édouard Ernest (1829–1915) MS	Prill.
Prillinger, H. (fl. 1991) M	Prillinger
Prima, V.M. (fl. 1973) S	Prima
Prime, Cecil Thomas (1909–1979) S	Prime
(Prinada, Basil P.)	
see Prynada, V.D.	Pryn.
Prinada, V.D. A	Prinada
Prince, Arthur Reginald (1900–1969) P	A.Prince
Prince, J.S. A	J.S.Prince
Prince, William (1766–1842) S	Prince
Prince, William Robert (1795–1869) S	W.R.Prince
Pring, George Henry ('Harry') (1885–1974) S	Pring
Pringle, Cyrus Guernsey (1838–1911) MS	Pringle
Pringle, James Scott (1937–) S	J.S.Pringle
Pringsheim, Ernst G. (1881–1970) AM	E.G.Pringsh.
Pringsheim, Nathanael (Nathaniel) (1823–1894) AMS	Pringsh.
Pringsheim, Olga A	O.Pringsh.
Prins, Ben A	Ben Prins
Prins, Bernard A	Bern.Prins
Prins, M. (fl. 1987) S	M.Prins
Prinsen Geerligs, Hendrik Cönraad (1864–) M	Prins.Geerl.
Prinsloo, Helene E. (fl. 1962) M	Prinsloo
Printz, Karl Henrik Oppegaard (1888–1978) AS	Printz
Prinz, William Alfred Joseph (1857–1910) A	Prinz
(Prinz zu Hohenlohe-Schillingsfürst, Egon Viktor Moritz Karl Maria von Ratibor)	
see West, James	J.West
Prior, Christopher (fl. 1977) M	C.Prior
Prior, Richard Chandler Alexander (1809–1902) S	Prior
Priou, J.P. (fl. 1986) M	Priou
Priszter, Szaniszló (1917–) S	Priszter
Pritchard, Andrew (1804–1882) A	A.Pritch.
Pritchard, Frederick John (1874–1931) M	F.J.Pritch.
Pritchard, Graham George S	G.G.Pritch.
Pritchard, Noël Marshall (1933–) S	N.M.Pritch.
Pritchard, Stephen F. (fl. 1836) PS	S.Pritch.
Pritzel, Ernst Georg (1875–1946) PS	E.Pritz.
Pritzel, George August (1815–1874) S	Pritz.
Privalova, L.A. (1919–) S	Privalova

Probatova, N.S. (1939–) S	**Prob.**
Probst, Ruldolph (1885–1940) M	**Probst**
(Procenko, A.E.)	
see Protsenko, A.E.	**Prots.**
Prochacki, H. (fl. 1974) M	**Prochacki**
Procházka, Frantisek (1939–) S	**F.Proch.**
Procházka, Jan Svatopluk (1891–1933) F	**Proch.**
Procopianu–Procopovici, Aurel (1862–1918) S	**Procop.**
Proctor, George Richardson (1920–) PS	**Proctor**
Proctor, Michael Charles Faraday (1929–) S	**M.Proctor**
Proctor, Vernon Willard (1927–) A	**V.W.Proctor**
Procupiu S	**Procupiu**
Prodán, Iuliu (Julius) (1875–1959) S	**Prodán**
Profice, Sheila Regina (1948–) S	**Profice**
Progel, August (1829–1889) BS	**Progel**
Prokhanov, Jaroslav Ivanovic (Yaroslav Ivanovich) (1902–1964) S	**Prokh.**
Prokhorov, V.P. (fl. 1990) M	**Prokhorov**
Prökschl, H. (fl. 1953) M	**Prökschl**
Prokudin, Juri Nikolajevi (1911–) S	**Prokudin**
Prola, G. (fl. 1983) S	**Prola**
Proll, Alois (fl. 1839) S	**Proll**
Prollius, Friedrich (fl. 1882) S	**Prollius**
Prolongo y García, Pablo (1806–1885) S	**Prolongo**
Prône, Michele (fl. 1967) M	**Prône**
Pronville, Auguste de (fl. 1818) S	**Pronville**
Propach–Giesler, Charlotte S	**Prop.-Giesl.**
Proschowsky S	**Prosch.**
Proshkina–Lavrenko, Anastasia Ivanovna (1891–1977) A	**Proshk.-Lavr.**
Proskauer, Johannes Max (1923–1970) ABS	**Prosk.**
Proskoriakov, Eugeny I. (1895–) S	**Proskor.**
(Proskoriakova, G.M.)	
see Proskuryakova, G.M.	**Proskur.**
(Proskorjakov, Eugeny I.)	
see Proskoriakov, Eugeny I.	**Proskor.**
Proskuryakova, G.M. (fl. 1976) S	**Proskur.**
Prosniakova, L.V. A	**Prosn.**
Prost, T.C. (–1848) BM	**Prost**
Prostakova, Zh.G. (fl. 1966) M	**Prostak.**
Prostoserdov, N.N. (fl. 1933) M	**Prostos.**
Proszyński, Konstanty (1859–1936) S	**Proszyński**
Protíc, Georg (Gjorgje) (1864–) AC	**Protíc**
Proton, M. (fl. 1913) M	**Proton**
Protsenko, A.E. (fl. 1941) M	**Prots.**
Proust, Louis (1878–1959) S	**Proust**
Provancher, Léon (1820–1892) S	**Prov.**
Provasoli, Luigi (1908–) A	**Provasoli**
Prowazek, Stanislaus von (1875–1915) A	**Prowazek**
Prowse, Gerald Albert (1916–1977) AM	**Prowse**
Prud'homme van Reine, W.F. (1941–) A	**Prud'homme**

Prudent, Paul A	**Prudent**
Prudhomme, J. (fl. 1986) S	**J.Prudhomme**
Prunet, A. (fl. 1897) M	**Prunet**
Pruski, John Francis (1955–) S	**Pruski**
Prynada, V.D. (1897–) BF	**Pryn.**
Pryor, Alfred Reginald (1839–1881) S	**Pryor**
Pryor, Lindsey Dixon (1915–) S	**L.D.Pryor**
Prytz, Lars Johan (1789–1823) S	**Prytz**
Przesmycki, Adam Maryan A	**Przesm.**
Przewalski, Nikolai Michailowicz (1839–1888) S	**Przew.**
(Przhevalsky, Nikolay Mikhaylovich)	
see Przewalski, Nikolai Michailowicz	**Przew.**
Psareva, E.N. (1893–) S	**Psareva**
Pshenin, L.N. A	**Pshenin**
Psomadakis, A. (fl. 1982) S	**Psomad.**
Pu, Fa Ting (Ding) (1936–) S	**F.T.Pu**
Pucci, Angiolo (1851–1935) S	**Pucci**
Puccinelli, Benedetto Luigi (1808–1850) S	**Puccin.**
Puccini, Giuliano (1914–1980) S	**Puccini**
Puech, Hippolyte (fl. 1893–1945) S	**Puech**
Puech, Suzette (fl. 1982) S	**S.Puech**
Puel, Timothée (1812–1890) S	**Puel**
Puente, Emilio (1957–) S	**Puente**
Puerari, Marc Nicolas (1766–1845) S	**Puerari**
Puerta y Ródenas, Gabriel de la (fl. 1891) S	**Puerta**
(Puerto, Osvaldo del)	
see Del Puerto, Osvaldo	**Del Puerto**
Pueschel, Curt M. (1950–) AM	**Pueschel**
Puff, Christian (1949–) S	**Puff**
Puget, E. (fl. 1954) M	**E.Puget**
Puget, François (1829–1880) S	**Puget**
Puget, Louis de S	**L.Puget**
Pugh, G.J.F. (fl. 1964) M	**G.J.F.Pugh**
Pugh, Grace Odel (later Wineland, G.O.) (1888–) M	**Pugh**
Pugliese, Alfredo S	**Pugliese**
Puglisi, A. (fl. 1927) M	**Puglisi**
Pugsley, Herbert William (1868–1947) S	**Pugsley**
Puig, Sr.Hermelina A	**Puig**
Puigdullés, Emilio Maffei (fl. 1895) S	**Puigd.**
Puiggari, Juan Ignacio (1823–1900) BCS	**Puigg.**
Puihn, Johann Georg (–1793) S	**Puihn**
Puissant, Pierre A. (1831–1911) S	**Puiss.**
Pujadas I Ferrer, J. (fl. 1981) S	**J.Pujadas**
Pujadas Salvá, Antonio (c.1950–) S	**A.Pujadas**
Pujals, Carmen (1916–) A	**Pujals**
Pulević, V. (1938–) S	**Pulević**
Pulle, August Adriaan (1878–1955) S	**Pulle**
Pullen, A.B. (fl. 1987) S	**A.B.Pullen**
Pullen, Thomas Marion (1919–) S	**Pullen**

Pulley, Jean M. (fl. 1973) S	**Pulley**
Pulliat, Victor (1827–1866) S	**Pulliat**
Pullinger, B.Davidine (fl. 1918) M	**Pullinger**
Pulteney, Richard (1730–1801) S	**Pult.**
Puncag, T. (fl. 1966) M	**Puncag**
Punčochárová, Marcela (1941–) A	**Punčoch.**
Punetha, N. (fl. 1985) P	**Punetha**
Punithalingam, Eliyathamby (1935–) M	**Punith.**
Punsola, L. (fl. 1986) M	**Punsola**
Punt, Wim (1929–) S	**Punt**
Puntillo, Domenico (1950–) M	**Puntillo**
Puntoni, V. (fl. 1935) M	**Puntoni**
Punugu, A. (fl. 1980) M	**Punugu**
Puolanne, Mielo E. (1877–1941) S	**Puol.**
Puppi, Giovanna (1948–) S	**Puppi**
Purchas, William Henry (1823–1903) S	**Purchas**
Purdie, Andrew William (1940–1989) S	**A.W.Purdie**
Purdie, William (c.1817–1857) S	**Purdie**
Purdom, M.A. S	**M.A.Purdom**
Purdom, William (1880–1921) S	**Purdom**
Purdy, Carlton Elmer (1861–1945) S	**Purdy**
Puri, H.S. (fl. 1985) S	**Puri**
Puring, Nicolai J. (1865–1904) S	**Puring**
Purkayastha, C.S. S	**Purkayastha**
Purkayasthra, R.P. (fl. 1974) M	**Purkay.**
Purkinje, J.E. (1787–1869) S	**Purkinje**
Purkyně, Emanuel von (1832–1882) S	**Purk.**
Purohit, D.K. (fl. 1972) M	**Purohit**
Purohit, K.M. (fl. 1979) S	**K.M.Purohit**
Puroshothaman, D. (fl. 1970) M	**Purosh.**
Purpus, Carl (Karl) Albert (1851–1941) S	**Purpus**
Purpus, Joseph Anton (1860–1932) S	**J.A.Purpus**
(Pursch, Friedrich Traugott)	
see Pursh, Frederick Traugott	**Pursh**
Pursell, Ronald Arling (1930–) B	**Pursell**
Pursh, Frederick Traugott (1774–1820) MPS	**Pursh**
Purss, G.S. (fl. 1957) M	**Purss**
Purton, Thomas (1768–1833) M	**Purton**
Purushothama, K.B. (fl. 1987) M	**Purush.**
Purvis, W.O. (1959–) M	**Purvis**
Puschkarew, B.M. A	**Puschk.**
Pushkaran, M. (fl. 1976) M	**Pushkaran**
Pushpavathy, K.K. (fl. 1979) M	**Pushpav.**
Pussard, Marc A	**Pussard**
Puttemans, Arsène (1873–1937) M	**Puttemans**
Putterill, K.M. (fl. 1954) M	**K.M.Putterill**
Putterill, V.A. (fl. 1919) M	**V.A.Putterill**
Putterlick, Alois (Aloys) (1810–1845) BS	**Putt.**
Puttock, C.F. A	**Puttock**

Putzeys, Jules Antoine Adolph Henri (1809–1882) S   **Putz.**
Puvilland (fl. 1879) S   **Puvill.**
(Puydt, Emile De)
   see De Puydt, Emile   **De Puydt**
(Puyfol, Jordan de)
   see Jordan de Puyfol   **Jord.Puyf.**
Puymaly, André Henri Laurent de (1883–) AM   **Puym.**
(Pylaie, Auguste Jean Marie Bachelot de la)
   see Bachelot de la Pylaie, Auguste Jean Marie   **Bach.Pyl.**
Pynaert, Charles (–1936) PS   **C.Pynaert**
Pynaert, Édouard–Christophe (1835–1900) S   **Pynaert**
(Pynaert–van Geert, Édouard–Christophe)
   see Pynaert, Édouard–Christophe   **Pynaert**
Pyrah, G.L. (fl. 1983) S   **Pyrah**

Qadri, Syed Shamsuddin A   **Qadri**
Qaiser, Mohammad (1946–) S   **Qaiser**
Qazilbash, Nawazish Ali S   **Qazilb.**
Qi, Cheng Jing (1932–) S   **C.J.Qi**
Qi, Hui Rong (fl. 1988) S   **H.R.Qi**
Qi, Yu Zao A   **Y.Z.Qi**
Qi, Zu Tong (1926–) M   **Z.T.Qi**
Qian, Hong (fl. 1988) S   **H.Qian**
Qian, Hong Jiang (fl. 1983) S   **H.J.Qian**
Qian, Shi Xin (fl. 1985) S   **S.X.Qian**
Qian, Xiao Hu (fl. 1984) S   **X.H.Qian**
Qian, Yi Yong (fl. 1989) S   **Y.Y.Qian**
Qian, Zhi Guang (1953–) M   **Z.G.Qian**
Qiao, Chuan Zhuo (fl. 1987) S   **C.Z.Qiao**
Qin, De Hai (fl. 1988) S   **D.H.Qin**
Qin, Jia Zhong (fl. 1980) M   **J.Z.Qin**
(Qin, Ren Chang)
   see Ching, Ren Chang   **Ching**
Qin, Y. (fl. 1982) M   **Y.Qin**
Qin, Zi Sheng (fl. 1981) S   **Z.S.Qin**
Qiu, Jin Xing (fl. 1989) S   **J.X.Qiu**
Qiu, Shu Hua (fl. 1979) S   **S.H.Qiu**
Qosja, Xh. (fl. 1983) S   **Qosja**
Qu, Shi Zeng (fl. 1989) S   **S.Z.Qu**
Quadraccia, Livio (1958–) M   **Quadr.**
Quaintance, Altus Lacy (1870–1958) S   **Quaint.**
Quandt, Christian (1720–after 1807) S   **Quandt**
Quansah, Nathaniel ('Nat') (fl. 1986) PS   **Quansah**
Quarin, Camilo Luis (1943–) S   **Quarin**
Quartin–Dillon, Richard (–1841) S   **Quart.–Dill.**
(Quatre, B.D.De)
   see De Quatre, B.D.   **De Quatre**
Quehl, Leopold (1849–1922) S   **Quehl**
Queiroz, Lusinete Aciole (fl. 1970) M   **L.A.Queiroz**
Queiroz, Luciano Paganucci de (1958–) S   **L.P.Queiroz**

Quekett, Edwin John (1808–1847) AMS   **E.J.Quekett**
Quekett, John Thomas (1815–1861) S   **J.T.Quekett**
Quélet, Lucien (1832–1899) M   **Quél.**
Quelle, Ferdinand Friedrich Hermann (1876–1963) B   **Quelle**
Quennerstedt, Nils   A   **Quenn.**
Quenstedt, Friedrich August von (1809–1889) A   **Quenst.**
Quer y Martínez, José (1695–1764) S   **Quer**
(Quer, Pio (Pius) Font)
   see Font Quer, Pio (Pius)   **Font Quer**
Quero, H.J. (fl. 1980) S   **H.J.Quero**
Quero, M.J. (fl. 1982) S   **M.J.Quero**
(Quesnay, M.C.du)
   see duQuesnay, M.C.   **duQuesnay**
Quesné, François Alexandre (1752–1820) S   **Quesné**
Quételet, Lambert Adolphe Jacques (Adolph Jacob) (1796–1874) S   **Quételet**
Queva, Charles (fl. 1894) S   **Queva**
Queyrat, Louis (fl. 1909) M   **Queyrat**
Quezada, Max (1936–) S   **Quezada**
Quézel, Pierre Ambrunaz (1926–) S   **Quézel**
Quian, Yi Yong   S   **Y.Y.Quian**
Quick, Clarence Roy (1902–) S   **Quick**
Quick, J.A. (fl. 1974) M   **J.A.Quick**
Quimio, Tricita H. (1939–) M   **Quimio**
Quincy, Charles (fl. 1900–1911) S   **Quincy**
Quinet, R.I. (fl. 1928) M   **Quinet**
Quinlan, M.S. (fl. 1958) M   **Quinlan**
Quinn, Christopher John (1936–) S   **Quinn**
Quinquaud, E. (fl. 1868) M   **Quinq.**
Quintanilha, Aurélio Periera da Silva (1892–) S   **Quint.**
(Quintanilla, J.A.)
   see Quintanilla Sáez, J.A.   **Quintan.**
Quintanilla Sáez, J.A. (fl. 1981) M   **Quintan.**
Quirk, Helen Mary (1953–1982) P   **H.M.Quirk**
Quirk, R. (fl. 1911–1916) S   **Quirk**
Quisumbing y Argüelles, Eduardo (1895–1986) S   **Quisumb.**
Quoy, Joy Rene Constant (1790–1869) A   **Quoy**
Quraishi, M.S. (fl. 1960) M   **Quraishi**
Qureshi, Rizwana Aleem (1950–) S   **Qureshi**

Raab, Christian Wilhelm? (1788–1835) BS   **Raab**
Raab, H. (fl. 1955) M   **H.Raab**
Raab, Ludwig (fl. 1900) S   **L.Raab**
Raabe, Henryk   A   **H.Raabe**
Raabe, Hildegard   A   **Raabe**
Raabe, Robert Donald (1924–) M   **R.D.Raabe**
Raabe, Zdzislaw (1909–1972) A   **Z.Raabe**
Raadts, Edith Marie (1914–) S   **Raadts**
Raap   S   **Raap**
Raatikainen, Terttu K. (1934–) S   **Raatik.**

Raatz, Georg Victor (fl. 1937) BF	**Raatz**
Raatz, Wilhelm (1864–1918) S	**W.Raatz**
Rabanus, Adolf (1890–) A	**Rabanus**
Rabaté, Jacques (–1941) A	**Rabaté**
Rabeler, Richard Kevin (1953–) S	**Rabeler**
Rabenau, (Benno Carl August) Hugo von (1845–1921) S	**Rabenau**
Rabenhorst, Gottlob (Gottlieb) Ludwig (1806–1881) ABMPS	**Rabenh.**
Raber, Oran Lee (1893–1940) S	**Raber**
Rabesa, Z.A. (fl. 1982) S	**Z.A.Rabesa**
Rabevohitra, Raymond (1946–) S	**R.Rabev.**
Rabie, C.J. (fl. 1966) M	**Rabie**
Rabinowitsch, Lydia (1871–1935) M	**Rabin.**
(Rabinowitsch–Kempner, Lydia)	
see Rabinowitsch, Lydia	**Rabin.**
Rach, Louis Theodor (1821–1859) FS	**Rach**
Raciborski, Marjan (Maryan, Marian, Maryjan) (1863–1917) ABFMPS	**Racib.**
(Raciborsky, Marjan (Maryan, Marian, Maryjan))	
see Raciborski, Marjan (Maryan, Marian, Maryjan)	**Racib.**
Racine, Rudolf (fl. 1889) S	**Racine**
Rački, Ranka A	**Rački**
Raclaru, Petru (1925–) S	**Raclaru**
(Racoviţă, Andrei)	
see Racovitza, Andrei	**Racov.**
Racovitza, Andrei (1911–) M	**Racov.**
Racovitza, Angela (1909–) M	**A.Racov.**
Rácz, László A	**Rácz**
Radais, Maxime Pierre François (1861–) S	**Radais**
(Radčenko, Georgii Pavlovich)	
see Radczenko, Georgii Pavlovich	**Radcz.**
(Radćenko, M.I.)	
see Radczenko, M.I.	**M.I.Radcz.**
(Radchenko, Georgii Pavlovich)	
see Radczenko, Georgii Pavlovich	**Radcz.**
Radcliffe–Smith, Alan (1938–) S	**Radcl.–Sm.**
Radczenko, Georgii Pavlovich (fl. 1933) ABFS	**Radcz.**
Radczenko, M.I. S	**M.I.Radcz.**
Radde, Gustav Ferdinand Richard Johannes von (1831–1903) S	**Radde**
Radde–Fomina, Olga S	**Radde-Fom.**
Raddi, Giuseppe (1770–1829) ABMPS	**Raddi**
Raddin, Charles Salisbury (1863–1930) S	**Raddin**
Rade, J. A	**Rade**
Radencova S	**Radencova**
Rader, William Ernest (1916–) M	**Rader**
Radermacher, Jacobus Cornelius Matthaeus (1741–1783) S	**Raderm.**
Radford, A.L. S	**A.L.Radford**
Radford, Albert Ernest (1918–) S	**Radford**
Radian, Simeon Stefan (1871–1958) B	**Radian**
Radić, Jure (1920–1990) S	**Radić**
Radius, Justus Wilhelm Martin (1797–1884) S	**Radius**

Radl, Florian (fl. 1896) S	**Radl**
Radley, F. S	**Radley**
Radlkofer, Ludwig Adolph Timotheus (1829–1927) AMS	**Radlk.**
Radloff, Fredrik Wilhelm (1766–1838) S	**Radloff**
Radoicić, Rajka A	**Radoicić**
Radugin, K.V. A	**Radugin**
Rădulescu, Eugen (1904–) M	**E.Rădul.**
Rădulescu, Ion M. (fl. 1932) M	**Rădul.**
Radulovic, S. (fl. 1980) S	**Radulovic**
Radzhi, A.D. (1936–) S	**Radzhi**
Rae, S.J. (fl. 1947) PS	**Rae**
Raenko, I. S	**Raenko**
Raeuschel, Ernst Adolf (fl. 1772–1797) BS	**Raeusch.**
Rafarin (fl. 1866) S	**Rafarin**
Raffaelli, Mauro (1944–) S	**Raffaelli**
(Raffeneau–Delile, Alire)	
see Delile, Alire Raffeneau	**Delile**
Raffill, Charles Percival (1876–1951) S	**Raffill**
Raffles, Thomas Stamford Bingley (1781–1826) S	**Raffles**
Rafinesque, Constantine Samuel (1783–1840) ABMPS	**Raf.**
(Rafinesque–Schmaltz, Constantine Samuel)	
see Rafinesque, Constantine Samuel	**Raf.**
Rafn, Carl Gottlob (1769–1808) S	**Rafn**
Ragab, Mohamed Ali (1925–) M	**Ragab**
Raghukumar, S. (fl. 1970) M	**Raghuk.**
Raghunathan, A.N. (fl. 1965) M	**Raghun.**
Ragionieri, Attilio (1856–1933) S	**Ragion.**
Ragonese, Ana Maria (1928–) F	**A.M.Ragonese**
Ragonese, Arturo Enrique (1909–) S	**Ragonese**
Ragot, Jules (fl. 1902) S	**Ragot**
Ragunath, T. (fl. 1963) M	**Ragunath**
Rahat, M. A	**Rahat**
Rahayu, G. (fl. 1991) M	**Rahayu**
Rahm, E.von (fl. 1958) M	**Rahm**
Rahman, M.U. (fl. 1981) M	**M.U.Rahman**
Rahman, Mohammad Matiur (1948–) S	**M.M.Rahman**
Rahman, S.M.A. (fl. 1962) M	**Rahman**
Rahn, Ann Worley (fl. 1984) M	**A.W.Rahn**
Rahn, Knud (1928–) S	**Rahn**
Rai, A.N. (fl. 1986) M	**A.N.Rai**
Rai, Bharat (fl. 1968) M	**B.Rai**
Rai, B.K. (fl. 1991) M	**B.K.Rai**
Rai, H. A	**H.Rai**
Rai, J.N. (fl. 1970) M	**J.N.Rai**
Rai, M.K. (fl. 1989) M	**M.K.Rai**
Rai, T.R.N. (fl. 1981) M	**T.R.N.Rai**
Raikova, Ilariya Alexeevna (1896–1981) S	**Raikova**
Raikwar, S.K.Singh A	**Raikwar**
Raillo, A.I. (fl. 1950) M	**Raillo**
Railonsala, Artturi Nikodemus (1902–1982) S	**Rail.**

Railyan, A.F. (1946–   ) S   **Railyan**
Raimann, Rudolf (1863–1896) PS   **Raim.**
Raimondi, Antonio (1826–1890) S   **Raimondi**
Raimondo, Francesco Maria (1944–   ) S   **Raimondo**
Rainer, Moriz (Moritz) von undzu Haarbach (1793–1847) S   **Rainer**
Raineri, Rita (1896–1980) AF   **Raineri**
Rainio, Aarne Jakob Pehr (1898–1943) M   **Rainio**
Rainville, Frédéric (–1779) S   **Rainv.**
Raistrick, Harold (fl. 1942) M   **Raistrick**
Raithelhuber, Jög (fl. 1969) M   **Raithelh.**
Raitviir, Ain (G.) (1938–   ) M   **Raitv.**
Raizada, Mukat Behari (1907–   ) PS   **Raizada**
Rajagopal, Tirunilai (1938–   ) S   **Rajagopal**
Rajagopalan, C. (fl. 1967) M   **C.Rajagop.**
Rajagopalan, Koran (1922–   ) M   **K.Rajagop.**
Rajak, P.K. (fl. 1983) M   **P.K.Rajak**
Rajak, R.C. (fl. 1978) M   **R.C.Rajak**
Rajak, R.K. (fl. 1983) M   **R.K.Rajak**
Rajam, R.V. (fl. 1965) M   **Rajam**
Rajan, R. (1950–   ) S   **Rajan**
Rajapaksa, N. (fl. 1964) M   **Rajap.**
Rajasab, A.H. (fl. 1981) M   **Rajasab**
Rajashekhar, M. (fl. 1991) M   **Rajash.**
Rajbhandari, K.R. (fl. 1988) MS   **Rajbh.**
Rajchenberg, Mario (1953–   ) M   **Rajchenb.**
Rajderkar, N.R. (fl. 1964) M   **Rajd.**
Rajendran, C. (fl. 1980) M   **Rajendran**
Rajendran, N.R.   A   **N.R.Rajendran**
Rajendren, R.B. (fl. 1966) M   **Rajendren**
Rajenko, L.M. (1957–   ) S   **Rajenko**
Rajhathy, Tibor   S   **Rajhathy**
Rajput, M.T.M.   S   **Rajput**
Raju, N.B. (fl. 1986) M   **N.B.Raju**
Raju, Vatsavaya S. (fl. 1985) MS   **V.S.Raju**
Rakestraw, Lulu (fl. 1948) B   **Rakestraw**
Rakhmankulov, U. (fl. 1982) S   **Rakhm.**
Rakotoudrainibe, F. (fl. 1988–1989) P   **Rakotoudr.**
Rakotozafy, Armand (1932–   ) S   **Rakot.**
Rakshit (fl. 1961) S   **Rakshit**
Rald, E. (fl. 1991) M   **Rald**
Ralfs, John (1807–1890) A   **Ralfs**
Rall, Gloria (fl. 1964) M   **Rall**
Rallet, Louis (1897–1969) S   **Rallet**
Ralph, Thomas Shearman (1813–1891) S   **Ralph**
Ralston, Barbara E.   S   **Ralston**
Ram, Asha (fl. 1972) M   **A.Ram**
Ram, C.S.Venkata (fl. 1961) M   **C.S.V.Ram**
Ram, Chatthoo (fl. 1967) M   **C.Ram**
(Ram Dayal)
     see Dayal, Ram   **Dayal**

(Ram Dular)
  see Dular, Ram   **Dular**
(Rama Rao, Lakshmeswar)
  see Rao, Lakshmeswar Rama   **L.R.Rao**
Rama Rao, Muttada (1865–) S   **Rama Rao**
Rama Rao, P. (fl. 1960) M   **P.Rama Rao**
Ramachandra Chary, Srimattirumala T. (fl. 1981) S   **Ram.Chary**
(Ramachandrachary, Srimattirumala T.)
  see Ramachandra Chary, Srimattirumala T.   **Ram.Chary**
Ramachandran, Kamala (1932–) S   **K.Ramach.**
Ramachandran, V.S. (fl. 1982) S   **V.S.Ramach.**
Ramachar, P. (fl. 1956) M   **Ramachar**
Ramain, P.I. (fl. 1955) M   **Ramain**
Ramakers, P.M.J. (fl. 1979) M   **Ramakers**
Ramakrishna, K. (1920–) M   **Ramakrishna**
Ramakrishna, T.M. (fl. 1981) S   **T.M.Ramakrishna**
Ramakrishnan, K. (fl. 1949) M   **K.Ramakr.**
Ramakrishnan, Taracad Subromania (fl. 1928) M   **T.S.Ramakr.**
Ramakrishnan, Venkataswami (1926–) S   **Ramakr.**
Ramaley, Annette W. (fl. 1987) M   **A.W.Ramaley**
Ramaley, Francis (1870–1942) S   **Ramaley**
Ramalingam, A. (fl. 1981) M   **Ramal.**
Ramamoorthy, T.P. (1945–) S   **Ramamoorthy**
Ramamurthi, B. (fl. 1965) M   **Ramamurthi**
Ramamurthi, C.S. (fl. 1957) M   **C.S.Ramamurthi**
Ramamurthy, Kandasamy (1933–) S   **Ramam.**
Raman, N. (fl. 1980) M   **Raman**
Ramana, S.V. (fl. 1978) M   **Ramana**
Ramanathan, K.R. (1910–) A   **Ramanathan**
Ramanujam, C.G.K. (fl. 1933) FM   **Ramanujam**
(Ramarao, P.)
  see Rama Rao, P.   **P.Rama Rao**
Ramaswami, Madabusi Srinivasa (fl. 1913) S   **Ramaswami**
Ramaswamy, Sengodagouder (1942–) S   **Ramaswamy**
Ramatuelle, Thomas Albin Joseph d'Audibert de (1750–1794) S   **Ramat.**
Ramayya, Nannegari (1929–) S   **Ramayya**
Ramazanov, E. (fl. 1963) S   **Ramazanov**
Ramazanova, S.S. (fl. 1986) M   **Ramaz.**
(Ramazanow, E.)
  see Ramazanov, E.   **Ramazanov**
Rambelli, Angelo (1932–) M   **Rambelli**
Rambelli, Antonella (fl. 1962) M   **Ant.Rambelli**
Ramberg, Lars   A   **Ramberg**
Rambert, Eugène (1810–1886) S   **Rambert**
Rambo, Balduino (1905–1961) S   **Rambo**
Rambold, Gerhard Walter (1956–) M   **Rambold**
Rambosson, Jean Pierre (1827–1886) S   **Rambosson**
Ramesh, Ch. (fl. 1986) M   **C.Ramesh**
Ramesh, S.R. (fl. 1986) S   **S.R.Ramesh**

(Ramìrez, Carlos)	
see Ramírez Gómez, Carlos	**Ram.Gómez**
Ramírez de Carnevali, Ivón (fl. 1987) S	**I.Ramírez**
Ramírez Espindola, Augusto (fl. 1983) S	**A.Ramírez**
Ramírez Gómez, Carlos (fl. 1952) M	**C.Ramírez**
Ramírez Goyena, Miguel (1857–1927) S	**Ram.Goyena**
(Ramírez, Ivón)	
see Ramìrez de Carnevali, Ivón	**I.Ramírez**
Ramírez, José (1852–1904) S	**Ramírez**
Ramírez, Nelson (1952–) S	**N.Ramírez**
Ramis, Aly Ibraham (1875–1928) S	**Ramis**
Ramis y Ramis, Juan (1746–1819) S	**Ramis y Ramis**
Ramisch, Franz Xaver (1798–1859) S	**Ramisch**
Ramm, Etienne (fl. 1989) M	**Ramm**
Rammeloo, Jan (1946–) M	**Rammeloo**
Ramond de Carbonnière, Louis François Elisabeth (1753–1827) MS	**Ramond**
Ramos, Clara Hilda S	**Ramos**
(Ramos de Moura, Nilsa)	
see Moura, Nilsa Ramos de	**N.R.Moura**
(Ramos Lopes, Maria Helena)	
see Lopes, M. Helena Ramos	**R.Lopes**
Ramos Núñez, A. (fl. 1983) S	**A.Ramos**
Ramovs, Anton A	**Ramovs**
Rampi, Leopoldo A	**Rampi**
Ramsay, A.T.S. A	**A.T.S.Ramsay**
Ramsay, Helen Patricia (1928–) B	**H.P.Ramsay**
Ramsay, James (1812–1888) B	**Ramsay**
Ramsbottom, John (1885–1974) M	**Ramsb.**
Ramsey, Glen Blaine (1889–) M	**Ramsey**
Ramsfjell, Einar A	**Ramsfjell**
(Ranalli de Cinto, Marìa E.)	
see Ranalli, María E.	**Ranalli**
Ranalli, María E. (1940–) M	**Ranalli**
Rand, Edward Lothrop (1859–1924) AS	**E.L.Rand**
Rand, Edward Sprague (1834–1897) S	**E.S.Rand**
Rand, Frederick Vernon (1883–) M	**F.V.Rand**
Rand, Isaac (–1743) L	**Rand**
Randall, R.P. (fl. 1988) S	**Randall**
Rändel, Ursula (1941–) S	**Rändel**
Randell, Barbara Rae (1942–) S	**Randell**
Randeria, Aban J. (fl. 1960) S	**Randeria**
Randhawa, H.S. (fl. 1963) M	**H.S.Randhawa**
Randhawa, Mohinder Singh (1909–1986) A	**Randhawa**
Randjelović, Novica (fl. 1990) S	**Randjel.**
Randlane, Tiina (1953–) M	**Randlane**
Randolph, Lowell Fitz (1894–1980) S	**Randolph**
Randow, Friedrich A	**Randow**
Rands, Robert Delafield (1890–1970) M	**Rands**
Rane, Daya (fl. 1980) M	**D.Rane**
Rane, M.S. (fl. 1966) M	**Rane**

Rangachari, Kadambi (1868–1934) S	**Rang.**
Ranganathan, K. (fl. 1973) M	**Rangan.**
Rangaswami, G. (fl. 1948) M	**Rangaswami**
Rangaswamy, Nanjangud Sreekantaiah (1932–) M	**Rangasw.**
Range, Paul Theodor (1879–1952) S	**Range**
(Rangel, Antonia Bastos)	
see Bastos, Antonia Rangel	**A.R.Bastos**
(Rangel Ch., J.Orlando)	
see Rangel, J.Orlando	**J.O.Rangel**
Rangel, Eugenio dos Santos (1877–1953) MS	**Rangel**
Rangel, J.Orlando (1950–) S	**J.O.Rangel**
Rangiah, P. (fl. 1965) M	**Rangiah**
Rangkuti, D. (fl. 1975) M	**Rangkuti**
Rani, N. (1955–) S	**N.Rani**
Ranjitha Devi, K.A. A	**Ranjitha**
Ranker, Thomas A. (1952–) P	**Ranker**
Rankin, Josephine Margaret (fl. 1980s) PS	**J.M.Rankin**
(Rankin Rodríguez, Rosa)	
see Rankin, Rosa	**R.Rankin**
Rankin, Rosa (1958–) S	**R.Rankin**
Rankin, William Howard (1888–) M	**Rankin**
Rankine, B.C. (fl. 1964) M	**Rankine**
Ranojević, Nikola (fl. 1905) M	**Ranoj.**
(Ranojevitch, Nikola)	
see Ranojević, Nikola	**Ranoj.**
Rantio–Lehtimäki, A.H. (fl. 1985) M	**Rant.-Leht.**
Ranzoni, Francis Verne (1916–) M	**Ranzoni**
Rao, A.N.S. (fl. 1980) M	**A.N.S.Rao**
Rao, A.Nageswara (fl. 1985) S	**A.N.Rao**
Rao, A.Sudhakar (fl. 1984) M	**A.Sudh.Rao**
Rao, A.V.N. (fl. 1984) S	**A.V.N.Rao**
Rao, Ananda R. (1924–) ABFM	**Rao**
Rao, Aragula Sathyanarayana (1924–1983) S	**A.S.Rao**
Rao, B. (fl. 1974) M	**B.Rao**
Rao, B.R.J. A	**B.R.J.Rao**
Rao, C.Bhashyakarla A	**C.B.Rao**
(Rao, Chellapilla Surya Prakasa)	
see Prakasa Rao, Chellapilla Surya	**Prak.Rao**
Rao, D.Koteswara (fl. 1957) M	**D.K.Rao**
Rao, D.P.C. (fl. 1976) M	**D.P.C.Rao**
Rao, Dev (fl. 1978) M	**D.Rao**
Rao, G.Koteswara (fl. 1980) M	**G.K.Rao**
Rao, G.Narasimha (fl. 1977) M	**G.N.Rao**
(Rao, Gorti Venkata Subba)	
see Subba Rao, Gorti Venkata	**Subba Rao**
Rao, Guang Yuan (fl. 1988) S	**G.Y.Rao**
Rao, Gurunath Vasant (fl. 1970) M	**G.V.Rao**
Rao, H.S. (fl. 1943) FM	**H.S.Rao**
Rao, K.Niranjan (fl. 1987) M	**K.N.Rao**

Rao, K.P.R. (fl. 1973) M	**K.P.R.Rao**
Rao, K.Purnachandra (fl. 1973) M	**K.P.Rao**
Rao, K.Seshagiri (fl. 1988) S	**K.S.Rao**
Rao, K.Sripada A	**K.Sr.Rao**
Rao, K.Venugopal (fl. 1980) M	**K.V.Rao**
Rao, L.N. (fl. 1944) P	**L.N.Rao**
Rao, Lakshmeswar Rama (1896–1974) A	**L.R.Rao**
Rao, M.M. (fl. 1980) M	**M.M.Rao**
Rao, M.R. (fl. 1986) M	**M.R.Rao**
Rao, M.Umamaheswara A	**M.U.Rao**
Rao, Mallikarjuna (fl. 1966) M	**M.Rao**
Rao, N.Krishna (fl. 1988) M	**N.K.Rao**
Rao, P.Govinda (fl. 1957) M	**P.G.Rao**
Rao, P.N. (fl. 1954) M	**P.N.Rao**
Rao, P.Raguveer (fl. 1962) M	**P.Rag.Rao**
(Rao, P.Rama)	
see Rama Rao, P.	**P.Rama Rao**
Rao, P.Sreenivasa (–1981) A	**P.S.Rao**
Rao, R.Raghavendra (1945–) P	**R.R.Rao**
Rao, Ramchandra (fl. 1964) M	**R.Rao**
Rao, Rolla Seshagiri (1921–) S	**R.S.Rao**
Rao, S.R.Narayana A	**S.R.N.Rao**
Rao, T.Bhaskar A	**T.B.Rao**
Rao, V.G. (1937–) M	**V.G.Rao**
Raoul, Édouard Fiacre Louis (1815–1852) BPS	**Raoul**
Raoul, Édouard François Armand (1845–1898) AS	**E.F.A.Raoul**
Raoul, M. (fl. 1844) M	**M.Raoul**
Rapaics, Raymund (1885–1953) MS	**Rapaics**
(Rapaics von Ruhmwerth, Raymund)	
see Rapaics, Raymund	**Rapaics**
Raper, Kenneth Bryan (1908–) M	**Raper**
Rapin, Daniel (1799–1882) S	**Rapin**
Rapp, Arthur Roman (1854–after 1894) S	**Rapp**
Rapp, Severin (1853–1941) S	**S.Rapp**
Rapp, William F., Jr. (fl. 1947) P	**W.F.Rapp**
Rappa, F. (fl. 1950s) S	**Rappa**
Rappaz, François (fl. 1987) M	**Rappaz**
Räsänen, Veli Johannes Paavo Bartholomeus (1888–1953) MS	**Räsänen**
Rasbach, Helga (1924–) PS	**Rasbach**
Rasbach, Kurt (fl. 1977) PS	**K.Rasbach**
Raschle, Paul (fl. 1977) M	**Raschle**
Rascio, Nicoletta A	**Rascio**
Rasheed, A. (fl. 1990) M	**Rasheed**
Răsín, K. (fl. 1928) M	**Răsín**
Rasinš, A.P. S	**Rasinš**
(Rasinsch, A.P.)	
see Rasinš, A.P.	**Rasinš**
Rásky, Klara (1908–1971) AF	**Rásky**
Rasmussen, Finn Nygaard (1948–) S	**F.N.Rasm.**

Rasmussen, Rasmus (1871–1962) S   **Rasm.**
Raspail, François Vincent (1794–1878) MS   **Raspail**
Rassadina, Ksenia Aleksandrovna (1903–　) M   **Rass.**
Rasskazova, E.C. (fl. 1968) F   **Rassk.**
Rassulova, M.R. (1926–　) S   **Rassulova**
Rasul, S.M.　A   **Rasul**
(Rasulova, M.R.)
   see Rassulova, M.R.   **Rassulova**
Rasumov, A.　A   **Rasumov**
Rataj, Karel (1925–　) S   **Rataj**
Ratcliffe, Herbert (1901–　) A   **Ratcl.**
Rath, Andrew C. (fl. 1989) M   **A.C.Rath**
Rath, Francesco (fl. 1987) M   **F.Rath**
Rathaiah, Y. (fl. 1971) M   **Rathaiah**
Rathakrishnan, N.C. (1939–　) S   **Rathakr.**
Ráthay, Emerich (1845–1900) M   **Ráthay**
Rathke, Jens (1769–1855) S   **Rathke**
Rathod, M.M. (fl. 1986) M   **Rathod**
Rathore, R.S. (fl. 1980) M   **Rathore**
Rathore, S.R. (fl. 1984) S   **S.R.Rathore**
Rathschlag, Heinz (fl. 1930) M   **Rathschlag**
Ratiani, Gulnara Sh. (1930–　) M   **Ratiani**
Raţiu, Onoriu (1927–1986) S   **Raţiu**
Ratnasabapathy, M. (1929–　) A   **Ratnas.**
Rattan, R.S.　A   **R.S.Rattan**
Rattan, Sarjit S. (fl. 1967) M   **S.S.Rattan**
Rattan, Volney (1840–1915) S   **Rattan**
Ratte, F. (fl. 1886–1887) F   **Ratte**
Ratter, James Alexander (1934–　) S   **Ratter**
Rattke, Wilhelm (fl. 1884) S   **Rattke**
Rattray, James (fl. 1835–1880) S   **J.Rattray**
Rattray, James McFarlane (1907–1974) S   **J.M.Rattray**
Rattray, John (1858–1900) A   **Rattray**
Ratzeburg, Julius Theodor Christian (1801–1871) S   **Ratzeb.**
Rau, Ambrosius (1784–1830) S   **A.Rau**
Rau, Eugene Abraham (1848–1932) BM   **Rau**
Raub, Thomas J. (fl. 1979) M   **Raub**
Rauch, Friedrich (1867–　) S   **Rauch**
Rauh, Werner (1913–　) BPS   **Rauh**
Rauhala, A. (fl. 1971) M   **Rauhala**
Raulin, Victor Félix (1819–1905) S   **Raulin**
Raunkiaer, Christen Christiansen (1860–1938) MS   **Raunk.**
Raunsgaard Pedersen, Kaj　A   **Raunsg.Ped.**
Raup, Hugh Miller (1901–　) S   **Raup**
Raus, Thomas (1949–　) S   **Raus**
Rausch, Walter (1928–　) S   **Rausch**
(Räuschel, Ernst Adolf)
   see Raeuschel, Ernst Adolf   **Raeusch.**
Rauschenbach, W.　A   **Rauschenb.**
Rauscher, Robert (1806–1890) A   **Rauscher**

Rauschert, R. (fl. 1975) M	**R.Rauschert**
Rauschert, Stephan (1931–1986) MPS	**Rauschert**
Rauth, Franz (1874–) S	**Rauth**
Rauwenhoff, Nicolaas Willem Pieter (1826–1909) S	**Rauwenh.**
Rauwerdink, J.B. (fl. 1986) S	**Rauwerd.**
(Rauwolf, Leonhard	
see Rauwolff, Leonhart	**Rauwolff**
Rauwolff, Leonhart (1535–1596) L	**Rauwolff**
Ravanko, Orvokki Maija–Liisa (1939–) A	**Ravanko**
Ravano, Carla S	**Ravano**
Răvăruţ, Mihai (1907–1981) S	**Răvăruţ**
Ravaud, Louis Célestine Mure (1822–1898) S	**Ravaud**
Ravaz, Louis Etienne (1863–1937) M	**Ravaz**
Raven, John Earle (1914–1980) S	**Raven**
Raven, Peter Hamilton (1936–) S	**P.H.Raven**
Raven, Tamra Engelhorn (née Engelhorn, T.) (1945–) S	**T.E.Raven**
Ravenel, Henry William (1814–1887) BMS	**Ravenel**
Ravenna, Pedro Felix (Pierfelice, Pierre Félice) (1938–) S	**Ravenna**
Ravenscroft, Edward James (1816–1890) S	**Ravenscr.**
Ravenshaw, Thomas Fitzarthur Torin (1829–1882) S	**Ravenshaw**
Ravi, N. (fl. 1969–1979) P	**Ravi**
Ravikumar, D.R. (fl. 1991) M	**Ravik.**
Ravin, Eugène (fl. 1861) S	**Ravin**
Ravinder, E.John (fl. 1987) M	**Ravinder**
Ravindran, P.N. (fl. 1987) S	**Ravindran**
Raviv, Varda (1940–) A	**Raviv**
Ravn, Frederik Kølpin (1873–1920) S	**Ravn**
Ravnik, Vlado (1924–) S	**Ravnik**
Rawal, R.D. (fl. 1984) M	**Rawal**
Rawat, M.S. (fl. 1963) AM	**Rawat**
Rawe, Rolf S	**Rawe**
Rawitscher, Felix (1890–1957) S	**Rawitscher**
Rawla, G.S. (1934–) M	**Rawla**
Rawlings, G.B. (fl. 1956) M	**Rawlings**
Rawlinson, Christopher James (1942–) M	**Rawl.**
Rawson, Rawson William (1812–1899) P	**Rawson**
Rawton, Olivier de (fl. 1889) S	**Rawton**
Ray, Harendra Nath (1898–1969) A	**H.N.Ray**
Ray, J.B. (fl. 1985) M	**J.B.Ray**
Ray, James Davis, Jr. (1918–) S	**J.D.Ray**
Ray, John (1627–1705) ABFLS	**Ray**
Ray, S.M. (fl. 1966) M	**S.M.Ray**
Ray, William Winfield (1909–) M	**W.W.Ray**
Raybaud, L. (fl. 1921) M	**Raybaud**
Rayburn, William Reed (1940–) A	**Rayburn**
Raychaudhuri, Saibal A	**S.Raych.**
Raychaudhuri, Syam Prasaa (1915–) M	**Raych.**
Rayman (fl. 1891) M	**Rayman**
Raymond, Louis-Florent-Marcel (1915–1972) PS	**Raymond**

Raymond–Hamet (1890–1972) S	**Raym.-Hamet**
Raynal, Aline Marie (1937–) S	**A.Raynal**
Raynal, Jean (1933–1979) S	**J.Raynal**
(Raynal–Roques, Aline Marie)	
see Raynal, Aline Marie	**A.Raynal**
Raynaud, Christian (1939–) S	**Raynaud**
Raynaud, J.F. A	**J.F.Raynaud**
Raynaud, Maurice A	**M.Raynaud**
Rayner, John Frederick (1854–1947) S	**Rayner**
Rayner, Ronald William (1914–) M	**R.W.Rayner**
Rayner, Timothy Guy Johnson (1963–) S	**T.G.J.Rayner**
Rayns, David Geoffrey (1935–) A	**Rayns**
Rayss, Tscharna (1890–1965) AMS	**Rayss**
Razafindratsira, A. (fl. 1987) S	**Razaf.**
Razi, Basheer Ahmed (1916–) S	**Razi**
Re, Filippo (1763–1817) MS	**F.Re**
Re, Giovanni Francesco (1773–1833) MS	**Re**
Re, S. (fl. 1925) M	**S.Re**
Rea, Carleton (1861–1946) M	**Rea**
Rea, Paul Marshall (1878–1948) M	**P.M.Rea**
Read, Charles Brian (fl. 1939–1942) F	**C.B.Read**
Read, D.J. (fl. 1974) M	**D.J.Read**
Read, Robert William (1931–) S	**Read**
Reade, John Moore (1876–1937) MS	**J.M.Reade**
Reade, Oswald Alan (1848–1929) S	**Reade**
Reader, (Henry Charles Lyon) Peter (1840–1929) S	**P.Reader**
Reader, Felix Maximilian (1850–1911) S	**Reader**
Réaubourg, Gaston (fl. 1906) S	**Réaub.**
Reaugh, Ann G. A	**Reaugh**
Rebassa, Antoni (fl. 1991) P	**Rebassa**
Rebentisch, Johann Friedrich (1772–1810) BMS	**Rebent.**
Reber, Burkhardt (1848–1926) S	**Reber**
Reboud, Victor Constant (1821–1889) S	**Reboud**
Reboul, Eugène de (1781–1851) S	**Reboul**
Rebrikova, N.L. (fl. 1978) M	**Rebr.**
(Rebristaja, O.V.)	
see Rebristaya, O.V.	**Rebrist.**
Rebristaya, O.V. (1930–) S	**Rebrist.**
Rebut, P. (–1898) S	**Rebut**
Recca, J. (fl. 1952) M	**Recca**
Réchin, Jules (1853–1913) B	**Réchin**
Rechinger, Karl (1867–1952) BS	**Rech.**
Rechinger, Karl Heinz (1906–) APS	**Rech.f.**
Recocochea, M. (fl. 1977) M	**Recoc.**
Record, Samuel James (1881–1945) S	**Record**
Redaelli, Piero (1898–1955) AM	**Redaelli**
Reddi, Bommareddi Venkata (1941–) M	**Reddi**
Reddick, Donald (1883–1955) M	**Reddick**

Reddy, A.P. (fl. 1980) M	A.P.Reddy
Reddy, B.Satyanarayana (fl. 1970) M	B.S.Reddy
Reddy, C.Narayana (fl. 1979) M	C.N.Reddy
Reddy, G.S. (fl. 1952) M	G.S.Reddy
Reddy, J.R. (fl. 1980) M	J.R.Reddy
Reddy, K.Adinarayana (fl. 1980) M	K.A.Reddy
Reddy, K.R.Chandra (fl. 1968) M	K.R.C.Reddy
Reddy, M.S. (fl. 1975) M	M.S.Reddy
Reddy, P.G. (fl. 1990) M	P.G.Reddy
Reddy, P.Santosh (fl. 1978) M	P.S.Reddy
Reddy, S.M. (fl. 1977) M	S.M.Reddy
Reddy, S.R. (fl. 1978) M	S.R.Reddy
Reddy, S.S. (fl. 1977) M	S.S.Reddy
Reddy, T.K.Ramachandra (fl. 1962) M	T.K.R.Reddy
Reddy, V.R.Thulasi (fl. 1986) M	V.R.T.Reddy
Redecker, G. (1902–1975) S	Redecker
Redeke, Heinrich Carl (1873–1945) AS	Redeke
(Redesdale, Algernon Bertram Freeman, Lord)	
see Mitford, Algernon Bertram Freeman	Mitford
Redeuilh, G. (fl. 1978) M	Redeuilh
Redfearn, Paul Leslie (1926–) B	Redf.
Redfern, D.B. (fl. 1987) M	Redfern
Redfield, John Howard (1815–1895) AFS	Redfield
Redforth, N.W. (fl. 1958) M	Redforth
Redhead, J.F. (fl. 1979) M	J.F.Redhead
Redhead, Scott Alan (1950–) M	Redhead
Redi, Francesco (1626–1698) S	Redi
Redinger, Karl Martin (1907–1940) AM	Redinger
Redlin, Scott C. (1954–) M	Redlin
Redmond, Paul John Dominic (1901–) S	Redmond
(Redoffsky, Ivan Ivanovich)	
see Redowsky, Ivan Ivanovich	I.Redowsky
Redón Figueroa, Jorge (1936–) M	Redón
Redouté, Pierre Joseph (1759–1840) S	Redouté
Redowsky, D. (1804–) S	Redowsky
Redowsky, Ivan Ivanovich (1774–1807) S	I.Redowsky
Redslob, Julius (fl. 1863) S	Redslob
Redtenbacher, Joseph (1810–1870) S	Redtenb.
Reed, Clarence Arthur (1880–1950) S	C.A.Reed
Reed, Clyde Franklin (1918–) ABPS	C.F.Reed
Reed, Edward Looman (1878–1946) S	E.L.Reed
Reed, George Matthew (1878–1956) M	G.M.Reed
Reed, Howard Sprague (1876–1950) MS	H.S.Reed
Reed, L.E. (fl. 1984) M	L.E.Reed
Reed, Minnie A	M.Reed
Reeder, Charlotte Olive (née Goodding, C.O.) (1916–) S	C.Reeder
Reeder, John Raymond (1914–) S	Reeder

Reedman, D.J.   A   **Reedman**
Reeleder, R.D. (fl. 1988) M   **Reeleder**
Reenen–Hoekstra, E.S.van (fl. 1989) M   **Reenen**
Reer, U. (fl. 1986) S   **Reer**
Rees, Abraham (1743–1825) PS   **Rees**
Rees, Anthony J.J.   A   **A.J.J.Rees**
Rees, Bertha   S   **B.Rees**
Rees, R.G. (fl. 1976) M   **R.G.Rees**
Reese, Heinrich (fl. 1931–1939) S   **Reese**
Reese, R.N. (fl. 1984) S   **R.N.Reese**
Reese, William Dean (1928–   ) B   **W.D.Reese**
Reess, Maximilian (Friedrich Timotheus Ferdinand Maria) (1845–1901) MS   **Reess**
Reeve, Helen (c.1939–   ) S   **Reeve**
Reeve, T.M. (fl. 1987) S   **T.M.Reeve**
Reeves, Enoch Lloyd (1901–   ) M   **E.L.Reeves**
Reeves, Robert Gatlin (1898–   ) S   **Reeves**
Reeves, Timothy (1947–   ) P   **T.Reeves**
Regel, (Johann) Albert von (1845–1908) S   **A.Regel**
Regel, Constantin Andreas von (1890–1970) S   **C.Regel**
Regel, Eduard August von (1815–1892) BMPS   **Regel**
Regel, Robert E. (1867–1920) S   **R.E.Regel**
Regnault, A. (fl. 1974) M   **A.Regnault**
Regnault, Nicolas François (1746–   ) S   **Regnault**
Regnell, Anders Fredrik (1807–1884) S   **Regnell**
Reguis, J.Marius F. (1850–   ) S   **Reguis**
Reháková, Helena   A   **H.Reháková**
Reháková, Zdenka   A   **Z.Reháková**
Rehana, A.R. (fl. 1975) M   **Rehana**
Rehder, Alfred (1863–1949) S   **Rehder**
Rehfous, Laurent (1890–   ) A   **Rehfous**
Rehill, P.S. (fl. 1959) M   **Rehill**
Rehm, Heinrich (Simon Ludwig Friedrich Felix) (1828–1916) MS   **Rehm**
Rehm, Sigmund Eugen Adolf (1911–   ) S   **S.E.A.Rehm**
(Rehman, Anton)
  see Rehmann, Anton   **Rehmann**
Rehman, K.   A   **Rehman**
Rehmann, Anton (1840–1917) BS   **Rehmann**
Rehnelt, F. (1861–   ) S   **Rehnelt**
Rehner, Stephen A. (fl. 1988) M   **S.A.Rehner**
Rehsteiner, Hugo (1864–1947) S   **Rehst.**
Reichard, Johann Jacob (Jakob) (1743–1782) ABMPS   **Reichard**
Reichardt, Auguste   A   **A.Reichardt**
Reichardt, Erwin   A   **E.Reichardt**
Reichardt, Heinrich Wilhelm (1835–1885) MPS   **Reichardt**
Reichardt, J.   S   **J.Reichardt**
Reiche, Karl Friedrich (Carlos Federico) (1860–1929) S   **Reiche**
Reichel, Georg Christian (1721–1771) S   **Reichel**
Reichelt, Hugo (1857–   ) A   **Reichelt**

Reichelt, W. (fl. 1887) M	W.Reichelt
Reichenau, Wilhelm von (1847–post 1913) S	Reichenau
Reichenbach, (Heinrich Gottlieb) Ludwig (1793–1879) ABMPS	Rchb.
Reichenbach, Carl (Karl) Ludwig von (1788–1869) S	C.Rchb.
Reichenbach, F. (fl. 1896) S	F.Rchb.
Reichenbach, Heinrich Gustav (1824–1889) PS	Rchb.f.
Reichenbach-Klinke, Heinz Herman (fl. 1956) AM	Rchb.-Klinke
Reichenow, Eduard (1883–1960) A	Reichenow
Reichensperger, Gedenktage August (c.1877) M	Reichensp.
Reichert, Israel G. (1889–1975) M	Reichert
Reichgelt, B. (fl. 1939) S	B.Reichg.
Reichgelt, Theodorus Johannes (1903–1966) S	Reichg.
Reichle, R.E. (fl. 1980s) M	Reichle
Reichstein, R. (fl. 1981–1989) S	R.Reichst.
Reichstein, Tadeus (1897–) PS	Reichst.
Reid, Clement L. (1853–1916) ABFS	C.Reid
Reid, Derek A. (1927–) M	D.A.Reid
Reid, Eleanor Mary Wynne-Edwards (1860–1953) ABFS	E.Reid
Reid, Freda M.H. A	F.Reid
Reid, James (fl. 1961) M	J.Reid
Reid, Philip C. A	P.C.Reid
Reider, Jakob Ernst von (1784–1853) S	Reider
Reiersöl, Signy (fl. 1958) M	Reiersöl
Reif, Charles Braddock (1912–) A	Reif
Reifschneider, Francisco J.B. (fl. 1979) M	Reifschn.
Reijnders, W.J. (fl. 1959) M	Reijnders
Reilly, Jacqueline (fl. 1977) S	Reilly
Reimann, Bernhard E.F. (1922–) A	Reimann
Reimer, Charles Wilson (1923–) A	Reimer
Reimers, Hermann Johann O. (1893–1961) BMPS	Reimers
Rein, Johannes Justus (1835–1918) S	Rein
Reinaud de Fonvert, Alexandre Jean Baptiste (1797–1871) S	Rein.Fonv.
Reinbold, Theodor (1840–1918) A	Reinbold
Reinecke, Franz (1866–) PS	Reinecke
Reinecke, Karl Lorenz (1854–1934) S	K.Reinecke
Reinecke, Pandora A	P.Reinecke
Reinecke, W. (fl. 1886) S	W.Reinecke
Reiner, Joseph (Josef) (1766–1797) S	Reiner
Reinhard, Edward George (1899–) A	E.G.Reinhard
Reinhard, Hans R. (1919–) S	H.R.Reinhard
Reinhard, Ludwig (Vasilievič) (1846–1922) A	Reinhard
Reinhardt, Donald J. (fl. 1967) M	D.J.Reinh.
Reinhardt, Ludwig (fl. 1910) S	L.Reinh.
Reinhardt, Max Otto (1854–1935) M	M.O.Reinh.
Reinhardt, Otto Wilhelm Hermann (1838–1924) S	Reinh.
Reinhardt, Peter (fl. 1961) AB	P.Reinh.
Reinhold, Thomas (1890–1955) A	Reinhold
Reinisch, Olga A	Reinisch
Reinke, Johannes (1849–1931) AM	Reinke

Reinking, Mark (fl. 1981) S	**M.Reinking**
Reinking, Otto August (1890–1962) M	**Reinking**
Reinsch, (Edgar) Hugo (Emil) (1809–1884) S	**H.Reinsch**
Reinsch, Paul Friedrich (1836–1914) AMPS	**Reinsch**
Reinwardt, Caspar Georg Carl (1773–1854) BPS	**Reinw.**
Reis, Cónego M.Póvoa dos A	**C.M.P.Reis**
Reis, M.L.C. (fl. 1969) M	**M.L.C.Reis**
Reis, Manuel Póvoa dos A	**M.P.Reis**
Reis, Otto Maria (1862–1934) A	**Reis**
Reisaeter, Oddvin (1913–1983) S	**Reisaeter**
Reischer, Helen Simpson (fl. 1949) M	**Reischer**
Reiser, C. (fl. 1984) S	**Reiser**
Reisigl, Herbert (1929–) AS	**Reisigl**
Reisinger, Otto (fl. 1967) M	**Reisinger**
(Reissaeter, Oddvin)	
see Reisaeter, Oddvin	**Reisaeter**
(Reisseck, Siegfried)	
see Reissek, Siegfried	**Reissek**
Reissek, Siegfried (1819–1871) MS	**Reissek**
Reitschel, Siegfried A	**Reitschel**
Reitsma, J. (fl. 1950) M	**Reitsma**
Reitsma, J.M. (fl. 1984) S	**J.M.Reitsma**
Reitter, Johann Daniel von (1759–1811) S	**Reitter**
Reitz, P.Raulino (1919–1990) S	**Reitz**
Rejdali, M. (fl. 1988) S	**Rejdali**
Rejtlinger, E.A. A	**Rejtl.**
Reker, Carl C. A	**Reker**
Relhan, Richard (1754–1823) BMS	**Relhan**
Remak, Robert (fl. 1845) M	**Remak**
Remaudière, G. (fl. 1980) M	**Remaud.**
Remler, Paula (fl. 1979) M	**Remler**
Remsberg, Ruth Elizabeth ('Honey') (1906–) M	**Remsberg**
Remsen, Charles C. (1937–) A	**Remsen**
Remy, Esprit Alexandre (1826–1893) S	**Remy**
Rémy, Ezechiel Jules (1826–1893) PS	**J.Rémy**
Remy, L. (fl. 1964) M	**L.Remy**
Remy, R. (fl. 1975) F	**R.Remy**
Remy, W. (fl. 1975) F	**W.Remy**
Ren, Jin Zhou (fl. 1987) S	**J.Z.Ren**
Ren, Shou Qin (fl. 1990) F	**S.Q.Ren**
Ren, Wei (fl. 1987) M	**W.Ren**
Ren, Yi (1960–) S	**Y.Ren**
Renaudet, Georges Benjamin (1852–) S	**Renaudet**
Renauld, Ferdinand François Gabriel (1837–1910) B	**Renauld**
Renault, Bernard (1836–1904) ABFMS	**Renault**
Renault, Pierre Antoine (1750–1835) S	**P.Renault**
Renberg, Ingemar A	**Renberg**
Rendle, Alfred Barton (1865–1938) APS	**Rendle**

(Renealmus, Paul)	
see Reneaulme, Paul	**Reneaulme**
Reneaulme, Paul (1560–1624) L	**Reneaulme**
Renfro, B.L. (fl. 1966) M	**Renfro**
Rengarten, N.V. A	**Reng.**
Renier, Armand–Marie–Vincent–Joseph (1876–1951) F	**Renier**
Renjifo, Carlos (1884–) S	**Renjifo**
Renner, Otto (1883–1960) BMS	**Renner**
Renner, Richard A	**R.Renner**
Renner, Susanne S. (1954–) S	**S.S.Renner**
Rennerfelt, Erik (fl. 1937) M	**Rennerf.**
Rennie, Robert (–1820) S	**Rennie**
Renny, J. (fl. 1874) M	**Renny**
Renobales, Gustavo (1958–) M	**Renob.**
Rensch, Ilse (1902–) PS	**Rensch**
(Rensch–Maier, Ilse)	
see Rensch, Ilse	**Rensch**
Rensselaer, Maunsell van (fl. 1942) S	**Renss.**
Renuka, C. (1951–) S	**Renuka**
Renvoize, Stephen Andrew (1944–) S	**Renvoize**
Renz, Jany (1907–) S	**Renz**
Renzoni, Giovanni Cela (1931–) S	**Renzoni**
Reppenhagen, W. (fl. 1980) S	**Repp.**
Requien, Esprit (1788–1851) FMPS	**Req.**
Resch, W. A	**Resch**
Resende, Flávio P.de (1907–1967) S	**Resende**
Resende–Pinto, Manuel Cabral de (–1989) MP	**Res.-Pinto**
Reshetova, I.S. (fl. 1975) M	**Reshetova**
Resmeriţă, Ion (1907–1987) S	**Resm.**
Resvoll, Thekla Susanne Ragnhild (1871–1948) S	**Resvoll**
Resvoll–Holmsen, Hanna (Marie) (1873–1943) S	**Resv.-Holms.**
Retallack, Gregory J. (fl. 1977–1980) F	**Retallack**
Retief, Elizabeth (1947–) S	**Retief**
Rettger, Leo Frederick (1874–) AM	**Rettger**
Rettig, J.H. S	**Rettig**
(Retz, Bernard G.G.De)	
see De Retz, Bernard G.G.	**De Retz**
Retzdorff, Adolf Eduard Willy (1856–1910) S	**Retzd.**
Retzius, (Magnus) Gustav (1842–1919) S	**G.Retz.**
Retzius, Anders Jahan (1742–1821) ABMPS	**Retz.**
Reukauf, E. (fl. 1912) M	**Reukauf**
Reuling, F. A	**Reuling**
Reum, Johann Adam (1780–1839) S	**Reum**
Reumaux, Patrick (fl. 1982) M	**Reumaux**
Reusch, Erhart (1678–1740) S	**Reusch**
Reuss, August Emanuel (Rudolph) von (1811–1873) AS	**A.E.Reuss**
Reuss, Christian Friedrich (1745–1813) S	**Reuss**
Reuss, Georg Christian (fl. 1869) S	**G.C.Reuss**
Reuss, Gustáv (1818–1861) S	**G.Reuss**

Reuss, Jeremies David (1750–1837) S	**J.D.Reuss**
Reuss, Leopold (fl. 1831) S	**L.Reuss**
Reusser, F.A. (fl. 1964) M	**Reusser**
Reuter, George François (1805–1872) PS	**Reut.**
Reuthe, G. (fl. 1880s) S	**Reuthe**
Révay, Ágnes (1952–) M	**Révay**
Reveal, James Lauritz (1941–) FS	**Reveal**
Réveil, (Pierre) Oscar (1821–1865) S	**Réveil**
Revel, Joseph (1811–1887) S	**Revel**
Revelière, Eugène (1822–1892) S	**Revelière**
Reverchon, Elisée (1835–1914) S	**E.Rev.**
Reverchon, Julien (1837–1905) S	**J.Rev.**
Reverchon, P. (fl. 1878–1892) S	**P.Rev.**
Reverdatto, Viktor Vladimirovich (1891–1969) S	**Reverd.**
Reverdin, Louis (1894–) A	**Reverdin**
Revjakina, N.V. (1940–) S	**Revjakina**
Révoil, Georges (fl. 1878–1880) S	**Révoil**
Revol, J. (fl. 1928) S	**Revol**
Revuschkin, A.S. (1952–) S	**Revuschkin**
(Revyakina, N.V.)	
see Revjakina, N.V.	**Revjakina**
Rewbridge, A.G. (fl. 1929) M	**Rewbr.**
Rex, George Abraham (1845–1895) M	**Rex**
Rexhepi, F. (fl. 1986) S	**Rexhepi**
Rey–Pailhade, Constantin de (1844–1930) P	**Rey-Pailh.**
(Reyes Prosper, Eduardo)	
see Reyes y Prosper, Eduardo	**Reyes y Prosper**
Reyes y Prosper, Eduardo (1860–1921) AS	**Reyes y Prosper**
Reyger, Gottfried (1704–1788) S	**Reyger**
Reymond, M.L.C. (fl. 1854) S	**Reymond**
Reymond, Olivier A	**O.Reymond**
Reyneke, William Frederick (1945–) S	**Reyneke**
Reynier, Alfred (1845–1932) S	**A.Reyn.**
Reynier, Jean Louis Antoine (1762–1824) S	**Reyn.**
Reynolds, Bruce Dodson (1894–) A	**B.D.Reynolds**
Reynolds, Don Rupert (1938–) AM	**D.R.Reynolds**
Reynolds, Gilbert Westacott (1895–1967) S	**Reynolds**
Reynolds, Nial (fl. 1965) A	**N.Reynolds**
Reynolds, Sally T. (1932–) S	**S.T.Reynolds**
Reyssac, Josette A	**Reyssac**
Rezak, Richard (1920–) A	**Rezak**
(Rezende–Pinto, Manuel Cabral de)	
see Resende–Pinto, Manuel Cabral de	**Res.Pinto**
Reznicek, Anton Albert (1950–) S	**Reznicek**
Reznik, A. S	**Reznik**
Rheede tot Draakestein, Hendrik Adriaan von (1637–1691) L	**Rheede**
Rhein, G.F. (1858–) S	**Rhein**
Rhind, William (fl. 1830–1860) S	**Rhind**
Rhiner, Joseph (1830–1897) S	**Rhiner**

Rhoads, Arthur Stevens (1893–) M	**Rhoads**
Rhode, Johann Gottlieb (1762–1827) S	**Rhode**
Rhodes, Donald Gene (1933–) S	**D.G.Rhodes**
Rhodes, Landon H. (fl. 1981) M	**L.H.Rhodes**
Rhodes, Philip Grafton Mole (1885–1934) BM	**Rhodes**
Rhumbler, L. A	**Rhumbler**
Rhyne, Charles F. A	**Rhyne**
Riba, Ramón (1934–) P	**Riba**
Ribaldi, M. (fl. 1948) M	**Ribaldi**
Ribeiro de Souza, Abigail Freire (fl. 1985) M	**Rib.Souza**
Ribeiro, J.O. (fl. 1991) M	**Ribeiro**
(Ribeiro Nesio, Maria L.)	
see Nesio, Maria L. Ribeiro	**Nesio**
Ribeyro, R.E. (fl. 1918) M	**Ribeyro**
Riboni, G. (fl. 1879) M	**Riboni**
Ricard, Michel (1943–) A	**Ricard**
Ricardi Salinas, Mario (1921–) S	**Ricardi**
Ricasoli, Vincenzo (1814–1891) S	**Ricasoli**
Ricca, Luigi (–1879) S	**Ricca**
Ricca, Ubaldo (1872–) S	**U.Ricca**
Ricceri, Carlo (1933–) S	**Ricceri**
Ricci, Angelo Maria (1777–1850) S	**Ricci**
Ricci, Ignazio (1922–1986) S	**I.Ricci**
Ricci, Sandra (1956–) S	**S.Ricci**
Ricciardi, Massimo (1935–) S	**Ricciardi**
Riccobono, Vincenzo (1861–1943) S	**Riccob.**
Rice, Ellen L. S	**Rice**
Ricek, E.W. (fl. 1974) M	**Ricek**
Rich, Mark (1932–) A	**M.Rich**
Rich, Marvin A. (fl. 1958) M	**M.A.Rich**
Rich, Mary Florence (1865–1939) AS	**M.F.Rich**
Rich, Obadiah (1783–1850) S	**O.Rich**
Rich, Timothy Charles Guy (1961–) S	**T.C.G.Rich**
Rich, William (1800–1864) S	**W.Rich**
Rich, William Penn (1849–1930) S	**W.P.Rich**
Richard, (Jean Michel) Claude (1784–1868) S	**J.M.C.Rich.**
Richard, Achille (1794–1852) ABMPS	**A.Rich.**
Richard, Claude (fl. 1971) M	**C.Rich.**
Richard, Joseph Herve Pierre (1946–) A	**J.H.P.Rich.**
Richard, Louis Claude Marie (1754–1821) BP	**Rich.**
Richard, Olivier Jules (1836–1896) BM	**O.J.Rich.**
Richards, Adrian John (1943–) S	**A.J.Richards**
Richards, Edward L. (1927–) S	**E.L.Richards**
Richards, Herbert Maule (1871–1928) A	**H.Richards**
Richards, Merfyn (fl. 1956) M	**M.Richards**
Richards, Paul Westmacott (1908–) B	**P.W.Richards**
Richardson, Alfred Thomas (1930–) S	**A.T.Richardson**
Richardson, Frederick Reginald (1915–) S	**F.R.Richardson**
Richardson, Ian Bertram Kay (1940–) S	**I.Richardson**
Richardson, James W. (1937–) S	**J.W.Richardson**

Richardson, John (1787–1865) PS	**Richardson**
Richardson, Michael John (1938–) M	**M.J.Richardson**
Richardson, Richard (1663–1741) L	**R.Richardson**
Richatt, F.M. (fl. 1953) M	**Richatt**
Richel, Thierry (1955–) S	**Richel**
Richen, Gottfried (1863–) S	**Richen**
Richens, Richard Hook (1919–1984) S	**Richens**
Richer de Belleval, Pierre (1564–1632) LS	**Rich.Bell.**
Richerson, Peter James (1943–) A	**Richerson**
Richiţeanu, Anghel (1942–) M	**Richiţ.**
Richon, Charles Édouard (1820–1893) M	**Richon**
Richter, August Gottlieb (1742–1812) S	**A.G.Richt.**
Richter, Berthold (1834–) S	**B.Richt.**
Richter, H.G. (fl. 1987) S	**H.G.Richt.**
Richter, Harold (1902–) M	**H.Richt.**
Richter, Hermann Eberhard Friedrich (1808–1876) S	**Richt.**
Richter, Jules Adolphe (1821–1910) S	**J.A.Richt.**
Richter, Karl (Carl) (1855–1891) S	**K.Richt.**
Richter, Ludwig (Lajos) (1844–1917) S	**L.Richt.**
Richter, Oswald (1878–1955) A	**O.Richt.**
Richter, Paul Boguslav (1853–1911) F	**P.B.Richt.**
Richter, Paul Gerhard (1837–1913) A	**P.G.Richt.**
Richter, Ursula (1941–) A	**U.Richt.**
Richter, Vincenz Aladár (1868–1927) PS	**V.A.Richt.**
Rick, Johann (João Evangelista) (1869–1946) M	**Rick**
Ricken, Adalbert (1851–1921) M	**Ricken**
Ricker, Percy Leroy (1878–1973) MS	**Ricker**
Ricker, Robert Wallace A	**R.W.Ricker**
Rickert, Francis Brilon (1914–) A	**Rickert**
Rickett, Harold William (1896–1989) S	**Rickett**
Rickett, Theresa Cecil (née Bauchman, T.C.) (1902–) S	**T.Rickett**
Rickli, M.A. (1868–1951) S	**Rickli**
Rico, Enrique (1953–) S	**E.Rico**
(Rico Hernández, Enrique)	
see Rico, Enrique	**E.Rico**
Rico, Maria de Lourdes ('Lulu') (1955–) S	**L.Rico**
Rico, V.J. (fl. 1985) M	**V.J.Rico**
(Rico–Arce, Maria de Lourdes)	
see Rico, Maria de Lourdes	**L.Rico**
Riddell, John Leonard (1807–1865) S	**Riddell**
Riddelsdell, Harry Joseph (1866–1941) S	**Ridd.**
Ridder–Numan, J.W.A. (fl. 1985) S	**Ridd.-Num.**
Riddle, Lincoln Ware (1880–1921) M	**Riddle**
Ridings, W.H. (fl. 1978) M	**Ridings**
Ridley, Geoffrey S. (fl. 1988) M	**G.S.Ridl.**
Ridley, Henry Nicholas (1855–1956) MPS	**Ridl.**
Ridlon, Harry Cooper (fl. 1921–1922) P	**Ridlon**
Ridsdale, Colin Ernest (1944–) S	**Ridsdale**
Rieber, Xaver (1860–1906) M	**Rieber**

Riedel, Ludwig (1790–1861) S	**Riedel**
Riedel, W.R. (1927–) A	**W.R.Riedel**
Rieder, C.L. (fl. 1982) M	**Rieder**
Riedl, Harald (Harold) Udo von (1936–) MS	**Riedl**
Riedl–Dorn, Christa (1955–) S	**Riedl–Dorn**
Riegel, Walter A	**Riegel**
Riehl, Nicholas (1808–1852) S	**Riehl**
Riehmer, Ernst (1874–1966) BM	**Riehm.**
Riek, R. S	**Riek**
Riemann, Bo (1946–) B	**Riemann**
(Riemer–Gerhardt, Elizabeth)	
see Gerhardt, Elizabeth	**El.Gerhardt**
Ries, Heinrich (1871–1951) S	**Ries**
Rieseberg, Loren H. (1961–) S	**Rieseberg**
Riess, H. (1809–1878) MS	**Riess**
Rietema, Hybo A	**Rietema**
Rieth, Alfred (fl. 1950) AM	**Rieth**
Rietsch, Maximilian (fl. 1882) S	**Rietsch**
Rietschel, P. (fl. 1935) M	**Rietschel**
(Rietz, Gustaf Einar)	
see Du Rietz, Gustaf Einar	**Du Rietz**
Rieuf, P. (fl. 1953) M	**Rieuf**
Rifai, Mien Achmad (1940–) M	**Rifai**
Rigaud, Antoine (fl. 1877) S	**Rigaud**
Rigaux, A. (fl. 1877) B	**Rigaux**
Rigby, John Francis (1927–) FM	**Rigby**
Rigden, E.Josephine A	**Rigden**
Rigg, E. (fl. 1983) M	**E.Rigg**
Rigg, George Burton (1872–1961) S	**Rigg**
Righter, Francis Irving (1897–) S	**Righter**
Rigo, Gregorio (1841–1922) S	**Rigo**
Rigo, P. S	**P.Rigo**
Rigoni, Victor A. (fl. 1958) S	**Rigoni**
Rigual Magallón, Abelardo (1918–) S	**Rigual**
Ríha, Jan (1947–) S	**Ríha**
(Rijn, Anne Renée Ariette Görts van	
see Görts–van Rijn, Anne Renée Ariette	**Görts**
Riker, Albert Joyce (1894–) M	**Riker**
Rikhy, Madhu (fl. 1974) M	**Rikhy**
Rikli, Martin Albert (1868–1951) S	**Rikli**
Riley, E.A. (fl. 1952) M	**E.A.Riley**
Riley, Herbert Parkes (1904–1988) S	**H.P.Riley**
Riley, J.P. A	**J.P.Riley**
Riley, John (c.1796–1846) P	**Riley**
Riley, Lawrence Athelstan Molesworth (1889–1928) S	**L.Riley**
Riley, Leslie A. A	**L.A.Riley**
Rilstone, Francis (1881–1953) BM	**Rilstone**
Rimpau, R.H. (fl. 1962) M	**Rimpau**
Rinaldi, Michael G. (fl. 1987) M	**Rinaldi**

Rines, Jan E.B.   A   **Rines**
Ringier, Victor Abraham (fl. 1823) S   **Ringier**
Ringius, Gordon Stacey (1949–   ) S   **G.S.Ringius**
Ringius, Hans Henric (Henrik) (1808–1874) S   **Ringius**
Ringo, Sandra L. (fl. 1969) M   **Ringo**
Ringueberg, Eugene Nicholas Sylvester (1859–   ) A   **Ringueb.**
Rink, Henrik Johannes (1819–1893) S   **Rink**
Rintz, Richard E. (fl. 1960) S   **Rintz**
Rinz (fl. 1935) M   **Rinz**
Riocreux, A. (1820–1912) S   **Riocreux**
Riofrio, B.F. (fl. 1923) M   **Riofrio**
Rion, Alphonse (Chanoine) (1809–1856) S   **Rion**
(Rios, Miguel de los)
    see de los Rios, Miguel   **de los Rios**
Riousset, Gisèle (fl. 1976) M   **G.Riousset**
Riousset, L. (fl. 1963) M   **Riousset**
Rioux, Jean-Antoine (1925–   ) M   **Rioux**
Ripart, Jean Baptiste Marie Joseph Solange Eugène (1814–1878) AMS   **Ripart**
Ripley, Harry Dwight Dillon (1908–1973) S   **Ripley**
Rippa, Giovanni (fl. 1932) S   **Rippa**
Rippka, R.   A   **Rippka**
Risatti, J.Bruno   A   **Risatti**
Rischin, M. (fl. 1921) M   **Rischin**
Risler, Johan   S   **Risler**
Risse, Horst (fl. 1985) S   **Risse**
Risso, Joseph Antoine (1777–1845) MS   **Risso**
Rist, E. (fl. 1902) M   **Rist**
Ristanovič, B. (fl. 1976) M   **Ristan.**
Rita, Juan (fl. 1989) S   **Rita**
Ritchie, James Cunningham (1929–   ) S   **Ritchie**
Ritgen, Ferdinand August Maria Franz von (1787–1867) MP   **Ritgen**
Ritschl, Georg (Adolf) (1816–1866) MS   **Ritschl**
Rittener, T. (fl. 1887) S   **Rittener**
Ritter, Carl (1779–1859) S   **C.Ritter**
Ritter, Christian Wilhelm Jonathan (1765–1821) S   **C.W.J.Ritter**
Ritter, Friedrich (1898–1989) S   **F.Ritter**
Ritter, Johann Jacob (1714–1784) S   **J.Ritter**
Ritter-Studnička, Hilda (1911–1976) PS   **Ritter-Studn.**
Rittershausen, Paul (fl. 1892) S   **Rittersh.**
Ritzberger, Englebert (1868–1923) S   **Ritzb.**
Ritzema Bos, Jan (1850–1928) M   **Ritz.Bos**
Riva, Alfredo (fl. 1987) M   **A.Riva**
Riva, Domenico (c.1856–1895) S   **Riva**
Rivalier, E. (fl. 1954) M   **Rivalier**
Rivas Goday, Salvador (1905–1981) S   **Rivas Goday**
Rivas Martínez, Salvador (1935–   ) MPS   **Rivas Mart.**
Rivas Mateos, Marcelo (1875–1931) S   **Rivas Mateos**
Rivas Ponce, Maria Antonia (1941–   ) S   **Rivas Ponce**

(Rivas–Martínez, Salvador)
  see Rivas Martínez, Salvador   **Rivas Mart.**
(Rivera Campanile, Giulia)
  see Campanile, Giulia Rivera   **Campan.**
Rivera, J. (fl. 1980) S   **J.Rivera**
Rivera Núñez, Diego (1958–) S   **D.Rivera**
Rivera, Patricio (1942–) A   **P.Rivera**
(Rivera Ramírez, Patricio)
  see Rivera, Patricio   **P.Rivera**
Rivera, Vincenzo (1890–1967) M   **Rivera**
Rivers, Thomas (1798–1877) S   **Rivers**
Rivet, Charles Gabriel (–1884) A   **Rivet**
Rivière, Charles Marie (1845–) S   **C.Rivière**
Rivière, Marie Auguste (1821–1877) S   **Rivière**
Rivinus, Augustus Quirinus (Bachmann) (1652–1723) L   **Riv.**
Rivoire, Antoine (fl. 1921) S   **Rivoire**
Rivola, Milan (1933–) B   **Rivola**
Rivolta, Sebastiano (1832–1893) M   **Rivolta**
Rix, Edward Martin (1943–) S   **Rix**
Rixford, Emmet (fl. 1896) M   **Rixford**
Riza, Ali (fl. 1920) M   **Riza**
Riznyk, Raymond Z. (1942–) A   **Riznyk**
Rizvi, S.R.N. (fl. 1960) M   **Rizvi**
Rizwi, M.A. (fl. 1978) M   **Rizwi**
Rizzini, Carlos Toledo (1921–) MS   **Rizzini**
Rizzotto, Milena (1936–) S   **Rizzotto**
Rjabkova, L.S. (1926–) S   **Rjabkova**
Rjabov, J.N. (1897–) S   **Rjabov**
Rjachovsky, N. (fl. 1931) M   **Rjach.**
Roach, Archibald Wilson Kilbourne (1920–) S   **A.Roach**
Roach, Blanche Muriel Bristol (1888–) A   **Roach**
Roane, Martha K. (fl. 1988) MS   **Roane**
Roark, Eugene Washburn (1894–1918) M   **Roark**
Robak, Håkon (fl. 1932) M   **Robak**
Robatsch, K. (fl. 1988) S   **Robatsch**
Robbertse, Petrus Johannes (1932–) S   **Robbertse**
Robbins, Charles Albert Sumner (1874–1930) M   **Robbins**
Robbins, Guy Thomas (1916–1960) S   **G.T.Robbins**
Robbins, James Watson (1801–1879) S   **J.W.Robbins**
Robbins, William Jacob (1890–1978) S   **W.J.Robbins**
Robbrecht, Elmar (1946–) MS   **Robbr.**
(Robecchi Bricchetti, Luigi)
  see Robecchi–Bricchetti, Luigi   **Rob.–Bricch.**
Robecchi–Bricchetti, Luigi (1855–1926) S   **Rob.–Bricch.**
Roberg, Lars (Laurentius) (1664–1742) L   **Roberg**
Roberg, Max (fl. 1930) M   **M.Roberg**
Roberge, Michel (Michael) Robert (–1864) M   **Roberge**
Róbert, Andrei A   **A.Róbert**
Robert, Gaspard Nicolas (1776–1857) S   **Robert**
Robert, H. (fl. 1985) M   **H.Robert**

Robert, Nicolas (1610–1684) L	N.Robert
Robert–Passini, M.–F. (fl. 1981) S	Rob.-Pass.
Roberts, Daniel Altman (1922–) M	D.A.Roberts
Roberts, Donald W. (1933–) M	D.W.Roberts
Roberts, Evan Paul (Michigan) (1914–) S	E.P.Roberts
Roberts, Florence M. (fl. 1952) M	F.M.Roberts
Roberts, J. (1912–1960) S	J.Roberts
Roberts, John M. (fl. 1952) M	J.M.Roberts
Roberts, John William (1882–1957) M	Roberts
Roberts, Joyce E. (née Zavortink, J.E.) (fl. 1971) S	J.E.Roberts
Roberts, Margaret A	M.Roberts
Roberts, Peter (fl. 1992) M	P.Roberts
Roberts, R.G. (fl. 1984) M	R.G.Roberts
Roberts, Richard Henry (1910–) S	R.H.Roberts
Robertson, Alexander (fl. 1980) S	A.Robertson
Robertson, Colin Charles (–1946) S	C.Robertson
Robertson, David (1806–1896) A	Robertson
Robertson, Eddie B. (fl. 1976) B	E.B.Robertson
Robertson, Enid Lucy (1925–) AS	E.L.Robertson
Robertson, G.I. (fl. 1979) M	G.I.Robertson
Robertson, J.S. (fl. 1961) M	J.S.Robertson
Robertson, Jack Alex (1943–) M	J.A.Robertson
Robertson, Kenneth R. (1941–) S	K.R.Robertson
Robertson, Marian Esther Ropes (1934–1975) B	M.E.R.Robertson
Robertson, Muriel (1883–1973) A	M.Robertson
Robertson, Noel Farnie (1923–) M	N.F.Robertson
Roberty, Guy Edouard (1907–1971) S	Roberty
Robich, G. (fl. 1988) M	Robich
Robillard d'Argentelle, Louis Marc Antoine (1777–1828) S	Robill.
Robin, Charles Philippe (1821–1885) AMS	C.P.Robin
Robin, Claude Cesar (1750–) S	C.C.Robin
Robin, Jean (1550–1629) L	J.Robin
Robins, Peter A. A	Robins
Robinson, A. (fl. 1814) S	A.Rob.
Robinson, Benjamin Lincoln (1864–1935) MPS	B.L.Rob.
Robinson, C.J. (fl. 1985) S	C.J.Rob.
Robinson, Charles Budd (1871–1913) AS	C.B.Rob.
Robinson, Edward Armitage (1921–) S	E.A.Rob.
Robinson, Harold Ernest (1932–) ABS	H.Rob.
Robinson, James Fraser (1857–1927) S	J.F.Rob.
Robinson, John (1846–1925) AP	J.Rob.
Robinson, William (1838–1935) S	Rob.
Robinson, Winifred Josephine (1867–) PS	W.J.Rob.
Robinson–Jeffrey, Robena C. (fl. 1964) M	Rob.-Jeffr.
Robison, Barbara M. (fl. 1970) M	B.M.Robison
Robison, Coleman R. (1940–) B	Robison
Robley, Augusta J. (fl. 1840) S	Robley
Robnett, Christie J. (fl. 1986) M	Robnett
Robolsky, H. (1796–1849) S	Robolsky
Robson, Edward (1763–1813) S	E.Robson

Robson, Norman Keith Bonner (1928–) S   **N.Robson**
Robson, Stephen (1741–1779) S   **Robson**
Robyns, André Georges Marie Walter Albert (1935–) S   **A.Robyns**
Robyns, Frans Hubert Edouard Arthur Walter (1901–1986) BS   **Robyns**
Rocha Afonso, Maria da Luz de Oliveira Tavares Monteiro (1925–) S   **Rocha Afonso**
(Rocha da Torre, Antonio)
    see Torre, Antonio Rocha da   **Torre**
(Roche, Daniel de la)
    see Delaroche, Daniel   **D.Delaroche**
(Roche, François de la)
    see Delaroche, François   **F.Delaroche**
Rochebrune, Alphonse Trémeau de (1834–1912) S   **Rochebr.**
Rochel, Anton (1770–1847) S   **Rochel**
Rochet d'Héricourt, C.L.X. (fl. 1846) S   **Rochet**
Rock, Howard Francis Leonard (1925–1964) S   **H.Rock**
Rock, Joseph Francis Charles (Joseph Franz Karl) (1884–1962) AS   **Rock**
Rockhausen, M.   S   **Rockh.**
Rockley, Alicia Margaret Amherst Cecil (1865–1941) S   **Rockley**
Rodati, Aloysius (Luigi) (1762–1832) S   **Rodati**
Rode, H.R. (fl. 1936) M   **H.R.Rode**
Rode, K.P. (fl. 1933–1936) F   **K.P.Rode**
Rodegher, Emilio (1856–1922) S   **Rodegher**
Rodenburg, W.F.   S   **Rodenb.**
Rodenwaldt, E.   A   **Rodenw.**
Röder, K. (fl. 1937) M   **Röder**
Rodet, Henri Jean Antoine (1810–1875) S   **Rodet**
Rodgers, Andrew Denny (1900–1981) S   **A.D.Rodgers**
Rodgers, C.Leland (fl. 1956) S   **Rodgers**
Rodhain, F. (fl. 1985) M   **Rodhain**
Rodhain, J.   A   **J.Rodhain**
Rodig, Friedrich Weinhold (1770–1844) M   **Rodig**
Rodigas, Émile (1831–1902) S   **Rodigas**
Rodigin, Michail Nikolaevich (fl. 1928) M   **Rodigin**
Rodin, Hippolyte (1829–1886) S   **Rodin**
Rodin, Leonid Efimovic (1907–1966?) S   **L.E.Rodin**
Rodina, L.G. (1894–) S   **Rodina**
Rodio, Gaetano (1886–) MS   **Rodio**
Rodionenko, Georgi Ivanovich (1913–) S   **Rodion.**
Rödl-Linder, Gisela (1954–) P   **Rödl-Linder**
Rodman, James Eric (1945–) S   **Rodman**
Rodrigo Trigo, América del Pilar (fl. 1938) S   **Rodrigo**
Rodrigues, C.M.A. (fl. 1987) S   **C.M.A.Rodrigues**
Rodrigues de Miranda, Lennart (1925–) M   **Rodr.Mir.**
(Rodrigues, João Barbosa)
    see Barbosa Rodrigues, João   **Barb.Rodr.**
Rodrigues, José E.de Mesquita (1916–) A   **Rodrigues**
Rodrigues, Katia Ferreira (1957–) M   **K.F.Rodrigues**
Rodrigues, L.   S   **L.Rodrigues**
Rodrigues, William Antônio (1928–) S   **W.A.Rodrigues**

Rodríguez, A.E. (fl. 1964) M	A.E.Rodr.
Rodríguez Acosta, Maricela (fl. 1986) S	Rodr.Acosta
Rodríguez de la Rosa, Nerci (fl. 1989) M	N.Rodr.
Rodríguez de Rios, Nora S	N.Rodr.Rios
Rodríguez Fuentes, Alicia (fl. 1983) S	A.Rodr.
Rodríguez Hernández, Mateo (1889–1949) M	M.Rodr.
Rodríguez Hernández, Miguel (fl. 1980) M	Mig.Rodr.
Rodríguez Jiménez, Concepción (fl. 1984) S	C.Rodr.Jim.
Rodríguez, José Demetrio (1780–1846) S	Rodr.
Rodríguez, L.M. (fl. 1976) M	L.M.Rodr.
Rodríguez López–Neyra, C. A	C.Rodr.
(Rodríguez, Miguel)	
see Rodríguez Hernández, Miguel	Mig.Rodr.
Rodríguez, N.B. (fl. 1984) M	N.B.Rodr.
Rodríguez Oubiña, Juan (fl. 1990) S	Rodr.Oubiña
(Rodríguez, Rafael Lucas)	
see Lucas Rodríguez, Rafael	Lucas Rodr.
Rodríguez Rios, Roberto (1944–) P	R.Rodr.
Rodríguez y Femenías, Juan Joaquín (1839–1905) AS	J.J.Rodr.
Rodríguez–Carrasquero, Henry (fl. 1983) S	H.Rodr.
Rodríguez–Rodríguez, Héctor (1948–) S	Rodr.-Rodr.
Rodschied, Ernst Carl (Karl) (–1796) S	Rodschied
Rodway, Leonard (1853–1936) BMPS	Rodway
Roe, John Septimus (1797–1878) S	Roe
Roe, Keith Edward (1937–) S	K.E.Roe
Roe, Margaret James (fl. 1961) S	M.J.Roe
(Roechoudt, Van)	
see Van Roechoudt	Van Roech.
Røed, H. (fl. 1949) M	Røed
Roehl, Ernst (Karl Gustav Wilhelm) von (1825–1881) S	Roehl
(Roehling, Johann Christoph)	
see Röhling, Johann Christoph	Röhl.
Roehrich, Olivier S	Roehr.
Roeijmans, H.J. (fl. 1989) M	Roeijmans
Roekmowati–Hartono (fl. 1965) S	Roekm.
Roell, Julius (1846–1928) B	Roell
Roemer, (Friedrich) Adolph (1809–1869) AMPS	A.Roem.
Roemer, Carl (Karl) Ferdinand von (1818–1891) AP	F.Roem.
Roemer, Johann Jakob (1763–1819) PS	Roem.
Roemer, Max Joseph (1791–1849) S	M.Roem.
Roemer, R.de (fl. 1852) S	R.Roem.
Roemer, Stephen C. A	S.C.Roem.
(Roenning, Olaf Inge)	
see Rønning, Olaf Inge	Rønning
Roeper, Johannes August Christian (1801–1885) PS	Roep.
Roesch, Charles (fl. 1894) A	Roesch
Roeser, Paul A	Roeser
(Roesler, R.)	
see Rösler, R.	R.Rösler

(Roessig, Carl Gottlob)	
see Rössig, Carl Gottlob	**Rössig**
Roessler, Helmut (1926–) S	**Roessler**
Roeth, J. (fl. 1978–) S	**Roeth**
Roever, William Edward (1907–) S	**Roever**
Roezl, Benedikt (Benito) (1824–1885) S	**Roezl**
Roffavier, Georges (1775–1866) S	**Roffavier**
Roffey, John (1860–1927) S	**Roffey**
Roger, L. (fl. 1934) M	**Roger**
Rogers, A.L. (fl. 1962) M	**A.L.Rogers**
Rogers, A.M. (fl. 1928) M	**A.M.Rogers**
(Rogers, Charles Coltman)	
see Coltman–Rogers, Charles	**Coltm.-Rog.**
Rogers, Charles Gilbert (1864–1937) S	**C.G.Rogers**
Rogers, Claude Marvin (1919–) S	**C.M.Rogers**
Rogers, David James (1918–) S	**D.J.Rogers**
Rogers, Donald Philip (1908–) M	**D.P.Rogers**
Rogers, Frederick Arundel (1876–1944) S	**F.A.Rogers**
Rogers, George King (1952–) S	**G.K.Rogers**
Rogers, Henry Darwin (1809–1886) S	**H.D.Rogers**
Rogers, Jack David (1937–) M	**J.D.Rogers**
Rogers, Joanne K. (fl. 1971) M	**J.K.Rogers**
Rogers, Julia Ellen (1866–) S	**J.E.Rogers**
Rogers, Ken E. (fl. 1968) S	**K.E.Rogers**
Rogers, Richard Sanders (1862–1942) S	**R.S.Rogers**
Rogers, Roderick Westgarth (1944–) M	**R.W.Rogers**
Rogers, Walter E. (1890–1951) S	**W.E.Rogers**
Rogers, William Edwin (1936–) A	**Wm.E.Rogers**
Rogers, William Moyle (1835–1920) S	**W.M.Rogers**
Rogerson, Clark Thomas (1918–) M	**Rogerson**
Rogler, George Albert (1913–) S	**Rogler**
(Rogovitch, Athanasi Semenovich)	
see Rogowicz, Athanasi Semenovich	**Rogow.**
(Rogowič, Athanasi Semenovich)	
see Rogowicz, Athanasi Semenovich	**Rogow.**
Rogowicz, Athanasi Semenovich (1812–1878) S	**Rogow.**
Rogozinski, H. (fl. 1985) S	**Rogoz.**
Rohde, Michael (1782–1812) S	**Rohde**
Rohde, Theo (fl. 1936) M	**T.Rohde**
Rohlena, Josef (Joseph) (1874–1944) PS	**Rohlena**
Rohlfs, Gerhard (1831–1896) S	**Rohlfs**
Röhling, Johann Christoph (1757–1813) BMPS	**Röhl.**
Rohmeder, Ernst (1903–1972) S	**Rohmeder**
Rohr, Julius Bernard von (1686–1742) LS	**J.B.Rohr**
Rohr, Julius Philip Benjamin von (1737–1793) S	**Rohr**
Rohrbach, Carl S	**C.Rohrb.**
Rohrbach, Paul (1846–1871) S	**Rohrb.**
Rohrer, Joseph Raphael (1954–) B	**J.R.Rohrer**
Rohrer, Rudolph (1805–1839) S	**Rohrer**

Rohweder, Otto (1919–) S	**Rohweder**
Rohwer, J.G. (fl. 1985) S	**Rohwer**
Roig, Fidel Antonio (1922–) S	**F.A.Roig**
(Roíg, Juan Tomás)	
see Roíg y Mesa, Juan Tomás	**Roíg**
Roíg y Mesa, Juan Tomás (1878–) S	**Roíg**
Roivainen, Heikki (1900–1983) BMPS	**Roiv.**
Rojas Acosta, Nicolás (1873–1947) PS	**Rojas Acosta**
(Rojas Clemente, Simón de)	
see Clemente y Rubio, Simón de Roxas (Rojas)	**Clemente**
(Rojas E. de Mendiola, Blanca)	
see Mendiola, Blanca Rojas E. de	**B.R. Mendiola**
Rojas, Isabel (fl. 1988) M	**I.Rojas**
Rojas, Teodoro (1877–1954) S	**Rojas**
Rojo, Justo P. (1935–) S	**Rojo**
Rokde, B.G. (fl. 1964) M	**Rokde**
(Roland, Louis)	
see Rolland-Germain	**Roll.-Germ.**
Roland–Gosselin, Robert (1854–1925) S	**Rol.-Goss.**
Rolander, Daniel (1725–1793) S	**Rol.**
Roldán, A. (fl. 1987) M	**A.Roldán**
Roldan, Emiliano F. (fl. 1936) M	**Roldan**
Roldugin, I.I. (1926–) S	**Roldugin**
Rołed, H. (fl. 1949) M	**Rołed**
Rolfe, Robert Allen (1855–1921) S	**Rolfe**
Rolfs, Peter Henry (1865–1944) S	**Rolfs**
Röll, Julius (1846–1928) S	**Röll**
Roll, Yakiv V. (Ja.W.) (1887–1961) A	**Y.V.Roll**
Roll–Hansen, Finn (fl. 1948) M	**Roll–Hansen**
Roll–Hansen, H. (fl. 1979) M	**H.Roll–Hansen**
Rolland, Eugène (1846–1909) S	**E.Rolland**
Rolland, Léon Louis (1841–1912) M	**Rolland**
Rolland–Germain (1881–1972) S	**Roll.-Germ.**
Rolle, Friedrich (1827–1887) S	**Rolle**
Rolleri, Cristina H. (fl. 1970–1988) P	**Rolleri**
(Rolleri de Daugherti, Cristina H.)	
see Rolleri, Cristina H.	**Rolleri**
Rolli, Ettore (1818–1876) S	**Rolli**
Röllin, O. (fl. 1973) M	**Röllin**
Rollins, Reed Clark (1911–) S	**Rollins**
Roloff, Christian Ludwig (1726–1800) S	**Roloff**
Romagnesi, Henri Charles Louis (Lewis) (1912–) M	**Romagn.**
Romagnoli, Massimiliano (–c.1870) S	**Romagnoli**
Romain, Charles (fl. 1848–1868) S	**Romain**
Roman, Nicolae (1927–) S	**Roman**
Romanczuk, M.C. (fl. 1976–) S	**Romanczuk**
Romanes, M.F. A	**Romanes**
Romanet du Caillaud, Frédéric (fl. 1881–1888) S	**Rom.Caill.**
Romankova, Anna Grigorievna (1909–) M	**Romankova**
Romano, Giovanni Battista (1810–1877) S	**G.B.Romano**

Romano, Girolamo (Gerolamo) (1765–1841) S	**Romano**
Romanov, Michael Alexander A	**Romanov**
Romanova, Evfrosima Andreevna (1910–) F	**Romanova**
Romans, Bernard (c.1720–1784) S	**Romans**
Romariz, Carlos Mateus (1920–) PS	**Romariz**
Rombach, Michiel C. (1955–) M	**Rombach**
Rombouts, Johannes Gottfried Hendrik (–1889) S	**Rombouts**
Romein, A.J.T. A	**Romein**
Romell, Lars Gunnar Torgny (1891–) M	**Romell**
Romeo, Antonino (1899–1941) S	**Romeo**
Römer, Carl (1815–1881) S	**C.Römer**
Römer, Heinrich (fl. 1843) S	**H.Römer**
Römer, Julius (1848–1926) S	**J.Römer**
Romero, Andrea I. (1955–) M	**A.I.Romero**
Romero, Edgardo Juan (1936–) F	**E.J.Romero**
Romero García, Ana Teresa (1957–) S	**Romero García**
Romero, Gustavo A. (1955–) S	**G.A.Romero**
Romero Manrique, P. (fl. 1990) S	**P.Romero**
(Romero, Maria E.Múlgura de)	
see Múlgura de Romero, Maria E.	**Múlg.**
Romero Martinengo, Javier A	**J.Romero**
Romero, Rafael (1910–1973) S	**Romero**
Romero, Tomás (fl. 1987) S	**T.Romero**
Romero Zarco, Carlos (1954–) S	**Romero Zarco**
(Romero–Castañeda, Rafael)	
see Romero, Rafael	**Romero**
(Romero–Zarco, Carlos)	
see Romero Zarco, Carlos	**Romero Zarco**
Romieux, Henri Auguste (1857–1937) S	**Romieux**
Rominger, James McDonald (1928–) S	**Rominger**
Römmp, Herman (fl. 1928) S	**Römmp**
Romney, S.V. (fl. 1974) M	**Romney**
Romo, Angel María (1955–) S	**Romo**
(Romo Díez, Angel María)	
see Romo, Angel María	**Romo**
(Rompaey, Emiel Van)	
see Van Rompaey, Emiel	**Van Romp.**
Roms, O.G. (fl. 1967) BM	**Roms**
Roncali, D.B. (fl. 1896) M	**Roncali**
Ronceray, Paul–Louis (1875–1953?) S	**Ronceray**
Rondelet, Guillaume (1507–1566) L	**Rondelet**
Rondon, Yves (1914–) M	**Rondon**
Rong, F. (fl. 1985) M	**Rong**
Rönn, Hans Ludwig Karl (1886–) S	**Rönn**
Ronniger, Karl (Carl) (1871–1954) PS	**Ronniger**
(Rönning, Olaf Inge)	
see Rønning, Olaf Inge	**Rønning**
Rønning, Olaf Inge (1925–) BS	**Rønning**
Ronse Decraene, Louis–Philippe S	**Ronse Decr.**
Rood, Anthony P. A	**Rood**

Roof, James B. (1910–1983) S   **Roof**
Rooke, Hayman (c.1722–1806) S   **Rooke**
(Roon, Adrianus Cornelis de)
  see de Roon, Adrianus Cornelis   **de Roon**
Roos, Johan Carl (1745–1828) S   **Roos**
Roos, Marco C. (1955–) PS   **M.C.Roos**
(Rooy, Jacques van)
  see van Rooy, Jacques   **van Rooy**
Roper, Freeman Clarke Samuel (1819–1896) AS   **Roper**
Roper, Ida Mary (1865–1935) S   **I.Roper**
(Röper, Johannes August Christian)
  see Roeper, Johannes August Christian   **Roep.**
Roquebert, Marie-France (1940–) M   **Roquebert**
Roques, Joseph (1772–1850) MS   **Roques**
Rorer (fl. 1922) M   **Rorer**
Ros, R.M.   B   **Ros**
Rosa, Karel (1901–1976) A   **Rosa**
Rosander, (Karl) Hendrik Andreas (1873–1950) S   **Rosander**
(Rosanoff, Sergei Matveevič)
  see Rozanov, Sergei Matveevich   **Rozanov**
Rosas, Marcelo R. (fl. 1989) S   **Rosas**
Rosatti, Thomas James (1951–) S   **Rosatti**
Rosbach, Heinrich (fl. 1880) S   **Rosbach**
(Roschevictz, Roman Julievich)
  see Roshevitz, Roman Julievich   **Roshevitz**
Roscoe, Margaret (fl. 1830) S   **M.Roscoe**
Roscoe, William (1753–1831) S   **Roscoe**
Rose, August David Friedrich Carl (1821–1873) B   **A.Rose**
Rose, D. (fl. 1966) M   **D.Rose**
Rose, Dean Humboldt (1878–1963) M   **D.H.Rose**
Rose, Francis (1921–) BMS   **F.Rose**
Rose, Hugh (c.1717–1792) S   **H.Rose**
Rose, Joseph Nelson (1862–1928) S   **Rose**
Rose, Lewis Samuel (1893–1973) S   **L.Rose**
Rose, Maurice   A   **M.Rose**
Rose, Sharon L. (fl. 1979) M   **S.L.Rose**
Rosella, Etienne (fl. 1937) M   **Rosella**
Rosellini, Ferdinando (1817–1873) S   **Rosell.**
Roselt, G.   S   **Roselt**
Rosemberg, J.A. (fl. 1959) M   **Rosemberg**
Rosén, Eberhard (1714–1796) S   **E.Rosén**
Rosen, Felix (1863–1925) S   **F.Rosen**
Rosen, Harry Robert (1889–1962) M   **H.R.Rosen**
Rosén, Johan Peter (1788–1825 or later) S   **Rosén**
Rosén, Nils (later Rosenstein, N.) (1706–1773) S   **N.Rosén**
Rosenbaum, Joseph (1887–1925) M   **Rosenbaum**
Rosenberg, Caroline Friderike (1810–1902) AB   **C.Rosenb.**
Rosenberg, Gustaf Otto (1872–1948) S   **O.Rosenb.**
Rosenberg, Marie (1907–) A   **M.Rosenb.**
Rosenberg, Mary Elizabeth (1820–1914) S   **M.E.Rosenb.**

(Rosenblad, Eberhard)
  see Rosén, Eberhard   **E.Rosén**
(Rosenburgh, Cornelis Rugier Willem Karel Alderwerelt van)
  see Alderwerelt van Rosenburgh, Cornelis Rugier Willem Karel van   **Alderw.**
Rosencranz, Ronald J. S   **Rosencr.**
Rosendahl, Carl Otto (1875–1956) MS   **Rosend.**
Rosendahl, Friedrich (1881–1942) S   **F.Rosend.**
Rosendahl, Henrik Viktor (1855–1918) PS   **H.Rosend.**
Rosengurtt, Bernardo (1916–1985) S   **Roseng.**
(Rosengurtt–Gurvich, Bernardo)
  see Rosengurtt, Bernardo   **Roseng.**
Rosenstein, Nils (né Rosén, N.) (1706–1773) S   **Rosenstein**
Rosenstingl, Walter S   **Rosenstingl**
Rosenstock, Eduard (1856–1938) PS   **Rosenst.**
Rosenthal, David August (1821–1875) S   **Rosenthal**
Rosenthal, Käthe (Kaethe) Hoffmann (1883–) S   **K.Rosenthal**
Rosenthal, R.C. (fl. 1882–1888) S   **R.C.Rosenthal**
Rosenvinge, Janus Lauritz Andreas Kolderup (1858–1939) AS   **Rosenv.**
Roshevitz, Roman Julievich (1882–1949) S   **Roshev.**
Roshkova, Olga Ivanovna (1909–1989) S   **Roshkova**
Rosier, François S   **Rosier**
Rosillo de Velasco, Salvador (1905–1987) S   **S.Rosillo**
Rosillo, Rodriguez A   **R.Rosillo**
Roskin, Gr. A   **Roskin**
Roskov, Ju. (Yu.) R. (1964–) S   **Roskov**
Rösler, Carl August (1770–1858) S   **Rösler**
Rösler, R. S   **R.Rösler**
Rosny, Louis Léon Lucien Prunol de (1837–1916) S   **Rosny**
Rosowski, James R. A   **Rosowski**
Ross, Alexander Milton (1832–1897) S   **A.M.Ross**
Ross, David (c.1810–1881) S   **D.Ross**
Ross, Estelle M. (1952–) S   **E.M.Ross**
Ross, Herman (1863–1942) PS   **H.Ross**
Ross, James Clarke (1800–1862) S   **J.C.Ross**
Ross, James Henderson (1941–) S   **J.H.Ross**
Ross, John (1777–1856) S   **Ross**
Ross, Nils–Erik (fl. 1949) B   **N.-E.Ross**
Ross, Pontus Henry S   **P.H.Ross**
Ross, Robert (1912–) AS   **R.Ross**
Ross, Ronald A   **Ron.Ross**
Ross–Craig, Stella (1906–) S   **Ross-Craig**
Rossbach, George Bowyer (1910–) S   **Rossbach**
Rossbach, Ruth Peabody (1912–) S   **R.Rossbach**
Rossberg, G. S   **Rossberg**
Rosselló, Josep Antoni (fl. 1988) PS   **Rosselló**
Rosser, Effie Moira (1923–1987) S   **Rosser**
Rosset, C. A   **Rosset**
Rossetti, Corrado (1866–) S   **Rossetti**
Rosshirt, Karl (fl. 1888) S   **Rosshirt**

Rossi, Albert E. (fl. 1984) M	A.E.Rossi
Rossi, Giovanni Batista (fl. 1825) S	G.Rossi
(Rossi, Giuseppe de)	
see de Rossi, Giuseppe	de Rossi
Rossi, Ludwig (Ljudevit) (1850–1930) P	L.Rossi
Rossi, M.L. (1850–1932) S	M.L.Rossi
Rossi, Pietro (1738–1804) S	Rossi
Rossi, Pietro (1871–1950) S	P.Rossi
Rossi, Stefano (1851–1898) S	S.Rossi
Rossi, Walter (1946–) MS	W.Rossi
Rossi–Doria, Tullio (fl. 1891) M	Rossi–Dorio
Rössig, Carl Gottlob (1752–1806) S	Rössig
Rossignol, Louis S	Rossignol
Rossignol, Martine A	M.Rossignol
Rossinsky, V. S	Rossinsky
Rössler, Wilhelm (1909–) S	Rössler
Rossman, Amy Y. (1946–) M	Rossman
Rossmann, (Georg Wilhelm) Julius (1831–1866) MS	Rossmann
Rossmässler, Emil Adolph (1806–1867) S	Rossm.
Rossow, Ricardo A. (1956–) S	Rossow
Rossy–Valderrama, Carmen (fl. 1958) M	Rossy–Vald.
Rost, Ernest C. (fl. 1932) S	E.C.Rost
Rost, Georg (fl. 1902–1909) S	Rost
(Rostafinski, Jozef Thomas von)	
see Rostafińsky, Józef Thomasz	Rostaf.
Rostafińsky, Józef Thomasz (1850–1928) AM	Rostaf.
Rostam, S. (fl. 1984) M	Rostam
Rostan, Edouard (1826–1895) S	Rostan
Rostański, Krzysztof (1930–) S	Rostański
Roster, Giorgio (–1968) S	Roster
Rostius, Christopher (1620–1687) L	Rostius
(Rostkov, Friedrich Wilhelm Gottlieb)	
see Rostkovius, Friedrich Wilhelm Gottlieb Theophil	Rostk.
Rostkovius, Friedrich Wilhelm Gottlieb Theophil (1770–1848) MS	Rostk.
Rostock, Michael (1821–1893) S	Rostock
Rostovzev, Semen Ivanovich (c.1862–1916) M	Rostovzev
(Rostowzew, Semen Ivanovich)	
see Rostovzev, Semen Ivanovich	Rostovzev
Rostram, S. (fl. 1987) M	Rostram
Rostrup, (Frederik Georg) Emil (1831–1907) MS	Rostr.
Rostrup, Ove Georg Frederik (1864–1933) M	O.Rostr.
(Rosua Campos, José Luis)	
see Rosua, José Luis	Rosua
Rosua, José Luis (1954–) S	Rosua
Rot von Schreckenstein, Friedrich (1753–1808) S	Rot Schreck.
Rota, Lorenzo (1819–1855) B	Rota
Rota–Rossi, Guido (fl. 1907) M	Rota–Rossi
Rotem, J. (fl. 1966) M	Rotem

(Rotert, Vladislav Adolfovich)	
see Rothert, (Karol) Władisław Adolfovich	**Rothert**
Roth, Albrecht Wilhelm (1757–1834) ABMPS	**Roth**
Roth, Ernst (Carl Ferdinand) (1857–1918) S	**E.Roth**
Roth, Georg (1842–1915) B	**G.Roth**
Roth, I.L. (fl. 1981) M	**I.L.Roth**
Roth, Johannes Rudolph (1814–1858) S	**J.Roth**
Roth, Peter Hans A	**P.H.Roth**
Roth, V.D. (fl. 1981) S	**V.D.Roth**
Roth, Wilhelm (1819–1875) S	**W.Roth**
Rothe, Heinrich August (fl. 1890) AS	**Rothe**
Rother, Wilhelm Otto (1853–) S	**Rother**
Rotheray, Lister (fl. 1889–1932) S	**Rotheray**
Rothers, Boris (1890–) M	**Rothers**
Rothert, (Karol) Władisław Adolfovich (1863–1916) AMS	**Rothert**
Rothmaler, Werner Hugo Paul (1908–1962) APS	**Rothm.**
Rothman, Georg (Göran) (1739–1778) S	**Rothman**
Rothman, Johan Stensson (1684–1763) S	**J.Rothman**
Rothmayr, Julius (fl. 1910) M	**Rothmayr**
Rothpletz, (Friedrich) August (1853–1918) AS	**Rothpletz**
Rothrock, Joseph Trimble (1839–1922) MS	**Rothr.**
Rothschild, Jules (1838–) S	**Rothsch.**
Rothwell, Frederick M. (fl. 1979) M	**F.M.Rothwell**
Rothwell, Gar W. (1944–) M	**G.W.Rothwell**
Roti Michelozzi, Guidetta (1928–) S	**Roti Mich.**
Rotman, A.D. (fl. 1976) S	**Rotman**
Rotov, R.A. (1930–) S	**Rotov**
Rotschild, E. (fl. 1961) S	**Rotschild**
(Rottboell, Christen Friis)	
see Rottbøll, Christen Friis	**Rottb.**
Rottbøll, Christen Friis (1727–1797) S	**Rottb.**
Rottenbach, Heinrich (1835–1917) S	**Rottenb.**
(Rottenburg, Ludwig von)	
see Sarnthein, Ludwig von	**Sarnth.**
Rottler, Johan Peter (1749–1836) S	**Rottler**
Roubal, Jindřich A	**Roubal**
Roubaud, Émile (1882–1962) A	**Roubaud**
Roubos, J.C. (fl. 1974) M	**Roubos**
Roucel, François Antoine (1735–1831) B	**Roucel**
Rouch, Janine (fl. 1966) M	**Rouch**
(Rouchijajnen, M.I.)	
see Roukhiyajnen, M.I.	**Roukh.**
(Rouchiyainen, M.I.)	
see Roukhiyajnen, M.I.	**Roukh.**
Rouhier, Alexandre (fl. 1926) S	**Rouhier**
Rouillier, Charles (1814–1858) F	**Rouillier**
Rouillure S	**Rouill.**
Roukhiyajnen, M.I. A	**Roukh.**
Rouleau, (Joseph Albert) Ernest (1916–1991) S	**Rouleau**

Roumeguère, Casimir (1828–1892) BCM	**Roum.**
Round, Frank Eric (1927–) A	**Round**
Roupell, Arabella Elizabeth (1817–1914) S	**Roupell**
Rouppert, Kasimierz(s) Stefan (1885–1963) A	**Rouppert**
Roure, L.A. (fl. 1963) M	**Roure**
Rourke, John Patrick (1942–) S	**Rourke**
Rouse, Glenn Everett (1928–) BM	**Rouse**
Roush, Eva Myrtelle (née Fling, E.M.) (1886–) S	**Roush**
Rousi, Arne Henrik (1931–) S	**Rousi**
Roussard, Monique (fl. 1969) M	**Roussard**
Rousseau, (Joseph Jules Jean) Jacques (1905–1970) PS	**J.Rousseau**
Rousseau, Jean–Jacques (1712–1778) S	**Rousseau**
Rousseau, Marietta Hannon (1850–1926) M	**M.Rousseau**
Roussel, Alexandre Victor (1795–1874) S	**A.Roussel**
Roussel, Ernest (fl. 1860) S	**E.Roussel**
Roussel, Henri François Anne de (1747–1812) AMS	**Roussel**
Rousselin, Geneviève A	**Rousselin**
Roussine, Nathalie A. (formerly Desiatova-Shostenko, N.A.) (1889–1968) S	**Roussine**
Routien, John Broderick (1913–) M	**Routien**
Rouville, Paul Gervais de (1823–1907) FS	**Rouville**
Rouwenhorst, R.J. (fl. 1987) S	**Rouw.**
Roux, Alain A	**A.Roux**
Roux, Cecilia (1947–) M	**Cec.Roux**
Roux, Claude (1945–) M	**Cl.Roux**
Roux, Honoré (1812–1892) S	**H.Roux**
Roux, Jacobus Petrus (1954–) PS	**J.P.Roux**
Roux, Jacques (1773–1822) S	**Roux**
Roux, Jean (1876–1939) B	**J.Roux**
Roux, Nisius (1854–1923) S	**N.Roux**
Roux, W. (fl. 1887) M	**W.Roux**
Rouy, Georges C.Chr. (1851–1924) PS	**Rouy**
Rouzeau, Christian (fl. 1967) M	**Rouzeau**
Rovelli, Renato (1806–1880) S	**Rovelli**
(Röver, William Edward)	
see Roever, William Edward	**Roever**
Rovira, Ana María (1947–) S	**Rovira**
(Rovira López, Ana María)	
see Rovira, Ana María	**Rovira**
Rovirosa, José N. (1849–1901) PS	**Rovirosa**
Rowan, Kingsley Spencer (1918–) A	**Rowan**
Rowland, Verner Hawsbrook (1883–) S	**Rowland**
Rowlee, Willard Winfield (1861–1923) S	**Rowlee**
Rowley, Gordon Douglas (1921–) S	**G.D.Rowley**
Rowley, J.A. S	**J.A.Rowley**
Rowley, Willard Winfield (1861–1923) S	**Rowley**
Rowntree, Gertrude Ellen Lester (1879–1979) S	**Rowntree**
(Roxas Clemente y Rubio, Simón de)	
see Clemente y Rubio, Simón de Roxas (Rojas)	**Clemente**
Roxburgh, William (1751–1815) PS	**Roxb.**

Roxon, J.E. (fl. 1974) M   **Roxon**
Roy, A.J. (fl. 1967) M   **A.J.Roy**
Roy, A.K. (fl. 1963) M   **A.K.Roy**
Roy, Anjali (fl. 1979) M   **A.Roy**
Roy, B. (fl. 1969) S   **B.Roy**
Roy, C.S. (fl. 1983) M   **C.S.Roy**
Roy, G.P. (1939–) S   **G.P.Roy**
Roy, Harendranath   A   **H.Roy**
Roy, Jessie D. (fl. 1871–1878) B   **J.D.Roy**
Roy, John (1826–1893) A   **R.Roy**
Roy, K. (fl. 1977) M   **K.Roy**
Roy, R.Y. (fl. 1961) M   **R.Y.Roy**
Roy, Sisir Kumar (1928–) FP   **S.K.Roy**
Royama, T. (fl. 1990) M   **Royama**
Roychoudhury, K.N. (fl. 1969) M   **Roych.**
Royen, Adriaan van (1704–1779) S   **Royen**
Royen, David van (1727–1799) S   **D.Royen**
Royen, Pieter van (1923–) S   **P.Royen**
Royer, Charles Louis Alexis (1831–1883) S   **Royer**
Royle, John Forbes (1798–1858) FS   **Royle**
Rozanov, Sergei Matveevich (1840–1870) AS   **Rozanov**
Rozanova, Maria Aleksandrovna (Alexandrovna) (1885–1957) S   **Rozanova**
Roze, Ernest (1833–1900) ABMS   **Roze**
Rozeira, Arnaldo Deodata da Fonseca (1912–) AS   **Rozeira**
Rozen, Betsy   A   **Rozen**
(Rozevic, Roman Julievich)
   see Roshevitz, Roman Julievich   **Roshevitz**
(Rozhevits, Roman Julievich)
   see Roshevitz, Roman Julievich   **Roshevitz**
Rozier, François (Jean-François) (1734–1793) S   **Rozier**
Rozin, A. (fl. 1791) S   **Rozin**
Rozsypal, J. (fl. 1966) M   **Rozsypal**
Rtischczeva, A.I. (1937–) M   **Rtischcz.**
(Rtyscheva, A.I.)
   see Rtischczeva, A.I.   **Rtischcz.**
Rua, Gabriel H. (1961–) S   **G.H.Rua**
Ruan, X.(Y.) (fl. 1986) M   **X.Y.Ruan**
Ruan, Yun Zhen (fl. 1988) S   **Y.Z.Ruan**
Ruban, E.L. (fl. 1970) M   **Ruban**
Rubaschevskaja, M.K. (1902–) S   **Rubasch.**
(Rubaschevskaya, M.K.)
   see Rubaschevskaja, M.K.   **Rubasch.**
Rübel, Eduard August (1876–1960) B   **Rübel**
Rubel, Franz (fl. 1778) S   **F.Rubel**
(Rübel–Blass, Eduard August)
   see Rübel, Eduard August   **Rübel**
Rubers, Wilhelmus (Wim) Vincentius (1944–) B   **Rubers**
Rubina, N.V.   A   **Rubina**

Rubio, Carmen (fl. 1982) M	**C.Rubio**
Rubio, P.P. A	**Rubio**
Rubner, Konrad (1886–1974) S	**Rubner**
(Rubtzoff, Peter)	
see Rubtzov, Peter	**P.Rubtzov**
Rubtzov, G.A. (1887–1942) S	**Rubtzov**
Rubtzov, Nicolaj Ivanovich (1907–1988) S	**N.I.Rubtzov**
Rubtzov, Peter (1920–) S	**P.Rubtzov**
(Rubzov, Nicolai Ivanovich)	
see Rubtzov, Nicolaj Ivanovich	**N.I.Rubtzov**
Ruchinger, Guiseppe Maria (1809–1879) S	**G.M.Ruch.**
Ruchinger, Guiseppe (Joseph) (1761–1847) AS	**Ruch.**
Rückbrodt, Dietrich (fl. 1970) S	**D.Rückbr.**
Rückbrodt, Ursula (fl. 1970) S	**U.Rückbr.**
Rückert, Ernst Ferdinand (1794–1843) S	**Rückert**
Rudakov, O.L. (fl. 1959) M	**Rudakov**
Rudavskaja, V.A. A	**Rudavsk.**
Rudbeck, Johan Olof (1711–1790) S	**J.O.Rudbeck**
Rudbeck, Olaus (Olof) Johannis (1630–1702) L	**O.J.Rudbeck**
Rudbeck, Olaus (Olof) Olai (1660–1740) L	**O.O.Rudbeck**
Rudberg, August (1842–1912) S	**Rudberg**
Rudd, Velva Elaine (1910–) S	**Rudd**
Rüdenberg, Lily (fl. 1970) S	**Rüdenberg**
Rudge, Edward (1763–1846) PS	**Rudge**
Rudio, Franz (1811–1877) S	**Rudio**
Rudjiman (1944–) S	**Rudjiman**
Rudloff, B. A	**Rudloff**
(Rudmose Brown, Robert Neal)	
see Brown, Robert Neal Rudmose	**R.N.R.Br.**
Rudnicka-Jezierska, Wanda (1924–) M	**Rudn.-Jez.**
Rudolph, Bert Alexander (1889–1953) M	**B.A.Rudolph**
Rudolph, Emanuel David (1927–) M	**E.D.Rudolph**
Rudolph, Johann Heinrich (1744–1809) S	**Rudolph**
Rudolph, Karl (1881–1937) S	**K.Rudolph**
Rudolph, Ludwig (1813–1896) S	**L.Rudolph**
Rudolphi, (Israel) Karl Asmund (Carl Asmunt (Asmus)) (1771–1832) MS	**Rudolphi**
Rudolphi, Friedrich Karl Ludwig (1801–1849) AMS	**F.Rudolphi**
Rudolphy, F. (fl. 1981) S	**Rudolphy**
Rudsky, J.A. S	**Rudsky**
Ruedemann, Rudolf (1864–) A	**Rued.**
(Ruehle, George Dewey)	
see Rühle, George Dewey	**Rühle**
Rueling, Johann Philipp (1741–) S	**Ruel.**
(Ruemke, Christian Ludwig)	
see Rümke, Christian Ludwig	**Rümke**
(Ruempler, Karl Theodor)	
see Rümpler, Karl Theodor	**Rümpler**
Rueness, Jan (1938–) A	**Rueness**
Ruenge (fl. 1850) S	**Ruenge**
Rugayah (1956–) S	**Rugayah**

Ruge, Georg (fl. 1893) S	**Ruge**
Rugel, Ferdinand Ignatius Xavier (1806–1879) S	**Rugel**
Rugg, Harold Goddard (1883–1957) P	**Rugg**
Rüggeberg, Hermann Karl August (1886–1967) M	**Rüggeb.**
Ruggeri, Antonio S	**Ruggeri**
Ruggieri, Gaetana (fl. 1933) M	**Ruggieri**
Ruggiero, Livio (1940–) S	**Ruggiero**
Rúgolo de Agrasar, Sulma (Zulma) E. (1940–) S	**Rúgolo**
(Rúgolo, Sulma (Zulma) E.)	
see Rúgolo de Agrasar, Sulma (Zulma) E.	**Rúgolo**
Ruhland, Wilhelm (Willy) Otto Eugen (1878–1960) MS	**Ruhland**
Rühle, George Dewey (1898–1962) M	**Rühle**
Ruhmer, Gustav (Gustaf) Ferdinand (1853–1883) S	**Ruhmer**
(Ruhmwerth, Raumund Rapaics von)	
see Rapaics, Raymund	**Rapaics**
Ruijs, Johannes Marinus (fl. 1884) S	**Ruijs**
(Ruijsch, Frederik)	
see Ruysch, Frederik	**Ruysch**
Ruinen, Jakoba (fl. 1964) AM	**Ruinen**
Ruini, Sergio (fl. 1990) M	**Ruini**
Ruiz de Azúa, Justo (1903–1980) S	**Ruiz de Azúa**
Ruíz de Ciolfi, Elsa N. (fl. 1970) S	**E.N.Ruíz**
Ruíz de Clavijo, Emilio (fl. 1980–1990) S	**E.Ruíz**
Ruíz de la Torre, Juan (1927–) S	**Ruíz Torre**
Ruíz del Castillo, Jacobo (1936–) S	**Ruíz Cast.**
Ruíz Leal, Ramón Adrián (1898–) S	**Ruíz Leal**
Ruiz López, Hipólito (1754–1815) PS	**Ruiz**
(Ruíz, Manuel)	
see Ruíz Oronoz, Manuel	**Ruíz Oronoz**
Ruíz, Martamonica S	**M.Ruíz**
Ruíz Oronoz, Manuel (1909–) M	**Ruíz Oronoz**
Ruíz Rejón, Manuel Enrique (1950–) S	**Ruíz Rejón**
Ruíz, Sebastien Joseph López (fl. 1802) S	**S.Ruíz**
Ruíz, Thirza (1949–) S	**T.Ruíz**
(Ruíz Z., Thirza)	
see Ruíz, Thirza	**T.Ruíz**
Ruíz–Terán, Luis Enrique (1923–1979) S	**Ruíz-Terán**
Rulamort, M.de (fl. 1986) M	**Rulamort**
Rumbold, Caroline Thomas (1877–1949) M	**Rumbold**
Rumjantsev, S.D. (fl. 1987) S	**Rumjantsev**
Rümke, Christian Ludwig (1898–1964) S	**Rümke**
Rummelspacher, Jörg (fl. 1965) S	**Rummelsp.**
(Rumpf, Georg Eberhard)	
see Rumphius, Georg Eberhard	**Rumph.**
Rumphius, Georg Eberhard (1628–1702) LMS	**Rumph.**
Rümpler, Karl Theodor (1817–1891) S	**Rümpler**
(Rumyantsev, S.D.)	
see Rumjantsev, S.D.	**Rumjantsev**
Rundel, Philip Wilson (1943–) M	**Rundel**
Rundina, L.A. (1937–) A	**Rundina**

Rune, Nils Olof (1919–) S	**Rune**
Runemark, Hans (1927–) MPS	**Runemark**
Rungby, Svend (1893–) B	**Rungby**
Runge, Annemarie (fl. 1962) M	**A.Runge**
Runge, C. (fl. 1889) S	**Runge**
Runyon, H.Everett (fl. 1935) S	**Runyon**
Ruokola, Anna–Liisa (fl. 1968) M	**Ruokola**
Ruoss, Engelbert (1956–) M	**Ruoss**
Ruotsalainen, Juhani (fl. 1990) M	**Ruots.**
Rupin, Ernest (Jean Baptiste) (1845–1909) PS	**Rupin**
(Rupp, Heinrich Bernard)	
see Ruppius, Heinrich Bernard	**Ruppius**
Rupp, Herman Montague Rucker (1872–1956) S	**Rupp**
(Rüppel, (Wilhelm Peter) Eduard (Simon))	
see Rüppell, (Wilhelm Peter) Eduard (Simon)	**Rüppell**
Ruppel, Manfred A	**Ruppel**
Rüppell, (Wilhelm Peter) Eduard (Simon) (1794–1884) S	**Rüppell**
Ruppert, Joseph (1864–1935) S	**Ruppert**
Ruppius, Heinrich Bernard (1688–1719) BL	**Ruppius**
Rupprecht, Heinrich von (fl. 1958) M	**H.Ruppr.**
Rupprecht, Johann Baptist (1776–1846) S	**Ruppr.**
Ruprecht, Franz Josef (Ivanovich) (1814–1870) ABMPS	**Rupr.**
Ruprecht, P. (fl. 1936) M	**P.Rupr.**
Rury, Philip M. (fl. 1978) P	**Rury**
Rusby, Henry Hurd (1855–1940) AS	**Rusby**
Ruschenberger, William Samuel Waithman (fl. 1831) MS	**Ruschenb.**
Ruschi, Augusto (1915–1986) S	**Ruschi**
Ruscoe, Q.W. (fl. 1970) M	**Ruscoe**
Rush, R. (fl. 1982) P	**Rush**
Rush–Munro, F.M. (fl. 1970) M	**Rush–Munro**
Rushforth, K.D. (fl. 1983) S	**Rushforth**
Rushforth, S.R. (fl. 1967) F	**S.R.Rushforth**
Rushforth, Samuel (1945–) A	**S.Rushforth**
Russ, Georg Philip (fl. 1868) S	**Russ**
Russanov, Fedor Nikolaevitch (1895–1979) S	**Russanov**
Russegger, Joseph von (1802–1863) S	**Russegger**
Russell, A.B. (fl. 1958) M	**A.B.Russell**
Russell, Alexander (c.1715–1768) S	**Russell**
Russell, Anna Worsley (1807–1876) M	**A.W.Russell**
Russell, Dennis J. A	**D.J.Russell**
Russell, George Archie (1882–) S	**G.A.Russell**
Russell, George Frederick (1953–) S	**G.F.Russell**
Russell, John (1766–1839) S	**J.Russell**
Russell, John Lewis (1808–1873) BM	**J.L.Russell**
Russell, Norman Hudson (1921–) S	**N.H.Russell**
Russell, Patrick (1727–1805) S	**P.Russell**
Russell, Paul George (1889–1963) S	**P.G.Russell**
Russell, Ralph Clifford (1896–) M	**R.C.Russell**
Russell, Thomas Hawkes (1851–1913) B	**T.H.Russell**

Rüssmann, M. (fl. 1984) S	**Rüssmann**
Russo, A. (fl. 1974) M	**Russo**
Russow, Edmund (August Friedrich) (1841–1897) BPS	**Russow**
Rüst (fl. 1889–1899) S	**Rüst**
Rustan, Øyvind H. (1954–) S	**Rustan**
(Ruström, Carl Birger)	
see Rutström, Carl Birger	**Rutstr.**
Rutenberg, Diedrich Christian (1851–1878) S	**Rutenb.**
Rutgers, Abraham Arnold Loderwijk (1884–) M	**Rutgers**
Ruthe, (Johann Gustav) Rudolf (1823–1905) BS	**R.Ruthe**
Ruthe, Johannes Friedrich (1788–1859) S	**Ruthe**
Rutherford, Daniel (1749–1819) S	**Rutherf.**
Rutishauser, R. (1949–) S	**Rutish.**
Rutschmann, J. (fl. 1988) S	**Rutschm.**
Rutström, Carl Birger (1758–1826) MS	**Rutstr.**
Ruttner, Franz (1882–1961) A	**Ruttner**
Rutty, John (1697–1775) S	**Rutty**
Ruud, Birgithe A	**Ruud**
Ruys, Jan Daniel (1897–1931) S	**J.D.Ruys**
Ruys, Johannes (1856–1933) M	**J.Ruys**
(Ruys, Johannes Marinus)	
see Ruijs, Johannes Marinus	**Ruijs**
Ruysch, Frederik (1638–1731) L	**Ruysch**
Ruyters, P. (fl. 1981) S	**Ruyters**
Růžič, J. (fl. 1967) M	**Růžic**
Růžička, Jirí (1909–) AM	**Růžička**
Ryabinin, D.V. A	**Ryabinin**
(Ryakhovsky, N.)	
see Rjachovsky, N.	**Rjach.**
Ryan, Bruce D. (fl. 1985) M	**B.D.Ryan**
Ryan, Elling (1849–1905) B	**Ryan**
Ryan, K.G. A	**K.G.Ryan**
Ryan, Ruth Winifred (1899–) M	**R.W.Ryan**
Rybak, Mariusz A	**Rybak**
Rybářová, J. (fl. 1980) M	**Rybářová**
Rybin, Vladimir Alekseevich (1895–1979) S	**Rybin**
Rycroft, Hedley Brian (1918–1990) S	**Rycroft**
Rydberg, Per Axel (1860–1931) MPS	**Rydb.**
Ryding, Per Olof (1951–) S	**Ryding**
Rydlo, Jaroslav (1950–) S	**Rydlo**
Rye, Barbara Lynette (1952–) S	**Rye**
Ryker, Truman Clifton (1908–) M	**Ryker**
Rylands, Thomas Glazebrook (1818–1900) AP	**Rylands**
Ryman, Svengunnar (1946–) M	**Ryman**
Ryppowa, Halina w Kowalskich (1899–1927) A	**Ryppowa**
Rytz, August Rudolf Walther (1882–1966) M	**Rytz**
(Rytz-Miller, August Rudolf Walther)	
see Rytz, August Rudolf Walther	**Rytz**

Ryvarden, Leif (1935–) M   **Ryvarden**
Ryves, Thomas Bruno (1930–) S   **Ryves**
Rzazade, Rza Jakhja Ogly (1909–) S   **Rzazade**
Rzedowski, Jerzy (1926–) S   **Rzed.**
Ržonsnickaja, M.A.   S   **Ržonsn.**

Sa'ad, Fatima El–Zahra Mahmoud (1925–) S   **Sa'ad**
Saag, A. (fl. 1991) M   **Saag**
Saage, Martin Joseph (1803–) S   **Saage**
Saarsoo, Bernhard (1899–1969) S   **Saarsoo**
Sab, S.C.D. (fl. 1956) F   **Sab**
Sabato, Sergio (1941–) S   **Sabato**
Sabbati, Liberato (1714–) S   **Sabbati**
Sabet, K.A. (fl. 1963) M   **Sabet**
Sabine, Edward (1788–1883) S   **E.Sabine**
Sabine, Joseph (1770–1837) S   **Sabine**
Sabirov, B.Z. (fl. 1959) S   **Sabirov**
Sablina, B.P. (fl. 1970) S   **Sablina**
Sabnis, Sharad Dwarkanath (1933–) S   **Sabnis**
Sabo, R. (fl. 1963) M   **Sabo**
Sabouraud, Raymond (Jacques) (1864–1938) M   **Sabour.**
Sabransky, Heinrich (1864–1915) BP   **Sabr.**
Sabrazès, J. (fl. 1921) M   **Sabrazès**
Sabu, M. (fl. 1988) S   **M.Sabu**
Sabu, T. (fl. 1985) S   **T.Sabu**
Sacamano, Charles M. (fl. 1967) M   **Sacamano**
Saccardo, Domenico (1872–1952) M   **D.Sacc.**
Saccardo, Francesco (1869–1896) M   **F.Sacc.**
Saccardo, Pier Andrea (1845–1920) ABMPS   **Sacc.**
Saccas, A.M. (fl. 1948) M   **Saccas**
Sacchetti, M. (fl. 1933) M   **Sacch.**
Sacco, José da Costa (1930–) S   **Sacco**
Sacconi, S. (fl. 1977) M   **Sacconi**
Sachar, G.S. (fl. 1934) M   **Sachar**
Sachet, Marie–Hélène (1922–1986) PS   **Sachet**
Sachokia, Michail Fedorovič (1902–) S   **Sachokia**
Sachs, (Ferdinand Gustav) Julius von (1832–1897) S   **Sachs**
Sachse, Carl Traugott (1815–1863) S   **Sachse**
Sackett, Walter George (1880–) M   **Sackett**
Sackin, M.J. (fl. 1989) M   **Sackin**
Sacleux, Charles (1856–1943) S   **Sacleux**
Saddi, N. (fl. 1984) S   **Saddi**
Sadebeck, Alexander (1843–1879) B   **A.Sadeb.**
Sadebeck, Richard Emil Benjamin (1839–1905) MPS   **Sadeb.**
Sadler, John (1837–1882) BM   **J.Sadler**
Sadler, Joseph (1791–1849) PS   **Sadler**
Sadler, Michael (Michal) (fl. 1831) S   **M.Sadler**
Sadovsky, Otakar (1893–1990) S   **Sadovsky**
Sadychov, I.A.   S   **Sadychov**
Saebi, E. (fl. 1991) M   **Saebi**

Saelán, Anders Thiodolf (1834–1921) B	**Saelán**
Saelan, E.L. S	**E.L.Saelan**
Saenger, P. A	**Saenger**
Saenko (fl. 1947) M	**Saenko**
Sáenz, A.A. (fl. 1979) S	**A.A.Sáenz**
Sáenz de Rivas, Concepción (1935–) PS	**Sáenz de Rivas**
Sáenz, José Alberto (1929–) M	**J.A.Sáenz**
(Sáenz Laín, Concepción)	
see Sáenz de Rivas, Concepción	**Sáenz de Rivas**
(Sáenz R., José Alberto)	
see Sáenz, José Alberto	**J.A.Sáenz**
Saëz, Henri (fl. 1960) M	**Saëz**
Sáez Soto, Francisco (fl. 1992) S	**F.Sáez**
Safeeulla, Kunigal M. (1924–) M	**Safeeulla**
Safford, William Edwin ('Ned') (1859–1926) S	**Saff.**
(Safina, L.K.)	
see Saphina, L.K.	**Saphina**
Safonov, G.E. (1933–) S	**Safonov**
Safonova, T.A. A	**Safonova**
Safralieva, N.A. (fl. 1984) S	**Safral.**
Safui, B. (1940–) S	**Safui**
Saga, Naotsune (1950–) A	**Saga**
Sagara, Naohiko (1938–) M	**Sagara**
Sagástegui, Abundio (1932–) S	**Sagást.**
(Sagástegui Alva, Abundio)	
see Sagástegui, Abundio	**Sagást.**
(Sagdullaeva, M.Sh.)	
see Sagdullajeva, M.Sh.	**Sagdull.**
Sagdullajeva, M.Sh. (fl. 1973) M	**Sagdull.**
Sageret, Augustin (1763–1851) S	**Sageret**
Saggese, V. (fl. 1938) M	**Saggese**
Sagorski, Ernst Adolf (1847–1929) PS	**Sagorski**
Sagot, Paul Antoine (1821–1888) BS	**Sagot**
Sagra, Ramón de la (1798–1871) BCS	**Sagra**
Sagredo, Rufino (1899–) S	**Sagredo**
Sah, B.C.L. (fl. 1964) M	**Sah**
Sah, S.C.D. (fl. 1960) AM	**S.C.D.Sah**
Saha, J.C. (fl. 1964) S	**Saha**
Sahambi, H.S. (fl. 1981) M	**Sahambi**
Sahashi, Norio (fl. 1979) P	**Sahashi**
Sahlberg, Carl Reinhold (1779–1860) S	**Sahlb.**
Sahlén, Anders Johan (1822–1891) S	**Sahlén**
Sahlin, Carl Ingmar (1912–1987) S	**Sahlin**
Sahling, P.-H. A	**Sahling**
Sahni, Birbal (1891–1949) AFM	**Sahni**
Sahni, Kailash Chandra (1921–) S	**K.C.Sahni**
Sahni, V.P. (fl. 1964) M	**V.P.Sahni**
Saho, Haruyoshi (fl. 1962) M	**Saho**
Sahtiyanci, S. (fl. 1962) M	**Sahtiy.**
Sahu, T.R. (fl. 1982) S	**Sahu**

Sahut, Félix (1835–1904) S	**Sahut**
Saifullah, S.M. A	**Saifullah**
Saiki, Yasuhisa (fl. 1973) PS	**Saiki**
Saikia, U.N. (fl. 1985) M	**Saikia**
Sailer, Franz Seraphin (1792–1847) S	**Sailer**
Saini, G.G. (fl. 1981) M	**G.G.Saini**
Saini, S.S. (fl. 1967) M	**Saini**
Sainsbury, George Osborne King (1880–1957) B	**Sainsbury**
Saint-Amans, Jean Florimond Boudon de (1748–1831) AMS	**St.-Amans**
Saint-Cyr, Dominique Napoleon Deshayes (1826–1899) S	**St.-Cyr**
Saint-Gal, Marie Joseph (1841–1932) S	**St.-Gal**
Saint-Germain, J.J.de (fl. 1784) S	**St.-Germ.**
Saint-Hilaire, Auguste François César Prouvençal de (1779–1853) MPS	**A.St.-Hil.**
Saint-Hilaire, Jean Henri Jaume (1772–1845) ABS	**J.St.-Hil.**
Saint-Lager, Jean Baptiste (1825–1912) PS	**St.-Lag.**
Saint-Léger, Léon (1868–1912) S	**St.-Lég.**
Saint-Marcel, Gaspard Dejean de (1763–1842) B	**St.-Marcel**
Saint-Moulin, Vincentius Josephus de (1804–1837) S	**St.-Moul.**
(Saint-Pierre, Jacques Nicolas Ernest Germain de)	
see Germain de Saint-Pierre, Jacques Nicolas Ernest	**Germ.**
Saint-Simon, Maximilien Henri de (1720–1799) S	**St.-Simon**
Saint-Yves, Alfred (Marie Augustine) (1855–1933) S	**St.-Yves**
Saito, Hidesaku (fl. 1931) M	**H.Saito**
Saito, Kamezo (1947–) BS	**K.Saito**
Saito, Kendo (1878–) M	**Saito**
Saito, Minoru (1919–) B	**M.Saito**
Saito, Yuzuru (1930–) A	**Y.Saito**
Sajdakovskij, L.Ya. A	**Sajdak.**
Sakaguchi, K. (fl. 1944) M	**Sakag.**
Sakai, Yoshio (1923–) A	**Sakai**
Sakalo, Dmytro Ivanovyč (1904–1965) S	**Sakalo**
Sakamoto, Sadao (1930–) S	**Sakam.**
Sakata, T. (fl. 1938) S	**Sakata**
Sakisaka, Michiji (1895–) A	**Sakis.**
Saksena, D.N. A	**D.N.Saksena**
Saksena, H.K. (fl. 1956) M	**H.K.Saksena**
Saksena, Ram Kumar (1897–1970) M	**R.K.Saksena**
Saksena, S.B. (fl. 1953) M	**S.B.Saksena**
Saksena, S.D. (fl. 1952) F	**S.D.Saksena**
Saksena, Shivdayal S. (1910–) B	**Saksena**
Sakuma, T. (fl. 1982) M	**Sakuma**
Sakurai, Kyuichi (1889–1963) BP	**Sakurai**
Sakurai, Y. (fl. 1954) M	**Y.Sakurai**
Salah, M.M. A	**Salah**
Salam, A.M.Abdus (1929–) A	**A.Salam**
Salam, M.A. (fl. 1954) M	**M.A.Salam**
Salard-Cheboldaeff, Marguerite (fl. 1980) M	**Sal.-Cheb.**
(Salazar Chávez, Gerardo A.)	
see Salazar, Gerardo A.	**Salazar**
Salazar, Gerardo A. (1961–) S	**Salazar**

Saldanha, Cecil John (1926–   ) P — C.J.Saldanha
Saldanha da Gama, José de (1839–1905) S — Saldanha
Saleem, Muhammad A — Saleem
(Sales Petzi, Franz von)
  see Petzi, Franz von Sales — Petzi
Saliba, J. (fl. 1988) M — Saliba
Salieva, Ya.S. (fl. 1989) M — Salieva
Salim, K.M. A — Salim
Salimbeni, A. (fl. 1920) M — Salimbeni
Salis–Marschlins, Carl Ulysses Adalbert von (1795–1886) PS — Salis
Salisbury, Edward James (1886–1978) S — E.Salisb.
Salisbury, George (1912–   ) MS — G.Salisb.
Salisbury, James Henry (1823–1905) A — J.H.Salisb.
Salisbury, Richard Anthony (1761–1829) PS — Salisb.
Salisbury, Robert K. A — R.K.Salisb.
Salisbury, William (–1823) S — W.Salisb.
Salkin, Ira Fred (1941–   ) M — Salkin
Salleh, K.M. (fl. 1988) P — Salleh
Sallent y Gotés, Angel (1857–1934) S — Sallent
(Sallenti Gotés, Angel)
  see Sallent y Gotés, Angel — Sallent
Salm–Reifferscheid–Dyck, Joseph Franz Maria Anton Hubert Ignatz
  Fürst zu (1773–1861) S — Salm–Dyck
Salman, A.H.P.M. (fl. 1977) S — Salman
Salmenova, K.Z. S — Salmenova
Salmi, Vera (1917–   ) S — Salmi
Salmon, Charles Edgar (1872–1930) S — C.E.Salmon
Salmon, Ernest Stanley (1871–1959) BM — E.S.Salmon
Salmon, John Drew (1802–1859) S — Salmon
Salomon, Carl E. (1829–1899) PS — Salomon
Salonen, A. (fl. 1968) M — Salonen
Salt, G.A. (fl. 1974) M — G.A.Salt
Salt, Henry (1780–1827) S — Salt
Salt, Jonathan (1759–1810) S — J.Salt
Salter, John (1798–1874) S — J.Salter
Salter, John Henry (1862–1942) S — J.H.Salter
Salter, John William (1820–1869) A — J.W.Salter
Salter, Samuel James Augustus 1825–1897 S — Salter
Salter, Terence Macleane (1883–1969) S — T.M.Salter
Salter, Thomas Bell (1814–1858) S — T.B.Salter
Såltin, Holger Torsten (1912–1969) S — Såltin
Salujha, S.K. A — Salujha
Salvanet–Duval, A. (fl. 1935) M — Salv.–Duval
Salvat, P. (fl. 1922) M — Salvat
Salvi, G. (fl. 1913) M — Salvi
(Salvo, Ángel Enrique)
  see Salvo Tierra, Ángel Enrique — Salvo
Salvo Tierra, Ángel Enrique (1957–   ) P — Salvo
Salvoza, Felipe Modesto (1892–   ) S — Salvoza

Salwey, Thomas (1791–1878) M	**Salwey**
Salzer, Michael (fl. 1860) S	**Salzer**
Salzmann, Philipp (1781–1851) ABMS	**Salzm.**
Sámano Bishop, Amelia A	**Sámano**
Samant, S.S. (fl. 1990) P	**Samant**
Samaritani, Giovanni Battista (1821–1894) S	**Samarit.**
(Samaritini, Giovanni Battista)	
see Samaritani, Giovanni Battista	**Samarit.**
Sambandam, C.N. (fl. 1961) M	**Samb.**
(Sambe Gowda, S.)	
see Gowda, S.Sambe	**S.S.Gowda**
Sambo, Maria (1888–1939) M	**Sambo**
Sambuc, Camille (1859–) S	**Sambuc**
Sambuk, F.V. (1900–1942) S	**Sambuk**
Samejima, Junichiro (1926–) S	**J.Samej.**
Samejima, Kazuko (1927–) S	**K.Samej.**
Sameva, E.F. (fl. 1987) M	**Sameva**
Samgina, D.I. (fl. 1971) M	**Samgina**
Samojlova, R.B. A	**Samojlova**
Samojlovich, S.R. A	**Samojl.**
Samoutsevitch, M.M. (fl. 1927) M	**Samouts.**
Sampaio, Alberto José de (1881–1946) PS	**A.Samp.**
Sampaio, Gonçalo António da Silva Ferreira (1865–1937) AMPS	**Samp.**
Sampaio, Joaquim A	**J.Samp.**
Sampaio, J.P. (fl. 1991) M	**J.P.Samp.**
(Sampaio Martins, Enrico)	
see Martins, Enrico Sampaio	**E.S.Martins**
Sampath, V. (fl. 1981) M	**Sampath**
Sampson, Dexter Reid (1930–) S	**D.R.Sampson**
Sampson, Frederick Bruce (1937–) S	**F.B.Sampson**
Sampson, Kathleen (fl. 1932) M	**Sampson**
Samra, A.S. (fl. 1963) M	**Samra**
Samson, Robert Archibald (1946–) M	**Samson**
Samsonoff, W.A. (fl. 1982) M	**Sams.**
Samsonoff-Aruffo, Caterina A	**Sams.-Aruffo**
Samsuddin, Wahid bin (fl. 1991) M	**Samsuddin**
Samuel, Geoffrey (1898–) M	**Samuel**
Samuels, Gary Joseph (1944–) M	**Samuels**
Samuelson, D.A. (fl. 1985) M	**Samuelson**
Samuelsson, F. S	**F.Sam.**
Samuelsson, Gunnar (1885–1944) S	**Sam.**
Samutina, M.I. (1962–) S	**Samutina**
Samylina, V.A. S	**Samylina**
Samzelius, Abraham (1723–1773) S	**Samzelius**
San Felice, Francesco (fl. 1895) M	**San Felice**
San Georgio, Anna di (1803–1874) S	**A.San Georgio**
San Georgio, Paolo (1748–1816) S	**San Georgio**
(San Giorgio, Anna di)	
see San Georgio, Anna di	**A.San Georgio**

San Martin, C. (fl. 1983) S	**San Martin**
Sanadze, K.S. (fl. 1964) S	**Sanadze**
Sanborn, Ethel Ida (1883–1952) B	**Sanborn**
Sancetta, Constance Antonina (1949–) A	**Sancetta**
Sánchez, Apolinar (fl. 1966) M	**A.Sánchez**
Sánchez, Dario (1947–) S	**D.Sánchez**
Sánchez, Evangelina A. (1934–) S	**E.A.Sánchez**
Sánchez, M.J. (fl. 1988) S	**M.J.Sánchez**
(Sánchez, María Elena)	
see Sanchez-Rodriques María Elena	**Sánchez-Rodr.**
Sánchez Mata, Daniel (fl. 1986) S	**Sánchez Mata**
(Sánchez S., Dario)	
see Sánchez, Dario	**D.Sánchez**
Sánchez Sánchez, José (1942–) S	**J.Sánchez**
(Sánchez V., I.)	
see Sánchez Vega, I.	**Sánchez Vega**
Sánchez Vega, Isidoro (1940–) S	**Sánchez Vega**
Sánchez Villaverde, Carlos (fl. 1981–1989) P	**C.Sánchez**
Sánchez Vindas, Pablo E. (fl. 1986) S	**P.E.Sánchez**
Sánchez–Gómez, Pedro (fl. 1986) S	**Sánchez–Gómez**
Sánchez–Mejorada, Hernando (1926–1988) S	**Sánchez–Mej.**
(Sánchez–Mejorada R., Hernando)	
see Sánchez–Mejorada, Hernando	**Sánchez–Mej.**
Sánchez–Pinto, C. (fl. 1990) M	**C.Sánchez–Pinto**
Sánchez–Pinto, I. (fl. 1980) M	**Sánchez–Pinto**
Sánchez–Pinto, Lázaro M. (1950–) M	**L.M.Sánchez–Pinto**
Sánchez–Rodríguez, María Elena (1929–) A	**Sánchez–Rodr.**
Sanchir, Ch. (1940–) S	**Sanchir**
Sancho, L.G. (fl. 1978) MS	**Sancho**
Sand, Austin Westman William (fl. 1926) S	**A.W.W.Sand**
Sanda, V. (1937–) S	**Sanda**
Sandahl, Oskar Theodor (1829–1894) S	**Sandahl**
Sandberg, John Herman (1848–1917) S	**Sandberg**
Sandberger, Guido (1821–1880) A	**G.Sandb.**
Sandberger, Karl (Carl) Ludwig Fridolin von (1826–1898) AF	**Sandb.**
Sande Lacoste, Cornelius Marinus van der (1815–1887) B	**Sande Lac.**
Sandéen, Peter Fredrik (1839–1868) S	**Sandéen**
Sander, Henry Frederick Conrad (1847–1920) PS	**Sander**
Sanders, Elizabeth Percy (1902–) A	**Sanders**
Sanders, F.E. (fl. 1986) M	**F.E.Sanders**
Sanders, Roger William (1950–) S	**R.W.Sanders**
Sanderson, Arthur Rufus (1877–1932) S	**A.Sand.**
Sanderson, Frank Reynolds (1936–) M	**F.R.Sand.**
Sanderson, John (1820/21–1881) S	**Sand.**
Sanderson, S.C. (fl. 1987) S	**S.C.Sand.**
Sández, E. S	**Sández**
Sandford, E. (fl. 1882) P	**Sandford**
Sandho, R.S. (fl. 1963) M	**Sandho**
Sandhu, Dhanwant K. (fl. 1963) M	**D.K.Sandhu**

Sandhu, R.S. (fl. 1963) M	R.S.Sandhu
Sandi, Alessandro Francesco (1794–1849) S	Sandi
Sandifort, Gerard (1779–1848) S	Sandifort
Sandino–Groot, W. S	Sand.-Groot
Sandmark, Gudmund (fl. 1809) P	Sandmark
Sandon, H. A	Sandon
Sandor, ?J. (fl. 1860s) P	Sandor
Sándor, I. (1853–) S	I.Sándor
Sandor, Rudolf (fl. 1958) M	R.Sandor
Sandri, G. (fl. 1842) M	Sandri
Sandru, Gheorghe (fl. 1971) M	Sandru
Sands, Martin Jonathan Southgate (1938–) S	Sands
Sandstede, (Johann) Heinrich (1859–1951) BMS	Sandst.
Sandt, Walter (1891–) S	Sandt
Sandu, R.S. (fl. 1933) M	R.S.Sandu
Sandu-Ville, Costantin (1897–1969) M	Sandu
Sandwith, Cecil Ivry (1871–1961) S	C.I.Sandwith
Sandwith, Noel Yvri (1901–1965) S	Sandwith
Sane, P.V. (1937–) B	Sane
Sanford, Guthrie Brown (1890–) M	Sanford
Sanford, Robert B. A	R.B.Sanford
Sang, John A	Sang
Sangiorgi, Giuseppe A	Sangiorgi
(Sangiorgio, Paolo)	
see San Georgio, Paolo	San Georgio
Sanguinetti, Pietro (1802–1868) S	Sanguin.
Sanio, Carl (Karl) Gustav (1832–1891) BPS	Sanio
Sanjappa, M. (1951–) S	Sanjappa
Sankaran, K.V. (fl. 1984) M	Sankaran
Sankaran, V. A	V.Sankaran
Sankhla, H.C. (fl. 1966) M	Sankhla
Sannemann, Dietrich A	Sann.
Sanson, C.A. S	C.A.Sanson
Sanson, M. (fl. 1830s) S	Sanson
(Sant'Anna, Célia Leite)	
see Leite, Célia Romano	C.R.Leite
Santa María, Juan (1915–) M	Santa María
Santa, Sébastien (fl. 1951) S	Santa
(Santamaría del Campo, Sergio (Sergi))	
see Santamaría, Sergio	Santam.
(Santamaría i del Campo, Sergio)	
see Santamaría Sergio	Santam.
Santamaría, Sergio (fl. 1985) M	Santam.
(Santana C., Elvinia)	
see Santana, Elvinia	Santana
Santana, Elvinia (fl. 1989) S	Santana
Santapau, Hermenegild (1903–1970) S	Santapau
Santelices, Bernabé (1945–) A	Santel.
Santesson, Carl Gustaf (1862–1939) M	Sant.

Santesson, J. (fl. 1960s) M	**J.Sant.**
Santesson, Rolf (1916–) MS	**R.Sant.**
Sántha, László (Ladislaus) (1886–1954) M	**Sántha**
Santi, Flavio (1856–1939) S	**F.Santi**
Santi, Giorgio (1746–1822) MS	**Santi**
Santiago, Alejandro E. A	**A.E.Santiago**
Santiago, Axel Rodolpho (fl. 1975) M	**A.R.Santiago**
Santiago, G. (fl. 1984) M	**G.Santiago**
Santiago, Laura Jane Moreira (fl. 1988) S	**L.J.M.Santiago**
Santisuk, Thawatchai (1944–) S	**Santisuk**
Santore, U.J. A	**Santore**
Santori (fl. 1903) M	**Santori**
Santos, Aniceta C.dos (fl. 1954) MS	**A.C.Santos**
(Santos Arnoldo)	
see Santos Guerra, Arnoldo	**A.Santos**
Santos, Aymar da Silva A	**A.S.Santos**
Santos, Emília Albina Alves dos (1936–) S	**E.Santos**
Santos Guerra, Arnoldo (1948–) S	**A.Santos**
Santos, João Ubiratan Moreira dos (fl. 1982)	**J.U.Santos**
Santos, José Vera (1908–1987) S	**Santos**
Santra, S.C. A	**Santra**
Sanwal, B.D. (fl. 1951) M	**Sanwal**
(Sapeghin, Andrej Afanasievich)	
see Sapjegin, Andrej Afanasievich	**Sapjegin**
Sapetza, Josef (1829–1868) S	**Sapetza**
Saphina, L.K. (1961–) S	**Saphina**
Sapjegin, Andrej Afanesievich (1883–1946) BS	**Sapjegin**
Saporta, Louis Charles Joseph Gaston de (1823–1895) ABFS	**Saporta**
Saposhnikow, Vasili Vasilievich (1861–1924) S	**Saposhn.**
Sappa, Francesco (1915–1957) BM	**Sappa**
Sapre, S.N. A	**Sapre**
(Sapriagaev, F.L.)	
see Zaprjagaev, F.L.	**Zaprjag.**
Sarafis, V. A	**Sarafis**
Saralegui Boza, Hildelisa (1949–) S	**Saralegui**
Sarasin, (Carl) Karl Friedrich (1859–1942) S	**Sarasin**
Sarato, César (1830–1893) S	**Sarato**
Sarauw, Georg Frederik Ludvig (1862–1928) S	**Sarauw**
Sarbajna, K.K. (fl. 1990) M	**Sarbajna**
Sarbhoy, Asha (fl. 1975) M	**A.Sarbhoy**
Sarbhoy, Ashok Kumar (1939–) M	**A.K.Sarbhoy**
Sardiña, Juan Rodriguez (fl. 1929) M	**Sardiña**
Sarejanni, Jean Antoine (1898–1962) M	**Sarej.**
Sarfatti, Giacomino (1920–1987) BS	**Sarfatti**
Sargant, Ethel (1816–1918) S	**Sargant**
Sargent, Charles Sprague (1841–1927) S	**Sarg.**
Sargent, Frederick LeRoy (1863–1928) S	**F.L.Sarg.**
Sargent, Henry Winthrop (1810–1882) S	**H.W.Sarg.**
Sargent, Oswald Hewlett (1880–1952) S	**O.H.Sarg.**

Sarim, Fazli Malik A	**Sarim**
Sarjeant, William Antony Swithin (1935–) A	**Sarjeant**
Sarkar, A.C. A	**A.C.Sarkar**
Sarkar, Anjali K. (fl. 1959) MS	**A.K.Sarkar**
Sarkar, Nandita (fl. 1991) M	**N.Sarkar**
Sarkar, Priyabrata K. (1929–) S	**P.K.Sarkar**
Sarkaria, J.S. (fl. 1980) S	**Sarkaria**
Sarkisova, S.A. (fl. 1970) S	**Sarkisova**
Sarma, Pranjit A	**P.Sarma**
Sarma, Yeleswarapu Siva Rama Krishna (1922–) A	**Sarma**
Sarmah, P.C. (fl. 1947) M	**Sarmah**
Sarmiento, M.N.R.de B	**Sarmiento**
Sarnari, Mauro (fl. 1990) M	**Sarnari**
Sarnthein, Ludwig von (1861–1914) BMPS	**Sarnth.**
Sarode, P.T. A	**Sarode**
Saroja, T.L. (fl. 1961) S	**Saroja**
Sarr, T. (fl. 1982) S	**Sarr**
Sarrazin, F. (fl. 1887) M	**Sarrazin**
Sarsons, Thomas Dixon (1880–1951) S	**Sarsons**
Sartoni, Gianfranco (1943–) A	**Sartoni**
Sartorelli, Giovanni Battista (1780–1853) S	**Sart.**
Sartori, Franz (1782–1832) S	**F.Sartori**
Sartori, Joseph (1809–1885) S	**Sartori**
Sartoris, George Bartholomew (1896–1949) M	**Sartoris**
Sartorius, Carl (Carlos) (Christian Wilhelm) (1796–1872) S	**Sartorius**
Sartory, Auguste Theodore (1881–1950) M	**Sartory**
Sartory, René (fl. 1930) M	**R.Sartory**
Sartwell, Henry Parker (1792–1867) S	**Sartwell**
Saruchanyan, F.G. (fl. 1957) M	**Saruch.**
Sarvela, Jaakko (1914–) P	**Sarvela**
Sarwal, B.M. (fl. 1983) M	**Sarwal**
Sarwar, M. (fl. 1965) M	**Sarwar**
(Sás, Elisabeta)	
see Szász, Elisabeta	**Szász**
Sasakawa, M. (fl. 1922) M	**Sasak.**
Sasaki, Katsuhito (fl. 1977) M	**K.Sasaki**
Sasaki, Rintarô (fl. 1939) M	**R.Sasaki**
Sasaki, Shun–ichi (Syun'iti) (1888–1960) PS	**Sasaki**
Sasaki, Yoshiyuki (1926–1972) M	**Y.Sasaki**
Sasamura, S. S	**Sasam.**
Sasaoka, Hisahiko (fl. 1934) B	**Sasaoka**
Sasidharan, N. (1952–) S	**Sasidh.**
Sass, Arthur Ferdinand von (1837–1870) S	**Sass**
Sass, Michael (fl. 1989) M	**M.Sass**
Sassenfeld, Joseph (fl. 1884) S	**Sassenf.**
Sassi, Agostino (–1852) B	**Sassi**
Sasson, A. A	**Sasson**
Sastre, B. (fl. 1990) P	**B.Sastre**
Sastre, Claude Henri Léon (1938–) PS	**Sastre**

Sastre-de Jesus, Ines (1955–   ) B — **I.Sastre**
Sastri, V.V.   A — **Sastri**
Sastry, A.R.K.Ramakrishna (1938–   ) PS — **Sastry**
Sastry, K.S.M. (fl. 1969) M — **K.S.M.Sastry**
Sastry, M.V.A.   A — **M.V.A.Sastry**
Sata, Tyósyun (1907–   ) S — **Sata**
Satabié, Benoît (1942–   ) S — **Satabié**
Satake, Yoshisuke (1902–   ) PS — **Satake**
Satchuthanthavale, V. (fl. 1966) M — **Satchuth.**
Sathe, A.V. (1935–   ) M — **Sathe**
Sathe, P.G. (fl. 1971) M — **P.G.Sathe**
Sathish Kumar, C. (1957–   ) S — **Sath.Kumar**
Sathyanarayanachar, M.B.   S — **Sathyan.**
Sati, S.C. (fl. 1983) M — **Sati**
Satija, Chander K. (fl. 1981–1989) P — **Satija**
Satô, Kunihiko (1902–   ) MPS — **K.Satô**
Satô, Masami (1910–1984) BM — **M.Satô**
Sato, Shoji (1927–   ) M — **S.Sato**
Sato, Taeko (1935–   ) MS — **T.Sato**
Sato, Takashi (fl. 1984) S — **Tak.Sato**
Satomi, Nobuo (1922–   ) PS — **Satomi**
Satow, Ernest Mason (1843–1929) S — **Satow**
Satya, H.N. (fl. 1963) M — **Satya**
Satyanarayan, Y. (fl. 1964) S — **Y.Satyan.**
Satyanarayana, B.A.K. (fl. 1965) M — **Satyanar.**
Satyanarayana, P. (fl. 1983) S — **P.Satyanar.**
Satyaprasad, K.S. (fl. 1981) M — **Satyapr.**
Satzyperova, I.F. (1922–   ) S — **Satzyp.**
Sauck, Jane Rees   S — **Sauck**
Sauer, Friedrich (Fritz) Ludwig Ferdinand (1852–   ) S — **Sauer**
Sauer, Friedrich Wilhelm Heinrich (1803–1873) S — **F.W.H.Sauer**
Sauer, G. (fl. 1980) S — **G.Sauer**
Sauer, Jonathan Deininger (1918–   ) S — **J.D.Sauer**
Sauer, Wilhelm (1935–   ) S — **W.Sauer**
Sauerbeck, Friedrich Wilhelm (1801–1882) B — **Sauerb.**
(Sauget y Barbis, Joseph Sylvestre)
   see Léon, (Hermano) — **Léon**
(Sauget-Barbier, Joseph Sylvestre)
   see Léon, (Hermano) — **Léon**
Sauleda, R.P. (1947–   ) S — **Sauleda**
Saunders, Charles Francis (1859–1941) S — **C.F.Saunders**
Saunders, De Alton (1870–1940) A — **D.A.Saunders**
Saunders, Edith Rebecca (1865–1945) S — **E.R.Saunders**
Saunders, G.W.   A — **G.W.Saunders**
Saunders, James (1839–1925) S — **J.Saunders**
Saunders, R.D.   A — **R.D.Saunders**
Saunders, William (1836–1914) S — **W.Saunders bis**
Saunders, William (1822–1900) S — **W.Saunders**
Saunders, William Wilson (1809–1879) S — **Saunders**

Saupe, Karl Alwin (1861–) S	**Saupe**
Saussure, Horace Bénédict de (1740–1799) S	**Sauss.**
Saussure, Nicolas Théodore de (1767–1845) S	**N.T.Sauss.**
Sauter, Anton Eleutherius (1800–1881) BMPS	**Saut.**
Sauvage, Charles Philippe Félix (1909–1980) MS	**Sauvage**
Sauvage, Monique (fl. 1974) M	**M.Sauvage**
Sauvageau, Camille François (1861–1936) AMS	**Sauv.**
(Sauvages de la Croix, François Boissier de)	
see Sauvages, François Boissier de la Croix de	**Sauvages**
Sauvages, François Boissier de la Croix de (1706–1767) S	**Sauvages**
Sauvaigo, Émile (1856–1927) S	**Sauvaigo**
Sauvalle, Francisco Adolfo (1807–1879) S	**Sauvalle**
Sauveur, (Dieudonné) Jean Joseph (1797–1862) F	**Sauveur**
Sauzé, Jean Charles (1815–1889) PS	**Sauzé**
Sava, Roberto (fl. 1844) S	**Sava**
Savage, Spencer (1886–1966) S	**Savage**
Savastano, Luigi Salvatore (1853–) M	**Savast.**
Savatier, Paul Alexandre (1824–1886) S	**A.Sav.**
Savatier, Paul Amedée Ludovic (1830–1891) PS	**Sav.**
Säve, Carl (Fredrik) (1812–1876) S	**Säve**
Saveleva–Dolgova, A.Ya. (1896–1925) A	**Sav.–Dolg.**
Savelli, Martino (1894–1918) S	**Savelli**
Savi, Gaetano (1769–1844) BMS	**Savi**
Savi, Paolo (1798–1871) F	**Pa.Savi**
Savi, Pietro (1811–1871) AS	**Pi.Savi**
(Savich, Vsevolod Pavlovich)	
see Savicz, Vsevolod Pavlovicz	**Savicz**
(Savich–Lubitskaya, Lydia Ivanovna)	
see Savicz–Lubitskaya, Lydia Ivanovna	**L.I.Savicz**
(Savicz, Lydia Ivanovna)	
see Savicz–Lubitskaya, Lydia Ivanovna	**L.I.Savicz**
Savicz, Natalie Mikhailovna (1894–) S	**N.Savicz**
Savicz, Vsevolod Pavlovicz (1885–1972) M	**Savicz**
Savicz–Lubitskaya, Lydia Ivanovna (1886–1982) BM	**L.I.Savicz**
Savicz–Ryczegorski, V.M. (1885–) S	**Sav.–Rycz.**
Savignone, Francesco (1818–) S	**Savign.**
Savigny, Marie Jules César Lélorgne de (1777–1851) APS	**Savigny**
Savile, Douglas Barton Osborne (1909–) MS	**Savile**
Savinceva, Z.D. (fl. 1970) M	**Savinceva**
Savoure, B. A	**B.Savoure**
Savouré, Henri Saintange (1861–1921) S	**Savouré**
Săvulescu, Alice (1905–1970) M	**A.Săvul.**
Săvulescu, Olga–Constanţa (1914–1969) M	**O.Săvul.**
Săvulescu, Traian (Trajan) (1889–1963) MPS	**Săvul.**
Sawa, Takashi (1929–) A	**Sawa**
Sawada, J. (fl. 1968) M	**J.Sawada**
Sawada, Kaneyoshi (Kenkichi) (1888–1950) M	**Sawada**
Sawada, Taketarô (1899–1938) S	**T.Sawada**

Sawadogo, Abdoussalam (fl. 1990) M	**Sawadogo**
Sawant, R.S. (fl. 1991) M	**Sawant**
Sawatari, A. A	**Sawatari**
(Saweljewa–Dolgowa, A.Ya.)	
see Saveleva–Dolgova, A.Ya.	**Sav.-Dolgova**
Sawer, John Charles (–1904) S	**Sawer**
Sawiczewski, Florian (1797–1876) S	**Sawicz.**
Sax, Karl (1892–1973) S	**Sax**
Saxeena, P.N. A	**Saxeena**
Saxén, Uno Alfons (1863–1948) S	**Saxén**
Saxena, Arjun Kishore (1928–) M	**A.K.Saxena**
Saxena, Hari Om (1938–) S	**H.O.Saxena**
Saxena, R.K. (fl. 1986) M	**R.K.Saxena**
Saxena, S.B. (fl. 1981) M	**S.B.Saxena**
Saxena, Y.N. (fl. 1958) F	**Y.N.Saxena**
Say, Thomas (1787–1834) S	**Say**
Saya, Ömer (1949–) S	**Saya**
Saylor, W.R. S	**Saylor**
Sayre, Geneva (1911–) B	**Sayre**
Sazonova, L.V. (1933–) S	**Sazonova**
Sbarbaro, Camillo (1888–1967) BM	**Sbarbaro**
Scagel, Robert F. (1921–) A	**Scagel**
Scalia, Giuseppe (1876–) M	**Scalia**
Scaling, William (fl. 1863–1882) S	**Scaling**
Scammacca, Blasco (1934–) A	**Scamm.**
Scamman, Edith (1882–1967) PS	**Scamman**
Scanagatta, Jonso Josue (1773–1823) S	**Scan.**
Scapoli, Giovanni Antonio S	**Scapoli**
Scarabelli Gommi Flamini, Giuseppe S	**Scarab.**
Scaramella, P. (fl. 1929) M	**Scaram.**
Scarr, M.P. (fl. 1966) M	**Scarr**
Scates, Katherine (Catherine) (fl. 1980) M	**Scates**
(Ščegolev, A.K.)	
see Shchegolev, A.K.	**Shcheg.**
Scepin, Constantin (1727–) S	**Scepin**
Schaack, Clark G. (fl. 1987) S	**Schaack**
(Schaack, George Booth Van)	
see Van Schaack, George Booth	**Van Schaack**
Schäar, Edward (1842–1913) S	**Schäar**
Schaarschmidt, Friedemann (1934–) M	**F.Schaarschm.**
Schaarschmidt, H. (fl. 1983) S	**H.Schaarschm.**
Schaarschmidt, Julius (later Istvanfy, G.) (1860–1930) AMS	**Schaarschm.**
Schabel, A. (1792–1836) S	**Schabel**
Schachnazarov, A.J. S	**Schachn.**
Schachner, J. (fl. 1929) M	**Schachner**
Schacht, Hermann (1814–1864) S	**Schacht**
Schack, Hans (1878–1946) S	**Schack**
Schade, Friedrich Alwin (1881–1976) BM	**Schade**
Schade, H. (fl. 1892) S	**H.Schade**
Schaede, Reinhold (fl. 1934) M	**Schaede**

Schaefer, Hermann (fl. 1872) S	H.Schaef.
Schaefer, Johannes (Bernhard) (1864–) S	J.Schaef.
Schaefer, Michael (1790–1846) S	Schaef.
Schaefer, Zdeněk (1906–1966) M	Z.Schaef.
Schaeffer, Asa Arthur (1883–) A	A.A.Schaeff.
Schaeffer, Cäsar (1867–1944) S	C.Schaeff.
(Schaeffer, Gottlieb August)	
see Herrich-Schaeffer, Gottlieb August	Herr.-Schaeff.
Schaeffer, Jacob Christian (H.von) (1718–1790) MS	Schaeff.
(Schaeffer, Julius)	
see Schäffer, Julius	Jul.Schäff.
Schaeftlein, Hans (1886–) S	Schaeftl.
Schaepe, Annemarie (1952–) B	Schaepe
(Schaer, Edward)	
see Schäar, Edward	Schäar
Schaerer, E.A. (fl. 1900s) M	E.A.Schaer.
Schaerer, Ludwig Emanuel (Louis-Emmanuel) (1785–1853) M	Schaer.
Schafer, G. (fl. 1980) S	G.Schaf.
Schäfer, H.I. (fl. 1973) S	H.I.Schäf.
Schäfer, Peter Andreas (1944–) S	P.Schäf.
Schafer-Verwimp, Alfons (1950–) B	Schaf.-Verw.
(Schäffer, Jacob Christian (H.von))	
see Schaeffer, Jacob Christian (H.von)	Schaeff.
Schäffer, Julius (1882–1944) MS	Jul.Schäff.
Schaffner, Johann Wilhelm (Guillermo) (1830–1882) PS	W.Schaffn.
Schaffner, John Henry (1866–1939) APS	J.H.Schaffn.
Schaffnit, (Johannes Martin) Ernst (Christian Otto) (1878–) MS	Schaffnit
Schaffranek, A. (fl. 1883–1894) S	Schaffr.
Schafhäutl, (Karl Franz) Emil (1803–1890) AFS	Schafh.
Schaga, V.S. S	Schaga
Schagalina, L.M. (fl. 1965) S	Schagalina
Schagerström, Johan August (1818–1867) S	Schag.
Schaijes, Michel (fl. 1987) S	Schaijes
(Schaikin, I.M.)	
see Shaikin, I.M.	Shaikin
Schakryl, A.K. (fl. 1968) F	Schakryl
Schalkwijk, Joost (fl. 1982) M	Schalkw.
Schallert, Paul Otto (1879–1970) S	Schallert
Schalow, Emil (fl. 1938) S	Schalow
Schalyt, Michail S. (1904–1968) S	Schalyt
Schander, Richard (1873–1933) M	Schander
Schanen, Noel (fl. 1970) AS	Schanen
Schangin, Petr Ivanovich (1741–1816) S	Schangin
Schäperclaus, Wilhelm (1899–) A	Schäp.
Schapoval, I.I. (1941–) S	Schapoval
Scharaschova, V.S. (1941–) S	Scharasch.
Scharfetter, Rudolf (1880–1956) S	Scharf.
Scharif, G. (fl. 1959) M	Scharif
Scharipov, A. S	Scharipov

Scharipova, B.A. S	**B.A.Scharip.**
Scharipova, M.M. (1935–) S	**M.M.Scharip.**
Scharlock, (Carl) Julius (Adolf) (1809–1899) S	**Scharlock**
Schatz, Albert (1920–) A	**A.Schatz**
Schatz, George Edward (1953–) S	**G.E.Schatz**
Schatz, Jacob Wilhelm (1802–1867) S	**Schatz**
Schatz, Scott (fl. 1980) M	**S.Schatz**
Schatzl, S. (fl. 1979) S	**Schatzl**
Schau, (Hermann Ernst) Otto (1900–) S	**Schau**
Schau, J.K. S	**J.K.Schau**
Schaub, Hans A	**Schaub**
Schauer, Johannes Conrad (1813–1848) S	**Schauer**
Schauer, L. S	**L.Schauer**
Schauer, Sebastian (fl. 1847) PS	**S.Schauer**
Schauer, Thomas (1938–) BM	**T.Schauer**
Schaulo, D.N. S	**Schaulo**
Schaumann, K. (fl. 1973) M	**Schaumann**
Schauroth, Karl (Carl) von (1818–1893) A	**Schauroth**
Schaver, F. (fl. 1985) M	**Schaver**
Schawo, Michael (1850–1909) S	**Schawo**
Schchian, A.S. (1905–1990) S	**Schchian**
Schczedrova, B.I. (fl. 1964) M	**Schczedr.**
Schebalina, M.A. (1900–) S	**Scheb.**
Schechter, Yaakov S	**Schechter**
Scheckler, S.E. (fl. 1971) F	**Scheckler**
Scheele, E. S	**E.Scheele**
Scheele, George Heinrich Adolf (1808–1864) S	**Scheele**
Scheer, D. (fl. 1935) M	**D.Scheer**
Scheer, Friedrich (Frederick) (1792–1868) S	**Scheer**
Scheeren, Walter F. (fl. 1970) S	**Scheeren**
Scheetz, R.W. (fl. 1980) M	**Scheetz**
Scheffelt, E. A	**Scheffelt**
Scheffer, Jozef (1903–1949) S	**J.Scheff.**
Scheffer, R.J. (fl. 1984) M	**R.J.Scheff.**
Scheffer, Rudolph Herman Christiaan Carel (1844–1880) S	**Scheff.**
Scheffer, Theodore Comstock (1904–) S	**T.C.Scheff.**
Scheffler, W. A	**Scheffler**
Scheibe, Arnold (1901–1989) S	**Scheibe**
Scheidegger, C. (fl. 1985) M	**Scheid.**
Scheidweiler, Michael Joseph François (1799–1861) S	**Scheidw.**
Scheinpflug, H. (fl. 1958) M	**Scheinpflug**
Scheinvar, Leia (fl. 1974) S	**Scheinvar**
Scheit, Max (1858–1888) S	**Scheit**
Scheitz, Antal A	**Scheitz**
Schekhovtsov, A.G. (fl. 1981) M	**Schekh.**
(Schelkovnikov, Alexander Bebutovich)	
see Schelkownikow, Alexandr Bebutovicz	**Schelk.**
Schelkownikow, Alexandr Bebutovicz (1870–1933) S	**Schelk.**
Schell, W.A. (fl. 1983) M	**Schell**

Schelle, Ernst (1864–1945) S	**Schelle**
Schellenberg, Gustav August Ludwig David (1882–1963) S	**G.Schellenb.**
Schellenberg, Hans Conrad (1872–1923) AMS	**Schellenb.**
Schellmann, C. (fl. 1938) S	**Schellm.**
Scheloske, H.-W. (fl. 1969) M	**Scheloske**
Schelpe, Edmund André Charles Lois Eloi ('Ted') (1924–1985) BFPS	**Schelpe**
Schelver, Friedrich ('Franz') Joseph (1778–1832) S	**Schelver**
Schembel, S.J. (fl. 1913) M	**Schembel**
Schemmann, Wilhelm (1845–c.1920) S	**Schemmann**
Schenck, A. (fl. 1845) S	**A.Schenck**
Schenck, Carl Alwin (1868–1955) S	**C.A.Schenck**
Schenck, Johann Heinrich Rudolf (1860–1927) PS	**Schenck**
Schenck, Norman C. (fl. 1982) M	**N.C.Schenck**
Schenckel, J. (fl. 1847) S	**Schenckel**
Schengelia, E.M. (fl. 1953) S	**Scheng.**
Schenk, Alexander (1864–1924) FS	**A.Schenk**
Schenk, Bernhard (1833–1893) M	**B.Schenk**
Schenk, Ernst (1880–1965) S	**E.Schenk**
Schenk, Joseph August (1815–1891) FMS	**Schenk**
Scherbius, Johannes (1769–1813) BMS	**Scherb.**
Scherer, Johann (Baptista) Andreas (1755–1844) S	**Scherer**
Scherer, Reed P. A	**R.P.Scherer**
Scherfel, Aurel Wilhelm (Vilmos Aurél) (1835–1895) S	**Scherfel**
Scherffel, Aladár (1865–1938) AM	**Scherff.**
Schery, Robert Walter (1917–1987) S	**Schery**
Schesmer, Andrea A	**Schesmer**
Scheuchzer, Johann Jacob (Jakob) (1672–1733) L	**J.J.Scheuchzer**
Scheuchzer, Johannes (1738–1815) S	**Scheuchzer f.**
Scheuchzer, Johannes Gaspar (1684–1738) L	**J.Scheuchzer**
Scheuer, Christian (fl. 1986) M	**Scheuer**
Scheuermann, Richard (1873–1949) S	**Scheuerm.**
Scheutz, Nils Johan Wilhelm (1836–1889) BS	**Scheutz**
(Scheviakov, Vladimir)	
see Schewiakoff, Wladimir	**Schew.**
Schewe, O. (1892–) S	**Schewe**
Schewiakoff, Wladimir (1859–1930) A	**Schew.**
Scheygrond, Arie (1905–) S	**Scheygr.**
Schiavone, Maria Magdalena (1945–) B	**Schiavone**
(Schibkova, I.F.)	
see Shibkova, I.F.	**Shibkova**
Schibler, Wilhelm (1861–1931) S	**Schibler**
Schick, Carl (1881–1953) S	**Schick**
Schick, G. (fl. 1970) M	**G.Schick**
Schick, Rudolf (1905–) S	**R.Schick**
Schick-Freiburg, Carl (1881–1953) S	**Schick-Freib.**
Schidlay, Eugen (1911–) PS	**Schidlay**
Schiede, Christian Julius Wilhelm (1798–1836) S	**Schiede**
Schiedermayr, Karl (Carl) B. (1818–1895) BM	**Schied.**

Schieferdecker, K. (fl. 1954) M	**Schief.**
Schiel, David R. A	**Schiel**
Schiemann, Carl Christianus (1763–1835) S	**Schiem.**
Schiemann, Elisabeth (1881–1972) S	**E.Schiem.**
Schiemek, A. (fl. 1921) M	**Schiemek**
Schiffner, Victor Félix (1862–1944) ABMS	**Schiffn.**
Schikora, Friedrich (1859–1932) M	**Schikora**
Schilberszky, Károly (1863–1935) BMS	**Schilb.**
Schild, Edwin (fl. 1970) M	**Schild**
Schildknecht, J. (fl. 1863) S	**Schildkn.**
Schiller, Eduard (fl. 1886) S	**E.Schiller**
Schiller, Ivan (fl. 1958) S	**I.Schiller**
Schiller, Josef (1877–1960) A	**J.Schiller**
Schiller, Sigismund (1847–1920) S	**Schiller**
Schilling, Anthony D. (1931–) S	**A.D.Schill.**
Schilling, August Jacob (1865–) A	**A.J.Schill.**
Schilling, Edward E. (1953–) S	**E.E.Schill.**
Schilling, Friedrich von (1897–) M	**F.Schill.**
Schilling, Gottfried Wilhelm (fl. 1778) S	**Schill.**
Schiman-Czeika, Helene (1933–) MS	**Schiman-Czeika**
Schimanski, Hans A	**Schim.**
(Schimert, Johann Peter)	
see Schimmert, Johann Peter	**Schimmert**
Schimmert, Johann Peter (fl. 1776) S	**Schimmert**
Schimon, O. (fl. 1911) M	**Schimon**
Schimper, Andreas Franz Wilhelm (1856–1901) ABS	**A.Schimp.**
Schimper, Georg (Heinrich) Wilhelm (1804–1878) S	**G.W.Schimp.**
Schimper, Karl (Carl) Friedrich (1803–1867) BS	**K.F.Schimp.**
Schimper, Wilhelm Philipp (1808–1880) ABFMS	**Schimp.**
Schindehütte, G. (fl. 1907) F	**Schindeh.**
Schindler, Anton Karl (1879–1964) S	**Schindl.**
Schindler, Hans (fl. 1917–1925) S	**H.Schindl.**
Schindler, Johann (1881–) S	**J.Schindl.**
Schindler, S.S. S	**S.S.Schindl.**
Schingnitz von Boselager, Ada A	**Schingn.**
Schinini, Aurelio (1943–) S	**Schinini**
Schinne, Isaac Evert Cornelis van (1741–1819) S	**Schinne**
Schinner, Franz (1947–) M	**Schinner**
Schinnerl, Martin (1861–1950) B	**Schinnerl**
Schinz, Christoph Salomon (1764–1847) S	**C.S.Schinz**
Schinz, Hans (1858–1941) PS	**Schinz**
Schinz, Salomon (1734–1784) S	**S.Schinz**
Schiønning, Holger Ludwig (Ludvig) (1868–1942) M	**Schiønning**
(Schiötz, Ludvig Theodor)	
see Schiøtz, Ludvig Theodor	**Schiøtz**
Schiøtz, Ludvig Theodor (1821–1900) S	**Schiøtz**
Schipczinski, Nikolaj Valerianovich (1886–1955) S	**Schipcz.**
(Schipczinsky, Nikolaj Valerianovich)	
see Schipczinski, Nikolaj Valerianovich	**Schipcz.**

(Schipczinzky, Nikolaj Valerianovich)	
see Schipczinski, Nikolaj Valerianovich	**Schipcz.**
Schipp, William August (1891–1967) S	**Schipp**
Schipper, Margarita A.A. (1923–) M	**Schipper**
(Schirjaev, Grigorij (Gregory) Ivanović)	
see Širjaev, Grigorij (Gregory) Ivanović	**Širj.**
(Schirjev, Grigorij (Gregory) Ivanovich)	
see Širjaev, Grigorij (Gregory) Ivanović	**Širj.**
Schirmer, P. (fl. 1987) M	**Schirmer**
(Schirnina–Grischina, L.V.)	
see Shirnina–Grishina, L.V.	**Shirn.-Grish.**
Schirschoff, P.P. A	**Schirsch.**
Schischkin, Boris Konstantinovich (1886–1963) S	**Schischk.**
Schischkin, Ivan Kusmich (1897–1934) S	**I.Schischk.**
Schischkina, A.K. (fl. 1973) M	**Schischkina**
Schkorbatow, Leonid Andrejewitsch (1884–) AM	**Schkorb.**
Schkuhr, Christian (1741–1811) BCP	**Schkuhr**
Schkurenko, V.A. (fl. 1971) M	**Schkur.**
Schlachter (fl. c.1857) S	**Schlachter**
Schlagdenhauffen, Charles Frédéric (1830–1907) S	**Schlagd.**
Schlagintweit, Adolf (von) (1829–1857) S	**A.Schlag.**
Schlagintweit–Sakünlünski, Herman Alfred Rudolph von (1826–1882) S	**H.Schlag.**
Schlange, K.G. S	**Schlange**
(Schläpfer, Elizabeth)	
see Schläpfer–Bernhard, Elizabeth	**Schläpf.-Bernh.**
Schläpfer, Georg (1797–1835) S	**Schläpfer**
Schläpfer–Bernhard, Elizabeth (fl. 1968) M	**Schläpf.-Bernh.**
Schlatter, Theodor (1847–1918) S	**Schlatter**
Schlauder, J. A	**Schlauder**
Schlauter, (E.) August (1803–1849) S	**Schlauter**
Schlayer, Philipp (fl. 1837) S	**Schlayer**
Schlechtendal, Diederich Franz Leonhard von (1794–1866) MPS	**Schltdl.**
Schlechtendal, Diederich Friedrich Karl (Carl) von (1767–1842) S	**D.F.K.Schltdl.**
Schlechtendal, Diederich (Dietrich) Hermann Reinhard von (1834–1916) F	**D.H.R.Schltdl.**
Schlechter, Friedrich Richard Rudolf (1872–1925) S	**Schltr.**
Schleicher, Johann Christoph (1768–1834) ABMPS	**Schleich.**
Schleiden, Matthias Jacob (1804–1881) S	**Schleid.**
Schlenker, Georg (1847–1932) S	**Schlenker**
Schlesinger, Karl (1869–) S	**Schles.**
Schlesner, H. (fl. 1986) M	**Schlesner**
Schlessman, Mark A. (1952–) S	**Schlessman**
Schlich, William (Wilhelm) (Philip Daniel) (1840–1925) S	**Schlich**
Schlickum, August (1867–) B	**A.Schlick.**
Schlickum, Julius (1804–1884) S	**J.Schlick.**
Schlickum, Oskar (1836–1889) S	**O.Schlick.**
Schlieben, Hans–Joachim Eberhardt (1902–1975) S	**Schlieb.**
Schliephacke, Karl (1834–1913) B	**Schlieph.**
Schlim, Louis Joseph (fl. 1845) S	**Schlim**

Schlitter, Jakob (fl. 1950) S	**Schlitter**
Schljakov, Roman Nicolaevich (1912–) BS	**Schljakov**
Schloemer-Jaeger, Anna (fl. 1958) F	**Schloemer-Jaeger**
Schloesser, Robert E. A	**Schloesser**
Schlögl, Ludwig (Ludvík) (1846–1899) S	**Schlögl**
Schlösser, U. (fl. 1970) M	**U.Schlöss.**
Schlosser von Klekovski, Joseph Calasenz (1808–1882) PS	**Schloss.**
Schlotheim, Ernst Friedrich von (1764–1832) AFS	**Schloth.**
Schlothgauer, S.D. (1941–) S	**Schlothg.**
Schlotterbeck, Julius Otto (1865–1917) S	**Schlott.**
Schlotthauber, August Friedrich (–1872) S	**Schlotth.**
(Schlotthauber, Augusto Frederico)	
see Schlotthauber, August Friedrich	**Schlotth.**
Schlumbach, Friedrich Alexander von (1772–1835) S	**Schlumbach**
Schlumberger, Frederic (1823–1893) S	**Schlumb.**
Schlüter, Clemens August Joseph (1836–1906) AF	**Schlüter**
(Schlyakov, Roman Nicolayevich)	
see Schljakov, Roman Nicolaevich	**Schljakov**
(Schmalgausen, Ivan Fedorovich)	
see Schmalhausen, Johannes Theodor	**Schmalh.**
Schmalhausen, Johannes Theodor (1849–1894) FS	**Schmalh.**
Schmaltz, J.Paul (1844–) S	**Schmaltz**
Schmalz, Eduard (1801–1871) AMS	**Schmalz**
Schmarda, Ludwig K. A	**Schmarda**
Schmarse, Helmut (fl. 1933) S	**Schmarse**
Schmeil, Franz Otto (1860–1943) S	**Schmeil**
Schmeiss, Oskar (fl. 1906) S	**Schmeiss**
Schmid, (Ludwig) Bernhard (Ehregott) (1788–1857) S	**L.B.E.Schmid**
Schmid, Anne-Marie A	**A.M.Schmid**
Schmid, B. (fl. 1983) S	**B.Schmid**
Schmid, Eduard (fl. 1906) S	**Ed.Schmid**
Schmid, Emil (1891–) S	**Em.Schmid**
Schmid, Ernst Ehrhard Friedrich Wilhelm (1815–1885) S	**Schmid**
Schmid, Gerlinde A	**G.Schmid**
Schmid, Manfred E. A	**M.E.Schmid**
Schmid, Maurice (fl. 1991) S	**M.Schmid**
Schmid, Rudolf (1942–) S	**R.Schmid**
Schmid, Rupertus (–1804) PS	**Rup.Schmid**
Schmid, Wilhelm Gustav Günther (1888–1949) AS	**W.G.G.Schmid**
Schmid-Heckel, Helmuth (1956–) M	**Schmid-Heckel**
Schmidel, Casimir Christoph (1718–1792) ABMPS	**Schmidel**
Schmidely, Auguste Isaac Samuel (1838–1918) S	**Schmidely**
Schmidle, Wilhelm (1860–1951) A	**Schmidle**
Schmidlin, Eduard (1808?–1890) A	**Schmidlin**
Schmidt, Adolf Wilhelm Ferdinand (1812–1899) A	**A.W.F.Schmidt**
Schmidt, Albert (fl. 1970) M	**Alb.Schmidt**
Schmidt, Alexander Friedrich Wolfgang (1932–) MS	**A.F.W.Schmidt**
Schmidt, Alfred (1886–) M	**Alf.Schmidt**
Schmidt, Antal A	**Ant.Schmidt**

Schmidt, C.E. S	C.E.Schmidt
Schmidt, C.P. (fl. 1831) S	C.P.Schmidt
Schmidt, Carl (Karl) Friedrich (1811–1890) S	C.F.Schmidt
Schmidt, D. (fl. 1990) M	D.Schmidt
Schmidt, Eduard Oscar (1823–1886) A	E.O.Schmidt
Schmidt, Ernst (1834–1902) S	E.Schmidt
Schmidt, Ernst Johannes (1877–1933) AMS	E.J.Schmidt
Schmidt, Franz (1751–1834) S	Schmidt
Schmidt, Franz Wilibald (1764–1796) S	F.W.Schmidt
Schmidt, Friedrich (Karl) (Fedor Bogdanovich) (1832–1908) S	F.Schmidt
Schmidt, G.W. A	G.W.Schmidt
Schmidt, H. (of Freiburg) (fl. 1916) B	H.Schmidt
Schmidt, Heinrich (fl. 1907) S	Heinr.Schmidt
Schmidt, Heinrich Christian Friedrich (1819–1863) S	H.C.F.Schmidt
Schmidt, Hermann (1821/2–1905) S	H.Schmidt Elberf.
Schmidt, Hermann Rudolph (1814–1867) S	H.R.Schmidt
Schmidt, I. (fl. 1969) M	I.Schmidt
Schmidt, Johann Anton (1823–1905) S	J.A.Schmidt
Schmidt, Johann August Friedrich (fl. 1832) S	J.A.F.Schmidt
Schmidt, Johann Carl (Karl) (1793–1850) MS	J.C.Schmidt
Schmidt, Johann Joachim (fl. 1797) S	J.J.Schmidt
Schmidt, Justus J.H. (1851–1930) PS	J.J.H.Schmidt
Schmidt, Karl Max (1880–) S	M.Schmidt
(Schmidt, of St.Petersberg)	
see Schmidt, Friedrich (Karl) (Fedor Bogdanovich)	F.Schmidt
Schmidt, Oswald (Hermann Wilhelm) (1907–) S	O.Schmidt
Schmidt, Otto Christian (1900–1951) ACPS	O.C.Schmidt
Schmidt, Paul (1846–) A	P.Schmidt
Schmidt, Peter A. (1946–) S	P.A.Schmidt
Schmidt, R.G. S	R.G.Schmidt
Schmidt, Reinold (fl. 1925) M	Rein.Schmidt
Schmidt, Robert (1826–1890) S	R.Schmidt
Schmidt, Robert J. A	R.J.Schmidt
Schmidt, Ronald R. A	R.R.Schmidt
Schmidt, Wilhelm Ludwig Ewald (1804–1843) S	W.L.E.Schmidt
Schmidtke, Ernst (fl. 1934) B	Schmidtke
Schmiedeknecht, Martin (1927–) M	Schmied.
(Schmiedel, Casimir Christoph)	
see Schmidel, Casimir Christoph	Schmidel
(Schmiedely, Auguste Isaac Samuel)	
see Schmidely, Auguste Isaac Samuel	Schmidely
Schmitt, John Arvid (1925–) M	J.A.Schmitt
Schmitt, Joseph (1862–1915) S	Schmitt
Schmitt, Ursula (fl. 1968) B	U.Schmitt
Schmitter (fl. 1923) M	Schmitter
Schmittner, Stella M. A	Schmittner
Schmitz, (Carl Johann) Friedrich (1850–1895) AS	F.Schmitz
Schmitz, André (1920–) S	A.Schmitz
Schmitz, Joachim (fl. 1985) S	J.Schmitz

Schmitz, Johann Joseph (1813–1845) P	**J.J.Schmitz**
Schmoll, Carolina (fl. 1951) S	**C.Schmoll**
Schmoll, Ferdinand S	**F.Schmoll**
Schmoll, Hazel Marguerite (1891–1990) S	**Schmoll**
Schmoller, H. (fl. 1966) M	**Schmoller**
Schmoranzer, J. (fl. 1934) M	**Schmor.**
Schmotina, G.E. (fl. 1974) M	**Schmotina**
Schmucker, Theodor (1894–1970) S	**Schmucker**
Schmula A	**Schmula**
Schmutzler, Clarence L. S	**Schmutzler**
Schnabel, Hans (fl. 1971) S	**Schnabel**
Schnabl, Johann Nepomuk (1853–1899) M	**Schnabl**
Schnack, Benno Julio Christian (1910–1981) S	**Schnack**
Schnarf, Karl (1879–1947) S	**Schnarf**
Schnebling S	**Schnebl.**
Schneck, Jacob (1843–1906) S	**Schneck**
Schnee, Ludwig (1908–1975) S	**Schnee**
(Schneevoight, George Voorhelm)	
see Schneevoogt, George Voorhelm	**Schneev.**
Schneevoogt, George Voorhelm (1775–1850) S	**Schneev.**
Schnegg, Hans (1875–1950) M	**Schnegg**
Schneidegger, C. (fl. 1987) M	**Schneidegger**
Schneider, Albert (1863–1928) AM	**A.Schneid.**
Schneider, Alfred A	**Alf.Schneid.**
Schneider, Camillo Karl (1876–1951) S	**C.K.Schneid.**
Schneider, Craig William (1948–) A	**C.W.Schneid.**
Schneider, E.E. (fl. 1912) S	**E.E.Schneid.**
Schneider, Eduard Karl Ludwig (Pfeil) (1809–1889) S	**L.Schneid.**
Schneider, Edward L. (1947–) S	**E.L.Schneid.**
Schneider, Ferdinand (1834–1882) S	**Ferd.Schneid.**
Schneider, Frits (1926–) S	**Fr.Schneid.**
Schneider, Georg (1888–) S	**G.Schneid.bis**
Schneider, George (1848–1917) P	**G.Schneid.**
Schneider, Gotthard (fl. 1979) M	**Gotth.Schneid.**
Schneider, Günther (1904–) S	**Gün.Schneid.**
Schneider, Gustav (1834–1900) S	**Gus.Schneid.**
Schneider, Josef (–1885) S	**J.Schneid.**
Schneider, Joseph (Josef) (fl. 1890) S	**J.Schneid.bis**
Schneider, Karl Friedrich Robert (1798–1872) S	**K.F.R.Schneid.**
Schneider, Oskar (1841–1903) S	**O.Schneid.**
Schneider, P. (fl. 1985) S	**P.Schneid.**
Schneider, R.W. (fl. 1979) M	**R.W.Schneid.**
Schneider, Richard Conrad (1890–) S	**R.C.Schneid.**
Schneider, Roswitha (fl. 1954) M	**R.Schneid.**
Schneider, Ulrike (1936–) PS	**U.Schneid.**
Schneider, W. (fl. 1927) M	**W.Schneid.**
Schneider, Wilhelm Gottlieb (1814–1889) M	**W.G.Schneid.**
(Schneider (of Hessen Nassau), Georg)	
see Schneider, Georg	**G.Schneid.bis**

Schneider–Binder, E. (1942–) S	**Schneid.-Bind.**
Schneider–Orelli, Otto (1880–) M	**Schneid.-Or.**
Schnekker, Johannes Daniel (1794–) S	**Schnekker**
Schnell, Raymond Albert Alfred (1913–) S	**Schnell**
Schneller, Johann Jakob (1942–) MPS	**Schneller**
Schnepf, Eberhard (1931–) A	**Schnepf**
Schnetter, Reinhard (1936–) A	**Schnetter**
Schnetzler, Johann Balthasar (1823–1896) B	**Schnetzl.**
Schnittler, M. (fl. 1990) M	**Schnittler**
Schnittspahn, Georg Friedrich (1810–1865) S	**Schnittsp.**
(Schnitzlein, Adalbert Carl (Karl) Friedrich Hellwig Conrad)	
see Schnizlein, Adalbert Carl (Karl) Friedrich Hellwig Conrad	**Schnizl.**
Schnizlein, (Karl Friedrich Chrisoph) Wilhelm (1780–1856) S	**W.Schnizl.**
Schnizlein, Adalbert Carl (Karl) Friedrich Hellwig Conrad (1814–1868) M	**Schnizl.**
Schnyder, Albert (1856–1938) S	**A.Schnyd.**
Schnyder, Otto (fl. 1878) S	**O.Schnyd.**
Schobinger–Pfister, Josef (1810–1874) S	**Schob.-Pfist.**
Schoch, Emil (fl. 1903) S	**E.Schoch**
Schoch, Gottlieb (1853–1905) S	**Schoch**
Schoch–Etzensperger, Emil (1863–1945) S	**Schoch-Etz.**
Schodde, Richard (1936–) S	**Schodde**
Schoellhorn, M. (fl. 1920) M	**Schoellh.**
Schoeman, Ferdinand Reynold (1943–) A	**Schoeman**
Schoenberg, Maria Miranda (fl. 1977) S	**Schoenb.**
Schoene, (Friedrich) Kurt (1880–) S	**Schoene**
Schoenefeld, Wladimir de (1816–1875) P	**Schoenef.**
Schoener, Carol S. S	**Schoener**
Schoenfeld, G. (fl. 1924) F	**Schoenfeld**
(Schoenheit, Friedrich Christian Heinrich)	
see Schönheit, Friedrich Christian Heinrich	**Schönh.**
Schoenichen, Walther (1876–1956) AS	**Schoen.**
(Schoenlein, Johann Lucas)	
see Schönlein, Johann Lucas	**Schönl.**
Schoenlein, Philipp (1834–1856) S	**P.Schoenl.**
Schoepf, Johann David (1752–1800) S	**Schoepf**
Schofield, Joseph Robert (1868–1928) M	**Schofield**
Schofield, Wilfred Borden (1927–) B	**W.B.Schofield**
Schoknecht, Jean Donze (1943–) M	**Schokn.**
Schol–Schwarz, Marie Beatrice (1898–1969) M	**Schol-Schwarz**
Scholander, Per Fredrik (1905–1980) MS	**Schol.**
Scholler, Friedrich Adam (1718–1795) S	**Scholler**
Scholten, G. (fl. 1964) M	**Scholten**
Scholtz, Johann Eduard Heinrich (1812–1859) PS	**H.Scholtz**
Scholz, Eduard (1860–) S	**E.Scholz**
Scholz, Hildemar Wolfgang (1928–) MS	**H.Scholz**
Scholz, Ilse (fl. 1988) M	**I.Scholz**
Scholz, Joseph B. (1858–1915) S	**Scholz**
Scholz, S. (fl. 1990) S	**S.Scholz**
Scholz, U. (fl. 1981) S	**U.Scholz**

Schomburgk, Moritz Richard (1811–1891) AS	**M.R.Schomb.**
Schomburgk, Robert Hermann (1804–1865) S	**R.H.Schomb.**
Schonach, Hugo (1847–) S	**Schonach**
Schonau, Karl von (1885–1944) S	**Schonau**
Schönbeck–Temesy, E. (1930–) S	**Schönb.-Tem.**
Schönfeld, G. (fl. 1924–1947) F	**Schönfeld**
Schönfeldt, Hilmar von (fl. 1907–1913) S	**Schönf.**
Schönheit, Friedrich Christian Heinrich (1789–1870) S	**Schönh.**
Schönland, Selmar (1860–1940) S	**Schönland**
Schönlein, Johann Lucas (1793–1864) S	**Schönl.**
Schoonoord, M.P. A	**Schoon.**
Schopf, James Morton (1911–1978) F	**J.M.Schopf**
Schopf, James William (1941–) AM	**J.W.Schopf**
Schöpf, Johann David (1752–1800) S	**Schöpf**
(Schöpfer, (Edvard))	
see Brandes, Edvard	**Brandes**
Schöpfer, Franz Xaver (1777–1855) S	**Schöpfer**
Schopfer, William Henri (1900–1962) S	**W.H.Schopfer**
Schorler, Bernhard (1859–1920) A	**Schorler**
Schorn, D.P.K. (fl. 1975) S	**Schorn**
Schornherst, Ruth Olive (1905–1987) B	**Schornh.**
Schoschiaschvili, I. (fl. 1940) M	**Schosch.**
Schoser, Gustav (1924–) S	**Schoser**
Schostakowitsch, Wladimir (fl. 1895) M	**Schostak.**
Schotsman, Henriette Dorothea (1921–) S	**Schotsman**
Schott, Anton (1866–1945) S	**Ant.Schott**
Schott, Arthur Carl Victor (1814–1875) A	**A.Schott**
Schott, Heinrich Wilhelm (1794–1865) PS	**Schott**
Schotter, Georges (1922–1963) AM	**Schotter**
Schottky, Ernst Max (1888–1915) S	**Schottky**
Schoubert S	**Schoubert**
Schoulties, Calvin L. (fl. 1978) M	**Schoult.**
Schousboe, Peder Kofod Anker (1766–1832) AS	**Schousb.**
Schoute, Johannes Cornelis (1877–1942) S	**Schoute**
Schouteden, Henri (1881–) A	**Schout.**
Schouten, R.T.A. (fl. 1986) S	**R.T.A.Schouten**
Schouten, Y. (fl. 1985) S	**Y.Schouten**
Schouw, Joakim (Joachim) Frederik (1789–1852) S	**Schouw**
Schrader, Christian Friedrich (1740s–1816) S	**C.Schrad.**
Schrader, Frank Charles (1860–1940) S	**F.C.Schrad.**
Schrader, Hans–Joachim (1940–) A	**H.-J.Schrad.**
Schrader, Heinrich Adolph (1767–1836) BMPS	**Schrad.**
Schrader, Herman Friedrich A	**H.F.Schrad.**
Schrader, Johann Christian Carl (1762–1826) S	**J.C.C.Schrad.**
Schrader, Johann Eduard Julius (1809–1898) S	**J.E.J.Schrad.**
Schrader, Wilhelm (1818–1895) S	**W.Schrad.**
Schramm, Alphons (1823–1875) A	**Schramm**
Schramm, Johannes Augustin (1773–1849) S	**J.A.Schramm**

Schramm, Otto Christoph (1791–1863) S	O.C.Schramm
Schrank, Eckart A	E.Schrank
Schrank, Franz von Paula von (1747–1835) ABMPS	Schrank
Schrantz, J.P. (fl. 1961) M	Schrantz
Schreber, Johann Christian Daniel von (1739–1810) ABMPS	Schreb.
Schreck, Christophorus Jacobus (fl. 1753) S	Schreck
(Schreckenstein, Friedrich Rot von)	
see Rot von Schreckenstein, Friedrich	Rot Schreck.
Schreiber, Annelis (1927–) S	A.Schreib.
Schreiber, Beryl Olive (1911–) S	B.Schreib.
Schreiber, Franz (1839–) S	F.Schreib.
Schreiber, Hans (1859–1936) S	H.Schreib.
Schreiber, Her(r)mann Rudolph Ferdinand (1811–1853) S	Schreib.
Schreiber, L.L. (1895–) S	L.L.Schreib.
Schreiber, Marvin Mandel (1925–) S	M.M.Schreib.
Schreibers, Carl Franz Anton von (1775–1852) S	Schreibers
Schreiner, Ernst Jefferson (1902–) S	Schreiner
(Schreiner, R.)	
see Tracey, R.	Tracey
(Schrenck, Alexander Gustav von)	
see Schrenk, Alexander Gustav von	Schrenk
Schrenk, Alexander Gustav von (1816–1876) S	Schrenk
Schrenk, Hermann von (1873–1953) M	H.Schrenk
Schrenk, Joseph (von) (1842–1890) S	J.Schrenk
Schrenk, W.J. (fl. 1975) S	W.J.Schrenk
Schreurs, J. (fl. 1978) M	Schreurs
Schrire, Brian David (1953–) S	Schrire
Schröder, (Theodor) Julius (Reinhold) von (1843–1895) S	J.Schröd.
Schröder, Julius Ludwig Bruno (1867–1928) ABS	Schröd.
Schröder, Richard Iwanowitch (1822–1903) S	R.I.Schröd.
Schrödinger, Rudolf (1857–1919) S	Schrödinger
Schroeder, Alfred Oskar Julius (1873–) S	A.Schroed.
Schroeder, Fred–Günter (1930–) S	F.G.Schroed.
(Schroeder, Julius Ludwig Bruno)	
see Schröder, Julius Ludwig Bruno	Schröd.
(Schroeder, Richard Iwanowitch)	
see Schröder, Richard Iwanowitch	R.I.Schröd.
(Schroedinger, Rudolf)	
see Schrödinger, Rudolf	Schrödinger
(Schroeter, Carl Joseph)	
see Schröter, Carl Joseph	Schröt.
Schroeter, Heinrich Ernst (–1910) S	H.E.Schroet.
Schroeter, Hilde (fl. 1936) S	H.Schroet.
Schröter, Carl Joseph (1855–1939) AS	Schröt.
(Schröter, Heinrich Ernst)	
see Schroeter, Heinrich Ernst	H.E.Schroet.
Schröter, Joseph (1837–1894) M	J.Schröt.
Schrüfer, Theodor (1836–1908) F	Schrüfer

(Schtscheglov, N.P.)	
see Schtscheglow, N.P.	**Schtscheglow**
Schtscheglow, N.P. S	**Schtscheglow**
Schtschenkova, M.S. (1899–) S	**Schtschenk.**
Schtscherbina, A. (fl. 1963) S	**Schtscherb.**
Schube, Theodor (1860–1934) PS	**Schube**
Schübeler, Frederik (Fritz) Christian (1815–1892) S	**Schübeler**
Schubert, Bernice Giduz (1913–) S	**B.G.Schub.**
Schubert, Carl (fl. 1820) M	**C.Schub.**
Schubert, Gotthilf Heinrich von (1780–1860) S	**Schub.**
Schubert, L.Elliot A	**L.E.Schub.**
Schubert, Michael (1787–1860) S	**M.Schub.**
Schubert, Richard Johann (1876–) AM	**R.J.Schub.**
Schubert, Rudolph (1927–) M	**R.Schub.**
Schubert, T.S. (fl. 1989) M	**T.S.Schub.**
Schübler, Gustav (1787–1834) S	**Schübl.**
Schuchardt, (Conrad Gideon) Theodor (1829–1892) S	**Schuch.**
(Schuebler, Gustav)	
see Schübler, Gustav	**Schübl.**
Schüep, Otto (1888–) S	**Schüep**
Schüepp, Hannes (fl. 1956) M	**Schüepp**
(Schüepp, Otto)	
see Schüep, Otto	**Schüep**
Schuette, Gretchen A	**G.Schuette**
Schuette, Joachim Heinrich (1821–1908) S	**Schuette**
(Schuetz, Bogumil)	
see Schütz, Bogumil	**Schütz**
Schuez, (Georg) Emil (Carl Christoph) (1828–1877) S	**Schuez**
Schuh, Richard Edwin (1860–) A	**Schuh**
Schuiteman, A. (fl. 1982) S	**Schuit.**
Schuldt, H. (fl. 1937) S	**Schuldt**
Schüler, G. (fl. 1982) M	**G.Schüler**
Schuler, Johann Alois Ernst (1853–1946) M	**Schuler**
Schüler, Johannes A	**J.Schüler**
Schulkina, T.V. (1934–) S	**Schulkina**
Schulmann, O.von (fl. 1955) M	**Schulm.**
Schultes, H. (fl. 1852) S	**H.Schult.**
Schultes, Josef (Joseph) August (1773–1831) BS	**Schult.**
Schultes, Julius Hermann (1804–1840) S	**Schult.f.**
Schultes, Julius Hermann (1820–1887) S	**J.H.Schult.bis**
Schultes, Richard Evans (1915–) S	**R.E.Schult.**
Schultz, Arthur (1838–1915) S	**A.Schultz**
(Schultz, Benjamin)	
see Shultz, Benjamin	**Shultz**
Schultz, Carl (Karl) Friedrich (1765/6–1837) BMPS	**Schultz**
Schultz, Carl (Karl) Heinrich 'Bipontinus' (1805–1867) S	**Sch.Bip.**
Schultz, Carl (Karl) Heinrich 'Schultzenstein' (1798–1871) MPS	**Schultz Sch.**
Schultz, Eugene S. (1884–) M	**E.S.Schultz**
Schultz, Franz Johann (fl. 1785–1847) S	**F.J.Schultz**

Schultz, Friedrich Wilhelm (1804–1876) PS   **F.W.Schultz**
Schultz, G.E. (fl. 1960) S   **G.E.Schultz**
(Schultz, Karl Friedrich)
   see Schultz, Carl Friedrich   **Schultz**
Schultz, Leila M. (1946–) S   **L.M.Schultz**
Schultz, Richard (1858–1936) S   **R.Schultz**
(Schultz–Schultzenstein, Carl (Karl) Heinrich)
   see Schultz, Carl (Karl) Heinrich 'Schultzenstein'   **Schultz Sch.**
Schultze, A.G.R. (fl. 1840) S   **A.Schultze**
Schultze, Christian Friedrich (1730–1755) F   **C.Schultze**
Schultze, Edwin A.   A   **E.A.Schultze**
Schultze, Johannes Dominik (Johann Dominicus) (1752–1790) S   **J.D.Schultze**
Schultze, Maximilian Johann Siegmund (Max Sigmund) (1825–1874) A   **M.Schultze**
Schultze, Wilhelm (1881–) S   **W.Schultze**
Schultze–Motel, Jürgen (1930–) S   **J.Schultze-Motel**
Schultze–Motel, Wolfram (1934–) BS   **W.Schultze-Motel**
Schulz, A. (fl. 1852) A   **A.Schulz**
Schulz, August Albert Heinrich (1862–1922) S   **A.A.H.Schulz**
Schulz, August Gustavo (1899–) S   **A.G.Schulz**
Schulz, Dieter   A   **D.Schulz**
Schulz, Dorothea L. (1931–) S   **D.L.Schulz**
Schulz, Ellen Dorothy (1892–1970) S   **E.D.Schulz**
Schulz, Ernst   A   **E.Schulz**
Schulz, Franz (fl. 1869) S   **F.Schulz**
Schulz, Hermann (1882–1970) S   **H.Schulz**
Schulz, Johann Heinrich (1799–1870) S   **J.H.Schulz**
Schulz, Otto Eugen (1874–1936) S   **O.E.Schulz**
Schulz, Paul Franz Ferdinand (1872–1919) S   **P.F.F.Schulz**
Schulz, Richard (1904–) S   **Rich.Schulz**
Schulz, Roman (1873–1926) M   **R.Schulz**
Schulz–Danzig, Paul (fl. 1926–1935) A   **Schulz-Danzig**
Schulz–Korth, Karl (1906–1931) M   **Schulz-Korth**
Schulz–Weddigen, I.(H.) (fl. 1985) M   **Schulz-Wedd.**
Schulze, (Carolus Otto) Rudolfus (1870–) S   **R.Schulze**
Schulze, (Johann Ernst) Ferdinand (fl. 1788) S   **J.E.F.Schulze**
Schulze, Arnold Edward (1914–) S   **A.E.Schulze**
Schulze, Bruno   A   **B.Schulze**
Schulze, Carl (1865–) S   **C.Schulze**
Schulze, Carl Theodor Maximilian (1841–1915) S   **M.Schulze**
Schulze, Erwin (1861–1931) S   **E.Schulze**
Schulze, Franz Eilhard   A   **F.E.Schulze**
Schulze, Georg Martin (1906–1985) S   **G.M.Schulze**
Schulze, Joachim? (fl. 1971) S   **J.Schulze**
Schulze, Johann Heinrich (1687–1744) L   **J.H.Schulze**
Schulze, U. (fl. 1982) M   **U.Schulze**
Schulze, Walther (1875–) S   **Walt.Schulze**
Schulze, Werner (1930–) S   **Wern.Schulze**
Schulze, Wilhelm (fl. 1899) S   **Wilh.Schulze**
Schulze–Menz, Georg Karl Wilhelm (1908–1978) S   **Schulze-Menz**

Schulzer von Müggenburg, Stephan V.M. (1802–1892) M	**Schulzer**
Schumacher, Albert (1893–1975) BPS	**A.Schumach.**
Schumacher, George John (1919–) A	**G.J.Schumach.**
Schumacher, Heinrich Christian Friedrich (1757–1830) MPS	**Schumach.**
Schumacher, Th.E. (1757–1830) S	**T.E.Schumach.**
Schumacher, Trond (1949–) M	**T.Schumach.**
Schumacker, René Leonard A. (1937–) B	**Schumacker**
Schumann, Eva (1889–) PS	**E.Schum.**
Schumann, I.H. (fl. 1966) M	**I.H.Schum.**
Schumann, Julius (Heinrich Karl) (1810–1868) A	**Schum.**
Schumann, Karl Moritz (1851–1904) ABPS	**K.Schum.**
Schummel, Theodor Emil (1785–1848) S	**Schummel**
Schur, Philipp Johann Ferdinand (1799–1878) BPS	**Schur**
Schürhoff, Paul Norbert (1878–1939) S	**Schürhoff**
Schüssler, Hermann (1889–1916) A	**Schüssler**
Schussnig, Bruno (1892–) A	**Schussnig**
Schuster, Curt (1860–1935) S	**C.Schust.**
Schuster, Frederick Lee (1934–) A	**F.L.Schust.**
Schuster, Johann Constantin (Christian) (1777–1839) S	**J.C.Schust.**
Schuster, Julius (1886–1949) AMS	**J.Schust.**
Schuster, R. (1935–) S	**R.Schust.**
Schuster, Rudolf Mathias (1921–) B	**R.M.Schust.**
Schuster, S. S	**S.Schust.**
Schustler, Frantisek (1893–1925) S	**Schustler**
Schütt, Bruno (1876–1967) S	**B.Schütt**
Schütt, Franz (1859–1921) A	**F.Schütt**
Schutte, Anne Lise (1962–) S	**A.L.Schutte**
Schutte, Anne Louise (1943–) M	**Schutte**
Schütz, Bogumil (1903–) S	**Schütz**
Schutzman, B. (fl. 1988) S	**Schutzman**
Schuurman, Johanna F.M. A	**Schuurman**
Schuurmans Stekhoven, Jacobus Hermanus (Herman) (1792–1855) S	**Schuurm.Stekh.**
Schuyler, Alfred Ernest (1935–) S	**Schuyler**
Schvedtschikova, N.K. (fl. 1982) S	**Shvedtsch.**
Schwab, A.J. (fl. 1986) M	**A.J.Schwab**
Schwab, G. (fl. 1978) B	**G.Schwab**
Schwabe, Samuel Heinrich (1799–1875) ABM	**Schwabe**
Schwacke, Carl August Wilhelm (1848–1904) PS	**Schwacke**
(Schwaegrichen, Christian Friedrich)	
see Schwägrichen, Christian Friedrich	**Schwägr.**
Schwägrichen, Christian Friedrich (1775–1853) BS	**Schwägr.**
Schwaighofer, Anton (fl. 1892) S	**Schwaigh.**
Schwalb, Karl Josef W. (1842–) M	**Schwalb**
Schwanecke, Carl (1821–1916) S	**Schwan.**
Schwann, Theodor (1810–1882) S	**Schwann**
Schwantes, Martin Heinrich Gustav (Georg) (1891–1960) S	**Schwantes**
Schwappach, Adam F. (1851–1932) S	**Schwapp.**
Schwartz, Ernest Justus (1869–1939) M	**Schwartz**
Schwartz, Oskar (1901–1945) S	**O.Schwartz**

Schwarz, (Erich) Frank (1857–1928) BS   **F.Schwarz**
Schwarz, August Friedrich (1852–1915) AS   **A.F.Schwarz**
Schwarz, Cornel (1818–1860) B   **Schwarz**
Schwarz, Elizabeth de Araujo (fl. 1981) S   **E.A.Schwarz**
Schwarz, Elisabeth A   **E.Schwarz**
Schwarz, Ernest Hubert Lewis (1873–1928) A   **E.H.L.Schwarz**
Schwarz, Fritz (fl. 1849) S   **Fritz Schwarz**
Schwarz, K. A   **K.Schwarz**
Schwarz, Marie Beatrice (1898–) M   **M.B.Schwarz**
Schwarz, N.V. (fl. 1927) M   **N.V.Schwarz**
Schwarz, Otto Karl Anton (1900–1983) PS   **O.Schwarz**
Schwarze, Carl Alois (1886–1956) M   **Schwarze**
Schwarzenbach, Fritz (1894–) S   **F.Schwarzenb.**
Schwarzenbach, Marthe (1900–) S   **M.Schwarzenb.**
Schwarzer, Carl Ferdinand (1829–1870) S   **Schwarzer**
Schwarzman, S.R. (fl. 1964) M   **Schwarzman**
(Schwarzmann, S.R.)
    see Schwarzman, S.R.   **Schwarzman**
Schwarzová, Terézia (1938–) S   **Schwarzová**
Schwatka, Frederick (1849–1892) S   **Schwatka**
Schweers, A.C.S. (fl. 1940) M   **Schweers**
Schwegler, H.-W. (1929–) S   **Schwegler**
Schwegler, J. (fl. 1978) M   **J.Schwegler**
Schwegman, J.E. (fl. 1982) S   **Schwegman**
Schweickerdt, Herold Georg Wilhelm Johannes (1903–1977) S   **Schweick.**
Schweiger, Hans George (1927–1986) A   **Schweiger**
Schweiger, M. A   **M.Schweiger**
Schweigger, August Friedrich (1783–1821) AS   **Schweigg.**
Schweik S   **Schweik**
Schweinfurth, Charles (1890–1970) S   **C.Schweinf.**
Schweinfurth, Georg August (1836–1925) PS   **Schweinf.**
Schweinitz, Lewis (Ludwig) David von (1780–1834) BMS   **Schwein.**
Schweizer, G. (fl. 1923) M   **G.Schweiz.**
Schweizer, Jakob (1885–) S   **Schweiz.**
Schweizer, K. (fl. 1921) M   **K.Schweiz.**
Schwencke, Martinus Wilhelmus (1707–1785) S   **Schwencke**
Schwendener, Simon (1829–1919) M   **Schwend.**
Schwenkel, Hans (1886–1957) S   **Schwenkel**
Schwere, Siegfried (1964–) S   **Schwere**
Schwerin, Fritz Kurt Alexander von (1856–1934) S   **Schwer.**
Schwerin, Karl H. S   **K.H.Schwer.**
Schwers, Henri A   **Schwers**
Schwertschlager, Joseph (1853–1924) S   **Schwertschl.**
Schweykert, J.M. (fl. 1791) S   **Schweyk.**
Schwickerath, Mathias Friederich (1892–1974) S   **Schwick.**
Schwier, Heinz (1881–1955) S   **Schwier**
Schwilgué, Charles Joseph Antoine (1774–1808) S   **Schwilgué**

Schwimmer, J. (1879–1959) S   **Schwimmer**
Schwinn, F.J. (fl. 1974) M   **Schwinn**
Schwöbel, H. (fl. 1969) M   **Schwöbel**
Schychowsky, Iwan (fl. 1832) S   **Schych.**
Sclavo, J.P. (fl. 1989) S   **Sclavo**
Scoffern, John (1814–1882) S   **Scoffern**
Scofield, Carl Schurz (1875–1965) S   **Scofield**
Scoggan, Homer John (1911–1986) P   **Scoggan**
Scopoli, Joannes Antonius (Giovanni Antonio) (1723–1788) AMPS   **Scop.**
Scora, Rainer Walter (1928–) S   **Scora**
Scortechini, Benedetto (1845–1886) S   **Scort.**
Scott, Andrew John (1950–) S   **A.J.Scott**
Scott, Arthur Moreland (1888–1963) AS   **A.M.Scott**
Scott, Charles Leslie (1913–) S   **C.L.Scott**
Scott, De B. (fl. 1976) M   **D.B.Scott**
Scott, Dukinfield Henry (1854–1934) FS   **D.H.Scott**
Scott, Edith Bohrer (1928–) B   **E.B.Scott**
(Scott Elliot, George Francis)
    see Scott-Elliot, George Francis   **Scott-Elliot**
Scott, Fiona J. A   **F.J.Scott**
Scott, George Anderson Macdonald (1933–) B   **G.A.M.Scott**
Scott, H.H. (fl. 1926) FM   **H.H.Scott**
Scott, John (1838–1880) PS   **J.Scott**
Scott, Kenneth John (1933–) F   **K.J.Scott**
Scott, Munro Briggs (1887–1917) S   **M.B.Scott**
Scott, Peter John (1948–) S   **P.J.Scott**
Scott, Richard A. (fl. 1956) M   **R.A.Scott**
Scott, Robert Robinson (1827–1877) PS   **R.R.Scott**
Scott, Robert W. (1936–) A   **R.W.Scott**
Scott, William Wallace (1920–) M   **W.W.Scott**
Scott-Elliot, George Francis (1862–1934) MS   **Scott-Elliot**
Scotti, Gilberto (1818–1880) S   **Scotti**
Scouler, John (1804–1871) S   **Scouler**
Scriba, Julius Karl (1848–1905) FPS   **J.Scriba**
Scriba, Ludwig Philipp Karl (1847–1933) M   **L.Scriba**
Scribner, Frank Lamson– (1851–1938) MS   **Scribn.**
Scrivenor, Robert (1848–1918) S   **Scriv.**
Scrugli, Antonio (fl. 1984) S   **Scrugli**
Sculczewski, J.W. (fl. 1952) M   **Sculcz.**
Scully, Reginald William (1858–1935) S   **Scully**
Scurti, J.C. (fl. 1958) M   **Scurti**
Scutari, N.C. (fl. 1991) M   **Scutari**
Sdobnina, L.I. (fl. 1973) S   **Sdobnina**
Seagrief, Stanley Charles (1927–) A   **Seagrief**
Sealy, Joseph Robert (1907–) S   **Sealy**
Seaman, Frederick C. (1948–) S   **Seaman**
Searle, G.O. (fl. 1921) M   **Searle**
Searles, Richard Brownlee (1936–) A   **Searles**
Sears, James R. A   **Sears**

Seaton, Henry Eliason (1869–1893) PS	**Seaton**
Seaver, Fred Jay (1877–1970) M	**Seaver**
Sebald, Oskar (1929–) S	**Sebald**
Sebastiani, Francesco Antonio (1782–1821) S	**Sebast.**
Sebastine, Kunju Mathew (1918–1967) S	**Sebastine**
Šebek, Svatopluk (fl. 1953) M	**Šebek**
Sebeók de Szent–Miklós, Alexander (Sándor) (fl. 1780) S	**Sebeók**
Seberg, Ole (1952–) S	**Seberg**
Sebert, Hippolyte (1839–1930) S	**Sebert**
Sebestyén, Olga A	**Sebestyén**
Sebille, René Léon (1851–1938) B	**Sebille**
Sébire, Albert (1863–1936) S	**Sébire**
Seboth, Joseph (1816–1883) S	**Seboth**
Sebring, Mary M. A	**Sebring**
Sebsebe Demissew (1953–) S	**Sebsebe**
Secall y Inda, José (1853–1918) S	**Secall**
Secco, R.de S. (fl. 1985) S	**Secco**
Sechet, M. (fl. 1953) M	**Sechet**
Séchier, P. (fl. 1835) M	**Séchier**
Sechkina, T.V. A	**Sechkina**
Seckt, Heinrich Karl Felix Hans (1879–1953) AS	**Seckt**
Secondat, Jean Baptiste de (1716–1796) S	**Secondat**
Secretan, (Gabriel–Abraam–Samuel–Jean) Louis (1758–1839) M	**Secr.**
Sedelnikova, N.V. (fl. 1984) M	**Sedeln.**
Sedgwick, Leonard John (1883–1925) BMS	**Sedgw.**
Seefried, F. (fl. 1908) S	**Seefried**
Seeland, Hermann (1868–1954) S	**Seeland**
Seeler, Edgar Viguers (1908–) M	**Seeler**
Seeliger, Rudolf Heinrich (1889–1943) S	**Seeliger**
Seeligmann, Peter (1923–) S	**Seeligm.**
Seelos, Gustav (1831–1911) P	**Seelos**
Seemann, Berthold Carl (1825–1871) BPS	**Seem.**
Seemann, Wilhelm Eduard Gottfried (–1868) S	**W.Seem.**
Seemen, Karl Otto von (1838–1910) S	**Seemen**
Seenayya, G. A	**Seenayya**
Seenus, Josef von (fl. 1799–1805) S	**Seenus**
Seethalakshmi, V. (fl. 1953) M	**Seethal.**
Seetharam, Y.N. (fl. 1982) S	**Seeth.**
Seetzen, Ulrich Jaspar (1767–1811) S	**Seetzen**
Segadas–Vianna, Fernando S	**Seg.-Vianna**
Segal, Ronald Henry (1940–) M	**Segal**
Segar, E.C.Margaret A	**Segar**
Segawa, Sokichi (1904–1960) A	**Segawa**
Segedin, Barbara P. (1923–) M	**Segedin**
Segelberg, Ivar (1914–) S	**Segelb.**
Segerstedt, Per Simeon Julius (1866–c.1930) S	**Segerst.**
Segi, Toshio (1914–1979) A	**Segi**
Segonzac, Geneviève A	**Segonzac**
Segret, L. (1867–1949) S	**Segret**

Segretain, G. (fl. 1959) M — **Segretain**
Segroves, Kenneth Lee (1938–) A — **Segroves**
Séguier, Jean François (1703–1784) MPS — **Ség.**
Segura, José Carmen (1846–1906) S — **Segura**
Segura Zubizarreta, Antonio (1921–) S — **A.Segura**
Sehgal, H.S. (fl. 1964) M — **H.S.Sehgal**
Sehgal, S.P. (fl. 1966) M — **S.P.Sehgal**
Sehlmeyer, Johann Friedrich (1788–1856) BMS — **Sehlm.**
Sehnem, Aloysio (1912–1981) BP — **Sehnem**
Seibert, Russell Jacob (1914–) S — **Seibert**
Seibert, Z. (fl. 1988) P — **Z.Seibert**
Seibt, D. (fl. 1988) M — **Seibt**
Seidel, Carl A.J. (1858–) S — **C.A.Seidel**
Seidel, Christoph Friedrich (fl. 1869) S — **C.F.Seidel**
Seidel, Johann Heinrich (fl. 1779) S — **Seidel**
Seidel, O.M. (1841–1917) S — **O.M.Seidel**
Seidel, Otto (fl. 1890) S — **O.Seidel**
Seidel, Traugott (Jacob Hermann) (1833–1896) S — **T.Seidel**
(Seidel, Wenzel Benno)
   see Seidl, Wenzel Benno — **Seidl**
Seidenfaden, Gunnar (1908–) AS — **Seidenf.**
Seidenschnur, Christiane Eva (later Anderson, C.E.) (1944–) S — **Seid.**
Seidensticker, Ute (fl. 1988) S — **Seidenst.**
Seidl, Michelle T. (fl. 1989) M — **M.T.Seidl**
Seidl, Wenzel Benno (1773–1842) MS — **Seidl**
Seidlitz, Nicolai Karlovič (Karl Samuel) (1831–1930) S — **Seidlitz**
Seifert, Keith A. (fl. 1979) M — **Seifert**
Seifriz, William Ernest (1888–1955) S — **Seifriz**
Seigle-Murandi, Françoise (fl. 1976) M — **Seigle-Mur.**
Seigler, David Stanley (1940–) S — **Seigler**
Seiler, Jean (1878–1939) S — **Seiler**
Seithe, Almut (fl. 1962) S — **Seithe**
Seits, Tobias Anton (1772–1833) S — **Seits**
Seitz, Ludwig (1792–1866) S — **Seitz**
Seitz, W. (1940–) P — **W.Seitz**
Sejfulin, E.M. (1936–) S — **Sejfulin**
Sekar, G. (fl. 1986) M — **Sekar**
Sekera, Wenzel Johann Vaclav Jan (1815–1875) S — **Sekera**
Seki, Tarow (1934–) B — **Seki**
Sekunova, V.N. (fl. 1960) M — **Sekunova**
Selander, Nils Sten Edvard (1891–1957) PS — **Selander**
Selbstherr, Carl (fl. 1832) S — **Selbsth.**
Selby, Augustine Dawson (1859–1924) M — **A.Selby**
Selby, Prideaux John (1788–1867) S — **P.Selby**
Seleban de Cattani, Maria   S — **Seleb.Catt.**
Seler, Eduard Georg (1849–1922) S — **Seler**
Selett, J.W. (fl. 1941) P — **Selett**
Selig, Christoph Wilhelmus (fl. 1802) S — **Selig**
Seliger, A. (fl. 1987) S — **A.Seliger**

Seliger, Ignaz (1752–1812) BC	**Seliger**
Seligmann, Johann Michael (1720–1762) S	**Seligm.**
Seligo, Arthur A	**Seligo**
Selin, Gustaf (1836–1862) S	**Selin**
Selivanova-Gorodkova, Elena Alexandrovna (1902–) S	**Seliv.-Gor.**
Seljak, G. (fl. 1988) M	**Seljak**
Selk, Heinrich (fl. 1908) A	**Selk**
Selkirk, D.R. (fl. 1972) M	**Selkirk**
Sell, Yves (fl. 1962–) S	**Y.Sell**
Sell, Peter Derek (1929–) S	**P.D.Sell**
Selland, Sjur Knutsen (1867–1920) S	**Selland**
Sellar, P.W. (fl. 1966) M	**Sellar**
Selling, Olof Hugo (1917–) FPS	**Selling**
(Sello, Friedrich)	
see Sellow, Friedrich	**Sellow**
Sello, Hermann Ludwig (1800–1876) S	**Sello**
Sellow, Friedrich (1789–1831) S	**Sellow**
Selma Fernández, Caridad (fl. 1990) S	**Selma**
Sélys-Longchamps, Michel Edmond de (1813–1900) S	**Sél.-Longch.**
Semadeni, Francesco Ottavio (fl. 1904) M	**Semadeni**
Semecnik, A. (fl. 1967) M	**Semecnik**
Semen, E.O. (fl. 1968) M	**E.O.Semen**
Semeniuk, George (1910–) M	**Semeniuk**
Semenov, B.B. (fl. 1921) B	**Semenov**
Semenov-Tjan-Schansky, Peter Petrovich von (1827–1914) S	**Semen.**
Semenova, M.N. S	**Semenova**
Semenova-Tjan-Schanskaya, Neonila Zenonovna (1906–1960) S	**N.Semenova**
(Semenow-Tjan-Shansky, Peter Petrowitsch von)	
see Semenov-Tjan-Schansky, Peter Petrovich von	**Semen.**
(Semenowa-Tjan-Shanskaja, Neonila Zenonovna)	
see Semenova-Tjan-Schanskaya, Neonila Zenonovna	**N.Semenova**
Semer, C.R. (fl. 1987) M	**Semer**
Sëmina, G.I. (H.J.) A	**Sëmina**
(Semiotrocheva, N.L.)	
see Semiotroczeva, N.L.	**Semiotr.**
Semiotroczeva, N.L. (1928–) S	**Semiotr.**
Semir, João (1937–) AS	**Semir**
Semler, C.L. (1875–1955) S	**Semler**
Semon, H.C. (fl. 1922) M	**Semon**
Semple, John Cameron (1947–) S	**Semple**
(Sen Gupta, G.)	
see Sengupta, G.	**Sengupta**
Sen, Tuhinsri (1942–) PS	**T.Sen**
Sen, Udayananda (1933–) PS	**U.Sen**
Senaratna, S.D.J.E. (fl. 1956) S	**Senaratna**
Senay, Pierre (1892–1954) S	**Senay**
Sencke, Ferdinand (fl. 1830–1866) S	**Sencke**
Senckenberg, Johann Christian (1707–1772) S	**Senck.**
Sendtner, G. S	**G.Sendtn.**

Sendtner, Otto (1813–1859) BS — **Sendtn.**
Sendulsky, Tatiana (1922–) S — **Send.**
Senebier, Jean (1742–1809) S — **Seneb.**
Senesse, S. (fl. 1988) S — **Senesse**
Senéz, A. (fl. 1913) M — **Senéz**
Senft, (Philip) Emanuel (1870–1922) M — **E.Senft**
Senft, Christian Carl Friedrich Ferdinand (1810–1893) S — **Senft**
Senghas, Karlheinz (1928–) S — **Senghas**
Sengupta, G. (1935–) S — **Sengupta**
(Senilis, Johannes)
  see Nelson, John — **J.Nelson**
Senior, Robert Michael (1881–) S — **Senior**
Senjaninova-Korczagina, M.v. (1900–1966) S — **Senjan.-Korcz.**
Senn, Gustav Alfred (1875–1945) AS — **Senn**
Senn, Harold Archie (1912–) S — **H.Senn**
Senn–Irlet, Beatrice (1954–) M — **Senn–Irlet**
Senna, Pedro Américo Cabral   A — **Senna**
Sennen, frère (1861–1937) PS — **Sennen**
Sennholz, Gustav (1850–1895) S — **Sennholz**
Senni, Lorenzo (1879–1954) S — **Senni**
Senoner, Adolf (1806–1895) S — **Senoner**
Seoane-Camba, Juan A. (1933–) AS — **Seoane-Camba**
Seppelt, Rodney David (1945–) BS — **Seppelt**
Sequeira, Eduardo (1861–) B — **Seq.**
Serafimov, V.S.   S — **Seraf.**
Serafino, Edoardo   S — **Serafino**
Şerbanescu, Ioan (1903–) AS — **Şerb.**
Şerbănescu, Mariya   A — **M.Şerb.**
(Serbinov, J.S.)
  see Serbinow, Ivan L'rovich — **Serbinow**
Serbinow, Ivan L'rovich (1872–1950) M — **Serbinow**
Serdyukova, L.B. (fl. 1973) S — **Serdyuk.**
Serea, Constantin (fl. 1962) M — **Serea**
Serebrianikow, J. (fl. 1912) M — **Serebrian.**
Serebrjakova, T.J. (1893–) S — **Serebr.**
Seredin, R.M. (1912–) S — **Seredin**
Sergeev, Marguérite (1877–) S — **Sergeev**
Sergeeva, Klavdia Spiridonovna (1909–) M — **Sergeeva**
(Sergejeva, Klavdia Spiridonovna)
  see Sergeeva, Klavdia Spiridonovna — **Sergeeva**
Sergent, Étienne Louis Marie Edmond (1876–) A — **Sergent**
Sergienko, L.A. (1950–) S — **L.A.Sergienko**
Sergienko, V.G. (1947–) S — **V.G.Sergienko**
Sergievskaja, Lydia Palladievna (1897–1970) S — **Serg.**
Sergievskaja, Ye.V. (1926–1985) S — **Ye.V.Serg.**
Sergio Costa Gomes, Cecília Loff Pereira (1942–) B — **Sergio**
(Serguéeff, Marguérite)
  see Sergeev, Marguérite — **Sergeev**
(Serguiejeva, Klavdia Spiridonovna)
  see Sergeeva, Klavdia Spiridonovna — **Sergeeva**

Serieyssol, Karen K. A	**Serieyssol**
Seringe, Nicolas Charles (1776–1858) S	**Ser.**
Seriu, Yoshihiro (fl. 1987) M	**Seriu**
Serizawa, Shunsuke (1948–) PS	**Seriz.**
Sériziat, Charles Victor Émile (1835–) S	**Sériziat**
(Serna, Irene E. La)	
see La Serna Ramos, Irene E.	**La Serna**
Sernander, Johan Rutger (1866–1944) MS	**Sern.**
Sernari, Mauro (fl. 1987) M	**Sernari**
Serov, V.P. (1957–) S	**Serov**
Serpette, Maurice A	**Serpette**
Serpukhova, Vera Ivanovna (1896–) S	**Serp.**
Serrano, Felicisimo B. (fl. 1940) M	**Serrano**
Serrão, Alves S	**A.Serrão**
Serrato–Valenti, Giorgina (1930–) S	**Serrato**
Serres, (Pierre) Marcel (Toussaint) de (1780/83–1862) S	**M.Serres**
Serres, Jean Jacques (1790–1858) S	**J.Serres**
Serrurier, Lindor (1846–1901) S	**Serrurier**
Sérusiaux, Emmannuël (1953–) M	**Sérus.**
Servais, Gaspard Joseph de (1735–1807) S	**Servais**
Servazzi, Ottone (fl. 1934) M	**Servazzi**
Serve, L. (fl. 1971) S	**Serve**
Servettaz, Camille (1870–1947) S	**Servett.**
Service, Michael William (fl. 1950) M	**Service**
Servít, Miroslav (1886–1959) BM	**Servít**
Seshadri, V.S. (fl. 1965) M	**Seshadri**
Sesler, Leonard (–1785) S	**Sesl.**
Sesma, Begoña (fl. 1978) M	**Sesma**
(Sessé, Martín)	
see Sessé y Lacasta, Martín	**Sessé**
Sessé y Lacasta, Martín (1751–1808) PS	**Sessé**
Sestini, Domenico (1750–1832) S	**Sest.**
Setchell, William Albert (1864–1943) ABMS	**Setch.**
Setchkarev, B.I. (1904–) S	**Setchk.**
Seth, Christopher J. (fl. 1984) S	**C.J.Seth**
Seth, H.K. (fl. 1971) M	**Seth**
Sethy, P.K. (fl. 1987) M	**Sethy**
Setliff, E.C. (fl. 1972) M	**Setliff**
Šetlík, Ivan A	**Šetlík**
Seto, Ryozo (1918–) A	**Seto**
Sette, Vincenzo (1785–1827) S	**Sette**
Setten, A.K.van (fl. 1985) S	**Setten**
Setty, M.G.A.P. (fl. 1986) M	**Setty**
Seubert, Moritz August (1818–1878) ABPS	**Seub.**
Seurat, Léon Gaston (1872–) S	**Seurat**
Sevastianov, A.F. (1782–1821) S	**Sevast.**
Severin, Viorel (fl. 1955) M	**Severin**
Severini, Guiseppe (1878–1918) M	**Severini**
Severo, Branca Maria Aimi (fl. 1982) S	**Severo**
Seward, Albert Charles (1863–1941) ABFP	**Seward**

Seyani, Jameson Henry (1948–) S	**Seyani**
Seybold, Siegmund Gerhard (1939–) S	**Seybold**
Seydler, Friedrich Wilhelm (1810–1897) S	**Seydler**
Seyfarth, H. (fl. 1980) M	**Seyfarth**
Seymann, Wilhelm (1887–1915) PS	**Seymann**
Seymour, Arthur Bliss (1859–1933) CM	**Seym.**
Seymour, C.P. (fl. 1963) M	**C.P.Seym.**
Seymour, Edward Loomis Davenport (1888–1956) S	**E.L.D.Seym.**
Seymour, Frank Conkling (1895–1985) PS	**F.Seym.**
Seymour, Roland Lee (1939–) M	**R.L.Seym.**
(Seynes, Jules De)	
see De Seynes, Jules	**De Seynes**
Seyot, Pierre (1876–1942) M	**Seyot**
Sha, Shi Gui (fl. 1981) S	**S.G.Sha**
Sha, Wen Lan (fl. 1981) S	**W.L.Sha**
Shabbir Ali, S. A	**Shabbir Ali**
Shabetai, Joseph Raphael (1895–) S	**Shab.**
Shadbolt, George (1817–1901) A	**Shadbolt**
Shadomy, Jean H. (fl. 1960s) M	**Shadomy**
Shafer, John Adolph (1863–1918) PS	**Shafer**
Shaffer, Robert Lynn (1929–) M	**Shaffer**
Shäffer-Fehre, Monika (1937–) S	**Shäffer-Fehre**
Shafik, Samir A	**Shafik**
Shagufta, S. (fl. 1979) M	**Shagufta**
Shah, G.L. (fl. 1969) S	**G.L.Shah**
Shah, Jehandar (1948–) S	**J.Shah**
Shah, M.B. S	**Shah**
Shahori (fl. 1980) S	**Shahori**
Shaikin, I.M. A	**Shaikin**
Shain, Louis (fl. 1988) M	**Shain**
Shaji, C. A	**Shaji**
Shakhmedov, I.Sh. (fl. 1977) S	**Shakhm.**
Shakirova, R.Yu. (fl. 1982) M	**Shakirova**
Shakya, P.R. (fl. 1970) S	**Shakya**
Shaler, Nathaniel Southgate (1841–1906) S	**Shaler**
Shallom, L.J. (fl. 1969) F	**Shallom**
Shalyt, Mikhail Solomonovich (1904–) S	**Shalyt**
Shambi, H.S. (fl. 1983) M	**Shambi**
Shameel, Mustafa A	**Shameel**
Shamraj, I.A. A	**Shamraj**
Shamsi, S. (fl. 1986) M	**Shamsi**
Shamsiev, S.Sh. (fl. 1982) M	**Shamsiev**
Shan, Ren Hwa (1909–) S	**R.H.Shan**
Shang, Chi(h) Bei (1935–) S	**C.B.Shang**
Shang, Yan Zhong (fl. 1987) M	**Y.Z.Shang**
Shang, Zong Yan (fl. 1985) S	**Z.Y.Shang**
Shangguan, Tie Liang (fl. 1984) M	**Shangguan**
(Shangin, Petr Ivanovich)	
see Schangin, Petr Ivanovich	**Schangin**
Shankar, Guri (fl. 1971) M	**Shankar**

Shankarnarayan, K.A. (fl. 1964) S	**Shank.**
Shannon, William Cummings (1851–1905) S	**Shannon**
Shanor, Leland (1914–) AM	**Shanor**
Shantz, Homer LeRoy (1876–1958) AS	**Shantz**
Shao, Jian Zhang (fl. 1986) MS	**J.Z.Shao**
Shaparenko, K.K. (1908–1941) S	**Shap.**
Shapiro, I.A. (fl. 1987) M	**Shapiro**
Sharapov, V.M. (fl. 1974) M	**Sharapov**
Sharda, R.M. (fl. 1984) M	**Sharda**
Sharipov, A. S	**Sharipov**
Sharipova, B.A. (fl. 1978) S	**Sharipova**
Sharma, A.D. (fl. 1978) M	**A.D.Sharma**
Sharma, Arun Kumar (1924–) M	**A.K.Sharma**
Sharma, B.M. (fl. 1985) M	**B.M.Sharma**
Sharma, Brahma Dutta (1935–) FMS	**B.D.Sharma**
Sharma, H.C. A	**H.C.Sharma**
Sharma, J.K. (fl. 1985) M	**J.K.Sharma**
Sharma, K.R. (fl. 1973) M	**K.R.Sharma**
Sharma, Lokendra R. (fl. 1990) M	**L.R.Sharma**
Sharma, M.B. (fl. 1982) M	**M.B.Sharma**
Sharma, Mahendra P. (1946–) M	**M.P.Sharma**
Sharma, Mithilesh (fl. 1979) S	**M.Sharma**
Sharma, N.D. (fl. 1974) M	**N.D.Sharma**
Sharma, O.P. (fl. 1957) M	**O.P.Sharma**
Sharma, R.M. (fl. 1985) M	**R.M.Sharma**
Sharma, Raghunandan (fl. 1981) M	**R.Sharma**
Sharma, S.K. (fl. 1985) S	**S.K.Sharma**
Sharma, S.L. (fl. 1972) M	**S.L.Sharma**
Sharma, S.P. A	**S.P.Sharma**
Sharma, V.S. (1931–) S	**V.S.Sharma**
Sharma, Veena (fl. 1987) M	**V.Sharma**
Sharman, Percy J. (fl. 1916) S	**Sharman**
Sharp, Aaron John ('Jack') (1904–) BMS	**Sharp**
Sharp, Seymour Sereno (1893–) S	**S.S.Sharp**
Sharp, Ward McClintic (1904–) S	**W.M.Sharp**
Sharpe, Philip Ridley (1915–) S	**Sharpe**
Sharples, Arnold (1887–1937) M	**Sharples**
Sharsmith, Carl William (1903–) S	**Sharsm.**
Sharsmith, Helen Katherine (1905–1982) S	**H.Sharsm.**
Shastry, S.V.S. (fl. 1961) S	**Shastry**
Shaulo, D.N. (1954–) S	**Shaulo**
Shaver, Jesse Milton (1888–1961) PS	**Shaver**
Shaw, Arthur Jonathan (fl. 1982) BS	**A.J.Shaw**
Shaw, Charles Gardner (1917–) M	**C.G.Shaw**
Shaw, Dorothy E. (fl. 1952) M	**D.E.Shaw**
Shaw, Elizabeth Anne (1938–) S	**E.A.Shaw**
Shaw, F.W.A. (fl. 1927) M	**F.W.A.Shaw**
Shaw, Frederick John Freshwater (1886–1936) M	**F.J.F.Shaw**
Shaw, George (1751–1813) S	**G.Shaw**

Shaw, George Russell (1848–1937) S   **Shaw**
Shaw, Henry (1800–1889) S   **H.Shaw**
(Shaw, Herbert Kenneth Airy)
    see Airy Shaw, Herbert Kenneth   **Airy Shaw**
Shaw, John (1837–1890) B   **J.Shaw**
Shaw, L.A. (fl. 1988) M   **L.A.Shaw**
Shaw, R.B. (fl. 1984) S   **R.B.Shaw**
Shaw, Richard J. (1923–) S   **R.J.Shaw**
Shaw, Walter Robert (1871–) A   **W.Shaw**
Shawhan, Fae McClung (1903–) A   **Shawhan**
Shayam, K. (fl. 1972) M   **Shayam**
(Shchegleev, Serge S.)
    see Stscheglejew, Serge S.   **Stschegl.**
Shchegolev, A.K. S   **Shcheg.**
Shchelokova, I.F. (fl. 1979) M   **Shchelok.**
Shcherbak, V.P. (fl. 1977) M   **Shcherbak**
Shcherban, E.P. (fl. 1977) M   **Shcherban**
Shcherbin-Parfenenko, A.L. (fl. 1953) M   **Shcherb.-Parf.**
Shcherbina, M.L. S   **Shcherbina**
Shchukin, G. (fl. 1985) M   **Shchukin**
Shchumenko, S.I. A   **Shchum.**
Shear, Cornelius Lott (1865–1956) MS   **Shear**
Sheard, John Wilson (1940–) M   **Sheard**
Shearer, Carol Ann (1941–) M   **Shearer**
Sheath, Robert G. (1950–) A   **Sheath**
Shebalina, M.A. (fl. 1985) S   **Shebalina**
Shecut, John (Linnaeus Edward Whitridge) (1770–1836) S   **Shecut**
Sheeba, T. (fl. 1990) M   **Sheeba**
Sheffy, M.V. (fl. 1970) M   **Sheffy**
Sheh, Men Lan (fl. 1986) S   **M.L.Sheh**
Shehata, A.M.El-Tabey (fl. 1956) M   **Shehata**
Sheikh, M.T. (fl. 1971) MS   **M.T.Sheikh**
Sheikh, Muhammad Yusuf (1939–) S   **M.Y.Sheikh**
Sheldon, Edmund Perry (1869–1947) S   **E.Sheld.**
Sheldon, John Lewis (1865–1947) MS   **J.Sheld.**
Shemberg, M.A. (fl. 1985) S   **Shemberg**
Shen, Bao An (fl. 1988) S   **B.A.Shen**
Shen, Kuan Mien (fl. 1970) S   **K.M.Shen**
Shen, Lian Dai (Tai) (fl. 1970) S   **L.D.Shen**
(Shen, Qi Yi)
    see Chen, C.I.   **C.I.Chen**
Shen, Ren Hwa (fl. 1986) S   **R.H.Shen**
Shen, Xian Sheng (fl. 1989) S   **X.S.Shen**
Shen, Yan Zhang (fl. 1980) M   **Y.Z.Shen**
Shen, Yin Wu (fl. 1989) A   **Y.W.Shen**
Shen, Yu Feng A   **Y.F.Shen**
Shen, Zu An (fl. 1984) S   **Z.A.Shen**
Shende, D.V. (fl. 1962) P   **Shende**

Sheng, Guo Ying (1936–) S	**G.Y.Sheng**
Sheng, Shi Fa (fl. 1990) M	**S.F.Sheng**
Shepeleva, E.D.　A	**Shepeleva**
Shephard, K.S. (fl. 1946) M	**Shephard**
Shepherd, Audrey M. (fl. 1955) M	**A.M.Sheph.**
Shepherd, Henry (c.1783–1858) S	**H.Sheph.**
Shepherd, J.D. (fl. 1981) S	**J.D.Sheph.**
Shepherd, John (1764–1836) S	**Sheph.**
Shepherd, K. (fl. 1964) S	**K.Sheph.**
Shepherd, Thomas (1779–1835) S	**T.Sheph.**
Shepley, E.Ann　A	**Shepley**
Sheppard, (Mrs.) (fl. 1783–1867) S	**Sheppard**
Sherard, William (1659–1728) L	**Sherard**
Sherbakoff, Constantine Demetry (Dmitriev) (1878–1965) M	**Sherb.**
Sherff, Earl Edward (1886–1966) S	**Sherff**
Sheridan, John Edmund (1937–) M	**Sheridan**
Sherif, A.S. (1950–) S	**Sherif**
Sherley, James L.　A	**Sherley**
Sherrard, Elizabeth M. (1904–) B	**Sherrard**
Sherrin, William Robert (1871–1955) B	**Sherrin**
Sherwin, Helen Shedd (fl. 1948) M	**Sherwin**
Sherwood, Martha Allen (fl. 1985) M	**Sherwood**
Sherwood, Ronald W.　A	**R.W.Sherwood**
(Sherwood–Pike, Martha Allen)	
see Sherwood, Martha Allen	**Sherwood**
Sheshegova, L.I. (fl. 1975) A	**Shesh.**
Sheshukova–Poretzkaja, V.S.　A	**Shesh.–Por.**
(Sheshukova–Poretzkaya, V.S.)	
see Sheshukova–Poretzkaja, V.S.	**Shesh.–Por.**
Shete, R.H. (fl. 1978) FM	**Shete**
Shete, S.G.　A	**S.G.Shete**
Shetler, Stanwyn Gerald (1933–) S	**Shetler**
Shetty, Brahmavar Vishwanath (1931–) S	**B.V.Shetty**
Shetty, H.S. (fl. 1980) M	**H.S.Shetty**
Shetty, K.Shivappa (fl. 1966) M	**K.S.Shetty**
Sheviak, Charles J. (1947–) S	**Sheviak**
Shevock, James R. (1950–) S	**Shevock**
Shi, Zhi Xin (1940–) A	**Z.X.Shi**
Shi, Zhi Yong (fl. 1984) M	**Z.Y.Shi**
Shibaki, Hideomi (fl. 1988) M	**Shibaki**
Shibasaki, Y. (fl. 1934) M	**Shibas.**
Shibata, Keita (1877–1949) S	**Shibata**
Shibkova, I.F. (fl. 1957) S	**Shibkova**
Shibkova, K.G.　A	**K.G.Shibkova**
Shibuichi, Heinai (fl. 1970) M	**Shibuichi**
Shibukawa, K. (fl. 1954) M	**Shibuk.**
Shieh, Wang Chueng (1929–) PS	**W.C.Shieh**
Shifrine, M. (fl. 1956) M	**Shifrine**
Shih, Chu (1932–) S	**C.Shih**

Shih, Wen Liang S	**W.L.Shih**
Shih, You Kuang (fl. 1936) M	**Y.K.Shih**
Shikovskii, I. S	**Shik.**
Shiller, Ivan (1895–) S	**Shiller**
Shim, Phyau Soon (fl. 1982) S	**Shim**
Shimabukuro, Shunichi (fl. 1957) M	**Shimab.**
Shimada, Shoichi (fl. 1930) M	**Shimada**
Shimadzu, Tadashige (fl. 1921) S	**Shimadzu**
Shimazaki, Togo A	**Shimaz.**
Shimazu, Mitsuaki (fl. 1988) M	**Shimazu**
Shimek, Bohumil (1861–1937) PS	**Shimek**
Shimizu, Daisuke (1915–) BM	**Shimizu**
Shimizu, Tatemi (1932–) S	**T.Shimizu**
Shimotomai, Naomasa (1899–) S	**Shimot.**
Shimura, Yoshio (fl. 1979) P	**Shimura**
Shin, D.S. (fl. 1988) M	**D.S.Shin**
Shin, Hyeon Dong (fl. 1988) M	**H.D.Shin**
Shin, Toshio (1917–1982) BS	**Shin**
(Shing, Gung Hsia)	
see Shing, Kung Hsieh (Hsia)	**K.H.Shing**
Shing, Kung Hsieh (Hsia) (1929–) P	**K.H.Shing**
Shing, S.Z. (fl. 1975) M	**S.Z.Shing**
Shingh, K.H. S	**Shingh**
Shinners, Lloyd Herbert (1918–1971) S	**Shinners**
Shinoda, T. (fl. 1967) M	**Shinoda**
(Shipczinski, Nikolaj Valerianovič)	
see Schipczinski, Nikolaj Valerianovich	**Schipcz.**
Shipley, Arthur Everett (1861–1927) M	**Shipley**
Shipman, Jill S. (fl. 1979) M	**Shipman**
Shipton, Warren A. (1940–) M	**Shipton**
Shirai, Mitsutarô ('Kotaro') (1863–1932) MS	**Shirai**
Shirasawa, Homi (Yasuyoshi) (1868–1947) S	**Shiras.**
(Shiriaev, Grigorij (Gregory) Ivanovich)	
see Širjaev, Grigorij (Gregory) Ivanović	**Širj.**
Shirley, John F. (1849–1922) FM	**Shirley**
Shirnina–Grishina, L.V. (fl. 1976) M	**Shirn.-Grish.**
(Shishkin, Boris Konstantinovich)	
see Schischkin, Boris Konstantinovich	**Schischk.**
(Shishkina, A.K.)	
see Schischkina, A.K.	**Schischkina**
(Shiskina, A.K.)	
see Schischkina, A.K.	**Schischkina**
Shivas, Mary Grant (later Walker, M.G.) (1926–) PS	**Shivas**
Shivas, R.G. (fl. 1983) M	**R.G.Shivas**
Shkarupa, A.G. (fl. 1977) M	**Shkarupa**
(Shkorbatov, Leonid Andrejewitsch)	
see Schkorbatow, Leonid Andrejewitsch	**Schkorb.**
Shlotgauer, S.D. (fl. 1972) S	**Shlotg.**
(Shlyakov, Roman Nicolayevich)	
see Schljakov, Roman Nicolaevich	**Schljakov**

Shmida, Avi (1945–) S	**Shmida**
Shockley, William Hillman (1855–1925) S	**Shockley**
Shoemaker, Robert Alan (1928–) M	**Shoemaker**
Shome, S.K. (fl. 1963) M	**S.K.Shome**
Shome, Usha (fl. 1964) M	**U.Shome**
Shoolbred, William Andrew (1852–1928) S	**Shoolbred**
Shope, Paul Franklin (1894–1986) M	**Shope**
Short, Charles Wilkins (1794–1863) S	**Short**
Short, John W. (fl. 1983) P	**J.W.Short**
Short, M. (fl. 1982) S	**M.Short**
Short, Philip Sydney (1955–) S	**P.S.Short**
Shortt, H.E. (fl. 1923) M	**Shortt**
(Shostenko–Dessiatova, Nathalie A.)	
see Desjatova–Shostenko, Nathalie A.	**Desj.-Shost.**
Shotter, G. (fl. 1959) M	**Shotter**
Showalter, Amos Martin (1891–1968) S	**Showalter**
Showers, David W. (fl. 1980) B	**Showers**
Shperk, Gustav (1840s–1870) A	**Shperk**
Shreemali, J.L. (fl. 1973) M	**Shreem.**
Shrestha, T.B. S	**Shrestha**
Shreve, Forrest (1878–1950) S	**Shreve**
Shrivastava, J.N. (fl. 1989) M	**Shrivast.**
Shriver, Howard (1824–1901) S	**Shriver**
Shrock, Robert Rakes (1904–) A	**Shrock**
Shrubsole, William Hobbs (1837–1927) A	**Shrubsole**
Shu, Lon Rhan S	**L.R.Shu**
Shu, Ting Chi S	**T.C.Shu**
Shu, Tsi Yun (fl. 1980) S	**T.Y.Shu**
(Shuchobodskii, B.A.)	
see Shukhobodsky, B.A.	**Shukhob.**
Shue, Long Zhan (fl. 1985) S	**L.Z.Shue**
Shuey, Allen G. S	**Shuey**
Shukhobodsky, B.A. (1920–) S	**Shukhob.**
Shukla, A.N. (fl. 1984) M	**A.N.Shukla**
Shukla, D.N. (fl. 1982) M	**D.N.Shukla**
Shukla, R.S. (fl. 1985) M	**R.S.Shukla**
Shukla, V.B. (fl. 1941–1953) AF	**Shukla**
Shull, Charles Albert (1879–1962) S	**C.Shull**
Shull, George Harrison (1874–1954) S	**Shull**
Shults, G.E. S	**Shults**
Shultz, Benjamin (1772–1814) S	**Shultz**
Shultz, John S. (fl. 1974) S	**J.S.Shultz**
Shultz, Leila M. (1946–) S	**L.M.Shultz**
Shultz, V.A. (1952–) S	**V.A.Shultz**
Shumaker–Lambry, J. (fl. 1966) F	**Shum.-Lambry**
Shumenko, Stanislav Ivanovich A	**Shumenko**
Shurly, Ernest William (1888–1963) S	**Shurly**
Shushan, Sam (1922–) M	**Shushan**
Shuttleworth, Robert James (1810–1874) AP	**Shuttlew.**

Shuvashov, B.I. A	**Shuvashov**
Shvartsman, Sofia Ruvimovna (1912–1975) M	**Shvartsman**
(Shvarzman, Sofia Ruvimovna)	
see Shvartsman, Sofia Ruvimovna	**Shvartsman**
(Shvedchikova, N.K.)	
see Schvedtschikova, N.K.	**Shvedtsch.**
Shyam, R. A	**Shyam**
Shylaja, M. (fl. 1986) S	**Shylaja**
Sibasaki, Y. (fl. 1930) M	**Sibas.**
Sibilia, Cesare (1895–) M	**Sibilia**
Sibree, James (1836–1929) S	**Sibree**
Sibthorp, Humphrey Waldo (1713–1797) S	**H.Sibth.**
Sibthorp, John (1758–1796) BMS	**Sibth.**
Sicard, Guillaume (1829–1886) M	**Sicard**
Sickenberger, Ernst (1831–1895) S	**Sickenb.**
Sickmann, Johann Rudolph (1779–1849) S	**Sickmann**
Siddiqi, M.A. (1924–) S	**Siddiqi**
Sidebotham, Joseph (1822–1884) A	**Sideboth.**
Sideris, Christos Plutarchos (1891–1965) M	**Sideris**
Sidibe, U. (fl. 1974) M	**Sidibe**
Sidorenko, G.T. (fl. 1968) S	**Sidorenko**
Sidorova, I.I. (fl. 1964) M	**Sidorova**
Siebenlist S	**Siebenl.**
Sieber, Franz(e) Wilhelm (1789–1844) BP	**Sieber**
Sieber, Johann (–1880) F	**J.Sieber**
Siebert, Arthur E. A	**A.E.Siebert**
Siebert, August (1854–1923) S	**Siebert**
Siebert, Russell Jacob (1914–) S	**R.J.Siebert**
Siebold, Carl Theodor Ernst von (1804–1885) S	**C.Siebold**
Siebold, Philipp Franz (Balthasar) von (1796–1866) S	**Siebold**
Sieburth, John McNeill (1927–) A	**Sieburth**
Siegel, John (fl. 1909) M	**Siegel**
Siegerist, E.S. (1925–) S	**Siegerist**
Siegers, Jean (1842–post 1904) S	**Siegers**
Siegert, Gottlob (1789–1868) S	**Siegert**
Siegesbeck, Johann Georg (1686–1755) PS	**Siegesb.**
Siegfried, Hans (1837–1903) S	**Siegfr.**
Siegler, Eugene Alfred (1891–) M	**Siegler**
Siegmund, Wilhelm (1821–1897) S	**Siegm.**
Siehe, Walter (1859–1928) S	**Siehe**
Siemaszko, J. (fl. 1928) M	**J.Siemaszko**
Siemaszko, Wincenty (Vincent) (1887–1943) M	**Siemaszko**
Siemers, C.L. S	**Siemers**
Siemińska, Jadwiga (1922–) A	**Siemińska**
Siemoni, Giancarlo (1838–1912) S	**Siemoni**
Siemssen, C.H. (fl. 1795) S	**Siemssen**
Siepmann, R. (fl. 1959) M	**Siepmann**
Sieren, David J. (1941–) S	**Sieren**
Sierk, Herbert Allen (1932–) M	**Sierk**

Sierotin, T. (fl. 1961) M	**Sierotin**
(Sierra Calzado, Jorge)	
see Sierra, Jorge	**J.Sierra**
Sierra, Eugeni (1919–) S	**Sierra**
Sierra, Jorge (fl. 1989) S	**J.Sierra**
(Sierra Rafols, Eugeni)	
see Sierra, Eugeni	**Sierra**
Sierra Rico, G. (fl. 1956) M	**G.Sierra**
Sierstorpff, Kaspar Heinrich von (1750–1842) S	**Sierst.**
Siesmayer (fl. 1888) S	**Siesm.**
Sieurin, Johan Magnus (1813–1846) S	**Sieurin**
Sieverding, Ewald (fl. 1984) M	**Sieverd.**
Sievers, Johann Erasmus (–1795) S	**Siev.**
Sievers, Johann Friedrich Ernst (1880–) S	**J.F.E.Siev.**
Sievers, Wilhelm (1860–1921) S	**W.Siev.**
Sigal, Lorene L (fl. 1975) M	**Sigal**
Sigaldi, de (fl. 1964) S	**Sigaldi**
Siggers, Paul Victor (1889–) M	**Sigg.**
Sigler, Lynne (1949–) M	**Sigler**
Sigmond, George Gabriel (fl. 1840) S	**Sigmond**
Signorello, Pietro (1939–) S	**Signor.**
Sigot, A. (fl. 1931) M	**Sigot**
Sigriansky, Alexandre (1882–) MS	**Sigr.**
Sigunov, A. (fl. 1977) S	**Sigunov**
Siim-Jensen, Johannes (1870–1921) S	**Siim-Jensen**
Siino, H. (fl. 1940) M	**Siino**
Sijazov, M.M. (fl. 1902) S	**Sijazov**
Sikdar, J.K. (fl. 1980) S	**Sikdar**
Sikora, Josef (–1837) B	**Sikora**
Sikstel, T.A. S	**Sikstel**
Sikura, Iosif Iosifovich (1932–) S	**Sikura**
Silba, John (fl. 1984) S	**Silba**
Silberschmidt, W. (fl. 1899) M	**Silb.**
Silfvenius, Antti Johannes A	**Silfv.**
Šilić, S.Čedomil (1932–) S	**Šilić**
Silipranti, Giovanni (fl. 1887) S	**Silipr.**
Sillans, Roger (fl. 1952) S	**Sillans**
Sillén, Olof Leopold (1813–1894) B	**Sillén**
Silliman, Benjamin (1779–1864) PS	**Silliman**
Sillinger, Pavel (1905–1938) S	**Sill.**
Silow, Ronald Alfred (1908–) S	**Silow**
Silva, A.A. (fl. 1964) M	**A.A.Silva**
Silva, António Rodrigo Pinto da (1912–) PS	**P.Silva**
Silva, Estela de Sousa e A	**E.S.Silva**
Silva, Herman A	**H.Silva**
Silva, Inês Machline (1957–) S	**I.M.Silva**
Silva, José Nery da (fl. 1950) M	**J.Silva**
(Silva, Júlia Dames e)	
see Dames e Silva, Júlia	**Dames e Silva**
Silva Lacaz, C.da (fl. 1940) M	**Silva Lacaz**

Silva Manso, António Luiz Patricio da (1788–1818) S   **Silva Manso**
Silva Manso, José da   S   **J.Silva Manso**
(Silva, Manuel Augusto Pirajá da)
    see Pirajá da Silva, Manuel Augusto   **Pirajá**
Silva, Manuel da (1916–) S   **M.Silva**
Silva, Margarita (fl. 1952) M   **Marg.Silva**
(Silva Maria Pita de Beauclair)
    see Guimarães, Silva Maria Pita de Beauclair   **S.M.Guim.**
Silva, Marlene Freitas da (1937–) S   **M.F.Silva**
(Silva, Nilda Marquete Ferreira da)
    see Marquete Ferreira da Silva, Nilda   **Marquete**
Silva Pando, Francisco Javier (1955–) S   **Silva Pando**
Silva, Paul Claude (1922–) A   **P.C.Silva**
Silva, Quitéria Jesus G. Pinto da (1911–) S   **Q.J.P.Silva**
(Silva, Rudolfo Albino Dias da)
    see Dias da Silva, Rodolfo Albino   **Dias da Silva**
(Silva Santos, Aymar da)
    see Santos, Aymar da Silva   **A.S.Santos**
Silva Tarouca, Ernst Emmanuel (1860–1936) S   **Silva Tar.**
Silva Vidal, R.da   S   **Silva Vidal**
Silveira, Alvaro Astolpho da (1867–1945) PS   **Silveira**
Silveira, Fernando (fl. 1925) S   **F.Silveira**
Silveira, G.C. (fl. 1959) M   **G.C.Silveira**
Silveira, Jarbas S. (fl. 1959) M   **J.S.Silveira**
Silveira, Nelson Jorge E. (1959–) S   **N.Silveira**
Silveiro Grillo, H.V.da (fl. 1934) M   **Silveiro**
Silvén, Nils Olaf Valdemar (1880–) S   **Silvén**
Silver, Alexandre (–1865) S   **Silver**
Silverside, Alan James (1947–) S   **Silverside**
Silvestre Domingo, Santiago (1944–) S   **Silvestre**
(Silvestre, Santiago)
    see Silvestre Domingo, Santiago   **Silvestre**
Silvestri, Antonio de (1836–) S   **A.Silvestri**
Silvestri, Filippo (1873–1949) S   **F.Silvestri**
Silveus, William Arents (1875–1953) S   **Silveus**
(Silvio Campos, T.C.)
    see Campos, Silvio T.C.   **Campos**
Sim, John (1824–1901) B   **J.Sim**
Sim, Robert (1791–1878) PS   **R.Sim**
Sim, Thomas Robertson (1856–1938) BPS   **Sim**
Simagin, V.S. (fl. 1986) S   **Simagin**
Simeria, G. (fl. 1983) M   **Simeria**
Simione, F.P. (fl. 1977) M   **Simione**
Simionescu, Ioan Th. (1873–) A   **Simion.**
Simizu, Hideo   S   **Simizu**
(Simkovics, Lájos Philipp von)
    see Simonkai, Lájos von   **Simonk.**
Simler, Rudolf Theodor (1833–1874) S   **Simler**
Simmler, E.   S   **E.Simmler**
Simmler, Gudrun (1884–) S   **Simmler**

Simmonds, Arthur (1892–1968) S	**A.Simmonds**
Simmonds, John Howard (fl. 1965) M	**J.H.Simmonds**
Simmonds, Joseph Henry (1845–1936) S	**Simmonds**
Simmonds, Norman Willison (1922–) S	**N.W.Simmonds**
Simmonds, Peter Lund (1814–1897) S	**P.Simmonds**
Simmons, Emory Guy (1920–) M	**E.G.Simmons**
Simmons, Herman George (1866–1943) ABS	**Simmons**
(Simmons, Peter Lund)	
see Simmonds, Peter Lund	**P.Simmonds**
Simmons, Robert B. (fl. 1990) M	**R.B.Simmons**
Simms, J. (fl. 1955) M	**Simms**
Simo, M. (fl. 1936) M	**Simo**
Simões Barbosa, F.A. (fl. 1941) M	**Simões**
Simon, (Wilhelm Otto) Friedrich (1868–) S	**F.Simon**
Simon, Bryan Kenneth (1943–) S	**B.K.Simon**
Simon, Charles (1908–1987) S	**C.Simon**
Simon, Eugène (1848–1924) S	**Simon**
Simon, Eugène Ernest (1871–1967) S	**E.Simon**
Simon, Jean–Pierre J. (1938–) S	**J.Simon**
Simon, Marie–France A	**M.-F.Simon**
Simon, Siegfried Veit (1877–1934) S	**S.V.Simon**
Simon, Tibor (1926–) S	**T.Simon**
Simon, William (fl. 1968) S	**W.Simon**
Simon–Louis, Léon L. (1834–1913) S	**Simon–Louis**
(Simon–Studer, Charles)	
see Simon, Charles	**C.Simon**
Simonds, Arthur Beaman (1867–) S	**Simonds**
Simonet, Marc (1899–1965) S	**Simonet**
(Simonian, Seda A.)	
see Simonyan, Seda A.	**Simonyan**
Simonis, J.E. (fl. 1981) S	**Simonis**
(Simonivich, L.G.)	
see Simonovicz, L.G.	**Simonov.**
(Simonjian, Seda A.)	
see Simonyan, Seda A.	**Simonyan**
Simonkai, Lájos von (1851–1910) BS	**Simonk.**
Simonneua, P. (fl. 1951) S	**Simonn.**
Simonova, E.F. (1927–) S	**E.F.Simonova**
Simonova, O.A. (1930–) S	**O.A.Simonova**
Simonovicz, L.G. (1926–) S	**Simonov.**
Simons, Jan (1940–) A	**J.Simons**
Simons, R. (fl. 1983) M	**R.Simons**
Simons, Richard Harold (1928–) A	**Simons**
Simonsen, Reimer (1931–) A	**Simonsen**
Simonyan, Seda A. (1929–) M	**Simonyan**
Simpson, Beryl Britnall (née Vuilleumier) (1942–) S	**B.B.Simpson**
Simpson, Charles Torrey (1826–1932) S	**Simpson**
Simpson, David Alan (1955–) S	**D.A.Simpson**

Simpson, Donald Ray (1932–   ) S   **D.R.Simpson**
Simpson, George (1880–1952) S   **G.Simpson**
Simpson, Graham Miller (1931–   ) S   **G.M.Simpson**
Simpson, Jack A. (1941–   ) M   **J.A.Simpson**
Simpson, Margaret Jane Annand (1920–   ) S **M.J.A.Simpson**
Simpson, Norman Douglas (1890–1974) S   **N.D.Simpson**
Simpson, Phillip D.   A   **P.D.Simpson**
Sims, J. (fl. 1964) M   **J.Sims**
Sims, John (1749–1831) S   **Sims**
Sims, Patricia Anne (1932–   ) A   **P.A.Sims**
Simson, Augustus (1836–1918) S   **Simson**
Simson-Scharold, E.   S   **Simson-Schar.**
Sinadskii, Yu.V. (fl. 1962) M   **Sinadskii**
(Sinadskij, Ju.V.)
    see Sinadskii, Yu.V.   **Sinadskii**
Sinclair, Andrew (1796–1861) S   **A.Sinclair**
Sinclair, George (1786–1834) S   **G.Sinclair**
Sinclair, Isabella (1842–1900) S   **I.Sinclair**
Sinclair, James (1913–1968) S   **J.Sinclair**
Sinclair, James Burton (1927–   ) M   **J.B.Sinclair**
Sinclair, John (1754–1835) S   **Sinclair**
Sinclair, Robert C. (fl. 1979) M   **R.C.Sinclair**
Sinden, James Whaples (1902–   ) M   **Sinden**
Singer, Jakob (1834–1901) S   **J.Singer**
Singer, Max (fl. 1885) S   **M.Singer**
Singer, Rolf (1906–   ) MS   **Singer**
Singh, A.D.   A   **A.D.Singh**
Singh, A.N. (1943–   ) S   **A.N.Singh**
Singh, Ajay (fl. 1980) M   **Ajay Singh**
Singh, Alakhnadan Prasad (fl. 1962) F   **A.P.Singh**
Singh, Amar (1921–   ) M   **Amar Singh**
Singh, Anil K. (fl. 1987) M   **A.K.Singh**
Singh, B.B. (fl. 1979) M   **B.B.Singh**
Singh, B.N. (fl. 1947) M   **B.N.Singh**
Singh, C.S.   A   **C.S.Singh**
Singh, D.N.   S   **D.N.Singh**
Singh, D.V. (fl. 1968) M   **D.V.Singh**
Singh, G.N. (fl. 1969) M   **G.N.Singh**
Singh, G.P. (fl. 1954) M   **G.P.Singh**
Singh, Gurcharan (fl. 1970) FS   **G.Singh**
Singh, H.P. (fl. 1986) AM   **H.P.Singh**
Singh, Harnek (fl. 1965) M   **H.Singh**
Singh, I.D. (fl. 1976) M   **I.D.Singh**
Singh, Jactar (fl. 1934) M   **Jactar Singh**
Singh, Jiwan (fl. 1930) M   **Jiwan Singh**
Singh, K.D. (fl. 1978) M   **K.D.Singh**
Singh, Kamala P.   A   **Kam.P.Singh**
Singh, Krishna Pal (fl. 1972) M   **Kr.P.Singh**
Singh, Kunwar Suresh   A   **K.S.Singh**

Singh, Lal Behari (1921–) M	**L.Singh**
Singh, M. (fl. 1986) M	**M.Singh**
Singh, N. (fl. 1978) M	**N.Singh**
Singh, N.B. (fl. 1985) S	**N.B.Singh**
Singh, N.I. (fl. 1981) M	**N.I.Singh**
Singh, Nedra Pal (1941–) S	**N.P.Singh**
Singh, P. (1956–) S	**P.Singh**
Singh, P.N. (fl. 1977) M	**P.N.Singh**
Singh, Pritam (fl. 1959) M	**Pr.Singh**
Singh, R.A. (fl. 1966) M	**R.A.Singh**
Singh, R.B. (fl. 1978) AM	**R.B.Singh**
Singh, R.S. (fl. 1956) M	**R.S.Singh**
Singh, R.Y. A	**R.Y.Singh**
Singh, Ram Dhan (fl. 1966) M	**R.D.Singh**
Singh, Ram Nath (1921–) M	**Ram N.Singh**
Singh, Ram Pratar (1930–) M	**R.P.Singh**
Singh, Rama Nagina (1915–) A	**Rama N.Singh**
Singh, S.D. (fl. 1980) M	**S.D.Singh**
Singh, S.J. A	**S.J.Singh**
Singh, S.K. (fl. 1982) M	**S.K.Singh**
Singh, S.L. (fl. 1977) M	**S.L.Singh**
Singh, S.M. (fl. 1972) M	**S.M.Singh**
Singh, S.S. (fl. 1972) M	**S.S.Singh**
Singh, Sarjeet (fl. 1980) M	**Sarj.Singh**
Singh, Sarnam (fl. 1989) P	**Sarn.Singh**
Singh, Satys Pal (1926–) A	**S.P.Singh**
Singh, Shri Ram (fl. 1973) M	**S.R.Singh**
Singh, Sujan (fl. 1987) M	**Suj.Singh**
Singh, Thakur K.S. (fl. 1962) M	**T.K.S.Singh**
Singh, Udai Pratap (1945–) M	**U.P.Singh**
Singh, V. (fl. 1984) S	**V.Singh**
Singh, Vijay Pratap (1917–) A	**V.P.Singh**
Singhai, L.C. (fl. 1958) BFM	**Singhai**
Sinha, A.K. (fl. 1981) S	**A.K.Sinha**
Sinha, A.R.P. (fl. 1989) M	**A.R.P.Sinha**
Sinha, B.M.B. (fl. 1970–1988) P	**B.M.B.Sinha**
Sinha, G.P. (fl. 1988) M	**G.P.Sinha**
Sinha, J.P. A	**J.P.Sinha**
Sinha, R.K. (fl. 1989) M	**R.K.Sinha**
Sinha, R.L. (fl. 1979) S	**R.L.Sinha**
Sinha, Saligram (1913–) MS	**S.Sinha**
Sinha, Veena (1943–) A	**V.Sinha**
Sinigaglia, Georg (fl. 1913) M	**Sinig.**
Siniscalco Gigliano, Gesualdo (1948–) S	**Sinisc.Gigl.**
Sinkora, Doris Martha (1927–) A	**Sinkora**
Sinkova, G.M. S	**Sinkova**
Sinning, Wilhelm (1792–1874) S	**Sinning**
Sinnott, Edmund Ware (1888–1968) S	**Sinnott**
Sinnott, Quinn P. (1954–) S	**Q.P.Sinnott**

Sinotô, Yosito (1895–) A   **Sinotô**
(Sinskaja, Eugenija Nikolayevna)
  see Sinskaya, Eugeniya Nikolayevna   **Sinskaya**
Sinskaya, Eugeniya Nikolayevna (1889–1965) S   **Sinskaya**
Sintenis, Paul Ernst Emil (1847–1907) S   **Sint.**
Sipe, F.P. (fl. 1946) M   **Sipe**
Sipkes (fl. 1921) S   **Sipkes**
(Siplivinskij, Vladimir N.)
  see Siplivinsky, Vladimir N.   **Sipliv.**
Siplivinsky, Vladimir N. (1937–) PS   **Sipliv.**
Sipman, Henricus Johannes Maria (1945–) BM   **Sipman**
Siqueira, Josafá Carlos de (1953–) S   **J.C.Siqueira**
Siqueira, Mauro Wanderley de (fl. 1960) M   **Siqueira**
Siradhana, B.S. (fl. 1980) M   **Siradhana**
Sirard, L. (fl. 1984) M   **Sirard**
Sirirugsa, Puangpen (Phuangpen) (1938–) S   **Sirirugsa**
Širjaev, Grigorij (Gregory) Ivanović (1882–1954) S   **Širj.**
Sirkar, Nagendranath (fl. 1929) B   **Sirkar**
Sirko, A.V. (fl. 1976) M   **Sirko**
Sirks, Marius Jakob (Jabob) (1889–1966) S   **Sirks**
Sirleo, L. (fl. 1895) M   **Sirleo**
Sirna, Giuseppe A   **Sirna**
Sirodot, Simon (1825–1903) AS   **Sirodot**
(Šiškin, Boris Konstantinovich)
  see Schischkin, Boris Konstantinovich   **Schischk.**
(Šiškina, A.K.)
  see Schischkina, A.K.   **Schischkina**
Sismonda, Eugenio (1815–1870) S   **Sismonda**
Sissingh, W. A   **Sissingh**
Sisson, W.R. (fl. 1917) M   **Sisson**
Sisterna, Marina N. (fl. 1989) M   **Sisterna**
Sitgreaves S   **Sitgr.**
Sitholey, Verma Rajendra (fl. 1943–1979) AF   **Sitholey**
Sittiracha, T. (fl. 1977) M   **Sittir.**
(Siuzev, Paul V.)
  see Siuzew, Paul W.   **Siuzew**
Siuzew, Paul W. S   **Siuzew**
Sivadasan, M. (1948–) S   **Sivad.**
Sivanesan, Asaipillai (1931–) M   **Sivan.**
Sivarajan, V.V. (fl. 1970) S   **Sivar.**
Sivasithamparam, K. (fl. 1975) M   **Sivasith.**
Siver, Peter A. A   **Siver**
Sivertsen, Sigmund (1929–) M   **Sivertsen**
Siwasan, C. (fl. 1971) M   **Siwasan**
Sixtel, (fl. 1960) F   **Sixtel**
Sizova, T.P. (fl. 1957) M   **Sizova**
(Sjoestrand, Magnus Gustaf)
  see Sjöstrand, Magnus Gustaf   **Sjöstr.**
Sjors, Hugo Mattias (1915–) B   **Sjors**

Sjöstedt, Bo Jonny (1937–) S   **Sjöstedt**
Sjöstedt, Lars Gunnar (1894–1975) AS   **G.Sjöstedt**
Sjöstrand, Magnus Gustaf (1807–1880) S   **Sjöstr.**
(Sjuzew, Paul W.)
   see Siuzew, Paul W.   **Siuzew**
Skabichevskaja, N.A.   A   **Skabich.**
Skabichevskij, A.P.   A   **Skab.**
Skalická, Anna (nëe Klasková, A.) (1932–) S   **Skalická**
Skalický, Vladimír (1930–) MS   **Skalický**
Skalińska, Maria (1890–1977) S   **Skalińska**
Skan, Sidney Alfred (1870–1939) S   **Skan**
Skarlato, O.A.   A   **Skarlato**
Skarupke, E.   S   **Skarupke**
Skean, James Daniel (1958–) S   **Skean**
Skeels, Homer Collar (1873–1934) S   **Skeels**
Skelly, R.J. (fl. 1988) S   **Skelly**
Skene, Macgregor (1889–1973) S   **Skene**
Skepper, Edmund (1825–1867) S   **Skepper**
Skifte, Ola (1923–) M   **Skifte**
Skinner, Charles Edward (1897–) M   **C.E.Skinner**
Skinner, F.A. (fl. 1951) M   **F.A.Skinner**
Skinner, Frank Leith (1882–1967) S   **F.L.Skinner**
Skinner, George Ure (1804–1867) S   **Skinner**
Skinner, S.G. (fl. 1982) S   **S.G.Skinner**
Skinner, Stephen (1622–1667) ALS   **S.Skinner**
Skipwith, G.R. (fl. 1929) S   **Skipwith**
Skipworth, John Peyton (1934–) S   **Skipw.**
Skirgiełło, Alina (1911–) MS   **Skirg.**
Skofitz, Alexander (1822–1892) S   **Skofitz**
Skog, Judith Ellen (1944–) FPS   **J.E.Skog**
Skog, Laurence Edgar (1943–) S   **L.E.Skog**
Skogsberg, (Karl Jonas) Tage (1887–) A   **Skogsb.**
Skolka, V.H.   A   **Skolka**
Skolko, Arthur John (1912–) M   **Skolko**
Skorepa, Allen Charles (1941–) M   **Skorepa**
Škorič, Vladimir (1890–1947) M   **Škorič**
Skorikow, A.S.   A   **Skorikow**
Skoropad, W.P. (fl. 1982) M   **Skoropad**
Skottsberg, Carl Johan Fredrik (1880–1963) ABMPS   **Skottsb.**
Skou, J.P.S. (fl. 1972) M   **Skou**
Skovsted, Aage Thorsen (1903–1983) MS   **Skovst.**
Skrine, Pattie Morfydd   A   **Skrine**
Skuja, Heinrich Leonhards (1892–1972) A   **Skuja**
Skulberg, Olav M.   A   **Skulberg**
Skult, H. (fl. 1984) M   **Skult**
Skupieński, Frantiszek (François–Xavier) (1888–) M   **Skup.**
Skutch, Alexander Frank (1904–) S   **Skutch**
Skvortsov, Alexei Konstantinovich (1920–) S   **A.K.Skvortsov**

Skvortsov, Boris Vassilievich (1890–1980) AMS   **Skvortsov**
Skvortsova, N.T. S   **Skvortsova**
(Skvortzov, Boris Vassilievich)
    see Skvortsov, Boris Vassilievich   **Skvortsov**
(Skvortzow, Boris Wassilievich)
    see Skvortsov, Boris Vassilievich   **Skvortsov**
Slaba (fl. 1985) S   **Slaba**
Slagg, Charles Mervyn (1890–) M   **Slagg**
Slater, Daniel W. (fl. 1946) M   **D.W.Slater**
Slater, Matthew B. (c.1829–1918) S   **M.Slater**
Slavík, Bohdan (1924–) S   **Slavík**
Slavík, Bohumil (1935–) S   **B.Slavík**
Slavíková, Elena (fl. 1976) M   **E.Slavíková**
Slavíková, Kvetoslava A   **K.Slavíková**
Slavíková, Zdeňka (1935–) S   **Slavíková**
Slavin, Arthur Daniel (1903–) S   **A.D.Slavin**
Slavin, Bernard Henry (1873–1960) S   **Slavin**
Sledge, William Arthur (1904–1991) PS   **Sledge**
Sleep, Anne (1939–) PS   **Sleep**
Sleesen, Ellen H.L.van der (fl. 1958) S   **Sleesen**
Śléndziński, Aleksander Jan (1849–1881) S   **Śleńdz.**
Sleumer, Hermann Otto (1906–) S   **Sleumer**
Slis, W. S   **Slis**
(Sljakov, R.N.)
    see Schljakov, Roman Nicolaevich   **Schljakov**
Sljussarenko, L.P. (1931–) S   **Sljuss.**
Sloane, Boyd Lincoln (1885–) S   **B.Sloane**
Sloane, Hans (1660–1753) L   **Sloane**
Sloboda, Daniel (1809–1888) S   **Sloboda**
Slobodov, A.A. S   **Slobodov**
Sloff, Jan G. (1892–1979) S   **Sloff**
Slogteren, Egbertus van (1888–) M   **Slogt.**
Slooff, Wilhelmina Ch. (fl. 1966) M   **Slooff**
Slooten, Dirk Fok van (1891–1953) S   **Slooten**
Slootweg, A.F.G. A   **Slootweg**
(Sloover, J.R.De)
    see De Sloover, J.R.   **J.R.De Sloover**
(Sloover, Jean Louis J.A.De)
    see De Sloover, Jean Louis J.A.   **De Sloover**
Slope, D.B. (fl. 1977) M   **Slope**
Slosson, Margaret (1872–) PS   **Sloss.**
Slysh, Anton R. (fl. 1960) M   **Slysh**
Small, Ernest (1940–) S   **E.Small**
Small, James (1889–1955) S   **J.Small**
Small, John Kunkel (1869–1938) BPS   **Small**
Small, William (fl. 1924) M   **W.Small**
Smalley, Eugene Byron (1926–) M   **Smalley**
Smarda, František (1902–1976) M   **F.Smarda**
Šmarda, Jan (1904–1968) B   **Šmarda**
Smarods, Julius (1884–1956) M   **Smarods**

Smeathman, Henry (–1786) S	**Smeathman**
Smedegård-Petersen, V. (1933–) M	**Smed.-Pet.**
Smejkal, Miroslav (1927–) S	**Smejkal**
(Smelowsky, Timotheus)	
see Smielowsky, Timotheus	**Smiel.**
Smelror, Morten A	**Smelror**
Smerlis, E. (fl. 1962) M	**Smerlis**
(Smet, Louis De)	
see De Smet, Louis	**De Smet**
(Smet, S.De)	
see De Smet, S.	**S.De Smet**
(Smidt, F.P.G.De)	
see De Smidt, F.P.G.	**De Smidt**
Smidth, Jens Hansen (1769–1847) S	**Smidth**
Smielowsky, Timotheus (1769–1815) S	**Smiel.**
(Smik, Ljudmila V.)	
see Smỹk, Ljudmila V.	**Smỹk**
Smiley, Frank Jason (1880–1969) S	**Smiley**
Smiley, M.J. (fl. 1974) M	**M.J.Smiley**
(Smirnoff, Michael Nikolajèwitsh)	
see Smirnov, Michael Nikolajèwitsh	**M.Smirn.**
(Smirnoff, Sergius)	
see Smirnov, Sergius	**S.Smirn.**
(Smirnoff, W.A.)	
see Smirnov, W.A.	**W.A.Smirn.**
Smirnov, Michael Nikolajèwitsh (1849–1889) S	**M.Smirn.**
Smirnov, Pavel Aleksandrovich (1896–1980) S	**P.A.Smirn.**
Smirnov, Sergei Michailowitsch (fl. 1870) S	**Smirn.**
Smirnov, Sergius A	**S.Smirn.**
Smirnov, W.A. A	**W.A.Smirn.**
Smirnova, Alexandrea Dmitrievna (1904–) S	**A.D.Smirnova**
Smirnova, W.A. (1941–) S	**W.A.Smirnova**
Smirnova, Zoya (Zoë) Nikolayevna (1898–1979) B	**Smirnova**
(Smirnow, Michael Nikolajèwitsh)	
see Smirnov, Michael Nikolajèwitsh	**M.Smirn.**
(Smirnow, Pavel Aleksandrovich)	
see Smirnov, Pavel Aleksandrovich	**P.A.Smirn.**
(Smirnow, Sergei Michailowitsch)	
see Smirnov, Sergei Michailowitsch	**Smirn.**
(Smirnowa, W.A.)	
see Smirnova, W.A.	**W.A.Smirnova**
Smit, A. (fl. 1971) S	**A.Smit**
Smit, J. (fl. 1934) M	**Smit**
Smit, Petra G. S	**P.G.Smit**
(Smith, Alan Radcliffe)	
see Radcliffe–Smith, Alan	**Radcl.-Sm.**
Smith, Alan Reid (1943–) FPS	**A.R.Sm.**
Smith, Albert Charles (1906–) PS	**A.C.Sm.**
Smith, Alexander Hanchett (1904–1986) M	**A.H.Sm.**

Smith, Anna Petronella Maria (1856–1946) S	**A.P.M.Sm.**
Smith, Annie Elizabeth Morrill (1856–1946) B	**A.M.Sm.**
Smith, Annie Lorrain (1854–1937) BM	**A.L.Sm.**
Smith, Anthony John Edwin (1935–) B	**A.J.E.Sm.**
Smith, Benjamin Harrison (1892–) A	**B.Harr.Sm.**
Smith, Benjamin Hayes (1841–1918) S	**B.H.Sm.**
Smith, Charles Eastwick (1820–1900) S	**C.E.Sm.**
Smith, Charles Leonard (1866–1923) M	**C.L.Sm.**
Smith, Charles Piper (1877–1955) S	**C.P.Sm.**
Smith, Christen (1785–1816) BS	**C.Sm.**
Smith, Christo Albertyn (1898–1956) S	**C.A.Sm.**
Smith, Christopher Parker (1835–1892) B	**Chr.P.Sm.**
Smith, Claude Earle, Jr. (1922–1987) S	**Earle Sm.**
Smith, Clayton Orville (1871–) M	**C.O.Sm.**
Smith, Clifford W. (1938–) M	**C.W.Sm.**
Smith, Clifton F. (1920–) S	**C.F.Sm.**
Smith, D.N. (fl. 1991) S	**D.N.Sm.**
Smith, Dale Metz (1928–) PS	**D.M.Sm.**
Smith, Dianne Mary Noelle (fl. 1983) S	**D.M.N.Sm.**
Smith, Donald J. (fl. 1941) M	**D.J.Sm.**
Smith, Doris Alme (later Goy, D.A.) (1912–) PS	**D.A.Sm.**
Smith, Douglas Roane (1930–) B	**D.R.Sm.**
Smith, Edwin Burnell (1936–) S	**E.B.Sm.**
Smith, Elizabeth Hight (1877–1933) M	**E.H.Sm.**
Smith, Elmer William (1920–1981) S	**E.W.Sm.**
Smith, Erwin Frink (1854–1927) M	**E.F.Sm.**
Smith, Eugene Allen (1841–1927) S	**E.A.Sm.**
(Smith, F.Donnell)	
see Donnell Smith, F.	**F.Donn.Sm.**
Smith, F.E.V. (fl. 1924) M	**F.E.V.Sm.**
Smith, F.J. (fl. 1989) S	**F.J.Sm.**
Smith, Frederick Porter (1833–1888) S	**F.P.Sm.**
Smith, Frederick William (1797–1835) S	**F.W.Sm.**
Smith, G.S. (fl. 1985) M	**G.S.Sm.**
Smith, Gary Lane (1939–) BS	**G.L.Sm.**
Smith, George (1895–1967) M	**G.Sm.**
Smith, Gerald Graham (1892–1976) S	**G.G.Sm.**
Smith, Gerard Edwards (1804–1881) S	**G.E.Sm.**
Smith, Gilbert Morgan (1885–1959) A	**G.M.Sm.**
Smith, Hamilton Lanphere (1819–1903) A	**H.L.Sm.**
(Smith, Harry)	
see Smith, Karl August Harald ('Harry')	**Harry Sm.**
Smith, Helen Vandervort (1909–) M	**H.V.Sm.**
Smith, Henry (1786–1868) S	**H.Sm.**
Smith, Henry (fl. 1857) S	**H.Sm.bis**
Smith, Henry George (1852–1924) S	**H.G.Sm.**
Smith, Herbert Huntingdon (1852–1919) S	**Herb.H.Sm.**
Smith, Huron Herbert (1883–1933) S	**Hur.H.Sm.**
Smith, J.C. (1900–) A	**J.C.Sm.**
Smith, J.Drew (fl. 1979) M	**J.D.Sm.**

Smith, J.M.B. (fl. 1963) M	J.M.B.Sm.
Smith, James (1760–1840) APS	Js.Sm.
Smith, James Edward (1759–1828) BMPS	Sm.
Smith, Jared Gage (1866–1925) PS	J.G.Sm.
Smith, Johannes Jacobus (1867–1947) S	J.J.Sm.
Smith, John (1798–1888) P	J.Sm.
(Smith, John Donnell)	
see Donnell Smith, John	Donn.Sm.
Smith, John Jay (1798–1881) S	J.Jay Sm.
Smith, Karl August Harald ('Harry') (1889–1971) S	Harry Sm.
Smith, Kenneth Manley (1892–) M	K.M.Sm.
Smith, Lee A. A	L.A.Sm.
Smith, Lindsay Stewart (1917–1970) S	L.S.Sm.
Smith, Loren Bartlett (1890–) S	Lor.B.Sm.
Smith, Lucius Chambers (1853–1896) S	L.C.Sm.
Smith, Lyman Bradford (1904–) PS	L.B.Sm.
Smith, M.C. (fl. 1981) S	M.C.Sm.
Smith, Marion Ashton (1897–) M	M.A.Sm.
Smith, Matilda (1854–1926) S	M.Sm.
Smith, Maudy Th. (1940–) M	M.T.Sm.
Smith, P.J. (fl. 1986) S	P.J.Sm.
Smith, Paul G. (fl. 1958) S	P.G.Sm.
Smith, Paul Hamilton (1936–) M	P.H.Sm.
Smith, Pearl (1883–) S	P.Sm.
Smith, Philip Morgans (1941–) S	P.M.Sm.
Smith, R.B. (fl. 1981) M	R.B.Sm.
Smith, R.J. (fl. 1973) M	R.J.Sm.
Smith, Ralph Elliott (Eliot) (1874–1953) MS	R.E.Sm.
Smith, Raymond Vaughan (1923–) S	R.V.Sm.
Smith, Rebecca Harris (1967–) S	R.H.Sm.
Smith, Rebecca Merritt (1832–) S	Reb.M.Sm.
Smith, Richard L. A	R.L.Sm.
Smith, Robert R. (1934–) S	R.R.Sm.
Smith, Rosemary Margaret (1933–) S	R.M.Sm.
Smith, Stanley Jay (1916–1977) S	S.J.Sm.
Smith, T.J. (fl. 1989) S	T.J.Sm.
Smith, Thomas (1857–1955) S	T.Sm.
Smith, William (1808–1857) A	W.Sm.
Smith, William Gardner (1866–1928) M	Wm.G.Sm.
Smith, William Wright (1875–1956) S	W.W.Sm.
Smith, Worthington George (1835–1917) M	W.G.Sm.
Smitinand, Tem (1920–) S	Smitinand
Smitka, C. (fl. 1981) M	Smitka
Smitska, M.F. (fl. 1972) M	Smitska
Smitt, Johan Wilhelm (1821–1904) M	Smitt
Smoljaninova, L.A. (1904–) S	Smoljan.
Smolsky, Nikolaï Vladislanovich (1905–) S	Smolsky
(Smolyaninova, L.A.)	
see Smoljaninova, L.A.	Smoljan.

Smoot, Edith L. (fl. 1985) BF — Smoot
Smotlacha, F. (fl. 1921) M — Smotl.
Smundin, Luciana   A — Smundin
Smuts, Jan Christiaan (1870–1950) S — Smuts
Smỹk, Ljudmila V. (1937–   ) M — Smỹk
Smyth, Bernard Bryan (1843–1913) S — Smyth
Smythies, Bertram E. (1912–   ) S — Smythies
Smyttère, Philippe Joseph Emmanuel de (1800–   ) S — Smyttère
Snarskis, Povilas Petrovich (1889–1969) S — Snarskis
Sneddon, Barry Victor (1942–   ) S — Sneddon
Snelders, H.C. (fl. 1981) S — Snelders
Snell, Karl (1881–1956) S — K.Snell
Snell, Walter Henry (1889–1980) M — Snell
Snelling, Lilian (1879–1972) S — Snelling
Snethlage, Emil Heinrich (1868–1929) S — Snethl.
Snetselaar, K.M. (fl. 1989) M — Snets.
Snider, Jerry Allen (1937–   ) B — Snider
Snijman, Deidré A. (1949–   ) S — Snijman
Snogerup, Britt (1934–   ) S — B.Snogerup
Snogerup, Sven E. (1929–   ) S — Snogerup
Snooke, William Drew (1787–1857) S — Snooke
Snow, Charles Henry (1863–1957) S — C.Snow
Snow, Julia Warner (1863–1927) A — J.Snow
Snowden, Joseph Davenport (1886–1973) MS — Snowden
Snuverink, J.H. (fl. 1981) S — Snuv.
Snyder, L.A.   S — L.A.Snyder
Snyder, Leon Carlton (1908–   ) M — Snyder
Snyder, William Cowperthwaite (1904–1980) M — W.C.Snyder
Soares da Cunha, Narciso   S — Soares da Cunha
Soares, G.G. (fl. 1986) M — Soares
Soares, Leo de Oliveira   A — L.O.Soares
Soares Nunes, José Manoel (fl. 1982) S — Soares Nunes
Soave, Marco   S — Soave
Sobers, Edward K. (fl. 1963) M — Sobers
Sobko, V.G.   S — Sobko
Sobolevskaja, Kira Arkadjevna (1911–   ) S — Sobolevsk.
(Sobolewski, Gregor Fedorovitch (Grigoriy Fedorowich))
    see Sobolewsky, Gregor Fedorovitch (Grigoriy Fedorowich) — Sobol.
Sobolewsky, Gregor Fedorovitch (Grigoriy Fedorowich) (1741–1807) S — Sobol.
Sobotka, Jaroslav (fl. 1941) M — Sobotka
Sobral, Marcos (1960–   ) S — Sobral
Sobrinho, Luis (1907–1969) S — Sobr.
(Sochadze, M.E.)
    see Sokhadze, M.E. — Sokhadze
Søchting, Ulrik (1947–   ) M — Søchting
Socias Amorós, A. (fl. 1954) M — Socias
Socorro Abreu, Oswaldo (1949–   ) S — Socorro
Soczava, V.B. (1905–   ) S — Soczava

(Soczawa, V.B.)	
see Soczava, V.B.	**Soczava**
Söderberg, Daniel Henric (1750–1781) S	**Söderb.**
Söderberg, Erik Sigurd (1890–1959) S	**E.S.Söderb.**
Söderström, B.E. (fl. 1978) M	**B.E.Söderstr.**
Söderström, Johan A	**J.Söderstr.**
Soderstrom, Thomas Robert (1936–1987) S	**Soderstr.**
Sodiro, Luis (Aloysius, Luigi) (1836–1909) PS	**Sodiro**
Soeder, Carl J. A	**Soeder**
Soegeng–Reksodihardjo, Wertit (1935–) S	**Soegeng**
Soehner, Ert (fl. 1921) M	**Soehner**
Soehns, Franz (fl. 1897) S	**Soehns**
Soejarto, Djaja Djendoel (1939–) S	**Soejarto**
Soejima (fl. 1987) S	**Soejima**
Soenarko, Soejatmi (later Dransfield, S.) (1939–) S	**Soenarko**
Soepadmo, Engkik (1937–) S	**Soepadmo**
Soest, Johannes Leendert van (1898–1983) S	**Soest**
Sofiekova, T.M. S	**Sofiekova**
Sofieva, Rafiga Mohamed Kysy (1927–) S	**Sofieva**
(Sofijeva, Rafiga Mohamed Kysy)	
see Sofieva, Rafiga Mohamed Kysy	**Sofieva**
Sohi, H.S. (fl. 1972) M	**Sohi**
Sohmer, Seymour Hans (1941–) S	**Sohmer**
Sohns, Ernest Reeves (1917–) S	**Sohns**
Söhrens, Johannes (fl. 1900) S	**Söhrens**
(Soie, Gaspard Abdon)	
see De la Soie, Gaspard Abdon	**De la Soie**
Soják, Jirí (1936–) S	**Soják**
Sokać, Branko A	**Sokać**
Sokhadze, M.E. (1909–) S	**Sokhadze**
Sokoloff, Demétrio A	**Sokoloff**
(Sokolov, Demétrio)	
see Sokoloff, Demétrio	**Sokoloff**
Sokolova, E.A. (fl. 1986) S	**E.A.Sokolova**
Sokolova, I.V. (1963–) S	**I.V.Sokolova**
(Sokolova, T.G.Derviz)	
see Derviz–Sokolova, T.G.	**Derv.-Sok.**
(Sokolovskaja, Alexandra Pavlova)	
see Sokolovskaya, Alexandra Pavlova	**Sokolovsk.**
Sokolovskaya, Alexandra Pavlova (1905–) S	**Sokolovsk.**
Solacolu, Théodor (1876–1940) MS	**Solacolu**
Soladoye, Michael O. (1945–) S	**Soladoye**
Solander, Daniel Carl (1733–1782) AS	**Sol.**
Solankure, R.T. (fl. 1970) M	**Solank.**
Soláns, Maria José (fl. 1984) M	**Soláns**
Solari, Silvia Susana (1944–) B	**Solari**
Solbrig, Otto Thomas (1930–) S	**Solbrig**
Sölch, A. (1932–) S	**Sölch**
Soldano, Adriano (fl. 1986) S	**Soldano**

Sole, William (1741–1802) S	**Sole**
Soleirol, Henry Augustin (1792–) S	**Soleirol**
Solel, Z. (fl. 1987) M	**Solel**
Solemacher–Antweiler, Johanne Viktor Ludwig Andreas Georg (1889–) S	**Solem.**
Soler, Ana (1940–) S	**A.Soler**
(Soler–Hernando, Ana)	
see Soler, Ana	**A.Soler**
Solereder, Hans (1860–1920) S	**Soler.**
Solheim, Halvor (fl. 1977) M	**H.Solheim**
Solheim, Wilhelm Gerhard (Gerard) (1898–1978) M	**Solheim**
Solier, Antoine Joseph Jean (1792–1851) A	**Solier**
Solkina, A.F. (fl. 1928) M	**Solkina**
Solla, Rüdiger Felix (Ruggero Felice) (1859–) AC	**Solla**
Sollers, Basil (1853–1909) S	**Sollers**
(Sollman, August)	
see Sollmann, August	**Sollm.**
Sollmann, August (fl. 1862) M	**Sollm.**
Solms–Laubach, Hermann Maximilian Carl Ludwig Friedrich zu (1842–1915) ABFMPS	**Solms**
(Soloviev, F.A.)	
see Solovjev, F.A.	**Solovjev**
Solovjev, F.A. (fl. 1930s) M	**Solovjev**
Solovjeva, L.V. (fl. 1980) S	**Solovjeva**
Soltoković, Marie (fl. 1901) S	**Soltok.**
Soltwedel, Friedrich (Wilhelm Otto) (1858–1890) S	**Soltw.**
Solum, Ingrid A	**Solum**
Solymosy, Sigmond L. (1906–1974) S	**Solymosy**
Somal, B.S. (fl. 1974) M	**Somal**
Somani, R.B. (fl. 1977) M	**Somani**
Somer, Paul (fl. 1975) P	**Somer**
Somers, Carl (1963–) S	**Somers**
Somerville, Alexander (1842–1907) S	**Somerv.**
Somes, Melvin Philip (–1928) S	**Somes**
Sommer, Friedrich Wilhelm A	**F.W.Sommer**
Sommer, Gustav (fl. 1886) S	**Sommer**
Sommer, Noel Frederick (1920–) M	**N.F.Sommer**
Sommerauer, Ignatius (–1854) S	**Sommerauer**
Sommerfeld, Milton R. A	**Sommerfeld**
Sommerfelt, Søren Christian (Severin Christianus) (1794–1838) ABMPS	**Sommerf.**
Sommerstorff, Hermann (1889–1913) M	**Sommerst.**
Sommier, Carlo Pietro Stefano (Stephen) (1848–1922) BMPS	**Sommier**
Somner, Genise Vieira (fl. 1988) S	**Somner**
Sonck, C.E. (1905–) MS	**Sonck**
Sondén, Carl Marten (1846–1913) S	**Sondén**
Sonder, Christoph (fl. 1890) S	**C.Sond.**
Sonder, Otto Wilhelm (1812–1881) ABS	**Sond.**
Soneda, Masami (1928–) AM	**Soneda**

Sonet, Ernest (fl. 1872) S	**Sonet**
Song, Rui Qing (fl. 1988) S	**R.Q.Song**
Song, Xiang Hou (fl. 1984) S	**X.H.Song**
Songeon, André (1826–1905) S	**Songeon**
Soni, K.K. (fl. 1978) M	**Soni**
Sonké, B. (fl. 1990) S	**Sonké**
Sonnad, G.R. A	**Sonnad**
Sonnerat, Pierre (1748–1814) S	**Sonn.**
Sönnerberg, Jacob (1770–1847) S	**Sönnerb.**
(Sonnet, Ernest)	
see Sonet, Ernest	**Sonet**
Sonnini de Manoncour, Charles Nicolas Sigisbert (1751–1812) S	**Sonnini**
Sonntag, C.O. (fl. 1894) S	**Sonntag**
(Soó, Károly Rezsö)	
see Soó von Bere, Károly Rezsö	**Soó**
Soó von Bere, Károly Rezsö (1903–1980) PS	**Soó**
Soong, Tse Pu (1926–) S	**Soong**
Soong, Z.P. (fl. 1984) S	**Z.P.Soong**
Soop, Karl (fl. 1987) M	**Soop**
Soper, James Herbert (1916–) S	**Soper**
Soper, Richard S. (fl. 1974) M	**R.S.Soper**
Sopp, Olav Johan (né Johan–Olsen, O.J.) (1860–1931) M	**Sopp**
Soppitt, Henry Thomas (1858–1899) M	**Soppitt**
Soprunov, F.F. (fl. 1951) M	**Soprunov**
Soran, V. (1828–) S	**Soran**
Sorarú, Stella Beatriz S	**Sorarú**
Sorauer, Paul Carl Moritz (1839–1916) M	**Sorauer**
Sorda, Francesco Saverio (1793–1885) S	**Sorda**
Sordelli, Ferdinando (1837–1916) S	**Sordelli**
Sordi, M. (fl. 1958) M	**Sordi**
Soreng, Robert J. (fl. 1991) S	**Soreng**
Sørensen, Henrik Lauritz (1842–1903) S	**Sørensen**
Sørensen, Marten (1952–) S	**M.Sørensen**
Sorensen, Paul Davidson (1934–) S	**P.D.Sorensen**
Sørensen, Thorwald (Thorvald) Julius (1902–1973) S	**T.J.Sørensen**
Sorenson, W.G. (fl. 1990) M	**Sorenson**
Sörgel, G. (fl. 1961) M	**Sörgel**
Sorger, F. (fl. 1979) S	**Sorger**
Soriano, Alberto (1920–) S	**A.Soriano**
Soriano, Carlos (1947–) S	**C.Soriano**
Soriano, Joventino Duque (1920–) A	**J.D.Soriano**
(Soriano Martín, Carlos)	
see Soriano, Carlos	**C.Soriano**
Sornay, Pierre de (fl. 1913) S	**Sornay**
Sorokin, Nicolai Vasilevitch (Vasilevich) (1846–1909) M	**Sorokin**
Sortino, Mario (1936–) S	**Sortino**
Sosa, Victoria (1952–) S	**Sosa**
Sosef, M.S.M. (1960–) S	**Sosef**

Sosin, Pavel Egrafovich (1895–) M   **Sosin**
Soska, Theodor (1876–1948) S   **Soska**
Soskov, Yurij (G.)D. (1930–) S   **Soskov**
(Sosnovskii, Dmitrii Ivanovich)
    see Sosnowsky, Dmitrii Ivanovich   **Sosn.**
Sosnowska, J.   A   **Sosnowska**
Sosnowsky, Dmitrii Ivanovich (1885–1952) S   **Sosn.**
(Sota, Elías Ramón de la)
    see de la Sota, Elías Ramón   **de la Sota**
Soto Arenas, Miquel Angel (1963–) S   **Soto Arenas**
Soto N., José C. (fl. 1989) S   **J.C.Soto**
Souaya, Fernand Joseph   A   **Souaya**
Soubeiran, Jean Léon (1827–1892) AS   **Soub.**
Souché, Baptiste (1846–1915) S   **Souché**
Součková-Tomková, Milada (fl. 1958) M   **Součk.-Tomk.**
Souèges, Etienne Charles René (1876–1967) S   **Souèges**
Sougnez, N.   S   **Sougnez**
Soukup, Jaroslav Jeníček (1902–1989) S   **Soukup**
Soukup, Victor G. (1924–) S   **V.G.Soukup**
Soulange-Bodin, Étienne (1774–1846) S   **Soul.-Bod.**
Soulat-Ribette (fl. 1892) S   **Soulat-Rib.**
Soulavie, Jean Louis Giraud (1752–1813) S   **Soulavie**
Soulié, Jean André (1858–1905) S   **J.Soulié**
Soulié, Joseph Auguste (1868–1930) S   **Soulié**
Soulsby, Basil Harrington (1864–1933) S   **Soulsby**
Soumini, C.K. (fl. 1947) M   **Soumini**
Sourek, Josef (1891–1968) S   **Sourek**
Šourková, Michaela (1944–) S   **Šourkova**
Sournia, Alain   A   **Sournia**
Sousa Costa, Nuno Maria (fl. 1978) S   **Sousa Costa**
Sousa da Câmara, Manuel Emmanuele de (1871–1955) M   **Sousa da Câmara**
(Sousa Dias, Maria Rosalia)
    see Dias, Maria Rosalia de Sousa   **Dias**
(Sousa e Silva, Estela de)
    see Silva, Estela de Sousa e   **E.S.Silva**
Sousa, Ester Concucão Pereira de (1907–) S   **E.C.Sousa**
(Sousa, Magdalena Peña de)
    see Peña de Sousa, Magdalena   **Peña de Sousa**
Sousa, Octavio E. (fl. 1985) M   **O.E.Sousa**
Sousa Sánchez, Mario (1940–) S   **M.Sousa**
Souster, John Eustace Sirett (1912–) S   **Souster**
South, Graham Robin (1940–) AM   **South**
Southall, Russell M. (fl. 1970) S   **Southall**
Southworth, Effie Almira (1860–1947) M   **Southw.**
(Souza Barreiros, Humberto de)
    see Barreiros, H. de Souza   **Barreiros**
Souza Brito, Ezequiel Candido (1860–1922) S   **Souza Brito**
(Souza Mello Netto, Ladislau de)
    see Netto, Ladislau de Souza Mello   **Netto**
Souza, Simone de (fl. 1979) S   **Souza**

Sovereign, H.E. (-1965) A   **Sovereign**
Sow bin Tandang (1911-1971) S   **Sow**
Sowerby, Charles Edward (1795-1842) S   **C.E.Sowerby**
Sowerby, James (1757-1822) AMPS   **Sowerby**
Sowerby, James (1815-1834) M   **J.Sowerby**
Sowerby, James DeCarle (1787-1871) ABMS   **J.C.Sowerby**
Sowerby, John Edward (1825-1870) S   **J.E.Sowerby**
Sowter, Frederick Archibald (1899-1972) BM   **Sowter**
Soyaux, Hermann (1852-   ) S   **Soyaux**
Soyer-Willemet, Hubert Félix (1791-1867) S   **Soy.-Will.**
Söyrinki, Niilo (1907-   ) S   **Söyrinki**
Spaaij, F. (fl. 1989) M   **Spaaij**
Spach, Édouard (1801-1879) PS   **Spach**
Spackman, William (1919-   ) S   **Spackman**
Spadoni, Paolo (1764-1826) S   **Spadoni**
Spae, Dieudonné (1819-1858) S   **Spae**
Spaendonck, Gerrit (1746-1822) S   **Spaend.**
(Spaeth, Franz Ludwig)
   see Späth, Franz Ludwig   **Späth**
(Spaeth, Louis)
   see Späth, Louis   **L.Späth**
Spain, Joyce Lance (fl. 1984) M   **Spain**
Spalding, Volney Morgan (1849-1918) S   **Spalding**
Spalla, C. (fl. 1973) M   **Spalla**
Spallanzani, Lazaro (1729-1799) S   **Spall.**
Spalowsky, Joachim Johann Nepomuk (1752-1797) S   **Spalowsky**
Spamer, Albert (fl. 1887) S   **Spamer**
Spampinato, Giovanni (1958-   ) S   **Spamp.**
Spannagel, Candidus (-1956) S   **Spann.**
Spanoghe, Johan Baptist (1798-1838) S   **Span.**
Spanowski, Wolfram (1929-   ) S   **Spanowski**
(Spanowsky, Wolfram)
   see Spanowski, Wolfram   **Spanowski**
Spargo, Mildred Webster (1888-   ) A   **Spargo**
Sparks, A.K. (fl. 1981) M   **Sparks**
Sparling, Shirley Ray (1929-   ) A   **Sparling**
Sparre, Benkt (Bengt) (1918-1986) S   **Sparre**
Sparrman, Anders (1748-1820) S   **Sparrm.**
Sparrow, Frederick Kroeber (1903-1977) M   **Sparrow**
Spasskaja, N.A. (1930-   ) S   **Spasskaja**
Späth, Franz Ludwig (1838-1913) S   **Späth**
Späth, Hellmut Ludwig (1885-1945) S   **H.L.Späth**
Späth, Louis   S   **L.Späth**
Spatzier, Johann (Jan) (1806-1883) S   **Spatzier**
Spaulding, Perley (1878-1960) M   **Spauld.**
Speare, Alden True (1885-   ) M   **Speare**
(Spechnew, Nicolai Nicolaievich von)
   see Speschnew, Nicolai Nicolaievich von   **Speschnew**
Specht, Raymond Louis (1924-   ) S   **Specht**

Speer, Eberhard Otto von (1940–) M	**Speer**
Spegazzini, Carlo Luigi (Carlos Luis) (1858–1926) AFMS	**Speg.**
Speggazini, Rutile A. F	**Spegg.**
Speke, John Hanning (1827–1864) S	**Speke**
Spellenberg, Richard William (1940–) S	**Spellenb.**
Spellman, David L. (fl. 1976) S	**Spellman**
Spence, John R. (1956–) B	**J.R.Spence**
Spence, Magnus (1853–1919) S	**Spence**
Spencer, Edwin Rollin (1881–) M	**E.R.Spencer**
Spencer, Herbert (1820–1903) A	**Spencer**
Spencer, J.F.T. (fl. 1964) M	**J.F.T.Spencer**
Spencer, James A. (fl. 1972) M	**J.A.Spencer**
Spencer, Lorraine Barney (1924–) S	**L.B.Spencer**
Spencer, Roger David (1945–) S	**R.D.Spencer**
Spencer, W.I. A	**W.I.Spencer**
Spenner, Fridolin Carl Leopold (1798–1841) PS	**Spenn.**
Speranza, Maria (1949–) S	**Speranza**
Sperapani, Clare P. A	**Sperapani**
Sperk, Gustav (1845/6–1870) S	**Sperk**
Sperlich, Adolf (1879–1963) S	**Sperlich**
Spero, Howard J. A	**Spero**
Speroni, D. (fl. 1930) M	**Speroni**
Sperry, Omer Edison (1902–1975) S	**Sperry**
Speschnew, Nicolai Nicolaievich von (1844–1907) M	**Speschnew**
Spessa, Carolina (fl. 1910) M	**Spessa**
Speta, Franz (1941–) S	**Speta**
Spethmann, Wolfgang (1945–) S	**Spethmann**
Spevak, Virginia Sager (1941–) S	**Spevak**
Spicer, K.W. (fl. 1981) M	**K.W.Spicer**
Spicer, William Webb (c.1820–1879) S	**Spicer**
Spichiger, Rodolphe E. (1946–) S	**Spichiger**
Spieckermann, Albert (1871–) M	**Spieck.**
Spiegel, F.W. (fl. 1978) M	**Spiegel**
Spieler, Alexander Julius Theodor (1817–) S	**Spieler**
Spielman, Linda J. (1946–) M	**Spielman**
Spielmann, Jakob Reinhold (1722–1783) S	**Spielm.**
Spieshnev, Nikolai (1844–1907) M	**Spieshnev**
Spigai, Raffaele (1850–1895) S	**Spigai**
(Spigno, (Marquese di))	
see Spin, (Marquis de)	**Spin**
Spilger, Ludwig (1881–1941) S	**Spilger**
Spillman, William Jasper (1863–1931) S	**Spillman**
Spiltoir, Charles Francis (1918–) M	**Spiltoir**
Spin, (Marquis de) (fl. 1809) S	**Spin**
Spindler, M. (fl. 1886–1917) B	**Spindl.**
Spinedi, Horacio A. (1951–) M	**Spinedi**
Spingarn, Joel Elias (1875–1939) S	**Spingarn**
Spinner, E. A	**E.Spinner**
Spinner, Henri (1875–1962) S	**Spinner**

Spire, André (fl. 1903) S	A.Spire
Spire, Camille Joseph (fl. 1903) S	Spire
Spirlet, Marie-Louise (1918–1966) S	Spirlet
Spitzel, Anton von (1807–1853) S	Spitzel
Spitzner, Wenzel (Václav) (1852–1907) MS	Spitzn.
Spix, Johann Baptist von (1781–1826) S	Spix
Spjeldnaes, Nils (1926–) A	Spjeldn.
Splendore, Achille (1868–1928) AM	Splend.
Splitgerber, Frederik Louis (1801–1845) PS	Splitg.
Spode, F. A	Spode
Spongberg, Stephen Alex (1942–) S	Spongberg
Spooner, Brian Martin (1951–) M	Spooner
Sporleder, Friedrich Wilhelm (1787–1875) S	Sporl.
Sprague, Charles James (1823–1903) M	C.Sprague
Sprague, Elizabeth Fern (1911–) S	E.F.Sprague
Sprague, Isaac (1811–1895) S	I.Sprague
Sprague, Mary Letitia ('Manna') (née Green, M.L.) (1886–1978) S	M.L.Sprague
Sprague, Roderick (1901–1962) MS	R.Sprague
Sprague, Thomas Archibald (1877–1958) S	Sprague
Spratt, George (fl. 1830) S	G.Spratt
Spratt, Thomas Abel Brimage (1811–1888) S	T.Spratt
Sprecher von Bernegg, Andreas (fl. 1907–1930) S	Sprecher
Spreitzenhofer, G.C. (1835–1883) S	Spreitz.
Sprengel, Anton (1803–1851) AFP	A.Spreng.
Sprengel, Christian Konrad (Conrad) (1750–1816) S	C.K.Spreng.
Sprengel, Curt (Kurt, Curtius) Polycarp Joachim (1766–1833) ABMPS	Spreng.
Sprenger, Carl (Charles) Ludwig (1846–1917) S	Sprenger
Spribille, Franz Joseph (1841–1921) S	Sprib.
Spriestersbach, Julius A	Spriest.
Spring, Antoine Frédéric (1814–1872) PS	Spring
(Spring, Anton Friedrich)	
see Spring, Antoine Frédéric	Spring
Springael, R. (fl. 1977) M	Springael
Springer, Martha Edith (1916–) M	Springer
Spruce, Richard (1817–1893) BPS	Spruce
Spruner, Wilhelm von (1805–1874) S	Spruner
Sprung, Douglas C. A	Sprung
Sprygin, Ivan Ivanovic (1873–1942) S	Sprygin
Squier, Ephraim George (1812–1888) S	Squier
Squinabol, Senofonte (Xenofonte) (1861–) AF	Squinab.
Squivet de Carondelet, Joseph (1878–1966) B	Squivet
Srebrodolskaja, I.N. BS	Srebrod.
Sredinsky, Nicolia K.(C.) (1843–1908) MS	Sred.
Sreekumar, P.V. (fl. 1982) S	Sreek.
Sreelatha, P.M. A	Sreel.
Sreemadhavan, C.P. S	Sreem.
Sreenivasa, M.R. A	Sreen.
(Sreenivasa Rao, P.)	
see Rao, P.Sreenivasa	P.S.Rao

Sreeramulu, T. (fl. 1958) A   Sreer.
Sridhar, T.S. (fl. 1979) M   Sridhar
Sridharan, V.T. A   Sridharan
Srimanobhas, Vithya A   Sriman.
Srinath, K.V. (fl. 1965) M   Srinath
Srinivasalu, B.V. (fl. 1967) M   Srinivas.
Srinivasan, Kadaym Subbiah (1912–   ) A   K.S.Sriniv.
Srinivasan, Kattalaicheri Venkataraman (1917–   ) M   Sriniv.
Srinivasan, M.C. (fl. 1961) M   M.C.Sriniv.
Srinivasan, M.V. A   M.V.Sriniv.
Srinivasan, S.R. (1943–   ) S   S.R.Sriniv.
Sriskantha, A. (fl. 1980) M   Srisk.
Srisuko (fl. 1969) S   Srisuko
Srivastava, B.P. (fl. 1955) F   B.P.Srivast.
Srivastava, G. (fl. 1977) M   G.Srivast.
Srivastava, G.C. (fl. 1976) M   G.C.Srivast.
Srivastava, Gopal Krishna (1939–   ) P   G.K.Srivast.
Srivastava, H.C. (fl. 1951) M   H.C.Srivast.
Srivastava, H.P. (fl. 1972) M   H.P.Srivast.
Srivastava, K.J. (fl. 1981) M   K.J.Srivast.
Srivastava, K.P. (1920–   ) B   K.P.Srivast.
Srivastava, K.S. (fl. 1979) M   K.S.Srivast.
Srivastava, L.S. (fl. 1981) M   L.S.Srivast.
Srivastava, M.N. A   M.N.Srivast.
Srivastava, M.P. (fl. 1966) M   M.P.Srivast.
Srivastava, Madhavi A   M.Srivast.
Srivastava, O.N. (fl. 1989) M   O.N.Srivast.
Srivastava, Pratap Narain (fl. 1952–1965) A   P.N.Srivast.
Srivastava, Pushpa (fl. 1976) AMP   P.Srivast.
Srivastava, R.C. (fl. 1976) MS   R.C.Srivast.
Srivastava, S.K. (1957–   ) MS   S.K.Srivast.
Srivastava, S.L. (fl. 1966) M   S.L.Srivast.
Srivastava, Suresh Chandra (1940–   ) BS   S.C.Srivast.
Srivastava, Y.N. (fl. 1975) M   Y.N.Srivast.
Srivastra, P.N. (fl. 1952–1955) F   Srivastra
Sroëlov, R. (fl. 1965) S   Sroëlov
(Ssu, H.C.)
   see Sze, Hsing Chien   H.C.Sze
(St.-Amans, Jean Florimond Boudon de)
   see Saint-Amans, Jean Florimond Boudon de   St.-Amans
(St.-Léger, Léon)
   see Saint-Léger, Léon   St.-Leg.
St.Brody, Gustavus A.Ornano (1828–1901) S   St.Brody
St.Cloud, S.F. S   St.Cloud
St.John, Edward Porter (1866–1952) PS   E.P.St.John
St.John, Harold (1892–1991) PS   H.St.John
St.John, Robert Porter (1869–1960) PS   R.P.St.John
Stabile de Nucci, L. (fl. 1930) M   Stabile
Stabler, George (1839–1910) S   Stabler
Stabler, Robert Miller (1904–1985) A   R.M.Stabler

Staby, Ludwig Hermann Friedrich Wilhelm (1861–) S   **Staby**
Stace, Clive Anthony (1938–) PS   **Stace**
Stacey, John William (1871–1943) S   **Stacey**
Stach, E. (fl. 1957) M   **Stach**
Stache, Karl Heinrich Hector Guido (1833–1921) AFS   **Stache**
Stack, R.W. (fl. 1988) M   **Stack**
Stackhouse, John (1742–1819) AMS   **Stackh.**
Stacy, John E. (1918–) S   **Stacy**
Stadel, O. (fl. 1911) M   **Stadel**
Stadelman, F. (fl. 1975) M   **Stadelman**
Stadelmann, R.J. (fl. 1979) M   **Stadelmann**
Stadelmeyer, Ernst (–1840) S   **Stadelm.**
Stadler, Solomon (1842–1917) S   **Stadler**
Stadlmann, Josef (1881–) S   **Stadlm.**
Stadman (fl. 1810) S   **Stadman**
(Stadmann)
    see Stadman   **Stadman**
Stadtmann, Jean Frédéric (1762–1807) S   **Stadtm.**
Staes, Gustaaf (1863–1918) S   **Staes**
Stafleu, Frans Antonie (1921–) MS   **Stafleu**
Stahel, Gerold (1887–1955) M   **Stahel**
Stahl, Augustin (1842–1917) S   **A.Stahl**
Ståhl, Bertil (1957–) S   **B.Ståhl**
Stahl, Christian Ernst (1848–1919) AS   **Stahl**
Stahl, William (fl. 1966) M   **W.Stahl**
Stahlecker, Eugen (1867–) S   **Stahlecker**
Stainier, F. (fl. 1978) S   **Stainier**
Stair, Leslie Dalrymple (1876–) S   **Stair**
Stakman, Elvin Charles (1885–1979) M   **Stakman**
Stal, Lucas J.   A   **Stal**
Staley, John M. (fl. 1964) M   **Staley**
Stalpers, Joost A. (1947–) M   **Stalpers**
Stamets, P.E. (fl. 1980) M   **Stamets**
Stampfer, Cölestin (1823–1895) S   **Stampfer**
Standish, John (1814–1875) S   **Standish**
Standley, Lisa A. (1953–) S   **L.A.Standl.**
Standley, Paul Carpenter (1884–1963) PS   **Standl.**
Staněk, M. (fl. 1958) M   **M.Staněk**
Staněk, V.J. (fl. 1954) M   **V.J.Staněk**
Staner, Pierre (1901–1984) S   **Staner**
Stanfield, Dennis Percival (1903–1971) S   **Stanf.**
Stanford, Ernest Elwood (1888–) S   **Stanford**
Stange, Johann Carl (Karl) Thomas (1792–1854) S   **Stange**
Stanger, William (1811–1854) S   **Stanger**
Stanghellini, M.E. (fl. 1984) M   **Stangh.**
Stangl, Johann (fl. 1963) M   **Stangl**
Stanislavsky, Franz Antonivich (fl. 1964) BF   **Stanisl.**
(Stankevich, A.K.)
    see Stankevicz, A.K.   **Stank.**

Stankevicz, A.K. (1927–) S   **Stank.**
(Stankevij, A.K.)
  see Stankevicz, A.K.   **Stank.**
Stankov, I.T. S   **I.T.Stankov**
Stankov, Sergej Sergevich (1892–1962) S   **Stankov**
Stanley, Trevor Donald (1952–) S   **Stanley**
Stannard, Brian Leslie (1944–) S   **Stannard**
Stansbury, Howard (1806–1863) S   **Stansb.**
Stansfield, Abraham (1802–1880) PS   **Stansf.**
Stansfield, F.W. (fl. 1858–1880) P   **F.W.Stansf.**
Stansfield, Olive P. (fl. 1924) M   **O.P.Stansf.**
Stapf, Otto (1857–1933) BS   **Stapf**
Staples, George William (1953–) S   **Staples**
Staplin, Frank Lyons (1923–) A   **Staplin**
Stapp, Carl (1888–) M   **Stapp**
Star, Wim A   **Star**
Starback, Karl (1863–1931) M   **Starbäck**
Starchenko, V.M. (1950–) S   **Starch.**
Starcs, Karlis (1897–1953) S   **Starcs**
Staring, Winand Carel Hugo (1808–1877) S   **Staring**
Staritz, Richard (1851–1922) M   **Staritz**
Stark, Lloyd Ralph (1955–) BS   **L.R.Stark**
Stark, Peter (1888–1932) S   **P.Stark**
Stark, Robert Mackenzie (1815–1873) S   **Stark**
Starke, Johann Christian (1744–1808) ABS   **Starke**
Starker, Thurman James (1890–) S   **Starker**
Starling, B.N. (fl. 1985) S   **Starling**
Starmach, Karol (1900–1988) A   **Starmach**
Starmer, William T. (fl. 1976) M   **Starmer**
Starodubtzev, V.N. (1948–) S   **Starod.**
Starr, Mortimer Paul (1917–) M   **Starr**
Starr, Richard Cawthorn (1924–) A   **R.C.Starr**
Starý, Frantisek (1925–) S   **Starý**
Staszkiewicz, Jerzy (1929–) S   **Staszk.**
States, Jack S. (1941–) M   **States**
Statzell, Adele C. (fl. 1970) M   **Statzell**
Staub, Moriz (Móricz, Moritz) (1842–1904) F   **Staub**
Staub, W. (fl. 1921) M   **W.Staub**
Staubo, I. (fl. 1985) S   **Staubo**
Staude, Friedrich (–1861) M   **Staude**
Staudt, Günther (fl. 1958) S   **Staudt**
Stauffer, Clinton Raymond (1928–   ) A   **C.R.Stauffer**
Stauffer, Hans Ulrich (1929–1965) S   **Stauffer**
Staunton, George Leonard (1737–1801) S   **Staunton**
Stautz, Walter (fl. 1931) M   **Stautz**
Stavelova, E. (fl. 1986) S   **Stavelova**
Stavrinos, G.N. A   **Stavrinos**
Stawiński, Wieslaw A   **Stawiński**
(Stchegleev, Serge S.)
  see Stscheglejew, Serge S.   **Stschegl.**

Stead, J.W. (fl. 1984) S	**Stead**
Stearn, William Thomas (1911–) S	**Stearn**
Stearns, Elmer (fl. 1912) S	**E.Stearns**
Stearns, Samuel (1747–1819) S	**Stearns**
Stearns, Winfred Alden (1852–1909) S	**W.Stearns**
Stebbing, Edward Percy (1870–1960) S	**Stebbing**
Stebbins, George Ledyard (1906–) S	**Stebbins**
Stebler, Friedrich Gottlieb (1852–1935) S	**Stebler**
Stechmann, Johannes (Johann) Paul (fl. 1775) S	**Stechm.**
Steck, Abraham (fl. 1757) S	**Steck**
Stecki, Konstanty (1885–1978) S	**Stecki**
Stedje, Brita (1956–) S	**Stedje**
Stedman, John Gabriel (1744–1797) S	**Stedman**
Steedman, Ellen Constance (1859–1949) S	**Steedman**
Steedman, Henry (1866–1953) S	**H.Steedman**
Steeg, M.G.van der S	**Steeg**
Steele, Edward Strieby (1850–1942) PS	**E.S.Steele**
Steele, K.P. (fl. 1981) S	**K.P.Steele**
Steele, Richard L. A	**R.L.Steele**
Steele, William Edward (1816–1883) S	**Steele**
Steemann Nielsen, Einer (Halfdan) (1907–) A	**Steem.Niels.**
Steenhauer, Anna Johanna (fl. 1919) S	**Steenh.**
Steenis, Cornelis Gijsbert Gerrit Jan van (1901–1986) S	**Steenis**
Steenis–Kruseman, Maria Johanna van (1904–) S	**Steen.-Krus.**
Steenstrup, Johannes Japetus Smith (1813–1897) S	**Steenstr.**
Steere, William Campbell (1907–1989) B	**Steere**
Steetz, Joachim (1804–1862) S	**Steetz**
Stefani, Carlo de (1851–1924) S	**Stefani**
Stefanini, Guiseppe (1882–1938) S	**Stefan.**
Stefanoff, Boris (1894–1979) S	**Stef.**
(Stefanov, Boris)	
see Stefanoff, Boris	**Stef.**
Stefánsson, Stefán (Johann) (1863–1921) S	**Stefánsson**
Stefenelli, Silvio (1941–1983) S	**Stefen.**
Steffen, Hans (1891–) S	**Steffen**
Steffens, Henrich (Henrik) (1773–1845) S	**Steffens**
Ştefureac, Traian Ion (1908–1986) AB	**Ştefur.**
Stegenga, H. (1947–) AM	**Stegenga**
Stehlé, Henri (1909–1983) AS	**Stehlé**
Stehlé, M. (fl. 1964) S	**M.Stehlé**
Stehmann, J.R. (fl. 1987) S	**Stehmann**
Steidinger, Karen A. A	**Steid.**
Steifelhagen, H. (fl. 1910) S	**Steif.**
Steigbigel, Roy T. (fl. 1974) M	**Steigb.**
Steiger, Emil (1861–1927) S	**E.Steiger**
Steiger, Jakob Robert (1801–1862) S	**Steiger**
Steiger, Rudolf (1823–1908) S	**R.Steiger**
Steiman, R. (fl. 1985) M	**Steiman**
Stein, B.A. (fl. 1988) S	**B.A.Stein**

Stein, Berthold (1847–1899) MS	**Stein**
Stein, Friedrich (1818–1885) A	**F.Stein**
Stein, Janet Ruth (1930–) A	**J.R.Stein**
Stein, Jeffrey A. A	**J.A.Stein**
Steinberg, Carlo Hermann (1923–1981) S	**C.H.Steinb.**
Steinberg, Christian A	**C.Steinb.**
Steinberg, Elisabeth Ivanovna (1884–1963) S	**Steinb.**
Steinberg, K.M. (fl. 1936) S	**K.M.Steinb.**
Steinbroun, K.K. (fl. 1986) M	**Steinbr.**
Steindórsson, Steindór (Jónas) (1902–) S	**Steind.**
Steinecke, Fritz (1892–) AM	**Steinecke**
Steiner, Gotthold (1886–1961) M	**G.Steiner**
Steiner, J.M. (fl. 1924) M	**J.M.Steiner**
Steiner, Julius (1844–1918) M	**J.Steiner**
Steiner, Kim E. (1953–) S	**K.E.Steiner**
Steiner, Mariette E. (fl. 1989) S	**M.E.Steiner**
Steiner, Maximilian (1904–) M	**M.Steiner**
Steiner, Mona Lisa (1915–) S	**M.L.Steiner**
Steinhaus, Julius (fl. 1887) M	**Steinhaus**
Steinhauser, N. A	**Steinhauser**
Steinheil, Adolph(e) (1810–1839) S	**Steinh.**
Steininger, Hans (1856–1891) S	**Steininger**
Steinmann, (B.) Alfred (1892–) M	**B.A.Steinm.**
Steinmann, A. A	**A.Steinm.**
Steinmann, Johann Heinrich Conrad Gottfried Gustav (1856–1929) AFS	**Steinm.**
Steinvorth, Heinrich (1817–1905) S	**Steinv.**
Stejskal, M. (fl. 1974) M	**Stejskal**
Stelfox, Arthur Wilson (1883–1972) S	**Stelfox**
Stellati, Vincenzo (1780–1852) S	**Stellati**
Steller, Georg Wilhelm (1709–1746) S	**Steller**
Stellfeld, Carlos (1900–1970) S	**Stellfeld**
(Stellfield, Carlos)	
see Stellfeld, Carlos	**Stellfeld**
Stelling–Dekker, N.M. (fl. 1952) M	**Stell.–Dekk.**
Stemen, Thomas Ray (1892–1968) S	**Stemen**
Stemler, Johann Gottlieb (1788–) S	**Stemler**
Stemmerik, J.F. (fl. 1964) S	**Stemm.**
Stemmermann, L. (fl. 1981) S	**Stemmerm.**
Stenar, (Axel) Helge (Svensson) (1896–1971) S	**Stenar**
Stenfort, F. (fl. 1874) S	**Stenfort**
Stenhammar, Christian (1783–1866) MS	**Stenh.**
Stenholm, Carl (1862–1939) M	**Stenholm**
Stenina, N.P. (fl. 1968) M	**Stenina**
Stenroos, Karl Emil (later Kivirikko, K.E.) (1870–1947) S	**Stenroos**
Stenroos, Soili Kristina (1958–) M	**S.Stenroos**
Stenström, Karl Otto Edvard (1858–1901) S	**Stenstr.**
Stent, Sydney Margaret (1875–1942) S	**Stent**
Stenzel, Georg (fl. 1893) S	**G.Stenzel**
Stenzel, Karl Gustav Wilhelm (1826–1905) FS	**Stenzel**

Steova, Milka S	**Steova**
Step, Edward (1855–1931) S	**Step**
Stepanek, J. (fl. 1983) S	**Stepanek**
Stepanova, I.V. (fl. 1975) M	**Stepanova**
Stepanova, M.V. A	**M.V.Stepanova**
Stepanova-Kartavenko, N.T. (fl. 1967) M	**Step.-Kart.**
Stephan, Christian Friedrich (1757–1814) S	**Stephan**
Stephan, Paul (fl. 1930–1938) S	**P.Stephan**
Stephani, Franz (1842–1927) B	**Steph.**
Stephanitz, Alexander Ludwig (fl. 1838) S	**Stephanitz**
(Stephanov, Boris)	
see Stefanoff, Boris	**Stef.**
(Stephen, Christian Friedrich)	
see Stephan, Christian Friedrich	**Stephan**
Stephen, D. (fl. 1990) M	**Stephen**
(Stephen, Paul)	
see Stephan, Paul	**P.Stephan**
Stephens, Edith Layard (1884–1966) AMS	**Stephens**
Stephens, Frances L. (fl. 1939) M	**F.L.Stephens**
Stephens, Henry Oxley (1816–1881) A	**H.O.Stephens**
Stephens, Stanley George (1911–) S	**S.G.Stephens**
Stephenson, Anne A	**A.Stephenson**
Stephenson, John (fl. 1820–1830) S	**Stephenson**
Stephenson, Steven L. (1943–) M	**S.L.Stephenson**
Stephenson, Thomas (1865–1948) S	**T.Stephenson**
Stephenson, Thomas Alan (1898–1961) AS	**T.A.Stephenson**
Sterba, G. (fl. 1970) M	**Sterba**
(Sterbeeck, Frans (Franciscus, François) Van)	
see Van Sterbeeck, Frans (Franciscus, François)	**Van Sterbeeck**
Sterk, A.A. (1931–) S	**Sterk**
Sterler, Alois (Aloys) (1787–1831) S	**Sterler**
Sterling, Clarence (1919–) S	**Sterling**
Sterling, G.R. (fl. 1978) M	**G.R.Sterling**
Stermitz, Frank R. (1928–) S	**Stermitz**
Stern, Arthur M. (fl. 1958) M	**A.M.Stern**
Stern, Frederick Claude (1884–1967) S	**Stern**
Stern, Kingsley Roland (1927–) S	**K.R.Stern**
Stern, Kurt (1892–) S	**K.Stern**
Stern, William Louis (1926–) S	**W.L.Stern**
Sternberg, Caspar (Kaspar) Maria von (1761–1838) BFPS	**Sternb.**
Sterne, Ernst Ludwig Krause (Carus) (1839–1903) S	**Sterne**
Sterneck, Jakob (Daublebsky) von (1864–1941) S	**Sterneck**
Sterner, Ewald (1890–1940) S	**E.Sterner**
Sterner, Karl Rikard (1891–1956) S	**Sterner**
Sternheim, K. (1818–1850) S	**Sternh.**
Sternon, Fernand (1895–1945) M	**Sternon**
Sterns, Emerson Ellick (1846–1926) PS	**Sterns**
Sterrett, William Dent (1881–) S	**Sterrett**
Sterzel, Johann Traugott (1841–1914) AFS	**Sterzel**

Sterzing, J. (fl. 1860) S — **Sterzing**
Steuber, L. (fl. 1900) M — **Steuber**
Steudel, Ernst Gottlieb von (1783–1856) ABMPS — **Steud.**
Steudner, German (1822–1863) S — **Steudn.**
Steuer, Adolphe (Adolf) (1871–) A — **Steuer**
Stevels, J.M.C. (fl. 1988) S — **Stevels**
Steven, Christian von (1781–1863) S — **Steven**
Stevens, Frank Lincoln (1871–1934) M — **F.Stevens**
Stevens, George Thomas (1832–1921) S — **Stevens**
Stevens, George Walter (1868–) S — **G.W.Stevens**
Stevens, Gweneth Nell (1934–) M — **G.N.Stevens**
Stevens, Neil Everett (1887–1949) M — **N.E.Stevens**
Stevens, Orin Alva (1885–1979) S — **O.Stevens**
Stevens, Peter Francis (1944–) S — **P.F.Stevens**
Stevens, Warren Douglas (1944–) S — **W.D.Stevens**
Stevens, William Chase (1861–1955) S — **W.Stevens**
Stevenson, Dennis William (1942–) S — **D.W.Stev.**
Stevenson, Greta (fl. 1962) M — **G.Stev.**
Stevenson, John (1836–1903) M — **Stev.**
Stevenson, John Albert (1890–1979) M — **J.A.Stev.**
Stevenson, John Barr (1882–1950) S — **J.B.Stev.**
Stevenson, Robert N. A — **R.N.Stev.**
Stevenson, William Clark (1848–1919) M — **W.C.Stev.**
Steward, Albert Newton (1897–1959) S — **Steward**
Stewart, Alban N. (1875–) S — **A.Stewart**
Stewart, D.A. S — **D.A.Stewart**
Stewart, Dewey (1898–) M — **D.Stewart**
Stewart, Elwin L. (1940–) M — **E.L.Stewart**
Stewart, Fred Carlton (1868–1946) M — **F.C.Stewart**
Stewart, Harold Lyndon (1941–) M — **H.L.Stewart**
Stewart, Joan Godsil (1932–) A — **J.G.Stewart**
Stewart, John Lindsay (1832–1873) S — **J.Stewart**
Stewart, Joyce L. (1936–) S — **J.L.Stewart**
Stewart, Kenneth D. A — **K.D.Stewart**
Stewart, Margaret Gaylord (later Grover) (1911–) S — **M.G.Stewart**
Stewart, Ralph Randles (1890–) BPS — **R.R.Stewart**
Stewart, Robert (1811–1865) S — **Stewart**
Stewart, Robert B. (fl. 1957) M — **R.B.Stewart**
Stewart, Samuel Alexander (1826–1910) S — **S.Stewart**
Stewart, Sara R. (1913–) S — **S.R.Stewart**
Stewart, Vern Bonham (1888–1918) M — **V.B.Stewart**
Stewart, W.Alan A — **W.A.Stewart**
Stewart, William Sheldon (1914–) S — **W.S.Stewart**
Steyaert, René Léopold Alix Ghislain Jules (1905–1978) MS — **Steyaert**
Steyermark, Julian Alfred (1909–1988) PS — **Steyerm.**
Steyn, R.L. (fl. 1987) M — **Steyn**
Stidolph, Stuart R. A — **Stidolph**
Stiebel, Salomon Friedrich (1792–1868) A — **Stiebel**
Stieber, Michael Thomas (1943–) S — **Stieber**

Stiefelhagen, Heinz (fl. 1910) S	**Stiefelh.**
Stiehler, August Wilhelm (1797–1878) F	**Stiehler**
Stifler, Mary Cloyd Burnley (1876–) M	**Stifler**
Stiles, Charles Wardell (1867–1933) A	**Stiles**
Stiles, F.Gary (fl. 1979) S	**F.G.Stiles**
Stiles, George Whitford (1977–) M	**G.W.Stiles**
Stillingfleet, Benjamin (1702–1771) S	**Still.**
Stipanicic (fl. 1949) F	**Stip.**
Stipes, Roland Jay (1936–) M	**Stipes**
Stirling, James (1852–1909) BS	**J.Stirl.**
Stirling, James (1836–1916) B	**Stirl.**
Stirton, Charles Howard (1946–) S	**C.H.Stirt.**
Stirton, James (1833–1917) BM	**Stirt.**
Stizenberger, Ernst (1827–1895) AM	**Stizenb.**
Stockdale, Phyllis Margaret (1927–1989) M	**Stockdale**
Stocker, Otto (1888–1979) S	**Stocker**
Stockhouse, R.E. (fl. 1983) S	**Stockh.**
Stocking, Kenneth Morgan (1911–) S	**Stocking**
Stockmans, François (1904–1986) AF	**Stockmans**
Stockmayer, Siegfried (1868–1933) A	**Stockm.**
Stocks, John Ellerton (1822–1854) A	**Stocks**
Stockwell, William Palmer (1898–1950) S	**Stockw.**
Stoddart, A.B. (fl. 1916) M	**A.B.Stoddart**
Stoddart, J.A. (fl. 1989) S	**J.A.Stoddart**
Stoddart, J.L. (fl. 1916) M	**J.L.Stoddart**
Stodder, Charles (1808–1883) A	**Stodder**
Stoelzel, Victoria A. A	**Stoelzel**
(Stoerck, Anton von)	
see Störck, Anton von	**Störck**
Stoermer, Eugene Filmore (1934–) A	**Stoermer**
Stoeva, Milka P. S	**Stoeva**
Stoever, Dietrich Heinrich (1767–1822) S	**Stoever**
Stoffers, Anton Lambertus (1926–) S	**Stoffers**
Stöhr, Emil A	**E.Stöhr**
Stohr, Gerrit (1928–) S	**Stohr**
Stoianovitch, Carmen (fl. 1960) M	**Stoian.**
Stoitschkoff, Jord S	**Stoitschk.**
Stojanov, Nikolai Andreev (1883–1968) S	**Stoj.**
Stojko, S.M. (1920–) S	**Stojko**
Stoker, Fred (1878–1943) S	**Stoker**
Stokes, Alfred Cheatham (fl. 1893) A	**A.Stokes**
Stokes, Jonathan S. (1755–1831) PS	**Stokes**
Stokes, Susan Gabriella (1868–1954) S	**S.Stokes**
Stokes, Whitley (1763–1845) B	**W.Stokes**
Stokes, William Royal (fl. 1896) M	**W.R.Stokes**
Stokey, Alma Gracey (1877–1968) P	**Stokey**
Stoletova, E.A. (1889–1964) S	**Stolet.**
Stoliczka, Ferdinand (1838–1874) F	**Stoliczka**
Stolk, Amelia C. (fl. 1955) M	**Stolk**

Stoll, C. (fl. 1969) M	**C.Stoll**
Stoll, O. (fl. 1904) M	**Stoll**
Stolle, Emil (1868–1940) B	**Stolle**
(Stöller, Georg Wilhelm)	
see Steller, Georg Wilhelm	**Steller**
Stolley, Ernst (1869–) AF	**Stolley**
Stoloff, Leonard (1915–) A	**Stoloff**
Stolt, Karl Alrik Hugo (1885–1942) S	**Stolt**
Stoltenberg, Nicolaus Theodor Marcus (1844–) S	**Stoltenb.**
Stolterforth, Henry A	**Stolterf.**
Stoltz, Jean Louis (1804–1896) S	**Stoltz**
(Stoltz, Jean–Chrétien)	
see Stolz, Jean-Chrétien (Johann Christian)	**J.-C.Stolz**
Stolz, Adolf Ferdinand (1871–1917) S	**Stolz**
Stolz, Friedrich (1878–1899) BC	**F.Stolz**
Stolz, Jean-Chrétien (Johann Christian) (1764–1828) S	**J.-C.Stolz**
Stolze, Robert G. (1927–) PS	**Stolze**
Stomps, Theodoor Jan (1885–1973) S	**Stomps**
Stone, Benjamin Clemens Masterman (1933–) PS	**B.C.Stone**
Stone, C. (fl. 1987) M	**C.Stone**
Stone, Donald Eugene (1930–) S	**D.E.Stone**
Stone, George Edward (1860–1941) S	**G.Stone**
Stone, Herbert (1865–) S	**H.Stone**
Stone, Ilma Grace (1913–) B	**I.G.Stone**
Stone, J.L. S	**J.L.Stone**
Stone, Jeffrey K. (fl. 1986) M	**J.K.Stone**
Stone, Robert J. (1751–1829) A	**Stone**
Stone, Roland Elisha (1881–1939) M	**R.Stone**
Stone, Witmer (1866–1939) S	**W.Stone**
Stoneburner, Ann Hudson (1947–) B	**A.H.Stoneb.**
Stoneburner, Daniel Lee (1945–) A	**D.L.Stoneb.**
Stoneman, Bertha M. (1866–1943) M	**Stoneman**
Stopes, Marie Charlotte Carmichael (1880–1958) FMPS	**Stopes**
Stopp, Klaus Dieter (1926–) S	**Stopp**
Stoppel, Rose (fl. 1907) M	**Stoppel**
Storch, Franz de Paula (1812–) S	**Storch**
Störck, Anton von (1731–1803) S	**Störck**
Storie, James G. (fl. 1880s) S	**Storie**
Stork, Adélaïde Louise (1937–) S	**Stork**
Storm, P.K. (fl. 1967) M	**P.K.Storm**
Storm, Vilhelm (Ferdinand Johan) (1835–1913) S	**Storm**
Störmer, Fredrik Carl Mülerz (1874–) M	**F.C.M.Störmer**
Størmer, Per (1907–) B	**Størmer**
Storrie, John (1843–1901) S	**Storrie**
Story, Robert (1913–) S	**Story**
Stosch, Hans Adolf von (1908–1987) A	**Stosch**
Stotler, Raymond Eugene (1940–) B	**Stotler**
Stotler–Crandall, Barbara B	**Stotl.-Crand.**
Stouffer, David James (1900–) M	**Stouffer**

Stout, Arlow Burdette (1876–1957) S	**Stout**
Stout, Gilbert Leonidas (1898–1963) M	**G.L.Stout**
Stover, Lewis Eugene (1925–　) A	**Stover**
Stover, Robert Harry (1926–　) M	**R.H.Stover**
Stovold, G.E. (fl. 1986) M	**Stovold**
Stowell, Willard Allen (　–1929) P	**Stowell**
(Stoyanoff, Nikolai Andreev)	
see Stojanov, Nikolai Andreev	**Stoj.**
Strachan, Isles　A	**I.Strachan**
Strachan, Jeffrey L. (1956–　) S	**Strachan**
Strachey, Richard (1817–1908) S	**Strachey**
Strack, Dieter (fl. 1987) S	**Strack**
Stradner, Herbert　A	**Stradner**
(Straehler, Adolf)	
see Strähler, Adolf	**Strähler**
Strafforello, Ildefonso (1823–1899) S	**I.Straff.**
Strafforello, J. (　–1883?) A	**J.Straff.**
Strähler, Adolf (1829–1897) S	**Strähler**
Strail, Charles Antoine (1808–1893) S	**Strail**
Straley, Gerald B. (fl. 1977) S	**Straley**
Strampelli, Nazareno　S	**Stramp.**
Strand, Bengt Johan (fl. 1756) S	**Strand**
Strand, Embrik (1876–　) A	**E.Strand**
Strandhede, Sven Olof (1930–　) S	**Strandh.**
Strang, Harold Edgard (1921–　) S	**Strang**
(Strangways, William Thomas Horner Fox)	
see Fox–Strangways, William Thomas Horner	**Fox–Strangw.**
Strank, K.J. (fl. 1985)　S	**Strank**
Strasburger, Eduard Adolf (1844–1912) AMPS	**Strasb.**
Strasser, Pius (1843–1927) M	**Strasser**
Straszewski, H.von (1887–1944) PS	**Strasz.**
Stratingh, Gozewinus Acker (1804–1876) S	**Strat.**
Stratman, Mary Agatha (1894–　) S	**Stratman**
Stratton, Frederic (1840–1916) S	**Stratton**
Stratton, Robert (1883–　) M	**R.Stratton**
Straub, Ernst Wilhelm　A	**E.W.Straub**
Straub, J.　S	**Straub**
Straus, Adolf Paul Carl (1901–　) ABFM	**Straus**
Strausbaugh, Perry Daniel (1886–1965) S	**Strausb.**
Strauss, Friedrich Karl (Carl) Joseph von (1787–1855) M	**F.Strauss**
Strauss, Heinrich Christian (1850–1922) S	**Strauss**
Strauss, Theodor (1859–1911) S	**T.Strauss**
Straw, Richard Myron (1926–　) S	**Straw**
Streber, F.　A	**Streber**
Strecker, Wilhelm (1858–1934) S	**Strecker**
Strehler, Ludwig F. (fl. 1841) S	**Strehler**
Strehlow, Karl　A	**Strehlow**
Streiblová, Eva (fl. 1963) M	**Streibl.**
Streimann, Heinar (1938–　) BM	**Streimann**

Streintz, Joseph Anton (fl. 1843) S	**Streintz**
Streinz, Wenzel Matern (1792–1876) S	**Streinz**
Strelnikova, N.I. A	**Streln.**
Strempel, Johannes Karl (Carl) Friedrich (1800–1872) PS	**Strempel**
Strenz, W.M. (1792–1876) M	**Strenz**
Stresemann, Erwin (1889–1972) S	**Stresem.**
Stretton, Hellena M. (fl. 1967) M	**Stretton**
Strey, Rudolf Georg (1907–1988) S	**Strey**
Strgar, Vinko (fl. 1971) S	**Strgar**
Stríbrný, Václav (1853–1927) S	**Stríbrný**
Strickland, C. (fl. 1911) A	**Strickland**
Strickland, John Claiborne, Jr. (1915–) A	**J.C.Strickland**
Strid, Åke (1932–) M	**Å.Strid**
Strid, P. Arne K. (1943–) S	**Strid**
Strigl, F. (fl. 1979) S	**Strigl**
Stritch, L.R. (fl. 1982) S	**Stritch**
Strizhova, T.G. (1935–) S	**Strizhova**
Strobel, Gary A. (fl. 1990) M	**Strobel**
Strobl, P. Gabriel (1846–1925) PS	**Strobl**
Strödel, R. (fl. 1984) M	**Strödel**
(Stroem, Hans)	
see Strøm, Hans	**Strøm**
Stroh, Georg (1864–) S	**Stroh**
Strohecker, Jonas Rudolph (fl. 1869) S	**Strohecker**
Strohmeyer, Otto (August Karl) (1869–) S	**Strohm.**
Strohschneider, I. S	**Strohschn.**
Strøm, Hans (1726–1797) BS	**Strøm**
Strøm, Kaare Münster (1902–) A	**K.Strøm**
Strömbom, Nils Gustaf (1847–1897) M	**Strömbom**
Stromeyer, August Adolf Ludwig (fl. 1829) S	**A.Stromeyer**
Stromeyer, Friedrich (1776–1835) S	**F.Stromeyer**
Stromeyer, Johann Friedrich (1750–1830) S	**Stromeyer**
Strömfelt, Harald Fredrik Georg (1861–1890) AS	**Strömf.**
Strong, Asa B. (fl. 1850) S	**Strong**
Strong, Richard (1945–) S	**R.Strong**
Strongman, D.B. (fl. 1985) M	**Strongman**
Stropnik, Z. (fl. 1988) M	**Stropnik**
Štros, F. (fl. 1981) M	**Štros**
Ströse, Karl (1853–) S	**Ströse**
Strother, John Lance (1941–) S	**Strother**
Struck, M. (fl. 1986) S	**Struck**
Strugnell, Edmund Jardine (1903–) S	**Strugnell**
Strunk, H.F. (fl. 1903) M	**Strunk**
Strutt, Jacob George (fl. 1820–1850) S	**Strutt**
Struve, Curt (fl. 1872) S	**C.Struve**
Struve, Gustav Adolph (1811–1889) S	**Struve**
Strzeszewski, Bol. A	**Strzesz.**
(Stschegleew, Serge S.)	
see Stscheglejew, Serge S.	**Stschegl.**

Stscheglejew, Serge S. (fl. 1851) S	**Stschegl.**
Stschepotjev, F.L. S	**Stschep.**
(Stuart, Combertus Pieter Cohen)	
see Cohen–Stuart, Combertus Pieter	**Cohen–Stuart**
Stuart, D.C. (1940–) S	**D.C.Stuart**
Stuart, John, Earl of Bute (1713–1792) S	**Stuart**
Stubblefield, S.P. (fl. 1983) M	**Stubblef.**
Stübel, (Moritz) Alphons (1835–1904) S	**Stübel**
Stüben, H. (fl. 1939) M	**Stüben**
Stübing, Gerardo (1957–) S	**Stübing**
Stuchlik, Jaroslav (1890–1967) S	**Stuchlik**
Stuckenberg, Elisabeth K. (1883–) M	**Stuckenb.**
Stuckert, Teodoro (Theodor) Juan Vicente (1852–1932) S	**Stuck.**
Stuckey, Ronald Lewis (1938–) S	**Stuckey**
Studer–Steinhäuslin, Bernhard (1847–1910) M	**Stud.-Steinh.**
Studnička, Miloslav (1949–) S	**Studnička**
Studt, Werner (1894–) S	**Studt**
Stuessy, Tod Falor (1943–) S	**Stuessy**
Stuhlmann, Franz Ludwig (1863–1928) S	**Stuhlmann**
Stümcke, Martin (1853–1927) S	**Stümcke**
Stumpf, Johann Georg (1750–1798) S	**Stumpf**
Stuntz, Daniel Elliot (1909–1983) M	**D.E.Stuntz**
Stuntz, Stephen Conrad (1875–1918) BS	**Stuntz**
Stupper, Carl (Karl) (1808–1874) S	**Stupper**
Stur, Dionys Rudolf Josef (1827–1893) FS	**Stur**
(Štur, Dionýz)	
see Stur, Dionys Rudolf Josef	**Stur**
Sturch, H.H. A	**Sturch**
Sturgis, William Codman (1886–1942) CM	**Sturgis**
Sturhan, D. (fl. 1981) M	**Sturhan**
Sturm, Jacob (Jakob) W. (1771–1848) ABMPS	**Sturm**
Sturm, Johann Wilhelm (1808–1865) PS	**J.W.Sturm**
Sturm, Karl (1879–) S	**K.Sturm**
Sturt, Charles (1795–1869) S	**Sturt**
Sturt, G. (1860–1947) A	**G.Sturt**
Sturtevant, Edward Lewis (1842–1898) S	**Sturtev.**
Stutz, Howard Coombs (1918–) S	**Stutz**
Stützel, Thomas (1954–) S	**Stützel**
Stutzer, M.J. (fl. 1926) M	**M.J.Stutzer**
Stützer, Olga (fl. 1935) S	**O.Stützer**
Stüwe, Hermann Heinrich Wilhelm Theodor (1875–) AS	**Stüwe**
Stuxberg, Anton Julius (1849–1902) S	**Stuxberg**
Styles, Brian Thomas (1936–) S	**Styles**
Su, C.H. (fl. 1986) M	**C.H.Su**
Su, Ho Yi (1937–) S	**H.Y.Su**
(Su, Hong Ji)	
see Su, Horng Jye	**H.J.Su**
Su, Horng Jye (fl. 1978) MS	**H.J.Su**

(Su, Qing Hua)	
see Su, C.H.	**C.H.Su**
Su, Song Wang (fl. 1982) S	**S.W.Su**
Su, Zhi Yun (1936–) S	**Z.Y.Su**
Su, Zong Ming (fl. 1985) S	**Z.M.Su**
Suard, V. (fl. 1839) S	**Suard**
Suárez Peregrín, E. A	**Suarez**
Suarez, Verónica L. (fl. 1990) M	**V.L.Suarez**
Suárez–Cervera, M.A. (1943–) S	**Suárez–Cerv.**
Subba Rao, Gorti Venkata (1928–) S	**Subba Rao**
Subbalakshmi, G. (fl. 1987) M	**Subbal.**
(Subbaraju)	
see Subbaraju N.	**Subbaraju**
Subbaraju N. A	**Subbaraju**
Suberkropp, Keller (fl. 1990) M	**Suberkr.**
Subhedar, A.W. (fl. 1976) M	**Subhedar**
Subhedar, R.P. (fl. 1985) S	**R.P.Subhedar**
Subík, H. S	**Subík**
Subils, Rosa (1929–) S	**Subils**
Subrahmanyam, A. (fl. 1978) AM	**Subrahm.**
Subrahmanyan, R. (fl. 1943) A	**Subrahman.**
Subramaniam, Lekshminarayananpuram Subramania (1890–) M	**Subraman.**
Subramanian, B. A	**B.Subram.**
Subramanian, Chirayathumadom Venkatachalier (1924–) M	**Subram.**
Subramanian, Dubagunta (1931–) A	**D.Subram.**
Subramanian, G. A	**G.Subram.**
Subramanyam, Krishnaier (1915–1980) S	**Subr.**
Subramoniam, V. (fl. 1975) M	**Subramon.**
Suc, Louis (fl. 1912) S	**Suc**
Succow, Friedrich Wilhelm Ludwig (1770–1838) S	
see Suckow, Friedrich Wilhelm Ludgwig	**F.W.L.Suckow**
(Succow, Laurenz (Lorenz) Johann Daniel)	
see Suckow, Laurenz (Lorenz) Johann Daniel	**L.Suckow**
(Suchanova, I.N.)	
see Sukhanova, I.N.	**Sukhanova**
Suchlandt, Otto (1873–1947) AS	**Suchl.**
Suchovejeva, M.V. A	**Suchov.**
Suckow, Friedrich Wilhelm Ludwig (1770–1838) S	**F.W.L.Suckow**
Suckow, Georg Adolph (1751–1813) S	**Suckow**
Suckow, Laurenz (Lorenz) Johann Daniel (1722–1801) S	**L.Suckow**
Suckow, Sigismund (1845–) S	**S.Suckow**
Sucre Benjamin, Dimitri (fl. 1970) S	**Sucre**
Suctancar, Crisna A	**Suctancar**
Sudha, K. (fl. 1978) M	**Sudha**
Sudnitsyna, I.G. (fl. 1974) S	**Sudn.**
Sudre, Henri L. (1862–1918) PS	**Sudre**
Sudworth, George Bishop (1864–1927) S	**Sudw.**
Sudzuki, H.Fusa S	**Sudzuki**
Suessenguth, Karl (1893–1955) APS	**Suess.**
Suetin, S.O. (fl. 1981) M	**Suetin**

Suffren, François Palamède de (1753–1824) S   **Suffren**
Suga, Hidefumi   A   **Suga**
Sugano, Y. (fl. 1930) M   **Sugano**
Sugawara, Shigezo (fl. 1937) S   **Sugaw.**
Sugawara, T. (fl. 1985) S   **T.Sugaw.**
Sugaya, Sadao (1917–) S   **Sugaya**
Sugden, A.M. (fl. 1982) S   **Sugden**
Suggs, Edna G. (fl. 1974) M   **Suggs**
Sugimoto, Junichi (1901–) PS   **Sugim.**
Sugimoto, M. (fl. 1986) S   **M.Sugim.**
Sugiyama, Junta (1939–) M   **Sugiy.**
Sugiyama, Keiichi (fl. 1973) M   **K.Sugiy.**
Sugiyama, Y. (fl. 1981) M   **Y.Sugiy.**
Suh, Young Bae (fl. 1990) S   **Y.B.Suh**
Suhonen, Pentti (Veikko Yrjänä) (1888–1966) S   **Suhonen**
Suhr, Johannes (1882–) A   **J.Suhr**
Suhr, Johannes Nicolaus von (1792–1847) A   **Suhr**
Sujkowski, Zbigniew Leliwa (1898–1954) A   **Sujkowski**
(Sukachev, Vladimir Nikolajevich)
    see Sukaczev, Vladimir Nikolajevich   **Sukaczev**
Sukaczev, Vladimir Nikolajevich (1880–1967) AS   **Sukaczev**
Sukaczeva, H.I.   S   **Sukaczeva**
Sukapure, R.S. (fl. 1962) M   **Sukapure**
(Sukatschew, Vladimir Nikolajevich)
    see Sukaczev, Vladimir Nikolajevich   **Sukaczev**
Sukawa, Tschonoski (Chonosuke) (1841–1925) S   **Sukawa**
Sukh, Dev (fl. 1959) F   **Sukh**
Sukhanova, I.N.   A   **Sukhanova**
Sukhoveeva, M.V.   A   **Sukhov.**
Sukopp, Herbert (1930–) S   **Sukopp**
Suksdorf, Wilhelm Nikolaus (1850–1932) PS   **Suksd.**
Šula, Josef (1909–) A   **Šula**
Sulakadze, Tamara S. (1900–) S   **Sulak.**
Šulc, Karel (fl. 1910) M   **Šulc**
Sulek, Josef   A   **Sulek**
Sulistiarini, D. (1955–) S   **Sulist.**
Sullia, Shanker Bhat (1940–) M   **Sullia**
Sullivan, David (1836–1895) S   **Sullivan**
Sullivan, Frank R.   A   **F.R.Sullivan**
Sullivan, G.A. (fl. 1985) S   **G.A.Sullivan**
Sullivan, Janet R. (1955–) S   **J.R.Sullivan**
Sullivan, Michael J.   A   **M.J.Sullivan**
Sullivant, William Starling (1803–1873) AB   **Sull.**
Sulmont, Ph. (fl. 1969) M   **Sulmont**
Sultana, F.Kauser (fl. 1980) M   **Sultana**
Sultana, Tayyaba   A   **T.Sultana**
Sultanova, B.A. (1941–) S   **Sultanova**

Sulzberger, Robert (fl. 1888) S	**Sulzb.**
Sümbül, Hüseyin (1955–) S	**Sümbül**
Suminoe, Kinsi (fl. 1926) M	**Suminoe**
Summerbell, Richard C. (fl. 1983) M	**Summerb.**
Summerfelt, Sörren Christian (1794–1838) S	**Summerf.**
Summerhayes, Victor Samuel (1897–1974) S	**Summerh.**
Summers, Lucia (1839–1898) P	**Summers**
Sumner, D.R. (fl. 1971) M	**D.R.Sumner**
Sumner, George (1793–1855) S	**Sumner**
Sumnevicz, Georgji Prokopievič (1909–1947) PS	**Sumnev.**
(Sumnevitcz, Georgji Prokopievič)	
see Sumnevicz, Georgji Prokopievič	**Sumnev.**
Sumstine, David Ross (1870–1965) MS	**Sumst.**
Sun, Bi Sin (1921–) S	**B.S.Sun**
Sun, Chen Reng (fl. 1988) S	**C.R.Sun**
(Sun, Hsiang Chung)	
see Sun, Siang Chung	**S.C.Sun**
Sun, Ji Liang (1952–) S	**J.L.Sun**
Sun, S.H. (fl. 1955) M	**S.H.Sun**
Sun, San Sheng (fl. 1989) S	**S.S.Sun**
Sun, Shou Kung (fl. 1978) M	**S.K.Sun**
Sun, Siang Chung (1908–) S	**S.C.Sun**
Sun, Tai Yang (fl. 1979) S	**T.Y.Sun**
(Sun, Xiang Zhong)	
see Sun, Siang Chung	**S.C.Sun**
Sun, Yon Zai S	**Y.Z.Sun**
Sun, Zeng Mei (fl. 1974) MS	**Z.M.Sun**
(Sundara Raghavan, R.)	
see Sundararaghavan, R.	**Sundararagh.**
Sundaralingam, Vellore Somasundaram (1918–) A	**Sundaral.**
Sundaram, Nangavalli Venkatesan (1924–) M	**Sundaram**
Sundararaghavan, R. (fl. 1982) S	**Sundararagh.**
Sundararaj, Daniel (1919–) S	**Sundararaj**
Sundararajan, M. A	**Sundararajan**
Sundararaman, S. (fl. 1922) M	**Sundar.**
Sundberg, S.D. (fl. 1984) S	**S.D.Sundb.**
Sundberg, Walter J. (1939–) M	**Sundb.**
Sundell, Eric (1942–) S	**Sundell**
Sundene, Ove A	**Sundene**
Sündermann, Franz (1864–1946) S	**Sünd.**
Sundermann, Hans (fl. 1961–1982) S	**H.Sund.**
Sunding, Per (1938–1980) S	**Sunding**
Sundström, Bo G. A	**Sundström**
Suneson, Karl Svante (1904–) AS	**Suneson**
Sung, Sun Huang (fl. 1950) M	**S.H.Sung**
Sung, Wan Chih (fl. 1970) S	**W.C.Sung**
Sunhede, Stellan (1942–) M	**Sunhede**
Supplie, F. (fl. 1988) S	**Supplie**
Supprian, Karl Wilhelm (1871–1917) S	**Supprian**

Suprun, T.P. (fl. 1957) M	**Suprun**
Sur, Banani (fl. 1985) M	**B.Sur**
Sur, P.R. (fl. 1975) S	**Sur**
Surana, A.C. (fl. 1971) F	**Surana**
Surange, Krishna Rajaram (1920–) F	**Surange**
Sureau, P. (fl. 1959) M	**Sureau**
Suresh, C.R. (fl. 1988) S	**Suresh**
Surgis, Eugène (fl. 1921) S	**Surgis**
Surian, Joseph Donat (–1691) LS	**Surian**
Suringar, Gerard Conrad Bernard (1802–1874) S	**G.Suringar**
Suringar, Jan Valckenier (1864–1932) S	**J.V.Suringar**
Suringar, Willem Frederik Reinier (1832–1898) AS	**Suringar**
Suriray A	**Suriray**
Suryanarayana, G. A	**G.Suryan.**
Suryanarayana, M.C. S	**Suryan.**
Suryanarayana, K. (fl. 1954) F	**K.Suryan.**
Susaki, Chusuke (1866–1933) PS	**Susaki**
(Susanna de la Serna, Alfonso)	
see Susanna, Alfonso	**Susanna**
Susanna, Alfonso (1956–) S	**Susanna**
Suse, Theodor (fl. 1910) B	**Suse**
Suslova, T.A. (1944–) S	**Suslova**
(Süssenguth, Karl)	
see Suessenguth, Karl	**Suess.**
Süssli, Peter A	**Süssli**
Šutara, Josef (fl. 1987) M	**Šutara**
Sutejew, G. (fl. 1929) M	**Sutejew**
Suter, Johann Rudolf (1766–1827) S	**Suter**
Sutherland, George Kenneth (fl. 1915) M	**G.K.Sutherl.**
Sutherland, James (c.1639–1719) L	**Sutherl.**
Sutherland, Jean L. A	**J.L.Sutherl.**
Sutherland, Joan A	**J.Sutherl.**
Sutherland, William (1832–1920) S	**W.Sutherl.**
Sutherland-Campbell, H. (fl. 1934) M	**Suth.-Campb.**
Sutô, Tiharu (1910–) S	**Sutô**
Suto, Y. (fl. 1980) M	**Y.Suto**
Sutorý, Karel (1947–) S	**Sutorý**
Sutra, G. (fl. 1972) M	**Sutra**
Sutton, Brian Charles (1938–) M	**B.Sutton**
Sutton, Charles (1756–1846) S	**Sutton**
Sutton, David A. (1952–) S	**D.A.Sutton**
Sutton, John Mayhew (1867–) S	**J.M.Sutton**
Sutulov, A.N. (fl. 1914) S	**Sutulov**
(Sutulow, A.N.)	
see Sutulov, A.N.	**Sutulov**
Suvatabandhu, Kasin (1916–) S	**Suvat.**
Suvorov, Vladimir Vasilevich (1902–) S	**Suvorov**
Suvorova, T.N. (1904–1967) S	**Suvorova**
Suxena, Mahendra Raj (1914–) A	**Suxena**

Suza, Jindrich (1890–1951) BMS	**Suza**
Suzuki, Hyoji (1915–) BPS	**H.Suzuki**
Suzuki, Kiyoshi (fl. 1982) PS	**K.Suzuki**
Suzuki, Motofumi (fl. 1987) MS	**M.Suzuki**
Suzuki, Sadao S	**Sad.Suzuki**
Suzuki, Shigetaka S	**Shig.Suzuki**
Suzuki, Shizuo (fl. 1962) M	**S.Suzuki**
Suzuki, Sigeyosi (1894–1937) S	**Suzuki**
Suzuki, Tadashi (fl. 1982) B	**Tad.Suzuki**
Suzuki, Tokio (1911–) PS	**T.Suzuki**
Svanidze, T.V. (fl. 1984) M	**Svanidze**
Svanlund, (Johan) Fredrik (Friedrich) (Eberhard) (1832–1902) S	**Svanlund**
Svarchevskii, Boris Aleksandrovich (1872–1930) A	**Svarch.**
(Svartz, Olof (Peter))	
see Swartz, Olof (Peter)	**Sw.**
Švecov, M.S. A	**Švecov**
Svedelius, Nils (Eberhard) (1873–1960) A	**Sved.**
Svendsen, Per A	**Svendsen**
Svenson, Henry Knute (Knut) (1897–1986) PS	**Svenson**
Svensson, Harry Gustaf (1894–1959) S	**H.G.Svenss.**
Svensson, Nicolaus Alexis (1871–) S	**N.Svenss.**
Svensson, Per (1839–1917) S	**Svenss.**
Sventenius, Eric R.Svensson (1910–1973) S	**Svent.**
Sveshnikov, P.de (fl. 1898) B	**Sveshn.**
Sveshnikova, I.N. (fl. 1967) F	**Sveshnik.**
Svestka, Frantisek (1898–) S	**Svestka**
Svihla, Ruth Isobel Dowell (1897–1974) B	**Svihla**
Svilvinyi, A.von (fl. 1936) M	**Svilv.**
Svirenko, D.O. A	**Svirenko**
Svoboda, Pravdomil (1908–) S	**Svoboda**
(Svobodová, Yvonne)	
see Svobodová-Poláková, Yvonne	**Svob.-Pol.**
Svobodová-Poláková, Yvonne (fl. 1959) M	**Svob.-Pol.**
Svrček, Mirko (1925–) M	**Svrček**
Swahari Sasibushan (Sasibhushan) (fl. 1982) S	**Swahari**
Swails, Lawrence F. (1932–) B	**Swails**
Swainson, William (1789–1855) S	**Swainson**
Swale, Erica M.F. (1929–) A	**Swale**
Swallen, Jason Richard (1903–1991) BS	**Swallen**
Swallow, George Clinton (1817–1899) S	**Swallow**
Swaminathan, Monkombu Sambasivan (1941–) S	**Swamin.**
Swamy, B.G.L. (fl. 1948) S	**Swamy**
Swamy, S.N.R. S	**S.N.R.Swamy**
Swank, George (fl. 1953) M	**Swank**
Swanlund, Claes Magnus Julius Richard (1875–1952) S	**Swanlund**
Swanson, J.R. (fl. 1986) S	**Swanson**
Swanton, Ernest William (Brockton) (1870–1958) S	**Swanton**
Swart, Haring ('Harry') Johannes (1922–) M	**H.J.Swart**
Swart, Jan Johannes (1901–1974) S	**Swart**

(Swarts, Olof (Peter))	
see Swartz, Olof (Peter)	Sw.
Swartz, Olof (Peter) (1760–1818) ABMPS	Sw.
Swarup, J. (fl. 1966) M	Swarup
Swarupanandan, Kundil (1952–) S	Swarupan.
(Swarz, Olof (Peter))	
see Swartz, Olof (Peter)	Sw.
Swatman, Cecil Charles (1884?–1958) A	Swatman
Swayne, George (1746–1827) S	Swayne
Swedelius, N. (1873–) S	Swedel.
Swederus, Magnus Bernhard (1840–1911) S	Swederus
Sweeney, A.W. (fl. 1975) M	A.W.Sweeney
Sweeney, Beatrice Marcy (1914–1989) A	Sweeney
Sweet, Herman Royden (1909–) S	H.R.Sweet
Sweet, Robert (1783–1835) PS	Sweet
Sweetser, Albert Raddin (1861–1940) S	Sweetser
Sweigger, August Friedrich (1783–1821) S	Sweigger
Swellengrebel, Nicholaas Hendrik (1885–) A	Swell.
Swete, Edmund Horace (1827–1912) S	Swete
Swett, Keene (1932–) A	Swett
Swezey, Goodwin Deloss (1851–1934) S	Swezey
Swezy, Olive (1878–) A	Swezy
Swift, Elijah V. (1938–) A	E.V.Swift
Swift, Marjorie Elizabeth (1905–) M	Swift
Swinbourne, Robert F.G. S	Swinb.
Swindell, Nellie A	Swindell
Swingle, Deane Bret (1879–1944) M	D.B.Swingle
Swingle, Leroy Dey (1881–) A	L.D.Swingle
(Swingle, Maud Kellerman)	
see Kellerman, Maud	M.Kellerm.
Swingle, Walter Tennyson (1871–1952) MS	Swingle
Swinhoe, Robert (1836–1877) S	Swinhoe
Swinscow, Thomas Douglas Victor (1917–) M	Swinscow
Swynnerton, Charles Francis Massey (1877–1938) S	Swynn.
Sydow, Hans (1879–1946) MS	Syd.
Sydow, Paul (1851–1925) BM	P.Syd.
Syed, Hadiuzzaman (fl. 1973) B	Syed
Syeda, S.T. (fl. 1980) S	Syeda
Sykes, William Russell (1927–) S	Sykes
(Sykora, Josef)	
see Sikora, Josef	Sikora
Sylvén, Nils Olof Valdemar (1880–1969) S	Sylvén
Syme, John Thomas Irvine Boswell (1822–1888) PS	Syme
Symes, Michael S	Symes
Symington, Colin Fraser (1905–1943) S	Symington
Symoens, Jean–Jacques A. (1927–) AS	Symoens
Symon, David Eric (1920–) S	Symon
Symons, Jelinger (1778–1851) PS	Symons
Symons–Jeune, Bertram Hamner Bunbury (fl. 1907–1953) S	Sym.–Jeune

Syozi, Yositika S	**Syozi**
Syrach–Larsen, Carl (1898–) S	**Syrach**
(Syreischikov, Dimitrii Petrovich)	
see Syreisczikov, Dimitrii Petrovich	**Syr.**
(Syreischtschikou, Dimitrii Petrovich)	
see Syreisczikov, Dimitrii Petrovich	**Syr.**
Syreisczikov, Dimitrii Petrovich (1868–1932) S	**Syr.**
(Syreitschikow, Dimitrii Petrovich)	
see Syreisczikov, Dimitrii Petrovich	**Syr.**
Syrgij, P.G. (fl. 1964) M	**Syrgij**
Sytin, A.K. (1952–) S	**Sytin**
Syvertsen, Erik E. A	**Syvertsen**
Szabados, András A	**A.Szabados**
Szabados, Margit A	**M.Szabados**
Szabó, Arpad (1884–1914) S	**A.Szabó**
Szabó, Attila T. (1941–) S	**A.T.Szabó**
Szabo, Emerich (fl. 1894) S	**E.Szabo**
Szabó, József (1822–1894) S	**J.Szabó**
Szabó, Zoltán von (1882–1944) MS	**Szabó**
Szafer, Władysław (1886–1970) AS	**Szafer**
Szafran, Bronisław (1897–1968) B	**Szafran**
Szajnocha, Władisław (1857–) F	**Szajnocha**
Szakien, B. (fl. 1927) M	**Szakien**
Szalai, István A	**Szalai**
Szász, Elisabeta (fl. 1959) M	**Szász**
Szatala, Ödön (1889–1958) M	**Szatala**
Szathmáry, S. (fl. 1966) M	**Szathmáry**
Szchieschang, G. (fl. 1984) M	**Szchiesch.**
Szczawinski, Adam Franciszek (1913–) S	**Szczaw.**
Sze, Hsing Chien (1902–) F	**H.C.Sze**
Szelubsky, R. S	**Szel.**
Szembel, Stefan Julianovič (1888–1934) M	**Szembel**
Szembelya, S. (fl. 1915) M	**Szemb.**
Szemere, László (1884–) M	**Szem.**
Szemes, G. A	**Szemes**
(Szentélek, Julius Kováts von)	
see Kováts von Szent-Lélek, Julius	**Kováts**
Szép, E. (fl. 1963) M	**E.Szép**
Szepesfalvy, Janos (1882–1959) B	**Szep.**
Szépligeti, Gyözö (1855–1915) S	**Szépl.**
Szijarto, E. (fl. 1986) M	**Szijarto**
(Szilvinyi, A.)	
see Svilvinyi, A.von	**Svilv.**
Szlachetko, Dariusz L. (1961–) S	**Szlach.**
Szmajda, Piotr (1945–) B	**Szmajda**
Szöllö, István S	**Szöllösi**
Szontágh, Nicolaus (Miklós) von (1843–1899) S	**Szontágh**
Szovits, A.J. (–1830) S	**Szov.**

Szweykowski, Jerzy (1925–) BS	**Szweyk.**
Szymański, Franz (1853–) A	**Szym.**
Szyszyłowicz, Ignaz (Ignacy) von (1857–1910) BPS	**Szyszył.**
't Hart, E. (fl. 1980) S	**E.'t Hart**
't Hart, Henk (1944–) S	**'t Hart**
't Mannetje, Len (fl. 1977) S	**'t Mannetje**
Taasen, Jens Petter A	**Taasen**
Tabor, Richard John (1875–1958) M	**Tabor**
Taborda de Morais, Antonio Artur (1900–1959) S	**Tab.Morais**
(Tachtadshjan, Armen Leonovich)	
see Takhtajan, Armen Leonovich	**Takht.**
Tacik, Tadeusz (1926–) S	**Tacik**
Täckholm, Gunnar Vilhelm (1891–1933) S	**G.Täckh.**
Täckholm, Vivi (1898–1978) S	**Täckh.**
Tada (fl. 1939) M	**Tada**
Taddei, Roberto (1939–) A	**Taddei**
(Tadesse, Mesfin)	
see Mesfin Tadesse	**Mesfin**
Tadulingam, Mudaliar C. (1878–) S	**Tadul.**
Tadulingham, Chinnakanavan (fl. 1932) S	**Tadulingham**
Tafalla, Juan José (1755–1811) S	**Tafalla**
Taft, Clarence Egbert (1906–1986) AS	**Taft**
Tafur, Issac A. A	**Tafur**
Tagaev, I.U. (1953–) S	**Tagaev**
Tagawa, Motozi (1908–1977) PS	**Tagawa**
Tagg, Harry Frank (1874–1933) S	**Tagg**
Taggart, Ralph E. (1941–) A	**Taggart**
Tagliabue, Giuseppe (fl. 1816) S	**Tagl.**
Täglich, Ulla (fl. 1991) M	**Täglich**
Tahir, H.M. (fl. 1977) M	**Tahir**
Tahourdin, Charles Baynard (1872–1942) S	**Tahourdin**
Tai, Fan Tsien (fl. 1988) S	**F.T.Tai**
Tai, Fung (Fang) Lan (L.Y.) (1893–1973) M	**F.L.Tai**
Tai, Li–Sun A	**L.S.Tai**
Tai, Ming Cheih (fl. 1946) M	**M.C.Tai**
Takács, Janos (fl. 1839) S	**Takács**
Takada, Masaki (fl. 1969) M	**Takada**
Takahashi, Eiji (1929–) A	**E.Takah.**
Takahashi, Genzo (fl. 1937) B	**G.Takah.**
Takahashi, Haruki (fl. 1973) M	**Har.Takah.**
Takahashi, Hideo S	**Hid.Takah.**
Takahashi, Hiroshi (fl. 1940) B	**Hir.Takah.**
Takahashi, I. (fl. 1972) M	**I.Takah.**
Takahashi, Kiyoshi A	**K.Takah.**
Takahashi, M. (fl. 1965) M	**M.Takah.**
Takahashi, T. (fl. 1934) M	**T.Takah.**
Takahashi, Yoshinao (–1914) MS	**Takah.**

Takahasi, Eitaro (1911–) B   **Takahasi**
Takahata, Naohiro A   **Takahata**
Takaki, Noriwo (1915–) B   **Takaki**
Takamatsu, Masahiko (1899–) A   **Takam.**
Takamine, N. (fl. 1921) P   **Takamine**
Takamiya, Atusi (fl. 1957) M   **Takamiya**
Takano, Hideaki (1927–) A   **Takano**
Takano, Ichiro (fl. 1969) M   **I.Takano**
Takase, S. (fl. 1966) M   **Takase**
Takashima, T. (fl. 1956) M   **Takash.**
Takashio, M. (fl. 1973) M   **Takashio**
Takatsuki, S. (fl. 1932) M   **Takats.**
Takayama, Toshiaki A   **Takay.**
Takeda, Hideo (fl. 1910) S   **H.Takeda**
Takeda, Hisayoshi (1883–1972) APS   **Takeda**
Takeda, Y. (fl. 1965) M   **Y.Takeda**
Takematsu, Akiko (fl. 1990) M   **Takem.**
Takenouch, (fl. 1982) S   **Takenouch**
Takenouchi, Makoto (1894–) S   **Taken.**
Takeuchi, H. (fl. 1929) M   **Takeuchi**
Takeuchi, Masayuki (fl. 1960) PS   **M.Takeuchi**
Takeuchi, Wayne N. (1952–) S   **W.N.Takeuchi**
Takeya, Minoru (fl. 1925) M   **Takeya**
(Takhtadjan, Armen Leonovich)
   see Takhtajan, Armen Leonovich   **Takht.**
(Takhtadzhian, Armen Leonovich)
   see Takhtajan, Armen Leonovich   **Takht.**
Takhtajan, Armen Leonovich (1910–) PS   **Takht.**
Takiguchi, K. (fl. 1979) P   **Takig.**
Takimoto, A. (fl. 1934) M   **A.Takim.**
Takimoto, Seito (fl. 1922) M   **S.Takim.**
(Talavera Lozano, Salvador)
   see Talavera, Salvador   **Talavera**
Talavera, Salvador (1945–) S   **Talavera**
Talbot, Patrick Henry Brabazon (1919–1979) M   **P.H.B.Talbot**
Talbot, W.S. S   **W.S.Talbot**
Talbot, William Alexander (1847–1917) S   **Talbot**
(Talbot, William Henry Fox)
   see Fox Talbot, William Henry   **Fox Talbot**
Talde, U.K. (fl. 1971) M   **Talde**
Talice, R.V. (fl. 1929) M   **Talice**
Taliev, Valerij Ivanovich (1872–1932) S   **Taliev**
Taligoola, H.K. (fl. 1977) M   **Talig.**
Tallman, Adele Statzell (fl. 1973) M   **Tallman**
Talou, A.de (fl. 1858–1866) S   **Talou**
Talpasayi, E.R.S. A   **Talp.**
Taltasse (fl. 1960) M   **Taltasse**
Tam, Pui Cheung (1921–) S   **P.C.Tam**

(Tamamschan, Sophia G.)
  see Tamamschjan, Sophia G.   **Tamamsch.**
(Tamamschian, Sophia G.)
  see Tamamschjan, Sophia G.   **Tamamsch.**
Tamamschjan, Sophia G. (1900–1981) S   **Tamamsch.**
Tamanian, Kamilla G. (1936–) S   **Tamanian**
(Tamanjan, Kamilla G.)
  see Tamanian, Kamilla G.   **Tamanian**
(Tamanyan, Kamilla G.)
  see Tamanian, Kamilla G.   **Tamanian**
Tamás, Alberto   A   **A.Tamás**
Tamás, Gizella   A   **G.Tamás**
Tamássy, Gé   S   **Tamássy**
Tamayo, Francisco (1902–1985) S   **Tamayo**
(Tamayo, Roberto González)
  see González Tamayo, Roberto   **R.González**
Tamburlini, Francesco (1859–) M   **Tamb.**
Tamin, R. (fl. 1986) S   **Tamin**
Tamiya, Hiroshi (1903–) M   **Tamiya**
Tamlander, Zacharias (fl. 1791) S   **Taml.**
Tamm, Yu. (fl. 1971) S   **Tamm**
Tammaro, Fernando (1942–) S   **Tammaro**
Tammes, Tine (Jantine) (1871–1941) S   **Tammes**
Tamsalu, Aleksander (1891–1960) S   **Tamsalu**
Tamura, Michio (1927–) MS   **Tamura**
Tan, Benito C. (1947–) BPS   **B.C.Tan**
Tan, C.S. (fl. 1989) M   **C.S.Tan**
Tan, Chung Ming   S   **C.M.Tan**
Tan, Hui Ci (1932–) M   **H.C.Tan**
Tan, Kiat W. (1943–) S   **K.W.Tan**
(Tan, Kit)
  see Kit Tan   **Kit Tan**
Tan, Shi Xian (fl. 1987) S   **S.X.Tan**
Tan, Sin Hok   A   **S.H.Tan**
Tan, T.K. (fl. 1990) M   **T.K.Tan**
Tan, Wei Ye (fl. 1979) M   **W.Y.Tan**
Tan, Zhong Ming (fl. 1989) S   **Z.M.Tan**
Tanabe, K. (fl. 1931) S   **Tanabe**
Tanabe, Misao   A   **M.Tanabe**
Tanai, T. (fl. 1987) M   **Tanai**
Tanaka, I. (fl. 1940) M   **I.Tanaka**
Tanaka, Jiro (1950–) A   **Ji.Tanaka**
Tanaka, Junko (fl. 1931) M   **J.Tanaka**
Tanaka, Kiyoshi (1943–) M   **K.Tanaka**
Tanaka, Shōichi (fl. 1964) M   **S.Tanaka**
Tanaka, Takesi (1907–) A   **Tak.Tanaka**
Tanaka, Tuyoshi   A   **Tuy.Tanaka**
Tanaka, Tyôzaburô (1885–) MS   **Tanaka**
Tanaka, Yoshio (1838–1916) S   **Y.Tanaka**

Tanaka, Yuichiro (1900–) S	**Yu.Tanaka**
Tanda, Seinosuke (fl. 1976) M	**Tanda**
Tandon, I.N. (fl. 1951) M	**I.N.Tandon**
Tandon, M.P. (fl. 1975) M	**M.P.Tandon**
Tandon, R.N. (fl. 1960) M	**Tandon**
Tandy, Geoffrey (1900–) A	**Tandy**
Tandy, P.A. (fl. 1975) M	**P.A.Tandy**
Tanfani, Enrico (1848–1892) S	**Tanfani**
Tanfiljew, Gavriel Ivanowitsch (Gavriil Ivanovič) (1857–1928) S	**Tanfil.**
Tang, Chang Lin (fl. 1986) S	**C.L.Tang**
Tang, Chen (Zhen) Zi (fl. 1982) S	**C.Z.Tang**
Tang, D.Z. (fl. 1984) M	**D.Z.Tang**
Tang, Geng Guo (1950–) S	**G.G.Tang**
Tang, Hune Cheung (1918–) S	**H.C.Tang**
Tang, Rong Guan (fl. 1991) M	**R.G.Tang**
Tang, Si Hua (fl. 1982) S	**S.H.Tang**
Tang, T. (Chin) (1897–1984) S	**T.Tang**
Tang, Tsin (1900–) S	**Ts.Tang**
Tang, Ya (fl. 1988) S	**Y.Tang**
Tang, Yan Cheng (1926–) S	**Y.C.Tang**
Tang, Yang Ping (fl. 1988) M	**Y.P.Tang**
(Tang, Yen Chen)	
see Tang, Yan Cheng	**Y.C.Tang**
Tang, Zhi Jie A	**Z.J.Tang**
Tangen, Karl A	**Tangen**
Tanger, Louise Forney Arnold (1889–) P	**Tanger**
Tangl, Eduard (Josef) (1848–1905) AS	**Tangl**
Tani, Toshikazu (fl. 1966) M	**Tani**
Taniguchi, Y. (fl. 1926) M	**Tanig.**
Tanimoto, Takeo (1940–) S	**Tanim.**
Tanner, Christopher E. A	**Tanner**
Tanner–Füllemann, M. A	**Tanner–Füll.**
Tanowitz, B.D. (fl. 1983) S	**Tanowitz**
Tansey, Michael R. (fl. 1975) M	**Tansey**
Tansley, Arthur George (1871–1955) A	**Tansley**
Tansley, S.A. (fl. 1984) S	**S.A.Tansley**
Tao, De Ding (1937–) S	**D.D.Tao**
Tao, Guang Fu (fl. 1985) S	**G.F.Tao**
Tao, Guo Da (1939–) S	**G.D.Tao**
Tao, Jia Feng (1926–) M	**J.F.Tao**
Tao, Jun Rong (fl. 1986) S	**J.R.Tao**
Tao, Kai (fl. 1985) M	**K.Tao**
Tao, Xiu Liu (fl. 1982) S	**X.L.Tao**
Taoda, Hiroshi (1945–) B	**Taoda**
Taparia, S.C. (fl. 1974) M	**Taparia**
Tapia de Fossaert, G. (fl. 1980) M	**Tap.Foss.**
Tapke, Victor Ferdinand (1890–) M	**Tapke**
Tappan, Helen (fl. 1961) AM	**Tappan**
Tappeiner, Franz (1816–1902) S	**Tapp.**

Taránek, Karel J. (1855–1888) S	**Taránek**
Taranto, Emmanuele (Emmanuello) (fl. 1845) S	**Taranto**
Tarasevich, V.F. (fl. 1987) S	**Taras.**
(Tarasov, A.O.)	
see Tarassov, A.O.	**A.O.Tarassov**
Tarassov, A.O. (1933–) S	**A.O.Tarassov**
Tarassov, R.P. (1917–) S	**Tarassov**
Tardelli, Marcello (1944–) S	**Tardelli**
Tardent, Charles (fl. 1841) S	**Tardent**
Tardieu, Marie Laure (1902–) PS	**Tardieu**
(Tardieu–Blot, Marie–Laure)	
see Tardieu, Marie Laure	**Tardieu**
Targé, A. (fl. 1965) M	**Targé**
(Targioni, Giovanni)	
see Targioni Tozzetti, Giovanni	**G.Targ.Tozz.**
Targioni Tozzetti, Adolfo (1823–1902) S	**Ad.Targ.Tozz.**
Targioni Tozzetti, Antonio (1785–1856) S	**Ant.Targ.Tozz.**
Targioni Tozzetti, Giovanni (1712–1783) AS	**Targ.Tozz.**
Targioni Tozzetti, Ottaviano (1755–1829) S	**O.Targ.Tozz.**
Tarjan, A.C. (fl. 1961) M	**Tarjan**
Tarn, T.R. (fl. 1988) S	**T.R.Tarn**
Tarnavschi, Ion (Teofil) (1904–1989) ABS	**Tarn.**
Tarnogradskii, D.A. A	**Tarnogr.**
Tärnström, Christopher (1703–1746) L	**Tärnström**
Taroda, Neusa (1952–) S	**Taroda**
Tarr, S.A.J. (1918–) M	**Tarr**
Tarrade, Adrien (1843–1889) M	**Tarrade**
Tarrega Bellver, I. (fl. 1984) S	**I.Tarrega**
Tarrega Bellver, S. (fl. 1985) S	**S.Tarrega**
Tartenova, M.A. (fl. 1957) M	**Tartenova**
Tasch, Paul A	**Tasch**
Taschenberg, Ernst Ludwig (1818–1898) S	**Taschenb.**
Taschereau, Pierre Michel (1939–) S	**Tascher.**
Taschner, Christian Friedrich (1817–) PS	**Taschner**
Tasenkevitsch, L.A. (1948–) S	**Tasenk.**
Tashiro, M. (fl. 1986) S	**M.Tash.**
Tashiro, Yasusada (1856–1928) S	**Tash.**
Tashiro, Zentarô (1872–1947) S	**Z.Tash.**
(Taslacktian, M.G.)	
see Taslakhchyan, M.G.	**Tasl.**
(Taslakhchiyan, M.G.)	
see Taslakhchyan, M.G.	**Tasl.**
Taslakhchyan, M.G. (1937–) M	**Tasl.**
Tassi, Attilio (1820–1905) S	**A.Tassi**
Tassi, Flaminio (1851–) M	**Tassi**
Tasugi, H. (fl. 1940) M	**Tasugi**
Tatar, Mathias (fl. 1878) S	**Tatar**
Tatara, S. (fl. 1973) M	**Tatara**

(Tatarinov, Alexander Alexejevitch)
 see Tatarinow, Alexander Alexejevitch   **Tat.**
Tatarinow, Alexander Alexejevitch (1817–1886) S   **Tat.**
Tate, Donald E. (fl. 1961) P   **D.E.Tate**
Tate, George (1805–1871) F   **G.Tate**
Tate, George Henry Hamilton (1894–1953) S   **G.H.Tate**
Tate, George Ralph (1835–1874) S   **G.R.Tate**
Tate, P. (fl. 1927) M   **P.Tate**
Tate, Ralph (1840–1901) S   **Tate**
Tateishi, Yoichi (1948–) S   **Tateishi**
Tatem, T.G. A   **Tatem**
Tateoka, Tsuguo (1931–) S   **Tateoka**
Tatewaki, Misao (1899–) PS   **Tatew.**
Tatnall, Edward (1818–1898) S   **Tatnall**
Taton, Auguste (1914–1989) PS   **Taton**
Tatsumi, C. (fl. 1950) M   **Tatsumi**
Taub, S. (fl. 1965) S   **S.Taub**
Taube, (Daniel) Johann (1727–1799) S   **Taube**
Taubenhaus, Jacob Joseph (1885–1937) M   **Taubenh.**
Taubenheim, Gerd (fl. 1975) S   **Taubenheim**
Taubert, Paul Hermann Wilhelm (1862–1897) PS   **Taub.**
Taugourdeau, Philippe A   **Taug.**
Taugourdeau-Lantz, J. A   **Taug.-Lantz**
Taupenot, J.M. (fl. 1851) S   **Taupenot**
Tausch, Ignaz Friedrich (1793–1848) PS   **Tausch**
Tauschanoff (fl. 1933) M   **Tauschan.**
Tauscher, Gyula (Julius August) (1832–1882) S   **Tauscher**
Täuscher, Lothar A   **L.Täuscher**
Taussig, Marguerite (fl. 1956) M   **Taussig**
Tavares, Carlos das Neves (1914–1972) BM   **Tav.**
Tavares, Isabelle I. (1921–) M   **I.I.Tav.**
Tavares, Joaquim (Joachim) da Silva (1866–1931) S   **J.Tav.**
Tavares, Sérgio (1931–) S   **S.Tav.**
Tavasiev, R.A. (1947–) S   **Tavasiev**
Tavel, (Rudolf) Franz von (1863–1941) MP   **Tavel**
Tavel, Catherine von (1898–) MS   **C.Tavel**
Tawada, Shinjun (1907–) S   **Tawada**
Tawan, Cheksum bt. (fl. 1984) S   **Tawan**
Tawfic (fl. 1939) M   **Tawfic**
Taylor, A.R.A. (1921–) A   **A.R.A.Taylor**
Taylor, Adam (fl. 1769) S   **A.Taylor**
Taylor, B. (fl. 1979) M   **B.Taylor**
Taylor, C.E.S. (fl. 1983) S   **C.E.S.Taylor**
Taylor, Charlotte M. (1955–) S   **C.M.Taylor**
Taylor, D.L. A   **D.L.Taylor**
Taylor, F.E. (fl. 1922) M   **F.E.Taylor**
Taylor, F.J.R. A   **F.J.R.Taylor**
Taylor, Frederick Beatson (1851–1931) S   **F.Taylor**
Taylor, G.Marie (1930–) M   **G.M.Taylor**

Taylor, George (1904–) S	**G.Taylor**
Taylor, George Crosbie (1901–1962) S	**G.C.Taylor**
Taylor, J.B. (fl. 1977) M	**J.B.Taylor**
Taylor, Jane (1924–) BS	**J.Taylor**
Taylor, Joan M. (1929–) S	**J.M.Taylor**
Taylor, John Ellor (1837–1895) S	**J.E.Taylor**
Taylor, John J. (fl. 1970) M	**J.J.Taylor**
Taylor, Joseph (c.1762–1844) S	**Jos.Taylor**
Taylor, M.W. (fl. 1927) M	**M.W.Taylor**
Taylor, Mary Ruth Fussel Jackson (1908–) S	**M.Taylor**
Taylor, Nigel Paul (1956–) S	**N.P.Taylor**
Taylor, Norman (1883–1967) S	**N.Taylor**
Taylor, P.A. (fl. 1985) M	**P.A.Taylor**
Taylor, Peter Geoffrey (1926–) S	**P.Taylor**
Taylor, Raymond Leech (1901–) S	**R.L.Taylor**
Taylor, Robert Hibbert S	**R.H.Taylor**
Taylor, Ronald J. (1932–) S	**R.J.Taylor**
Taylor, Thomas (1775–1848) BMS	**Taylor**
Taylor, Thomas (1820–1910) M	**T.Taylor**
Taylor, Thomas Mayne Cunninghame (1904–1983) PS	**T.M.C.Taylor**
Taylor, Thomas N. (fl. 1969) BFM	**T.N.Taylor**
Taylor, W. Carl (1946–) PS	**W.C.Taylor**
Taylor, William Randolph (1895–1990) A	**W.R.Taylor**
Tcheremissinoff (fl. 1936) M	**Tcherem.**
Tcherntzoff, I.A. (fl. 1934) M	**Tcherntz.**
Tchichatscheff, Petr Aleksandrovich (1812–1890) S	**Tchich.**
(Tchihatcheff, Petr Aleksandrovich)	
see Tchichatscheff, Petr Aleksandrovich	**Tchich.**
(Tchihatscheff, Pierre de)	
see Tchichatscheff, Petr Aleksandrovich	**Tchich.**
Tchoumé, M. (fl. 1967) S	**Tchoumé**
Tchuvashov, B.I. A	**Tchuv.**
Teakle, D.S. (fl. 1959) M	**Teakle**
Teasca, G. (fl. 1970) M	**Teasca**
Tébar, Javier (fl. 1988) S	**Tébar**
Tebbs, Margaret C. (1948–) S	**Tebbs**
Techet, Karl (Carl) (1877–1919) S	**Techet**
Tedeschi, C. (fl. 1930) M	**Tedeschi**
Tedin, Hans (1860–1930) S	**Tedin**
Tedlie, Henry (1792–1818) S	**Tedlie**
Teesdale, Robert (c. 1740–1804) S	**Teesd.**
Tehler, Anders G. (1947–) M	**Tehler**
Tehon, Leo Roy (1895–1954) M	**Tehon**
Teich, C.A. (fl. 1934) M	**Teich**
Teichert, Julius (–1873) S	**Teichert**
Teichmeyer, Hermann Friedrich (1685–1744) L	**Teichm.**
Teijsmann, Johannes Elias (1809–1882) PS	**Teijsm.**
Teiling, Einar (Johan Sigurd) (1888–1974) AMS	**Teiling**
Teirlinck, Isidoor (1851–1934) S	**Teirl.**

Teissier, George (1900–) A	**Teissier**
Teixeira, Alcides Ribeiro (1918–) MS	**Teixeira**
(Teixeira, Beulah Coe)	
see Coe Teixeira, Beulah	**Coe Teix.**
Teixeira, Carlos (fl. 1948) BF	**C.Teixeira**
Teixeira, Clovis A	**Cl.Teixeira**
Teixeira, Cyro Gonçalves (1922–) M	**C.G.Teixeira**
Tejada, R. (fl. 1913) S	**Tejada**
(Tejera, Esperanza Beltran)	
see Beltrán–Tejera, Esperanza	**Beltrán–Tej.**
Tekunaga (fl. 1935) M	**Tekun.**
Tekutjev, G. S	**Tekutjev**
Teles, Antonio do Nascimento (1925–) S	**Teles**
Telfair, Charles (1778–1833) S	**Telfair**
Telford, Ian R.H. (1941–) S	**I.Telford**
Telford, Sam R. (1932–) A	**Telford**
Tell, Guillermo A	**Tell**
Tellam, Richard Vercoe (1826–1908) S	**Tellam**
Tellería, Maria Teresa (1950–) M	**Tellería**
Téllez Valdés, Oswaldo (1953–) S	**O.Téllez**
Téllez–Alvarado, Carolina (fl. 1987) S	**C.Téllez**
(Temesy, E.Schönbeck)	
see Schönbeck–Temesy, E.	**Schönb.–Tem.**
Temme, F. (fl. 1887) M	**Temme**
Temminck, Coenraad Jacob (1778–1858) S	**Temminck**
Tempère, G. (fl. 1947) M	**G.Temp.**
Tempère, Joannes Albert (Jean Clodius) (1847–1926) A	**Temp.**
Temple, Augusta A. (fl. 1907) S	**A.Temple**
Temple, F.L. (fl. 1885) S	**F.L.Temple**
Templeton, Bonnie Carolyn (1906–) S	**B.C.Templeton**
Templeton, G.E. (fl. 1973) M	**G.E.Templeton**
Templeton, John (1766–1825) S	**Templeton**
Templeton, Robert A	**R.Templeton**
Templeton, William (fl. 1802) S	**W.Templeton**
Temu, Ruwa–Aichi Pius Cosmos (1955–) S	**Temu**
Tendulkar, J.S. (fl. 1971) M	**Tend.**
Teng, Shu Chün (1902–1970) M	**Teng**
Tengström, Johan Magnus af (1793–1856) S	**Tengstr.**
Tengwall, Tor Ake (Åke) (1892–1946) M	**Tengwall**
Tenison–Woods, Julian Edmund (1832–1889) FPS	**Ten.–Woods**
Tennant, James Robert (1928–) S	**Tennant**
Tennent, James Emerson (1804–1869) S	**Tennent**
Tenore, Michele (1780–1861) PS	**Ten.**
Tenore, Vincenzo (1825–1886) S	**V.Ten.**
Tenzel, Franz Bernhard Richard (1790–) S	**Tenzel**
Teodoresco, Emanoil Constantin (1866–1949) A	**Teodor.**
(Teodorescu, Emmanuel)	
see Teodoresco, Emanoil Constantin	**Teodor.**
Teodoro, Irmá Luis (1904–) S	**I.L.Teodoro**
Teodoro, Ramón Malagarriga Heras S	**Teodoro**

Teodoro y Gregorio, Nicanor Gonzalo (1890–) S   **N.G.Teodoro**
(Teodorowicz, F.von)
  see Theodorowicz, F.von   **Theodor.**
Teodoru, Ion (fl. 1948) M   **Teodoru**
(Teplakova, T.E.)
  see Teplyakova, T.E.   **Teplyak.**
Teplouchow, Fedor (Theodor) Alexandrowitsch (1845–1905) S   **Tepl.**
Teplyakova, T.E. (1954–) S   **Teplyak.**
Tepper, Johann Gottlieb Otto (1841–1923) M   **Tepper**
Teppner, Herwig (1941–) S   **Teppner**
Ter-Chatschaturova, S.Yakovlevna (1902–) S   **Ter-Chatsch.**
Terada, Katsuyuki (fl. 1978) M   **Terada**
Teramoto, T. (fl. 1951) M   **Teram.**
Terán, Manuel de Mier y (–1852) S   **Terán**
Terao, H. (fl. 1977) S   **Terao**
Terashita, Takakiyo (fl. 1968) M   **Terash.**
Terazaki, Tomekichi   S   **Teraz.**
Terechov, A.F. (1890–1974) S   **Terechov**
Tereg, Elinor   A   **Tereg**
(Terekhin, E.S.)
  see Teryokhin, E.S.   **Teryokhin**
Terjajeva, I.G. (fl. 1962) M   **Terjajeva**
Terkelsen, Frede (fl. 1956) M   **Terk.**
Terletzki, Paul (1863–) S   **Terl.**
Termier, Geneviève   A   **G.Termier**
Termier, Henri François Emile (1897–) A   **Termier**
Termo, M.B. (fl. 1837) S   **Termo**
Ternetz, Charlotte (fl. 1900) S   **Ternetz**
Terni, C. (fl. 1894) M   **Terni**
Ternisien, Théophile Antoine Timoléon (1804–1879) S   **Ternis.**
Ternovsky, M.F. (1888–) S   **Tern.**
(Ternström, Christopher)
  see Tärnström, Christopher   **Tärnström**
Terpó, A. (1925–) S   **Terpó**
Terra, Paule (fl. 1953) M   **Terra**
Terracciano, Achille (1861–1917) PS   **A.Terracc.**
Terracciano, Nicola (1837–1921) PS   **N.Terracc.**
Terracino, A.   S   **Terracino**
Terraneo, Lorenzo (1676–1714) L   **Terraneo**
Terrasi, Maria-Carmen (1947–) S   **Terrasi**
Terrell, Edward E. (1923–) S   **Terrell**
(Terri, Jean Joseph)
  see Therry, Jean Joseph   **Therry**
Terrier, Charles A. (fl. 1950) M   **Terrier**
Terry, Emily (Hitchcock) (1837–1921) P   **Terry**
Ters, Mireille   A   **Ters**
Terscheck (fl. c.1840) S   **Terscheck**
Terui, Mutsuo (fl. 1930) M   **Terui**
Teryokhin, E.S. (1932–) S   **Teryokhin**

Teschner, Hans (1894–) S	**Teschner**
Teschner, Walter Paul (1927–) S	**W.P.Teschner**
Tessendorff, (Konrad) Ferdinand (1879–1924) S	**Tessend.**
Tesseron, Yves Augustin (1831–1925) S	**Tess.**
Tessier, Henri Alexandre (1741–1837) S	**Tessier**
Tessmann, Günther (fl. 1904–1926) S	**Tessmann**
Tester, L.S. A	**Tester**
Tetenyi, P. (fl. 1987) S	**Tetenyi**
Teterevnikova, D.N. (fl. 1929) M	**Teterevn.**
(Teterevnikova–Babayan, D.N.)	
see Teterevnikova, D.N.	**Teterevn.**
Tetrik, Robert M. (fl. 1949) P	**Tetrik**
Tettelbach, (Ernst Gustav) Moritz (1794–1870) S	**Tettelb.**
Teunisson, Dorothea J. (fl. 1961) M	**Teun.**
Teuscher, Heinrich (Henry) (1891–1984) S	**Teusch.**
Teutschell S	**Teutsch.**
Tewari, A. A	**A.Tewari**
Tewari, Ishwari (fl. 1968) M	**I.Tewari**
Tewari, J.P. (fl. 1975) M	**J.P.Tewari**
Tewari, Vijay Pratap (1931–) M	**V.P.Tewari**
Tewary, P.K. (fl. 1987) S	**Tewary**
(Texidor, Juan)	
see Texidor y Cos, Juan	**Texidor**
Texidor y Cos, Juan (1836–1885) S	**Texidor**
Teyber, Alois (1846–1913) S	**Teyber**
(Teysmann, Johannes Elias)	
see Teijsmann, Johannes Elias	**Teijsm.**
Thaer, (Konrad Wilhelm) Albrecht (1828–1906) S	**Thaer**
Thaisz, Lájos von (1867–1937) S	**Thaisz**
Thaithong, Obchant (fl. 1978) B	**Thaithong**
Thakur, Ji (fl. 1969) M	**J.Thakur**
Thakur, S.B. (fl. 1973) M	**S.B.Thakur**
Thal, Johannes (1542/3–1583) L	**Thal**
Thaler, Aurelius Ant. (1796–1843) S	**Thaler**
Thamavit, W. (fl. 1977) M	**Thamavit**
Than, Dinh Dai (fl. 1985) S	**Than**
Tharp, Benjamin Carroll (1885–1964) MS	**Tharp**
Thate, R. (fl. 1974) M	**Thate**
Thaung, M.M. (fl. 1973) M	**Thaung**
Thaxter, Roland (1858–1932) M	**Thaxt.**
Thayer, Emma (1842–1908) S	**Thayer**
Thayer, Lewis A. (1903–) A	**L.A.Thayer**
Thays, Carlos (Charles) (1849–1934) S	**Thays**
Thaysen, A.C. (fl. 1943) M	**Thaysen**
Thedenius, Carl Gustaf Hugo (1843–1929) S	**C.G.H.Thed.**
Thedenius, Knut Fredrik (1814–1894) B	**Thed.**
Theel, Jakob Gustaf Gösta (1846–1885) S	**Theel**
Theile, Michael A	**Theile**
Théis, Alexandre (Étienne Guillaume) de (1765–1842) S	**Théis**

Theiss, T. (fl. 1952) M	T.Theiss
Theissen, Ferdinand (1877–1919) MS	Theiss.
(Theiszen, Ferdinand)	
see Theissen, Ferdinand	Theiss.
Thellung, Albert (1881–1928) PS	Thell.
Thenen, Salvator (fl. 1911) S	Thenen
Thénint, André S	Thénint
Theobald, Gottfried Ludwig (1810–1869) S	Theob.
Theobald, William (1829–1908) S	W.Theob.
Theobald, William Louis (1936–) S	W.L.Theob.
Theodorescu, Georgeta (fl. 1989) M	Theod.
Theodorou, Michael K. (fl. 1991) M	Theodorou
(Theodorov, Alexander Alexandrovich)	
see Fedorov, Alexander Alexandrovich	Al.Fed.
(Theodorov, Andrej Aleksandrovich)	
see Fedorov, Andrej Aleksandrovich	Fed.
Theodorowicz, F.von (fl. 1934) M	Theodor.
Theohari, I. (fl. 1977) M	Theoh.
Theophileo, Rodolpho S	Theoph.
Theophrastus (4th–3rd Century BC) LMS	Theophr.
Theorin, Gustav Robert Alfons (1841–1881) S	G.Theorin
Theorin, Pehr Gustaf Emanuel (1842–1916) S	Theorin
Therese, Charlotte Marianne Augusta von Bayern (1850–1925) S	Therese
Thérézien, Y. A	Thérézien
Thériot, (Marie Hypolite) Irénée (1859–1947) BS	Thér.
Theriot, Edward C. (1953–) A	E.C.Ther.
Theron, Johannes Jacobus (1905–) S	Theron
Therry, Jean Joseph (1833–1888) MS	Therry
Thesleff, Arthur (1871–1920) M	Thesleff
Theune, Erich (1884–) S	Theune
Theunissen, J. (fl. 1977) S	J.Theun.
Theunissen, S. (fl. 1981) S	S.Theun.
Theuss, Theodor (fl. 1811) S	Theuss
Thévenau, Antonin Victor (1815–1876) S	Thévenau
Thibaud, ?E. (fl. 1785) S	Thibaud
Thibault de Chanvalon, Jean Baptiste S	Thib.Chanv.
Thibaut, Monique (fl. 1971) M	Thibaut
(Thiébaud de Berneaud, Arsenne (Arsène))	
see Thiébaut–de–Berneaud, Arsenne (Arsène)	Thiéb.–Bern.
Thiébaut, Charles (1837–1884) S	C.Thiébaut
Thiébaut, Jean (–1953) S	Thiébaut
Thiébaut, M.Joseph (fl. 1921–1955) S	J.Thiébaut
Thiébaut–de–Berneaud, Arsenne (Arsène) (1777–1850) S	Thiéb.–Bern.
Thiel, (C.E.) Hugo (1839–1918) S	Thiel
(Thiel, P.H.Van)	
see Van Thiel, P.H.	Van Thiel
Thielau, Friedrich Joachim Siegismund von (1796–) S	Thielau
Thiele, Friedrich Leopold (–1841) S	Thiele
Thiele, Kevin R. (fl. 1988) S	K.R.Thiele

Thielens, Armand (1833–1874) S	**Thielens**
Thien, Leonard B. (1938–　) S	**Thien**
Thienemann, (Friedrich August) Ludwig (1793–1858) AS	**L.Thienem.**
Thienemann, F.W. (fl. 1839) M	**Thienem.**
Thieret, John William (1926–　) PS	**Thieret**
Thierry (fl. 1882) M	**Thierry**
Thierry, Jacques　A	**J.Thierry**
Thiers, Harry Delbert (1919–　) M	**Thiers**
Thierstein, Hans R.　A	**Thierst.**
Thiéry de Ménonville, Nicolas Joseph (1739–1780) S	**Thiéry Mén.**
Thijsse, G. (fl. 1984) S	**G.Thijsse**
Thijsse, Jacobus Pieter (1863–1945) S	**Thijsse**
Thin, Nguen N. (fl. 1983) S	**Thin**
Thind, I.P.S. (fl. 1983) M	**I.P.S.Thind**
Thind, Kartar Singh (1917–　) M	**K.S.Thind**
Thiollière, Victor (1801–1859) A	**Thioll.**
Thiriart, T.Fr. (fl. 1806) S	**Thiriart**
Thiroux　A	**Thiroux**
Thirring, E. (fl. 1962) M	**Thirring**
Thirumalachar, Mandayani Jeersannidhi (1914–　) M	**Thirum.**
(Thiselton-Dyer, William Turner)	
see Dyer, William Turner Thiselton (Thistleton)	**Dyer**
Thisquen (fl. 1853–1876) S	**Thisquen**
(Thistleton-Dyer, William Turner)	
see Dyer, William Turner Thiselton (Thistleton)	**Dyer**
Thite, A.N. (fl. 1970) M	**Thite**
Thivend, S. (fl. 1981) Simone	**Thivend**
Thivy, Francesca　A	**Thivy**
Tho, Nguyen Quang (fl. 1972) M	**Tho**
Thoday, David (1883–1964) S	**Thoday**
Thode, (Hans) Justus (1859–1932) S	**Thode**
Thoen, D. (fl. 1969) M	**Thoen**
Thoizon, G. (fl. 1967) M	**Thoizon**
Thom, Charles (1872–1956) M	**Thom**
Thom, J.E.　S	**J.E.Thom**
Thomae, Carl (Karl) Th. (1808–1885) S	**Thomae**
Thomae, Karl Friedrich (1863–　) S	**K.Thomae**
Thomann, Anton (fl. 1859) S	**Thomann**
Thomas, Abraham Louis Emmanuel (Emanuel) (1788–1859) S	**Thomas**
Thomas, A.V.　S	**A.V.Thomas**
Thomas, Barry A. (1940–　) B	**B.A.Thomas**
Thomas, Benjamin Walden (fl. 1886) A	**B.W.Thomas**
Thomas, Berthold Ernst Friedrich (1910–　) S	**B.Thomas**
Thomas, Charles Arden (fl. 1938–1973) S	**C.A.Thomas**
Thomas, Cyrus (1825–1910) S	**C.Thomas**
Thomas, D.W. (fl. 1986) S	**D.W.Thomas**
Thomas, David (1776–1859) S	**D.Thomas**
Thomas, Eugen A. (1912–　) ACM	**E.A.Thomas**
Thomas, Friedrich August Wilhelm (1840–1918) M	**F.Thomas**

Thomas, Harold Earl (1900–) M	**H.E.Thomas**
Thomas, Hugh Hamshaw (1885–1962) FS	**H.H.Thomas**
Thomas, J.C. A	**J.C.Thomas**
Thomas, Joab Langston (1933–) S	**J.L.Thomas**
Thomas, Joanna E. A	**J.E.Thomas**
Thomas, John Hunter (1928–) S	**J.H.Thomas**
Thomas, Joseph Marie Paul (1872–) AS	**J.Thomas**
Thomas, K.S. (fl. 1930) M	**K.S.Thomas**
Thomas, Margarete A	**M.Thomas**
Thomas, Mason Blanchard (1866–1912) S	**M.B.Thomas**
Thomas, Owen (1843–1923) S	**O.Thomas**
Thomas, P.C. A	**P.C.Thomas**
Thomas, R.Dale (1936–) S	**R.D.Thomas**
Thomas, Roy Curtis (1887–) M	**R.C.Thomas**
Thomas, S.M. (fl. 1983) S	**S.M.Thomas**
Thomas, William Sturgis (1871–1941) M	**W.S.Thomas**
Thomas, William Wayt (1951–) S	**W.W.Thomas**
Thomasson, Kuno (1923–) A	**Thomasson**
Thomé, Otto Wilhelm (1840–1925) M	**Thomé**
Thommen, Edouard (1880–1961) S	**Thommen**
Thompson, Aldworth William ('Tommy') (1900–1982) M	**A.W.Thomps.**
Thompson, Arnold (1876–1959) B	**A.Thomps.**
Thompson, B.L. (fl. 1985) M	**B.L.Thomps.**
Thompson, Bertha Emogene (fl. 1903) M	**B.E.Thomps.**
Thompson, Charles Henry (1870–1931) MS	**C.H.Thomps.**
Thompson, David M. (1957–) S	**D.M.Thomps.**
Thompson, George Edward (1903–) M	**G.E.Thomps.**
Thompson, Harold Stuart (1870–1940) S	**H.S.Thomps.**
Thompson, Henry Joseph (1921–) S	**H.J.Thomps.**
Thompson, John (fl. 1798) S	**J.Thomps.**
Thompson, John Vaughan (1779–1847) S	**J.V.Thomps.**
Thompson, John William (1890–) S	**J.W.Thomps.**
Thompson, Joy (née Garden, J.) (1923–) S	**Joy Thomps.**
Thompson, Mary Fraser (1941–1982) S	**M.F.Thomps.**
Thompson, Rufus Henney (1908–1980) ABPS	**R.H.Thomps.**
Thompson, William (1823–1903) AS	**W.Thomps.**
Thompson, Zodock (1796–1856) S	**Z.Thomps.**
Thoms, Hermann (Friedrich Maria) (1859–1931) S	**Thoms**
Thomsen, Christen (1822–1874) S	**Thomsen**
Thomsen, Helge Abildhauge A	**H.A.Thomsen**
Thomsen, R. A	**R.Thomsen**
Thomson, Anthony Todd (1778–1849) S	**A.T.Thomson**
Thomson, Charles Wyville (1830–1882) A	**C.W.Thomson**
Thomson, George Malcolm (1849–1933) PS	**G.M.Thomson**
Thomson, John Scott (1882–1943) S	**J.S.Thomson**
Thomson, John Walter (1913–) M	**J.W.Thomson**
Thomson, Joseph (1858–1895) S	**J.Thomson**
Thomson, Paul William (1892–1957) B	**P.W.Thomson**
Thomson, Robert Boyd (1870–1947) S	**R.B.Thomson**

Thomson, Spencer (c.1817–1886) S	S.Thomson
Thomson, Thomas (1817–1878) S	Thomson
Thongpukdee, A. (fl. 1988) S	Thongp.
Thonner, Franz (1863–1928) S	Thonner
Thonning, Peter (1775–1848) S	Thonn.
Thor, Göran (1953–) MS	G.Thor
Thor, Sig (fl. 1930) M	Thor
Thore, Jean (1762–1823) AMPS	Thore
Thorel, Clovis (1833–1911) S	Thorel
Thorenaar, Adriaan (1894–1977) S	Thorenaar
Thormeyer, Paul (1878–) S	Thorm.
Thorn, R.Greg (fl. 1981) M	Thorn
Thornber, John James (1872–1962) S	Thornber
Thornberry, Halbert Houston (1902–) M	Thornb.
Thorne, Kaye H. (1939–) S	K.H.Thorne
Thorne, Robert Folger (1920–) S	Thorne
Thornton, Robert John (1768?–1837) S	Thornton
Thoroddsen, Thorvaldur (1855–1921) S	Thoroddsen
Thorold, Charles Aubrey (1906–) M	Thorold
Thorrington-Smith, Margaret A	Thorr.-Sm.
Thorsrud, Arne (1895–1964) S	Thorsrud
Thorstenson, Georg Leonhard (1851–1918) S	Thorst.
Thory, Claude Antoine (1759–1827) S	Thory
Thothathri, K. (1929–) S	Thoth.
Thouars, Abel Aubert du Petit- (1793–1864) S	A.Thouars
Thouars, Louis Marie Aubert du Petit (1758–1831) PS	Thouars
Thouin, André (1747–1824) PS	Thouin
Thouvenin, Maurice (–François) (1857–) S	Thouv.
Thozet, Anthelme (1826–1878) S	Thozet
Threlfall, S. (fl. 1983) S	Threlfall
Threlkeld, Caleb (1676–1728) L	Threlkeld
Throndsen, Jahn A	Throndsen
Thrower, L.B. (fl. 1954) M	Thrower
Thrower, Stella L. (1925–) S	S.L.Thrower
Thuan, Nguyén Van (–1987) S	Thuan
Thuillier, Jean Louis (1757–1822) ABS	Thuill.
Thulin, Mats (1948–) S	Thulin
Thümen, Felix (Karl Albert Ernst Joachim) von (1839–1892) MS	Thüm.
(Thümen–Gräfendorf, Felix (Karl Albert Ernst Joachim) von)	
see Thümen, Felix (Karl Albert Ernst Joachim) von	Thüm.
Thunberg, Carl Peter (1743–1828) ABMPS	Thunb.
Thunmark, Sven A	Thunmark
Thurber, George (1821–1890) S	Thurb.
Thurén, Alfred Fredrik Abraham (1840–) S	Thurén
Thuret, Gustave Adolphe (1817–1875) AMS	Thur.
Thurmann, Jules (1804–1855) AS	Thurm.
Thurn, Everard Ferdinand im (1852–1932) PS	Thurn
Thurston, Charles Orion (1857–1933) S	C.Thurst.
Thurston, Edgar (1855–1935) S	E.Thurst.

Thurston, Henry Winfred (1893–) M	**Thurst.**
Thurston, John Bates (1836–1897) S	**J.Thurst.**
Thury, (Jean) Marc (Antoine) (1822–1905) S	**Thury**
Thusu, Bindra A	**Thusu**
Thwaites, George Henry Kendrick (1812–1882) ABMPS	**Thwaites**
Thyr, B.D. (fl. 1964) M	**Thyr**
Thyssen, Paul (1891–1974) BS	**Thyssen**
Tiagi, B. (fl. 1979) S	**Tiagi**
Tian, Cun Zeng A	**C.Z.Tian**
Tian, Hong (fl. 1985) S	**H.Tian**
Tian, Xing Jun (fl. 1988) S	**X.J.Tian**
Tibell, Leif (1944–) M	**Tibell**
Tichelaar, G.M. (fl. 1972) M	**Tichelaar**
(Tichomirov, Vadim Nikolaevich)	
see Tikhomirov, Vadim Nikolaevich	**V.N.Tikhom.**
(Tichomirov, Vladimir A.)	
see Tikhomirov, Vladimir A.	**Tikhom.**
(Tichomirow, Wadim Nikolaevich)	
see Tikhomirov, Vadim Nikolaevich	**V.N.Tikhom.**
(Tichomirow, Wladimir A.)	
see Tikhomirov, Vladimir A.	**Tikhom.**
Tidestrom, Ivar (Frederick) (1864–1956) PS	**Tidestr.**
Tidyman, Philipp (Philip) (–1850) FS	**Tidyman**
Tiegel, E. (–1936) S	**Tiegel**
Tieghem, Phillippe Édouard Léon van (1839–1914) AMS	**Tiegh.**
Tiep, N.V. (fl. 1980) S	**N.V.Tiep**
Tiep, Nguyen Anh (fl. 1965) S	**Tiep**
Tiesenhausen, Manfred B.von (1875–) M	**Tiesenh.**
(Tietz, Johann Daniel)	
see Titius, Johann Daniel	**Titius**
Tietz, Solveig (fl. 1988) S	**Tietz**
Tiffany, Lewis Hanford (1894–1965) S	**Tiffany**
Tiffany, Lois C. (fl. 1952) M	**L.C.Tiffany**
Tiffany, Lois H. (1924–) M	**L.H.Tiffany**
Tikhomirov, Boris A. (1909–1976) S	**B.A.Tikhom.**
Tikhomirov, Vadim Nikolaevich (1932–) S	**V.N.Tikhom.**
Tikhomirov, Vladimir A. (fl. 1868) M	**Tikhom.**
Tikhonenko, Jurij Ja. (1954–) M	**Tikhon.**
Tikovsky, H. S	**Tikovsky**
(Til–landz, Elias Erici)	
see Tillandz, Elias Erici	**Tillandz**
Tilak, S.T. (fl. 1958) M	**Tilak**
Tilden, Josephine Elizabeth (1869–1957) A	**Tilden**
Tilesius von Tilenau, Wilhelm Gottlieb (1769–1857) S	**Tilesius**
Tilford, Paul Edward (1900–) M	**Tilford**
Tiling, G. S	**G.Tiling**
Tiling, Heinrich Sylvester Theodor (1818–1871) S	**Tiling**
Till, Hans (fl. 1982) S	**H.Till**
Till, Walter (1956–) S	**W.Till**

(Tillander, Elias Erici)
 see Tillandz, Elias Erici   **Tillandz**
Tillandz, Elias Erici (1640–1693) L   **Tillandz**
Tillett, Stephen S. (1930–) S   **Tillett**
(Tilling, Heinrich Sylvester Theodor)
 see Tiling, Heinrich Sylvester Theodor **Tiling**
Tilloch (fl. 1823) S   **Tilloch**
Tilney, P.M. (fl. 1987) S   **Tilney**
Tim, Stephen K.-M. (1937–) M   **Tim**
Timbal–Lagrave, Pierre Marguérite Édouard (1819–1888) S **Timb.-Lagr.**
Timbrook, S. (fl. 1986) S   **Timbrook**
Timdal, Einar (1957–) M   **Timdal**
Timeroy (fl. 1947) S   **Timeroy**
Timkó, György (1876–1945) M   **Timkó**
Timler, Friedrich Karl (1914–) S   **Timler**
Timm, Carl Theodor (1824–1907) S   **C.Timm**
Timm, Gisela S   **G.Timm**
Timm, Joachim Christian (1734–1805) BS   **Timm**
Timm, Rudolph (1859–1936) B   **R.Timm**
Timmer, L.W. (fl. 1970) M   **Timmer**
Timmerman, A. (fl. 1984) S   **Timmerman**
Timmermans, Adrianna J. (fl. 1941) M   **Timmerm.**
Timofeev, Boris V. A   **B.V.Timofeev**
Timofeev, E.V. A   **E.V.Timofeev**
(Timofejew, Boris W.)
 see Timofeev, Boris V.   **B.V.Timofeev**
Timokhina, S.A. (1935–) S   **Timokhina**
Timonin, M.I. (fl. 1940) M   **Timonin**
Timothy, D.H. (fl. 1981) S   **Timothy**
Timpano, Peter (1948–1987) A   **Timpano**
Timperman, Jules (1937–) S   **Timp.**
Timpko, V.A. (1919–) S   **Timpko**
Tims, Eugene Chapel (1894–) M   **Tims**
Tinant, François Auguste (1803–1853) S   **Tinant**
Tindale, Mary Douglas (1920–) PS   **Tindale**
Tindall, Donald R. A   **D.R.Tind.**
Tindall, Isabella (Ella) Mary (1850–1928) B   **Tind.**
Tineo, Giuseppe (1757–1812) S   **G.Tineo**
Tineo, Vincenzo (1791–1856) PS   **Tineo**
(Tineo–Ragusa, Giuseppe)
 see Tineo, Giuseppe   **G.Tineo**
Ting, Chih Chi S   **C.C.Ting**
Ting, Chih Tsun (fl. 1982) S   **C.T.Ting**
Ting, Kwang Chi (fl. 1963) S   **K.C.Ting**
Ting, W.P. (fl. 1964) M   **W.P.Ting**
Tiong, S.K.K. (fl. 1984) S   **Tiong**
Tiraboschi, Carlo (fl. 1905) M   **Tirab.**
Tiraby, G. (fl. 1990) M   **Tiraby**
Tirel, Christiane (1939–) S   **Tirel**

(Tirel–Roudet, Christiane)
  see Tirel, Christiane   **Tirel**
Tirilly, Y. (fl. 1976) M   **Tirilly**
Tirvengadum, D.D. (fl. 1986) S   **Tirveng.**
Tischer, Arthur (1895–) S   **Tischer**
Tischler, Georg (Friedrich Leopold) (1878–1955) S   **Tischler**
Tisdale, William Burleigh (1890–) M   **Tisdale**
Tiselius, Gustaf August (1833–1904) S   **Tiselius**
Tish, R.V. (fl. 1985) M   **Tish**
Tison, (Eugène) Édouard (Augustin) (1842–) S   **Tison**
Tison, Adrien (fl. 1900–1909) MS   **A.Tison**
Tisserant, Charles (1886–1962) S   **Tisser.**
Tissière, Pierre Germain (1828–1868) S   **Tissière**
Tita, Antonius (–1729) L   **Tita**
Titford, William Jowit (Jowett) (1784–1823/7) S   **Titford**
Titius, Johann Daniel (1729–1796) S   **Titius**
Titius, Pius (Pio) Vendel (1801–1884) AS   **P.Titius**
Titorenko, T.N. A   **Titor.**
Titov, A.N. (1959–) M   **Titov**
Titov, V.S. S   **V.S.Titov**
Titova, Yu.A. (fl. 1983) M   **Titova**
Tittel, C. (fl. 1986) S   **Tittel**
Tittley, Ian (1945–) A   **Tittley**
Tittmann, (Friedrich) Hermann (1863–) S   **H.Tittmann**
Tittmann, Johann August (1774–1840) S   **Tittmann**
Titz, W. (1941–) S   **Titz**
Tiu, Danny (fl. 1984) S   **D.Tiu**
Tivoli, B. (fl. 1988) M   **Tivoli**
Tiwari, D.N. A   **D.N.Tiwari**
Tiwari, D.P. (fl. 1977) M   **D.P.Tiwari**
Tiwari, G.L. A   **G.L.Tiwari**
Tiwari, Nutan (fl. 1992) M   **N.Tiwari**
Tiwari, R.S. A   **R.S.Tiwari**
Tixier, Pierre (1918–) BS   **Tixier**
Tixier–Durivault, A. (fl. 1961) S   **Tix.–Dur.**
Tjaden, William Louis (1913–) S   **Tjaden**
Tjallingii–Beukers, G. (fl. 1983) M   **Tjall.–Beuk.**
(Tjean, Shian–Shong)
  see Tzean, Shean–Shung   **Tzean**
(Tkachenko, V.I.)
  see Tkatschenko, V.I.   **Tkatsch.**
Tkany, František (1851–) S   **F.Tkany**
Tkany, Wilhelm (1792–1863) S   **Tkany**
Tkatschenko, V.I. (fl. 1962) S   **Tkatsch.**
Tobe, Hiroshi (fl. 1987) S   **Tobe**
Tobias, A.V. (1955–) M   **Tobias**
Tobisch, Julius (fl. 1934) M   **Tobisch**
Tobler, Friedrich (1879–1957) MS   **Tobler**
Tobler, Gertrud (Paula) (1877–1948) MS   **G.Tobler**

Tobolewski, Zygmunt (1927–1988) BM	**Tobol.**
Tochinai, Yoshiniko (1893–) M	**Tochinai**
Tocl, Karel (Karl) C. (1870–1910) S	**Tocl**
Todaro, Agostino (1818–1892) PS	**Tod.**
Todd, R.L. (fl. 1936) M	**Todd**
Todd, S.Rammohan A	**S.R.Todd**
Tode, Heinrich Julius (1733–1797) MS	**Tode**
Todor, Ioan (1914–1981) S	**Todor**
Todsen, Thomas A. (fl. 1971) S	**T.A.Todsen**
Todsen, Thomas K. (1918–) S	**Todsen**
Todzia, C.A. (fl. 1985) S	**Todzia**
Toelken, Helmut R. (1939–) S	**Toelken**
Toepffer, Adolph (1853–1931) S	**Toepff.**
Tofsrud, Robert B. S	**Tofsrud**
Togashi, Kogo (1895–1952) M	**Togashi**
Togliani, Franco (fl. 1951) M	**Togliani**
Tognini, Filippo (1868–1896) M	**Tognini**
Toha, Moehamad (fl. 1972) S	**Toha**
Toilliez–Genoud, J. (fl. 1960) S	**Toill.–Gen.**
(Toit, C.A.du)	
see du Toit, C.A.	**C.A.du Toit**
(Toit, J.W.du)	
see du Toit, J.W.	**du Toit**
Toivonen, Heikki (1947–) S	**Toivonen**
Toki, S. (fl. 1954) M	**Toki**
Tokida, Jun (fl. 1948) AM	**Tokida**
Tokio, I. (fl. 1984) M	**Tokio**
Tokumasu, Seiji (fl. 1983) M	**Tokum.**
Tokunaga, Y. (fl. 1931) M	**Tokun.**
Tokuoka, Keiko (fl. 1991) M	**Tokuoka**
Tokura, Ryoichi (fl. 1982) M	**Tokura**
Tole, C. (fl. 1981–1989) S	**Tole**
Toledo, Joaquim Franco de (1905–1952) S	**Toledo**
(Toledo, Laura S. Domínguez de)	
see Domínguez de Toledo, Laura S.	**L.S.Domínguez**
Toledo Manzur, C.A. (fl. 1984) S	**C.A.Toledo**
Tolentino, Danielo B. (fl. 1987) P	**Tolentino**
Tolf, K.Robert (1849–1903) S	**Tolf**
(Tölken, Helmut R.)	
see Toelken, Helmut R.	**Toelken**
Tollard, Claude (fl. 1805) S	**Tollard**
Tollemache, Stanhope (fl. 1901) S	**Tollem.**
(Tolmacev, Alexandr Innokentevich)	
see Tolmatchew, Alexandr Innokentevich	**Tolm.**
(Tolmachev, Alexandr Innokentevich)	
see Tolmatchew, Alexandr Innokentevich	**Tolm.**
Tolmatchew, Alexandr Innokentevich (1903–1979) PS	**Tolm.**
Tolmie, William Fraser (1812–1886) S	**Tolmie**
Tolomio, C. A	**Tolomio**
Tolsma, Jan (fl. 1978) S	**Tolsma**

Toma, Mihai T. (1934–) M	**M.T.Toma**
Toma, Nicolae (1934–) M	**N.Toma**
Toman, Jan (1933–) S	**J.Toman**
Toman, Miloslav (1932–) AS	**M.Toman**
Tomaschek, Antonín (Anton) (1826–1891) MS	**Tomaschek**
Tomaselli, Marcello (1949–) S	**M.Tomas.**
Tomaselli, Ruggero (1920–1982) MS	**Tomas.**
Tomaševič, Milica (fl. 1957) M	**Tomaševič**
Tomb, Andrew Spencer (1943–) S	**Tomb**
Tombe, Frans Andries des (1884–1926) S	**Tombe**
Tomida, Mikio (fl. 1973) S	**Tomida**
Tomilin, B.A. (fl. 1965) M	**Tomilin**
Tomin, Mikhail Petrovich (1883–1967) M	**Tomin**
Tominaga, K. (fl. 1986) M	**K.Tominaga**
Tominaga, Tokitô (fl. 1965) M	**Tominaga**
Tominz, Raimondo (fl. 1879) S	**Tominz**
Tomiya (fl. 1947) S	**Tomiya**
Tomkovich, L.P. (fl. 1981) S	**Tomk.**
Tomlinson, J.A. (fl. 1952) M	**J.A.Toml.**
Tomlinson, Philip Barry (1932–) S	**Toml.**
Tommaselli, Guiseppe (1733–1818) S	**Tommas.**
Tommasini, Muzio Guiseppe Spirito de (Mutius Joseph Spiritus) (1794–1879) BS	**Tomm.**
Tommerup, Inez C. (fl. 1970) M	**Tommerup**
Tomooka, H. S	**Tomooka**
Tömösváry, Ö. A	**Tömösváry**
Tomoyasu, Ryokichi (fl. 1925) M	**Tomoy.**
Tompkins, C.M. (fl. 1941) M	**Tompkins**
Tomsovic, P. (fl. 1986) S	**Tomsovic**
Tonduz, Adolphe (1862–1921) S	**Tonduz**
Tong, Cheng Ren (fl. 1985) S	**C.R.Tong**
Tong, Koe Yang (1896–) S	**K.Y.Tong**
Tong, Shao Quan (1935–) S	**S.Q.Tong**
Tongiorgi, Ezio (1913–1987) BPS	**Tongiorgi**
Tongiorgi, Marco A	**M.Tongiorgi**
Tonglet, August (François Marie Antoine) (1864–1936) M	**Tonglet**
(Toni, Ettore (Hector) De)	
see De Toni, Ettore (Hector)	**E.De Toni**
(Toni, Giovanni Batista De)	
see De Toni, Giovanni Batista	**De Toni**
(Toni, Giuseppe De)	
see De Toni, Giuseppe	**G.De Toni**
Tonini, Carlo (1803–1877) P	**Tonini**
Tonjan, U.R. (1924–) S	**Tonjan**
Tonning, Henrik (Henrich, Heinrich) (1732–1796) S	**Tonning**
Tonolo, A. (fl. 1967) M	**Tonolo**
Tønsberg, Tor (1948–) M	**Tønsberg**
(Tonyan, U.R.)	
see Tonjan, U.R.	**Tonjan**

Toole, Eben Richard (1913–) M   **Toole**
Toomey, Donald Francis (1927–) A   **Toomey**
Toovey, F.W. (fl. 1949) M   **Toovey**
Tooyama, A. (fl. 1979) M   **Tooyama**
Top, W.G. (1824–1896) S   **Top**
Topa, A. S   **A.Topa**
Ţopa, Emilian (1900–1987) S   **Ţopa**
Topal, R. (fl. 1981) M   **Topal**
Topali, S. (1900–1944) S   **Topali**
Topf, Alfred (fl. 1850) S   **Topf**
Topham, Paul (1904–1979) S   **Topham**
Topitz, Anton (1857–1948) S   **Topitz**
Topsent, E. (fl. 1892) M   **Topsent**
Torell, Otto Martin (1828–1900) S   **Torell**
Toren, D.R. (fl. 1975) M   **D.R.Toren**
Torén, Olof (1718–1753) S   **Torén**
Torgård, Salomon Svenson (1885–) S   **Torgård**
Torges, (Karl) Emil (Wilhelm) (1831–1917) S   **Torges**
Toriumi, Saburo A   **Toriumi**
Torka, Valentin (1867–1952) AB   **Torka**
Torkelsen, Anna–Elise (1938–) M   **Tork.**
(Torkelson, Anna–Elise)
   see Torkelsen, Anna–Elise   **Tork.**
Tornabene, Francesco (1813–1897) MS   **Tornab.**
Torner, Erik (fl. 1756) S   **E.Torner**
Törner, Samuel (1762–) S   **Törner**
Toro, Rafael Andres (J.) (1897–) M   **Toro**
Toro, Silvia (fl. 1985) M   **S.Toro**
(Toro T., Silvia)
   see Toro, Silvia   **S.Toro**
Török, Katalin (1954–) M   **Török**
Torralbas, José Ildefonso (1842–1903) S   **Torralbas**
Torre, Antonio Rocha da (1904–) S   **Torre**
(Torre, Juan Ruíz de la)
   see Ruíz de la Torre, Juan   **Ruíz Torre**
Torre, Margarita de la (1940–) M   **M.Torre**
Torrend, Camille (1875–1961) M   **Torrend**
Torrente, P. (fl. 1989) M   **Torrente**
Torres, Amelia Maria (1931–) S   **Torres**
Torres, Andrew M. (fl. 1972) S   **A.M.Torres**
Torres Colín, Rafael (fl. 1988) S   **R.Torres**
Torres, Jorge Hernán (1935–) S   **J.H.Torres**
Torres Juan, J. (fl. 1965) M   **J.Torres**
Torres, M.F. (fl. 1985) M   **M.F.Torres**
Torres, Nestor (fl. 1990) S   **N.Torres**
(Torres–Romero, Jorge Hernán)
   see Torres, Jorge Hernán   **J.H.Torres**
Torrey, George Safford (1891–) S   **G.S.Torr.**
Torrey, John (1796–1873) BMPS   **Torr.**

Torrey, Ray Ethan (1887–1956) S	R.E.Torr.
Torrey, Raymond Hezekiah (1880–1938) S	R.H.Torr.
Torssander, Axel (Gustav Abraham) (1843–1905) S	Torssander
Torssell, Gustav (1811–1849) M	Torss.
Tortič, M.(V.) (fl. 1974) M	Tortič
Tortolero, Omar (fl. 1987) M	Tort.
Tortosa, Roberto D. (1946–) S	Tortosa
Toscano, Antonio Luiz Vieira (1957–) S	Toscano
(Toscano de Brito, Antonio Luiz Vieira)	
see Toscano, Antonio Luiz Vieira	Toscano
Tosco, Uberto (1915–) B	Tosco
Tosi, G. (fl. 1981) S	Tosi
Tosquinet, Pierre Jules (1825–1902) S	Tosq.
Tóth, Sándor (fl. 1959) M	Tóth
Totschilina, L.A. (1930–) S	Totsch.
Totten, Henry Roland (1892–1974) MS	Totten
Toula, Franz (1845–1920) A	Toula
Toumey, James William (1865–1932) MS	Toumey
Toumikoski, R. (fl. 1953) M	Toumik.
Touraine-Desvaux, F. (fl. 1923) M	Tour.-Desv.
Tourlet, Ernest Henry (1843–1907) FS	Tourlet
Tournay, Roland Louis Jules Alfred (1925–1972) S	Tournay
Tournefort, Joseph Pitton de (1656–1708) LMPS	Tourn.
Tournon, Dominique Jérôme (1758–post 1827) S	Tournon
Tourret, Eugène Gilbert (1881–1914) B	Tourret
Tourrette, Claret de la (1729–1793) A	Tourr.
Toursarkissian, Martin (fl. 1958) S	Toursark.
Toussaint, Anatole (1863–1943) S	Touss.
Toussaint, Leon (fl. 1953) S	L.Touss.
Toussoun, T.A. (fl. 1968) M	Toussoun
Touton, Karl (1858–1934) S	Touton
Touw, Andries (1935–) B	Touw
Tovar, Oscar (1923–) S	Tovar
(Tovar Serpa, Oscar)	
see Tovar, Oscar	Tovar
Tovey, James Richard (1873–1922) S	Tovey
Towe, Kenneth M. (1935–) A	Towe
Townely, Patricia J. (fl. 1961) M	Townely
Towner, Howard Frost (1943–) S	Towner
Townrow, Jocelyn Elizabeth Suzanne (1932–) S	J.Townrow
Townrow, John A. (1927–) BF	Townrow
Townsend, Charles Orvin (1863–1937) M	C.O.Towns.
Townsend, Clifford Charles (1926–) BS	C.C.Towns.
Townsend, Frederick (1822–1905) S	F.Towns.
Townsend, John Kirk (1809–1851) S	J.K.Towns.
Townsend, Joseph (1739–1816) S	Towns.
Townsend, Roberta A. A	R.A.Towns.
Townson, R. (fl. 1797) S	R.Townson
Townson, William (1850–1926) S	Townson

Towpasz, Krystyna (1943–) S	**Towpasz**
Toyama, Reizo (1913–1947) B	**Toyama**
Toyama, Shirô A	**S.Toyama**
Toyazaki, Noristsuna (fl. 1986) M	**Toyaz.**
Toyoda, Takeshi (1933–) S	**Toyoda**
Toyokuni, Hideo (1932–) PS	**Toyok.**
Toyoshima, Masami (fl. 1972) S	**Toyosh.**
Tozzi, Bruno (1656–1743) LS	**Tozzi**
Traaen, Alf E. (fl. 1914) M	**Traaen**
Trabucco, Giacomo (1845–) A	**Trabucco**
Trabut, Louis (Charles) (1853–1929) BMPS	**Trab.**
Tracanna, Beatriz C. (1949–) A	**Tracanna**
(Tracanna de Albornoz, Beatriz C.)	
see Tracanna, Beatriz C.	**Tracanna**
Tracey, John Geoffrey (1930–) S	**Tracey**
Tracey, R. (1951–) S	**R.Tracey**
Trachsel, Kaspar (Caspar) (1788–1832) S	**Trachsel**
Tracy, Clarissa (1918–1905) S	**C.Tracy**
Tracy, Cyrus Mason (1824–1891) S	**C.M.Tracy**
Tracy, Joseph Prince (1879–1953) S	**J.P.Tracy**
Tracy, Samuel Mills (1847–1920) MS	**Tracy**
Tradescant, John (1608–1662) L	**Trad.**
Trafvenfelt, Eric Carl (1774–1835) S	**Trafv.**
(Trafvenveldt, Eric Carl)	
see Trafvenfelt, Eric Carl	**Trafv.**
Trail, James William Helenus (1851–1919) MS	**Trail**
Traill, Catherine Parr (1802–1899) S	**C.Traill**
Traill, George William (1836–1897) S	**Traill**
(Traill, James William Helenus)	
see Trail, James William Helenus	**Trail**
Trainor, Francis R. (1929–) A	**Trainor**
Tralau, Hans (1932–1977) S	**Tralau**
Tramier, R. (fl. 1960) M	**Tramier**
Tran, Cong Khanh (1936–) S	**C.K.Tran**
Tran, Dinh Ly (fl. 1986) S	**D.L.Tran**
Trana, Erik Andreas (1847–1933) S	**Trana**
Transeau, Edgar Nelson (1875–1960) A	**Transeau**
(Transhel, Woldemar (Andrejevitch))	
see Tranzschel, Woldemar (Andrejevitch)	**Tranzschel**
Tranzschel, Woldemar (Andrejevitch) (1868–1942) MS	**Tranzschel**
Trappe, James M. (1931–) MS	**Trappe**
Trappen, Johannes (Jan) Everhardus van der (fl. 1834–1849) S	**Trappen**
Trask, B.J. A	**Trask**
Trass, Hans (Kh.K.) (1928–) M	**Trass**
Tratnik, B. (fl. 1988) M	**Tratnik**
Trattinnick, Leopold (1764–1849) MS	**Tratt.**
Traub, Hamilton Paul (1890–1983) S	**Traub**
Traunfellner, Aloys (1782–1840) S	**Traunf.**
Traunsteiner, Joseph (1798–1850) S	**Traunst.**

Trauth, Friedrich (1883–) A	**Trauth**
Trautmann, Robert (1873–1953) S	**Trautm.**
Trautvetter, Ernst Rudolf von (1809–1889) PS	**Trautv.**
Trautwein, Johannes (Carl) (1858–) S	**Trautwein**
Travares, J.S. S	**Travares**
Traversi, Leopoldo S	**Traversi**
Traverso, Giovanni Battista (1878–1955) M	**Traverso**
Travis, Bernard V. (1907–) A	**B.V.Travis**
Travis, William Gladstone (1877–1958) MS	**Travis**
Treboux, Octave (1876–) AM	**Treboux**
Trécul, Auguste (Adolphe Lucien) (1818–1896) S	**Trécul**
Tredici, V. (fl. 1937) M	**Tredici**
Tredick, Joanne (fl. 1986) M	**Tredick**
Tredick–Kline, J. (fl. 1987) M	**Tred.-Kline**
Treffer, Georg (1847–1902) S	**Treffer**
Treffner, Eduard (fl. 1881) S	**Treffner**
Treichel, Alexander (Johann August) (1837–1901) S	**Treich.**
Trejos, A. (fl. 1954) M	**Trejos**
Trelease, William (1857–1945) AMPS	**Trel.**
Tremaut S	**Tremaut**
(Trémeau de Rochebrune, Alphonse)	
see Rochebrune, Alphonse Trémeau de	**Rochebr.**
Trémols y Borrell, Federico (1831–1900) S	**Trémols**
Trench, Robert Kent (1940–) A	**Trench**
Trentepohl, Johann Friedrich (1784–1806) ABMS	**Trentep.**
Treschew, Cecil (fl. 1940) M	**Treschew**
(Treschow, Cecil)	
see Treschew, Cecil	**Treschew**
Trescol, F. (fl. 1986) M	**Trescol**
Tresner, Homer D. (fl. 1953) M	**Tresner**
Trespalacios, F. (fl. 1959) M	**Trespal.**
Tressens, Sara Graciela (1944–) S	**Tressens**
Tretiu, T. (1911–1969) S	**Tretiu**
Treub, Melchior (1851–1910) APS	**Treub**
Treuinfels, Leo M. (1848–) S	**Treuinf.**
Treumann, Karl (fl. 1880) S	**Treumann**
Trevelyan, Walter Caverly (1797–1879) S	**Trevelyan**
Treviranus, Gottfried Reinhold (1776–1837) S	**G.Trevir.**
Treviranus, Ludolf Christian (1779–1864) PS	**Trevir.**
Trevisan de Saint-Léon, Vittore Benedetto Antonio (1818–1897) ACMPS	**Trevis.**
Trew, Christoph Jakob (1695–1769) S	**Trew**
Triana, José Jéronimo (1834–1890) PS	**Triana**
Tribe, H.T. (fl. 1977) M	**Tribe**
Tribondeau (fl. 1899) M	**Trib.**
Tricker, Charles William Bret (1852–1916) S	**Tricker**
Trickett, R.S. S	**Trickett**
Triebel, Dagmar (1957–) M	**Triebel**
Triebner, Wilhelm (1883–1957) S	**Triebner**
Triemer, Richard E. A	**Triemer**

Trier, Georg (1884–1944) S   **Trier**
Triest, Ludwig J. (1957–) S   **Triest**
Trifonova, V.I. (fl. 1988) S   **Trifonova**
Trigaux, G. (fl. 1985) M   **Trigaux**
(Trigo, América del Pilar Rodrigo)
   see Rodrigo Trigo, América del Pilar   **Rodrigo**
Trikha, Chander K. (fl. 1968–1979) P   **Trikha**
Trimbach, Jacques (1933–) M   **Trimbach**
Trimboli, D. (fl. 1986) M   **Trimboli**
Trimen, Henry (1843–1896) PS   **Trimen**
Trimen, Roland (1840–1916) S   **R.Trimen**
Trimmer, Kirby (1804–1887) S   **Trimmer**
Trinajstic, Ivo (fl. 1985) S   **Trinajstic**
Trinchieri, Giulio (fl. 1911) M   **Trinchieri**
Trinchinetti, Augusto (1813–1847) S   **Trinch.**
Trindade, I.B. (fl. 1970) S   **Trindade**
Trinius, Carl Bernhard von (1778–1844) S   **Trin.**
Trinkwalter, Leopold (fl. 1913) S   **Trinkw.**
(Trinta, Elza Fromm)
   see Fromm Trinta, Elza   **Fromm**
Triolo, E. (fl. 1976) M   **Triolo**
Tripathi, A.K. (fl. 1986) P   **A.K.Tripathi**
Tripathi, R.C. (fl. 1966) M   **Tripathi**
Tripodi, Giacomo (1938–) A   **Tripodi**
Tripp, Frances E (fl. 1868) B   **Tripp**
Trist, Philip John Owen (1908–) S   **Trist**
Tristan, Jules Marie Claude de (1776–1861) S   **Tristan**
Tristani, A. (fl. 1988) M   **Tristani**
Tristram, Henry Baker (1822–1906) S   **Tristram**
Trivedi, B.S. (fl. 1970) FMP   **Trivedi**
Trivelli, Piera (1933–) M   **Trivelli**
Trochain, Jean–Louis (1903–1976) S   **Troch.**
Trochain–Marquès, Yvonne (fl. 1940–1959) S   **Troch.-Marq.**
Troeltzsch, Georgius Christianus (fl. 1751) S   **Troeltzsch**
Trofimovskaya, A.Ya (1905–) S   **Trofim.**
Trog, Jakob Gabriel (1781–1865) M   **Trog**
Troickij S   **Troickij**
Troilius, Adolf Magnus (1838–1909) S   **Troilius**
Troitskaya, O.V. A   **O.V.Troitsk.**
Troitskaya, Z.F. S   **Z.F.Troitsk.**
Troitsky, N.A. S   **Troitsky**
(Troitzkaja, Z.F.)
   see Troitskaya, Z.F.   **Z.F.Troitsk.**
Trojan, Johannes (1837–1925) S   **Trojan**
Troll, (Julius Georg Hubertus) Wilhelm (1897–1978) APS   **Troll**
Troll, Carl (Karl) (1899–1975) S   **C.Troll**
Trommer, Ernst Emil (fl. 1881) S   **Trommer**
Tron, E.Z. S   **Tron**
Tronchet, Antonin Benoît Joseph (1902–) S   **Tronchet**

(Troncoso de Burkart, Nélida Sara)
  see Troncoso, Nélida Sara   **Tronc.**
Troncoso, Nélida Sara (1914–1988) S   **Tronc.**
Tröndle, Arthur (1881–1920) S   **Tröndle**
Trono, Gavino C. (1931–) A   **Trono**
Tropea, Caledonio (fl. 1910) S   **Tropea**
Tropova, A.T. (fl. 1933) M   **Tropova**
Troschel, (Franz Ernst) Innocenz (1858–) S   **Troschel**
Trotsenko, R.S. (fl. 1981) M   **R.S.Trots.**
Trotsenko, Yu.A. (fl. 1974) M   **Y.A.Trots.**
Trotter, Alessandro (1874–1967) MPS   **Trotter**
(Trotzky, Petrus (Peter) Kornuch)
  see Kornuch–Trotzky, Petrus (Peter)   **Korn.-Trotzky**
Trouette, Édouard (1855–) S   **Trouette**
Trouillard, Charles (1821–1888) S   **Trouill.**
Troup, Robert Scott (1874–1939) S   **Troup**
Troupin, Georges M.D.J. (1923–) S   **Troupin**
Trow, Albert Howard (1863–1939) M   **Trow**
Trozelius, Clas Blechert (1719–1794) S   **Trozel.**
Trpin, D. (fl. 1965) S   **Trpin**
Truan y Luard, Alfredo (1837–1890) A   **Truan**
Trudell, Harry William (1884–) P   **Trudell**
Trudgen, Malcolm Eric (1951–) S   **Trudgen**
True, E.Y. (fl. 1919) M   **E.Y.True**
True, Gordon H.   S   **G.H.True**
True, Rodney Howard (1866–1940) S   **True**
Trueblood, Ellen (fl. 1972) M   **Trueblood**
Truffaut, Albert (1844–1925) S   **Truff.**
Truffaut, Georges (fl. 1912) S   **G.Truff.**
Truffi, M. (fl. 1901) M   **Truffi**
Trujillo, Baltazar (1927–) S   **Trujillo**
Trujillo, Eduardo E. (fl. 1963) M   **E.E.Trujillo**
Trukhaleva, N.A. (1927–) S   **Trukh.**
Trülzsch, (Paul) Otto (1889–) S   **Trülzsch**
Trumbull, James Hammond (1821–1897) S   **Trumbull**
Trummer, Franz Xaver (–1858) S   **Trummer**
Truncová, Eva   A   **Truncová**
Trüper, E. (fl. 1928) M   **Trüper**
Truscott, John Henry Lloyd (1905–) M   **Truscott**
Trusov, B.A. (1932–) S   **Trusov**
Trusova, N. (fl. 1915) M   **Trusova**
Truszkowska, (I.) Wanda (fl. 1958) M   **Truszk.**
Tryon, Alice Faber (1920–) PS   **A.F.Tryon**
Tryon, Henry (1856–1943) S   **Tryon**
Tryon, Rolla Milton (1916–) PS   **R.M.Tryon**
Trzebiński, Józef (Joseph) (1867–) M   **Trzeb.**
Tsagolova, V.G. (fl. 1970) S   **Tsag.**
Tsai, Hse Tao (1911–1981) S   **H.T.Tsai**
Tsai, Shao Lan   S   **S.L.Tsai**

Tsanava, N.I. (fl. 1973) M	**Tsanava**
Tsao, G.C. (fl. 1965) M	**G.C.Tsao**
(Tschechow, W.P.)	
see Czechov, V.P.	**Czechov**
(Tscheremissinoff, N.A.)	
see Cheremisinov, N.A.	**Cheremis.**
(Tschermak, Elizabeth)	
see Tschermak–Woess, Elizabeth	**Tscherm.-Woess**
Tschermak–Seysenegg, Erich von (1871–1962) S	**Tscherm.-Seys.**
Tschermak–Woess, Elizabeth (fl. 1976) AM	**Tscherm.-Woess**
(Tschernaiew, Vassili Matveievitch)	
see Czernajew, Vassiliĭ Matveievitch	**Czern.**
Tschernetzkaja, S.S. (fl. 1929) M	**Tschern.**
Tscherneva, O.V. (1929) S	**Tscherneva**
Tscherning, Johannes Wilhelm (1856–1890) S	**Tscherning**
Tschernov, V.K. A	**Tschernov**
Tscheuschner, Irmgard T. (fl. 1956) M	**Tscheuschner**
(Tschewrenidi, S.Kh.)	
see Czevrenidi, S.Kh.	**Czevr.**
(Tschichatscheff, Petr Aleksandrovich)	
see Tchichatscheff, Petr Aleksandrovich	**Tchich.**
Tschirch, (Wilhelm Oswald) Alexander (1856–1939) MS	**Tschirch**
Tscholokaschvili, N.B. (1932–) S	**Tscholok.**
Tschourina, Olga A	**Tschourina**
Tschuchrukidze, A.N. (1940–) S	**Tschuchr.**
Tschudovskaia, Inna (fl. 1928) M	**Tschud.**
Tschudy, R.H. (fl. 1937) M	**Tschudy**
Tseng, Chang Jiang (fl. 1965) S	**C.J.Tseng**
Tseng, Charles Chiao (1932–) S	**C.C.Tseng**
Tseng, Cheng Kwei (Kuei) A	**C.K.Tseng**
Tseng, Ling Chang (fl. 1981) S	**L.C.Tseng**
Tseng, Yong Qian (fl. 1978) S	**Y.Q.Tseng**
Tseng, Yung Chien (1930–) S	**Y.C.Tseng**
Tsi, Zhan Huo (1937–) S	**Z.H.Tsi**
Tsiang, Ying (1898–1982) S	**Tsiang**
Tsien, Cho Po (1916–) S	**C.P.Tsien**
Tsiklinsky, P. (fl. 1899) M	**Tsikl.**
Tsikov, D.K. (fl. 1973) S	**Tsikov**
(Tsinovskis, Raĭmond Ekabovich)	
see Cinovskis, Raĭmond Ekabovich	**Cinovskis**
Tsintsadze, K.V. (fl. 1976) M	**Tsints.**
Tsitsin, Nikolai Vasiljevich (1898–1980) S	**Tsitsin**
Tsitsvidze, A.T. (fl. 1960) S	**Tsitsv.**
Tso, Ching Lieh S	**C.L.Tso**
Tsoong, Chi Hsin (fl. 1963) S	**C.H.Tsoong**
Tsoong, Kuan Kwang S	**K.K.Tsoong**
Tsoong, Pu Chiu (1906–) S	**P.C.Tsoong**
Tsou, C.S. (fl. 1977) M	**C.S.Tsou**
Tsubo, Yoshihiro (1926–) A	**Tsubo**

Tsubouchi, Haruo (fl. 1986) M	**Tsub.**
Tsuchiya, S. (fl. 1986) M	**S.Tsuchiya**
Tsuchiya, Takeshi (1904–1989) M	**Tsuchiya**
Tsuda, Mitsuya (fl. 1978) M	**Tsuda**
Tsuda, Roy Toshio (1939–) A	**R.T.Tsuda**
(Tsueh, Yu Wan)	
see Tsui, You Wen	**Y.W.Tsui**
Tsui, Hung Pin (fl. 1980s) S	**H.P.Tsui**
Tsui, Yon Wen (1907–) S	**Tsui**
Tsui, You Wen (fl. 1985) S	**Y.W.Tsui**
Tsuji, R. (fl. 1919) M	**Tsuji**
Tsukamoto, F. (fl. 1953) M	**Tsukam.**
(Tsukervanik, T.I.)	
see Tzukervanik, T.I.	**Tzukerv.**
Tsumura, Kohei (1913–) A	**Tsumura**
Tsuneda, Akihiko (1946–) M	**Tsuneda**
(Tsvelev, Nikolai Nikolaievich)	
see Tzvelev, Nikolai Nikolaievich	**Tzvelev**
(Tsvelov, Nikolai Nikolaievich)	
see Tzvelev, Nikolai Nikolaievich	**Tzvelev**
(Tsvelyov, Nikolai Nikolaievich)	
see Tzvelev, Nikolai Nikolaievich	**Tzvelev**
Tsvetkova, N.L. A	**Tsvetkova**
Tswett, Michael (Michel) Semenovich (1872–1919) S	**Tswett**
Tsyrenova, D.Yu. (fl. 1985) S	**Tsyren.**
Tsyrina, T.S. S	**Tsyrina**
Tu, C.C. (fl. 1969) M	**C.C.Tu**
Tu, V.G. (fl. 1980s) S	**V.G.Tu**
Tu, Vu (Vy) Nguyen (fl. 1979) P	**V.N.Tu**
Tu, Yu Lin (fl. 1984) S	**Y.L.Tu**
Tuan, Pei Chin (fl. 1978) AS	**P.C.Tuan**
Tubaki, Keisuke (1924–) AM	**Tubaki**
Tubergen, Cornelis Gerrit van (1844–1919) S	**Tubergen**
Tubeuf, Carl (Karl) von (1862–1941) MS	**Tubeuf**
Tubilla, Tomás Andrés y (1859–1882) S	**Tubilla**
Tucker, Bruce E. (fl. 1981) M	**B.E.Tucker**
Tucker, Clarence Mitchell (1897–1954) M	**Tucker**
Tucker, Ethelyn (Daliaette) Maria (1871–) S	**E.M.Tucker**
Tucker, Gordon C. (1957–) S	**G.C.Tucker**
Tucker, John Maurice (1916–) S	**J.M.Tucker**
Tucker, R. (fl. 1986) S	**R.Tucker**
Tucker, Shirley Cotter (1927–) M	**S.C.Tucker**
Tuckerman, Edward (1817–1886) MPS	**Tuck.**
Tuckey, James Hingston (1776–1816) S	**Tuckey**
Tucović, A. (fl. 1972) S	**Tucović**
Tudosescu-Bănescu, Veronica (1932–) M	**Tud.-Băn.**
Tuisl, Gerhard (1942–) S	**Tuisl**
Tukuoka, Keiko (fl. 1987) M	**Tukuoka**
Tulasne, Charles (1816–1884) M	**C.Tul.**

Turner

Tulasne, Louis René ('Edmond') (1815–1885) MS | Tul.
(Tuljaganova, M.T.)
  see Tulyaganova, M.T. | Tulyag.
Tullberg, Sven Axel Teodor (1852–1886) S | Tullb.
Tullis, Edgar Cecil (1901–) M | Tullis
Tulloch, A.P. (fl. 1970) M | A.P.Tulloch
Tulloch, Margaret C. (fl. 1972) M | M.C.Tulloch
Tulloss, Rodham E. (fl. 1984) M | Tulloss
Tully, William (1785–1859) S | Tully
Tulyaganova, M.T. (1936–) S | Tulyag.
Tumadzanov, I.I. (1910–) S | Tumadz.
(Tumadzhanov, I.I.)
  see Tumadzanov, I.I. | Tumadz.
Tumanian, M.G. S | Tumanian
Tumanow, Ivan I. (1894–) S | Tumanow
Tung, Shi Lin (fl. 1988) S | S.L.Tung
Tunkina, T.V. (fl. 1955) M | Tunkina
Tunmann, Otto (1867–1919) S | Tunmann
Tunnela, E. (fl. 1978) M | Tunnela
Tunstall, A.C. (1903–) M | Tunstall
Tuntas, Basilios (1871–) S | Tuntas
Tuomikoski, Risto Kalevi (1911–1989) B | Tuom.
Tupa, Dianna D. A | Tupa
Tur, Nuncia Maria (fl. 1975) S | Tur
Turakulov, I. (fl. 1986) S | Turak.
Türckheim, Hans von (1853–1920) S | Türckh.
Turconi, Malusio (1879–1929) M | Turconi
Turczaninow, Porphir Kiril Nicolai Stepanowitsch (1796–1863) PS | Turcz.
Turenschi, Eugen (1922–) S | Turenschi
Turesson, Göte Wilhelm (1892–1970) MS | Turesson
Turfitt, G.E. (fl. 1939) M | Turfitt
Turhan, G. (fl. 1973) M | Turhan
Turian, G. (fl. 1969) M | Turian
Turio, Bernardino (1779–1854) S | Turio
Turkensteen, L.J. (fl. 1986) M | Turkenst.
Turkevicz, S.Ju. S | Turkev.
Turnau, Katarzyna (1954–) M | Turnau
Turnbull, J.R. (fl. 1988) S | J.R.Turnbull
Turnbull, Robert (c.1813–1891) S | Turnbull
Turner, Billie Lee (1925–) S | B.L.Turner
Turner, Dawson (1775–1858) ABMS | Turner
Turner, Edward Phillips (1865–1937) S | E.Turner
Turner, Elizabeth M. (fl. 1940) M | E.M.Turner
Turner, Frederick (1852–1939) S | F.Turner
Turner, Hubert (fl. 1992) S | H.Turner
Turner, Melvin D. (fl. 1988) P | M.D.Turner
Turner, Samuel S | S.Turner
Turner, Wilhelm (1860–) A | W.Turner
Turner, William Barwell (1845–1917) A | W.B.Turner
Turner Ettlinger, D.M. (fl. 1990) S | Turner Ettl.

Türpe, Anna Maria (1946–) S	**Türpe**
Turpin, P.G. (fl. 1982) S	**P.G.Turpin**
Turpin, Pierre Jean François (1775–1840) AMS	**Turpin**
Turquet, Jean (fl. 1910) S	**Turquet**
(Turquier de Longchamp, Joseph Alexandre de)	
see Le Turquier de Longchamp, Joseph Alexandre	**Le Turq.**
Turra, Antonio (1730–1796) AS	**Turra**
Turrell, E.E. S	**Turrell**
Turrettini, G. S	**Turrett.**
Turrill, William Bertram (1890–1961) S	**Turrill**
Turton, William (1762–1835) A	**Turton**
Tuset, J.J. (fl. 1989) M	**J.J.Tuset**
Tuset, M.T. (fl. 1989) M	**M.T.Tuset**
Tussac, François Richard de (1751–1837) S	**Tussac**
Tutcher, William James (1867–1920) PS	**Tutcher**
Tuthill, Dorothy E. (fl. 1986) M	**Tuthill**
Tutin, Thomas Gaskell (1908–1987) AFS	**Tutin**
Tutunaru, Vasile (1922–) MS	**Tutunaru**
Tüxen, Reinhold (1899–1980) S	**Tüxen**
Tuyama, Takasi (1910–) PS	**Tuyama**
Tuyn, P. (fl. 1960) S	**Tuyn**
Tuz, A.S. (1919–) S	**Tuz**
Tuzet, Odette (fl. 1951) AM	**Tuzet**
Tuzibe, Masanobu (1915–1944) B	**Tuzibe**
Tuzlaci, E. (fl. 1983) S	**Tuzlaci**
Tuzson, János (1870–1943) PS	**Tuzson**
Tuzson, Johann von (1870–1952) S	**J.Tuzson**
(Twamley, Louisa Anne)	
see Meredith, Louisa Anne	**Meredith**
Tweed, Ronald Duncan (1900–1989) A	**Tweed**
Tweedie, John ('James') (1775–1862) S	**Tweedie**
Tweedy, Frank (1854–1937) S	**Tweedy**
Twenhofel, William Henry (1875–1957) A	**Twenhofel**
Twentyman, J.D. S	**Twentyman**
Twining, Alfred (1853–1922) S	**A.Twining**
Twining, Elizabeth (1805–1889) S	**Twining**
Twiss, Wilfred Charles (1868–) S	**Twiss**
Twisselmann, Ernest C. (1917–1972) S	**Twisselm.**
Tyagi, A.K. (fl. 1978) M	**A.K.Tyagi**
Tyagi, R.N.S. (fl. 1962) M	**Tyagi**
Tyas, Robert (1811–1879) S	**Tyas**
Tyler, Ansel Augustus (1869–1922) S	**Tyler**
Tyler, Peter Alfred (1936–) A	**P.A.Tyler**
Tyler, Stanley A. A	**S.A.Tyler**
(Tylitki, Edmund E.)	
see Tylutki, Edmund E.	**Tylutki**
Tylutki, Edmund E. (1926–) M	**Tylutki**
Tynan, Eugene J. (1924–) A	**Tynan**
Tynni, Risto (1924–) A	**Tynni**

Tyrl, Ronald Jay (1943–) S   **Tyrl**
Tyroff, Helmut   A   **Tyroff**
Tyrrell, D. (fl. 1966) M   **Tyrrell**
Tyson, William (1851–1920) A   **Tyson**
Tyteca, B. (fl. 1981) S   **B.Tyteca**
Tyteca, D. (fl. 1981) S   **D.Tyteca**
Tyzzer, Ernest Edward (1875–) A   **Tyzzer**
Tzagolova, V.G.   S   **Tzagolova**
Tzanoudakis, D.B. (1950–) S   **Tzanoud.**
Tzavella–Klonari, K. (fl. 1974) M   **Tzav.-Klon.**
Tzean, Shean–Shung (1944–) M   **Tzean**
(Tzenkovskii, Lev Semenovich)
   see Cienkowski, Leo de   **Cienk.**
Tzukervanik, T.I. (fl. 1983) S   **Tzukerv.**
Tzvelev, Nikolai Nikolaievich (1925–) S   **Tzvelev**

Ubaldi, D. (fl. 1983) M   **Ubaldi**
Ubera Jimenez, José Luis (1950–) S   **Ubera**
(Ubera, José Luis)
   see Ubera Jimenez, José Luis   **Ubera**
Ubrizsy, G. (1919–1973) M   **Ubrizsy**
Uchida, J.Y. (fl. 1979) MS   **J.Y.Uchida**
Uchida, Masahiro (fl. 1991) M   **M.Uchida**
Uchida, S. (fl. 1960) M   **S.Uchida**
Uchida, Shigetaro (1885–) S   **Uchida**
Uchiyama, Shigeru (fl. 1987) M   **Uchiy.**
Ucria, Bernardino da (1739–1796) S   **Ucria**
Udachin, R.A. (fl. 1972) S   **Udachin**
Udagawa, Shun–ichi (1931–) M   **Udagawa**
Udar, Ram (1926–1985) B   **Udar**
Udayalakshmi, G.P. (fl. 1978) M   **Udayal.**
Uden, N.van (1921–1991) M   **Uden**
Udipi, P.A. (fl. 1976) M   **Udipi**
Uebel, E.   A   **Uebel**
Uebelmesser, Esther–Ruth (fl. 1956) M   **Uebelm.**
Uechtritz, Maximilian (Max) Friedrich Sigismund von (1785–1851) S   **Uechtr.**
Uechtritz, Rudolf (Karl (Carl)) Friedrich von (1838–1886) S   **R.Uechtr.**
Uecker, Francis August (1930–) M   **Uecker**
Ueda, Kunihiko (fl. 1980) S   **K.Ueda**
Ueda, Saburo (1898–) A   **Ueda**
Ueda, Seichi (Seiichi) (fl. 1988) M   **S.Ueda**
Uehlinger, Arthur Bernhard (1896–) S   **Uehl.**
Ueki, Robert (fl. 1973) M   **Ueki**
Uematsu, Y. (fl. 1991) M   **Uematsu**
Uemura, Y. (fl. 1932) M   **Uemura**
Ueno (fl. 1971) M   **Ueno**
Uexküll–Gyllenbrand, Margarete von (fl. 1901) S   **Uexküll**
Ueyama, Akinori (fl. 1978) M   **Ueyama**

Ugadim, Yumiko (1937–) A	**Ugadim**
Ugborogho, R.E. (fl. 1972) S	**Ugbor.**
Ugent, Donald (1933–) S	**Ugent**
(Uglitskikh, Alexander N.)	
see Uglitzkich, Alexander N.	**Uglitzk.**
Uglitzkich, Alexander N. (1876–) S	**Uglitzk.**
Ugolini, Ugolino (1856–1942) PS	**Ugolini**
Ugrinsky, K.A. (fl. 1920) S	**Ugr.**
Uherkovich, Gábor (Gabriel) (1912–) A	**Uherk.**
Uhl, Charles Harrison (1918–) S	**C.H.Uhl**
Uhl, Natalie Whitford (1919–) S	**N.W.Uhl**
Uhlik, David J. A	**Uhlik**
Uhlmann, D. A	**Uhlmann**
Uhlworm, Oskar (1849–1929) S	**Uhlworm**
(Uhrová, Anezka)	
see Hrabětova–Uhrová, Anezka	**Hrabětova**
Ui, T. (fl. 1986) M	**Ui**
Uildriks, Frederike Johanna van (fl. 1898) S	**Uildriks**
Uilkens, Theodorus Frederik (1812–1891) S	**Uilkens**
Uitewaal, Antonius Josephus Adrianus (1899–1963) S	**Uitewaal**
Uittien, Hendrik (1898–1944) S	**Uittien**
Ujhelyi, József (1910–1979) S	**Ujhelyi**
Ujvárosi, M. (1913–) S	**Ujvárosi**
(Ukedem, J.d')	
see d'Ukedem, J.	**d'Ukedem**
Ukkelberg, H.G. (fl. 1935) M	**Ukkelberg**
Ulander, Axel Petrus (1874–) S	**Ulander**
Ulanova, K.P. (fl. 1981) S	**Ulanova**
Ulbrich, Oskar Eberhard (1879–1952) MPS	**Ulbr.**
Ule, Ernst Heinrich Georg (1854–1915) BMS	**Ule**
Úlehla, Vladimír A	**Úlehla**
Ulibarri, Emilio A. (1946–) S	**Ulibarri**
Uline, Edwin Burton (1867–1933) S	**Uline**
Ulitzsch, Carl August (fl. 1796) S	**Ulitzsch**
Uljanishchev, V.I. (fl. 1938) M	**Uljan.**
Uljanova, T.N. (1926–) S	**Uljanova**
Uljé, C.B. (fl. 1985) M	**Uljé**
Ulke, Titus (1866–1961) AS	**Ulke**
Ulken, Annemarie (fl. 1965) M	**Ulken**
Ullasa, B.A. (fl. 1969) M	**Ullasa**
Ulle, Z.G. (1944–) S	**Ulle**
Ullepitsch, Joseph (1827–1896) S	**Ullep.**
Ullgren, Olof Matthias (1785–1819) S	**Ullgren**
Ullmann, I. (fl. 1987) S	**I.Ullmann**
Ullmann, J. (fl. 1990) S	**J.Ullmann**
Ulloa, M. (fl. 1978) M	**Ulloa**
Ullrich, Bernd (fl. 1992) S	**B.Ullrich**
Ullrich, Hans von (1913–) M	**Ullrich**
Ullstrup, Arnold John (1907–) M	**Ullstrup**
Ulomskii, S.N. A	**Ulomskii**

Uloth, Wilhelm (Ludwig Heinrich) (1833–1895) S	**Uloth**
Ulpiani, Celso S	**Ulpiani**
Ulrich, Edward Oscar (1857–1944) A	**E.O.Ulrich**
Ulrich, Wilhelm (fl. 1870–1890) S	**Ulrich**
Ulsamer, A. (fl. 1895) S	**Ulsamer**
Ulvinen, (Eero) Arvi (1897–) S	**Ulvinen**
Ulvinen, Tauno (1930–) M	**T.Ulvinen**
(Ulyanishchev, V.I.)	
see Uljanishchev, V.I.	**Uljan.**
(Ulziikutag, N.)	
see Ulziykhutag, N.	**N.Ulziykh.**
Ulziykhutag, N. (fl. 1987) S	**N.Ulziykh.**
Ulziykhutag, Reginald E. (1934–) S	**R.E.Ulziykh.**
Umamaheswara Rao, M. A	**Umam.Rao**
Umezaki, Isamu (1925–) A	**Umezaki**
Umezawa, H. (fl. 1954) M	**Umezawa**
Umezu, Y. (fl. 1987) S	**Umezu**
Umphlett, Clyde J. (fl. 1956) M	**Umphlett**
(Unamuno, Luis M.)	
see Unamuno Yrigoyen, Luis M.	**Unamuno**
Unamuno Yrigoyen, Luis Mariano (1873–1943) MS	**Unamuno**
Unanue, José Hipólito (1755–1853) S	**Unanue**
Underwood, John (–1834) S	**J.Underw.**
Underwood, Judson Kemp (1898–) S	**J.K.Underw.**
Underwood, Lucien Marcus (1853–1907) BMPS	**Underw.**
Ungar, Karl (1869–) S	**Ungar**
Unger, Franz (Joseph Andreas Nicolaus) (1800–1870) ABFMS	**Unger**
Unger, Gottfried (fl. 1971) S	**G.Unger**
Unger, Michael (fl. 1806) S	**M.Unger**
Ungern–Sternberg, Franz (1808–1885) S	**Ung.-Sternb.**
Uniyal, B.P. (1943–) S	**Uniyal**
Unna, P.G. (fl. 1895) M	**Unna**
Uno, Kakuo (fl. 1940) B	**Uno**
Unruh, Martin S	**Unruh**
Untawale, A.G. A	**Untawale**
Unverricht, Carl (1809–1883) S	**Unverr.**
Unwin, Arthur Harold (1878–) S	**A.Unwin**
Unwin, William Charles (1811–1887) S	**Unwin**
Uotila, Pertti Johannes (1944–) S	**Uotila**
Upadhyay, H.B.P. (fl. 1964) M	**H.B.P.Upadhyay**
Upadhyay, Harbansh Prasad (fl. 1966) M	**H.P.Upadhyay**
Upadhyay, J.M. (fl. 1984) M	**J.M.Upadhyay**
Upadhyay, Janeshwar (fl. 1966) M	**J.Upadhyay**
Upadhyay, R.S. (fl. 1982) M	**R.S.Upadhyay**
Upadhyay, Rajeev K. (fl. 1990) M	**R.K.Upadhyay**
Upham, Warren (1850–1934) S	**Upham**
Uphof, Johannes Cornelius Theodorus (Theodoor) (1886–) AS	**Uphof**
Uppal, Badri Nath (fl. 1936) M	**Uppal**
Upreti, D.K. (1958–) M	**Upreti**

Upton, Walter T. (fl. 1967) S	**Upton**
Ura, T. (fl. 1956) M	**Ura**
Urakami, T. (fl. 1977) M	**Urakami**
Urasawa, Y. (fl. 1951) M	**Urasawa**
Urban, Ignatz (1848–1931) ABPS	**Urb.**
Urban, J.B. A	**J.B.Urb.**
Urban, Otto (fl. 1934) S	**O.Urb.**
Urban, Zdeněk (fl. 1956) M	**Z.Urb.**
Urbani, Malvina (1958–) S	**Urbani**
Urbatsch, Lowell Edward (1942–) S	**Urbatsch**
Urbina y Altamirano, Manuel (1843–1906) S	**Urbina**
Urbonas, Vincas (1940–) M	**Urbonas**
Ureña Plaza, Juan Fernando (fl. 1986) S	**Ureña**
Uribe, (Antonio) Lorenzo Uribe (1901–1980) S	**L.Uribe**
Uribe, Cesar A	**C.Uribe**
(Uribe Echebarria Díaz, Pedro María)	
see Uribe-Echebarria Díaz, Pedro María	**Uribe-Ech.**
Uribe-Echebarria Díaz, Pedro María (1953–) S	**Uribe-Ech.**
(Uribe-Echebarria, Pedro María)	
see Uribe-Echebarria Díaz, Pedro María	**Uribe-Ech.**
(Uribe-Uribe, (Antonio) Lorenzo)	
see Uribe, (Antonio) Lorenzo Uribe	**L.Uribe**
Uriburu, Julio Vincente (fl. 1909) M	**Uriburu**
(Urmi-Koenig, Katherina)	
see Urmi-König, Katherina	**Urmi-König**
Urmi-König, Katherina (fl. 1975) S	**Urmi-König**
Urquiola, A. (fl. 1987) S	**Urquiola**
Urries de Azara, M.J.de (fl. 1932) M	**Urries**
Ursch, Eugène (1882–1962) S	**Ursch**
Ursprung, Alfred (1876–1952) S	**Ursprung**
Urton, N.R. (fl. 1986) S	**Urton**
Urumoff, Ivan Kiroff (1857–1937) S	**Urum.**
(Urumov, Ivan Kirov)	
see Urumoff, Ivan Kiroff	**Urum.**
(Urusov, V.M.)	
see Urussov, V.M.	**Urussov**
Urussov, V.M. (1943–) S	**Urussov**
(Urville, Jules Sébastian César Dumont d')	
see d'Urville, Jules Sébastian César Dumont	**d'Urv.**
Usharani, P.(U.) (fl. 1981) M	**Ushar.**
Usherwood, M. (fl. 1985) M	**Usherw.**
Uslar, Johann Julius von (1762–1838) S	**Uslar**
Uspenskaja, G.D. (fl. 1974) M	**Uspensk.**
Uspenskaja, M.S. (fl. 1987) S	**M.S.Uspensk.**
(Uspenskaya, G.D.)	
see Uspenskaja, G.D.	**Uspensk.**
Usteri, Alfred (1869–1948) PS	**A.Usteri**
Usteri, Paul (1768–1831) PS	**Usteri**
Ustjuzhanina, L.A. A	**Ustjuzh.**

Usui K. (fl. 1974) M	**Usui**
Utatsu, Ikuyo (fl. 1979) M	**Utatsu**
Utenkow, M. (fl. 1929) M	**Utenkow**
Utermöhl, Hans A	**Utermöhl**
Utinet (fl. 1839) S	**Utinet**
Utkin, Leonid Antonovich (1884–1964) S	**Utkin**
Utkin, M.S. S	**M.S.Utkin**
Utley, John F. (1944–) S	**Utley**
Utsch, Jacob (1824–1901) S	**Utsch**
Uttal, Leonard J. (fl. 1962) S	**Uttal**
Uttangi, J.C. A	**Uttangi**
Utz, J.P. (fl. 1963) M	**Utz**
(Üxip, Albert Jakovlevič)	
see Juxip, Albert Jakovlevič	**Juxip**
Uyeda, Y. (fl. 1905) M	**Uyeda**
Uyeki, Homiki (1882–) S	**Uyeki**
Uzonyi–Latkoczky, A. (fl. 1962) M	**Uz.-Latk.**
(V'ulf, Eugenii Vladimirowitsch)	
see Wulff, Eugenii Vladimirowitsch	**E.Wulff**
Vaarama, (Otto) Antero (1912–1975) BS	**Vaar.**
Vaartaja, O. (fl. 1960) M	**Vaartaja**
Vaccaneo, Roberto (1905–) S	**Vaccaneo**
Vaccari, Antonio (1868–) BS	**A.Vacc.**
Vaccari, Enrico (1912–1944) S	**E.Vacc.**
Vaccari, Lino (1873–1951) PS	**Vacc.**
Vacek, V. (fl. 1949) M	**Vacek**
Vachard, Daniel A	**Vachard**
Vachell, Eleanor (1879–1948) S	**Vachell**
Vacherot, M. S	**Vacherot**
Vachey, G. A	**Vachey**
Váczy, C. (1913–) S	**Váczy**
Vadas, Jenö (Eugen, Eugène) (1857–) S	**Vadas**
Vaga, August (1893–1960) S	**Vaga**
Vágner, L. (1815–1888) S	**Vágner**
Vaheeduddin, Syed (fl. 1955) M	**Vaheed.**
Vahl, Jens Laurentius (Lorenz) Moestue (1796–1854) S	**J.Vahl**
Vahl, Martin (Henrichsen) (1749–1804) ABMPS	**Vahl**
Vahl, Martin (II) (1869–1946) MPS	**M.Vahl**
Vahrameev, V.A. S	**Vahram.**
Vaidehi, B.K. (fl. 1969) M	**Vaidehi**
Vaidya, J.G. (fl. 1976) M	**Vaidya**
Vail, Anna Murray (1863–1955) S	**Vail**
Vailionis, Liudas (1886–1939) A	**Vailionis**
Vaillant, Léon (Louis) (1834–1914) S	**L.Vaill.**
Vaillant, Sébastien (Sebastian) (1669–1722) LM	**Vaill.**
Vainberg, T.I. S	**Vainberg**
Vainio, Edvard (Edward) August (1853–1929) BMS	**Vain.**
Vaizey, John Reynolds (1862–1889) B	**Vaiz.**

663

Vajda, Laszló (1890–) BS	**Vajda**
(Vajdan, Laszló)	
see Vajda, Laszló	**Vajda**
Vajravelu, E. (1936–) S	**Vajr.**
Vakhrameev, V.A. (fl. 1958) AF	**Vakhram.**
Vakhrusheva, T.E. (fl. 1974) M	**Vakhrush.**
Vakulin, D.Ja. S	**Vakulin**
Vala, D.G. (fl. 1984) M	**Vala**
Valbusa, Ubaldo (1874–1939) S	**Valbusa**
(Valckenier Suringar, Jan)	
see Suringar, Jan Valckenier	**J.V.Suringar**
Valdebenito, H.A. (fl. 1986) S	**Valdeb.**
(Valdes, A. Barreto)	
see Barreto Valdés, A.	**A.Barreto**
Valdés, Benito (1942–) S	**Valdés**
Valdés Bermejo, Enrique (1945–) S	**Valdés Berm.**
(Valdés Castrillón, Benito)	
see Valdés, Benito	**Valdés**
Valdés, Javier (fl. 1962) S	**J.Valdés**
(Valdés, Jésus)	
see Valdés–Reyna, Jésus	**Valdés–Reyna**
Valdés–Reyna, Jesús (1948–) S	**Valdés–Reyna**
Valencia, Juan I. S	**Valencia**
Valensi, Lionel A	**Valensi**
Valenta, Vlk (fl. 1948) M	**Valenta**
Valente, Antonio (fl. 1803) S	**Valente**
Valente, Marie da Conceição (1938–) S	**C.Valente**
(Valenti, Giorgina Serrato)	
see Serrato–Valenti, Giorgina	**Serrato**
Valenti–Serini, Francesco (1795–1872) MS	**Valenti**
Valentine, David Henriques (1912–1987) PS	**Valentine**
Valentine, William (–1884) B	**W.Valentine**
Valenzuela, F. (fl. 1920s) M	**Valenz.**
Valenzuela, R. (fl. 1986) M	**R.Valenz.**
Valenzuela, Salvador A	**S.Valenz.**
(Valéra, Máirín de)	
see de Valéra, Máirín	**de Valéra**
Vales, M.A. (fl. 1986) S	**Vales**
Valet, (August) Friedrich (1811–1889) S	**Valet**
Valet, Gabriel A	**G.Valet**
Valeton, Theodoric (1855–1929) S	**Valeton**
Vălev, S.T. (1910–1974) S	**Vălev**
Valiante, R. A	**Valiante**
Valkanov, Alexander (fl. 1929) AM	**Valkanov**
(Valkenburg, Shirley D.Van)	
see Van Valkenburg, Shirley D.	**Van Valk.**
Valla, G. (fl. 1972) M	**Valla**
Valldosera, M. (fl. 1987) M	**Valldos.**

(Valle, Cipriano J.)	
see Valle Gutiérrez, Cipriano J.	**C.J.Valle**
Valle Gutiérrez, Cipriano J. (fl. 1984) S	**C.J.Valle**
Valle, R.C. (fl. 1961) M	**Valle**
Valle Tendero, Francisco (1953–) S	**F.Valle**
Valleau, William Dorney (1891–) M	**Valleau**
Vallée d'Alfort, H. (fl. 1903) M	**Vallée Alf.**
Valleggi, M. (fl. 1933) M	**Valleggi**
Vallentin, Elinor Frances (1873–1924) S	**Vallentin**
Vallès–Xirau, Joan (1959–) S	**Vallès–Xirau**
Vallin, H. (fl. 1953) M	**Vallin**
Vallino, Filippo S	**Vallino**
Vallot, Jean Nicolas (fl. 1840) M	**J.N.Vallot**
Vallot, Joseph (1854–1925) S	**Vallot**
Valls, José Francisco Montenegro (1945–) S	**Valls**
Valmaseda, M. (fl. 1986) M	**Valmaseda**
Valmayor, H.L. (fl. 1984) S	**Valmayor**
Valmont de Bomare, Jacques Christophe (1731–1807) S	**Valmont**
Valsecchi, Franca (1931–) S	**Vals.**
Valta, Akseli (1911–1982) S	**Valta**
(Valva, Vincenzo la)	
see la Valva, Vincenzo	**la Valva**
Vampola, Petr (fl. 1991) M	**Vampola**
(van Alderwerelt van Rosenburgh, Cornelis Rugier Willem Karel)	
see Alderwerelt van Rosenburgh, Cornelis Rugier Willem Karel van	**Alderw.**
Van Baalen, Chase (1925–1986) A	**Van Baalen**
(van Balgooy, Max Michael Josephus)	
see Balgooy, Max Michael Josephus van	**Balgooy**
Van Bambeke, Charles Eugène Marie (1829–1918) S	**Van Bamb.**
Van Bastelaer, Désiré Alexandre (Henri) (1823–1907) S	**Van Bast.**
Van Beem, A.P. A	**Van Beem**
(van Benthem, J.)	
see Benthem, J.van	**Benthem**
(van Beusekom, C.F.)	
see Beusekom, C.F.van	**Beusekom**
(van Beverwijk, A.L.)	
see Beverwijk, A.L.van	**Beverw.**
(van Beyma, J.F.H.)	
see Beyma, J.F.H.van	**J.F.H.Beyma**
(van Beyma, T.H.)	
see Beyma, T.H.van	**T.H.Beyma**
Van Bockstal, Liliane (1937–) S	**Van Bockstal**
Van Boekel, Norma M.da Costa A	**Van Boekel**
(van Borssum Waalkes, Jan)	
see Borssum Waalkes, Jan van	**Borss.Waalk.**
(van Breda de Haar, Jacob Gijsbert Samuel)	
see Breda, Jacob Gijsbert Samuel van	**Breda**
(van Breemen, Pieter Johan)	
see Breemen, Pieter Johan van	**P.J.Breemen**

(van Bruggen, A.C.)	
see Bruggen, A.C.van	**A.C.Bruggen**
(van Bruggen, Heinrich Wilhelm Eduard)	
see Bruggen, Heinrich Wilhelm Eduard van	**H.W.E.Bruggen**
(van Bruggen, Theodore)	
see Bruggen, Theodore van	**Bruggen**
(van Brummelen, J.)	
see Brummelen, J.van	**Brumm.**
Van Campo, Madeleine (1920–) S	**Van Campo**
(Van Campo–Duplan, Madeleine)	
see Van Campo, Madeleine	**Van Campo**
Van Criekinge, Louis (fl. 1986) S	**Van Criek.**
(van Dam, Herman)	
see Dam, Herman van	**Dam**
(van de Beeke, A.)	
see Beeke, A.van de	**Beeke**
(van de Bogart, Fred)	
see Bogart, Fred van de	**Bogart**
(van de Klashorst, G.)	
see Klashorst, G.van de	**Klashorst**
(van de Meerendonk, J.P.M.)	
see Meerendonk, J.P.M.van de	**Meerend.**
(van de Meerssche, J.)	
see Meerssche, J.van de	**Meerssche**
Van de Vyvere, Ernest (1811–1853) S	**Van de Vyvere**
(van de Weyer, W.)	
see Weyer, W.van de	**Weyer**
Van den Assem, J. (fl. 1953) S	**Van den Assem**
(van den Berg, Maria Elizabeth)	
see Berg, Maria Elizabeth van den	**M.E.Berg**
(van den Bosch, Roelof Benjamin)	
see Bosch, Roelof Benjamin van den	**Bosch**
(van den Brink, Reinier Cornelis Bakhuizen)	
see Bakhuizen van den Brink, Reinier Cornelis	**Bakh.f.**
Van den Broecke, R. (fl. 1936) M	**Van den Broecke**
(van den Ende, Willem Pieter)	
see Ende, Willem Pieter van den	**Ende**
(van den Heede, Adolphe)	
see Heede, Adolphe van den	**Heede**
(van der Aa, H.A.)	
see Aa, H.A.van der	**Aa**
Van der Ben, Dick A	**Van der Ben**
(Van der Bijl, Paul Andries)	
see Van der Byl, Paul Andries	**Van der Byl**
(van der Burgh, Johannes)	
see Burgh, Johannes van der	**Burgh**
Van der Byl, Paul Andries (1888–1939) M	**Van der Byl**

(van der Gronde, Keympe)	
see Gronde, Keympe van der	**Gronde**
(van der Ham, R.W.J.M.)	
see Ham, R.W.J.M.van der	**R.W.Ham**
(van der Hammen, T.)	
see Hammen, T.van der	**Hammen**
(van der Heijden, E.)	
see Heyden, E.van der	**Heyden**
(van der Heyden, E.)	
see Heyden, E.van der	**Heyden**
(van der Hoeven, E.P.)	
see Hoeven, E.P.van der	**E.P.Hoeven**
(van der Hoeven, Jan)	
see Hoeven, Jan van der	**Hoeven**
(van der Klift, W.C.)	
see Klift, W.C.van der	**Klift**
(van der Laan, F.M.)	
see Laan, F.M.van der	**Laan**
Van der Linde, E.J. (fl. 1991) M	**Van der Linde**
(van der Linden, B.L.)	
see Linden, B.L.van der	**B.L.Linden**
(van der Maesen, L.J.G.)	
see Maesen, L.J.G.van der	**Maesen**
(Van der Meer, John P.)	
see Vandermeer, John P.	**Vanderm.**
(van der Meijden, R.)	
see Meijden, R.van	**Meijden**
Van der Merwe, Frederick Ziervogel (1894–1968) S	**Van der Merwe**
Van der Merwe, Jacoba Johanna Maria(?) (1946–) S	**J.J.M.van der Merwe**
Van der Merwe, W.J.J. (fl. 1990) M	**W.J.J.van der Merwe**
(van der Meyden, R.)	
see Meijden, R.van	**Meijden**
(van der Moezel, P.G.)	
see Moezel, P.G.van der	**Moezel**
(van der Pijl, Leendert)	
see Pijl, Leendert van der	**Pijl**
(van der Plaäts–Niterink, A.J.)	
see Plaäts–Niterink, A.J.van der	**Plaäts–Nit.**
(van der Plas, F.)	
see Plas, F.van der	**Plas**
(van der Ploeg, D.T.E.)	
see Ploeg, D.T.E.van der	**D.T.E.Ploeg**
(van der Ploeg, J.)	
see Ploeg, J.van der	**J.Ploeg**
(van der Sleesen, Ellen H.L.)	
see Sleesen, Ellen H.L.van der	**Sleesen**
(van der Steeg, M.G.)	
see Steeg, M.G.van der	**Steeg**
Van der Veken, Paul A.J.B. (1928–) MS	**Van der Veken**

Van der Walt, J.P. (fl. 1956) M	**Van der Walt**
Van der Walt, Johannes Jacobus Adriaan (1938–) S	**J.J.A.van der Walt**
van der Werff, Henk (fl. 1980) PS	**van der Werff**
Van der Westhuizen, G.C.A. (fl. 1956) M	**Van der Westh.**
Van der Westhuizen, Suzelle (fl. 1988) S	**S.van der Westh.**
(van der Wolk, P.C.)	
see Wolk, P.C.van der	**Wolk**
Van Dersal, William Richard (1907–) S	**Van Dersal**
(van Dijk, D.Eduard)	
see Dijk, D.Eduard van	**Dijk**
(van Donselaar, Johannes)	
see Donselaar, Johannes van	**Donsel.**
(van Doorn–Hoekman, H.)	
see Doorn–Hoekman, H.van	**Doorn-Hoekm.**
Van Dover, Cindy Lee (fl. 1986) M	**Van Dover**
(van Druten, Denise)	
see Druten, Denise van	**Druten**
(van Duin, W.)	
see Duin, W.van	**Duin**
Van Dyke, C.Gerald (fl. 1988) M	**Van Dyke**
(van Eck–Boorsboom, M.H.J.)	
see Eck–Boorsboom, M.H.J.van	**Eck-Boorsb.**
(van Eeden, Frederik Willem)	
see Eeden, Frederik Willem van	**Eeden**
(van Eek, Th.)	
see Eek, Th.van	**Eek**
Van Eel, Ludo A	**Van Eel**
(van Eijk, G.W.)	
see Eijk, G.W.van	**Eijk**
(van Emden, J.H.)	
see Emden, J.H.van	**Emden**
Van Erve, A.W. A	**Van Erve**
Van Eseltine, Glen Parker (1888–1938) PS	**Van Eselt.**
(van Eyndhoven, G.L.)	
see Eyndhoven, G.L.van	**Eyndh.**
Van Geel, Pierre Corneille (Petrus Cornelius) (1796–1836) S	**Van Geel**
Van Geert, August(e) (1888–1938) PS	**Van Geert**
(van Geuns, Steven Jan)	
see Geuns, Steven Jan van	**S.Geuns**
(van Goor, Andreas Cornelis Joseph)	
see Goor, Andreas Cornelis Joseph van	**Goor**
(van Grinsven, A.M.)	
see Grinsven, A.M.van	**Grinsven**
Van Haesendonck, (Gérard) Constant (1810–1881) S	**Van Haes.**
(van Hall, Constant Johann Jakob)	
see Hall, Constant Johann Jakob van	**C.J.J.Hall**
(van Hall, Herman (Hermanus) Christiaan)	
see Hall, Herman (Hermanus) Christiaan van	**H.C.Hall**

(van Hall, Hermann)	
see Hall, Herman van	**Herm.Hall**
(van Haluwyn, C.)	
see Haluwyn, C.van	**Haluwyn**
(van Hasselt, Arend Ludolf)	
see Hasselt, Arend Ludolf van	**A.Hasselt**
(van Hasselt, Johan Coenraad)	
see Hasselt, Johan Coenraad van	**Hasselt**
Van Heek, Werner (fl. 1982) S	**Van Heek**
(van Heerdt, P.François)	
see Heerdt, P.François van	**Heerdt**
Van Hest, J.J. (fl. 1917) M	**Van Hest**
Van Heurck, Henri Ferdinand (1838–1909) APS	**Van Heurck**
(van Heurn, Willem Cornelis)	
see Heurn, Willem Cornelis van	**Heurn**
(van Heusden, E.C.H.)	
see Heusden, E.C.H.van	**Heusden**
Van Hoek, L. (fl. 1979) P	**Van Hoek**
(van Hoof, H.A.)	
see Hoof, H.A.van	**Hoof**
(van Hook, James Mon)	
see Hook, James Mon van	**J.M.Hook**
Van Hoorebeke, Charles Joseph (1790–1821) PS	**Van Hooreb.**
Van Horn, Gene Stanley (1940–) S	**Van Horn**
Van Houtte, Louis (1898–1952) S	**L.Van Houtte**
Van Houtte, Louis Benoît (1810–1876) PS	**Van Houtte**
(van Hoven, Frederick Johan Jacob Slingsbij)	
see Hoven, Frederick Johan Jacob Slingsbij van	**Hoven**
(van Huyssteen, Dorothea Christina)	
see Huyssteen, D.C.	**Huysst.**
Van Jaarsveld, Ernst Jacobus (1953–) S	**Van Jaarsv.**
(van Joenckema, Rembert)	
see Dodoens, Rembert	**Dodoens**
(van Keppel, Johannes Cornelius)	
see Keppel, Johannes Cornelius van	**Keppel**
(van Kerken, Amelia E.)	
see Kerken, Amelia E.van	**Kerken**
(van Kesteren, H.A.)	
see Kesteren, H.A.van	**Kesteren**
(van Kleef, C.)	
see Kleef, C.van	**Kleef**
Van Laer, H. (fl. 1921) M	**Van Laer**
(Van Landingham, Sam L.)	
see VanLandingham, Sam L.	**VanLand.**
(van Leeuwen, Betsy Louise Jacoba)	
see Leeuwen, Betsy Louise Jacoba van	**Leeuwen**
(van Looken, H.)	
see Looken, H.van	**Looken**

(van Luijk, Abraham)	
see Luijk, Abraham van	**Luijk**
(van Marum, Martin (Martinus))	
see Marum, Martin (Martinus) van	**Marum**
(van Melle, Peter Jacobus)	
see Melle, Peter Jacobus van	**Melle**
(van Niel, C.B.)	
see Niel, C.B.van	**C.B.Niel**
(van Nooten, Bertha Hoola)	
see Hoola van Nooten, Bertha	**Hoola van Nooten**
Van Oers, F. A	**Van Oers**
(van Olden, E.)	
see Olden, E.van	**Olden**
(van Ommering, G.)	
see Ommering, G.van	**Ommering**
(van Oorschot, Connie A.N.)	
see Oorschot, Connie A.N.van	**Oorschot**
(van Oosten, M.W.B)	
see Oosten, M.W.B van	**Oosten**
(van Ooststroom, Simon Jan)	
see Ooststroom, Simon Jan van	**Ooststr.**
(van Oven, E.)	
see Oven, E.van	**Oven**
(van Overeem, Casper)	
· see Overeem, Casper van	**Overeem**
(van Overeem, D.)	
see Overeem, D.van	**D.Overeem**
(van Overeem de Haas, Casper)	
see Overeem, Casper van	**Overeem**
(van Pelt Lechner, A.A.)	
see Lechner, A.A.van Pelt	**Lechner**
(van Phelsum, Murk (Murck, Mark))	
see Phelsum, Murk (Murck, Mark) van	**Phelsum**
(van Reenen–Hoekstra, E.S.)	
see Reenen–Hoekstra, E.S.van	**Reenen**
(van Rensselaer, Maunsell)	
see Rensselaer, Maunsell van	**Renss.**
(van Rijn, Anne Renée Ariette Görts)	
see Görts–van Rijn, Anne Renée Ariette	**Görts**
Van Roechoudt (fl. 1932) S	**Van Roech.**
Van Rompaey, Emiel (1895–1975) S	**Van Romp.**
van Rooy, Jacques (1953–) B	**van Rooy**
(van Rosenburgh, Cornelis Rugier Willem Karel Alderwerelt)	
see Alderwerelt van Rosenburgh, Cornelis Rugier Willem Karel van	**Alderw.**
(van Royen, Adriaan)	
see Royen, Adriaan van	**Royen**
(van Royen, David)	
see Royen, David van	**D.Royen**

(van Royen, Pieter)	
see Royen, Pieter van	**P.Royen**
Van Schaack, George Booth (1903–1983) S	**Van Schaack**
(van Schinne, Isaac Evert Cornelis)	
see Schinne, Isaac Evert Cornelis van	**Schinne**
(van Setten, A.K.)	
see Setten, A.K.van	**Setten**
(van Slogteren, Egbertus)	
see Slogteren, Egbertus van	**Slogt.**
(van Slooten, Dirk Fok)	
see Slooten, Dirk Fok van	**Slooten**
(van Soest, Johannes Leendert)	
see Soest, Johannes Leendert van	**Soest**
(van Steenis, Cornelis Gijsbert Gerrit Jan)	
see Steenis, Cornelis Gijsbert Gerrit Jan van	**Steenis**
(van Steenis–Kruseman, Maria Johanna)	
see Steenis–Kruseman, Maria Johanna van	**Steen.-Krus.**
Van Sterbeeck, Frans (Franciscus, François) (1630–1693) L	**Van Sterbeeck**
(Van T.Cotter, H.)	
see Cotter, H.Van T.	**Cotter**
Van Thiel, P.H. (fl. 1954) M	**Van Thiel**
(van Tieghem, Phillippe Édouard Léon)	
see Tieghem, Phillippe Édouard Léon van	**Tiegh.**
(van Tubergen, Cornelis Gerrit)	
see Tubergen, Cornelis Gerrit van	**Tubergen**
(van Uden, N.)	
see Uden, N.van	**Uden**
(van Uildriks, Frederike Johanna)	
see Uildriks, Frederike Johanna van	**Uildriks**
Van, V.M. (1950–) S	**Van**
Van Valkenburg, Shirley D. (1925–) A	**Van Valk.**
(van Vierssen, W.)	
see Vierssen, W.van	**Vierssen**
(van Vliet, D.J.)	
see Vliet, D.J.van	**Vliet**
(van Vloten, H.)	
see Vloten, H.van	**Vloten**
(van Vuure, M.)	
see Vuure, M.van	**Vuure**
(van Wachendorff, Evert Jacob)	
see Wachendorff, Evert Jacob van	**Wach.**
Van Warmelo, Konrad T. (fl. 1967) M	**Van Warmelo**
(van Welzen, P.C.)	
see Welzen, P.C.van	**Welzen**
(van Wijk, Hugo Leonardus Gerth)	
see Gerth van Wijk, Hugo Leonardus	**Gerth**
(van Wisselingh, Cornelius)	
see Wisselingh, Cornelius van	**Wissel.**

(van Wissen, M.J.)	
see Wissen, M.J.van	**Wissen**
van Wyk, Abraham Erasmus (1952–) S	**A.E.van Wyk**
van Wyk, Ben–Erik (fl. 1987) S	**B.-E.van Wyk**
van Wyk, P.S. (fl. 1975) M	**P.S.van Wyk**
(van Zaayen, A.)	
see Zaayen, A.van	**Zaayen**
(van Zanten, Bennard Otto)	
see Zanten, Bennard Otto van	**Zanten**
(van Zijp, Coenraad)	
see Zijp, Coenraad van	**Zijp**
(van Zinderen–Bakker, Edward Meine)	
see Zinderen–Bakker, Edward Meine van	**Zind.-Bakker**
Vána, Jiri (1940–) B	**Vána**
Vanalstyne, Kathryn L. A	**Vanalst.**
Vanbreuseghem, R. (fl. 1950) M	**Vanbreus.**
Vandamme, Henri (1803–) S	**Vandamme**
Vandas, Karel (Karl) (1861–1923) S	**Vandas**
Vandelaar, Jan (1738–) S	**Vandel.**
Vandelli, Domingo (Domingos, Domenico) (1735–1816) S	**Vand.**
Vanden Berghen, Constant (1914–) S	**Vanden Berghen**
Vandenbranden, J. A	**Vandenbr.**
Vandendries, René François Prosper (1874–1952) M	**Vandendr.**
Vandercolme, Édouard (fl. 1780) S	**Vanderc.**
Vandermeer, John P. (1943–) AM	**Vanderm.**
Vanderwalle, R. (fl. 1945) M	**Vanderw.**
Vanderyst, Hyacinthe Julien Robert (1860–1934) S	**Vanderyst**
Vandevelde, Albert (Jacob Josef) (1871–) S	**Vandev.**
Vanek, Rudolf (1899–1953) B	**Vanek**
Vanev, Simeon (G.) (fl. 1963) M	**Vanev**
Vang, J. (fl. 1945) M	**Vang**
Vanguestaine, Michel A	**Vanguest.**
Vaňha, Johann J. (fl. 1903) M	**Vaňha**
Vanhöffen, Ernst (1858–1918) A	**Vanhöffen**
Vanin, Stephen Ivanovich (1890–) M	**Vanin**
Vaniot, Eugène (–1913) S	**Vaniot**
Vánky, Kálmán (1930–) M	**Vánky**
Vánky, Lia (1930–) M	**L.Vánky**
VanLandingham, Sam L. (1935–) A	**VanLand.**
Vannacci, G. (fl. 1976) M	**Vannacci**
(Vanner Corley, Martin Francis)	
see Corley, Martin Francis Vanner	**M.F.V.Corley**
Vanni, Ricardo Oscar (1954–) S	**Vanni**
Váňová, Marie (fl. 1968) M	**Váňová**
Vänskä, Heino (1943–) M	**Vänskä**
Vanterpool, Thomas Clifford (1898–) M	**Vanterp.**
Vanucchi (fl. 1838) S	**Vanucchi**
(Vanucci)	
see Vanucchi	**Vanucchi**

Varadarajan, G.S. (fl. 1983) S	**G.S.Varad.**
Varadarajan, P.D. (fl. 1958) M	**Varad.**
Varadpande, D.G. (fl. 1981) M	**Varadp.**
Varano, Lorenzo (1942–) A	**Varano**
Vardhanabhuti, Sman (fl. 1959) M	**Vardhan.**
Varela de Vega, Alida (fl. 1986) S	**Varela**
Vareschi, Volkmar (1906–1991) BMPS	**Vareschi**
Vargas Calderón, Julio César (1907–1960) S	**Vargas**
Vargas, P. (fl. 1986) S	**P.Vargas**
Varghese, K.I.Mani (fl. 1977) M	**Varghese**
Varitchak, B. (fl. 1933) M	**Varitchak**
Varma, A. (fl. 1986) M	**A.Varma**
Varma, Ajit K. A	**A.K.Varma**
Varma, Chandra Prakash (fl. 1950) AFM	**C.P.Varma**
Varma, R.Prasanna (1925–1970) A	**R.P.Varma**
Varma, Y.N.R. (fl. 1985) M	**Y.N.R.Varma**
Varo Alcala, Juan (fl. 1972) S	**Varo**
Varoczky, Edith Cs. (fl. 1963) P	**Varoczky**
Varossieau, W.W. S	**Vaross.**
Varsavsky, Edith (fl. 1965) M	**Varsavsky**
Varshney, J.L. (fl. 1981) M	**Varshney**
Vartak, Vaman Dattatraya (1925–) S	**Vartak**
Vartapetov, S.G. (fl. 1976) M	**Vartap.**
Vartapetyan, V.V. (fl. 1980) S	**Vartapetyan**
Vasak, Vladimir (fl. 1972) S	**Vasak**
Vasconcellos, João de Carvalho e (1897–1972) PS	**Vasc.**
Vasconcelos, Augusto Teixeira de (fl. 1952) M	**A.T.Vasconc.**
Vasconcelos, C.T.de (fl. 1966) M	**C.T.Vasconc.**
Vasconcelos, Ilo (fl. 1966) M	**I.Vasconc.**
Vasek, Frank Charles (1927–) S	**Vasek**
Vasey, George (1822–1893) S	**Vasey**
Vasey, M.C. (fl. 1985) S	**M.C.Vasey**
Vasic, O. (fl. 1986) S	**Vasic**
(Vasilchenko, I.T.)	
see Vassilczenko, I.T.	**Vassilcz.**
Vasilenko, S.V. A	**Vasilenko**
(Vasilev, P.)	
see Vassiljev, P.	**P.Vassil.**
(Vasilev, Viktor Nikolayevich)	
see Vassiljev, Viktor Nikolayevich	**V.N.Vassil.**
(Vasileva, A.N.)	
see Vassiljeva, A.N.	**A.N.Vassiljeva**
(Vasileva, L.I.)	
see Vassiljeva, L.I.	**L.I.Vassiljeva**
(Vasileva, M.G.)	
see Vassiljeva, M.G.	**M.G.Vassiljeva**
(Vasileva–Pupysheva, L.I.)	
see Vassiljeva, L.I.	**L.I.Vassiljeva**
(Vasiljeva, Larissa N.)	
see Vassiljeva, Larissa N.	**Lar.N.Vassiljeva**

(Vasiljevskij, N.I.)	
see Vassiljevsky, N.I.	**Vassiljevsky**
(Vasilyeva, Larissa N.)	
see Vassiljeva, Larissa N.	**Lar.N.Vassiljeva**
Vasinger, Antonina Vasilievna (1892–1940) S	**Vasinger**
(Vasinger–Alektorova, Antonina Vasilievna)	
see Vasinger, Antonina Vasilievna	**Vasinger**
Vasishta, P.C. A	**Vasishta**
(Vásquez Ch., Roberto)	
see Vásquez, Roberto	**R.Vásquez**
Vásquez, Roberto (1942–) S	**R.Vásquez**
Vassal, J. (1932–) S	**Vassal**
Vassal, J.-J. A	**J.-J.Vassal**
(Vasser, Solomon P.)	
see Wasser, Solomon P.	**Wasser**
Vassilczenko, I.T. (1903–) S	**Vassilcz.**
(Vassileva, Larissa N.)	
see Vassiljeva, Larissa N.	**Lar.N.Vassiljeva**
(Vassileva, Ljubov Nikolaevna)	
see Vassiljeva, Ljubov Nikolaevna	**Lj.N.Vassiljeva**
(Vassilevskii, N.I.)	
see Vassiljevsky, N.I.	**Vassiljevsky**
(Vassilieva, A.N.)	
see Vassiljeva, A.N.	**A.N.Vassiljeva**
(Vassilieva, Larissa N.)	
see Vassiljeva, Larissa N.	**Lar.N.Vassiljeva**
Vassiljev, A.V. (1902–) S	**A.V.Vassil.**
Vassiljev, Igor V. (1921–) S	**I.V.Vassil.**
Vassiljev, Ja.Ja. (fl. 1940) S	**J.J.Vassil.**
Vassiljev, P. (fl. 1973) S	**P.Vassil.**
Vassiljev, V.F. S	**V.F.Vassil.**
Vassiljev, Viktor Nikolayevich (1890–1987) PS	**V.N.Vassil.**
Vassiljeva, A.N. (fl. 1969) S	**A.N.Vassiljeva**
Vassiljeva, I.M. (1955–) S	**I.M.Vassiljeva**
Vassiljeva, L.I. (1923–) MS	**L.I.Vassiljeva**
Vassiljeva, Larissa N. (fl. 1950–1976) MS	**Lar.N.Vassiljeva**
Vassiljeva, Ljubov Nikolaevna (1901–) MS	**Lj.N.Vassiljeva**
Vassiljeva, M.G. (1953–) S	**M.G.Vassiljeva**
(Vassiljeva–Pupysheva, L.I.)	
see Vassiljeva, L.I.	**L.I.Vassiljeva**
Vassiljevsky, N.I. (1884–1950) M	**Vassiljevsky**
Vassiljevsky, W. (fl. 1929) M	**W.Vassiljevsky**
Vassilkov, B.P. (fl. 1953) M	**Vassilkov**
Vassilkovskaja, A.P. S	**Vassilkovsk.**
Vasudeva, R.S. (fl. 1953) M	**Vasudeva**
Vasudeva Rao, M.K. (fl. 1979) S	**Vasudeva Rao**
Vasudevan Nair, R. (fl. 1987) S	**Vasud.Nair**
Vasuki, S. (fl. 1989) M	**Vasuki**
Vasyagina, M.P. (fl. 1957) M	**Vasyag.**

Vater, Henrich (fl. 1884) F	**Vater**
Vatke, (Georg (George) Carl) Wilhelm (1849–1889) S	**Vatke**
Vàtova, Aristocle (1897–) A	**Vàtova**
Vatsala, P. (fl. 1981) S	**Vatsala**
Vattimo, Ítalo de (1928–) BS	**Vattimo**
Vattimo–Gil, Ida de (fl. 1980s) BS	**Vattimo-Gil**
Vaucher, Jean Pierre Étienne (1763–1841) AMPS	**Vaucher**
Vaughan, Alice Esther (fl. 1939) S	**A.E.Vaughan**
Vaughan, Duncan A. (fl. 1990) S	**D.A.Vaughan**
Vaughan, John (1855–1922) S	**Vaughan**
Vaughan Martini, Ann (1948–) S	**Vaughan Mart.**
Vaughan, Reginald Edward (1895–1987) S	**R.E.Vaughan**
Vaughan, Richard English (1884–1952) S	**Rich.Vaughan**
Vaughn, R.H. (fl. 1942) M	**Vaughn**
Vaulina, E.N. A	**Vaulina**
(Vaumartoise, François Victor)	
see Mérat de Vaumartoise, François Victor	**Mérat**
Vaupel, D. S	**D.Vaupel**
Vaupel, Friedrich Karl Johann (1876–1927) S	**Vaupel**
Vaupell, Christian Theodor (1821–1862) AS	**Vaupell**
Vauquelin, Louis Nicolas (1763–1829) S	**Vauquelin**
Vauras, Jukka (1946–) M	**Vauras**
Vautier, Simone (1908–1971) S	**Vautier**
Vauvel, Léopold Eugène (1848–1915) S	**Vauvel**
Vavasseur, Pierre Henri Louis Dominique (1797–) S	**Vavass.**
Vavilov, Nikolaj Ivanovich (1887–1943) S	**Vavilov**
Vavra, J. (fl. 1966) M	**Vavra**
Vavrdová, Milada A	**Vavrdová**
Vayreda y Vila, Estanislao (1848–1901) S	**Vayr.**
Vaz, Angela M. Studart da Fonseca (fl. 1981) S	**Vaz**
Vazquez, Antonio (fl. 1989) S	**A.Vazquez**
Vázquez Avila, M.D. (fl. 1981) S	**Vázq.Avila**
Vázquez Torres, Mario (fl. 1980) S	**Vázq.Torres**
Vázquez, V.M. (fl. 1982) M	**Vázquez**
(Vázquez–Torres, Mario)	
see Vázquez Torres, Mario	**Vázq.Torres**
Vecchi, Octavio S	**Vecchi**
Ved Prakash (fl. 1979) S	**Ved Prakash**
Vedajanani, K. A	**Vedaj.**
Veenbaas–Rijks, Johanna W. (fl. 1970) M	**Veenb.-Rijks**
Veer, Jacob van der A	**Veer**
Veeraraghavan, J. (fl. 1965) M	**Veerar.**
Veerkamp, J. (fl. 1983) M	**Veerkamp**
Veesenmeyer, Karl Gustav (1814–1901) S	**Veesenm.**
Vegh, I. (fl. 1974) M	**Vegh**
Vegni, G. (fl. 1963) M	**Vegni**
(Veiga, A.da)	
see Da Veiga, A.	**Da Veiga**
Veilex, J. (fl. 1962) S	**Veilex**

Veillard (fl. 1800) S	**Veill.**
Veillon, Jean–Marie (fl. 1982) S	**Veillon**
Veitch, Harry James (1840–1924) PS	**H.J.Veitch**
Veitch, James Herbert (1868–1907) S	**J.H.Veitch**
Veitch, James James (1815–1869) S	**J.J.Veitch**
Veitch, John Gould (1839–1870) S	**Veitch**
(Vej, Tzjan–czunj)	
see Wei, Jiang Chun	**J.C.Wei**
(Veken, Paul A.J.B.van der)	
see Van der Veken, Paul A.J.B.	**Van der Veken**
Vekshina, V.N. A	**Vekshina**
(Velarde Nuñez, Octavio)	
see Velarde, Octavio	**Velarde**
Velarde, Octavio (fl. 1945–1959) S	**Velarde**
(Velarde–Nuñez, Octavio)	
see Velarde, Octavio	**Velarde**
Velasco, M. (fl. 1986) S	**M.Velasco**
Velasco Negueruela, Arturo (1944–) S	**A.Velasco**
Velásquez, Dilia (1943–) S	**D.Velásquez**
Velásquez, Justiniano (1936–) S	**Velásquez**
(Velásquez de Orsini, Dilia)	
see Valásquez, Dilia	**D.Velásquez**
(Velayos, Mauricio)	
see Velayos Rodríguez, Mauricio	**Velayos**
Velayos Rodríguez, Mauricio (1955–) S	**Velayos**
Velazquez, Gregorio Tiongson (1901–) A	**Velazquez**
Velchev, Velcho Ivanov (1928–) S	**Velchev**
Veldeman, R. (fl. 1971) M	**Veldeman**
Veldkamp, Jan Frederik (1941–) S	**Veldkamp**
Velenovský, Josef (Joseph) (1858–1949) BFMPS	**Velen.**
Velez (fl. 1913) A	**Velez**
Velez, C.G. A	**C.G.Velez**
Velez, M.C. (fl. 1981) S	**M.C.Velez**
Velić, Ivo A	**Velić**
Velican, Vasile (1904–) S	**Velican**
Velichkin, E.M. (1942–) S	**Velichkin**
Velley, Thomas (1749–1806) A	**Velley**
Vellez, Th. S	**Vellez**
Vellinga, Else C. (fl. 1986) M	**Vellinga**
Velloso de Miranda, Joaquim (1733–1815) S	**Velloso**
(Velloso, José Mariano da Conceição)	
see Vellozo, José Mariano da Conceição	**Vell.**
Vellozo, José Mariano da Conceição (1742–1811) PS	**Vell.**
Veloira del Rosario, Nieva E. (fl. 1959) B	**Veloira**
(Veloira, Nieva E.)	
see Veloira del Rosario, Nieva E.	**Veloira**
(Veloso, José Mariano da Conceição)	
see Vellozo, José Mariano da Conceição	**Vell.**
Velu, H. (fl. 1924) M	**Velu**

Venanzi, Guiseppe (1851–) S	**Venanzi**
Vendrely, (François) Xavier (1837–1908) S	**Vendrely**
Vendriés, Albert (–1916) P	**Vendriés**
Venetz, Ignaz (1788–1859) S	**Venetz**
Venitz, Herbert (fl. 1934) B	**Venitz**
Venkatachala, Bangalore S. (fl. 1969) AFM	**Venkatach.**
Venkatakrishnaiah, N.S. (fl. 1952) M	**Venkatakr.**
Venkataraman, Gopalasamudram Sitaraman (1930–) A	**Venkataram.**
Venkataramani, K.S. (fl. 1961) M	**Venkatar.**
Venkatarayan, S.V. (fl. 1949) M	**Venkatarayan**
Venkateswarlu, Jillella (1912–1978) S	**Venkat.**
Venkateswarlu, V. A	**V.Venkat.**
Vent, Walter (1920–) S	**W.Vent**
Vente, M. (fl. 1982) S	**Vente**
Ventenat, Étienne Pierre (1757–1808) MPS	**Vent.**
Venter, Hendrik Johannes Tjaart (1938–) S	**Venter**
Venter, S. (fl. 1988) S	**S.Venter**
Ventrice, Maria Rosa A	**Ventrice**
(Ventslavovich, F.S.)	
see Wenzlawowicz, F.S.	**Wenzlaw.**
Venturelli, M. (fl. 1982) S	**Ventur.**
Venturi, Antonio (1805–1864) M	**A.Venturi**
Venturi, Carl Antonio S	**C.A.Venturi**
Venturi, Gustavo (1830–1898) B	**Venturi**
Venturi, Santiago (fl. 1910) S	**S.Venturi**
Venzo, Sebastiano (1815–1876) S	**Venzo**
Vera de la Puente, Ma. Luisa (1954–) S	**Vera**
Verachtert, H. (fl. 1971) M	**Veracht.**
Verbeek, J.W. A	**Verbeek**
Verbunt, J.A. (fl. 1938) M	**Verbunt**
Verdam, H.D. A	**Verdam**
Verdcourt, Bernard (1925–) PS	**Verdc.**
Verdier, Jean-Pierre A	**Verdier**
Verdon, Douglas (1921–) M	**Verdon**
Verdoorn, Frans (1906–1984) BPS	**Verd.**
Verdoorn, Inez Clare (1896–1989) S	**I.Verd.**
Verdun, P. (fl. 1909) M	**Verdun**
Verduyn, G.P. (fl. 1987) P	**Verduyn**
Vereitinov, I.A. (1878–1922) M	**Vereit.**
Vergara y Perez de Arandu, Mariano (1833–) S	**Vergara**
Verguin, Louis (1868–1936) S	**Verg.**
Verhey, C.J. (1917–) PS	**Verhey**
Verheyen, K. (fl. 1969) M	**Verheyen**
Verhoeff, Carl (Karl) Wilhelm (1867–1944) S	**Verhoeff**
Verhoek-Williams, Susan Elizabeth (1942–) S	**Verh.–Will.**
Verhoeven, Adriana A. (fl. 1972) M	**Verh.**
Verhoeven, R.L. (1945–) S	**R.L.Verh.**
Verity, David S. (1930–) S	**Verity**
Verlaque, M. A	**M.Verlaque**

Verlaque, R. (fl. 1978) S	Verlaque
Verlot, (Pierre) Bernard (Lazare) (1836–1897) S	B.Verl.
Verlot, Jean Baptiste (1825–1891) S	Verl.
Verma, B.L. (fl. 1987) M	B.L.Verma
Verma, B.N. A	B.N.Verma
Verma, C.L. (fl. 1970) M	C.L.Verma
Verma, D.M. (1937–) S	D.M.Verma
Verma, G.S. (fl. 1940) M	G.S.Verma
Verma, J.K. (fl. 1956–1958) F	J.K.Verma
Verma, M.P. A	M.P.Verma
Verma, R.A.B. (fl. 1970) M	R.A.B.Verma
Verma, R.K. (fl. 1988) M	R.K.Verma
Verma, R.P. (fl. 1986) M	R.P.Verma
Verma, R.V. (fl. 1986) M	R.V.Verma
Verma, Sandeep (fl. 1990) P	S.Verma
Verma, Satish Chander (1931–) MP	S.C.Verma
Verma, T. (fl. 1990) M	T.Verma
Vermeulen, Jaap J. (1955–) S	J.J.Verm.
Vermeulen, Pieter (1899–1981) S	Verm.
Vermoesen, (François Marie) Camille (1882–1922) MS	Vermoesen
Vernet (fl. 1904) S	Vernet
Verona, O. (fl. 1933) M	Verona
Verplancke, Germain (fl. 1926) M	Verpl.
Verrall, Arthur Frederic (1905–) M	Verrall
Verschaffelt, Ambroise Colette Alexandre (1825–1886) S	Verschaff.
Verschaffelt, Charles S	C.Verschaff.
Verseghy, Klára (fl. 1956) M	Verseghy
Verses, Patricia A. A	Verses
Versluys, W. (fl. 1977) M	Versluys
Vervoorst, Frederico B. (1923–) S	Verv.
Verwoerd, Len (fl. 1924) M	Verwoerd
Veselsky, Bedřich (Friedrich) (1813–1866) P	Veselsky
Veselský, J. (fl. 1975) M	J.Veselský
Vesely, R. (fl. 1932) M	Vesely
Vesonder, R.F. (fl. 1974) M	Vesonder
Vesque, Julien (Joseph) (1848–1895) S	Vesque
Vesselovskaya, M.A. (1898–) S	Vessel.
Vest, Lorenz Chrysanth von (1776–1840) S	Vest
Vestal, Arthur Gibson (1888–1964) S	Vestal
Vestergren, (Jacob) Tycho (Conrad) (1875–1930) MS	Vestergr.
Vesterholt, Jan (fl. 1989) M	Vesterh.
Vesterlund, (Per) Otto (1857–1953) S	Vesterl.
Veszprémi, D. (fl. 1907) M	Veszprémi
Vetrova, Zinaida I. (1931–) A	Vetrova
Vetter, J. (1865–1945) S	J.Vetter
Vetter, Jean Jacques (1826–1913) S	Vetter
Veulliot, Charles (1829–1890) M	Veull.
Vey, A. (fl. 1965) M	Vey
Veyret, Y. (fl. 1978) S	Veyret
Vězda, Antonín (fl. 1963) BM	Vězda

Viala, G. A	**G.Viala**
Viala, Pierre (1859–1936) MS	**Viala**
Viallanes, (Jacques Joseph) Alfred (1828–1899) S	**Viall.**
Viana, Gaspar (fl. 1913) M	**Viana**
Viand–Grand–Marais, A. (fl. 1892) M	**Viand–Grand–Marais**
Viane, Ronald L.L. (1951–) PS	**Viane**
Vianna, G. (fl. 1913) M	**Vianna**
Viano, Josette S	**Viano**
Viaud–Grand–Marais, Ambroise (1833–1913) S	**Viaud**
Vibert, J.P. (fl. 1824) S	**Vibert**
Viborg, Erik Nissen (1759–1822) S	**Viborg**
Vicary, Nathaniel S	**Vicary**
Vicat, Phillipe Rodolphe (1720–1783) S	**Vicat**
Vicherek, Jirí (1929–) S	**Vicherek**
(Vicioso, Benito)	
see Vicioso Trigo, Benito	**Vicioso**
Vicioso Martínez, Carlos (1886/7–1968) S	**C.Vicioso**
Vicioso Trigo, Benito (1850–1929) S	**Vicioso**
Vickerman, Keith A	**Vick.**
Vickers, Anna (1852–1906) A	**Vickers**
Vickery, Albert Roy (1947–) S	**A.R.Vickery**
Vickery, Joyce Winifred (1908–1979) S	**Vickery**
Vickery, R.S. A	**R.S.Vickery**
Vickery, Robert Kingston (1922–) S	**R.K.Vickery**
Vicol, E.C. (1936–) S	**Vicol**
Vicotin (fl. 1945) P	**Vicotin**
Vicq, (Léon–Bonaventure) Éloy de (1810–1886) S	**Vicq**
Victor, B.J. (fl. 1984) M	**Victor**
Victorin, Joseph Louis Conrad Marie– (1885–1944) PS	**Vict.**
Vida, Gábor (1935–) PS	**Vida**
Vidal, António José (Rodrigo) (1808–1879) S	**Vidal**
Vidal, Domingo (–1878) S	**D.Vidal**
Vidal, G. S	**G.Vidal**
Vidal, Jules Eugène (1914–) S	**J.E.Vidal**
Vidal, Louis (fl. 1900–1906) PS	**L.Vidal**
Vidal, Luis Mariano (1842–1922) S	**L.M.Vidal**
(Vidal, R.da Silva)	
see Silva Vidal, R.da	**Silva Vidal**
Vidal, Sebastian (1842–1889) PS	**S.Vidal**
(Vidal y Carreras, Luis Mariano	
see Vidal, Luis Mariano	**L.M.Vidal**
(Vidal y Soler, Domingo)	
see Vidal, Domingo	**D.Vidal**
(Vidal y Soler, Sebastian)	
see Vidal, Sebastian	**S.Vidal**
Vidal–Leiria, Manuela (fl. 1966) M	**Vidal–Leir.**
Vidhyasekaran, P. (fl. 1973) M	**Vidhyas.**
Vido, Aloysii (fl. 1879) M	**Vido**
Viégas, Ahmés Pinto (1905–1986) MS	**Viégas**
Viegi, Lucia (1947–) S	**Viegi**

Vieillard, Eugène (Deplanche Émile) (1819–1896) PS   **Vieill.**
Vieira, J.R. (fl. 1959) M   **Vieira**
(Vieira Toscano de Brito, Antonio Luiz)
  see Toscano, Antonio Luiz   **Toscano**
Viennot–Bourgin, Georges (1906–) M   **Vienn.-Bourg.**
Viera y Clavija, José de (1731–1813) S   **Viera y Clavija**
Viereck, L.A. (1930–) B   **Viereck**
Vierhapper, Friedrich (1844–1903) S   **F.Vierh.**
Vierhapper, Friedrich (Karl Max) (1876–1932) S   **Vierh.**
Vierssen, W.van (fl. 1982) S   **Vierssen**
Viethen, Birgit (fl. 1987) M   **Viethen**
Vietz, Ferdinand Bernhard (1772–1815) S   **Vietz**
Vigineix, Guillaume (–1877) S   **Vigin.**
Vigna, María Susana (1949–) A   **Vigna**
Vignal, Victor William Montgomery (1852–) S   **Vignal**
(Vigne, Gislain François De la)
  see Delavigne, Gislain François   **Delavigne**
Vignet, A.von (fl. 1795) S   **Vignet**
Vigneux, A. (fl. 1812) S   **Vigneux**
Vignier, A. (fl. 1914) M   **Vignier**
Vignolo–Lutati, Ferdinando (1878–1965) PS   **Vignolo**
Vigo Bonada, Josep (1937–) PS   **Vigo**
(Vigo i Bonada, Josep)
  see Vigo Bonada, Josep   **Vigo**
Viguié, M.-Th. (fl. 1960) S   **Viguié**
Viguier, A. (fl. 1914) M   **A.Vig.**
Viguier, L.G.Alexandre (1790–1867) S   **Vig.**
Viguier, Pierre S   **P.Vig.**
Viguier, René (1880–1931) FMS   **R.Vig.**
Vijay Kumar, B.K. (fl. 1983) S   **Vij.Kumar**
Vikulova, N.V. (fl. 1939) S   **Vikulova**
Vilchez, Oscar (1933–) S   **Vilchez**
(Vilensky, D.G.)
  see Wilensky, D.G.   **Wilensky**
Vilgalys, Rytas (1958–) M   **Vilgalys**
Vilhelm, Jan (1876–1931) AB   **Vilh.**
Vill, August (1851–1930) M   **Vill**
Villa Carenzo, Martín (fl. 1960) S   **Villa**
Villafañe Lastra, T.de (fl. 1939) M   **Villafañe**
Villagrán, Carolina (fl. 1971) P   **Villagrán**
Villalobos, Francisco (fl. 1784–1789) S   **Villalobos**
Villani, Armando (1875–1930) S   **Villani**
(Villar, Celastino Fernández)
  see Fernández–Villar, Celestino   **Fern.-Vill.**
(Villar, Domínique)
  see Villars, Domínique   **Vill.**

Villar y Serratacó, Emile (Emilio) Huguet del (1871–1951) S	**Villar**
Villar Peréz, Luis (1946–) S	**L.Villar**
Villard S	**Villard**
Villareal, Tracy A. A	**Villareal**
(Villarreal Quintanilla, José Angel)	
see Villarreal–Quintanilla, José Angel	**Villarreal**
Villarreal–Quintanilla, José Angel (1956–) S	**Villarreal**
Villars, Domínique (1745–1814) AMPS	**Vill.**
Villaseñor, José Luis (1954–) S	**Villaseñor**
Villatte, Juliette A	**Villatte**
(Ville, Jean Baptiste De)	
see De Ville, Jean Baptiste	**De Ville**
(Ville, Nicolas De)	
see De Ville, Nicolas	**N.De Ville**
(Villegas de G., Marina)	
see Villegas, Marina	**Villegas**
Villegas, Marina (1940–) M	**Villegas**
Villeret, Serge A	**Villeret**
Villiers, J.F. (1943–) S	**Villiers**
(Villiers, J.J.R.De)	
see De Villiers, J.J.R.	**De Villiers**
Villiers, J.P. S	**J.P.Villiers**
Villot, J.P. A	**Villot**
Villouta, E. A	**Villouta**
Vilmorin, (Auguste Louis) Maurice (Lévêque) de (1849–1918) S	**M.Vilm.**
Vilmorin, (Charles Philippe) Henry (Lévêque) de (1843–1899) S	**H.Vilm.**
Vilmorin, (Joseph Marie) Philippe (Lévêque) de (1872–1917) S	**P.Vilm.**
Vilmorin, (Philippe) Victoire (Lévêque) de (1746–1840) S	**V.Vilm.**
Vilmorin, (Pierre Philippe) André (Lévêque) de (1776–1862) S	**A.Vilm.**
Vilmorin, (Pierre) Louis (François Lévêque) de (1816–1860) S	**Vilm.**
Vilmorin, Jacques Lévêque de (1882–1933) S	**J.Vilm.**
Vilmorin, Roger (Marie Vincent Philippe Lévêque) de (1905–1980) S	**R.Vilm.**
Vilmos, Syemann S	**Vilmos**
Vilyasoo, L. (fl. 1980s) S	**Vilyasoo**
Vimal, K.P. A	**Vimal**
Vimala Bai, B. A	**Vimala Bai**
Vimba, Edgars (1930–) M	**Vimba**
Vinall, Harry Nelson (1880–1937) S	**Vinall**
Vinatzer, Georg A	**Vinatzer**
Vincens, François (1880–1925) M	**Vincens**
Vincent, Michael A. (1955–) M	**Vincent**
Vindt, Jacques (1915–) S	**Vindt**
Vines, Sydney Howard (1849–1934) S	**Vines**
Vinha, Sérgio Guimarães da (fl. 1970) S	**Vinha**
Viniklář, L. S	**Viniklář**
Vink, Willem (Willen) (1931–) S	**Vink**
Vinniková, Alena A	**Vinniková**
Vinogradova, K.L. (1937–) A	**K.L.Vinogr.**
Vinogradova, V.M. (1937–) S	**V.M.Vinogr.**

Vinson, (Jean François Dominique) Émile (fl. 1855–1870) S	**Vinson**
Vinyard, William Corwin (1922–) A	**Vinyard**
Viola, S. (fl. 1963) M	**Viola**
Vipper, Pavel Borisovich (1920–) S	**Vipper**
Virdi, S.S. (fl. 1986) M	**Virdi**
Viret, Louis (1875–1928) A	**Viret**
Virey, Jules–Joseph (1775–1846) S	**Virey**
Virginio, Giovanni Vincenzo S	**Virginio**
Virieux, Joseph–Jean–Marie (1890–1915) A	**Virieux**
Virot, Robert (1915–) S	**Virot**
Virskaia, I.Yu. A	**Virskaia**
(Vis, Charles Walter De)	
see De Vis, Charles Walter	**De Vis**
Visalakshmi, V. A	**Visal.**
Vischer, Wilhelm (1890–1960) AS	**Vischer**
Vischnevskaja, Z.A. (fl. 1955) M	**Vischn.**
Visher, Stephen Sargent (1887–1967) S	**Visher**
(Vishnevskaia, Z.A.)	
see Vischnevskaja, Z.A.	**Vischn.**
Vishniac, Helen S. (fl. 1955) M	**Vishniac**
Vishwarathan, M.B. (fl. 1988) M	**Vishwar.**
Visiani, Roberto de (1800–1878) PS	**Vis.**
Visnadi, Sandra Regina (1965–) B	**Visnadi**
Viswanathan, M.N. (fl. 1989) P	**M.N.Viswan.**
Viswanathan, M.Venkatesan (1942–) S	**M.V.Viswan.**
Viswanathan, T.S. (fl. 1957) M	**T.S.Viswan.**
(Visyulina, E.D.)	
see Wissjulina, E.D.	**E.D.Wissjul.**
Vital, A.Fernandes (fl. 1953) M	**A.F.Vital**
Vital, Daniel Moreira (1924–) B	**Vital**
Vitalariu, G. (fl. 1982) S	**Vitalariu**
Vitasey, Joanna S	**Vitasey**
Vitek, Ernst (1953–) S	**Vitek**
Vitez, I. (fl. 1964) M	**Vitez**
Vitikainen, Orvo (1949–) M	**Vitik.**
Vitkovskiya, V.L. (1928–) S	**Vitk.**
Vitman, Fulgenzio (1728–1806) PS	**Vitman**
Vitt, Dale Hadley (1944–) BS	**Vitt**
Vittadini, Carlo (1800–1865) M	**Vittad.**
Vittal, Balumuri Pandu Ranga (1944–) M	**Vittal**
Vittal, M. A	**M.Vittal**
Vittet, Nelly (fl. 1954) S	**Vittet**
Vivaldi, José L. (1948–) S	**Vivaldi**
Vivant, J. (fl. 1956) MS	**Vivant**
Vivanti, Anna A	**Vivanti**
Vivekananthan, K. (1938–) S	**Vivek.**
Vives, José C. (1931–) S	**Vives**
Viviand–Morel, Joseph Victor (1843–1915) P	**Viv.–Morel**
Viviani, Domenico (1772–1840) MPS	**Viv.**

Vivoli, D. (fl. 1932) M	**Vivoli**
Vize, John Edward (1831–1916) M	**Vize**
Vizioli, José (1896–) M	**Vizioli**
Vizkelety, Eva A	**Vizk.**
Vizoso, M.T. (fl. 1991) M	**Vizoso**
Vlad, Elena (fl. 1962) M	**Vlad**
Vlădescu, Mihai (fl. 1888) M	**Vlădescu**
Vladimirović, V.P. AS	**Vladim.**
Vlasák, J. (fl. 1990) M	**Vlasák**
(Vlasova, E.)	
see Vlassova, E.	**Vlassova**
(Vlasova, N.V.)	
see Vlassova, N.V.	**N.V.Vlassova**
Vlassova, E. (fl. 1968) M	**Vlassova**
Vlassova, N.V. (fl. 1980) S	**N.V.Vlassova**
Vlémincq, Aimé (fl. 1937) S	**Vlémincq**
Vleugel, Jens Schanke (1854–1927) M	**Vleugel**
Vliet, D.J.van (fl. 1969) S	**Vliet**
Vlk, Wladimir H. A	**Vlk**
Vlok, Jan H.J. (1957–) S	**Vlok**
Vloten, H.van (fl. 1953) M	**Vloten**
Vobis, G. (fl. 1980) M	**Vobis**
Vobogdin, A.G. A	**Vobogdin**
Vöchting, Hermann (1847–1917) S	**Vöcht.**
Vocke, Adolf (1821–1901) S	**Vocke**
Vodeničarov, Dimiter G. A	**Voden.**
Vodopianova, N.S. (1932–) S	**Vodop.**
(Vodopyanova, N.S.)	
see Vodopianova, N.S.	**Vodop.**
Voeltzkow, Alfred (1860–1946) AB	**Voeltzk.**
Vogel, (Julius Rudolph) Theodor (1812–1841) S	**Vogel**
Vogel, A. (fl. 1982) S	**A.Vogel**
Vogel, Benedict Christian (1745–1825) S	**B.Vogel**
(Vogel, Eduard Ferdinand de)	
see de Vogel, Eduard Ferdinand	**de Vogel**
Vogel, Heinrich (fl. 1875) S	**H.Vogel**
Vogel, Rudolph Augustin (1724–1775) S	**R.Vogel**
Vogel, Stefan (1925–) AM	**S.Vogel**
Vogeli–Zuber, M. (fl. 1952) M	**Vog.-Zuber**
Vogellehner, D. S	**Vogell.**
Vogelsberger, Albert (fl. 1893) S	**Vogelsb.**
Voggenreiter, Volker (fl. 1975) S	**Voggenr.**
Vogl, August Emil (1833–1909) S	**Vogl**
Vogl, Balthasar (fl. 1888) S	**B.Vogl**
Vogler, Johann Andreas (fl. 1781) PS	**J.A.Vogler**
Vogler, Johann Philipp (1746–1816) S	**. Vogler**
Vogler, Paul (1875–) A	**P.Vogler**
Voglino, Enrico S	**E.Voglino**
Voglino, Pietro (Piero) (1864–1933) M	**Voglino**

Vogt, Robert M. (1957–) S	**Vogt**
Vogtherr, Max (Adalbert Theodor Eugen) (1850–1915) S	**Vogtherr**
Vohra, Jitinder Nath (1934–) B	**Vohra**
(Voight, Friedrich Sigismund)	
see Voigt, Friedrich Siegmund	**F.Voigt**
Voigt, (Julius) Alfred (1864–1935) S	**Alf.Voigt**
Voigt, Albert (1858–) S	**Alb.Voigt**
Voigt, Friedrich Siegmund (1781–1850) S	**F.Voigt**
Voigt, Joachim (Johann) Otto (1798–1843) PS	**Voigt**
Voigt, Manfred A	**M.Voigt**
Voigt, Max (1888–) A	**Max Voigt**
Voigt, W.E.Alwin (1852–c.1927) S	**Alw.Voigt**
Voit, Johann Gottlob Wilhelm (1787–1813) B	**Voit**
Voith, Ignaz von (1759–1848) S	**Voith**
(Vol, Charles Edward De)	
see De Vol, Charles Edward	**De Vol**
Volcinschi, Adrian (1927–) M	**Volc.**
Volgin, Serjej A. (fl. 1981) S	**Volgin**
Volgunov, D.K. (fl. 1940) S	**Volgunov**
Voliotis, Dimitrios T. (1933–) S	**Voliotis**
Volk, Otto Heinrich (1903–) BS	**O.H.Volk**
Volk, Richard (1849?–1911) A	**Volk**
Volkart, Albert (1873–1951) MS	**Volkart**
Volkens, Georg Ludwig August (1855–1917) S	**Volkens**
Volkmann–Kohlmeyer, B. (fl. 1987) M	**Volkm.-Kohlm.**
Volkov, Luka Illarionovich (1886–1963) A	**Volkov**
Volkova, E.V. (1910–) MS	**Volkova**
Voll, Otto (–1959) S	**Voll**
Vollesen, Kaj Børge (1946–) S	**Vollesen**
Vollmann, Franz (1858–1917) PS	**Vollm.**
Vollmer, Albert Michael (1896–1977) S	**Vollmer**
Vollmer, C. A	**C.Vollmer**
Vollrath, H. (fl. 1967) S	**Vollrath**
Vologdin, Aleksandr Grigorevich (1896–) AF	**Vologdin**
Volponi, C.R. (fl. 1986) S	**Volponi**
Volschenk, B. (fl. 1982) S	**Volschenk**
Vondracek, Miloslav (1922–) B	**Vondr.**
(Voogd, W.B.de)	
see de Voogd, W.B.	**de Voogd**
Voorhees, Richard Kenneth (1907–) M	**Voorhees**
Voorhelm, George (1711/2–1787) S	**Voorhelm**
Voorhoeve, A.G. (1934–) S	**Voorh.**
Voos, Jane R. (fl. 1968) M	**Voos**
Vorderman, A.G. (fl. 1893) M	**Vord.**
Vorel, Jaromír (fl. 1973) S	**Vorel**
(Vorobev, D.P.)	
see Vorobiev, D.P.	**Vorob.**
Vorobiev, D.P. (1906–1984) S	**Vorob.**
Vorobik, Linda A. (1955–) S	**Vorobik**

(Voronikhin, Nikolai Nikolaevich)
   see Woronichin, Nikolai Nikolaevich **Woron.**
Voronin, L.V. (fl. 1986) M **Voronin**
(Voronov, Georg Gjurij Nikolaewitch)
   see Woronow, Georg Jurij Nikolaewitch **Woronow**
Vörös, J. (fl. 1956) M **Vörös**
Vörös, Lajos A **L.Vörös**
Vörös–Felkai, G. (fl. 1961) M **Vörös-Felkai**
Voroschilov, Vladimir N. (1908–) S **Vorosch.**
(Voroshilov, Vladimir N.)
   see Voroschilov, Vladimir N. **Vorosch.**
Vöröss, L.Zs. (fl. 1984) S **Vöröss**
Vorster, Pieter Johannes (1945–) PS **Vorster**
(Vos, André (Pascal Alexandre) de)
   see de Vos, André (Pascal Alexandre) **A.de Vos**
(Vos, Anna Petronella Cornelia de)
   see de Vos, Anna Petronella Cornelia **A.P.C.de Vos**
(Vos, Cornelis De)
   see de Vos, Cornelis **de Vos**
(Vos, Miriam Phoebe de)
   see de Vos, Miriam Phoebe **M.P.de Vos**
Voskresenskaya, O.A. (1904–1949) S **Voskr.**
Voss, Andreas (1857–1924) S **Voss**
Voss, Edward Groesbeck (1929–) S **E.G.Voss**
Voss, John William (1907–) S **J.W.Voss**
Voss, Wilhelm (Guglielmo) (1849–1895) M **W.Voss**
Vosseler, J. (fl. 1902) M **Vosseler**
Votava, F. (fl. 1982) S **Votava**
Voth, W. (fl. 1980) S **Voth**
Votsch (fl. 1904) S **Votsch**
Vouaux, Léon (1870–1914) M **Vouaux**
Vouk, Vale (Valentin) (1886–1962) A **Vouk**
Vouyeas, H. (fl. 1963) M **H.Vouyeas**
Vouyeas, V. (fl. 1963) M **V.Vouyeas**
Vovides, Andrés P. (1944–) S **Vovides**
Vozzhennikova, Tamara Fedorovna A **Vozzhenn.**
Vozzhinskaja, V.B. A **Vozzhinsk.**
(Vries, Bernhard W.L.de)
   see de Vries, Bernhard W.L. **B.de Vries**
(Vries, Gerardus Albertus de)
   see de Vries, Gerardus Albertus **G.A.de Vries**
(Vries, Hugo de)
   see de Vries, Hugo **de Vries**
(Vries, N.F.de)
   see de Vries, N.F. **N.F.de Vries**
(Vriese, Willem Hendrik de)
   see de Vriese, Willem Hendrik **de Vriese**
Vrishcz, D.L. (1939–) S **Vrishcz**
(Vroey, C.De)
   see De Vroey, C. **De Vroey**

Vrolik, Gerardus (Gerard) (1775–1859) S   **Vrolik**
Vroman, M. (1927–) A   **Vroman**
Vrugtman, Freek (1927–) S   **Vrugtman**
(Vuarambon, Roger Jacques)
  see Jacques–Vuarambon, Roger   **Jacq.-Vuar.**
Vucan (Vu Van Can) (fl. 1991) S   **Vucan**
Vuikj, Jacobus (1910–) S   **Vuijk**
Vuillemin, (Jean) Paul (1861–1932) M   **Vuill.**
Vuillet, J. (fl. 1914) S   **Vuillet**
Vuilleumier, Beryl Britnall (later Simpson) (1942–) S   **Vuilleum.**
Vukićević, E. (fl. 1972) S   **Vukić.**
Vukotinović, Ljudevit Farkaš (1813–1893) PS   **Vuk.**
Vulev, S. (1910–) S   **Vulev**
Vural, M. (fl. 1983–) S   **Vural**
Vustin, M.(M.) (fl. 1981) M   **Vustin**
Vuure, M.van (fl. 1957) M   **Vuure**
Vuyck, Laurens (1862–1931) S   **Vuyck**
(Vvedenskii, Aleksei Ivanovich)
  see Vvedensky, Aleksei Ivanovich   **Vved.**
Vvedensky, Aleksei Ivanovich (1898–1972) S   **Vved.**
Vyas, B.L. (fl. 1984) S   **B.L.Vyas**
Vyas, K.M. (fl. 1990) M   **K.M.Vyas**
Vyas, N.L. (fl. 1972) M   **N.L.Vyas**
Vyawahare, S.V. (fl. 1985) M   **Vyaw.**
Vyschin, I.B. (1951–) S   **Vyschin**
(Vyshin, I.B.)
  see Vyschin, I.B.   **Vyschin**
Vysokoostrovskaja, I.B. (fl. 1950) S   **Vysok.**
Vysotskij, A.V. A   **Vysotskij**
(Vyvere, Ernest Van de)
  see Van de Vyvere, Ernest   **Van de Vyvere**
Vyverman, Wim (1963–) A   **Vyverman**
Vzkulieva, V.E. A   **Vzkul.**

(Waalkes, Jan van Borssum)
  see Borssum Waalkes, Jan van   **Borss.Waalk.**
Wachendorff, Evert Jacob van (1702–1758) S   **Wach.**
Wächter, Wilhelm (1870–1928) MS   **W.Wächt**
Wachter, Willem Hendrik (1882–1946) BPS   **Wacht.**
Wacker, Hermann (–1899) S   **Wacker**
Wada, Koichiro (1911–1981) S   **Wada**
Wada, Masahiko (fl. 1978) M   **M.Wada**
Waddell, Coslett Herbert (1858–1919) S   **Waddell**
Waddell, David R. (fl. 1984) M   **D.R.Waddell**
Wade, Arthur Edwin (1895–1989) MS   **A.E.Wade**
Wade, Mary E. A   **M.E.Wade**
Wade, Walter (1760–1825) S   **Wade**
Wade, Wilbert Ernest (1922–) A   **W.E.Wade**

Wadhwa, Brij Mohan (1933–) BS   **Wadhwa**
Wadhwani, K. (fl. 1975) M   **Wadhwani**
Wadkins, R.F. (fl. 1931) M   **Wadkins**
Wadley, Bryce N. (fl. 1952) M   **Wadley**
Wadmond, Samuel Christensen (1871–) S   **Wadmond**
Wadood Khan, M.A (fl. 1982) S   **Wad.Khan**
Waern, Mats (1912–) A   **Waern**
Waga, August Ja. (1893–1960) S   **A.J.Waga**
Waga, Jakub (Jakob) Ignacy (1800–1872) S   **Waga**
Wagabayashi, Michio (1942–) S   **Wagab.**
(Wagenaar Hummelinck, Pieter)
    see Hummelinck, Pieter Wagenaar   **Hummel.**
Wagener, Philipp Christian (fl. 1798) S   **Wagener**
Wagener, Willis Westlake (1892–1969) M   **W.W.Wagener**
Wagenitz, Gerhard (1927–) S   **Wagenitz**
Wagenknecht, Burdette Lewis (1925–) S   **B.Wagenkn.**
Wagenknecht, Rodolfo (fl. 1955) S   **Wagenkn.**
Wager, Horace Athelstan (1876–1951) B   **Wager**
Wager, Vincent Athelstan (1904–) MS   **V.A.Wager**
Waghorne, Arthur Charles (1851–1900) S   **Waghorne**
Wagner, Adolf (1869–1940) S   **A.Wagner**
Wagner, D. (1800–) S   **D.Wagner**
Wagner, D.T.S. (fl. 1970) M   **D.T.S.Wagner**
Wagner, Edward (1864–) S   **E.Wagner**
Wagner, Florence Signaigo (1919–) AP   **F.S.Wagner**
Wagner, Fritz (fl. 1950) MP   **F.Wagner**
Wagner, Hermann (of Hadamar) (1824–1904) S   **H.Wagner bis**
Wagner, Hermann (of Weissenfels) (1824–1879) S   **H.Wagner**
Wagner, János (Johannes) (1870–1955) S   **J.Wagner**
Wagner, Josef (fl. 1825) M   **Wagner**
Wagner, Kenneth A. (1919–) B   **K.A.Wagner**
Wagner, Klaus (1935–1987) S   **K.Wagner**
Wagner, Moritz Friedrich (1813–1887) S   **M.Wagner**
Wagner, Paul Raymond (1910–) S   **P.R.Wagner**
Wagner, Rudolf (1872–) MS   **R.Wagner**
Wagner, Rudolf Eduard (1842–1913) S   **R.E.Wagner**
Wagner, Warren Herbert (1920–) PS   **W.H.Wagner**
Wagner, Warren L. (1950–) S   **W.L.Wagner**
Wagnon, Harvey Keith (1916–) S   **Wagnon**
Wahl, Carl Georg von (1869–) S   **C.Wahl**
Wahl, Herbert Alexander (1900–1975) S   **Wahl**
Wahl, I. (fl. 1966) M   **I.Wahl**
Wahlberg, J. (1810–1856) S   **J.Wahlb.**
Wahlberg, Pehr Fredrik (1800–1877) S   **Wahlb.**
Wahlbom, Johan Gustav (1724–1808) S   **Wahlbom**
Wahlenberg, Georg (Göran) (1780–1851) AMPS   **Wahlenb.**
Wahlstedt, Lars Johan (1836–1917) AS   **Wahlst.**
Wahlström, Johan Erik (1821–1892) S   **Wahlstr.**

Wahlström, Rolf (fl. 1982) S	R.Wahlstr.
Wahnschaffe, (Gustav Albert Bruno) Felix (1851–1914) S	Wahnsch.
Wahrlich, Woldemar Karlowitsch (1859–1923) MS	Wahrlich
Wahul, M.A. A	Wahul
Wailes, George H. (1862–1945) A	Wailes
Waines, R.H. A	Waines
(Wainio, Edvard (Edward) August)	
see Vainio, Edvard (Edward) August	Vain.
Wainwright, M.Ruth A	Wainwr.
Waisbecker, Anton (1835–1916) PS	Waisb.
Waisel S	Waisel
Waite, Merton Benway (1865–1945) M	Waite
Waitz, Friedrich August Carl (1798–1882) S	Waitz
Waitz, Karl Friedrich (1774–1848) S	K.F.Waitz
Wakabayashi, Michio (1942–) S	Wakab.
Wakefield, Elsie Maud (1886–1972) M	Wakef.
Wakefield, Norman Arthur (1918–1972) PS	N.A.Wakef.
Wakefield, Priscilla (1751–1832) S	P.Wakef.
Wakelin, P.O. A	Wakelin
Wakker, Jan Hendrik (1859–1927) M	Wakker
Waksman, Selman Abraham (1888–) M	Waksman
Walcott, Charles Doolittle (1850–1927) AFM	C.Walcott
Walcott, John (1754–1831) S	Walcott
Walcott, Mary Vaux (1860–1940) S	M.Walcott
Walcz, I. (fl. 1983) M	Walcz
Waldenburg, Ilse (1908–) S	Waldenb.
Waldheim, Stig (Gunnar Anton) (1911–1976) BM	Waldh.
Waldner, Heinrich (fl. 1880) S	H.Waldner
Waldner, Martin (fl. 1887) S	M.Waldner
Waldschmidt, J. (fl. 1865) S	Waldschm.
Waldstein, Franz de Paula Adam von (1759–1823) S	Waldst.
(Waldstein-Wartemburg, Franz de Paula Adam von)	
see Waldstein, Franz de Paula Adam von	Waldst.
Wale, Royden Samuel (–1952) S	Wale
Walgate, Marion Meason (1914–) S	Walgate
Waliyar, F. (fl. 1977) M	Waliyar
Walker, A.J. (fl. 1938) M	A.J.Walker
(Walker Arnott, George Arnott)	
see Arnott, George Arnott Walker	Arn.
Walker, C.C. (fl. 1981) S	C.C.Walker
Walker, Christopher (fl. 1979) M	C.Walker
Walker, Egbert Hamilton (1899–1991) BS	E.Walker
Walker, F.J. (fl. 1985) M	F.J.Walker
Walker, F.S. (1924–) S	F.S.Walker
Walker, James Willard (1943–) S	J.W.Walker
Walker, Jane (fl. 1983) M	Jane Walker
Walker, John (1731–1803/4) S	Walker
Walker, John (1930–) M	J.Walker
Walker, John Charles (1893–) M	J.C.Walker

Walker, Leva Belle (1878–1970) M	**L.B.Walker**
Walker, Mary Grant (née Shivas, M.G.) (1926–) P	**M.G.Walker**
Walker, Richard (1791–1870) S	**R.Walker**
Walker, Stanley (1924–1985) PS	**S.Walker**
Walker, Trevor George (1927–) PS	**T.G.Walker**
Walker, W.C. A	**W.C.Walker**
Walkington, David Leo (1930–) S	**Walk.**
Walkom, Arthur Bache (fl. 1915–1925) FS	**Walkom**
Wall, Arnold (1869–1966) S	**A.Wall**
Wall, David A	**D.Wall**
Wall, George (1821–1894) P	**G.Wall**
Wall, J.T. (fl. 1934) S	**J.T.Wall**
Wall, Mary E. A	**M.E.Wall**
Wall, Wilhelm August (1813–1861) S	**W.Wall**
Wallace, Alexander (1829–1899) S	**A.Wallace**
Wallace, Alfred Russel (1823–1913) S	**Wallace**
Wallace, Benjamin John (1947–) S	**B.J.Wallace**
Wallace, Franklin Gerhard (1909–) A	**F.G.Wallace**
Wallace, Gary D. (1946–) S	**G.D.Wallace**
Wallace, John Hume (1918–) A	**J.H.Wallace**
Wallace, Robert Whistler (1867–1955) S	**R.W.Wallace**
Wallace, Thomas Jennings (1912–) M	**T.J.Wallace**
Wallays, Antoine Charles François (1812–1888) M	**Wallays**
Waller, Floyd R. S	**F.R.Waller**
Waller, James Martin (1938–) M	**J.M.Waller**
Waller, S. (fl. 1951) M	**S.Waller**
Wallerius, Johan Gottschalk (1709–1785) S	**Wallerius**
Wallich, George Charles (1815–1899) A	**G.C.Wall.**
Wallich, Nathaniel (1786–1854) PS	**Wall.**
Wallis, Gustav (1830–1878) S	**Wallis**
Wallis, John (1714–1793) S	**J.Wallis**
Wallis, M. (fl. 1866) S	**M.Wallis**
Wallman, Johan Haquin (1792–1853) AS	**Wallman**
Wallner, J. A	**Wallner**
Wallnöfer, Anton (fl. 1888) B	**Walln.**
Wallnöfer, Bruno (1960–) S	**B.Walln.**
Wallroth, Carl (Karl) Friedrich Wilhelm (1792–1857) AMPS	**Wallr.**
Wallwork, H. (fl. 1988) M	**Wallwork**
Walpers, Wilhelm Gerhard (1816–1853) S	**Walp.**
Walpert, H. (fl. 1852–1855) S	**Walpert**
Walpole, Branson Alva (1890–1952) S	**Walpole**
Walpole, R. S	**R.Walpole**
Walraven, Wesley Clifton (1936–) S	**Walraven**
Walsemann (fl. 1987) S	**Walseman**
Walsh, Neville Grant (1956–) S	**N.G.Walsh**
Walsh, Robert (1772–1852) S	**Walsh**
Walsh-Held, M.E. (1881–1973) S	**Walsh-Held**
(Walt, J.P.Van der)	
see Van der Walt, J.P.	**Van der Walt**

(Walt, Johannes Jacobus Adriaan van der)	
see Van der Walt, Johannes Jacobus Adriaan	**J.J.A.van der Walt**
Walter, Charles (1831–1907) S	**C.Walter**
Walter, Émile (1873–1953) PS	**E.Walter**
Walter, Hans Paul Heinrich (1882–) S	**H.Walter**
Walter, Heinrich (Karl) (1898–) S	**H.K.Walter**
Walter, Heinrich Henry Fraser (1822–1893) S	**H.H.F.Walter**
Walter, James M. (fl. 1952) M	**J.M.Walter**
Walter, Johan Ernst Christian (1799–1860) S	**J.E.C.Walter**
Walter, Kerry Scott (1950–) P	**K.S.Walter**
Walter, Thomas (1740–1789) MPS	**Walter**
Walters, Maurice B. (fl. 1943) M	**M.B.Walters**
Walters, Neville E.M. (fl. 1956) M	**N.Walters**
Walters, Stuart Max (1920–) S	**Walters**
Walters, Terrence W. (1955–) S	**T.W.Walters**
Walther, (Edward) Eric (1892–1959) S	**E.Walther**
Walther, Alexander Wilhelm Hannibal Franz (1813–1890) B	**A.W.H.Walther**
Walther, Augustin Friedrich (1688–1746) L	**A.Walther**
Walther, Friedrich Ludwig (1759–1824) S	**Walther**
Walther, Kurt Herbert (1910–) B	**K.H.Walther**
Walton, D.W.H. (1945–) S	**D.W.H.Walton**
Walton, Elijah (1832–1880) S	**E.Walton**
Walton, Frederick Arthur (1853–1922) S	**Walton**
Walton, John (1895–1971) F	**J.Walton**
Walton, Lee Baker (1871–1937) A	**L.Walton**
Walz, A. (fl. 1987) M	**A.Walz**
Walz, Jacob Jacoblevič (1840–1904) AM	**J.Walz**
Walz, Lajos (Ludwig) (1845–1914) S	**L.Walz**
Wan, Shao Bin (fl. 1983) S	**S.B.Wan**
Wan, Thung Ling (1907–) B	**T.L.Wan**
Wan, Yu (fl. 1985) S	**Y.Wan**
Wanderley, Maria das Graças Lapa (1947–) S	**Wand.**
Wang, A.Q. (fl. 1987) M	**A.Q.Wang**
Wang, Bi Nong (fl. 1986) S	**B.N.Wang**
Wang, Bin (fl. 1989) S	**B.Wang**
Wang, C.R. (fl. 1983) MP	**C.R.Wang**
Wang, C.S. (fl. 1955) M	**C.S.Wang**
Wang, Ce Zhen (fl. 1989) A	**C.Z.Wang**
Wang, Chang S	**C.Wang**
Wang, Chang Yong (fl. 1984) PS	**C.Y.Wang**
Wang, Chao Jiang (fl. 1991) M	**Chao J.Wang**
Wang, Chen Hwa (1908–) S	**Chen H.Wang**
Wang, Chen Ju S	**Chen J.Wang**
Wang, Cheng Ping (fl. 1982) S	**C.P.Wang**
Wang, Chi Wu (1913–) S	**C.W.Wang**
Wang, Chia Chi A	**Chia C.Wang**
Wang, Chin Chih (fl. 1973) M	**Chin C.Wang**
Wang, Ching Jui (1928–) MS	**Ching J.Wang**
Wang, Chu Chia A	**Chu C.Wang**

Wang, Chu Hao (1923–) PS   **Chu H.Wang**
Wang, Chun Juan Kao (1928–) M   **C.J.K.Wang**
Wang, Chung Hsin (1923–) PS   **Chung H.Wang**
Wang, Chung Kuei   B   **C.K.Wang**
Wang, Cong Jiao (fl. 1988) S   **C.J.Wang**
Wang, Da Ming (fl. 1981) S   **D.M.Wang**
Wang, De Qun (fl. 1989) S   **D.Q.Wang**
Wang, De Yin (fl. 1982) S   **D.Y.Wang**
Wang, Dung (Dong) Sheng (fl. 1990) M   **D.S.Wang**
Wang, Fa Tsuan (1929–) S   **F.T.Wang**
Wang, Fan Lung   S   **F.L.Wang**
Wang, Fu Xing   A   **F.X.Wang**
Wang, G.C. (fl. 1984) S   **G.C.Wang**
Wang, Gan Jin (fl. 1984) S   **G.J.Wang**
Wang, Gui Zhen (fl. 1981) M   **G.Z.Wang**
Wang, H.H. (fl. 1971) M   **H.H.Wang**
Wang, Han Jin (1929–) S   **H.J.Wang**
Wang, Hsien Chih   S   **H.C.Wang**
Wang, Hui Qin (1934–) S   **H.Q.Wang**
Wang, J.Z. (fl. 1989) P   **J.Z.Wang**
(Wang, Jan Ru)
  see Wang, Chen Ju   **Chen J.Wang**
Wang, Jen Li (fl. 1967) S   **J.L.Wang**
Wang, Ji E (fl. 1979) M   **J.E.Wang**
Wang, Jin Bang (fl. 1981) M   **J.B.Wang**
Wang, Jin Wu (1928–) S   **J.W.Wang**
Wang, Jing Quan (1938–) S   **J.Q.Wang**
Wang, Jing Xiang (fl. 1981) S   **J.X.Wang**
Wang, Kai Yun (fl. 1989) S   **K.Y.Wang**
Wang, Ke Rong (fl. 1987) M   **K.R.Wang**
Wang, L.Z. (fl. 1983) S   **L.Z.Wang**
Wang, Lan Chow (fl. 1986) S   **L.C.Wang**
Wang, Liang Min (fl. 1984) S   **L.M.Wang**
Wang, Liu Ying (fl. 1985) S   **L.Y.Wang**
Wang, M.Y. (fl. 1989) M   **M.Y.Wang**
Wang, Mao (fl. 1988) S   **M.Wang**
Wang, Mei Zhi (1948–) B   **M.Z.Wang**
Wang, Ming Chang (fl. 1988) S   **M.Chang Wang**
Wang, Ming Chin (Jin) (fl. 1954) S   **M.C.Wang**
Wang, Ming Kin   S   **M.K.Wang**
Wang, Ning (fl. 1989) S   **N.Wang**
Wang, Ning Zhu (fl. 1985) S   **N.Z.Wang**
Wang, P.T. (fl. 1961) S   **P.T.Wang**
Wang, P.X. (fl. 1988) M   **P.X.Wang**
Wang, Pei Shan (1936–) PS   **P.S.Wang**
Wang, Qi (fl. 1989) M   **Q.Wang**
Wang, Qian Sheng (fl. 1987) S   **Q.S.Wang**
Wang, Qing Jiang (1916–) S   **Q.J.Wang**
Wang, Qing Li (fl. 1984) S   **Q.L.Wang**

Wang, Ren Shi (fl. 1985) S	**R.S.Wang**
Wang, S.H. (fl. 1979) S	**S.H.Wang**
Wang, Shu Fen (fl. 1988) S	**S.F.Wang**
Wang, Shu Song (fl. 1987) A	**S.S.Wang**
Wang, Siang Chung (fl. 1981) S	**S.C.Wang**
Wang, Song (fl. 1989) S	**S.Wang**
Wang, Su Chuan (Juan) A	**Su C.Wang**
Wang, Sui Yi (fl. 1989) S	**S.Y.Wang**
Wang, Tieh Seng (fl. 1986) S	**T.S.Wang**
Wang, Ting Fen (fl. 1965) S	**T.F.Wang**
Wang, Tong Kun (fl. 1989) S	**T.K.Wang**
Wang, Tso Pin (1904–) S	**T.P.Wang**
Wang, Tsung Hsuin S	**T.H.Wang**
Wang, Wan Xian (fl. 1984) S	**W.X.Wang**
Wang, Wei (1909–) S	**W.Wang**
Wang, Wei Yi (fl. 1982) S	**W.Y.Wang**
Wang, Wen Tsai (1926–) S	**W.T.Wang**
Wang, Xiao Ming (fl. 1987) M	**X.M.Wang**
Wang, Xue Wen (fl. 1986) S	**X.W.Wang**
Wang, Ya Jin (fl. 1979) A	**Y.J.Wang**
Wang, Yan Lin (fl. 1987) M	**Y.L.Wang**
Wang, Ying (fl. 1988) S	**Ying Wang**
Wang, Ying Xiang (fl. 1986) M	**Ying X.Wang**
Wang, Yong Xiao (fl. 1982) S	**Y.X.Wang**
Wang, You Fang (1953–) B	**Y.F.Wang**
Wang, Yu Sheng (fl. 1985) S	**Y.S.Wang**
Wang, Yun (1941–) M	**Y.Wang**
Wang, Yun Chang (1906–) M	**Y.C.Wang**
Wang, Yun Zhang (fl. 1980) M	**Y.Z.Wang**
Wang, Yung Chuan A	**Yung C.Wang**
Wang, Z.B. (fl. 1987) P	**Z.B.Wang**
Wang, Zhan (1911–) S	**Z.Wang**
Wang, Zhang Ping (fl. 1989) S	**Z.P.Wang**
Wang, Zhi Min (fl. 1980) S	**Z.M.Wang**
Wang, Zhong Ren (1939–) P	**Z.R.Wang**
Wang, Zhu Hao S	**Z.H.Wang**
Wang-Wei (fl. 1958) P	**Wang-Wei**
Wang-Yang, Jen Rong (1939–) BM	**Wang-Yang**
(Wange, Nicolaus Alexis)	
see Svensson, Nicolaus Alexis	**N.Svenss.**
Wangenheim, Friedrich Adam Julius von (1749–1800) S	**Wangenh.**
Wangerin, Walther (Leonhard) (1884–1938) S	**Wangerin**
Wangikar, B.P. (fl. 1984) M	**B.P.Wangikar**
Wangikar, P.D. (fl. 1977) M	**Wangikar**
Wani, D.D. (fl. 1968) M	**Wani**
Wankow, Iwan Wasiljewitsch (fl. 1928) S	**Wankow**
Wann, Frank Burkett (1892–1954) S	**Wann**
Wannan, B.S. (fl. 1985) S	**Wannan**
Wanner, Atreus (1852–) A	**Wanner**
Wanner, G. A	**G.Wanner**

Wanner, Stefan J. (fl. 1885) S	S.J.Wanner
Wanntorp, Hans–Erik (1940–) APS	Wanntorp
Waraitch, K.S. (fl. 1964) M	Waraitch
Warburg, Edmund Frederic (1908–1966) PS	E.F.Warb.
Warburg, Oscar Emanuel (1876–1937) S	O.E.Warb.
Warburg, Otto (1859–1938) PS	Warb.
Warcup, J.H. (1921–) M	Warcup
Ward, Daniel Bertram (1928–) S	D.B.Ward
Ward, F.S. (fl. 1929) M	F.S.Ward
Ward, George Henry (1916–) S	G.H.Ward
Ward, Harry Marshall (1854–1906) CMS	H.M.Ward
Ward, Henry Baldwin (1865–) A	H.B.Ward
Ward, Henry William (1840–1916) S	H.W.Ward
Ward, John E. (fl. 1965) M	J.E.Ward
Ward, Lester Frank (1841–1913) FS	Ward
Ward, Nathaniel Bagshaw (1791–1868) S	N.B.Ward
Warder, John Aston (1812–1883) S	Warder
Wardle, Peter (1931–) S	Wardle
Ware, S. (fl. 1967) S	S.Ware
Ware, William Melville (fl. 1925) MS	Ware
Warén, Harry Ilmari (1893–1973) AS	Warén
Warfa, Ahmed Mumin (fl. 1988) S	Warfa
Warion, (Jean Pierre) Adrien (1837–1880) S	Warion
(Warmelo, Konrad T.van)	
see Van Warmelo, Konrad T.	Van Warmelo
Warming, Johannes Eugen(ius) Bülow (1841–1924) AMS	Warm.
Wärn, Mats (fl. 1912) S	Wärn
Warncke, Esbern (1939–) B	E.Warncke
Warncke, K. (1937–) S	K.Warncke
Warner, Gloria M. (fl. 1970) M	G.M.Warner
Warner, Richard (1711/13?–1775) S	Warner
Warner, Robert (c.1815–1896) S	R.Warner
Warnock, Barton Holland (1911–) S	Warnock
Warnock, Michael J. (1956–) S	M.J.Warnock
Warnstorf, Carl (Friedrich E.) (1837–1921) BPS	Warnst.
Warren, Fred Adelbert (1902–) S	F.A.Warren
Warren, John Byrne Leicester (1835–1895) S	Warren
Warren, John R. (fl. 1948) M	J.R.Warren
Warren, Tyler B. (fl. 1973) M	T.B.Warren
(Warscewicz, Josef von)	
see Warszewicz, Joseph von Rawicz	Warsz.
Warsi, Akhlaq A. A	Warsi
Warszewicz, Joseph von Rawicz (1812–1866) S	Warsz.
Wartenberg, Arnold A	Wartenb.
Wartenweiler, Alfred (fl. 1917) M	Wartenw.
Wartmann, Friedrich Bernhard (1830–1902) MS	Wartm.
Wartmann, Jakob (1803–1873) S	J.Wartm.
Wasscher, Jacob (1911–1966) S	Wasscher
Wasser, Solomon P. (1946–) M	Wasser

Wasshausen, Dieter Carl (1938–) S   **Wassh.**
(Wassiliew, V.F.)
  see Vassiljev, V.F.   **V.F.Vassil.**
(Wassiliewski, N.I.)
  see Vassiljevsky, N.I.   **Vassiljevsky**
(Wassilijew, V.F.)
  see Vassiljev, V.F.   **V.F.Vassil.**
(Wassillijew, V.F.)
  see Vassiljev, V.F.   **V.F.Vassil.**
Wassner, Ludwig (1861–1929) S   **Wassner**
Wastler, Franz (1837–1936) S   **Wastler**
Watanabe, Atsuki (fl. 1935) M   **A.Watan.**
Watanabe, H. (fl. 1936) M   **H.Watan.**
Watanabe, Kiyohiko (1900–) S   **Watan.**
Watanabe, Masayuki (1941–) A   **M.Watan.**
Watanabe, Noboru (fl. 1935) M   **N.Watan.**
Watanabe, Ryozo   B   **R.Watan.**
Watanabe, Sadamoto (1934–) S   **S.Watan.**
Watanabe, Shin (1948–) A   **Shin Watan.**
Watanabe, Tatsuwo (fl. 1952) M   **T.Watan.**
Watanabe, Tsuneo (fl. 1989) M   **Ts.Watan.**
Watelet, (Jean–François) Adolphe (1811–1879) FP   **Watelet**
Waterbury, J.   A   **Waterbury**
Waterfall, Umaldy Theodore (1910–1971) S   **Waterf.**
Waterhouse, Benjamin (1754–1846) S   **Waterh.**
Waterhouse, Grace Marion (1906–) M   **G.M.Waterh.**
Waterhouse, John Teast (1924–1983) S   **J.T.Waterh.**
Waterhouse, W.L. (fl. 1951) M   **W.L.Waterh.**
Waterman, Alma May (1893–) M   **Waterman**
Waterman, P.G. (fl. 1975) S   **P.G.Waterman**
Waters, Campbell Easter (1872–1955) P   **Waters**
Waters, Josephine   A   **J.Waters**
Waters, S.D. (fl. 1989) M   **S.D.Waters**
Waterston, John MacLaren (1911–) M   **Waterston**
Watkins, T.P.   A   **Watkins**
Watling, Roy (1938–) M   **Watling**
Watson, A.K. (fl. 1988) M   **A.K.Watson**
(Watson, Alexander)
  see Beatson, Alexander   **Beatson**
Watson, Alice Marie Johnson (1898–) M   **A.M.J.Watson**
Watson, E.A. (1879–) A   **E.A.Watson**
Watson, E.M. (fl. 1987) S   **E.M.Watson**
Watson, Elba Emanuel (1871–1936) S   **E.Watson**
Watson, Hewett Cottrell (1804–1881) PS   **H.C.Watson**
Watson, Joan   A   **J.Watson**
Watson, John Forbes (1827–1892) S   **J.F.Watson**
Watson, Leslie (1938–) S   **L.Watson**
Watson, Peter William (1761–1830) S   **P.Watson**
Watson, R.J. (fl. 1987) S   **R.J.Watson**
Watson, S.W. (fl. 1957) M   **S.W.Watson**

Watson, Sereno (1826–1892) PS	S.Watson
Watson, Wade Ralph (1903–) M	W.R.Watson
Watson, Walter (1872–1960) BMS	Walt.Watson
Watson, William (1858–1925) S	W.Watson
Watson, William (1715–1787) S	Watson
Watson, William Charles Richard (1885–1954) S	W.C.R.Watson
Watt, David Allan Poe (1830–1917) PS	Watt
Watt, George (1851–1930) S	G.Watt
Watt, John Mitchell (1892–1980) S	J.M.Watt
Watts, William Walter (1856–1920) BPS	Watts
Watzel, Kajetán (Cajetán) (1812–1885) BPS	Watzel
Watzl, Bruno (1886–1945) S	Watzl
Wauer, Brent (fl. 1990) S	Wauer
Waugh, Frank Albert (1869–1947) S	Waugh
Wawra, Heinrich (1831–1887) PS	Wawra
(Wawra von Fernsee, Heinrich)	
see Wawra, Heinrich	Wawra
Wawrik, Friedericke (fl. 1952) AM	Wawrik
Wear, Sylvanus (1858–1920) S	Wear
Weatherby, Charles Alfred (1875–1949) PS	Weath.
Weathers, John (1867–1928) S	Weathers
Weatherwax, Paul (1888–1976) S	Weatherwax
Weaver, Richard P. (1943–) S	Weaver
Webb, Colin James (1949–) S	C.J.Webb
Webb, David Allardice (1912–) S	D.A.Webb
Webb, Jill (fl. 1991) M	J.Webb
Webb, Philip Barker (1793–1854) APS	Webb
Webb, Robert Holden (1805–1880) S	R.Webb
Webber, Herbert John (1865–1946) MS	Webber
Webber, John Milton (1897–) S	J.M.Webber
Weber, A. Alois (1852–1942) S	A.A.Webber
Weber, Anton (1947–) S	A.Weber
Weber, Carl Albert (1856–1931) MS	C.A.Weber
Weber, Carl Otto (fl. 1851) F	C.O.Weber
Weber, Claude (fl. 1968) S	C.Weber
Weber, Cornelius I. (1927–) A	C.I.Weber
Weber, Daniel (fl. 1973) S	D.Weber
Weber, E.H. (fl. 1963) M	E.H.Weber
Weber, F.E. S	F.E.Weber
Weber, Ferdinand (1903–) S	Ferd.Weber
Weber, Frédéric Albert Constantin (1830–1903) S	F.A.C.Weber
Weber, Friedrich (1781–1823) ABMPS	F.Weber
Weber, G. (fl. 1989) M	G.Weber
Weber, George Frederick (1894–) M	G.F.Weber
Weber, George Heinrich (1752–1828) AMS	Weber
Weber, Hans (1911–) S	H.Weber
Weber, Heinrich E. (fl. 1988) S	H.E.Weber
Weber, J.B. (fl. 1874) S	J.B.Weber
Weber, Jean–Germaine Claude (1922–) S	J.G.C.Weber
Weber, Joseph (1856–1908) S	J.Weber

Weber, Joseph Carl (Karl) (1801–1875) S	**J.C.Weber**
Weber, Joseph Zvonko (1930–) S	**J.Z.Weber**
Weber, Mihály A	**M.Weber**
Weber, Nancy S. (1943–) M	**N.S.Weber**
Weber, Ulrich (fl. 1922) P	**U.Weber**
Weber, W.W. (–1823) S	**W.W.Weber**
Weber, Wilhelm (fl. 1979) S	**W.Weber**
Weber, William Alfred (1918–) MS	**W.A.Weber**
Weber-van Bosse, Anna Antoinette (1852–1942) A	**Weber Bosse**
Weberbauer, August (1871–1948) BS	**Weberb.**
Weberbauer, Otto (1846–1881) MS	**O.Wederb.**
Weberling, Focko H.E. (1926–) S	**Weberling**
Webr, Karel Mirko (fl. 1974) S	**Webr**
Webster, Angus Duncan (fl. 1886–1920) S	**Webster**
Webster, Anne M. A	**A.M.Webster**
Webster, Clement Lyon (1859–) A	**C.L.Webster**
Webster, Grady Linder (1927–) S	**G.L.Webster**
Webster, John (1925–) M	**J.Webster**
Webster, Neri (fl. 1989) F	**N.Webster**
Webster, Robert D. (1950–) S	**R.D.Webster**
Webster, Robert K. (fl. 1987) M	**R.K.Webster**
Webster, S.D. (1959–) S	**S.D.Webster**
Wechuysen, Ronny (1953–) S	**Wech.**
Weddell, Hugh Algernon (1819–1877) MS	**Wedd.**
Wedemayer, Gary J. (1955–) A	**G.J.Wedem.**
Wedemeyer (fl. pre 1803) S	**Wedem.**
Wedin, Mats (fl. 1991) M	**Wedin**
Wee, James L. A	**Wee**
Weeber, Gustav (1857–1943) S	**Weeber**
Weed, Clarence Moores (1864–1943) S	**Weed**
Weeda, E.J. (1952–) S	**Weeda**
Weedon, Amy Gertrude (fl. 1927) M	**Weedon**
Weeks, R.A. (fl. 1979) M	**R.A.Weeks**
Weeks, R.J. (fl. 1985) M	**R.J.Weeks**
Weese, Josef Karl (1888–1962) M	**Weese**
Weevers, Theodorus (1875–1952) S	**Weevers**
Wegelin, (Antonius) Theodoor (fl. 1837) S	**Wegelin**
Wegelin, Heinrich (1853–1940) MS	**H.Wegelin**
Wegener, H. (fl. 1894) M	**H.Wegener**
Wegener, Th. (fl. 1844) S	**Wegener**
Wehbe, Jorge A. (fl. 1986) S	**Wehbe**
Wehlburg, Cornelis (1903–) S	**Wehlb.**
Wehmer, Carl (Friedrich Wilhelm) (1858–1935) MS	**Wehmer**
Wehmeyer, Lewis Edgar (1897–1971) M	**Wehm.**
Weholt, Ø. (fl. 1983) M	**Weholt**
Wehrhahn, Heinrich Rudolf (1887–1940) S	**H.R.Wehrh.**
Wehrhahn, Herbert Fuller (1879–) S	**H.F.Wehrh.**
Wehrhahn, Wilhelm (1857–1926) B	**Wehrh.**
Wehrli, M. S	**Wehrli**
Wei, C.T. (fl. 1933) M	**C.T.Wei**

Wei, Chao Fen (1934–) S	**C.F.Wei**
Wei, Fa Nan (1941–) S	**F.N.Wei**
Wei, Hong Tu (fl. 1982) S	**H.T.Wei**
Wei, Jiang Chun (1932–) M	**J.C.Wei**
Wei, Liu Gen (fl. 1979) M	**L.G.Wei**
Wei, Shu Xia (fl. 1980) M	**S.X.Wei**
Wei, Si Qi (fl. 1984) S	**S.Q.Wei**
Wei, Song Ji (fl. 1986) S	**S.J.Wei**
Wei, X.W. (fl. 1982) S	**X.W.Wei**
Wei, Yin Xin (Shin) (1939–) A	**Y.X.Wei**
Wei, Yu(e) Tsung (1936–) S	**Y.T.Wei**
Wei, Yun (fl. 1984) PS	**Y.Wei**
Wei, Zhi (fl. 1985) S	**Z.Wei**
Weibel, Raymond (1905–) S	**Weibel**
Weibull, Claes A	**Weibull**
Weick, Alphonse (fl. 1863) S	**Weick**
Weide, Heinz (1933–) S	**Weide**
Weidemann, G.J. (fl. 1982) M	**Weid.**
Weidlich, E. (fl. 1928) S	**Weidlich**
Weidman, F.D. (fl. 1945) M	**Weidman**
Weidmann, Antonín (1850–1915) B	**Weidmann**
Weigel, Christian Ehrenfried (von) (1748–1831) MPS	**Weigel**
Weigel, Johann Adam Valentin (1740–1806) S	**J.Weigel**
Weigelt, Christoph (–1828) S	**Weigelt**
Weihe, Carl Ernst August (1779–1834) S	**Weihe**
Weijman, A.C.M. (fl. 1979) M	**Weijman**
Weik, Kenneth L. A	**Weik**
Weilbacher, B.F. (fl. 1970) M	**Weilb.**
Weiler, Helmut A	**H.Weiler**
Weiler, J.H. (fl. 1962) S	**Weiler**
Weill, A. (fl. 1932) M	**A.Weill**
Weill, Jean S	**J.Weill**
Weill, P.–E. (fl. 1919) M	**Weill**
Weill, R. (fl. 1932) M	**R.Weill**
Weiller, A. S	**A.Weiller**
Weiller, Marc (1880–1945) PS	**Weiller**
Weimarck, August Henning (1903–1980) S	**Weim.**
Weimarck, Karl Gunnar Henning (1936–) S	**G.Weim.**
Weimer, James Leroy (1887–) M	**Weimer**
Wein, Kurt (1883–1968) S	**Wein**
Weinberg, Barbara (fl. 1985) S	**B.Weinberg**
Weinberg, F. S	**Weinberg**
Weindlmayr, J.von (fl. 1964) M	**Weindlm.**
Weinert, Erich (1931–) S	**Weinert**
Weingart, Wilhelm (1856–1936) S	**Weing.**
Weinhart, Max (1824–1905) CS	**Weinh.**
Weinhold A	**Weinhold**
Weinmann, Johann Anton (1782–1858) MS	**Weinm.**
Weinmann, Johann Georg (1764–1769) S	**J.G.Weinm.**

Weinmann, Johann Wilhelm (1683–1741) L	**J.W.Weinm.**
Weintal, C. (fl. 1987) M	**Weintal**
Weintroub, D. S	**Weintroub**
Weir, James Robert (1882–1943) M	**Weir**
(Weis, Friedrich Wilhelm G.)	
see Weiss, Friedrich Wilhelm G.	**F.W.Weiss**
Weis, Ludwig S	**L.Weis**
Weisbord, Norman (1901–) A	**Weisbord**
Weise, Johann Gottlob Christoph (1762–1840) S	**Weise**
Weiser, Jaroslav (fl. 1965) AM	**Weiser**
Weismann, August (1834–1914) S	**Weism.**
Weiss, (Christian) Ernst (1833–1890) F	**C.E.Weiss**
Weiss, (Joseph Gustav) Adolf (1837–1894) S	**G.A.Weiss**
Weiss, Emanuel (1837–1870) S	**E.Weiss**
Weiss, Frederick Ernest (1865–1953) MS	**F.E.Weiss**
Weiss, Freeman Albert (1892–) M	**F.A.Weiss**
Weiss, Friedrich Wilhelm G. (1744–1826) ABMPS	**Weiss**
Weiss, Heinrich (–1912) S	**H.Weiss**
Weiss, Johann Evangelista (1850–1918) AS	**J.Weiss**
Weiss, P. (fl. 1930) M	**P.Weiss**
Weisse, Arthur (Friedrich Hermann) (1861–1939) S	**A.Weisse**
Weisse, Johann Friedrich von (1792–1869) A	**Weisse**
Weissenberg, Julius Richard (1882–) A	**J.R.Weissenb.**
Weissenberg, Richard (fl. 1912) M	**Weissenb.**
(Weissmann–Kollmann, Fania)	
see Kollmann, Fania Weissmann	**Kollmann**
Weitenweber, Wilhelm Rudolf (1804–1870) S	**Weitenw.**
Weitzman, Anna Lisa (1958–) S	**A.L.Weitzman**
Weitzman, Irene (fl. 1962) M	**Weitzman**
Welch, Donald Stuart (1894–1972) M	**Welch**
Welch, Marcus Baldwin (1895–1942) S	**M.B.Welch**
Welch, Winona Hazel (1896–1990) B	**W.H.Welch**
Welden, Arthur L. (1927–) M	**A.L.Welden**
Welden, Franz Ludwig von (1782–1853) S	**Welden**
Weldt, Eduardo (fl. 1970) S	**Weldt**
Well, Johann Jacob von (1725–1787) S	**Well**
Welles, Colin Gilchrist (fl. 1922) M	**Welles**
Wellheim A	**Wellheim**
Welling, Christian Friedrich von (fl. 1791) S	**Welling**
Wellman, A.M. (fl. 1975) M	**A.M.Wellman**
Wellman, Frederick Lovejoy (1897–) M	**Wellman**
Wells, Bertram Whittier (1884–) S	**Wells**
Wells, Doreen E. (fl. 1954) M	**D.E.Wells**
Wells, James Ray (1932–) S	**J.R.Wells**
Wells, Kenneth (1927–) M	**K.Wells**
Wells, P. A	**P.Wells**
Wells, Philipp Vincent (1928–) S	**P.V.Wells**
Wells, Theodore W. (fl. 1970) S	**T.W.Wells**
Wells, Virginia L. (fl. 1968) M	**V.L.Wells**

Welsford, Evelyn J. A	**Welsford**
Welsh, Henry (1906–1967) A	**Welsh**
Welsh, Stanley Larson (1928–) S	**S.L.Welsh**
Weltner, W. A	**Weltner**
Weltzien, H.C. (fl. 1963) M	**Weltzien**
Welwitsch, Friedrich Martin Josef (1806–1872) MPS	**Welw.**
Welzen, P.C. van (1958–) S	**Welzen**
Wemple, Don Kimberley (1929–) S	**Wemple**
Wen, Du Su (fl. 1982) S	**D.S.Wen**
Wen, He Qun (1957–) S	**H.Q.Wen**
Wen, Hua An (1949–) M	**H.A.Wen**
Wen, Tai Hui (1924–) S	**T.H.Wen**
Wen, Xuan Kai (fl. 1982) S	**X.K.Wen**
Wendelberger, Gustav (1915–) S	**Wendelb.**
Wendelbo, Per Erland Berg (1927–1981) S	**Wendelbo**
Wenderoth, Georg Wilhelm Franz (1774–1861) S	**Wender.**
Wendisch, Ernst (fl. 1892–1905) M	**Wendisch**
Wendland, Heinrich Ludolph (1792–1869) S	**H.L.Wendl.**
(Wendland, "H.A.")	
see Wendland, Hermann	**H.Wendl.**
Wendland, Hermann (1825–1903) S	**H.Wendl.**
Wendland, Johann Christoph (1755–1828) S	**J.C.Wendl.**
Wendt, Albert (1887–1958) S	**A.Wendt**
Wendt, Georg Friedrich Carl (fl. 1804) S	**Wendt**
Wendt, Johann Christian Wilhelm (1778–1838) S	**J.Wendt**
Wendt, Thomas Leighton (1950–) P	**T.Wendt**
Wengenmayr, Xaver (fl. 1930) S	**Weng.**
Weniger, H.L. (fl. 1969) S	**Weniger**
Wenner, John Jeremaiah (1885–) A	**Wenner**
Wenrich, David H. (1885–1968) A	**Wenrich**
Went, Friedrich August Ferdinand Christian (1863–1935) MS	**Went**
Went, Frits Warmolt (1903–1990) S	**F.W.Went**
Went, J.C. S	**J.C.Went**
Wentz, William Alan (1946–) S	**Wentz**
Wenyon, Charles Morely (1878–1948) A	**Wenyon**
Wenzig, (Johann) Theodor (1824–1892) S	**Wenz.**
Wenzlaff, Franz (1810–) S	**Wenzlaff**
(Wenzlavovitsch, F.S.)	
see Wenzlawowicz, F.S.	**Wenzlaw.**
Wenzlawowicz, F.S. (1904–) S	**Wenzlaw.**
Wercklé, Karl (Carl) (1860–1924) PS	**Wercklé**
Werdermann, Erich (1892–1959) MS	**Werderm.**
Weresub, Luella K. (1918–1979) M	**Weresub**
(Werff, Henk van der)	
see van der Werff, Henk	**van der Werff**
Wermel, E. A	**Wermel**
Werneck, H.L. (1890–) S	**H.L.Werneck**
Werneck, Ludwig Friedrich Franz von (fl. 1791) S	**Werneck**
Wernekinck, Franz (1764–1839) AS	**Wernek.**
Werner, Alexander S	**A.Werner**

Werner, H. (1839–) S	**H.Werner**
Werner, H.M. (fl. 1970) M	**H.M.Werner**
Werner, Klaus (1928–) S	**K.Werner**
Werner, Roger–Guy (1901–1977) ABCM	**Werner**
Wernham, Clifford Charles (1903–) M	**C.C.Wernham**
Wernham, Herbert Fuller (1879–1941) S	**Wernham**
Wernicke, Albrecht Ludwig Agathon (fl. 1892) M	**Wernicke**
Wernischeck, Johann Jacob (Jakob) (1743–1804) S	**Wernisch.**
Werth, Emil Albert Karl August (1869–1961) AS	**Werth**
Werz, Günther A	**Werz**
(Weselsky, Bedřich (Friedrich))	
see Veselsky, Bedřich (Friedrich)	**Veselsky**
Wesenburg–Lund, Carl Jørgen (1867–1955) A	**Wesenb.-Lund**
Weskamp, W. (fl. 1979) S	**Weskamp**
Wesmacl, Alfred (1832–1905) S	**Wesm.**
Wessel, (Paul) Philipp (Friedrich) (1825–1855) S	**P.Wessel**
Wessel, A.W. (fl. 1858) S	**A.Wessel**
Wessel, Otto (fl. 1874) S	**O.Wessel**
Wessels Boer, Jan Gerard (1936–) PS	**Wess.Boer**
Wessels, Dirk C.J. (1950–) M	**Wessels**
Wessely, I. (1814–1898) S	**Wessely**
Wessén, Carl Johan (1812–1843) S	**Wessén**
Wessner, Wilhelm (1904–1903) S	**Wessner**
West, Cyril (1887–1986) MS	**C.West**
West, Erdman (1894–) M	**E.West**
West, George Stephen (1876–1919) A	**G.S.West**
West, Hans (1758–1811) S	**H.West**
West, James (1875–1939) S	**J.West**
West, John A. A	**J.A.West**
West, Judith Gay (1949–) S	**J.G.West**
West, Keith R. S	**K.R.West**
West, Louise (1910–) S	**L.West**
West, Tuffen (1823–1891) A	**T.West**
West, William (1848–1914) A	**West**
West, William (1875–1901) A	**W.West**
Westberg, E.H.V. S	**E.H.V.Westb.**
Westberg, G.F. (1872–) S	**G.Westb.**
Westcott, Frederic (–1861) S	**Westc.**
Weste, Gretna Margaret (1917–) M	**Weste**
Westendorp, Gérard Daniel (1813–1869) CM	**Westend.**
Wester, Peter Jansen (1877–1931) S	**Wester**
Westerdijk, Johanna (1883–1961) M	**Westerd.**
Westerhoff, Rembertus (1801–1874) S	**Westerh.**
Westerlund, Carl Agardh (1831–1908) S	**Westerl.**
Westerlund, Carl Gustaf (1864–1914) S	**G.C.Westerl.**
Westermaier, Maximilian (1852–1903) S	**Westerm.**
Westermann, Diedrich Hermann (1875–1956) S	**Westermann**
Westhoff, Friedrich (Fritz) (1867–1896) S	**Westh.**
(Westhuizen, G.C.A.van der)	
see Van der Westhuizen, G.C.A.	**Van der Westh.**

(Westhuizen, S.van der)	
see Van der Westhuizen, S.	**S.Van der Westh.**
Westling, (Per) Richard (1868–1942) MS	**Westling**
Weston, Arthur Stewart (1932–) S	**A.S.Weston**
Weston, Peter Henry (1956–) S	**P.H.Weston**
Weston, Richard (1733–1806) S	**Weston**
Weston, William Henry (1890–1978) M	**W.Weston**
Westphal, E. S	**Westphal**
Westra, Lübbert Ybele Theodoor (1932–) S	**Westra**
Westring, Johan Peter (1753–1833) M	**Westr.**
Westwood, L.O. (fl. 1931) M	**Westwood**
Westwood, N.J. (fl. 1977) M	**N.J.Westwood**
(Wet, Johannes Martenis Jacob de)	
see de Wet, Johannes Martenis Jacob	**de Wet**
Wetherbee, Richard (1945–) A	**Wetherbee**
Wethered, Edward A	**Wethered**
Wetmore, Clifford M. (1934–) M	**Wetmore**
Wettstein, Fritz (Friedrich) (1895–1945) ABMPS	**F.Wettst.**
Wettstein, Richard (1863–1931) AMPS	**Wettst.**
(Wettstein von Westersheim, Fritz (Friedrich))	
see Wettstein, Fritz (Friedrich)	**F.Wettst.**
(Wettstein von Westersheim, Richard von)	
see Wettstein, Richard	**Wettst.**
Wetzel, Otto Christian August (1891–1971) A	**Wetzel**
Wetzel, W. (1887–1978) A	**W.Wetzel**
Weyer, W.van de (fl. 1920) S	**Weyer**
Weyhe, (M.) F. (1775–1846) S	**Weyhe**
Weyl, Carl Friedrich (1792–1872) S	**Weyl**
Weyland, Hermann (Gerhard) (1888–1974) AF	**Weyland**
Weymayr, Thassilo (fl. 1867) S	**Weymayr**
Weymouth, William Anderson (1842–1928) B	**Weymouth**
Weyrich, Heinrich (1828–1863) S	**Weyr.**
Whalen, Michael Dennis (1950–1985) S	**Whalen**
Whalley, Anthony J.S. (fl. 1978) M	**Whalley**
Whedon, Walter Forest (1904–) A	**Whedon**
Wheeler, Charles Fay (1842–1910) S	**C.F.Wheeler**
Wheeler, George Montague (1842–) P	**G.M.Wheeler**
Wheeler, Gerald A. (1940–) S	**G.A.Wheeler**
Wheeler, Helen–Mar. (fl. 1935) S	**H.-M.Wheeler**
Wheeler, James (fl. 1763) S	**Wheeler**
Wheeler, James Lowe (fl. 1830–1870) S	**J.L.Wheeler**
Wheeler, Judith Roderick (1944–) S	**J.R.Wheeler**
Wheeler, Louis Cutter (1910–1980) S	**L.C.Wheeler**
Wheeler, William Archie (1876–1968) MS	**W.A.Wheeler**
Wheelock, William Efner (1852–1926) S	**Wheelock**
Whelden, Roy M. (fl. 1935) AM	**Whelden**
Wheldon, James Alfred (1862–1924) MS	**Wheldon**
Wherry, Edgar Theodore (1885–1982) PS	**Wherry**
Whetstone, R. David (1949–) S	**Whetstone**

Whetzel, Herbert Hice (1877–1944) M	**Whetzel**
Whibley, David John Edward (1936–) S	**Whibley**
Whiffen, Alma Joslyn (1916–) M	**Whiffen**
(Whiffen–Barksdale, Alma Joslyn)	
see Whiffen, Alma Joslyn	**Whiffen**
Whiffin, Trevor Paul (1947–) S	**Whiffin**
Whipple, Amiel Weeks (1816–1863) B	**Whipple**
Whipple, George Chandler (1866–) A	**G.C.Whipple**
Whisler, Howard C. (fl. 1963) M	**Whisler**
Whistler, W.Arthur (1944–) S	**Whistler**
Whistling, Christian Gottfried (1757–1807) S	**Whistling**
Whitaker, Thomas Wallace (1905–) S	**Whitaker**
Whitcombe, R.P. (fl. 1983) S	**Whitc.**
White, (Charles) David (1862–1935) AF	**C.D.White**
White, Alain Campbell (1880–) S	**A.C.White**
White, Cyril Tenison (1890–1950) PS	**C.T.White**
White, Donald (c.1892–) S	**D.White**
White, Edward Albert (1872–1943) S	**E.A.White**
White, F.Joy (fl. 1987) M	**F.J.White**
White, Francis Buchanan (1842–1894) MS	**F.B.White**
White, Frank (1927–) S	**F.White**
White, George P. (fl. 1985) M	**G.P.White**
White, Gilbert (1720–1793) S	**G.White**
White, Harold Everett (1899–) M	**H.E.White**
White, Henry Hoply (1790–1876) A	**H.H.White**
White, James F. (fl. 1985) M	**J.F.White**
White, James Walter (1846–1932) S	**J.W.White**
White, Jean (c.1750–1832) S	**White**
White, Jean (fl. 1911) S	**Jean White**
White, John (c.1760–1837) of Dublin S	**J.White Dubl.**
White, John (1756?–1832) of Royal Navy S	**J.White R.N.**
White, N.H. (–1991) M	**N.H.White**
White, Richard Alan (1935–) P	**R.A.White**
White, Richard Peregrine (1896–) M	**R.P.White**
White, Stephen S. (1909–) S	**S.S.White**
White, Theodore Greely (1872–1901) S	**T.G.White**
White, Violetta Susan Elizabeth (1875–1949) M	**V.S.White**
White, William Lawrence (1908–1952) M	**W.L.White**
Whitehead, F.H. (1913–) S	**Whitehead**
Whitehead, Jack (fl. 1943) S	**J.Whitehead**
Whitehead, Marvin D. (fl. 1951) M	**M.D.Whitehead**
Whitehouse, Eula (1892–1974) BS	**Whitehouse**
Whitehouse, Harold Leslie Keer (1917–) B	**H.Whitehouse**
Whitelegge, Thomas (1850–1927) B	**Whitel.**
Whiteside, J.O. (fl. 1972) M	**Whiteside**
Whitfield, Robert Parr (1828–1910) AF	**Whitfield**
Whitford, A.C. (fl. 1916) M	**A.C.Whitford**
Whitford, Harry Nichols (1872–1941) S	**Whitford**
Whitford, Larry Alston (1902–) A	**L.A.Whitford**

Whiting, Marian Muriel (1881–) S	**Whiting**
Whiting, Mark C. A	**M.C.Whiting**
Whiting, R.E. (fl. 1982) S	**R.E.Whiting**
Whitmore, Eugene Randolph (1874–) A	**E.R.Whitmore**
Whitmore, H.C. S	**H.C.Whitmore**
Whitmore, S.A. (fl. 1983) S	**S.A.Whitmore**
Whitmore, Timothy Charles (1935–) S	**Whitmore**
Whitney, Hugh S. (1935–) M	**H.S.Whitney**
Whitney, Kenneth D. (fl. 1978) M	**K.D.Whitney**
Whitney, Leo David (1908–1937) S	**Whitney**
Whittaker, R.H. A	**Whittaker**
Whittard, W.F. A	**Whittard**
Whittemore, A.T. (fl. 1988) S	**Whittem.**
Whitten, W.M. (fl. 1988) S	**Whitten**
Whittick, Alan (fl. 1972) AM	**Whittick**
Whittier, Barbara B	**B.Whittier**
Whittier, Henry O. (1937–) B	**H.Whittier**
Whitting, Frances G. A	**Whitting**
Whitwell, William (1839–1920) PS	**Whitw.**
Wibel, August Wilhelm Eberhard Christoph (1775–1814) MS	**Wibel**
Wibiral, Erich (1878–1950) S	**Wibiral**
Wicander, (Edwin) Reed (1946–) A	**Wicander**
Wichansky, Evžen (fl. 1958) M	**Wichansky**
(Wichers, Friedrich Heinrich (Fridrich Hindrich))	
see Wiggers, Friedrich Heinrich (Fridrich Hindrich)	**F.H.Wigg.**
Wichmann, Lucia A	**Wichmann**
Wichura, Max Ernst (1817–1866) S	**Wich.**
Wickens, Gerald Ernest (1927–) S	**Wickens**
Wickerham, Lynford J. (fl. 1951) M	**Wick.**
Wicklow, D.T. (fl. 1983) M	**Wicklow**
Wickström, J.E. (1789–1856) S	**Wickstr.**
Widden, Paul (fl. 1988) M	**Widden**
Widder, Felix Joseph (1892–1974) MS	**Widder**
Widdowson, Thomas B. A	**Widd.**
Widén, Karl (Carl) Johann (1935–) PS	**Widén**
Widjaja, Elizabeth A. (1951–) S	**Widjaja**
Widmer, Elisabeth (1862–1952) S	**Widmer**
Widnmann, Friedrich (1765–1848) S	**Widnmann**
Wiebe, Gustav A. (1899–) S	**Wiebe**
Wiebel, C.A. (fl. 1799) M	**Wiebel**
Wieben, M. (fl. 1927) M	**Wieben**
Wied-Neuwied, Maximilian Alexander Philipp zu (1782–1867) S	**Wied-Neuw.**
Wiedemann, Ferdinand (Johannes) (1805–1887) S	**Wiedem.**
Wieden, M.A. (fl. 1986) M	**Wieden**
Wiedling, Sten (1909–) A	**Wiedling**
Wieffering, J.H. (fl. 1964) P	**Wieff.**
Wiegand, Karl McKay (1873–1942) S	**Wiegand**
Wiegleb, G. (1948–) S	**Wiegleb**
Wiegmann, Arend (Joachim) Friedrich (1770–1853) S	**Wiegmann**

Wiehe, Paul Octave (–1975) MS	**Wiehe**
Wiehler, Hans Joachim (1930–) S	**Wiehler**
Wieland, George Reber (1865–1953) AF	**Wieland**
Wieland, H. (fl. 1934) M	**H.Wieland**
Wiens, Delbert (1932–) S	**Wiens**
Wier, D.B. (fl. 1877) S	**Wier**
Wieringer, K.T. (fl. 1956) M	**Wieringer**
Wiersema, John H. (1950–) S	**Wiersema**
Wierzbicki, Antal S	**A.Wierzb.**
Wierzbicki, Piotr Pawlus (Peter (Petrus) Paulus) (1794–1847) PS	**Wierzb.**
Wiesbaur, Johann Baptist (1836–1906) S	**Wiesb.**
Wieslander, Albert Everett (1890–) S	**Wiesl.**
Wiesner, H. A	**H.Wiesner**
Wiesner, Julius von (1838–1916) S	**Wiesner**
Wiest, Anton (–1835) S	**Wiest**
Wigand, (Julius Wilhelm) Albert (1821–1886) AMS	**Wigand**
Wiggers, (Heinrich) August (Ludwig) (1803–1880) S	**A.Wigg.**
Wiggers, Friedrich Heinrich (Fridrich Hindrich) (1746–1811) AMS	**F.H.Wigg.**
Wiggers, Heinrich August (1752–1828) M	**H.A.Wigg.**
Wiggins, Ira Loren (1899–1987) S	**Wiggins**
Wiggins, Virgil Dale (1931–) A	**V.D.Wiggins**
Wight, Robert (1796–1872) S	**Wight**
Wight, William Franklin (1874–1954) S	**W.Wight**
Wight, William W. A	**W.W.Wight**
Wigman, H.J. (fl. 1894) S	**Wigman**
Wiinstedt, A.K. S	**A.K.Wiinst.**
Wiinstedt, Knud Jørgen Frederik (1878–1964) S	**Wiinst.**
Wijk, Roelof J.van der (1895–1981) BS	**Wijk**
Wijn, M. (fl. 1986) S	**Wijn**
Wijnands, D. Onno (1945–) S	**Wijnands**
Wijsman, H.J.W. (fl. 1982) S	**Wijsman**
Wiklund, Annette (1953–) S	**Wiklund**
Wikström, Johan Emanuel (1789–1856) PS	**Wikstr.**
Wilberforce, Peter William Charles (1935–) M	**Wilberf.**
Wilbrand, Johann Bernhard (1779–1846) S	**Wilbr.**
Wilbur, Robert Lynch (1925–) S	**Wilbur**
Wilce, Joan Hubbell (1931–) P	**J.H.Wilce**
Wilce, Robert Thayer (1924–) A	**R.T.Wilce**
Wilcke, Georg Wilhelm Constantin von (fl. 1788) S	**G.Wilcke**
Wilcke, Samuel Gustav (–1791) S	**S.Wilcke**
Wilcock, Christopher C. (1946–) S	**Wilcock**
Wilcox, Edwin (Forrest) Mead (1876–1931) MS	**Wilcox**
Wilcox, H.E. (fl. 1971) M	**H.E.Wilcox**
Wilcox, Lee W. A	**L.W.Wilcox**
Wilcox, Marguerite Statira (1902–) M	**M.S.Wilcox**
Wilcox, Raymond Boorman (1889–1949) M	**R.B.Wilcox**
Wilcoxon, James A. A	**Wilcoxon**
Wilczek, Ernst (1867–1948) MPS	**Wilczek**
Wilczek, Rudolf (1903–1984) S	**R.Wilczek**

Wilczyński, T. (fl. 1911) M — **Wilcz.**
Wild, Aloysius   A — **A.Wild**
Wild, Hiram (1917–1982) S — **Wild**
Wilde, Earle Irving (1888–1949) S — **E.Wilde**
(Wilde, Jan Jacobus Friedrich Egmond de)
   see de Wilde, Jan Jacobus Friedrich Egmond — **J.J.de Wilde**
Wilde, Julius (1864–1947) S — **J.Wilde**
(Wilde, Willem Jan Jacobus Oswald de)
   see de Wilde, Willem Jan Jacobus Oswald — **W.J.de Wilde**
(Wilde-Duyfjes, Brigitta E.E.de)
   see Duyfjes, Brigitta E.E. — **Duyfjes**
(Wildeman, Émile August(e) Joseph De)
   see De Wildeman, Émile Auguste(e) Joseph — **De Wild.**
Wilder, Gerrit Parmile (1863–1935) S — **Wilder**
Wildhaber, Othmar J. (1908–1976) S — **Wildh.**
Wilding, N. (fl. 1984) M — **Wilding**
Wildman, Robert D.   A — **Wildman**
(Wildpret de la Torre, Wolfredo)
   see Wildpret, Wolfredo de la Torre — **Wildpret**
Wildpret, Wolfredo de la Torre (1933–   ) S — **Wildpret**
Wildt, Albin (1845–1927) S — **A.Wildt**
Wilensky, D.G. (1892–1959) S — **Wilensky**
Wiley, Bonnie J. (fl. 1971) M — **B.J.Wiley**
Wiley, Farida Anna (1887–1927) S — **Wiley**
Wiley, Henry (1824–1907) M — **H.Wiley**
Wilhelm, Christian   A — **C.Wilh.**
Wilhelm, Gerould (1948–   ) M — **G.Wilh.**
Wilhelm, Gottlieb Tobias (1758–1811) S — **Wilh.**
Wilhelm, Karl (Carl) (Adolf) (1848–1933) MS — **K.Wilh.**
Wilhelm, Rudolph (1857–   ) S — **R.Wilh.**
Wilhelm, Stephen (fl. 1956) M — **S.Wilh.**
Wilke, Fritz (Wilhelm) (1888–   ) S — **Wilke**
Wilke, Hermann   A — **H.Wilke**
Wilken, Dieter H. (1944–   ) S — **Wilken**
Wilkes, Charles (1798–1877) PS — **Wilkes**
Wilkie, David (–1961) S — **Wilkie**
Wilkins, George Hubert (1888–1958) S — **Wilkins**
Wilkinson, Caroline Catharine (1822–1881) S — **Wilk.**
Wilkinson, Hazel Patricia (1932–   ) S — **H.P.Wilk.**
Wilkinson, Michael J. (fl. 1991) S — **M.J.Wilk.**
Will, H. (fl. 1916) M — **H.Will**
Willd, H.B.   S — **H.B.Willd**
Willdenow, Carl Ludwig von (1765–1812) AMPS — **Willd.**
Wille, Johan Nordal Fischer (1858–1924) AMS — **Wille**
Wille, Wolfgang   A — **W.Wille**
Willemet, (Pierre) Remi (1735–1807) S — **Willemet**
(Willemet, Hubert Félix Soyer)
   see Soyer-Willemet, Hubert Félix — **Soy.-Will.**
Willemet, Pierre Rémi François de Paule (1762–1790) PS — **P.Willemet**

Willems, G. (fl. 1990) M	**G.Willems**
Willems, Liliane (fl. 1956) S	**Willems**
Willemse, L.P.M. (fl. 1982) S	**L.P.M.Willemse**
Willemse, R.H. (fl. 1979) S	**R.H.Willemse**
Willén, Torbjörn A	**Willén**
Willer, K.H. (1933–) S	**Willer**
Willey, Henry (1824–1907) M	**Willey**
Willey, Ruth Lippitt (1928–) A	**R.L.Willey**
Williams, Benjamin Samuel (1824–1890) S	**B.S.Williams**
Williams, David Mervyn (1954–) A	**D.M.Williams**
Williams, E.B. (fl. 1957) M	**E.B.Williams**
Williams, F.J. (fl. 1978) M	**F.J.Williams**
Williams, Frederic Newton (1862–1923) S	**F.N.Williams**
Williams, Graham Lee A	**G.L.Williams**
Williams, Iolo Aneurin (1890–1962) S	**I.A.Williams**
Williams, Ion James Muirhead (1912–) S	**I.Williams**
Williams, J.E. (fl. 1982) S	**J.E.Williams**
Williams, J.Trevor (fl. 1977) S	**J.T.Williams**
Williams, John Beaumont (1932–) S	**J.B.Williams**
Williams, Leonard Howard John (1915–) S	**L.H.J.Williams**
Williams, Llewelyn (1901–1984) S	**Ll.Williams**
Williams, Louis G. A	**L.G.Williams**
Williams, Louis Otho (Otto) (1908–1991) BS	**L.O.Williams**
Williams, Margot (fl. 1984) S	**M.Williams**
Williams, Marvin C. (fl. 1984) M	**M.C.Williams**
Williams, Mary B. A	**M.B.Williams**
Williams, Norris H. (1943–) S	**N.H.Williams**
Williams, O.B. (fl. 1952) M	**O.B.Williams**
Williams, R.D. B	**R.D.Williams**
Williams, R.M. (fl. 1943) S	**R.M.Williams**
Williams, Reginald George (1935–) P	**R.G.Williams**
Williams, Robert Orchard (1891–1967) S	**R.O.Williams**
Williams, Robert Statham (1859–1945) BS	**R.S.Williams**
Williams, Samuel (c.1898–1965) P	**S.Williams**
Williams, Samuel (1743–1817) S	**Williams**
(Williams, Susan Elizabeth Verhoek) see Verhoek–Williams, Susan Elizabeth	**Verh.-Will.**
Williams, Thomas Albert (1865–1900) S	**T.A.Williams**
Williams, Yolande N. (fl. 1966) M	**Y.N.Williams**
Williamson, David B. A	**D.B.Will.**
Williamson, Graham (1932–) S	**G.Will.**
Williamson, H.S. (fl. 1923) M	**H.S.Will.**
Williamson, Herbert Bennett (1860–1931) S	**H.B.Will.**
Williamson, John (1839–1884) S	**J.Will.**
Williamson, Margaret A. (fl. 1988) M	**M.A.Will.**
Williamson, P.M. (fl. 1991) M	**P.M.Will.**
Williamson, Phyllis Alison (1925–) S	**P.A.Will.**
Williamson, William Crawford (1816–1895) AFM	**Will.**
Willich, Christian Ludwig (1718–1873) S	**Willich**
Willière, Yvonne A	**Willière**

(Willière–Stockmans, Yvonne)
   see Willière, Yvonne   **Willière**
Willing, Barbara (fl. 1985) S   **B.Willing**
Willing, Eckhard (fl. 1985) S   **E.Willing**
Willis, Barbara   B   **B.Willis**
Willis, J.L. (fl. 1956) S   **J.L.Willis**
Willis, James Hamlyn (1910–) BMS   **J.H.Willis**
Willis, John Christopher (1868–1958) S   **Willis**
Willis, Oliver Rivington (1815–1902) S   **O.R.Willis**
Willkomm, Heinrich Moritz (1821–1895) MPS   **Willk.**
Willmott, Ellen Ann (1860–1934) S   **E.Willm.**
Willmott, John (1775–1834) S   **J.Willm.**
Willoughby, L.G. (fl. 1956) M   **Willoughby**
Wilmot-Dear, Christine Melanie (1952–) S   **Wilmot-Dear**
Wilmott, Alfred James (1888–1950) S   **Wilmott**
Wilms, Friedrich (1848–1919) BS   **F.Wilms**
Wilms, Friedrich Heinrich (1811–1880) P   **F.H.Wilms**
Wilpert, Hubert (1901–) S   **Wilpert**
Wilson, Albert (1862–1949) BM   **A.Wilson**
Wilson, Alexander Stephen (1827–1893) M   **A.S.Wilson**
Wilson, Carl Louis (1897–) PS   **C.L.Wilson**
Wilson, Charles M. (fl. 1954) M   **C.M.Wilson**
Wilson, Doreen E. (fl. 1967) M   **D.E.Wilson**
Wilson, E.C.   A   **E.C.Wilson**
Wilson, Edward Elmer (1900–) M   **E.E.Wilson**
Wilson, Ernest Henry (1876–1930) S   **E.H.Wilson**
Wilson, F.D. (fl. 1963) S   **F.D.Wilson**
Wilson, Francis Robert Muter (1832–1903) M   **F.Wilson**
Wilson, Frank C.   S   **F.C.Wilson**
Wilson, G.B. (fl. 1954) M   **G.B.Wilson**
Wilson, George Fox (1896–1951) S   **G.F.Wilson**
Wilson, Graeme J.   A   **G.J.Wilson**
Wilson, Guy West (1877–1956) M   **G.W.Wilson**
Wilson, Harriet L.   A   **H.L.Wilson**
Wilson, Hugh Dale (1945–) S   **H.D.Wilson**
Wilson, Irene M. (fl. 1954) M   **I.M.Wilson**
Wilson, James Stewart (1932–) S   **J.S.Wilson**
Wilson, John (1696–1751) L   **J.Wilson**
Wilson, John Bracebridge (1828–1895) A   **J.B.Wilson**
Wilson, K.I. (fl. 1960) M   **K.I.Wilson**
Wilson, Karen Louise (1950–) S   **K.L.Wilson**
Wilson, Kenneth Allen (1928–) PS   **K.A.Wilson**
Wilson, Leonard Richard (1906–) AFMP   **L.R.Wilson**
Wilson, Malcolm (1882–1960) M   **M.Wilson**
Wilson, Marcia C. (fl. 1979) S   **M.C.Wilson**
Wilson, Nathaniel (1809–1874) S   **N.Wilson**
Wilson, Paul Graham (1928–) S   **Paul G.Wilson**
Wilson, Paul S. (1965–) B   **P.S.Wilson**
Wilson, Percy (1879–1944) S   **P.Wilson**

Wilson, Perry William (1902–) S	P.W.Wilson
Wilson, Peter Gordon (1950–) S	Peter G.Wilson
Wilson, Robert Gardner (1911–) S	R.G.Wilson
Wilson, T.R.S. A	T.R.S.Wilson
Wilson, Thomas Braidwood (1792–1843) A	T.B.Wilson
Wilson, W.C. (fl. 1956) S	W.C.Wilson
Wilson, William M. (1799–1871) BS	Wilson
Wilton, A.C. (fl. 1964) S	Wilton
Wiltshear, Felix Gilbert (1882–1917) S	Wiltshear
Wiltshire, Samuel Paul (1891–1967) M	Wiltshire
Wimmer, Christian Friedrich Heinrich (1803–1868) MPS	Wimm.
Wimmer, Franz Elfried (1881–1961) S	E.Wimm.
Wimpenny, R.S. A	Wimpenny
Winch, Nathaniel John (1768–1838) S	Winch
Winchell, Newton Horace (1839–1914) S	Winchell
(Winckler, Constantin (Konstantin) Georg Alexander)	
see Winkler, Constantin (Konstantin) Georg Alexander	C.G.A.Winkl.
Winckler, Emil Leonhard Wilhelm (1824–c.1871) S	Winckler
Wind, F.H. A	Wind
Winder, Daniel Knode (fl. 1871) M	Winder
Windham, Michael D. (1954–) PS	Windham
Windisch, Paulo Guenter (Günther) (1948–) PS	P.G.Windisch
Windisch, S. (fl. 1952) M	Windisch
Windler, Donald Richard (1940–) S	Windler
Windsor, John (1787–1868) S	Windsor
Wineland, Grace Odel (née Pugh, G.O.) (1889–) M	Wineland
Wing, Bruce Larry (1938–) A	Wing
Wingard, Samuel Andrew (1895–) M	Wingard
Wingate, Harold (1852–1926) M	Wingate
Winge, Øjvind (1886–1964) MS	Winge
Wingfield, Michael J. (fl. 1985) M	M.J.Wingf.
Wingfield, Robert C. (1936–) S	Wingf.
Winkelmann, Johannes (1842–1921) S	Wink.
Winkler, (Karl Gustav) Adolf (1810–1893) S	A.Winkl.
Winkler, (Wilhelm) William (1842–1927) S	W.Winkl.
Winkler, Constantin (Konstantin) Georg Alexander (1848–1900) S	C.Winkl.
Winkler, Eduard (1799–) S	Winkl.
Winkler, H. (fl. 1868) S	H.Winkl.
Winkler, Hans Karl Albert (1877–1945) S	H.K.A.Winkl.
Winkler, Hubert J.P. (1875–1941) S	H.J.P.Winkl.
Winkler, Moritz (Mauritz) (1812–1889) S	M.Winkl.
Winkler, Sieghard (fl. 1965) MS	S.Winkl.
Winks, Barbara L. (fl. 1966) M	Winks
Winogradsky, Sergey Nikolayevich (1856–1953) A	Winogr.
Winslow, Andreas Peter (Petersson) (1835–1900) S	Winslow
Winslow, Charles Edward Amory (1877–) M	C.E.A.Winslow
Winslow, Charles Frederick (1811–1877) S	C.Winslow
Winslow, Evelyn James (1870–1949) PS	E.J.Winslow

Winslow, Marcia R. A	**M.R.Winslow**
Winstead, Nash N. (fl. 1964) M	**Winstead**
(Winter, Bernard de)	
see De Winter, Bernard	**De Winter**
Winter, F.B. (1795–1869) S	**F.B.Winter**
Winter, F.W. (fl. 1907) A	**F.W.Winter**
Winter, Ferdinand (1835–1888) S	**F.Winter**
Winter, Heinrich Georg (1848–1887) M	**G.Winter**
Winter, John Mack (1899–1964) S	**J.M.Winter**
Winter, L. S	**L.Winter**
Winter, N.A. (1898–1934) S	**N.A.Winter**
Winterbottom, J.M. B	**J.M.Winterb.**
Winterbottom, James Edward (1803–1854) S	**Winterb.**
Winterhoff, W. (fl. 1980) M	**Winterh.**
Winterl, Jacob Joseph (1739–1809) S	**Winterl**
Winterschmidt, Johann Samuel (1760–1824/29?) S	**Winterschm.**
Winward, A.H. (fl. 1985) S	**Winward**
Wipff, J.K. (1962–) S	**Wipff**
Wirawan, Nengah (1941–) S	**Wirawan**
Wiriadinata, H. (1949–) S	**Wiriad.**
(Wirinow, Georg Jurij Nikolaewitch)	
see Woronow, Georg Jurij Nikolaewitch	**Woronow**
Wirsing, Adam Ludwig (1734–1797) S	**Wirsing**
Wirtgen, Ferdinand Paul (1848–1924) PS	**F.Wirtg.**
Wirtgen, Phillip Wilhelm (1806–1870) MPS	**Wirtg.**
Wirth, Carl (1883–) S	**Wirth**
Wirth, Michael (fl. 1959) MS	**M.Wirth**
Wirth, Volkmar (1943–) M	**V.Wirth**
Wirz, Johannes (1850–1915) S	**Wirz**
(Wirzbicki, Piotr Pawlus (Peter (Petrus) Paulus))	
see Wierzbicki, Piotr Pawlus (Peter (Petrus) Paulus)	**Wierzb.**
Wirzén, Johan Ernst Adhemar (1812–1857) S	**Wirzén**
Wise, Frederick Clunie (1884–1962) A	**Wise**
Wise, Sherwood W. A	**S.W.Wise**
Wislizenus, Friedrich (Frederick) Adolph (1810–1889) S	**Wisl.**
Wisłouch, Stanislav (Michaelovic) (1875–1927) AS	**Wisłouch**
Wiśniewski, Jerzy (fl. 1974) M	**J.Wiśn.**
Wiśniewski, Tadeusz (1905–1943) B	**Wiśn.**
Wisselingh, Cornelius van (fl. 1892–1899) A	**Wissel.**
Wissen, M.J.van A	**Wissen**
Wissjulina, E. (1898–1972) S	**Wissjul.**
Wissjulina, E.D. (1902–) S	**E.D.Wissjul.**
Wissotsky, A.V. A	**Wissotsky**
Wiström, Johan Alfred (1830–1896) S	**Wiström**
Wiström, Per Wilhelm (1865–1926) S	**P.Wiström**
(Wit, Hendrik Cornelius Dirk de)	
see de Wit, Hendrik Cornelius Dirk	**de Wit**
(Wit, R.de)	
see de Wit, R.	**R.de Wit**

Wita, Irena   A — **Wita**
Witasek, Johanna A. (1865–1910) S — **Witasek**
Wite, David   S — **Wite**
Witham, Henry Thomas Maire Silvertop (1779–1844) F — **Witham**
Withering, William (1741–1799) ABMPS — **With.**
Withner, Carl Leslie (1918–   ) S — **Withner**
Witt, N.H. (fl. 1902) S — **Witt**
Witt, Otto N. (1852?–1915) A — **O.N.Witt**
Witte, (Bror Otto) Hernfrid (1877–1945) S — **H.Witte**
Witte, Eduard Theodor (1865–1936) S — **E.Witte**
Witte, Heinrich (1825–1917) S — **Witte**
Wittern, I. (fl. 1986) M — **Wittern**
Wittlake, Eugene B.   BF — **Wittlake**
Wittmack, (Marx Carl) Ludwig (Ludewig) (1839–1929) MS — **Wittm.**
Wittmann, H. (fl. 1985) S — **Wittmann**
Wittrock, Veit Brecher (1839–1914) AS — **Wittr.**
Wittstein, Georg Christian (1810–1887) S — **Wittst.**
Wium–Andersen, S. (fl. 1986) S — **Wium–And.**
Wize, Casimir (fl. 1905) M — **Wize**
Wlangali, A.   S — **Wlangali**
Wobst, Karl (Carl) August (1842–1914) S — **Wobst**
Wocke, Erich (1863–1941) S — **Wocke**
Wodehouse, Roger Philip (1889–1978) S — **Wodehouse**
Woelkerling, William J. (1941–   ) A — **Woelk.**
Woenig, Franz (1851–1899) S — **Woenig**
Woerlein, Georg (1848–1899) PS — **Woerl.**
Wofford, B.Eugene (1943–   ) S — **Wofford**
(Wohlbach, S.B.)
   see Wolbach, S.B. — **Wohlbach**
Wohlfarth, Rudolf (1830–1888) S — **Wohlf.**
Wohlleben, Johann Friedrich (–1796) S — **Wohll.**
Wohlschlager, M. (fl. 1988) S — **Wohlschl.**
Wohltmann, Ferdinand (Friedrich Wilhelm) (1857–1919) S — **Wohltm.**
Wojciechowski, E. (fl. 1948) M — **Wojc.**
Wojewoda, Władysław (1932–   ) M — **Wojewoda**
Wojinowić, Wenislaw P. (1864–1892) B — **Wojinowić**
Wojnowski, Wieslaw   S — **Wojn.**
Wojtowicz, A. (fl. 1935) M — **Wojt.**
Wolbach, S.B. (fl. 1917) M — **Wohlbach**
Wolde, E.R. (fl. 1987) S — **Wolde**
Wolf, Carl Brandt (1905–1974) S — **C.B.Wolf**
Wolf, E. (fl. 1954) M — **E.Wolf**
Wolf, Egbert Ludwigowitsch (1860–1931) S — **E.L.Wolf**
Wolf, Ferdinand Otto (1838–1906) S — **F.O.Wolf**
Wolf, Franz Theodor (1841–1921) S — **Th.Wolf**
Wolf, Frederick Adolph (1885–1975) M — **F.A.Wolf**
Wolf, Frederick Taylor (1915–   ) M — **F.T.Wolf**
(Wolf, Gordon Parker De)
   see DeWolf, Gordon Parker — **DeWolf**

(Wolf, Johann Philipp)
   see Wolff, Johann Philipp **J.P.Wolff**
Wolf, Nathanael Matthaeus von (1724–1784) S **Wolf**
Wolf, Richard (fl. 1986) S **R.Wolf**
Wolf, Steven J. (fl. 1979) S **S.J.Wolf**
Wolf, Wolfgang (1875–1950) S **W.Wolf**
Wolfard, A. A **Wolfard**
Wolfe, Carl B. (1945–) M **Wolfe**
Wölfel, G. (fl. 1984) M **Wölfel**
Wolff, (Jacobus Otto) Reinhold (1845–) MS **R.Wolff**
Wolff, Caspar Friedrich (1735–1794) S **C.F.Wolff**
Wolff, D. S **D.Wolff**
Wolff, G. (1811–1893) S **G.Wolff**
Wolff, Johann Friedrich (1778–1806) S **J.F.Wolff**
Wolff, Johann Philipp (1743–1825) S **J.P.Wolff**
Wolff, Julius (1844–1921) S **J.Wolff**
Wolff, Karl Friedrich August Hermann (1866–1929) MS **H.Wolff**
Wolff, Manfred (fl. 1989) S **M.Wolff**
(Wolff, Nathaniel)
   see Wallich, Nathaniel **Wall.**
Wolff, S.L. (fl. 1987) S **S.L.Wolff**
(Wolffius, Johann Philipp)
   see Wolff, Johann Philipp **J.P.Wolff**
Wolfgang, Johann Friedrich (1776–1859) S **Wolfg.**
(Wolfius, Johann Philipp)
   see Wolff, Johann Philipp **J.P.Wolff**
Wolfner, Wilhelm (fl. 1858) S **Wolfner**
Wolfram, S. (fl. 1933) M **Wolfram**
Wolk, P.C.van der (fl. 1913) M **Wolk**
Wollaston, Elise Margaretta (1922–) A **E.M.Woll.**
Wollaston, George Buchanan (1814–1899) PS **Woll.**
Wolle, Francis (1817–1893) A **Wolle**
Wollenweber, Eckhard (fl. 1990) P **E.Wollenw.**
Wollenweber, Hans Wilhelm (1879–1949) AMP **Wollenw.**
Wolley–Dod, Anthony Hurt (1861–1948) S **Wolley–Dod**
Wollman, Constance E. (fl. 1968) M **Wollman**
Wollny, Robert A **Wollny**
Wolny, Andreas Raphael (–1829) S **Wolny**
Wołoszczak, Eustach (1835–1918) S **Woł.**
Wołoszyńska, Jadwiga (1882–1951) A **Wołosz.**
Wolpert, Josef (Johann Baptist) (1881–) S **Wolpert**
Womersley, Hugh Bryan Spencer (1922–) A **Womersley**
Womersley, John Spencer (1920–1985) S **J.S.Womersley**
Wong, H.D. S **H.D.Wong**
Wong, K.K. (fl. 1932) PS **K.K.Wong**
Wong, Khoon Meng (1954–) S **K.M.Wong**
Wong, Percy T.W. (fl. 1984) **MP.Wong**
Wong Sui, Tak Ping (fl. 1990) S **Wong Sui**
Wonisch, Franz (fl. 1909) S **Wonisch**
Wood, Alan A **A.Wood**

Wood, Alec E. (1933–) M	**A.E.Wood**
Wood, Alfred S	**Alf.Wood**
Wood, Alphonso W. (1810–1881) PS	**A.W.Wood**
Wood, Bertha S	**B.Wood**
Wood, Carroll E. (1921–) S	**C.E.Wood**
Wood, David (1939–) S	**D.Wood**
Wood, Edward James Ferguson (1904–) A	**E.J.F.Wood**
Wood, F.C. (fl. 1957) M	**F.C.Wood**
Wood, Fae Donat (1906–) A	**F.D.Wood**
Wood, Geoffrey H.S. (1927–) S	**G.H.S.Wood**
Wood, Gordon D. A	**G.D.Wood**
Wood, Horatio Charles (1841–1920) AF	**H.C Wood**
Wood, Jeffrey James (1952–) S	**J.J.Wood**
Wood, John Bland (1813–1890) B	**J.B.Wood**
Wood, John Medley (1827–1915) S	**J.M.Wood**
Wood, John Richard Ironside (1944–) S	**J.R.I.Wood**
Wood, Joseph Garnett (1900–1959) S	**J.G.Wood**
Wood, Mark W. (fl. 1973) S	**M.W.Wood**
Wood, Richard Dawson (1918–1977) A	**R.D.Wood**
Wood, S.N. (fl. 1983) M	**S.N.Wood**
Wood, Thomas Fanning (1841–1892) S	**T.F.Wood**
Wood, William (1745–1808) S	**Wood**
Woodall, Edward H.W. (1843–1937) S	**Woodall**
Woodcock, H.M. A	**H.M.Woodcock**
Woodcock, Hubert Bayley Drysdale (1867–1957) S	**Woodcock**
Wooden, Helen A. S	**Wooden**
Woodforde, James (1771–1837) S	**Woodf.**
Woodhead, Norman (1903–1978) A	**Woodhead**
Woodhead, Thomas William (1863–1940) S	**T.W.Woodhead**
Woodrow, G.M. (fl. 1898) S	**Woodrow**
Woodruff, H.B. (fl. 1941) M	**Woodruff**
Woodruffe–Peacock, Edward Adrian (1858–1922) S	**Woodr.-Peac.**
Woods, Albert Fred (1866–1948) A	**A.F.Woods**
Woods, Joseph (1776–1864) S	**Woods**
(Woods, Julian Edmund Tenison)	
see Tenison–Woods, Julian Edmund	**Ten.-Woods**
Woods, Patrick James Blythe (1932–) S	**P.Woods**
Woods, T.A.D. (fl. 1985) M	**T.A.D.Woods**
Woodson, Robert Everard (1904–1963) S	**Woodson**
Woodville, William (1752–1805) S	**Woodv.**
Woodward, John B. A	**J.B.Woodw.**
Woodward, Robert (1877–1915) S	**R.Woodw.**
Woodward, Thomas Jenkinson (1745–1820) AS	**Woodw.**
Woollam, R. A	**Woollam**
Woolley, Robert Vernon Giffard (fl. 1921–1926) S	**Woolley**
Woolls, William (1814–1893) S	**Woolls**
Woolsey, Theodore Salisbury (1880–1933) S	**Woolsey**
Woolson, Grace A. (1856–1911) PS	**Woolson**
Woolward, Florence Helen (1854–1930) S	**Woolward**

Wooster, David (c.1824–1888) S — **Wooster**
Wooten, Jean W. (fl. 1971) S — **Wooten**
Wooton, Elmer Ottis (1865–1945) S — **Wooton**
Worley, Anna C. (fl. 1977) M — **Worley**
Wormald, Hugh (1879–1955) M — **Wormald**
Wormley, T.G. A — **Wormley**
(Wormskiöld, Morten)
    see Wormskjöld, Martin — **Wormsk.**
Wormskjöld, Martin (1783–1845) AMS — **Wormsk.**
Wornardt, Walter W. A — **Wornardt**
Woronichin, Nikolai Nikolaevich (1882–   ) AM — **Woron.**
Woronin, Michael Stepanovitch (1838–1903) AMS — **Woronin**
Woronow, Georg Jurij Nikolaewitch (1874–1931) BMS — **Woronow**
(Woroschilov, Wladimir N.)
    see Voroschilov, Vladimir N. — **Vorosch.**
Worsdell, Wilson Crosfield (1867–1957) S — **Worsd.**
Worsley, Arthington (1861–1944) S — **Worsley**
Worsley, Thomas R. (1942–   ) A — **T.R.Worsley**
Worster, Pieter Johannes (1945–   ) S — **Worster**
Worthington, Richard D. (1941–   ) S — **Worth.**
Wortmann, Julius (1856–1925) S — **Wortmann**
Wossidlo, Paul (1836–   ) S — **Wossidlo**
Wóycicki, Zygmunt (1871–1941) S — **Wóycicki**
Woynar, Heinrich Karl (1865–1917) PS — **Woyn.**
Wraber, T. (fl. 1982) S — **Wraber**
Wrangel, Fredrik Anton (1786–1842) A — **Wrangel**
(Wray, John)
    see Ray, John — **Ray**
Wray, John L. (1925–   ) A — **Wray**
Wrede, Ernst Christian Conrad S — **Wrede**
Wreden, Robert (fl. 1874) M — **Wreden**
Wredow, Johann Christian Ludwig (Ludewig) (1773–1823) S — **Wredow**
Wretschko, Matthias (1834–1914) S — **Wretschko**
Wright, Albert Allen (1846–1905) S — **A.A.Wright**
Wright, Anthony Ernest (1954–   ) S — **A.E.Wright**
Wright, Charles Henry (1864–1941) APS — **C.H.Wright**
Wright, Charles (Carlos) (1811–1885) PS — **C.Wright**
Wright, Edward Perceval (1834–1910) AMS — **E.P.Wright**
Wright, Elaine F. (fl. 1968) M — **E.F.Wright**
Wright, Ernest (fl. 1935) M — **E.Wright**
Wright, Greg (fl. 1989) M — **G.Wright**
Wright, James (fl. 1928) M — **J.Wright**
Wright, Joan M. (fl. 1966) S — **J.M.Wright**
Wright, John (1811–1846) S — **Wright**
Wright, Jorge Eduardo (1922–   ) M — **J.E.Wright**
Wright, Samuel Hart (1825–1905) S — **S.H.Wright**
Wright, William (1735–1819) S — **W.Wright**
Wright, Ysabel (fl. 1937) S — **Y.Wright**

Wrigley, Fenella Ann (1936–) S	**Wrigley**
Wróblewski, Anton(i) (–1944) MS	**Wróbl.**
Wrońska, B. (fl. 1986) M	**Wrońska**
Wu, Bo Tang A	**B.T.Wu**
Wu, C.N. (fl. 1983) S	**C.N.Wu**
Wu, Chang Chun (fl. 1964) S	**C.C.Wu**
Wu, Cheng Yih (1916–) S	**C.Y.Wu**
Wu, Chi Guang (fl. 1987) M	**C.G.Wu**
Wu, Ching Ju (1934–) S	**C.J.Wu**
Wu, Chung Luen S	**C.L.Wu**
Wu, Han (fl. 1986) S	**H.Wu**
Wu, Hong Qi (fl. 1986) S	**H.Q.Wu**
Wu, Ji Nong (1922–) M	**J.N.Wu**
Wu, Jia Kun (fl. 1987) S	**J.K.Wu**
Wu, Jia Lin (fl. 1989) S	**J.L.Wu**
Wu, Jiunn Tzong (1949–) AS	**J.T.Wu**
Wu, Kuo Fang (1930–) S	**K.F.Wu**
Wu, Ming Hsiang S	**M.H.Wu**
Wu, Pan(g) Cheng (1935–) BS	**P.C.Wu**
Wu, Qi Xin (fl. 1985) S	**Q.X.Wu**
Wu, S.X. (fl. 1983) M	**S.X.Wu**
Wu, Sheng Hua (fl. 1990) M	**Sheng H.Wu**
Wu, Shi Fu (fl. 1990) P	**S.F.Wu**
Wu, Shiew Hung (fl. 1974) PS	**S.H.Wu**
(Wu, Shu Gong)	
see Wu, Su Kung	**S.K.Wu**
Wu, Su Kung (1935–) PS	**S.K.Wu**
Wu, T.C. (fl. 1978) S	**T.C.Wu**
Wu, Te Lin(g) (1934–) S	**T.L.Wu**
Wu, Tie Hang (fl. 1991) M	**T.H.Wu**
Wu, Wen Chen(g) S	**W.C.Wu**
Wu, X.B. (fl. 1987) M	**X.B.Wu**
Wu, X.Y. (fl. 1985) S	**X.Y.Wu**
Wu, Y. (fl. 1983) M	**Y.Wu**
Wu, Y.J. S	**Y.J.Wu**
Wu, Yeng Fen (fl. 1963) S	**Y.F.Wu**
Wu, Yin Chan (Ch'An) (fl. 1932) PS	**Y.C.Wu**
Wu, Ying Siang (fl. 1980) S	**Y.S.Wu**
Wu, Yu Ting (fl. 1970s) S	**Y.T.Wu**
Wu, Zhen Hai (1964–) S	**Z.H.Wu**
Wu, Zhen Lan (1939–) S	**Z.L.Wu**
(Wu, Zheng Yi)	
see Wu, Cheng Yih	**C.Y.Wu**
Wu, Zhi Min (fl. 1988) S	**Z.M.Wu**
Wuensche, Johann Georg (fl. 1804) S	**J.Wuensche**
Wuest, J. (fl. 1969) M	**Wuest**
Wuilbaut, J.J. (fl. 1986) M	**Wuilb.**
Wuitner, Émile (1865–1946) A	**Wuitner**
Wujek, Daniel Everett (1939–) AS	**Wujek**

Wujek, Mildred G.   A   **M.G.Wujek**
Wulf, A. (fl. 1988) M   **Wulf**
(Wulf, Eugenii Vladimirowitsch)
  see Wulff, Eugenii Vladimirowitsch   **E.Wulff**
Wulfen, Franz Xavier von (1728–1805) AMPS   **Wulfen**
Wulff, Alfred   A   **A.Wulff**
Wulff, Eugenii Vladimirowitsch (1885–1941) S   **E.Wulff**
Wulff, Heinz Diedrich (1910–) S   **H.Wulff**
Wulff, Johann Christoph (–1767) S   **Wulff**
(Wulff, Nathan)
  see Wallich, Nathaniel   **Wall.**
Wulff, Thorild (1877–1917) S   **T.Wulff**
Wulfsberg, Nils Gregers Ingvald (1847–1888) B   **Wulfsb.**
Wullschlaegel, Heinrich Rudolph (1805–1864) S   **Wullschl.**
Wunder, Helmut (1940–) M   **Wunder**
Wunderlich, Rosalie (fl. 1971) S   **Wund.**
Wunderlin, Richard P. (1939–) S   **Wunderlin**
Wundsch, H.H. (fl. 1929) M   **Wundsch**
Wünsche, Friedrich Otto (1839–1905) MPS   **Wünsche**
Wunschmann, Ernst (1848–) S   **Wunschm.**
Wurdack, John Julius (1921–) S   **Wurdack**
Wurm, Franz (1845–1922) S   **Wurm**
Wurmb, Friedrich von (–1781) S   **Wurmb**
Wurstle, B. (fl. 1986) S   **Wurstle**
Wurth, Theophil (1875–1922) M   **Wurth**
Württemberg, Prince Friedrich Paul Wilhelm von (1797–1860) S   **Württemb.**
Würtz, Alfred (1918–1964) A   **Würtz**
Wussow, James R. (fl. 1979) S   **Wussow**
Wüstnei, Karl Georg Gustav (1810–1858) S   **Wüstnei**
Wuzhi (fl. 1979) S   **Wuzhi**
Wyatt, Mary (–c.1950) A   **Wyatt**
Wyatt, Robert Edward (1950–) B   **R.E.Wyatt**
Wyatt–Smith, John (1917–) S   **Wyatt–Sm.**
Wydler, Heinrich (1800–1883) S   **Wydler**
(Wyk, Abraham Erasmus van)
  see Van Wyk, Abraham Erasmus   **A.E.van Wyk**
(Wyk, Ben–Erik van)
  see Van Wyk, Ben–Erik   **B–E.van Wyk**
(Wyk, P.S.van)
  see Van Wyk, P.S.   **P.S.van Wyk**
Wyley, Andrew (1820–) S   **Wyley**
Wylie, Robert Bradford (1870–1959) S   **Wylie**
Wyman, A.Phelps (1870–1947) S   **Wyman**
Wynd, Frederick Lyle (1904–) S   **F.L.Wynd**
Wynne, E.S. (fl. 1956) M   **E.S.Wynne**
Wynne, Frances Elizabeth (1916–) B   **Wynne**
Wynne, Michael James (1940–) A   **M.J.Wynne**
(Wynne–Edwards, Eleanor Mary)
  see Reid, Eleanor Mary Wynne–Edwards   **E.Reid**

Wyse Jackson, Michael B. (1962–) S   **M.B.Wyse Jacks.**
Wyse Jackson, Peter S. (1955–) S   **P.S.Wyse Jacks.**
Wyss–Chodat, F. (fl. 1927) M   **Wyss–Chod.**
(Wyssotsky, A.V.)
  see Vysotskij, A.V.   **Vysotskij**
Wyssotzky, G.H. (1899–) S   **Wyss.**

Xavier Filho, Lauro (fl. 1962) M   **L.Xavier**
Xavier, Miriam Borges (1949–) A   **M.B.Xavier**
Xeldencar, Vamona A   **Xeld.**
(Xena de Enrech, Nereida)
  see Xena, Nereida   **Xena**
Xena, Nereida (fl. 1983) S   **Xena**
Xi, Jing Qing (fl. 1987) S   **J.Q.Xi**
Xi, You Wei (fl. 1989) M   **Y.W.Xi**
Xia, Bang Mei A   **B.M.Xia**
Xia, De Yun (fl. 1986) S   **D.Y.Xia**
Xia, En Zhan A   **E.Z.Xia**
Xia, Qun (Quan) (1957–) P   **Q.Xia**
Xia, Yong Mei (fl. 1987) S   **Y.M.Xia**
Xian, Ting (fl. 1980s) M   **T.Xian**
Xiang, Cun Ti (Di) (1933–) S   **C.T.Xiang**
Xiang, Mei Mei (fl. 1988) M   **M.M.Xiang**
Xiang, Qiu Yun (1962–) S   **Q.Y.Xiang**
Xiang, Xian Zheng (fl. 1986) S   **X.Z.Xiang**
Xiao, Mian Yun (fl. 1982) S   **M.Y.Xiao**
Xiao, Pei Gen (fl. 1985) S   **P.G.Xiao**
Xiao, Sheng Rong (fl. 1984) M   **S.R.Xiao**
Xiao, Shun Chung (fl. 1988) S   **S.C.Xiao**
Xiao, Y.G. (fl. 1988) M   **Y.G.Xiao**
Xiao, Z.M. (fl. 1984) M   **Z.M.Xiao**
Xie, Bao Gui (fl. 1990) A   **B.G.Xie**
Xie, De Zi (fl. 1987) M   **D.Z.Xie**
Xie, Li Shan (fl. 1987) S   **L.S.Xie**
Xie, Quan Zheng (fl. 1981) S   **Q.Z.Xie**
Xie, Shu Qi (fl. 1985) A   **S.Q.Xie**
Xie, Yin Tong (Tang) (1929–) PS   **Y.T.Xie**
Xie, Zhi Xi (1939–) M   **Z.X.Xie**
Xie, Zong Wan (1924–) S   **Z.W.Xie**
Xifreda, Cecilia C. (1942–) S   **Xifreda**
(Xing, Gong Xia)
  see Shing, Kung Hsieh (Hsia)   **K.H.Shing**
Xing, Ji Qing (fl. 1987) S   **J.Q.Xing**
Xing, Wu Fu (fl. 1990) S   **W.F.Xing**
Xiong, Ruo Li (1929–1985) B   **R.L.Xiong**
Xiong, Zhi Ting (fl. 1985) S   **Z.T.Xiong**
Xu, C.Q. (fl. 1980) S   **C.Q.Xu**
Xu, Chang (fl. 1979) S   **C.Xu**
Xu, J.L. (fl. 1989) M   **J.L.Xu**

Xu, Jie Mei (fl. 1985) S ... **J.M.Xu**
Xu, Kun Yi (fl. 1986) M ... **K.Y.Xu**
Xu, Li Guo (1936–) S ... **L.G.Xu**
(Xu, Lian Wang)
  see Hsu, Lian Wang ... **L.W.Hsu**
Xu, Ling Chuan (fl. 1989) S ... **L.C.Xu**
Xu, Ming De (fl. 1989) S ... **M.D.Xu**
Xu, Shun Mei (fl. 1989) S ... **S.M.Xu**
Xu, Ting Zhi (fl. 1985) S ... **T.Z.Xu**
Xu, Tong (fl. 1983) M ... **T.Xu**
Xu, Wen Xuan (1919–1985) B ... **W.X.Xu**
Xu, Yi Wei (fl. 1990) M ... **Y.W.Xu**
Xu, Yin (fl. 1982) S ... **Y.Xu**
Xu, Zhao Ran (1957–) S ... **Z.R.Xu**
Xu, Zhu (fl. 1982) S ... **Z.Xu**
Xuan, Y. (fl. 1985) M ... **Y.Xuan**
Xuarez, Gaspar (1731–1804) S ... **Xuarez**
Xue, Ji Ru (1921–) S ... **J.R.Xue**
Xue, Jia Rong (fl. 1991) S ... **Jia R.Xue**
Xue, Xiang Ji (fl. 1983) S ... **X.J.Xue**
Xue, Zhao Wen (fl. 1986) S ... **Z.W.Xue**

Yabe, Hisakatsu (1878–1969) AF ... **H.Yabe**
Yabe, K. (fl. 1902) M ... **K.Yabe**
Yabe, Yoshitaka (1876–1931) PS ... **Y.Yabe**
Yabu, Hiroshi (1929–) A ... **Yabu**
Yabuno, Tomosaburo (1924–) S ... **Yabuno**
Yacubson, Sara A ... **Yacubson**
Yadav, A.K. (fl. 1979) P ... **A.K.Yadav**
Yadav, A.S. (fl. 1956) M ... **Yadav**
Yadav, B.R.Dayaker (fl. 1977) M ... **B.R.D.Yadav**
Yadav, R.R. (fl. 1979) M ... **R.R.Yadav**
Yadav, S.R. (fl. 1982) S ... **S.R.Yadav**
Yadava, Raj Nath A ... **Yadava**
Yaegashi, H. (fl. 1978) M ... **Yaegashi**
Yagi, Shigeichi (fl. 1973) AS ... **Yagi**
Yahara, Tetsukazu (1954–) S ... **Yahara**
Yahata, P.S. (fl. 1988) M ... **Yahata**
Yakimov, V.L. A ... **Yakimov**
Yakovlev, G.P. (1938–) S ... **Yakovlev**
Yakubtsiner, M.M. (fl. 1971) S ... **Yakubts.**
Yakushiji, E. (fl. 1930) M ... **Yakush.**
Yaltırık, Faik (1930–) S ... **Yalt.**
Yamada, Gentaro (–1943) M ... **G.Yamada**
Yamada, Iemasa (1935–) A ... **I.Yamada**
Yamada, Kohsaku (1934–) M ... **K.Yamada**
Yamada, Yukio (1900–1975) A ... **Yamada**
Yamada, Yuzo (fl. 1987) M ... **Y.Yamada**
Yamagishi, Takaaki (1923–) A ... **Yamagishi**

Yamagiwa, Suewo (fl. 1934) M	**Yamagiwa**
Yamaguchi, Hisanao A	**Yamag.**
Yamaguchi, T. B	**T.Yamag.**
(Yamaguishi, Noemy)	
see Yamaguishi–Tomita, Noemy	**Yam.–Tomita**
Yamaguishi–Tomita, Noemy (1935–) A	**Yam.–Tomita**
Yamamoto, Hirotoshi (1937–) A	**H.Yamam.**
Yamamoto, Hiroyuki (fl. 1966) M	**Hiroy.Yamam.**
Yamamoto, Masao (fl. 1986) M	**M.Yamam.**
Yamamoto, Shōji (fl. 1964) M	**S.Yamam.**
Yamamoto, Wataro (fl. 1955) M	**W.Yamam.**
Yamamoto, Yoshimatsu (1893–1947) PS	**Yamam.**
Yamamoto, Yukinori (fl. 1984) M	**Y.Yamam.**
Yamamura, M. (fl. 1978) M	**Yamamura**
Yamanaka, Kei (fl. 1984) M	**K.Yamanaka**
Yamanaka, Tsugiwo (1925–) S	**T.Yamanaka**
Yamano, Y. (fl. 1931) M	**Yamano**
Yamanouchi, Shigée (1878–) A	**Yaman.**
Yamaoka, Yuichi (fl. 1990) M	**Yamaoka**
Yamashiro, M. (fl. 1934) M	**Yamash.**
Yamatoya, Kaori (fl. 1989) M	**Yamat.**
Yamauchi, S. (fl. 1960) M	**Yamauchi**
Yamazaki, H. (fl. 1918) M	**H.Yamaz.**
Yamazaki, Takasi (Takashi) (1921–) S	**T.Yamaz.**
Yan, Ji Zhe (fl. 1981) S	**J.Z.Yan**
Yan, Jian Ping (fl. 1986) S	**J.P.Yan**
Yan, Min Shen(g) (1950–) MS	**M.S.Yan**
Yan, Su Zhu (fl. 1981) S	**S.Z.Yan**
Yanagawa, M. (fl. 1986) M	**Yanagawa**
Yanagihara, Masayuki (fl. c.1941) S	**Yanagih.**
Yanagita, K. (fl. 1989) M	**K.Yanagita**
Yanagita, Yoshizo (1872–1945) S	**Yanagita**
Yang, Bao Min (1928–) S	**B.M.Yang**
Yang, Chen Yuan (fl. 1982–1984) P	**Chen Y.Yang**
Yang, Chien Chow S	**C.C.Yang**
Yang, Chin S. (fl. 1985) MS	**Chin S.Yang**
Yang, Chun Shu (fl. 1983) S	**C.S.Yang**
Yang, Chun Yu (fl. 1984) S	**C.Y.Yang**
Yang, D.Q. (fl. 1986) M	**D.Q.Yang**
Yang, Da Rong (fl. 1990) M	**D.R.Yang**
Yang, Guang Hui (fl. 1987) S	**G.H.Yang**
Yang, H.Z. (fl. 1987) S	**H.Z.Yang**
Yang, Han Pi (1930–) S	**H.P.Yang**
Yang, Hsi (Xi) Ling (1931–) S	**H.L.Yang**
Yang, Ji (1961–) S	**J.Yang**
Yang, Jia Cheng (fl. 1980s) M	**J.C.Yang**
Yang, Jing Jing A	**J.J.Yang**
Yang, Jun Liang (fl. 1988) S	**J.L.Yang**
(Yang, Kuang Hui)	
see Yang, Guang Hui	**G.H.Yang**

Yang, Lang (fl. 1984) S	**L.Yang**
Yang, Ming Jin (fl. 1988) S	**M.J.Yang**
Yang, Ming Quan (fl. 1979) S	**M.Q.Yang**
Yang, Ming Zhu (fl. 1988) S	**M.Z.Yang**
Yang, Ping Hou (fl. 1988) S	**P.H.Yang**
Yang, Qian (fl. 1983) S	**Q.Yang**
Yang, Qiu Sheng (fl. 1989) S	**Q.S.Yang**
Yang, Ren Jun (fl. 1984) S	**R.J.Yang**
Yang, Shao Cheng (fl. 1988) S	**S.C.Yang**
Yang, Shao Zeng (fl. 1981) S	**S.Z.Yang**
Yang, Si Yuan (1964–) S	**S.Y.Yang**
Yang, Tsai Yeong (fl. 1967) S	**T.Y.Yang**
Yang, Ya Ling (1933–) S	**Y.L.Yang**
Yang, Yen Chin (1913–1984) S	**Yen C.Yang**
Yang, Yong Kang (fl. 1987) S	**Y.K.Yang**
Yang, Yu Pei (fl. 1981) S	**Yu P.Yang**
Yang, Yuan You (fl. 1978) S	**Y.Y.Yang**
Yang, Yuen Po (1943–) S	**Yuen P.Yang**
Yang, Yung Chang (1927–) S	**Y.C.Yang**
Yang, Zen Hong (fl. 1988) S	**Z.H.Yang**
Yang, Zheng Lu (fl. 1987) S	**Z.L.Yang**
Yano, Koji (1912–) B	**Yano**
Yao, Akira A	**A.Yao**
Yao, Chang Yu (fl. 1980) S	**C.Y.Yao**
Yao, G. (fl. 1987) P	**G.Yao**
Yao, Kan (fl. 1987) S	**K.Yao**
Yao, Q.W. (fl. 1988) S	**Q.W.Yao**
Yao, Te Shen (fl. 1988) S	**T.S.Yao**
Yao, Y.Y. (fl. 1987) S	**Y.Y.Yao**
Yao, Yi Jian (fl. 1989) M	**Y.J.Yao**
Yap, K.F. (fl. 1972) S	**Yap**
Yapp, Richard Henry (1871–1929) PS	**Yapp**
Yarchuk, T.A. (fl. 1979) S	**Yarchuk**
Yarish, Charles (1948–) A	**Yarish**
(Yarmolenko, A.V.)	
see Jarmolenko, A.V.	**Jarm.**
Yarrow, David (1935–) M	**Yarrow**
Yasnitskii, V. A	**Yasn.**
Yasuda, A. (1868–1924) M	**Yasuda**
Yasuda, M. BMS	**M.Yasuda**
Yasui, Kono (1880–) B	**Yasui**
Yatabe, Ryôkichi (Ruôkichi) (1851–1899) APS	**Yatabe**
Yatel, P. (fl. 1938) M	**Yatel**
Yates, Harris Oliver (1934–) S	**H.O.Yates**
Yates, Henry Stanley (1886–1938) M	**H.S.Yates**
Yates, James (1789–1871) S	**J.Yates**
Yates, Lorenzo Gordin (1837–1909) AFP	**Yates**
Yatindra A	**Yatindra**
(Yatskievich, George A.)	
see Yatskievych, George A.	**Yatsk.**

Yatskievych, George A. (1957–) PS	**Yatsk.**
Yaug, C.Y. (fl. 1988) P	**Yaug**
Yavin, Ziva (1936–) S	**Yavin**
Yawale, N.R. (fl. 1978) M	**Yawale**
Yazdani, S.S. (fl. 1983) M	**Yazdani**
Ye, Guang Han (1938–) S	**G.H.Ye**
Ye, Jia Song A	**J.S.Ye**
Ye, Neng Gan (fl. 1988) S	**N.G.Ye**
Ye, Wen Cai (fl. 1989) S	**W.C.Ye**
Ye, Yin Min S	**Y.M.Ye**
Ye, Yu Qing (fl. 1987) S	**Y.Q.Ye**
Yeh, Chen Tsung (fl. 1959) M	**C.T.Yeh**
Yeh, K.W. (fl. 1985) M	**K.W.Yeh**
Yeh, Z.Y. (fl. 1985) M	**Z.Y.Yeh**
Yen, Chi (fl. 1983) S	**C.Yen**
Yen, D.F. (fl. 1962) M	**D.F.Yen**
Yen, H.C. (fl. 1957) M	**H.C.Yen**
Yen, Jing Chu (fl. 1966) M	**J.C.Yen**
Yen, Jo Min (fl. 1978) M	**J.M.Yen**
Yen, Tzu Cheng (fl. 1966) M	**T.C.Yen**
(Yen, Wen Yu)	
see Yen, Jo Min	**J.M.Yen**
Yendo, Kichisaburo (1874–1921) A	**Yendo**
Yendo, Y. (fl. 1932) M	**Y.Yendo**
Yeo, Peter Frederick (1929–) S	**Yeo**
Yesodharan, K. (fl. 1987) M	**Yesodh.**
Yi, Lu Mei (fl. 1989) M	**L.M.Yi**
Yi, Tong Pei (fl. 1980) S	**T.P.Yi**
Yildirimli, Şinasi (1949–) S	**Yild.**
Yıldız, Beyram (1946–) S	**Yıldız**
Yin, Gong Yi (1912–) M	**G.Y.Yin**
Yin, Li S	**L.Yin**
Yin, Shu Fen (fl. 1983) S	**S.F.Yin**
Yin, Zu (Tsu) Tang (1931–) S	**Z.T.Yin**
(Ying, Gong Yi)	
see Yin, Gong Yi	**G.Y.Yin**
Ying, Jian Zhe (fl. 1980) M	**J.Z.Ying**
Ying, Shao Shun (fl. 1970) S	**S.S.Ying**
Ying, Tsun Shen (1933–) S	**T.S.Ying**
Yip, Hin Yuen (fl. 1986) M	**H.Y.Yip**
Yoder–Williams, M. (fl. 1984) S	**Yoder–Will.**
Yoganarasimhan, Sunkam Narayana Iyengar (1944–) S	**Yogan.**
Yohem, K.H. (fl. 1985) M	**Yohem**
Yokogi, K. (fl. 1955) M	**Yokogi**
Yokota, Masatsugu (1955–) S	**Yokota**
Yokote, Y. (fl. 1976) M	**Yokote**
Yokoyama, Kazumasa (fl. 1976) M	**K.Yokoy.**
Yokoyama, S. A	**S.Yokoy.**
Yokoyama, Tatsuo (fl. 1959) M	**T.Yokoy.**

Yokutsuka, I. (fl. 1955) M · · · · · · · · · · · · · · · · Yokuts.
Yoneda, Yûichi (1907–1977) A · · · · · · · · · · · · · · Yoneda
Yoneshigue, Yocie (1939–   ) A · · · · · · · · · · · · · Yonesh.
Yoneyama, Minoru (fl. 1959) M · · · · · · · · · · · · · Yoney.
Yonezawa, Nobumichi (fl. 1986) S · · · · · · · · · · · N.Yonez.
Yonezawa, W. (fl. 1957) M · · · · · · · · · · · · · · · · Yonez.
Yong, Ling   S · L.Yong
Yong, Shi Peng (1933–   ) S · · · · · · · · · · · · · · S.P.Yong
Yoo, Soon Ae   A · S.A.Yoo
Yoon, Ha Yong   A · H.Y.Yoon
York, Harlan Harvey (1875–   ) S · · · · · · · · · · · · · York
Yoshida, Meiko (1929–   ) A · · · · · · · · · · · · · M.Yoshida
Yoshida, Tadao (1933–   ) A · · · · · · · · · · · · · T.Yoshida
Yoshii, Hazime (1900–   ) M · · · · · · · · · · · · · · · Yoshii
Yoshikawa, N. (fl. 1985) P · · · · · · · · · · · · · · · · Yoshik.
Yoshima, S. (fl. 1979) M · · · · · · · · · · · · · · · · · Yoshima
Yoshimura, Isao (1933–   ) M · · · · · · · · · · · · · · Yoshim.
Yoshinaga, Torama (1871–1946) MS · · · · · · · · · · · Yoshin.
Yoshino, Kiichi (fl. 1905) M · · · · · · · · · · · · · · · Yoshino
Yoshizaki, Makoto (1943–   ) A · · · · · · · · · · · · · Yoshiz.
You, Cheng Xia (fl. 1985) S · · · · · · · · · · · · · · · C.X.You
Young, A.M. (fl. 1973) M · · · · · · · · · · · · · · · A.M.Young
Young, Aaron (1819–1898) S · · · · · · · · · · · · · · A.Young
Young, B.E. (fl. 1985) S · · · · · · · · · · · · · · · · B.E.Young
Young, David N.   A · · · · · · · · · · · · · · · · · · D.N.Young
Young, Donald Peter (1917–1972) S · · · · · · · · · · D.P.Young
Young, Esther (1893–   ) M · · · · · · · · · · · · · · · E.Young
Young, Herbert Andrew (1857–1894) S · · · · · · · · H.A.Young
Young, Joyce Redemske   A · · · · · · · · · · · · · · · J.R.Young
Young, Kavina (fl. 1924) M · · · · · · · · · · · · · · · K.Young
Young, Mary Sophie (1872–1919) S · · · · · · · · · · M.S.Young
Young, Maurice (fl. 1872) S · · · · · · · · · · · · · · M.Young
Young, Paul Allen (1898–   ) M · · · · · · · · · · · · P.A.Young
Young, Ralph George Norwood (1904–1979) S · · · · R.G.N.Young
Young, Robert Armstrong (1876–1963) S · · · · · · · R.A.Young
Young, Roy A. (fl. 1957) M · · · · · · · · · · · · · · Roy A.Young
Young, S.B. (fl. 1970) S · · · · · · · · · · · · · · · · S.B.Young
Young, Thomas (1773–1829) S · · · · · · · · · · · · · · Young
Young, Thomas W.K. (fl. 1969) M · · · · · · · · · · T.W.K.Young
Young, William (1742–1785) S · · · · · · · · · · · · · W.Young
Young, William (1865–1947) of Kirkcaldy S · · · · W.Young Kirkc.
Youngberg, Alv Dan (fl. 1972) S · · · · · · · · · · · · Youngberg
Youngken, Heber Williamson (1885–1963) S · · · · · · Youngken
Ysabeau, Alexandre Victor Frédéric (1793–1873) S · · · Ysabeau
Yu, C.Y. (fl. 1984) S · · · · · · · · · · · · · · · · · · · C.Y.Yu
Yu, Cheng Hung (fl. 1970) S · · · · · · · · · · · · · · · C.H.Yu
Yu, Guo Dian (fl. 1985) S · · · · · · · · · · · · · · · · G.D.Yu
Yu, H.T. (fl. 1986) P · · · · · · · · · · · · · · · · · · · H.T.Yu

Yu, Ling Long (fl. 1988) S   **L.L.Yu**
Yu, Qing Zhu (fl. 1986) S   **Q.Z.Yu**
Yu, Ta Fuh (1902–) MS   **T.F.Yu**
Yu, Te Tsun (fl. 1988) S   **Te T.Yu**
Yu, Tong Nian (fl. 1986) M   **T.N.Yu**
Yu, Tse Tsun (1908–1986) S   **T.T.Yu**
Yu, Xiang Yun (fl. 1989) S   **X.Y.Yu**
Yu, Yong Nian (1923–) M   **Y.N.Yu**
Yu, Yong Xin (fl. 1980) M   **Y.X.Yu**
(Yü, Yung Nien)
 see Yu, Yong Nian   **Y.N.Yu**
Yu, Ze Jun (fl. 1980) S   **Z.J.Yu**
Yu, Zhao Ying (fl. 1987) S   **Z.Y.Yu**
Yu, Zhi Xiong (fl. 1988) S   **Z.X.Yu**
Yuan, Chang Chi (fl. 1980s) S   **C.C.Yuan**
Yuan, Chang Qi (1934–) S   **C.Q.Yuan**
Yuan, G.F. (fl. 1984) M   **G.F.Yuan**
Yuan, Jian Yu (fl. 1985) M   **J.Y.Yuan**
Yuan, Pi Gang (fl. 1984) M   **P.G.Yuan**
Yuan, Xiao Ying (fl. 1985) S   **X.Y.Yuan**
Yuan, Zhi Wen (fl. 1986) M   **Z.W.Yuan**
Yuan, Zi Qing (1956–) M   **Z.Q.Yuan**
Yuasa, Hiroshi (fl. 1975) S   **Yuasa**
(Yue, Chong Xi)
 see Yueh, Chung Hsi   **C.H.Yueh**
Yue, Jing Zhu (Chu) (1935–) M   **J.Z.Yue**
Yue, Zong Shu (fl. 1987) S   **Z.S.Yue**
Yueh, Chung Hsi (fl. 1970) S   **C.H.Yueh**
Yuhki, Yoshimi (1904–) S   **Yuhki**
Yuill, Edward (fl. 1938) M   **E.Yuill**
Yuill, John L. (fl. 1938) M   **J.L.Yuill**
Yukawa, Matao (fl. 1912) M   **Yukawa**
Yukawa, Y. (fl. 1954) M   **Y.Yukawa**
Yukhananov, D.K. (fl. 1973) S   **Yukhan.**
Yuksip, A. (fl. 1978) S   **Yuksip**
Yun, Hye–Su A   **Yun**
Yuncker, Truman George (1891–1964) PS   **Yunck.**
Yung, Ying–Kit A   **Yung**
Yunitskaya, F.A. (fl. 1916) M   **Yunitsk.**
Yunusov, S.Yu. (fl. 1978) S   **Yunusov**
(Yurtsev, B.A.)
 see Jurtzev, B.A.   **Jurtzev**
(Yurtzew, B.A.)
 see Jurtzev, B.A.   **Jurtzev**
Yusef, Hasan M. (fl. 1966) M   **Yusef**
Yushev, A.A. (fl. 1978) S   **Yushev**

Z (Herr) (fl. 1774) S   **Z**
Zaayen, A.van (fl. 1982) M   **Zaayen**

Zabel, Hermann (1832–1912) PS	**Zabel**
Zabelina, M.M. A	**Zabelina**
Zaberzhinskav, Z.B. A	**Zaberzh.**
Zablackis, Earl A	**Zablackis**
Záborský, Ján (1928–) S	**Záborský**
Zabriskie, Jeremiah Lott (fl. 1891) M	**Zabriskie**
Zach, Franz (fl. 1939) M	**Zach**
Zachariah, S. (fl. 1981) M	**Zachariah**
Zacharias, (Emil) Otto (1846–1916) AS	**O.Zacharias**
Zacharias, Edward (1852–1911) S	**E.Zacharias**
Zacharof, E. (fl. 1980) S	**Zacharof**
Zachleder, Vilém A	**Zachleder**
Zachlederová, Milada A	**Zachled.**
Zachos, D.G. (fl. 1960) M	**Zachos**
Zadoks, J.C. (fl. 1955) S	**Zadoks**
Zaffran, J. (1935–) PS	**Zaffran**
Zafra Valverde, Ma.Luisa (1942–) S	**Zafra**
Zaganiaris, D. (1903–1940) S	**Zagan.**
Zagareli, Pavel Petrovich (1897–1937) S	**Zagar.**
Zagolin, A. S	**Zagolin**
Zagorenko, G.F. A	**Zagorenko**
Zagorodskich, P.F. (1901–1942) S	**Zagor.**
Zahariadi, Constantine (1901–1985) S	**Zahar.**
Zaharof, Eugenia (1950–) S	**Zaharof**
Zahl, P.A. A	**Zahl**
Zahlbruckner, Alexander (1860–1938) MS	**Zahlbr.**
Zahlbruckner, Johann Baptist (1782–1851) S	**J.Zahlbr.**
Zahn, Karl Hermann (1865–1940) S	**Zahn**
Zahn, Rudolf A	**R.Zahn**
Zähner, H. (fl. 1957) M	**Zähner**
Zahradnikova, K. (fl. 1976) S	**Zahradn.**
(Zaikonnikova, T.I.)	
see Zaikonnikowa, T.I.	**Zaik.**
Zaikonnikowa, T.I. (1929–) S	**Zaik.**
Zaim, M. (fl. 1979) M	**Zaim**
Zainal, A. (fl. 1988) M	**Zainal**
Zainuddin, Hardaniah (fl. 1989) M	**Zainuddin**
Zaitsev (fl. 1983) S	**Zaitsev**
Zając, Adam (1940–) S	**Zając**
Zájara Jiménez, J. (fl. 1958) M	**Zájara**
Zak, Bratislav (1919–) M	**Zak**
Zakaljabina, L.G. (fl. 1973) S	**Zakal.**
(Zakalyabina, L.G.)	
see Zakaljabina, L.G.	**Zakal.**
Zakharchenko, V.O. (fl. 1973) M	**Zakharch.**
Zakharyeva, O.I. (1929–) S	**Zakhar.**
Zakia, Banu (fl. 1962) M	**Zakia**
Zakirov, K.Z. (1906–) S	**Zakirov**
Zakryś, Bozena (1953–) A	**Zakryś**

Zalasky, H. (fl. 1964) M	**Zalasky**
(Zaleski, Karol M.)	
see Zalessky, Karol M.	**K.M.Zalessky**
(Zaleski, W.)	
see Zalessky, W.	**W.Zalessky**
(Zaleskij, Karol M.)	
see Zalessky, Karol M.	**K.M.Zalessky**
Zalessky, Karol M. (1890–) MS	**K.M.Zalessky**
Zalessky, Mikhail (Dmitrievich) (1877–1946) ABF	**Zalessky**
Zalessky, W. (fl. 1927) M	**W.Zalessky**
Zalewski, Aleksander (1854–1906) MS	**Zalewski**
Zalkind, F.L. (1901–) S	**Zalkind**
(Zalleskij, Mikhail (Dmitrievich))	
see Zalessky, Mikhail (Dmitrievich)	**Zalessky**
(Zalocar de Demitrović, Yolanda)	
see Zalocar, Yolanda	**Zalocar**
Zalocar, Yolanda A	**Zalocar**
(Zalyesskii, Mikhail (Dmitrievich))	
see Zalessky, Mikhail (Dmitrievich)	**Zalessky**
Zalzmann, P. S	**Zalzmann**
Zambettakis, Charalambos (fl. 1951) M	**Zambett.**
Zämelis, Aleksander (1897–1943) S	**Zämelis**
(Zämels, Aleksander)	
see Zamelis, Aleksander	**Zämelis**
Zammuto (fl. 1965) M	**Zammuto**
Zamora, Nelson (fl. 1988) S	**N.Zamora**
Zamora, Prescillano M. (1933–) P	**P.M.Zamora**
Zamora, R. (fl. 1973) M	**R.Zamora**
(Zamora V., Nelson)	
see Zamora, Nelson	**N.Zamora**
Zämuda, Antoni Józef (1889–1916) S	**Zämuda**
Zamudio Ruíz, Sergio (1953–) S	**Zamudio**
Zanardini, Giovanni (Antonio Maria) (1804–1878) AS	**Zanardini**
Zander, Richard Henry (1941–) B	**R.H.Zander**
Zander, Robert (1892–1969) S	**Zander**
Zaneveld, Jacques Simon (1909–) A	**Zaneveld**
Zanfrogini, Carlo (1866–) M	**Zanfr.**
Zang, Mu (1930–) BM	**M.Zang**
Zang, X.Q. (fl. 1979) M	**X.Q.Zang**
Zangheri, Pietro (1889–1983) S	**Zangh.**
(Zanichelli, Giovanni Gerolamo (Gian Girolamo))	
see Zannichelli, Giovanni Gerolamo (Gian Girolamo)	**Zannich.**
Zanin, Ana (fl. 1990) S	**A.Zanin**
Zanin Buri, C. A	**Zanin Buri**
Zannichelli, Giovanni Gerolamo (Gian Girolamo) (1662–1729) LM	**Zannich.**
Zanon, Vito (1875–1949) A	**Zanon**
Zanoschi, Valeriu (1934–) A	**Zanoschi**
Zantedeschi, August S	**A.Zanted.**
Zantedeschi, Giovanni (1773–1846) S	**Zanted.**
Zanten, Bennard Otto van (1927–) B	**Zanten**

Zantner, Alfred S	**Zantner**
(Zapałowich, Hugo)	
see Zapałowicz, Hugo	**Zapał.**
Zapałowicz, Hugo (1852–1917) PS	**Zapał.**
Zapater, Bernardo (1823–1907) S	**Zapater**
(Zapater y Marconell, Bernardo)	
see Zapater, Bernardo	**Zapater**
Zaprjagaev, F.L. (1903–1942) S	**Zaprjag.**
Zaprometov, Nikolai Georgievich (fl. 1926) M	**Zaprom.**
(Zárate P., Sergio)	
see Zárate Pedroche, Sergio	**Zárate**
Zárate Pedroche, Sergio (1952–) S	**Zárate**
Zardetto de Toledo, O. (fl. 1954) M	**Zardetto**
Zardini, Elsa Matilde (1949–) S	**Zardini**
Zareh, M. (fl. 1989) S	**Zareh**
Zarucchi, James Lee (1952–) S	**Zarucchi**
Zate, B.R. (fl. 1982) S	**Zate**
Zauer, L.M. A	**Zauer**
Zauschner, Johann Baptista Josef (1737–1799) S	**Zauschn.**
Zavaro Pérez, Carlos (fl. 1991) S	**Zavaro**
Zaverucha, Boris V. (1927–) S	**Zaver.**
Zavortink, Joyce E. (1930–) S	**Zavort.**
Zavřek, H. (fl. 1968) M	**Zavřek**
Zawadski, Aleksander (Jan Antoni) (1798–1868) S	**Zaw.**
(Zawadzki, Aleksander (Jan Antoni))	
see Zawadski, Aleksander (Jan Antoni)	**Zaw.**
(Zawadzsky, Aleksander (Jan Antoni))	
see Zawadski, Aleksander (Jan Antoni)	**Zaw.**
Zdárek, Robert (fl. 1881–1900) S	**Zdárek**
(Zdoroveva, E.N.)	
see Zdorovjeva, E.N.	**Zdor.**
Zdorovjeva, E.N. (1941–) S	**Zdor.**
Zea, Francisco Antonio (1770–1822) S	**Zea**
Zea, M. (fl. 1991) M	**M.Zea**
Zeberzhinskaya, E.B. A	**Zeberzh.**
Zebrowski, George (fl. 1936) M	**Zebrowski**
Zecher, E. (fl. 1974) S	**Zecher**
Zederbauer, Emmerich (1877–1950) S	**Zederb.**
Zefirov, B.M. (1915–1957) S	**Zefir.**
Zeh, W. A	**Zeh**
Zehetleitner, Gerda (fl. 1978) M	**Zehetl.**
Zeil, A.E. (fl. 1964) S	**Zeil**
Zeile, Elsie May (fl. 1895) S	**Zeile**
Zeiller, Charles René (1847–1915) FPS	**Zeiller**
Zeinalova, S.A. (1932–) S	**Zein.**
Zeissold, H. (fl. 1895) S	**Zeiss.**
Zeitler, Ilse A	**Zeitler**
Zeitlin, A. (fl. 1929) M	**Zeitlin**
(Zejnalova, S.A.)	
see Zeinalova, S.A.	**Zein.**

Zelenetzky, Nikolaj Michailowitsch (1859–1923) S	**Zelen.**
Zelený, Václav (1936–) S	**Zelený**
Zeliff, Clarke Courson (1897–) A	**Zeliff**
Zelig, Y. (fl. 1988) M	**Zelig**
Zeller, G. A	**G.Zeller**
Zeller, Sanford Myron (1885–1948) AM	**Zeller**
Zemann, Margarete (1883–) S	**Zemann**
(Zen, S.H.)	
see Zen, Shu Ying	**S.Y.Zen**
Zen, Shu Ying B	**S.Y.Zen**
Zenari, Silvia (1896–1956) S	**Zenari**
Zender, J. (fl. 1926) AM	**Zender**
Zeng, Cheng Kui (? Tseng Cheng Kuei) A	**C.K.Zeng**
Zeng, Fan An (fl. 1985) S	**F.A.Zeng**
Zeng, S.H. B	**S.H.Zeng**
Zeng, Shu Ying (1940–) B	**S.Y.Zeng**
Zeng, Wan Zhang (fl. 1986) S	**W.Z.Zeng**
Zeng, Wei (fl. 1989) M	**W.Zeng**
Zeng, X. (fl. 1978) M	**X.Zeng**
Zeng, X.L. (fl. 1987) M	**X.L.Zeng**
Zeng, Yi Quan (fl. 1979) S	**Y.Q.Zeng**
Zenitani, B. (fl. 1953) M	**Zenitani**
Zenker, Friderich Albert von (1825–1898) S	**F.Zenker**
Zenker, Georg August (1855–1922) S	**G.Zenker**
Zenker, Jonathan Carl (Karl) (1799–1837) ABMPS	**Zenker**
Zenkert, Charles Anthony (1886–1972) S	**Zenkert**
Zenoni, C. (fl. 1912) M	**Zenoni**
Zentmyer, George Aubrey (1913–) M	**Zentmyer**
Zenyuk, T.I. A	**Zenyuk**
Zepernick, Bernhard (1926–) S	**Zepern.**
Zerafa, Stephano (1791–1871) S	**Zerafa**
Žerbele, I.Ya. (fl. 1972) M	**Žerbele**
Zerbst, Kurt-Jürgen (1941–1975) S	**Zerbst**
Zeretelli, I. (fl. 1924) M	**Zeret.**
Zermann, Chrysostomus Alexander (fl. 1893–1895) S	**Zermann**
Zerov, Dmitriy Konstantinovich (1895–1971) MS	**Zerov**
Zerova, Marija Ja. (Mariya Ya.) (1902–) M	**Zerova**
Zersi, Elia (1818–1880) P	**Zersi**
Žertová, Anna (1930–) (later Chrtková, Anna) S	**Žertová**
Zetterstedt, Johan Emanuel (1828–1880) S	**J.E.Zetterst.**
Zetterstedt, Johan Wilhelm (1785–1874) S	**J.W.Zetterst.**
Zeven, A.C. (fl. 1973) S	**Zeven**
Zevenbergen, H.A. (1943–) S	**Zevenb.**
Zeyher, Carl Ludwig Philip(p) (1799–1858) S	**Zeyh.**
Zeyher, K.Johann Michael (1770–1843) S	**J.Zeyh.**
Zhang, Bin Cheng (1956–) M	**B.C.Zhang**
Zhang, Chao Fang (1923–) P	**C.F.Zhang**
Zhang, De Rui A	**D.R.Zhang**
Zhang, Ding Cheng (fl. 1986) S	**D.C.Zhang**

Zhang, F.M. (fl. 1984) S   **F.M.Zhang**
Zhang, Guang Chu (1940–) B   **G.C.Zhang**
Zhang, Gui Zhen (fl. 1984) S   **G.Z.Zhang**
Zhang, Guo Chun (fl. 1989) M   **Guo C.Zhang**
Zhang, Hai Dao (fl. 1965) S   **H.D.Zhang**
Zhang, Hai Yan (fl. 1985) S   **H.Y.Zhang**
Zhang, Han Jie (fl. 1985) S   **H.J.Zhang**
Zhang, J.Z. (fl. 1984) M   **J.Z.Zhang**
Zhang, Ji Lin (fl. 1981) S   **J.L.Zhang**
Zhang, Ji Zu (fl. 1985) S   **Ji Zu Zhang**
Zhang, Jia Yan (fl. 1989) S   **J.Y.Zhang**
Zhang, Jian Hou (fl. 1983) S   **J.H.Zhang**
Zhang, Jing Chun (fl. 1989) MS   **J.C.Zhang**
Zhang, Jing Rong A   **J.R.Zhang**
Zhang, Jun Fu A   **J.F.Zhang**
Zhang, K.Q. (fl. 1987) M   **K.Q.Zhang**
Zhang, Kai Yi (fl. 1989) M   **K.Y.Zhang**
Zhang, Lai Fa (fl. 1990) P   **L.F.Zhang**
Zhang, Li Bin (fl. 1989) S   **L.B.Zhang**
Zhang, Man Xiang (1934–) B   **M.X.Zhang**
Zhang, Mu (fl. 1985) M   **M.Zhang**
Zhang, Pei Xin (fl. 1985) S   **P.X.Zhang**
Zhang, Peng Yong (Yun) (1920–) S   **P.Y.Zhang**
Zhang, Qi Hong (fl. 1987) S   **Q.H.Zhang**
Zhang, Qi Tai (fl. 1988) S   **Q.T.Zhang**
Zhang, Qing (fl. 1989) S   **Q.Zhang**
Zhang, Qing Bin (fl. 1991) S   **Q.B.Zhang**
Zhang, Qiu Gen (fl. 1983) S   **Q.G.Zhang**
Zhang, Ru Song (fl. 1983) S   **R.S.Zhang**
Zhang, S. (fl. 1980) M   **S.Zhang**
Zhang, S.D. (fl. 1985) S   **S.D.Zhang**
Zhang, Shi Cai (fl. 1989) S   **S.C.Zhang**
Zhang, Shu Fan (fl. 1983) S   **S.F.Zhang**
Zhang, Shu Xian (fl. 1987) S   **S.X.Zhang**
(Zhang, T.)
  see Zhang, Guang Chu   **G.C.Zhang**
Zhang, Tian Yu (1937–) M   **T.Y.Zhang**
Zhang, Ting Zhen (fl. 1987) S   **T.Z.Zhang**
Zhang, W. (fl. 1986) M   **W.Zhang**
Zhang, Wei Ping (1963–) S   **W.P.Zhang**
Zhang, Weng Hui (fl. 1984) S   **W.H.Zhang**
(Zhang, Wi Pien)
  see Zhang, Wei Ping   **W.P.Zhang**
Zhang, Xian Nu (fl. 1984) S   **X.N.Zhang**
Zhang, Xiao Ling (1926–) M   **X.L.Zhang**
Zhang, Xiao Ping (fl. 1987) S   **X.P.Zhang**
Zhang, Xiao Qing (1951–) M   **X.Q.Zhang**
Zhang, Xiao Zen (fl. 1979) M   **X.Z.Zhang**
Zhang, Xing Xiang (fl. 1986) S   **X.X.Zhang**

(Zhang, Xui Shi)
   see Chang, Siu Shih   **S.S.Chang**
Zhang, Yao Jia (fl. 1984) S   **Y.J.Zhang**
Zhang, Yu Hua (fl. 1986) S   **Y.H.Zhang**
Zhang, Yuan (fl. 1989) S   **Y.Zhang**
Zhang, Z.F. (fl. 1986) M   **Z.F.Zhang**
(Zhang, Zhi Yu)
   see Chang, Chi(h) Yu   **Chi Yu Chang**
Zhang, Zhi Yun (1950–) S   **Zhi Y.Zhang**
Zhang, Zhong Yi (fl. 1980) M   **Z.Y.Zhang**
Zhang, Zi An   A   **Z.A.Zhang**
Zhao, Da Zhen (fl. 1991) M   **D.Z.Zhao**
Zhao, Guan Cai (fl. 1990) M   **G.C.Zhao**
Zhao, Hui Ru (fl. 1985) S   **H.R.Zhao**
Zhao, Ji Ding (1916–) MS   **J.D.Zhao**
Zhao, Jing Zhou (fl. 1987) M   **J.Z.Zhao**
Zhao, Mei Hua (fl. 1979) S   **M.H.Zhao**
Zhao, Neng (fl. 1980) S   **N.Zhao**
Zhao, Qing Sheng (fl. 1988) S   **Q.S.Zhao**
Zhao, Ru Neng (fl. 1987) S   **R.N.Zhao**
Zhao, S.Y. (fl. 1985) S   **S.Y.Zhao**
Zhao, Shi Dong (1941–) S   **S.D.Zhao**
Zhao, W.   A   **W.Zhao**
Zhao, Wei Liang (fl. 1987) S   **W.L.Zhao**
Zhao, Wen Xia (fl. 1989) M   **W.X.Zhao**
Zhao, Yi Zhi (1939–) S   **Y.Z.Zhao**
Zhao, Yu Tang (fl. 1982) S   **Y.T.Zhao**
Zhao, Zhen Ju (fl. 1987) S   **Z.J.Zhao**
Zhao, Zheng Yu (1928–) M   **Z.Y.Zhao**
Zhdanova, N.N. (fl. 1957) M   **Zhdanova**
Zhebrak, A.R. (1901–) S   **Zhebrak**
Zheng, Bai Lin   A   **B.L.Zheng**
Zheng, Bao Fu   A   **B.F.Zheng**
Zheng, Chao Zong (1939–) S   **C.Z.Zheng**
Zheng, Cheng Jin (fl. 1982) S   **C.J.Zheng**
Zheng, De Rong (fl. 1988) M   **D.R.Zheng**
Zheng, Guo Yang (fl. 1984) M   **G.Y.Zheng**
Zheng, Qing Fang (fl. 1985) S   **Q.F.Zheng**
Zheng, Ru Yong (1931–) M   **R.Y.Zheng**
Zheng, Ruo Xian (fl. 1986) S   **R.X.Zheng**
Zheng, Shu Lian (1935–) M   **S.L.Zheng**
Zheng, W.K. (fl. 1985) M   **W.K.Zheng**
Zheng, Xiu Ju (fl. 1987) S   **X.J.Zheng**
(Zherbele, I.Ya.)
   see Žerbele, I.Ya.   **Žerbele**
Zhilina, Z.A. (fl. 1963) M   **Zhilina**
Zhirov, E.G. (fl. 1980) S   **Zhirov**
Zhiteneva, N.E. (1900–) S   **Zhit.**
Zhmylev, P.Yu. (fl. 1985) S   **Zhmylev**

Zhogoleva, E.P. (1938–) S	**Zhogoleva**
Zhong, Guo Rong (fl. 1988) S	**G.R.Zhong**
Zhong, Jian (fl. 1981) S	**J.Zhong**
Zhong, Ming Jin (fl. 1986) S	**M.J.Zhong**
Zhong, Shi Li (1926–) S	**S.L.Zhong**
Zhong, Shi Qiang (fl. 1984) S	**S.Q.Zhong**
Zhong, Ye Cong (fl. 1981) S	**Y.C.Zhong**
Zhong, Zhao Kang (fl. 1988) M	**Z.K.Zhong**
Zhou, Han Qiu A	**H.Q.Zhou**
Zhou, Hong (fl. 1985) S	**H.Zhou**
Zhou, Jing Hua A	**J.H.Zhou**
Zhou, Li Hua (1934–) S	**L.H.Zhou**
Zhou, Li Jiang (fl. 1983) S	**L.J.Zhou**
Zhou, Lin (fl. 1981) S	**L.Zhou**
Zhou, Ling Yun (1918–) S	**L.Y.Zhou**
Zhou, Ren Zhang (fl. 1987) S	**R.Z.Zhou**
Zhou, Rui Chang (fl. 1981) S	**R.C.Zhou**
(Zhou, Shi Quan)	
see Zhou, Si Quang	**S.Q.Zhou**
Zhou, Si Quang (1937–) S	**S.Q.Zhou**
Zhou, Tong Xin (fl. 1987) M	**T.X.Zhou**
Zhou, W. (fl. 1987) M	**W.Zhou**
Zhou, Yin Sou (fl. 1985) S	**Y.S.Zhou**
Zhou, Z.C. (fl. 1984) S	**Z.C.Chou**
Zhou, Z.H. (fl. 1978) M	**Z.H.Zhou**
Zhou, Zhen Ying A	**Z.Y.Zhou**
Zhu, Da Yue (fl. 1990) A	**D.Y.Zhu**
(Zhu, Ge Lin(g))	
see Chu, Ge Lin(g)	**G.L.Chu**
Zhu, Hua (fl. 1988) A	**H.Zhu**
Zhu, Hui Zhong (1930–) A	**H.Z.Zhu**
Zhu, Pei Liang (fl. 1991) M	**P.L.Zhu**
Zhu, Wei Ming (fl. 1979) P	**W.M.Zhu**
Zhu, Wen (Wan) Jia (1925–) A	**W.J.Zhu**
Zhu, Xin Ting (fl. 1989) M	**X.T.Zhu**
Zhu, Yu Jian (fl. 1986) S	**Y.J.Zhu**
Zhu, Zheng Yin (fl. 1982) S	**Z.Y.Zhu**
Zhu, Zong Yuan (fl. 1989) S	**Zong Y.Zhu**
Zhuang, Jian Yun (1944–) M	**J.Y.Zhuang**
Zhuang, Ti De (1937–) S	**T.D.Zhuang**
Zhuang, Wen Ying (1948–) M	**W.Y.Zhuang**
(Zhuang, Xuan)	
see Chuang, Hsuan	**H.Chuang**
Zhukovsky, O. S	**O.Zhuk.**
Zhukovsky, Peter Mikhailovich (1888–1975) S	**Zhuk.**
Zhuo, Li Huan (fl. 1983) S	**L.H.Zhuo**
Zhurbenko, M.P. (1958–) M	**Zhurb.**
(Zhuze, A.P.)	
see Jousé, A.P.	**Jousé**

Zickendrath, Ernst (1846–1903) B   **Zick.**
Zickler, H. (fl. 1934) M   **Zickler**
Ziegenspeck, Hermann (Robert Theodor) (1891–1959) AS   **Ziegensp.**
Ziegler, A.W. (fl. 1950) M   **Ziegler**
Zieliński, Jerzy (1943–) S   **Ziel.**
Zielonkowski, W. (1940–) S   **Zielonk.**
Ziemann, Hans   A   **Ziemann**
Ziembińska–Tworzydlo, Maria   A   **Ziemb.-Tworz.**
Ziesenhenne, Rudolf Christian (1911–) S   **Ziesenh.**
Zigno, Achille de (1813–1892) AFM   **Zigno**
Zigra, Johannes Hermann (1775–1857) S   **Zigra**
Zijp, Coenraad van (1879–) S   **Zijp**
Zikes, H. (fl. 1906) M   **Zikes**
(Zilah, Endre Kiss von)
   see Kiss von Zilah, Endre   **E.Kiss**
Ziling, M.K. (fl. 1929) M   **Ziling**
Ziliotto   S   **Ziliotto**
Ziller, Wolf Gunther (1930–) M   **Ziller**
Zillig, Hermann (1893–1952) M   **Zillig**
Zimina, V.G.   S   **Zimina**
Zimmer, Brigitte (1943–) P   **B.Zimmer**
Zimmer, R.C. (fl. 1964) M   **R.C.Zimmer**
Zimmerman, Dale A. (fl. 1972) S   **Zimmerman**
Zimmermann, (Philipp William) Albrecht (1860–1931) S   **A.Zimm.**
Zimmermann, Abraham (1787–1850) S   **Ab.Zimm.**
Zimmermann, Albrecht Wilhelm Phillip (1860–1931) M   **Zimm.**
Zimmermann, Birthe   A   **B.Zimm.**
Zimmermann, C.   A   **C.Zimm.**
Zimmermann, Ferdinand J.von   S   **F.J.Zimm.**
Zimmermann, Friedrich (1855–1928) S   **F.Zimm.**
Zimmermann, G. (fl. 1978) M   **G.Zimm.**
Zimmermann, Hugo (1862–1933) MS   **H.Zimm.**
Zimmermann, Johann Georg von (1728–1795) S   **J.G.Zimm.**
Zimmermann, Martin Huldrych (1926–1980s) A   **M.H.Zimm.**
Zimmermann, Oscar Emil Reinhold (–1902) S   **O.E.R.Zimm.**
Zimmermann, Walter Max (1892–1980) APS   **W.Zimm.**
Zimmeter, Albert (1848–1897) S   **Zimmeter**
(Zinderen Bakker, Edward Meine van)
   see Zinderen–Bakker, Edward Meine van   **Zind.-Bakker**
Zinderen–Bakker, Edward Meine van (1907–) BM   **Zind.-Bakker**
Zinger, Nikola Wasiljevicz (1866–1923) S   **N.W.Zinger**
Zinger, Vasili Jakololewitsch (1836–1907) S   **V.J.Zinger**
Zingone, Adriana   A   **Zingone**
Zinn, Johann Gottfried (1727–1759) S   **Zinn**
Zinno, Y. (fl. 1972) M   **Zinno**
Zinova, A.D. (1903–1985) A   **Zinova**
Zinova, E.S.   A   **E.S.Zinova**
Zinserling, George   S   **G.Zinserl.**
Zinserling, Iurij Dmitrievitch (1894–1938) PS   **Zinserl.**

Zinssmeister, Carl Luther (fl. 1918) M	**Zinssm.**
Zippel, Hermann (fl. 1879–1885) S	**Zippel**
Zippelius, Alexander (1797–1828) MPS	**Zipp.**
Zippelius, H. (fl. 1923) M	**H.Zipp.**
Zittel, Karl Alfred von (1839–1904) S	**Zitt.**
Zitti, R. (1909–1974) S	**Zitti**
Ziz, Johann Baptist (1779–1829) S	**Ziz**
(Zizin, Nikolai Vasiljevich)	
see Tsitsin, Nikolai Vasiljevich	**Tsitsin**
Zizka, G. (fl. 1987) S	**Zizka**
Zlatník, Alios (1902–1979) S	**Zlatník**
Zlattner, L. (fl. 1983) M	**Zlattner**
Zlinska, J. (fl. 1987) S	**Zlinska**
Zlotina, G.D. (fl. 1969) M	**Zlotina**
Zlotsky, A. (fl. 1986) S	**Zlotsky**
Zobel, August (1861–1934) S	**A.Zobel**
Zobel, Johann Baptista (1812–1865) M	**Zobel**
Zobell, C.E. (fl. 1962) M	**Zobell**
Zodda, Giuseppe (1877–1968) S	**Zodda**
Zoëga, Johann (1742–1788) APS	**Zoëga**
Zoellner, Otto (fl. 1972) S	**Zoellner**
Zogg, Emil (1915–) P	**Zogg**
Zogg, Hans (fl. 1949) M	**H.Zogg**
Zohary, Daniel (1926–) S	**D.Zohary**
Zohary, Michael (1898–1983) S	**Zohary**
Zola, Alberto (1850–1942) S	**Zola**
Zolkiewicz, A.J. (fl. 1922) M	**Zolk.**
Zollikofer, Caspar Tobias (1774–1843) S	**Zollik.**
Zollikofer, Mor S	**M.Zollik.**
Zollinger, Heinrich (1818–1859) AMPS	**Zoll.**
Zolotuchin, N.I. (1952–) S	**Zolot.**
(Zolotukhin, N.I.)	
see Zolotuchin, N.I.	**Zolot.**
Zona, S. (fl. 1985) S	**Zona**
Zondag, J.L.P. (fl. 1929) M	**Zondag**
Zong, Y.C. (fl. 1987) P	**Y.C.Zong**
Zonneveld, B.J.M. (fl. 1981) S	**Zonn.**
Zopf, (Friederich) Wilhelm (1846–1909) AM	**Zopf**
Zorn, Helmut A	**H.Zorn**
Zorn, Johannes (1739–1799) S	**Zorn**
Zornow, Robert (1842–1873) S	**Zornow**
Zotov, Victor Dmitrievich (1908–1977) S	**Zotov**
Zotto, M. A	**Zotto**
Zou, Hui Yu (fl. 1984) S	**H.Y.Zou**
Zou, Shou Qing (fl. 1984) S	**S.Q.Zou**
Zoz, I.G. (1903–) S	**Zoz**
Zsák, Zoltán (1880–1966) S	**Zsák**
Zschacke, (Georg) Hermann (1867–1937) BMS	**Zschacke**

Zschieschang, Gerhard (fl. 1987) M — Zschiesch.
Zsolt, J. (fl. 1957) M — Zsolt
Zubcova, R.D. (fl. 1971) M — Zubcova
(Zubizarreta, Antonio Segura)
  see Segura Zubizarreta, Antonio — Segura
Zuccagni, Attilio (1754–1807) S — Zuccagni
Zuccarini, Joseph Gerhard (1797–1848) S — Zucc.
Zuccherella, A. (fl. 1987) M — Zuccher.
(Zucconi Galli Fonseca, Laura)
  see Zucconi, Laura — Zucconi
Zucconi, Laura (1960–) M — Zucconi
Zuchold, Ernst Amandus (–1867) S — Zuchold
Zuckerwanik, T.I. (1930–) S — Zuckerw.
Zuev, V.V. (1955–) S — Zuev
Zukal, Hugo (1845–1900) AMS — Zukal
(Zukervanik, T.I.)
  see Zuckerwanik, T.I. — Zuckerw.
Zukowski, Waldemar (1935–) S — Zukowski
Zuloaga, Fernando Omar (1951–) S — Zuloaga
Zumaglini, Antonio Maurizio (1804–1865) PS — Zumagl.
Zundel, George (Lorenzo Ingram) (1885–1950) M — Zundel
Zurett, S. (fl. 1938) M — Zurett
Zürn, L. (fl. 1986) M — Zürn
Zurowetz, J.E. S — Zurowetz
Zurzycka, A. (fl. 1963) M — Zurzycka
Zvára, Jaroslav I. (fl. 1922) M — Zvára
Zvirgzd, A.V. (1928–) S — Zvirgzd
Zwackh-Holzhausen, (Philipp Franz) Wilhelm von (1826–1903) M — Zwackh
Zwanziger, Gustaf Adolf (1839–1893) S — Zwanziger
Zweili, Fred A — Zweili
Zwicky, Henny (1889–) S — Zwicky
Zwillenberg, L.O. (fl. 1966) M — Zwillenb.
Żwodny, J. (fl. 1903) M — Żwodny
Zycha, Herbert (1903–) M — Zycha
Zykoff, W. A — Zykoff